Calculus

with analytic geometry

Calculus

with analytic geometry

HOWARD ANTON
Drexel University

SECOND EDITION

JOHN WILEY & SONS
New York Chichester Brisbane Toronto Singapore

To the Student Study guides and a solutions manual for this textbook are available through your college bookstore under the titles:

The Calculus Companion, volumes 1 and 2, to Accompany Calculus with Analytic Geometry, Second Edition by Howard Anton, prepared by William H. Barker and James Ward

Student's Solutions Manual to Accompany Calculus with Analytic Geometry, Second Edition by Howard Anton, prepared by Albert Herr

The study guide can help you with course material by acting as a tutorial, review, and study aid. The solutions manual contains detailed solutions to all odd-numbered exercises. If either the study guide or solutions manual is not in stock, ask the bookstore manager to order a copy for you.

Library of Congress Cataloging in Publication Data:

Anton, Howard.
  Calculus, with analytic geometry.

  Includes index.
  1. Calculus. I. Title.
QA303.A53 1984   516'.04   83-19778
ISBN 0-471-08271-6

Printed in the United States of America

10 9 8

To
My wife Pat
My children Brian, David, and Lauren
My mother Shirley

In memory of
Stephen Girard (1750-1831) benefactor

preface

This text is designed for a standard three-semester or four-quarter calculus course. My intent is to present the material in the clearest possible way. The primary effort has been devoted to *teaching* the subject matter with a level of rigor suitable for the mainstream calculus audience.

Anyone who has taught calculus knows that the course contains so much material that it is impossible for an instructor to spend an adequate amount of classroom time on every topic. For this reason a calculus textbook bears a major portion of the burden in the teaching process. I am hopeful that this book will be one that can be relied on for sound, clear, and complete explanations, thereby freeing valuable classroom time for the instructor.

FEATURES

Precalculus Material Because of the vast amount of material to be covered, it is desirable to spend as little time as possible on precalculus topics. However, it is a fact that freshmen have a wide variety of educational backgrounds and different levels of preparedness. Therefore, I have included an optional first chapter devoted to precalculus material. It is written in enough detail to enable the instructor to feel confident in moving quickly through these preliminaries.

Trigonometry Deficiencies in trigonometry plague many students throughout the entire calculus sequence. Therefore, I have included a substantial trigonometry review in the appendix. It is more detailed than most such reviews because I feel that students will appreciate having this material readily available. The review is broken into two units: the first to be mastered before reading Section 1.4 and the second before Section 3.3

Illustrations This calculus text is more heavily illustrated than most. It is my experience that beginners in mathematics frequently have difficulty extracting concepts from mathematical formulas, yet when the right picture is presented the concept becomes clear immediately. For this reason, I have chosen to take full advantage of modern two-color typography and the most up-to-date illustrative techniques. There are over 1200 illustrations designed to clarify and enhance the exposition. (Most calculus books have half this number.)

Pedagogy   I have devoted special effort to the explanations of more difficult concepts. In places where students traditionally have trouble, I have tried to give the reader some foothold on the problem. At times I have simply used artwork designed to focus the student's thinking process in the right direction; at other times I have paraphrased ideas informally in a way that cuts through the technical roadblocks; and at still other times I have simply broken the discussion into smaller, more understandable pieces than generally found in most calculus texts.

Rigor   Where possible, theory is presented precisely in a style tailored for freshmen. However, in those places where precision conflicts with clarity, I have presented informal intuitive discussions. Whenever this occurs, there is a clear indication that the arguments given are not intended as formal proof. I have tried to make a clear distinction between rigorous mathematics and informal developments. Theory involving $\delta\epsilon$ arguments has been placed in optional sections so it can be avoided if desired.

Flexibility   This book is designed to ensure maximum flexibility. There will be no difficulty in permuting chapters in any reasonable way.

Order of Presentation   The order in which topics are presented is fairly standard, with two exceptions: derivatives of trigonometric functions are introduced early (Chapter 3) and a discussion of first order separable and linear differential equations is given in the chapter on logarithmic and exponential functions (Chapter 7). This placement of differential equations material allows us to give some nice applications of logarithms and exponentials immediately, and also helps meet the needs of those engineering and science students who will encounter this material in other courses taken concurrently with calculus.

Applications   An abundance of applications to physics, engineering, biology, population growth, chemistry, and economics appear throughout the text.

Exercises   Each exercise set begins with routine drill problems and progresses gradually toward problems of greater difficulty. I have tried to construct well-balanced exercise sets with more variety than is available in most calculus texts. Answers, including art, are given for the odd-numbered exercises, and each chapter contains a set of supplementary exercises to help the student consolidate his or her mastery of the chapter.

Supplements There are two supplements available to students: a solutions manual containing worked-out solutions to odd-numbered exercises and a study guide for students who need additional help with text material.

Computer Graphics The use of computers to generate mathematical surfaces is fascinating and growing in importance. To help make the student aware of this tool, I have included a brief (optional) discussion of this topic at the end of Section 16.1.

Reviewing and Class Testing All of the material in this text has been refined and polished as a result of our extensive experience with the first edition. In addition to the hundreds of instructors who wrote to me with comments and constructive suggestions, *twenty-two* reviewers worked with me on the second edition to ensure mathematical accuracy and quality exposition.

CHANGES FOR THE SECOND EDITION

The second edition is a refinement of the first. Although some new material has been added, the major effort has been devoted to improving the exposition and the exercise sets.

A team of reviewers examined each exercise set to determine which ones needed more drill problems, more challenging problems, or more problems of specific types. The exercise sets were expanded accordingly. In addition, many exercise sets were modified to create a better balance between the even and odd exercises in order to give instructors more flexibility in assigning exercises with answers or without.

Throughout the text, material has been rewritten or reorganized to improve the exposition, and some new material has been added based on reviewer and user suggestions. The main changes are as follows:

- New material on graphing $y = ax^2 + bx + c$ was added to Chapter 1 to better prepare students for exercises in which such curves occur.

- The section on differentials was extensively rewritten and a discussion of linear approximation was added to it.

- The section on the chain rule was revised to give equal emphasis to the functional and differential versions of the formula. A rigorous proof of the chain rule was added to the appendix.

- Chapter 4 was totally reorganized as requested by many users.

- A new section on Newton's method was added.

- A review of logarithms was added to the chapter on the natural logarithm, the integral definition of $\ln x$ is now better motivated, and it is shown that $\ln x$ and $\log_e x$ are equivalent.

- The sections on inverse functions and inverse trigonometric functions were rewritten for greater clarity.

- New sections on surface integrals, Stokes' Theorem, and the Divergence Theorem in three dimensions were added.

HOWARD ANTON

acknowledgments

Reviewers It was my good fortune to have the advice and guidance of many fine reviewers. The knowledge and skills that they shared with me have greatly enhanced this book. For their contributions, I thank:

Dennis DeTurck, *University of Pennsylvania*
Garret J. Etgen, *University of Houston*
William R. Fuller, *Purdue University*
Douglas W. Hall, *Michigan State University*
Albert Herr, *Drexel University*
Robert Higgins, *Quantics Corporation*
Harvey B. Keynes, *University of Minnesota*
Leo Lampone, *Quantics Corporation*
Stanley M. Lukawecki, *Clemson University*
Melvin J. Maron, *University of Louisville*
Mark A. Pinsky, *Northwestern University*
Donald R. Sherbert, *University of Illinois*
William F. Trench, *Drexel University*
Richard C. Vile, *Eastern Michigan University*
Irving Drooyan, *Los Angeles Pierce College*
Charles Denlinger, *Millersville State College*
Edith Ainsworth, *University of Alabama*
Norton Starr, *Amherst College*
Katherine Franklin, *Los Angeles Pierce College*
Barbara Flajnik, *Virginia Military Institute*
Jacqueline Dewar, *Loyola Marymount University*
Hugh B. Easler, *College of William and Mary*
Paul Kumpel, *SUNY Stony Brook*
A.L. Deal, *Virginia Military Institute*
Michael Grossman, *University of Lowell*
David Bolen, *Virginia Military Institute*
David Armacost, *Amherst College*
Gary Grimes, *Mt. Hood Community College*
Joseph Meier, *Millersville State College*
Joseph M. Egar, *Cleveland State University*

John Brothers, *Indiana University*
Phil Locke, *University of Maine, Orono*
Richard Thompson, *University of Arizona*
David Cohen, *University of California, Los Angeles*
Marilyn Blockus, *San Jose State University*
Thomas McElligott, *University of Lowell*
Mark Bridger, *Northeastern University*
David Sandell, *U.S. Coast Guard Acdemy*

Exercises and Proofreading   I gratefully acknowledge the fine work of Melvin J. Maron who helped prepare the end-of-chapter supplementary exercises. Thanks are also due for the contribution of those who helped with the text exercises and proofreading: Albert Herr, Robert Higgins, Leo Lampone, Hal Schwalm, Andrew Galardi, Steven Fratini, Douglas McCloud, and Evelyn Weinstock.

Assistants   For their help with duplicating, paste-up, and collating, I thank my children: Brian Anton, David Anton, and Lauren Anton.

Computer Graphics   The excellent computer graphics were prepared by Robert Conley, James Warner, and Richard Yuskaitis of Precision Visuals, P.O. Box 1185, Boulder, Colorado, 80306.

Typing   The fine appearance of this text is due in part to the careful typing and formatting of the manuscript by Technitype, 304 Fries Lane, Cherry Hill, New Jersey, 08003. I am especially grateful to Kathleen R. McCabe, who personally handled the project and magically transformed my rough work into polished copy. A word of thanks is also due to Susan Gershuni who typed preliminary versions of the manuscript.

Wiley Staff   I give special thanks to the entire Wiley staff for the care and cooperation they have shown in the production and marketing of this edition of the text. In particular, I thank the following people with whom I had the good fortune to work personally:

Robert Pirtle, *Mathematics Editor*
Elaine Rauschal, *Production Manager*
Alejandra Longarini, *Production Supervisor*
Eugene T. Patti, *Supervising Editor*
Trumbull Rogers, *Manuscript Editor*
John Balbalis, *Consulting Artist*
Rafael H. Hernandez, *Assistant Design Director*
Carolyn Moore, *Marketing Manager*

Finally, I wish to acknowledge the major role of my former editor, Gary W. Ostedt, whose dedication to excellence contributed in large measure to the success of the first edition. I feel fortunate to have been the beneficiary of his talent for the past eleven years.

HOWARD ANTON

contents

introduction

Calculus is the mathematical tool used to analyze changes in physical quantities. It was developed in the seventeenth century to study four major classes of scientific and mathematical problems of the time:

1. Find the tangent to a curve at a point.
2. Find the length of a curve, the area of a region, and the volume of a solid.
3. Find the maximum or minimum value of a quantity—for example, the maximum and minimum distances of a planet from the sun, or the maximum range attainable for a projectile by varying its angle of fire.
4. Given a formula for the distance traveled by a body in any specified amount of time, find the velocity and acceleration of the body at any instant. Conversely, given a formula that specifies the acceleration or velocity at any instant, find the distance traveled by the body in a specified period of time.

These problems were attacked by the greatest minds of the seventeenth century, culminating in the crowning achievements of Gottfried Wilhelm Leibniz and Isaac Newton—the creation of calculus.

Gottfried Wilhelm Leibniz (1646–1716)
This gifted genius was one of the last people to have mastered most major fields of knowledge—an impossible accomplishment in our own era of specialization. He was an expert in law, religion, philosophy, literature, politics, geology, metaphysics, alchemy, history, and mathematics.

Leibniz was born in Leipzig, Germany. His father, a professor of moral philosophy at the University of Leipzig, died when Leibniz was six years old. The precocious boy then gained access to his father's library and began reading voraciously on a wide range of subjects, a habit that he maintained throughout his life. At age 15 he entered the University of Leipzig as a law student and by the age of 20 received a doctorate from the University of Altdorf. Subsequently, Leibniz followed a career in law and international politics, serving as counsel to kings and princes.

During his numerous foreign missions, Leibniz came in contact with outstanding mathematicians and scientists who stimulated his interest in mathematics—most notably, the physicist Christian Huygens. In mathematics Leibniz was self-taught, learning the subject by reading papers and journals. As a result of this fragmented mathematical education, Leibniz often duplicated the results of others, and this ultimately led to a raging conflict over the inventor of calculus—Leibniz or Newton? The argument over this question engulfed the scientific circles of England and Europe with most scientists on the continent supporting Leibniz and those in England supporting Newton. The conflict was unfortunate, and both sides suffered in the end. The continent lost the benefit of Newton's discoveries in astronomy and physics for more than 50 years, and for a long period England became a second-rate country mathematically because its mathematicians were hampered by Newton's inferior calculus notation. It is of interest to note that Newton and Leibniz never went to the lengths of vituperation of their advocates—both were sincere admirers of each other's work. The fact is that both men invented calculus independently. Leibniz invented it 10 years after Newton, in 1685, but he published his results 20 years before Newton published his own work on the subject.

Leibniz never married. He was moderate in his habits, quick tempered, but easily appeased, and charitable in his judgment of other people's work. In spite of his great achievements, Leibniz never received the honors show-

Gottfried Leibniz
(Culver Pictures)

Isaac Newton
(Culver Pictures)

ered on Newton, and he spent his final years as a lonely embittered man. At his funeral there was one mourner, his secretary. An eyewitness stated, "He was buried more like a robber than what he really was—an ornament of his country."

Isaac Newton (1642–1727)

Newton was born in the village of Woolsthorpe, England. His father died before he was born and his mother raised him on the family farm. As a youth he showed little evidence of his later brilliance, except for an unusual talent with mechanical devices—he apparently built a working water clock and a toy flour mill powered by a mouse. In 1661 he entered Trinity College in Cambridge with a deficiency in geometry. Fortunately, Newton caught the eye of Isaac Barrow, a gifted mathematician and teacher. Under Barrow's guidance Newton immersed himself in mathematics and science, but he graduated without any special distinction. Because the plague was spreading rapidly through London, Newton returned to his home in Woolsthorpe and stayed there during the years of 1665 and 1666. In those two momentous years the entire framework of modern science was miraculously created in Newton's mind—he discovered calculus, recognized the underlying principles of planetary motion and gravity, and determined that "white" sunlight was composed of all colors, red to violet. For some reason he kept his discoveries to himself. In 1667 he returned to Cambridge to obtain his Master's degree and upon graduation became a teacher at Trinity. Then in 1669 Newton succeeded his teacher, Isaac Barrow, to the Lucasian chair of mathematics at Trinity, one of the most honored chairs of mathematics in the world. Thereafter, brilliant discoveries flowed from Newton steadily. He formulated the law of gravitation and used it to explain the motion of the moon, the planets, and the tides; he formulated basic theories of light, thermodynamics, and hydrodynamics; and he devised and constructed the first modern reflecting telescope.

Throughout his life Newton was hesitant to publish his major discoveries, revealing them only to a select circle of friends, perhaps because of a fear of criticism or controversy. In 1687, only after intense coaxing by the astronomer, Edmond Halley (Halley's comet), did Newton publish his masterpiece, *Philosophae Naturalis Principia Mathematica* (The Mathematical Principles of Natural Philosophy). This work is generally considered to be the most important and influential scientific book ever written. In it Newton explained the workings of the solar system and formulated the basic laws of motion which to this day are fundamental in engineering and physics. However, not even the pleas of his friends could convince Newton to publish his discovery of calculus. Only after Leibniz published his results did Newton relent and publish his own work on calculus.

After 35 years as a professor, Newton suffered depression and a nervous breakdown. He gave up research in 1695 to accept a position as warden and later master of the London mint. During the 25 years that he worked at the mint, he did virtually no scientific or mathematical work. He was knighted in 1705 and on his death was buried in Westminster Abbey with all the honors his country could bestow. It is interesting to note that Newton was a learned theologian who viewed the primary value of his work to be its support of the existence of God. Throughout his life he worked passionately to date biblical events by relating them to astronomical phenomena. He was so consumed with this passion that he wasted years searching the Book of Daniel for clues to the end of the world and the geography of hell.

Newton described his brilliant accomplishments as follows, "I seem to have been only like a boy playing on the seashore and diverting myself in now and then finding a smoother pebble or prettier shell than ordinary, whilst the great ocean of truth lay all undiscovered before me."

1 coordinates, graphs, lines

1.1 REAL NUMBERS, SETS, AND INEQUALITIES (A REVIEW)

Since numbers and their properties play a fundamental role in calculus, we will begin by reviewing some terminology and facts about them.

The simplest numbers are the *integers:*

$$\ldots, \; -4, \; -3, \; -2, \; -1, \; 0, \; 1, \; 2, \; 3, \; 4, \; \ldots$$

With the exception that division by zero is ruled out, ratios of integers are called *rational numbers.* Examples are

$$\frac{2}{3}, \; \frac{0}{9}, \; \frac{6}{2}, \; \frac{17}{1000}, \; -\frac{5}{2}\left(=\frac{-5}{2}=\frac{5}{-2}\right)$$

Observe that every integer is also a rational number because an integer p can be written as the ratio

$$p = \frac{p}{1}$$

Division by zero is ruled out because we want to be able to express the relationship

$$y = \frac{p}{0}$$

in the alternative form

$$0 \cdot y = p$$

However, if p is different from zero, this equation is contradictory; and if p is equal to zero, this equation is satisfied by any number y, which means that the ratio $p/0$ does not have a unique value—a situation that is mathematically unsatisfactory. For these reasons such symbols as

$$\frac{p}{0} \quad \text{and} \quad \frac{0}{0}$$

are not assigned a value; they are said to be **undefined.**

The early Greeks believed that the size of every physical quantity could, in theory, be represented by a rational number. They reasoned that the size of a physical quantity must consist of a certain whole number of units plus some fraction m/n of an additional unit. This idea was shattered in the fifth century B.C. by Hippasus of Metapontum,* who demonstrated by geometric methods that the hypotenuse of the triangle in Figure 1.1.1 cannot be expressed as a ratio of integers. This dramatic discovery demonstrated the existence of **irrational numbers;** that is, numbers not expressible as ratios of integers. Other examples of irrational numbers are

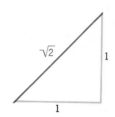

Figure 1.1.1

$$1 + \sqrt{2}, \quad \sqrt{3}, \quad \sqrt[3]{7}, \quad \pi, \quad \cos 19°$$

The proof that π is irrational is difficult and evaded mathematicians for centuries; it was finally proved in 1761 by J. H. Lambert.†

Rational and irrational numbers can be distinguished by their decimal representations. Rational numbers have **repeating decimals;** that is, from some point the decimal consists entirely of zeros or else a fixed finite string of digits that is repeated over and over. For example,

$$\frac{1}{2} = 0.5000 \ldots \quad \frac{8}{25} = 0.32000 \ldots \quad \frac{12}{4} = 3.000 \ldots$$
$$\frac{4}{3} = 1.3333 \ldots \quad \frac{3}{11} = 0.272727 \ldots \quad \frac{5}{7} = 0.714285714285714285 \ldots$$

Irrational numbers are represented by decimals that are not repeating.

▶ Example 1   Since π is irrational, the decimal

$$\pi = 3.14159265358979323846 \ldots$$

---

* HIPPASUS OF METAPONTUM (circa 500 B.C.). A Greek Pythagorean philosopher. According to legend, Hippasus made his discovery at sea and was thrown overboard by fanatic Pythagoreans because his result contradicted their doctrine. The discovery of Hippasus is one of the most fundamental in the entire history of science.

---

†JOHANN HEINRICH LAMBERT (1728–1777). A Swiss-German scientist, Lambert taught at the Berlin Academy of Sciences. In addition to his work on irrational numbers, he wrote landmark books on geometry, the theory of cartography, and perspective in art. His influential book on geometry foreshadowed the discovery of modern non-Euclidean geometry.

does not begin to repeat from some point. Since the decimal

$$0.101001000100001000001\ldots$$

is not repeating (the number of zeros between the ones keeps growing), it represents an irrational number.   ◄

In 1637 René Descartes* published a philosophical work called *Discourse on the Method of Rightly Conducting the Reason.* In the back of that book were three appendices that purported to show how the "method" could be applied to concrete examples. The first two appendices were minor works that endeavored to explain the behavior of lenses and the movement of shooting stars. The third appendix, however, was an inspired stroke of genius; it was described by the nineteenth century British philosopher John Stuart Mill as, "The greatest single step ever made in the progress of the exact sciences." In that appendix René Descartes linked together two branches of mathematics, algebra and geometry. Descartes' work evolved into a new subject called ***analytic geometry;*** it gave a way of describing algebraic formulas by means of geometric curves and, conversely, geometric curves by algebraic formulas.

In analytic geometry, the key step is to establish a correspondence between real numbers and points on a line. To accomplish this, we arbitrarily choose one direction along the line to be called ***positive*** and the other ***negative.*** It is usual to mark the positive direction with an arrowhead, as shown in Figure 1.1.2. Next, we choose an arbitrary reference point on the line to be called the ***origin,*** and select a unit of length for measuring distances.

Figure 1.1.2
     −   Origin   +

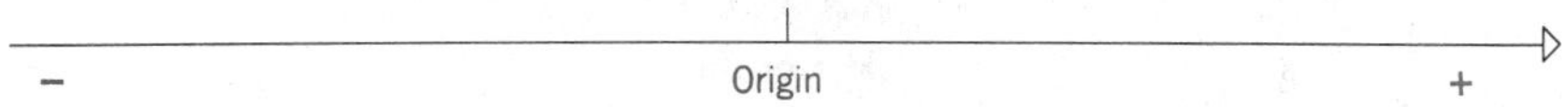

With each real number we can associate a point on the line as follows:

(a)   Associate with each positive number r the point that is a distance of r units in the positive direction from the origin.

---

*RENÉ DESCARTES (1596–1650). Descartes, a French aristocrat, was the son of a government official. He graduated from the University of Poitiers with a law degree at age 20. After a brief probe into the pleasures of Paris he became a military engineer, first for the Dutch Prince of Nassau and then for the German Duke of Bavaria. It was during his service as a soldier that Descartes began to pursue mathematics seriously and develop his analytic geometry. After the wars, he returned to Paris where he stalked the city as an eccentric, wearing a sword in his belt and a plumed hat. He lived in leisure, never arose before 11 A.M., and dabbled in the study of human physiology, philosophy, glaciers, meteors, and rainbows. He eventually moved to Holland, where he published his *Discourse on the Method,* and finally to Sweden where he died while serving as tutor to Queen Christina. Descartes is regarded as a genius of the first magnitude. In addition to major contributions in mathematics and philosophy, he is considered, along with William Harvey, to be a founder of modern physiology.

(b)   Associate with each negative number −r the point that is a distance of r units in the negative direction from the origin.

(c)   Associate the origin with the number 0.

The real number corresponding to a point on the line is called the ***coordinate*** of the point, and the line is called a ***coordinate line*** or sometimes the ***real line.***

▶ Example 2   In Figure 1.1.3 we have marked the approximate location of the points whose coordinates are −3, −1.75, −$\frac{1}{2}$, $\sqrt{2}$, π, and 4.   ◀

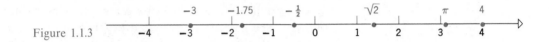

Figure 1.1.3

It is evident from the way in which real numbers and points on a coordinate line are related that each real number corresponds to a single point and each point corresponds to a single real number. To describe this fact we say that the real numbers and the points on a coordinate line are in ***one-to-one correspondence.***

Of particular importance to us is the fact that the real numbers are *ordered,* that is, given any two numbers a and b, exactly one of the following is true:

a is less than b
b is less than a
a is equal to b

The order symbols < (less than) and ≤ (less than or equal to) are defined as follows.

1.1.1   DEFINITION   If a and b are real numbers, then

$a < b$ means $b − a$ is positive
$a \leq b$ means $a < b$ or $a = b$

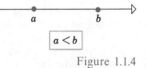

$a < b$

Figure 1.1.4

The inequality $a < b$, which is read *a is less than b*, can also be written as $b > a$, which is read *b is greater than a*. Geometrically, the inequality $a < b$ states that b falls to the right of a on a coordinate line (Figure 1.1.4). Similarly $a \leq b$, which is read *a is less than or equal to b*, can also be written as $b \geq a$, which is read, *b is greater than or equal to a.*

▶ Example 3   The following inequalities are correct:

$$3 < 8, \quad -7 < 1.5, \quad -12 \le -\pi, \quad 5 \le 5$$
$$8 \ge 3, \quad 1.5 > -7, \quad -\pi > -12, \quad 5 \ge 5 \qquad \blacktriangleleft$$

A number b will be called **nonnegative** if $0 \le b$, and **nonpositive** if $b \le 0$. The following properties of inequalities are frequently used in calculus. We omit the proofs.

1.1.2   THEOREM   *(a)   If $a < b$ and $b < c$, then $a < c$.*
(b)   If $a < b$, then $a + c < b + c$.
(c)   If $a < b$ and $c < d$, then $a + c < b + d$.
(d)   If $a < b$, then $ac < bc$ when c is positive and $ac > bc$ when c is negative.
(e)   If a and b are both positive or both negative and $a < b$, then $1/a > 1/b$.

REMARK.   These five properties remain true if $<$ and $>$ are replaced by $\le$ and $\ge$, respectively.

▶ Example 4   To paraphrase property (b), *an inequality remains true if the same number is added to both sides.*
   For example, adding 7 to both sides of the inequality

$$-2 < 6 \tag{1}$$

yields the valid inequality

$$5 < 13$$

Similarly, adding -8 to both sides of (1) yields the valid inequality

$$-10 < -2 \qquad \blacktriangleleft$$

▶ Example 5   To paraphrase property (c), *inequalities in the same direction can be added.*
   For example, adding

$$2 < 6 \quad \text{and} \quad -8 < 5$$

yields the valid inequality

$$-6 < 11 \qquad \blacktriangleleft$$

▶ Example 6   To paraphrase property (d), *if both sides of an inequality are multiplied by a positive number, the direction of the inequality remains the same, and if both sides of an inequality are multiplied by a negative number, the direction of the inequality is reversed.*

For example, multiplying the inequality

$$-4 < 7$$

by 2 yields

$$-8 < 14$$

and multiplying by -3 yields

$$12 > -21$$ ◄

▶ Example 7   To paraphrase property (*e*), *if both sides of an inequality have the same sign, then taking reciprocals reverses the direction of the inequality.*

For example, taking reciprocals on both sides of the inequality

$$3 < 7$$

yields

$$\frac{1}{3} > \frac{1}{7}$$

and taking reciprocals on both sides of the inequality

$$-8 < -6$$

yields

$$-\frac{1}{8} > -\frac{1}{6}$$ ◄

1.1.3   DEFINITION   If a, b, and c are real numbers, we will write

$$a < b < c$$

when $a < b$ and $b < c$.

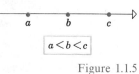

Figure 1.1.5

Geometrically, $a < b < c$ states that on a coordinate line, b is to the right of a and c is to the right of b (Figure 1.1.5).

The symbol $a < b \le c$ means $a < b$ and $b \le c$. We leave it to the reader to deduce the meanings of such symbols as $a \le b < c$, $a \le b \le c$, $a < b < c < d$, and so on.

1.1.4 DEFINITION · A *set* is a collection of objects; the objects are called the ***elements*** or ***members*** of the set.

In this text we will be concerned primarily with sets whose members are real numbers. One way of describing a set is to list its elements between braces. Thus, the set of all positive integers less than 5 can be written

$$\{1, 2, 3, 4\}$$

and the set of all positive even integers can be written

$$\{2, 4, 6, \ldots\}$$

where the dots are used to indicate that only a portion of the members have been explicitly listed and the rest of the members are obtained by continuing the pattern.

When it is inconvenient or impossible to list the members, it is sufficient to define a set by stating a property common only to its members. For example,

the set of all rational numbers
the set of all real numbers x such that $2x^2 - 4x + 1 = 0$
the set of all real numbers between 2 and 3

As an alternative to such verbal descriptions of sets, we can use the notation

$$\{x: \underline{\qquad}\}$$

which is read, "the set of all x such that." Where the blue line is placed, we state the property that specifies the set.

▶ Example 8
(a) $\{x: x$ is a rational number$\}$ is read, "the set of all x such that x is a rational number."
(b) $\{x: x$ is a real number satisfying $2x^2 - 4x + 1 = 0\}$ is read, "the set of all x such that x is a real number satisfying $2x^2 - 4x + 1 = 0$."
(c) $\{x: x$ is a real number between 2 and 3$\}$ is read, "the set of all x such that x is a real number between 2 and 3." ◀

REMARK. When it is clear that the members of a set are real numbers, we will omit the reference to this fact. Thus, the sets in Example 8 might be written

$$\{x: x \text{ is rational}\}$$
$$\{x: 2x^2 - 4x + 1 = 0\}$$
$$\{x: 2 < x < 3\}$$

To indicate that an element a is a member of a set A, we write

$$a \in A$$

which is read, "a is an element of A" or "a belongs to A." To indicate that the element a is *not* a member of the set A, we write

$$a \notin A$$

which is read, "a is not an element of A" or "a does not belong to A."

▶ **Example 9** Let $A = \{x : x \text{ is rational}\}$ and $B = \{x : 2 < x < 4\}$, then

$$\frac{3}{4} \in A, \quad -2 \in A, \quad \pi \notin A, \quad -\sqrt{2} \notin A$$
$$2.5 \in B, \quad \pi \in B, \quad -1 \notin B, \quad 4 \notin B \qquad \blacktriangleleft$$

Sometimes sets arise that have no members. For example,

$$\{x : x^2 < 0\}$$

A set with no members is called an ***empty set*** or a ***null set*** and is denoted by the symbol $\emptyset$. Thus,

$$\emptyset = \{x : x^2 < 0\}$$

Of special interest are certain sets of real numbers called ***intervals***. Geometrically, an interval is a line segment. For example, if $a < b$, then the ***closed interval*** from a to b is the set

$$\{x : a \leq x \leq b\}$$

and the ***open interval*** from a to b is the set

$$\{x : a < x < b\}$$

These sets are pictured in Figure 1.1.6.

REMARK. A closed interval includes both its endpoints (indicated by solid dots in Figure 1.1.6), while an open interval excludes both endpoints (indicated by open dots in Figure 1.1.6). Closed and open intervals are usually denoted by the symbols $[a, b]$ and (a, b), respectively, so

$$[a, b] = \{x : a \leq x \leq b\}$$
$$(a, b) = \{x : a < x < b\}$$

A square bracket [or] indicates that the endpoint is included, while a rounded bracket (or) indicates that the endpoint is excluded.

Figure 1.1.6

An interval can include one endpoint and exclude the other. Such intervals are called **half-open** (or sometimes **half-closed**). For example,

$$[a, b) = \{x : a \leq x < b\}$$
$$(a, b] = \{x : a < x \leq b\}$$

$\{x : x > a\}$

An interval can extend indefinitely in either the positive or negative direction (see Figure 1.1.7). The intervals in Figure 1.1.7 are denoted by

$$(a, +\infty) = \{x : x > a\}$$
$$(-\infty, b] = \{x : x \leq b\}$$

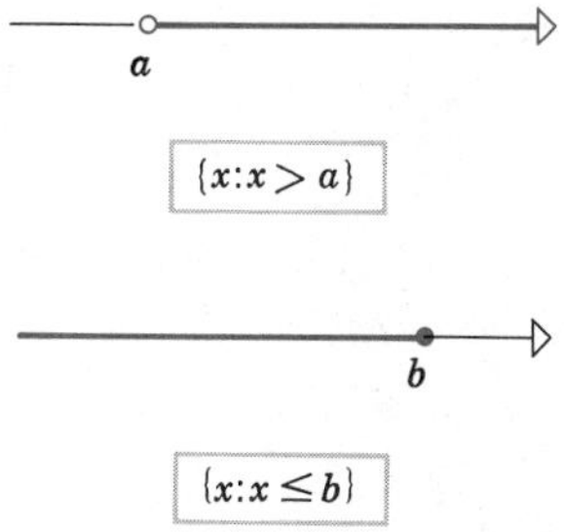

$\{x : x \leq b\}$

Figure 1.1.7

where the symbol $+\infty$ ("plus infinity") indicates the interval extends indefinitely in the positive direction and the symbol $-\infty$ ("minus infinity") indicates the interval extends indefinitely in the negative direction.

Table 1.1.1 gives a complete list of the kinds of intervals possible.

Table 1.1.1

INTERVAL NOTATION	SET NOTATION	GEOMETRIC PICTURE
$[a, b]$	$\{x : a \leq x \leq b\}$	
(a, b)	$\{x : a < x < b\}$	
$[a, b)$	$\{x : a \leq x < b\}$	
$(a, b]$	$\{x : a < x \leq b\}$	
$(-\infty, b]$	$\{x : x \leq b\}$	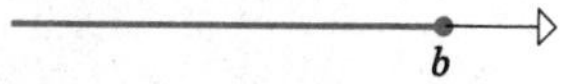
$(-\infty, b)$	$\{x : x < b\}$	
$[a, +\infty)$	$\{x : x \geq a\}$	
$(a, +\infty)$	$\{x : x > a\}$	
$(-\infty, +\infty)$	$\{x : x \text{ is a real number}\}$	

1.1.5 DEFINITION   Two sets A and B are said to be **equal** if they have the same elements, in which case we write $A = B$.

▶ Example 10

$$\{x: x^2 = 1\} = \{-1, 1\}, \qquad \{\pi, 0, 3\} = \{3, \pi, 0\}$$
$$\{x: x^2 < 9\} = \{x: -3 < x < 3\}$$
◀

1.1.6 DEFINITION If every member of a set A is also a member of set B, then we say A is a ***subset*** of B and write $A \subset B$.

In addition, we will agree that the empty set $\varnothing$ is a subset of every set.

▶ Example 11

$$\{-2, 4\} \subset \{-2, 1, 0, 4\}$$
$$\{x: x \text{ is rational}\} \subset \{x: x \text{ is a real number}\}$$
$$\varnothing \subset A \text{ (for every set } A)$$
◀

REMARK. If $A \subset B$ and $B \subset A$, then $A = B$. (Why?)

1.1.7 DEFINITION If A and B are two given sets, then the set of all elements belonging to both A and B is called the ***intersection*** of A and B; it is denoted by $A \cap B$.

1.1.8 DEFINITION If A and B are two given sets, then the set of all elements belonging to A or B or both is called the ***union*** of A and B; it is denoted by $A \cup B$.

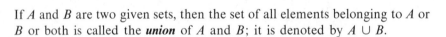

▶ Example 12   In Figure 1.1.8, A is the set of points inside the left circle and B the set of points inside the right circle. The sets $A \cap B$ and $A \cup B$ are shown as shaded regions.
◀

▶ Example 13   Sketch the sets:

(a) $[-4, 1) \cup (3, 6)$
(b) $(-3, 2] \cup (1, 7]$

Solution. See Figure 1.1.9.
◀

Figure 1.1.8

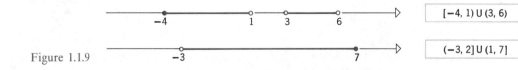

Figure 1.1.9

▶ **Example 14**   Sketch the set $(-4, 3] \cap (2, 6)$.

Solution.   The set $(-4, 3] \cap (2, 6)$ consists of all points common to the intervals $(-4, 3]$ and $(2, 6)$. As shown in Figure 1.1.10, the intersection is the interval $(2, 3]$.   ◀

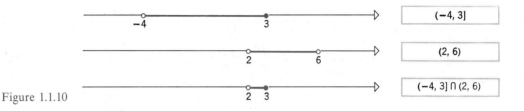

Figure 1.1.10

In the following examples we use Theorem 1.1.2 to solve some inequalities; the solutions are expressed in terms of intervals. The first example points out a subtle problem in logic that we shall encounter throughout the text.

▶ **Example 15**   Solve $2x - 3 < 7$. (That is, find all real numbers satisfying the inequality.)

Solution.   If x is any solution, then

$$
\begin{array}{ll}
2x - 3 < 7 & \text{[Given.]} \\
2x \quad\;\; < 10 & \text{[We added 3 to both sides.]} \\
x \quad\;\; < 5 & \text{[We multiplied both sides by } \tfrac{1}{2}.\text{]}
\end{array}
$$

At this point we are tempted to conclude that the solutions of $2x - 3 < 7$ consist of all x less than 5, that is, all x in the interval $(-\infty, 5)$. While this conclusion is correct, it is premature. To see why, let us examine the logic of our argument. We have proved: *If x is a solution, then $x < 5$.* In other words, if S denotes the set of solutions to $2x - 3 < 7$, we have shown

$$S \subset (-\infty, 5) \tag{2}$$

We have *not* yet shown that

$$S = (-\infty, 5) \tag{3}$$

If we can prove that

$$(-\infty, 5) \subset S \tag{4}$$

then by the remark after Example 11, (3) will follow from (2) and (4).

To prove (4) we must show that every member of $(-\infty, 5)$ is also a member of S. That is, we must assume $x < 5$ and deduce that $2x - 3 < 7$. We do this as follows:

$$
\begin{array}{lll}
x & < 5 & \text{[Given.]} \\
2x & < 10 & \text{[We multiplied both sides by 2.]} \\
2x - 3 & < 7 & \text{[We added } -3 \text{ to both sides.]}
\end{array}
$$

Thus, the solutions of $2x - 3 < 7$ form the interval $(-\infty, 5)$. ◄

REMARK. In the last example, compare the steps needed to prove (2) with the steps needed to prove (4):

$$
\begin{array}{cc}
\textit{Steps to} & \textit{Steps to} \\
\textit{Prove (2)} & \textit{Prove (4)} \\
\\
2x - 3 < 7 & x \quad\ < 5 \\
2x \quad\ < 10 & 2x \quad\ < 10 \\
x \quad\ < 5 & 2x - 3 < 7
\end{array}
$$

In the two parts the steps are the same, but in opposite orders. In any "two-stage" mathematical argument where the steps in the second stage are the same as those in the first stage, but in the opposite order, it is common to give only the steps for the first stage and then conclude the proof by stating that *the steps are reversible.*

▶ Example 16 Solve $3 + 7x \leq 2x - 9$.

Solution.

$$
\begin{array}{ll}
3 + 7x \leq 2x - 9 & \text{[Given.]} \\
7x \leq 2x - 12 & \text{[We added } -3 \text{ to both sides.]} \\
5x \leq -12 & \text{[We added } -2x \text{ to both sides.]} \\
x \leq -\tfrac{12}{5} & \text{[We multiplied both sides by } \tfrac{1}{5}.\text{]}
\end{array}
$$

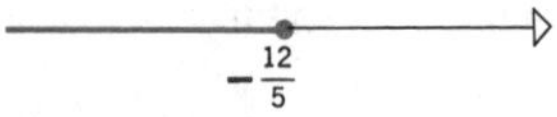

Figure 1.1.11

Since the steps are reversible, the set of solutions is $(-\infty, -\tfrac{12}{5}]$; this solution set is shown in Figure 1.1.11. ◄

▶ Example 17 Solve $7 \leq 2 - 5x < 9$.

Solution.

$$
\begin{array}{ll}
7 \leq 2 - 5x < 9 & \text{[Given.]} \\
5 \leq \quad\ -5x < 7 & \text{[We added } -2 \text{ to each member.]} \\
-1 \geq \quad\quad x > -\tfrac{7}{5} & \text{[We multiplied each member by } -\tfrac{1}{5}.\text{]}
\end{array}
$$

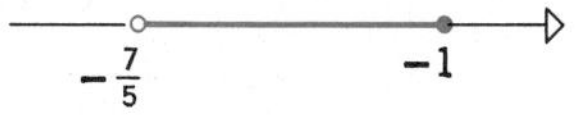

Figure 1.1.12

Since the steps are reversible, the set of solutions is $(-\frac{7}{5}, -1]$; this solution set is shown in Figure 1.1.12.   ◀

REMARK.   Observe that in Example 17 the inequalities were reversed when we multiplied by the negative quantity $-\frac{1}{5}$.

▶ **Example 18**   Solve $(x + 2)(x - 5) > 0$.

Solution. The solutions are those values of x for which the factors $(x + 2)$ and $(x - 5)$ have the same sign. From Figure 1.1.13 we see that the solutions are those x for which $x < -2$ or $x > 5$, since $(x + 2)(x - 5)$ must be positive (>0). Equivalently, the solution set is $(-\infty, -2) \cup (5, +\infty)$: see Figure 1.1.14.   ◀

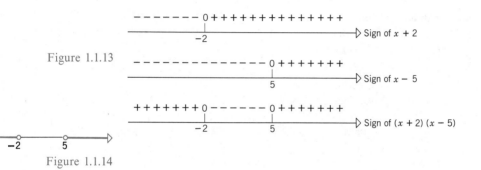

Figure 1.1.13

Figure 1.1.14

▶ **Example 19**   Solve

$$\frac{2x - 5}{x - 2} < 1$$

Solution.   Rewrite the inequality as

$$\frac{2x - 5}{x - 2} - 1 < 0 \qquad \text{[We subtracted 1 from both sides.]}$$

$$\frac{(2x - 5) - (x - 2)}{x - 2} < 0 \qquad \text{[We combined terms.]}$$

$$\frac{x - 3}{x - 2} < 0 \qquad \text{[We simplified.]}$$

The solutions are those values of x for which the quotient $(x - 3)/(x - 2)$ is negative, and this occurs if and only if the numerator and denominator have opposite signs. From Figure 1.1.15, we see that this is so if and only if x is in the interval $(2, 3)$: see Figure 1.1.16.   ◀

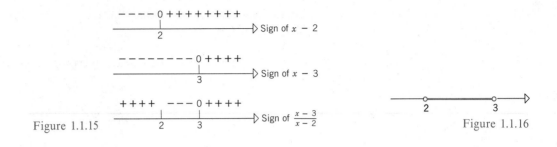

Figure 1.1.15   Figure 1.1.16

► Exercise Set 1.1

1. In each part, sketch on a coordinate line all values of x that satisfy the stated condition.
   (a) $x \leq 4$ (b) $x \geq -3$ (c) $-1 \leq x \leq 7$
   (d) $x^2 = 9$ (e) $x^2 \leq 9$ (f) $x^2 \geq 9$.

2. In parts (a)–(d), sketch on a coordinate line all values of x, if any, that satisfy the stated conditions.
   (a) $x > 4$ and $x \leq 8$
   (b) $x \leq 2$ or $x \geq 5$
   (c) $x > -2$ and $x \geq 3$
   (d) $x \leq 5$ and $x > 7$.

3. Among the terms, *integer, rational,* and *irrational,* which ones apply to the given number?
   (a) $-\frac{3}{4}$ (b) 0 (c) $\frac{24}{8}$
   (d) 0.25 (e) $-\sqrt{16}$ (f) $2^{1/2}$
   (g) $0.020202 \ldots$ (h) $7.000 \ldots$

4. Which of the terms, *integer, rational,* and *irrational,* apply to the given number?
   (a) $0.31311311131111 \ldots$ (b) $0.729999 \ldots$
   (c) $0.376237623762 \ldots$ (d) $17\frac{4}{5}$.

5. Which of the following are always correct if $a \leq b$?
   (a) $a - 3 \leq b - 3$ (b) $-a \leq -b$
   (c) $3 - a \leq 3 - b$ (d) $6a \leq 6b$
   (e) $a^2 \leq ab$ (f) $a^3 \leq a^2 b$.

6. Which of the following are always correct if $a \leq b$ and $c \leq d$?
   (a) $a + 2c \leq b + 2d$ (b) $a - 2c \leq b - 2d$
   (c) $a - 2c \geq b - 2d$.

7. The repeating decimal $0.137137137 \ldots$ can be expressed as a ratio of integers by writing

   $$x = 0.137137137 \ldots$$
   $$1000x = 137.137137137 \ldots$$

   and subtracting to obtain $999x = 137$ or $x = \frac{137}{999}$. Use this idea, where needed, to express the following decimals as ratios of integers.
   (a) $0.123123123 \ldots$ (b) $12.7777 \ldots$
   (c) $38.07818181 \ldots$ (d) $0.4296000 \ldots$

8. Show that the repeating decimal $0.99999 \ldots$ represents the number 1. Since $1.000 \ldots$ is also a decimal representation of 1, this problem shows that a real number can have two different decimal representations. [*Hint:* Use the technique of Exercise 7].

9. For what values of a are the following inequalities valid?
   (a) $a \leq a$ (b) $a < a$.

10. If $a \leq b$ and $b \leq a$, what can you say about a and b?

11. (a) If $a < b$ is true, does it follow that $a \leq b$ must also be true?
    (b) If $a \leq b$ is true, does it follow that $a < b$ must also be true?

12. In each part, list the elements in the set.
    (a) $\{x : x^2 - 5x = 0\}$
    (b) $\{x : x \text{ is an integer satisfying } -2 < x < 3\}$.

13. In each part, express the set in the notation $\{x : \underline{\hspace{1cm}}\}$.

(a) $\{1, 3, 5, 7, 9, \ldots\}$

(b) the set of even integers

(c) the set of irrational numbers

(d) $\{7, 8, 9, 10\}$.

14. Let $A = \{1, 2, 3\}$. Which of the following are equal to A?

(a) $\{0, 1, 2, 3\}$ (b) $\{3, 2, 1\}$

(c) $\{x : (x - 3)(x^2 - 3x + 2) = 0\}$.

15. Which of the following sets are empty?

(a) $\{x : x^2 + 1 = 0\}$

(b) $\{x : x^2 - 1 = 0\}$

(c) $\{x : x > 3 \text{ and } x < 3\}$

(d) $\{x : x \geq 3 \text{ and } x \leq 3\}$.

16. List all subsets of

(a) $\{a_1, a_2, a_3\}$ (b) $\varnothing$.

17. In Figure 1.1.17, let

S = the set of points inside the square

T = the set of points inside the triangle

C = the set of points inside the circle

and let a, b, and c be the points shown. Answer the following as true or false.

(a) $T \subset C$ (b) $T \subset S$ (c) $a \notin T$

(d) $a \notin S$

(e) $b \in T$ and $b \in C$

(f) $a \in C$ or $a \in T$

(g) $c \in T$ and $c \notin C$.

Figure 1.1.17

18. In each part, sketch the set on a coordinate line.

(a) $[-3, 2] \cup [1, 4]$

(b) $[4, 6] \cup [8, 11]$

(c) $(-4, 0) \cup (-5, 1)$

(d) $[2, 4) \cup (4, 7)$

(e) $(-2, 4) \cap (0, 5]$

(f) $[1, 2.3) \cup (1.4, \sqrt{2})$

(g) $(-\infty, -1) \cup (-3, +\infty)$

(h) $(-\infty, 5) \cap [0, +\infty)$.

In Exercises 19–40, solve the inequality and sketch the solution on a coordinate line.

19. $3x - 2 < 8$.

20. $\frac{1}{5}x + 6 \geq 14$.

21. $4 + 5x \leq 3x - 7$.

22. $2x - 1 > 11x + 9$.

23. $3 \leq 4 - 2x < 7$.

24. $-2 \geq 3 - 8x \geq -11$.

25. $\frac{x}{x - 3} < 4$.

26. $\frac{x}{8 - x} \geq -2$.

27. $\frac{3x + 1}{x - 2} < 1$.

28. $\frac{\frac{1}{2}x - 3}{4 + x} > 1$.

29. $\frac{4}{2 - x} \leq 1$.

30. $\frac{3}{x - 5} \leq 2$.

31. $x^2 > 9$.

32. $x^2 \leq 5$.

33. $(x - 4)(x + 2) > 0$.

34. $(x - 3)(x + 4) < 0$.

35. $x^2 - 9x + 20 \leq 0$.

36. $2 - 3x + x^2 \geq 0$.

37. $\frac{2}{x} < \frac{3}{x - 4}$.

38. $\frac{1}{x + 1} \geq \frac{3}{x - 2}$.

39. $x^3 - x^2 - x - 2 > 0$.

40. $x^3 - 3x + 2 \leq 0$.

41. Prove the following results about sums of rational and irrational numbers:

(a) rational + rational = rational

(b) rational + irrational = irrational.

42. Prove the following results about products of rational and irrational numbers:

(a) rational · rational = rational

(b) rational · irrational = irrational (provided the rational factor is nonzero).

43. Show that the sum or product of two irrational numbers can be rational or irrational.

44. Classify the following as rational or irrational and justify your conclusion.

(a) $3 + \pi$ (b) $\frac{3}{4}\sqrt{2}$

(c) $\sqrt{8}\,\sqrt{2}$ (d) $\sqrt{\pi}$.

(See Exercises 41 and 42.)

45. Prove: The average of two rational numbers is a rational number, but the average of two irrational numbers can be rational or irrational.

46. Can a rational number satisfy $10^z = 3$?

47. Solve: $8x^3 - 4x^2 - 2x + 1 < 0$.

48. Solve: $12x^3 - 20x^2 \geq -11x + 2$.

49. Prove: If a, b, c, and d are positive numbers such that $a < b$ and $c < d$, then $ac < bd$. (This result gives conditions under which inequalities can be "multiplied together.")

50. Show that rational numbers are represented by repeating decimals. [*Hint:* Examine the long division process that converts a ratio of integers into a decimal.]

1.2 ABSOLUTE VALUE

1.2.1 DEFINITION The ***absolute value*** or ***magnitude*** of a real number a is denoted by $|a|$ and is defined by

$$\begin{cases} |a| = a & \text{if} \quad a \geq 0 \\ |a| = -a & \text{if} \quad a < 0 \end{cases}$$

▶ Example 1

$$|5| = 5 \qquad \text{[since } 5 \geq 0\text{]}$$
$$|-\tfrac{4}{7}| = -(-\tfrac{4}{7}) = \tfrac{4}{7} \qquad \text{[since } -\tfrac{4}{7} < 0\text{]}$$
$$|0| = 0 \qquad \text{[since } 0 \geq 0\text{]} \qquad\qquad ◀$$

REMARK. Symbols such as $+a$ and $-a$ are deceptive, since it is tempting to conlude that $+a$ is positive and $-a$ is negative. However, this need not be so, since a itself can represent either a positive or negative number. In fact, if a itself is negative, then $-a$ is positive and $+a$ is negative. With this comment in mind, it should be evident that $|a| \geq 0$ for any number a.

Recall that a number whose square is a is called a ***square root*** of a. In algebra it is learned that every nonnegative real number has exactly one *nonnegative* square root; we denote this square root by $\sqrt{a}$. For example, the number 9 has two square roots, -3 and 3. Since 3 is the nonnegative square root, we have $\sqrt{9} = 3$.

REMARK. Readers who were previously taught to write $\sqrt{9} = \pm 3$ are advised to stop doing so, since it is incorrect.

It is another common error to write $\sqrt{a^2} = a$. Although this equality is correct when a is nonnegative, it is false for negative a. For example, if $a = -4$, then

$$\sqrt{a^2} = \sqrt{(-4)^2} = \sqrt{16} = 4 \neq a$$

A result that is correct for all a is given in the following theorem.

1.2.2 THEOREM *For any real number a,*

$$\sqrt{a^2} = |a|$$

Proof. Since $a^2 = (+a)^2 = (-a)^2$, the numbers $+a$ and $-a$ are square roots of a^2. If $a \geq 0$, then $+a$ is the nonnegative square root of a^2 and if $a < 0$, then $-a$ is the nonnegative square root of a^2. Since $\sqrt{a^2}$ denotes the nonnegative square root of a^2, we have

$$\begin{cases} \sqrt{a^2} = +a & \text{if} \quad a \geq 0 \\ \sqrt{a^2} = -a & \text{if} \quad a < 0 \end{cases}$$

That is, $\sqrt{a^2} = |a|$. ▮

Some basic properties of absolute value are listed in the following theorem.

1.2.3 THEOREM *If a and b are real numbers and n is an integer, then*

    (a) $-|a| \leq a \leq |a|$

    (b) $|ab| = |a|\,|b|$

    (c) $\left|\dfrac{a}{b}\right| = \dfrac{|a|}{|b|}$

    (d) $|a^n| = |a|^n$

We will prove parts *(a)* and *(b)* only.

Proof of (a). If $a \geq 0$, then we can write

$$-a \leq a \leq a \qquad \text{and} \qquad |a| = a$$

from which it follows that

$$-|a| \leq a \leq |a|$$

If $a < 0$, we can write

$$a \leq a \leq -a \qquad \text{and} \qquad -a = |a|$$

from which it again follows that

$$-|a| \leq a \leq |a|$$

Thus, statement *(a)* holds in all cases.

Proof (b). Using Theorem 1.2.2

$$|ab| = \sqrt{(ab)^2} = \sqrt{a^2b^2} = \sqrt{a^2}\sqrt{b^2} = |a|\,|b|. \quad ▊$$

REMARK. In words, parts (*b*) and (*c*) of this theorem state that *the absolute value of a product is the product of the absolute values and the absolute value of a ratio is the ratio of the absolute values.*

The notion of absolute value arises naturally in distance problems. On a coordinate line, let A and B be points with coordinates a and b. Because distance is always nonnegative, the distance d between A and B is

$$d = b - a$$

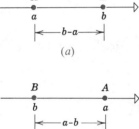

Figure 1.2.1

when B is to the right of A (Figure 1.2.1*a*), and

$$d = a - b$$

when B is to the left of A (Figure 1.2.1*b*).
In the first case, $b - a$ is positive, so we can write

$$d = b - a = |b - a|$$

and in the second case, $b - a$ is negative, so we can write

$$d = a - b = -(b - a) = |b - a|$$

Thus, regardless of whether B is to the right or left of A, the distance d between A and B is

$$d = |b - a|$$

This formula is useful when the relative positions of A and B are unknown.
For any real number b we can write

$$|b| = |b - 0|$$

Therefore, the absolute value of a number b can be interpreted geometrically as its distance from the origin on a coordinate line. For example, if $|b| = 5$, then b lies 5 units from the origin, that is, $b = 5$ or $b = -5$.

▶ Example 2   Solve $|x - 3| = 4$.

Solution 1 (geometric).   The solution consists of all x that are 4 units away from the point 3. There are two such x values, $x = 7$ and $x = -1$ (Figure 1.2.2).

Figure 1.2.2

Solution 2 (algebraic). Depending on whether $x - 3$ is positive or negative, the equation $|x - 3| = 4$ can be written as

$$x - 3 = 4 \quad \text{or} \quad -(x - 3) = 4$$

Solving these two equations gives

$$x = 7 \quad \text{and} \quad x = -1$$

which agrees with the solutions obtained geometrically. ◀

▶ **Example 3** Solve $|x - 3| < 4$.

Solution. The solution consists of all x whose distance from 3 is less than 4 units, that is, all x satisfying

$$-1 < x < 7$$

This is the interval $(-1, 7)$ shown in Figure 1.2.3. ◀

Figure 1.2.3

The following general result will be of importance in later sections.

1.2.4 THEOREM *For any real numbers x and a and any positive number k*

(a) $|x| < k$ *if and only if* $-k < x < k$

(b) $|x - a| < k$ *if and only if* $a - k < x < a + k$

Part (*a*) states that $|x| < k$ if and only if x is within k units of the origin (Figure 1.2.4*a*) and part (*b*) states that $|x - a| < k$ if and only if x is within k units of a (Figure 1.2.4*b*).

REMARK. Theorem 1.2.4 is also true if we replace $<$ by $\leq$.

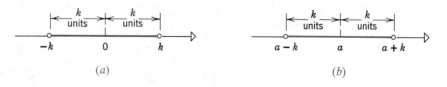

Figure 1.2.4   (*a*)   (*b*)

The following examples illustrate additional techniques for solving equations and inequalities involving absolute values.

▶ Example 4   Solve $|x + 4| > 2$.

Solution. The given inequality can be rewritten

$$|x - (-4)| > 2$$

Thus, the solution consists of all x whose distance from -4 is greater than 2 units. This is the set

$$(-\infty, -6) \cup (-2, +\infty)$$

shown in Figure 1.2.5.   ◀

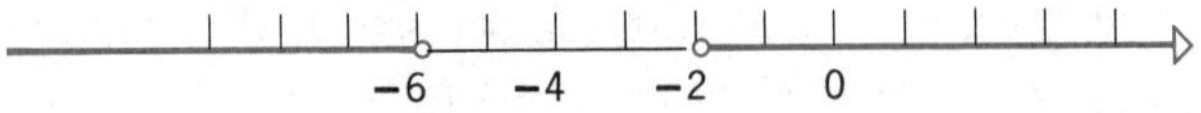

Figure 1.2.5

▶ Example 5   Solve $|3x - 2| = |5x + 4|$.

Solution. The equation will be satisfied if either

$$3x - 2 = 5x + 4 \quad \text{or} \quad 3x - 2 = -(5x + 4)$$

Solving the first equation gives $x = -3$ and solving the second equation gives $x = -\frac{1}{4}$. Thus, $x = -3$ and $x = -\frac{1}{4}$ are the solutions.   ◀

It is *not* generally true that $|a + b| = |a| + |b|$. For example, if $a = 2$ and $b = -3$, then $a + b = -1$, so that

$$|a + b| = |-1| = 1$$

whereas

$$|a| + |b| = |2| + |-3| = 2 + 3 = 5$$

The following theorem shows that *the absolute value of a sum is always less than or equal to the sum of the absolute values.*

1.2.5 THEOREM *For any real numbers, a and b,*
The Triangle Inequality

$$|a + b| \leq |a| + |b|$$

Proof. From Theorem 1.2.3a.

$$-|a| \leq a \leq |a| \quad \text{and} \quad -|b| \leq b \leq |b|$$

Adding these inequalities yields

$$-(|a| + |b|) \leq a + b \leq (|a| + |b|) \tag{1}$$

Using Theorem 1.2.4a with $\leq$ in place of $<$, with $x = a + b$, and with $k = |a| + |b|$, it follows from (1) that

$$|a + b| \leq |a| + |b| \quad \blacksquare$$

▶ **Exercise Set 1.2**

1. Compute $|x|$ if
   (a) $x = 7$ (b) $x = -\sqrt{2}$
   (c) $x = k^2$ (d) $x = -k^2$.

2. Rewrite $\sqrt{(x-6)^2}$ without using a square root or absolute value sign.

3. Verify $\sqrt{a^2} = |a|$ for $a = 7$ and $a = -7$.

4. Verify the inequalities $-|a| \leq a \leq |a|$ for $a = 2$ and $a = -5$.

5. Let A and B be points with coordinates a and b. In each part find the distance between A and B.
   (a) $a = 9, b = 7$ (b) $a = 2, b = 3$
   (c) $a = -8, b = 6$ (d) $a = \sqrt{2}, b = -3$
   (e) $a = -11, b = -4$ (f) $a = 0, b = -5$.

6. Is the equality $\sqrt{a^4} = a^2$ valid for all values of a? Explain.

7. Let A and B be points with coordinates a and b. In each part, use the given information to find b.
   (a) $a = -3$, B is to the left of A, and $|b - a| = 6$
   (b) $a = -2$, B is to the right of A, and $|b - a| = 9$
   (c) $a = 5$, $|b - a| = 7$, and $b > 0$.

8. Let E and F be points with coordinates e and f. In each part, determine whether E is to the left or to the right of F on a coordinate line.
   (a) $f - e = 4$ (b) $e - f = 4$
   (c) $f - e = -6$ (d) $e - f = -7$.

In Exercises 9–16 solve for x.

9. $|6x - 2| = 7$.

10. $|3 + 2x| = 11$.

11. $|6x - 7| = |3 + 2x|$.

12. $|4x + 5| = |8x - 3|$.

13. $|9x| - 11 = x$.

14. $2x - 7 = |x + 1|$.

15. $\left| \dfrac{x + 5}{2 - x} \right| = 6$.

16. $\left| \dfrac{x - 3}{x + 4} \right| = 5$.

In Exercises 17–28, solve for x and express the solution in terms of intervals.

17. $|x + 6| < 3$.

18. $|7 - x| \leq 5$.

19. $|5 - 2x| \geq 4$.

20. $|7x + 1| > 3$.

21. $|x + 3| < |x - 8|$.

22. $|3x| \leq |2x - 5|$.

23. $|4x| \geq |7 - 6x|$.

24. $|2x + 1| > |x - 5|$.

25. $\left| \dfrac{x - \frac{1}{2}}{x + \frac{1}{2}} \right| < 1$.

26. $\left| \dfrac{3 - 2x}{1 + x} \right| \leq 4$.

27. $\dfrac{1}{|x - 4|} < \dfrac{1}{|x + 7|}$.

28. $\dfrac{1}{|x - 3|} - \dfrac{1}{|x + 4|} \geq 0$.

29. For which values of α is $\sqrt{(\alpha^2 - 5\alpha + 6)^2} = \alpha^2 - 5\alpha + 6$?

30. Verify the triangle inequality $|a + b| \leq |a| + |b|$ (Theorem 1.2.5) for
    (a) $a = 3, b = 4$ (b) $a = -2, b = 6$
    (c) $a = -7, b = -8$ (d) $a = -4, b = 4$.

31. Under what conditions will equality hold in the triangle inequality, $|a + b| \leq |a| + |b|$ (Theorem 1.2.5)?

32. Solve the problem in Example 4 algebraically.

33. Prove: $|a - b| \leq |a| + |b|$.

34. Prove: $|a| - |b| \leq |a - b|$.

35. Prove: $||a| - |b|| \leq |a - b|$. [*Hint*: Use Exercise 34.]

36. Find the smallest value of M such that
    $$\left| \frac{1}{x} \right| \leq M \text{ for all } x \text{ in the interval } [2, 7].$$

37. Find the smallest value of M such that
    $$\left| \frac{1}{x + 7} \right| \leq M$$
    for all x in the interval $(-4, 2)$.

38. Use the triangle inequality to find a value of M such that

$$|x^3 - 2x + 1| \leq M$$

for all x in the interval $(-2, 3)$.

39. Find a value of M such that

$$\left| \frac{x + 3}{x - 3} \right| \leq M$$

for all x in the interval $[-\frac{3}{4}, \frac{1}{4}]$.

1.3   COORDINATE PLANES; DISTANCE; GRAPHS

Just as points on a line can be placed in one-to-one correspondence with the real numbers, so points in a plane can be placed in one-to-one correspondence with pairs of real numbers by using two perpendicular coordinate lines that intersect at their origins. For convenience, we make one of the lines horizontal with its positive direction to the right and the other line vertical with its positive direction up (Figure 1.3.1). The two lines are called *coordinate axes;* the horizontal line is called the *x-axis,* the vertical line is called the *y-axis,* and the coordinate axes together form what is called a *Cartesian coordinate system* or sometimes a *rectangular coordinate system.* The point of intersection of the coordinate axes is denoted by O and is called the *origin* of the coordinate system.

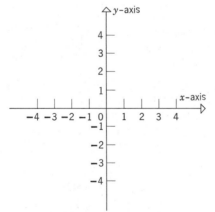

Figure 1.3.1

REMARK.   Throughout this text we will assume that the same unit of measurement is used on both coordinate axes (see Exercise 40).

A plane in which a rectangular coordinate system has been introduced is called a *coordinate plane* or an *xy-plane.* We will now show how to establish a

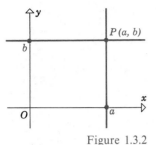

Figure 1.3.2

one-to-one correspondence between points in a coordinate plane and pairs of real numbers. If P is a point in a coordinate plane, then we draw two lines through P, one perpendicular to the x-axis and one perpendicular to the y-axis. If the first line intersects the x-axis at the point with coordinate a and the second line intersects the y-axis at the point with coordinate b, then we associate the pair (a, b) with the point P (Figure 1.3.2). The number a is called the **x-coordinate** or **abscissa** of P and the number b is called the **y-coordinate** or **ordinate** of P; we say that P is the point with **coordinates** (a, b) and denote the point by $P(a, b)$.

▶ Example 1 In Figure 1.3.3 we have located the points

$$P(2, 5), \qquad Q(-4, 3), \qquad R(-5, -2), \qquad S(4, -3) \qquad ◄$$

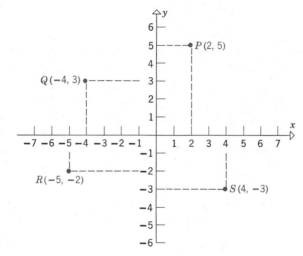

Figure 1.3.3

From the above construction, each point in a coordinate plane determines a unique pair of numbers. Conversely, starting with a pair of numbers (a, b) we can construct lines perpendicular to the x-axis and y-axis at the points with coordinates a and b, respectively; the intersection of these lines determines a unique point P in the plane whose coordinates are (a, b) (Figure 1.3.2). Thus, we have a one-to-one correspondence between pairs of real numbers and points in a coordinate plane.

REMARK. Since the order in which the members of a set are listed does not matter, the set $\{a, b\}$ and the set $\{b, a\}$ are the same. However, the pair of real numbers (a, b) and the pair of real numbers (b, a) represent different points (unless $a = b$), so that order is important. For this reason, a pair (a, b) of real numbers is often called an **ordered pair.**

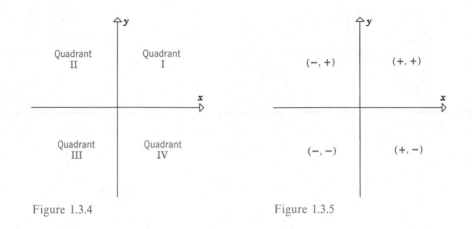

Figure 1.3.4   Figure 1.3.5

The coordinate axes divide the plane into four parts, called **quadrants.** These quadrants are numbered from one to four as shown in Figure 1.3.4. As shown in Figure 1.3.5 it is easy to determine the quadrant in which a point lies from the signs of its coordinates. A point with two positive coordinates $(+, +)$ lies in Quadrant I, a point with a negative x-coordinate and a positive y-coordinate $(-, +)$ lies in Quadrant II, and so on.

Occasionally, we will use letters other than x and y to label coordinate axes. Figure 1.3.6 shows a uv-plane and a tQ-plane. The first letter in the name of the plane refers to the horizontal axis and the second to the vertical axis.

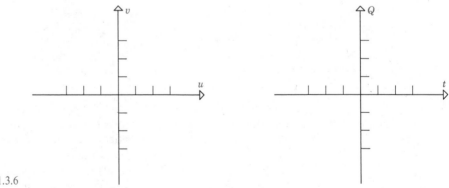

Figure 1.3.6

DISTANCE   In the last section we showed that the distance between points a and b on a coordinate line is $|b - a|$. It follows that the distance d between two points $A(x_1, y)$ and $B(x_2, y)$ on a horizontal line in the xy-plane is

$$d = |x_2 - x_1|$$

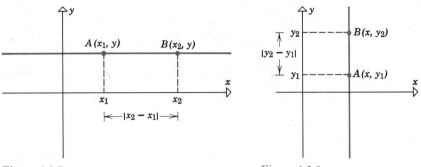

Figure 1.3.7   Figure 1.3.8

Figure 1.3.9

(Figure 1.3.7) and the distance d between two points $A(x, y_1)$ and $B(x, y_2)$ on a vertical line in the xy-plane is

$$d = |y_2 - y_1|$$

(See Figure 1.3.8.)

To find the distance d between any two points $P_1(x_1, y_1)$ and $P_2(x_2, y_2)$ in the xy-plane we apply the Theorem of Pythagoras to the triangle shown in Figure 1.3.9. This yields

$$d = \sqrt{|x_2 - x_1|^2 + |y_2 - y_1|^2}$$

Since $|x_2 - x_1|^2 = (x_2 - x_1)^2$ and $|y_2 - y_1|^2 = (y_2 - y_1)^2$, we are led to the following result.

1.3.1   THEOREM   *The distance d between two points (x_1, y_1) and (x_2, y_2) in a coordinate plane is given by*

$$d = \sqrt{(x_2 - x_1)^2 + (y_2 - y_1)^2} \tag{1}$$

▶ Example 2   The distance between $(-1, 2)$ and $(3, 4)$ is

$$d = \sqrt{(3 - (-1))^2 + (4 - 2)^2} = \sqrt{4^2 + 2^2} = \sqrt{20} \qquad ◀$$

If a and b are points on a coordinate line, then their average $\frac{1}{2}(a + b)$ is the midpoint of the line segment joining a and b (Figure 1.3.10a). In the xy-plane the midpoint of the line segment joining two points (x_1, y_1) and (x_2, y_2) is obtained by averaging the coordinates of the endpoints. More precisely:

$$\begin{bmatrix} \textit{The midpoint of the line segment} \\ \textit{joining } P_1(x_1, y_1) \textit{ and } P_2(x_2, y_2) \end{bmatrix} = \left(\frac{x_1 + x_2}{2}, \frac{y_1 + y_2}{2} \right) \qquad (2)$$

(See Figure 1.3.10*b*.) The proof is left as an exercise.

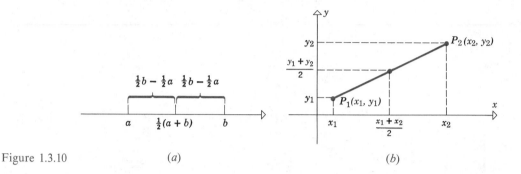

Figure 1.3.10   (*a*)   (*b*)

▶ **Example 3** From (2), the midpoint of the line segment joining the points $(3, -4)$ and $(7, 2)$ is

$$\left(\frac{3 + 7}{2}, \frac{-4 + 2}{2} \right) = (5, -1) \qquad ◀$$

GRAPHS   To see how rectangular coordinate systems enable us to describe equations geometrically, assume we have constructed a rectangular coordinate system and are given an equation involving only two variables, x and y; for example,

$$5xy = 2, \qquad x^2 + 2y^2 = 7, \qquad \text{or} \qquad y = \frac{1}{1 - x}$$

We define a **solution** of such an equation to be an ordered pair of real numbers (a, b) such that the equation is satisfied when we substitute $x = a$ and $y = b$.

▶ **Example 4** The pair $(3, 2)$ is a solution of

$$6x - 4y = 10$$

since the equation is satisfied when we substitute $x = 3$ and $y = 2$. However, the pair $(2, 0)$ is not a solution, since the equation is not satisfied when we substitute $x = 2$ and $y = 0$.   ◀

The set of all solutions of an equation is called its **solution set.** If we locate or **plot** all members of the solution set in a rectangular coordinate system, we obtain a set of points called the **graph** of the equation.

▶ Example 5   Sketch the graph of $y = x^2$.

Solution. The solution set of $y = x^2$ has infinitely many members, so that it is impossible to plot them all. However, some sample members of the solution set can be obtained by substituting some arbitrary x values into the right side of $y = x^2$ and solving for the associated values of y. Some typical computations are given in Table 1.3.1.

Table 1.3.1

x	0	1	2	3	-1	-2	-3
$y = x^2$	0	1	4	9	1	4	9
(x, y)	$(0, 0)$	$(1, 1)$	$(2, 4)$	$(3, 9)$	$(-1, 1)$	$(-2, 4)$	$(-3, 9)$

In Figure 1.3.11 the points listed in Table 1.3.1 are connected by a smooth curve. This curve approximates the graph of $y = x^2$.   ◀

REMARK. It is important to keep in mind that the sketch in Figure 1.3.11 is only an *approximation* to the graph of $y = x^2$ based on plotting a few points. It is conceivable that the smooth curve connecting the plotted points does not accurately describe the true graph. For example, isn't it logically possible that the graph of $y = x^2$ oscillates between the points we have plotted and looks like the dark blue curve in Figure 1.3.12? Although we might look for such oscillations by plotting additional points between those already obtained, we can never resolve the problem with certainty by point plotting, since we will never be sure how the true graph behaves *between* our plotted points. In later sections we will show how calculus can be used to determine the true shape of a graph.

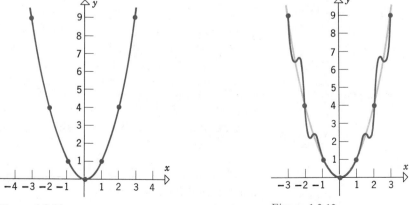

Figure 1.3.11   Figure 1.3.12

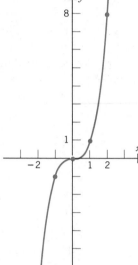

Figure 1.3.13

▶ Example 6   Sketch the graph of $y = x^3$.

Solution.   From Table 1.3.2 we obtain the curve in Figure 1.3.13.   ◀

Table 1.3.2

x	0	1	2	-1	-2
$y = x^3$	0	1	8	-1	-8
(x, y)	$(0, 0)$	$(1, 1)$	$(2, 8)$	$(-1, -1)$	$(-2, -8)$

▶ Example 7   Sketch the graph of $y = \sqrt{x}$.

Solution.   Since $\sqrt{x}$ is imaginary if $x < 0$, we can only plot points for which $x \geq 0$. From Table 1.3.3 we obtain the curve in Figure 1.3.14.   ◀

Table 1.3.3

x	0	1	2	3	4
$y = \sqrt{x}$	0	1	$\sqrt{2} \approx 1.4$	$\sqrt{3} \approx 1.7$	2
(x, y)	$(0, 0)$	$(1, 1)$	$(2, \sqrt{2})$	$(3, \sqrt{3})$	$(4, 2)$

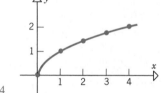

Figure 1.3.14

SYMMETRY   The work required to graph an equation can be reduced if the graph enjoys certain symmetry properties. A curve is called **symmetric about the x-axis** if for each point (x, y) on the curve, the point $(x, -y)$ is also on the curve (Figure 1.3.15a); geometrically, this makes the portion of curve below the x-axis the mirror image of the portion above. Similarly, a curve is **symmetric about the y-axis** if for each point (x, y) on the curve, the point $(-x, y)$ is also on the curve (Figure 1.3.15b). Finally, a curve is **symmetric about the origin** if for each point (x, y) on the curve, the point $(-x, -y)$ is also on the curve (Figure 1.3.15c); geometrically, this means that for each point P on the curve there is a companion point Q on the curve with origin bisecting the line segment joining P and Q.

Symmetries can often be detected from the equation of a curve. For example, the graph of

$$y = x^2 \tag{3}$$

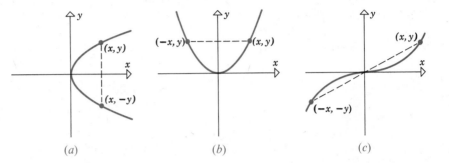

Figure 1.3.15

(a)   (b)   (c)

must be symmetric about the y-axis because for any point (x, y) whose coordinates satisfy (3), the coordinates of the point $(-x, y)$ also satisfy (3), since substituting these values in (3) yields

$$y = (-x)^2 \quad \text{or} \quad y = x^2.$$

(Figure 1.3.11). This discussion suggests the following symmetry tests.

1.3.2 THEOREM
Symmetry Tests

(a)   *A plane curve is symmetric about the y-axis if replacing x by $-x$ in its equation produces an equivalent equation.*

(b)   *A plane curve is symmetric about the x-axis if replacing y by $-y$ in its equation produces an equivalent equation.*

(c)   *A plane curve is symmetric about the origin if replacing x by $-x$ and y by $-y$ in its equation produces an equivalent equation.*

▶ **Example 8**   Sketch the graph of $x = y^2$.

Figure 1.3.16

Solution. If we solve $x = y^2$ for y in terms of x, we obtain two solutions, $y = \sqrt{x}$ and $y = -\sqrt{x}$. The graph of $y = \sqrt{x}$ is the portion of the curve $x = y^2$ that is above the x-axis (since $y = \sqrt{x} \geq 0$), and the graph of $y = -\sqrt{x}$ is the portion below (since $y = -\sqrt{x} \leq 0$). However, the curve $x = y^2$ is symmetric about the x-axis because substituting $-y$ for y yields $x = (-y)^2$, which is equivalent to the original equation. Thus, we need only graph $y = \sqrt{x}$ (see Figure 1.3.14) and reflect it about the x-axis to complete the graph (Figure 1.3.16). ◀

▶ Example 9 Sketch the graph of $y = 1/x$.

Solution. Because $1/x$ is undefined when x is zero, we can only plot points for which $x \neq 0$. From Table 1.3.4 we obtain the curve in Figure 1.3.17. Note that the curve is symmetric about the origin because replacing x by $-x$ and y by $-y$ yields

$$-y = \frac{1}{-x}$$

Table 1.3.4

x	$\frac{1}{3}$	$\frac{1}{2}$	1	2	3	$-\frac{1}{3}$	$-\frac{1}{2}$	-1	-2	-3
$y = 1/x$	3	2	1	$\frac{1}{2}$	$\frac{1}{3}$	-3	-2	-1	$-\frac{1}{2}$	$-\frac{1}{3}$
(x, y)	$(\frac{1}{3}, 3)$	$(\frac{1}{2}, 2)$	$(1, 1)$	$(2, \frac{1}{2})$	$(3, \frac{1}{3})$	$(-\frac{1}{3}, -3)$	$(-\frac{1}{2}, -2)$	$(-1, -1)$	$(-2, -\frac{1}{2})$	$(-3, -\frac{1}{3})$

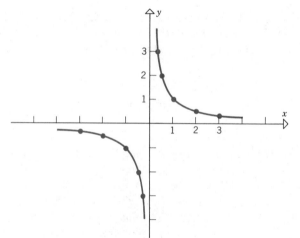

Figure 1.3.17

which is equivalent to $y = 1/x$. Thus, we could have plotted the points in the first quadrant and obtained those in the fourth quadrant by symmetry. ◀

We conclude this section with a catalog of basic curves (Figure 1.3.18) that we will encounter frequently. Some of these curves have been graphed in the examples in this section, while others have not. Try to convince yourself that the graphs are correct by plotting points in each case.

CATALOG OF BASIC CURVES

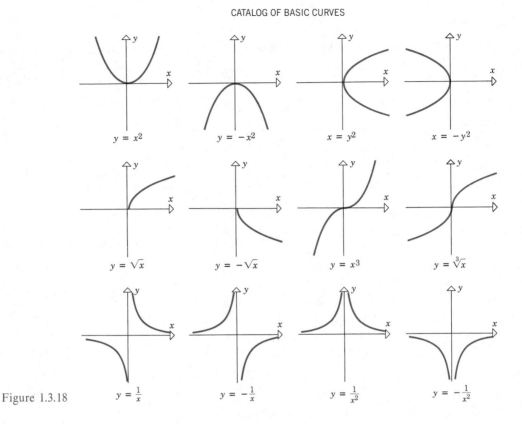

Figure 1.3.18

▶ Exercise Set 1.3

1. Draw a rectangular coordinate system and locate the points
   (a) $(3, 4)$ (b) $(-2, 5)$
   (c) $(-2.5, -3)$ (d) $(1.7, -2)$
   (e) $(0, -6)$ (f) $(4, 0)$.

In Exercises 2 and 3, draw a rectangular coordinate system and sketch the set of points whose coordinates (x, y) satisfy the given conditions.

2. (a) $x = 0$ (b) $y = 0$
   (c) $y < 0$ (d) $x \geq 1$ and $y \leq 2$
   (e) $x = 3$ (f) $|x| = 5$.

3. (a) $x = 2$ (b) $y = -3$ (c) $x \geq 0$
   (d) $y = x$ (e) $y \geq x$ (f) $|x| \geq 1$.

4. Find the center of the square whose vertices are $(1, 1)$, $(5, 1)$, $(5, -3)$, and $(1, -3)$.

5. Find the fourth vertex of the rectangle, three of whose vertices are $(-1, 4)$, $(6, 4)$, and $(-1, 9)$.

In Exercises 6 and 7, the points lie on a horizontal or vertical line. Determine whether the line is horizontal or vertical, and find the distance between the points.

6. (a) $A(9, 2)$, $B(7, 2)$
   (b) $A(2, -6)$, $B(3, -6)$
   (c) $A(-8, 1)$, $B(6, 1)$.

7. (a) $A(-4, \sqrt{2})$, $B(-4, -3)$
   (b) $A(3, -11)$, $B(3, -4)$
   (c) $A(0, 0)$, $B(0, -5)$.

In Exercises 8–11, find
(a) the distance between A and B
(b) the midpoint of the line segment joining A and B.

8. $A(2, 5)$, $B(-1, 1)$.

9. $A(7, 1)$, $B(1, 9)$.

10. $A(2, 0)$, $B(-3, 6)$.

11. $A(-2, -6)$, $B(-7, -4)$.

12. Show that the points $(-1, 7)$, $(5, 4)$, $(2, -2)$, and $(-4, 1)$ are vertices of a square.

13. Show that the triangle with vertices $(5, -2)$, $(6, 5)$, $(2, 2)$ is isosceles.

14. Show that $(1, 3)$, $(4, 2)$, and $(-2, -6)$ are vertices of a right triangle and specify the vertex at which the right angle occurs.

15. Show that $(0, -2)$, $(-4, 8)$, and $(3, 1)$ lie on a circle with center $(-2, 3)$.

16. Show that for all values of t the point $(t, 2t - 6)$ is equidistant from $(0, 4)$ and $(8, 0)$.

17. In each part determine if the given ordered pair (x, y) is a solution of $x^2 - 2x + y = 4$.
    (a) $(0, 4)$ (b) $(-3, 7)$
    (c) $(\frac{1}{2}, \frac{19}{4})$ (d) $(1 + \sqrt{5 - t}, t)$.

In Exercises 18 and 19, determine whether the graph is symmetric about the x-axis, the y-axis, or the origin.

18. (a) $x = 5y^2 + 9$ 19. (a) $x^4 = 2y^3 + y$
    (b) $x^2 - 2y^2 = 3$
    (c) $xy = 5$ (b) $y = \dfrac{x}{3 + x^2}$
                  (c) $y^2 = |x| - 5$.

In Exercises 20–29, sketch the graph of the equation. (A hand calculator will be helpful in some of these problems.)

20. $y = 2x - 3$.   21. $y = 6 - x$.
22. $y = 1 + x^2$.   23. $y = 4 - x^2$.
24. $y = -\sqrt{x + 1}$.   25. $y = \sqrt{x - 4}$.

26. $y = |x|$.   27. $y = |x - 3|$.
28. $xy = -1$.   29. $x^2y = 2$.

In Exercises 30 and 31, sketch the portion of the graph in the first quadrant, and use symmetry to complete the rest of the graph. (A hand calculator will be helpful.)

30. $9x^2 + 4y^2 = 36$.   31. $4x^2 + 16y^2 = 16$.

32. Sketch the graph of $y^2 = 3x$ and explain how this graph is related to the graphs of $y = \sqrt{3x}$ and $y = -\sqrt{3x}$.

33. Sketch the graph of $(x - y)(x + y) = 0$ and explain how it is related to the graphs of $x - y = 0$ and $x + y = 0$.

34. Graph $F = \frac{9}{5}C + 32$ in a CF-coordinate system.

35. Graph $u = 3v^2$ in a uv-coordinate system.

36. Graph $Y = 4X + 5$ in a YX-coordinate system.

37. Find an equation whose graph is the perpendicular bisector of the line segment connecting $(-2, 1)$ and $(4, -3)$.

38. Use the distance formula in Theorem 1.3.1 to prove that $(1, 1)$, $(-2, -8)$, and $(4, 10)$ lie on a straight line.

39. Find k, given that $(2, k)$ is equidistant from $(3, 7)$ and $(9, 1)$.

40. Where in this section did we use the fact that the same unit of measure was used on both coordinate axes?

41. Prove: The midpoint of the hypotenuse of a right triangle is equidistant from the three vertices. [*Hint:* Let a and b denote the lengths of the sides and introduce coordinate axes so that the vertices are $(0, 0)$, $(a, 0)$, and $(0, b)$.]

42. Use Theorem 1.3.2 to show that a graph which is symmetric about the x-axis and y-axis must be symmetric about the origin. Give an example to show that the converse is not true.

1.4 SLOPE OF A LINE

Before starting this section, readers who need to review trigonometry are advised to read Unit 1 of the trigonometry review in Appendix 1.

In this section we will discuss ways to measure the "steepness" or "slope" of a line in the plane. The ideas we develop here will be important when we discuss equations and graphs of straight lines.

Consider a particle moving left to right along a *nonvertical* line segment from a point $P_1(x_1, y_1)$ to a point $P_2(x_2, y_2)$. As shown in Figure 1.4.1, the particle moves $y_2 - y_1$ units in the y-direction as it travels $x_2 - x_1$ units in the x-direction. The vertical change $y_2 - y_1$ is called the **rise,** and the horizontal change $x_2 - x_1$ the **run.**

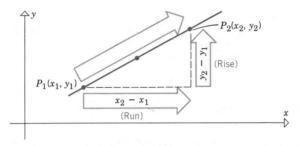

Figure 1.4.1

The rise over the run is called the *slope* of the line segment and is denoted by m.

1.4.1 DEFINITION   If $P_1(x_1, y_1)$ and $P_2(x_2, y_2)$ are the endpoints of a nonvertical line segment, then the **slope** m of the line segment is defined by

$$m = \frac{\text{rise}}{\text{run}} = \frac{y_2 - y_1}{x_2 - x_1} \qquad (1)$$

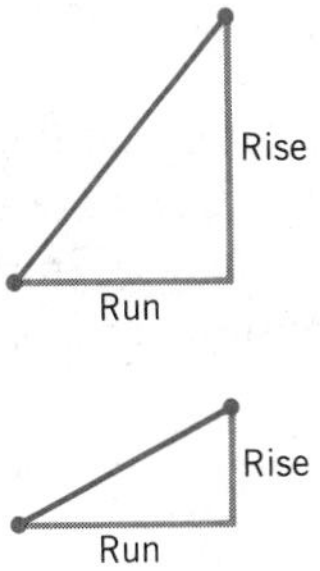

Rise

Run

Rise

Run

Figure 1.4.2

For two line segments with the same run, the more steeply inclined segment will have the greater rise and therefore the greater slope (Figure 1.4.2). Thus, the slope m is a numerical measure of steepness.

Note that Definition 1.4.1 does not apply to vertical line segments. For such segments we would have $x_2 = x_1$, so (1) would involve a division by zero. The slope of a vertical line segment is **undefined.**

REMARK.   In light of the equality

$$\frac{y_2 - y_1}{x_2 - x_1} = \frac{y_1 - y_2}{x_1 - x_2}$$

it does not matter which endpoint is labeled P_1 and which is labeled P_2 when we apply (1) to compute the slope of a line segment.

▶ **Example 1** Let L be the line segment connecting $(6, 2)$ and $(9, 8)$. Label $(6, 2)$ as P_1 and $(9, 8)$ as P_2; then from (1) the slope is

$$m = \frac{8 - 2}{9 - 6} = 2$$

Similarly, the slope of the line segment connecting the points $(2, 9)$ and $(4, 3)$ is

$$m = \frac{3 - 9}{4 - 2} = -3$$ ◀

It is clear geometrically that two line segments on the same straight line will be equally "steep," and consequently have the same slope. This is the content of the following theorem.

1.4.2 THEOREM *On a nonvertical line, all line segments have the same slope.*

Proof. Let $P_1(x_1, y_1)$ and $P_2(x_2, y_2)$ be distinct points on a nonvertical line L and let $P_1'(x_1', y_1')$ and $P_2'(x_2', y_2')$ be another pair of distinct points on L. We will show that the slope

$$m = \frac{y_2 - y_1}{x_2 - x_1}$$

of the line segment joining P_1 and P_2 is equal to the slope

$$m' = \frac{y_2' - y_1'}{x_2' - x_1'}$$

of the line segment joining P_1' and P_2'. (We will assume the points are ordered as in Figure 1.4.3. The proofs of the remaining cases are similar and will be omitted for brevity.) The triangles $P_1 Q P_2$ and $P_1' Q' P_2'$ in Figure 1.4.3 are similar, so that the lengths of corresponding sides are proportional. Therefore,

$$\frac{y_2 - y_1}{x_2 - x_1} = \frac{y_2' - y_1'}{x_2' - x_1'}$$

or, equivalently,

$$m = m' \quad \blacksquare$$

We define the **slope** of a nonvertical line L to be the common slope of all line segments on L. The slope of a vertical line is **undefined.** Speaking informally, some people say that a vertical line has **infinite slope.**

Figure 1.4.3

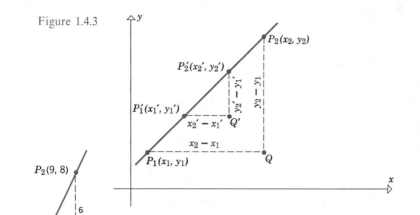

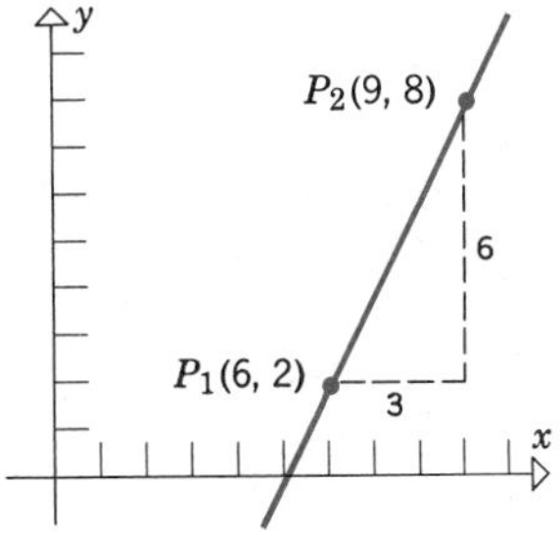

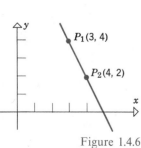

Figure 1.4.4

The slope of a line has a useful geometric interpretation, which can be seen by rewriting (1) as

$$y_2 - y_1 = m(x_2 - x_1) \tag{2}$$

It follows from (2) that as we travel left to right along a nonvertical line L, the rise or change in y is proportional to the run or change in x, and the slope m is the constant of proportionality. For this reason, the slope m is said to measure the *rate* at which y changes with x along L.

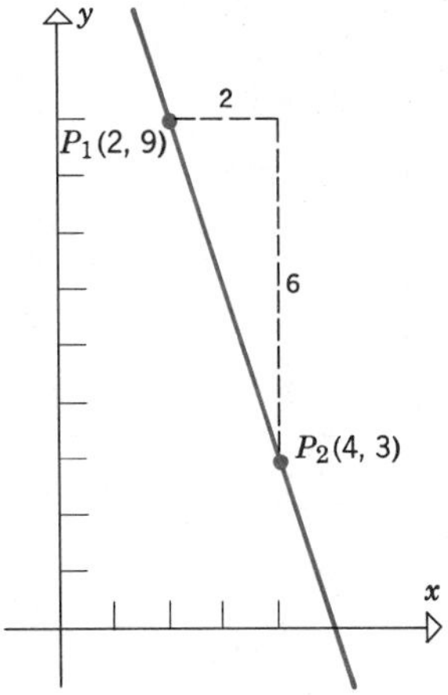

▶ **Example 2** In Example 1, we showed that the line determined by the points $P_1(6, 2)$ and $P_2(9, 8)$ has slope $m = 2$. This means that a particle traveling left to right along this line *rises* two units for every unit it moves in the positive x-direction (Figure 1.4.4). ◀

▶ **Example 3** In Example 1, we showed that the line determined by the points $P_1(2, 9)$ and $P_2(4, 3)$ has slope $m = -3$. This means that a particle traveling left to right along this line *falls* three units for every unit it moves in the positive x-direction (Figure 1.4.5). ◀

Figure 1.4.5

▶ **Example 4** Construct a line through $(3, 4)$ with slope -2.

Solution. A particle traveling left to right along the line *falls* 2 units for every unit it travels in the positive x-direction. Thus, if it begins at $P_1(3, 4)$ and travels so that x increases by one unit, it will end up at $P_2(4, 2)$. The line through P_1 and P_2 is the one we want (Figure 1.4.6). ◀

In Figure 1.4.7 we have sketched some lines with different slopes. Observe that lines with positive slope are inclined upward to the right, while lines with negative slope are inclined downward to the right.

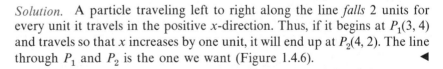

Figure 1.4.6

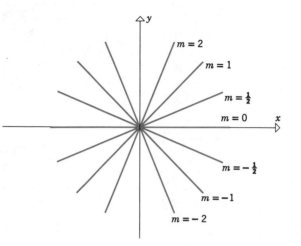

Figure 1.4.7

The slope of a line is related to the angle the line makes with the positive x-axis. To establish this relationship we need the following definition.

1.4.3   DEFINITION   For a line L not parallel to the x-axis, the ***angle of inclination*** is the smallest angle ϕ measured counterclockwise from the direction of the positive x-axis to L (Figure 1.4.8). For a line parallel the x-axis, we take $\phi = 0$.

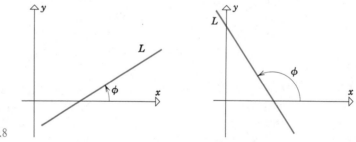

Figure 1.4.8

REMARK.   In degree measure the angle of inclination satisfies $0° \leq \phi < 180°$ and in radian measure it satisfies $0 \leq \phi < \pi$.

The following theorem relates the slope of a line to its angle of inclination.

1.4.4   THEOREM   *For a line not parallel to the y-axis, the slope and angle of inclination are related by*

$$m = \tan \phi \qquad\qquad\qquad (3)$$

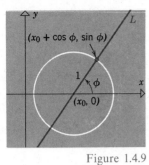

Figure 1.4.9

Proof. If the line is horizontal, then $m = 0$ and $\phi = 0°$. Thus,

$$\tan \phi = \tan 0 = 0 = m$$

so (3) holds in this case. If the line is not horizontal, let $(x_0, 0)$ be its point of intersection with the x-axis, and construct a circle of radius 1 centered at this point. Because the angle of inclination of the line is ϕ, the line will intersect this circle at the point $(x_0 + \cos \phi, \sin \phi)$ (Figure 1.4.9). (This follows from Theorem 3 in Unit 1 of the trigonometry review in Appendix 1.)

Since the points $(x_0, 0)$ and $(x_0 + \cos \phi, \sin \phi)$ lie on L, we can use them to compute the slope m. This gives

$$m = \frac{\sin \phi - 0}{(x_0 + \cos \phi) - x_0} = \frac{\sin \phi}{\cos \phi} = \tan \phi$$

which proves (3). ▨

REMARK. If the line L is parallel to the y-axis, then $\phi = \frac{1}{2}\pi$ and $\tan \phi$ in (3) is undefined. This agrees with the fact that the slope m is also undefined in this case.

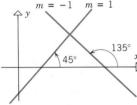

Figure 1.4.10

▶ **Example 5** Find the angle of inclination for a line of slope $m = 1$ and also for a line of slope $m = -1$.

Solution. If $m = 1$, then from (3), $\tan \phi = 1$ so that $\phi = \pi/4$ (or in degree measure $\phi = 45°$). If $m = -1$, then from (3), $\tan \phi = -1$; from this equality and the fact that $0 \leq \phi < \pi$ we obtain $\phi = 3\pi/4$ or, in degree measure, $\phi = 135°$ (Figure 1.4.10). ◀

As a consequence of Theorem 1.4.4, we obtain the following basic result.

1.4.5 THEOREM *Two nonvertical lines are parallel if and only if they have the same slope.*

Proof. If L_1 and L_2 are nonvertical parallel lines, then their angles of inclination ϕ_1 and ϕ_2 are equal, since two parallel lines cut by a transversal have equal corresponding angles (Figure 1.4.11). Thus,

$$\text{slope } L_1 = \tan \phi_1 = \tan \phi_2 = \text{slope } L_2$$

Conversely, if L_1 and L_2 have the same slope m, then

$$m = \tan \phi_1 = \tan \phi_2 \qquad (4)$$

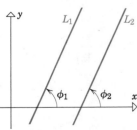

Figure 1.4.11

Because $0 \leq \phi_1 < \pi$ and $0 \leq \phi_2 < \pi$, it follows from (4) that ϕ_1 and ϕ_2 are equal. Thus, L_1 and L_2 are parallel (Figure 1.4.11). ▨

The next theorem shows how slopes can be used to determine whether two lines are perpendicular.

1.4.6 THEOREM *Two nonvertical lines are perpendicular if and only if the product of their slopes is* −1; *that is, lines with slopes m_1 and m_2 are perpendicular if and only if*

$$m_2 = -\frac{1}{m_1} \tag{5}$$

Proof. Suppose two nonvertical perpendicular lines, L_1 and L_2, have angles of inclination ϕ_1 and ϕ_2 and slopes m_1 and m_2, respectively. Assume L_1 has the smaller angle of inclination (Figure 1.4.12), so that

$$\phi_2 = \phi_1 + \frac{\pi}{2}$$

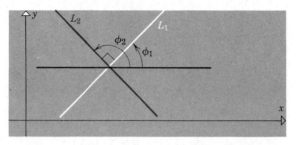

Figure 1.4.12

Thus,

$$m_2 = \tan \phi_2 = \tan\left(\phi_1 + \frac{\pi}{2}\right) = \frac{\sin\left(\phi_1 + \frac{\pi}{2}\right)}{\cos\left(\phi_1 + \frac{\pi}{2}\right)}$$

$$= \frac{\sin \phi_1 \cos \frac{\pi}{2} + \cos \phi_1 \sin \frac{\pi}{2}}{\cos \phi_1 \cos \frac{\pi}{2} - \sin \phi_1 \sin \frac{\pi}{2}}$$

$$= -\frac{\cos \phi_1}{\sin \phi_1} = -\frac{1}{\sin \phi_1 / \cos \phi_1} = -\frac{1}{\tan \phi_1} = -\frac{1}{m_1}$$

which establishes (5). The proof of the converse, that is, that L_1 and L_2 are perpendicular if (5) holds, is discussed in Exercise 29. ■

▶ **Example 6** Use slopes to show that the points $A(1, 3)$, $B(3, 7)$, and $C(7, 5)$ are vertices of a right triangle.

Solution. The line through A and B has slope

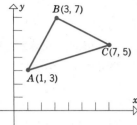

Figure 1.4.13

$$m_1 = \frac{7 - 3}{3 - 1} = 2$$

and the line through B and C has slope

$$m_2 = \frac{5 - 7}{7 - 3} = -\frac{1}{2}$$

Since $m_1 m_2 = -1$, the line through A and B is perpendicular to the line through B and C; thus, ABC is a right triangle (Figure 1.4.13). ◀

▶ Exercise Set 1.4

1. Find the slope of the line through
   (a) $(-1, 2)$ and $(3, 4)$
   (b) $(5, 3)$ and $(7, 1)$
   (c) $(4, \sqrt{2})$ and $(-3, \sqrt{2})$
   (d) $(-2, -6)$ and $(-2, 12)$.

In Exercises 2 and 3, find the slope of the line whose angle of inclination is given. (You should be able to solve these problems without tables or a calculator.)

2. (a) $45°$ (b) $\dfrac{2\pi}{3}$ (c) $30°$.

3. (a) $\dfrac{\pi}{6}$ (b) $135°$ (c) $60°$.

In Exercises 4 and 5, use a calculator or the trigonometric tables in Appendix 3, where necessary, to find (to the nearest degree) the angle of inclination of a line with the given slope.

4. (a) $m = \frac{1}{2}$ (b) $m = -1$ (c) $m = 2$
   (d) $m = -57$.

5. (a) $m = -\frac{1}{2}$ (b) $m = 1$ (c) $m = -2$
   (d) $m = 57$.

6. Draw the line through $(4, 2)$ with slope
   (a) $m = 3$ (b) $m = -2$ (c) $m = -\frac{3}{4}$.

7. Draw the line through $(-1, -2)$ with slope
   (a) $m = \frac{2}{5}$ (b) $m = -1$ (c) $m = \sqrt{2}$.

8. Let L be a line with slope $m = 2$. Determine whether the given line L' is parallel to L, perpendicular to L, or neither.
   (a) L' is the line through $(2, 4)$ and $(4, 8)$
   (b) L' is the line through $(2, 4)$ and $(6, 2)$
   (c) L' is the line through $(1, 5)$ and $(2, -3)$.

9. Let L be a line with slope $m = -3$. Determine whether the given line L' is parallel to L, perpendicular to L, or neither.
   (a) L' is the line through $(1, 8)$ and $(2, 5)$
   (b) L' is the line through $(6, 5)$ and $(3, 4)$
   (c) L' is the line through $(1, 0)$ and $(-2, 1)$.

10. A particle, initially at $(7, 5)$, moves along a line of slope $m = -2$ to a new position (x, y).
    (a) Find y if $x = 9$.   (b) Find x if $y = 12$.

11. A particle, initially at $(1, 2)$, moves along a line of slope $m = 3$ to a new position (x, y).
    (a) Find y if $x = 5$.   (b) Find x if $y = -2$.

12. Given that the point $(k, 4)$ is on the line through $(1, 5)$ and $(2, -3)$, find k.

13. Let the point $(3, k)$ lie on the line of slope $m = 5$ through $(-2, 4)$; find k.

14. Find the slopes of the sides of the triangle with vertices $(-1, 2)$, $(6, 5)$, and $(2, 7)$.

15. Use slopes to determine whether the given points lie on a line.
    (a)   (1, 1), (−2, −5), and (0, −1)
    (b)   (−2, 4), (0, 2), and (1, 5).

16. An equilateral triangle has one vertex at the origin, another on the x-axis, and the third in the first quadrant. Find the slopes of its sides.

17. Show that if (x, y) lies on the line of slope $m = 2$ passing through $(0, 3)$, then $y = 2x + 3$.

18. Use slopes to show that $(3, 1)$, $(6, 3)$, and $(2, 9)$ are vertices of a right triangle.

19. Use slopes to show that $(3, −1)$, $(6, 4)$, $(−3, 2)$, and $(−6, −3)$ are vertices of a parallelogram.

20. Find x and y if the line through $(0, 0)$ and (x, y) has slope $\frac{1}{2}$, and the line through (x, y) and $(7, 5)$ has slope 2.

21. Given two intersecting lines, let L_2 be the line with the larger angle of inclination ϕ_2, and let L_1 be the line with the smaller angle of inclination ϕ_1. We define the **angle θ between L_1 and L_2** by

    $$\theta = \phi_2 - \phi_1$$

    (a)   Prove geometrically that θ is the angle pictured in Figure 1.4.14. [*Remark:* The angle θ is the smallest positive angle through which L_1 can be rotated until it coincides with L_2.]

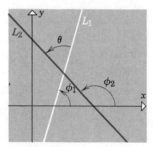

Figure 1.4.14

(b)   Prove: If L_1 and L_2 are not perpendicular, then

$$\tan \theta = \frac{m_2 - m_1}{1 + m_1 m_2}$$

22. Use the result of Exercise 21 to find the tangent of the angle between the lines whose slopes are
    (a)   1 and 3   (b)   −2 and 4
    (c) $-\frac{1}{2}$ and $-\frac{2}{3}$.

23. Use the result of Exercise 21 to find the tangent of the angle between the lines whose slopes are
    (a) $\frac{4}{5}$ and $\frac{1}{3}$ (b)   −0.7 and 5
    (c)   −6 and −2.

24. Use the result of Exercise 21 and a hand calculator or the trigonometric tables in Appendix 3 to find, to the nearest degree, the angle between the lines whose slopes are given in Exercise 22.

25. Repeat the directions of Exercise 24 for the lines whose slopes are given in Exercise 23.

26. Use the result of Exercise 21 to find, to the nearest degree, the interior angles of the triangle whose vertices are $(4, −3)$, $(−3, −1)$, and $(6, 6)$.

27. Use the result of Exercise 21 to find the slope of the line that bisects the angle A of the triangle whose vertices are $A(0, 2)$, $B(−8, 8)$, and $C(8, 6)$.

28. Let L be a line of slope $m = −2$. Use the result of Exercise 21 to find the slope of a line K such that the angle between K and L is $45°$. (Two solutions.)

29. Complete the proof of Theorem 1.4.6 by showing: If L_1 and L_2 are lines whose slopes satisfy $m_1 m_2 = −1$, then L_1 and L_2 are perpendicular. [*Hint:* One slope is positive and the other negative; if $m_1 > 0$ and $m_2 < 0$, then $0 < \phi_1 < \frac{1}{2}\pi$ and $\frac{1}{2}\pi < \phi_2 < \pi$. Use this and the identity $\tan(\phi_1 + \frac{1}{2}\pi) = −1/\tan \phi_1$.]

1.5 EQUATIONS OF STRAIGHT LINES

In this section we will be concerned with recognizing those equations whose graphs are straight lines and finding equations for lines specified geometrically.

VERTICAL LINES   A line parallel to the y-axis meets the x-axis at some point $(a, 0)$. Every point on this line has an x-coordinate of a and conversely every point with an x-coordinate of a lies on the line (Figure 1.5.1). Thus,

1.5.1 THEOREM   *The vertical line passing through $(a, 0)$ is represented by the equation*

$$x = a \tag{1}$$

▶ **Example 1**   The graph of $x = -2$ is the vertical line passing through $(-2, 0)$. ◀

LINES DETERMINED
BY POINT AND SLOPE

A nonvertical line can be determined by specifying a point on the line and the slope; the slope determines how the line is tilted and the point pins down the location. Let us try to find an equation for the line L passing through $P_1(x_1, y_1)$ and having slope m.

If $P(x, y)$ is any point on L, other than P_1, then the slope of L can be obtained from the points $P(x, y)$ and $P_1(x_1, y_1)$; this gives

$$m = \frac{y - y_1}{x - x_1}$$

which can be rewritten as

$$y - y_1 = m(x - x_1) \tag{2}$$

With the possible exception of (x_1, y_1), we have shown that every point on L satisfies (2). But $x = x_1, y = y_1$ satisfies (2), so that all points on L satisfy (2). We leave it as an exercise to show that every point satisfying (2) lies on L. To summarize:

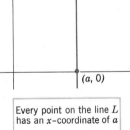

Every point on the line L has an x-coordinate of a

Figure 1.5.1

1.5.2 THEOREM   *The line passing through $P_1(x_1, y_1)$ and having slope m is given by the equation*

$$y - y_1 = m(x - x_1) \tag{3}$$

*This is called the **point-slope form** of the line.*

▶ **Example 2** Find the point-slope form of the line through $(4, -3)$ with slope 5.

Solution. Substituting the values $x_1 = 4$, $y_1 = -3$, and $m = 5$ in (3) yields the point-slope form $y + 3 = 5(x - 4)$. ◀

LINES DETERMINED
BY SLOPE AND
y-INTERCEPT

A nonvertical line crosses the y-axis at some point $(0, b)$. If we use this point in the point-slope form of its equation, we obtain

$$y - b = m(x - 0)$$

which we can rewrite as

$$y = mx + b$$

The number b in this equation is called the **y-intercept** of the line; it is the y-coordinate of the point where the line crosses the y-axis.

To summarize:

1.5.3 THEOREM *The line with y-intercept b and slope m is given by the equation*

$$y = mx + b \tag{4}$$

*This is called the **slope-intercept form** of the line.*

REMARK. The slope-intercept form is of special importance because it expresses y in terms of x.

▶ **Example 3** Comparing $y = 4x + 7$ to (4), we have $m = 4$ and $b = 7$, so that the equation represents a line crossing the y-axis at $(0, 7)$ with slope 4. ◀

HORIZONTAL LINES By definition, the angle of inclination of a horizontal line is $\phi = 0$. Thus, the slope of a horizontal line is

$$m = \tan \phi = \tan 0 = 0$$

Substituting $m = 0$ in (4) yields the following result.

1.5.4 THEOREM *The horizontal line passing through $(0, b)$ is represented by the equation*

$$y = b \tag{5}$$

▶ **Example 4** The equation $y = -7$ represents the horizontal line passing through $(0, -7)$. ◀

LINES DETERMINED
BY TWO POINTS

If $P_1(x_1, y_1)$ and $P_2(x_2, y_2)$ are distinct points on a nonvertical line, then the slope of the line is

$$m = \frac{y_2 - y_1}{x_2 - x_1}$$

Substituting this expression in (3) we obtain the following result.

1.5.5 THEOREM *The nonvertical line determined by the points $P_1(x_1, y_1)$ and $P_2(x_2, y_2)$ can be represented by the equation*

$$y - y_1 = \frac{y_2 - y_1}{x_2 - x_1}(x - x_1) \tag{6}$$

*This is called the **two-point form** of the line.*

▶ **Example 5** Find the slope-intercept form of the line passing through $(3, 4)$ and $(2, -5)$.

Solution. Letting $(x_1, y_1) = (3, 4)$ and $(x_2, y_2) = (2, -5)$ and substituting in (6), we obtain the two-point form

$$y - 4 = \frac{-5 - 4}{2 - 3}(x - 3)$$

which can be written $y - 4 = 9(x - 3)$. Solving for y yields the slope-intercept form

$$y = 9x - 23$$ ◀

THE GENERAL
EQUATION OF A LINE

An equation expressible in the form

$$Ax + By + C = 0 \tag{7}$$

where A, B, and C are constants and A and B are not both zero, is called a ***first-degree equation*** in x and y.

▶ **Example 6** Comparing $7x + 2y - 3 = 0$ to (7), we see that this is a first-degree equation in x and y with $A = 7$, $B = 2$, and $C = -3$. ◀

The following theorem shows that the first-degree equations in x and y are precisely the equations whose graphs in the xy-plane are straight lines.

1.5.6 THEOREM *Every first-degree equation in x and y has a straight line as its graph and, conversely, every straight line can be represented by a first-degree equation in x and y.*

Proof. Let

$$Ax + By + C = 0 \tag{8}$$

be any first-degree equation in x and y. To prove that the graph of this equation is a straight line, we distinguish between two cases, $B = 0$ and $B \neq 0$.

If $B \neq 0$, we can solve (8) for y in terms of x to obtain

$$y = -\frac{A}{B}x - \frac{C}{B}$$

But this is the slope-intercept form (4) of the line with

$$m = -\frac{A}{B} \quad \text{and} \quad b = -\frac{C}{B}$$

If $B = 0$, then $A \neq 0$, since A and B are not both zero in a first-degree equation in x and y. Thus, (8) reduces to $Ax + C = 0$, which can be rewritten as

$$x = -\frac{C}{A}$$

But this is the equation of a line parallel to the y-axis. Thus, in either case, the graph of (8) is a straight line.

Conversely, consider any straight line in the xy-plane. If the line is vertical, then it has an equation of the form $x = a$. This can be rewritten as $x - a = 0$, which is of form (7) with

$$A = 1, \quad B = 0, \quad \text{and } C = -a$$

If the line is not vertical, it can be expressed in slope-intercept form

$$y = mx + b$$

This can be rewritten as

$$mx - y + b = 0$$

which is of form (7) with

$$A = m, \qquad B = -1, \qquad \text{and } C = b$$

Because every straight line has a first-degree equation in x and y and every first-degree equation in x and y has a straight line for its graph, (7) is sometimes called the **general equation** of a line or a **linear equation** in x and y.

▶ Example 7   Graph the equation

$$3x - 4y + 12 = 0 \tag{9}$$

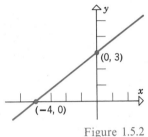

Figure 1.5.2

Solution.   Since this is a linear equation in x and y, its graph is a straight line. Thus, to sketch the graph we need only plot two points on the graph and draw the line through them. It is particularly convenient to plot the points where the line crosses the coordinate axes. The line crosses the y-axis when $x = 0$ and crosses the x-axis when $y = 0$. Substituting $x = 0$ in (9) and solving for y, we obtain the intersection $(0, 3)$ with the y-axis. Substituting $y = 0$ in (9) and solving for x, we obtain the intersection $(-4, 0)$ with the x-axis. The graph of (9) is given in Figure 1.5.2.   ◀

▶ Exercise Set 1.5

1. Graph the equations
(a) $2x + 5y = 15$
(b) $x = 3$
(c) $y = -2$
(d) $y = 2x - 7$.

2. Graph the equations
(a) $\dfrac{x}{3} - \dfrac{y}{4} = 1$
(b) $x = -8$
(c) $y = 0$
(d) $x = 3y + 2$.

3. Graph the equations
(a) $y = 2x - 1$
(b) $y = 3$
(c) $y = -2x$.

4. Graph the equations
(a) $y = 2 - 3x$
(b) $y = \frac{1}{4}x$
(c) $y = -\sqrt{3}$.

5. Find the slope and y-intercept of
(a) $y = 3x + 2$
(b) $y = 3 - \frac{1}{4}x$
(c) $3x + 5y = 8$
(d) $y = 1$
(e) $\dfrac{x}{a} + \dfrac{y}{b} = 1$.

6. Find the slope and y-intercept of
(a) $y = -4x + 2$
(b) $x = 3y + 2$
(c) $\dfrac{x}{2} + \dfrac{y}{3} = 1$
(d) $y - 3 = 0$
(e) $a_0 x + a_1 y = 0$.

7. To the nearest degree, find the angle of inclination of
(a) $y = \sqrt{3}x + 2$
(b) $y + 2x + 5 = 0$.

8. To the nearest degree, find the angle of inclination of
(a) $3y = 2 - \sqrt{3}x$
(b) $y - 4x + 7 = 0$.

In Exercises 9–20, find the slope-intercept form of the line satisfying the given conditions.

9. Slope $= -2$, y-intercept $= 4$.

10. $m = 5$, $b = -3$.

11. The line is parallel to $y = 4x - 2$ and has y-intercept 7.

12. The line is perpendicular to $y = 5x + 9$ and has y-intercept 6.

13. The line passes through $(2, 4)$ and $(1, -7)$.

14. The line passes through $(-3, 6)$ and $(-2, 1)$.

15. The line has angle of inclination $\phi = \frac{1}{6}\pi$ and y-intercept -3.

16. The line has angle of inclination $\phi = \frac{2}{3}\pi$ and passes through the point $(1, 2)$.

17. The y-intercept is 2 and the x-intercept is -4.

18. The y-intercept is b and the x-intercept is a.

19. The line is perpendicular to the y-axis and passes through $(-4, 1)$.

20. The line is parallel to $y = -5$ and passes through $(-1, -8)$.

21. Find an equation for the line that passes through $(5, -2)$ and has angle of inclination $\phi = \frac{1}{2}\pi$.

22. Find an equation for the line along the y-axis.

23. In each part, classify the lines as parallel, perpendicular, or neither.
    (a) $y = 4x - 7$ and $y = 4x + 9$
    (b) $y = 2x - 3$ and $y = 7 - \frac{1}{2}x$
    (c) $5x - 3y + 6 = 0$ and $10x - 6y + 7 = 0$
    (d) $Ax + By + C = 0$ and $Bx - Ay + D = 0$
    (e) $y - 2 = 4(x - 3)$ and $y - 7 = \frac{1}{4}(x - 3)$.

24. In each part, classify the lines as parallel, perpendicular, or neither.
    (a) $y = -5x + 1$ and $y = 3 - 5x$
    (b) $y - 1 = 2(x - 3)$ and $y - 4 = -\frac{1}{2}(x + 7)$
    (c) $4x + 5y + 7 = 0$ and $5x - 4y + 9 = 0$
    (d) $Ax + By + C = 0$ and $Ax + By + D = 0$
    (e) $y = \frac{1}{2}x$ and $x = \frac{1}{2}y$.

25. In each part, find the point of intersection of the lines.
    (a) $2x + 3y = 5$ and $y = -1$
    (b) $4x + 3y = -2$ and $5x - 2y = 9$.

26. In each part, find the point of intersection of the lines.
    (a) $6x - 9y = 7$ and $x = -\frac{2}{3}$
    (b) $6x - 2y = -3$ and $-8x + 3y = 5$.

27. Find the distance from the point $(2, 1)$ to the line $4x - 3y + 10 = 0$. [*Hint:* Find the foot of the perpendicular dropped from the point to the line.]

28. Find the distance from the point $(8, 4)$ to the line $5x + 12y - 36 = 0$. [*Hint:* See the hint in Exercise 27.]

29. Use the method described in Exercise 27 to prove that the distance d from (x_0, y_0) to the line $Ax + By + C = 0$ is

$$d = \frac{|Ax_0 + By_0 + C|}{\sqrt{A^2 + B^2}}$$

30. Use the formula in Exercise 29 to solve Exercise 27.

31. Use the formula in Exercise 29 to solve Exercise 28.

32. Is the graph of

$$\frac{y}{x} = 1$$

a line? Explain.

33. Prove: The midpoint of the line segment joining (a_1, b_1) and (a_2, b_2) is the point

$$\left(\frac{a_1 + a_2}{2}, \frac{b_1 + b_2}{2} \right)$$

34. Use the result in Exercise 33 to find
    (a) the midpoint of the line segment joining $(2, 8)$ and $(-4, 6)$
    (b) the equation of the perpendicular bisector of the line segment in part (a).

35. Repeat Exercise 34 for the line segment joining $(5, -1)$ and $(4, 8)$.

36. Prove: For any triangle, the perpendicular bisectors of the sides meet at a point. [*Hint:* See Exercises 33 and 34; also position the triangle with one vertex on the y-axis and the opposite side on the x-axis, so that the vertices are $(0, a)$, $(b, 0)$, and $(c, 0)$.]

37. Find the point on the line $4x - 2y + 3 = 0$ that is equidistant from $(3, 3)$ and $(7, -3)$.

38. In physical problems linear equations involving variables other than x and y often arise. In parts (a)–(g) determine whether the equation is linear.
    (a) $3\alpha - 2\beta = 5$
    (b) $A = 2000(1 + 0.06t)$
    (c) $A = \pi r^2$

(d) $E = mc^2$ (*c* constant)
(e) $V = C(1 - rt)$ (*r* and *C* constant)
(f) $V = \frac{1}{3}\pi r^2 h$ (*r* constant)
(g) $V = \frac{1}{3}\pi r^2 h$ (*h* constant).

39. There are two common systems for measuring temperature, Celsius and Fahrenheit. Water freezes at 0° Celsius and 32° Fahrenheit; it boils at 100° Celsius and 212° Fahrenheit.

(a) Assuming that the Celsius temperature *C* and the Fahrenheit temperature *F* are related by a linear equation, find the equation.

(b) What is the slope of the line relating *F* and *C* if *F* is plotted on the horizontal axis?

40. Prove: If (x, y) satisfies Equation (2), then the point $P(x, y)$ lies on the line with slope *m* passing through $P_1(x_1, y_1)$.

1.6 CIRCLES AND EQUATIONS OF THE FORM $y = ax^2 + bx + c$

In this section we will study equations of circles and graphs of equations of the form $y = ax^2 + bx + c$.

CIRCLES   If (x_0, y_0) is a fixed point in the plane, then the circle of radius *r* centered at (x_0, y_0) is the set of all points in the plane whose distance from (x_0, y_0) is *r* (Figure 1.6.1). Thus, a point (x, y) will lie on this circle if and only if

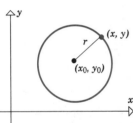

$$\sqrt{(x - x_0)^2 + (y - y_0)^2} = r$$

or equivalently

$$(x - x_0)^2 + (y - y_0)^2 = r^2 \qquad (1)$$

Figure 1.6.1   This is called the **standard form of the equation of a circle**.

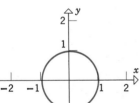

▶ **Example 1**   Sketch the graph of $x^2 + y^2 = 1$.

Solution. This equation results when $x_0 = y_0 = 0$ and $r = 1$ in (1). Thus, the graph is a circle of radius 1 centered at the origin (Figure 1.6.2). This is often called the **unit circle.** ◀

Figure 1.6.2

▶ **Example 2**   Find an equation for the circle of radius 5 centered at $(-3, 2)$.

Solution. From (1) with $x_0 = -3$, $y_0 = 2$, and $r = 5$ we obtain

$$(x + 3)^2 + (y - 2)^2 = 25$$

or, if preferred, we can square out the terms to obtain

$$x^2 + y^2 + 6x - 4y - 12 = 0$$ ◀

An alternate version of Equation (1) can be obtained by squaring out the terms and simplifying. This yields an equation of the form

$$x^2 + y^2 + dx + ey + f = 0 \tag{2}$$

Still another version of the equation of a circle can be obtained by multiplying both sides of Equation (2) by a nonzero constant A. This yields an equation of the form

$$Ax^2 + Ay^2 + Dx + Ey + F = 0 \tag{3}$$

where A, D, E, and F are constants and $A \neq 0$.

If the equation of a circle is given in form (2) or (3), then the center and radius can be determined by rewriting the equation in standard form and then reading off the center (x_0, y_0) and radius r from this equation. The following example illustrates how this is done using the algebraic technique of *completing the square*.

▶ Example 3   Find the center and radius of the circle with equation

$$x^2 + y^2 - 8x + 2y + 8 = 0$$

Solution.   First, group the x-terms, group the y-terms, and take the constant to the right side:

$$(x^2 - 8x) + (y^2 + 2y) = -8$$

Next add the appropriate constant within each set of parentheses to complete the square, and add the same constants to the right side to maintain equality:

$$(x^2 - 8x + 16) + (y^2 + 2y + 1) = -8 + 16 + 1$$

or

$$(x - 4)^2 + (y + 1)^2 = 9$$

Thus from (1), the circle has center $(4, -1)$ and radius 3.   ◀

REMARK.   To find the center and radius of a circle whose equation is of form (3) with $A \neq 1$, we can first divide both sides by A to put the equation in form (2), and then proceed as in Example 3.

An equation of form (3) does not always have a circle as its graph. To see why, suppose we divide both sides of (3) by A and complete the squares to obtain

$$(x - x_0)^2 + (y - y_0)^2 = k$$

If $k > 0$, then the graph is a circle with center (x_0, y_0) and radius $\sqrt{k}$.
If $k = 0$, the only solution of the equation is $x = x_0$, $y = y_0$; thus, the graph is the single point (x_0, y_0).

If $k < 0$, then the equation has no solutions and there is no graph. In summary, we have the following result.

1.6.1   THEOREM   *An equation of the form*

$$Ax^2 + Ay^2 + Dx + Ey + F = 0$$

where $A \neq 0$, represents a circle, or a point, or else has no graph.

The last two cases in Theorem 1.6.1 are called the **degenerate cases.**

THE GRAPH OF
$y = ax^2 + bx + c$

An equation of the form

$$y = ax^2 + bx + c \qquad (a \neq 0) \tag{4}$$

is called a **quadratic equation in x.** Its graph is a curve called a *parabola*. If a is positive, the parabola will open up as shown in Figure 1.6.3a, and if a is negative, it will open down as in Figure 1.6.3b. In both cases the parabola is symmetric about a vertical line parallel to the y-axis. This line of symmetry cuts the parabola at a point called the *vertex*. The vertex is the low point on the curve if $a > 0$ and the high point if $a < 0$.

In the exercises we will help the reader show that the x-coordinate of the vertex is given by the formula

$$x = -\frac{b}{2a} \tag{5}$$

With the aid of this formula, a reasonably accurate graph of a quadratic equation in x can be obtained by plotting the vertex and one or two points on each side of it.

▶ **Example 4**   Sketch the graph of $y = x^2 - 4x + 5$.

Solution.   The given equation is of form (4), with $a = 1$, $b = -4$, and $c = 5$, so by (5) the x-coordinate of the vertex is

$$x = -\frac{b}{2a} = 2$$

Using this value and two additional values on each side, we obtain Table 1.6.1.

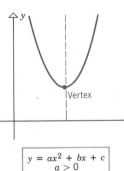

$y = ax^2 + bx + c$
$a > 0$

(a)

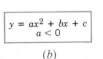

$y = ax^2 + bx + c$
$a < 0$

(b)

Figure 1.6.3

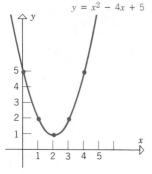

$y = x^2 - 4x + 5$

5
4
3
2
1

1 2 3 4 5

Figure 1.6.4

Table 1.6.1

x	0	1	2	3	4
$y = x^2 - 4x + 5$	5	2	1	2	5

By plotting the points in this table, we obtain the curve in Figure 1.6.4.
◄

Sometimes it is desirable to find the points where the parabola intersects the coordinate axes. The x-coordinates of the intersections with the x-axis are obtained by setting $y = 0$ and solving for x; and the y-coordinates of the intersections with the y-axis are obtained by setting $x = 0$ and solving for y.

► **Example 5** From Figure 1.6.4, we see that the parabola $y = x^2 - 4x + 5$ does not intersect the x-axis. This can also be seen algebraically by setting $y = 0$ and solving for x. We obtain

$$x^2 - 4x + 5 = 0$$

By the quadratic formula

$$x = \frac{-b \pm \sqrt{b^2 - 4ac}}{2a} = \frac{4 \pm \sqrt{16 - 20}}{2} = 2 \pm \frac{\sqrt{-4}}{2}$$

so the solutions are imaginary, which tells us that there are no intersections with the x-axis.
◄

► **Example 6** Sketch the graph of $y = -x^2 + 2x + 2$

Solution. This equation is of form (4), with $a = -1$, $b = 2$, and $c = 2$, so by (5) the x-coordinate of the vertex is

$$x = -\frac{b}{2a} = 1$$

Using this value and two additional values on each side, we obtain Table 1.6.2.

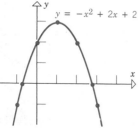

$y = -x^2 + 2x + 2$

Figure 1.6.5

Table 1.6.2

x	-1	0	1	2	3
$y = -x^2 + 2x + 2$	-1	2	3	2	-1

Plotting the points in this table (Figure 1.6.5) shows that the graph intersects the x-axis at two points. To obtain a more accurate sketch, it will be helpful to locate these intersections. Setting $y = 0$ yields

$$-x^2 + 2x + 2 = 0$$

and solving by the quadratic formula gives

$$x = \frac{-b \pm \sqrt{b^2 - 4ac}}{2a} = \frac{-2 \pm \sqrt{12}}{-2} = 1 \pm \sqrt{3}$$

Thus, the intersections with the x-axis occur at

$$x = 1 + \sqrt{3} \approx 2.7 \quad \text{and} \quad x = 1 - \sqrt{3} \approx -.7$$

Using these values and Table 1.6.2, we obtain the graph in Figure 1.6.5.

◀

REMARK. If x and y are interchanged in (4), the resulting equation,

$$x = ay^2 + by + c$$

is called a **quadratic equation in y.** The graph of such an equation is a parabola with its line of symmetry parallel to the x-axis and its vertex at the point with y-coordinate $y = -b/2a$. Some problems relating to such equations appear in the exercises.

▶ Exercise Set 1.6

In Exercises 1 and 2, find the center and radius of each circle.

1. (a) $x^2 + y^2 = 25$
   (b) $(x - 1)^2 + (y - 4)^2 = 16$
   (c) $(x + 1)^2 + (y + 3)^2 = 5$
   (d) $x^2 + (y + 2)^2 = 1$.

2. (a) $x^2 + y^2 = 9$
   (b) $(x - 3)^2 + (y - 5)^2 = 36$
   (c) $(x + 4)^2 + (y + 1)^2 = 8$
   (d) $(x + 1)^2 + y^2 = 1$.

In Exercises 3–10, find the standard equation of the circle satisfying the given conditions.

3. Center $(3, -2)$; radius $= 4$.

4. Center $(1, 0)$; diameter $= \sqrt{8}$.

5. Center $(-4, 8)$; circle is tangent to the x-axis.

6. Center $(5, 8)$; circle is tangent to the y-axis.

7. Center $(-3, -4)$; circle passes through the origin.

8. Center $(4, -5)$; circle passes through $(1, 3)$.

9. A diameter has endpoints $(2, 0)$ and $(0, 2)$.

10. A diameter has endpoints $(6, 1)$ and $(-2, 3)$.

In Exercises 11–22, determine whether the equation represents a circle, a point, or no graph. If the equation represents a circle, find the center and radius.

11. $x^2 + y^2 - 2x - 4y - 11 = 0$.

12. $x^2 + y^2 + 8x + 8 = 0$.

13. $2x^2 + 2y^2 + 4x - 4y = 0$.

14. $6x^2 + 6y^2 - 6x + 6y = 3$.

15. $x^2 + y^2 + 2x + 2y + 2 = 0$.

16. $x^2 + y^2 - 4x - 6y + 13 = 0$.

17. $9x^2 + 9y^2 = 1$.

18. $\dfrac{x^2}{4} + \dfrac{y^2}{4} = 1$.

19. $x^2 + y^2 + 10y + 26 = 0.$

20. $x^2 + y^2 - 10x - 2y + 29 = 0.$

21. $16x^2 + 16y^2 + 40x + 16y - 7 = 0.$

22. $4x^2 + 4y^2 - 16x - 24y = 9.$

In Exercises 23–36, graph the parabola and label the coordinates of the vertex and the intersections with the coordinate axes.

23. $y = x^2 + 2.$

24. $y = x^2 - 3.$

25. $y = x^2 + 2x - 3.$

26. $y = x^2 - 3x - 4.$

27. $y = -x^2 + 4x + 5.$

28. $y = -x^2 + x.$

29. $y = (x - 2)^2.$

30. $y = (3 + x)^2.$

31. $x^2 - 2x + y = 0.$

32. $x^2 + 8x + 8y = 0.$

33. $y = 3x^2 - 2x + 1.$

34. $y = x^2 + x + 2.$

35. $x = -y^2 + 2y + 2.$

36. $x = y^2 - 4y + 5.$

37.   If a ball is thrown straight up with an initial velocity of 32 ft/sec, then after t seconds, the distance s above its starting height is given by $s = 32t - 16t^2$.
   (a)   Graph this equation in an st-coordinate system with the t-axis horizontal.
   (b)   At what time t will the ball be at its highest point, and how high will it rise?

38.   A point (x, y) moves so that its distance to $(2, 0)$ is $\sqrt{2}$ times its distance to $(0, 1)$.
   (a)   Show that the point moves along a circle.
   (b)   Find the center and radius.

39.   A point (x, y) moves so that the sum of the squares of its distances from $(4, 1)$ and $(2, -5)$ is 45.
   (a)   Show that the point moves along a circle.
   (b)   Find the center and radius.

40.   (a)   By completing the square, show that $y = ax^2 + bx + c$ can be rewritten as

$$y = a\left(x + \frac{b}{2a}\right)^2 + \left(c - \frac{b^2}{4a}\right)$$

   if $a \neq 0$.
   (b)   Use the result in (a) to show that the graph of $y = ax^2 + bx + c$ has its high point at $x = -b/2a$ if $a < 0$ and its low point there if $a > 0$.

▶ SUPPLEMENTARY EXERCISES

In Exercises 1–5, use interval notation to describe the set of all values of x (if any) that satisfy the given inequalities.

1.   (a) $-3 < x \leq 5$
     (b) $-1 < x^2 \leq 9$
     (c) $x^2 \geq \frac{1}{4}.$

2.   (a) $|2x + 1| > 5$
     (b) $|x^2 - 9| \geq 7$
     (c) $1 \leq |x| \leq 3.$

3.   (a) $2x^2 - 5x > 3$
     (b) $x^2 - 5x + 4 \leq 0.$

4.   (a) $\dfrac{x}{1 - x} \geq 3$

     (b) $\dfrac{2x + 3}{x} \geq x.$

5.   (a) $\dfrac{|x| - 1}{|x| - 2} \leq 0$

     (b) $|x - 1| \leq 2|x + 2|.$

6.   Among the terms integer, rational, irrational, which ones apply to the given number?
     (a) $\sqrt{4/9}$ (b) 2^{-2}
     (c) $-4^{1/3}$ (d)   0.87
     (e) $-4^{1/2}$ (f)   0.1010010001 . . .
     (g)   3.222   (h) $3/(-1).$

7.   (a)   Find values of a and b such that $a < b$, but $a^2 > b^2$.

(b) If $a < b$, what additional assumptions on a and b are required to ensure that $a^2 < b^2$?

8. Which of the following are true for all sets A and B?
   (a) $A \subset (A \cap B)$ (b) $(A \cap B) \subset A$
   (c) $\phi \subset A$ (d) $A \subset (A \cup B)$
   (e) $(A \cap B) \subset (A \cup B)$
   (f) either $A \subset B$ or $B \subset A$
   (g) $A \in B$.

9. Prove: $|x| \le \sqrt{x^2 + y^2}$ and $|y| \le \sqrt{x^2 + y^2}$. Interpret this geometrically.

In Exercises 10–14, draw a rectangular coordinate system and sketch the set of points whose coordinates (x, y) satisfy the given conditions.

10. (a) $y = 0$ and $x > 0$
    (b) $2x - y \le 3$.

11. (a) $xy = x^2$
    (b) $y(x - 1) = x^2 - 1$.

12. (a) $y = (x^3 - 1)/(x - 1)$
    (b) $y > x^2 - 9$.

13. (a) $y^2 - 6y + x^2 - 2x - 6 \ge 0$
    (b) $x + |y - 2| = 1$.

14. (a) $|x| + |y| = 4$
    (b) $|x| - |y| = 4$.

15. Where does the parabola $y = x^2$ intersect the line $y - 2 = x$?

In Exercises 16–19, sketch the graph of the given equation.

16. $xy + 4 = 0$.

17. $y = |x - 2|$.

18. $y = \sqrt{4 - x^2}$.

19. $y = x(x - 2)$.

In Exercises 20–24, find the standard equation for the circle satisfying the given conditions.

20. The circle centered at $(3, -2)$ and tangent to the line $y = 1$.

21. The circle centered at $(1, 2)$ and passing through the point $(4, -2)$.

22. The circle centered on the line $x = 2$ and passing through the points $(1, 3)$ and $(3, -11)$.

23. The circle of radius 5 tangent to the lines $y = 7$ and $x = 6$.

24. The circle of radius 13 that passes through the origin and the point $(0, -24)$.

In Exercises 25–28, determine whether the equation represents a circle, a point, or has no graph. If it represents a circle, find the center and radius.

25. $x^2 + y^2 + 4x + 2y + 5 = 0$.

26. $4x^2 + 4y^2 - 4x + 8y + 1 = 0$.

27. $x^2 + y^2 - 3x + 2y + 4 = 0$.

28. $3x^2 + 3y^2 - 5x + 7y + 3 = 0$.

29. In each part, find an equation for the line through A and B, the distance between A and B, and the coordinates of the midpoint of the line segment joining A and B.
    (a) $A(3, 4)$, $B(-3, -4)$
    (b) $A(3, 4)$, $B(3, -4)$
    (c) $A(3, 4)$, $B(-3, 4)$
    (d) $A(3, 4)$, $B(4, 3)$.

30. Show that the point $(8, 1)$ is *not* on the line through the points $(-3, -2)$ and $(1, -1)$.

31. Where does the circle of radius 5 centered at the origin intersect the line of slope $-3/4$ through the origin?

32. Fahrenheit and Celsius temperatures are related by $F - 32 = 9C/5$. What temperature is the same in both Fahrenheit and Celsius?

33. Find the inclination angle of the line whose equation is
    (a) $x = 3$ (b) $y = -2$
    (c) $2x + 2y = 1$ (d) $\sqrt{3}x - y = 4$.

34. Find the slope of the line whose angle of inclination is
    (a) $30°$ (b) $120°$ (c) $90°$.

In Exercises 35–37, find the slope-intercept form of the line satisfying the stated conditions.

35. The line through $(2, -3)$ and $(4, -3)$.

36. The line with x-intercept -2 and angle of inclination $\phi = 45°$.

37. The line parallel to $x + 2y = 3$ that passes through the origin.

38. Find the equation of the perpendicular bisector of the line segment joining $A(-2, -3)$ and $B(1, 1)$.

In Exercises 39–41, find equations of the lines L and L' and determine their point of intersection.

39. L passes through $(1, 0)$ and $(-1, 4)$.
L' is perpendicular to L and has y-intercept -3.

40. L passes through $(-2, 0)$ and $(-2, 3)$.
L' passes through $(-1, 4)$ and is perpendicular to L.

41. L has slope $2/5$ and passes through $(3, 1)$.
L' has x-intercept $-8/3$ and y-intercept -4.

42. Consider the triangle with vertices $A(5, 2)$, $B(1, -3)$, and $C(-3, 4)$. Find the point-slope form of the line containing:
(a) the median from C to AB
(b) the altitude from C to AB.

43. Use slopes to show that the points $(5, 6)$, $(-4, 3)$, $(-3, -2)$, and $(6, 1)$ are vertices of a parallelogram. Is it a rectangle?

44. For what value of k (if any) will the line $2x + ky = 3k$ satisfy the stated condition?
(a) have slope 3
(b) have y-intercept 3
(c) be parallel to the x-axis
(d) pass through $(1, 2)$.

2 functions and limits

2.1 FUNCTIONS

In this section we consider one of the most fundamental concepts in all of mathematics, the notion of a *function*. Historically, the term "function" was first used by Leibniz in 1673 to denote the dependence of one quantity on another. To illustrate:

1. The area A of a circle depends on its radius r by the equation $A = \pi r^2$; we say that "A is a function of r."

2. The velocity of a ball dropped from a height increases with time until it hits the ground. Thus, the velocity v depends on the time t and we say that "v is a function of t."

3. At a fixed point on earth, the wind speed w varies with the time t. Thus, "w is a function of t."

4. In a bacteria culture, the number of bacteria present after one hour of growth depends on the number present initially; we say that "the size of the bacteria population after one hour is a function of the initial population size."

In order to discuss functions or relationships between quantities without stating specific formulas, the Swiss mathematician, Leonhard Euler* (see page 56), developed the ingenious idea of using a letter of the alphabet such as f to denote a function or relationship. Thus, by writing

$$y = f(x)$$

(read "y equals f of x"), we convey the idea that y is a function of x, that is, the value of y depends on the value of x. To indicate that the wind speed w

at a fixed point on earth depends on the time t at which it is measured, we can write

$$w = f(t) \tag{1}$$

There is nothing special about the letter f; any symbol can be used to denote a function; thus, (1) could have been written

$$w = F(t), \qquad w = f_1(t), \qquad w = g(t), \qquad \text{or} \qquad w = \phi(t)$$

By having a variety of symbols available to denote functions we are able to distinguish between different possible relationships. For example, at a point in Alaska the wind speed w will have a certain dependence on the time t, say, $w = g_1(t)$, while at a point in Florida the wind speed w will have a different dependence on the time t, say $w = g_2(t)$.

Note that in (1) the letters w and t denote measurable *physical* quantities, wind speed and time. The letter f, however, does not denote a physical quantity; it stands for a *dependence* of one physical quantity on another; this is a completely *abstract* idea.

Since the time of Euler and Leibniz the original idea of a function has evolved into the following more precise and more general mathematical concept:

---

*LEONHARD EULER (1707–1783). Euler was probably the most prolific mathematician who ever lived. It has been said that, "Euler wrote mathematics as effortlessly as most men breathe." He was born in Basel, Switzerland, and was the son of a Protestant minister who had himself studied mathematics. Euler's genius developed early. He attended the University of Basel, where by age 16 he obtained both a Bachelor of Arts degree and a Master's degree in philosophy. While at Basel, Euler had the good fortune to be tutored one day a week in mathematics by a distinguished mathematician, Johann Bernoulli. At the urging of his father, Euler then began to study theology. The lure of mathematics was too great, however, and by age 18 Euler had begun to do mathematical research. Nevertheless, the influence of his father and his theological studies remained, and throughout his life Euler was a deeply religious, unaffected person. At various times Euler taught at St. Petersburg Academy of Sciences (in Russia), the University of Basel, and the Berlin Academy of Sciences. Euler's energy and capacity for work were virtually boundless. His collected works form about 60 to 80 quarto sized volumes and it is believed that much of his work has been lost. What is particularly astonishing is that Euler was blind for the last 17 years of his life, and this was one of his most productive periods! Euler's flawless memory was phenomenal. Early in his life he memorized the entire *Aeneid* by Virgil and at age 70 could not only recite the entire work, but could also state the first and last sentence on each page of the book from which he memorized the work. His ability to solve problems in his head was beyond belief. He worked out in his head major problems of lunar motion that baffled Isaac Newton and once did a complicated calculation in his head to settle an argument between two students whose computations differed in the fiftieth decimal place.

Euler's main contribution was the systematization of mathematics. Following the development of calculus by Leibniz and Newton, results in mathematics developed rapidly in a disorganized way. Euler's genius gave coherence to the mathematical landscape. He was the first mathematician to bring the full power of calculus to bear on problems from physics. He made major contributions to virtually every branch of mathematics as well as to the theory of optics, planetary motion, electricity, magnetism, and general mechanics.

2.1.1 DEFINITION A ***function*** is a rule that assigns to each element in a set A one and only one element in a set B.

In general the sets A and B need not be sets of real numbers; however, for the time being we will only be concerned with functions for which A and B are both subsets of the real numbers.

To see how Definition 2.1.1 relates to the historical concept of a function, consider the relationship

$$w = f(t) \tag{2}$$

where w denotes the wind speed at a point and t denotes the time of measurement. Let A be the set of all possible t values and B be the set of all possible w values. With each instant t in time there is associated by (2) a unique wind speed w; that is, the function f in (2) assigns to each element t in A one and only one element w in B. This is precisely the definition of a function as stated in Definition 2.1.1.

The set A in Definition 2.1.1 is called the ***domain*** of the function. If x is an element in the domain of a function f, then the element that f associates with x is denoted by the symbol $f(x)$ (read "f of x") and is called the ***image of x under f*** or the ***value of f at x*** (Figure 2.1.1). The set of all possible values of $f(x)$ as x varies over the domain is called the ***range*** of f.

Most often the rule for obtaining the value of a function is given by means of a formula.

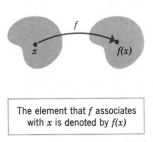

The element that f associates with x is denoted by $f(x)$

Figure 2.1.1

▶ **Example 1** The formula $f(x) = x^3$ tells us that the value of f at x is x^3. Thus,

$$f(2) = 2^3 = 8, \qquad f(-4) = (-4)^3 = -64, \qquad \text{and} \qquad f(0) = 0^3 = 0$$
◄

▶ **Example 2** If $\phi(x) = 2x^2 - 1$, then

$$\phi(4) = 2(4)^2 - 1 = 31, \qquad \phi(t) = 2t^2 - 1$$
$$\phi(k + 1) = 2(k + 1)^2 - 1 = 2k^2 + 4k + 1$$
◄

REMARK. If x is an element in the domain of a function f, the definition of a function (Definition 2.1.1) requires that f assign one and only one value to x. This means that a function cannot be "multiple-valued." For example, the expression $\pm \sqrt{x}$ does not define a function of x, since it assigns two values to each positive x.

REMARK. If the rule for evaluating a function is given by a formula, and if there is no mention of the domain, it is understood that the domain consists of all real numbers for which the formula makes sense and yields a real value.

▶ **Example 3** Let

$$g(x) = \sqrt{x - 1}$$

When $x < 1$, $g(x)$ is imaginary; thus the domain of g is $[1, +\infty)$. As x varies over this interval, $g(x)$ varies from 0 to $+\infty$; therefore, the range of g is $[0, +\infty)$. ◀

▶ **Example 4** Let

$$h(x) = \frac{1}{(x - 1)(x - 3)}$$

When $x = 1$ or $x = 3$, $h(x)$ is undefined since division by zero is not allowed; otherwise, $h(x)$ yields real values. Thus, the domain of h consists of all real numbers other than $x = 1$ and $x = 3$. In interval notation the domain is $(-\infty, 1) \cup (1, 3) \cup (3, +\infty)$. (See Figure 2.1.2.) ◀

Figure 2.1.2

Sometimes it is necessary to specify restrictions on the domain, even though the formula for the function makes sense elsewhere.

▶ **Example 5** Assume that it costs 12 cents to manufacture a certain kind of resistor, and let $f(x)$ be the cost in dollars for manufacturing x such resistors. The formula for $f(x)$ is

$$f(x) = 0.12x \tag{3}$$

If we were to specify no restrictions on the domain of this function, then the domain would be $(-\infty, +\infty)$ since (3) makes sense for all real x. Physically, however, x must be a nonnegative integer, so we must restrict the domain to the set $\{0, 1, 2, \ldots\}$. We do this by writing

$$f(x) = 0.12x, \qquad x = 0, 1, 2, \ldots \tag{4}$$

From (4) it is to be understood that the formula $f(x) = 0.12x$ only applies when $x = 0, 1, 2, \ldots$; for other values of x, $f(x)$ is undefined. Thus,

$$f(3) = 0.12(3) = 0.36 \qquad \text{and} \qquad f(\tfrac{3}{2}) \text{ is undefined} \qquad ◀$$

The next example shows how functions can be specified by formulas that have been "pieced together."

▶ **Example 6** The cost of a taxi ride in a certain metropolitan area is 75 cents for any ride up to and including one mile. After one mile the rider pays an additional amount at the rate of 50 cents per mile. If $f(x)$ is the total cost in dollars for a ride of x miles, then the value of $f(x)$ is

$$0.75 \quad \text{if} \quad 0 < x \leq 1 \qquad \left(\begin{matrix} \text{\$0.75 for a ride up to} \\ \text{and including one mile} \end{matrix} \right)$$

and

$$0.75 + 0.50(x - 1) \quad \text{if} \quad x > 1 \qquad \left(\begin{matrix} \text{\$0.75 for the first} \\ \text{mile plus \$0.50 a mile} \\ \text{for each mile after the} \\ \text{first.} \end{matrix} \right)$$

We express this by writing

$$f(x) = \begin{cases} 0.75, & 0 < x \leq 1 \\ 0.75 + 0.50(x - 1), & 1 < x \end{cases} \qquad \blacktriangleleft$$

Although most functions are specified by formulas, this is not the only possibility. Any description that tells us what values f assigns to the points in its domain will suffice. One possibility is to use a table relating the values of x and $f(x)$.

▶ Example 7 At 10 A.M. on a weekday, a 2-minute call is made between two stations in New Jersey. Table 2.1.1, obtained from a New Jersey telephone directory, shows how the cost of such a call varies with the distance between the stations. If $f(x)$ is the cost in dollars for a call between stations x miles apart, then the function f is completely specified by Table 2.1.1. For example,

$$f(22) = 0.25, \quad f(114) = 0.60, \quad \text{and} \quad f(64) = 0.40 \qquad \blacktriangleleft$$

Table 2.1.1

DISTANCE IN MILES	more than	0	10	15	20	25	32	48	64	80	96	112
	up to and including	10	15	20	25	32	48	64	80	96	112	more
COST IN DOLLARS		0.10	0.15	0.20	0.25	0.30	0.35	0.40	0.45	0.50	0.55	0.60

If x is an element in the domain of a function f, then by forming the ordered pair of numbers

$$(x, f(x))$$

we match each x with the value of f at x. For the function $f(x) = x^3$, a few such ordered pairs would be

$$(2, 8), \quad (-3, -27), \quad (0, 0), \quad \text{and} \quad (\tfrac{1}{2}, \tfrac{1}{8})$$

As illustrated in the next example, a function can be specified by giving the set of all such possible ordered pairs.

▶ Example 8 Let f be a function with domain $\{1, 2, 3, 4, 5\}$ and let the set C of all ordered pairs $(x, f(x))$ be

$$C = \{(1, 0), (2, -1), (3, 0), (4, 7), (5, \tfrac{1}{7})\}$$

The set C completely specifies the function, since it tells us that

$$f(1) = 0, \quad f(2) = -1, \quad f(3) = 0, \quad f(4) = 7, \quad \text{and} \quad f(5) = \tfrac{1}{7}$$

◀

In the symbol $f(x)$, we can think of x as a quantity that can be varied arbitrarily over the domain of f. As x varies, so does the number $f(x)$, but with one difference; where the value of x can be varied arbitrarily within the domain of f, the value of $f(x)$ is determined once the value of x is specified. For this reason, x is sometimes called the **independent variable** and $f(x)$ the **dependent variable.** In many problems it is convenient to introduce a single letter such as y to stand for the dependent variable. Thus, if we write

$$y = f(x) \tag{5}$$

then x is the independent variable and y is the dependent variable.

▶ Example 9 In the equation $y = 9x^3 - 2x + 7$, x is the independent variable and y the dependent variable for the function $f(x) = 9x^3 - 2x + 7$.

◀

The following definition will enable us to study functions geometrically.

2.1.2 DEFINITION We define the **graph of a function f** to be the graph of the equation $y = f(x)$.

▶ Example 10 Sketch the graph of $f(x) = x + 2$.

Solution. By definition, the graph of $f(x) = x + 2$ is the graph of the equation $y = x + 2$, which is a line of slope 1 and y-intercept 2 (Figure 2.1.3).

◀

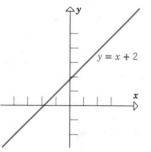

$y = x + 2$

Figure 2.1.3

▶ Example 11 Sketch the graph of $f(x) = |x|$.

Solution. By definition, the graph of $f(x) = |x|$ is the graph of the equation

$$y = |x|$$

or equivalently

$$y = \begin{cases} x, & x \geq 0 \\ -x, & x < 0 \end{cases}$$

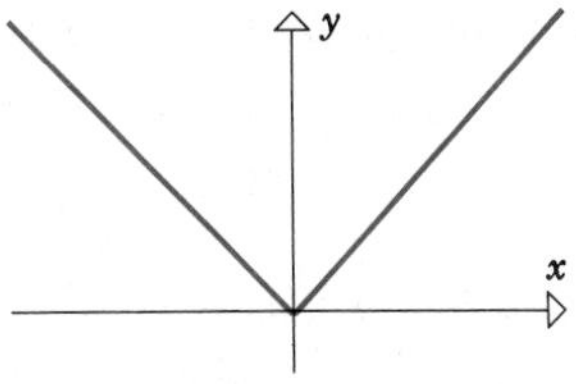

The graph coincides with the line $y = x$ for $x \geq 0$ and with the line $y = -x$ for $x < 0$ (Figure 2.1.4). ◀

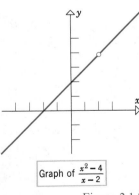

Graph of $|x|$

Figure 2.1.4

▶ Example 12   Sketch the graph of

$$h(x) = \frac{x^2 - 4}{x - 2} \tag{6}$$

Solution.   The function h can be rewritten as

$$h(x) = \frac{(x - 2)(x + 2)}{(x - 2)}$$

Algebraically, it is desirable to cancel the $x - 2$ factors, but we must exercise some care. If we simply cancel the factors and write

$$h(x) = x + 2 \tag{7}$$

then we would erroneously alter the domain of h. To see this, observe that the domain of the function h in (6) consists of all x other than $x = 2$, whereas the domain of the function h in (7) consists of all x. In fact, when $x = 2$, the function h in (7) has the value

$$h(2) = 2 + 2 = 4$$

whereas for the original function h in (6),

$$h(2) = \frac{2^2 - 4}{2 - 2} = \frac{0}{0}$$

which is undefined.

To cancel the $x - 2$ factors in (6) and not alter the domain of h, we must restrict the domain in (7) and write

$$h(x) = x + 2, \qquad x \neq 2 \tag{8}$$

The function h in (8) is identical to the function f in Example 10, except that h is undefined at $x = 2$. Thus, the graph of h is identical to the graph of f (Figure 2.1.3), except that the graph of h has a hole in it above $x = 2$ (Figure 2.1.5). ◀

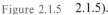

Graph of $\frac{x^2 - 4}{x - 2}$

Figure 2.1.5

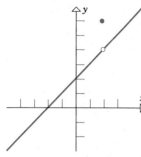

$$y = \begin{cases} x + 2, & x \neq 2 \\ 6, & x = 2 \end{cases}$$

Figure 2.1.6

▶ Example 13   Sketch the graph of

$$\phi(x) = \begin{cases} x + 2, & x \neq 2 \\ 6, & x = 2 \end{cases}$$

Solution. The function ϕ is identical to the function f in Example 10, except at $x = 2$, where we have

$$\phi(2) = 6 \quad \text{and} \quad f(2) = 4$$

Thus, the graph of f (Figure 2.1.3) is identical to the graph of ϕ (Figure 2.1.6), except that the graph of ϕ has a point separated from the line at $x = 2$. ◀

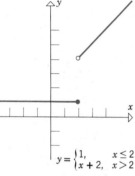

$$y = \begin{cases} 1, & x \leq 2 \\ x + 2, & x > 2 \end{cases}$$

Figure 2.1.7

▶ Example 14   Sketch the graph of

$$g(x) = \begin{cases} 1, & x \leq 2 \\ x + 2, & x > 2 \end{cases}$$

Solution. By definition, the graph of g is the graph of

$$y = \begin{cases} 1, & x \leq 2 \\ x + 2, & x > 2 \end{cases}$$

Thus, for $x \leq 2$, the value of y is always 1 and for $x > 2$, y is given by $y = x + 2$, which is the straight line graphed in Example 10. The graph of g is shown in Figure 2.1.7. ◀

REMARK. In Figure 2.1.7, we used the heavy dot and open circle above $x = 2$ to emphasize that the value $g(2) = 1$ lies on the horizontal line and not on the inclined line.

Sometimes the graph of a function can be obtained by translating the graph of a "simpler" function. In Table 2.1.2 we have listed four operations whose effect is to translate the graph of a function f.

▶ Example 15   Sketch the graph of $g(x) = \sqrt{x + 3}$.

Solution. The function $g(x) = \sqrt{x + 3}$ results when x is replaced by $x + 3$ in the formula $\sqrt{x}$. Thus, from the fourth entry in Table 2.1.2 the graph of g is obtained by translating the graph of $y = \sqrt{x}$ left 3 units (Figure 2.1.8). ◀

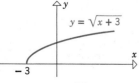

$y = \sqrt{x + 3}$

-3

Figure 2.1.8

▶ Example 16   Sketch the graph of $g(x) = x^2 - 4x + 5$.

$y = x^2 - 4x + 5$

Solution. Complete the square on the first two terms:

$$g(x) = (x^2 - 4x + 4) + 1 = (x - 2)^2 + 1$$

In this form we see that the graph can be obtained by translating the graph of x^2 right 2 units because of the $x - 2$, and up one unit because of the $+1$ (Figure 2.1.9).

Alternate Solution. Use the procedure in Example 4 of Section 1.6. ◄

(2,1)

Figure 2.1.9

Table 2.1.2

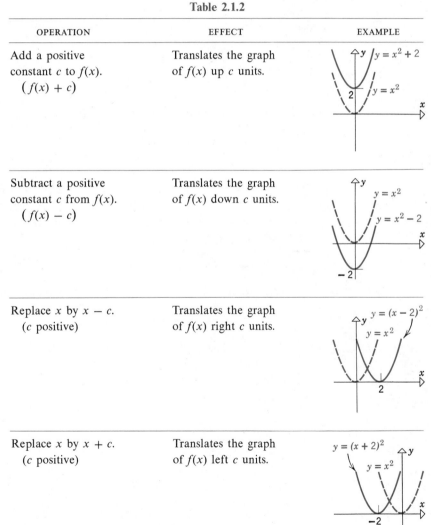

OPERATION	EFFECT	EXAMPLE
Add a positive constant c to $f(x)$. $(f(x) + c)$	Translates the graph of $f(x)$ up c units.	$y = x^2 + 2$, $y = x^2$
Subtract a positive constant c from $f(x)$. $(f(x) - c)$	Translates the graph of $f(x)$ down c units.	$y = x^2$, $y = x^2 - 2$
Replace x by $x - c$. (c positive)	Translates the graph of $f(x)$ right c units.	$y = (x - 2)^2$, $y = x^2$
Replace x by $x + c$. (c positive)	Translates the graph of $f(x)$ left c units.	$y = (x + 2)^2$, $y = x^2$

In each of the last seven examples we found the graph of a *given* function. We shall now consider the converse problem.

2.1.3 PROBLEM *Given a curve in the xy-plane, does there exist a function f whose graph is the given curve?*

▶ Example 17 Show that the graph of

$$3x^2 - 2y = 1 \tag{9}$$

is also the graph of $f(x)$ for some function f.

Solution. Equation (9) can be rewritten in the equivalent form

$$y = \tfrac{1}{2}(3x^2 - 1)$$

so that the graph of (9) is also the graph of the function

$$f(x) = \tfrac{1}{2}(3x^2 - 1) \qquad\qquad\blacktriangleleft$$

▶ Example 18 The curve in Figure 2.1.10*a* cannot be the graph of any function of x. To see why, consider the vertical line in Figure 2.1.10*b*; this line intersects the curve at the two points (a, b) and (a, c). If the curve were the graph of

$$y = f(x) \tag{10}$$

for some function f, then, since (a, b) and (a, c) lie on the curve, (10) would imply that

$$b = f(a) \qquad \text{and} \qquad c = f(a)$$

But this is impossible since f cannot assign two different values to a. Thus, there is no function f whose graph is the curve in Figure 2.1.10*a*. ◀

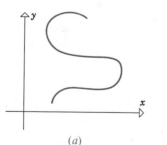

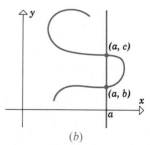

Figure 2.1.10   (*a*)   (*b*)

This example illustrates the following general result, which we will call the **vertical line test.**

2.1.4 THE VERTICAL LINE TEST   *A curve is the graph of f(x) for some function f if and only if no vertical line intersects the curve more than once.*

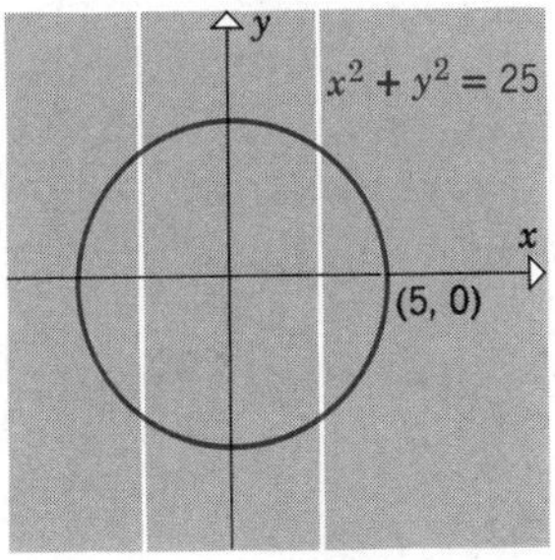

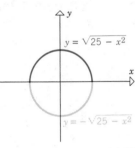

Figure 2.1.11

▶ Example 19   Is the graph of the circle

$$x^2 + y^2 = 25 \tag{11}$$

also the graph of $f(x)$ for some function f?

Solution. Since some vertical lines intersect the circle more than once (see Figure 2.1.11), the circle is not the graph of any function.

We can also deduce this result algebraically by observing that (11) can be written as

$$y = \pm \sqrt{25 - x^2} \tag{12}$$

But the right side of (12) is not a function of x since it is "multiple-valued." Thus, (11) is not equivalent to an equation of the form $y = f(x)$.   ◀

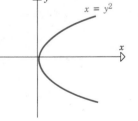

Figure 2.1.12

REMARK. Although the circle $x^2 + y^2 = 25$ is not the graph of any single function f, it follows from (12) that it is the union of the graphs of the *two* functions, $\sqrt{25 - x^2}$ and $-\sqrt{25 - x^2}$. The graph of $y = \sqrt{25 - x^2}$ is the upper semicircle, and the graph of $y = -\sqrt{25 - x^2}$ is the lower semicircle (Figure 2.1.12).

Sometimes it is desirable to reverse the roles of x and y, treating y as the independent variable and x as the dependent variable. Thus, we define the graph of the function $g(y)$ to be the graph of the equation

$$x = g(y)$$

▶ Example 20   Sketch the graph of the function $g(y) = y^2$.

Solution. By definition, the graph of $g(y) = y^2$ is the graph of the equation $x = y^2$ (Figure 2.1.13).   ◀

Figure 2.1.13

Just as the graph of a function of x is cut at most once by any vertical line, so the graph of a function of y is cut at most once by any horizontal line. (See Figure 2.1.13, for example.)

FUNCTIONS DEFINED BY EQUATIONS

Sometimes functions are defined by equations rather than formulas. If the graph of an equation in x and y is cut at most once by any vertical line, then the graph of the equation is also the graph of some function of x, and we say that the equation defines *y as a function of x.* Similarly, an equation is said to define *x as a function of y* if its graph is cut at most once by any horizontal line.

▶ Example 21 The graph of the equation

$$2x + 3y = 6 \tag{13}$$

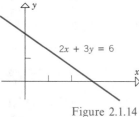

2x + 3y = 6

Figure 2.1.14

is cut at most once by any vertical line and cut at most once by any horizontal line (Figure 2.1.14). Thus, the equation defines *y* as a function of *x* and *x* as a function of *y*. This can also be seen algebraically by solving (13) for *y* and *x*. Solving for *y* yields

$$y = \tfrac{1}{3}(6 - 2x) \tag{14}$$

and solving for *x* yields

$$x = \tfrac{1}{2}(6 - 3y) \tag{15}$$

From (14), we see that (13) is the graph of the function $f(x) = \tfrac{1}{3}(6 - 2x)$, and from (15) we see that (13) is the graph of the function $g(y) = \tfrac{1}{2}(6 - 3y)$. ◀

▶ Example 22 The equation

$$3x^2 - 2y = 1 \tag{16}$$

defines *y* as a function of *x* because it can be rewritten as

$$y = \tfrac{1}{2}(3x^2 - 1) \tag{17}$$

which has the same graph as the function $f(x) = \tfrac{1}{2}(3x^2 - 1)$. However, the equation does not define *x* as a function of *y* because solving (16) for *x* yields

$$x = \pm \sqrt{\tfrac{1}{3}(1 + 2y)}$$

which is "multiple-valued" and therefore does not define a function of *y*.

We leave it for the reader to verify these conclusions geometrically by showing that the graph of (16) is cut at most once by any vertical line, but more than once by some horizontal line. ◀

We conclude this section with some useful terminology. Because the function $f(x)$ stands alone on one side of the equation

$$y = f(x)$$

this equation is said to define *y* **explicitly** as a function of *x*. If an equation defines *y* as a function of *x*, but is written in some form other than $y = f(x)$, then the equation is said to define *y* **implicitly** as a function of *x*. Thus, in

Example 22, formula (16) defines y implicitly as a function of x, and (17) defines y explicitly as a function of x.

Similarly, a function of y can be defined explicitly by an equation $x = g(y)$ or implicitly by an equation of some different form.

▶ Exercise Set 2.1

1. Given that $f(x) = 3x^2 + 2$, find
   (a) $f(-2)$ (b) $f(4)$ (c) $f(0)$
   (d) $f(-\sqrt{3})$ (e) $f(t)$.

2. Given that $g(x) = \dfrac{x + 1}{x - 1}$, find
   (a) $g(1.1)$ (b) $g(\frac{1}{4})$ (c) $g(\sqrt[3]{5} + 1)$
   (d) $g(\pi)$ (e) $g(a - 1)$.

3. Given that
   $$\phi(x) = \begin{cases} \dfrac{1}{x}, & x \geq 3 \\ 2x, & x < 3 \end{cases}$$
   find
   (a) $\phi(2)$ (b) $\phi(-4)$ (c) $\phi(3)$
   (d) $\phi(3.1)$ (e) $\phi(2.9)$.

4. Given that $f(x) = \sqrt{4x - 3}$, find
   (a) $f(\frac{3}{4})$ (b) $f(-\frac{3}{4})$ (c) $f(0)$
   (d) $f(3)$ (e) $f(5.1)$.

5. Given that $H(x) = x/|x|$, find
   (a) $H(3)$ (b) $H(7.1)$ (c) $H(-6)$
   (d) $H(-\pi)$ (e) $H(0)$.

In Exercises 6–17, find the domain of the given function.

6. $f(x) = \dfrac{1}{x - 3}$.

7. $f(x) = \dfrac{1}{5x + 7}$.

8. $g(x) = \sqrt{x^2 - 3}$.

9. $g(x) = \sqrt{(x - 1)(x + 2)}$.

10. $\phi(x) = \sqrt{x^2 + 3}$.

11. $\phi(x) = \dfrac{x}{\sqrt{|x| + 1}}$.

12. $H(x) = x^2, \; x < -3$.

13. $H(x) = \sqrt{x}, \; x \geq 5$.

14. $F(x) = \begin{cases} \sqrt{x}, & x \geq 2 \\ \dfrac{1}{x - 2}, & x < 2. \end{cases}$

15. $F(x) = \begin{cases} \sqrt{-x}, & x \leq -3 \\ 0, & x \geq 2. \end{cases}$

16. $f(x) = \dfrac{(x - 2)(x - 1)}{x - 1}$.

17. $f(x) = \dfrac{(x - 1)(x - 3)}{x - 3}$.

In Exercises 18–37, sketch the graph of the function.

18. $f(x) = 2x + 1$.

19. $f(x) = 3x - 2$.

20. $G(x) = x, \quad 1 \leq x \leq 2$.

21. $G(x) = x - 2, \quad -1 \leq x \leq 1$.

22. $h(x) = x^2 - 3$.

23. $h(x) = (x - 2)^2$.

24. $\phi(x) = \dfrac{x}{|x|}$.

25. $\phi(x) = \dfrac{|x - 2|}{x - 2}$.

26. $g(x) = \begin{cases} x^2, & x \neq 4 \\ 0, & x = 4. \end{cases}$

27. $g(x) = \begin{cases} x - 1, & x \neq 1 \\ 3, & x = 1. \end{cases}$

28. $F(x) = \sqrt{x(x - 1)}$.
29. $F(x) = \sqrt{x^2 + 3x}$.

30. $f(y) = 4y - 2$.

31. $f(y) = 6 - 3y$.

32. $f(y) = \sqrt{y}$.

33. $f(y) = |y|$.

34. $f(x) = \begin{cases} x + 2, & x \leq 3 \\ x + 4, & x > 3. \end{cases}$

35. $f(x) = \begin{cases} x^2, & x > 1 \\ 2, & x \leq 1. \end{cases}$

36. $h(x) = \begin{cases} 1, & 0 < x \leq 1 \\ 3, & 1 < x \leq 2 \\ -1, & 2 < x \leq 3 \\ 0, & \text{elsewhere.} \end{cases}$

37. $h(x) = \begin{cases} -2, & -2 \leq x < -1 \\ 1, & -1 \leq x < 0 \\ 2, & 0 \leq x < 1 \\ 0, & \text{elsewhere.} \end{cases}$

38. Let $g(x) = x - 3$ and let

$$f(x) = \begin{cases} \dfrac{x^2 - 9}{x + 3}, & x \neq -3 \\ k, & x = -3 \end{cases}$$

Find k so that $f(x) = g(x)$ for all x.

39. Let $g(x) = x - 4$ and

$$f(x) = \begin{cases} \dfrac{x^2 - 5x + 4}{x - 1}, & x \neq 1 \\ k, & x = 1 \end{cases}$$

Find k so that $f(x) = g(x)$ for all x.

40. Use Table 2.1.2 and the graph of $y = |x|$ to graph the following:
(a) $y = |x - 4|$ (b) $y = |x| + 4$
(c) $y = |x - 4| + 4$ (d) $y = |x + 5| - 2$.

41. Use Table 2.1.2 and the graph of $y = \sqrt{x}$ to graph the following:
(a) $y = \sqrt{x - 3}$ (b) $y = \sqrt{x} + 3$
(c) $y = \sqrt{x - 3} + 3$ (d) $y = \sqrt{x + 1} - 2$.

In Exercises 42 and 43, determine whether the equation defines y as a function of x, or x as a function of y, or both, or neither.

42. (a) $3x - 4y = 12$ (b) $xy^2 = 1$

(c) $x^2 + y^2 = 1$ (d) $\dfrac{1 + x}{1 - y} = 2$.

43. (a) $4x + 2y = -8$ (b) $x^2 y^3 = 1$

(c) $3x^2 + 4y^2 = 12$ (d) $\dfrac{xy}{1 - xy} = 1$.

44. In each part express x explicitly as a function of y.
(a) $xy - x = 1$ (b) $y = \dfrac{x}{1 + x}$
(c) $x^2 + 2xy + y^2 = 0$

45. In each part express y explicitly as a function of x.
(a) $x^2 y - 1 = 0$ (b) $x = \dfrac{1 - y}{1 + y}$
(c) $y^2 + 2xy + x^2 = 0$.

46. Show that $y^2 + 3xy + x^2 = 0$ does not define y implicitly as a function of x.

47. Show that $y^2 + 4xy + 1 = 0$ does not define y implicitly as a function of x.

48. Find two functions, the union of whose graphs is the graph of the equation in Exercise 47.

49. Criticize the following statement: The function

$$\frac{1 - (1/x)}{1 + (1/x)}$$

can be simplified by multiplying numerator and denominator by x to obtain

$$\frac{1 - (1/x)}{1 + (1/x)} = \frac{x - 1}{x + 1}$$

How would you write the statement to make it accurate?

In Exercises 50 and 51, simplify the function by canceling factors, but be sure not to alter the domain.

50. (a) $f(x) = \dfrac{x^2 - 4}{x + 2}$

(b) $f(x) = \dfrac{(x + 2)(x^2 - 1)}{(x + 2)(x + 1)}$

(c) $f(x) = \dfrac{x^2 + x}{x}$

(d) $f(x) = \dfrac{x + 1 + \sqrt{x + 1}}{\sqrt{x + 1}}$

51. (a) $f(x) = \dfrac{x^2 - 9}{x - 3}$

(b) $f(x) = \dfrac{x^3 + 2x^2 - 3x}{(x - 1)(x + 3)}$

(c) $f(x) = \dfrac{x + \sqrt{x}}{\sqrt{x}}$.

52. Consider the function f graphed in Figure 2.1.15. In each part, find all values of x satisfying the given condition.

(a) $f(x) = 0$ (b) $f(x) = 3$ (c) $f(x) \geq 0$.

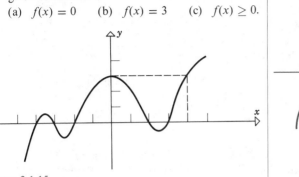

Figure 2.1.15

53. In Figure 2.1.16, determine whether the curve is the graph of a function of x, a function of y, both, or neither.

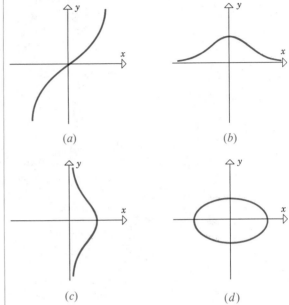

(a)

(b)

(c)

(d)

Figure 2.1.16

2.2   OPERATIONS ON FUNCTIONS; CLASSIFYING FUNCTIONS

Just as numbers can be added, subtracted, multiplied, and divided to produce other numbers, there is a useful way of adding, subtracting, multiplying, and dividing functions to produce other functions. These operations are defined as follows:

2.2.1 DEFINITION   Given functions f and g, their *sum $f + g$*, *difference $f - g$*, *product $f \cdot g$*, and *quotient f/g*, are defined by

$$(f + g)(x) = f(x) + g(x)$$
$$(f - g)(x) = f(x) - g(x)$$
$$(f \cdot g)(x) = f(x) \cdot g(x)$$
$$(f/g)(x) = f(x)/g(x)$$

For the functions $f + g$, $f - g$, and $f \cdot g$ the domain is defined to be the intersection of the domains of f and g, and for f/g the domain is this intersection with the points where $g(x) = 0$ excluded.

▶ Example 1   Let f and g be the functions $f(x) = \sqrt{5 - x}$ and $g(x) = \sqrt{x - 3}$. Then the formulas for $f + g, f - g, f \cdot g$, and f/g are

$$(f + g)(x) = f(x) + g(x) = \sqrt{5 - x} + \sqrt{x - 3} \tag{1}$$
$$(f - g)(x) = f(x) - g(x) = \sqrt{5 - x} - \sqrt{x - 3} \tag{2}$$
$$(f \cdot g)(x) \quad = f(x) \cdot g(x) \quad = \sqrt{5 - x}\sqrt{x - 3} \tag{3}$$
$$(f/g)(x) \quad = f(x)/g(x) \quad = \sqrt{5 - x}/\sqrt{x - 3} \tag{4}$$

Since the domain of f is $(-\infty, 5]$ and the domain of g is $[3, +\infty)$, the domain of $f + g, f - g$, and $f \cdot g$ is

$$[3, 5] \tag{5}$$

because this is the intersection of $(-\infty, 5]$ and $[3, +\infty)$. Since $g(x) = 0$ when $x = 3$, we must exclude this point to obtain the domain of f/g; thus, the domain of f/g is

$$(3, 5] \tag{6}$$

◀

REMARK.   In the last section we agreed that when a function is given by a formula and there is no mention of the domain, it is understood that the domain consists of all real numbers where the formula makes sense and yields real values. Since the set in (5) is precisely where formulas (1), (2), and (3) make sense and yield real values, the formulas themselves specify the domain for us. Similarly, the formula in (4) specifies that (6) is the domain of f/g since this is precisely where the formula makes sense and yields real values. However, as the next example shows, the formula for $f + g, f - g, f \cdot g$, or f/g does not always suffice to specify the domain.

▶ Example 2   Let $f(x) = 3\sqrt{x}$ and $g(x) = \sqrt{x}$. Find $(f \cdot g)(x)$.

Solution.   Because the domain of f is $[0, +\infty)$ and the domain of g is $[0, +\infty)$, the domain of $f \cdot g$ is also $[0, +\infty)$, since this is the intersection of the domains of f and g. The formula for $f \cdot g$ is

$$(f \cdot g)(x) = f(x) \cdot g(x) = (3\sqrt{x}) \cdot (\sqrt{x}) = 3x \tag{7}$$

Because the formula in (7) makes sense and yields real values for all x, the formula alone will not correctly describe the domain of $f \cdot g$ for us; we must write

$$(f \cdot g)(x) = 3x, \qquad x \geq 0$$

◀

Sometimes we will write f^2 to denote the product $f \cdot f$. For example, if $f(x) = 8x$, then

$$f^2(x) = (f \cdot f)(x) = f(x) \cdot f(x) = (8x) \cdot (8x) = 64x^2$$

Similarly, we will denote $f^2 \cdot f$ by f^3, $f^3 \cdot f$ by f^4, and so on.

Sometimes a given function has the same effect as two simpler functions evaluated in succession. For example, consider the function h given by

$$h(x) = (x + 1)^2$$

To evaluate $h(x)$ we first compute $x + 1$ and then square the result. In other words, if we consider the functions g and f given by

$$g(x) = x + 1 \quad \text{and} \quad f(x) = x^2$$

then

$$h(x) = (x + 1)^2 = [g(x)]^2 = f(g(x))$$

Thus, h has the same effect as g and f evaluated successively. Loosely speaking, h is "composed" of the two functions f and g. This idea is formalized in the following definition.

2.2.2 DEFINITION   Given two functions f and g, the ***composition of f with g,*** denoted by $f \circ g$, is the function defined by

$$(f \circ g)(x) = f(g(x))$$

where the domain of $f \circ g$ consists of all x in the domain of g for which $g(x)$ is in the domain of f.

▶ **Example 3**   Find $(f \circ g)(x)$ if $f(x) = x^2 + 3$ and $g(x) = \sqrt{x}$.

Solution.   Since the domain of g is $[0, +\infty)$ and the domain of f is $(-\infty, +\infty)$, the domain of $f \circ g$ consists of all x in $[0, +\infty)$ such that $g(x) = \sqrt{x}$ lies in $(-\infty, +\infty)$; thus, the domain of $f \circ g$ is $[0, +\infty)$. Since

$$f(g(x)) = [g(x)]^2 + 3 = (\sqrt{x})^2 + 3 = x + 3$$

the composition of f with g is given by

$$(f \circ g)(x) = x + 3, \quad x \geq 0 \qquad \blacktriangleleft$$

▶ **Example 4**   Let $f(x) = x - 1$ and $g(x) = \sqrt{x}$. Find

(a) $(f \circ g)(x)$ (b) $(g \circ f)(x)$

Solution (a). Since the domain of g is $[0, +\infty)$ and the domain of f is $(-\infty, +\infty)$, the domain of $f \circ g$ consists of all x in $[0, +\infty)$ such that $g(x) = \sqrt{x}$ lies in $(-\infty, +\infty)$, that is, all x in $[0, +\infty)$. Since

$$f(g(x)) = g(x) - 1 = \sqrt{x} - 1$$

we obtain

$$(f \circ g)(x) = \sqrt{x} - 1 \tag{8}$$

In (8) there is no need to indicate that the domain is $[0, +\infty)$ since this is precisely the set where $\sqrt{x} - 1$ is defined and yields real values.

Solution (b). The domain of $g \circ f$ consists of all x in the domain of f such that $f(x)$ lies in the domain of g. Since the domain of f is $(-\infty, +\infty)$ and the domain of g is $[0, +\infty)$, the domain of $g \circ f$ consists of all x in $(-\infty, +\infty)$ such that $f(x) = x - 1$ lies in $[0, +\infty)$; thus, the domain is $[1, +\infty)$. Since

$$g(f(x)) = \sqrt{f(x)} = \sqrt{x - 1}$$

we have

$$(g \circ f)(x) = \sqrt{x - 1} \tag{9}$$

[As in part (a), there is no need to indicate in (9) that the domain is $[1, +\infty)$, since this is precisely the set where $\sqrt{x - 1}$ is defined and yields real values.] ◄

► **Example 5** Express $f(x) = (x - 4)^5$ as a composition of two functions.

Solution. To evaluate $f(x)$ for a given value of x we would first compute $x - 4$ and then raise the result to the fifth power. Therefore, if

$$g(x) = x^5 \qquad \text{and} \qquad h(x) = x - 4$$

then

$$g(h(x)) = [h(x)]^5 = (x - 4)^5$$

so $f = g \circ h$. ◄

CLASSIFICATION OF FUNCTIONS

A function that assigns the same value to every member of its domain is called a ***constant function.***

► **Example 6** The function f given by

$$f(x) = 3$$

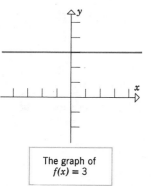

The graph of
$f(x) = 3$

Figure 2.2.1

is constant since it assigns the value 3 to every real number x. By definition, the graph of f is the graph of the equation $y = f(x)$; that is, $y = 3$. (See Figure 2.2.1.) ◀

REMARK. The constant function that assigns the value c to each real number is sometimes called *the constant function c.* Thus, the function f in Example 6 is *the constant function* 3.

A function of the form cx^n, where c is a constant and n is a nonnegative integer, is called a *monomial in x.* Examples are:

$$2x^3, \qquad \pi x^7, \qquad 4x^0 \,(= 4), \qquad -6x, \qquad \text{and} \qquad x^{17}$$

The functions $4x^{1/2}$ and x^{-3} are *not* monomials because the powers of x are not nonnegative integers.

A function that is expressible as the sum of finitely many monomials in x is called a *polynomial in x.* Examples are

$$x^3 + 4x + 7, \qquad 17 - \tfrac{2}{3}x, \qquad 9, \qquad \text{and} \qquad x^5$$

Also,

$$(x^2 - 4)^3$$

is a polynomial in x because it is expressible as a sum of monomials (by carrying out the cubing operation). In general, f is a polynomial in x if it is expressible in the form

$$f(x) = a_0 + a_1 x + a_2 x^2 + \cdots + a_n x^n$$

where n is a nonnegative integer and $a_0, a_1, a_2, \ldots, a_n$ are real constants. The domain of a polynomial is $(-\infty, +\infty)$.

A polynomial is called *linear* if it has the form

$$a_0 + a_1 x \qquad \text{(where } a_1 \neq 0\text{)}$$

quadratic if it has the form

$$a_0 + a_1 x + a_2 x^2 \qquad \text{(where } a_2 \neq 0\text{)}$$

and *cubic* if it has the form

$$a_0 + a_1 x + a_2 x^2 + a_3 x^3 \qquad \text{(where } a_3 \neq 0\text{)}$$

A function that is expressible as a ratio of two polynomials is called a *rational function.* Examples are:

$$\frac{x^5 - 2x^2 + 1}{x^2 - 4}, \qquad \frac{x}{x + 1}, \qquad \frac{1}{x^5}$$

In general, f is a rational function if it is expressible in the form

$$f(x) = \frac{a_0 + a_1 x + \cdots + a_n x^n}{b_0 + b_1 x + \cdots + b_m x^m}$$

The domain of f consists of all x where the denominator differs from zero.

An **explicit algebraic function** is a function that can be evaluated using finitely many additions, subtractions, multiplications, divisions, and root extractions. For example,

$$f(x) = x^{2/3} = (\sqrt[3]{x})^2 \quad \text{and} \quad g(x) = \frac{(x-3)\sqrt[4]{x}}{x^5 + \sqrt{x^2 + 1}}$$

define explicit algebraic functions of x. Also, polynomials and rational functions are explicit algebraic functions.

All remaining functions fall into two categories, **implicit algebraic functions** and **transcendental functions**. We shall not define these terms, but instead refer the interested reader to a classic book in calculus, G. H. Hardy,* *A Course of Pure Mathematics,* Cambridge Press, 1958 (10th edition). Among the transcendental functions already familiar to the reader are the trigonometric functions.

---

* G. H. HARDY (1877–1947). Hardy was a world-renowned British mathematician. He taught at Cambridge and Oxford universities and was a prolific researcher who produced over 300 research papers. He received numerous medals and honorary degrees for his accomplishments. Hardy's book, *A Course of Pure Mathematics,* which gave the first rigorous English exposition of functions and limits for the college undergraduate, had a great impact on university teaching. Hardy had a rebellious spirit; he once listed among his most ardent wishes: (1) to prove the Riemann hypothesis (a famous unsolved mathematical problem), (2) to make a brilliant play in a crucial cricket match, (3) to prove the nonexistence of God, and (4) to murder Mussolini.

▶ Exercise Set 2.2

1. Let $f(x) = x^2 + 1$. Find
   (a) $f(t)$
   (b) $f(t + 2)$
   (c) $f(x + 2)$
   (d) $f\left(\dfrac{1}{x}\right)$
   (e) $f(x + h)$
   (f) $f(-x)$
   (g) $f(\sqrt{x})$
   (h) $f(3x)$.

2. Let $g(x) = \sqrt{x}$. Find
   (a) $g(5s + 2)$
   (b) $g(\sqrt{x} + 2)$
   (c) $3g(5x)$
   (d) $\dfrac{1}{g(x)}$
   (e) $g(g(x))$
   (f) $g^2(x)$

   (g) $g\left(\dfrac{1}{\sqrt{x}}\right)$
   (h) $g((x - 1)^2)$.

In Exercises 3–8, find
   (a) $f + g$
   (b) $f - g$
   (c) $f \cdot g$
   (d) f/g
   (e) $f \circ g$
   (f) $g \circ f$
   (g) kf

where k is a constant.

3. $f(x) = 2x$, $g(x) = x^2 + 1$.
4. $f(x) = 3x - 2$, $g(x) = |x|$.
5. $f(x) = \sqrt{x} + 1$, $g(x) = x - 2$.
6. $f(x) = \dfrac{x}{1 + x^2}$, $g(x) = \dfrac{1}{x}$.

7. $f(x) = \sqrt{x - 2}$, $g(x) = \sqrt{x - 3}$.

8. $f(x) = x^3$, $g(x) = \dfrac{1}{\sqrt[3]{x}}$.

9. Let h be defined by $h(x) = 2x - 5$. Find
    (a) $h \circ h$ (b) h^2.

In Exercises 10–15, express f as a composition of two functions, that is, find functions g and h such that $f = g \circ h$. (Each exercise has more than one correct solution.)

10. $f(x) = x^2 + 1$.

11. $f(x) = \sqrt{x + 2}$.

12. $f(x) = \dfrac{1}{x - 3}$.

13. $f(x) = (x - 5)^7$.

14. $f(x) = a + bx$.

15. $f(x) = |x^2 - 3x + 5|$.

16. Is it ever true that $f \circ g = g \circ f$? Is it always true that $f \circ g = g \circ f$?

17. Let $f(x) = \dfrac{1}{x}$.

    (a) If $g(x) = x^2 + 1$, show that $f \circ g$ is defined for all x even though f is not defined when $x = 0$.
    (b) Can you find a different function g such that $f \circ g$ is defined for all x?
    (c) What property must a function g have in order for $f \circ g$ to be defined for all x?

For the functions in Exercises 18–21, state which of the following terms apply: monomial, polynomial, rational function, explicit algebraic function.

18. (a) $3x^7$ (b) $4x^{1/7}$

    (c) $\dfrac{1}{5x^6}$ (d) $2x^3 - 1$.

19. (a) $2x^{1/3} + 1$ (b) x^{-2}

    (c) $x^{-1/2}$ (d) $(x - 3)^{12}$.

20. (a) $x^2 \sqrt{x^3 - 3}$ (b) $\dfrac{x + 1}{x + 2}$

    (c) $\pi^{-2} + 1$ (d) $|x|$.

21. (a) $\sqrt{x^2 + \sqrt{x}}$ (b) $\dfrac{x^3 - 2x + 1}{x^2 - 9}$

    (c) $\sqrt{\pi} - 3$ (d) $|x - 2|$.

22. A function f is called **even** if $f(-x) = f(x)$ for each x in the domain of f, and **odd** if $f(-x) = -f(x)$ for each such x. In each part, classify the function as even, odd, or neither.

    (a) $f(x) = x^2$ (b) $f(x) = x^3$
    (c) $f(x) = |x|$ (d) $f(x) = x + 1$
    (e) $f(x) = \dfrac{x^5 - x}{1 + x^2}$ (f) $f(x) = 2$.

23. In Figure 2.2.2 we have sketched part of the graph of a function f (the part to the right of the y-axis.) Complete the graph assuming
    (a) f is an even function
    (b) f is an odd function.
    (See Exercise 22 for the definitions of even and odd functions.)

Figure 2.2.2

24. Classify the functions graphed in Figure 2.2.3 as even, odd, or neither. (See Exercise 22 for the definitions of even and odd.)

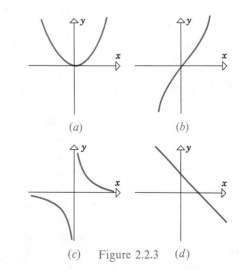

(a)   (b)

(c)   Figure 2.2.3   (d)

25. Can a function be both even and odd? (See Exercise 22 for the definitions of even and odd functions.)

26. Let a be a constant and suppose $f(a - x) = f(a + x)$ for all x. What geometric property must the graph of f have?

27. Find $(f \circ g)(x)$ if

$$f(x) = \sqrt{x - 4}$$

and

$$g(x) = \frac{1}{2}x + 1, \qquad x \geq 6$$

28. Let

$$f(x) = \begin{cases} 5x, & x \leq 0 \\ -x, & 0 < x \leq 8 \\ \sqrt{x}, & x > 8 \end{cases} \quad \text{and} \quad g(x) = x^3$$

Find $(f \circ g)(x)$.

29. Prove or disprove: For any three functions f, g, and h, $f \circ (g \circ h) = (f \circ g) \circ h$.

2.3 INTRODUCTION TO CALCULUS: TANGENTS AND VELOCITY

Calculus centers around the following two fundamental problems.

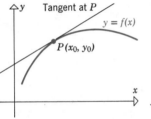

Figure 2.3.1

THE TANGENT PROBLEM.   Given a function f and a point $P(x_0, y_0)$ on its graph, find the equation of the tangent to the graph at P (Figure 2.3.1).

THE AREA PROBLEM.   Given a function f, find the area between the graph of f and an interval $[a, b]$ on the x-axis (Figure 2.3.2).

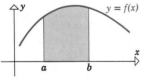

Figure 2.3.2

Traditionally, that portion of calculus arising from the tangent problem is called **differential calculus** and that portion arising from the area problem is called **integral calculus.** As we shall see, however, the tangent and area problems are closely related, so that the distinction between differential calculus and integral calculus is often hard to discern. For now, we will focus on the tangent problem.

In plane geometry, a line is called **tangent** to a circle if it meets the circle at precisely one point (Figure 2.3.3a). However, this definition is not satisfactory for other kinds of curves. In Figure 2.3.3b the line meets the curve exactly once, yet is not a tangent, and in Figure 2.3.3c the line is tangent yet meets the curve more than once. It is important to define mathematically the concept of a tangent so that the definition applies not only to circles, but to other curves as well. To motivate the appropriate definition, we must approach the tangent concept in a different way. Consider the tangent to a circle at a point P and consider a secant line from P to another point Q on the circle. If, as shown in Figure 2.3.4, we move the point Q along the circle toward P, then the position of the secant line moves closer and closer to the

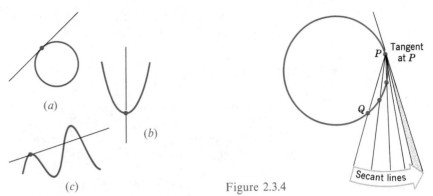

Figure 2.3.3

(a)

(b)

(c)

Figure 2.3.4

Secant lines

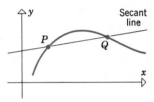

Secant line

Figure 2.3.5

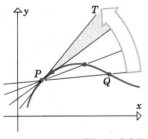

Figure 2.3.6

position of the tangent at P. In other words, the secant lines "approach" the tangent line as Q approaches P along the circle. Using this idea we will be able to define the notion of a tangent so that it applies to curves other than circles.

Consider a point P on a curve in the xy-plane. To define the tangent to the curve at P it suffices to define the *slope* of the tangent at P since the slope and the point P together can be used to obtain the equation of the tangent (from the point-slope form of a line). If Q is any point on the curve different from P, the line through P and Q is called a **secant line** for the curve (Figure 2.3.5). As with the circle, intuition suggests that if we move the point Q along the curve toward P, the secant line will rotate toward a "limiting" position (Figure 2.3.6). The line T occupying this limiting position we define to be the **tangent line** at P.

With this *intuitive* understanding of the tangent we can now discuss how to determine the slope of a tangent line. Let us suppose that $P(x_0, y_0)$ is a point on the graph of a function f, and assume we want to obtain the slope $m_{\tan}$ of the tangent to the graph at P. If we let $Q(x_1, y_1)$ be any point on the graph of f, different from P, then as suggested by Figure 2.3.7 the slope $m_{\sec}$ of the secant line joining P and Q is

$$m_{\sec} = \text{slope of } PQ = \frac{y_1 - y_0}{x_1 - x_0} \tag{1}$$

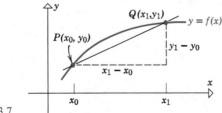

Figure 2.3.7

But (x_0, y_0) and (x_1, y_1) lie on the graph of f so that $y_0 = f(x_0)$ and $y_1 = f(x_1)$. Thus, (1) can be rewritten as

$$m_{\text{sec}} = \frac{f(x_1) - f(x_0)}{x_1 - x_0} \tag{2}$$

As Q approaches P along the graph of f, or equivalently as x_1 gets closer and closer to x_0, the secant line through P and Q approaches the tangent line at P, so the slope m_{sec} of the secant line approaches the slope m_{tan} of the tangent line. Therefore, (2) suggests the definition

$$m_{\text{tan}} = \begin{array}{c} \text{limiting value} \\ \text{as } x_1 \text{ approaches } x_0 \\ \text{of} \end{array} \frac{f(x_1) - f(x_0)}{x_1 - x_0} \tag{3}$$

This discussion would be complete if it were not for the fact that the notion of a "limiting value" is an intuitive idea, not yet mathematically defined. One of our primary objectives is to give a precise definition of the term "limit" and thereby complete the definition of a tangent. However, an intuitive understanding of these notions will suffice in the following example.

▶ **Example 1** Find the slope and the equation of the tangent line to the graph of $f(x) = x^2$ at the point $P(3, 9)$. (See Figure 2.3.8.)

Solution. From (2) with $x_0 = 3$,

$$m_{\text{sec}} = \frac{f(x_1) - f(3)}{x_1 - 3}$$

or, since $f(x) = x^2$,

$$m_{\text{sec}} = \frac{x_1^2 - 9}{x_1 - 3} = \frac{(x_1 + 3)(x_1 - 3)}{x_1 - 3}$$

Therefore,

$$m_{\text{sec}} = x_1 + 3, \qquad x_1 \neq 3 \tag{4}$$

If we now let x_1 approach 3, then the value of m_{sec} in (4) will approach 6. Thus, the slope of the tangent line at $P(3, 9)$ is

$$m_{\text{tan}} = 6$$

Using this slope and the point $P(3, 9)$, the point-slope form of the tangent line is

$$(y - 9) = 6(x - 3)$$

and the slope-intercept form is $y = 6x - 9$. ◀

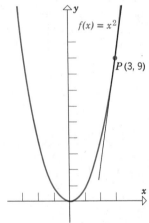

$f(x) = x^2$

$P(3, 9)$

Figure 2.3.8

While the tangent problem is of interest as a matter of pure geometry, much of the impetus for studying this problem arose in the seventeenth century when scientists recognized that many problems involving objects moving with variable velocity could be reduced to problems involving tangents. To see why this is so, we need to examine critically the meaning of the word "velocity."

If a car travels 75 miles over a straight road in a 3-hour period, then we say that the average velocity of the car is 25 miles per hour (25 mi/hr). More generally, the *average velocity* of an object moving in *one direction* along a line is

$$\text{average velocity} = \frac{\text{distance traveled}}{\text{time elapsed}}$$

Obviously, if an object travels with an average velocity of 25 mi/hr during a 3-hour trip, it need not travel at a fixed velocity of 25 mi/hr; sometimes it may speed up and sometimes it may slow down.

While average velocity is useful for some purposes, it is not always significant in physical problems. For example, if a moving car strikes a tree, the damage sustained is not determined by the average velocity up to the time of impact, but rather the instantaneous velocity at the precise moment of impact.

A clear understanding of instantaneous velocity evaded scientists until the advent of calculus in the seventeenth century. The subtlety of this concept was nicely described by Morris Kline* who wrote,

> In contrasting average velocity with instantaneous velocity we implicitly utilize a distinction between interval and instant. . . . An average velocity is one that concerns what happens over an interval of time—3 hours, 5 seconds, one-half second, and so forth. The interval may be small or large, but it does represent the passage of a definite amount of time. We use the word instant, however, to state the fact that something happens so fast that no time elapses. The event is momentary. When we say, for example, that it is 3 o'clock, we refer to an instant, a precise moment. If the lapse of time is pictured by length along a line, then an interval (of time) is represented by a line segment, whereas an instant corresponds to a point. The notion of an instant, although it is used in everyday life, is strictly a mathematical idealization.
>
> Our ways of thinking about real events cause us to speak in terms of

---

*MORRIS KLINE (1908–   ) American mathematician, scholar, and educator. Kline has made numerous contributions to mathematical thought, written extensively on education, especially mathematics education, and has taught, lectured, and served as a consultant throughout his very active career. He is the author of many popular books including *Mathematical Thought from Ancient to Modern Times* and *Why Johnny Can't Add: The Failure of the New Mathematics.* I wish to thank him for permission to use the above quotation, which is taken from *Calculus: An Intuitive and Physical Approach,* Wiley, New York, 1977, p. 17.

instants and velocity at an instant, but closer examination shows that the concept of velocity at an instant presents difficulties. Average velocity, which is simply the distance traveled during some interval of time divided by that amount of time, is easily calculated. Suppose, however, that we try to carry over this process to instantaneous velocity. The distance an automobile travels in one instant is 0 and the time that elapses during one instant is also 0. Hence the distance divided by the time is 0/0, which is meaningless. Thus, although instantaneous velocity is a physical reality, there seems to be a difficulty in calculating it, and unless we can calculate it, we cannot work with it mathematically.

In order to understand the concept of instantaneous velocity and work with it mathematically we must think in terms of approximating instantaneous velocity by average velocity. For example, suppose we are interested in the instantaneous velocity of a car at a certain instant of time, say, exactly 5 sec after it starts to move along a straight road. Although the velocity of the car may be changing, it is evident that over a short interval of time, say, 0.1 sec, the velocity does not vary much. Thus, we make only a small error if we approximate the instantaneous velocity after 5 sec by the *average velocity* over the *interval* from 5 to 5.1 sec. This average velocity can be calculated by measuring the distance traveled during the time interval between 5 and 5.1 sec and then dividing by the time elapsed, which is 0.1 sec. Thus, even though the instantaneous velocity after 5 sec cannot be calculated directly, it can be approximated by an average velocity that can be determined from physical measurements.

The error that results from approximating instantaneous velocity by average velocity can be reduced by shrinking the time interval over which the average velocity is computed. For example, we would expect that the average velocity between 5 and 5.01 sec better approximates the instantaneous velocity after 5 sec than does the average velocity between 5 and 5.1 sec, because the time interval is smaller. Thus, by computing average velocity over smaller and smaller time intervals, we expect to get closer and closer to the instantaneous velocity.

We stated earlier that there is a relationship between the tangent problem and physical problems involving objects moving with varying velocity. To illustrate this relationship we will consider the simplest kind of motion, a particle moving in one direction along a line. For example, the particle might idealize a rock dropped toward earth from a height, a billiard ball moving across a table, or a car traveling on a straight track.

For convenience, introduce a coordinate line along the path of the particle. Let the positive direction be the direction of motion, and imagine that a clock is keeping track of the elapsed time t, starting with $t = 0$ initially (Figure 2.3.9). After t units of time, the particle will be at a certain distance s from the origin. As t changes, so does s; thus, s is a function of t, say,

$$s = f(t) \tag{5}$$

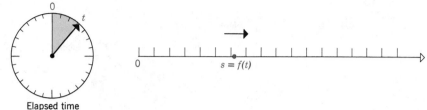

Figure 2.3.9   Elapsed time

For example, if a rock, initially at rest, is dropped from a height not too far above the surface of the earth, we will see later that the relationship between s and t is approximately

$$s = 16t^2 \tag{6}$$

where t is the number of seconds elapsed from the time the rock is released and s is the distance in feet from the starting point.

If we graph (5) with the t-axis horizontal and the s-axis vertical, we obtain a *position versus time curve* (Figure 2.3.10). Using this curve, we can interpret average velocity and instantaneous velocity geometrically. For example, suppose we are interested in the average velocity of the particle over the time interval from t_0 to t_1. At time t_0 the particle is at some distance s_0 from the origin and at time t_1 it is at some distance s_1. Over the time interval from t_0 to t_1 the distance traveled is

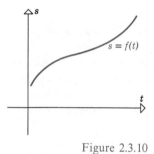

Figure 2.3.10

$$s_1 - s_0$$

and the time elapsed is

$$t_1 - t_0$$

so that the average velocity during the interval is given by

$$\text{average velocity} = \frac{s_1 - s_0}{t_1 - t_0} \tag{7}$$

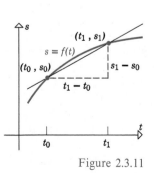

Figure 2.3.11

However, the points (t_0, s_0) and (t_1, s_1) lie on the position versus time curve, so that expression (7) is also the slope of the secant line connecting these points (Figure 2.3.11). In other words:

> **Geometric Interpretation of Average Velocity**
> For a particle moving in one direction on a straight line, the average velocity between time t_0 and t_1 is represented geometrically by the slope of the secant line connecting (t_0, s_0) and (t_1, s_1) on the position versus time curve.

If we choose t_1 close to t_0, then the average velocity between times t_0 and t_1 closely approximates the instantaneous velocity at time t_0; moreover, as

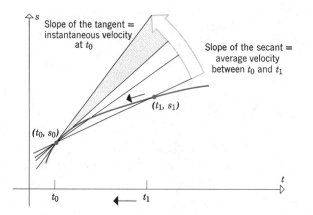

Figure 2.3.12

we move t_1 closer and closer to t_0, intuition suggests that these approximations will approach the exact value of the instantaneous velocity at t_0. However, as t_1 moves toward t_0, the point (t_1, s_1) moves toward (t_0, s_0) on the position versus time curve (Figure 2.3.12), so that the slope of the secant line between (t_0, s_0) and (t_1, s_1) approaches the slope of the tangent line at (t_0, s_0). Since the slope of the secant line represents the average velocity between times t_0 and t_1, we conclude that:

> **Geometric Interpretation of Instantaneous Velocity**
> For a particle moving in one direction on a straight line, the instantaneous velocity at time t_0 is represented geometrically by the slope of the tangent line at (t_0, s_0) on the position versus time curve.

▶ **Example 2** If a rock is dropped from a height not too far above the surface of the earth, then it is shown in physics that the relationship between the distance traveled the time elapsed is approximately

$$s = 16t^2 \tag{8}$$

where t is the number of seconds elapsed from the time the rock is released and s is its distance in feet from the starting point.
(a) Find the instantaneous velocity of the rock after t_0 sec.
(b) Find the instantaneous velocity of the rock after 3 sec.

Solution (a). Let s_0 be the distance traveled after t_0 sec and s_1 the distance traveled after t_1 sec. From (8),

$$s_0 = 16t_0{}^2 \quad \text{and} \quad s_1 = 16t_1{}^2$$

so that the average velocity in the time interval from t_0 to t_1 sec is

$$\text{average velocity} = \frac{s_1 - s_0}{t_1 - t_0} = \frac{16t_1{}^2 - 16t_0{}^2}{t_1 - t_0} \tag{9}$$

Factoring (9) yields

$$\text{average velocity} = \frac{16(t_1 - t_0)(t_1 + t_0)}{t_1 - t_0}$$

or equivalently

$$\text{average velocity} = 16(t_1 + t_0), \qquad t_1 \neq t_0 \tag{10}$$

As t_1 approaches t_0, the average velocity between t_0 and t_1 approaches the instantaneous velocity at t_0. But as t_1 approaches t_0, the quantity $16(t_1 + t_0)$ approaches $16(t_0 + t_0) = 32t_0$, so

$$\text{instantaneous velocity} = 32t_0 \tag{11}$$

Since (8) assumes that distance is measured in feet and time in seconds, (11) tells us that the instantaneous velocity after t_0 sec is $32t_0$ feet per second (ft/sec).

Solution (b).   Letting $t_0 = 3$ in (11), we find that the instantaneous velocity of the rock after 3 sec is $32(3) = 96$ ft/sec.   ◄

▶ Exercise Set 2.3

1. Let $f(x) = \frac{1}{2}x^2$.
    (a) Find the slope of the secant line between those points on the graph of f for which $x = 3$ and $x = 4$.
    (b) Use the method of Example 1 to find the slope and equation for the tangent to the graph of f at the point where $x = 3$.
    (c) Sketch the graph of f together with the secant and tangent lines from (a) and (b).

2. Let $f(x) = x^3$.
    (a) Find the slope of the secant line between those points on the graph of f for which $x = 1$ and $x = 2$.
    (b) Use the method of Example 1 to find the

slope and equation for the tangent to the graph of f at the point where $x = 1$.
    (c) Sketch the graph of f together with the secant and tangent lines from (a) and (b).

3. Let $f(x) = \dfrac{1}{x}$.
    (a) Find the slope of the secant line between those points on the graph of f for which $x = 2$ and $x = 3$.
    (b) Use the method of Example 1 to find the slope and equation for the tangent to the graph of f at the point where $x = 2$.
    (c) Sketch the graph of f together with the secant and tangent lines from (a) and (b).

4. Let $f(x) = \dfrac{1}{x^2}$.

(a) Find the slope of the secant line between those points on the graph of f at which $x = 1$ and $x = 2$.

(b) Use the method of Example 1 to find the slope and equation for the tangent to the graph of f at the point where $x = 1$.

(c) Sketch the graph of f together with the secant and tangent lines from (a) and (b).

5. Let $f(x) = x^3$.

(a) Use the method of Example 1 to show that the slope of the tangent to the graph of f at the point where $x = x_0$ is $3x_0{}^2$.

(b) Use the result in (a) to find the equation of the tangent to the graph of f at the point where $x = 5$.

(c) Use the result in (a) to find the equation of the tangent to the graph of f at the point where $x = x_0$.

6. Let $f(x) = \dfrac{1}{x}$.

(a) Use the method of Example 1 to show that the slope of the tangent to the graph of f at the point where $x = x_0$ is $-1/x_0{}^2$.

(b) Use the result in (a) to find the equation of the tangent to the graph of f at the point where $x = -7$.

(c) Use the result in (a) to find the equation of the tangent to the graph of f at the point where $x = x_0$.

7. Let $f(x) = x^2 + x$.

(a) Use the method of Example 1 to find the slope of the tangent to the graph of f at the point where $x = x_0$.

(b) Use the result in (a) to find the equation of the tangent to the graph of f at the point where $x = 2$.

(c) Use the result in (a) to find the equation of the tangent to the graph of f at the point where $x = x_0$.

8. Follow the directions of Exercise 7 for the function $f(x) = x^2 + 3x + 2$.

9. Figure 2.3.13 shows the position versus time curve for a certain particle moving on a straight line.

(a) Is the particle moving faster at time t_0 or time t_2? Explain.

(b) At the origin, the tangent is horizontal. What does this tell us about the initial velocity of the particle?

(c) Is the particle speeding up or slowing down in the interval $[t_0, t_1]$? Explain.

(d) Is the particle speeding up or slowing down in the interval $[t_1, t_2]$? Explain.

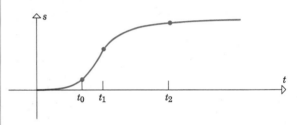

Figure 2.3.13

10. An automobile, initially at rest, begins to move along a straight track. The velocity increases continually until suddenly the driver sees a concrete barrier in the road and applies the brakes sharply at time t_0. The car decelerates rapidly, but it is too late—the car crashes into the barrier at time t_1 and instantaneously comes to rest. Sketch a position versus time curve that might represent the motion of the car.

11. If a particle moves at constant velocity, what can you say about its position versus time curve?

12. A rock is dropped from a height of 576 ft and falls toward earth in a straight line. In t sec the rock drops $16t^2$ ft.

(a) How long does it take for the rock to hit the ground?

(b) What is the average velocity of the rock during the time it is falling?

(c) What is the average velocity of the rock for the first 3 sec?

(d) What is the velocity of the rock when it hits the ground?

13. During the first 40 sec of a rocket flight, the rocket is propelled straight up so that in t sec it reaches a height of $5t^3$ ft.
  (a) How high does the rocket travel in 40 sec?
  (b) What is the average velocity of the rocket during the first 40 sec?
  (c) What is the average velocity of the rocket during the first 135 ft of its flight?
  (d) What is the velocity of the rocket at the end of 40 sec?

14. A particle moves on a line away from its initial position so that after t hr it is $s = 3t^2 + t$ mi from its initial position.
  (a) Find the average velocity of the particle over the interval [1, 3].
  (b) Find the instantaneous velocity at $t = 1$.

15. A particle moves in one direction along a straight line so that after t min its distance is $s = 6t^4$ ft from the origin.
  (a) Find the average velocity of the particle over the interval [2, 4].
  (b) Find the instantaneous velocity at $t = 2$.

16. A car is traveling on a straight road that is 120 mi long. For the first 100 mi the car travels at an average velocity of 50 mi/hr. Show that no matter how fast the car travels for the final 20 mi it cannot bring the average velocity up to 60 mi/hr for the entire trip.

17. Let $f(x) = x^2$. If we approximate the slope of the tangent at the point $(x_0, f(x_0))$ by the slope of the secant line between $(x_0, f(x_0))$ and $(x_1, f(x_1))$, show that the error is $|x_1 - x_0|$. (By error we mean $|m_{tan} - m_{sec}|$, where m_{tan} is the slope of the tangent line and m_{sec} the slope of the secant line.)

2.4 LIMITS (AN INTUITIVE INTRODUCTION)

In the last section we saw that the concepts of tangent and instantaneous velocity ultimately rest on the notion of a "limit" or "value approached by" a function. In this section as well as the next few we will investigate the notion of limit in more detail. Our development of limits in this text proceeds in three stages:

1. First we discuss limits intuitively.

2. Then we discuss methods for computing limits.

3. Finally, we give a precise mathematical discussion of limits.

Limits are used to describe how a function behaves as the independent variable moves toward a certain value. To illustrate, consider the function

$$f(x) = \frac{\sin x}{x}$$

where x is in radians. Although this function is not defined at $x = 0$, it still makes sense to ask what happens to the values of $f(x)$ as x moves along the x-axis toward $x = 0$. To answer this question, we used a calculator to obtain

Table 2.4.1

x	$\dfrac{\sin x}{x}$ (x radians)
1.0	0.84147
0.9	0.87036
0.8	0.89670
0.7	0.92031
0.6	0.94107
0.5	0.95885
0.4	0.97355
0.3	0.98507
0.2	0.99335
0.1	0.99833
0.01	0.99998

Table 2.4.2

x	$\dfrac{\sin x}{x}$ (x in radians)
−1.0	0.84147
−0.9	0.87036
−0.8	0.89670
−0.7	0.92031
−0.6	0.94107
−0.5	0.95885
−0.4	0.97355
−0.3	0.98507
−0.2	0.99335
−0.1	0.99833
−0.01	0.99998

values of $f(x)$ at a succession of points moving along the *positive* x-axis toward $x = 0$. The results, which appear in Table 2.4.1, suggest that the values of $f(x)$ approach 1. We call the number 1 the **limit** of

$$\frac{\sin x}{x}$$

as x approaches 0 from the right side, and we write

$$\lim_{x \to 0^+} \frac{\sin x}{x} = 1 \tag{1}$$

In this expression, "lim" tells us that we are computing a limit; the symbol $x \to 0$ tells us that we are letting x approach 0; and the label "$+$" on $x \to 0$ tells us that x is approaching zero from the *right* side. We can also ask what happens to the values of

$$f(x) = \frac{\sin x}{x}$$

as x approaches 0 from the left side. From Table 2.4.2 [or from the fact that $f(-x) = f(x)$] it is evident that the values of $f(x)$ again approach 1. We denote this by writing

$$\lim_{x \to 0^-} \frac{\sin x}{x} = 1 \tag{2}$$

In (2), the label "$-$" on $x \to 0$ tells us that we are computing the limit as x approaches zero from the *left* side.

REMARK. It is important to keep in mind that the limits given in (1) and (2) are really just guesses based on numerical evidence. It is conceivable that if we continue the calculator computations in Tables 2.4.1 and 2.4.2 for values of x still closer to 0, the pattern of values for $(\sin x)/x$ might change and approach some value different from 1 or perhaps approach no number at all. Moreover, no matter how much we enlarge our tables, this problem will persist since we can never be certain what happens for values of x not included in the table. To be certain that the limits in (1) and (2) are really correct, we need mathematical proof, and to give a mathematical proof, we need a precise mathematical definition of a limit. These questions will be taken up later. However, in this section we will rely completely on our intuition to obtain limits. Our main objective at this time is to introduce the mathematical notation associated with limits and to develop our intuitive understanding of limits geometrically.

If the value of $f(x)$ approaches the number L_1 as x approaches x_0 from the right side, we write

$$\lim_{x \to x_0^+} f(x) = L_1 \tag{3}$$

If the value of $f(x)$ approaches the number L_2 as x approaches x_0 from the left side, we write

$$\lim_{x \to x_0^-} f(x) = L_2 \tag{4}$$

Expression (3) is read, "the limit of $f(x)$ as x approaches x_0 from the right equals L_1" and expression (4) is read, "the limit of $f(x)$ as x approaches x_0 from the left equals L_2."

If the limit from the left side is the same as the limit from the right side, say

$$\lim_{x \to x_0^-} f(x) = \lim_{x \to x_0^+} f(x) = L$$

then we write

$$\lim_{x \to x_0} f(x) = L \tag{5}$$

Equation (5) is read, "the limit of $f(x)$ as x approaches x_0 equals L."

▶ Example 1   From (1) and (2)

$$\lim_{x \to 0^+} \frac{\sin x}{x} = \lim_{x \to 0^-} \frac{\sin x}{x} = 1$$

Therefore,

$$\lim_{x \to 0} \frac{\sin x}{x} = 1 \qquad\qquad ◀$$

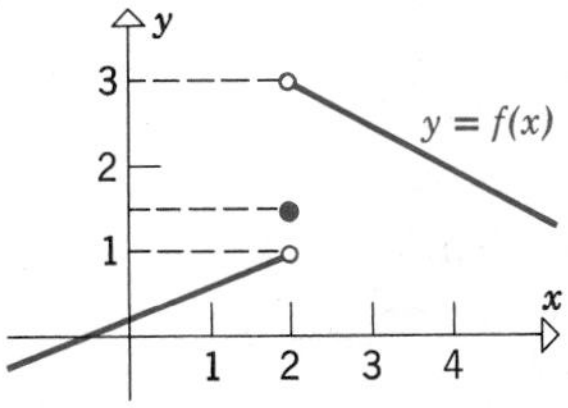

The following examples illustrate limits geometrically.

Figure 2.4.1

▶ Example 2   Let f be the function whose graph is shown in Figure 2.4.1. As x approaches 2 from the left, $f(x)$ approaches 1, so

$$\lim_{x \to 2^-} f(x) = 1$$

As x approaches 2 from the right, $f(x)$ approaches 3, so that

$$\lim_{x \to 2^+} f(x) = 3$$ ◀

Since $f(2) = 1.5$, this example shows that the value of a function at a point, and the left- and right-hand limits at the point can all be different.

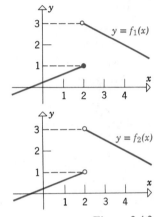

▶ **Example 3** Let f_1 and f_2 be the functions with graphs as shown in Figure 2.4.2.

The functions f_1 and f_2 are identical to the function f in Example 2, except at the point $x = 2$; there we have $f(2) = 1.5$ and $f_1(2) = 1$, while $f_2(2)$ is undefined. However, even though the functions $f, f_1,$ and f_2 behave differently *at* $x = 2$, their limits as x approaches 2 are the same, that is,

$$\lim_{x \to 2^-} f(x) = \lim_{x \to 2^-} f_1(x) = \lim_{x \to 2^-} f_2(x) = 1$$

and

$$\lim_{x \to 2^+} f(x) = \lim_{x \to 2^+} f_1(x) = \lim_{x \to 2^+} f_2(x) = 3$$ ◀

Figure 2.4.2

This example illustrates a general principle about limits that can be stated loosely as follows:

> The limit of a function as the independent variable approaches a point does not depend on the value of the function *at* the point.

Thus, if we alter the value of a function f only at $x = x_0$, then we do not affect

$$\lim_{x \to x_0^-} f(x), \qquad \lim_{x \to x_0^+} f(x), \qquad \text{or} \qquad \lim_{x \to x_0} f(x)$$

Function f_2 in Example 3 illustrates another idea: namely, that the left- and right-hand limits can have values at a point where the function itself is undefined.

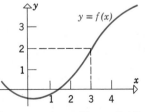

▶ **Example 4** Let f be the function whose graph is shown in Figure 2.4.3. From the graph we see that

$$\lim_{x \to 3^-} f(x) = 2 \qquad \text{and} \qquad \lim_{x \to 3^+} f(x) = 2$$

Figure 2.4.3   so that

$$\lim_{x \to 3} f(x) = 2$$

The graph also shows that $f(3) = 2$. Thus, the value of a function at a point and the limits as x approaches that point may be equal. ◄

▶ **Example 5** Let f be the function whose graph is shown in Figure 2.4.4. This graph is intended to convey the idea that as x approaches 6 from the right side, $f(x)$ oscillates between -3 and 3 infinitely often and with increasing frequency. Since $f(x)$ approaches no fixed value as x approaches 6 from the right, we say that

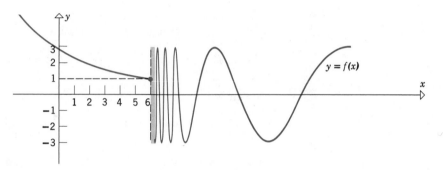

Figure 2.4.4

$$\lim_{x \to 6^+} f(x) \ \textit{does not exist}$$

On the other hand, as x approaches 6 from the left side, $f(x)$ approaches 1, so that

$$\lim_{x \to 6^-} f(x) = 1$$ ◄

▶ **Example 6** Let f be the function whose graph is shown in Figure 2.4.5. As x approaches 0 from the right side, $f(x)$ gets larger and larger without bound and consequently approaches no fixed value. Thus,

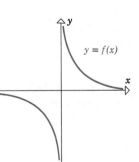

$$\lim_{x \to 0^+} f(x)$$

does not exist. In this case, we would write

$$\lim_{x \to 0^+} f(x) = +\infty \tag{6}$$

to indicate that the limit fails to exist because $f(x)$ is increasing without bound.

Figure 2.4.5

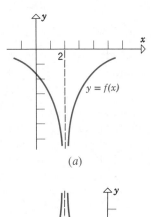

(a)

(b)

Figure 2.4.6

As x approaches 0 from the left side, $f(x)$ becomes more and more negative without bound and consequently approaches no fixed value. Thus,

$$\lim_{x \to 0^-} f(x)$$

does not exist. In this case, we would write

$$\lim_{x \to 0^-} f(x) = -\infty \qquad\qquad (7)$$

to indicate that the limit fails to exist because $f(x)$ is decreasing without bound. ◄

▶ **Example 7** Let f and g be the functions with graphs as shown in Figure 2.4.6. From the graph of f we see that

$$\lim_{x \to 2^-} f(x) = -\infty \qquad \text{and} \qquad \lim_{x \to 2^+} f(x) = -\infty$$

We can indicate these two results more succinctly by writing

$$\lim_{x \to 2} f(x) = -\infty$$

From the graph of g we see that

$$\lim_{x \to -3^-} g(x) = +\infty \qquad \text{and} \qquad \lim_{x \to -3^+} g(x) = +\infty$$

More briefly, we will write

$$\lim_{x \to -3} g(x) = +\infty \qquad\qquad\qquad ◄$$

So far we have used limits to describe how a function behaves as the independent variable approaches a fixed point on the x-axis. However, limits can also be used to describe how a function behaves as the independent variable moves "indefinitely far" from the origin along the x-axis. If x is allowed to increase without bound, then we write $x \to +\infty$ (read, "x approaches plus infinity"), and if x is allowed to decrease without bound then we write $x \to -\infty$ (read, "x approaches minus infinity").

▶ **Example 8** Let f be the function with the graph shown in Figure 2.4.7. As $x \to +\infty$ the graph of f approaches the line $y = 4$ so that the value of $f(x)$ approaches 4. We denote this by writing

$$\lim_{x \to +\infty} f(x) = 4$$

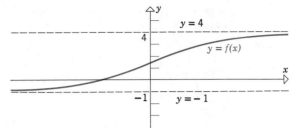

Figure 2.4.7

As $x \to -\infty$ the graph of f tends toward the line $y = -1$ and so the value of $f(x)$ approaches -1. We denote this by writing

$$\lim_{x \to -\infty} f(x) = -1$$ ◄

▶ **Example 9** Let f be the function with the graph shown in Figure 2.4.8. As x approaches $-\infty$, the value of $f(x)$ increases without bound and consequently approaches no fixed value. Thus

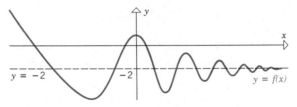

Figure 2.4.8

$$\lim_{x \to -\infty} f(x)$$

does not exist. In this case, we would write

$$\lim_{x \to -\infty} f(x) = +\infty$$

to indicate that the limit fails to exist because $f(x)$ is increasing without bound.

As x approaches $+\infty$, the graph of f oscillates but tends toward the line $y = -2$. Thus, the value of $f(x)$ approaches -2 as x approaches $+\infty$ and we write

$$\lim_{x \to +\infty} f(x) = -2$$

▶ Example 10 Let f be the function with graph as shown in Figure 2.4.9. As x approaches $-\infty$, the value of $f(x)$ decreases without bound and consequently approaches no fixed value. Thus,

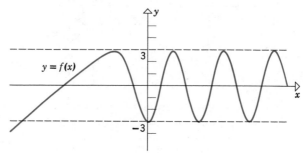

Figure 2.4.9

$$\lim_{x \to -\infty} f(x)$$

does not exist. In this case, we would write

$$\lim_{x \to -\infty} f(x) = -\infty$$

to indicate that the limit fails to exist because $f(x)$ is decreasing without bound.

As x approaches $+\infty$, the value of $f(x)$ oscillates between -3 and $+3$ and consequently approaches no fixed value. Thus,

$$\lim_{x \to +\infty} f(x)$$

does not exist. There is no special notation used to describe limits that fail to exist because of oscillation. ◀

We conclude this section with some comments on terminology. The expressions

$$\lim_{x \to x_0^-} f(x) \quad \text{and} \quad \lim_{x \to x_0^+} f(x)$$

are called the **one-sided limits of $f(x)$ at x_0** and the expression

$$\lim_{x \to x_0} f(x)$$

is called the **two-sided limit of $f(x)$ at x_0**. As noted previously, the two-sided limit exists only when the one-sided limits exist and are equal. The two-sided limit does not exist if either of the one-sided limits does not exist or if the one-sided limits exist but are unequal.

▶ Exercise Set 2.4

1. For the function f graphed below, find
(a) $\lim\limits_{x \to 3^-} f(x)$ (b) $\lim\limits_{x \to 3^+} f(x)$

(c) $\lim\limits_{x \to 3} f(x)$ (d) $f(3)$

(e) $\lim\limits_{x \to -\infty} f(x)$ (f) $\lim\limits_{x \to +\infty} f(x)$. •

4. For the function g graphed below, find
(a) $\lim\limits_{x \to 0^-} g(x)$ (b) $\lim\limits_{x \to 0^+} g(x)$

(c) $\lim\limits_{x \to 0} g(x)$ (d) $g(0)$

(e) $\lim\limits_{x \to -\infty} g(x)$ (f) $\lim\limits_{x \to +\infty} g(x)$.

2. For the function f graphed below, find
(a) $\lim\limits_{x \to 2^-} f(x)$ (b) $\lim\limits_{x \to 2^+} f(x)$

(c) $\lim\limits_{x \to 2} f(x)$ (d) $f(2)$

(e) $\lim\limits_{x \to -\infty} f(x)$ (f) $\lim\limits_{x \to +\infty} f(x)$.

5. For the function F graphed below, find
(a) $\lim\limits_{x \to -2^-} F(x)$ (b) $\lim\limits_{x \to -2^+} F(x)$

(c) $\lim\limits_{x \to -2} F(x)$ (d) $F(-2)$

(e) $\lim\limits_{x \to -\infty} F(x)$ (f) $\lim\limits_{x \to +\infty} F(x)$.

3. For the function g graphed below, find
(a) $\lim\limits_{x \to 4^-} g(x)$ (b) $\lim\limits_{x \to 4^+} g(x)$

(c) $\lim\limits_{x \to 4} g(x)$ (d) $g(4)$

(e) $\lim\limits_{x \to -\infty} g(x)$ (f) $\lim\limits_{x \to +\infty} g(x)$.

6. For the function F graphed below, find
(a) $\lim\limits_{x \to 3^-} F(x)$ (b) $\lim\limits_{x \to 3^+} F(x)$

(c) $\lim\limits_{x \to 3} F(x)$ (d) $F(3)$

(e) $\lim\limits_{x \to -\infty} F(x)$ (f) $\lim\limits_{x \to +\infty} F(x)$.

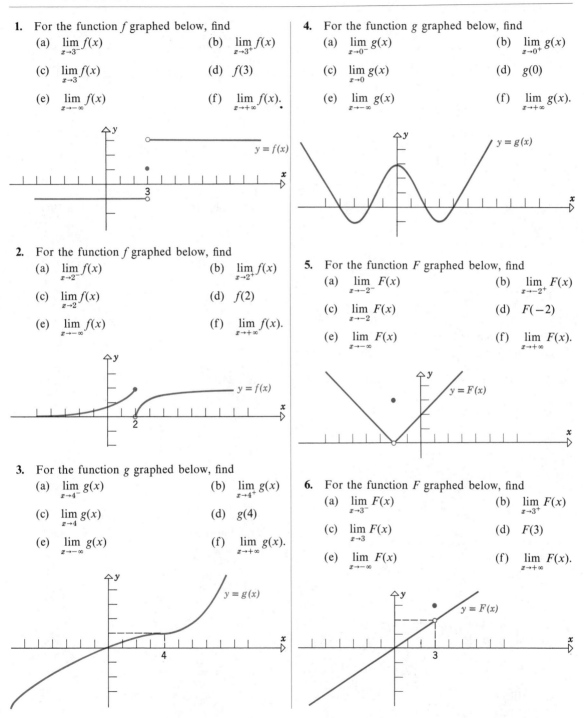

7. For the function ϕ graphed below, find

(a) $\lim\limits_{x \to -2^-} \phi(x)$

(b) $\lim\limits_{x \to -2^+} \phi(x)$

(c) $\lim\limits_{x \to -2} \phi(x)$

(d) $\phi(-2)$

(e) $\lim\limits_{x \to -\infty} \phi(x)$

(f) $\lim\limits_{x \to +\infty} \phi(x)$.

8. For the function ϕ graphed below, find

(a) $\lim\limits_{x \to 4^+} \phi(x)$

(b) $\lim\limits_{x \to 4^-} \phi(x)$

(c) $\lim\limits_{x \to 4} \phi(x)$

(d) $\phi(4)$

(e) $\lim\limits_{x \to -\infty} \phi(x)$

(f) $\lim\limits_{x \to +\infty} \phi(x)$.

9. For the function f graphed below, find

(a) $\lim\limits_{x \to 3^-} f(x)$

(b) $\lim\limits_{x \to 3^+} f(x)$

(c) $\lim\limits_{x \to 3} f(x)$

(d) $f(3)$

(e) $\lim\limits_{x \to -\infty} f(x)$

(f) $\lim\limits_{x \to +\infty} f(x)$.

10. For the function f graphed below, find

(a) $\lim\limits_{x \to 0^-} f(x)$

(b) $\lim\limits_{x \to 0^+} f(x)$

(c) $\lim\limits_{x \to 0} f(x)$

(d) $f(0)$

(e) $\lim\limits_{x \to -\infty} f(x)$

(f) $\lim\limits_{x \to +\infty} f(x)$.

11. For the function G graphed below, find

(a) $\lim\limits_{x \to 0^-} G(x)$

(b) $\lim\limits_{x \to 0^+} G(x)$

(c) $\lim\limits_{x \to 0} G(x)$

(d) $G(0)$

(e) $\lim\limits_{x \to -\infty} G(x)$

(f) $\lim\limits_{x \to +\infty} G(x)$.

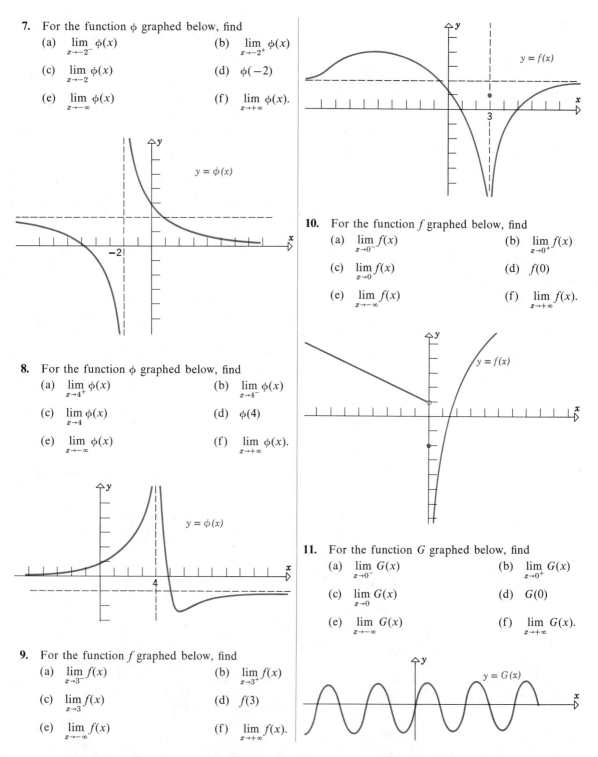

12. For the function G graphed below, find

    (a) $\lim\limits_{x \to 0^-} G(x)$ (b) $\lim\limits_{x \to 0^+} G(x)$

    (c) $\lim\limits_{x \to 0} G(x)$ (d) $G(0)$

    (e) $\lim\limits_{x \to -\infty} G(x)$ (f) $\lim\limits_{x \to +\infty} G(x)$.

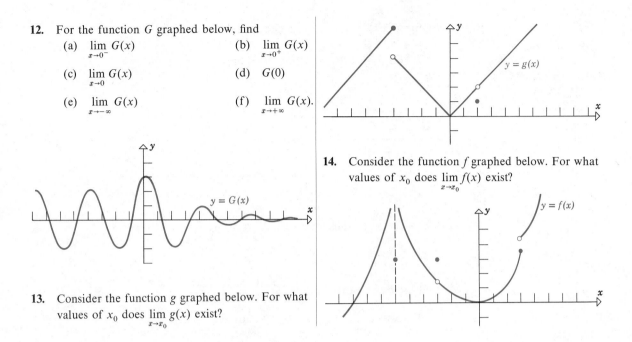

14. Consider the function f graphed below. For what values of x_0 does $\lim\limits_{x \to x_0} f(x)$ exist?

13. Consider the function g graphed below. For what values of x_0 does $\lim\limits_{x \to x_0} g(x)$ exist?

2.5 LIMITS (COMPUTATIONAL TECHNIQUES)

In the last section we concentrated on the graphical interpretation of limits. In this section we will discuss techniques for finding limits directly from the formula for a function. Again our results will be based on intuition. Mathematical proofs of the results here will have to wait until we have a precise mathematical definition of a limit.

The limits discussed in our first four examples are the basic tools used to solve more complicated limit problems.

▶ **Example 1** Since a constant function $f(x) = k$ has the same value k everywhere, it follows that at each point a

$$\lim_{x \to a} k = k$$

and also

$$\lim_{x \to +\infty} k = \lim_{x \to -\infty} k = k$$

(Figure 2.5.1). These results can be summarized by the statement:

> The limit of a constant function k at any point is k. ◀

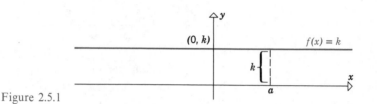

Figure 2.5.1

▶ **Example 2** The limit

$$\lim_{x \to a} x = a$$

is self-evident. For example,

$$\lim_{x \to 5} x = 5, \qquad \lim_{x \to 0} x = 0, \qquad \lim_{x \to -2} x = -2 \qquad ◀$$

▶ **Example 3** Find the limits

$$\lim_{x \to 0^+} \frac{1}{x}, \qquad \lim_{x \to 0^-} \frac{1}{x}, \qquad \lim_{x \to 0} \frac{1}{x}$$

Solution. It will be evident from some simple calculations that none of these limits exist. As x approaches 0 from the right, the value of $1/x$ gets larger and larger without bound. For example, if x successively assumes values

$$x = 1, \frac{1}{10}, \frac{1}{100}, \frac{1}{1000}, \frac{1}{10,000}, \ldots$$

then $1/x$ successively assumes values

$$\frac{1}{x} = 1, 10, 100, 1000, 10,000, \ldots$$

Thus,

$$\lim_{x \to 0^+} \frac{1}{x} = +\infty$$

As x approaches 0 from the left, the value of $1/x$ decreases (that is, becomes more and more negative) without bound. For example, if x successively assumes values

$$x = -1, \ -\frac{1}{10}, \ -\frac{1}{100}, \ -\frac{1}{1000}, \ -\frac{1}{10,000}, \ \ldots$$

then $1/x$ successively assumes values

$$\frac{1}{x} = -1, \ -10, \ -100, \ -1000, \ -10,000, \ \ldots$$

Thus,

$$\lim_{x \to 0^-} \frac{1}{x} = -\infty$$

It follows that

$$\lim_{x \to 0} \frac{1}{x} \text{ does not exist}$$

since the one-sided limits as x approaches zero do not exist. Moreover, we cannot even write

$$\lim_{x \to 0} \frac{1}{x} = +\infty \qquad \text{or} \qquad \lim_{x \to 0} \frac{1}{x} = -\infty$$

(Why?)

The reader should check that the results in this example are consistent with the graph of $1/x$ (see Figure 1.3.18).   ◀

▶ Example 4   Find the limits

$$\lim_{x \to +\infty} \frac{1}{x}, \qquad \lim_{x \to -\infty} \frac{1}{x}$$

Solution.   These limits are also evident from some sample calculations. To investigate the case $x \to +\infty$, assume x takes on successive values

$$x = 1, \ 10, \ 100, \ 1000, \ 10,000, \ \ldots$$

Then $1/x$ successively assumes values

$$\frac{1}{x} = 1, \ \frac{1}{10}, \ \frac{1}{100}, \ \frac{1}{1000}, \ \frac{1}{10,000}, \ \ldots$$

It is clear from these values of $1/x$ that

$$\lim_{x \to +\infty} \frac{1}{x} = 0$$

To investigate the case $x \to -\infty$, assume x takes on successive values

$$x = -1, \ -10, \ -100, \ -1000, \ -10,\!000, \ \ldots$$

Then $1/x$ successively assumes values

$$\frac{1}{x} = -1, \ -\frac{1}{10}, \ -\frac{1}{100}, \ -\frac{1}{1000}, \ -\frac{1}{10,\!000}, \ \ldots$$

It is clear from these values of $1/x$ that

$$\lim_{x \to -\infty} \frac{1}{x} = 0$$

The reader should check that the results in this example are consistent with the graph of $1/x$ (see Figure 1.3.18). ◀

The results in our first four examples together with the fundamental properties of limits given in the following theorem can be used to help solve more complicated limit problems. Parts of this theorem are proved in Appendix 2.

2.5.1   THEOREM   *Let* $\lim$ *stand for one of the limits* $\lim\limits_{x \to a}$, $\lim\limits_{x \to a^-}$, $\lim\limits_{x \to a^+}$, $\lim\limits_{x \to +\infty}$, *or* $\lim\limits_{x \to -\infty}$. *If* $L_1 = \lim f(x)$ *and* $L_2 = \lim g(x)$ *both exist, then*

(a) $\lim [f(x) + g(x)] = \lim f(x) + \lim g(x) = L_1 + L_2$.
(b) $\lim [f(x) - g(x)] = \lim f(x) - \lim g(x) = L_1 - L_2$.
(c) $\lim [f(x)g(x)] = \lim f(x) \lim g(x) = L_1 L_2$.
(d) $\lim \dfrac{f(x)}{g(x)} = \dfrac{\lim f(x)}{\lim g(x)} = \dfrac{L_1}{L_2}$ *if* $L_2 \neq 0$.
(e) $\lim \sqrt[n]{f(x)} = \sqrt[n]{\lim f(x)} = \sqrt[n]{L_1}$ *provided* $L_1 > 0$ *if* n *is even.*

In other words, this theorem tells us:

(a)   *The limit of a sum is the sum of the limits.*
(b)   *The limit of a difference is the difference of the limits.*
(c)   *The limit of a product is the product of the limits.*
(d)   *The limit of a quotient is the quotient of the limits provided the limit of the denominator is nonzero.*
(e)   *The limit of an nth root is the nth root of the limit.*

REMARK.   Although results (a) and (c) are stated for two functions f and g, these results hold as well for any finite number of functions; that is, if

$$\lim f_1(x), \ \lim f_2(x), \ldots, \lim f_n(x)$$

all exist, then

$$\lim \left[f_1(x) + f_2(x) + \cdots + f_n(x) \right] = \lim f_1(x) + \lim f_2(x) + \cdots + \lim f_n(x)$$

and

$$\lim \left[f_1(x) f_2(x) \cdots f_n(x) \right] = \lim f_1(x) \lim f_2(x) \cdots \lim f_n(x) \qquad (1)$$

In particular, if $f_1, f_2, \ldots, f_n$ are all the same function f, then (1) reduces to

$$\lim \left[f(x) \right]^n = \left[\lim f(x) \right]^n \qquad (2)$$

From (2) we obtain the useful result

$$\lim_{x \to a} x^n = \left[\lim_{x \to a} x \right]^n = a^n \qquad (3)$$

For example,

$$\lim_{x \to 3} x^4 = 3^4 = 81$$

Another useful result follows from Example 1 and part (c) of Theorem 2.5.1. If k is a constant, then

$$\lim k f(x) = \lim k \lim f(x) = k \lim f(x) \qquad (4)$$

The two end expressions in (4) tell us:

A constant factor can be moved through a limit sign.

▶ Example 5   Find $\lim\limits_{x \to 5} (x^2 - 4x + 3)$ and justify each step.

Solution.

$$
\begin{aligned}
\lim_{x \to 5} (x^2 - 4x + 3) &= \lim_{x \to 5} x^2 - \lim_{x \to 5} 4x + \lim_{x \to 5} 3 && \text{[Theorem 2.5.1}(a), (b)\text{]}\\
&= \lim_{x \to 5} x^2 - 4 \lim_{x \to 5} x + \lim_{x \to 5} 3 && \text{[Equation (4)]}\\
&= 5^2 - 4(5) + 3 && \text{[Equation (3)]}\\
&= 8
\end{aligned}
$$

◀

▶ Example 6   Show that for any polynomial

$$p(x) = c_0 + c_1 x + \cdots + c_n x^n$$

and any real number a,

$$\lim_{x \to a} p(x) = c_0 + c_1 a + \cdots + c_n a^n = p(a)$$

Solution.

$$\lim_{x \to a} p(x) = \lim_{x \to a} (c_0 + c_1 x + \cdots + c_n x^n)$$

$$= \lim_{x \to a} c_0 + \lim_{x \to a} c_1 x + \cdots + \lim_{x \to a} c_n x^n$$

$$= \lim_{x \to a} c_0 + c_1 \lim_{x \to a} x + \cdots + c_n \lim_{x \to a} x^n$$

$$= c_0 + c_1 a + \cdots + c_n a^n = p(a) \qquad \blacktriangleleft$$

REMARK. Example 6 states that for a polynomial $p(x)$, the limit as x approaches a is equal to the value of the polynomial at a. If we apply this result to the limit problem in Example 5, we can bypass the intermediate steps and write immediately

$$\lim_{x \to 5} (x^2 - 4x + 3) = 5^2 - 4(5) + 3 = 8$$

▶ Example 7 Find

$$\lim_{x \to 2} \frac{5x^3 + 4}{x - 3}$$

Solution. Using the result of Example 6 and part (*d*) of Theorem 2.5.1, we obtain

$$\lim_{x \to 2} \frac{5x^3 + 4}{x - 3} = \frac{\lim_{x \to 2} (5x^3 + 4)}{\lim_{x \to 2} (x - 3)} = \frac{5 \cdot 2^3 + 4}{2 - 3} = -44 \qquad \blacktriangleleft$$

▶ Example 8 Find

$$\lim_{x \to 2} \frac{x^2 - 4}{x - 2}$$

Solution. The method used in the previous example cannot be used here since part (*d*) of Theorem 2.5.1 does not apply (the limit of the denominator is 0). However, by factoring the numerator and canceling, we obtain

$$\lim_{x \to 2} \frac{x^2 - 4}{x - 2} = \lim_{x \to 2} \frac{(x - 2)(x + 2)}{x - 2} = \lim_{x \to 2} (x + 2) = 4 \qquad \blacktriangleleft$$

Though correct, the second equality in this calculation needs some justification. We pointed out in Example 12 of Section 2.1 that to cancel the $x - 2$ factors from

$$h(x) = \frac{x^2 - 4}{x - 2} = \frac{(x - 2)(x + 2)}{(x - 2)}$$

we must write

$$h(x) = x + 2, \qquad x \neq 2$$

in order not to alter the domain of h. However, the functions

$$f(x) = x + 2 \qquad \text{and} \qquad h(x) = x + 2, \qquad x \neq 2$$

differ only at $x = 2$, so that their limits as x approaches 2 are the same. Thus,

$$\lim_{x \to 2} \frac{(x - 2)(x + 2)}{x - 2} = \lim_{x \to 2} (x + 2)$$

even though

$$\frac{(x - 2)(x + 2)}{x - 2} \qquad \text{and} \qquad x + 2$$

are different functions.

▶ Example 9 Find

$$\lim_{x \to 4} \frac{2 - x}{(x - 4)(x + 2)}$$

Solution. Again, part (*d*) of Theorem 2.5.1 cannot be applied since the limit of the denominator is zero. There are no common factors in the numerator and denominator as there were in Example 8, so we must use a different approach. This kind of problem is best attacked by looking at the one-sided limits.

First consider

$$\lim_{x \to 4^+} \frac{2 - x}{(x - 4)(x + 2)}$$

As x approaches 4 from the right, the numerator is a negative quantity approaching -2 and the denominator a positive quantity approaching 0. Consequently the ratio is a negative quantity that decreases without bound. That is,

$$\lim_{x \to 4^+} \frac{2 - x}{(x - 4)(x + 2)} = -\infty \tag{5}$$

Now consider

$$\lim_{x \to 4^-} \frac{2 - x}{(x - 4)(x + 2)}$$

As x approaches 4 from the left, the numerator is eventually a negative quantity approaching -2 and the denominator is a negative quantity approaching 0. Consequently the ratio is a positive quantity that increases without bound; that is,

$$\lim_{x \to 4^-} \frac{2 - x}{(x - 4)(x + 2)} = +\infty \qquad (6)$$

From (5) and (6)

$$\lim_{x \to 4} \frac{2 - x}{(x - 4)(x + 2)}$$

does not exist. ◄

Next we will consider methods for finding limits of rational functions as $x \to +\infty$ or $x \to -\infty$. But first we need some preliminary results.

In Example 4 we obtained the limits

$$\lim_{x \to +\infty} \frac{1}{x} = \lim_{x \to -\infty} \frac{1}{x} = 0$$

Consequently, for any positive integer n,

$$\lim_{x \to +\infty} \frac{1}{x^n} = \left(\lim_{x \to +\infty} \frac{1}{x} \right)^n = 0 \qquad (7)$$

and

$$\lim_{x \to -\infty} \frac{1}{x^n} = \left(\lim_{x \to -\infty} \frac{1}{x} \right)^n = 0 \qquad (8)$$

Results (7) and (8) are also evident from the graph of $1/x^n$, which has one of the two general shapes shown in Fig. 2.5.2, depending on whether n is even or odd.

If we divide numerator and denominator of a rational function by the highest power of x that occurs in the function, then all the powers of x become constants or powers of $1/x$. The next two examples show how this observation together with (7) and (8) can be used to find $\lim_{x \to +\infty}$ or $\lim_{x \to -\infty}$ for rational functions.

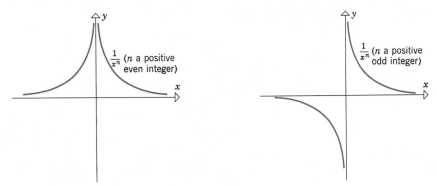

Figure 2.5.2

▶ Example 10   Find

$$\lim_{x \to +\infty} \frac{3x + 5}{6x - 8}$$

Solution. Divide the numerator and denominator by the highest power of x that occurs; this is $x^1 = x$. We obtain

$$\lim_{x \to +\infty} \frac{3x + 5}{6x - 8} = \lim_{x \to +\infty} \frac{3 + 5/x}{6 - 8/x} = \frac{\lim_{x \to +\infty} (3 + 5/x)}{\lim_{x \to +\infty} (6 - 8/x)}$$

$$= \frac{\lim_{x \to +\infty} 3 + \lim_{x \to +\infty} 5/x}{\lim_{x \to +\infty} 6 - \lim_{x \to +\infty} 8/x} = \frac{3 + 5 \lim_{x \to +\infty} 1/x}{6 - 8 \lim_{x \to +\infty} 1/x}$$

$$= \frac{3 + 5 \cdot 0}{6 - 8 \cdot 0} = \frac{1}{2} \qquad ◀$$

▶ Example 11   Find

$$\lim_{x \to -\infty} \frac{4x^2 - x}{2x^3 - 5}$$

Solution. Divide numerator and denominator by the highest power of x that occurs, namely x^3. We obtain

$$\lim_{x \to -\infty} \frac{4x^2 - x}{2x^3 - 5} = \lim_{x \to -\infty} \frac{4/x - 1/x^2}{2 - 5/x^3} = \frac{\lim_{x \to -\infty} (4/x - 1/x^2)}{\lim_{x \to -\infty} (2 - 5/x^3)}$$

$$= \frac{\lim_{x \to -\infty} 4/x - \lim_{x \to -\infty} 1/x^2}{\lim_{x \to -\infty} 2 - \lim_{x \to -\infty} 5/x^3} = \frac{4 \lim_{x \to -\infty} 1/x - \lim_{x \to -\infty} 1/x^2}{2 - 5 \lim_{x \to -\infty} 1/x^3}$$

$$= \frac{4 \cdot 0 - 0}{2 - 5 \cdot 0} = \frac{0}{2} = 0 \qquad ◀$$

▶ Example 12 Find

$$\lim_{x \to +\infty} \sqrt[3]{\frac{3x + 5}{6x - 8}}$$

Solution.

$$\lim_{x \to +\infty} \sqrt[3]{\frac{3x + 5}{6x - 8}} = \sqrt[3]{\lim_{x \to +\infty} \frac{3x + 5}{6x - 8}} \qquad \text{[Theorem 2.5.1, part } (e)\text{]}$$

$$= \sqrt[3]{\frac{1}{2}} \qquad \text{[Example 10]} \qquad \blacktriangleleft$$

▶ Example 13 Find

$$\lim_{x \to +\infty} \frac{3 - 2x^4}{x + 1}$$

Solution. Dividing numerator and denominator by x^4, we obtain

$$\lim_{x \to +\infty} \frac{3 - 2x^4}{x + 1} = \lim_{x \to +\infty} \frac{3/x^4 - 2}{1/x^3 + 1/x^4} \qquad (9)$$

Since the limit of the denominator in the right side of (9) is zero, we cannot apply Theorem 2.5.1. However, we can argue as follows. In the numerator, $3/x^4$ approaches 0 as $x \to +\infty$ so that

$$\frac{3}{x^4} - 2$$

is eventually negative and approaches -2. As $x \to +\infty$, the denominator is a positive quantity approaching 0. Consequently, as $x \to +\infty$, the ratio is eventually a negative quantity that decreases without bound. That is,

$$\lim_{x \to +\infty} \frac{3 - 2x^4}{x + 1} = \lim_{x \to +\infty} \frac{3/x^4 - 2}{1/x^3 + 1/x^4} = -\infty \qquad \blacktriangleleft$$

▶ Example 14 Find

$$\lim_{x \to +\infty} \frac{\sqrt{x^2 + 2}}{3x - 6}$$

Solution. Since we are letting $x \to +\infty$, it would be helpful to manipulate the function so that the powers of x become powers of $1/x$. To achieve this, we divide numerator and denominator by x. In the numerator, we write x as

$$x = \sqrt{x^2}$$

This is justified because x eventually assumes positive values as $x \to +\infty$, so that

$$\sqrt{x^2} = |x| = x$$

Thus,

$$\lim_{x \to +\infty} \frac{\sqrt{x^2 + 2}}{3x - 6} = \lim_{x \to +\infty} \frac{\sqrt{x^2 + 2}/\sqrt{x^2}}{(3x - 6)/x} = \lim_{x \to +\infty} \frac{\sqrt{1 + 2/x^2}}{3 - 6/x}$$

$$= \frac{\lim_{x \to +\infty} \sqrt{1 + 2/x^2}}{\lim_{x \to +\infty} (3 - 6/x)} = \frac{\sqrt{\lim_{x \to +\infty} (1 + 2/x^2)}}{\lim_{x \to +\infty} (3 - 6/x)}$$

$$= \frac{\sqrt{\lim_{x \to +\infty} 1 + 2 \lim_{x \to +\infty} 1/x^2}}{\lim_{x \to +\infty} 3 - 6 \lim_{x \to +\infty} 1/x} = \frac{\sqrt{1 + 2 \cdot 0}}{3 - 6 \cdot 0} = \frac{1}{3} \qquad \blacktriangleleft$$

► Example 15 Find

$$\lim_{x \to -\infty} \frac{\sqrt{x^2 + 2}}{3x - 6}$$

Solution. The function here is the same as in Example 14. However, in this problem we want $\lim_{x \to -\infty}$ rather than $\lim_{x \to +\infty}$. As in Example 14, we divide numerator and denominator by x, but in the numerator we write x as

$$x = -\sqrt{x^2}$$

This is justified because x eventually assumes negative values as $x \to -\infty$, so that

$$\sqrt{x^2} = |x| = -x$$

We obtain

$$\lim_{x \to -\infty} \frac{\sqrt{x^2 + 2}}{3x - 6} = \lim_{x \to -\infty} \frac{\sqrt{x^2 + 2}/(-\sqrt{x^2})}{3 - (6/x)}$$

$$= \lim_{x \to -\infty} \frac{-\sqrt{1 + 2/x^2}}{3 - (6/x)} = -\frac{1}{3} \qquad \blacktriangleleft$$

► Example 16 Find $\lim_{x \to 3} f(x)$ for

$$f(x) = \begin{cases} x^2 - 5, & x \leq 3 \\ \sqrt{x + 13}, & x > 3 \end{cases}$$

Solution. As x approaches 3 from the left, the formula for f is

$$f(x) = x^2 - 5$$

so that

$$\lim_{x \to 3^-} f(x) = \lim_{x \to 3^-} (x^2 - 5) = 3^2 - 5 = 4$$

As x approaches 3 from the right, the formula for f is

$$f(x) = \sqrt{x + 13}$$

so that

$$\lim_{x \to 3^+} f(x) = \lim_{x \to 3^+} \sqrt{x + 13} = \sqrt{\lim_{x \to 3^+} (x + 13)} = \sqrt{16} = 4$$

Therefore,

$$\lim_{x \to 3} f(x) = 4$$

since the one-sided limits are equal. ◀

▶ Exercise Set 2.5

Find the limits in Exercises 1–48.

1. $\lim\limits_{x \to 8} 7$.

2. $\lim\limits_{x \to -\infty} (-3)$.

3. $\lim\limits_{x \to 0^+} \pi$.

4. $\lim\limits_{x \to -2} 3x$.

5. $\lim\limits_{y \to 3^+} 12y$.

6. $\lim\limits_{h \to +\infty} (-2h)$.

7. $\lim\limits_{x \to 5} \sqrt{x^3 - 3x - 1}$.

8. $\lim\limits_{x \to 0^-} (x^4 + 12x^3 - 17x + 2)$.

9. $\lim\limits_{y \to -1} (y^6 - 12y + 1)$.

10. $\lim\limits_{x \to 3} \dfrac{x^2 - 2x}{x + 1}$.

11. $\lim\limits_{y \to 2^-} \dfrac{(y - 1)(y - 2)}{y + 1}$.

12. $\lim\limits_{x \to 0} \dfrac{6x - 9}{x^3 - 12x + 3}$.

13. $\lim\limits_{x \to 4} \dfrac{x^2 - 16}{x - 4}$.

14. $\lim\limits_{t \to -2} \dfrac{t^3 + 8}{t + 2}$.

15. $\lim\limits_{x \to 1^+} \dfrac{x^4 - 1}{x - 1}$.

16. $\lim\limits_{x \to 2} \dfrac{x^2 - 4x + 4}{x^2 + x - 6}$.

17. $\lim\limits_{x \to -1} \dfrac{x^2 + 6x + 5}{x^2 - 3x - 4}$.

18. $\lim\limits_{t \to 1} \dfrac{t^3 + t^2 - 5t + 3}{t^3 - 3t + 2}$.

19. $\lim\limits_{x \to +\infty} \dfrac{3x + 1}{2x - 5}$.

20. $\lim\limits_{x \to +\infty} \dfrac{1}{x - 12}$.

21. $\lim\limits_{y \to -\infty} \dfrac{3}{y + 4}$.

22. $\lim\limits_{x \to +\infty} \dfrac{5x^2 + 7}{3x^2 - x}$.

23. $\lim\limits_{x \to -\infty} \dfrac{x - 2}{x^2 + 2x + 1}$.

24. $\lim\limits_{s \to +\infty} \sqrt[3]{\dfrac{3s^7 - 4s^5}{2s^7 + 1}}$.

25. $\lim\limits_{x \to -\infty} \dfrac{\sqrt{5x^2 - 2}}{x + 3}$.

26. $\lim\limits_{x \to +\infty} \dfrac{\sqrt{5x^2 - 2}}{x + 3}$.

27. $\lim\limits_{y \to -\infty} \dfrac{2 - y}{\sqrt{7 + 6y^2}}$.

28. $\lim\limits_{y \to +\infty} \dfrac{2 - y}{\sqrt{7 + 6y^2}}$.

29. $\lim\limits_{x \to -\infty} \dfrac{\sqrt{3x^4 + x}}{x^2 - 8}$.

30. $\lim\limits_{x \to +\infty} \dfrac{\sqrt{3x^4 + x}}{x^2 - 8}$.

31. $\lim\limits_{x \to 3^+} \dfrac{x}{x - 3}$.

32. $\lim\limits_{x \to 3^-} \dfrac{x}{x - 3}$.

33. $\lim\limits_{x \to 3} \dfrac{x}{x - 3}$.

34. $\lim\limits_{x \to 2^+} \dfrac{x}{x^2 - 4}$.

35. $\lim\limits_{x \to 2^-} \dfrac{x}{x^2 - 4}$.

36. $\lim\limits_{x \to 2} \dfrac{x}{x^2 - 4}$.

37. $\lim\limits_{y \to 6^+} \dfrac{y + 6}{y^2 - 36}$.

38. $\lim\limits_{y \to 6^-} \dfrac{y + 6}{y^2 - 36}$.

39. $\lim\limits_{y \to 6} \dfrac{y + 6}{y^2 - 36}$.

40. $\lim\limits_{x \to 4^+} \dfrac{3 - x}{x^2 - 2x - 8}$.

41. $\lim\limits_{x \to 4^-} \dfrac{3 - x}{x^2 - 2x - 8}$.

42. $\lim\limits_{x \to 4} \dfrac{3 - x}{x^2 - 2x - 8}$.

43. $\lim\limits_{x \to +\infty} \dfrac{7 - 6x^5}{x + 3}$.

44. $\lim\limits_{t \to -\infty} \dfrac{5 - 2t^3}{t^2 + 1}$.

45. $\lim\limits_{t \to +\infty} \dfrac{6 - t^3}{7t^3 + 3}$.

46. $\lim\limits_{x \to 0^+} \dfrac{x}{|x|}$.

47. $\lim\limits_{x \to 0^-} \dfrac{x}{|x|}$.

48. $\lim\limits_{x \to 3^-} \dfrac{1}{|x - 3|}$.

49. $\lim\limits_{x \to 9} \dfrac{x - 9}{\sqrt{x} - 3}$.

50. $\lim\limits_{y \to 4} \dfrac{4 - y}{2 - \sqrt{y}}$.

51. Let

$$f(x) = \begin{cases} x - 1, & x \le 3 \\ 3x - 7, & x > 3 \end{cases}$$

Find:

(a) $\lim\limits_{x \to 3^-} f(x)$ (b) $\lim\limits_{x \to 3^+} f(x)$ (c) $\lim\limits_{x \to 3} f(x)$.

52. Let

$$g(t) = \begin{cases} t^2, & t \ge 0 \\ t - 2, & t < 0 \end{cases}$$

Find:

(a) $\lim\limits_{t \to 0^-} g(t)$ (b) $\lim\limits_{t \to 0^+} g(t)$ (c) $\lim\limits_{t \to 0} g(t)$.

53. Find $\lim\limits_{x \to 3} h(x)$ given that

$$h(x) = \begin{cases} x^2 - 2x + 1, & x \ne 3 \\ 7, & x = 3 \end{cases}$$

54. Let

$$F(x) = \begin{cases} \dfrac{x^2 - 9}{x + 3}, & x \ne -3 \\ k, & x = -3 \end{cases}$$

(a) Find k so that $F(-3) = \lim\limits_{x \to -3} F(x)$.

(b) With k assigned the value $\lim\limits_{x \to -3} F(x)$, show that $F(x)$ can be expressed as a polynomial.

55. (a) Explain why the following calculation is incorrect.

$$\lim\limits_{x \to 0^+} \left(\frac{1}{x} - \frac{1}{x^2} \right) = \lim\limits_{x \to 0^+} \frac{1}{x} - \lim\limits_{x \to 0^+} \frac{1}{x^2}$$
$$= +\infty - (+\infty) = 0$$

(b) Show that $\lim\limits_{x \to 0^+} \left(\dfrac{1}{x} - \dfrac{1}{x^2} \right) = -\infty$.

56. Find

(a) $\lim\limits_{x \to 0} \dfrac{\sqrt{x + 3} - \sqrt{3}}{x}$

(b) $\lim\limits_{x \to +\infty} \dfrac{\sin x}{x}$

(c) $\lim\limits_{x \to 0} x^2 \sin \dfrac{1}{x}$.

57. Let $r(x)$ be a rational function. Under what conditions is it true that $\lim\limits_{x \to a} r(x) = r(a)$?

58. Find

$$\lim\limits_{x \to +\infty} \frac{c_0 + c_1 x + \cdots + c_n x^n}{d_0 + d_1 x + \cdots + d_m x^m}$$

where $c_n \ne 0$ and $d_m \ne 0$.
[*Hint:* Your answer will depend on whether $m < n$, $m = n$, or $m > n$.]

2.6 LIMITS: A RIGOROUS APPROACH (OPTIONAL)

This section gives a rigorous treatment of two-sided limits. Readers interested in pursuing the theory of one-sided and infinite limits can turn to Appendix 2, where we give a rigorous treatment of these topics and also prove some of the basic limit theorems.

Let a and L be real numbers. When we write

$$\lim_{x \to a} f(x) = L \tag{1}$$

we mean intuitively that the value of $f(x)$ approaches L as x approaches a from either side. To capture this concept mathematically, we must clarify the meanings of the phrases "x approaches a from either side" and "$f(x)$ approaches L."

The limit in (1) is intended to describe the behavior of f when x is near but *different from a*. The value of f at a is irrelevant to the limit. In fact, we have seen examples such as

$$\lim_{x \to 0} \frac{\sin x}{x} = 1$$

where the function f is not even defined at the point a. Thus, when we say "x approaches a from either side," we mean that x assumes values arbitrarily close to a, on one side or the other, but does not actually take on the value a. On the other hand, when we say that "$f(x)$ approaches L," we want to allow the possibility that $f(x)$ may actually take on the value L.

Let us see if we can phrase these ideas more precisely. To say that $f(x)$ becomes arbitrarily close to L as x approaches a means that if we pick any positive number ϵ (no matter how small), eventually the difference between $f(x)$ and L will be less than ϵ as x approaches a. That is, eventually $f(x)$ will satisfy the condition

$$|f(x) - L| < \epsilon$$

Figure 2.6.1 illustrates this idea. For the curve in that figure we have

$$\lim_{x \to a} f(x) = L$$

where a and L are the points shown. We picked an arbitrary positive number ϵ and indicated the interval

$$(L - \epsilon, L + \epsilon)$$

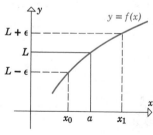

Figure 2.6.1

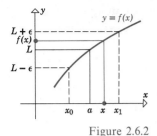

Figure 2.6.2

on the y-axis. As x approaches a, x will eventually lie between the points x_0 and x_1; when this occurs, $f(x)$ will lie in the interval $(L - \epsilon, L + \epsilon)$, that is, $f(x)$ will satisfy

$$L - \epsilon < f(x) < L + \epsilon$$

or equivalently

$$|f(x) - L| < \epsilon \tag{2}$$

(Figure 2.6.2). It is evident geometrically that no matter how small we make ϵ, eventually $f(x)$ will satisfy (2) as x approaches a.

For the curve in Figure 2.6.1, we have

$$f(a) = \lim_{x \to a} f(x) = L$$

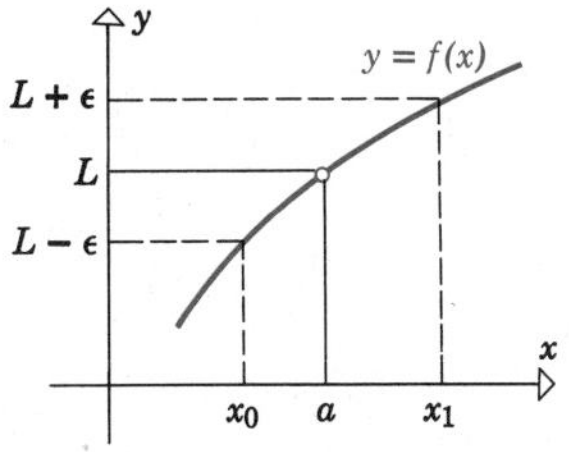

But the equality of $f(a)$ and L is irrelevant to the limit discussion. As x approaches a, x is not allowed to take on the value a. Thus, if $f(a)$ is undefined, as in Figure 2.6.3, or if $f(a)$ is defined, but unequal to L, as in Figure 2.6.4, it is still true that for an arbitrary positive value of ϵ, $f(x)$ will eventually satisfy (2) as x approaches a. In brief, the condition $|f(x) - L| < \epsilon$ need only be satisfied for x in the set $(x_0, a) \cup (a, x_1)$.

Before we can give a precise definition of limit, we need a preliminary observation. If the condition

$$|f(x) - L| < \epsilon \tag{3}$$

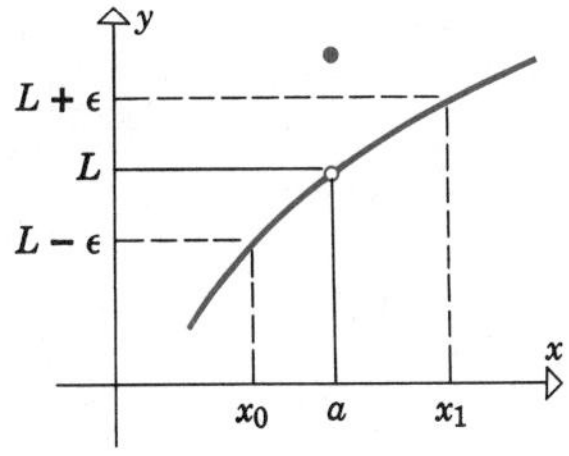

holds for all x in the set

$$(x_0, a) \cup (a, x_1) \tag{4}$$

then this condition holds as well for all x in any *subset* of (4). In particular, if we let δ be any positive number that is smaller than both

$$a - x_0 \quad \text{and} \quad x_1 - a$$

then

$$(a - \delta, a) \cup (a, a + \delta) \tag{5}$$

will be a subset of (4) (Figure 2.6.5) and condition (3) will hold for all x in this subset.

Since the set in (5) consists of all x satisfying

$$0 < |x - a| < \delta$$

we are led to the following definition.

Figure 2.6.3

Figure 2.6.4

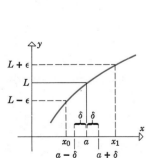

Figure 2.6.5

2.6.1 DEFINITION Let $f(x)$ be defined for all x in some open interval containing the number a, with the possible exception that $f(x)$ may or may not be defined at a. We will write

$$\lim_{x \to a} f(x) = L$$

if given any number $\epsilon > 0$, we can find a number $\delta > 0$ such that $f(x)$ satisfies

$$|f(x) - L| < \epsilon \qquad \text{whenever } x \text{ satisfies} \qquad 0 < |x - a| < \delta$$

▶ Example 1 Use Definition 2.6.1 to prove

$$\lim_{x \to 2} (3x - 5) = 1$$

Solution. We must show that given any positive number ϵ, we can find a positive number δ such that $f(x) = 3x - 5$ satisfies

$$|(3x - 5) - 1| < \epsilon \tag{6}$$

whenever x satisfies

$$0 < |x - 2| < \delta \tag{7}$$

To find δ, we can rewrite (6) as

$$|3x - 6| < \epsilon$$

or equivalently

$$3|x - 2| < \epsilon$$

or

$$|x - 2| < \frac{\epsilon}{3} \tag{8}$$

We must choose δ so that (8) is satisfied whenever (7) is satisfied. We can do this by taking

$$\delta = \frac{\epsilon}{3}$$

To prove that this choice for δ works, assume x satisfies (7). Since we are letting $\delta = \epsilon/3$, it follows from (7) that x satisfies

$$0 < |x - 2| < \frac{\epsilon}{3} \tag{9}$$

Thus, (8) is satisfied since it is just the right-hand inequality in (9). This proves that $\lim_{x \to 2} (3x - 5) = 1$. ◀

As illustrated in Example 1, a limit proof proceeds as follows: We *assume* that an arbitrary positive number ϵ is given to us. Then we try to *find* a positive number δ such that the conditions in the definition are fulfilled. The idea is to find a formula for δ in terms of ϵ so that no matter what ϵ is chosen, we automatically obtain from the formula the required δ. (In Example 1, the formula was $\delta = \epsilon/3$.)

Example 1 is about as easy as a limit proof can get; most limit proofs require a little more algebraic and logical ingenuity. The reader who finds "δ-ϵ" discussions hard going should not become discouraged. The concepts and techniques are intrinsically difficult. In fact, a precise understanding of limits evaded the finest mathematical minds for centuries.

Note that the value of δ in Definition 2.6.1 is not unique. Once one value of δ is found that fulfills the requirements of the definition, any *smaller* positive value for δ will also fulfill these requirements. To see why, assume we have found a value of δ such that

$$|f(x) - L| < \epsilon \tag{10}$$

whenever x satisfies

$$0 < |x - a| < \delta \tag{11}$$

Let δ_1 be any positive number smaller than δ. If x satisfies

$$0 < |x - a| < \delta_1 \tag{12}$$

then we have

$$0 < |x - a| < \delta_1 < \delta$$

Thus, x satisfies (11) and consequently $f(x)$ satisfies (10). Therefore, δ_1 meets the requirements of Definition 2.6.1 also. For example, we found in Example 1 that $\delta = \epsilon/3$ fulfills the requirements of Definition 2.6.1. Consequently, any smaller value for δ, such as $\delta = \epsilon/4$, $\delta = \epsilon/5$, or $\delta = \epsilon/6$, does also.

▶ **Example 2** Prove that

$$\lim_{x \to 3} x^2 = 9$$

Solution. We must show that given any positive number ϵ we can find a positive number δ such that

$$|x^2 - 9| < \epsilon \tag{13}$$

whenever x satisfies

$$0 < |x - 3| < \delta \tag{14}$$

Because $|x - 3|$ occurs in (14), it will be helpful to rewrite (13) so that $|x - 3|$ appears as a factor on the left side. Therefore, we will rewrite (13) as

$$|x + 3|\,|x - 3| < \epsilon \tag{15}$$

If we can somehow assure that when x satisfies (14) the factor $|x + 3|$ remains less than some positive constant, say,

$$|x + 3| < k \tag{16}$$

then on choosing

$$\delta = \frac{\epsilon}{k} \tag{17}$$

it will follow from (14) that

$$0 < |x - 3| < \frac{\epsilon}{k}$$

or

$$0 < k|x - 3| < \epsilon \tag{18}$$

From (16) and the right-hand inequality in (18) we will then have

$$|x + 3|\,|x - 3| < k|x - 3| < \epsilon$$

so that (15) will be satisfied, and the proof will be complete. As we shall now explain, condition (16) can be obtained by restricting the size of δ. We remarked in the discussion preceding this example that δ is not unique; once a value of δ is found, any *smaller* positive value for δ can also be used. At this point we do not have a formula relating δ and ϵ. However, let us agree in advance that if, for a given ϵ, δ turns out to exceed 1, we will choose the smaller value $\delta = 1$ instead. Thus, we can assume in our discussion that δ satisfies

$$0 < \delta \leq 1 \tag{19}$$

Assume that x satisfies (14). From (19) and the right side of (14) we obtain

$$|x - 3| < \delta \leq 1$$

so that

$$|x - 3| < 1$$

or equivalently

$$2 < x < 4$$

so

$$5 < x + 3 < 7$$

Therefore,

$$|x + 3| < 7$$

Comparing the last inequality to (16) suggests $k = 7$; and from (17)

$$\delta = \frac{\epsilon}{k} = \frac{\epsilon}{7}$$

In summary, given $\epsilon > 0$, we choose

$$\delta = \frac{\epsilon}{7}$$

provided $\epsilon/7$ does not exceed 1. If $\epsilon/7$ exceeds 1, we choose

$$\delta = 1$$

In other words, we take δ to be the minimum of the numbers $\epsilon/7$ and 1. This is sometimes written

$$\delta = \min\left(\frac{\epsilon}{7}, 1\right)$$ ◄

   In this last example the reader may have wondered how we knew to make the restriction $\delta \leq 1$ as opposed to some other restriction such as $\delta \leq \frac{1}{2}$ or $\delta \leq 5$. Actually, our selection was completely arbitrary; any other restriction of the form $\delta \leq c$ would have worked equally well (Exercise 27).

► Example 3   Prove that

$$\lim_{x \to 1/2} \frac{1}{x} = 2$$

Solution. We must show that given $\epsilon > 0$, there exists a $\delta > 0$ such that

$$\left| \frac{1}{x} - 2 \right| < \epsilon \tag{20}$$

whenever x satisfies

$$0 < \left| x - \frac{1}{2} \right| < \delta \tag{21}$$

Because $|x - \frac{1}{2}|$ occurs in (21), it will be helpful to rewrite (20) so that $|x - \frac{1}{2}|$ appears as a factor on the left side. We do this as follows:

$$\left| \frac{1}{x} - 2 \right| = \left| \frac{2}{x} \left(\frac{1}{2} - x \right) \right| = \left| \frac{2}{x} \right| \left| \frac{1}{2} - x \right| = \left| \frac{2}{x} \right| \left| x - \frac{1}{2} \right|$$

Thus, (20) is equivalent to

$$\left| \frac{2}{x} \right| \left| x - \frac{1}{2} \right| < \epsilon \tag{22}$$

From here on, the procedure is similar to that used in the last example. By restricting δ we will try to make the factor $|2/x|$ satisfy

$$\left| \frac{2}{x} \right| < k \tag{23}$$

for some constant k when x satisfies (21). Then on choosing

$$\delta = \frac{\epsilon}{k} \tag{24}$$

we will have, from (21),

$$0 < \left| x - \frac{1}{2} \right| < \frac{\epsilon}{k}$$

or

$$0 < k \left| x - \frac{1}{2} \right| < \epsilon \tag{25}$$

From (23) and the right-hand inequality in (25) it will then follow that

$$\left| \frac{2}{x} \right| \left| x - \frac{1}{2} \right| < k \left| x - \frac{1}{2} \right| < \epsilon$$

so that (22) will hold and the proof will be complete. Therefore, to finish the proof, we must show that δ can be restricted so that (23) holds for some k when x satisfies (21).

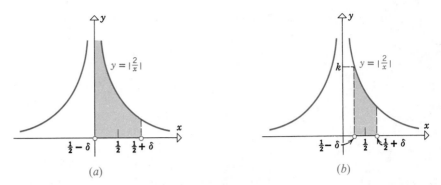

Figure 2.6.6
(a)
(b)

In Figure 2.6.6 we have sketched the graph of $|2/x|$ and marked the set of x values satisfying (21). In part (a) of the figure we took $\delta = \frac{1}{2}$ and in part (b) of the figure we took $\delta < \frac{1}{2}$. Part (a) of this figure makes it clear that if $\delta = \frac{1}{2}$ (or $\delta > \frac{1}{2}$) then the values of $|2/x|$ will have no upper bound, making it impossible to satisfy (23). However, part (b) shows that when $\delta < \frac{1}{2}$, there is an upper bound k to the values of $|2/x|$. Therefore, the arbitrary restriction

$$\delta \leq \frac{1}{4} \tag{26}$$

will be satisfactory to assure that $|2/x|$ has an upper bound.

Assume that x satisfies (21). From (26) and the right side of (21), we obtain

$$\left| x - \frac{1}{2} \right| < \delta \leq \frac{1}{4}$$

so that

$$\left| x - \frac{1}{2} \right| < \frac{1}{4}$$

or equivalently

$$\frac{1}{4} < x < \frac{3}{4} \tag{27}$$

To estimate the size of $|2/x|$ we take reciprocals in (27) and multiply through by 2, which gives

$$8 > \frac{2}{x} > \frac{8}{3}$$

Thus,

$$\left| \frac{2}{x} \right| < 8$$

Comparing the last inequality to (23) suggests $k = 8$; and from (24)

$$\delta = \frac{\epsilon}{k} = \frac{\epsilon}{8}$$

In summary, given $\epsilon > 0$, we can choose

$$\delta = \frac{\epsilon}{8}$$

provided this choice does not violate $\delta \leq \frac{1}{4}$. If it does, we choose the value $\delta = \frac{1}{4}$ instead. In other words,

$$\delta = \min\left(\frac{\epsilon}{8}, \frac{1}{4}\right)$$ ◀

▶ Example 4   Let

$$f(x) = \begin{cases} 1, & x > 0 \\ -1, & x < 0 \end{cases}$$

Prove that $\lim_{x \to 0} f(x)$ does not exist.

Solution.   We will assume that there is a limit and obtain a contradiction. Assume there is a number L such that

$$\lim_{x \to 0} f(x) = L$$

Then given any $\epsilon > 0$, there exists a $\delta > 0$ such that

$$|f(x) - L| < \epsilon \qquad \text{whenever} \qquad 0 < |x - 0| < \delta$$

In particular, if we take $\epsilon = 1$, there is a $\delta > 0$ such that

$$|f(x) - L| < 1$$

whenever

$$0 < |x - 0| < \delta \tag{28}$$

But $x = \delta/2$ and $x = -\delta/2$ both satisfy (28) so that

$$\left| f\left(\frac{\delta}{2}\right) - L \right| < 1 \qquad \text{and} \qquad \left| f\left(-\frac{\delta}{2}\right) - L \right| < 1 \tag{29}$$

However, $\delta/2$ is positive and $-\delta/2$ is negative, so

$$f\left(\frac{\delta}{2}\right) = 1 \quad \text{and} \quad f\left(-\frac{\delta}{2}\right) = -1$$

Thus, (29) states that

$$|1 - L| < 1 \quad \text{and} \quad |-1 - L| < 1$$

or equivalently

$$0 < L < 2 \quad \text{and} \quad -2 < L < 0$$

But this is a contradiction, since no number L can satisfy these two conditions. ◄

▶ Exercise Set 2.6

In Exercises 1–10, we are told that $\lim\limits_{x \to a} f(x) = L$ and we are given a value of ϵ. In each exercise, find a number δ such that $|f(x) - L| < \epsilon$ whenever $0 < |x - a| < \delta$.

1. $\lim\limits_{x \to 4} 2x = 8$; $\epsilon = 0.1$.

2. $\lim\limits_{x \to -2} \frac{1}{2}x = -1$; $\epsilon = 0.1$.

3. $\lim\limits_{x \to -1} (7x + 5) = -2$; $\epsilon = 0.01$.

4. $\lim\limits_{x \to 3} (5x - 2) = 13$; $\epsilon = 0.01$.

5. $\lim\limits_{x \to 2} \frac{x^2 - 4}{x - 2} = 4$; $\epsilon = 0.05$.

6. $\lim\limits_{x \to -1} \frac{x^2 - 1}{x + 1} = -2$; $\epsilon = 0.05$.

7. $\lim\limits_{x \to 4} x^2 = 16$; $\epsilon = 0.001$.

8. $\lim\limits_{x \to 9} \sqrt{x} = 3$; $\epsilon = 0.001$.

9. $\lim\limits_{x \to 5} \frac{1}{x} = \frac{1}{5}$; $\epsilon = 0.05$.

10. $\lim\limits_{x \to 0} |x| = 0$; $\epsilon = 0.05$.

In Exercises 11–22, use Definition 2.6.1 to prove that the given limit statement is correct.

11. $\lim\limits_{x \to 5} 3x = 15$.

12. $\lim\limits_{x \to 3} (4x - 5) = 7$.

13. $\lim\limits_{x \to 0} \frac{x^2 + x}{x} = 1$.

14. $\lim\limits_{x \to -3} \frac{x^2 - 9}{x + 3} = -6$.

15. $\lim\limits_{x \to 1} 2x^2 = 2$.

16. $\lim\limits_{x \to 3} (x^2 - 5) = 4$.

17. $\lim\limits_{x \to 1/3} \frac{1}{x} = 3$.

18. $\lim\limits_{x \to -2} \frac{1}{x + 1} = -1$.

19. $\lim\limits_{x \to 4} \sqrt{x} = 2$.

20. $\lim\limits_{x \to 6} \sqrt{x + 3} = 3$.

21. $\lim\limits_{x \to 1} f(x) = 3$, where $f(x) = \begin{cases} x + 2, & x \neq 1 \\ 10, & x = 1. \end{cases}$

22. $\lim\limits_{x \to 2} (x^2 + 3x - 1) = 9$.

23. Let $f(x) = \begin{cases} \dfrac{1}{8}, & x > 0 \\[2mm] -\dfrac{1}{8}, & x < 0. \end{cases}$

Use the method of Example 4 to prove that $\lim\limits_{x \to 0} f(x)$ does not exist.

24. Let $g(x) = \begin{cases} 1 + x, & x > 0 \\ x - 1, & x < 0. \end{cases}$ Prove that $\lim\limits_{x \to 0} g(x)$ does not exist.

25. Prove that $\lim\limits_{x \to 1} \dfrac{1}{x - 1}$ does not exist.

26. (a) In Definition 2.6.1 there is a condition requiring that $f(x)$ be defined for every x in an open interval containing a, except possibly at a itself. What is the purpose of this requirement?

(b) Why is $\lim\limits_{x \to 0} \sqrt{x} = 0$ an incorrect statement?

(c) Is $\lim\limits_{x \to 0.01} \sqrt{x} = 0.1$ a correct statement?

27. Prove the result in Example 2 under the assumption that $\delta \le 2$ rather than $\delta \le 1$.

▶ SUPPLEMENTARY EXERCISES

In Exercises 1–5, find the domain of f and then evaluate f (if defined) at the given values of x.

1. $f(x) = \sqrt{4 - x^2}$; $x = -\sqrt{2}, 0, \sqrt{3}$.

2. $f(x) = 1/\sqrt{(x - 1)^3}$; $x = 0, 1, 2$.

3. $f(x) = (x - 1)/(x^2 + x - 2)$; $x = 0, 1, 2$.

4. $f(x) = \sqrt{|x| - 2}$; $x = -3, 0, 2$.

5. $f(x) = \begin{cases} x^2 - 1, & x \le 2 \\ \sqrt{x - 1}, & x > 2 \end{cases}$; $x = 0, 2, 4$.

In Exercises 6 and 7, find:

  (a) $f(x^2) - (f(x))^2$
  (b) $f(x + 3) - [f(x) + f(3)]$
  (c) $f(1/x) - 1/f(x)$
  (d) $(f \circ f)(x)$.

6. $f(x) = \sqrt{3 - x}$.

7. $f(x) = \dfrac{3 - x}{x}$.

In Exercises 8–15, sketch the graph of f and find its domain and range.

8. $f(x) = (x - 2)^2$.

9. $f(x) = -\pi$.

10. $f(x) = |2 - 4x|$.

11. $f(x) = \dfrac{x^2 - 4}{2x + 4}$.

12. $f(x) = \sqrt{-2x}$.

13. $f(x) = -\sqrt{3x + 1}$.

14. $f(x) = 2 - |x|$.

15. $f(x) = \dfrac{2x - 4}{x^2 - 4}$.

16. In each part, complete the square, and then find the range of f.
  (a) $f(x) = x^2 - 5x + 6$
  (b) $f(x) = -3x^2 + 12x - 7$.

17. Express $f(x)$ as a composite function $(g \circ h)(x)$ in two different ways.
  (a) $f(x) = x^6 + 3$
  (b) $f(x) = \sqrt{x^2 + 1}$
  (c) $f(x) = \sin(3x + 2)$.

In Exercises 18–20, find the equation of the tangent line at x_0.

18. $f(x) = x^2 - 6x$; $x_0 = -1$.

19. $f(x) = 3x^3 - 1$; $x_0 = 0$.

20. $f(x) = \sqrt{x}$; $x_0 = \frac{1}{4}$.
  [*Hint:* $\sqrt{a} - \sqrt{b} = (a - b)/(\sqrt{a} + \sqrt{b})$]

In Exercises 21 and 22, $s = f(t)$ is the distance from the origin at time t of a particle moving along a coordinate line. Find the instantaneous velocity of the particle at time t_0.

21. $s(t) = 5t + 8$; $t_0 = 2$.

22. $s(t) = \dfrac{3}{(t + 1)^2}$; $t_0 = 0$.

In Exercises 23 and 24, sketch the graph of f and find the indicated limits (if they exist).

23.
$$f(x) = \begin{cases} 1/x, & x < 0 \\ x^2, & 0 \le x < 1 \\ 2, & x = 1 \\ 2 - x, & x > 1 \end{cases}$$

(a) as $x \to -1$ (b) as $x \to 0$
(c) as $x \to 1$ (d) as $x \to 0^+$
(e) as $x \to 0^-$ (f) as $x \to 2^+$
(g) as $x \to -\infty$ (h) as $x \to +\infty$.

24.
$$f(x) = \begin{cases} 2, & x \le -1 \\ -x, & -1 < x < 0 \\ x/(2 - x), & 0 < x < 2 \\ 1, & x \ge 2 \end{cases}$$

(a) as $x \to -1^+$ (b) as $x \to -1^-$
(c) as $x \to -1$ (d) as $x \to 0$

(e) as $x \to 2^+$ (f) as $x \to 2^-$
(g) as $x \to 2$ (h) as $x \to -\infty$.

In Exercises 25–28, find $\lim\limits_{x \to a} f(x)$ (if it exists).

25. $f(x) = \sqrt{2 - x}$;
$a = -2, 1, 2^-, 2^+, -\infty, +\infty.$

26. $f(x) = \begin{cases} (x - 2)/|x - 2|, & x \ne 2 \\ 0, & x = 2 \end{cases}$
$a = 0, 2^-, 2^+, 2, -\infty, +\infty.$

27. $f(x) = (x^2 - 25)/(x - 5)$;
$a = 0, 5^+, -5^-, 5, -5, -\infty, +\infty.$

28. $f(x) = (x + 5)/(x^2 - 25)$;
$a = 0, 5^+, -5^-, -5, 5, -\infty, +\infty.$

3 differentiation

3.1 THE DERIVATIVE

Many physical phenomena involve changing quantities—the speed of a rocket, the inflation of currency, the number of bacteria in a culture, the shock intensity of an earthquake, the voltage of an electrical signal, and so forth. In this chapter we will develop the notion of a *derivative*, which is the basic mathematical tool for studying rates of change. We will also see that there is a close relationship between rates of change and tangent lines to curves.

Let us review some ideas encountered earlier. If $P(x_0, y_0)$ and $Q(x_1, y_1)$ are points on the graph of a function f then the secant line connecting P and Q has slope

$$m_{\text{sec}} = \frac{y_1 - y_0}{x_1 - x_0}$$

or since $y_0 = f(x_0)$ and $y_1 = f(x_1)$

$$m_{\text{sec}} = \frac{f(x_1) - f(x_0)}{x_1 - x_0} \qquad (1)$$

(See Figure 3.1.1a.) If we let x_1 approach x_0, then Q will approach P along the graph of f, and the secant line through P and Q will approach the tangent line at P. Thus, the slope m_{sec} of the secant line will approach the slope m_{tan} of the tangent line as x_1 approaches x_0; so from (1)

$$m_{\text{tan}} = \lim_{x_1 \to x_0} \frac{f(x_1) - f(x_0)}{x_1 - x_0} \qquad (2)$$

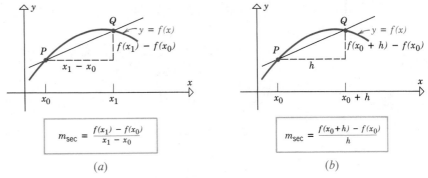

Figure 3.1.1   (a)   (b)

Alternatively, if we let

$$h = x_1 - x_0$$

then $x_1 = x_0 + h$ and $h \to 0$ as $x_1 \to x_0$. Thus, (2) can also be written as

$$m_{\text{tan}} = \lim_{h \to 0} \frac{f(x_0 + h) - f(x_0)}{h} \tag{3}$$

(See Figure 3.1.1b.)

The following definition summarizes this discussion.

3.1.1   DEFINITION   If $P(x_0, y_0)$ is a point on the graph of a function f, then the **tangent line** to the graph of f at P is defined to be the line through P with slope

$$m_{\text{tan}} = \lim_{h \to 0} \frac{f(x_0 + h) - f(x_0)}{h} \tag{4}$$

provided the limit exists.

Sometimes we will abbreviate the terminology and refer to the tangent line at $P(x_0, y_0)$ as the **tangent line at $x = x_0$**.

The limit in (4) occurs so frequently that it has a special notation. We denote it by

$$f'(x_0) = \lim_{h \to 0} \frac{f(x_0 + h) - f(x_0)}{h} \tag{5}$$

where $f'(x_0)$ is read, "f prime of x_0."

If we go one step further and drop the subscript on x_0 in (5), we obtain one of the most important functions in mathematics, *the derivative function*.

3.1.2 DEFINITION The **derivative** of a function f is the function f' defined by

$$f'(x) = \lim_{h \to 0} \frac{f(x + h) - f(x)}{h} \qquad (6)$$

The domain of f' consists of all x where this limit exists.

REMARK. If x_0 lies in the domain of f', then it follows from (6) and (4) that

$$f'(x_0) = \lim_{h \to 0} \frac{f(x_0 + h) - f(x_0)}{h} = m_{\text{tan}}$$

In other words, the derivative of f is a function whose value at $x = x_0$ is the slope of the tangent line to $y = f(x)$ at $x = x_0$. The domain of the derivative function is the set of x values at which $y = f(x)$ has a tangent line.

▶ Example 1 Find the derivative of $f(x) = x^2 + 1$.

Solution. From Definition 3.1.2,

$$f'(x) = \lim_{h \to 0} \frac{f(x + h) - f(x)}{h}$$

$$= \lim_{h \to 0} \frac{[(x + h)^2 + 1] - [x^2 + 1]}{h}$$

$$= \lim_{h \to 0} \frac{x^2 + 2xh + h^2 + 1 - x^2 - 1}{h}$$

$$= \lim_{h \to 0} \frac{2xh + h^2}{h} = \lim_{h \to 0} (2x + h) = 2x \qquad ◀$$

▶ Example 2 Use the derivative obtained in Example 1 to find the slope of the tangent line to $y = x^2 + 1$ at $x = 2$, $x = 0$, and $x = -2$.

Solution. The slope of the tangent line at any point x is

$$f'(x) = 2x$$

Thus, the slope of the tangent line at $x = 2$ is

$$f'(2) = 2(2) = 4$$

The slope at $x = -2$ is

$$f'(-2) = 2(-2) = -4$$

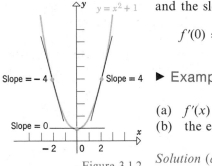

Figure 3.1.2

and the slope at $x = 0$ is

$$f'(0) = 2(0) = 0 \qquad \text{(See Figure 3.1.2.)}$$ ◄

▶ **Example 3** For the function $f(x) = \sqrt{x}$ find:

(a) $f'(x)$
(b) the equation of the tangent line to $y = \sqrt{x}$ at the point where $x = 4$.

Solution (a). From Definition 3.1.2,

$$f'(x) = \lim_{h \to 0} \frac{f(x + h) - f(x)}{h} = \lim_{h \to 0} \frac{\sqrt{x + h} - \sqrt{x}}{h}$$

$$= \lim_{h \to 0} \frac{(\sqrt{x + h} - \sqrt{x})(\sqrt{x + h} + \sqrt{x})}{h(\sqrt{x + h} + \sqrt{x})} = \lim_{h \to 0} \frac{(x + h) - x}{h(\sqrt{x + h} + \sqrt{x})}$$

$$= \lim_{h \to 0} \frac{h}{h(\sqrt{x + h} + \sqrt{x})} = \lim_{h \to 0} \frac{1}{\sqrt{x + h} + \sqrt{x}}$$

$$= \frac{1}{\sqrt{x} + \sqrt{x}} = \frac{1}{2\sqrt{x}}$$

Solution (b). At $x = 4$, the corresponding y value on the curve $y = \sqrt{x}$ is $y = 2$, so the tangent line passes through the point $(4, 2)$. From (a), the slope of the tangent line at this point is

$$f'(4) = \frac{1}{2\sqrt{4}} = \frac{1}{4}$$

Thus, the point-slope form of the tangent line is

$$y - 2 = \tfrac{1}{4}(x - 4)$$

or after simplifying

$$y = \tfrac{1}{4}x + 1$$ ◄

If x_0 lies in the domain of f', then we say that f is **differentiable at x_0**. There are three common ways in which a function f can *fail* to be differentiable at a point. Loosely phrased, these may be classified as:

1. Breaks.
2. Corners.
3. Vertical tangents.

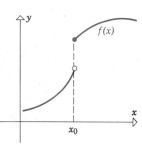

Figure 3.1.3

It is intuitively clear that if the graph of a function f has a "break" at $x = x_0$ (Figure 3.1.3), then the function cannot have a tangent at x_0. We shall prove this to be true after we make the term "break" more precise.

The graph of a function f has a "corner" at a point $P(x_0, f(x_0))$ if the graph of f is unbroken at P and the limiting position for the secant line joining P and Q depends on whether Q approaches P from the left or the right (Figure 3.1.4). At corners a tangent line does not exist since the slopes of the secant lines do not have a (two-sided) limit.

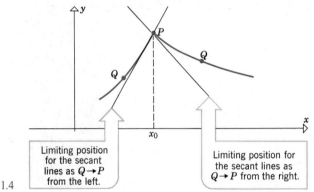

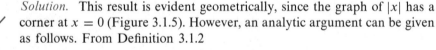

Figure 3.1.4

▶ **Example 4** Show that $f(x) = |x|$ is not differentiable at $x = 0$.

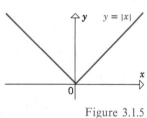

Figure 3.1.5

Solution. This result is evident geometrically, since the graph of $|x|$ has a corner at $x = 0$ (Figure 3.1.5). However, an analytic argument can be given as follows. From Definition 3.1.2

$$f'(0) = \lim_{h \to 0} \frac{f(0 + h) - f(0)}{h} = \lim_{h \to 0} \frac{f(h) - f(0)}{h}$$

$$= \lim_{h \to 0} \frac{|h| - |0|}{h} = \lim_{h \to 0} \frac{|h|}{h}$$

But

$$\frac{|h|}{h} = \begin{cases} 1, & h > 0 \\ -1, & h < 0 \end{cases}$$

so that

$$\lim_{h \to 0^-} \frac{|h|}{h} = -1 \quad \text{and} \quad \lim_{h \to 0^+} \frac{|h|}{h} = 1$$

Thus,

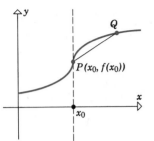

The slope of the secant line tends toward $+\infty$ as $Q \to P$

(a)

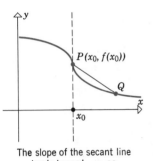

The slope of the secant line tends toward $-\infty$ as $Q \to P$

(b)

Figure 3.1.6

$$f'(0) = \lim_{h \to 0} \frac{|h|}{h}$$

does not exist since the one-sided limits are not equal. Consequently, $f(x) = |x|$ is not differentiable at $x = 0$. ◀

If the slope of the secant line joining P and Q tends toward $+\infty$ or $-\infty$ as Q approaches P along the graph of f, then f is not differentiable at x_0. Geometrically, such points occur where the secant lines tend toward a vertical limiting position (Figure 3.1.6).

The process of finding a derivative is called **differentiation.** It is often useful to think of differentiation as an operation which, when applied to a function f, produces a new function f'. In the case where the independent variable is x, the differentiation operation is often denoted by the symbol

$$\frac{d}{dx}[\;\;]$$

which is read, **the derivative with respect to x of.**

For example,

$$\frac{d}{dx}[x^2 + 1]$$

stands for the derivative with respect to x of $x^2 + 1$.

From Example 1,

$$\frac{d}{dx}[x^2 + 1] = 2x \tag{7}$$

The d/dx notation allows us to write derivatives without using any function symbols like f or f'. This notation also makes it easy to write derivatives of functions involving variables other than x. For example, if we use u as the variable in (7) instead of x, we can write

$$\frac{d}{du}[u^2 + 1] = 2u$$

(The symbol $d/du [\;\;]$ is read, "*the derivative with respect to u of*".)

With the d/dx notation it is awkward to denote the value of the derivative at a particular point; the prime notation is better for this purpose. As an illustration, we found in Example 2 that if

$$f(x) = x^2 + 1$$

then $f'(x_0) = 2x_0$. To express this in d/dx notation, we must write

$$\frac{d}{dx}[x^2 + 1]\Big|_{x=x_0} = 2x_0$$

which is rather clumsy. In general,

$$\frac{d}{dx}[\ \ ]\Big|_{x=x_0}$$

denotes the value of the derivative with respect to x at $x = x_0$.

When we use a dependent variable

$$y = f(x)$$

we will usually denote the derivative as

$$\frac{dy}{dx}$$

and the value of the derivative at a point $x = x_0$ as

$$\frac{dy}{dx}\Big|_{x=x_0}$$

For example, if $y = x^2 + 1$ then

$$\frac{dy}{dx} = 2x \quad \text{and} \quad \frac{dy}{dx}\Big|_{x=x_0} = 2x_0$$

and if $y = u^2 + 1$, then

$$\frac{dy}{du} = 2u \quad \text{and} \quad \frac{dy}{du}\Big|_{u=u_0} = 2u_0.$$

Later the symbols dy and dx will be defined separately. However, for the time being, dy/dx should not be regarded as a ratio; rather, it should be considered as a single symbol denoting the derivative.

REMARK. Instead of writing

$$\frac{d}{dx}[\ \ ]$$

to denote the differentiation operation, some writers use $D_x[\ \ ]$. We will not use this notation, however.

RATES OF CHANGE   Up to now we have interpreted the derivative dy/dx as the slope of a tangent line to the curve $y = f(x)$. Derivatives also have interpretations as rates of change. To illustrate, consider the curve $y = f(x)$ in Figure 3.1.7. As we travel along that curve from P to Q, y increases slowly at first, then more rapidly and then more slowly again. However, at the end of the trip from P to Q, y has increased by 6 units and x by 8 units. Thus, "on the average" y increases $\frac{6}{8} = \frac{3}{4}$ of a unit for each one unit increase in x between P and Q. Note that this average rate of change is just the slope of the secant line connecting P and Q. This suggests the following definition.

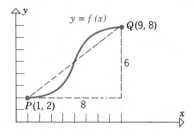

Figure 3.1.7

3.1.3   DEFINITION   If (x_0, y_0) and (x_1, y_1) are points on the graph of $y = f(x)$, then we define

$$m_{\text{sec}} = \frac{y_1 - y_0}{x_1 - x_0}$$

to be the **average rate at which y changes with x** over the interval $[x_0, x_1]$.

Just as we distinguished between the average velocity of a moving particle over a time interval and the instantaneous velocity at a single point in time, so we distinguish between the average rate of change of y with x over an interval and the instantaneous rate of change of y with x at a point. Motivated by our discussion of instantaneous velocity, we make the following definition.

3.1.4   DEFINITION   If $y = f(x)$ and f is differentiable at x_0, then we define

$$m_{\text{tan}} = \frac{dy}{dx}\bigg|_{x=x_0}$$

to be the **instantaneous rate at which y changes with x** at x_0.

▶ Example 5   For the curve $y = x^2 + 1$ find

(a)   the average rate of change of y with x over the interval $[3, 5]$;

(b)   the instantaneous rate of change of y with x at the point $x = 3$.

Solution (a).   We will apply Definition 3.1.3 with $x_0 = 3$ and $x_1 = 5$. The y values corresponding to these x values are $y_0 = (3)^2 + 1 = 10$ and $y_1 = (5)^2 + 1 = 26$, so

$$\begin{array}{l} \text{ave. rate of} \\ \text{change of } y \\ \text{over } [3, 5] \end{array} = \frac{y_1 - y_0}{x_1 - x_0} = \frac{26 - 10}{5 - 3} = \frac{16}{2} = 8$$

Thus, on the average, y increases 8 units for each unit increase in x over the interval $[3, 5]$.

Solution (b). We will apply Definition 3.1.4 with $x_0 = 3$. In Example 1 we found

$$\frac{dy}{dx} = f'(x) = 2x$$

so

$$\begin{array}{l} \text{inst. rate of} \\ \text{change of } y \\ \text{at } x = 3 \end{array} = \frac{dy}{dx}\bigg|_{x=3} = 2x\bigg|_{x=3} = 6$$

Thus, at the point $x = 3$, y is increasing 6 times as fast as x.   ◄

▶ **Example 6**   For the line $y = mx + b$,

$$\frac{dy}{dx} = \lim_{h \to 0} \frac{f(x + h) - f(x)}{h}$$

$$= \lim_{h \to 0} \frac{[m(x + h) + b] - (mx + b)}{h}$$

$$= \lim_{h \to 0} \frac{mx + mh + b - mx - b}{h}$$

$$= \lim_{h \to 0} \frac{mh}{h} = \lim_{h \to 0} m = m$$

Thus, $dy/dx = m$ is constant, which tells us that the instantaneous rate at which y varies with x along a straight line is the same at each point of the line. This constant rate of change is the slope of the line.   ◄

▶ Exercise Set 3.1

In Exercises 1–10, use Definition 3.1.2 to find $f'(x)$.

1. $f(x) = 3x^2$.

2. $f(x) = x^2 - x$.

3. $f(x) = x^3$.

4. $f(x) = 2x^3 + 1$.

5. $f(x) = \sqrt{x + 1}$.

6. $f(x) = x^4$.

7. $f(x) = \dfrac{1}{x}$.

8. $f(x) = \dfrac{1}{x^2}$.

9. $f(x) = ax^2 + b$ (a, b constants).

10. $f(x) = \dfrac{1}{x + 1}$.

In Exercises 11–16, find $f'(a)$ and the equation of the tangent to the graph of f at the point where $x = a$.

11. f is the function in Exercise 1; $a = 3$.

12. f is the function in Exercise 2; $a = 2$.

13. f is the function in Exercise 3; $a = 0$.

14. f is the function in Exercise 4; $a = -1$.

15. f is the function in Exercise 5; $a = 8$.

16. f is the function in Exercise 6; $a = -2$.

17. Let $y = 4x^2 + 2$. Find

(a) $\dfrac{dy}{dx}$ (b) $\dfrac{dy}{dx}\Big|_{x=1}$.

18. Let $y = \dfrac{5}{x} + 1$. Find

(a) $\dfrac{dy}{dx}$ (b) $\dfrac{dy}{dx}\Big|_{x=-2}$.

In Exercises 19–22, use Definition 3.1.2 (with the appropriate change in notation) to obtain the derivative requested.

19. Find $f'(t)$ if $f(t) = 4t^2 + t$.

20. Find $g'(u)$ if $g(u) = 5u + 3$.

21. Find $\dfrac{dA}{d\lambda}$ if $A = 3\lambda^2 - \lambda$.

22. Find $\dfrac{dV}{dr}$ if $V = \frac{4}{3}\pi r^3$.

23. Let $y = 2x^2 - 1$.
(a) Find the average rate at which y changes with x over the interval $[1, 4]$.
(b) Find the instantaneous rate at which y changes with x at the point $x = 1$.

24. Let $y = \dfrac{1}{x^2 + 1}$.
(a) Find the average rate at which y changes with x over the interval $[-1, 2]$.
(b) Find the instantaneous rate at which y changes with x at the point $x = -1$.

25. Use the formula $A = \pi r^2$ for the area of a circle to find:
(a) the average rate at which the area of a circle changes with r as the radius increases from $r = 1$ to $r = 2$;

(b) the instantaneous rate at which the area changes with r when $r = 2$.

26. Use the formula $V = l^3$ for the volume of a cube of side l to find:
(a) the average rate at which the volume of a cube changes with l as l increases from $l = 2$ to $l = 4$;
(b) the instantaneous rate at which the volume of a cube changes with l when $l = 5$.

In Exercises 27–34, sketch the graph of the derivative of the function whose graph is shown.

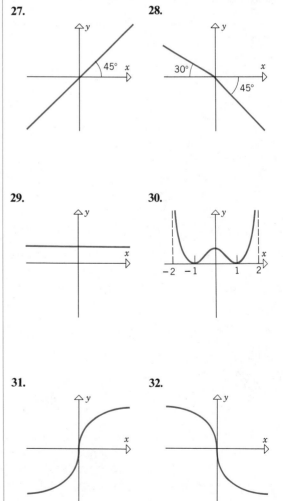

27.   **28.**

29.   **30.**

31.   **32.**

33.

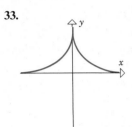

34.

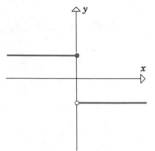

35. Show that $f(x) = \sqrt[3]{x - 2}$ is not differentiable at $x = 2$. Sketch the graph of f.

36. Show that

$$f(x) = \begin{cases} x^2 + 1, & x \le 1 \\ 2x, & x > 1 \end{cases}$$

is differentiable at 1, but that

$$g(x) = \begin{cases} x^2 + 2, & x \le 1 \\ 3x, & x > 1 \end{cases}$$

is not differentiable at 1. Sketch the graphs of f and g. [*Remark:* Geometrically, the results in this problem tell us that the graph of g has a "corner" at $x = 1$, but the graph of f does not.]

37. Use Definition 3.1.2 to find $f'(x)$ if

$$f(x) = \frac{1}{\sqrt{x}}$$

38. Use Definition 3.1.2 to find $f'(x)$ if

$$f(x) = x^{1/3}$$

3.2 TECHNIQUES OF DIFFERENTIATION

Up to now we have obtained derivatives directly from the definition. In this section we will develop some theorems and formulas that provide more efficient methods.

3.2.1 THEOREM *If f is a constant function, say $f(x) = c$ for all x, then $f'(x) = 0$.*

Proof.

$$f'(x) = \lim_{h \to 0} \frac{f(x + h) - f(x)}{h} = \lim_{h \to 0} \frac{c - c}{h} = \lim_{h \to 0} 0 = 0 \quad \blacksquare$$

REMARK. The result in Theorem 3.2.1 is obvious geometrically. If $f(x) = c$ is a constant function, then the graph of f is a line parallel to the x-axis; consequently, the tangent at each point is a horizontal line. Therefore, $f'(x) = m_{\tan} = 0$, since a horizontal line has slope 0.

In another notation, Theorem 3.2.1 states

$$\frac{d}{dx}[c] = 0$$

▶ **Example 1**   If $f(x) = 5$ for all x, then $f'(x) = 0$ for all x; that is,

$$\frac{d}{dx}[5] = 0 \qquad\qquad\qquad ◀$$

3.2.2   THEOREM
The Power Rule

If n is a positive integer, then

$$\frac{d}{dx}[x^n] = nx^{n-1}$$

Proof.   Let $f(x) = x^n$. Then

$$f'(x) = \lim_{h \to 0} \frac{f(x + h) - f(x)}{h} = \lim_{h \to 0} \frac{(x + h)^n - x^n}{h}$$

Expanding $(x + h)^n$ by the binomial theorem, we obtain

$$f'(x) = \lim_{h \to 0} \frac{\left[x^n + nx^{n-1}h + \dfrac{n(n-1)}{2!}x^{n-2}h^2 + \cdots + nxh^{n-1} + h^n\right] - x^n}{h}$$

$$= \lim_{h \to 0} \frac{nx^{n-1}h + \dfrac{n(n-1)}{2!}x^{n-2}h^2 + \cdots + nxh^{n-1} + h^n}{h}$$

Canceling a factor of h we obtain

$$f'(x) = \lim_{h \to 0} \left[nx^{n-1} + \frac{n(n-1)}{2!}x^{n-2}h + \cdots + nxh^{n-2} + h^{n-1}\right]$$

Every term but the first has a factor of h, and so approaches zero as $h \to 0$. Therefore,

$$f'(x) = nx^{n-1} \quad \blacksquare$$

REMARK.   In words, *to differentiate x to a positive integer power, take the power and multiply it by x to the next lower integer power.*

► Example 2

$$\frac{d}{dx}[x^5] = 5x^4, \qquad \frac{d}{dx}[x] = 1 \cdot x^0 = 1, \qquad \frac{d}{dx}[x^{12}] = 12x^{11} \qquad ◄$$

3.2.3 THEOREM *Let c be a constant. If f is differentiable at x, then so is cf, and*

$$\frac{d}{dx}[cf(x)] = c\frac{d}{dx}[f(x)]$$

Proof.

$$\frac{d}{dx}[cf(x)] = \lim_{h \to 0} \frac{cf(x + h) - cf(x)}{h}$$

$$= \lim_{h \to 0} c\left[\frac{f(x + h) - f(x)}{h}\right]$$

$$= c\lim_{h \to 0} \frac{f(x + h) - f(x)}{h} \qquad \begin{bmatrix} \text{A constant factor can be} \\ \text{moved through a} \\ \text{limit sign.} \end{bmatrix}$$

$$= c\frac{d}{dx}[f(x)] \quad ▮$$

In another notation, Theorem 3.2.3 states

$$(cf)'(x) = cf'(x)$$

REMARK. In words, *a constant factor c can be moved through a derivative sign.*

► Example 3

$$\frac{d}{dx}[4x^8] = 4\frac{d}{dx}[x^8] = 4[8x^7] = 32x^7$$

$$\frac{d}{dx}[-x^{12}] = (-1)\frac{d}{dx}[x^{12}] = -12x^{11} \qquad ◄$$

3.2.4 THEOREM *If f and g are differentiable at x, then so is f + g, and*

$$\frac{d}{dx}[f(x) + g(x)] = \frac{d}{dx}[f(x)] + \frac{d}{dx}[g(x)]$$

Proof.

$$\frac{d}{dx}[f(x) + g(x)] = \lim_{h \to 0} \frac{[f(x + h) + g(x + h)] - [f(x) + g(x)]}{h}$$

$$= \lim_{h \to 0} \frac{[f(x + h) - f(x)] + [g(x + h) - g(x)]}{h}$$

$$= \lim_{h \to 0} \frac{f(x + h) - f(x)}{h} + \lim_{h \to 0} \frac{g(x + h) - g(x)}{h} \quad \left[\begin{array}{l} \text{The limit of a sum} \\ \text{is the sum of} \\ \text{the limits.} \end{array}\right]$$

$$= \frac{d}{dx}[f(x)] + \frac{d}{dx}[g(x)] \quad \blacksquare$$

Theorem 3.2.4 can be written in the alternate notation,

$$(f + g)'(x) = f'(x) + g'(x)$$

By writing $f - g = f + (-1)g$ and then applying Theorems 3.2.3 and 3.2.4 it follows that

$$\frac{d}{dx}[f(x) - g(x)] = \frac{d}{dx}[f(x)] - \frac{d}{dx}[g(x)]$$

REMARK. In words, *the derivative of a sum equals the sum of the derivatives,* and *the derivative of a difference equals the difference of the derivatives.*

▶Example 4

$$\frac{d}{dx}[x^4 + x^2] = \frac{d}{dx}[x^4] + \frac{d}{dx}[x^2] = 4x^3 + 2x$$

$$\frac{d}{dx}[x^4 - x^2] = \frac{d}{dx}[x^4] - \frac{d}{dx}[x^2] = 4x^3 - 2x$$

$$\frac{d}{dx}[6x^{11} + 9] = \frac{d}{dx}[6x^{11}] + \frac{d}{dx}[9] = 66x^{10} + 0 = 66x^{10} \quad \blacktriangleleft$$

The result in Theorem 3.2.4 can be extended to any finite number of functions. More precisely, if the functions $f_1, f_2, \ldots, f_n$ are all differentiable at x, then their sum is differentiable at x and

$$\frac{d}{dx}[f_1(x) + f_2(x) + \cdots + f_n(x)] = \frac{d}{dx}[f_1(x)] + \frac{d}{dx}[f_2(x)] + \cdots + \frac{d}{dx}[f_n(x)]$$

▶ Example 5

$$\frac{d}{dx}[3x^8 - 2x^5 + 6x + 1] = \frac{d}{dx}[3x^8] + \frac{d}{dx}[-2x^5] + \frac{d}{dx}[6x] + \frac{d}{dx}[1]$$

$$= 24x^7 - 10x^4 + 6 \quad \blacktriangleleft$$

3.2.5   THEOREM   *If f and g are differentiable at x, then so is the product f · g, and*
The Product Rule

$$\frac{d}{dx}[f(x)g(x)] = f(x)\frac{d}{dx}[g(x)] + g(x)\frac{d}{dx}[f(x)]$$

Proof.

$$\frac{d}{dx}[f(x)g(x)] = \lim_{h \to 0} \frac{f(x+h) \cdot g(x+h) - f(x) \cdot g(x)}{h}$$

If we add and subtract $f(x+h) \cdot g(x)$ in the numerator, we obtain

$$\frac{d}{dx}[f(x)g(x)] = \lim_{h \to 0} \frac{f(x+h)g(x+h) - f(x+h)g(x) + f(x+h)g(x) - f(x)g(x)}{h}$$

$$= \lim_{h \to 0} \left[f(x+h) \cdot \frac{g(x+h) - g(x)}{h} + g(x) \cdot \frac{f(x+h) - f(x)}{h} \right]$$

$$= \lim_{h \to 0} f(x+h) \cdot \lim_{h \to 0} \frac{g(x+h) - g(x)}{h} + \lim_{h \to 0} g(x) \cdot \lim_{h \to 0} \frac{f(x+h) - f(x)}{h}$$

$$= [\lim_{h \to 0} f(x+h)]\frac{d}{dx}[g(x)] + [\lim_{h \to 0} g(x)]\frac{d}{dx}[f(x)] \qquad (1)$$

Consider the limit

$$\lim_{h \to 0} g(x)$$

occurring in the second term of (1). The expression $g(x)$ does not involve h so that $g(x)$ remains *constant* as $h \to 0$. Therefore,

$$\lim_{h \to 0} g(x) = g(x) \qquad (2)$$

Next consider the limit

$$\lim_{h \to 0} f(x+h)$$

which appears in the first term of (1). We can rewrite this limit as

$$\lim_{h \to 0} f(x+h) = \lim_{h \to 0} \left[\frac{f(x+h) - f(x)}{h} \cdot h + f(x) \right]$$

$$= \lim_{h \to 0} \frac{f(x+h) - f(x)}{h} \cdot \lim_{h \to 0} h + \lim_{h \to 0} f(x)$$

$$= \frac{d}{dx}[f(x)] \cdot 0 + \lim_{h \to 0} f(x)$$

Since $f(x)$ does not involve h

$$\lim_{h \to 0} f(x) = f(x)$$

so that

$$\lim_{h \to 0} f(x + h) = f(x) \tag{3}$$

Substituting (2) and (3) into (1) yields

$$\frac{d}{dx}[f(x)g(x)] = f(x)\frac{d}{dx}[g(x)] + g(x)\frac{d}{dx}[f(x)] \quad \blacksquare$$

The product rule can be written in the alternate notation,

$$(f \cdot g)'(x) = f(x)g'(x) + g(x)f'(x)$$

REMARK. In words, *the derivative of a product of two functions equals the first function times the derivative of the second plus the second function times the derivative of the first.*

▶ Example 6   Find $\dfrac{dy}{dx}$ if $y = (4x^2 - 1)(7x^3 + x)$.

Solution. There are two methods. We can either use the product rule or we can multiply out the factors in y and then differentiate. We will give both methods.

Method I.   (*Using the Product Rule.*)

$$\frac{dy}{dx} = \frac{d}{dx}[(4x^2 - 1)(7x^3 + x)]$$

$$= (4x^2 - 1)\frac{d}{dx}[7x^3 + x] + (7x^3 + x)\frac{d}{dx}[4x^2 - 1]$$

$$= (4x^2 - 1)(21x^2 + 1) + (7x^3 + x)(8x)$$

If desired, we can simplify this expression to obtain

$$\frac{dy}{dx} = 140x^4 - 9x^2 - 1$$

Method II. *(Multiplying First.)*

$$y = (4x^2 - 1)(7x^3 + x) = 28x^5 - 3x^3 - x$$

Thus,

$$\frac{dy}{dx} = \frac{d}{dx}[28x^5 - 3x^3 - x] = 140x^4 - 9x^2 - 1$$

which agrees with the result obtained using the product rule. ◀

3.2.6 THEOREM *If f and g are differentiable at x and $g(x) \neq 0$, then f/g is differentiable at x*
The Quotient Rule *and*

$$\frac{d}{dx}\left[\frac{f(x)}{g(x)}\right] = \frac{g(x)\frac{d}{dx}[f(x)] - f(x)\frac{d}{dx}[g(x)]}{[g(x)]^2}$$

Proof.

$$\frac{d}{dx}\left[\frac{f(x)}{g(x)}\right] = \lim_{h \to 0} \frac{\dfrac{f(x+h)}{g(x+h)} - \dfrac{f(x)}{g(x)}}{h}$$

$$= \lim_{h \to 0} \frac{f(x+h) \cdot g(x) - f(x) \cdot g(x+h)}{h \cdot g(x) \cdot g(x+h)}$$

Adding and subtracting $f(x) \cdot g(x)$ in the numerator yields

$$\frac{d}{dx}\left[\frac{f(x)}{g(x)}\right] = \lim_{h \to 0} \frac{f(x+h) \cdot g(x) - f(x) \cdot g(x) - f(x) \cdot g(x+h) + f(x) \cdot g(x)}{h \cdot g(x) \cdot g(x+h)}$$

$$= \lim_{h \to 0} \frac{\left[g(x) \cdot \dfrac{f(x+h) - f(x)}{h}\right] - \left[f(x) \cdot \dfrac{g(x+h) - g(x)}{h}\right]}{g(x) \cdot g(x+h)}$$

$$= \frac{\displaystyle\lim_{h \to 0} g(x) \cdot \lim_{h \to 0} \frac{f(x+h) - f(x)}{h} - \lim_{h \to 0} f(x) \cdot \lim_{h \to 0} \frac{g(x+h) - g(x)}{h}}{\displaystyle\lim_{h \to 0} g(x) \cdot \lim_{h \to 0} g(x+h)}$$

$$= \frac{[\displaystyle\lim_{h \to 0} g(x)] \cdot \frac{d}{dx}[f(x)] - [\displaystyle\lim_{h \to 0} f(x)] \cdot \frac{d}{dx}[g(x)]}{\displaystyle\lim_{h \to 0} g(x) \cdot \lim_{h \to 0} g(x+h)} \qquad (4)$$

Since the expressions $f(x)$ and $g(x)$ do not involve h,

$$\lim_{h \to 0} g(x) = g(x) \qquad \text{and} \qquad \lim_{h \to 0} f(x) = f(x) \tag{5}$$

By an argument like that used to derive (3) in the previous theorem, the reader will be able to show that

$$\lim_{h \to 0} g(x + h) = g(x) \tag{6}$$

Substituting (5) and (6) into (4) yields

$$\frac{d}{dx}\left[\frac{f(x)}{g(x)}\right] = \frac{g(x)\dfrac{d}{dx}[f(x)] - f(x)\dfrac{d}{dx}[g(x)]}{[g(x)]^2} \quad \blacksquare$$

The quotient rule can be written in the alternate notation,

$$\left(\frac{f}{g}\right)'(x) = \frac{g(x)f'(x) - f(x)g'(x)}{[g(x)]^2}$$

REMARK. In words, *the derivative of a quotient of two functions is the denominator times the derivative of the numerator minus the numerator times the derivative of the denominator, all divided by the denominator squared.*

▶ Example 7 Find dy/dx if

$$y = \frac{5x^4 + x^2}{x^3 + 2}$$

Solution.

$$\frac{dy}{dx} = \frac{d}{dx}\left[\frac{5x^4 + x^2}{x^3 + 2}\right] = \frac{(x^3 + 2)\dfrac{d}{dx}[5x^4 + x^2] - (5x^4 + x^2)\dfrac{d}{dx}[x^3 + 2]}{(x^3 + 2)^2}$$

$$= \frac{(x^3 + 2)(20x^3 + 2x) - (5x^4 + x^2)(3x^2)}{(x^3 + 2)^2}$$

If necessary, the numerator can be simplified. ◀

In Theorem 3.2.2 we established the formula

$$\frac{d}{dx}[x^n] = nx^{n-1}$$

for *positive* integer values of n. The following theorem extends this formula so that it applies to *all* integer values of n. Eventually, we will show that this formula is valid for all real exponents.

3.2.7   THEOREM   *If n is any integer, then*

$$\frac{d}{dx}[x^n] = nx^{n-1} \tag{7}$$

Proof. The result has already been established in the case where $n > 0$. If $n < 0$, then let

$$m = -n$$

and let

$$f(x) = x^{-m} = \frac{1}{x^m}$$

From Theorem 3.2.6,

$$f'(x) = \frac{d}{dx}\left[\frac{1}{x^m}\right] = \frac{x^m \cdot \frac{d}{dx}[1] - 1 \cdot \frac{d}{dx}[x^m]}{(x^m)^2}$$

Since $n < 0$, it follows that $m > 0$, so x^m can be differentiated using Theorem 3.2.2. Thus,

$$f'(x) = \frac{x^m \cdot 0 - 1 \cdot mx^{m-1}}{x^{2m}} = -mx^{m-1-2m}$$

$$= -mx^{-m-1} = nx^{n-1}$$

Formula (7) also holds in the case $n = 0$ since the formula reduces to

$$\frac{d}{dx}[1] = 0 \cdot x^{-1} = 0$$

which is correct by Theorem 3.2.1.   ∎

▶ Example 8

$$\frac{d}{dx}[x^{-9}] = -9x^{-9-1} = -9x^{-10}$$

$$\frac{d}{dx}\left[\frac{1}{x}\right] = \frac{d}{dx}[x^{-1}] = (-1)x^{-1-1} = -x^{-2} = -\frac{1}{x^2}$$

◀

HIGHER DERIVATIVES If the derivative f' of a function f is itself differentiable, then the derivative of f' is denoted by f'' and is called the **second derivative** of f. As long as we have differentiability, we can continue the process of differentiating derivatives to obtain third, fourth, fifth, and even higher derivatives of f. The successive derivatives of f are denoted by

$$f'$$ (the first derivative of f)
$$f'' = (f')'$$ (the second derivative of f)
$$f''' = (f'')'$$ (the third derivative of f)
$$f^{(4)} = (f''')'$$ (the fourth derivative of f)
$$f^{(5)} = (f^{(4)})'$$ (the fifth derivative of f)
$$\vdots \qquad \vdots \qquad \qquad \vdots$$

Beyond the third derivative, it is too clumsy to continue using primes, so we switch from primes to integers in parentheses to denote the **order** of the derivative. In this notation it is easy to denote a derivative of arbitrary order by writing

$$f^{(n)}$$ (the nth derivative of f)

The significance of the derivatives of order 2 and higher will be discussed later.

▶ Example 9 If

$$f(x) = 3x^4 - 2x^3 + x^2 - 4x + 2$$

then

$$f'(x) = 12x^3 - 6x^2 + 2x - 4$$
$$f''(x) = 36x^2 - 12x + 2$$
$$f'''(x) = 72x - 12$$
$$f^{(4)}(x) = 72$$
$$f^{(5)}(x) = 0$$
$$\vdots$$
$$f^{(n)}(x) = 0 \qquad (n \geq 5)$$ ◀

In the case where f is a function of x, successive applications of the derivative operation are denoted as follows:

$$f'(x) = \frac{d}{dx}[f(x)]$$

$$f''(x) = \frac{d}{dx}\left[\frac{d}{dx}[f(x)]\right] = \frac{d^2}{dx^2}[f(x)]$$

$$f'''(x) = \frac{d}{dx}\left[\frac{d}{dx}\left[\frac{d}{dx}[f(x)]\right]\right] = \frac{d^3}{dx^3}[f(x)]$$

$$\vdots \qquad \qquad \qquad \vdots$$
etc. $\qquad \qquad \qquad$ etc.

In general, we write

$$f^{(n)}(x) = \frac{d^n}{dx^n}[f(x)]$$

which is read, *the nth derivative of f with respect to x.*
When a dependent variable is involved, say,

$$y = f(x)$$

then successive derivatives can be denoted by writing

$$y', \quad y'', \quad y''', \quad y^{(4)}, \dots, y^{(n)}, \dots$$

or

$$\frac{dy}{dx}, \quad \frac{d^2y}{dx^2}, \quad \frac{d^3y}{dx^3}, \quad \frac{d^4y}{dx^4}, \dots, \frac{d^ny}{dx^n}, \dots$$

The following symbols denote values of derivatives at a particular point x_0; their meanings should be self-evident.

$$y''(x_0), \quad f^{(4)}(x_0), \quad \frac{d^3y}{dx^3}\bigg|_{x=x_0}, \quad \frac{d^2}{dx^2}[x^7 - x]\bigg|_{x=x_0}$$

▶ Example 10

$$\frac{d^2}{dx^2}[x^5] = \frac{d}{dx}\left[\frac{d}{dx}(x^5)\right] = \frac{d}{dx}[5x^4] = 20x^3$$

Thus,

$$\frac{d^2}{dx^2}[x^5]\bigg|_{x=2} = 20 \cdot 8 = 160 \qquad \blacktriangleleft$$

▶ Exercise Set 3.2

In Exercises 1–28, use the results of this section to find dy/dx.

1. $y = 4x^7$.

2. $y = -3x^{12}$.

3. $y = 3x^8 + 2x + 1$.

4. $y = \frac{1}{2}(x^4 + 7)$.

5. $y = \pi^3$.

6. $y = \sqrt{2}x + \dfrac{1}{\sqrt{2}}$.

7. $y = -\dfrac{1}{3}(x^7 + 2x - 9)$.

8. $y = \dfrac{x^2 + 1}{5}$.

9. $y = ax^3 + bx^2 + cx + d$
   (a, b, c, d constant).

10. $y = \frac{1}{a}\left(x^2 + \frac{1}{b}x + c\right)$
    (a, b, c constant).

11. $y = -3x^{-8}$.

12. $y = 7x^{-6}$.

13. $y = x^{-3} + \frac{1}{x^7}$.

14. $y = x + \frac{1}{x}$.

15. $y = (3x^2 + 6)\left(2x - \frac{1}{4}\right)$.

16. $y = (2 - x - 3x^3)(7 + x^5)$.

17. $y = (x^3 + 7x^2 - 8)(2x^{-3} + x^{-4})$.

18. $y = \left(\frac{1}{x} + \frac{1}{x^2}\right)(3x^3 + 27)$.

19. $y = (x^5 + 2x)^2$.

20. $y = (x^2 + x)^{-2}$.

21. $y = \frac{3x}{2x + 1}$.

22. $y = \frac{x^2 + 1}{3x}$.

23. $y = \frac{3x^8 - 6x^2 + 1}{2x^7 + 4x + 2}$.

24. $y = \frac{x^3 + 7x + 1}{x^3 - x}$.

25. $y = \left(\frac{3x + 2}{x}\right)(x^{-5} + 1)$.

26. $y = (2x^7 - x^2)\left(\frac{x - 1}{x + 1}\right)$.

27. $y = 3 + \frac{1}{4}\left(\frac{x}{x + 1}\right)(x^{-2} + 5)$.

28. $y = \frac{1}{5} + \frac{1}{3}\left(\frac{x - 1}{x + 1}\right)(1 + x^{-2})$.

In Exercises 29–34, the functions involve independent variables other than x. Use the results in this section to find the indicated derivative.

29. Find $\frac{d}{dt}[16t^2]$.

30. $c = 2\pi r$; find $\frac{dc}{dr}$.

31. $V(r) = \pi r^3$; find $V'(r)$.

32. Find $\frac{d}{d\alpha}[2\alpha^{-1} + \alpha]$.

33. $s = \frac{t}{t^3 + 7}$; find $\frac{ds}{dt}$.

34. Find $\frac{d}{d\lambda}\left[\frac{\lambda\lambda_0 + \lambda^6}{2 - \lambda_0}\right]$ (λ_0 is constant).

35. Newton's law of gravitation states that the magnitude of the force F exerted by a body of mass M on a body of mass m is

$$F = \frac{GmM}{r^2}$$

where G is a constant and r is the distance between the bodies. Assuming that the bodies are moving, find a formula for the instantaneous rate of change of F with r.

36. The volume of a sphere is $V = \frac{4}{3}\pi r^3$. Assuming that the radius is changing, find a formula for the instantaneous rate of change of V with r.

37. Find d^2y/dx^2.
    (a) $y = 7x^3 - 5x^2 + x$
    (b) $y = 12x^2 - 2x + 3$
    (c) $y = \frac{x + 1}{x}$
    (d) $y = (5x^2 - 3)(7x^3 + x)$.

38. Find y''.
    (a) $y = 4x^7 - 5x^3 + 2x$
    (b) $y = 3x + 2$
    (c) $y = \frac{3x - 2}{5x}$
    (d) $y = (x^3 - 5)(2x + 3)$.

39. Find y'''.
    (a) $y = x^{-5} + x^5$
    (b) $y = \frac{1}{x}$
    (c) $y = ax^3 + bx + c$
        (a, b, c constant).

40. Find $\frac{d^3y}{dx^3}$.
    (a) $y = 5x^2 - 4x + 7$
    (b) $y = 3x^{-2} + 4x^{-1} + x$
    (c) $y = ax^4 + bx^2 + c$
        (a, b, c constant).

41. Find

(a) $f'''(2)$, where $f(x) = 3x^2 - 2$

(b) $\dfrac{d^2y}{dx^2}\Big|_{x=1}$, where $y = 6x^5 - 4x^2$

(c) $\dfrac{d^4}{dx^4}[x^{-3}]\Big|_{x=1}$

42. Find

(a) $y'''(0)$, where $y = 4x^4 + 2x^3 + 3$

(b) $\dfrac{d^4y}{dx^4}\Big|_{x=1}$, where $y = \dfrac{6}{x^4}$.

43. Show that $y = x^3 + 3x + 1$ satisfies the equation $y''' + xy'' - 2y' = 0$.

44. Show that $y = 1/x$ satisfies the equation $x^3y'' + x^2y' - xy = 0$.

45. At which point(s) does the graph of $y = \frac{1}{3}x^3 - \frac{3}{2}x^2 + 2x$ have a horizontal tangent?

46. Find an equation for the line that is tangent to $y = (1 - x)/(1 + x)$ at the point where $x = 2$.

47. Find the values of a and b if the tangent to $y = ax^2 + bx$ at $(1, 5)$ has slope $m_{\tan} = 8$.

48. Find k if the curve $y = x^2 + k$ is tangent to the line $y = 2x$.

49. Use Theorems 3.2.3 and 3.2.4 to prove: If the functions f and g are differentiable at x, then $f - g$ is differentiable at x and $(f - g)' = f' - g'$.

50. (a) Let the functions f, g, and h be differentiable at x. By applying Theorem 3.2.5 twice, show that the product $f \cdot g \cdot h$ is differentiable at x and

$$(f \cdot g \cdot h)'(x) = f(x)g(x)h'(x) + f(x)g'(x)h(x) + f'(x)g(x)h(x)$$

(b) State a formula for differentiating a product of n functions.

51. Use the results of Exercise 50 to find:

(a) $\dfrac{d}{dx}\left[(2x + 1)\left(1 + \dfrac{1}{x}\right)(x^{-3} + 7)\right]$

(b) $\dfrac{d}{dx}[x^{-5}(x^2 + 2x)(4 - 3x)(2x^9 + 1)]$

(c) $\dfrac{d}{dx}[(x^7 + 2x - 3)^3]$

(d) $\dfrac{d}{dx}[(x^2 + 1)^{50}]$.

52. Prove: If f is differentiable at x and $f(x) \neq 0$, then $1/f(x)$ is differentiable at x and

$$\frac{d}{dx}\left[\frac{1}{f(x)}\right] = -\frac{f'(x)}{[f(x)]^2}$$

53. (a) Find $f^{(n)}(x)$ if $f(x) = x^n$.

(b) Find $f^{(n)}(x)$ if $f(x) = x^k$ and $n > k$, where k is a positive integer.

(c) Find $f^{(n)}(x)$ if $f(x) = a_0 + a_1x + a_2x^2 + \cdots + a_nx^n$.

54. In each part compute f', f'', f''' and then state the formula for $f^{(n)}$.

(a) $f(x) = \dfrac{1}{x}$

(b) $f(x) = \dfrac{1}{x^2}$.

[*Hint:* The value of $(-1)^n$ is 1 if n is even and -1 if n is odd. Use this expression in your answer.]

55. (a) Prove:

$$\frac{d^2}{dx^2}[cf(x)] = c\frac{d^2}{dx^2}[f(x)]$$

$$\frac{d^2}{dx^2}[f(x) + g(x)] = \frac{d^2}{dx^2}[f(x)] + \frac{d^2}{dx^2}[g(x)]$$

(b) Do the results in part (a) generalize to nth derivatives? Justify your answer.

56. Prove:

$$(f \cdot g)''(x) = f''(x) \cdot g(x) + 2f'(x) \cdot g'(x) + f(x) \cdot g''(x)$$

57. (a) In our proof of Theorem 3.2.5, we showed that

$$\lim_{h \to 0} f(x + h) = f(x)$$

[see Equation (3)]. The argument given used the hypothesis that f is differentiable at x. Find the fallacy in the following proof that makes no assumptions about f. As h approaches 0, $x + h$ approaches x; consequently, $f(x + h)$ approaches $f(x)$; that is,

$$\lim_{h \to 0} f(x + h) = f(x)$$

(b) Let

$$f(x) = \begin{cases} x, & x \neq 1 \\ 3, & x = 1 \end{cases}$$

Show that

$$\lim_{h \to 0} f(x + h) \neq f(x)$$

when $x = 1$

58. Let $f(x) = x^8 - 2x + 3$ and $x_0 = 2$; find

$$\lim_{h \to 0} \frac{f'(x_0 + h) - f'(x_0)}{h}$$

3.3   DERIVATIVES OF TRIGONOMETRIC FUNCTIONS

Readers who need to review trigonometry should read Unit II of the trigonometry review in Appendix 1 before beginning this section.

The main objective of this section is to find the derivatives of the trigonometric functions: $\sin x$, $\cos x$, $\tan x$, $\cot x$, $\sec x$, and $\csc x$. We assume throughout that *all angles are in radian measure*.

Let us first consider the problem of differentiating $\sin x$. From the definition of a derivative,

$$\frac{d}{dx}[\sin x] = \lim_{h \to 0} \frac{\sin (x + h) - \sin x}{h}$$

$$= \lim_{h \to 0} \frac{\sin x \cos h + \cos x \sin h - \sin x}{h}$$

$$= \lim_{h \to 0} \left[\sin x \left(\frac{\cos h - 1}{h} \right) + \cos x \left(\frac{\sin h}{h} \right) \right]$$

Since $\sin x$ and $\cos x$ do not involve h, they remain constant as $h \to 0$; thus,

$$\lim_{h \to 0} (\sin x) = \sin x \qquad \text{and} \qquad \lim_{h \to 0} (\cos x) = \cos x$$

Consequently,

$$\frac{d}{dx}[\sin x] = \sin x \cdot \lim_{h \to 0} \left(\frac{\cos h - 1}{h} \right) + \cos x \cdot \lim_{h \to 0} \left(\frac{\sin h}{h} \right) \tag{1}$$

In Theorems 3.3.2 and 3.3.4 below we will show that

$$\lim_{h \to 0} \frac{\cos h - 1}{h} = 0 \qquad \text{and} \qquad \lim_{h \to 0} \frac{\sin h}{h} = 1 \tag{2}$$

so that (1) reduces to

$$\frac{d}{dx}[\sin x] = \cos x \tag{3}$$

The derivative of $\cos x$ is obtained similarly:

$$\frac{d}{dx}[\cos x] = \lim_{h \to 0} \frac{\cos (x + h) - \cos x}{h}$$

$$= \lim_{h \to 0} \frac{\cos x \cos h - \sin x \sin h - \cos x}{h}$$

$$= \lim_{h \to 0} \left[\cos x \cdot \left(\frac{\cos h - 1}{h} \right) - \sin x \cdot \left(\frac{\sin h}{h} \right) \right]$$

$$= \cos x \cdot \lim_{h \to 0} \left(\frac{\cos h - 1}{h} \right) - \sin x \cdot \lim_{h \to 0} \left(\frac{\sin h}{h} \right)$$

Thus, from (2)

$$\frac{d}{dx}[\cos x] = -\sin x \tag{4}$$

The derivatives of the remaining trigonometric functions can be obtained using the relationships

$$\tan x = \frac{\sin x}{\cos x} \qquad \cot x = \frac{\cos x}{\sin x} \qquad \sec x = \frac{1}{\cos x} \qquad \csc x = \frac{1}{\sin x}$$

For example,

$$\frac{d}{dx}[\tan x] = \frac{d}{dx} \left[\frac{\sin x}{\cos x} \right]$$

$$= \frac{\cos x \cdot \frac{d}{dx}[\sin x] - \sin x \cdot \frac{d}{dx}[\cos x]}{\cos^2 x}$$

$$= \frac{\cos x \cdot \cos x - \sin x \cdot (-\sin x)}{\cos^2 x}$$

$$= \frac{\cos^2 x + \sin^2 x}{\cos^2 x} = \frac{1}{\cos^2 x} = \sec^2 x$$

Thus,

$$\frac{d}{dx}[\tan x] = \sec^2 x \tag{5}$$

We leave the remaining formulas for the exercises:

$$\frac{d}{dx}[\cot x] = -\csc^2 x \qquad (6)$$

$$\frac{d}{dx}[\sec x] = \sec x \tan x \qquad (7)$$

$$\frac{d}{dx}[\csc x] = -\csc x \cot x \qquad (8)$$

REMARK. The derivative formulas for the trigonometric functions should be memorized. An easy way of doing this is discussed in Exercise 52.

▶ Example 1 Find $f'(x)$ if $f(x) = x^2 \tan x$.

Solution. Using the product rule and formula (5), we obtain

$$f'(x) = x^2 \cdot \frac{d}{dx}[\tan x] + \tan x \cdot \frac{d}{dx}[x^2]$$

$$= x^2 \sec^2 x + 2x \tan x \qquad \blacktriangleleft$$

▶ Example 2 Find dy/dx if

$$y = \frac{\sin x}{1 + \cos x}$$

Solution. Using the quotient rule together with formulas (3) and (4) we obtain

$$\frac{dy}{dx} = \frac{(1 + \cos x) \cdot \dfrac{d}{dx}[\sin x] - \sin x \cdot \dfrac{d}{dx}[1 + \cos x]}{(1 + \cos x)^2}$$

$$= \frac{(1 + \cos x)(\cos x) - (\sin x)(-\sin x)}{(1 + \cos x)^2}$$

$$= \frac{\cos x + \cos^2 x + \sin^2 x}{(1 + \cos x)^2}$$

$$= \frac{\cos x + 1}{(1 + \cos x)^2} = \frac{1}{1 + \cos x} \qquad \blacktriangleleft$$

As we have seen, the derivative formulas for the trigonometric functions ultimately rest on the two limits stated in (2). We will conclude this section by proving these two fundamental results. Unfortunately, the values of these limits cannot be obtained using the methods available so far. In fact, it is not even clear that these limits exist. To see the difficulty, consider the limit

$$\lim_{h \to 0} \frac{\cos h - 1}{h}$$

Intuition tells us that as $h \to 0$ the numerator and denominator both approach 0. As a result, there are two conflicting influences on the ratio. The numerator approaching 0 drives the magnitude of the ratio toward zero, while the denominator approaching 0 drives the magnitude of the ratio toward $+\infty$.

The precise way in which these influences offset one another determine whether the limit exists and what its value is. In a limit problem where the numerator and denominator both approach 0, it is often possible to obtain the limit by some algebraic manipulations. Unfortunately, no such algebraic manipulations will work for the limits in (2); we must develop other techniques.

When it is difficult to find the limit of a function directly, it is sometimes possible to obtain the limit indirectly by "squeezing" the function between simpler functions whose limits are known. For example, suppose we are unable to calculate directly

$$\lim_{x \to 0} f(x)$$

but we are able to "squeeze" $f(x)$ between $g(x) = 1 - 2x^2$ and $h(x) = 1 + x^2$ by means of the inequalities

$$1 - 2x^2 \leq f(x) \leq 1 + x^2 \tag{9}$$

Since $1 + x^2$ and $1 - 2x^2$ both approach 1 as $x \to 0$, and since the value of $f(x)$ is always between $1 + x^2$ and $1 - 2x^2$, it is intuitively evident that $f(x)$ must also approach 1 as $x \to 0$. Thus,

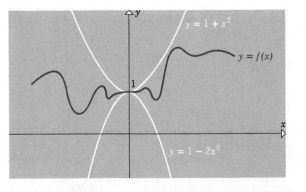

Figure 3.3.1

$$\lim_{x \to 0} f(x) = 1$$

This result is also clear geometrically, since the inequalities in (9) imply that the graph of f is "squeezed" or "pinched" between the graphs of $g(x) = 1 - 2x^2$ and $h(x) = 1 + x^2$ (Figure 3.3.1).

This idea is formalized in the following theorem, which is called the **Pinching Theorem** or sometimes the **Squeezing Theorem.** We omit the proof.

3.3.1   THEOREM

The Pinching Theorem

Let f, g, and h be functions satisfying

$$g(x) \leq f(x) \leq h(x)$$

for all x in some open interval containing the point a, with the possible exception that the inequalities need not hold at a. If g and h have the same limit as x approaches a, say

$$\lim_{x \to a} g(x) = \lim_{x \to a} h(x) = L$$

then f also has this limit as x approaches a, that is,

$$\lim_{x \to a} f(x) = L$$

REMARK.   The Pinching Theorem remains true if $\lim\limits_{x \to a}$ is replaced by $\lim\limits_{x \to a^+}$ or $\lim\limits_{x \to a^-}$. Moreover, for $\lim\limits_{x \to a^+}$ the condition

$$g(x) \leq f(x) \leq h(x)$$

need only hold on an open interval extending to the right from a, and for $\lim\limits_{x \to a^-}$ the condition need only hold on an open interval extending to the left from a. The Pinching Theorem can also be extended to limits of the form $\lim\limits_{x \to -\infty}$ and $\lim\limits_{x \to +\infty}$.

▶ Example 3   Use the Pinching Theorem to evaluate the limit

$$\lim_{x \to 0} x^2 \sin^2 \frac{1}{x}$$

Solution.   If $x \neq 0$, we can write

$$0 \leq \sin^2 \frac{1}{x} \leq 1$$

Multiplying through by x^2 yields

$$0 \leq x^2 \sin^2 \frac{1}{x} \leq x^2$$

if $x \neq 0$. But

$$\lim_{x \to 0} 0 = \lim_{x \to 0} x^2 = 0$$

so by the Pinching Theorem

$$\lim_{x \to 0} x^2 \sin^2 \frac{1}{x} = 0 \qquad \blacktriangleleft$$

3.3.2 THEOREM

$$\lim_{h \to 0} \frac{\cos h - 1}{h} = 0$$

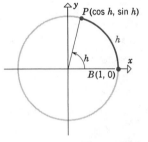

Figure 3.3.2

Proof.* Construct the angle h in standard position (Figure 3.3.2). Since the terminal side of h intersects the unit circle at the point $P(\cos h, \sin h)$ and since h is measured in radians, the *signed* arc length along the unit circle from $B(1, 0)$ to $P(\cos h, \sin h)$ is h. (See Definition 4 of the trigonometry review in Appendix 1.)

Thus, the arc length along the circle from B to P is $|h|$. Since the straight-line distance between $B(1, 0)$ and $P(\cos h, \sin h)$ does not exceed the arc length along the circle between these points, we have

$$0 \leq \sqrt{(1 - \cos h)^2 + (0 - \sin h)^2} \leq |h|$$

On squaring and simplifying we get

$$0 \leq 2 - 2 \cos h \leq h^2 \tag{10}$$

If h is *positive,* then on division by $2h$, (10) becomes

$$0 \leq \frac{1 - \cos h}{h} \leq \frac{1}{2}h$$

But

$$\lim_{h \to 0^+} \frac{1}{2}h = 0 \qquad \text{and} \qquad \lim_{h \to 0^+} 0 = 0$$

so that

---

*This proof, due to Stephen Hoffman, was published in the *American Mathematical Monthly,* Vol. 67 (1960), pp. 671–672.

$$\lim_{h \to 0^+} \frac{1 - \cos h}{h} = 0 \tag{11}$$

by the Pinching Theorem. If h is *negative*, then on dividing (10) by $2h$ we obtain

$$\frac{1}{2}h \le \frac{1 - \cos h}{h} \le 0$$

But

$$\lim_{h \to 0^-} \frac{1}{2}h = 0 \qquad \text{and} \qquad \lim_{h \to 0^-} 0 = 0$$

so that

$$\lim_{h \to 0^-} \frac{1 - \cos h}{h} = 0 \tag{12}$$

by the Pinching Theorem. From (11) and (12) we can conclude

$$\lim_{h \to 0} \frac{1 - \cos h}{h} = 0$$

from which it follows that

$$\lim_{h \to 0} \frac{\cos h - 1}{h} = \lim_{h \to 0} \left[-\left(\frac{1 - \cos h}{h} \right) \right] = 0 \quad \blacksquare$$

As a consequence of Theorem 3.3.2 we obtain the following important result.

3.3.3 COROLLARY $\lim_{h \to 0} \cos h = 1$

Proof.

$$\lim_{h \to 0} \cos h = \lim_{h \to 0} \left(1 + \frac{\cos h - 1}{h} \cdot h \right)$$

$$= \lim_{h \to 0} 1 + \left(\lim_{h \to 0} \frac{\cos h - 1}{h} \cdot \lim_{h \to 0} h \right) = 1 + (0 \cdot 0) = 1 \quad \blacksquare$$

Before proceeding to the last theorem we must develop the formula for the area of a circular sector. Consider a sector with radius r and a central angle of θ radians (Figure 3.3.3). If $\theta = 2\pi$, the sector is the full circle that has area πr^2. Thus, if we assume that the area A of an arbitrary sector is proportional to its central angle θ, we obtain

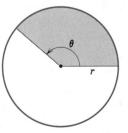

Figure 3.3.3

$$\frac{\text{area of sector}}{\text{area of circle}} = \frac{\text{central angle of sector}}{\text{central angle of circle}}$$

or

$$\frac{A}{\pi r^2} = \frac{\theta}{2\pi}$$

Solving for A we obtain the following formula.

> **Area of a sector with angle θ and radius r**
>
> $$A = \frac{1}{2}r^2\theta \qquad\qquad (13)$$

3.3.4 THEOREM

$$\lim_{h\to 0}\frac{\sin h}{h} = 1$$

Proof. Assume that h satisfies $0 < h < \pi/2$ and construct the angle h in standard position. The terminal side of h intersects the unit circle at $P(\cos h, \sin h)$ and intersects the vertical line through $B(1, 0)$ at the point $Q(1, \tan h)$ (Figure 3.3.4). From the figure

$$0 < \text{area of } \triangle\, OBP < \text{area of sector } OBP < \text{area of } \triangle\, OBQ$$

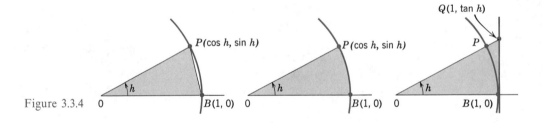

Figure 3.3.4

But

$$\text{area } \triangle\, OBP = \tfrac{1}{2}\text{ base}\cdot\text{altitude} = \tfrac{1}{2}\cdot 1\cdot\sin h = \tfrac{1}{2}\sin h$$
$$\text{area sector } OBP = \tfrac{1}{2}(1)^2\cdot h = \tfrac{1}{2}h \qquad\qquad [\text{Equation (13)}]$$
$$\text{area } \triangle\, OBQ = \tfrac{1}{2}\text{ base}\cdot\text{altitude} = \tfrac{1}{2}\cdot 1\cdot\tan h = \tfrac{1}{2}\tan h$$

Therefore,

$$0 < \tfrac{1}{2}\sin h < \tfrac{1}{2}h < \tfrac{1}{2}\tan h$$

Multiplying through by $2/\sin h$ yields

$$1 < \frac{h}{\sin h} < \frac{1}{\cos h}$$

and taking reciprocals yields

$$\cos h < \frac{\sin h}{h} < 1 \tag{14}$$

We have derived (14) under the assumption that $0 < h < \pi/2$. However, (14) is also valid if $-\pi/2 < h < 0$ (Exercise 51) so that (14) holds for all h in the interval $(-\pi/2, \pi/2)$ except $h = 0$.

From Corollary 3.3.3 we have

$$\lim_{h \to 0} \cos h = 1$$

and since

$$\lim_{h \to 0} 1 = 1$$

we can apply the Pinching Theorem to (14) and conclude

$$\lim_{h \to 0} \frac{\sin h}{h} = 1 \quad \blacksquare$$

▶ Example 4   Evaluate $\lim\limits_{\theta \to 0} \dfrac{\sin 2\theta}{\theta}$.

Solution.   Make the substitution $\phi = 2\theta$. Since $\phi \to 0$ as $\theta \to 0$, we can write

$$\lim_{\theta \to 0} \frac{\sin 2\theta}{\theta} = \lim_{\phi \to 0} \frac{\sin \phi}{\tfrac{1}{2}\phi} = 2 \lim_{\phi \to 0} \frac{\sin \phi}{\phi} = 2(1) = 2 \qquad ◀$$

▶ Exercise Set 3.3

In Exercises 1–18, find $f'(x)$.

1. $f(x) = 2 \cos x - 3 \sin x$.

2. $f(x) = \sin x \cos x$.

3. $f(x) = \dfrac{\sin x}{x}$.

4. $f(x) = x^2 \cos x$.

5. $f(x) = x^3 \sin x - 5 \cos x$.

6. $f(x) = \dfrac{\cos x}{x \sin x}$.

7. $f(x) = \sec x - \sqrt{2} \tan x$.

8. $f(x) = (x^2 + 1) \sec x$.

9. $f(x) = \sec x \tan x$.

10. $f(x) = \dfrac{\sec x}{1 + \tan x}$.

11. $f(x) = x - 4 \csc x + 2 \cot x$.

12. $f(x) = \csc x \cot x$.

13. $f(x) = \dfrac{\cot x}{1 + \csc x}$.

14. $f(x) = \dfrac{\csc x}{\tan x}$.

15. $f(x) = \sin^2 x + \cos^2 x$.

16. $f(x) = \dfrac{1}{\cot x}$.

17. $f(x) = \dfrac{\sin x \sec x}{1 + x \tan x}$.

18. $f(x) = \dfrac{(x^2 + 1)\cot x}{3 - \cos x \csc x}$.

In Exercises 19–23, find d^2y/dx^2.

19. $y = \sec x$.

20. $y = \csc x$.

21. $y = x \sin x - 3 \cos x$.

22. $y = x^2 \cos x + 4 \sin x$.

23. $y = \sin x \cos x$.

In Exercises 24–35, find the limit if it exists.

24. $\lim\limits_{h \to 0} \dfrac{\sin h}{2h}$.

25. $\lim\limits_{\theta \to 0} \dfrac{\sin 3\theta}{\theta}$.

26. $\lim\limits_{x \to 0} \dfrac{\sin 6x}{\sin 8x}$.

27. $\lim\limits_{x \to 0} \dfrac{\tan 7x}{\sin 3x}$.

28. $\lim\limits_{\theta \to 0} \dfrac{\sin^2 \theta}{\theta}$.

29. $\lim\limits_{h \to 0} \dfrac{h}{\tan h}$.

30. $\lim\limits_{h \to 0} \dfrac{\sin h}{1 - \cos h}$.

31. $\lim\limits_{\theta \to 0} \dfrac{\theta^2}{1 - \cos \theta}$.

32. $\lim\limits_{x \to 0} \dfrac{x}{\cos \left(\frac{1}{2}\pi - x \right)}$.

33. $\lim\limits_{\theta \to 0} \dfrac{\theta}{\cos \theta}$.

34. $\lim\limits_{t \to 0} \dfrac{t^2}{1 - \cos^2 t}$.

35. $\lim\limits_{h \to 0} \dfrac{1 - \cos 5h}{\cos 7h - 1}$.

In Exercises 36 and 37, find all points where the graph of f has a horizontal tangent.

36. (a) $f(x) = \sin x$ (b) $f(x) = \tan x$
    (c) $f(x) = \sec x$.

37. (a) $f(x) = \cos x$ (b) $f(x) = \cot x$
    (c) $f(x) = \csc x$.

38. Find the equation of the tangent to the graph of $\sin x$ at the point where:
    (a) $x = 0$ (b) $x = \pi$

    (c) $x = \dfrac{\pi}{4}$.

39. Find the equation of the tangent to the graph of $\tan x$ at the point where:
    (a) $x = 0$ (b) $x = \dfrac{\pi}{4}$ (c) $x = -\dfrac{\pi}{4}$.

40. (a) Show that $y = \cos x$ and $y = \sin x$ are solutions of the equation $y'' + y = 0$.
    (b) Show that $y = A \sin x + B \cos x$ is a solution for all constants A and B.

41. In each part, determine where f is differentiable.
    (a) $f(x) = \sin x$ (b) $f(x) = \cos x$
    (c) $f(x) = \tan x$ (d) $f(x) = \cot x$
    (e) $f(x) = \sec x$ (f) $f(x) = \csc x$
    (g) $f(x) = \dfrac{1}{1 + \cos x}$ (h) $f(x) = \dfrac{1}{\sin x \cos x}$

    (i) $f(x) = \dfrac{\cos x}{2 - \sin x}$.

42. Derive the formulas
    (a) $\dfrac{d}{dx}[\cot x] = -\csc^2 x$

    (b) $\dfrac{d}{dx}[\sec x] = \sec x \tan x$

    (c) $\dfrac{d}{dx}[\csc x] = -\csc x \cot x$.

43. Use Theorem 3.3.4 to prove that $\lim\limits_{h \to 0} \sin h = 0$.

44. (a) Show that $\lim\limits_{h \to 0} \dfrac{\tan h}{h} = 1$.

    (b) Use the result in (a) to help derive the formula $(d/dx)[\tan x] = \sec^2 x$ directly from the definition of a derivative.

45. In each part, find the limit by making the indicated substitution.
    (a) $\lim\limits_{x \to +\infty} x \sin \dfrac{1}{x}$. $\left[\text{Let } t = \dfrac{1}{x}. \right]$

    (b) $\lim\limits_{x \to -\infty} x \left(1 - \cos \dfrac{1}{x} \right)$. $\left[\text{Let } t = \dfrac{1}{x}. \right]$

    (c) $\lim\limits_{x \to \pi} \dfrac{\pi - x}{\sin x}$.   [Let $t = \pi - x$.]

46. Find $\lim\limits_{x \to 2} \dfrac{\cos (\pi/x)}{x - 2}$ by making appropriate substitutions.

47. Let f be a function that satisfies

    $1 - x^2 \le f(x) \le \cos x$

for all x in $(-\pi/2, \pi/2)$. Does $\lim\limits_{x \to 0} f(x)$ exist? If so, find the limit. If not, explain why.

48. Let

$$f(x) = \begin{cases} 1 & \text{if } x \text{ is a rational number} \\ 0 & \text{if } x \text{ is an irrational number} \end{cases}$$

Use the Pinching Theorem to prove $\lim_{x \to 0} xf(x) = 0$.

49. Prove: If there are constants L and M such that

$$L \le f(x) \le M$$

for all x in some open interval containing 0, with the possible exception that the inequalities may not hold at 0, then

$$\lim_{x \to 0} xf(x) = 0$$

50. State versions of the Pinching Theorem that apply to the limits $\lim_{x \to +\infty} f(x)$ and $\lim_{x \to -\infty} f(x)$. Draw some pictures to illustrate these results.

51. Prove that (14) holds if $-\pi/2 < h < 0$ by substituting $-h$ for h in (14).

52. Let us agree to call the functions $\cos x$, $\cot x$, and $\csc x$ the **cofunctions** of $\sin x$, $\tan x$, and $\sec x$, respectively. Convince yourself that the derivative of any cofunction can be obtained from the derivative of the corresponding function by introducing a minus sign and replacing each function in the derivative by its cofunction. Memorize the derivatives of $\sin x$, $\tan x$, and $\sec x$ and then use the above observation to deduce the derivatives of the cofunctions.

53. The derivative formulas for $\sin x$, $\cos x$, $\tan x$, $\cot x$, $\sec x$, and $\csc x$ were obtained under the assumption that x is measured in radians. This exercise shows that different (more complicated) formulas result if x is measured in degrees. Prove that if h and x are degree measures, then

(a) $\lim_{h \to 0} \dfrac{\cos h - 1}{h} = 0$

(b) $\lim_{h \to 0} \dfrac{\sin h}{h} = \dfrac{\pi}{180}$

(c) $\dfrac{d}{dx}[\sin x] = \dfrac{\pi}{180} \cos x.$

54. Evaluate $\lim_{x \to \pi/4} \dfrac{\tan x - 1}{x - \pi/4}.$

3.4 Δ-NOTATION; DIFFERENTIALS

Often the key to solving or understanding a problem is to introduce the right notation. In this section, we pause in our development of differentiation methods to discuss some powerful notational tools.

If a variable changes from one value to another, then its final value minus its initial value is called the **increment** in the variable. It is traditional in calculus to denote an increment in a variable x by the symbol Δx ("delta x"). In this notation, "Δx" is not a product of "Δ" and "x." Rather, Δx is a single symbol representing the *change* in the value of x. Similarly, Δy, Δt, $\Delta \theta$, and so forth, denote increments in the variables y, t, and θ.

If $y = f(x)$, and x changes from an initial value x_0 to a final value x_1, then there is corresponding change in the value of y from $y_0 = f(x_0)$ to $y_1 = f(x_1)$. Stated another way, the increment

$$\Delta x = x_1 - x_0 \tag{1}$$

produces a corresponding increment

$$\Delta y = y_1 - y_0 = f(x_1) - f(x_0) \tag{2}$$

in y (Figure 3.4.1). Increments can be either positive or negative depending on the relative positions of the initial and final points. In Figure 3.4.1, Δx is positive since the final point x_1 is to the right of the initial point x_0. If x_1 were to the left of x_0, then Δx would be negative. Relation (1) can be rewritten

$$x_1 = x_0 + \Delta x$$

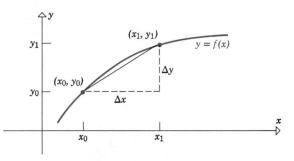

Figure 3.4.1

which states that the final value of x is the initial value of x plus the increment in x. With this notation, (2) can be rewritten as

$$\Delta y = f(x_0 + \Delta x) - f(x_0) \tag{3}$$

Sometimes it is convenient to drop the zero subscript on x_0 and use x to denote the initial value as well as the name of the variable. When this is done, $x + \Delta x$ represents the final value of the variable. Similarly, the initial and final values of y may be denoted by y and $y + \Delta y$ rather than y_0 and y_1 (Figure 3.4.2). Moreover, (3) becomes

$$\Delta y = f(x + \Delta x) - f(x) \tag{4}$$

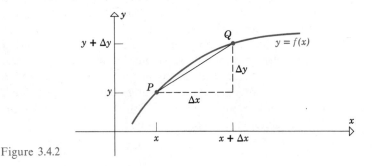

Figure 3.4.2

The Δ-notation provides a compact way of rewriting the derivative definition,

$$\frac{dy}{dx} = \lim_{h \to 0} \frac{f(x + h) - f(x)}{h} \tag{5}$$

Referring to Figure 3.4.2, the slope of the secant line joining the points P and Q is $\Delta y / \Delta x$. As $\Delta x \to 0$, Q tends toward P along the curve $y = f(x)$, so that the slope of the secant line approaches the slope of the tangent line at P; that is,

$$\frac{dy}{dx} = \lim_{\Delta x \to 0} \frac{\Delta y}{\Delta x} \tag{6}$$

This result is not unexpected since $\Delta y = f(x + \Delta x) - f(x)$, so (6) can be written

$$\frac{dy}{dx} = \lim_{\Delta x \to 0} \frac{f(x + \Delta x) - f(x)}{\Delta x} \tag{7}$$

which is simply a restatement of (5) with Δx used in place of h.

If it is undesirable to introduce a dependent variable y, then we write

$$\Delta f = f(x + \Delta x) - f(x)$$

and (6) as

$$f'(x) = \lim_{\Delta x \to 0} \frac{\Delta f}{\Delta x} \tag{8}$$

Expressions (5), (6), (7), and (8) provide four ways of writing the definition of a derivative. All are commonly used.

DIFFERENTIALS   Up to now we have been viewing the expression dy/dx as a single symbol for the derivative. The symbols "dy" and "dx" in this expression, which are called **differentials**, have had no meaning by themselves. We will now show how to interpret these symbols so that dy/dx can be regarded as the ratio of dy and dx.

Regard x as fixed and *define dx* to be an independent variable that can be assigned an arbitrary value. If f is differentiable at x, then we *define dy* by the formula

$$dy = f'(x) \, dx \tag{9}$$

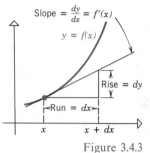

Slope $= \frac{dy}{dx} = f'(x)$

$y = f(x)$

Rise $= dy$

Run $= dx$

$x \quad x + dx$

Figure 3.4.3

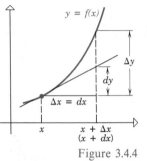

$y = f(x)$

Δy

dy

$\Delta x = dx$

$x \quad x + \Delta x$
$\quad (x + dx)$

Figure 3.4.4

If $dx \neq 0$, then we can divide both sides of (9) by dx to obtain

$$\frac{dy}{dx} = f'(x)$$

Thus, we have achieved our goal of defining dy and dx so their ratio is $f'(x)$. Because

$$\frac{dy}{dx} = f'(x) = m_{\tan}$$

where $m_{\tan}$ is the slope of the tangent to $y = f(x)$ at x, the differentials dy and dx can be interpreted as a corresponding rise and run of this tangent line (Figure 3.4.3).

It is important to understand the distinction between the increment Δy and the differential dy. To see the difference, let us assign the independent variables dx and Δx the same value, so $dx = \Delta x$. Then Δy represents the change in y that occurs when we start at x and travel *along the curve* $y = f(x)$ until we have moved $\Delta x \, (= dx)$ units in the x-direction, while dy represents the change in y that occurs if we start at x and travel *along the tangent* line until we have moved $dx \, (= \Delta x)$ units in the x-direction (Figure 3.4.4).

▶ **Example 1** If $y = x^2$, then the relation $dy/dx = 2x$ can be written in the *differential form*

$$dy = 2x \, dx$$

When $x = 3$, this becomes

$$dy = 6 \, dx$$

This tells us that if we travel along the tangent to the curve $y = x^2$ at $x = 3$, then a change of dx units in x produces a change of $6 \, dx$ units in y. For example, if the change in x is

$$dx = 4$$

then the change in y along the tangent is

$$dy = 6(4) = 24 \quad \text{units} \qquad ◀$$

▶ **Example 2** Let $y = \sqrt{x}$. Find dy and Δy if $x = 4$ and $dx = \Delta x = 3$. Then make a sketch of $y = \sqrt{x}$, showing dy and Δy in the picture.

Solution. From (4) with $f(x) = \sqrt{x}$,

$$\Delta y = \sqrt{x + \Delta x} - \sqrt{x} = \sqrt{7} - \sqrt{4} \approx .65$$

It follows from Example 3 in Section 3.1 that if $y = \sqrt{x}$ then

$$\frac{dy}{dx} = \frac{1}{2\sqrt{x}}$$

Thus,

$$dy = \frac{1}{2\sqrt{x}}\,dx = \frac{1}{2\sqrt{4}}(3) = .75$$

Figure 3.4.5 shows the curve $y = \sqrt{x}$ together with dy and Δy. ◀

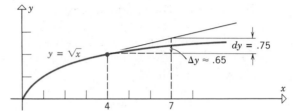

Figure 3.4.5

TANGENT LINE APPROXIMATIONS

Figure 3.4.6 suggests that if f is differentiable at x_0, then the tangent line to the curve $y = f(x)$ at x_0 is a reasonably good approximation to the curve $y = f(x)$ near x_0. Since the tangent line passes through the point $\left(x_0, f(x_0)\right)$ and has slope $f'(x_0)$, the point-slope form of its equation is

$$y - f(x_0) = f'(x_0)(x - x_0)$$

or

$$y = f(x_0) + f'(x_0)(x - x_0)$$

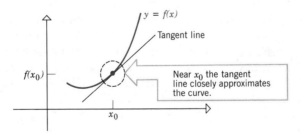

Figure 3.4.6

For values of x close to x_0, the height y of this tangent line will closely approximate the height $f(x)$ of the curve, which yields the approximation

$$f(x) \approx f(x_0) + f'(x_0)(x - x_0) \tag{10}$$

for x near x_0. If we let $\Delta x = x - x_0$, so $x = x_0 + \Delta x$, then (10) can be written in the alternate form

$$f(x_0 + \Delta x) \approx f(x_0) + f'(x_0)\,\Delta x \qquad (11)$$

which is a good approximation for Δx near zero. This result is called the *linear approximation of f near x_0*.

In the event that $f(x_0 + \Delta x)$ is tedious to calculate, but $f(x_0)$ and $f'(x_0)$ are not, then this formula enables us to use the values of $f(x_0)$ and $f'(x_0)$ to approximate $f(x_0 + \Delta x)$.

▶ **Example 3** Use (11) to approximate $\sqrt{1.1}$.

Solution. If we let $f(x) = \sqrt{x}$, then the problem is to approximate $f(1.1)$. But $f'(x) = 1/(2\sqrt{x})$, so that $f(1)$ and $f'(1)$ are easy to compute. Thus, we will apply (11) with

$$x_0 + \Delta x = 1.1 \qquad \text{and} \qquad x_0 = 1$$

It follows that $\Delta x = .1$, so (11) yields

$$f(1.1) \approx f(1) + f'(1)(.1)$$

or

$$\sqrt{1.1} \approx \sqrt{1} + \frac{1}{2\sqrt{1}}(.1) = 1.05. \qquad \blacktriangleleft$$

▶ **Example 4** Use (11) to approximate $\cos 62°$.

Solution. We will take advantage of the fact that $62°$ is close to $60°$, at which point the trigonometric functions are easy to evaluate. However, to apply (11), we must first convert to radian measure because the derivative formulas for the trigonometric functions are based on the assumption that x is in radians. If we let $f(x) = \cos x$ and note that $62° = 31\pi/90$ (radians), then the problem is to approximate $f(31\pi/90)$. Thus, we will apply (11) with

$$x_0 + \Delta x = \frac{31\pi}{90} \qquad \text{and} \qquad x_0 = \frac{\pi}{3}(=60°)$$

It follows that $\Delta x = \pi/90$, so (11) yields

$$f\left(\frac{31\pi}{90}\right) \approx f\left(\frac{\pi}{3}\right) + f'\left(\frac{\pi}{3}\right)\left(\frac{\pi}{90}\right)$$

$$\cos \frac{31\pi}{90} \approx \cos \frac{\pi}{3} - \left(\sin \frac{\pi}{3}\right)\left(\frac{\pi}{90}\right)$$

$$\cos \frac{31\pi}{90} \approx \frac{1}{2} - \frac{\sqrt{3}}{2}\left(\frac{\pi}{90}\right) \approx 0.5 - 0.0302299 = 0.4697701$$

To seven digits, the value of $\cos 62°$ is 0.4694715, so that the error is less than 0.0003. ◄

Formula (11) has a useful alternate form, which we will now derive. Subtract $f(x_0)$ from both sides to obtain

$$f(x_0 + \Delta x) - f(x_0) \approx f'(x_0)\,\Delta x$$

and use (3) to rewrite this as

$$\Delta y \approx f'(x_0)\,\Delta x$$

If we drop the subscript on x_0 and assume that $\Delta x = dx$, then from (9) this can be rewritten as

$$\Delta y \approx dy \tag{12}$$

This formula has applications in the study of **error propagation.** Suppose a researcher measures a physical quantity. Because of limitations in the instrumentation and other factors, the researcher will not usually obtain the exact value x of the quantity, but rather will obtain $x + \Delta x$, where Δx is a measurement error. This recorded value may then be used to calculate some other quantity y. In this way the measurement error Δx propagates to produce an error Δy in the calculated value of y.

▶ Example 5   The radius of a sphere is measured to be 50 in. with a possible measurement error of ± 0.02 in. Estimate the possible error in the computed volume of the sphere.

Solution.   The volume of the sphere is

$$V = \tfrac{4}{3}\pi r^3 \tag{13}$$

We are given that the error in the radius is $\Delta r = \pm 0.02$, and we want to find the error ΔV in V. If we consider Δr to be small and we let $dr = \Delta r$, then ΔV can be approximated by dV. Thus, from (13),

$$\Delta V \approx dV = 4\pi r^2\,dr \tag{14}$$

Substituting $r = 50$ and $dr = \pm 0.02$ in (14), we obtain

$$\Delta V \approx 4\pi(2500)(\pm 0.02) \approx \pm 628.32$$

Therefore, the possible error in the volume is approximately ± 628.32 cubic inches (in.^{3}). ◀

REMARK. In (14), r represents the exact value of the radius. Since the exact value of r was unknown, we substituted the measured value $r = 50$ to obtain ΔV. This is reasonable since the error Δr was assumed to be small.

If the exact value of a quantity is q and a measurement or calculation results in an error Δq, then $\Delta q/q$ is called the **relative error** in the measurement or calculation; when expressed as a percentage, $\Delta q/q$ is called the **percentage error.** As a practical matter, the exact value q is usually unknown, so that the measured or calculated value of q is used instead; and the relative error is approximated by dq/q.

▶ Example 6 For the sphere in Example 5,

$$\text{relative error in } r \approx \frac{dr}{r} = \frac{\pm 0.02}{50} = \pm 0.0004$$

$$\text{relative error in } V \approx \frac{dV}{V} = \frac{4\pi r^2\, dr}{\frac{4}{3}\pi r^3} = 3\frac{dr}{r} = \pm 0.0012$$

Thus, the percentage error in the radius is approximately $\pm 0.04\%$, and the percentage error in the volume is approximately $\pm 0.12\%$. ◀

▶ Example 7 The side of a square is measured with a possible percentage error of $\pm 5\%$. Use differentials to estimate the possible percentage error in the area of the square.

Solution. The area of a square with side x is given by

$$A = x^2 \tag{15}$$

and the relative errors in A and x are approximately dA/A and dx/x, respectively. From (15),

$$dA = 2x\, dx$$

so

$$\frac{dA}{A} = \frac{2x\,dx}{A} = \frac{2x\,dx}{x^2} = 2\frac{dx}{x}$$

We are given that $dx/x \approx \pm 0.05$, so that

$$\frac{dA}{A} \approx \pm 2(0.05) = \pm 0.1$$

Thus, there is a possible percentage error of $\pm 10\%$ in the area of the square.

◀

DIFFERENTIAL
FORMULAS

Just as

$$\frac{d}{dx}[\ \ ]$$

denotes the derivative of the expression inside the brackets, so

$$d[\ \ ]$$

denotes the differential of the expression in the brackets. For example,

$$d[x^3] = 3x^2\,dx$$
$$d[f(x)] = f'(x)\,dx$$

The basic rules of differentiation can be expressed in terms of differentials. In the following table, the differential formulas on the right result when the derivative formulas on the left are multiplied through by dx.

DERIVATIVE FORMULA	DIFFERENTIAL FORMULA
$\dfrac{d}{dx}[c] = 0$	$d[c] = 0$
$\dfrac{d}{dx}[cf] = c\dfrac{df}{dx}$	$d[cf] = c\,df$
$\dfrac{d}{dx}[f + g] = \dfrac{df}{dx} + \dfrac{dg}{dx}$	$d[f + g] = df + dg$
$\dfrac{d}{dx}[fg] = f\dfrac{dg}{dx} + g\dfrac{df}{dx}$	$d[fg] = f\,dg + g\,df$
$\dfrac{d}{dx}\left[\dfrac{f}{g}\right] = \dfrac{g\dfrac{df}{dx} - f\dfrac{dg}{dx}}{g^2}$	$d\left[\dfrac{f}{g}\right] = \dfrac{g\,df - f\,dg}{g^2}$

▶ Example 8   Find dy if $y = x \sin x$.

Solution.

$$dy = d[x \sin x] = x \, d[\sin x] + (\sin x) \, d[x]$$
$$= x(\cos x) \, dx + (\sin x) \, dx \qquad [\text{since } d[x] = 1 \cdot dx = dx]$$
$$= (x \cos x + \sin x) \, dx \qquad\qquad ◀$$

▶ Exercise Set 3.4

1. Let $y = x^2$.
    (a) Find Δy if $\Delta x = 1$ and the initial value of x is $x = 2$.
    (b) Find dy if $dx = 1$ and the initial value of x is $x = 2$.
    (c) Make a sketch of $y = x^2$ and show Δy and dy in the picture.

2. Repeat Exercise 1 with $x = 2$ as the initial value again, but $\Delta x = -1$ and $dx = -1$.

3. Let $y = \dfrac{1}{x}$.
    (a) Find Δy if $\Delta x = 0.5$ and the initial value of x is $x = 1$.
    (b) Find dy if $dx = 0.5$ and the initial value of x is $x = 1$.
    (c) Make a sketch of $y = 1/x$ and show Δy and dy in the picture.

4. Repeat Exercise 3 with $x = 1$ as the initial value again, but $\Delta x = -0.5$ and $dx = -0.5$.

In Exercises 5–8, find general formulas for dy and Δy.

5. $y = x^3$.

6. $y = 8x - 4$.

7. $y = x^2 - 2x + 1$.

8. $y = \sin x$.

In Exercises 9–12, find dy.

9. $y = 4x^3 - 7x^2 + 2x - 1$.

10. $y = \dfrac{1}{x^3 - 1}$.

11. $y = x \cos x$.

12. $y = \dfrac{1 - x^3}{2 - x}$.

In Exercises 13–16, find the limit by expressing it as a derivative.

13. $\displaystyle \lim_{\Delta x \to 0} \frac{(x + \Delta x)^2 - x^2}{\Delta x}$.

14. $\displaystyle \lim_{\Delta x \to 0} \frac{(3 + \Delta x)^2 - 3^2}{\Delta x}$.

15. $\displaystyle \lim_{\Delta x \to 0} \frac{\sin (\pi + \Delta x) - \sin \pi}{\Delta x}$.

16. $\displaystyle \lim_{\Delta x \to 0} \frac{5(2 + \Delta x)^4 - 5(2)^4}{\Delta x}$.

In Exercises 17–22, use a linear approximation [formula (11)] to estimate the value of the given quantity. Where needed, use the result of Example 3, Section 3.1.

17. $\sqrt{65}$.
18. $\sqrt{26}$.

19. $\sqrt{80.9}$.
20. $\sin 59°$.

21. $\cos 31°$.
22. $\sin 44°$.

23. The side of a square is measured to be 10 ft, with a possible error of ± 0.1 ft.
    (a) Use differentials to estimate the error in the calculated area.
    (b) Estimate the percentage errors in the side and the area.

24. The side of a cube is measured to be 25 cm, with a possible error of ± 1 cm.
    (a) Use differentials to estimate the error in the calculated volume.
    (b) Estimate the percentage errors in the side and volume.

25. The hypotenuse of a right triangle is known to be 10″ exactly, and one of the acute angles is measured to be 30°, with a possible error of $\pm 1°$.
    (a) Use differentials to estimate the errors in the sides opposite and adjacent to the measured angle.
    (b) Estimate the percentage errors in the sides.

26. One side of a right triangle is known to be 25 cm exactly. The angle opposite to this side is measured to be 60°, with a possible error of $\pm 0.5°$.
    (a) Use differentials to estimate the errors in the adjacent side and the hypotenuse.
    (b) Estimate the percentage errors in the adjacent side and hypotenuse.

27. The electrical resistance R of a certain wire is given by $R = k/r^2$, where k is a constant and r is the radius of the wire. Assuming that the radius r has a possible error of $\pm 5\%$, use differentials to estimate the percentage error in R. (Assume k is exact.)

28. The side of a square is measured with a possible percentage error of $\pm 1\%$. Use differentials to estimate the percentage error in the area.

29. The side of a cube is measured with a possible percentage error of $\pm 2\%$. Use differentials to estimate the percentage error in the volume.

30. The volume of a sphere is to be computed from a measured value of its radius. Estimate the maximum permissible percentage error in the measurement if the percentage error in the volume must be kept within $\pm 3\%$. [$V = \frac{4}{3}\pi r^3$ is the volume of a sphere of radius r.]

31. The area of a circle is to be computed from a measured value of its diameter. Estimate the maximum permissible percentage error in the measurement if the percentage error in the area must be kept within $\pm 1\%$.

32. A steel cube with 1 in. sides is coated with .01 in. of copper. Use differentials to estimate the volume of copper in the coating. [*Hint:* Let ΔV be the change in the volume of the cube.]

33. A metal rod 15 cm long and 5 cm in diameter is to be covered (except for the ends) with insulation that is .001 cm thick. Use differentials to estimate the volume of insulation. [*Hint:* Let ΔV be the change in volume of the rod.]

34. The time required for one complete oscillation of a pendulum is called its **period.** If the length L of the pendulum is measured in feet and the period P in seconds, then the period is given by $P = 2\pi\sqrt{L/g}$, where g is a constant. Use differentials to show that the percentage error in P is approximately half the percentage error in L. [*Hint:* Use the result of Example 3 in Section 3.1.]

3.5 THE CHAIN RULE

In this section we consider the following problem.

3.5.1 PROBLEM If we know the derivatives of f and g, how can we use this information to find the derivative of the composition $f \circ g$?

For example, we know how to differentiate

$$f(x) = \sin x \quad \text{and} \quad g(x) = x^2 - 1$$

How can we use this knowledge to obtain the derivative of

$$(f \circ g)(x) = f(g(x)) = \sin{(x^2 - 1)}?$$

The idea is as follows. Introduce dependent variables

$$y = \sin{(x^2 - 1)} \quad \text{and} \quad u = x^2 - 1 \tag{1}$$

so that

$$y = \sin{u}$$

We know that

$$\frac{dy}{du} = \frac{d}{du}[\sin{u}] = \cos{u} \tag{2}$$

and

$$\frac{du}{dx} = \frac{d}{dx}[x^2 - 1] = 2x \tag{3}$$

We want to find

$$\frac{dy}{dx} = \frac{d}{dx}[\sin{(x^2 - 1)}]$$

If we view a derivative as a rate of change, then intuition suggests

$$\frac{dy}{du} \cdot \frac{du}{dx} = \frac{dy}{dx} \tag{4}$$

(For example, if y changes at 4 times the rate of u and u changes at 2 times the rate of x, then y changes at $4 \times 2 = 8$ times the rate of x.) Substituting (2) and (3) in (4) yields

$$\frac{dy}{dx} = (\cos{u})(2x)$$

which may be expressed in terms of x alone using (1)

$$\frac{dy}{dx} = [\cos{(x^2 - 1)}]\,2x = 2x\cos{(x^2 - 1)}$$

The ideas in this example are formalized in the following theorem.

3.5.2 THEOREM
The Chain Rule

If g is differentiable at the point x and f is differentiable at the point g(x), then the composition f ∘ g is differentiable at the point x. Moreover, if

$$y = f(g(x)) \quad \text{and} \quad u = g(x)$$

then

$$\frac{dy}{dx} = \frac{dy}{du} \cdot \frac{du}{dx} \tag{5}$$

The proof of this result, which is optional, is given in part III of Appendix 2.

▶ **Example 1** Find dy/dx if

$$y = 4\cos(x^3)$$

Solution. Let $u = x^3$ so that

$$y = 4\cos u$$

By the chain rule,

$$\frac{dy}{dx} = \frac{dy}{du} \cdot \frac{du}{dx} = \frac{d}{du}[4\cos u] \cdot \frac{d}{dx}[x^3]$$

$$= (-4\sin u) \cdot (3x^2) = -12x^2 \sin(x^3) \qquad ◀$$

Formula (5) is easy to remember because the left side is exactly what results if we "cancel" the du's on the right side. This "canceling" device provides a good way to remember the chain rule when variables other than x, y, and u are used.

▶ **Example 2** Suppose

$$w = \tan x \quad \text{and} \quad x = 4t^3 + t$$

Find dw/dt.

Solution. In this case, the chain rule takes the form

$$\frac{dw}{dt} = \frac{dw}{dx} \cdot \frac{dx}{dt} = \frac{d}{dx}[\tan x] \cdot \frac{d}{dt}[4t^3 + t]$$

$$= (\sec^2 x)(12t^2 + 1) = (12t^2 + 1)\sec^2(4t^3 + t) \qquad ◀$$

There is another version of the chain rule that is useful when it is undesirable to use dependent variables. To derive this result, suppose $y = f(g(x))$ and $u = g(x)$, so that $y = f(u)$. Then

$$\frac{dy}{dx} = \frac{d}{dx}[f(g(x))]$$

$$\frac{dy}{du} = f'(u) = f'[g(x)]$$

$$\frac{du}{dx} = g'(x)$$

Substituting these in (5) yields the following version of the chain rule:

$$\frac{d}{dx}[f(g(x))] = f'[g(x)]g'(x) \qquad (6)$$

▶ Example 3 Find

$$\frac{d}{dx}[(x^3 + 7x + 1)^{35}]$$

Solution. Let

$$f(x) = x^{35} \qquad \text{and} \qquad g(x) = x^3 + 7x + 1$$

Thus,

$$\frac{d}{dx}[(x^3 + 7x + 1)^{35}] = \frac{d}{dx}[f(g(x))] = f'[g(x)]g'(x) \qquad (7)$$

But, $f'(x) = 35x^{34}$, so

$$f'[g(x)] = 35(x^3 + 7x + 1)^{34} \qquad \text{and} \qquad g'(x) = 3x^2 + 7$$

Substituting in (7) yields

$$\frac{d}{dx}[(x^3 + 7x + 1)^{35}] = 35(x^3 + 7x + 1)^{34}(3x^2 + 7) \qquad ◀$$

The last example illustrates the four steps required to apply Formula (6):

Step 1. Express the function to be differentiated in the form $f(g(x))$, where f and g are functions you know how to differentiate.

Step 2. Compute $f'(x)$, and then replace x by $g(x)$ to obtain $f'[g(x)]$.

Step 3. Compute $g'(x)$.

Step 4. Multiply $f'[g(x)]$ and $g'(x)$.

▶ Example 4   Find $\dfrac{d}{dx}[\sin 5x]$.

Solution.   Let

$$f(x) = \sin x \quad \text{and} \quad g(x) = 5x$$

Then,

$$\frac{d}{dx}[\sin 5x] = \frac{d}{dx}[f(g(x))] = f'[g(x)]g'(x)$$

But $f'(x) = \cos x$, so

$$f'[g(x)] = \cos 5x \quad \text{and} \quad g'(x) = 5$$

Therefore,

$$\frac{d}{dx}[\sin 5x] = (\cos 5x) \cdot 5 = 5\cos 5x \qquad\qquad ◀$$

Eventually, the reader should be able to apply the chain rule without explicitly writing down $f(x)$ and $g(x)$. To reach this goal, it is helpful to express the formula

$$\frac{d}{dx}[f(g(x))] = f'[g(x)]g'(x)$$

in words. For this purpose call f the "outside function" and g the "inside function." Thus, *the derivative of $f(g(x))$ is the derivative of the outside function evaluated at the inside function times the derivative of the inside function.*

Note that in the expression $f[g(x)]$, the inside function is evaluated first and the outside second. For example, to evaluate $\sin(x^3)$, we first compute x^3 and then sine, so $g(x) = x^3$ is the inside function and $f(x) = \sin x$ is the outside function. Some more examples are given in the following table.

FUNCTION	INSIDE	OUTSIDE
$(x^2 + 1)^{10}$	$x^2 + 1$	x^{10}
$\sin^3 x$	$\sin x$	x^3
$\tan(x^5)$	x^5	$\tan x$
$\sqrt{4 - 3x}$	$4 - 3x$	$\sqrt{x}$

▶ Example 5   Find

$$\frac{d}{dx}[\cos(x^2 + 9)]$$

Solution. The inside function is $x^2 + 9$ and the outside function is $\cos x$, so

$$\frac{d}{dx}[\cos (x^2 + 9)] = \underbrace{-\sin (x^2 + 9) \cdot}_{\substack{\text{derivative of} \\ \text{the outside} \\ \text{evaluated at} \\ \text{the inside}}} \underbrace{2x}_{\substack{\text{derivative} \\ \text{of the} \\ \text{inside}}} \quad \blacktriangleleft$$

▶ **Example 6** Find

$$\frac{d}{dx}[\tan^2 x]$$

Solution. The inside function is $\tan x$ and the outside function is x^2, so

$$\frac{d}{dx}[\tan^2 x] = \underbrace{(2 \tan x) \cdot}_{\substack{\text{derivative of} \\ \text{the outside} \\ \text{evaluated at} \\ \text{the inside}}} \underbrace{(\sec^2 x)}_{\substack{\text{derivative} \\ \text{of the} \\ \text{inside}}} = 2 \tan x \sec^2 x \quad \blacktriangleleft$$

▶ **Example 7** Find dy/dx if

$$y = \frac{1}{2x^4 - x^2 + 8}$$

Solution. First rewrite y as

$$y = (2x^4 - x^2 + 8)^{-1}$$

so the inside function is $2x^4 - x^2 + 8$ and the outside function is x^{-1}. Thus,

$$\frac{dy}{dx} = \frac{d}{dx}[(2x^4 - x^2 + 8)^{-1}] = -(2x^4 - x^2 + 8)^{-2} \cdot (8x^3 - 2x)$$

$$= \frac{2x - 8x^3}{(2x^4 - x^2 + 8)^2}$$

[Note that this problem can be solved without the chain rule by treating y as a quotient and applying Theorem 3.2.6.] $\quad \blacktriangleleft$

In the following example the chain rule must be applied more than once.

▶ **Example 8** Find

$$\frac{d}{dx}[\sin^3 (9x + 1)]$$

Solution.

$$\frac{d}{dx}[\sin^3(9x+1)] = 3\sin^2(9x+1) \cdot \frac{d}{dx}[\sin(9x+1)]$$

$$= 3\sin^2(9x+1) \cdot \cos(9x+1) \cdot \frac{d}{dx}(9x+1)$$

$$= 3\sin^2(9x+1) \cdot \cos(9x+1) \cdot 9$$

$$= 27\sin^2(9x+1) \cdot \cos(9x+1) \qquad \blacktriangleleft$$

GENERALIZED DERIVATIVE FORMULAS

If we let $u = g(x)$ in (6), then that formula becomes

$$\frac{d}{dx}[f(u)] = f'(u)\frac{du}{dx} \qquad (8)$$

This version of the chain rule is useful for generalizing some familiar derivative formulas. For example, if $f(x) = \sin x$, then it follows from (8) that

$$\frac{d}{dx}[\sin u] = \cos u \frac{du}{dx}$$

With this formula, we can differentiate the sine of any differentiable function of x directly. For example,

$$\frac{d}{dx}\left[\sin\left(\frac{1}{x}\right)\right] = \cos\left(\frac{1}{x}\right) \cdot \frac{d}{dx}\left[\frac{1}{x}\right] = \cos\left(\frac{1}{x}\right) \cdot \left(-\frac{1}{x^2}\right)$$

$$= -\frac{1}{x^2}\cos\left(\frac{1}{x}\right)$$

The generalized derivative formulas in the following list are all consequences of (8).

GENERALIZED DERIVATIVE FORMULAS

$\dfrac{d}{dx}[\sin u] = \cos u \dfrac{du}{dx}$	$\dfrac{d}{dx}[\sec u] = \sec u \tan u \dfrac{du}{dx}$
$\dfrac{d}{dx}[\cos u] = -\sin u \dfrac{du}{dx}$	$\dfrac{d}{dx}[\csc u] = -\csc u \cot u \dfrac{du}{dx}$
$\dfrac{d}{dx}[\tan u] = \sec^2 u \dfrac{du}{dx}$	$\dfrac{d}{dx}[u^n] = nu^{n-1} \dfrac{du}{dx}$ (n an integer)
$\dfrac{d}{dx}[\cot u] = -\csc^2 u \dfrac{du}{dx}$	

► Example 9

$$\frac{d}{dx}[\tan 3x] = \sec^2 3x \cdot \frac{d}{dx}[3x] = 3\sec^2 3x$$

$$\frac{d}{dx}\left[\left(\frac{x}{1+x}\right)^3\right] = 3\left(\frac{x}{1+x}\right)^2 \cdot \frac{d}{dx}\left[\frac{x}{1+x}\right]$$

$$= 3\left(\frac{x}{1+x}\right)^2 \cdot \frac{1}{(1+x)^2} = \frac{3x^2}{(1+x)^4} \qquad ◄$$

► Exercise Set 3.5

In Exercises 1–25, find $f'(x)$.

1. $f(x) = (x^3 + 2x)^{37}$.

2. $f(x) = (3x^2 + 2x - 1)^6$.

3. $f(x) = \left(x^3 - \dfrac{7}{x}\right)^{-2}$.

4. $f(x) = \dfrac{1}{(x^5 - x + 1)^9}$.

5. $f(x) = \dfrac{4}{(3x^2 - 2x + 1)^3}$.

6. $f(x) = \sin^3 x$.

7. $f(x) = \tan(4x^2)$.

8. $f(x) = 3\cot^4 x$.

9. $f(x) = 4\cos^5 x$.

10. $f(x) = \csc(x^3)$.

11. $f(x) = \sin\left(\dfrac{1}{x^2}\right)$.

12. $f(x) = \tan^4(x^3)$.

13. $f(x) = 2\sec^2(x^7)$.

14. $f(x) = \cos^3\left(\dfrac{x}{x+1}\right)$.

15. $f(x) = [x + \csc(x^3 + 3)]^{-3}$.

16. $f(x) = [x^4 - \sec(4x^2 - 2)]^{-4}$.

17. $f(x) = x^5 \sec\left(\dfrac{1}{x}\right)$.

18. $f(x) = \dfrac{\sin x}{\sec(3x + 1)}$.

19. $f(x) = \cos(\cos x)$.

20. $f(x) = \dfrac{1 + \csc(x^2)}{1 - \cot(x^2)}$.

21. $f(x) = (5x + 8)^{13}(x^3 + 7x)^{12}$.

22. $f(x) = \left(\dfrac{1 + x^2}{1 - x^2}\right)^{17}$.

23. $f(x) = \dfrac{(4x^2 - 1)^{-8}}{(2x + 1)^{-3}}$.

24. $f(x) = [1 + \sin^3(x^5)]^{12}$.

25. $f(x) = [x \sin 2x + \tan^4(x^7)]^5$.

In Exercises 26–28, find d^2y/dx^2.

26. $y = \sin(3x^2)$.

27. $y = x\cos(5x) - \sin^2 x$.

28. $y = x\tan\left(\dfrac{1}{x}\right)$.

In Exercises 29–32, find an equation for the tangent to the graph at the specified point.

29. $y = x\cos 3x$, $x = \pi$.

30. $y = \sin(1 + x^3)$, $x = -3$.

31. $y = \sec^3\left(\dfrac{\pi}{2} - x\right)$, $x = -\dfrac{\pi}{2}$.

32. $y = \left(x - \dfrac{1}{x}\right)^3$, $x = 2$.

In Exercises 33–36, find dy/dx in two ways: first by using the chain rule and then by expressing y in terms of x and differentiating directly.

33. $x = 5t + 2$, $y = t^2$.

34. $x = \dfrac{1}{2}\pi - \theta$, $y = \sec \theta$.

35. $x = u^{-1}$, $y = 3 \sin^3 u$.

36. $x = \dfrac{\lambda}{1 - \lambda}$, $y = (1 + \lambda)^{12}$.

In Exercises 37–40, find the indicated derivative.

37. $y = \cot^3(\pi - \theta)$; find $\dfrac{dy}{d\theta}$.

38. $\lambda = \left(\dfrac{au + b}{cu + d}\right)^6$; find $\dfrac{d\lambda}{du}$ (a, b, c, d constants).

39. $\dfrac{d}{d\omega}[a \cos^2 \pi\omega + b \sin^2 \pi\omega]$ (a, b constants).

40. $x = \csc^2\left(\dfrac{\pi}{3} - y\right)$; find $\dfrac{dx}{dy}$.

41. (a) Show that
$$\frac{d}{dx}(|x|) = \begin{cases} 1, & x > 0 \\ -1, & x < 0 \end{cases}$$

(b) Use the result in (a) and the chain rule to find
$$\frac{d}{dx}(|\sin x|) \quad \text{and} \quad \frac{d}{dx}(\sin |x|)$$
for nonzero x in the interval $(-\pi, \pi)$.

42. Use the identity
$$\cos x = \sin\left(\frac{\pi}{2} - x\right)$$
and the derivative formula for $\sin x$ to obtain the derivative formula for $\cos x$.

43. Find a formula for
$$\frac{d}{dx}[f(g(h(x)))]$$

44. Find $f'(x^2)$ if $\dfrac{d}{dx}[f(x^2)] = x^2$.

45. Given that $f(0) = 1$, $f'(0) = 2$, $g(0) = 0$, and $g'(0) = 3$, find $(f \circ g)'(0)$.

46. Let $y = f_1(u)$, $u = f_2(v)$, $v = f_3(w)$, and $w = f_4(x)$. Express dy/dx in terms of dy/du, dw/dx, du/dv, and dv/dw.

3.6 IMPLICIT DIFFERENTIATION

In previous sections we found tangent lines to curves of the form $y = f(x)$. In this section we will show how to find tangent lines to curves whose equations do not express y explicitly as a function of x.

Consider the equation

$$xy = 1 \tag{1}$$

One way to obtain dy/dx is to rewrite this equation as

$$y = \frac{1}{x} \tag{2}$$

which implies that

$$\frac{dy}{dx} = \frac{d}{dx}\left[\frac{1}{x}\right] = -\frac{1}{x^2}$$

However, there is another possibility. We can differentiate both sides of (1) *before* solving for y in terms of x, treating y as a (temporarily unspecified) differentiable function of x. With this approach we obtain

$$\frac{d}{dx}[xy] = \frac{d}{dx}[1]$$

$$x\frac{d}{dx}[y] + y\frac{d}{dx}[x] = 0$$

$$x\frac{dy}{dx} + y = 0$$

$$\frac{dy}{dx} = -\frac{y}{x}$$

If we now substitute (2) into the last expression, we obtain

$$\frac{dy}{dx} = -\frac{1}{x^2}$$

which agrees with the previous result. This method of obtaining derivatives is called ***implicit differentiation.*** It is especially useful when it is inconvenient or impossible to solve explicitly for y in terms of x.

▶ **Example 1** By implicit differentiation find dy/dx if $5y^2 + \sin y = x^2$.

Solution. Differentiating both sides with respect to x and treating y as a differentiable function of x, we obtain

$$\frac{d}{dx}[5y^2 + \sin y] = \frac{d}{dx}[x^2]$$

$$5\frac{d}{dx}[y^2] + \frac{d}{dx}[\sin y] = 2x$$

$$5\left(2y\frac{dy}{dx}\right) + (\cos y)\frac{dy}{dx} = 2x$$

$$10y\frac{dy}{dx} + (\cos y)\frac{dy}{dx} = 2x$$

Solving for dy/dx, we obtain

$$\frac{dy}{dx} = \frac{2x}{10y + \cos y} \tag{3}$$

Note that this formula for dy/dx involves the variables x and y. In order to obtain a formula involving x alone we would have to solve the original equation for y in terms of x and substitute in (3). However, it is impossible to do this, so the formula for dy/dx must be left in terms of x and y. ◀

In problems where dy/dx is used to calculate the slope of a tangent line at a point on a curve, the x and y coordinates of the point are often both known, so that there is no problem using a formula for dy/dx that involves both x and y.

▶ Example 2 Find the slope of the tangent line at $(4, 0)$ to the graph of

$$7y^4 + x^3y + x = 4. \tag{4}$$

[Note that $x = 4, y = 0$ satisfies the equation so that $(4, 0)$ is actually a point on the graph.]

Solution. It is difficult to solve (4) for y in terms of x, so we will differentiate implicitly. We obtain

$$\frac{d}{dx}[7y^4 + x^3y + x] = \frac{d}{dx}[4]$$

$$\frac{d}{dx}[7y^4] + \frac{d}{dx}[x^3y] + \frac{d}{dx}[x] = 0$$

$$\frac{d}{dx}[7y^4] + \left(x^3 \frac{dy}{dx} + y\frac{d}{dx}[x^3]\right) + \frac{d}{dx}[x] = 0$$

$$28y^3 \frac{dy}{dx} + x^3 \frac{dy}{dx} + 3yx^2 + 1 = 0$$

Solving for dy/dx yields

$$\frac{dy}{dx} = -\frac{3yx^2 + 1}{28y^3 + x^3} \tag{5}$$

At the point $(4, 0)$ we have $x = 4$ and $y = 0$, so (5) yields

$$m_{\text{tan}} = \frac{dy}{dx}\bigg|_{\substack{x=4 \\ y=0}} = -\frac{1}{64} \qquad ◀$$

▶ Example 3 Use implicit differentiation to find d^2y/dx^2 if $4x^2 - 2y^2 = 9$.

Solution. Differentiating both sides of $4x^2 - 2y^2 = 9$ implicitly yields

$$8x - 4y\frac{dy}{dx} = 0$$

from which we obtain

$$\frac{dy}{dx} = \frac{2x}{y} \tag{6}$$

Differentiating both sides of (6) implicitly yields

$$\frac{d^2y}{dx^2} = \frac{(y)(2) - (2x)(dy/dx)}{y^2} \tag{7}$$

Substituting (6) into (7) and simplifying, we obtain

$$\frac{d^2y}{dx^2} = \frac{2y - 2x(2x/y)}{y^2} = \frac{2y^2 - 4x^2}{y^3}$$

Finally, using the original equation to simplify further, we obtain

$$\frac{d^2y}{dx^2} = \frac{(-9)}{y^3} = -\frac{9}{y^3} \qquad \blacktriangleleft$$

DERIVATIVES OF RATIONAL POWERS OF x

Earlier, we obtained the formula

$$\frac{d}{dx}[x^n] = nx^{n-1} \tag{8}$$

for integer values of n. By using implicit differentiation we will show that this formula holds for any rational power of x. More precisely, we will show that if r is a rational number, then

$$\frac{d}{dx}[x^r] = rx^{r-1} \tag{9}$$

whenever x^r and x^{r-1} are defined. In our computations we will assume that x^r is differentiable; the justification for this assumption will be considered later.

Let $y = x^r$. Since r is a rational number, it can be expressed as a ratio of integers $r = m/n$.

Thus,

$$y = x^r = x^{m/n}$$

can be written as

$$y^n = x^m$$

so that

$$\frac{d}{dx}[y^n] = \frac{d}{dx}[x^m]$$

On differentiating implicitly with respect to x and using (8), we obtain

$$ny^{n-1}\frac{dy}{dx} = mx^{m-1} \tag{10}$$

But

$$y^{n-1} = [x^{m/n}]^{n-1} = x^{m-(m/n)}$$

Thus, (10) can be written as

$$nx^{m-(m/n)}\frac{dy}{dx} = mx^{m-1}$$

so that

$$\frac{dy}{dx} = \frac{m}{n}x^{(m/n)-1} = rx^{r-1}$$

which establishes (9). ▊

▶ Example 4   From (9)

$$\frac{d}{dx}[x^{4/5}] = \frac{4}{5}x^{(4/5)-1} = \frac{4}{5}x^{-1/5}$$

and

$$\frac{d}{dx}[(x^2 + 1)^{-3/4}] = -\frac{3}{4}(x^2 + 1)^{-(3/4)-1}(2x) = -\frac{3}{2}x(x^2 + 1)^{-7/4} \quad ◀$$

If $r = \frac{1}{2}$, then (9) states that

$$\frac{d}{dx}[x^{1/2}] = \frac{1}{2}x^{-1/2}$$

or in square root notation

$$\frac{d}{dx}[\sqrt{x}] = \frac{1}{2\sqrt{x}}$$

▶ Example 5

$$\frac{d}{dx}\left[\frac{x}{1+\sqrt{x}}\right] = \frac{(1+\sqrt{x})\frac{d}{dx}[x] - x\frac{d}{dx}[1+\sqrt{x}]}{(1+\sqrt{x})^2}$$

$$= \frac{(1+\sqrt{x}) - x\left(\frac{1}{2\sqrt{x}}\right)}{(1+\sqrt{x})^2} = \frac{1+\frac{1}{2}\sqrt{x}}{(1+\sqrt{x})^2} \qquad \blacktriangleleft$$

▶ Example 6

$$\frac{d}{dx}[x^2 - x + 2]^{3/4} = \frac{3}{4}(x^2 - x + 2)^{-1/4} \cdot \frac{d}{dx}[x^2 - x + 2]$$

$$= \frac{3}{4}(x^2 - x + 2)^{-1/4}(2x - 1)$$

$$\frac{d}{dx}[\sin\sqrt{x^3 + 2}] = \cos\sqrt{x^3 + 2} \cdot \frac{d}{dx}(\sqrt{x^3 + 2})$$

$$= (\cos\sqrt{x^3 + 2}) \cdot \frac{1}{2\sqrt{x^3 + 2}} \cdot \frac{d}{dx}(x^3 + 2)$$

$$= \frac{3x^2}{2\sqrt{x^3 + 2}}\cos\sqrt{x^3 + 2} \qquad \blacktriangleleft$$

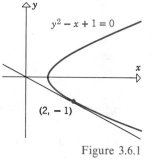

$y^2 - x + 1 = 0$

$(2, -1)$

Figure 3.6.1

▶ Example 7   Find the slope of the tangent line at $(2, -1)$ to $y^2 - x + 1 = 0$ (Figure 3.6.1).

Solution.   Differentiating implicitly yields

$$\frac{d}{dx}[y^2 - x + 1] = \frac{d}{dx}[0]$$

$$\frac{d}{dx}[y^2] - \frac{d}{dx}[x] + \frac{d}{dx}[1] = \frac{d}{dx}[0]$$

$$2y\frac{dy}{dx} - 1 = 0$$

Thus,

$$\frac{dy}{dx} = \frac{1}{2y}$$

At $(2, -1)$ we have $y = -1$, so the slope of the tangent line there is

$$m_{\text{tan}} = \frac{dy}{dx}\Bigg|_{y=-1} = -\frac{1}{2}$$

Alternate Solution.   If we solve $y^2 - x + 1 = 0$ for y in terms of x, we obtain

$$y = \sqrt{x - 1} \qquad \text{and} \qquad y = -\sqrt{x - 1}$$

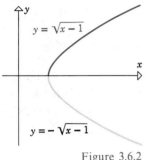

As shown in Figure 3.6.2, the graph of the first equation is the upper half of the curve $y^2 - x + 1 = 0$ (since $y \geq 0$), and the graph of the second is the lower half (since $y \leq 0$).

Since $(2, -1)$ lies on the lower half of the curve, the slope m_{tan} of the tangent line at this point can be obtained by evaluating the derivative of

$$y = -\sqrt{x - 1}$$

when $x = 2$; thus,

Figure 3.6.2

$$\frac{dy}{dx} = \frac{d}{dx}[-\sqrt{x - 1}] = -\frac{1}{2\sqrt{x - 1}}$$

and

$$m_{\text{tan}} = \frac{dy}{dx}\Bigg|_{x=2} = -\frac{1}{2}$$

which agrees with solution obtained by differentiating implicitly.   ◀

IMPLICIT FUNCTIONS

We conclude this section with some observations about the mathematical assumptions that underlie the method of implicit differentiation.

When differentiating implicitly, it is assumed that y represents a differentiable function of x. If this is not so, the resulting calculations may be nonsense. For example, if we implicitly differentiate the equation

$$x^2 + y^2 + 1 = 0 \tag{11}$$

we obtain

$$2x + 2y\frac{dy}{dx} = 0$$

or

$$\frac{dy}{dx} = -\frac{x}{y} \tag{12}$$

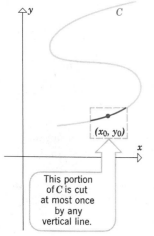

This portion
of C is cut
at most once
by any
vertical line.

Figure 3.6.3

However, the derivative in (12) is meaningless because no function satisfies (11). (The left side of the equation is always greater than zero.)

To help understand when implicit differentiation is justified, we introduce some terminology.

Suppose that a curve C is the graph of an equation in x and y and $P(x_0, y_0)$ is any point on C. We will say that the equation defines y as an *implicit function of x in a neighborhood of (x_0, y_0)* if we can find some rectangle centered at (x_0, y_0), with sides parallel to the coordinate axes, and such that the portion of C contained within the rectangle is cut at most once by any vertical line (Figure 3.6.3). This portion of the curve C is the graph of a function f. If this function is differentiable at x_0, then we can calculate dy/dx at x_0 by implicit differentiation.

Unfortunately, it can be difficult to determine whether an equation defines y as a function of x in a neighborhood of a point and if so, whether the function is differentiable. We leave such questions for a course in advanced calculus. However, the following example should clarify the basic idea.

▶ **Example 8** The equation of the unit circle $x^2 + y^2 = 1$ does not define y as a function of x in a neighborhood of the point $P(1, 0)$, because the portion of the circle within any rectangle centered at P is cut twice by some vertical line (Figure 3.6.4a). The equation does, however, define y as a function of x in a neighborhood of the point $(1/\sqrt{2}, 1/\sqrt{2})$ (Figure 3.6.4b).

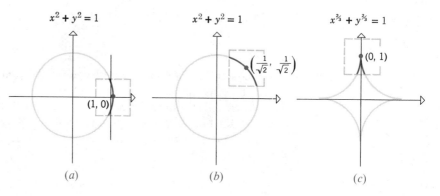

Figure 3.6.4   (a)   (b)   (c)

The graph of the equation

$$x^{2/3} + y^{2/3} = 1$$

is called a *four-cusped hypocycloid* (Figure 3.6.4c). This equation defines y as a function of x in a neighborhood of the point $(0, 1)$; however, the function is not differentiable at $x = 0$, because of the corner or "cusp" at $(0, 1)$.   ◀

▶ Exercise Set 3.6

In Exercises 1–10, find dy/dx.

1. $y = 2x^{3/4}$.

2. $y = (4x^2 - 1)^{-2/3}$.

3. $y = \sqrt{1 + x^3}$.

4. $y = (\cos x^2)^{-5/8}$.

5. $y = \tan \sqrt{x}$.

6. $y = \sqrt{1 + \cos^2 x}$.

7. $y = (x + 1/x)^{-1/7}$.

8. $y = \sqrt{1 + \sqrt{x}}$.

9. $y = \sqrt{x} + \sqrt[3]{3x} + \sqrt[4]{5x}$.

10. $y = [\sec(x^{-2/3} + 1)]^{4/5}$.

In Exercises 11–27, find dy/dx by implicit differentiation.

11. $x^2 + y^2 = 100$.

12. $x^3 - y^3 = 6xy$.

13. $x^2y + 3xy^3 - x = 3$.

14. $x^3y^2 - 5x^2y + x = 1$.

15. $\dfrac{1}{y} + \dfrac{1}{x} = 1$.

16. $x^2 = \dfrac{x + y}{x - y}$.

17. $\sqrt{x} + \sqrt{y} = 8$.

18. $\sqrt{xy} + 1 = y$.

19. $(x^2 + 3y^2)^{35} = x$.

20. $xy^{2/3} + yx^{2/3} = x^2$.

21. $3xy = (x^3 + y^2)^{3/2}$.

22. $\cos xy = y$.

23. $\sin(x^2y^2) = x$.

24. $x^2 = \dfrac{\cot y}{1 + \csc y}$.

25. $\tan^3(xy^2 + y) = x$.

26. $\dfrac{xy^3}{1 + \sec y} = 1 + y^4$.

27. $\sqrt{1 + \sin^3(xy^2)} = y$.

In Exercises 28–32, use implicit differentiation to find the slope of the tangent to the given curve at the specified point.

28. $x^2y - 5xy^2 + 6 = 0$; (3, 1).

29. $x^3y + y^3x = 10$; (1, 2).

30. $\sin xy = y$; $\left(\dfrac{\pi}{2}, 1\right)$.

31. $x^{2/3} - y^{2/3} - y = 1$; (1, −1).

32. $\sqrt{3 + \tan xy} - 2 = 0$; ($\pi/12$, 3).

In Exercises 33–37, find the value of dy/dx at the specified point in two ways: first by solving for y in terms of x and then by implicit differentiation. (See Example 7.)

33. $xy = 8$; (2, 4).

34. $y^2 - x + 1 = 0$; (10, 3).

35. $x^2 + y^2 = 1$; $\left(\dfrac{1}{\sqrt{2}}, -\dfrac{1}{\sqrt{2}}\right)$.

36. $\dfrac{1 - y}{1 + y} = x$; (0, 1).

37. $y^2 - 3xy + 2x^2 = 4$; (3, 2).

In Exercises 38–43, find d^2y/dx^2 by implicit differentiation.

38. $3x^2 - 4y^2 = 7$.

39. $x^3 + y^3 = 1$.

40. $x^3y^3 - 4 = 0$.

41. $2xy - y^2 = 3$.

42. $y + \sin y = x$.

43. $x \cos y = y$.

In Exercises 44–47, use implicit differentiation to find the specified derivative.

44. $\sqrt{u} + \sqrt{v} = 5$; du/dv.

45. $a^4 - t^4 = 6a^2t$; da/dt.

46. $y = \sin x$; dx/dy.

47. $a^2\omega^2 + b^2\lambda^2 = 1$ (a, b constants); $d\omega/d\lambda$.

48. Use implicit differentiation to show that the equation of the tangent to the curve $y^2 = kx$ at (x_0, y_0) is

$$y_0y = \dfrac{k}{2}(x + x_0)$$

49. Find dy/dx if

$$2y^3t + t^3y = 1 \quad \text{and} \quad \dfrac{dt}{dx} = \dfrac{1}{\cos t}$$

50. At what point(s) is the tangent to the curve

$$y^2 = 2x^3$$

perpendicular to the line $4x - 3y + 1 = 0$?

51. Find equations for two lines through the origin that are tangent to the curve

$$x^2 - 4x + y^2 + 3 = 0$$

52. (a) Use implicit differentiation to find the slope of the tangent to the four-cusped hypocycloid $x^{2/3} + y^{2/3} = 1$ at $P(-\frac{1}{4}\sqrt{2}, \frac{1}{4}\sqrt{2})$. (See Figure 3.6.4c.)

(b) Do part (a) by solving explicitly for y as a function of x and then differentiating.

(c) Trace the curve in Figure 3.6.4c and sketch the tangent line at P.

(d) Where does $x^{2/3} + y^{2/3} = 1$ *not* define y as an implicit function of x?

53. The graph of $8(x^2 + y^2)^2 = 100(x^2 - y^2)$, shown in Figure 3.6.5, is called a **lemniscate.**

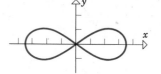

Figure 3.6.5

(a) Where does this equation *not* define y as an implicit function of x?

(b) Use implicit differentiation to find the equation of the tangent to the curve at the point $(3, 1)$.

(c) Trace the curve in Figure 3.6.5 and sketch the tangent at $(3, 1)$.

54. (a) Show that $f(x) = x^{4/3}$ is differentiable at 0, but not twice differentiable at 0.

(b) Show that $f(x) = x^{7/3}$ is twice differentiable at 0, but not three times differentiable at 0.

(c) Find an exponent k such that $f(x) = x^k$ is $(n-1)$-times differentiable at 0, but not n-times differentiable at 0.

3.7 CONTINUITY

A moving physical object cannot vanish at some point and reappear someplace else to continue its motion. Thus, we perceive the path of a moving object as a single, unbroken curve without gaps, jumps, or holes. Such curves can be described as "continuous." In this section we will express this intuitive idea mathematically and develop some properties of continuous curves.

Before we give any formal definitions, let us consider some of the ways in which curves can be "discontinuous." In Figure 3.7.1 we have graphed some curves which, because of their behavior at the point c, fail to be continuous.

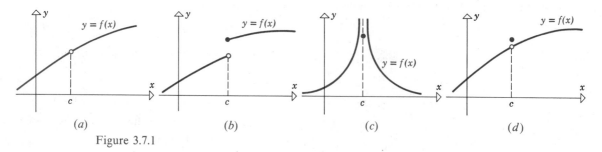

Figure 3.7.1

The curve in Figure 3.7.1a has a hole at the point c because the function f is undefined there. For the curves in Figures 3.7.1b and 3.7.1c, the function f is defined at c, but

$$\lim_{x \to c} f(x)$$

does not exist, thereby causing a break in the graph. For the curve in Figure 3.7.1*d*, the function f is defined at c, and the limit

$$\lim_{x \to c} f(x)$$

exists, yet the graph still has a break at the point c because

$$\lim_{x \to c} f(x) \neq f(c)$$

Based on this discussion, we see that there is a break or discontinuity in the graph of $y = f(x)$ at a point $x = c$ if any of the following conditions occur:

(i) The function f is undefined at c.

(ii) The limit $\lim_{x \to c} f(x)$ does not exist.

(iii) The function f is defined at c and the limit $\lim_{x \to c} f(x)$ exists, but the value of the function at c and the value of the limit at c are different.

This suggests the following definition:

3.7.1 DEFINITION A function f is said to be ***continuous at a point c*** if the following conditions are satisfied:

1. $f(c)$ is defined.

2. $\lim_{x \to c} f(x)$ exists.

3. $\lim_{x \to c} f(x) = f(c)$.

If one or more of the conditions in this definition fails to hold, then f is called ***discontinuous at c*** and c is called a ***point of discontinuity*** of f. If f is continuous at all points of an open interval (a, b), then f is said to be ***continuous on (a, b).*** A function that is continuous on $(-\infty, +\infty)$ is said to be ***continuous everywhere*** or simply ***continuous.***

▶ Example 1 Let

$$f(x) = \frac{x^2 - 4}{x - 2} \quad \text{and} \quad g(x) = \begin{cases} \dfrac{x^2 - 4}{x - 2}, & x \neq 2 \\ 3, & x = 2 \end{cases}$$

Both f and g are discontinuous at 2; the function f because $f(2)$ is undefined, and the function g because $g(2) = 3$, while

$$\lim_{x \to 2} g(x) = \lim_{x \to 2} \frac{x^2 - 4}{x - 2} = \lim_{x \to 2} (x + 2) = 4$$

so that

$$\lim_{x \to 2} g(x) \neq g(2)$$ ◀

▶ **Example 2** Show that $f(x) = \cos x$ is continuous at $x = 0$.

Solution. From Corollary 3.3.3 we obtain

$$\lim_{x \to 0} \cos x = 1$$

Since $\cos (0) = 1$, it follows that $\cos x$ is continuous at $x = 0$. ◀

3.7.2 THEOREM *Polynomials are continuous functions.*

Proof. If p is a polynomial and c is any real number, then by Example 6 in Section 2.5

$$\lim_{x \to c} p(x) = p(c)$$

which proves the continuity of p at c. Since c is an arbitrary real number, p is continuous everywhere. ▮

The following basic result is an immediate consequence of Theorem 2.5.1.

3.7.3 THEOREM *If the functions f and g are continuous at c, then*

(a) *$f + g$ is continuous at c*
(b) *$f - g$ is continuous at c*
(c) *$f \cdot g$ is continuous at c*
(d) *f/g is continuous at c if $g(c) \neq 0$, and is discontinuous at c if $g(c) = 0$.*

We will prove part (*d*) and leave the remaining proofs as exercises.

Proof of (d). If $g(c) = 0$, then f/g is discontinuous at c because $f(c)/g(c)$ is undefined.

Assume $g(c) \neq 0$. We must show that

$$\lim_{x \to c} \frac{f(x)}{g(x)} = \frac{f(c)}{g(c)} \tag{1}$$

Since f and g are continuous at c,

$$\lim_{x \to c} f(x) = f(c) \qquad \text{and} \qquad \lim_{x \to c} g(x) = g(c)$$

Thus, by Theorem 2.5.1(d)

$$\lim_{x \to c} \frac{f(x)}{g(x)} = \frac{\lim_{x \to c} f(x)}{\lim_{x \to c} g(x)} = \frac{f(c)}{g(c)}$$

which proves (1). ▮

▶ Example 3 Where is

$$h(x) = \frac{x^2 - 9}{x^2 - 5x + 6}$$

continuous?

Solution. The numerator and denominator of h are polynomials and, therefore, continuous everywhere by Theorem 3.7.2. Thus, by Theorem 3.7.3(d), the ratio is continuous everywhere except at the points where the denominator is zero. Since the solutions of

$$x^2 - 5x + 6 = 0$$

are $x = 2$ and $x = 3$, $h(x)$ is continuous everywhere except at these points. ◀

The result in Example 3 is a special case of the following general theorem. We leave the proof as an exercise.

3.7.4 THEOREM *A rational function is continuous everywhere except at the points where the denominator is zero.*

The following major theorem relates the notions of continuity and differentiability.

3.7.5 THEOREM *If f is differentiable at a point c, then f is also continuous at c.*

Proof. We want to show that

$$\lim_{x \to c} f(x) = f(c)$$

or equivalently,

$$\lim_{x \to c} [f(x) - f(c)] = 0$$

To prove this, let $h = x - c$, so $x = c + h$ and $h \to 0$ as $x \to c$. Thus,

$$\lim_{x \to c} [f(x) - f(c)] = \lim_{h \to 0} [f(c + h) - f(c)]$$

$$= \lim_{h \to 0} \left[\frac{f(c + h) - f(c)}{h} \cdot h \right]$$

$$= \lim_{h \to 0} \frac{f(c + h) - f(c)}{h} \cdot \lim_{h \to 0} h$$

$$= f'(c) \cdot 0 = 0 \quad \blacksquare$$

▶ **Example 4**   Since the derivatives

$$\frac{d}{dx} [\sin x] = \cos x$$

$$\frac{d}{dx} [\cos x] = -\sin x$$

exist for all x, the functions $\sin x$ and $\cos x$ are differentiable everywhere; thus, by Theorem 3.7.5, they are also continuous everywhere.   ◀

▶ **Example 5**   Where is $\tan x$ continuous?

Solution.   We can write

$$\tan x = \frac{\sin x}{\cos x}$$

Since the numerator and denominator are continuous everywhere by Example 4, the ratio will be continuous everywhere except at the points where the denominator is zero. But the solutions of $\cos x = 0$ are

$$x = \pm \frac{\pi}{2}, \pm \frac{3\pi}{2}, \pm \frac{5\pi}{2}, \dots$$

so that $\tan x$ is continuous everywhere except at these points.   ◀

Theorem 3.7.5 shows that differentiability at a point implies continuity at that point. The converse, however, is false—*a function may be continuous at a point but not differentiable there*. Whenever the graph of a function has a corner at a point, but no break or gap there, we have a point where the function is continuous, but not differentiable. For example, the function

$$f(x) = |x|$$

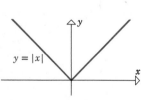

Figure 3.7.2   is continuous, but not differentiable at $x = 0$. (Figure 3.7.2.)

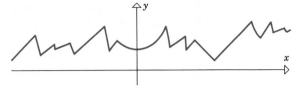

Figure 3.7.3

The relationship between continuity and differentiability was of great historical significance in the development of calculus. In the early nineteenth century mathematicians believed that the graph of a continuous function could not have too many points of nondifferentiability bunched up. They felt that if a continuous function had many points of nondifferentiability, these points, like the tips of a sawblade, would have to be separated from each other and joined by smooth curve segments (Figure 3.7.3). This misconception was shattered by a series of discoveries beginning in 1834. In that year a Bohemian priest, philosopher, and mathematician named Bernhard Bolzano* discovered a procedure for constructing a continuous function that is not differentiable at any point. Later, in 1860, the great German mathematician, Karl Weierstrass,** (see p. 186) produced the first formula for such a function. The graphs of such functions are impossible to draw; it is as if they oscillate so wildly that a corner occurs everywhere. Although these pathological functions rarely occur in practical problems,† (see p. 186) their discovery was of major importance because it made mathematicians distrustful of their geometric intuition and more reliant on precise mathematical proof.

3.7.6 THEOREM *Let* $\lim$ *stand for one of the limits* $\lim\limits_{x \to c}$, $\lim\limits_{x \to c^-}$, $\lim\limits_{x \to c^+}$, $\lim\limits_{x \to +\infty}$, *or* $\lim\limits_{x \to -\infty}$. *If* $\lim g(x) = L$ *and if the function f is continuous at L, then* $\lim f(g(x)) = f(L)$. *That is,* $\lim f(g(x)) = f(\lim g(x))$.

For those who have read Section 2.6, the optional proof of this result appears at the end of this section.

---

*BERNHARD BOLZANO (1781–1848). Bolzano, the son of an art dealer, was born in Prague, Bohemia (Czechoslovakia). He was educated at the University of Prague, and eventually won enough mathematical fame to be recommended for a mathematics chair there. However, Bolzano became an ordained Roman Catholic priest, and in 1805 he was appointed to a chair of Philosophy at the University of Prague. Bolzano was a man of great human compassion; he spoke out for educational reform, he voiced the right of individual conscience over government demands, and he lectured on the absurdity of war and militarism. His views so disenchanted Emperor Franz I of Austria that the emperor pressed the Archbishop of Prague to have Bolzano recant his statements. Bolzano refused and was then forced to retire in 1824 on a small pension. Bolzano's main contribution to mathematics was philosophical. His work helped convince mathematicians that sound mathematics must ultimately rest on rigorous proof rather than intuition. In addition to his work in mathematics, Bolzano investigated problems concerning space, force, and wave propagation.

REMARK. In words, this theorem states that the limit symbol can be moved through a function sign provided the limit of the expression inside the function sign exists and the function is continuous at this limit.

▶ Example 6 Since the functions $\sin x$ and $\cos x$ are continuous everywhere, it follows from Theorem 3.7.6 that we can write

$$\lim [\sin (g(x))] = \sin [\lim g(x)]$$

and

$$\lim [\cos (g(x))] = \cos [\lim g(x)]$$

if $\lim g(x)$ exists. For example,

$$\lim_{x \to \pi} \left[\sin \left(\frac{x^2}{\pi + x} \right) \right] = \sin \left[\lim_{x \to \pi} \left(\frac{x^2}{\pi + x} \right) \right] = \sin \frac{\pi}{2} = 1$$

$$\lim_{x \to +\infty} \left[\cos \left(\frac{\pi x^2 + 1}{x^2 + 3} \right) \right] = \cos \left[\lim_{x \to +\infty} \left(\frac{\pi x^2 + 1}{x^2 + 3} \right) \right] = \cos \left[\lim_{x \to +\infty} \left(\frac{\pi + (1/x^2)}{1 + (3/x^2)} \right) \right]$$

$$= \cos \pi = -1 \qquad \blacktriangleleft$$

---

*KARL WEIERSTRASS (1815–1897). Weierstrass, the son of a customs officer, was born in Ostenfelde, Germany. As a youth Weierstrass showed outstanding skills in languages and mathematics. However, at the urging of his dominant father, Weierstrass entered the law and commerce program at the University of Bonn. To the chagrin of his family, the rugged and congenial young man concentrated instead on fencing and beer drinking. Four years later he returned home without a degree. In 1839 Weierstrass entered the Academy of Münster to study for a career in secondary education, and he met and studied under an excellent mathematician named Christof Gudermann. Gudermann's ideas greatly influenced the work of Weierstrass. After receiving his teaching certificate, Weierstrass spent the next 15 years in secondary education teaching German, geography, and mathematics. In addition he taught handwriting to small children. During this period much of Weierstrass's mathematical work was ignored because he was a secondary school teacher and not a college professor. Then in 1854 he published a paper of major importance, which created a sensation in the mathematics world and catapulted him to international fame overnight. He was immediately given an honorary Doctorate at the University of Königsberg and began a new career in college teaching at the University of Berlin in 1856. In 1859 the strain of his mathematical research caused a temporary nervous breakdown and led to spells of dizziness that plagued him for the rest of his life. Weierstrass was a brilliant teacher and his classes overflowed with multitudes of auditors. In spite of his fame, he never lost his early beer-drinking congeniality and was always in the company of students, both ordinary and brilliant. Weierstrass was acknowledged as the leading mathematical analyst in the world. He and his students opened the door to the modern school of mathematical analysis.

†Important applications of these functions have been discovered recently. See, Bruce Schecter, "A New Geometry of Nature," *Discover Magazine* (June 1982), pp. 66–68.

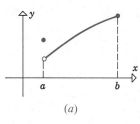

(a)

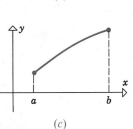

(b)

(c)

Figure 3.7.4

▶ **Example 7** Since $|x|$ is continuous everywhere (Exercise 24), Theorem 3.7.6 implies that

$$\lim |g(x)| = |\lim g(x)|$$

if $\lim g(x)$ exists. For example,

$$\lim_{x \to 3} |5 - x^2| = |\lim_{x \to 3} (5 - x^2)| = |-4| = 4 \qquad \blacktriangleleft$$

Figure 3.7.4 shows the graphs of three functions defined only on a closed interval $[a, b]$. Obviously the function shown in Figure 3.7.4a should be regarded as discontinuous at the left-hand endpoint a, the function in Figure 3.7.4b should be regarded as discontinuous at the right-hand endpoint b, and the function in Figure 3.7.4c should be regarded as continuous at both endpoints. However, the definition of continuity (Definition 3.7.1) does not apply at the endpoints, since the two-sided limits appearing in parts 2 and 3 of the definition make no sense. At the left-hand endpoint the only sensible limit is the one-sided limit

$$\lim_{x \to a^+} f(x)$$

and at the right-hand endpoint the only sensible limit is the one-sided limit

$$\lim_{x \to b^-} f(x)$$

This suggests the following definitions.

3.7.7   DEFINITION   A function f is called *continuous from the left at the point c* if the conditions in the left column below are satisfied, and is called *continuous from the right at the point c* if the conditions in the right column are satisfied.

1. $f(c)$ is defined.
2. $\lim_{x \to c^-} f(x)$ exists.
3. $\lim_{x \to c^-} f(x) = f(c)$.

1'. $f(c)$ is defined.
2'. $\lim_{x \to c^+} f(x)$ exists.
3'. $\lim_{x \to c^+} f(x) = f(c)$.

▶ **Example 8** If f denotes the function graphed in Figure 3.7.4a, then

$$f(a) \neq \lim_{x \to a^+} f(x) \qquad \text{and} \qquad f(b) = \lim_{x \to b^-} f(x)$$

Thus, f is continuous from the left at b, but is not continuous from the right at a. ◀

When we state that a function f is *continuous on a closed interval* $[a, b]$, we will mean:

(i) f is continuous on (a, b),
(ii) f is continuous from the right at a,
(iii) f is continuous from the left at b.

The reader should have no trouble deducing the appropriate definitions of continuity on intervals of the form $[a, +\infty)$, $(-\infty, b]$, $[a, b)$, and $(a, b]$.

▶ **Example 9** Show that $f(x) = \sqrt{9 - x^2}$ is continuous on the closed interval $[-3, 3]$.

Solution. Observe that the domain of f is the interval $[-3, 3]$. For c in the interval $(-3, 3)$ we have by Theorem 3.7.6

$$\lim_{x \to c} f(x) = \lim_{x \to c} \sqrt{9 - x^2} = \sqrt{9 - c^2} = f(c)$$

so that f is continuous on $(-3, 3)$. Also,

$$\lim_{x \to 3^-} f(x) = \lim_{x \to 3^-} \sqrt{9 - x^2} = 0 = f(3)$$

and

$$\lim_{x \to -3^+} f(x) = \lim_{x \to -3^+} \sqrt{9 - x^2} = 0 = f(-3)$$

so that f is continuous on $[-3, 3]$. ◀

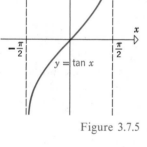

Figure 3.7.5

▶ **Example 10** Example 5 implies that $f(x) = \tan x$ is continuous on the open interval $(-\pi/2, \pi/2)$. However,

$$f\left(-\frac{\pi}{2}\right) = \tan\left(-\frac{\pi}{2}\right) \quad \text{and} \quad f\left(\frac{\pi}{2}\right) = \tan\frac{\pi}{2}$$

are undefined, so $\tan x$ is not continuous on the closed interval $[-\pi/2, \pi/2]$. (See Figure 3.7.5.) ◀

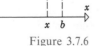

Figure 3.7.6

If f is continuous on a closed interval $[a, b]$, and we draw a horizontal line crossing the y-axis between the numbers $f(a)$ and $f(b)$ (Figure 3.7.6), then it is geometrically obvious that this line will cross the curve $y = f(x)$ at least once over the interval $[a, b]$. Stated another way, a continuous function must assume every possible value between $f(a)$ and $f(b)$ as x varies from a to b. This idea is formalized in the following theorem.

3.7.8 THEOREM
Intermediate Value Theorem

If f is continuous on a closed interval $[a, b]$ and C is any number between $f(a)$ and $f(b)$, inclusive, then there is at least one number x in the interval $[a, b]$ such that $f(x) = C$.

(See Figure 3.7.6.) This theorem, though intuitively obvious, is not easy to prove. The proof can be found in most advanced calculus texts.

The following useful result is an immediate consequence of the Intermediate Value Theorem.

3.7.9 THEOREM

If f is continuous on $[a, b]$, and if $f(a)$ and $f(b)$ have opposite signs, then there is at least one solution of the equation $f(x) = 0$ in the interval (a, b).

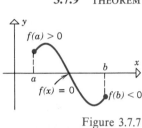

Figure 3.7.7

The proof is left as an exercise, but the result is illustrated in Figure 3.7.7 in the case where $f(a) > 0$ and $f(b) < 0$.

▶ **Example 11** The equation

$$x^3 - x - 1 = 0$$

cannot be solved by factoring because the left side has no simple factors. However, from Figure 3.7.8, which was generated on a microcomputer, we see that there is a solution in the interval $(1, 2)$. This can also be seen from Theorem 3.7.9 by letting $f(x) = x^3 - x - 1$, and noting that $f(1) = -1$ and $f(2) = 5$ have opposite signs.

To pinpoint the solution more exactly, we can divide the interval $(1, 2)$ into 10 equal parts and evaluate $f(x)$ at each point of subdivision using a calculator (Table 3.7.1).

Table 3.7.1

x	1	1.1	1.2	1.3	1.4	1.5	1.6	1.7	1.8	1.9	2
$f(x)$	-1	$-.77$	$-.47$	$-.10$	$.34$	$.88$	1.5	2.2	3.0	3.9	5

$y = x^3 - x - 1$

Figure 3.7.8

From the table we see that $f(1.3)$ and $f(1.4)$ have opposite signs, so there is a solution in the interval $(1.3, 1.4)$. If we use the midpoint 1.35 as an approximation to the solution, then our error is at most half the interval length or $.05$. If desired, we could obtain even greater accuracy by subdividing the interval $(1.3, 1.4)$ into 10 parts and repeating the sign analysis process. With sufficiently many subdivisions, the solution can be determined to any degree of accuracy. ◀

OPTIONAL

We conclude this section with a proof of Theorem 3.7.6 for those who have read Section 2.6. We will consider only the case where lim is the two-sided limit $\lim_{x \to c}$.

Proof. To prove that

$$\lim_{x \to c} f(g(x)) = f(L)$$

we must show that for every $\epsilon > 0$ there is a $\delta > 0$ such that $f(g(x))$ satisfies

$$|f(g(x)) - f(L)| < \epsilon \tag{2}$$

whenever x satisfies

$$0 < |x - c| < \delta \tag{3}$$

Since f is continuous at L,

$$\lim_{x \to L} f(x) = f(L)$$

If we use u as the variable rather than x, then this limit can be written

$$\lim_{u \to L} f(u) = f(L)$$

Thus, given $\epsilon > 0$ we can find a $\delta_1 > 0$ such that $f(u)$ satisfies

$$|f(u) - f(L)| < \epsilon \tag{4}$$

whenever u satisfies

$$|u - L| < \delta_1 \tag{5}$$

Since

$$\lim_{x \to c} g(x) = L$$

there exists a $\delta > 0$ such that

$$|g(x) - L| < \delta_1 \tag{6}$$

whenever x satisfies (3). To complete the proof let

$$u = g(x)$$

and assume that x satisfies (3). This implies that $g(x)$ satisfies (6), which implies that u satisfies (5). Therefore, $f(u)$ satisfies (4), which implies that $f(g(x))$ satisfies (2). Thus, (2) is satisfied whenever (3) is satisfied. ∎

▶ Exercise Set 3.7

In Exercises 1–4, let f be the function whose graph is shown. On which of the following intervals, if any, is f continuous?

(a) [1, 3] (b) (1, 3)
(c) [1, 2] (d) (1, 2)
(e) [2, 3] (f) (2, 3).

On those intervals where f is discontinuous, state where the discontinuities occur.

1.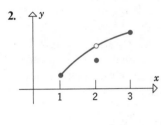

2.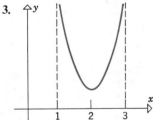

3.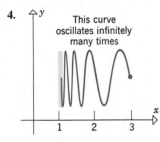

4.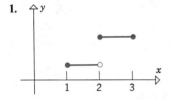

This curve oscillates infinitely many times

In Exercises 5–20, find the points of discontinuity, if any.

5. $f(x) = x^3 - 2x + 3$.

6. $f(x) = (x - 5)^{17}$.

7. $f(x) = \dfrac{x}{x^2 + 1}$.

8. $f(x) = \dfrac{x}{x^2 - 1}$.

9. $f(x) = \dfrac{x - 4}{x^2 - 16}$.

10. $f(x) = \dfrac{3x + 1}{x^2 + 7x - 2}$.

11. $f(x) = \sin(x^2 - 2)$.

12. $f(x) = \cos\left(\dfrac{x}{x - \pi}\right)$.

13. $f(x) = \cot x$.

14. $f(x) = \sec x$.

15. $f(x) = \csc x$.

16. $f(x) = \dfrac{1}{1 + \sin^2 x}$.

17. $f(x) = |\cos x|$.

18. $f(x) = \dfrac{x}{|x|}$.

19. $f(x) = \sqrt[3]{\dfrac{1}{x} - \dfrac{1}{x - 1}}$.

20. $f(x) = \sqrt{2 + \tan^2 x}$.

21. Find a value for the constant k that will make the function continuous.

(a) $f(x) = \begin{cases} 7x - 2, & x \le 1 \\ kx^2, & x > 1 \end{cases}$

(b) $f(x) = \begin{cases} \dfrac{\sin x}{x}, & x \ne 0 \\ k, & x = 0. \end{cases}$

22. On which of the following intervals is

$$f(x) = \dfrac{1}{\sqrt{x - 2}}$$

continuous?

(a) $[2, +\infty)$ (b) $(-\infty, +\infty)$
(c) $(2, +\infty)$ (d) $[1, 2)$.

23. Prove that $f(x) = \sqrt{x}$ is continuous on $[0, +\infty)$.

24. Prove that $f(x) = |x|$ is continuous everywere.

25. Prove that $\sin(g(x))$ is continuous at every point where $g(x)$ is continuous.

26. Prove: If g is continuous at c and f is continuous at $g(c)$, then $f \circ g$ is continuous at the point c.

27. Use Exercise 26 to prove that the composition of continuous functions is continuous.

28. Use Exercise 27 to prove that the following functions are continuous:
(a) $\sin(x^3 + 7x + 1)$ (b) $|\cos x|$
(c) $\cos^3(x + 1)$.

29. Prove:
(a) part (a) of Theorem 3.7.3
(b) part (b) of Theorem 3.7.3
(c) part (c) of Theorem 3.7.3.

30. Prove Theorem 3.7.4.

31. Let f and g be discontinuous at c. Give examples to show:
(a) $f + g$ can be continuous or discontinuous at c
(b) $f \cdot g$ can be continuous or discontinuous at c.

32. **(For students who have read Section 2.6.)** Let f be defined at c. Prove that f is continuous at c if and only if, given $\epsilon > 0$ there exists a $\delta > 0$ such that $|f(x) - f(c)| < \epsilon$ whenever $|x - c| < \delta$.

In Exercises 33–35, you will have to determine whether a function f is differentiable at a point x_0 where the formula for f changes. Use the following result:

Theorem. *Let f be continuous at x_0 and suppose that*

$$\lim_{x \to x_0^+} f'(x) \quad \text{and} \quad \lim_{x \to x_0^-} f'(x)$$

exist. Then f is differentiable at x_0 if and only if these limits are equal. Moreover, in the case of equality,

$$f'(x_0) = \lim_{x \to x_0^+} f'(x) = \lim_{x \to x_0^-} f'(x)$$

33. Let
$$f(x) = \begin{cases} x^2, & x \leq 1 \\ \sqrt{x}, & x > 1 \end{cases}$$

Determine whether f is differentiable at $x = 1$. If so, find the value of the derivative there.

34. Let
$$f(x) = \begin{cases} x^3 + \dfrac{1}{16}, & x < \dfrac{1}{2} \\ \dfrac{3}{4}x^2, & x \geq \dfrac{1}{2} \end{cases}$$

Determine whether f is differentiable at $x = \frac{1}{2}$. If so, find the value of the derivative there.

35. Find all points where f fails to be differentiable. Justify your answer.
(a) $f(x) = |3x - 2|$
(b) $f(x) = |x^2 - 4|$.

In Exercises 36 and 37, show that the equation has at least one solution in the given interval.

36. (a) $x^3 - 4x + 1 = 0$; $[1, 2]$
(b) $x = \cos x$; $[0, \pi/2]$.

37. (a) $x^3 + x^2 - 2x = 1$; $[-1, 1]$
(b) $x + \sin x = 1$; $[0, \pi/6]$.

38. Figure 3.7.9 shows the graph of $y = x^4 + x - 1$ generated on a microcomputer. Use the method of Example 11 to approximate the real solutions of $x^4 + x - 1 = 0$ with an error of at most .05.

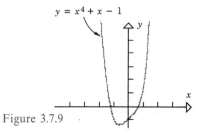

$y = x^4 + x - 1$

Figure 3.7.9

39. Figure 3.7.10 shows the graph of $y = 5 - x - x^4$ generated on a microcomputer. Use the method of Example 11 to approximate the real solutions of $5 - x - x^4 = 0$ with an error of at most .05.

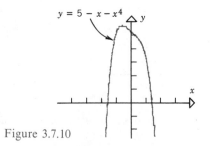

$y = 5 - x - x^4$

Figure 3.7.10

40. For the equation $x^3 - x - 1 = 0$ discussed in Example 11, show that the real solution is approximately 1.325, with an error of at most .005.

41. Use the fact that $\sqrt{5}$ is a solution of $x^2 - 5 = 0$ to approximate $\sqrt{5}$ with an error of at most
(a) .05 (b) .005.

42. Use Theorem 3.7.8 to prove Theorem 3.7.9.

43. (a) Prove: If $f''(x)$ exists for each x in (a, b), then both f and f' are continuous on (a, b).
(b) What can be said about the continuity of f and its derivatives if $f^{(n)}(x)$ exists for each x in (a,b)?

▶ SUPPLEMENTARY EXERCISES

In Exercises 1–4, use Definition 3.1.2 to find $f'(x)$.

1. $f(x) = kx$ (k constant).

2. $f(x) = (x - a)^2$ (a constant).

3. $f(x) = \sqrt{9 - 4x}$.

4. $f(x) = \dfrac{x}{x + 1}$.

5. Use Definition 3.1.2 to find $\dfrac{d}{dx}[|x|^3]\Big|_{x=0}$.

6. Suppose $f(x) = \begin{cases} x^2 - 1, & x \leq 1 \\ k(x - 1), & x > 1. \end{cases}$
For what values of k is f
(a) continuous
(b) differentiable?

7. Suppose $f(3) = -1$ and $f'(3) = 5$. Find an equation for the tangent line to the graph of f at $x = 3$.

8. Let $f(x) = x^2$. Show that for any distinct values of a and b, the slope of the tangent line to $y = f(x)$ at $x = \frac{1}{2}(a + b)$ is equal to the slope of the secant line through the points (a, a^2) and (b, b^2).

9. Given the following table of values at $x = 1$ and $x = -2$, find the indicated derivatives in (a)–(l).

x	$f(x)$	$f'(x)$	$g(x)$	$g'(x)$
1	1	3	-2	-1
-2	-2	-5	1	7

(a) $\dfrac{d}{dx}[f^2(x) - 3g(x^2)]\Big|_{x=1}$

(b) $\dfrac{d}{dx}[f(x)g(x)]\Big|_{x=1}$

(c) $\dfrac{d}{dx}\left[\dfrac{f(x)}{g(x)}\right]\Big|_{x=-2}$

(d) $\dfrac{d}{dx}\left[\dfrac{g(x)}{f(x)}\right]\Big|_{x=-2}$

(e) $\dfrac{d}{dx}[f(g(x))]\Big|_{x=1}$

(f) $\dfrac{d}{dx}[f(g(x))]\Big|_{x=-2}$

(g) $\dfrac{d}{dx}[g(f(x))]\Big|_{x=-2}$

(h) $\dfrac{d}{dx}[g(g(x))]\Big|_{x=-2}$

(i) $\dfrac{d}{dx}[f(g(4 - 6x))]\Big|_{x=1}$

(j) $\dfrac{d}{dx}[g^3(x)]\Big|_{x=1}$

(k) $\dfrac{d}{dx}[\sqrt{f(x)}]\Big|_{x=1}$

(l) $\dfrac{d}{dx}[f(-\frac{1}{2}x)]\Big|_{x=-2}$

In Exercises 10–15, find $f'(x)$ and determine those values of x for which $f'(x) = 0$.

10. $f(x) = (2x + 7)^6(x - 2)^5$.

11. $f(x) = \dfrac{(x - 3)^4}{x^2 + 2x}$.

12. $f(x) = \sqrt{3x + 1}\,(x - 1)^2$.

13. $f(x) = \left(\dfrac{3x + 1}{x^2}\right)^3$.

14. $f(x) = \dfrac{3(5x - 1)^{1/3}}{3x - 5}$.

15. $f(x) = \sqrt{x}\,\sqrt[3]{x^2 + x + 1}$.

16. Suppose that $f'(x) = 1/x$ for all $x \neq 0$.
   (a) Use the chain rule to show that for any non-zero constant a, $d(f(ax))/dx = d(f(x))/dx$.
   (b) If $y = f(\sin x)$ and $v = f(1/x)$, find dy/dx and dv/dx.

In Exercises 17–26, find the indicated derivatives.

17. $\dfrac{d}{dx}\left(\dfrac{\sqrt{2}}{x^2} - \dfrac{2}{5x}\right)$.

18. $\dfrac{dy}{dx}$ if $y = \dfrac{3x^2 + 7}{x^2 - 1}$.

19. $\dfrac{dz}{dr}\bigg|_{r=\frac{\pi}{6}}$ if $z = 4\sin^2 r \cos^2 r$.

20. $g'(2)$ if $g(x) = 1/\sqrt{2x}$.

21. $\dfrac{du}{dx}$ if $u = \left(\dfrac{x}{x-1}\right)^{-2}$.

22. $\dfrac{dw}{dv}$ if $w = \sqrt[5]{v^3} - \sqrt[4]{v}$.

23. $d(\sec^2 x - \tan^2 x)/dx$.

24. $\dfrac{dy}{dx}\bigg|_{x=\frac{\pi}{4}}$ if $y = \tan t$ and $t = \cos(2x)$.

25. $F'(x)$ if $F(x) = \dfrac{(1/x) + 2x}{\frac{1}{2}(1/x^2) + 1}$.

26. $\Phi'(x)$ if $\Phi(x) = \dfrac{x^2 - 4x}{5\sqrt{x}}$.

27. Find all values of x for which the tangent to $f(x) = x - (1/x)$ is parallel to the line $2x - y = 5$.

28. Find all values of x for which the tangent to $f(x) = 2x^3 - x^2$ is perpendicular to the line $x + 4y = 10$.

29. Find all values of x for which the tangent to $f(x) = (x + 2)^2$ passes through the origin.

30. Find all values of x for which the tangent to $f(x) = x - \sin 2x$ is horizontal.

31. Find all values of x for which the tangent to $f(x) = 3x - \tan x$ is parallel to the line $y - x = 2$.

In Exercises 32–34, find Δx, Δy, and dy.

32. $y = 1/(x - 1)$; x decreases from 2 to 1.5.

33. $y = \tan x$; x increases from $-\pi/4$ to 0.

34. $y = \sqrt{25 - x^2}$; x increases from 0 to 3.

35. Use a differential to approximate (a) $\sqrt[3]{-8.25}$; (b) $\cot 46°$.

36. The area of a right triangle whose hypotenuse is H centimeters is calculated using the formula $A = \frac{1}{4}H^2 \sin 2\theta$, where θ is one of the acute angles. Use differentials to approximate the error in calculating A if $H = 4$ cm (exactly) and $\theta = 30° \pm 15'$.

37. A 12-ft ladder leaning against a wall makes an angle θ with the floor. If the top of the ladder is h feet up the wall, express h in terms of θ and then use dh to estimate the change in h if θ changes from $60°$ to $59°$.

38. Let V and S denote the volume and surface area of a cube. Find the rate of change of V with respect to S.

39. The amount of water in a tank t minutes after it has started to drain is given by $W = 100(t - 15)^2$ (gal).
   (a) At what rate is the water running out at the end of 5 min?
   (b) What is the average rate at which the water flows out during the first 5 min?

In Exercises 40–42, find dy/dx by implicit differentiation and use it to find the equation of the tangent line at the indicated points.

40. $(x + y)^3 + 3xy = -7$; $(-2, 1)$.

41. $xy^2 = \sin(x + 2y)$; $(0, 0)$.

42. $(x + y)^3 - 5x + y = 1$; at the point where the curve intersects the line $x + y = 1$.

43. Show that the curves whose equations are $2x^2 + 3y^2 = 5$ and $y^2 = x^3$ intersect at the point $(1, 1)$ and that their tangent lines are perpendicular there.

44. Show that for any point $P_0(x_0, y_0)$ on the circle $x^2 + y^2 = r^2$, the tangent line at P_0 is perpendicular to the radial line from the origin to P_0.

In Exercises 45–48, find the indicated limit if it exists.

45. $\displaystyle\lim_{x \to 0} \dfrac{\tan ax}{\sin bx}$ ($a \neq 0, b \neq 0$).

46. $\lim\limits_{x\to 0} \dfrac{\sin 3x}{\tan 3x}$.

47. $\lim\limits_{\theta\to 0} \dfrac{\sin 2\theta}{\theta^2}$.

48. $\lim\limits_{x\to 0} \dfrac{x \sin x}{1 - \cos x}$.

49. Given that the sine and cosine functions are continuous at $x = 0$, use a suitable trigonometric identity to deduce that for any x, $\lim\limits_{h\to 0} \sin (x + h) = \sin x$. What have you proved?

50. Find d^2y/dx^2 implicitly if (a) $y^3 + 3x^2 = 4y$; (b) $\sin y + \cos x = 1$.

51. Verify that $y = \cos x - 3 \sin x$ satisfies $y''' + y'' + y' + y = 0$.

4 applications of differentiation

4.1 RELATED RATES

In this section we will study *related rates* problems. In such problems one tries to find the rate at which some quantity is changing by relating it to other quantities whose rates of change are known.

▶ Example 1   Assume that oil spilled from a ruptured tanker spreads in a circular pattern whose radius increases at a constant rate of 2 ft/sec. How fast is the area of the spill increasing when the radius of the spill is 60 ft?

Solution.   Let

 t = number of seconds elapsed from the time of the spill
 r = radius of the spill in feet after t seconds
 A = area of the spill in square feet after t seconds

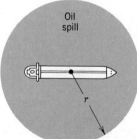

(See Figure 4.1.1.) At each instant the rate at which the radius is increasing with time is dr/dt, and the rate at which the area is increasing with time is dA/dt. We want to find

$$\left. \frac{dA}{dt} \right|_{r=60}$$

which is the rate at which the area of the spill is increasing at the instant when $r = 60$.

Figure 4.1.1

Since the radius increases at the constant rate of 2 ft/sec, we know

$$\frac{dr}{dt} = 2 \tag{1}$$

for all t.

From the formula for the area of a circle we obtain

$$A = \pi r^2 \tag{2}$$

Because A and r are functions of t, we can differentiate both sides of (2) with respect to t to obtain

$$\frac{dA}{dt} = 2\pi r \frac{dr}{dt}$$

Substituting (1) yields

$$\frac{dA}{dt} = 2\pi r(2) = 4\pi r$$

Thus, when $r = 60$ the area of the spill is increasing at the rate

$$\left. \frac{dA}{dt} \right|_{r=60} = 4\pi(60) = 240\pi \ \text{ft}^2/\text{sec}$$

or approximately 754 ft²/sec. ◄

With only minor variations, the method used in Example 1 can be used to solve a variety of related rate problems. The method consists of four steps:

> **Step 1.** Draw a figure and label the quantities that vary.
>
> **Step 2.** Find an equation relating the quantity with the un-known rate of change to quantities whose rates of change are known.
>
> **Step 3.** Differentiate both sides of this equation with respect to time and solve for the derivative that will give the unknown rate of change.
>
> **Step 4.** Evaluate this derivative at the appropriate point.

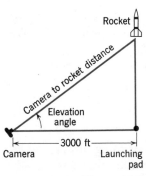

Figure 4.1.2

In Figure 4.1.2 we have shown a camera mounted at a point 3000 ft from the base of a rocket launching pad. Let us assume that rocket rises vertically and the camera is to take a series of photographs. Because the rocket will be rising, the elevation angle of the camera will have to vary at just the right rate to keep the rocket in sight. Moreover, because the camera-to-rocket distance will be changing constantly, the camera focusing mechanism will also have to vary at just the right rate to keep the picture sharp. Our next two examples are concerned with these problems.

▶ **Example 2** If the rocket shown in Figure 4.1.2 is rising vertically at 880 ft/sec when it is 4000 ft up, how fast is the camera-to-rocket distance changing at that instant?

Solution. Let
  t = number of seconds elapsed from the time of launch.
  y = camera-to-rocket distance in feet after t seconds
  x = height of rocket in feet after t seconds

At each instant the rate at which the camera-to-rocket distance is changing is dy/dt, and the rate at which the rocket is rising is dx/dt. We want to find

$$\frac{dy}{dt}\bigg|_{x=4000}$$

which is the rate the camera-to-rocket distance is changing at the instant when $x = 4000$.

From the Theorem of Pythagoras and Figure 4.1.3 we have

$$y^2 = x^2 + (3000)^2 \tag{3}$$

Because x and y are functions of t, we can differentiate both sides of (3) with respect to t to obtain

$$2y\frac{dy}{dt} = 2x\frac{dx}{dt}$$

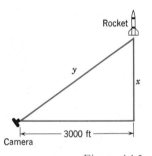

Figure 4.1.3

or

$$\frac{dy}{dt} = \frac{x}{y}\frac{dx}{dt} \tag{4}$$

When $x = 4000$, it follows from (3) that $y = 5000$; moreover, we are given that $dx/dt = 880$ when $x = 4000$ so that from (4)

$$\frac{dy}{dt}\bigg|_{x=4000} = \frac{4000}{5000}(880) = 704 \text{ ft/sec} \qquad ◀$$

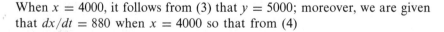

Figure 4.1.4

▶ **Example 3** If the rocket shown in Figure 4.1.4 is rising vertically at 880 ft/sec when it is 4000 ft up, how fast must the camera elevation angle change at that instant to keep the rocket in sight?

Solution.   Let

t = number of seconds elapsed from the time of launch

ϕ = camera elevation angle in radians after t seconds

x = height of rocket in feet after t seconds

(See Figure 4.1.4.) At each instant the rate at which the camera elevation angle must change is $d\phi/dt$, and the rate at which the rocket is rising is dx/dt. We want to find

$$\frac{d\phi}{dt}\bigg|_{x=4000}$$

which is the rate the camera elevation angle must change at the instant when $x = 4000$.

From Figure 4.1.4 we see that

$$\tan \phi = \frac{x}{3000} \tag{5}$$

Because ϕ and x are functions of t, we can differentiate both sides of (5) with respect to t to obtain

$$(\sec^2 \phi)\frac{d\phi}{dt} = \frac{1}{3000}\frac{dx}{dt}$$

or

$$\frac{d\phi}{dt} = \frac{1}{3000 \sec^2 \phi}\frac{dx}{dt} \tag{6}$$

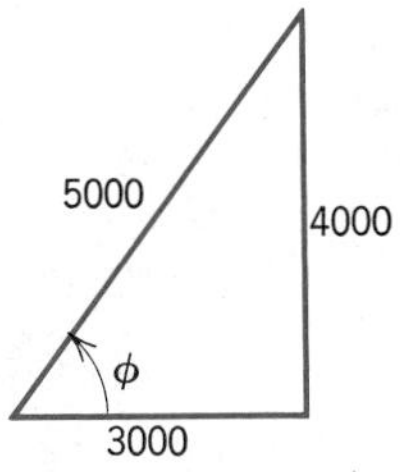

When $x = 4000$, it follows that

$$\sec \phi = \frac{5000}{3000} = \frac{5}{3}$$

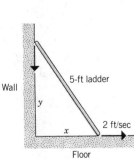

Figure 4.1.5

(See Figure 4.1.5.) Moreover, we are given that $dx/dt = 880$ when $x = 4000$ so that from (6)

$$\frac{d\phi}{dt}\bigg|_{x=4000} = \frac{1}{3000(\frac{5}{3})^2}\cdot 880 = \frac{66}{625} \approx 0.11 \text{ radian/sec} \quad \blacktriangleleft$$

▶ **Example 4** A 5-ft ladder, leaning against a wall (Figure 4.1.6), slips so that its base moves away from the wall at a rate of 2 ft/sec. How fast will the top of the ladder be moving down the wall when the base is 4 ft from the wall?

Figure 4.1.6

Solution. Let

> t = number of seconds after the ladder starts to slip
> x = distance in feet from the base of the ladder to the wall
> y = distance in feet from the top of the ladder to the floor

(See Figure 4.1.6.) At each instant the rate at which the base moves is dx/dt, and the rate at which the top moves is dy/dt. We want to find

$$\frac{dy}{dt}\bigg|_{x=4}$$

which is the rate at which the top is moving at the instant the base is 4 ft from the wall.

From the Theorem of Pythagoras we have

$$x^2 + y^2 = 25 \tag{7}$$

Differentiating both sides of this equation with respect to t yields

$$2x\frac{dx}{dt} + 2y\frac{dy}{dt} = 0$$

or

$$\frac{dy}{dt} = -\frac{x}{y}\frac{dx}{dt} \tag{8}$$

When $x = 4$, it follows from (7) that $y = 3$; moreover, we are given that $dx/dt = 2$, so that (8) yields

$$\frac{dy}{dt}\bigg|_{x=4} = -\frac{4}{3}(2) = -\frac{8}{3}\ \text{ft/sec}$$

The negative sign in the answer tells us that y is decreasing, which makes sense physically, since the top of the ladder is moving *down* the wall. ◀

▶ **Example 5** Suppose a liquid is to be cleared of sediment by pouring it through a cone-shaped filter as shown in Figure 4.1.7*a*. Assume the height of the cone is 16 in. and the radius at the base of the cone is 4 in. (Figure 4.1.7*b*.) If the liquid is flowing out of the cone at a constant rate of 2 in.^{3}/min, how fast is the depth of the liquid decreasing when the level is 8 in. deep?

Solution. Let

> t = time elapsed from the initial observation (min)
> V = volume of liquid in the cone at time t (in.^{3})
> y = depth of the liquid in the cone at time t (in.)
> x = radius of the liquid surface at time t (in.)

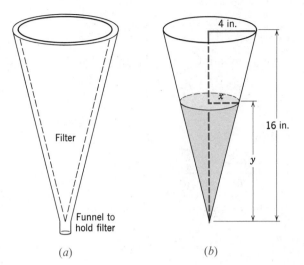

Figure 4.1.7   (*a*)   (*b*)

(See Figure 4.1.7*b*.) At each instant the rate at which the volume of liquid is changing is dV/dt, and the rate at which the depth is changing is dy/dt. We want to find

$$\left.\frac{dy}{dt}\right|_{y=8}$$

which is the rate at which the depth is changing at the instant the depth is 8 in.

From the formula for the volume of a cone, the volume V and the depth y are related by

$$V = \tfrac{1}{3}\pi x^2 y \tag{9}$$

Since we are given that the volume of liquid decreases at a constant rate of 2 in.^{3}/min, we have

$$\frac{dV}{dt} = -2 \tag{10}$$

(We must use a minus sign here because V *decreases* as t increases.) If we differentiate both sides of (9) with respect to t, the right side will involve the quantity dx/dt. Since we have no direct information about dx/dt, it is desirable to eliminate x from (9) before differentiating. This can be done using similar triangles. From Figure 4.1.7*b*, we see that

$$\frac{x}{y} = \frac{4}{16} \qquad \text{or} \qquad x = \frac{1}{4}y$$

Substituting this expression in (9) gives

$$V = \frac{\pi}{48} y^3 \tag{11}$$

Differentiating both sides of (11) with respect to t we obtain

$$\frac{dV}{dt} = \frac{\pi}{48}\left(3y^2 \frac{dy}{dt}\right)$$

or

$$\frac{dy}{dt} = \frac{16}{\pi y^2}\frac{dV}{dt} \tag{12}$$

Substituting (10) into (12), and letting $y = 8$ yields

$$\left.\frac{dy}{dt}\right|_{y=8} = \frac{16}{\pi(8)^2}(-2) = -\frac{1}{2\pi} \approx -0.16 \text{ in./min}$$

Thus, when $y = 8$ in., the depth of the liquid is decreasing at an approximate rate of 0.16 in./min. (The minus sign simply means that the depth y is decreasing as t is increasing.) ◀

▶ Exercise Set 4.1

1. Let A be the area of a square whose sides have length x, and assume that x varies with time.
   (a) How are dA/dt and dx/dt related?
   (b) At a certain instant the sides are 3 ft long and growing at a rate of 2 ft/min. How fast is the area growing at that instant?

2. Let A be the area of a circle whose radius is r, and assume that r varies with time.
   (a) How are dA/dt and dr/dt related?
   (b) At a certain instant the radius is 5 in. and is increasing at a rate of 2 in./sec. How fast is the area of the circle increasing at that instant?

3. Let V be the volume of a cylinder having height h and radius r, and assume that h and r vary with time.
   (a) How are dV/dt, dh/dt, and dr/dt related?
   (b) At a certain instant, the height is 6 in. and increasing at 1 in./sec, while the radius is 10 in. and decreasing at 1 in./sec. How fast is the volume changing at that instant? Is the volume increasing or decreasing at that instant?

4. Let l be the length of a diagonal of a rectangle whose sides have lengths x and y, and assume that x and y vary with time.
   (a) How are dl/dt, dx/dt, and dy/dt related?

(b) If x increases at a constant rate of $\frac{1}{2}$ ft/sec and y decreases at a constant rate of $\frac{1}{4}$ ft/sec, how fast is the size of the diagonal changing when $x = 3$ ft and $y = 4$ ft? Is the diagonal increasing or decreasing at that instant?

5. Let θ (in radians) be an acute angle in a right triangle, and let x and y, respectively, be the lengths of the sides adjacent and opposite θ. Suppose also that x and y vary with time.
   (a) How are $d\theta/dt$, dx/dt, and dy/dt related?
   (b) At a certain instant, $x = 2$ units and is increasing at 1 unit/sec, while $y = 2$ units and is decreasing at $\frac{1}{4}$ unit/sec. How fast is θ changing at that instant? Is θ increasing or decreasing at that instant?

6. A stone dropped into a still pond sends out a circular ripple whose radius increases at a constant rate of 3 ft/sec. How rapidly is the area enclosed by the ripple increasing at the end of 10 sec?

7. Oil spilled from a ruptured tanker spreads in a circle whose area increases at a constant rate of 6 mi²/hr. How fast is the radius of the spill increasing when the area is 9 mi²?

8. A spherical balloon is inflated so that its volume is increasing at the rate of 3 ft³/min. How fast is the diameter of the balloon increasing when the radius is 1 ft?

9. A spherical balloon is to be deflated so that its radius decreases at a constant rate of 15 cm/min. At what rate must air be removed when the radius is 9 cm?

10. A 17-ft ladder is leaning against a wall. If the bottom of the ladder is pulled along the ground away from the wall at a constant rate of 5 ft/sec, how fast will the top of the ladder be moving down the wall when it is 8 ft above the ground?

11. A 13-ft ladder is leaning against a wall. If the top of the ladder slips down the wall at a rate of 2 ft/sec, how fast will the foot be moving away from the wall when the top is 5 ft above the ground?

12. A rocket, rising vertically, is tracked by a radar station that is on the ground 5 mi from the launch pad. How fast is the rocket rising when it is 4 mi high and its distance from the radar station is increasing at a rate of 2000 mi/hr?

13. For the camera and rocket shown in Figure 4.1.2, at what rate is the elevation angle changing when the rocket is 3000 ft up and rising vertically at 500 ft/sec?

14. For the camera and rocket shown in Figure 4.1.2, at what rate is the rocket rising when the elevation angle is $\pi/4$ radians and increasing at a rate of .2 radian/sec?

15. A conical water tank with vertex down has a radius of 10 ft at the top and is 24 ft high. If water flows into the tank at a rate of 20 ft³/min, how fast is the depth of the water increasing when the water is 16 ft deep?

16. Grain pouring from a chute at the rate of 8 ft³/min forms a conical pile whose altitude is always twice its radius. How fast is the altitude of the pile increasing at the instant when the pile is 6 ft high?

17. Sand pouring from a chute forms a conical pile whose height and diameter are always the same. If the height increases at a constant rate of 5 ft/min, at what rate is sand pouring from the chute when the pile is 10 ft high?

18. Wheat is poured through a chute at the rate of 10 ft³/min, and falls in a conical pile whose bottom radius is always half the altitude. How fast will the circumference of the base be increasing when the pile is 8 ft high?

19. An aircraft is climbing at a 30° angle to the horizontal. How fast is the aircraft gaining altitude if its speed is 500 mi/hr?

20. A boat is pulled into a dock by means of a rope attached to a pulley on the dock. The rope is attached to the bow of the boat at a point 10 ft below the pulley. If the rope is pulled through the pulley at a rate of 20 ft/min, at what rate will the boat be approaching the dock when 125 ft of rope is out?

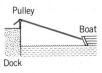

21. For the boat in Exercise 20, how fast must the rope be pulled if we want the boat to approach the dock at a rate of 12 ft/min?

22. A man 6 ft tall is walking at the rate of 3 ft/sec toward a street light 18 ft tall.
   (a) At what rate is the size of his shadow changing?
   (b) How fast is the tip of his shadow moving?

23. A beacon which makes one revolution every 10 sec is located on a ship 4 kilometers (km) from a straight shoreline. How fast is the beam moving along the shoreline when it makes an angle of 45° with the shore?

24. An aircraft is flying at a constant altitude with a constant speed of 600 mi/hr. An antiaircraft missile is fired on a straight line perpendicular to the flight path of the aircraft so it will hit the aircraft at a point P. At the instant the aircraft is 2 mi from the impact point P the missile is 4 mi from P and flying at 1200 mi/hr. At that instant, how rapidly is the distance between missile and aircraft decreasing?

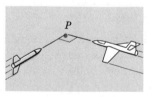

25. Solve Exercise 24 under the assumption that the angle between the flight paths is 120° instead of the assumption that the paths are perpendicular. [*Hint:* Use the law of cosines.]

26. A police helicopter is flying due north at 100 mi/hr, and at a constant altitude of $\frac{1}{2}$ mi. Below, a car is traveling west on a highway at 75 mi/hr. At the moment the helicopter crosses over the highway the car is 2 mi east of the helicopter.

   (a) How fast is the distance between the car and helicopter changing at the moment the helicopter crosses the highway?
   (b) Is the distance between car and helicopter increasing or decreasing at that moment?

27. A particle is moving along the curve whose equation is

$$\frac{xy^3}{1 + y^2} = \frac{8}{5}$$

Assume that the x-coordinate is increasing at the rate of 6 units/sec when the particle is at the point $(1, 2)$.
   (a) At what rate is the y-coordinate of the point changing at that instant?
   (b) Is the particle rising or falling at that instant?

28. Water is stored in a cone-shaped reservoir (vertex down). Assuming that the water evaporates at a rate proportional to the surface area exposed to the air, show the depth of the water will decrease at a constant rate that does not depend on the dimensions of the reservoir.

29. A meteorite enters the earth's atmosphere and burns up at a rate that, at each instant, is proportional to its surface area. Assuming that the meteorite is always spherical, show that the radius decreases at a constant rate.

30. On a certain clock the minute hand is 4 in. long and the hour hand is 3 in. long. How fast is the distance between the tips of the hands changing at 9 o'clock?

31. Coffee is poured at a uniform rate of 2 cm³/sec into a cup whose bowl is shaped like a truncated cone. If the upper and lower radii of the cup are 4 cm and 2 cm and the height of the cup is 6 cm, how fast will the coffee level be rising when the coffee is halfway up? [*Hint:* The volume of a truncated cone with height h and upper and lower radii of r_1 and r_2 is $V = \frac{1}{3}\pi(r_1{}^2 + r_1 r_2 + r_2{}^2)h$.]

4.2   INTERVALS OF INCREASE AND DECREASE; CONCAVITY

Although point plotting is useful for determining the general shape of a graph, it only provides an approximation. For no matter how many points are plotted, we can only guess at the shape of the graph *between* these points. For example, between two successive plotted points we are often forced to guess whether the graph exhibits a downward curvature (Figure 4.2.1*a*), an upward curvature (Figure 4.2.1*b*), or both (Figure 4.2.1*c*). In this section we will show how the derivative can be used to resolve such ambiguities.

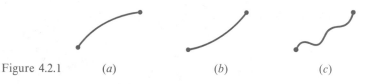

Figure 4.2.1   (*a*)   (*b*)   (*c*)

The terms *increasing* and *decreasing* are used to describe the behavior of a function as we travel *left to right* along its graph. For example, the function graphed in Figure 4.2.2 can be described as increasing on the interval $(-\infty, 1)$, decreasing on the interval $(1, 2)$, and increasing again on the interval $(2, +\infty)$.

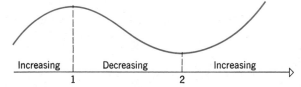

Figure 4.2.2

The following definition expresses these intuitive ideas precisely.

4.2.1   DEFINITION   Let f be defined on an interval, and let x_1 and x_2 be any points in the interval with $x_1 < x_2$. Then f is ***increasing*** on the interval if $f(x_1) < f(x_2)$ and is ***decreasing*** on the interval if $f(x_2) < f(x_1)$. (See Figure 4.2.3.)

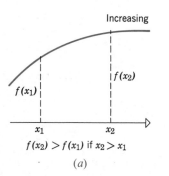

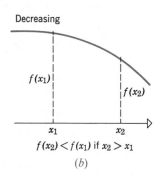

Figure 4.2.3   (*a*)   (*b*)

A function is said to be **strictly monotone** on a given interval if it is either increasing or decreasing on the interval.

Figure 4.2.4 suggests that if the graph of a function has tangent lines with positive slopes over an interval, then the function is increasing on that interval; and similarly, if the graph has tangent lines with negative slopes, the function is decreasing. This intuitive observation suggests the following fundamental theorem whose proof will be deferred until Section 4.11.

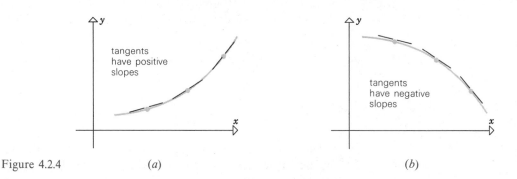

Figure 4.2.4   (a)   (b)

4.2.2   THEOREM   (a)   *If $f'(x) > 0$ on an open interval (a, b), then f is increasing on (a, b).*
(b)   *If $f'(x) < 0$ on an open interval (a, b), then f is decreasing on (a, b).*

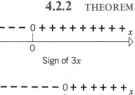

Sign of $3x$

Sign of $(x - 2)$

Sign of $3x (x - 2)$

Figure 4.2.5

▶ **Example 1**   On which intervals is the function

$$f(x) = x^3 - 3x^2 + 1$$

increasing? Decreasing?

Solution.   Differentiating f, we obtain

$$f'(x) = 3x^2 - 6x = 3x(x - 2)$$

Thus, f will be increasing when

$$3x(x - 2) > 0$$

and decreasing when

$$3x(x - 2) < 0$$

From the sign analysis in Figure 4.2.5 we see that f is

increasing if $x < 0$
decreasing if $0 < x < 2$
increasing if $x > 2$

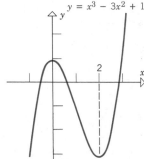

$y = x^3 - 3x^2 + 1$

Figure 4.2.6   The graph of f is shown in Figure 4.2.6. ·   ◄

CONCAVITY

Although the derivative of a function f can tell us where the graph of f is increasing or decreasing, it does not reveal where the graph has a downward curvature (Figure 4.2.7a) or an upward curvature (Figure 4.2.7b). To investigate this question, we must study the behavior of the tangent lines shown in the figure.

(a)   (b)

Figure 4.2.7

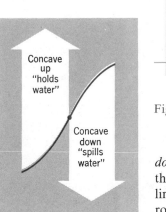

Figure 4.2.8

The curve in part (a) lies below its tangent lines and is called *concave down*. As we travel left to right along this curve, the tangent lines rotate so their slopes *decrease*. In contrast, the curve in part (b) lies above its tangent lines and is called *concave up*. As we travel left to right, its tangent lines rotate so their slopes *increase*. Since f' is the slope of a tangent line to the graph of f, we are led to the following definition.

4.2.3   DEFINITION   A function f is called ***concave up*** on a given interval if f' is increasing on the interval, and is called ***concave down*** if f' is decreasing on the interval.

REMARK.   Informally, a curve that is concave up "holds water" and one that is concave down "spills water" (Figure 4.2.8).

Since f'' is the derivative of f', it follows from Theorem 4.2.2 that f' is increasing on an open interval (a, b) if $f''(x) > 0$ for all x in (a, b), and f' is decreasing on (a, b) if $f''(x) < 0$ for all x in (a, b). Thus, we have the following result.

4.2.4   THEOREM   (a)   If $f''(x) > 0$ on an open interval (a, b), then f is concave up on (a, b).
(b)   If $f''(x) < 0$ on an open interval (a, b), then f is concave down on (a, b).

▶ Example 2   Where is the function

$$f(x) = x^3 - 3x^2 + 1$$

concave up? Concave down?

Solution. Calculating the second derivative, we obtain

$$f'(x) = 3x^2 - 6x$$
$$f''(x) = 6x - 6$$

Thus, the curve is concave up where

$$6x - 6 > 0 \tag{1}$$

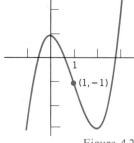

y $y = x^3 - 3x^2 + 1$

x

1

$(1, -1)$

Figure 4.2.9

and concave down where

$$6x - 6 < 0 \tag{2}$$

Since (1) holds if $x > 1$ and (2) holds if $x < 1$, we conclude:

f is concave up if $x > 1$
f is concave down if $x < 1$

The reader can check that these results agree with the graph of f shown in Figure 4.2.9. ◀

The point $x = 1$ in Example 2 is of particular interest because it is the point where the graph changes the direction of its concavity. There is a name for such points.

4.2.5 DEFINITION   If f is continuous on an open interval containing x_0, and if f changes the direction of its concavity at x_0, then the point $(x_0, f(x_0))$ on the graph of f is called an ***inflection point*** of f, and we say that f has an ***inflection point at x_0***. (See Figure 4.2.10.)

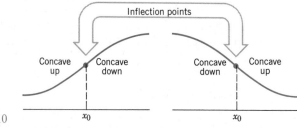

Inflection points

Concave up   Concave down   Concave down   Concave up

x_0 x_0

Figure 4.2.10

▶   **Example 3**   In Example 2 we showed that $f(x) = x^3 - 3x^2 + 1$ is concave down when $x < 1$ and concave up when $x > 1$. Thus, f has an inflection point at $x = 1$. Since $f(1) = -1$, the inflection point is $(1, -1)$. (See Figure 4.2.9.) ◀

▶ Exercise Set 4.2

In Exercises 1–15, determine open intervals on which f is (a) increasing, (b) decreasing, (c) concave up, (d) concave down, and (e) find the location of all inflection points.

1. $f(x) = x^2 - 5x + 6$.

2. $f(x) = 4 - 3x - x^2$.

3. $f(x) = (x + 2)^3$.

4. $f(x) = 5 + 12x - x^3$.

5. $f(x) = 3x^3 - 4x + 3$.

6. $f(x) = x^4 - 8x^2 + 16$.

7. $f(x) = 3x^4 - 4x^3$.

8. $f(x) = \dfrac{x}{x^2 + 2}$.

9. $f(x) = \cos x$, $0 < x < 2\pi$.

10. $f(x) = \sin^2 2x$, $0 < x < \pi$.

11. $f(x) = \tan x$, $-\pi/2 < x < \pi/2$.

12. $f(x) = x^{2/3}$.

13. $f(x) = \sqrt[3]{x + 2}$.

14. $f(x) = x^{4/3} - x^{1/3}$.

15. $f(x) = x^{1/3}(x + 4)$.

16. In each part sketch a continuous curve $y = f(x)$ with the stated properties.
   (a) $f(2) = 4$, $f'(2) = 0$, $f''(x) < 0$ for all x
   (b) $f(2) = 4$, $f'(2) = 0$, $f''(x) > 0$ for $x < 2$, $f''(x) < 0$ for $x > 2$
   (c) $f(2) = 4$, $f''(x) > 0$ for $x > 2$, $f''(x) > 0$ for $x < 2$ and

$$\lim_{x \to 2^+} f'(x) = -\infty, \ \lim_{x \to 2^-} f'(x) = +\infty$$

17. In each part sketch a continuous curve $y = f(x)$ with the stated properties.
   (a) $f(2) = 4$, $f'(2) = 0$, $f''(x) > 0$ for all x
   (b) $f(2) = 4$, $f'(2) = 0$, $f''(x) < 0$ for $x < 2$, $f''(x) > 0$ for $x > 2$
   (c) $f(2) = 4$, $f''(x) < 0$ for $x > 2$, $f''(x) < 0$ for $x < 2$ and

$$\lim_{x \to 2^+} f'(x) = +\infty, \ \lim_{x \to 2^-} f'(x) = -\infty$$

18. In each part sketch a continuous curve $y = f(x)$ with the stated properties.
   (a) $f(2) = 4$, $f'(2) = 1$, $f''(x) < 0$ if $x < 2$, $f''(x) > 0$ if $x > 2$
   (b) $f(2) = 4$, $f''(x) > 0$ if $x < 2$, $f''(x) < 0$ if $x > 2$ and

$$\lim_{x \to 2^-} f'(x) = +\infty, \ \lim_{x \to 2^+} f'(x) = +\infty$$

   (c) $f(2) = 4$, $f''(x) < 0$ if $x < 2$, $f''(x) < 0$ if $x > 2$ and

$$\lim_{x \to 2^-} f'(x) = 1, \ \lim_{x \to 2^+} f'(x) = -1$$

19. In each part sketch a continuous curve $y = f(x)$ with the stated properties.
   (a) $f(2) = 4$, $f''(x) < 0$ if $x < 2$, $f''(x) < 0$ if $x > 2$ and

$$\lim_{x \to 2^-} f'(x) = 0, \ \lim_{x \to 2^+} f'(x) = +\infty$$

   (b) $f(0) = 0$, $f(2) = 4$, $f''(x) = 0$ for all x.

20. Find the inflection points, if any, of the function $f(x) = (x - a)^3$.

21. Find the inflection points, if any, of the function $f(x) = (x - a)^4$.

22. Find the inflection points, if any, of the function $f(x) = ax^2 + bx + c$ ($a \neq 0$).

23. Prove that a third-degree polynomial $f(x) = ax^3 + bx^2 + cx + d$ ($a \neq 0$) has exactly one inflection point.

24. Prove that an nth-degree polynomial $f(x) = a_0 x^n + a_1 x^{n-1} + \cdots + a_n$ ($a_0 \neq 0$) has at most $n - 2$ inflection points.

4.3   RELATIVE EXTREMA; FIRST AND SECOND DERIVATIVE TESTS

In this section we will develop techniques for finding the highest and lowest points on the graph of a function or, equivalently, the largest and smallest values of the function. The methods that we develop here will have important applications in later sections.

The graphs of many functions form hills and valleys. The tops of the hills are called *relative maxima* and the bottoms of the valleys are called *relative minima* (Figure 4.3.1). Just as the top of a hill on the earth's terrain need not be the highest point on earth, so a relative maximum need not be the highest point on the entire graph. However, relative maxima, like tops of hills, are the high points in their *immediate vicinity,* and relative minima, like valley bottoms, are the low points in their *immediate vicinity.* These ideas are captured in the following definitions.

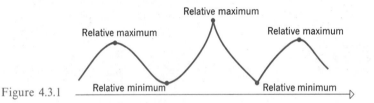

Figure 4.3.1

4.3.1   DEFINITION   A function f is said to have a ***relative maximum*** at x_0 if $f(x_0) \geq f(x)$ for all x in some open interval containing x_0.

4.3.2   DEFINITION   A function f is said to have a ***relative minimum*** at x_0 if $f(x_0) \leq f(x)$ for all x in some open interval containing x_0.

4.3.3   DEFINITION   A function f is said to have a ***relative extremum*** at x_0 if it has either a relative maximum or a relative minimum at x_0.

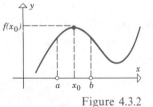

Figure 4.3.2

▶ Example 1   The function f graphed in Figure 4.3.2 has a relative maximum at x_0 because there is an open interval containing x_0 on which $f(x_0) \geq f(x)$ holds. (See interval (a, b) in the figure.)   ◀

Relative extrema can be viewed as the transition points that separate the regions where a graph is rising from those where it is falling. As suggested by Figure 4.3.3, the relative extrema of a function f occur either at points where the graph of f has a horizontal tangent or at points where f is not differentiable. This is the content of the following theorem (whose proof is given at the end of this section.)

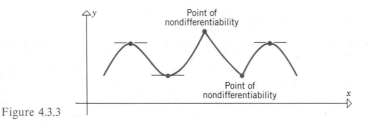

Figure 4.3.3

4.3.4 THEOREM *If f has a relative extremum at x_0, then either $f'(x_0) = 0$ or f is not differentiable at x_0.*

Theorem 4.3.4 is of such importance that there is some terminology associated with it.

4.3.5 DEFINITION A ***critical point*** for a function f is any value of x in the domain of f at which $f'(x) = 0$ or f is not differentiable; the critical points where $f'(x) = 0$ are called ***stationary points*** of f.

▶ Example 2 For each of the functions graphed in Figure 4.3.4, x_0 is a critical point. In parts (*a*), (*b*), (*c*), and (*d*), x_0 is a stationary point because there is a horizontal tangent at x_0, and in the remaining parts x_0 is a critical point, but not a stationary point because the derivative does not exist there. ◀

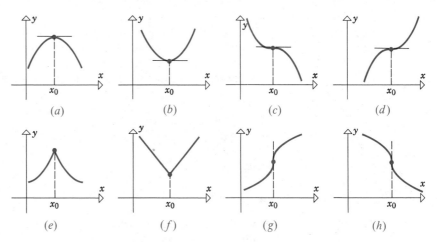

Figure 4.3.4   (*e*)   (*f*)   (*g*)   (*h*)

With the terminology of Definition 4.3.5, Theorem 4.3.4 states that *the relative extrema of a function occur at its critical points.* We emphasize, however, that there need not be a relative extremum at every critical point.

Indeed, in parts (c), (d), (g), and (h) of Figure 4.3.4, x_0 is a critical point at which there is no relative extremum. We will conclude this section with two tests that can be used to determine which critical points correspond to relative extrema and which do not.

FIRST AND SECOND
DERIVATIVE TESTS

The nature of a critical point x_0 can often be determined by studying the rise and fall of the curve near x_0. For example, if the curve is increasing to the left of x_0 and decreasing to the right, then there is a relative maximum at x_0 (Figures 4.3.4a, e); if the curve is decreasing to left of x_0 and increasing to the right, then there is a relative minimum at x_0 (Figures 4.3.4b, f); and if the curve is decreasing on both sides of x_0 or increasing on both sides of x_0, then there is no relative extremum at x_0 (Figures 4.3.4c, d, g, h).

These ideas are stated precisely in the following theorem.

4.3.6   THEOREM
First Derivative Test

Suppose f is continuous at a critical point x_0.
(a)   *If $f'(x) > 0$ on an open interval extending left from x_0 and $f'(x) < 0$ on an open interval extending right from x_0, then f has a relative maximum at x_0.*
(b)   *If $f'(x) < 0$ on an open interval extending left from x_0 and $f'(x) > 0$ on an open interval extending right from x_0, then f has a relative minimum at x_0.*
(c)   *If $f'(x)$ has the same sign [either $f'(x) > 0$ or $f'(x) < 0$] on an open interval extending left from x_0 and on an open interval extending right from x_0, then f does not have a relative extremum at x_0.*

REMARK.   To paraphrase this theorem, the relative extrema of a continuous function f occur at those critical points where f' changes sign.

A formal proof of the theorem is given in Section 4.11.

▶ Example 3   Locate the relative extrema of $f(x) = 3x^{5/3} - 15x^{2/3}$.

Solution.

$$f'(x) = 5x^{2/3} - 10x^{-1/3} = 5x^{-1/3}(x - 2)$$

Since $f'(x)$ does not exist when $x = 0$, and $f'(x) = 0$ when $x = 2$, the critical points of f are 0 and 2.

As shown in Figure 4.3.5a, the sign of f' changes from positive to negative at 0 and from negative to positive at 2. Thus, there is a relative maximum at 0 and a relative minimum at 2. Although not needed for the solution, the graph of f is shown in Figure 4.3.5b.   ◄

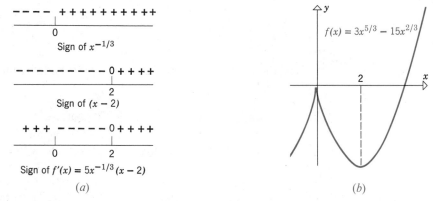

Figure 4.3.5   (a)   (b)

▶ **Example 4**   Locate the relative extrema of $f(x) = x^3 - 3x^2 + 3x - 1$.

Solution.   Differentiating yields

$$f'(x) = 3x^2 - 6x + 3 = 3(x - 1)^2$$

Solving $f'(x) = 0$ yields $x = 1$ as the only critical point. Since $3(x - 1)^2 \geq 0$ for all x, $f'(x)$ does not change sign at $x = 1$, so f does not have a relative extremum at $x = 1$. Thus, f has no relative extrema.   ◀

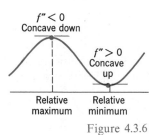

Figure 4.3.6

There is another test for relative extrema that is often easier to apply than the first derivative test. It is based on the geometric observation that a function has a relative maximum where its graph has a horizontal tangent line and is concave down (Figure 4.3.6), and a function has a relative minimum where its graph has a horizontal tangent line and is concave up. (See Figure 4.3.6.)

4.3.7   THEOREM
Second Derivative Test

Suppose f is twice differentiable at a stationary point x_0.
(a)   If $f''(x_0) > 0$, then f has a relative minimum at x_0.
(b)   If $f''(x_0) < 0$, then f has a relative maximum at x_0.

A formal proof of this result is given in Section 4.11.

▶ **Example 5**   Locate and describe the relative extrema of $f(x) = x^4 - 2x^2$.

Solution.

$$f'(x) = 4x^3 - 4x = 4x(x - 1)(x + 1)$$
$$f''(x) = 12x^2 - 4$$

Solving $f'(x) = 0$ yields the stationary points $x = 0$, $x = 1$, and $x = -1$. Since

$$f''(0) = -4 < 0$$
$$f''(1) = 8 > 0$$
$$f''(-1) = 8 > 0$$

there is a relative maximum at $x = 0$ and there are relative minima at $x = 1$ and $x = -1$. ◀

REMARK. If f is not twice differentiable at the critical point x_0, or if $f''(x_0) = 0$, then we must either rely on the first derivative test or devise more imaginative techniques appropriate to the problem.

REMARK. In some problems the first derivative test is easier to apply, and in others the second derivative test is easier. Thus, in each problem the reader should think carefully before selecting the test.

OPTIONAL

Proof of Theorem 4.3.4. We will prove this theorem in the case where there is a relative maximum at x_0. The proof in the case of a relative minimum is similar and is left to the reader.

There are two possibilities—either f is differentiable at x_0 or it is not. If it is not, then x_0 is a critical point for f and we are done. If f is differentiable at x_0, then we must show that $f'(x_0) = 0$. We will do this by showing that $f'(x_0) \geq 0$ and $f'(x_0) \leq 0$, from which it follows that $f'(x_0) = 0$. From the definition of a derivative we have

$$f'(x_0) = \lim_{h \to 0} \frac{f(x_0 + h) - f(x_0)}{h}$$

so that

$$f'(x_0) = \lim_{h \to 0^+} \frac{f(x_0 + h) - f(x_0)}{h} \tag{1}$$

and

$$f'(x_0) = \lim_{h \to 0^-} \frac{f(x_0 + h) - f(x_0)}{h} \tag{2}$$

Because f has a relative maximum at x_0, there is an open interval (a, b) containing x_0 on which $f(x) \leq f(x_0)$ for all x in (a, b).

Assume that h is sufficiently small so that $x_0 + h$ lies in the interval (a, b). Thus,

$$f(x_0 + h) \leq f(x_0)$$

or equivalently

$$f(x_0 + h) - f(x_0) \leq 0$$

Thus, if h is negative

$$\frac{f(x_0 + h) - f(x_0)}{h} \geq 0 \tag{3}$$

and if h is positive

$$\frac{f(x_0 + h) - f(x_0)}{h} \leq 0 \tag{4}$$

But an expression that never assumes negative values cannot approach a negative limit and an expression that never assumes positive values cannot approach a positive limit, so that from (1), (2), (3), and (4) it follows that

$$f'(x_0) = \lim_{h \to 0^-} \frac{f(x_0 + h) - f(x_0)}{h} \geq 0$$

and

$$f'(x_0) = \lim_{h \to 0^+} \frac{f(x_0 + h) - f(x_0)}{h} \leq 0$$

Since $f'(x_0) \geq 0$ and $f'(x_0) \leq 0$, it must be that $f'(x_0) = 0$. ◼

▶ Exercise Set 4.3

In Exercises 1–16, locate the critical points and classify them as stationary points or points of nondifferentiability.

1. $f(x) = x^2 - 5x + 6$.

2. $f(x) = 4x^2 + 2x - 5$.

3. $f(x) = x^3 + 3x^2 - 9x + 1$.

4. $f(x) = 2x^3 - 6x + 7$.

5. $f(x) = x^4 - 6x^2 - 3$.

6. $f(x) = 3x^4 - 4x^3$.

7. $f(x) = \dfrac{x}{x^2 + 2}$.

8. $f(x) = \dfrac{x^2 - 3}{x^2 + 1}$.

9. $f(x) = x^{2/3}$.

10. $f(x) = \sqrt[3]{x + 2}$.

11. $f(x) = \cos 3x$.

12. $f(x) = x \tan x$, $-\pi/2 < x < \pi/2$.

13. $f(x) = \sin^2 2x$, $0 < x < 2\pi$.

14. $f(x) = |\sin x|$.

15. $f(x) = x^{1/3}(x + 4)$.

16. $f(x) = x^{4/3} - 6x^{1/3}$.

In Exercises 17–20, find the relative extrema using (a) the first derivative test; (b) the second derivative test.

17. $f(x) = 1 - 4x - x^2$.

18. $f(x) = 2x^3 - 9x^2 + 12x$.

19. $f(x) = \sin^2 x$, $0 < x < 2\pi$.

20. $f(x) = \frac{1}{2}x - \sin x$, $0 < x < 2\pi$.

In Exercises 21–38, use any method to find the relative extrema.

21. $f(x) = x^3 + 5x - 2$. **22.** $f(x) = x^4 - 2x^2 + 7$.

23. $f(x) = x(x - 1)^2$. **24.** $f(x) = x^4 + 2x^3$.

25. $f(x) = 2x^2 - x^4$. **26.** $f(x) = (2x - 1)^5$.

27. $f(x) = x^{4/5}$. **28.** $f(x) = 2x + x^{2/3}$.

29. $f(x) = \dfrac{x^2}{x^2 + 1}$. **30.** $f(x) = \dfrac{x}{x + 2}$.

31. $f(x) = |x^2 - 4|$.

32. $f(x) = \begin{cases} 9 - x, & x \le 3 \\ x^2 - 3, & x > 3 \end{cases}$

33. $f(x) = \cos^2 x$.

34. $f(x) = \sqrt{3}x + 2\sin x$, $0 < x < 2\pi$.

35. $f(x) = \tan(x^2 + 1)$.

36. $f(x) = \dfrac{\sin x}{2 + \cos x}$, $0 < x < 2\pi$.

37. $f(x) = |\sin 2x|$, $0 < x < 2\pi$.

38. $f(x) = \cos 4x + 2\sin 2x$, $0 < x < \pi$.

39. Recall that the second derivative test (Theorem 4.3.7) does not apply if $f''(x_0) = 0$. Give examples to show that a function f can have a relative maximum at x_0, a relative minimum at x_0, or neither if $f''(x_0) = 0$. [*Hint:* Try functions of the form $f(x) = x^n$.]

40. Let h and g have relative maxima at x_0. Prove or disprove:
(a) $h + g$ has a relative maximum at x_0
(b) $h - g$ has a relative maximum at x_0.

41. Sketch some curves that show that the three parts of the first derivative test (Theorem 4.3.6) can be false without the assumption that f is continuous at x_0.

4.4 SKETCHING GRAPHS OF POLYNOMIALS AND RATIONAL FUNCTIONS

In this section we will show how to use the tools developed in the previous section to obtain graphs of polynomials and rational functions.

Polynomials are among the simplest functions to graph; the information obtained in the following steps is usually sufficient to obtain an accurate graph.

> **How to Graph a Polynomial P(x)**
> **Step 1.** Calculate $P'(x)$ and $P''(x)$.
> **Step 2.** From $P'(x)$ determine the stationary points and the intervals where P is increasing and decreasing.
> **Step 3.** From $P''(x)$ determine the inflection points and the intervals where P is concave up and concave down.
> **Step 4.** Plot a few well-chosen points to provide any additional information needed to complete the shape of the graph. (For example, the points where the graph crosses the axes.)

▶ Example 1   Sketch the graph of

$$y = x^3 - 3x + 2$$

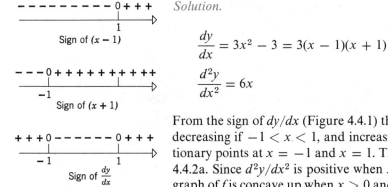

Figure 4.4.1

Solution.

$$\frac{dy}{dx} = 3x^2 - 3 = 3(x - 1)(x + 1)$$

$$\frac{d^2y}{dx^2} = 6x$$

From the sign of dy/dx (Figure 4.4.1) the graph of f is increasing if $x < -1$, decreasing if $-1 < x < 1$, and increasing again when $x > 1$; there are stationary points at $x = -1$ and $x = 1$. This information is recorded in Figure 4.2.2a. Since d^2y/dx^2 is positive when $x > 0$ and negative when $x < 0$, the graph of f is concave up when $x > 0$ and concave down when $x < 0$; there is an inflection point at $x = 0$, since f changes concavity there. The concavity information is also recorded in Figure 4.4.2a. From the information obtained thus far we can deduce the basic shape of the graph (Figure 4.4.2b).

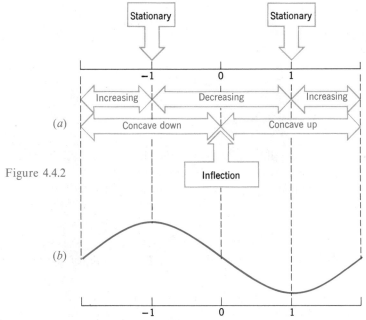

Figure 4.4.2

All that remains is to refine this sketch by plotting some key points. In this case, the stationary points, the inflection point, and two additional points suffice to give an accurate sketch of the graph. From Table 4.4.1 and Figure 4.4.2 we obtain the final graph shown in Figure 4.4.3. ◀

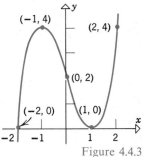

Figure 4.4.3

Table 4.4.1

x	-2	-1	0	1	2
$y = x^3 - 3x + 2$	0	4	2	0	4

GRAPHS OF
RATIONAL
FUNCTIONS

Recall that if $P(x)$ and $Q(x)$ are polynomials, then their ratio

$$f(x) = \frac{P(x)}{Q(x)}$$

is called a rational function of x. Graphing rational functions is complicated by the fact that discontinuities occur at points where $Q(x) = 0$. In Figure 4.4.4 we have sketched the graph of

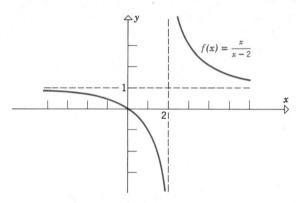

Figure 4.4.4

$$f(x) = \frac{x}{x - 2}$$

This figure illustrates most of the characteristics that are typical of rational functions. At $x = 2$ the denominator of $f(x)$ is zero, so that a discontinuity occurs in the graph at this point. Approaching the point $x = 2$, we have

$$\lim_{x \to 2^+} \frac{x}{x - 2} = +\infty \quad \text{and} \quad \lim_{x \to 2^-} \frac{x}{x - 2} = -\infty$$

The line $x = 2$ is called a *vertical asymptote* for the graph. Also

$$\lim_{x \to +\infty} \frac{x}{x - 2} = 1 \quad \text{and} \quad \lim_{x \to -\infty} \frac{x}{x - 2} = 1$$

so that the graph tends toward the line $y = 1$ as $x \to +\infty$ and as $x \to -\infty$. The line $y = 1$ is called a *horizontal asymptote* for the graph. For reference, let us formalize these ideas.

4.4.1   DEFINITION   A line $x = x_0$ is called a **vertical asymptote** for the graph of a function f if $f(x) \to +\infty$ or $f(x) \to -\infty$ as x approaches x_0 from the right or from the left. A line $y = L$ is called a **horizontal asymptote** for the graph of a function f if $f(x) \to L$ as $x \to +\infty$ or $x \to -\infty$.

In our discussion of graphing rational functions, we will assume for simplicity that the numerator and denominator have no common factors. For such functions, the information obtained in the following steps is usually sufficient to obtain an accurate sketch of the graph.

How to Graph a Rational Function $f(x) = \dfrac{P(x)}{Q(x)}$

Step 1. Find the values of x where $P(x) = 0$. At these values we have $f(x) = 0$, so that the graph intersects the x-axis at these points.

Step 2. Find the values of x where $Q(x) = 0$. At these values, $f(x)$ approaches $+\infty$ or $-\infty$, and the graph has a vertical asymptote.

Step 3. Compute $\lim\limits_{x \to +\infty} f(x)$ and $\lim\limits_{x \to -\infty} f(x)$. If either limit has a finite value L, then the line $y = L$ is a horizontal asymptote.

Step 4. The only places where $f(x)$ can change sign are at the points where the graph intersects the x-axis or has a vertical asymptote. Calculate a sample value of $f(x)$ in each of the open intervals determined by these points to see whether the graph is above or below the x-axis over the interval.

Step 5. From $f'(x)$ and $f''(x)$ determine the stationary points, inflection points, intervals of increase, decrease, upward concavity, and downward concavity.

Step 6. If needed, plot a few well-chosen points and determine whether the graph crosses any of the horizontal asymptotes.

▶ Example 2 Sketch the graph of

$$f(x) = \frac{x^2 - 1}{x^3}$$

Solution. Setting $P(x) = x^2 - 1$ equal to zero yields $x = 1$ and $x = -1$ as the points where the curve intersects the x-axis.

Setting $Q(x) = x^3$ equal to zero yields the line $x = 0$ as the only vertical asymptote.

From the limits

$$\lim_{x \to +\infty} \frac{x^2 - 1}{x^3} = \lim_{x \to +\infty} \left(\frac{1}{x} - \frac{1}{x^3} \right) = 0 \qquad (1)$$

and

$$\lim_{x \to -\infty} \frac{x^2 - 1}{x^3} = \lim_{x \to -\infty} \left(\frac{1}{x} - \frac{1}{x^3}\right) = 0 \tag{2}$$

it follows that the line $y = 0$ is the only horizontal asymptote.

If we mark on the x-axis, the location of the vertical asymptote and the places where the curve intersects the axis, then these points separate the x-axis into the open intervals

$$(-\infty, -1) \qquad (-1, 0) \qquad (0, 1) \qquad (1, +\infty)$$

Over each interval the graph remains above or below the x-axis. For each interval, we can determine which is the case by finding the sign of $f(x)$ at an arbitrary sample point in the interval.

At this stage we can make a rough sketch of the graph (Figure 4.4.5). Curve segments A and B in the figure were constructed by using Table 4.4.2 together with limits (1) and (2). Since the curve intersects the x-axis at -1 and 1, curve segments C and D follow from Table 4.4.2. Finally, since the y-axis is a vertical asymptote, curve segments E and F follow from Table 4.4.2.

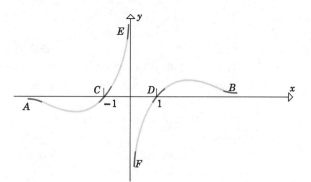

Figure 4.4.5

Table 4.4.2

INTERVAL	SAMPLE POINT	$f(x)$	CONCLUSION
$(-\infty, -1)$	$x = -2$	$f(-2) = -\dfrac{3}{8} < 0$	The graph is below the x-axis.
$(-1, 0)$	$x = -\dfrac{1}{2}$	$f\left(-\dfrac{1}{2}\right) = 6 > 0$	The graph is above the x-axis.
$(0, 1)$	$x = \dfrac{1}{2}$	$f\left(\dfrac{1}{2}\right) = -6 < 0$	The graph is below the x-axis.
$(1, +\infty)$	$x = 2$	$f(2) = \dfrac{3}{8} > 0$	The graph is above the x-axis.

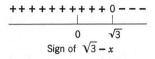

It is conceivable that our rough sketch may have missed some inflection points or stationary points, but we can investigate this possibility using the first and second derivatives:

$$f'(x) = \frac{x^3(2x) - (x^2 - 1)(3x^2)}{(x^3)^2} = \frac{3 - x^2}{x^4}$$

$$f''(x) = \frac{x^4(-2x) - (3 - x^2)(4x^3)}{(x^4)^2} = \frac{2(x^2 - 6)}{x^5}$$

Since the denominator of $f'(x)$ cannot be negative, the sign of $f'(x)$ is determined by the sign of the numerator. A sign analysis of $f'(x)$ is shown in Figure 4.4.6, and a sign analysis of $f''(x)$ is shown in Figure 4.4.7. From Figures 4.4.6 and 4.4.7 we deduce the results recorded in Figure 4.4.8. For our final sketch (Figure 4.4.9) we used our earlier rough sketch, and the data in Figure 4.4.8. We also plotted the stationary points and the inflection points. ◀

Figure 4.4.6

Figure 4.4.7

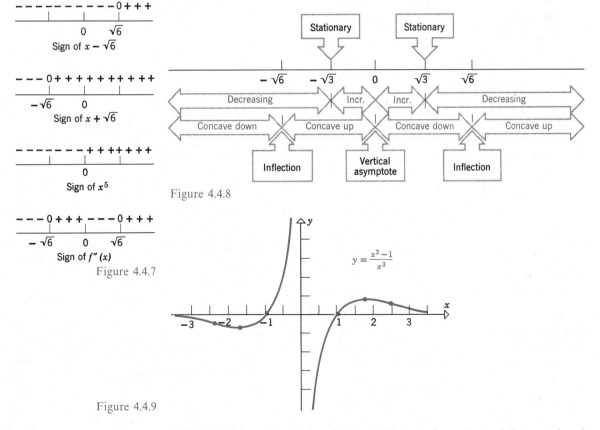

Figure 4.4.8

$$y = \frac{x^2 - 1}{x^3}$$

Figure 4.4.9

REMARK. Sometimes the work required to graph a polynomial or a rational function can be reduced by taking advantage of its symmetry properties. To illustrate, the curve

$$y = \frac{x^2 - 1}{x^3}$$

in Example 2 is symmetric about the origin, since replacing y by $-y$ and x by $-x$ yields

$$-y = \frac{(-x)^2 - 1}{(-x)^3}$$

or

$$-y = -\frac{x^2 - 1}{x^3}$$

which reduces to the original equation on multiplying by -1. Thus, we could have proceeded by sketching the portion of the graph to the right of the y-axis, and then constructing the graph in the second and third quadrants by symmetry (Figure 4.4.9).

▶ Exercise Set 4.4

In Exercises 1–14, use the techniques illustrated in this section to sketch the graph of the given polynomial. Plot the stationary points and the inflection points.

1. $x^2 - 2x - 3$.
2. $1 + x - x^2$.
3. $x^3 - 3x + 1$.
4. $2x^3 - 6x + 4$.
5. $x^3 + 3x^2 + 5$.
6. $x^2 - x^3$.
7. $2x^3 - 3x^2 + 12x + 9$.
8. $x^3 - 3x^2 + 3$.
9. $(x - 1)^4$.
10. $(x - 1)^5$.
11. $x^4 + 2x^3 - 1$.
12. $x^4 - 2x^2 - 12$.
13. $x^4 - 3x^3 + 3x^2 + 1$.
14. $x^5 - 4x^4 + 4x^3$.

In Exercises 15–23, use the techniques illustrated in this section to sketch the graph of the given rational function. Show any horizontal and vertical asymptotes, and plot the stationary points and inflection points.

15. $\dfrac{2x}{x - 3}$.
16. $\dfrac{x}{x^2 - 1}$.
17. $\dfrac{x^2}{x^2 - 1}$.
18. $\dfrac{1}{(x - 1)^2}$.
19. $\dfrac{x}{1 + x^2}$.
20. $1 - \dfrac{1}{x}$.
21. $\dfrac{x - 1}{x - 2}$.
22. $\dfrac{1}{x^2 + 1}$.
23. $x^2 - \dfrac{1}{x}$.

24. (*Oblique Asymptotes*) If a rational function $P(x)/Q(x)$ is such that the degree of the numerator exceeds the degree of the denominator by *one*, then the graph of $P(x)/Q(x)$ will have an *oblique asymptote*, that is, an asymptote that is neither vertical nor horizontal. To see why, we perform the division of $P(x)$ by $Q(x)$ to obtain

$$\frac{P(x)}{Q(x)} = (ax + b) + \frac{R(x)}{Q(x)}$$

where $ax + b$ is the quotient and $R(x)$ is the remainder. Use the fact that the degree of the remainder $R(x)$ is less than the degree of the divisor $Q(x)$ to help prove:

$$\lim_{x \to +\infty} \left| \frac{P(x)}{Q(x)} - (ax + b) \right| = 0$$

$$\lim_{x \to -\infty} \left| \frac{P(x)}{Q(x)} - (ax + b) \right| = 0$$

These results tell us that the graph of the equation $y = P(x)/Q(x)$ "tends" toward the line (oblique asymptote) $y = ax + b$ as $x \to +\infty$ or $x \to -\infty$ (Figure 4.4.10).

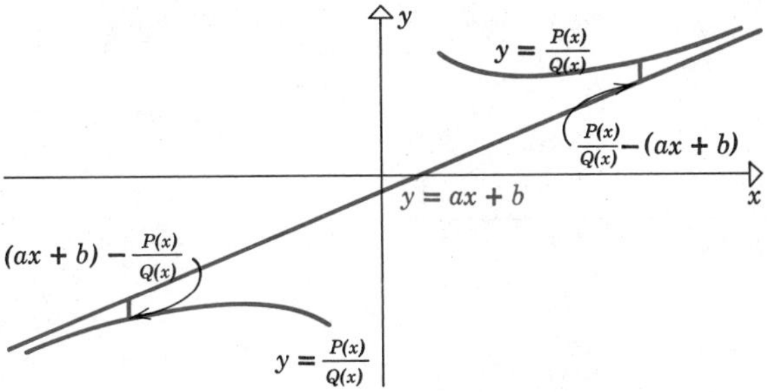

In Exercises 25–27, sketch the graph of the rational function. Show all vertical, horizontal, and oblique asymptotes (see Exercise 24).

25. $\dfrac{x^2 - 2}{x}$.

26. $\dfrac{x^2 - 2x - 3}{x + 2}$.

27. $\dfrac{(x - 2)^3}{x^2}$.

Figure 4.4.10

4.5 OTHER GRAPHING PROBLEMS

In this section we consider functions whose graphs have characteristics not found in the graphs of polynomials and rational functions.

4.5.1 DEFINITION The graph of a function f is said to have a ***vertical tangent line*** at x_0 if f is continuous at x_0 and $|f'(x)|$ approaches $+\infty$ as $x \to x_0$.

▶ **Example 1** Sketch the graph of $f(x) = \sqrt[3]{x}$.

Solution. The function $f(x) = \sqrt[3]{x} = x^{1/3}$ is continuous and

$$f'(x) = \frac{1}{3}x^{-2/3} = \frac{1}{3x^{2/3}}$$

$$f''(x) = \frac{1}{3}\left(-\frac{2}{3}\right)x^{-5/3} = -\frac{2}{9x^{5/3}}$$

The first derivative is positive for all x except $x = 0$, so that $f(x)$ is increasing on the intervals $(-\infty, 0)$ and $(0, +\infty)$. There are no stationary points since $f'(x) \neq 0$ for any x. However, $x = 0$ is a critical point since $f'(0)$ is undefined. At this point

$$\lim_{x \to 0} |f'(x)| = \lim_{x \to 0} \left| \frac{1}{3x^{2/3}} \right| = +\infty$$

so the graph of f has a vertical tangent line at $x = 0$.

The second derivative of f is negative if $x > 0$ and positive if $x < 0$, so that the graph of f is concave down on the interval $(0, +\infty)$ and concave up on the interval $(-\infty, 0)$ (Figure 4.5.1a).

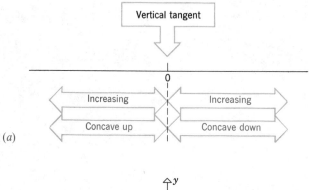

(a)

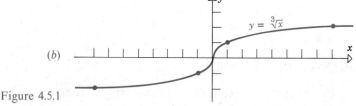

(b)

Figure 4.5.1

Since

$$\lim_{x \to +\infty} f(x) = \lim_{x \to +\infty} \sqrt[3]{x} = +\infty$$

$$\lim_{x \to -\infty} f(x) = \lim_{x \to -\infty} \sqrt[3]{x} = -\infty$$

the graph of f has no horizontal asymptotes. Thus, from Figure 4.5.1a and the points plotted from the following table, we obtain the sketch in Figure 4.5.1b.

x	-8	-1	0	1	8
$\sqrt[3]{x}$	-2	-1	0	1	2

◀

4.5.2 DEFINITION   The graph of a function f is said to have a **cusp** at x_0 if f is continuous at x_0 and $f'(x) \to +\infty$ as x approaches x_0 from one side, while $f'(x) \to -\infty$ as x approaches x_0 from the other side (Figure 4.5.2).

At a cusp there is a vertical tangent since $|f'(x)| \to +\infty$ as $x \to x_0$ from either side.

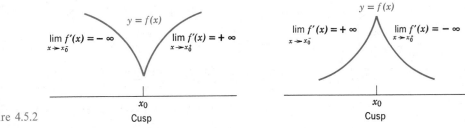

Figure 4.5.2

▶ Example 2   Sketch the graph of $f(x) = (x - 4)^{2/3}$.

Solution.   The function f is continuous and

$$f'(x) = \frac{2}{3}(x - 4)^{-1/3} = \frac{2}{3(x - 4)^{1/3}}$$

$$f''(x) = -\frac{2}{9}(x - 4)^{-4/3} = -\frac{2}{9(x - 4)^{4/3}}$$

Since $f'(x) > 0$ when $x > 4$ and $f'(x) < 0$ when $x < 4$, the graph of f is increasing when $x > 4$ and decreasing when $x < 4$. There are no stationary points since $f'(x) \neq 0$ for any value of x. However, $x = 4$ is a critical point since $f'(4)$ does not exist. At this point

$$\lim_{x \to 4^-} f'(x) = \lim_{x \to 4^-} \frac{2}{3(x - 4)^{1/3}} = -\infty$$

$$\lim_{x \to 4^+} f'(x) = \lim_{x \to 4^+} \frac{2}{3(x - 4)^{1/3}} = +\infty$$

so that the graph of f has a cusp at $x = 4$ (Figure 4.5.3).

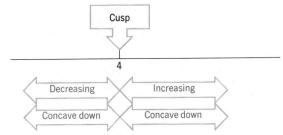

Figure 4.5.3

With the exception of the point $x = 4$, where $f''(x)$ is undefined, $f''(x)$ is negative. Thus, the graph of f is concave down if $x < 4$ and also concave down if $x > 4$ (Figure 4.5.3).

Since

$$\lim_{x \to -\infty} f(x) = \lim_{x \to -\infty} (x - 4)^{2/3} = +\infty$$

$$\lim_{x \to +\infty} f(x) = \lim_{x \to +\infty} (x - 4)^{2/3} = +\infty$$

the graph of f has no horizontal asymptotes. Thus, from Figure 4.5.3 and the points plotted from the following table we obtain the graph in Figure 4.5.4. ◄

x	0	3	4	5	8
$y = (x - 4)^{2/3}$	$\sqrt[3]{16} \approx 2.52$	1	0	1	$\sqrt[3]{16} \approx 2.52$

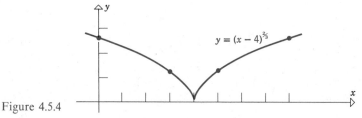

Figure 4.5.4

► Example 3   Sketch the graph of $f(x) = \sqrt{1 - x}$.

Solution.   Since $f(x)$ is undefined when $x > 1$, the graph of f extends only over the interval $-\infty < x \le 1$. Over this interval $f(x) = \sqrt{1 - x} = (1 - x)^{1/2}$ is continuous and

$$f'(x) = -\frac{1}{2}(1 - x)^{-1/2} = -\frac{1}{2\sqrt{1 - x}}$$

$$f''(x) = -\frac{1}{4}(1 - x)^{-3/2} = -\frac{1}{4(\sqrt{1 - x})^3}$$

For $x < 1$ both $f'(x)$ and $f''(x)$ are negative so that the graph of f is decreasing and concave down over the interval $(-\infty, 1)$. Also,

$$\lim_{x \to 1^-} f'(x) = \lim_{x \to 1^-} \left(-\frac{1}{2\sqrt{1 - x}} \right) = -\infty$$

so that the tangent lines tend toward a vertical position as x approaches 1 from the left.

From the above information and by plotting the points in the following table we are led to the graph in Figure 4.5.5. ◄

x	1	0	-3	-8
$y = \sqrt{1-x}$	0	1	2	3

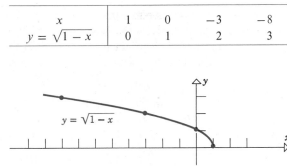

$y = \sqrt{1-x}$

Figure 4.5.5

▶ Exercise Set 4.5

In Exercises 1–18, use the techniques illustrated in this section to sketch the graph of the given function.

1. $(x-2)^{1/3}$.

2. $x^{1/4}$.

3. $x^{1/5}$.

4. $x^{2/5}$.

5. $x^{4/3}$.

6. $x^{-1/3}$.

7. $1 - x^{2/3}$.

8. $\sqrt{x+2}$.

9. $\sqrt{x^2-1}$.

10. $\sqrt[3]{x^2-4}$.

11. $2x + 3x^{2/3}$.

12. $4x - 3x^{4/3}$.

13. $x - \cos x$.

14. $x + \tan x$.

15. $\sin x + \cos x$.

16. $\sqrt{3}\cos x + \sin x$.

17. $\sin^2 x$, $0 \le x \le 2\pi$. **18.** $x\tan x$, $-\frac{\pi}{2} < x < \frac{\pi}{2}$.

19. (a) How can one obtain the graph of $y = |f(x)|$ from the graph of $y = f(x)$?
(b) Use the graph of $\sin x$ to obtain the graph of $|\sin x|$.

20. (a) How can one obtain the graph of $y = f(|x|)$ from the graph of $y = f(x)$?
(b) Use the graph of $\sin x$ to obtain the graph of $\sin |x|$.

4.6 MAXIMUM AND MINIMUM VALUES OF A FUNCTION

Problems concerned with finding the "best" way to perform a task are called *optimization* problems. A large class of optimization problems can be reduced to finding the largest or smallest value of a function and determining where this value occurs. In this section we will develop some mathematical tools for solving such problems, and in the next section we will use these tools to solve some applied problems.

Recall that a relative maximum (or minimum) of a function *f* is the largest (or smallest) value of *f* in the *immediate vicinity* of a point. However, we will often be interested in the largest or smallest value of a function *f* over its *entire domain*. This leads us to the following definitions.

4.6.1   DEFINITION   If $f(x_0) \geq f(x)$ for all x in the domain of f, then $f(x_0)$ is called the *maximum value* or *absolute maximum value* of f.

4.6.2   DEFINITION   If $f(x_0) \leq f(x)$ for all x in the domain of f, then $f(x_0)$ is called the *minimum value* or *absolute minimum value* of f.

4.6.3   DEFINITION   A number that is either the maximum or the minimum value of a function f is called an *extreme value* or *absolute extreme value* of f. Sometimes the terms *extremum* or *absolute extremum* are also used.

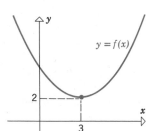

Figure 4.6.1

Often we will be concerned with the extreme values of f on some specified interval, rather than on the entire domain of f. The meaning of such terms as: *maximum value of f on [a, b]* or *minimum value of f on (a, b)* should be clear.

► **Example 1**   The function f graphed in Figure 4.6.1 has no maximum value. Its minimum value is 2, and the minimum occurs when $x = 3$.   ◄

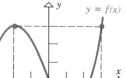

Figure 4.6.2

► **Example 2**   The function graphed in Figure 4.6.2 has neither a maximum nor a minimum value. However, on the interval $[-3, 3]$ it has both. The maximum value on $[-3, 3]$ is $f(x) = 3$, which occurs when $x = -2$ and $x = 3$. The minimum value on $[-3, 3]$ is $f(x) = -2$, which occurs when $x = 1$.   ◄

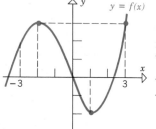

Figure 4.6.3

► **Example 3**   The function $f(x) = 2x + 1$ graphed in Figure 4.6.3 has a minimum, but no maximum value on $[0, 3)$. The minimum value is 1, and this occurs when $x = 0$. The reason why $f(x)$ has no maximum value is subtle, but important to understand. If we had been considering the interval $[0, 3]$ rather than $[0, 3)$, then $f(x)$ would have had a maximum value of 7 occurring at $x = 3$. However, the point $x = 3$ does not lie in the interval $[0, 3)$, so that 7 is *not* the maximum value on $[0, 3)$. But no number *less* than 7 can be the maximum either, for if M is any number less than 7, there are values of x in $[0, 3)$ where $f(x) > M$ (see Figure 4.6.4). Thus, $f(x) = 2x + 1$ has no maximum value on $[0, 3)$.   ◄

Given a function f, there are various questions we can ask about its maximum and minimum values; for example:

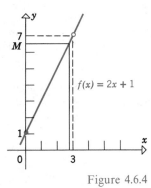

Figure 4.6.4

1. Does $f(x)$ have a maximum value?
2. If $f(x)$ has a maximum value, what is it?
3. If $f(x)$ has a maximum value, where does it occur?

These same questions could, of course, be asked about the minimum value of f. We will now obtain some results that will help us to answer such questions.

The following theorem gives conditions under which the existence of maximum and minimum values of a function is assured.

4.6.4 THEOREM
Extreme-Value Theorem

If a function f is continuous on a closed interval [a, b], *then f has both a maximum value and a minimum value on* [a, b].

The proof of this theorem is surprisingly difficult, and will be omitted. However, the result is intuitively obvious if we imagine a particle moving along the graph of a continuous function over a closed interval [a, b]; during the trip the particle will have to pass through a highest point and a lowest point (Figure 4.6.5).

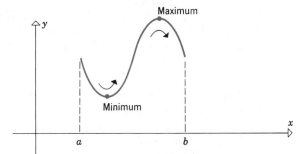

Figure 4.6.5

In the Extreme-Value Theorem the hypotheses that f is continuous and the interval is closed are essential. If either hypothesis is violated, the existence of maximum or minimum values cannot be guaranteed; this is shown in the next two examples.

▶ **Example 4** Since $f(x) = 2x + 1$ is a polynomial, it is continuous everywhere. In particular, it is continuous on the interval [0, 3). However, as shown in Example 3, the function f does not have a maximum value on [0, 3). Thus, the closed-interval hypothesis in Theorem 4.6.4 is essential. ◀

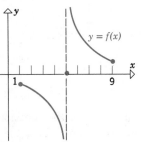

Figure 4.6.6

▶ **Example 5** The function f graphed in Figure 4.6.6 is defined everywhere on the closed interval [1, 9], yet it has neither a maximum nor minimum value on the interval. Since f has a point of discontinuity on the interval $[1, 9]$, this example shows that the continuity hypothesis in Theorem 4.6.4 is essential.   ◄

▶ **Example 6**   The polynomial $f(x) = 2x^3 - 15x^2 + 36x$ is continuous everywhere. In particular, it is continuous on the closed interval $[1, 5]$. Thus, the Extreme-Value Theorem tells us that f has both maximum and minimum values on $[1, 5]$. However, this theorem does not tell us what the maximum and minimum values are or where they occur.   ◄

The Extreme-Value Theorem is an example of what mathematicians call an *existence theorem;* it states conditions under which something exists, in this case maximum and minimum values for a function f. However, finding these maximum and minimum values is a separate problem.

The following theorem is the key tool for finding the extreme values of a function.

4.6.5   THEOREM   *If a function f has an extreme value (either a maximum or a minimum) on an open interval (a, b), then the extreme value occurs at a critical point for f.*

Proof.   If f has a maximum value on (a, b) at x_0, then $f(x_0)$ is the largest value of f on (a, b) and therefore the largest value of f in the immediate vicinity of x_0. Thus, f has a relative maximum at x_0 and by Theorem 4.3.4, x_0 is a critical point for f. The proof in the case of a minimum value is similar.   ▪

This theorem can be used to locate the extreme values of a continuous function f on a *closed* interval $[a, b]$. For example, consider the possible locations of the maximum. Either the maximum occurs at an endpoint (Figure 4.6.7a) or it occurs in the open interval (a, b), in which case it occurs at a critical point (Figures 4.6.7b and 4.6.7c.) This suggests the following procedure:

How to Find the Extreme Values of a Continuous Function f on a Closed Interval $[a, b]$.

Step 1.   Find the critical points of f in (a, b).
Step 2.   Evaluate f at the critical points and the endpoints a and b.
Step 3.   The largest of the values in Step 2 is the maximum and the smallest is the minimum.

▶ **Example 7**   Find the maximum and minimum values of the function $f(x) = 2x^3 - 15x^2 + 36x$ on the interval $[1, 5]$, and determine where they occur.

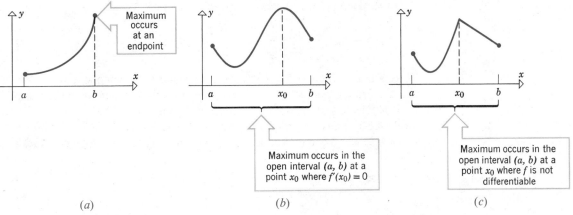

(a)

(b)

(c)

Figure 4.6.7

Solution.   Since polynomials are differentiable everywhere, f is differentiable everywhere on the interval $(1, 5)$. Thus, if an extreme value occurs on the open interval $(1, 5)$, it must occur at a point where the derivative is zero. Since

$$f'(x) = 6x^2 - 30x + 36$$

the equation $f'(x) = 0$ becomes

$$6x^2 - 30x + 36 = 0$$

or

$$x^2 - 5x + 6 = 0$$

or

$$(x - 3)(x - 2) = 0$$

Thus, there are two points in the interval $(1, 5)$ where $f'(x) = 0$, namely $x = 2$ and $x = 3$. Evaluating f at these points and the endpoints, we have

$$f(1) = 2(1)^3 - 15(1)^2 + 36(1) = 23$$
$$f(2) = 2(2)^3 - 15(2)^2 + 36(2) = 28$$
$$f(3) = 2(3)^3 - 15(3)^2 + 36(3) = 27$$
$$f(5) = 2(5)^3 - 15(5)^2 + 36(5) = 55$$

Thus, the minimum value is 23 and the maximum value is 55. The minimum occurs at $x = 1$ and the maximum occurs at $x = 5$.   ◄

► **Example 8**   Find the extreme values of $f(x) = 6x^{4/3} - 3x^{1/3}$ on the interval $[-1, 1]$ and determine where these values occur.

Solution.   Differentiating, we obtain

$$f'(x) = 8x^{1/3} - x^{-2/3} = x^{-2/3}(8x - 1) = \frac{8x - 1}{x^{2/3}}$$

Thus, $f'(x) = 0$ when $x = \frac{1}{8}$ and $f'(x)$ does not exist when $x = 0$. It follows that the critical points of f are $x = 0$, $x = \frac{1}{8}$, both of which lie in the interval $[-1, 1]$. Evaluating f at these critical points and the endpoints we obtain Table 4.6.1. Thus, the minimum value of f on $[-1, 1]$ is $-\frac{9}{8}$, which occurs when $x = \frac{1}{8}$ and the maximum value of f on $[-1, 1]$ is 9, which occurs when $x = -1$.   ◄

Table 4.6.1

x	-1	0	$\dfrac{1}{8}$	1
$f(x)$	9	0	$-\dfrac{9}{8}$	3

For a continuous function f on a closed interval $[a, b]$, the Extreme-Value Theorem assures that f has both a maximum and minimum value; and the procedure for finding these values is mechanical—we simply evaluate f at the critical points and the endpoints. For a continuous function on an open interval, half open interval, or infinite interval, the problem of finding maximum and minimum values is more complicated for two reasons:

1.   The function may not have a maximum or minimum value on such an interval.

2.   If the function does have maximum or minimum values, it often requires some ingenuity to find them.

It is often possible to determine whether a function has extrema by graphing it. Once it is determined that extrema exist, Theorem 4.6.5 can be applied to help locate them.

► Example 9   Determine whether the function $f(x) = 3x^4 + 4x^3$ has maximum or minimum values on $(-\infty, +\infty)$ and, if so, find them.

Solution.   Using the methods of Section 4.4, the reader should be able to obtain the graph of f shown in Figure 4.6.8. From this graph we see that the function has a minimum value but no maximum. By Theorem 4.6.5 this minimum must occur at a critical point. To locate the critical points, we set $f'(x)$ equal to zero to obtain

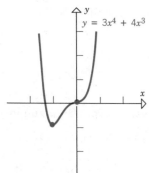

Figure 4.6.8

$$12x^3 + 12x^2 = 0$$

or

$$12x^2(x + 1) = 0$$

Thus, the critical points are $x = 0$ and $x = -1$. At $x = 0$ we have an inflection point, and at $x = -1$ we have the desired minimum. Substituting $x = -1$ in $f(x) = 3x^4 + 4x^3$ yields $f(-1) = -1$, which is the minimum value of f on $(-\infty, +\infty)$. ◀

The following theorem, when it applies, makes it possible to answer questions about the extrema of a function without constructing its graph.

4.6.6 THEOREM *Let f be continuous on an interval I and assume that f has exactly one relative extremum on I, say at x_0.*
(a) If f has a relative minimum at x_0, then $f(x_0)$ is the minimum value of f on the interval I.
(b) If f has a relative maximum at x_0, then $f(x_0)$ is the maximum value of f on the interval I.

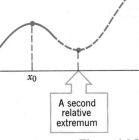

A second relative extremum

Figure 4.6.9

Although we will omit the proof, the result is easy to visualize. For example, if f has a relative maximum at x_0 and if $f(x_0)$ is *not* the maximum value of f on I, then the graph of f must make an upward turn somewhere on the interval I, thereby introducing a second relative extremum (Figure 4.6.9). Thus, if there is only one relative extremum on I, an absolute maximum value for f must also occur at x_0.

Theorem 4.6.6 reduces the problem of finding extrema to one of finding relative extrema.

▶ **Example 10** Find the maximum and minimum values, if any, of

$$f(x) = x + \frac{1}{x}$$

on $(0, +\infty)$.

Solution. The function f is continuous on $(0, +\infty)$ because the only discontinuity is at 0, which is outside this interval. We have

$$f'(x) = 1 - \frac{1}{x^2}$$

so that $f'(x) = 0$ yields the critical points $x = 1$ and $x = -1$. Thus, $x = 1$ is the only critical point in the interval $(0, +\infty)$. Since

$$f''(x) = \frac{2}{x^3}$$

it follows that

$$f''(1) = 2 > 0$$

Thus, a relative minimum occurs at $x = 1$ by the second derivative test. Since this is the only relative extremum on $(0, +\infty)$, it follows from Theorem 4.6.6 that the minimum value for f on $(0, +\infty)$ is

$$f(1) = 2$$

Since the maximum value of f on $(0, +\infty)$ would have to occur at a critical point, and since we have shown that a minimum occurs at the only critical point in $(0, +\infty)$, there is no maximum.   ◀

▶ Exercise Set 4.6

In Exercises 1–14, find the maximum and minimum values of f on the given closed interval and state where these values occur.

1. $f(x) = 4x^2 - 4x + 1$; $[0, 1]$.

2. $f(x) = 8x - x^2$; $[0, 6]$.

3. $f(x) = (x - 1)^3$; $[0, 4]$.

4. $f(x) = 2x^3 - 3x^2 - 12x$; $[-2, 3]$.

5. $f(x) = \dfrac{3x}{\sqrt{4x^2 + 1}}$; $[-1, 1]$.

6. $f(x) = \dfrac{x}{x^2 + 2}$; $[-1, 4]$.

7. $f(x) = x^{2/3}(20 - x)$; $[-1, 20]$.

8. $f(x) = (x^2 + x)^{2/3}$; $[-2, 3]$.

9. $f(x) = x - \tan x$; $[-\pi/4, \pi/4]$.

10. $f(x) = \sin x - \cos x$; $[0, \pi]$.

11. $f(x) = 2 \sec x - \tan x$; $[0, \pi/4]$.

12. $f(x) = \sin^2 x + \cos x$; $[-\pi, \pi]$.

13. $f(x) = 1 + |9 - x^2|$; $[-5, 1]$.

14. $f(x) = |6 - 4x|$; $[-3, 3]$.

In Exercises 15–26, find the maximum and minimum values of f on the given interval, if they exist.

15. $f(x) = x^2 - 3x - 1$; $(-\infty, +\infty)$.

16. $f(x) = 3 - 4x - 2x^2$; $(-\infty, +\infty)$.

17. $f(x) = 4x^3 - 3x^4$; $(-\infty, +\infty)$.

18. $f(x) = x^4 + 4x$; $(-\infty, +\infty)$.

19. $f(x) = (x^2 - 1)^2$; $(-\infty, +\infty)$.

20. $f(x) = (x - 1)^2(x + 2)^2$; $(-\infty, +\infty)$.

21. $f(x) = x^3 - 3x - 2$; $(-\infty, +\infty)$.

22. $f(x) = x^3 - 9x + 1$; $(-\infty, +\infty)$.

23. $f(x) = 1 + \dfrac{1}{x}$; $(0, +\infty)$.

24. $f(x) = \dfrac{x}{x^2 + 1}$; $[0, +\infty)$.

25. $f(x) = \dfrac{x^2}{x + 1}$; $(-5, -1)$.

26. $f(x) = \dfrac{x + 3}{x - 3}$; $[-5, 5]$.

27. Find the maximum and minimum values of

$$f(x) = 2 \sin 2x + \sin 4x$$

and state where the maximum and minimum values occur. [*Hint:* Since f is periodic, this problem can be solved by first finding the maximum and minimum values on an appropriate closed interval.]

28. Find the maximum and minimum values of

$$f(x) = 3 \cos \frac{x}{3} + 2 \cos \frac{x}{2}$$

[See the hint in the previous problem.]

29. Find the maximum and minimum values of

$$f(x) = \begin{cases} 4x - 2, & x < 1 \\ (x - 2)(x - 3), & x \geq 1 \end{cases}$$

on $[\frac{1}{2}, \frac{7}{2}]$.

30. Let $f(x) = x^2 + px + q$. Find values of p and q such that $f(1) = 3$ is an extreme value of f on $[0, 2]$. Is this value a maximum or minimum?

31. Let $f(x) = (x - a)^p$, where p is an integer greater than 1 and a is any real number. Find the relative extrema of f if
   (a) p is even
   (b) p is odd.

32. What is the smallest possible slope for a tangent to $y = x^3 - 3x^2 + 5x$?

33. (a) Show that

$$f(x) = \frac{64}{\sin x} + \frac{27}{\cos x}$$

   has a minimum value, but no maximum value on the interval $(0, \pi/2)$.
   (b) Find the minimum value.

34. Prove: For every positive value of t the function $f(x) = x + t/x$ has a minimum value but no maximum value on $(0, +\infty)$.

35. Find a_0, a_1, and a_2 such that the graph of $f(x) = a_0 + a_1 x + a_2 x^2$ passes through the point $(0, 9)$ and f has a minimum value of 1 that occurs when $x = 2$.

36. Prove that $\sin x \leq x$ for all x in the interval $[0, 2\pi]$. [*Hint:* Find the minimum value of $x - \sin x$ on the interval.]

37. Prove that

$$1 - \frac{x^2}{2} \leq \cos x$$

for all x in the interval $[0, 2\pi]$. [See the hint in the previous problem.]

38. (a) Sketch the graph of a function f that is continuous on the open interval $(-1, 1)$ and has both maximum and minimum values on the interval.
   (b) Sketch the graph of a function f that is defined everywhere on the open interval $(-1, 1)$ and is not continuous everywhere on $(-1, 1)$, yet has both maximum and minimum values on $(-1, 1)$.
   (c) Do the functions in parts (a) and (b) violate the Extreme Value Theorem? Explain.

39. Prove: If $ax^2 + bx + c = 0$ has two distinct real roots, then the midpoint between these roots is a stationary point for $f(x) = ax^2 + bx + c$.

40. Prove Theorem 4.6.5 in the case where the extreme value is a minimum.

41. Let $f(x) = ax^2 + bx + c$, where $a > 0$. Prove that $f(x) \geq 0$ for all x if and only if $b^2 - 4ac \leq 0$. [*Hint:* Find the minimum of $f(x)$.]

42. Prove that the minimum value of

$$f(x) = x^2 + \frac{16x^2}{(8 - x)^2}, \quad x > 8$$

occurs at $x = 4(2 + \sqrt[3]{2})$.

4.7 APPLIED MAXIMUM AND MINIMUM PROBLEMS

In this section we will show how the methods developed in the previous section can be used to solve some applied optimization problems.

▶ **Example 1** Find the dimensions of a rectangle with perimeter 100 ft whose area is as large as possible.

Solution. Let

$$x = \text{length of the rectangle}$$
$$y = \text{width of the rectangle}$$
$$A = \text{area of the rectangle}$$

Then

$$A = xy \tag{1}$$

Since the perimeter of the rectangle is 100 ft, the variables x and y are related by the equation

$$2x + 2y = 100$$

or

$$y = 50 - x \tag{2}$$

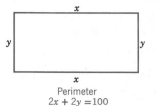

x

y y

x

Perimeter
$2x + 2y = 100$

Figure 4.7.1

(See Figure 4.7.1.) Substituting (2) in (1) yields

$$A = x(50 - x) = 50x - x^2 \tag{3}$$

Because x represents a length it cannot be negative, and because the two sides of length x cannot have a combined length exceeding the total perimeter of 100 ft, the variable x must satisfy

$$0 \leq x \leq 50 \tag{4}$$

Thus, we have reduced the problem to that of finding the value (or values) of x in [0, 50] for which (3) is maximum. Since A is a polynomial in x, it is continuous on [0, 50] and so the maximum must occur at an endpoint of this interval or at a critical point.

From (3) we obtain

$$\frac{dA}{dx} = 50 - 2x$$

Setting $dA/dx = 0$ we obtain

$$50 - 2x = 0$$

or $x = 25$. Thus, the maximum occurs at one of the points

$$x = 0, \qquad x = 25, \qquad x = 50$$

Substituting these values in (3) yields Table 4.7.1, which tells us that the maximum area of 625 ft^2 occurs when $x = 25$. From (2) the corresponding

Table 4.7.1

x	0	25	50
A	0	625	0

value of y is $y = 25$, so the rectangle of perimeter 100 ft with greatest area is a square with sides of length 25 ft.   ◀

REMARK. In (4) we included $x = 0$ and $x = 50$ as possible values for x. Because $x = 50$ implies $y = 0$ from (2), these x values correspond to rectangles with two sides of length zero. One can argue that these x values should not be allowed because a "true" rectangle cannot have sides of length zero.

Actually, it is just a matter of the assumptions we choose to make. If we view Example 1 as a purely mathematical problem, then there is nothing wrong with allowing sides of length zero. However, if we view it as a practical problem in which the rectangle is to be constructed from physical material, we would not want to allow $x = 0$ or $x = 50$ and (4) should be replaced by $0 < x < 50$. In this case we no longer have a closed interval to work with and the problem has to be solved by other methods (considered in the next section).

In this section we will consider only optimization problems over finite closed intervals. In the next section we will discuss other types of applied optimization problems.

Example 1 illustrates the following 5-step procedure for solving many applied maximum and minimum problems.

> **Step 1.** Label the quantities relevant to the problem.
>
> **Step 2.** Find a formula for the quantity to be maximized or minimized.
>
> **Step 3.** Using the conditions stated in the problem to eliminate variables, express the quantity to be maximized or minimized as a function of one variable.
>
> **Step 4.** Find the interval of possible values for this variable from the physical restrictions in the problem.
>
> **Step 5.** If applicable, use the techniques of the previous section to obtain the maximum or minimum.

▶ Example 2 An open box is to be made from a 16 in. by 30 in. piece of cardboard by cutting out squares of equal size from the four corners and bending up the sides (Figure 4.7.2). What size should the squares be to obtain a box with largest possible volume?

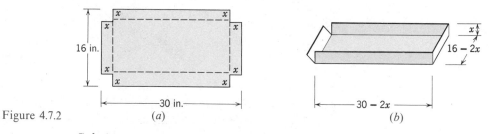

Figure 4.7.2 (a) (b)

Solution. Let

x = length (in inches) of the sides of the squares to be cut out
V = volume (in cubic inches) of the resulting box

Because we are removing a square of side x from each corner, the resulting box will have dimensions $16 - 2x$ by $30 - 2x$ by x (Figure 4.7.2*b*). Since the volume of a box is the product of its dimensions, we have

$$V = (16 - 2x)(30 - 2x)x = 480x - 92x^2 + 4x^3 \qquad (5)$$

The variable x in this expression is subject to certain restrictions. Because x represents a length it cannot be negative and because the width of the cardboard is 16 in., we cannot cut out squares whose sides are more than 8 in. long. Thus, the variable x in (5) must satisfy

$$0 \le x \le 8$$

We have thus reduced our problem to finding the value (or values) of x in the interval $[0, 8]$ for which (5) is maximum. Since the right side of (5) is a polynomial in x, it is continuous on the closed interval $[0, 8]$ and consequently we can use the methods developed in the previous section to find the maximum.

From (5) we obtain

$$\frac{dV}{dx} = 480 - 184x + 12x^2 = 4(120 - 46x + 3x^2)$$

Setting $dV/dx = 0$ yields

$$120 - 46x + 3x^2 = 0$$

which can be solved by the quadratic formula to obtain the critical points

$$x = \tfrac{10}{3} \qquad \text{and} \qquad x = 12$$

Since $x = 12$ falls outside the interval $[0, 8]$, the maximum value of V must occur either at the critical point $x = \tfrac{10}{3}$ or at one of the endpoints $x = 0$, $x = 8$. Substituting these values in (5) yields Table 4.7.2, which tells us that the greatest possible volume $V = \tfrac{19600}{27}$ in.³ ≈ 726 in.³ occurs when we cut out squares whose sides have length $\tfrac{10}{3}$ in. ◀

Table 4.7.2

x	0	$\dfrac{10}{3}$	8
V	0	$\dfrac{19600}{27}$	0

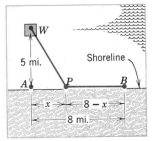

Figure 4.7.3

▶ **Example 3**   An offshore oil well is located in the ocean at a point W, which is 5 mi from the closest shorepoint A on a straight shoreline (Figure 4.7.3). The oil is to be piped to a shorepoint B that is 8 mi from A by piping it on a straight line underwater from W to some shorepoint P between A and B and then on to B via a pipe along the shoreline. If the cost of laying pipe is $100,000 per mile underwater and $75,000 per mile over land, where should the point P be located to minimize the cost of laying the pipe?

REMARK.   Since the shortest distance between two points is a straight line, a pipeline directly from W to B uses the least amount of pipe. However, the pipe, being completely under water, would be expensive to lay. At the other extreme, a pipeline from W to A to B uses the least amount of expensive underwater pipe, but uses the greatest total amount of pipe. Thus, it seems plausible that by piping to some point P between A and B, one might incur less total cost than by piping to either of the extreme locations.

Solution.   Let

x = distance (in miles) between A and P
c = cost (in thousands of dollars) for the entire pipeline

From Figure 4.7.3 the length of pipe underwater is the distance between W and P. By the Theorem of Pythagoras, that length is

$$\sqrt{x^2 + 25} \text{ mi} \tag{6}$$

Also from Figure 4.7.3, the length of pipe over land is the distance between P and B, which is

$$8 - x \text{ miles} \tag{7}$$

From (6) and (7) it follows that the total cost c (in thousands of dollars) for the pipeline is

$$c = 100\sqrt{x^2 + 25} + 75(8 - x) \tag{8}$$

Because the distance between A and B is 8 mi, the distance x between A and P must satisfy

$$0 \leq x \leq 8$$

We have thus reduced our problem to finding the value (or values) of x in the interval $[0, 8]$ for which (8) is a minimum. Since c is a continuous function of x on the closed interval $[0, 8]$, we can use the methods developed in the previous section to find the minimum.

From (8) we obtain

$$\frac{dc}{dx} = \frac{100x}{\sqrt{x^2 + 25}} - 75 = 25\left(\frac{4x}{\sqrt{x^2 + 25}} - 3\right)$$

Setting $dc/dx = 0$ yields

$$\frac{4x}{\sqrt{x^2 + 25}} - 3 = 0 \qquad\qquad (9)$$

or

$$4x = 3\sqrt{x^2 + 25}$$
$$16x^2 = 9(x^2 + 25)$$
$$7x^2 = 225$$
$$x = \pm\frac{15}{\sqrt{7}}$$

The number $-15/\sqrt{7}$ is not a solution of (9) and must be discarded, leaving $x = 15/\sqrt{7}$ as the only critical point. Since this point lies in the interval $[0, 8]$, the minimum must occur at one of the points:

$$x = 0, \qquad x = 15/\sqrt{7}, \qquad x = 8$$

Substituting these values in (8) yields Table 4.7.3, which tells us that the least possible cost for the pipeline, approximately $c = 930.719$ (thousand dollars) $= \$930,719$ occurs when the point P is located at a distance of $15/\sqrt{7} \approx 5.67$ mi from A. ◀

Table 4.7.3

x	0	$\dfrac{15}{\sqrt{7}}$	8
c	1100	$600 + 125\sqrt{7} \approx 930.719$	$100\sqrt{89} \approx 943.398$

▶ Example 4   Find the radius and height of the right-circular cylinder of largest volume that can be inscribed in a right-circular cone with radius 6 in. and height 10 in. (Figure 4.7.4a).

Solution.   Let

r = radius (in inches) of the cylinder
h = height (in inches) of the cylinder
V = volume (in cubic inches) of the cylinder

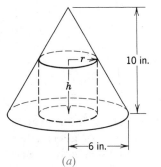

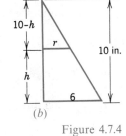

(a)

(b)

Figure 4.7.4

The formula for the volume of the inscribed cylinder is

$$V = \pi r^2 h \tag{10}$$

To eliminate one of the variables in (10) we need a relationship between r and h. Using similar triangles (Figure 4.7.4b) we obtain

$$\frac{10 - h}{r} = \frac{10}{6}$$

or

$$h = 10 - \tfrac{5}{3}r \tag{11}$$

Substituting (11) into (10) we obtain

$$V = \pi r^2 (10 - \tfrac{5}{3}r) = 10\pi r^2 - \tfrac{5}{3}\pi r^3 \tag{12}$$

which expresses V in terms of r alone. Because r represents a radius it cannot be negative, and because the radius of the inscribed cylinder cannot exceed the radius of the cone, the variable r must satisfy

$$0 \le r \le 6$$

Thus, we have reduced the problem to that of finding the value (or values) of r in [0, 6] for which (12) is a maximum. Since V is a continuous function of r on [0, 6], the methods developed in the previous section apply.

From (12) we obtain

$$\frac{dV}{dr} = 20\pi r - 5\pi r^2 = 5\pi r(4 - r)$$

Setting $dV/dr = 0$ gives

$$5\pi r(4 - r) = 0$$

so that

$$r = 0 \quad \text{and} \quad r = 4$$

are critical points. Since these lie in the interval [0, 6], the maximum must occur at one of the points

$$r = 0, \quad r = 4, \quad r = 6$$

Substituting these values in (12) yields Table 4.7.4, which tells us that the maximum volume $V = \tfrac{160}{3}\pi \approx 168$ in.^{3} occurs when the inscribed cylinder has radius 4 in. When $r = 4$ it follows from (11) that $h = \tfrac{10}{3}$. Therefore, the inscribed cone of largest volume has radius $r = 4$ in. and height $h = \tfrac{10}{3}$ in. ◄

Table 4.7.4

r	0	4	6
V	0	$\dfrac{160}{3}\pi$	0

MARGINAL ANALYSIS (AN APPLICATION TO ECONOMICS)

Economists and business people are interested in how changes in such variables as inventory, production, supply, advertising, and price affect other variables such as profit, revenue, demand, inflation, and employment. Such problems are studied using **marginal analysis.** The word, **marginal,** is the economist's term for a rate of change or derivative.

Three functions of importance to an economist or a manufacturer are:

$C(x)$ = total cost of producing x units of a product during some time period.

$R(x)$ = total revenue received from selling x units of the product during the time period.

$P(x)$ = total profit obtained by selling x units of the product during the time period.

These are called, respectively, the **cost function, revenue function,** and **profit function.** If all units produced are sold, then these are related by

$$P(x) = R(x) - C(x)$$
$$[\text{profit}] = [\text{revenue}] - [\text{cost}]$$

(13)

The derivatives $C'(x)$, $R'(x)$, and $P'(x)$ are called the **marginal cost, marginal revenue,** and **marginal profit.** If all units produced are sold, then a relationship between them is obtained by differentiating (13),

$$P'(x) = R'(x) - C'(x)$$
$$\begin{bmatrix}\text{marginal} \\ \text{profit}\end{bmatrix} = \begin{bmatrix}\text{marginal} \\ \text{revenue}\end{bmatrix} - \begin{bmatrix}\text{marginal} \\ \text{cost}\end{bmatrix}$$

The quantities $P'(x)$, $R'(x)$, and $C'(x)$ represent the instantaneous rates of change of profit, revenue, and cost with respect to x, where x is the amount of the product produced and sold.

In practice, $C'(x)$ is frequently interpreted as the cost of manufacturing the $(x + 1)$-th unit. Although this is not exact, it is usually a good approximation. The justification for this interpretation is based on the fact that x is usually large, so $\Delta x = 1$ can be considered close to zero by comparison. Thus,

$$C'(x) = \lim_{\Delta x \to 0} \frac{C(x + \Delta x) - C(x)}{\Delta x} \approx \frac{C(x + 1) - C(x)}{1}$$

$$= C(x + 1) - C(x)$$

Since $C(x + 1)$ is the cost of producing $x + 1$ units and $C(x)$ is the cost of producing x units, it follows that $C'(x) \approx C(x + 1) - C(x)$ is the approximate cost of producing the $(x + 1)$-th unit. Similarly, $R'(x)$ is the approximate revenue received from selling the $(x + 1)$-th unit and $P'(x)$ is the approximate profit from selling the $(x + 1)$-th unit.

The total cost $C(x)$ of producing x units can be expressed as a sum

$$C(x) = a + M(x) \tag{14}$$

where a is a constant, called **overhead,** and $M(x)$ is a function representing **manufacturing cost.** The overhead, which includes such fixed costs as rent and insurance, does not depend on x; it must be paid even if nothing is produced. On the other hand, the manufacturing cost $M(x)$, which includes such items as cost of materials and labor, depends on the number of items manufactured. It is shown in economics that with suitable simplifying assumptions, $M(x)$ can be expressed in the form

$$M(x) = bx + cx^2$$

Substituting this in (14) yields

$$C(x) = a + bx + cx^2 \tag{15}$$

▶ Example 5   A paint manufacturer determines that the total cost in dollars of producing x gallons of paint per day is

$$C(x) = 5000 + x + 0.001x^2$$

(a)   Find the marginal cost when the production level is 500 gal per day.
(b)   Use the marginal cost to approximate the cost of producing the 501st gallon.
(c)   Find the exact cost of producing the 501st gallon.

Solution.
(a)   The marginal cost is $C'(x) = 1 + 0.002x$, so

$$C'(500) = 1 + (0.002)(500) = 2.$$

(b)   Since $C'(500) = 2$, the cost of producing the 501st gallon is approximately $2.00.
(c)   The total cost of producing 501 gal is

$$C(501) = 5000 + 501 + 0.001(501)^2 = \$5752.001$$

and the total cost of producing 500 gal is

$$C(500) = 5000 + 500 + 0.001(500)^2 = \$5750$$

so the exact cost of producing the 501st gallon is

$$C(501) - C(500) = \$2.001 \qquad \blacktriangleleft$$

If a manufacturing firm can sell all the items it produces for p dollars apiece, then its total revenue $R(x)$ (in dollars) will be

$$R(x) = px \qquad (16)$$

and its total profit $P(x)$ (in dollars) will be

$$P(x) = [\text{total revenue}] - [\text{total cost}] = R(x) - C(x) = px - C(x)$$

Thus, if the cost function is given by (15),

$$P(x) = px - (a + bx + cx^2) \qquad (17)$$

Depending on such factors as number of employees, amount of machinery available, economic conditions, and competition, there will be some upper limit l on the number of items a manufacturer is capable of producing and selling. Thus, during a fixed time period the variable x in (17) will satisfy

$$0 \leq x \leq l$$

By determining the value or values of x in $[0, l]$ that maximize (17), the firm can determine how many units of its product must be manufactured and sold to yield the greatest profit. This is illustrated in the following numerical example.

▶ Example 6   A liquid form of penicillin manufactured by a pharmaceutical firm is sold in bulk at a price of \$200 per unit. If the total production cost (in dollars) for x units is

$$C(x) = 500{,}000 + 80x + 0.003x^2$$

and if the production capacity of the firm is at most 30,000 units in a specified time, how many units of penicillin must be sold in that time to maximize the profit?

Solution.   Since the total revenue for selling x units is $R(x) = 200x$, the profit $P(x)$ on x units will be

$$P(x) = R(x) - C(x) = 200x - (500{,}000 + 80x + 0.003x^2) \qquad (18)$$

Since the production capacity is at most 30,000 units, x must lie in the interval $[0, 30,000]$. From (18)

$$\frac{dP}{dx} = 200 - (80 + 0.006x) = 120 - 0.006x$$

Setting $dP/dx = 0$ gives

$$120 - 0.006x = 0$$

or

$$x = 20,000$$

Since this critical point lies in the interval $[0, 30,000]$, the maximum profit must occur at one of the points

$$x = 0, \qquad x = 20,000, \qquad \text{or} \qquad x = 30,000$$

Substituting these values in (18) yields Table 4.7.5, which tells us that the maximum profit $P = \$700,000$ occurs when $x = 20,000$ units are manufactured and sold in the specified time. ◄

Table 4.7.5

x	0	20,000	30,000
$P(x)$	$-500,000$	700,000	400,000

► Exercise Set 4.7

1. Express the number 10 as a sum of two nonnegative terms whose product is as large as possible.

2. How should two nonnegative numbers be chosen so that their sum is 1 and the sum of their squares is:
   (a) as large as possible?
   (b) as small as possible?

3. Find a number in the closed interval $[\frac{1}{2}, \frac{3}{2}]$ such that the sum of the number and its reciprocal is:
   (a) as small as possible;
   (b) as large as possible.

4. A rectangular field is to be bounded by a fence on three sides and by a straight stream on the fourth side. Find the dimensions of the field with maxi-

mum area that can be enclosed with 1000 feet of fence.

5. A rectangular plot of land is to be fenced in using two kinds of fencing. Two opposite sides will use heavy-duty fencing selling for $3 a foot, while the remaining two sides will use standard fencing selling for $2 a foot. What are the dimensions of the rectangular plot of greatest area that can be fenced in at a cost of $6000?

6. Show that among all rectangles with perimeter p, the square with side $p/4$ has the maximum area.

7. Find the dimensions of the rectangle with maximum area that can be inscribed in a circle of radius 10.

8. A rectangle is to be inscribed in a right triangle having sides of length 6 in., 8 in., and 10 in. Find the dimensions of the rectangle with greatest area assuming the rectangle is positioned as in Figure 4.7.5a.

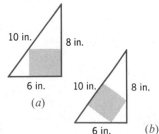

(a)

Figure 4.7.5

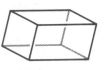

(b)

9. Solve the problem in Exercise 8 assuming the rectangle is positioned as in Figure 4.7.5b.

10. A rectangle has its two lower corners on the x-axis and its two upper corners on the curve $y = 16 - x^2$. For all such rectangles, what are the dimensions of the one with largest area?

11. A sheet of cardboard 12 in. square is used to make an open box by cutting squares of equal size from the corners and folding up the sides. What size squares should be cut to obtain a box with largest possible volume?

12. A square sheet of cardboard of side k is used to make an open box by cutting squares of equal size from the corners and folding up the sides. What size squares should be cut from the corners to obtain a box with largest possible volume?

13. A wire of length 12 in. can be bent into a circle, bent into a square, or cut into two pieces to make both a circle and a square. How much wire should be used for the circle if the total area enclosed by the figure(s) is to be:
(a) a maximum?
(b) a minimum?

14. Assume that the operating cost of a certain truck (excluding driver's wages) is $12 + x/6$ cents per mile when the truck travels at x mi/hr. If the driver earns $6 per hour, what is the most economical speed to operate the truck on a 400-mi turnpike where the minimum speed is 40 mi/hr and the maximum speed is 70 mi/hr?

15. Find the lengths of the sides of the isosceles triangle with perimeter 12 and maximum area.

16. A triangle is inscribed in a semicircle of radius 10 so that one side is along the diameter (Figure 4.7.6). Find the dimensions of the triangle with maximum area.

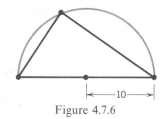

Figure 4.7.6

17. A box-shaped wire frame consists of two identical wire squares whose vertices are connected by four straight wires of equal length (Figure 4.7.7). If the frame is to be made from a wire of length L, what should the dimensions be to obtain a box of greatest volume?

Figure 4.7.7

18. If air resistance is neglected, then the range R of a cannonball fired from a cannon whose barrel makes an angle θ with the horizontal is

$$R = \left(\frac{v_0^2}{g}\right) \sin 2\theta$$

where the constants v_0 and g are the initial velocity and the acceleration due to gravity. Show that the maximum range is achieved when $\theta = 45°$.

19. (a) A chemical manufacturer sells sulphuric acid in bulk at a price of $100 per unit. If the daily total production cost in dollars for x units is

$$C(x) = 100,000 + 50x + 0.0025x^2$$

and if the daily production capacity is at most 7000 units, how many units of sulphuric acid must be manufactured and sold daily to maximize the profit?

(b) Would it benefit the manufacturer to expand the daily production capacity?

20. Find the dimensions of the rectangle of greatest area that can be inscribed in a semicircle of radius R.

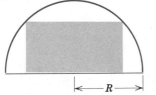

21. Find the height and radius of the cone of slant height L whose volume is as large as possible.

22. Find the dimensions of the right-circular cylinder of largest volume that can be inscribed in a sphere of radius R.

23. Find the dimensions of the right-circular cylinder of greatest surface area that can be inscribed in a sphere of radius R.

24. Show that the right-circular cylinder of greatest volume that can be inscribed in a right-circular cone has volume that is $\frac{4}{9}$ the volume of the cone (Figure 4.7.8).

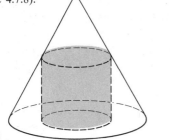

Figure 4.7.8

25. A drainage channel is to be made so that its cross section is a trapezoid with equally sloping sides (Figure 4.7.9). If the sides and bottom all have a length of 5 ft, how should the angle θ be chosen to yield the greatest cross-sectional area? Assume $0 \leq \theta \leq \pi/2$.

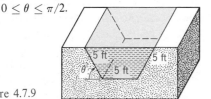

Figure 4.7.9

26. A commercial cattle ranch currently allows 20 steers per acre of grazing land; on the average its steers weigh 2000 lb at market. Estimates by the Agriculture Department indicate that the average market weight per steer will be reduced by 50 lb for each additional steer added per acre of grazing land. How many steers per acre should be allowed in order for the ranch to get the largest possible total market weight for its cattle?

27. A cone is made from a circular sheet of radius R by cutting out a sector and gluing the cut edges of the remaining piece together (Figure 4.7.10). What is the maximum volume attainable for the cone?

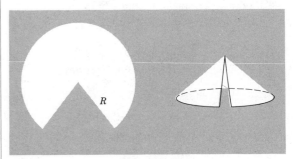

Figure 4.7.10

28. A man is on the bank of a river that is 1 mi wide. He wants to travel to a town on the opposite bank, but 1 mi upstream. He intends to row on a straight line to some point P on the opposite bank and then walk the remaining distance along the bank (Figure 4.7.11). To what point should he row in order to reach his destination in the least time if:
(a) he can walk 5 mi/hr and row 3 mi/hr?
(b) he can walk 5 mi/hr and row 4 mi/hr?

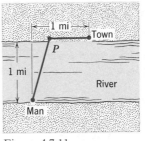

Figure 4.7.11

29. The total cost of producing x units of a commodity per week is

$$C(x) = 200 + 4x + 0.1x^2$$

(a) Find the marginal cost when the production level is 100 units.

(b) Use the marginal cost to approximate the cost of producng the 101st unit.

(c) Find the exact cost of producing the 101st unit.

(d) Assuming that the commodity is sold for $10 per unit, find the marginal revenue and marginal profit functions.

30. A firm determines that x units of its product can be sold daily at p dollars per unit, where

$$x = 1000 - p$$

The cost of producing x units per day is

$$C(x) = 3000 + 20x$$

(a) Find the revenue function $R(x)$.

(b) Find the profit function $P(x)$.

(c) Assuming that the production capacity is at most 500 units per day, determine how many units the company must produce and sell each day to maximize the profit.

(d) Find the maximum profit.

(e) What price per unit must be charged to obtain the maximum profit?

31. Prove that $(1, 0)$ is the closest point on the curve $x^2 + y^2 = 1$ to $(2, 0)$.

32. We are given a straight fence 100 ft long and we wish to use 200 ft of additional fence to form a rectangular enclosure whose boundary contains all of the original fence. How should we lay out the additional fence to obtain an enclosure of maximum area? (This problem, due to V. L. Klee, Jr., appeared in *Selected Papers on Calculus,* Mathematical Association of America (1969), p. 246.)

33. (a) Find the smallest value of M such that

$$|x^2 - 3x + 2| \leq M$$

for all x in the interval $[1, \frac{5}{2}]$.

(b) Find the largest value of m such that

$$|x^2 - 3x + 2| \geq m$$

for all x in the interval $[\frac{3}{2}, \frac{7}{4}]$.

34. At what point(s) in the interval $[0, \pi]$ are the graphs of

$$y = \tfrac{1}{2}x \quad \text{and} \quad y = \sin x$$

farthest apart?

35. Fermat's* principle in optics states that light traveling from one point to another follows that path for which the total travel time is minimum. In a uniform medium, the paths of "minimum time" and "shortest distance" turn out to be the same, so that light, if unobstructed, travels along a straight line. Assume we have a light source, a flat mirror, and an observer in a uniform medium. If a light ray leaves the source, bounces off the mirror, and travels on to the observer, then its path will consist of two line segments, as shown in Figure 4.7.12. According to Fermat's principle, the path will be such that the total travel time t is minimum or, since the medium is uniform, the path will be such that the total distance traveled from A to P to B is as small as possible. Assuming that the minimum occurs when $dt/dx = 0$, show that the light ray will strike the mirror at the point P where the "angle of incidence" θ_1 equals the "angle of reflection" θ_2.

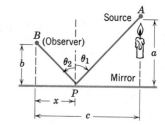

Figure 4.7.12

36. Fermat's* principle (Exercise 35) also explains why light rays traveling between air and water undergo bending or refraction. Imagine we have two uniform media (such as air and water) and a light ray traveling from a source A in one medium to an observer B in the other medium (Figure 4.7.13). It is known that light travels at a constant speed in a uniform medium, but more slowly in a dense medium (such as water) than in a thin medium (such as air.) Consequently, the path of shortest time from A to B is not necessarily a straight line, but rather some broken line path A to P to B allowing the light to take greatest advantage of its higher speed through the thin medium. Snell's† law of refraction states that the path of the light ray will be such that

$$\frac{\sin \theta_1}{v_1} = \frac{\sin \theta_2}{v_2}$$

where v_1 is the speed of light in the first medium, v_2 is the speed of light in the second medium, and θ_1 and θ_2 are the angles shown in Figure 4.7.13.

Show that this follows from the assumption that the path of minimum time occurs when $dt/dx = 0$.

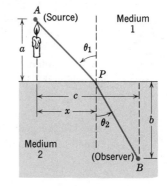

Figure 4.7.13

---

*PIERRE DE FERMAT (1601–1665). Fermat, the son of a successful French leather merchant, was a lawyer who practiced mathematics as a hobby. He received a Bachelor of Civil Laws degree from the University of Orleans in 1631 and subsequently held various government positions, including a post as councillor to the Toulouse parliament. Although he was apparently financially successful, confidential documents of that time suggest that his performance in office and as a lawyer was poor, perhaps because he devoted so much time to mathematics. Throughout his life, Fermat fought all efforts to have his mathematical results published. He had the unfortunate habit of scribbling his work in the margins of books and often sent his results to friends without keeping copies for himself. As a result, he never received credit for many major achievements until his name was raised from obscurity in the mid-nineteenth century. It is now known that Fermat, simultaneously and independently of Descartes, developed analytic geometry. Unfortunately, Descartes and Fermat argued bitterly over various problems so that there was never any real cooperation between these two great geniuses.

Fermat solved many fundamental calculus problems. He obtained the first procedure for differentiating polynomials, and solved many important maximization, minimization, area, and tangent problems. His work served to inspire Isaac Newton. Fermat is best known for his work in number theory, the study of properties and relationships between whole numbers. He was the first mathematician to make substantial contributions to this field after the ancient Greek mathematician Diophantus. Unfortunately, none of Fermat's contemporaries appreciated his work in this area, a fact that eventually pushed Fermat into isolation and obscurity in later life.

In addition to his work in calculus and number theory Fermat was one of the founders of probability theory and made major contributions to the theory of optics. Outside mathematics, Fermat was a classical scholar of some note, was fluent in French, Italian, Spanish, Latin, and Greek, and he composed a considerable amount of Latin poetry.

One of the great mysteries of mathematics is shrouded in Fermat's work in number theory. In the margin of a book by Diophantus, Fermat scribbled that for values of n greater than 2, the equation $x^n + y^n = z^n$ has no nonzero integer solutions for x, y, and z. He stated, "I have discovered a truly marvelous proof of this, which however the margin is not large enough to contain." For the past 300 years the greatest mathematical geniuses have been unable to prove this result, even though it seems to be true. This result is now known as "Fermat's last theorem." In 1908 a prize of 100,000 German marks was offered for its solution. Although the prize is worthless because of post–World War I inflation, it has never been won. Whether or not Fermat really proved this theorem is a mystery to this day.

NOTE. A major breakthrough in Fermat's last theorem has just been made by a young West German mathematician named Gerd Faltings (See *Newsweek,* Aug. 1, 1983, p. 66).

37. A farmer wants to walk at a constant rate from her barn to a straight river, fill her pail, and carry it to her house in the least time.

   (a) Explain how this problem relates to Fermat's principle and the light-reflection problem in Exercise 35.

   (b) Use the result of Exercise 35 to describe geometrically the best path for the farmer to take.

   (c) Use part (b) to determine where the farmer should fill her pail if her house and barn are located as in Figure 4.7.14.

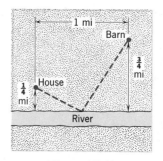

Figure 4.7.14

---

†WILLEBRORD VAN ROIJEN SNELL (1591–1626) Dutch mathematician. Snell, who succeeded his father to the post of Professor of Mathematics at the University of Leiden in 1613, is most famous for the result on light refraction that bears his name. Although this phenomenon was studied as far back as the ancient Greek astronomer Ptolemy, until Snell's work the relationship was incorrectly thought to be $\theta_1/v_1 = \theta_2/v_2$. Snell's law was published by Descartes in 1638 without giving proper credit to Snell. Snell also discovered a method for determining distances by triangulation that founded the modern technique of mapmaking.

4.8   MORE APPLIED MAXIMUM AND MINIMUM PROBLEMS

In the previous section we discussed optimization problems that reduced to maximizing or minimizing a continuous function over a closed interval $[a, b]$. For such problems, the Extreme Value Theorem guarantees that a solution exists. In this section we will consider optimization problems that reduce to maximizing or minimizing continuous functions over open intervals or infinite intervals. For such problems, there is no general guarantee that a solution exists. Thus, part of the problem is to determine whether a solution exists and then find it if it does.

▶ **Example 1**   A closed cylindrical can is to hold 1 liter (1000 cm³) of liquid. How should we choose the height and radius to minimize the amount of material needed to manufacture the can?

Solution.   Let

$$h = \text{height (in cm) of the can}$$
$$r = \text{radius (in cm) of the can}$$
$$S = \text{surface area (in cm}^2\text{) of the can}$$

Assuming there is no waste or overlap, the amount of material needed for manufacture will be the same as the surface area of the can. Since the can can be made from two circular disks of radius r and a rectangular sheet with dimensions h by $2\pi r$ (Figure 4.8.1), the surface area will be

$$S = 2\pi r^2 + 2\pi rh \qquad (1)$$

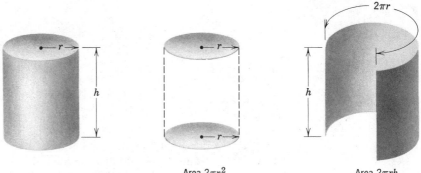

Figure 4.8.1

Area $2\pi r^2$

Area $2\pi rh$

We will now eliminate one of the variables in (1) so that S will be expressed as a function of one variable. Since the volume of the can is 1000 cm³, it follows from the formula

$$V = \pi r^2 h$$

for the volume of a cylinder that

$$1000 = \pi r^2 h \tag{2}$$

or

$$h = \frac{1000}{\pi r^2} \tag{3}$$

Substituting (3) in (1) yields

$$S = 2\pi r^2 + \frac{2000}{r} \tag{4}$$

Because r represents a radius it must be positive;* thus the variable r in (4) must satisfy

$$0 < r < +\infty$$

We have now reduced the problem to that of finding a value of r in $(0, +\infty)$ for which (4) is minimum (provided such a value exists). From (4) we obtain

$$\frac{dS}{dr} = 4\pi r - \frac{2000}{r^2} \tag{5}$$

Setting $dS/dr = 0$ gives

$$4\pi r - \frac{2000}{r^2} = 0$$

---

*The value $r = 0$ must be excluded, otherwise (2) cannot be satisfied.

or

$$4\pi r^3 = 2000$$

or

$$r = \frac{10}{\sqrt[3]{2\pi}} \tag{6}$$

To see that a minimum occurs at this value of r, it suffices to show that a relative minimum occurs at r because this is the only critical point (see Theorem 4.6.6). This can be done using either the first or second derivative test. To use the first derivative test, we rewrite (5) as

$$\frac{dS}{dr} = \frac{4\pi}{r^2}\left(r^3 - \frac{1000}{2\pi}\right)$$

Thus, on $(0, +\infty)$, S is decreasing $(dS/dr < 0)$ for $r < 10/\sqrt[3]{2\pi}$ and S is increasing $(dS/dr > 0)$ for $r > 10/\sqrt[3]{2\pi}$, which tells us that a minimum on $(0, +\infty)$ occurs at the value of r in (6). Thus, the radius of the can of least surface area is

$$r = \frac{10}{\sqrt[3]{2\pi}}$$

and from (3) the corresponding height is

$$h = \frac{1000}{\pi(10/\sqrt[3]{2\pi})^2} = \frac{20}{\sqrt[3]{2\pi}}$$

Second Solution. The conclusion that a minimum occurs at the value of r in (6) can be deduced from the second derivative test by noting that

$$\frac{d^2S}{dr^2} = 4\pi + \frac{4000}{r^3}$$

so that

$$\left.\frac{d^2S}{dr^2}\right|_{r=10/\sqrt[3]{2\pi}} = 4\pi + \frac{4000}{(10/\sqrt[3]{2\pi})^3} = 12\pi$$

Since the second derivative is positive, a relative minimum, and therefore a minimum, occurs at the critical point $r = 10/\sqrt[3]{2\pi}$.

Third Solution. Instead of using the first or second derivative tests, this problem can be solved by graphing (4). Using the methods of Section 4.4,

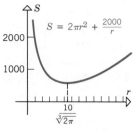

$$S = 2\pi r^2 + \frac{2000}{r}$$

Figure 4.8.2

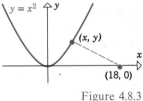

$y = x^2$

(x, y)

$(18, 0)$

Figure 4.8.3

the reader should be able to obtain the graph in Figure 4.8.2. From this graph we see that S has a minimum, and this minimum occurs at the critical point $r = 10/\sqrt[3]{2\pi}$ calculated in our first solution of this problem [see (6)].

◄

REMARK. Note that S has no maximum on $(0, +\infty)$. Thus, had we asked for the dimensions of the can requiring the maximum amount of material for its manufacture, there would have been no solution to the problem. Optimization problems with no solution are sometimes called **ill posed.**

▶ **Example 2** Find a point on the curve $y = x^2$ that is closest to the point $(18, 0)$.

Solution. The distance L between $(18, 0)$ and an arbitrary point (x, y) on the curve $y = x^2$ (Figure 4.8.3) is given by

$$L = \sqrt{(x - 18)^2 + (y - 0)^2}$$

Since (x, y) lies on the curve, x and y satisfy $y = x^2$; thus,

$$L = \sqrt{(x - 18)^2 + x^4} \tag{7}$$

Because there are no restrictions on x, the problem reduces to finding a value of x in $(-\infty, +\infty)$ for which (7) is minimum, provided such a value exists.

In problems of minimizing or maximizing a distance, there is a trick that is helpful for simplifying the computations. It is based on the observation that the distance and the square of the distance have their maximum or minimum at the same point (see Exercise 22). Thus, the minimum value of L in (7) and the minimum value of

$$S = L^2 = (x - 18)^2 + x^4 \tag{8}$$

occur at the same x value.

From (8),

$$\frac{dS}{dx} = 2(x - 18) + 4x^3 = 4x^3 + 2x - 36 \tag{9}$$

so that the critical points satisfy $4x^3 + 2x - 36 = 0$ or equivalently

$$2x^3 + x - 18 = 0 \tag{10}$$

To solve for x we will begin by searching for integer solutions. This task can be simplified by using the fact that all integer solutions, if there are any, to a polynomial equation with integer coefficients

$$a_0 + a_1 x + a_2 x^2 + \cdots + a_n x^n = 0$$

must be divisors of the constant term a_0. This result is usually proved in algebra courses. Thus, the only possible integer solutions of (10) are the divisors of -18: $\pm 1, \pm 2, \pm 3, \pm 6, \pm 9, \pm 18$. Successively substituting these values in (10) we find that $x = 2$ is a solution; therefore, $x - 2$ is a factor of the left side of (10). After dividing by the factor $x - 2$ we can rewrite (10) as

$$(x - 2)(2x^2 + 4x + 9) = 0$$

Thus, the remaining solutions of (10) satisfy the quadratic equation

$$2x^2 + 4x + 9 = 0$$

But these solutions are complex numbers (use the quadratic formula) so that $x = 2$ is the only real solution of (10) and consequently the only critical point. To determine the nature of this critical point we will use the second derivative test. From (9),

$$\frac{d^2S}{dx^2} = 12x^2 + 2$$

so

$$\left. \frac{d^2S}{dx^2} \right|_{x=2} = 50 > 0$$

which shows that a relative minimum occurs at $x = 2$. Since $x = 2$ is the only relative extremum for L, it follows from Theorem 4.6.6 that an absolute minimum value of L also occurs at $x = 2$. Thus, the point on the curve $y = x^2$ closest to $(18, 0)$ is

$$(x, y) = (x, x^2) = (2, 4) \qquad \blacktriangleleft$$

▶ Exercise Set 4.8
_____

1. Find two numbers whose sum is 20 and whose product is
    (a) maximum
    (b) minimum.

2. Find the dimensions of the rectangle of area A whose perimeter is
    (a) minimum
    (b) maximum.

3. A rectangular area of 1600 ft² is to be fenced off. Two opposite sides will use fencing costing $1 per foot and the remaining sides will use fencing cost-

ing $2 per foot. Find the dimensions of the rectangle of least cost.

4. A closed rectangular container with a square base is to have a volume of 3000 in.³ The material for the top and bottom of the container will cost $2 per in.², and the material for the sides will cost $3 per in.² Find the dimensions of the container of least cost.

5. A closed rectangular container with a square base is to have a volume of 1000 cm³. It costs twice as much per square centimeter for the top and bot-

tom as it does for the sides. Find the dimensions of the container of least cost.

6. A container with square base, vertical sides, and open top is to be made from 1000 ft² of material. Find the dimensions of the container with greatest volume.

7. A cylindrical can, open at the top, is to hold 500 cm³ of liquid. Find the height and radius that minimize the amount of material needed to manufacture the can.

8. A rectangular sheet of paper is to contain 72 in.² of printed matter, with 2-in. margins at top and bottom and 1-in. margins on each side. What dimensions for the sheet will use the least paper?

9. A cone-shaped paper drinking cup is to hold 10 cm³ of water. Find the height and radius of the cup that will require the least amount of paper.

10. A church window consisting of a rectangle topped by a semicircle is to have a perimeter p. Find the radius of the semicircle if the area of the window is to be maximum.

11. Find all points on the curve $x^2 - y^2 = 1$ closest to $(0, 2)$.

12. Find a point on the curve $x = 2y^2$ closest to $(0, 9)$.

13. Suppose a line L of variable slope passes through $(1, 3)$ and intersects the coordinate axes at the points $(a, 0)$ and $(0, b)$ where a and b are positive. Find the slope of L for which the area of the triangle with vertices $(a, 0)$, $(0, b)$, and $(0, 0)$ is
    (a) maximum
    (b) minimum.

14. In a certain chemical manufacturing process, the daily weight y of defective chemical output depends on the total weight x of all output according to the empirical formula

    $$y = 0.01x + 0.00003x^2$$

    where x and y are in pounds. If the profit is $100 per pound of nondefective chemical produced and the loss is $20 per pound of defective chemical produced, how many pounds of chemical should be produced daily to maximize the profit?

15. Find the points on the circle $x^2 + y^2 = 80$ that are closest to and farthest from $(1, 2)$.

16. A lamp is suspended above the center of a round table of radius r. How high above the table should the lamp be placed to achieve maximum illumination at the edge of the table? [Assume that the illumination I is directly proportional to the cosine of the angle of incidence ϕ of the light rays and inversely proportional to the square of the distance l from the light source (Figure 4.8.4).]

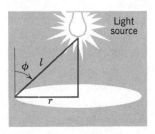

Figure 4.8.4

17. Where on the curve

    $$y = \frac{1}{1 + x^2}$$

    does the tangent line have the greatest slope?

18. Find the dimensions of the isosceles triangle of least area that can be circumscribed about a circle of radius R.

19. Find the height and radius of the right circular cone with least volume that can be circumscribed about a sphere of radius R.

20. If an unknown physical quantity x is measured n times, the measurements $x_1, x_2, \ldots, x_n$ often vary because of uncontrollable factors such as temperature, atmospheric pressure, and so forth. Thus, a scientist is often faced with the problem of using n different observed measurements to obtain an estimate $\bar{x}$ of an unknown quantity x. One method for making such an estimate is based on the *least squares principle*, which states that the estimate $\bar{x}$ should be chosen to minimize

    $$s = (x_1 - \bar{x})^2 + (x_2 - \bar{x})^2 + \cdots + (x_n - \bar{x})^2$$

    which is the sum of the squares of the deviations between the estimate $\bar{x}$ and the measured values. Show that the estimate resulting from the least squares principle is

$$\bar{x} = \frac{1}{n}(x_1 + x_2 + \cdots + x_n)$$

that is, $\bar{x}$ is the arithmetic average of the observed values.

21. A pipe of negligible thickness is to be carried horizontally around a corner from a hallway 8 ft wide into a hallway 4 ft wide (Figure 4.8.5). What is the maximum length that the pipe can have? [An interesting discussion of this problem in the case where the thickness of the pipe is not neglected is given by Norman Miller in the *American Mathematical Monthly*, vol. 56 (1949), pp. 177–179.]

22. Prove: If $f(x) \geq 0$ on an interval I and if $f(x)$ has a maximum value on I at x_0, then $\sqrt{f(x)}$ also has a maximum value at x_0. Similarly, for minimum values. [*Hint:* Use the fact that $\sqrt{x}$ is an increasing function on the interval $[0, +\infty)$.]

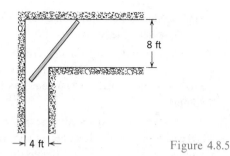

Figure 4.8.5

23. Prove: If $P(x_1, y_1)$ is a fixed point, then the minimum distance between P and a point on the line $ax + by + c = 0$ is

$$\frac{|ax_1 + by_1 + c|}{\sqrt{a^2 + b^2}}$$

4.9 NEWTON'S METHOD

In Section 3.7 we showed how to use the Intermediate-Value Theorem to approximate solutions of an equation $f(x) = 0$ to any degree of accuracy (see Example 11 in that section.) In this section we will develop *Newton's Method,* which provides a more efficient technique for approximating such solutions.

We note first that the solutions of $f(x) = 0$ are the values of x where the graph of f crosses the x-axis. Suppose that $x = r$ is the solution we are seeking. Even if we can't find the value of r exactly, it is usually possible to approximate it by graphing f and applying Theorem 3.7.9 to estimate where the graph crosses the x-axis. If we let x_1 denote our initial approximation to r, then we can generally improve on this approximation by moving along the tangent line to $y = f(x)$ at x_1 until we meet the x-axis at a point x_2 (Figure 4.9.1). Usually, x_2 will be closer to r than x_1. To improve the approximation further, we can repeat the process by moving along the tangent line to $y = f(x)$ at x_2 until we meet the x-axis at a point x_3. Continuing in this way we can generate a succession of values $x_1, x_2, x_3, x_4, \ldots$ that will usually get closer and closer to r. This procedure for approximating r is called *Newton's Method.*

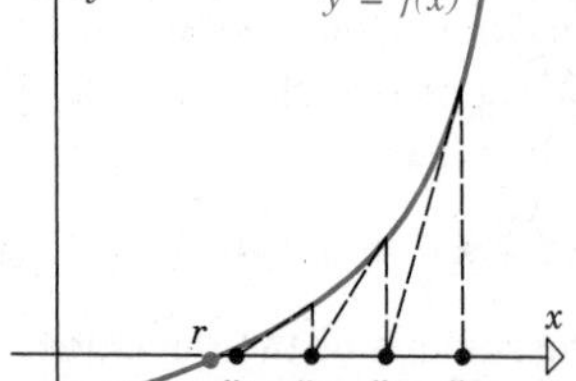

To implement Newton's Method analytically, we must derive a formula that will tell us how to calculate each improved approximation from the

Figure 4.9.1

preceding approximation. For this purpose, we note that the point-slope form of the tangent line to $y = f(x)$ at the initial approximation x_1 is

$$y - f(x_1) = f'(x_1)(x - x_1) \qquad (1)$$

If $f'(x_1) \neq 0$, then this line is not parallel to the x-axis and consequently it crosses the x-axis at some point $(x_2, 0)$. Substituting the coordinates of this point in (1) yields

$$-f(x_1) = f'(x_1)(x_2 - x_1)$$

Solving for x_2 we obtain

$$x_2 - x_1 = -\frac{f(x_1)}{f'(x_1)}$$

or

$$x_2 = x_1 - \frac{f(x_1)}{f'(x_1)} \qquad (2)$$

The next approximation can be obtained more easily. If we view x_2 as the starting approximation and x_3 the new approximation, we can simply apply (2) with x_2 in place of x_1 and x_3 in place of x_2. This yields

$$x_3 = x_2 - \frac{f(x_2)}{f'(x_2)} \qquad (3)$$

provided $f'(x_2) \neq 0$. In general, if x_n is the nth approximation, then it is evident from the pattern in (2) and (3) that the improved approximation x_{n+1} is given by:

> **Newton's Method**
>
> $$x_{n+1} = x_n - \frac{f(x_n)}{f'(x_n)} \qquad n = 1, 2, 3, \ldots \qquad (4)$$

REMARK. If $f'(x_n) = 0$ for some n, then this formula does not work because there is a division by zero. However, this is to be expected because the tangent line to $y = f(x)$ is parallel to the x-axis where $f'(x_n) = 0$, and so this tangent line does not cross the x-axis to generate the next approximation (Figure 4.9.2).

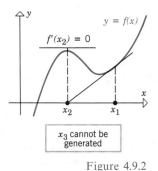

Figure 4.9.2

▶ **Example 1** Use Newton's Method to approximate the real solutions of

$$x^3 - x - 1 = 0.$$

Solution. Let $f(x) = x^3 - x - 1$, so $f'(x) = 3x^2 - 1$ and (4) becomes

$$x_{n+1} = x_n - \frac{x_n^3 - x_n - 1}{3x_n^2 - 1}$$

or after combining terms and simplifying

$$x_{n+1} = \frac{2x_n^3 + 1}{3x_n^2 - 1} \tag{5}$$

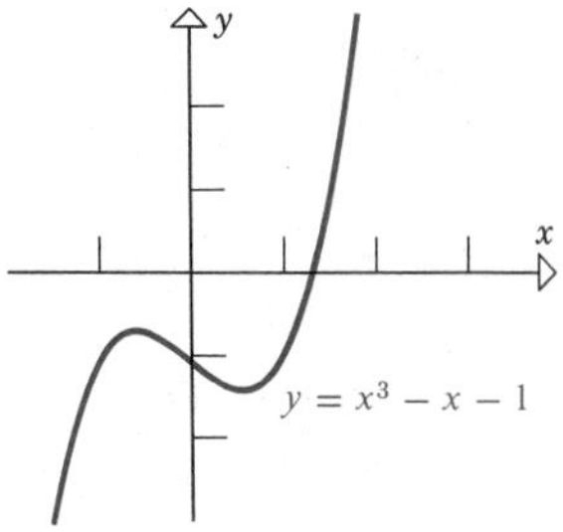

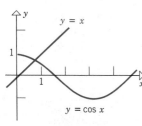

$y = x^3 - x - 1$

Figure 4.9.3

From the graph of f in Figure 4.9.3, we see that the given equation has only one real solution. This solution lies between 1 and 2 because $f(1) = -1 < 0$ and $f(2) = 5 > 0$. We will use $x_1 = 1.5$ as our first approximation ($x_1 = 1$ or $x_1 = 2$ would also be reasonable choices.)

Letting $n = 1$ in (5) and substituting $x_1 = 1.5$ yields

$$x_2 = \frac{2(1.5)^3 + 1}{3(1.5)^2 - 1} = 1.34782609$$

(We used a calculator that displays nine digits.) Next, we let $n = 2$ in (5) and substitute $x_2 = 1.34782609$ to obtain

$$x_3 = \frac{2(1.34782609)^3 + 1}{3(1.34782609)^2 - 1} = 1.32520040$$

If we continue this process until two identical approximations are generated in succession, we will obtain:

$$x_1 = 1.5$$
$$x_2 = 1.34782609$$
$$x_3 = 1.32520040$$
$$x_4 = 1.32471817$$
$$x_5 = 1.32471796$$
$$x_6 = 1.32471796$$

At this stage there is no need to continue further because we have reached the accuracy limit of our calculator, and all subsequent approximations that it generates will be the same. Thus, the solution is approximately $x \approx 1.32471796$ ◀

$y = x$

1

1

$y = \cos x$

Figure 4.9.4

▶ **Example 2** It is evident from Figure 4.9.4 that if x is in radians, then the equation

$$\cos x = x$$

has a solution between 0 and 1. Use Newton's Method to approximate it.

Solution. Rewrite the equation as

$$x - \cos x = 0$$

and apply (4) with $f(x) = x - \cos x$. Since $f'(x) = 1 + \sin x$, (4) becomes

$$x_{n+1} = x_n - \frac{x_n - \cos x_n}{1 + \sin x_n}$$

or after combining terms and simplifying

$$x_{n+1} = \frac{x_n \sin x_n + \cos x_n}{1 + \sin x_n} \tag{6}$$

From Figure 4.9.4, the solution seems closer to $x = 1$ than $x = 0$, so we will use $x_1 = 1$ (radian) as our initial approximation. Letting $n = 1$ in (6) and substituting $x_1 = 1$ yields

$$x_2 = \frac{1 \cdot \sin 1 + \cos 1}{1 + \sin 1} = .750363868$$

Next, letting $n = 2$ in (6) and substituting this value of x_2 yields

$$x_3 = \frac{(.750363868) \sin (.750363868) + \cos (.750363868)}{1 + \sin (.750363868)} = .739112891$$

If we continue this process until two identical approximations are generated in succession we obtain:

$$x_1 = 1$$
$$x_2 = .750363868$$
$$x_3 = .739112891$$
$$x_4 = .739085133$$
$$x_5 = .739085133$$

Thus, to the accuracy limit of our calculator, the solution is $x \approx .739085133$. ◀

REMARK. It is possible to construct examples where the approximations generated by Newton's Method do not get closer and closer to a solution of $f(x) = 0$. (See Exercise 18.) For a statement of conditions under which Newton's Method works and a discussion of error questions, the reader should consult a book on numerical analysis, for example, Peter Henrici, *Elements of Numerical Analysis,* Wiley, New York, 1964.

► Exercise Set 4.9

In this exercise set, use a hand calculator or microcomputer, and keep as many decimal places as your machine displays.

1. Approximate $\sqrt{2}$ by applying Newton's Method to the equation $x^2 - 2 = 0$.

2. Approximate $\sqrt{7}$ by applying Newton's Method to the equation $x^2 - 7 = 0$.

3. Approximate $\sqrt[3]{6}$ by applying Newton's Method to the equation $x^3 - 6 = 0$.

In Exercises 4–7, the equation has one real solution. Approximate it by Newton's Method.

4. $x^3 + x - 1 = 0$.

5. $x^3 - x + 3 = 0$.

6. $x^5 - x + 1 = 0$.

7. $x^5 + x^4 - 5 = 0$.

In Exercises 8–15, the equation has one solution satisfying the given conditions. Approximate it by Newton's Method.

8. $2x^2 + 4x - 3 = 0$; $x < 0$.

9. $2x^2 + 4x - 3 = 0$; $x > 0$.

10. $x^4 + x - 3 = 0$; $x > 0$.

11. $x^4 + x - 3 = 0$; $x < 0$.

12. $x^5 - 5x^3 = 0$; $x > 0$.

13. $2 \sin x = x$; $x > 0$.

14. $\sin x = x^2$; $x > 0$.

15. $x - \tan x = 0$; $\frac{\pi}{2} < x < \frac{3\pi}{2}$.

16. Many hand calculators compute reciprocals using the approximation $1/a \approx x_{n+1}$ where

$$x_{n+1} = x_n(2 - ax_n) \qquad n = 1, 2, 3, \ldots$$

and x_1 is an initial approximation to $1/a$. This formula makes it possible to use multiplications and subtractions (which can be done quickly) to perform divisions that would be slow to obtain directly.

(a) Use Newton's Method to derive this approximation.

(b) Use the formula to approximate $\frac{1}{17}$.

17. The *mechanic's rule* for approximating square roots states that $\sqrt{a} \approx x_{n+1}$ where

$$x_{n+1} = \frac{1}{2}\left(x_n + \frac{a}{x_n}\right) \qquad n = 1, 2, 3, \ldots$$

and x_1 is any positive approximation to $\sqrt{a}$.

(a) Use Newton's Method to derive the mechanic's rule.

(b) Use the mechanic's rule to approximate $\sqrt{10}$.

18. (a) Show that Newton's Method does not work for the equation $x^{1/3} = 0$ if $x_1 \neq 0$.

(b) Sketch the graph of $f(x) = x^{1/3}$ and illustrate the difficulty geometrically.

4.10 ROLLE'S THEOREM; MEAN-VALUE THEOREM

In this section we will discuss a result called *the Mean-Value Theorem*. There are so many major consequences of this theorem that it is regarded as one of the most fundamental results in calculus.

We will begin with a special case of the Mean-Value Theorem, called **Rolle's* Theorem**. Geometrically, Rolle's Theorem states that between any two points where a "well-behaved" curve $y = f(x)$ crosses the x-axis, there must be at least one place where the tangent line to the curve is horizontal (Figure 4.10.1). The precise statement of this result is as follows.

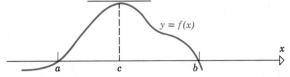

Figure 4.10.1

4.10.1 THEOREM

Rolle's Theorem

Let f be differentiable on (a, b) and continuous on $[a, b]$. If $f(a) = f(b) = 0$, then there is at least one point c in (a, b) where $f'(c) = 0$

Proof. Either $f(x)$ is equal to zero for all x in $[a, b]$ or it is not. If it is, then $f'(x) = 0$ for all x in (a, b), since f is constant on (a, b). Thus, for any c in (a, b)

$$f'(c) = 0$$

If $f(x)$ is not equal to zero for all x in $[a, b]$, then there must be a point x in (a, b) where $f(x) > 0$ or $f(x) < 0$. We will consider the first case and leave the second as an exercise.

Since f is continuous on $[a, b]$, it follows from the Extreme-Value Theorem that f has a maximum value at some point c in $[a, b]$. Since $f(a) = f(b) = 0$ and $f(x) > 0$ at some point in (a, b), the point c cannot be an endpoint; it must lie in (a, b). By hypothesis, f is differentiable everywhere on (a, b). In particular, it is differentiable at c so that

$$f'(c) = 0$$

by Theorem 4.6.5. ∎

*MICHEL ROLLE (1652–1719) French mathematician. Rolle, the son of a shopkeeper, received only an elementary education. He married early and as a young man struggled hard to support his family on the meager wages of a transcriber for notaries and attorneys. In spite of his financial problems and minimal education, Rolle studied algebra and Diophantine analysis (a branch of number theory) on his own. Rolle's fortune changed dramatically in 1682 when he published an elegant solution of a difficult, unsolved problem in Diophantine analysis. The public recognition of his achievement led to a patronage under minister Louvois, a job as an elementary mathematics teacher, and eventually to a short-termed administrative post in the Ministry of War. In 1685 he joined the Académie des Sciences in a very low-level position for which he received no regular salary until 1699. He remained there until he died of apoplexy in 1719.

While Rolle's forté was always Diophantine analysis, his most important work was a book on the algebra of equations, called *Traité d'algèbre,* published in 1690. In that book Rolle firmly established the notation $\sqrt[n]{a}$ [earlier written as $\sqrt{\textcircled{n} a}$] for the nth root of a, and proved a polynomial version of the theorem that today bears his name. (Rolle's Theorem was named by Giusto Bellavitis in 1846.) Ironically, Rolle was one of the most vocal early antagonists of calculus. He strove intently to demonstrate that it gave erroneous results and was based on unsound reasoning. He quarreled so vigorously on the subject that the Académie des Sciences was forced to intervene on several occasions. Among his several achievements, Rolle helped advance the currently accepted size order for negative numbers. Descartes, for example, viewed -2 as smaller than -5. Rolle preceded most of his contemporaries by adopting the current convention in 1691.

▶ Example 1   Let

$$f(x) = \sin x$$

The function $\sin x$ is both continuous and differentiable everywhere, so that $f(x) = \sin x$ is continuous on $[0, 2\pi]$ and differentiable on $(0, 2\pi)$. Moreover,

$$f(0) = \sin 0 = 0 \quad \text{and} \quad f(2\pi) = \sin 2\pi = 0$$

so that f satisfies the hypotheses of Rolle's Theorem on the interval $[0, 2\pi]$. Thus, we are guaranteed that there is at least one point c in $(0, 2\pi)$ satisfying

$$f'(c) = \cos c = 0 \tag{1}$$

Because (1) is so simple, we can actually find the values of c; they are

$$c_1 = \frac{\pi}{2} \quad \text{and} \quad c_2 = \frac{3\pi}{2}$$

Sometimes, however, the function f is sufficiently complicated that it is difficult or impossible to find values for c.   ◀

Rolle's Theorem is a special case of a more general result, which states that between any two points A and B on a "well-behaved" curve $y = f(x)$, there must be at least one place where the tangent line to the curve is parallel to the secant line joining A and B (Figure 4.10.2).

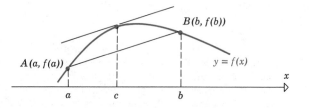

Figure 4.10.2

This result is called the **Mean-Value Theorem.** Noting that the slope of the secant line joining $A(a, f(a))$ and $B(b, f(b))$ is

$$\frac{f(b) - f(a)}{b - a}$$

and the slope of the tangent at c is $f'(c)$, the Mean-Value Theorem can be stated precisely as follows.

4.10.2   THEOREM
Mean-Value Theorem   *Let f be differentiable on (a, b) and continuous on $[a, b]$. Then there is at least one point c in (a, b) where*

$$f'(c) = \frac{f(b) - f(a)}{b - a} \tag{2}$$

The key to proving this theorem is the intuitive observation that the tangent line will be parallel to the secant line at the points where the vertical distance between the curve $y = f(x)$ and the secant line is greatest (Figure 4.10.2). Since the two-point form of the secant line joining $(a, f(a))$ and $(b, f(b))$ is

$$y - f(a) = \frac{f(b) - f(a)}{b - a}(x - a)$$

or equivalently

$$y = \frac{f(b) - f(a)}{b - a}(x - a) + f(a)$$

the difference $v(x)$ between the height of the graph of f and the height of the secant line is

$$v(x) = f(x) - \left[\frac{f(b) - f(a)}{b - a}(x - a) + f(a)\right] \tag{3}$$

Since the maximum vertical distance between $y = f(x)$ and the secant line is an extreme value for $v(x)$, we are led naturally to investigate the stationary points of v. Let c be any point where $v'(c) = 0$. From (3)

$$v'(x) = f'(x) - \frac{f(b) - f(a)}{b - a}$$

so that

$$v'(c) = f'(c) - \frac{f(b) - f(a)}{b - a}$$

or since $v'(c) = 0$,

$$f'(c) = \frac{f(b) - f(a)}{b - a}$$

Thus, we have shown that (2) holds at any point c where $v'(c) = 0$. Therefore, to prove the Mean-Value Theorem we need only show that there is a point c in the open interval (a, b) where $v'(c) = 0$. This can be done as follows.

Proof of Theorem 4.10.2. Since $f(x)$ is continuous on $[a, b]$ and differentiable on (a, b), so is the function $v(x)$ in (3). It is easy to check that

$$v(a) = 0 \quad \text{and} \quad v(b) = 0$$

so that $v(x)$ satisfies the hypotheses of Rolle's Theorem on the interval $[a, b]$. Thus, there is a point c in (a, b) such that $v'(c) = 0$. ∎

▶ Example 2   Let $f(x) = x^3 + 1$. Show that f satisfies the hypotheses of the Mean-Value Theorem on the interval $[1, 2]$ and find all values of c in this interval whose existence is guaranteed by the theorem.

Solution.   Because f is a polynomial, f is continuous and differentiable everywhere. In particular, f is continuous on $[1, 2]$ and differentiable on $(1, 2)$, so that the hypotheses of the Mean-Value Theorem are satisfied.
   If we take $a = 1$ and $b = 2$, then

$$f(a) = f(1) = 2, \qquad f(b) = f(2) = 9$$
$$f'(x) = 3x^2, \qquad f'(c) = 3c^2$$

so that the equation

$$f'(c) = \frac{f(b) - f(a)}{b - a}$$

becomes

$$3c^2 = 7$$

Of the two solutions,

$$c = \sqrt{7/3} \qquad \text{and} \qquad c = -\sqrt{7/3}$$

only the first is in the interval $(1, 2)$ so that $c = \sqrt{7/3}$ is the only number whose existence is guaranteed by the Mean-Value Theorem.   ◀

   We stated earlier that the Mean-Value Theorem is the starting point for many important results in calculus. We will conclude this section with two such results.
   We know that the derivative of a constant function is zero. Using the Mean-Value Theorem we will now prove the converse is also true.

4.10.3   THEOREM   *If $f'(x) = 0$ for all x in an interval, then f is constant on the interval.*

Proof.   Assume $f'(x) = 0$ for all x in an interval I. To prove that f is constant on I it suffices to show that f has the same value at any two points in I. Let a and b be arbitrary points in I with $a < b$. Since f is differentiable on I, it is continuous on I and consequently the hypotheses of the Mean-Value Theorem hold on the interval $[a, b]$. Thus, there is a number c in (a, b) such that

$$f'(c) = \frac{f(b) - f(a)}{b - a}$$

But $f'(c) = 0$ by hypothesis so that $f(b) = f(a)$, which proves that f has the same value at any two points in I.   ▪

4.10.4 COROLLARY   *If $f'(x) = g'(x)$ for all x in an interval I, then f and g differ by a constant on I, that is, there is a constant k such that $f(x) - g(x) = k$ for all x in I.*

Proof. Let $h(x) = f(x) - g(x)$. Then for every x in I

$$h'(x) = f'(x) - g'(x) = 0$$

Thus, $h(x) = f(x) - g(x)$ is constant on I by Theorem 4.10.3. ∎

This corollary has a useful geometric interpretation. It states that two functions with the same derivative at each point of an interval have "parallel" graphs over the interval (Figure 4.10.3).

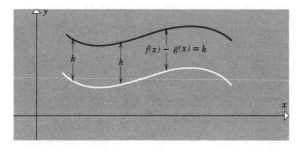

Figure 4.10.3

▶ Exercise Set 4.10

In Exercises 1–4, verify that the hypotheses of Rolle's Theorem are satisfied on the given interval and find all values of c that satisfy the conclusion of the theorem.

1. $f(x) = x^2 - 6x + 8$; [2, 4].

2. $f(x) = x^3 - 3x^2 + 2x$; [0, 2].

3. $f(x) = \cos x$; $\left[\dfrac{\pi}{2}, \dfrac{3\pi}{2}\right]$.

4. $f(x) = \dfrac{x^2 - 1}{x - 2}$; [−1, 1].

In Exercises 5–8, verify that the hypotheses of the Mean-Value Theorem are satisfied on the given interval and find all values of c that satisfy the conclusion of the theorem.

5. $f(x) = x^2 + x$; [−4, 6].

6. $f(x) = x^3 + x - 4$; [−1, 2].

7. $f(x) = \sqrt{x + 1}$; [0, 3].

8. $f(x) = x + \dfrac{1}{x}$; [3, 4].

9. Let $f(x) = \tan x$.
    (a) Show that there is no point c in $(0, \pi)$ such that $f'(c) = 0$, even though $f(0) = f(\pi) = 0$.
    (b) Explain why the result in part (a) does not violate Rolle's Theorem.

10. Let $f(x) = x^{2/3}$, $a = -1$, and $b = 8$.
    (a) Show that there is no point c in (a, b) such that

    $$f'(c) = \frac{f(b) - f(a)}{b - a}$$

    (b) Explain why the result in part (a) does not violate the Mean-Value Theorem.

11. Use the Mean-Value Theorem to prove that

    $$|\sin x - \sin y| \leq |x - y|$$

    holds for all real values of x and y.

12. Use the Mean-Value Theorem to prove that

$$|\tan x + \tan y| \geq |x + y|$$

holds for all real values of x and y in the interval $(-\pi/2, \pi/2)$.

13. Use Rolle's Theorem to prove that

$$4x^3 + 9x^2 - 4x - 2 = 0$$

has a solution in the interval $(0, 1)$.

$$\left[Hint: \frac{d}{dx}[x^4 + 3x^3 - 2x^2 - 2x] \right.$$

$$\left. = 4x^3 + 9x^2 - 4x - 2. \right]$$

14. Use Rolle's Theorem to prove that

$$x^3 + 4x - 1 = 0$$

does not have more than one real root.

15. Let f and g be continuous on $[a, b]$ and differentiable on (a, b). Prove: If $f(a) = g(a)$ and $f(b) = g(b)$, then there is a point c in (a, b) where $f'(c) = g'(c)$.

16. Prove: If f is continuous on $[a, b]$ and $f'(x) = 0$ for all x in (a, b), then f is constant on $[a, b]$.

17. Prove: If f and g are continuous on $[a, b]$ and $f'(x) = g'(x)$ for all x in (a, b), then f and g differ by a constant on $[a, b]$.

18. (a) Prove: If $f''(x) > 0$ for all x in (a, b), then $f'(x) = 0$ at most once in (a, b).
    (b) Give a geometric interpretation of the result in part (a).

19. Prove: If f is continuous on $[a, b]$ and differentiable on (a, b), and if $f(a) = f(b)$, then there is a point c in (a, b) where $f'(c) = 0$.

20. Let $s = f(t)$ be the position versus time curve for a particle moving in the positive direction along a coordinate line. Prove: If f satisfies the hypotheses of the Mean-Value Theorem on a time interval $[a, b]$, then there will be an instant t_0 in (a, b) where the instantaneous velocity at time t_0 equals the average velocity over the interval $[a, b]$.

21. Complete the proof of Rolle's Theorem by considering the case where $f(x) < 0$ at some point x in (a, b).

22. Use the Mean-Value Theorem to prove:

$$1.5 < \sqrt{3} < 1.75$$

[Hint: Let $f(x) = \sqrt{x}$, $a = 3$, and $b = 4$ in the Mean-Value Theorem.]

23. Theorem 4.2.2 gives conditions under which a function is increasing or decreasing on an *open* interval. Prove the following result concerned with *closed* intervals: Let the function f be *continuous* on the closed interval $[a, b]$ and differentiable on the open interval (a, b):
    (a) if $f'(x) > 0$ for all x in (a, b), then f is increasing on $[a, b]$;
    (b) if $f'(x) < 0$ for all x in (a, b), then f is decreasing on $[a, b]$.

4.11   PROOFS OF KEY RESULTS USING THE MEAN-VALUE THEOREM (OPTIONAL)

In this section we will prove three results that were stated without proof in earlier sections: Theorem 4.2.2, Theorem 4.3.6 (the first derivative test), and Theorem 4.3.7 (the second derivative test). As we shall see, the proofs of all these important results rest on the Mean-Value Theorem.

4.2.2   THEOREM   (a) *If $f'(x) > 0$ on an open interval (a, b), then f is increasing on (a, b).*
(b) *If $f'(x) < 0$ on an open interval (a, b), then f is decreasing on (a, b).*

Proof of (a).   Let x_1 and x_2 be points in (a, b) such that $x_1 < x_2$. We must show that $f(x_1) < f(x_2)$. Since f is differentiable on (a, b), it is continuous on (a, b). Therefore, f is continuous on $[x_1, x_2]$ and differentiable on (x_1, x_2), since the interval $[x_1, x_2]$ is contained inside the interval (a, b). Thus, we can apply the Mean-Value Theorem over the interval $[x_1, x_2]$ and conclude that there is a number c in the interval (x_1, x_2) such that

$$\frac{f(x_2) - f(x_1)}{x_2 - x_1} = f'(c)$$

or equivalently

$$f(x_2) - f(x_1) = f'(c)(x_2 - x_1) \tag{1}$$

But $x_2 - x_1 > 0$ since $x_1 < x_2$, and $f'(c) > 0$ follows from the hypothesis; thus, from (1)

$$f(x_2) - f(x_1) > 0$$

or equivalently

$$f(x_1) < f(x_2) \quad \blacksquare$$

The proof of part (b) is similar and is left as an exercise.

4.3.6   THEOREM   *Suppose f is continuous at a critical point x_0.*
(a)   If $f'(x) > 0$ on an open interval extending left from x_0 and $f'(x) < 0$ on an open interval extending right from x_0, then f has a relative maximum at x_0.
(b)   If $f'(x) < 0$ on an open interval extending left from x_0 and $f'(x) > 0$ on an open interval extending right from x_0, then f has a relative minimum at x_0.
(c)   If $f'(x)$ has the same sign [either $f'(x) > 0$ or $f'(x) < 0$] on an open interval extending left from x_0 and on an open interval extending right from x_0, then f does not have a relative extremum at x_0.

Proof of (a).   In accordance with the hypothesis, assume that $f'(x) > 0$ on the interval (a, x_0) and $f'(x) < 0$ on the interval (x_0, b). To prove that f has a relative maximum at x_0, we will show that

$$f(x_0) \geq f(x) \tag{2}$$

for all x in (a, b). First let x be any point in the interval (a, x_0). Because f is differentiable on the interval (a, x_0) and because of the hypothesis that f is continuous at x_0, it follows that the hypotheses of the Mean-Value Theorem are satisfied on the interval $[x, x_0]$. Thus, there is a point c in the interval (x, x_0) such that

$$\frac{f(x_0) - f(x)}{x_0 - x} = f'(c)$$

or equivalently

$$f(x_0) - f(x) = (x_0 - x)f'(c) \tag{3}$$

But $x_0 - x > 0$ since $x_0 > x$, and $f'(c) > 0$ holds since f' is positive everywhere on the interval (a, x_0). Thus, from (3)

$$f(x_0) - f(x) > 0$$

which shows that (2) holds for all x in the interval (a, x_0).

Next, let x be any point in the interval (x_0, b). As above, the hypotheses of the Mean-Value Theorem hold on the interval $[x_0, x]$, so there is a point c in the interval (x_0, x) such that

$$\frac{f(x) - f(x_0)}{x - x_0} = f'(c)$$

or equivalently

$$f(x) - f(x_0) = (x - x_0)f'(c) \tag{4}$$

But $x - x_0 > 0$ since $x > x_0$, and $f'(c) < 0$ holds since f' is negative everywhere on the interval (x_0, b). Thus, from (4)

$$f(x) - f(x_0) < 0$$

which shows that (2) holds for all x in the interval (x_0, b).

Since (2) obviously holds when $x = x_0$, we have shown that it holds for all x in (a, b). ▉

The proofs of parts (b) and (c) are left as exercises.

4.3.7 THEOREM *Suppose f is twice differentiable at a stationary point x_0.*
(a) If $f''(x_0) > 0$, then f has a relative minimum at x_0.
(b) If $f''(x_0) < 0$, then f has a relative maximum at x_0.

Proof of (a). Using the definition of a derivative we can write

$$f''(x_0) = \lim_{h \to 0} \frac{f'(x_0 + h) - f'(x_0)}{h} \tag{5}$$

By hypothesis, $f''(x_0) > 0$ so that we can use $\epsilon = \frac{1}{2} f''(x_0)$ in the definition of a limit and deduce from (5) that there exists a $\delta > 0$ such that

$$\left| \frac{f'(x_0 + h) - f'(x_0)}{h} - f''(x_0) \right| < \frac{1}{2} f''(x_0) \tag{6}$$

whenever h satisfies

$$0 < |h| < \delta \tag{7}$$

To prove that f has a relative minimum at x_0, we will show that

$$f'(x) > 0 \qquad \text{for all } x \text{ in } (x_0, x_0 + \delta) \tag{8}$$

and

$$f'(x) < 0 \qquad \text{for all } x \text{ in } (x_0 - \delta, x_0) \tag{9}$$

It will then follow from the first derivative test that f has a relative minimum at x_0.

From (6) and (7) it follows that

$$\frac{1}{2}f''(x_0) < \frac{f'(x_0 + h) - f'(x_0)}{h} < \frac{3}{2}f''(x_0) \tag{10}$$

whenever h satisfies

$$0 < |h| < \delta$$

By hypothesis, x_0 is a stationary point for f, so that $f'(x_0) = 0$. From this, the fact that $f''(x_0) > 0$, and the left-hand inequality in (10), it follows that

$$0 < \frac{f'(x_0 + h)}{h} \tag{11}$$

whenever h satisfies (7).

To prove (8), let x be any point in $(x_0, x_0 + \delta)$. If we let

$$h = x - x_0 \tag{12}$$

then $h > 0$ and h satisfies (7), so that from (11)

$$0 < \frac{f'(x_0 + h)}{h}$$

or on multiplying by h

$$0 < f'(x_0 + h) \tag{13}$$

Substituting (12) in (13) yields $0 < f'(x)$, which establishes (8). To obtain (9) let x be any point in $(x_0 - \delta, x_0)$. If we let

$$h = x - x_0 \tag{14}$$

then $h < 0$ and h satisfies (7) so that from (11)

$$0 < \frac{f'(x_0 + h)}{h}$$

or on multiplying by h

$$f'(x_0 + h) < 0 \tag{15}$$

Substituting (14) in (15) yields $f'(x) < 0$, which establishes (9). ▌

The proof of part (b) is similar and is left as an exercise.

▶ Exercise Set 4.11

1. Prove part (b) of Theorem 4.2.2.

2. Prove part (b) of Theorem 4.3.6.

3. Prove part (c) of Theorem 4.3.6.

4. Prove part (b) of Theorem 4.3.7.

5. A function f is called *nondecreasing* on a given interval if

   $$f(x_2) \geq f(x_1) \text{ whenever } x_2 > x_1$$

   where x_1 and x_2 are points in the interval. Similarly, f is called *nonincreasing* on the interval if

   $$f(x_2) \leq f(x_1) \text{ whenever } x_2 > x_1$$

   (a) Sketch the graph of a function that is nondecreasing, yet is not increasing.
   (b) Sketch the graph of a function that is nonincreasing, yet is not decreasing.
   (c) Prove: If $f'(x) \geq 0$ on an open interval (a, b), then f is nondecreasing on (a, b).
   (d) Prove: If $f'(x) \leq 0$ on an open interval (a, b), then f is nonincreasing on (a, b).

   [REMARK. Unfortunately, terminology is not always consistent in the mathematical literature. Some writers use the terms "strictly increasing" and "strictly decreasing" where we have used the terms increasing and decreasing. These writers then use the terms "increasing" and "decreasing" where we have used the terms nondecreasing and nonincreasing.]

6. Prove: if $f'(x) < g'(x)$ for all x in (a, b), then for

all points x_1, x_2 in (a, b), where $x_2 > x_1$,

$$f(x_2) - f(x_1) < g(x_2) - g(x_1)$$

7. (a) Prove: If f is continuous on (a, b) and $f'(x) > 0$, except at a single point x_0 in (a, b), then f is increasing on (a, b).
   (b) Does this result remain true if the condition $f'(x) > 0$ fails to hold at any finite number of points in (a, b)? Justify your answer.
   (c) Use part (a) to show that $f(x) = x^3$ is increasing on $(-\infty, +\infty)$.

8. (a) Prove: If $f'(x_0) > 0$ and f' is continuous at x_0, then there is an open interval containing x_0 on which f is increasing.
   (b) Prove: If $f'(x_0) < 0$ and f' is continuous at x_0, then there is an open interval containing x_0 on which f is decreasing.

9. (a) Use Exercise 8(a) to prove: If $f''(x_0) > 0$ and f'' is continuous at x_0, then there is an open interval containing x_0 on which f is concave up.
   (b) Use Exercise 8(b) to prove: If $f''(x_0) < 0$ and f'' is continuous at x_0, then there is an open interval containing x_0 on which f is concave down.

10. Prove: If f is differentiable at each point of the interval $[a, b]$ and f' is continuous at a and b, and $f'(a)f'(b) < 0$, then there is a number c in (a, b) such that $f'(c) = 0$. [*Hint:* Use the results in Exercise 8.]

▶ SUPPLEMENTARY EXERCISES

1. For the hollow cylinder shown, assume that R and r are increasing at a rate of 2 meters/sec, and h is decreasing at a rate of 3 meters/sec. At what rate is the volume changing when $R = 7$ m, $r = 4$ m, and $h = 5$ m?

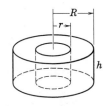

2. The vessel shown is filled at the rate of 4 ft³/min. How fast is the fluid level rising when the level is 1 ft?

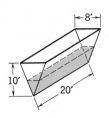

3. A ball is dropped from a point 10 ft away from a light at the top of a 48-ft pole as shown. When the ball has dropped 16 ft, its velocity (downward) is 32 ft/sec. At what rate is its shadow moving along the ground at that instant?

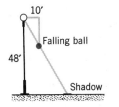

In Exercises 4–9, find the minimum value m and the maximum value M of f on the indicated interval (if they exist) and state where these extreme values occur.

4. $f(x) = 1/x$; $[-2, -1]$.

5. $f(x) = x^3 - x^4$; $[-1, \frac{3}{2}]$.

6. $f(x) = x^2(x - 2)^{1/3}$; $(0, 3]$.

7. $f(x) = 2x/(x^2 + 3)$; $(0, 2]$.

8. $f(x) = 2x^5 - 5x^4 + 7$; $(-1, 3)$.

9. $f(x) = -|x^2 - 2x|$; $[1, 3]$.

10. Use Newton's Method to approximate the smallest positive solution of $\sin x + \cos x = 0$.

11. Use Newton's Method to approximate all three solutions of $x^3 - 4x + 1 = 0$.

In Exercises 12–19, sketch the graph of f. Use symmetry, where possible, and show all relative extrema, inflection points, and asymptotes.

12. $f(x) = (x^2 - 3)^2$.

13. $f(x) = \dfrac{1}{1 + x^2}$.

14. $f(x) = \dfrac{2x}{1 + x}$.

15. $f(x) = \dfrac{x^3 - 2}{x}$.

16. $f(x) = (1 + x)^{2/3}(3 - x)^{1/3}$.

17. $f(x) = 2\cos^2 x$, $0 \le x \le \pi$.

18. $f(x) = x - \tan x$, $0 \le x \le 2\pi$.

19. $f(x) = \dfrac{3x}{(x + 8)^2}$.

20. Use implicit differentiation to show that a function defined implicitly by $\sin x + \cos y = 2y$ has a critical point whenever $\cos x = 0$. Then use either the first or second derivative test to classify these critical points as relative maxima or minima.

21. Find the equations of the tangent lines at all inflection points of the graph of
$$f(x) = x^4 - 6x^3 + 12x^2 - 8x + 3.$$

In Exercises 22–24, find all critical points and use the first derivative test to classify them.

22. $f(x) = x^{1/3}(x - 7)^2$.

23. $f(x) = 2\sin x - \cos 2x$, $0 \le x \le 2\pi$.

24. $f(x) = 3x - (x - 1)^{3/2}$.

In Exercises 25–27, find all critical points and use the second derivative test (if possible) to classify them.

25. $f(x) = x^{-1/2} + \frac{1}{9}x^{1/2}$.

26. $f(x) = x^2 + 8/x$.

27. $f(x) = \sin^2 x - \cos x$, $0 \le x \le 2\pi$.

28. Find two nonnegative numbers whose sum is 20 and such that: (a) the sum of their squares is a maximum, and (b) the product of the square of one and the cube of the other is a maximum.

29. Find the dimensions of the rectangle of maximum area that can be inscribed inside the ellipse $(x/4)^2 + (y/3)^2 = 1$.

30. Find the coordinates of the point on the curve $2y^2 = 5(x + 1)$ that is nearest to the origin. [Note that all points $P(x, y)$ on the curve satisfy $x \ge -1$.]

31. A church window consists of a blue semicircular section surmounting a clear rectangular section as shown. The blue glass lets through half as much light per unit area as the clear glass. Find the radius r of the window that admits the most light if the perimeter of the entire window is to be P feet.

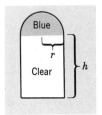

Blue

r

Clear h

32. The cost c (in dollars per hour) to run an ocean liner at a constant speed v (in miles per hour) is given by $c = a + bv^n$ where a, b, and n are positive constants with $n > 1$. Find the speed needed to make the cheapest 3000-mi run.

33. A soup can in the shape of a right-circular cylinder of radius r and height h is to have a prescribed volume V. The top and bottom are cut from squares as shown. If the shaded corners are wasted, but there is no other waste, find the ratio r/h for the can requiring the least material (including waste).

r

34. If a calculator factory produces x calculators per day, the total daily cost (in dollars) incurred is $0.25x^2 + 35x + 25$. If they are sold for $50 - \frac{1}{2}x$ dollars each, find the value of x that maximizes the daily profit.

In Exercises 35–37, determine if all hypotheses of Rolle's Theorem are satisfied on the stated interval. If not, state which hypotheses fail; if so, find all values of c guaranteed in the conclusion of the theorem.

35. $f(x) = \sqrt{4 - x^2}$ on $[-2, 2]$.

36. $f(x) = x^{2/3} - 1$ on $[-1, 1]$.

37. $f(x) = \sin(x^2)$ on $[0, \sqrt{\pi}]$.

In Exercises 38–41, determine if all hypotheses of the Mean-Value Theorem are satisfied on the stated interval. If not, state which hypotheses fail; if so, find all values of c guaranteed in the conclusion of the theorem.

38. $f(x) = |x - 1|$ on $[-2, 2]$.

39. $f(x) = \sqrt{x}$ on $[0, 4]$.

40. $f(x) = \dfrac{x + 1}{x - 1}$ on $[2, 3]$.

41. $f(x) = \begin{cases} 3 - x^2 & \text{if } x \le 1 \\ 2/x & \text{if } x > 1 \end{cases}$ on $[0, 2]$

5 integration

5.1 INTRODUCTION

In this chapter we will study the second major problem of calculus:

THE AREA PROBLEM. Given a function f that is continuous and nonnegative on an interval $[a, b]$, find the area between the graph of f and the interval $[a, b]$ on the x-axis (Figure 5.1.1).

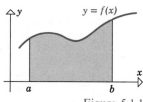

Figure 5.1.1

Area formulas for basic geometric figures such as rectangles, polygons, and circles date back to the earliest written records of mathematics. The first real advance beyond the elementary level of area computation was made by the Greek mathematician, Archimedes,* (see p. 274) who devised an ingenius, but cumbersome method for obtaining areas, called the *method of exhaustion*. Using this technique Archimedes was able to obtain areas of parabolic regions and spirals. By the early seventeenth century several mathematicians had learned to obtain such areas more simply by calculating appropriate limits. However, both the method of exhaustian and its successor lacked generality. For each different problem one had to devise special procedures or formulas that worked, and more often than not these were difficult or impossible to obtain. The major breakthrough in solving the general area problem was made independently by Newton and Leibniz when they discovered that areas could be obtained by reversing the process of differentiation. This major discovery, which marked the real beginning of calculus, was circulated by Newton in 1669 and then published in 1711 in a paper entitled, *De Analysi per Aequationes Numero Terminorum Infinitas* (*On Analysis by Means of Equations with Infinitely Many Terms.*) Independently, Leibniz discovered the same result around 1673 and stated it in an unpublished manuscript dated November 11, 1675.

*ARCHIMEDES (287 B.C.–212 B.C.) Greek mathematician and scientist. Born in Syracuse, Sicily, Archimedes was the son of the astronomer Pheidias and possibly related to Heiron II, king of Syracuse. Most of the facts about his life come from the Roman biographer, Plutarch, who inserted a few tantalizing pages about him in the massive biography of the Roman soldier, Marcellus. In the words of one writer, "the account of Archimedes is slipped like a tissue-thin shaving of ham in a bull-choking sandwich."

Archimedes ranks with Newton and Gauss as one of the three greatest mathematicians who ever lived, and he is certainly the greatest mathematician of antiquity. His mathematical work is so modern in spirit and technique that it is barely distinguishable from that of a seventeenth-century mathematician, yet it was all done without benefit of algebra or a convenient number system. Among his mathematical achievements, Archimedes developed a general method (exhaustian) for finding areas and volumes, and he used the method to find areas bounded by parabolas and spirals and to find volumes of cylinders, paraboloids, and segments of spheres. He gave a procedure for approximating π and bounded its value between $3\frac{1}{7}$ and $3\frac{10}{71}$. In spite of the limitations of the Greek numbering system, he devised methods for finding square roots and invented a method based on the Greek myriad (10,000) for representing numbers as large as 1 followed by 80 million billion zeros.

Of all his mathematical work, Archimedes was most proud of his discovery of the method for finding the volume of a sphere—he showed that the volume of a sphere is two thirds the volume of the smallest cylinder that can contain it. At his request, the figure of a sphere and cylinder was engraved on his tombstone.

In addition to mathematics, Archimedes worked extensively in mechanics and hydrostatics. Nearly every school child knows Archimedes as the absent-minded scientist who, on realizing that a floating object displaces its weight of liquid, leaped from his bath and ran naked through the streets of Syracuse shouting, "Eureka, Eureka!"—(meaning, "I have found it!"). Archimedes actually created the discipline of hydrostatics and used it to find equilibrium positions for various floating bodies, he laid down the fundamental postulates of mechanics, discovered the laws of levers, and calculated centers of gravity for various flat surfaces and solids. In the excitement of discovering the mathematical laws of the lever, he is said to have declared, "Give me a place to stand and I will move the earth."

Although Archimedes was apparently more interested in pure mathematics than its applications, he was an engineering genius. During the second Punic war, when Syracuse was attacked by the Roman fleet under the command of Marcellus, it was reported by Plutarch that Archimedes' military inventions held the fleet at bay for three years. He invented super catapults that showered the Romans with rocks weighing a quarter ton or more, and fearsome mechanical devices with iron "beaks and claws" that reached over the city walls, grasped the ships, and spun them against the rocks. After the first repulse, Marcellus called Archimedes a "geometrical Briareus (a hundred-armed mythological monster) who uses our ships like cups to ladle water from the sea."

Eventually the Roman army was victorious and contrary to Marcellus' specific orders the 75-year-old Archimedes was killed by a Roman soldier. According to one report of the incident, the soldier cast a shadow across the sand in which Archimedes was working on a mathematical problem. When the annoyed Archimedes yelled, "Don't disturb my circles," the soldier flew into a rage and cut the old man down.

With his death the Greek gift of mathematics passed into oblivion, not to be fully resurrected again until the sixteenth century. Unfortunately, there is no known accurate likeness or statue of this great man.

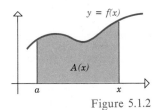

Figure 5.1.2

To obtain a better understanding of this major discovery, consider a continuous curve $y = f(x)$ lying above the x-axis, and let $[a, x]$ be an interval on the x-axis with a fixed left-hand endpoint a and a variable right-hand endpoint. As we move the right-hand endpoint, the area between $y = f(x)$ and the interval $[a, x]$ changes and is therefore a function of x. We denote it by $A(x)$ (Figure 5.1.2). Newton and Leibniz showed that the derivative of the area function $A(x)$ is $f(x)$; that is,

$$A'(x) = f(x) \tag{1}$$

Thus, finding the area function reduces to "reversing" the differentiation process and recovering $A(x)$ from its known derivative $A'(x)$. Once $A(x)$ is found, the area under $y = f(x)$ over a specific interval $[a, b]$ can be obtained by evaluating $A(x)$ at $x = b$.

The following example illustrates the method.

▶ Example 1   Find the area under the graph of $f(x) = x^2$ over the interval $[0, 1]$.

Solution.   Let $A(x)$ denote the area under the graph of $f(x) = x^2$ over the interval $[0, x]$. We are interested in finding the value $A(1)$. Since $f(x) = x^2$, it follows from (1) that

$$A'(x) = x^2 \tag{2}$$

Thus, to find $A(x)$ we must look for a function whose derivative is x^2. This is called an ***antidifferentiation*** problem because we are trying to find $A(x)$ by "undoing" a differentiation.

By simply guessing we see that

$$A(x) = \frac{1}{3}x^3$$

is one solution to (2). But this is not the only solution, since

$$A(x) = \frac{1}{3}x^3 + C \tag{3}$$

also satisfies (2) for any constant C. However, there is one additional condition that we can use to single out the appropriate choice for $A(x)$. If $x = 0$, then the interval $[0, x]$ reduces to a single point and consequently the area under the graph of f above this interval is zero. In other words,

$$A(x) = 0 \quad \text{if} \quad x = 0$$

It therefore follows from (3) with $x = 0$ that

$$A(0) = 0 + C = 0$$

or

$$C = 0$$

Substituting this value in (3) gives

$$A(x) = \frac{1}{3}x^3$$

Thus, the area under the graph of $f(x) = x^2$ over the interval $[0, 1]$ is

$$A(1) = \frac{1}{3}$$ ◀

This example shows how certain areas that are impossible to obtain using formulas from plane geometry can be determined using calculus. However, the solution hinged on our ability to *guess* at a function $A(x)$ satisfying $A'(x) = x^2$.

In this chapter we will develop a systematic way of finding functions from their derivatives and we will use the results to calculate areas.

5.2 ANTIDERIVATIVES; THE INDEFINITE INTEGRAL

The problem of finding a function whose derivative is specified arose in the last section. In this section we will discuss this problem in more detail.

5.2.1   DEFINITION   A function F is called an ***antiderivative*** of a function f if the derivative of F is f.

▶ Example 1   The functions

$$\frac{1}{3}x^3, \qquad \frac{1}{3}x^3 + 2, \qquad \frac{1}{3}x^3 - \pi$$

are antiderivatives of $f(x) = x^2$ since

$$\frac{d}{dx}\left[\frac{1}{3}x^3\right] = \frac{d}{dx}\left[\frac{1}{3}x^3 + 2\right] = \frac{d}{dx}\left[\frac{1}{3}x^3 - \pi\right] = x^2$$ ◀

This example shows that a function can have many antiderivatives. In fact, if $F(x)$ is any antiderivative of $f(x)$ and C is any constant, then

$$F(x) + C$$

is also an antiderivative of $f(x)$ since

$$\frac{d}{dx}[F(x) + C] = \frac{d}{dx}[F(x)] + \frac{d}{dx}[C] = f(x) + 0 = f(x)$$

It is reasonable to ask if there are other antiderivatives of f that cannot be obtained by adding a constant to F. Provided we consider only values of x in an interval I, the answer is *no*. To see this, let $G(x)$ be any other antiderivative of $f(x)$; then

$$\frac{d}{dx}[F(x)] = \frac{d}{dx}[G(x)] = f(x)$$

so that by Corollary 4.10.4, F and G differ only by some constant C on I, that is,

$$G(x) = F(x) + C$$

for x in I. The following theorem summarizes these observations.

5.2.2 THEOREM *If $F(x)$ is any antiderivative of $f(x)$, then for any value of C*

$$F(x) + C$$

is also an antiderivative of $f(x)$; moreover on any interval, every antiderivative of $f(x)$ is expressible in the form $F(x)$ plus a constant.

▶ Example 2 For all x,

$$\frac{d}{dx}[\sin x] = \cos x$$

so that $F(x) = \sin x$ is an antiderivative of $f(x) = \cos x$ on the interval $(-\infty, +\infty)$; thus, every antiderivative of $\cos x$ on this interval is expressible in the form

$$\sin x + C$$

where C is a constant. ◀

The process of finding antiderivatives is called **antidifferentiation** or **integration.** If

$$\frac{d}{dx}[F(x)] = f(x)$$

then the functions of the form $F(x) + C$ are the antiderivatives of $f(x)$. We denote this by writing

$$\int f(x)\, dx = F(x) + C \tag{1}$$

The symbol $\int$ is called an ***integral sign,*** and statement (1) is read, "the indefinite integral of $f(x)$ equals $F(x)$ plus C." The adjective "indefinite" is used because the right side of (1) is not a definite function, but rather a whole set of possible functions; the constant C is called the ***constant of integration.***

REMARK.   The symbol dx in the differentiation operation

$$\frac{d}{dx}[\ \ ]$$

and in the antidifferentiation operation*

$$\int[\ \ ]\,dx$$

serves to identify the independent variable. For example, the symbol

$$\int f(t)\,dt$$

denotes a function of t whose derivative with respect to t is $f(t)$.

The equation

$$\frac{d}{dx}[f(x)] = f'(x)$$

is equivalent to

$$\int f'(x)\,dx = f(x) + C \tag{2}$$

▶ Example 3

$$\frac{d}{dx}[x^3] = 3x^2 \qquad \text{is equivalent to} \qquad \int 3x^2\,dx = x^3 + C$$

$$\frac{d}{dt}[\tan t] = \sec^2 t \qquad \text{is equivalent to} \qquad \int \sec^2 t\,dt = \tan t + C$$

$$\frac{d}{du}[u^{3/2}] = \frac{3}{2}u^{1/2} \qquad \text{is equivalent to} \qquad \int \frac{3}{2}u^{1/2}\,du = u^{3/2} + C \quad ◀$$

---

*This notation was devised by Leibniz. In his early papers Leibniz used the notation "omn." (an abbreviation for the Latin word "omnes") to denote integration. Then on October 29, 1675 he wrote, "It will be useful to write $\int$ for omn., thus $\int \ell$ for omn. ℓ . . . ." Two or three weeks later he refined the notation further and wrote $\int[\ \ ]\,dx$ rather than $\int$ alone. This notation is so useful and so powerful that its development by Leibniz must be regarded as a major milestone in the history of mathematics and science.

Antidifferentiation is guesswork. By looking only at the derivative of a function we try to guess the function itself. It simplifies this guessing process if we keep in mind that every differentiation formula produces a companion integration formula. For example, the differentiation formula

$$\frac{d}{dx}[\sin x] = \cos x$$

gives rise to the integration formula

$$\int \cos x \, dx = \sin x + C$$

Similarly,

$$\frac{d}{dx}[-\cos x] = \sin x \qquad \text{yields} \qquad \int \sin x \, dx = -\cos x + C$$

and

$$\frac{d}{dx}\left[\frac{x^{r+1}}{r+1}\right] = x^r \qquad \text{yields} \qquad \int x^r \, dx = \frac{x^{r+1}}{r+1} + C \qquad (r \neq -1)$$

Some basic integration formulas are given in Table 5.2.1.

Table 5.2.1

DIFFERENTIATION FORMULA	INTEGRATION FORMULA
1. $\dfrac{d}{dx}[x] = 1$	$\displaystyle\int 1 \, dx = x + C$
2. $\dfrac{d}{dx}\left[\dfrac{x^{r+1}}{r+1}\right] = x^r \; (r \neq -1)$	$\displaystyle\int x^r \, dx = \dfrac{x^{r+1}}{r+1} + C \; (r \neq -1)$
3. $\dfrac{d}{dx}[\sin x] = \cos x$	$\displaystyle\int \cos x \, dx = \sin x + C$
4. $\dfrac{d}{dx}[-\cos x] = \sin x$	$\displaystyle\int \sin x \, dx = -\cos x + C$
5. $\dfrac{d}{dx}[\tan x] = \sec^2 x$	$\displaystyle\int \sec^2 x \, dx = \tan x + C$
6. $\dfrac{d}{dx}[-\cot x] = \csc^2 x$	$\displaystyle\int \csc^2 x \, dx = -\cot x + C$
7. $\dfrac{d}{dx}[\sec x] = \sec x \tan x$	$\displaystyle\int \sec x \tan x \, dx = \sec x + C$
8. $\dfrac{d}{dx}[-\csc x] = \csc x \cot x$	$\displaystyle\int \csc x \cot x \, dx = -\csc x + C$

▶ Example 4   From the second integration formula in Table 5.2.1 we obtain

$$\int x^2 \, dx = \frac{x^3}{3} + C \qquad (r = 2)$$

$$\int x^3 \, dx = \frac{x^4}{4} + C \qquad (r = 3)$$

$$\int \frac{1}{x^5} \, dx = \int x^{-5} \, dx = \frac{x^{-5+1}}{-5+1} = -\frac{1}{4x^4} + C \qquad (r = -5)$$

$$\int \sqrt{x} \, dx = \int x^{\frac{1}{2}} \, dx = \frac{x^{\frac{1}{2}+1}}{\frac{1}{2}+1} + C = \frac{2}{3}x^{\frac{3}{2}} + C$$

$$= \frac{2}{3}(\sqrt{x})^3 + C \qquad (r = \frac{1}{2}) \qquad ◀$$

If we differentiate an antiderivative of $f(x)$, we obtain $f(x)$ back again. Thus,

$$\frac{d}{dx} \left[\int f(x) \, dx \right] = f(x) \tag{3}$$

This result is helpful for proving the following basic properties of antiderivatives.

5.2.3   THEOREM   (a)   *A constant factor can be moved through an integral sign; that is,*

$$\int cf(x) \, dx = c \int f(x) \, dx$$

(b)   *An antiderivative of a sum is the sum of the antiderivatives; that is,*

$$\int [f(x) + g(x)] \, dx = \int f(x) \, dx + \int g(x) \, dx$$

Proof.   To prove formula (a) we must show that $c \int f(x) \, dx$ is an antiderivative of $cf(x)$, and to prove (b) we must show that $\int f(x) \, dx + \int g(x) \, dx$ is an antiderivative of $f(x) + g(x)$. But, these conclusions follow from (3):

$$\frac{d}{dx} \left[c \int f(x) \, dx \right] = c \frac{d}{dx} \left[\int f(x) \, dx \right] = cf(x)$$

$$\frac{d}{dx} \left[\int f(x) \, dx + \int g(x) \, dx \right] = \frac{d}{dx} \left[\int f(x) \, dx \right] + \frac{d}{dx} \left[\int g(x) \, dx \right]$$

$$= f(x) + g(x) \qquad ∎$$

▶ Example 5 Evaluate

(a) $\displaystyle\int 4\cos x\,dx$

(b) $\displaystyle\int (x + x^2)\,dx$

Solution (a).

$$\int 4\cos x\,dx = 4\int \cos x\,dx \qquad\qquad \text{[Theorem 5.2.3}a\text{]}$$

$$= 4\,(\sin x + C) \qquad\qquad \text{[Table 5.2.1]}$$

$$= 4\sin x + 4C$$

Since C is an arbitrary constant, so is $4C$. However, no purpose is served by keeping the arbitrary constant in this form, so we will replace it by a single letter, say $K = 4C$, and write

$$\int 4\cos x\,dx = 4\sin x + K$$

Solution (b).

$$\int (x + x^2)\,dx = \int x\,dx + \int x^2\,dx \qquad\qquad \text{[Theorem 5.2.3}b\text{]}$$

$$= \left[\frac{x^2}{2} + C_1\right] + \left[\frac{x^3}{3} + C_2\right] \qquad\qquad \text{[Table 5.2.1]}$$

$$= \frac{x^2}{2} + \frac{x^3}{3} + C_1 + C_2$$

Since C_1 and C_2 are arbitrary constants, so is $C_1 + C_2$. If we denote this arbitrary constant by the single letter C, we obtain

$$\int (x + x^2)\,dx = \frac{x^2}{2} + \frac{x^3}{3} + C \qquad\qquad ◀$$

Part (*b*) of Theorem 5.2.3 can be extended to include more than two terms. More precisely,

$$\int [f_1(x) + f_2(x) + \cdots + f_n(x)]\,dx$$

$$= \int f_1(x)\,dx + \int f_2(x)\,dx + \cdots + \int f_n(x)\,dx$$

In addition, we have left it as an exercise to show:

$$\int [f(x) - g(x)]\, dx = \int f(x)\, dx - \int g(x)\, dx$$

► Example 6

$$\int (3x^6 - 2x^2 + 7x + 1)\, dx = 3\int x^6\, dx - 2\int x^2\, dx + 7\int x\, dx + \int 1\, dx$$

$$= \frac{3x^7}{7} - \frac{2x^3}{3} + \frac{7x^2}{2} + x + C \qquad ◄$$

REMARK. The function $f(x)$ in the expression $\int f(x)\, dx$ is called the *integrand*. Sometimes, for compactness of notation, the dx is incorporated into the integrand. For example,

$$\int 1\, dx \qquad \text{can be written as} \qquad \int dx$$

and

$$\int \frac{1}{x^2}\, dx \qquad \text{can be written as} \qquad \int \frac{dx}{x^2}$$

Sometimes an integrand must be rewritten in a different form before the integration can be performed.

► Example 7   Evaluate

$$\int \frac{\cos x}{\sin^2 x}\, dx$$

Solution.

$$\int \frac{\cos x}{\sin^2 x}\, dx = \int \frac{1}{\sin x}\frac{\cos x}{\sin x}\, dx$$

$$= \int \csc x \cot x\, dx$$

$$= -\csc x + C \quad \text{[Formula 8 in Table 5.2.1]} \qquad ◄$$

► Example 8   Evaluate

$$\int \frac{t^2 - 2t^4}{t^4}\, dt$$

Solution.

$$\int \frac{t^2 - 2t^4}{t^4}\, dt = \int \left(\frac{1}{t^2} - 2\right) dt = \int (t^{-2} - 2)\, dt$$

$$= \frac{t^{-1}}{-1} - 2t + C = -\frac{1}{t} - 2t + C \qquad \blacktriangleleft$$

▶ Exercise Set 5.2

In Exercises 1–28, evaluate the integrals and check your results by differentiating the answers.

1. $\int x^8\, dx.$

2. $\int \frac{1}{x^6}\, dx.$

3. $\int x^{5/7}\, dx.$

4. $\int \sqrt[3]{x^2}\, dx.$

5. $\int \frac{4}{\sqrt{t}}\, dt.$

6. $\int \frac{dx}{x^{-2}}.$

7. $\int x^3 \sqrt{x}\, dx.$

8. $\int (u^3 - 2u + 7)\, du.$

9. $\int \left(x^{-3} + \sqrt{x} - 3x^{1/4} + \frac{1}{x^{-2}}\right) dx.$

10. $\int (x^{2/3} - 4x^{-1/5} + 4)\, dx.$

11. $\int \left(\frac{7}{y^{3/4}} - \sqrt[3]{y} + \frac{4}{y^{-1/2}}\right) dy.$

12. $\int (2 + y^2)^2\, dy.$

13. $\int x(1 + x^3)\, dx.$

14. $\int (1 + x^2)(2 - x)\, dx.$

15. $\int x^{1/3}(2 - x)^2\, dx.$

16. $\int \left[\frac{1}{t^2} - \cos t\right] dt.$

17. $\int [4 \sin x + 2 \cos x]\, dx.$

18. $\int [4 \sec^2 x + \csc x \cot x]\, dx.$

19. $\int \sec x(\sec x + \tan x)\, dx.$

20. $\int [\sqrt{\theta} - \csc^2 \theta]\, d\theta.$

21. $\int \sec x(\tan x + \cos x)\, dx.$

22. $\int \frac{dy}{\csc y}.$

23. $\int \frac{\sin x}{\cos^2 x}\, dx.$

24. $\int \frac{\sin 2x}{\cos x}\, dx.$

25. $\int [1 + \sin^2 \theta \csc \theta]\, d\theta.$

26. $\int \left[\phi + \frac{2}{\sin^2 \phi}\right] d\phi.$

27. $\int \frac{x^5 + 2x^2 - 1}{x^4}\, dx.$

28. $\int \frac{x^2 \sin x + 2 \sin x}{2 + x^2}\, dx.$

29. Find the antiderivative F of $f(x) = \sqrt[3]{x}$ that satisfies $F(1) = 2$.

30. Find a function f such that

$$f'(x) + \sin x = 0$$

and

$$f(0) = 2$$

31. Find the general form of a function whose second derivative is $\sqrt{x}$. [*Hint:* Solve the equation $f''(x) = \sqrt{x}$ for $f(x)$ by integrating both sides twice.]

32. Find a function f such that

$$f''(x) = x + \cos x$$

and such that $f(0) = 1$ and $f'(0) = 2$.

[*Hint:* Integrate both sides of the equation twice.]

33. Prove: $\int [f(x) - g(x)] dx = \int f(x) dx - \int g(x) dx.$

34. (a) Show that

$$F(x) = \begin{cases} x, & x > 0 \\ -x, & x < 0 \end{cases}$$

and

$$F_1(x) = \begin{cases} x + 2, & x > 0 \\ -x + 3, & x < 0 \end{cases}$$

are both antiderivatives of

$$f(x) = \begin{cases} 1, & x > 0 \\ -1, & x < 0 \end{cases}$$

but that $F_1(x) \neq F(x)$ plus a constant.

(b) Does this violate Theorem 5.2.2? Explain.

5.3 INTEGRATION BY u-SUBSTITUTION

In this section we will discuss a technique, called **u-substitution,** which can often be used to transform complicated integration problems into simpler ones. The method hinges on the following formula in which u stands for a differentiable function of x.

$$\int \left[f(u) \frac{du}{dx} \right] dx = \int f(u)\, du \tag{1}$$

To justify this formula, let F be an antiderivative of f, so that

$$\frac{d}{du}[F(u)] = f(u)$$

or, equivalently,

$$\int f(u)\, du = F(u) + C \tag{2}$$

If u is a differentiable function of x, the chain rule implies that

$$\frac{d}{dx}[F(u)] = \frac{d}{du}[F(u)] \cdot \frac{du}{dx} = f(u) \frac{du}{dx}$$

or, equivalently,

$$\int \left[f(u) \frac{du}{dx} \right] dx = F(u) + C \tag{3}$$

Formula (1) follows from (2) and (3). ▮

Suppose we are interested in evaluating

$$\int h(x)\,dx$$

It follows from (1) that if we can express this integral in the form

$$\int h(x)\,dx = \int f(g(x))g'(x)\,dx$$

then the substitution $u = g(x)$ and $du/dx = g'(x)$ will yield

$$\int h(x)\,dx = \int \left[f(u)\frac{du}{dx} \right] dx = \int f(u)\,du$$

With a "good" choice of $u = g(x)$, the integral on the right will be easier to evaluate than the original.

In practice, this substitution process is carried out as follows:

Step 1. Make a choice for u, say $u = g(x)$.
Step 2. Compute $du/dx = g'(x)$.
Step 3. Make the substitution

$$u = g(x), \qquad du = g'(x)\,dx$$

At this stage, the *entire* integral must be in terms of u; no x's should remain. If this is not the case, try a different choice of u.

Step 4. Evaluate the resulting integral.
Step 5. Replace u by $g(x)$, so the final answer is in terms of x.

▶ Example 1 Evaluate

$$\int (x^2 + 1)^{50} \cdot 2x\,dx$$

Solution. If we let

$$u = x^2 + 1$$

then $du/dx = 2x$, so $du = 2x\,dx$. Therefore

$$\int (x^2 + 1)^{50} \cdot 2x\,dx = \int u^{50}\,du = \frac{u^{51}}{51} + C = \frac{(x^2 + 1)^{51}}{51} + C \qquad ◀$$

▶ Example 2   Evaluate

$$\int \sin^2 x \cos x \, dx$$

Solution.   If we let

$$u = \sin x$$

then $du/dx = \cos x$, so $du = \cos x \, dx$. Thus,

$$\int \sin^2 x \cos x \, dx = \int u^2 \, du = \frac{u^3}{3} + C = \frac{\sin^3 x}{3} + C$$

◀

▶ Example 3   Evaluate

$$\int \frac{\cos \sqrt{x}}{2\sqrt{x}} \, dx$$

Solution.   If we let

$$u = \sqrt{x}$$

then

$$\frac{du}{dx} = \frac{1}{2\sqrt{x}} \qquad \text{so} \qquad du = \frac{1}{2\sqrt{x}} dx$$

Thus,

$$\int \frac{\cos \sqrt{x}}{2\sqrt{x}} \, dx = \int \cos u \, du = \sin u + C = \sin \sqrt{x} + C$$

◀

▶ Example 4   Evaluate

$$\int 3x^2 \sqrt{x^3 + 1} \, dx$$

Solution.   If we let

$$u = x^3 + 1$$

then $du/dx = 3x^2$, so $du = 3x^2 \, dx$. Thus,

$$\int 3x^2 \sqrt{x^3 + 1} \, dx = \int \sqrt{u} \, du = \int u^{1/2} \, du$$

$$= \frac{u^{3/2}}{3/2} + C$$

$$= \frac{2}{3}(x^3 + 1)^{3/2} + C \qquad \blacktriangleleft$$

▶ **Example 5** The easiest *u*-substitutions occur when the integrand is the derivative of a known function, except for a constant added to or subtracted from the independent variable. For example,

$$\int \sin{(x + 9)} \, dx = \int \sin u \, du \qquad \begin{bmatrix} u = x + 9 \\ du = 1 \cdot dx = dx \end{bmatrix}$$

$$= -\cos u + C$$

$$= -\cos{(x + 9)} + C$$

$$\int (x - 8)^{23} \, dx = \int u^{23} \, du \qquad \begin{bmatrix} u = x - 8 \\ du = 1 \cdot dx = dx \end{bmatrix}$$

$$= \frac{u^{24}}{24} + C$$

$$= \frac{(x - 8)^{24}}{24} + C \qquad \blacktriangleleft$$

Another easy *u*-substitution occurs when the integrand is the derivative of a known function, except for a constant that multiplies or divides the independent variable. The following example illustrates two ways to evaluate such integrals.

▶ **Example 6** Evaluate

$$\int \cos 5x \, dx.$$

First Solution.

$$\int \cos 5x \, dx = \int (\cos u) \cdot \frac{1}{5} \, du \qquad \begin{bmatrix} u = 5x \\ du = 5 \, dx \text{ or } dx = \frac{1}{5} du \end{bmatrix}$$

$$= \frac{1}{5} \int \cos u \, du$$

$$= \frac{1}{5} \sin u + C = \frac{1}{5} \sin 5x + C$$

Second Solution. Rather than solve $du = 5 \, dx$ for dx as above, we can replace dx by $5 \, dx$ in the original integrand and compensate by putting a factor of $\frac{1}{5}$ in front of the integral. This yields

$$\int \cos 5x \, dx = \frac{1}{5} \int \cos 5x \cdot 5 \, dx$$

$$= \frac{1}{5} \int \cos u \, du \qquad \begin{bmatrix} u = 5x \\ du = 5 \, dx \end{bmatrix}$$

$$= \frac{1}{5} \sin u + C$$

$$= \frac{1}{5} \sin 5x + C \qquad \blacktriangleleft$$

► Example 7

$$\int \frac{dx}{(\frac{1}{3}x - 8)^5} = \int \frac{3 \, du}{u^5} \qquad \begin{bmatrix} u = \frac{1}{3}x - 8 \\ du = \frac{1}{3} dx \text{ or } dx = 3 \, du \end{bmatrix}$$

$$= 3 \int u^{-5} \, du$$

$$= -\frac{3}{4} u^{-4} + C$$

$$= -\frac{3}{4} \left(\frac{1}{3}x - 8 \right)^{-4} + C \qquad \blacktriangleleft$$

► **Example 8** With the help of Theorem 5.2.3, a complicated integral can sometimes be computed by expressing it as a sum of simpler integrals. For example,

$$\int (x + \sec^2 \pi x) \, dx = \int x \, dx + \int \sec^2 \pi x \, dx$$

$$= \frac{x^2}{2} + \int \sec^2 \pi x \, dx$$

$$= \frac{x^2}{2} + \frac{1}{\pi} \int \sec^2 u \, du \qquad \begin{bmatrix} u = \pi x \\ du = \pi dx \text{ or } dx = \frac{1}{\pi} du \end{bmatrix}$$

$$= \frac{x^2}{2} + \frac{1}{\pi} \tan u + C$$

$$= \frac{x^2}{2} + \frac{1}{\pi} \tan \pi x + C \qquad \blacktriangleleft$$

► **Example 9** Evaluate

$$\int t^4 \sqrt[3]{3 - 5t^5} \, dt$$

Solution. After some possible false starts most readers would eventually hit on the following substitution:

$$\int t^4 \sqrt[3]{3 - 5t^5}\, dt = -\frac{1}{25}\int \sqrt[3]{u}\, du \quad \left[\begin{array}{c} u = 3 - 5t^5 \\ du = -25t^4\, dt \text{ or } -\frac{1}{25}\, du = t^4\, dt \end{array}\right]$$

$$= -\frac{1}{25}\int u^{1/3}\, du$$

$$= -\frac{1}{25}\frac{u^{4/3}}{4/3} + C$$

$$= -\frac{3}{100}(3 - 5t^5)^{4/3} + C \qquad \blacktriangleleft$$

▶ Example 10 Evaluate

$$\int x^2 \sqrt{x - 1}\, dx$$

Solution. Let

$$u = x - 1 \tag{4}$$

so that

$$du = dx$$

From (4)

$$x^2 = (u + 1)^2 = u^2 + 2u + 1$$

so that

$$\int x^2 \sqrt{x - 1}\, dx = \int (u^2 + 2u + 1)\sqrt{u}\, du$$

$$= \int (u^{5/2} + 2u^{3/2} + u^{1/2})\, du$$

$$= \frac{2}{7}u^{7/2} + \frac{4}{5}u^{5/2} + \frac{2}{3}u^{3/2} + C$$

$$= \frac{2}{7}(x - 1)^{7/2} + \frac{4}{5}(x - 1)^{5/2} + \frac{2}{3}(x - 1)^{3/2} + C \qquad \blacktriangleleft$$

REMARK. If you find integration by substitution hard, don't despair—it is not an easy topic. Work lots of problems.

▶ Exercise Set 5.3

1. Evaluate the integrals by making the indicated substitutions.

(a) $\int 2x(x^2 + 1)^{23} \, dx; \; u = x^2 + 1$

(b) $\int \cos^3 x \sin x \, dx; \; u = \cos x$

(c) $\int \frac{1}{\sqrt{x}} \sin \sqrt{x} \, dx; \; u = \sqrt{x}$

(d) $\int \frac{3x \, dx}{\sqrt{4x^2 + 5}}; \; u = 4x^2 + 5.$

2. Evaluate the integrals by making the indicated substitutions.

(a) $\int \sec^2 (4x + 1) \, dx; \; u = 4x + 1$

(b) $\int y \sqrt{1 + 2y^2} \, dy; \; u = 1 + 2y^2$

(c) $\int \sqrt{\sin \pi\theta} \cos \pi\theta \, d\theta; \; u = \sin \pi\theta$

(d) $\int (2x + 7)(x^2 + 7x + 3)^{4/5} \, dx;$
$\qquad u = x^2 + 7x + 3.$

3. Evaluate the integrals by making the indicated substitutions.

(a) $\int \cot x \csc^2 x \, dx; \; u = \cot x$

(b) $\int (1 + \sin t)^9 \cos t \, dt; \; u = 1 + \sin t$

(c) $\int x^2 \sqrt{1 + x} \, dx; \; u = 1 + x$

(d) $\int [\csc (\sin x)]^2 \cos x \, dx; \; u = \sin x.$

In Exercises 4–25, evaluate the integrals.

4. $\int 4x^5 \, dx.$

5. $\int (1 + \cos x) \, dx.$

6. $\int \sin (x + 4) \, dx.$

7. $\int (x - 2)^{17} \, dx.$

8. $\int (7t^5 - t^3 + 2t + 1) \, dt.$

9. $\int 5 \cos (\pi x - 3) \, dx.$

10. $\int \left(\frac{1}{x^2} + \frac{1}{x^3} \right) dx.$

11. $\int \frac{ds}{\sqrt{3s + 1}}.$

12. $\int x \sqrt{7x^2 + 12} \, dx.$

13. $\int \sqrt[n]{a + bx} \, dx \qquad (b \neq 0).$

14. $\int 6x^2(x^3 + 1)^{19} \, dx.$

15. $\int \sin^6 x \cos x \, dx.$

16. $\int x^{2/3} \sin (x^{5/3}) \, dx.$

17. $\int \frac{x^2 \, dx}{\sqrt{x^3 + 1}}.$

18. $\int \csc^2 (\pi\theta + 1) \, d\theta.$

19. $\int \sqrt{\tan x} \sec^2 x \, dx.$

20. $\int (3u^2 + 1)(u^3 + u - 2)^{3/4} \, du.$

21. $\int \tan^2 x \sec^2 x \, dx.$

22. $\int \tan x \sec^3 x \, dx.$

23. $\int [\sin (\sin \theta)] \cos \theta \, d\theta.$

24. $\int \frac{\sin v \, dv}{(5 - \cos v)^{12}}.$

25. $\int \sin^n (a + bx) \cos (a + bx) \, dx \qquad (n > 0, b \neq 0).$

In Exercises 26–34, evaluate the integrals. These are a little trickier than those in the preceding exercises.

26. $\displaystyle\int \frac{t^2 - 1}{\sqrt{t}}\, dt.$

27. $\displaystyle\int x \sqrt[3]{(x^2 + 5)^2}\, dx.$

28. $\displaystyle\int (x^2 - 6x + 9)^{3/5}\, dx.$

29. $\displaystyle\int \frac{y\, dy}{\sqrt{y + 1}}.$

30. $\displaystyle\int x^2 \sqrt{2 - x}\, dx.$

31. $\displaystyle\int \sin^3 x\, dx.$

32. $\displaystyle\int \frac{1 + \cos \theta}{\sin^2 \theta}\, d\theta.$

33. $\displaystyle\int \tan^2 x\, dx.$

34. $\displaystyle\int (x^2 - 3)^3\, dx.$

35. Evaluate the integral in Example 10 by making the substitution $u = \sqrt{x - 1}$. Check your answer against the one obtained in the example.

36. (a) Evaluate $\int \sin x \cos x\, dx$ by two methods: first, let $u = \sin x$; then, let $u = \cos x$.
   (b) Explain why the two apparently different answers obtained in part (a) are really equivalent.

5.4 RECTILINEAR MOTION (AN APPLICATION OF THE INDEFINITE INTEGRAL)

In Section 2.3 we studied the simplest kind of motion, a particle moving in one direction along a straight line. In this section we will consider a more complicated kind of motion, a particle moving back and forth on a line; this is called *rectilinear motion*. Examples are: a piston moving up and down in a cylinder, a rock tossed straight up and returning to earth straight down, back and forth vibrations of a spring, and so forth.

To study the motion of a particle along a line, it is usually desirable to coordinatize the line by selecting an arbitrary origin, a positive direction, and a unit of length. As the particle moves along the coordinate line its coordinate s will vary as a function of time t. This function, denoted by $s(t)$, is called the *position function* of the particle. The rate at which the particle's coordinate changes with time is called the *velocity* of the particle, and the magnitude or absolute value of the velocity is called the *speed* of the particle. More precisely:

5.4.1 DEFINITION   If $s(t)$ is the position function of a particle moving on a coordinate line, then the *instantaneous velocity* at time t is defined by

$$v(t) = s'(t) = \frac{ds}{dt}$$

and the *instantaneous speed* at time t is defined by

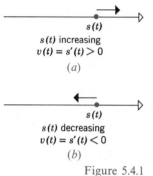

s(t) increasing
$v(t) = s'(t) > 0$

(a)

s(t) decreasing
$v(t) = s'(t) < 0$

(b)

Figure 5.4.1

$$\text{speed at time } t = |v(t)| = |s'(t)| = \left|\frac{ds}{dt}\right|$$

If $v(t) > 0$ at a given time t, then the coordinate s of the particle is increasing at that instant, which means that the particle is moving in the positive direction along the line; similarly if $v(t) < 0$, then s is decreasing at time t and the particle is moving in the negative direction (Figures 5.4.1a and 5.4.1b). The speed of a particle is always nonnegative; it tells how fast the particle is moving, but provides no information about the direction of motion.

▶ Example 1   Let $s = t^4 - 8t^2$ be the position function of a particle moving on a coordinate line, where t is in seconds and s is in feet. The instantaneous velocity and instantaneous speed of the particle are given by

$$v(t) = \frac{ds}{dt} = 4t^3 - 16t$$

$$|v(t)| = \left|\frac{ds}{dt}\right| = |4t^3 - 16t|$$

Thus, when $t = 1$ sec the particle's velocity is $v(1) = -12$ ft/sec, which means the particle is moving in the negative direction with a speed of $|v(1)| = 12$ ft/sec. When $t = 3$ sec the velocity is $v(3) = 60$ ft/sec, which means that the particle is moving in the positive direction with a speed of $|v(3)| = 60$ ft/sec. When $t = 2$ sec the velocity is $v(2) = 0$ ft/sec, which means that the particle has momentarily stopped. ◀

If a particle moves along a coordinate line, the rate at which its velocity changes with time is called the *acceleration* of the particle; more precisely:

5.4.2   DEFINITION   If $v(t)$ is the velocity at time t of a particle moving on a coordinate line, then its ***instantaneous acceleration*** at time t is defined by

$$a(t) = v'(t) = \frac{dv}{dt}$$

Since $v(t) = s'(t) = ds/dt$, the formula in this definition can also be written

$$a(t) = s''(t) = \frac{d^2s}{dt^2}$$

Since acceleration is the rate at which velocity changes with time, it is expressed in units of velocity per unit of time. For example, if t is in seconds

and s is in feet, then velocity units are feet per second (ft/sec), and acceleration units are feet per second per second, which would be written

(ft/sec)/sec   or   ft/sec²

▶ Example 2 Let $s = t^3 - 6t^2$, where t is in seconds and s is in feet. Then

$$v(t) = \frac{ds}{dt} = 3t^2 - 12t$$

and

$$a(t) = \frac{dv}{dt} = 6t - 12$$

At $t = 1$, the acceleration is $a(1) = -6$ ft/sec², which means that the velocity is decreasing at the rate of 6 ft/sec². At $t = 2$ the acceleration is $a(2) = 0$
ft/sec², which means that the velocity is not changing at that instant. At
$t = 4$ the acceleration is $a(4) = 12$ ft/sec², which means that the velocity is
increasing at the rate of 12 ft/sec².   ◀

REMARK. We leave it as an exercise to show that the speed of a particle is
increasing if its velocity and acceleration have the same sign, and the speed
is decreasing if they have opposite signs (Exercise 32). Thus, a particle moving in the positive direction $[v(t) > 0]$ is speeding up when its acceleration is
positive and slowing down when its acceleration is negative, whereas a particle moving in the negative direction $[v(t) < 0]$ is speeding up when its acceleration is negative and slowing down when its acceleration is positive.

▶ Example 3 The position function of a particle moving on a coordinate
line is given by $s(t) = 2t^3 - 21t^2 + 60t + 3$, where s is in feet and t is in
seconds. Describe the motion of the particle for $t \geq 0$.

Solution. The velocity and acceleration at time t are

$$v(t) = \frac{ds}{dt} = 6t^2 - 42t + 60 = 6(t - 2)(t - 5)$$

$$a(t) = \frac{dv}{dt} = 12t - 42 = 12\left(t - \frac{7}{2}\right)$$

At each instant we can determine the direction of motion from the sign of
$v(t)$ and whether the particle is speeding up or slowing down from the signs
of $v(t)$ and $a(t)$ together (Figures 5.4.2a and 5.4.2b). The motion of the
particle is described schematically by the curved line in Figure 5.4.3. At time
$t = 0$ the particle is at the point $s(0) = 3$ moving right with velocity
$v(0) = 60$ ft/sec, but slowing down with acceleration $a(0) = -42$ ft/sec².

The particle continues moving right until time $t = 2$, when it stops at the point $s(2) = 55$, reverses direction, and begins to speed up with an acceleration of $a(2) = -18$ ft/sec². At time $t = \frac{7}{2}$ the particle begins to slow down, but continues moving left until time $t = 5$, when it stops at the point $s(5) = 28$, reverses direction again, and begins to speed up with acceleration $a(5) = 18$ ft/sec². The particle then continues moving right thereafter with increasing speed. ◄

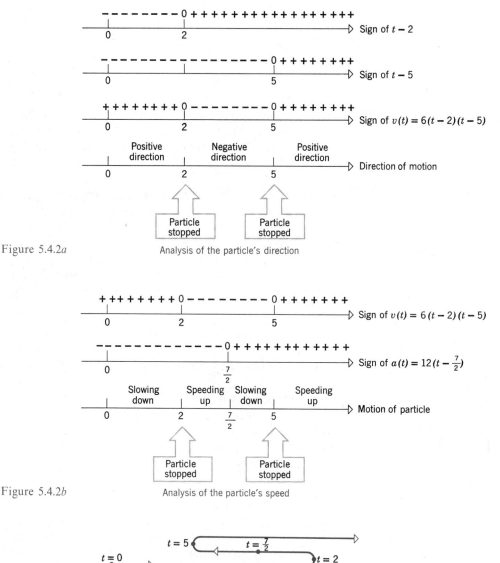

Figure 5.4.2a

Analysis of the particle's direction

Figure 5.4.2b

Analysis of the particle's speed

Figure 5.4.3

REMARK. The curved line in Figure 5.4.3 is descriptive only. The actual path of the particle is back and forth on the axis.

The formulas

$$v(t) = \frac{ds}{dt}$$

$$a(t) = \frac{dv}{dt}$$

can be written in the alternative forms,

$$s(t) = \int v(t)\, dt$$

$$v(t) = \int a(t)\, dt$$

Thus, if we have sufficient information to determine the constants of integration, the velocity can be determined from the acceleration, and the position can be determined from the velocity.

▶ Example 4 Find the position function of a particle moving with velocity $v(t) = \cos \pi t$ along a straight line, assuming the particle is at $s = 4$ when $t = 0$.

Solution. The position function is

$$s(t) = \int v(t)\, dt = \int \cos \pi t\, dt = \frac{1}{\pi} \sin \pi t + C$$

Since $s = 4$ when $t = 0$, it follows that

$$4 = s(0) = \frac{1}{\pi} \sin 0 + C = C$$

Thus,

$$s(t) = \frac{1}{\pi} \sin \pi t + 4 \qquad\qquad ◀$$

MOTION NEAR THE EARTH'S SURFACE

It is a fact of physics that an object moving on a vertical line near the earth's surface and subject only to the force of gravity moves with constant acceleration.* This constant, denoted by the letter g, is approximately 32 ft/sec^2 or 9.8 m/sec^2, depending on whether distance is measured in feet or meters.

---

* Strictly speaking, the acceleration changes with the distance from the earth's center. However, near the surface of the earth this distance is approximately constant.

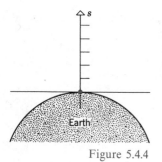

Figure 5.4.4

From this simple principle it is possible to describe completely the vertical motion of a particle moving freely near the earth's surface, provided the initial position and velocity of the particle are known. To see how, assume the line of motion has been coordinatized with the origin at the surface of the earth and the positive direction upward (Figure 5.4.4). Moreover, let us assume that time is measured in seconds, distance is measured in feet or meters, and that:

the position at time $t = 0$ is s_0
the velocity at time $t = 0$ is v_0

where the initial displacement s_0 and initial velocity v_0 are known. Recall that a particle speeds up when its velocity and acceleration have the same sign and slows down when they have opposite signs. Because we have chosen the positive direction to be up, and because a particle moving up (positive velocity) slows down under the force of gravity and a particle moving down (negative velocity) speeds up under the force of gravity, it follows that the acceleration a of the particle must always be negative, that is,

$$a(t) = -g$$

Since the acceleration of the particle is constant, we obtain

$$v(t) = \int a(t)\, dt = \int -g\, dt = -gt + C_1 \tag{1}$$

To determine the constant of integration we use the fact that the velocity is v_0 when $t = 0$. Thus,

$$v_0 = v(0) = -g \cdot 0 + C_1 = C_1$$

Substituting this in (1) yields

$$v(t) = -gt + v_0 \tag{2}$$

Since v_0 is a constant, it follows that

$$s(t) = \int v(t)\, dt = \int (-gt + v_0)\, dt = -\frac{1}{2}gt^2 + v_0 t + C_2 \tag{3}$$

To determine the constant C_2 we use the fact that the position is s_0 when $t = 0$. Thus,

$$s_0 = s(0) = -\frac{1}{2}g \cdot 0 + v_0 \cdot 0 + C_2 = C_2$$

Substituting this in (3) yields

$$s(t) = -\frac{1}{2}gt^2 + v_0 t + s_0 \tag{4}$$

▶ Example 5 A rock, initially at rest, is dropped from a height of 400 ft. Assuming gravity is the only force acting, how long does it take for the rock to hit the ground and what is its speed at the time of impact?

Solution. Since distance is in feet, we take $g = 32$ ft/sec². Initially, we have $s_0 = 400$ and $v_0 = 0$, so that from (4)

$$s(t) = -16t^2 + 400$$

Impact occurs when $s(t) = 0$. Solving this equation for t, we obtain

$$-16t^2 + 400 = 0$$
$$t^2 = 25$$
$$t = \pm 5$$

Since the rock is released at $t = 0$, the time of impact is positive, so that we can discard the negative solution and conclude that it takes 5 sec for the rock to hit the ground. Substituting $t = 5$ and $v_0 = 0$ in (2), we obtain the velocity at time of impact,

$$v(5) = -32(5) + 0 = -160 \text{ ft/sec}$$

Thus, the speed at impact is

$$|v(5)| = 160 \text{ ft/sec} \qquad \blacktriangleleft$$

▶ Example 6 A ball is thrown directly upward from a point 8 meters above the ground with an initial velocity of 49 meters/sec. Assuming gravity is the only force acting on the ball after its release, how high will the ball travel?

Solution. Since distance is in meters, we take $g = 9.8$ meters/sec². Initially, $s_0 = 8$ and $v_0 = 49$, so that from (2) and (4)

$$v(t) = -9.8t + 49$$
$$s(t) = -4.9t^2 + 49t + 8$$

The ball will rise until $v(t) = 0$, that is, until $-9.8t + 49 = 0$ or $t = 5$. At this instant the height will be

$$s(5) = -4.9(5)^2 + 49(5) + 8 = 130.5 \text{ meters} \qquad \blacktriangleleft$$

▶ Exercise Set 5.4

In Exercises 1–15, $s(t)$, $v(t)$, and $a(t)$ are the position, velocity, and acceleration functions for a particle moving on a coordinate line; distance is measured in feet and time in seconds.

1. (a) Let $s(t) = t^3 - 6t^2$. Make a table showing the position, velocity, speed, and acceleration at times $t = 1$, $t = 2$, $t = 3$, $t = 4$, and $t = 5$.
    (b) At each of these times specify the direction of motion, if any, and whether the particle is speeding up, slowing down, or neither.

In Exercises 2–7, use the given information to find the position function of the particle.

2. $v(t) = 3t^2$; $s(0) = 0$.

3. $v(t) = t^3 - 2t^2 + 1$; $s(0) = 1$.

4. $v(t) = 1 + \sin t$; $s(0) = -3$.

5. $a(t) = 4$; $v(0) = 1$, $s(0) = 0$.

6. $a(t) = t^2 - 3t + 1$; $v(0) = 0$; $s(0) = 0$.

7. $a(t) = 4 \cos 2t$; $v(0) = -1$; $s(0) = -3$.

8. In each part use the given information to find the position, velocity, speed, and acceleration at time $t = 1$.
    (a) $v = \sin \frac{1}{2}\pi t$; $s = 0$ when $t = 0$.
    (b) $a = -3t$; $s = 1$ and $v = 0$ when $t = 0$.

9. Let $s = 5t^2 - 22t$.
    (a) Find the maximum speed of the particle during the time interval $1 \le t \le 3$.
    (b) When, during the time interval $1 \le t \le 3$, is the particle farthest from the origin? What is its position at that instant?

In Exercises 10–15, describe the motion of the particle for $t \ge 0$ (as in Example 3) and make a sketch as in Figure 5.4.3.

10. $s = -3t + 2$.

11. $s = 1 + 6t - t^2$.

12. $s = t^3 - 6t^2 + 9t + 1$.

13. $s = t^3 - 9t^2 + 24t$.

14. $s = t + \dfrac{9}{t+1}$.

15. $s = \begin{cases} \cos t, & 0 \le t \le 2\pi \\ 1, & t > 2\pi. \end{cases}$

16. A car traveling 60 mi/hr along a straight road decelerates at a constant rate of 10 ft/sec^2.
    (a) How long will it take until the speed is 45 mi/hr? [*Note:* 60 mi/hr = 88 ft/sec.]
    (b) How far will the car travel before coming to a stop?

In Exercises 17–25, assume that the only force acting is the earth's gravity, and apply formulas (2) and (4) to solve the problems.

17. A projectile is launched vertically upward from ground level with an initial velocity of 112 ft/sec.
    (a) Find the velocity at $t = 3$ and $t = 5$.
    (b) How high will the projectile rise?
    (c) Find the speed of the projectile when it hits the ground.

18. A projectile fired downward from a height of 112 ft reaches the ground in 2 sec. What is its initial velocity?

19. A projectile is fired vertically upward from ground level with an initial velocity of 16 ft/sec.
    (a) How long will it take for the projectile to hit the ground?
    (b) How long will the projectile be moving upward?

20. A rock is dropped from the top of the Washington Monument, which is 555 ft high.
    (a) How long will it take for the rock to hit the ground?
    (b) What is the speed of the rock at impact?

21. A helicopter pilot drops a package when the helicopter is 200 ft above the ground and rising at a speed of 20 ft/sec.
    (a) How long will it take for the package to hit the ground?
    (b) What will be its speed at impact?

22. A stone is thrown downward with an initial velocity of -96 ft/sec from a height of 112 ft.
    (a) How long will it take for the stone to hit the ground?
    (b) What will be its speed at impact?

23. A projectile is fired vertically upward with an initial velocity of 49 meters/sec from a tower 150 meters high.

(a) How long will it take for the projectile to reach its maximum height?

(b) What is the maximum height?

(c) How long will it take for the projectile to pass its starting point on the way down?

(d) What is the velocity when it passes the starting point on the way down?

(e) How long will it take for the projectile to hit the ground?

(f) What will be its speed at impact?

24. A man drops a stone from a bridge. How high is the bridge if:

(a) the stone hits the water 4 sec later?

(b) the sound of the splash reaches the man 4 sec later? (Take 1080 ft/sec as the speed of sound.)

25. A projectile fired upward from ground level is to reach a height of 1000 ft. What must its initial velocity be?

26. A car traveling 60 mi/hr skids 180 ft after its brakes are applied. Find the acceleration of the car, assuming that it is constant.

27. Use the chain rule to show that for a particle with rectilinear motion

$$a = v \frac{dv}{ds}$$

28. A stone is released from rest from a point 40 ft above the ground.

(a) Find a formula for v in terms of s.

(b) Use the result in part (a) to find v and dv/ds when the stone is 15 ft above the ground, and verify that these values satisfy the equation in Exercise 27.

29. The position function of a particle moving on a coordinate line is shown in Figure 5.4.5.

(a) Is the particle moving in the negative direction or positive direction at time t_0?

(b) Is the acceleration positive or negative at time t_0?

(c) Is the particle speeding up or slowing down at time t_0?

(d) Is the particle speeding up or slowing down at time t_1?

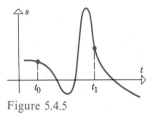

Figure 5.4.5

30. For a particle moving on a coordinate line, the *average velocity* v_{ave} and *average acceleration* a_{ave} over a time interval $[t_0, t_1]$ are defined by:

$$v_{ave} = \frac{\text{change in position}}{\text{time elapsed}} = \frac{s(t_1) - s(t_0)}{t_1 - t_0}$$

$$a_{ave} = \frac{\text{change in velocity}}{\text{time elapsed}} = \frac{v(t_1) - v(t_0)}{t_1 - t_0}$$

(a) Interpret v_{ave} geometrically on the graph of $s(t)$.

(b) Interpret a_{ave} geometrically on the graph of $v(t)$.

(c) Find the average velocity and acceleration over the time interval $2 \le t \le 4$ for a particle with position function $s(t) = t^3 - 3t^2$.

31. (a) Show that the average velocity v_{ave} over $[t_0, t_1]$ approaches the instantaneous velocity $v(t_0)$ as t_1 approaches t_0. [See Exercise 30 for definitions of v_{ave} and a_{ave}.]

(b) Show that the average acceleration a_{ave} over $[t_0, t_1]$ approaches the instantaneous acceleration $a(t_0)$ as t_1 approaches t_0.

32. Let $r(t) = |v(t)|$ be the speed of a particle moving on a coordinate line. Prove:

(a) $r'(t_0) > 0$ if $a(t_0) > 0$ and $v(t_0) > 0$ or $a(t_0) < 0$ and $v(t_0) < 0$

(b) $r'(t_0) < 0$ if $a(t_0) > 0$ and $v(t_0) < 0$ or $a(t_0) < 0$ and $v(t_0) > 0$.

(To paraphrase: The particle is speeding up if velocity and acceleration have the same sign, and slowing down if they have opposite signs.)

5.5 SIGMA NOTATION

In this section we digress from the main theme of this chapter to introduce a notation that can be used to write lengthy sums in a compact form. This material will be helpful when we discuss areas in the next section. The notation uses the uppercase Greek letter Σ (sigma) and is called *sigma notation* or *summation notation.*

To illustrate the idea, consider the sum

$$1^2 + 2^2 + 3^2 + 4^2 + 5^2$$

Each term in the sum is of the form k^2, where k is one of the integers from 1 to 5. In sigma notation this sum can be written

$$\sum_{k=1}^{5} k^2$$

which is read, "the summation of k^2, where k runs from 1 to 5." The notation tells us to form the sum of the terms that result when we substitute successive integers for k in the expression k^2, starting with $k = 1$ and ending with $k = 5$.

More generally, if $f(k)$ is a function of k, and a and b are integers, then

$$\sum_{k=a}^{b} f(k) \tag{1}$$

denotes the sum of the terms that result when we substitute successive integers for k, starting with $k = a$ and ending with $k = b$.

▶ Example 1

$$\sum_{k=4}^{8} k^3 = 4^3 + 5^3 + 6^3 + 7^3 + 8^3$$

$$\sum_{k=1}^{5} 2k = 2 \cdot 1 + 2 \cdot 2 + 2 \cdot 3 + 2 \cdot 4 + 2 \cdot 5 = 2 + 4 + 6 + 8 + 10$$

$$\sum_{k=0}^{5} (2k + 1) = 1 + 3 + 5 + 7 + 9 + 11$$

$$\sum_{k=0}^{5} (-1)^k (2k + 1) = 1 - 3 + 5 - 7 + 9 - 11$$

$$\sum_{k=-3}^{1} k^3 = (-3)^3 + (-2)^3 + (-1)^3 + 0^3 + 1^3 = -27 - 8 - 1 + 0 + 1$$

$$\sum_{k=1}^{3} k \sin\left(\frac{k\pi}{5}\right) = \sin\frac{\pi}{5} + 2\sin\frac{2\pi}{5} + 3\sin\frac{3\pi}{5}$$ ◀

The numbers a and b in (1) are called, respectively, the **lower** and **upper** **limits of summation;** and the letter k is called the **index of summation.** It is not essential to use k as the index of summation; any letter will do. For example,

$$\sum_{i=1}^{6} \frac{1}{i}, \qquad \sum_{j=1}^{6} \frac{1}{j}, \qquad \text{and} \qquad \sum_{n=1}^{6} \frac{1}{n}$$

all denote the sum

$$1 + \frac{1}{2} + \frac{1}{3} + \frac{1}{4} + \frac{1}{5} + \frac{1}{6}$$

If the upper and lower limits of summation are the same, then the "sum" in (1) reduces to one term. For example,

$$\sum_{k=2}^{2} k^3 = 2^3 \qquad \text{and} \qquad \sum_{i=1}^{1} \frac{1}{i+2} = \frac{1}{1+2} = \frac{1}{3}$$

In the sums

$$\sum_{i=1}^{5} 2, \qquad \sum_{k=3}^{6} 7, \qquad \text{and} \qquad \sum_{j=0}^{2} x^3$$

the expression to the right of the Σ sign does not involve the index of summation. In such cases, we take all the terms in the sum to be the same, with one term for each allowable value of the summation index. Thus,

$$\sum_{i=1}^{5} 2 = 2 + 2 + 2 + 2 + 2$$

$$\sum_{k=3}^{6} 7 = 7 + 7 + 7 + 7$$

$$\sum_{j=0}^{2} x^3 = x^3 + x^3 + x^3$$

A sum can be written in more than one way with sigma notation, depending on the limits of summation. For example, the sum of the first five positive even integers can be written as

$$\sum_{k=1}^{5} 2k = 2 + 4 + 6 + 8 + 10$$

or

$$\sum_{k=0}^{4} (2k + 2) = 2 + 4 + 6 + 8 + 10$$

or

$$\sum_{k=2}^{6} (2k - 2) = 2 + 4 + 6 + 8 + 10$$

On occasion we will want to change the sigma notation for a given sum to a sigma notation with different summation limits. The following example illustrates a method for doing this.

▶ Example 2 Express

$$\sum_{k=3}^{7} 5^{k-2}$$

in sigma notation so that the lower limit of summation is 0 rather than 3.

Solution. If we define a new summation index j by means of the formula

$$j = k - 3 \tag{2}$$

then j runs from 0 up to 4 as k runs from 3 up to 7. From (2), $k = j + 3$, so that

$$\sum_{k=3}^{7} 5^{k-2} = \sum_{j=0}^{4} 5^{(j+3)-2} = \sum_{j=0}^{4} 5^{j+1}$$

As a check, the reader can verify that

$$\sum_{j=0}^{4} 5^{j+1} \qquad \text{and} \qquad \sum_{k=3}^{7} 5^{k-2}$$

both denote the sum $5 + 5^2 + 5^3 + 5^4 + 5^5$. ◀

REMARK. In this solution the summation index was changed from k to j. If it is desirable to keep the same symbol for the summation index, we can change the j back to k *at the very end* and express the final result as

$$\sum_{k=0}^{4} 5^{k+1} \qquad \text{instead of} \qquad \sum_{j=0}^{4} 5^{j+1}$$

When we want to represent a general sum we will use letters with subscripts. For example, a general sum with five terms might be written as

$$a_1 + a_2 + a_3 + a_4 + a_5$$

or in sigma notation as

$$\sum_{k=1}^{5} a_k, \qquad \sum_{j=1}^{5} a_j, \qquad \text{or} \qquad \sum_{m=1}^{5} a_m$$

A general sum with n terms might be written as

$$b_1 + b_2 + \cdots + b_n$$

or in sigma notation as

$$\sum_{k=1}^{n} b_k, \qquad \sum_{j=1}^{n} b_j, \qquad \text{or} \qquad \sum_{m=1}^{n} b_m$$

The following algebraic properties of sigma notation help to manipulate sums.

5.5.1 THEOREM   (a) $\displaystyle\sum_{k=1}^{n} (a_k + b_k) = \sum_{k=1}^{n} a_k + \sum_{k=1}^{n} b_k$

(b) $\displaystyle\sum_{k=1}^{n} (a_k - b_k) = \sum_{k=1}^{n} a_k - \sum_{k=1}^{n} b_k$

(c) $\displaystyle\sum_{k=1}^{n} c a_k = c \sum_{k=1}^{n} a_k$

We will prove parts (a) and (c) and leave (b) as an exercise.

Proof of (a).

$$\sum_{k=1}^{n} (a_k + b_k) = (a_1 + b_1) + (a_2 + b_2) + \cdots + (a_n + b_n)$$
$$= (a_1 + a_2 + \cdots + a_n) + (b_1 + b_2 + \cdots + b_n)$$
$$= \sum_{k=1}^{n} a_k + \sum_{k=1}^{n} b_k$$

Proof of (c).

$$\sum_{k=1}^{n} c a_k = c a_1 + c a_2 + \cdots + c a_n = c(a_1 + a_2 + \cdots + a_n) = c \sum_{k=1}^{n} a_k$$

REMARK.   Loosely phrased, this theorem states: *Sigma of a sum equals the sum of the sigmas; sigma of a difference equals the difference of the sigmas; and a constant factor can be moved through a sigma sign.*

The following formulas will be used in our later work.

5.5.2   THEOREM   (a) $\displaystyle\sum_{k=1}^{n} k = 1 + 2 + 3 + \cdots + n = \frac{n(n+1)}{2}$

(b) $\displaystyle\sum_{k=1}^{n} k^2 = 1^2 + 2^2 + 3^2 + \cdots + n^2 = \frac{n(n+1)(2n+1)}{6}$

(c) $\displaystyle\sum_{k=1}^{n} k^3 = 1^3 + 2^3 + 3^3 + \cdots + n^3 = \left[\frac{n(n+1)}{2}\right]^2$

We will prove parts (a) and (b) and leave part (c) as an exercise.

Proof of (a).   If we write the terms of

$$\sum_{k=1}^{n} k = 1 + 2 + 3 + \cdots + (n-2) + (n-1) + n \qquad (3)$$

in the opposite order, we obtain

$$\sum_{k=1}^{n} k = n + (n-1) + (n-2) + \cdots + 3 + 2 + 1 \qquad (4)$$

Adding (3) and (4) term by term yields

$$2\sum_{k=1}^{n} k = \underbrace{(n+1) + (n+1) + (n+1) + \cdots + (n+1)}_{n \text{ terms}} = n(n+1)$$

Thus,

$$\sum_{k=1}^{n} k = \frac{n(n+1)}{2}$$

Proof of (b). This proof begins with a trick. Since

$$(k + 1)^3 - k^3 = k^3 + 3k^2 + 3k + 1 - k^3 = 3k^2 + 3k + 1$$

we obtain

$$\sum_{k=1}^{n} [(k + 1)^3 - k^3] = \sum_{k=1}^{n} (3k^2 + 3k + 1) \tag{5}$$

Writing out the left side of (5) yields

$$[2^3 - 1^3] + [3^3 - 2^3] + [4^3 - 3^3] + \cdots + [(n + 1)^3 - n^3] \tag{6}$$

Observe that each term in (6) cancels part of the next term, so that the entire sum collapses like a folding telescope, leaving only $-1^3 + (n + 1)^3$. Thus, (5) can be rewritten as

$$-1 + (n + 1)^3 = \sum_{k=1}^{n} (3k^2 + 3k + 1) \tag{7}$$

or, from Theorem 5.5.1,

$$-1 + (n + 1)^3 = 3 \sum_{k=1}^{n} k^2 + 3 \sum_{k=1}^{n} k + \sum_{k=1}^{n} 1 \tag{8}$$

But

$$\sum_{k=1}^{n} 1 = \underbrace{1 + 1 + \cdots + 1}_{n \text{ terms}} = n$$

and by part (*a*) of this theorem

$$\sum_{k=1}^{n} k = \frac{n(n + 1)}{2}$$

Thus, (8) can be written

$$-1 + (n + 1)^3 = 3 \sum_{k=1}^{n} k^2 + 3\frac{n(n + 1)}{2} + n$$

Therefore,

$$\sum_{k=1}^{n} k^2 = \frac{1}{3}\left[(n + 1)^3 - 3\frac{n(n + 1)}{2} - (n + 1)\right]$$

$$= \frac{n+1}{6}[2(n+1)^2 - 3n - 2]$$

$$= \frac{n+1}{6}(2n^2 + n) = \frac{n(n+1)(2n+1)}{6} \quad \blacksquare$$

▶ Example 3 Evaluate

$$\sum_{k=1}^{30} k(k+1)$$

Solution.

$$\sum_{k=1}^{30} k(k+1) = \sum_{k=1}^{30} (k^2 + k)$$

$$= \sum_{k=1}^{30} k^2 + \sum_{k=1}^{30} k$$

$$= \frac{30(31)(61)}{6} + \frac{30(31)}{2} \qquad \text{[Theorem 5.5.2a,b]}$$

$$= 9920 \qquad\qquad\qquad ◀$$

REMARK. In formulas such as

$$\sum_{k=1}^{n} k^2 = \frac{n(n+1)(2n+1)}{6}$$

or

$$1^2 + 2^2 + \cdots + n^2 = \frac{n(n+1)(2n+1)}{6}$$

the left side of the equality is said to express the sum in **open form** and the right side is said to express it in **closed form;** the open form just indicates the terms to be added, while the closed form gives their total.

▶ Example 4 Express

$$\sum_{k=1}^{n} (3+k)^2$$

in closed form.

Solution.

$$\sum_{k=1}^{n} (3 + k)^2 = \sum_{k=1}^{n} (9 + 6k + k^2) = \sum_{k=1}^{n} 9 + 6 \sum_{k=1}^{n} k + \sum_{k=1}^{n} k^2$$

$$= 9n + 6 \frac{n(n + 1)}{2} + \frac{n(n + 1)(2n + 1)}{6}$$

$$= \frac{1}{3} n^3 + \frac{7}{2} n^2 + \frac{73}{6} n \qquad \blacktriangleleft$$

▶ Exercise Set 5.5

1. Evaluate

(a) $\displaystyle\sum_{k=1}^{3} k^3$

(b) $\displaystyle\sum_{j=2}^{6} (3j - 1)$

(c) $\displaystyle\sum_{i=-4}^{1} (i^2 - i)$

(d) $\displaystyle\sum_{n=0}^{5} 1.$

2. Evaluate

(a) $\displaystyle\sum_{k=1}^{4} k \sin \frac{k\pi}{2}$

(b) $\displaystyle\sum_{j=0}^{5} (-1)^j$

(c) $\displaystyle\sum_{i=7}^{20} \pi$

(d) $\displaystyle\sum_{m=3}^{5} 2^{m+1}.$

In Exercises 3–16, express in sigma notation, but do not evaluate.

3. $1 + 2 + 3 + \cdots + 10.$

4. $3 \cdot 1 + 3 \cdot 2 + 3 \cdot 3 + \cdots + 3 \cdot 20.$

5. $1 \cdot 2 + 2 \cdot 3 + 3 \cdot 4 + \cdots + 49 \cdot 50.$

6. $1 + 2 + 2^2 + 2^3 + 2^4.$

7. $2 + 4 + 6 + 8 + \cdots + 20.$

8. $1 + 3 + 5 + 7 + \cdots + 15.$

9. $1 - 3 + 5 - 7 + 9 - 11.$

10. $1 - \dfrac{1}{2} + \dfrac{1}{3} - \dfrac{1}{4} + \dfrac{1}{5}.$

11. $-1 + \dfrac{1}{2} - \dfrac{1}{3} + \dfrac{1}{4} - \dfrac{1}{5}.$

12. $1 + \cos \dfrac{\pi}{7} + \cos \dfrac{2\pi}{7} + \cos \dfrac{3\pi}{7}.$

13. $\sin \dfrac{\pi}{8} + \sin \dfrac{3\pi}{8} + \sin \dfrac{5\pi}{8} + \sin \dfrac{7\pi}{8}.$

14. $2 + 4 + 8 + 16 + 32.$

15. $\dfrac{1}{2} + \dfrac{2}{3} + \dfrac{3}{4} + \dfrac{4}{5} + \dfrac{5}{6}.$

16. $15 + 24 + 35 + \cdots + (n^2 - 1).$

17. Express in sigma notation.

    (a) $a_1 - a_2 + a_3 - a_4 + a_5$

    (b) $-b_0 + b_1 - b_2 + b_3 - b_4 + b_5$

    (c) $a_0 + a_1 x + a_2 x^2 + \cdots + a_n x^n$

    (d) $a^5 + a^4 b + a^3 b^2 + a^2 b^3 + ab^4 + b^5.$

In Exercises 18–25, use Theorem 5.5.2 to evaluate the sums.

18. $\displaystyle\sum_{k=1}^{100} k.$

19. $\displaystyle\sum_{k=3}^{100} k.$

20. $\displaystyle\sum_{k=1}^{100} (7k + 1).$

21. $\displaystyle\sum_{k=1}^{20} k^2.$

22. $\displaystyle\sum_{k=4}^{20} k^2.$

23. $\displaystyle\sum_{k=1}^{6} (4k^3 - 2k + 1).$

24. $\displaystyle\sum_{k=1}^{6} (k - k^3).$

25. $\displaystyle\sum_{k=1}^{30} k(k - 2)(k + 2).$

When each term of a sum cancels part of the next term, leaving only portions of the first and last terms at the end, the sum is said to *telescope*. In Exercises 26–31, evaluate the telescoping sum.

26. $\displaystyle\sum_{k=1}^{50}\left(\frac{1}{k}-\frac{1}{k+1}\right).$ 　　27. $\displaystyle\sum_{k=5}^{17}(3^k-3^{k-1}).$

28. $\displaystyle\sum_{k=1}^{100}(2^{k+1}-2^k).$ 　　29. $\displaystyle\sum_{k=2}^{20}\left(\frac{1}{k^2}-\frac{1}{(k-1)^2}\right).$

30. $\displaystyle\sum_{k=1}^{n}(a_k-a_{k+1}).$ 　　31. $\displaystyle\sum_{k=1}^{n}(a_k-a_{k-1}).$

32. Evaluate

(a) $\displaystyle\sum_{j=0}^{m}m$ 　　(b) $\displaystyle\sum_{n=4}^{4}5$

(c) $\displaystyle\sum_{k=1}^{n}x$ 　　(d) $\displaystyle\sum_{i=1}^{n}i^2c.$

33. Evaluate

(a) $\displaystyle\sum_{k=1}^{n}n$ 　　(b) $\displaystyle\sum_{i=0}^{0}(-3)$

(c) $\displaystyle\sum_{k=1}^{n}kx$ 　　(d) $\displaystyle\sum_{k=m}^{n}c\ (n\geq m).$

34. Express $1+2+2^2+2^3+2^4+2^5$ in sigma notation with:
   (a) $j=0$ as the lower limit of summation
   (b) $j=1$ as the lower limit of summation
   (c) $j=2$ as the lower limit of summation.

35. Express

$$\sum_{k=4}^{18}k(k-3)$$

in sigma notation with:
   (a) $k=0$ as the lower limit of summation;
   (b) $k=5$ as the lower limit of summation.

36. Express

$$\sum_{k=5}^{9}k2^{k+4}$$

in sigma notation with:
   (a) $k=1$ as the lower limit of summation;
   (b) $k=13$ as the upper limit of summation.

37. Simplify

$$\sum_{k=11}^{28}(k-10)\sin\left(\frac{\pi}{k-10}\right)$$

by changing the limits of summation.

38. Which of the following are valid identities?

(a) $\displaystyle\sum_{i=1}^{n}a_ib_i=\sum_{i=1}^{n}a_i\sum_{i=1}^{n}b_i$

(b) $\displaystyle\sum_{i=1}^{n}\frac{a_i}{b_i}=\sum_{i=1}^{n}a_i\Big/\sum_{i=1}^{n}b_i$

(c) $\displaystyle\sum_{i=1}^{n}a_i^2=\left(\sum_{i=1}^{n}a_i\right)^2.$

39. By writing out the sums, determine whether the following are valid identities.

(a) $\displaystyle\int\left[\sum_{i=1}^{n}f_i(x)\right]dx=\sum_{i=1}^{n}\left[\int f_i(x)\,dx\right]$

(b) $\displaystyle\frac{d}{dx}\left[\sum_{i=1}^{n}f_i(x)\right]=\sum_{i=1}^{n}\left[\frac{d}{dx}[f_i(x)]\right].$

40. (a) Evaluate

$$\sum_{k=0}^{n}ar^k-r\sum_{k=0}^{n}ar^k$$

(b) Use the result in part (a) to prove that

$$\sum_{k=0}^{n}ar^k=a+ar+ar^2+\cdots+ar^n$$

$$=\frac{a-ar^{n+1}}{1-r}\qquad(r\neq1)$$

(A sum of this form is called a *geometric sum.*)

41. Use Exercise 40 to evaluate

(a) $\displaystyle\sum_{k=1}^{20}3^k$ 　　(b) $\displaystyle\sum_{k=5}^{30}2^k$

(c) $\displaystyle\sum_{k=0}^{100}(-1)^{k+1}\frac{1}{2^k}.$

42. Express the following sums in closed form:

(a) $\displaystyle\sum_{k=1}^{n}(k-\sin^k\theta)$ 　　(b) $\displaystyle\sum_{k=1}^{m-3}k^3.$

[*Hint:* Exercise 40 will help in part (a).]

43. Evaluate

$$\sum_{i=1}^{4}\left(\sum_{j=1}^{5}(i+j)\right)$$

44. Let $\bar{x}$ denote the arithmetic average of the n numbers $x_1, x_2, \ldots, x_n$. Use Theorem 5.5.1 to prove

$$\sum_{i=1}^{n} (x_i - \bar{x}) = 0$$

45. Prove part (b) of Theorem 5.5.1.

46. Prove part (c) of Theorem 5.5.2. [*Hint:* Begin with the difference $(k+1)^4 - k^4$ and follow the steps used to prove part (b) of the theorem.]

5.6 AREAS AS LIMITS

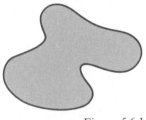

Figure 5.6.1

Until now we have used the term "area" freely, assuming that its meaning is intuitively clear. However, we must ultimately replace this intuitive concept of area with a precise mathematical definition. It is natural to hope that we might be able to devise a definition that would specify the area of an arbitrary region in the plane (Figure 5.6.1). Surprisingly, there is no satisfactory way to do this! It became clear in the late nineteenth century that there are regions in the plane of such complexity that any attempt to assign them areas would ultimately lead to mathematical inconsistencies. While we will not be concerned with such complicated regions, it is important to know that they exist. In this section we will show how to obtain areas of certain regions using limits. The ideas we develop here will form the basis for a precise definition of area in the next section.

5.6.1 PROBLEM Find the area of a region R bounded below by the x-axis, on the sides by the lines $x = a$ and $x = b$, and above by a curve $y = f(x)$, where f is continuous on $[a, b]$ and $f(x) \geq 0$ for all x in $[a, b]$ (Figure 5.6.2).

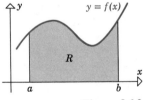

Figure 5.6.2

In Section 5.1 we showed how to obtain the area of this type of region using antiderivatives. However, we may also proceed as follows. Choose an arbitrary positive integer n and divide the interval $[a, b]$ into n subintervals of width $(b - a)/n$ by introducing points

$$x_1, x_2, \ldots, x_{n-1}$$

equally spaced between a and b (Figure 5.6.3). Next, draw vertical lines through the points $a, x_1, x_2, \ldots, x_{n-1}, b$ to divide the region R into n strips of uniform width. If we approximate each of these strips by a rectangle

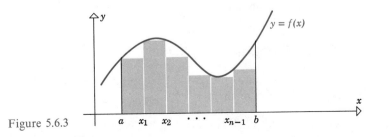

Figure 5.6.3

inscribed under the curve $y = f(x)$ (Figure 5.6.3), then the union of these rectangles will form a region R_n, which we can view as an approximation to the entire region R. The area of this approximating region can be calculated by adding the areas of its component rectangles. Moreover, if we allow n to increase, the widths of the rectangles will get smaller, so that the approximation of R by R_n will get better as the smaller rectangles fill in more of the gaps under the curve (Figure 5.6.4). Thus, we can define the *exact* area of R as the limit of the areas of the approximating regions as n goes to plus infinity; that is,

$$A = \text{area } (R) = \lim_{n \to +\infty} [\text{area } (R_n)] \tag{1}$$

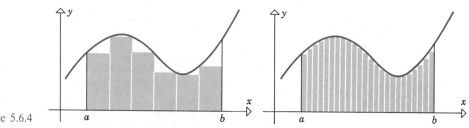

Figure 5.6.4

REMARK. There is a difference between writing $\lim_{n \to +\infty}$ and writing $\lim_{x \to +\infty}$, where n represents a positive integer, and x has no such restriction. Later we will study limits of the type $\lim_{n \to +\infty}$ in detail, but for now let it suffice to say that the computational techniques we have used for limits of the type $\lim_{x \to +\infty}$ will also work for $\lim_{n \to +\infty}$; in fact, if $\lim_{x \to +\infty} f(x) = L$, then $\lim_{n \to +\infty} f(n)$ is also L.

For computational purposes, (1) can be expressed in a more useful form. If we denote the heights of the inscribed rectangles by $h_1, h_2, \ldots, h_n$ and use the fact that each rectangle has a base of length $(b - a)/n$, then

$$\text{area } (R_n) = h_1 \cdot \frac{b - a}{n} + h_2 \cdot \frac{b - a}{n} + \cdots + h_n \cdot \frac{b - a}{n} \tag{2}$$

Because f is assumed to be continuous on $[a, b]$, it follows from the Extreme-Value Theorem that f assumes a minimum value on each of the n closed subintervals

$$[a, x_1], [x_1, x_2], \ldots, [x_{n-1}, b]$$

If these minimum values occur at the points $c_1, c_2, \ldots, c_n$, then the heights of the inscribed rectangles are

$$h_1 = f(c_1), h_2 = f(c_2), \ldots, h_n = f(c_n)$$

(Figure 5.6.5), so (2) can be written

$$\text{area } (R_n) = f(c_1) \cdot \frac{b-a}{n} + f(c_2) \cdot \frac{b-a}{n} + \cdots + f(c_n) \cdot \frac{b-a}{n} \tag{3}$$

Finally, it will be helpful to write

$$\Delta x = \frac{b-a}{n}$$

for the base dimension of the rectangles, so that (3) becomes

$$\text{area } (R_n) = f(c_1) \cdot \Delta x + f(c_2) \cdot \Delta x + \cdots + f(c_n) \cdot \Delta x$$

or, in sigma notation,

$$\text{area } (R_n) = \sum_{k=1}^{n} f(c_k) \cdot \Delta x$$

With this notation (1) becomes

$$A = \lim_{n \to +\infty} \sum_{k=1}^{n} f(c_k) \cdot \Delta x \tag{4}$$

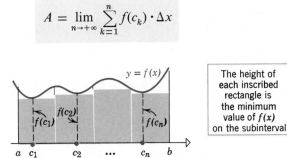

The height of each inscribed rectangle is the minimum value of $f(x)$ on the subinterval

Figure 5.6.5

▶ Example 1   Use inscribed rectangles to find the area under the line $y = x$ over the interval $[1, 2]$.

Solution.   If we subdivide the interval $[1, 2]$ into n equal parts (Figure 5.6.6), then each part will have length

$$\Delta x = \frac{b-a}{n} = \frac{2-1}{n} = \frac{1}{n}$$

and the points of subdivision will be

$$x_1 = 1 + \Delta x = 1 + \frac{1}{n}$$

$$x_2 = 1 + 2\,\Delta x = 1 + \frac{2}{n}$$

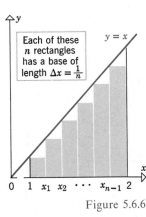

Each of these n rectangles has a base of length $\Delta x = \frac{1}{n}$

Figure 5.6.6

$$x_3 = 1 + 3\,\Delta x = 1 + \frac{3}{n}$$

$$\vdots$$

$$x_{n-1} = 1 + (n-1)\,\Delta x = 1 + \frac{n-1}{n}$$

Because $y = f(x) = x$ is increasing, the minimum value for $f(x)$ on each subinterval occurs at the left endpoint (Figure 5.6.6), so

$$c_1 = 1$$

$$c_2 = x_1 = 1 + \frac{1}{n}$$

$$c_3 = x_2 = 1 + \frac{2}{n}$$

$$c_4 = x_3 = 1 + \frac{3}{n}$$

$$\vdots$$

$$c_n = x_{n-1} = 1 + \frac{n-1}{n}$$

Thus, the inscribed rectangles have areas

$$f(c_1) \cdot \Delta x = c_1 \cdot \Delta x = 1 \cdot \frac{1}{n}$$

$$f(c_2) \cdot \Delta x = c_2 \cdot \Delta x = \left(1 + \frac{1}{n}\right) \cdot \frac{1}{n}$$

$$f(c_3) \cdot \Delta x = c_3 \cdot \Delta x = \left(1 + \frac{2}{n}\right) \cdot \frac{1}{n}$$

$$f(c_4) \cdot \Delta x = c_4 \cdot \Delta x = \left(1 + \frac{3}{n}\right) \cdot \frac{1}{n}$$

$$\vdots$$

$$f(c_n) \cdot \Delta x = c_n \cdot \Delta x = \left(1 + \frac{n-1}{n}\right) \cdot \frac{1}{n}$$

and the sum is

$$\sum_{k=1}^{n} f(c_k) \cdot \Delta x$$

$$= \left[1 + \left(1 + \frac{1}{n}\right) + \left(1 + \frac{2}{n}\right) + \left(1 + \frac{3}{n}\right) + \cdots + \left(1 + \frac{n-1}{n}\right)\right] \cdot \frac{1}{n}$$

$$= \left[n + \left(\frac{1}{n} + \frac{2}{n} + \frac{3}{n} + \cdots + \frac{n-1}{n}\right)\right] \cdot \frac{1}{n}$$

$$= 1 + \frac{1}{n^2}[1 + 2 + 3 + \cdots + (n-1)]$$

$$= 1 + \frac{1}{n^2} \cdot \frac{(n-1)(n)}{2} \qquad \text{[Theorem 5.5.2a with } n-1 \text{ substituted for } n\text{]}$$

$$= \frac{3}{2} - \frac{1}{2n}$$

Thus, from (4) the area is

$$A = \lim_{n \to +\infty} \sum_{k=1}^{n} f(c_k) \cdot \Delta x$$

$$= \lim_{n \to +\infty} \left(\frac{3}{2} - \frac{1}{2n} \right)$$

$$= \frac{3}{2} - 0 = \frac{3}{2}$$

The region whose area we have computed is a trapezoid with height $h = 1$ and bases $b_1 = 1$ and $b_2 = 2$. From plane geometry, the area of this trapezoid is $A = \frac{1}{2}h(b_1 + b_2) = \frac{1}{2}(1)(1 + 2) = \frac{3}{2}$, which agrees with the result obtained here.   ◄

▶ Example 2   Use inscribed rectangles to find the area under the curve $y = 9 - x^2$ over the interval $[0, 3]$.

Solution. If we divide the interval $[0, 3]$ into n subintervals of equal length, then each subinterval will have length

$$\Delta x = \frac{b - a}{n} = \frac{3 - 0}{n} = \frac{3}{n}$$

and the points of subdivision will be

$$x_1 = 0 + \Delta x = \frac{3}{n}$$

$$x_2 = 0 + 2\,\Delta x = 2\left(\frac{3}{n}\right)$$

$$x_3 = 0 + 3\,\Delta x = 3\left(\frac{3}{n}\right)$$

$$\vdots$$

$$x_{n-1} = 0 + (n-1)\,\Delta x = (n-1)\left(\frac{3}{n}\right)$$

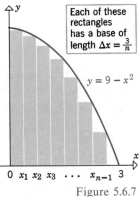

Each of these rectangles has a base of length $\Delta x = \frac{3}{n}$

$y = 9 - x^2$

0 $x_1\ x_2\ x_3$ $\cdots$ x_{n-1} 3

Figure 5.6.7

(Figure 5.6.7). Because $y = f(x) = 9 - x^2$ is decreasing on $[0, 3]$, the minimum value for $f(x)$ on each subinterval occurs at the right endpoint (Figure 5.6.7), so that

$$c_1 = x_1 = \frac{3}{n}$$

$$c_2 = x_2 = 2 \cdot \frac{3}{n}$$

$$c_3 = x_3 = 3 \cdot \frac{3}{n}$$

$$\vdots$$

$$c_{n-1} = x_{n-1} = (n-1) \cdot \frac{3}{n}$$

$$c_n = 3$$

In short, $c_k = k \cdot (3/n)$ for $k = 1, 2, \ldots, n$. Thus, the kth inscribed rectangle has area

$$f(c_k) \cdot \Delta x = (9 - c_k{}^2) \cdot \Delta x = \left[9 - k^2\left(\frac{9}{n^2}\right)\right] \cdot \frac{3}{n} = \frac{27}{n} - \frac{27k^2}{n^3}$$

and the sum of these areas is

$$\sum_{k=1}^{n} f(c_k) \cdot \Delta x = \sum_{k=1}^{n} \left(\frac{27}{n} - \frac{27k^2}{n^3}\right)$$

$$= \sum_{k=1}^{n} \frac{27}{n} - \sum_{k=1}^{n} \frac{27k^2}{n^3}$$

$$= n \cdot \frac{27}{n} - \frac{27}{n^3} \sum_{k=1}^{n} k^2$$

$$= 27 - \frac{27}{n^3} \cdot \frac{n(n+1)(2n+1)}{6} \qquad \text{[Theorem 5.5.2}b\text{]}$$

Thus, from (4)

$$A = \lim_{n \to +\infty} \sum_{k=1}^{n} f(c_k) \cdot \Delta x = \lim_{n \to +\infty} \left[27 - \frac{27}{n^3} \cdot \frac{n(n+1)(2n+1)}{6}\right]$$

$$= 27 - \lim_{n \to +\infty} \frac{27}{6}\left(1 + \frac{1}{n}\right)\left(2 + \frac{1}{n}\right) = 27 - \frac{27}{6}(1)(2) = 18 \qquad \blacktriangleleft$$

It may already have occurred to the reader that we could have used *circumscribed* rather than inscribed rectangles in these examples. Indeed, if the *maximum* values of $f(x)$ in the various subintervals occur at the points $d_1, d_2, \ldots, d_n$ then the sum

$$\sum_{k=1}^{n} f(d_k) \cdot \Delta x$$

is an approximation by circumscribed rectangles to the area under the curve (Figure 5.6.8), and the exact area A is

$$A = \lim_{n \to +\infty} \sum_{k=1}^{n} f(d_k) \cdot \Delta x \tag{5}$$

since the error (represented by the overlap) diminishes as the rectangles get thinner.

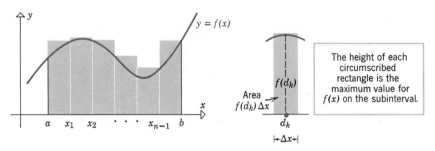

Figure 5.6.8

The height of each circumscribed rectangle is the maximum value for $f(x)$ on the subinterval.

▶ **Example 3**   Use circumscribed rectangles to find the area under the line $y = x$ over the interval $[1, 2]$.

Solution.   As in Example 1, the points

$$x_1 = 1 + \frac{1}{n}$$

$$x_2 = 1 + \frac{2}{n}$$

$$x_3 = 1 + \frac{3}{n}$$

$$\vdots$$

$$x_{n-1} = 1 + \frac{n-1}{n}$$

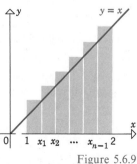

Figure 5.6.9

divide the interval $[1, 2]$ into n subintervals of length $\Delta x = 1/n$. Because $y = f(x) = x$ is increasing, the maximum value for $f(x)$ on each subinterval occurs at the right-hand endpoint (Figure 5.6.9), so that $d_1 = x_1 = 1 + 1/n$, $d_2 = x_2 = 1 + 2/n$, $d_3 = x_3 = 1 + 3/n, \ldots, d_n = 1 + n/n = 2$. Thus, the kth circumscribed rectangle has area

$$f(d_k) \cdot \Delta x = d_k \cdot \Delta x = \left(1 + \frac{k}{n}\right) \cdot \Delta x = \left(1 + \frac{k}{n}\right) \cdot \frac{1}{n}$$

and the sum of the areas of the rectangles is

$$\sum_{k=1}^{n} f(d_k) \cdot \Delta x = \sum_{k=1}^{n} \left[\left(1 + \frac{k}{n} \right) \cdot \frac{1}{n} \right]$$

$$= \frac{1}{n} \sum_{k=1}^{n} 1 + \frac{1}{n^2} \sum_{k=1}^{n} k$$

$$= \frac{1}{n} \cdot n + \frac{1}{n^2} \left[\frac{n(n + 1)}{2} \right] \qquad \text{[Theorem 5.5.2a]}$$

$$= \frac{3}{2} + \frac{1}{2n}$$

From (5), the area under the curve is

$$A = \lim_{n \to +\infty} \sum_{k=1}^{n} f(d_k) \cdot \Delta x = \lim_{n \to +\infty} \left(\frac{3}{2} + \frac{1}{2n} \right) = \frac{3}{2}$$

This result agrees with that obtained in Example 1 using inscribed rectangles. ◀

REMARK. Although we shall not do it, it can be proved that, in general, the method of inscribed rectangles [formula (4)] and the method of circumscribed rectangles [formula (5)] both yield the same value for A.

▶ Exercise Set 5.6

In Exercises 1–4, divide the interval $[a, b]$ into $n = 4$ subintervals of equal length and compute: (a) the sum of the areas of the inscribed rectangles, (b) the sum of the areas of the circumscribed rectangles.

1. $y = 3x + 1$; $a = 2$, $b = 6$.

2. $y = 1/x$; $a = 1$, $b = 9$.

3. $y = \cos x$; $a = -\frac{\pi}{2}$, $b = \frac{\pi}{2}$.

4. $y = 2x - x^2$; $a = 1$, $b = 2$.

In Exercises 5–10, use inscribed rectangles to find the area under the curve $y = f(x)$ over the interval $[a, b]$.

5. $y = \frac{1}{2}x$; $a = 1$, $b = 4$.

6. $y = -x + 5$; $a = 0$, $b = 5$.

7. $y = x^2$; $a = 0$, $b = 1$.

8. $y = 4 - \frac{1}{4}x^2$; $a = 0$, $b = 3$.

9. $y = x^3$; $a = 2$, $b = 6$.

10. $y = 1 - x^3$; $a = -3$, $b = -1$.

11. Use circumscribed rectangles to find the area of the region in Exercise 5.

12. Use circumscribed rectangles to find the area of the region in Exercise 6.

13. Use circumscribed rectangles to find the area of the region in Exercise 7.

14. Use inscribed rectangles to find the area under $y = mx$ over the interval $[a, b]$, where $m > 0$ and $a \geq 0$.

15. (a) Show that the area under $y = x^3$ over the interval $[0, b]$ is $b^4/4$.

    (b) Find a formula for the area under $y = x^3$ over the interval $[a, b]$, where $a \geq 0$.

16. Find the area between the curve $y = \sqrt{x}$ and the interval $0 \le y \le 1$ on the y-axis.

17. Assuming that $a > 0$, find the area under $y = x$ over the interval $[a, b]$ four different ways:
  (a) using the method of Section 5.1 (antidifferentiation)
  (b) using inscribed rectangles
  (c) using circumscribed rectangles
  (d) using an appropriate area formula from plane geometry.

5.7 THE DEFINITE INTEGRAL

In mathematics and science there are a variety of concepts, such as length, volume, density, probability, work, and others, whose properties are remarkably similar to properties of area. In this section we will introduce the concept of a *definite integral,* which is the unifying thread relating these diverse ideas.

In the last section we gave two equivalent ways of obtaining the area under a continuous curve $y = f(x)$ over an interval $[a, b]$:

$$A = \lim_{n \to +\infty} \sum_{k=1}^{n} f(c_k)\,\Delta x \qquad \text{(inscribed rectangles)} \qquad (1)$$

and

$$A = \lim_{n \to +\infty} \sum_{k=1}^{n} f(d_k)\,\Delta x \qquad \text{(circumscribed rectangles)} \qquad (2)$$

However, these are not the only possible formulas for the area A. Instead of choosing the height of the rectangle on the kth subinterval to be the minimum value or the maximum value of f, we can construct a rectangle of some intermediate height. For this purpose, let us select an *arbitrary* point in each subinterval; call these points

$$x_1^*, x_2^*, \ldots, x_n^*$$

Since $f(c_k)$ and $f(d_k)$ are, respectively, the smallest and largest values of f on the kth subinterval, it follows that

$$f(c_k) \le f(x_k^*) \le f(d_k)$$

and

$$f(c_k) \cdot \Delta x \le f(x_k^*) \cdot \Delta x \le f(d_k) \cdot \Delta x$$

so that

$$\sum_{k=1}^{n} f(c_k)\,\Delta x \le \sum_{k=1}^{n} f(x_k^*)\,\Delta x \le \sum_{k=1}^{n} f(d_k)\,\Delta x \qquad (3)$$

As $n \to +\infty$, the two outside sums approach A [see (1) and (2)], so the inside sum must also approach A, since it is "pinched" between the outer sums. Thus, for all possible choices of $x_1^*, x_2^*, \ldots, x_n^*$ we obtain

$$A = \lim_{n \to +\infty} \sum_{k=1}^{n} f(x_k^*) \Delta x$$

While rectangles with equal widths are convenient computationally, they are not essential; we can just as well express the area A as the limit of a sum of rectangular areas with different widths (Figure 5.7.1).

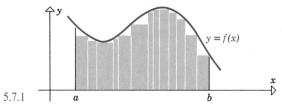

Figure 5.7.1

However, there is a complication here that did not occur before. For rectangles of equal width,

$$\Delta x = \frac{b - a}{n}$$

it follows that $\Delta x \to 0$ as $n \to +\infty$. In other words, we can ensure that the width of *every* rectangle decreases to zero by letting the number of rectangles increase to infinity. For rectangles whose widths vary in size this is not the case. For example, suppose we were to continually divide only the left half of the interval $[a, b]$ into more and more subintervals, but were always to leave the right half of the interval alone (Figure 5.7.2). With this construction the number of rectangles increases to infinity, but the sum of the areas of the rectangles does not approach the area under the curve because there is always an "error" on the right half of the interval. To remedy this problem, we need to ensure that the widths of *all* rectangles decrease to zero as the number of rectangles increases to infinity.

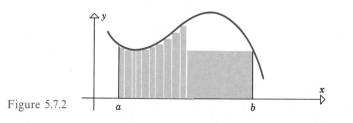

Figure 5.7.2

Let us suppose that the interval $[a, b]$ is divided into n subintervals whose widths are

$$\Delta x_1, \Delta x_2, \ldots, \Delta x_n$$

and let us denote the largest of these by the symbol

$$\max \Delta x_k$$

(read, "the maximum of the Δx_k's.")

If x_k^* is an arbitrary point in the kth subinterval, then

$$f(x_k^*) \, \Delta x_k$$

is the area of a rectangle of height $f(x_k^*)$ and width Δx_k, so

$$\sum_{k=1}^{n} f(x_k^*) \, \Delta x_k \tag{4}$$

is the sum of the shaded rectangular areas in Figure 5.7.3.

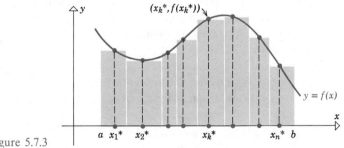

Figure 5.7.3

If we now increase n in such a way that

$$\max \Delta x_k \to 0$$

then the width of *every* rectangle tends to zero because no width exceeds the maximum. Thus, (4) approaches the exact area under the curve as $\max \Delta x_k \to 0$. We denote this by writing

$$A = \lim_{\max \Delta x_k \to 0} \sum_{k=1}^{n} f(x_k^*) \, \Delta x_k$$

We will give a precise definition of this limit at the end of this section. The preceding informal discussion suggests the following definition.

5.7.1 DEFINITION

Area

If the function f is continuous on $[a, b]$ and if $f(x) \geq 0$ for all x in $[a, b]$, then the **area** under the curve $y = f(x)$ over the interval $[a, b]$ is defined by

$$A = \lim_{\max \Delta x_k \to 0} \sum_{k=1}^{n} f(x_k^*) \, \Delta x_k$$

where $x_1^*, x_2^*, \ldots, x_n^*$ are arbitrary points in successive subintervals.

REMARK.   It is proved in advanced calculus that the limit in this definition exists, and that the area A is a nonnegative quantity.

Definition 5.7.1 applies only to *nonnegative* continuous functions. However, if f is continuous and $f(x) \leq 0$ for all x in $[a, b]$, then $-f$ is continuous and $-f(x) \geq 0$ for all x in $[a, b]$ (Figure 5.7.4). We define the area A between $y = f(x)$ and the interval $[a, b]$ to be the same as the area between $y = -f(x)$ and the interval $[a, b]$.

The limiting process used in Definition 5.7.1 to define area is of fundamental importance in a host of physical and mathematical applications that have no immediate connection with area. In many such applications the function f may have discontinuities and may assume *both* positive and negative values on $[a, b]$. To pave the way for these applications, we will retrace the three steps leading to Definition 5.7.1, but now we will drop the assumptions that f is continuous and nonnegative on $[a, b]$; we will assume only that f is defined on $[a, b]$.

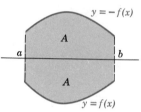

Figure 5.7.4

Step 1.   Divide the interval $[a, b]$ into n subintervals by choosing arbitrary points $x_1, x_2, \ldots, x_{n-1}$ satisfying

$$a < x_1 < x_2 < \cdots < x_{n-1} < b$$

The points $a, x_1, x_2, \ldots, x_{n-1}, b$ are said to form a **partition** of the interval. Let

$$\Delta x_1, \Delta x_2, \ldots, \Delta x_n$$

denote the lengths of successive subintervals formed by the partition. The largest of these lengths is called the **norm** of **mesh size** of the partition and is denoted by

$$\max \Delta x_k$$

Step 2.   Choose arbitrary points $x_1^*, x_2^*, \ldots, x_n^*$

in successive subintervals and form the sum

$$f(x_1^*) \, \Delta x_1 + f(x_2^*) \, \Delta x_2 + \cdots + f(x_n^*) \, \Delta x_n$$

$$= \sum_{k=1}^{n} f(x_k^*) \, \Delta x_k$$

> The sum in Step 2 is called a **Riemann* sum.**
>
> **Step 3.** Increase n, so that the mesh size of the partition approaches zero, and form the limit
>
> $$\lim_{\max \Delta x_k \to 0} \sum_{k=1}^{n} f(x_k^*)\, \Delta x_k$$
>
> if it exists.

The limit in Step 3 is of such importance that there is some notation and terminology associated with it.

5.7.2   DEFINITION   If a function f is defined on a closed interval $[a, b]$, then the **definite integral** of f from a to b, denoted by $\int_a^b f(x)\, dx$, is defined to be

$$\int_a^b f(x)\, dx = \lim_{\max \Delta x_k \to 0} \sum_{k=1}^{n} f(x_k^*)\, \Delta x_k \tag{5}$$

provided the limit exists.

The numbers a and b in the symbol $\int_a^b f(x)\, dx$ are called, respectively, the **lower** and **upper limits of integration,** and $f(x)$ is called the **integrand.** The

---

*GEORG FRIEDRICH BERNHARD RIEMANN (1826–1866) German mathematician. Bernard Riemann, as he is commonly known, was the son of a Protestant minister. He received his elementary education from his father and showed brilliance in arithmetic at an early age. In 1846 he enrolled at Göttingen University to study theology and philology, but he soon transferred to mathematics. He studied physics under W. E. Weber and mathematics under Karl Friedrich Gauss, whom some people consider to be the greatest mathematician who ever lived. In 1851 Riemann received his Ph.D. under Gauss, after which he remained at Göttingen to teach. In 1862, one month after his marriage, Riemann suffered an attack of pleuritis, and for the remainder of his life was a sick and dying man. He finally succumbed to tuberculosis in 1866 at age 39.

An interesting story surrounds Riemann's work in geometry. For his introductory lecture prior to becoming an associate professor, Riemann submitted three possible topics to Gauss. Gauss surprised Riemann by choosing the topic Riemann liked the least, the foundations of geometry. The lecture was like a scene from a movie. The old and failing Gauss, a giant in his day, watching intently as his brilliant and youthful protege skillfully pieced together portions of the old man's own work into a complete and beautiful system. Gauss is said to have gasped with delight as the lecture neared its end, and on the way home he marveled at his student's brilliance. Gauss died shortly thereafter. The results presented by Riemann that day eventually evolved into a fundamental tool that Einstein used some 50 years later to develop relativity theory.

In addition to his work in geometry, Riemann made major contributions to the theory of complex functions and mathematical physics. The notion of the definite integral, as it is presented in most basic calculus courses, is due to him. Riemann's early death was a great loss to mathematics, for his mathematical work was brilliant and of fundamental importance.

reason for the use of an integral sign in the above definition will become clear in the next section where we will establish a relationship between the definite integral defined by (5) and the indefinite integral studied earlier.

Because the definite integral is defined as a limit, the integral may or may not exist depending on the nature of the integrand f. If the integral does exist, then f is said to be **Riemann integrable** or simply **integrable** on $[a, b]$. The problem of determining precisely which functions are integrable is beyond the scope of this text. However, the following theorem, stated without proof, is useful to know.

5.7.3   THEOREM   (a)   *If f is continuous on $[a, b]$, then f is integrable on $[a, b]$.*
(b)   *If f has only finitely many points of discontinuity on $[a, b]$ and if there is a positive number M such that $-M \le f(x) \le M$ for all x in $[a, b]$, then f is integrable on $[a, b]$.*

REMARK.   In part (b), the condition that there exists a positive number M such that $-M \le f(x) \le M$ for all x in $[a, b]$ is described by stating that f is **bounded** on $[a, b]$. Geometrically, this means that over the interval $[a, b]$ the entire graph lies between two horizontal lines, $y = -M$ and $y = M$. A function that approaches $+\infty$ or $-\infty$ somewhere on the interval $[a, b]$ does not satisfy this condition. It can be proved that every Riemann integrable function on $[a, b]$ must be bounded on $[a, b]$.

The function f graphed in Figure 5.7.5 is integrable on $[a, b]$, since it is bounded on the interval $[a, b]$ and has only finitely many points of discontinuity on $[a, b]$.

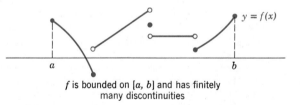

$y = f(x)$

f is bounded on $[a, b]$ and has finitely many discontinuities

Figure 5.7.5

The following theorem follows from Definitions 5.7.1 and 5.7.2.

5.7.4   THEOREM   *If f is nonnegative and continuous on $[a, b]$, then*

$$A = \begin{bmatrix} \text{area under} \\ y = f(x) \\ \text{over } [a, b] \end{bmatrix} = \int_a^b f(x)\, dx \qquad (6)$$

▶ Example 1   In Example 2 of Section 5.6 we showed the area under the curve $y = 9 - x^2$ over the interval $[0, 3]$ to be 18. Thus,

$$\int_0^3 (9 - x^2)\, dx = 18$$ ◀

If f is continuous and assumes both positive and negative values on $[a, b]$, then the definite integral can be interpreted as a *difference* of areas. To see why, consider a typical Riemann sum

$$\sum_{k=1}^{n} f(x_k^*)\, \Delta x_k$$

If $f(x_k^*)$ is nonnegative, then the term $f(x_k^*)\,\Delta x_k$ represents the area A_k of a rectangle with height $f(x_k^*)$ and base Δx_k. On the other hand, if $f(x_k^*)$ is negative, then $f(x_k^*)\,\Delta x_k$ is *not* the area of a rectangle, but rather the negative $-A_k$ of such an area. For example, in Figure 5.7.6

$$\sum_{k=1}^{6} f(x_k^*)\,\Delta x_k = f(x_1^*)\,\Delta x_1 + f(x_2^*)\,\Delta x_2 + \cdots + f(x_6^*)\,\Delta x_6$$

$$= A_1 + A_2 - A_3 - A_4 + A_5 + A_6$$
$$= (A_1 + A_2 + A_5 + A_6) - (A_3 + A_4)$$

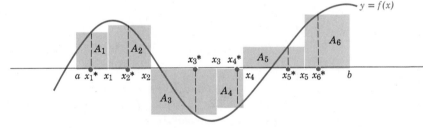

Figure 5.7.6

It follows that the definite integral

$$\int_a^b f(x)\, dx = \lim_{\max \Delta x_k \to 0} \sum_{k=1}^{n} f(x_k^*)\,\Delta x_k$$

can be interpreted as a difference of areas: the area above the x-axis between $y = f(x)$ and $[a, b]$ minus the area below the x-axis between $y = f(x)$ and $[a, b]$. As an illustration, the Riemann sums in Figure 5.7.7 tend toward $A_I - A_{II} + A_{III}$ as a limit, so that

$$\int_a^b f(x)\, dx = (A_I + A_{III}) - A_{II} = \begin{bmatrix} \text{area above} \\ x\text{-axis} \end{bmatrix} - \begin{bmatrix} \text{area below} \\ x\text{-axis} \end{bmatrix}$$

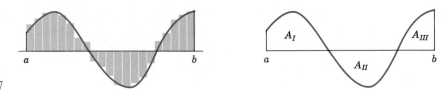

Figure 5.7.7

In the case where f is continuous and $f(x) \geq 0$ on $[a, b]$, the portion of area below the x-axis is zero so that $\int_a^b f(x)\, dx$ represents the area under $y = f(x)$ over $[a, b]$. On the other hand, if f is continuous and $f(x) \leq 0$ on $[a, b]$, then the portion of area above the x-axis is zero so that $\int_a^b f(x)\, dx$ represents the negative of the area above $y = f(x)$ under $[a, b]$. This agrees with our earlier results.

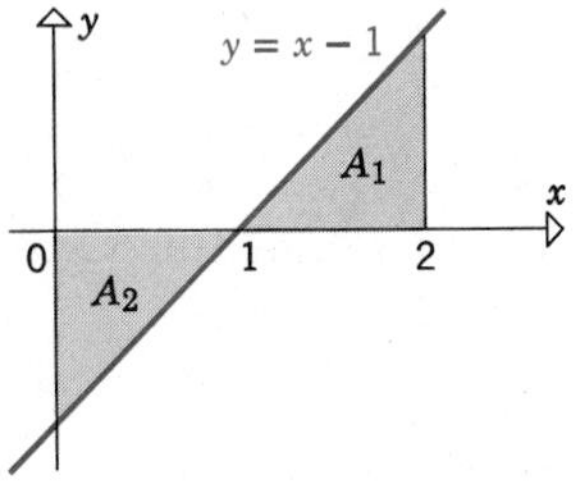

▶ **Example 2** Referring to Figure 5.7.8, the area A_1 is equal to the area A_2 so that

$$\int_0^2 (x - 1)\, dx = A_1 - A_2 = 0 \qquad\blacktriangleleft$$

Figure 5.7.8

OPTIONAL

The limits in Definitions 5.7.1 and 5.7.2 are different from those discussed in Chapter 2. Loosely phrased, the expression

$$\lim_{\max \Delta x_k \to 0} \sum_{k=1}^{n} f(x_k^*)\, \Delta x_k = L$$

is intended to convey the idea that we can force the Riemann sums to be as close as we please to L, regardless of how $x_1^*, x_2^*, \ldots, x_n^*$ are chosen, by making the mesh size of the partition sufficiently small. This idea is captured in the following definition.

5.7.5 DEFINITION We will write

$$\lim_{\max \Delta x_k \to 0} \sum_{k=1}^{n} f(x_k^*)\, \Delta x_k = L \tag{7}$$

if given any number $\epsilon > 0$, there is a number $\delta > 0$ such that

$$\left| \sum_{k=1}^{n} f(x_k^*)\, \Delta x_k - L \right| < \epsilon$$

whenever

$$\max \Delta x_k < \delta$$

and regardless of how $x_1^*, x_2^*, \ldots, x_n^*$ are chosen.

It can be shown that when a number L satisfying (7) exists, it is unique (Exercise 19). We denote this number by $\int_a^b f(x)\, dx$ and write

$$\lim_{\max \Delta x_k \to 0} \sum_{k=1}^n f(x_k^*)\, \Delta x_k = \int_a^b f(x)\, dx \qquad (8)$$

REMARK. Some writers use the symbol $\|\Delta\|$ rather than $\max \Delta x_k$ for the mesh size of the partition, in which case (8) would be written

$$\lim_{\|\Delta\| \to 0} \sum_{k=1}^n f(x_k^*)\, \Delta x_k = \int_a^b f(x)\, dx$$

▶ Exercise Set 5.7

In Exercises 1–4, find the value of:

(a) $\displaystyle\sum_{k=1}^n f(x_k^*)\, \Delta x_k$ (b) $\max \Delta x_k$.

1. $f(x) = x + 1;\quad a = 0,\quad b = 4;\quad x_1 = 1,\quad x_2 = 2;$
    $x_1^* = \frac{1}{3},\quad x_2^* = \frac{3}{2},\quad x_3^* = 3.$

2. $f(x) = \cos x;\qquad a = 0,\qquad b = 2\pi;\qquad x_1 = \pi/2,$
    $x_2 = 5\pi/4,\quad x_3 = 7\pi/4;\quad x_1^* = \pi/4,\quad x_2^* = \pi,$
    $x_3^* = 3\pi/2,\quad x_4^* = 7\pi/4.$

3. $f(x) = 4 - x^2; a = -3, b = 4; x_1 = -2, x_2 = 0,$
    $x_3 = 1;\quad x_1^* = -\frac{5}{2},\quad x_2^* = -1,\quad x_3^* = \frac{1}{4},\quad x_4^* = 3.$

4. $f(x) = x^3;\quad a = -3,\quad b = 3;\quad x_1 = -1,\quad x_2 = 0,$
    $x_3 = 1;\quad x_1^* = -2,\quad x_2^* = 0,\quad x_3^* = 0,\quad x_4^* = 2.$

5. Verify that relationship (3) holds when

    $$f(x) = \frac{1}{x};\quad a = 1,\quad b = 9;\quad x_1 = 3,\quad x_2 = 5,$$

    $$x_3 = 7;\quad x_1^* = 2,\quad x_2^* = 4,\quad x_3^* = 6,\quad x_4^* = 8.$$

In Exercises 6–8, use the given values of a and b to express the following limits as definite integrals. (Do not evaluate the integrals.)

6. $\displaystyle\lim_{\max \Delta x_k \to 0} \sum_{k=1}^n (x_k^*)^3\, \Delta x_k;\qquad a = 1, b = 2$

7. $\displaystyle\lim_{\max \Delta x_k \to 0} \sum_{k=1}^n 4x_k^*(1 - 3x_k^*)\, \Delta x_k;\qquad a = -3, b = 3$

8. $\displaystyle\lim_{\max \Delta x_k \to 0} \sum_{k=1}^n (\sin^2 x_k^*)\, \Delta x_k;\qquad a = 0, b = \pi/2.$

In Exercises 9–11, express the definite integrals as limits. (Do not evaluate.)

9. $\displaystyle\int_1^2 2x\, dx.$ 10. $\displaystyle\int_{-\pi/2}^{\pi/2} (1 + \cos x)\, dx.$

11. $\displaystyle\int_0^1 \frac{x}{x + 1}\, dx.$

12. Evaluate the following definite integrals by using appropriate area formulas from plane geometry.

    (a) $\displaystyle\int_0^3 x\, dx$ (b) $\displaystyle\int_{-2}^{-1} x\, dx$

    (c) $\displaystyle\int_{-1}^4 x\, dx$ (d) $\displaystyle\int_{-5}^5 x\, dx.$

13. Let

    $$f(x) = \begin{cases} -1, & x \geq 0 \\ -x, & x < 0 \end{cases}$$

    Evaluate the following definite integrals by using appropriate area formulas from plane geometry.

    (a) $\displaystyle\int_4^5 f(x)\, dx$ (b) $\displaystyle\int_{-3}^{-2} f(x)\, dx$

    (c) $\displaystyle\int_{-5}^2 f(x)\, dx$ (d) $\displaystyle\int_{-k}^k f(x)\, dx$ $(k > 0).$

14. Evaluate $\int_0^2 \sqrt{4 - x^2}\, dx$.

[*Hint:* Interpret the integral as an area.]

15. Use the hint in Exercise 14 to evaluate

$$\int_{-1}^1 \sqrt{1 - x^2}\, dx$$

16. Prove that the function

$$f(x) = \begin{cases} 1 & \text{if } x \text{ is rational} \\ 0 & \text{if } x \text{ is irrational} \end{cases}$$

is not integrable on any closed interval $[a, b]$.

17. Prove that the function

$$f(x) = \begin{cases} \dfrac{1}{x^2}, & x > 0 \\ 0, & x = 0 \end{cases}$$

is not integrable on the interval $[0, 1]$.

18. Use Theorem 5.7.3 to show that

$$f(x) = \begin{cases} \sin \dfrac{1}{x}, & x \neq 0 \\ 0, & x = 0 \end{cases}$$

is integrable on the interval $[-1, 1]$.

Exercises 19–21 are for readers who have read the optional material at the end of this section.

19. **(Optional)** Prove: If f is integrable on $[a, b]$, then the value of $\int_a^b f(x)\, dx$ is unique. [*Hint:* Show that if

$$\lim_{\max \Delta x_k \to 0} \sum_{k=1}^n f(x_k^*)\, \Delta x_k = L_1$$

and

$$\lim_{\max \Delta x_k \to 0} \sum_{k=1}^n f(x_k^*)\, \Delta x_k = L_2$$

then $L_1 = L_2$.]

20. **(Optional)** Prove: If f is integrable on $[a, b]$ and c is any constant, then cf is integrable on $[a, b]$ and

$$\int_a^b cf(x)\, dx = c \int_a^b f(x)\, dx$$

21. **(Optional)** Prove: If f and g are integrable on $[a, b]$, then $f + g$ is integrable on $[a, b]$ and

$$\int_a^b [f(x) + g(x)]\, dx = \int_a^b f(x)\, dx + \int_a^b g(x)\, dx$$

5.8   THE FIRST FUNDAMENTAL THEOREM OF CALCULUS

In Section 5.1 we saw that areas under curves can be obtained by antidifferentiation. In this section we will prove a result, called the *First Fundamental Theorem of Calculus,* which shows that the relationship between areas and antiderivatives is a special case of a more general relationship between definite integrals and antiderivatives.

5.8.1   THEOREM
The First Fundamental Theorem of Calculus

If f is continuous on $[a, b]$ and if F is an antiderivative of f on $[a, b]$, then

$$\int_a^b f(x)\, dx = F(b) - F(a) \tag{1}$$

Proof.   Let $x_1, x_2, \ldots, x_{n-1}$ be any points in $[a, b]$ such that

$$a < x_1 < x_2 < \cdots < x_{n-1} < b$$

These points divide $[a, b]$ into n subintervals

$$[a, x_1], [x_1, x_2], \ldots, [x_{n-1}, b] \tag{2}$$

whose lengths, as usual, we denote by

$$\Delta x_1, \Delta x_2, \ldots, \Delta x_n \tag{3}$$

Now we write $F(b) - F(a)$ as a telescoping sum:

$$\begin{aligned} F(b) - F(a) &= [F(x_1) - F(a)] + [F(x_2) - F(x_1)] + [F(x_3) - F(x_2)] \\ &\quad + \cdots + [F(b) - F(x_{n-1})] \end{aligned} \tag{4}$$

By hypothesis,

$$F'(x) = f(x) \tag{5}$$

for all x in $[a, b]$, so F satisfies the hypotheses of the Mean-Value Theorem (Theorem 4.10.2) on each subinterval in (2). Hence, (4) can be rewritten

$$\begin{aligned} F(b) - F(a) &= F'(x_1^*)(x_1 - a) + F'(x_2^*)(x_2 - x_1) + F'(x_3^*)(x_3 - x_2) \\ &\quad + \cdots + F'(x_n^*)(b - x_{n-1}) \end{aligned} \tag{6}$$

where $x_1^*, x_2^*, \ldots, x_n^*$ are points in successive subintervals.
 Using (3) and (5) we can rewrite (6) as

$$F(b) - F(a) = f(x_1^*) \Delta x_1 + f(x_2^*) \Delta x_2 + f(x_3^*) \Delta x_3 + \cdots + f(x_n^*) \Delta x_n$$

or in sigma notation,

$$F(b) - F(a) = \sum_{k=1}^{n} f(x_k^*) \Delta x_k \tag{7}$$

Let us now increase n in such a way that $\max \Delta x_k \to 0$. Since f is assumed to be continuous, the right side of (7) will approach $\int_a^b f(x)\, dx$, by Theorem 5.7.3a. However, the left side of (7) is a constant that is independent of n; thus,

$$F(b) - F(a) = \lim_{\max \Delta x_k \to 0} \sum_{k=1}^{n} f(x_k^*) \Delta x_k = \int_a^b f(x)\, dx \quad \blacksquare$$

The difference $F(b) - F(a)$ is commonly denoted by $F(x)]_a^b$ so that (1) can be written

$$\int_a^b f(x)\, dx = F(x) \Big]_a^b \tag{8}$$

REMARK. The symbols $[F(x)]_a^b$, $[F(x)]_{x=a}^{x=b}$, and $F(x)]_{x=a}^b$ are also used to denote $F(b) - F(a)$.

▶ Example 1 Evaluate

$$\int_1^2 x\, dx$$

Solution. The function

$$F(x) = \frac{1}{2}x^2$$

is an antiderivative of $f(x) = x$; thus, from (8)

$$\int_1^2 x\, dx = \frac{1}{2}x^2 \bigg]_1^2 = \frac{1}{2}(2)^2 - \frac{1}{2}(1)^2 = 2 - \frac{1}{2} = \frac{3}{2}$$ ◀

When applying the First Fundamental Theorem of Calculus, it does not matter which antiderivative of f is used, for if F is any antiderivative of f on $[a, b]$, then all others have the form

$$F(x) + C$$

(Theorem 5.2.2). Thus,

$$[F(x) + C]_a^b = [F(b) + C] - [F(a) + C]$$
$$= F(b) - F(a) = F(x)]_a^b$$
$$= \int_a^b f(x)\, dx \qquad (9)$$

which shows that all antiderivatives of f on $[a, b]$ yield the same value for $\int_a^b f(x)\, dx$.

Since

$$\int f(x)\, dx = F(x) + C$$

it follows from (9) that

$$\int_a^b f(x)\, dx = \left[\int f(x)\, dx \right]_a^b \qquad (10)$$

which relates the definite and indefinite integrals of f.

▶ **Example 2** Use the First Fundamental Theorem of Calculus to find the area under the curve $y = \cos x$ over the interval $[0, \pi/2]$.

Solution. Since $\cos x \geq 0$ for $0 \leq x \leq \pi/2$, the area is

$$A = \int_0^{\pi/2} \cos x \, dx$$

$$= \left[\int \cos x \, dx \right]_0^{\pi/2} \qquad \text{[from (10)]}$$

$$= \sin x \bigg]_0^{\pi/2} = \sin \frac{\pi}{2} - \sin 0 = 1 \qquad \blacktriangleleft$$

REMARK. In the third equality of this example we took the constant of integration for the indefinite integral to be $C = 0$. This is justified because we can select any antiderivative of f on $[a, b]$, in particular the one for which $C = 0$.

▶ **Example 3** Evaluate

$$\int_0^2 x^2 \sqrt{x^3 + 1} \, dx$$

Solution. We first evaluate the indefinite integral

$$\int x^2 \sqrt{x^3 + 1} \, dx$$

by a u-substitution. Let

$$u = x^3 + 1 \qquad \text{so that} \qquad \frac{du}{dx} = 3x^2 \qquad \text{or} \qquad \frac{1}{3} du = x^2 \, dx$$

Making this substitution gives

$$\int x^2 \sqrt{x^3 + 1} \, dx = \int \frac{1}{3} \sqrt{u} \, du = \frac{2}{9} u^{3/2} + C = \frac{2}{9}(x^3 + 1)^{3/2} + C$$

Therefore,

$$\int_0^2 x^2 \sqrt{x^3 + 1} \, dx = \frac{2}{9}(x^3 + 1)^{3/2} \bigg]_0^2 = \frac{2}{9}(27) - \frac{2}{9} = \frac{52}{9} \qquad \blacktriangleleft$$

The following properties of definite integrals follow from (10) and the corresponding properties for indefinite integrals. We will omit the proofs.

5.8.2 THEOREM   *If f and g are continuous on [a, b] and if c is a constant, then*

(a) $\displaystyle\int_a^b cf(x)\,dx = c \int_a^b f(x)\,dx$

(b) $\displaystyle\int_a^b [f(x) + g(x)]\,dx = \int_a^b f(x)\,dx + \int_a^b g(x)\,dx$

(c) $\displaystyle\int_a^b [f(x) - g(x)]\,dx = \int_a^b f(x)\,dx - \int_a^b g(x)\,dx$

Part (b) of Theorem 5.8.2 can be extended to include more than two functions. More precisely,

$$\int_a^b [f_1(x) + f_2(x) + \cdots + f_n(x)]\,dx \tag{11}$$

$$= \int_a^b f_1(x)\,dx + \int_a^b f_2(x)\,dx + \cdots + \int_a^b f_n(x)\,dx$$

REMARK.   The bracket notation $F(x)]_a^b$ has some properties in common with the definite integral. We leave it as an exercise to show:

$$[F(x) + G(x)]_a^b = F(x)]_a^b + G(x)]_a^b$$
$$[F(x) - G(x)]_a^b = F(x)]_a^b - G(x)]_a^b$$
$$[cF(x)]_a^b = c[F(x)]_a^b$$

▶ Example 4   Evaluate

$$\int_0^3 (x^3 - 4x + 1)\,dx$$

Solution.

$$\int_0^3 (x^3 - 4x + 1)\,dx = \int_0^3 x^3\,dx - 4\int_0^3 x\,dx + \int_0^3 dx$$

$$= \left[\frac{x^4}{4} - 4 \cdot \frac{x^2}{2} + x \right]_0^3$$

$$= \left(\frac{81}{4} - 18 + 3 \right) - (0) = \frac{21}{4} \qquad ◀$$

The following theorem is useful when it is necessary to compare the sizes of definite integrals.

5.8.3   THEOREM   (a)   *If f is continuous and $f(x) \geq 0$ for all x in $[a, b]$, then*

$$\int_a^b f(x)\, dx \geq 0$$

(b)   *If f and g are continuous and $f(x) \geq g(x)$ for all x in $[a, b]$, then*

$$\int_a^b f(x)\, dx \geq \int_a^b g(x)\, dx$$

Proof of (a).   Since f is continuous and nonnegative on $[a, b]$, the integral $\int_a^b f(x)\, dx$ represents the area under $y = f(x)$ over $[a, b]$. Thus, the integral is nonnegative by the remark following Definition 5.7.1.

Proof of (b).   It follows from the hypotheses that $f - g$ is continuous and $f(x) - g(x) \geq 0$ for all x in $[a, b]$. Thus by part *(a)*,

$$\int_a^b [f(x) - g(x)]\, dx = \int_a^b f(x)\, dx - \int_a^b g(x)\, dx \geq 0$$

from which it follows that

$$\int_a^b f(x)\, dx \geq \int_a^b g(x)\, dx$$

   In the case where $f(x) \geq g(x) \geq 0$ for all x in $[a, b]$, part *(b)* of this theorem states that the area under the "higher" curve $y = f(x)$ is at least as large as the area under the "lower" curve $y = g(x)$.

   Sometimes it is convenient to use a letter other than x for the variable of integration in a definite integral. For example,

$$\int_a^b f(t)\, dt, \qquad \int_a^b f(u)\, du, \qquad \int_a^b f(y)\, dy$$

It is an important fact that:

> *The value of a definite integral is unaffected if we change the letter used for the variable of integration, but do not change the limits of integration.*

As an illustration, let us compare the steps in the evaluation of $\int_1^3 x^2\, dx$ $\int_1^3 t^2\, dt$ and $\int_1^3 u^2\, du$. These integrals have the same limits of integration, but different variables of integration.

$$\int_1^3 x^2\, dx = \left. \frac{x^3}{3} \right]_{x=1}^{x=3} = \frac{27}{3} - \frac{1}{3} = \frac{26}{3}$$

$$\int_1^3 t^2 \, dt = \frac{t^3}{3} \Bigg]_{t=1}^{t=3} = \frac{27}{3} - \frac{1}{3} = \frac{26}{3}$$

$$\int_1^3 u^2 \, du = \frac{u^3}{3} \Bigg]_{u=1}^{u=3} = \frac{27}{3} - \frac{1}{3} = \frac{26}{3}$$

Because the letter used for the variable of integration has no effect on the final value of the definite integral, it is sometimes called a ***dummy variable.***

DISTANCE TRAVELED
IN RECTILINEAR
MOTION

We conclude this section with an application of the First Fundamental Theorem of Calculus to rectilinear motion.

In Section 5.4 we showed that for a particle with rectilinear motion the position function $s(t)$ and the velocity function $v(t)$ are related by

$$s(t) = \int v(t) \, dt$$

It follows from this and the First Fundamental Theorem of Calculus that

$$\int_{t_1}^{t_2} v(t) \, dt = s(t) \Bigg]_{t_1}^{t_2} = s(t_2) - s(t_1) \tag{12}$$

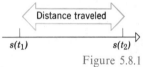

Figure 5.8.1

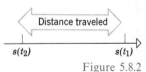

Figure 5.8.2

Since $s(t_1)$ is the position of the particle at time t_1 and $s(t_2)$ is the position at time t_2, the difference, $s(t_2) - s(t_1)$, is the change in position or ***displacement*** of the particle during the time interval $[t_1, t_2]$. A positive displacement means that the particle is farther to the right at time t_2 than at time t_1; a negative displacement means that the particle is farther to the left. In the case where $v(t) \geq 0$ throughout the time interval $[t_1, t_2]$, the particle moves in the positive direction only; thus, the displacement $s(t_2) - s(t_1)$ is the same as the distance traveled by the particle (Figure 5.8.1). In the case where $v(t) \leq 0$ throughout the time interval $[t_1, t_2]$, the particle moves in the negative direction only; thus, the displacement $s(t_2) - s(t_1)$ is the negative of the distance traveled by the particle (Figure 5.8.2). In the case where $v(t)$ assumes both positive and negative values during the time interval $[t_1, t_2]$, the particle moves back and forth and the displacement is the distance traveled in the positive direction minus the distance traveled in the negative direction. If we want to find the total distance traveled in this case (distance traveled in the positive direction *plus* the distance traveled in the negative direction), we must integrate the absolute value of the velocity function, that is,

$$\begin{bmatrix} \text{total distance} \\ \text{traveled during} \\ \text{time interval} \\ [t_1, t_2] \end{bmatrix} = \int_{t_1}^{t_2} |v(t)| \, dt \tag{13}$$

▶ **Example 5** A particle moves on a coordinate line so that its velocity at time t is $v(t) = t^2 - 2t$ meters/sec. Find

(a) the displacement of the particle during the time interval $0 \le t \le 3$;

(b) the distance traveled by the particle during the time interval $0 \le t \le 3$.

Solution (a). From (12) the displacement is

$$\int_0^3 v(t)\, dt = \int_0^3 (t^2 - 2t)\, dt = \left[\frac{t^3}{3} - t^2\right]_0^3 = 0$$

Thus, the particle is at the same position at time $t = 3$ as at $t = 0$.

Solution (b). The velocity can be written as $v(t) = t^2 - 2t = t(t - 2)$ from which it is evident that $v(t) \le 0$ for $0 \le t \le 2$ and $v(t) \ge 0$ for $2 \le t \le 3$. Thus, it follows from (13) that the distance traveled is

$$\int_0^3 |v(t)|\, dt = \int_0^2 -v(t)\, dt + \int_2^3 v(t)\, dt$$

$$= \int_0^2 -(t^2 - 2t)\, dt + \int_2^3 (t^2 - 2t)\, dt$$

$$= -\left[\frac{t^3}{3} - t^2\right]_0^2 + \left[\frac{t^3}{3} - t^2\right]_2^3$$

$$= \frac{4}{3} + \frac{4}{3} = \frac{8}{3} \text{ meters}$$ ◀

▶ **Exercise Set 5.8**

In Exercises 1–30, evaluate the definite integrals using the First Fundamental Theorem of Calculus.

1. $\displaystyle\int_2^3 x^3\, dx.$

2. $\displaystyle\int_{-1}^1 x^4\, dx.$

3. $\displaystyle\int_{-1}^2 x(1 + x^3)\, dx.$

4. $\displaystyle\int_{-3}^0 (x^2 - 4x + 7)\, dx.$

5. $\displaystyle\int_1^2 (t^2 - 2t + 8)\, dt.$

6. $\displaystyle\int_0^1 (x^5 - x^3 + 2x)\, dx.$

7. $\displaystyle\int_1^3 \frac{1}{x^2}\, dx.$

8. $\displaystyle\int_1^2 \frac{1}{x^6}\, dx.$

9. $\displaystyle\int_1^2 \left(\frac{1}{x^3} - \frac{2}{x^2} + x^{-4}\right) dx.$

10. $\displaystyle\int_{-2}^{-1} \left(u^{-4} + 3u^{-2} - \frac{1}{u^5}\right) du.$

11. $\displaystyle\int_1^9 \sqrt{x}\, dx.$

12. $\displaystyle\int_1^4 x^{-3/5}\, dx.$

13. $\displaystyle\int_4^9 2y\sqrt{y}\, dy.$

14. $\displaystyle\int_1^8 (5x^{2/3} - 4x^{-2})\, dx.$

15. $\displaystyle\int_1^4 \left(\frac{3}{\sqrt{x}} - 5\sqrt{x} - x^{-3/2}\right) dx.$

16. $\displaystyle\int_{4}^{9} (4y^{-1/2} + 2y^{1/2} + y^{-5/2})\, dy.$

17. $\displaystyle\int_{-\pi/2}^{\pi/2} \sin\theta\, d\theta.$

18. $\displaystyle\int_{0}^{\pi/4} \sec^2\theta\, d\theta.$

19. $\displaystyle\int_{-\pi/4}^{\pi/4} \frac{\sin 2x}{\cos x}\, dx.$

20. $\displaystyle\int_{0}^{1} (x - \sec x \tan x)\, dx.$

21. $\displaystyle\int_{\pi/6}^{\pi/2} \left(x + \frac{2}{\sin^2 x}\right) dx.$

22. $\displaystyle\int_{0}^{1} \sqrt[3]{a + bx}\, dx \ (b \neq 0).$

23. $\displaystyle\int_{0}^{1} \frac{du}{\sqrt{3u + 1}}.$

24. $\displaystyle\int_{-1}^{1} \pi \cos (x + 2)\, dx.$

25. $\displaystyle\int_{-3\pi/4}^{-\pi/4} \sin x \cos x\, dx.$

26. $\displaystyle\int_{0}^{\pi/4} \sqrt{\tan x}\, \sec^2 x\, dx.$

27. $\displaystyle\int_{-1}^{1} \frac{x^2\, dx}{\sqrt{x^3 + 9}}.$

28. $\displaystyle\int_{-1}^{0} 6t^2(t^3 + 1)^{19}\, dt.$

29. $\displaystyle\int_{a}^{4a} (a^{1/2} - x^{1/2})\, dx \ (a \text{ a positive constant}).$

30. $\displaystyle\int_{0}^{2\pi/t} t^2 \sin tx\, dx \ (t \text{ a positive constant}).$

31. Find the area under the curve $y = x^2 + 1$ over the interval $[0, 3]$. Make a sketch of the region.

32. Find the area above the x-axis, but below the curve $y = (1 - x)(x - 2)$. Make a sketch of the region.

33. Find the area under the curve $y = 3 \sin 2x$ over the interval $[0, \pi/8]$. Make a sketch of the region.

34. Find the area below the interval $[-2, -1]$, but above the curve $y = x^3$. Make a sketch of the region.

35. Find the total area that is between the curve $y = x^2 - 3x - 10$ and the interval $[-3, 8]$. Make a sketch of the region. [*Hint:* Find the portion of area above the interval and the portion of area below the interval separately.]

In Exercises 36–41, $v(t)$ is the velocity (meters/sec) of a particle moving on a coordinate line. Find the displacement and the distance traveled by the particle during the given time interval.

36. $v(t) = 2t - 4; \ 0 \leq t \leq 4.$

37. $v(t) = t^2 + t - 2; \ 0 \leq t \leq 2.$

38. $v(t) = |t - 3|; \ 0 \leq t \leq 5.$

39. $v(t) = \cos t; \ 0 \leq t \leq \pi.$

40. $v(t) = 3 \sin t; \ \pi/4 \leq t \leq \pi.$

41. $v(t) = t^3 - 3t^2 + 2t; \ 0 \leq t \leq 3.$

In Exercises 42–45, $a(t)$ is the acceleration (meters/sec²) of a particle moving on a coordinate line, and v_0 is its velocity at time $t = 0$. Find the displacement and the distance traveled by the particle during the given time interval.

42. $a(t) = -2; \ v_0 = 3; \ 1 \leq t \leq 4.$

43. $a(t) = t - 2; \ v_0 = 0; \ 1 \leq t \leq 5.$

44. $a(t) = \sin t; \ v_0 = 1; \ \dfrac{\pi}{4} \leq t \leq \dfrac{\pi}{2}.$

45. $a(t) = \dfrac{1}{\sqrt{5t + 1}}; \ v_0 = 2; \ 0 \leq t \leq 3.$

46. Express Equation (11) in sigma notation.

47. Prove: If f is continuous and $f(x) \leq 0$ for all x in $[a, b]$, then $\int_a^b f(x)\, dx \leq 0$.

48. Prove:
   (a) $[F(x) + G(x)]_a^b = F(x)]_a^b + G(x)]_a^b$
   (b) $[F(x) - G(x)]_a^b = F(x)]_a^b - G(x)]_a^b$
   (c) $[cF(x)]_a^b = c[F(x)]_a^b.$

49. Assuming the functions involved are all integrable on $[a, b]$, which of the following are always valid?
   (a) $\displaystyle\int_a^b f(x)g(x)\, dx = \int_a^b f(x)\, dx \int_a^b g(x)\, dx$

(b) $\displaystyle\int_a^b [c_1 f(x) + c_2 g(x)]\, dx$

$\qquad = c_1 \displaystyle\int_a^b f(x)\, dx + c_2 \int_a^b g(x)\, dx$

$\qquad (c_1, c_2 \text{ constant})$

(c) $\displaystyle\int_a^b \left(\sum_{k=1}^n c_k f_k(x) \right) dx = \sum_{k=1}^n \left[c_k \int_a^b f_k(x)\, dx \right]$

$(c_1, c_2, \ldots, c_k \text{ constant})$

(d) $\displaystyle\int_a^b [f(x)]^n\, dx = \left[\int_a^b f(x)\, dx \right]^n$

(e) $\displaystyle\int_a^b \sqrt{f(x)}\, dx = \sqrt{\int_a^b f(x)\, dx}.$

5.9   PROPERTIES OF THE DEFINITE INTEGRAL; AVERAGE VALUE

In this section we discuss additional properties of the definite integral and define the "average value" of a function.

Whenever we speak of the "closed interval $[a, b]$" it is assumed that $a < b$. Thus, our definition of the definite integral

$$\int_a^b f(x)\, dx$$

(Definition 5.7.2) does not allow for the possibilities $a = b$ or $b < a$. The following definition extends the notion of the definite integral to include these cases.

5.9.1   DEFINITION   (a)   If a is in the domain of f, we define

$$\int_a^a f(x)\, dx = 0$$

(b)   If $b < a$ and if f is integrable on $[b, a]$, then we define

$$\int_a^b f(x)\, dx = -\int_b^a f(x)\, dx$$

REMARK.   Geometrically, part (a) states that there is zero area between a curve $y = f(x)$ and a single point a on the x-axis.

▶ Example 1

(a) $\displaystyle\int_1^1 x^2\, dx = 0$

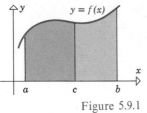

Figure 5.9.1

(b) $\displaystyle\int_{4}^{0} x\,dx = -\int_{0}^{4} x\,dx = -\frac{x^2}{2}\Big]_{0}^{4} = -8$ ◀

If f is continuous and nonnegative on $[a, b]$ and if c is a point between a and b, then it is evident that the area under $y = f(x)$ over the interval $[a, b]$ can be split into two parts, the area under curve from a to c plus the area under the curve from c to b (Figure 5.9.1), that is,

$$\int_{a}^{b} f(x)\,dx = \int_{a}^{c} f(x)\,dx + \int_{c}^{b} f(x)\,dx$$

This is a special case of the following theorem about definite integrals.

5.9.2 THEOREM *If f is integrable on a closed interval containing the three points a, b, and c, then*

$$\int_{a}^{b} f(x)\,dx = \int_{a}^{c} f(x)\,dx + \int_{c}^{b} f(x)\,dx \tag{1}$$

no matter how the points are ordered.

We omit the proof.

Theorem 5.9.2 is helpful when the formula for the integrand changes between the limits of integration.

▶ Example 2 Evaluate

$$\int_{0}^{6} f(x)\,dx \text{ if}$$

$$f(x) = \begin{cases} x^2, & x \le 2 \\ 3x - 2, & x \ge 2 \end{cases}$$

Solution.

$$\int_{0}^{6} f(x)\,dx = \int_{0}^{2} f(x)\,dx + \int_{2}^{6} f(x)\,dx$$

$$= \int_{0}^{2} x^2\,dx + \int_{2}^{6} (3x - 2)\,dx$$

$$= \frac{x^3}{3}\Big]_{0}^{2} + \left[\frac{3x^2}{2} - 2x\right]_{2}^{6}$$

$$= \left(\frac{8}{3} - 0\right) + (42 - 2) = \frac{128}{3}$$ ◀

▶ Example 3   Evaluate

$$\int_{-1}^{2} |x|\, dx$$

Solution.   Since $|x| = x$ when $x \geq 0$ and $|x| = -x$ when $x < 0$,

$$\int_{-1}^{2} |x|\, dx = \int_{-1}^{0} |x|\, dx + \int_{0}^{2} |x|\, dx$$

$$= \int_{-1}^{0} (-x)\, dx + \int_{0}^{2} x\, dx$$

$$= -\frac{x^2}{2}\bigg]_{-1}^{0} + \frac{x^2}{2}\bigg]_{0}^{2} = \frac{1}{2} + 2 = \frac{5}{2} \qquad\qquad ◀$$

EVALUATING DEFINITE INTEGRALS BY *u*-SUBSTITUTION

In Section 5.3 we studied the technique of *u*-substitution, whereby an indefinite integral of the form

$$\int f(g(x))g'(x)\, dx$$

can be transformed into an equivalent but simpler indefinite integral,

$$\int f(u)\, du$$

by making the substitution

$$u = g(x) \qquad \text{and} \qquad du = g'(x)\, dx$$

The technique of *u*-substitution can also be used to evaluate a definite integral of the form

$$\int_{a}^{b} f(g(x))g'(x)\, dx \qquad\qquad (2)$$

One way is to evaluate the indefinite integral

$$\int f(g(x))g'(x)\, dx$$

by *u*-substitution and then apply the First Fundamental Theorem of Calculus:

$$\int_{a}^{b} f(g(x))g'(x)\, dx = \left[\int f(g(x))g'(x)\, dx \right]_{a}^{b}$$

This was the method used in Example 3 of the last section.

Another possibility is to make the substitution

$$u = g(x) \quad \text{and} \quad du = g'(x)\,dx$$

directly into the definite integral (2). However, to do this we must change the x-limits of integration to corresponding u-limits of integration. Since $u = g(x)$, it follows that

$$u = g(a) \quad \text{when} \quad x = a$$

and

$$u = g(b) \quad \text{when} \quad x = b$$

Thus, if (2) is expressed in terms of u, we obtain

$$\int_a^b f(g(x))g'(x)\,dx = \int_{g(a)}^{g(b)} f(u)\,du$$

The following examples illustrate this idea.

▶ **Example 4** Evaluate

$$\int_0^2 2x(x^2 + 1)^3\,dx$$

Solution. Let

$$u = x^2 + 1 \quad \text{so that} \quad du = 2x\,dx$$

Since $u = 1$ when $x = 0$, and $u = 5$ when $x = 2$, we obtain

$$\int_0^2 2x(x^2 + 1)^3\,dx = \int_1^5 u^3\,du = \frac{u^4}{4}\Big]_1^5 = \frac{625}{4} - \frac{1}{4} = 156$$

As a check, let us compare this solution to the solution that results when we first evaluate the indefinite integral by u-substitution and then apply the First Fundamental Theorem of Calculus. We obtain

$$\int 2x(x^2 + 1)^3\,dx = \int u^3\,du = \frac{u^4}{4} + C = \frac{(x^2 + 1)^4}{4} + C$$

Thus,

$$\int_0^2 2x(x^2 + 1)^3\,dx = \left[\int 2x(x^2 + 1)^3\,dx\right]_0^2 = \frac{(x^2 + 1)^4}{4}\Big]_0^2$$

$$= \frac{625}{4} - \frac{1}{4} = 156$$

which agrees with our previous result. ◀

▶ **Example 5** Evaluate

$$\int_0^{\pi/4} \cos(\pi - x)\, dx$$

Solution. Let

$$u = \pi - x \qquad \text{so that} \qquad du = -dx$$

Since $u = \pi$ when $x = 0$ and $u = 3\pi/4$ when $x = \pi/4$,

$$\int_0^{\pi/4} \cos(\pi - x)\, dx = \int_\pi^{3\pi/4} \cos u\,(-du)$$

$$= -\int_\pi^{3\pi/4} \cos u\, du$$

$$= \int_{3\pi/4}^\pi \cos u\, du \qquad [\text{Definition 5.9.1}b]$$

$$= \sin u \Big]_{3\pi/4}^\pi$$

$$= \sin \pi - \sin(3\pi/4)$$

$$= 0 - \tfrac{1}{2}\sqrt{2} = -\tfrac{1}{2}\sqrt{2} \qquad ◀$$

The following theorem justifies the method of u-substitution illustrated in these examples.

5.9.3 THEOREM *If g' is continuous on $[a, b]$ and f is continuous and has an antiderivative on an interval containing the values of $g(x)$ for $a \le x \le b$, then*

$$\int_a^b f(g(x))g'(x)\, dx = \int_{g(a)}^{g(b)} f(u)\, du$$

provided the integrals exist.

Proof. Let $u = g(x)$ and let F be an antiderivative of f on an interval containing the values of $g(x)$ for $a \le x \le b$. Then by the chain rule

$$\frac{d}{dx} F(g(x)) = \frac{d}{dx} F(u) = \frac{dF}{du}\frac{du}{dx} = f(u)\frac{du}{dx} = f(g(x))g'(x)$$

for each x in $[a, b]$. Thus, $F(g(x))$ is an antiderivative of $f(g(x))g'(x)$ on $[a, b]$. Therefore, by the First Fundamental Theorem of Calculus (Theorem 5.8.1)

$$\int_a^b f(g(x))g'(x)\, dx = F(g(x)) \Big]_a^b$$

$$= F(g(b)) - F(g(a))$$

$$= \int_{g(a)}^{g(b)} f(u)\, du \quad \blacksquare$$

AVERAGE VALUE
OF A FUNCTION

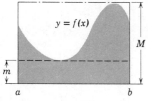

Figure 5.9.2

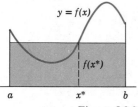

Figure 5.9.3

In order to see how the definite integral can be used to define the "average value" of continuous function, we must first develop a fundamental result, called the *Mean-Value Theorem for Integrals*.

Let f be a continuous nonnegative function on $[a, b]$, and let m and M be the minimum and maximum values of $f(x)$ on this interval. Consider the rectangles of heights m and M over the interval $[a, b]$ (Figure 5.9.2). It is clear geometrically that the area

$$A = \int_a^b f(x)\, dx$$

is at least as large as the area of the rectangle of height m, and no larger than the area of the rectangle of height M. It seems reasonable, therefore, that somewhere between m and M there is an appropriate height $f(x^*)$ such that the rectangle of this height over $[a, b]$ has precisely area A; that is,

$$\int_a^b f(x)\, dx = f(x^*)(b - a)$$

(Figure 5.9.3). This is a special case of the following result.

5.9.4   THEOREM
The Mean-Value Theorem for Integrals

If f is continuous on a closed interval $[a, b]$, then there is at least one number x^* in $[a, b]$ such that

$$\int_a^b f(x)\, dx = f(x^*)(b - a) \tag{3}$$

Proof. By the Extreme-Value Theorem (4.6.4), f assumes a maximum value M and a minimum value m on $[a, b]$. Thus, for all x in $[a, b]$,

$$m \le f(x) \le M \tag{4}$$

and from Theorem 5.8.3b

$$\int_a^b m\, dx \le \int_a^b f(x)\, dx \le \int_a^b M\, dx$$

or

$$m(b - a) \le \int_a^b f(x)\, dx \le M(b - a)$$

or

$$m \le \frac{1}{b-a} \int_a^b f(x)\,dx \le M \tag{5}$$

Since (5) states that

$$\frac{1}{b-a} \int_a^b f(x)\,dx \tag{6}$$

is a number between m and M, and since $f(x)$ assumes the values m and M on $[a, b]$, it follows from the Intermediate-Value Theorem (3.7.8) that $f(x)$ must assume the value (6) at some point x^* in $[a, b]$, that is,

$$\frac{1}{b-a} \int_a^b f(x)\,dx = f(x^*)$$

or

$$\int_a^b f(x)\,dx = f(x^*)(b-a)$$

▶ **Example 6** Since $f(x) = x^2$ is continuous on the interval $[1, 4]$, the Mean-Value Theorem for Integrals guarantees that there is a number x^* in $[1, 4]$ such that

$$\int_1^4 x^2\,dx = f(x^*)(4-1) = (x^*)^2(4-1) = 3(x^*)^2$$

But

$$\int_1^4 x^2\,dx = \frac{x^3}{3}\Big]_1^4 = 21$$

so that

$$3(x^*)^2 = 21 \quad \text{or} \quad (x^*)^2 = 7$$

Thus, $x^* = \sqrt{7} \approx 2.65$ is the number in the interval $[1, 4]$ whose existence is guaranteed by the Mean-Value Theorem for Integrals. ◄

The number $f(x^*)$ in Theorem 5.9.4 is closely related to the familiar notion of an *arithmetic average*. To see this, divide the interval $[a, b]$ into n subintervals of equal length

$$\Delta x = \frac{b-a}{n} \tag{7}$$

and choose arbitrary points $x_1^*, x_2^*, \ldots, x_n^*$ in successive subintervals. Then the arithmetic average of the numbers $f(x_1^*), f(x_2^*), \ldots, f(x_n^*)$ is

$$\text{ave} = \frac{1}{n}[f(x_1^*) + f(x_2^*) + \cdots + f(x_n^*)]$$

or from (7)

$$\text{ave} = \frac{1}{b-a}[f(x_1^*)\,\Delta x + f(x_2^*)\,\Delta x + \cdots + f(x_n^*)\,\Delta x]$$

$$= \frac{1}{b-a}\sum_{k=1}^{n} f(x_k^*)\,\Delta x$$

Taking the limit as $n \to +\infty$ yields

$$\lim_{n \to +\infty} \frac{1}{b-a}\sum_{k=1}^{n} f(x_k^*)\,\Delta x = \frac{1}{b-a}\int_a^b f(x)\,dx \tag{8}$$

Since this equation describes what happens when we average "more and more" values of $f(x)$, we are led to the following definition.

5.9.5 DEFINITION   If f is integrable on $[a, b]$, then the **average value** (or **mean value**) of f on $[a, b]$ is defined to be

$$f_{\text{ave}} = \frac{1}{b-a}\int_a^b f(x)\,dx$$

▶ **Example 7**   Find the average value of $f(x) = x^2$ on the interval $[1, 4]$.

Solution.

$$f_{\text{ave}} = \frac{1}{b-a}\int_a^b f(x)\,dx = \frac{1}{4-1}\int_1^4 x^2\,dx = \frac{1}{3}(21) = 7 \qquad ◀$$

REMARK.   In light of Definition 5.9.5, the quantity $f(x^*)$ in the Mean-Value Theorem for Integrals (5.9.4) is just the average value of f over $[a, b]$. Thus, the Mean-Value Theorem for Integrals can be stated in the form

$$\int_a^b f(x)\,dx = (b-a)f_{\text{ave}}$$

► Exercise Set 5.9

---

1. In each part express the integral in terms of the variable u, but do not evaluate.

(a) $\displaystyle\int_0^2 (x + 1)^7 \, dx; \; u = x + 1$

(b) $\displaystyle\int_{-1}^2 x\sqrt{8 - x^2} \, dx; \; u = 8 - x^2$

(c) $\displaystyle\int_{-1}^1 \sin(\pi\theta) \, d\theta; \; \theta = u/\pi$

(d) $\displaystyle\int_0^{\pi/4} \tan^2 x \sec^2 x \, dx; \; u = \tan x$

(e) $\displaystyle\int_0^1 x^3\sqrt{x^2 + 3} \, dx; \; u = x^2 + 3$

(f) $\displaystyle\int_0^3 (x + 2)(x - 3)^{20} \, dx; \; u = x - 3$

In Exercises 2–5, use Theorem 5.9.2 to evaluate the integrals.

2. $\displaystyle\int_1^5 |x - 2| \, dx.$

3. $\displaystyle\int_0^{3\pi/4} |\cos x| \, dx.$

4. $\displaystyle\int_{-2}^3 f(x) \, dx, \text{ where } f(x) = \begin{cases} -x, & x \geq 0 \\ x^2, & x < 0. \end{cases}$

5. $\displaystyle\int_0^9 g(x) \, dx, \text{ where } g(x) = \begin{cases} 1, & 0 \leq x < 1 \\ x^3, & 1 \leq x < 4 \\ \sqrt{x}, & 4 \leq x \leq 9. \end{cases}$

In Exercises 6–15, evaluate the integrals two ways: first by a u-substitution in the definite integral, and then by a u-substitution in the corresponding indefinite integral.

6. $\displaystyle\int_1^2 (4x - 2)^3 \, dx.$

7. $\displaystyle\int_0^1 (2x + 1)^4 \, dx.$

8. $\displaystyle\int_1^2 (4 - 3x)^8 \, dx.$

9. $\displaystyle\int_{-1}^0 (1 - 2x)^3 \, dx.$

10. $\displaystyle\int_{-5}^0 x\sqrt{4 - x} \, dx.$

11. $\displaystyle\int_0^8 x\sqrt{1 + x} \, dx.$

12. $\displaystyle\int_0^{\pi/6} 2\cos 3x \, dx.$

13. $\displaystyle\int_0^{\pi/2} 4\sin(x/2) \, dx.$

14. $\displaystyle\int_{1-\pi}^{1+\pi} \sec^2\left(\tfrac{1}{4}x - \tfrac{1}{4}\right) \, dx.$

15. $\displaystyle\int_{-2}^{-1} \frac{x}{(x^2 + 2)^3} \, dx.$

In Exercises 16–23, evaluate the integrals by any method.

16. $\displaystyle\int_{-1}^4 \frac{x \, dx}{\sqrt{5 + x}}.$

17. $\displaystyle\int_0^1 \frac{y^2 \, dy}{\sqrt{4 - 3y}}.$

18. $\displaystyle\int_{-\pi/4}^{\pi} \sin\theta\cos\theta \, d\theta.$

19. $\displaystyle\int_0^{\sqrt{\pi}} 5x\cos(x^2) \, dx.$

20. $\displaystyle\int_{-2}^2 \sqrt{2 + |x|} \, dx.$

21. $\displaystyle\int_1^3 \frac{x + 2}{\sqrt{x^2 + 4x + 7}} \, dx.$

22. $\displaystyle\int_1^2 \frac{dx}{x^2 - 6x + 9}.$

23. $\displaystyle\int_{\pi^2}^{4\pi^2} \frac{1}{\sqrt{x}} \sin\sqrt{x} \, dx.$

24. (a) Find f_{ave} of $f(x) = 2x$ over $[0, 4]$.
(b) Find a point x^* in $[0, 4]$ such that $f(x^*) = f_{\text{ave}}$.
(c) Sketch the graph of $f(x) = 2x$ over $[0, 4]$ and construct a rectangle over the interval whose area is the same as the area under the graph of f over the interval.

25. (a) Find f_{ave} of $f(x) = x^2$ over $[0, 2]$.
(b) Find a point x^* in $[0, 2]$ such that $f(x^*) = f_{\text{ave}}.$

(c) Sketch the graph of $f(x) = x^2$ over $[0, 2]$ and construct a rectangle over the interval whose area is the same as the area under the graph of f over the interval.

In Exercises 26–29, find the average value of $f(x)$ over the interval and find all values of x^* described in the Mean-Value Theorem for Integrals.

26. $f(x) = x^3$; $[1, 2]$.

27. $f(x) = \sqrt{x}$; $[0, 9]$.

28. $f(x) = \sin x$; $[-\pi, \pi]$.

29. $f(x) = \alpha x + \beta$; $[x_0, x_1]$.

30. Let $s(t)$ be the position function of a particle moving on a coordinate line. In Section 5.4 (Exercise 30) we defined the average velocity v_{ave} over a time interval $[t_0, t_1]$ by

$$v_{ave} = \frac{\text{change in position}}{\text{time elapsed}} = \frac{s(t_1) - s(t_0)}{t_1 - t_0}$$

(a) Show that

$$v_{ave} = \frac{1}{t_1 - t_0} \int_{t_0}^{t_1} v(t)\, dt$$

so that v_{ave} can also be interpreted as the average value of the velocity function over the time interval $[t_0, t_1]$.

(b) Use the result in (a) to compute the average velocity over the time interval $[0, 5]$ for a particle moving on a coordinate line with velocity function $v(t) = 32t$.

31. For a particle moving on a coordinate line, show that the average acceleration a_{ave} over a time interval $[t_0, t_1]$ (as defined in Exercise 30 of Section 5.4) can be written as

$$a_{ave} = \frac{1}{t_1 - t_0} \int_{t_0}^{t_1} a(t)\, dt$$

which is the average of the acceleration function over the time interval $[t_0, t_1]$.

32. Consider a particle moving on a coordinate line. Use Exercises 30 and 31 to find:

(a) v_{ave} for $1 \leq t \leq 4$ if $v(t) = 3t^3 + 2$
(b) a_{ave} for $2 \leq t \leq 9$ if $a(t) = t^{1/2}$
(c) v_{ave} for $t_0 \leq t \leq t_1$ if $v(t) = 32t + v_0$
                                 (v_0 constant).

33. Water is run at a constant rate of 1 ft^3/min to fill a cylindrical tank of radius 3 ft and height 5 ft. Assuming the tank is initially empty, find the average force on the bottom over the time period required to fill the tank (density of water $= 62.4$ lb/ft^3).

34. Prove: If $f(x) = k$ is constant on $[a, b]$, then $f_{ave} = k$ on $[a, b]$.

35. (a) Prove: If f is continuous on $[a, b]$, then

$$\int_a^b [f(x) - f_{ave}]\, dx = 0$$

(b) Does there exist a constant $c \neq f_{ave}$ such that

$$\int_a^b [f(x) - c]\, dx = 0?$$

36. Prove: If m and n are positive integers, then

$$\int_0^1 x^m (1 - x)^n\, dx = \int_0^1 x^n (1 - x)^m\, dx$$

[*Hint:* Do not evaluate; use a substitution.]

37. Prove: If n is a positive integer, then

$$\int_0^{\pi/2} \sin^n x\, dx = \int_0^{\pi/2} \cos^n x\, dx.$$

[*Hint:* Do not evaluate; use a trigonometric identity and a substitution.]

38. (a) Prove: If $f(-x) = -f(x)$ for all x in $[-a, a]$ and if f is continuous on $[-a, a]$, then $\int_{-a}^a f(x)\, dx = 0$.

(b) Give a geometric explanation of the result in part (a).

5.10 THE SECOND FUNDAMENTAL THEOREM OF CALCULUS

The First Fundamental Theorem of Calculus (5.8.1) tells us that if f is continuous on $[a, b]$ and *if* we can find an antiderivative F of f on $[a, b]$, then

$$\int_a^b f(x)\, dx = F(b) - F(a) \tag{1}$$

Notice that we have set the word "if" in italics. This is because the First Fundamental Theorem of Calculus does not guarantee that f actually has an antiderivative F. It simply says that if there is one, then formula (1) applies. In this section we will discuss the *Second Fundamental Theorem of Calculus,* which is a major theorem concerned with the existence of antiderivatives.

In the work to follow we will need to consider definite integrals of the form

$$\int_a^x \underline{\hspace{3cm}}$$

where the upper limit x is allowed to vary. For such integrals we will use a letter different from x (often t) for the variable of integration; thus, we would write

$$\int_a^x f(t)\, dt$$

rather than

$$\int_a^x f(x)\, dx \tag{2}$$

We do this because x is used in two different ways in (2), as a variable of integration and as a limit of integration. This can cause errors.

▶ Example 1 Evaluate

$$\int_2^x t^2\, dt$$

Solution.

$$\int_2^x t^2\, dt = \frac{t^3}{3}\Bigg]_{t=2}^{t=x} = \frac{x^3}{3} - \frac{8}{3} \qquad\qquad ◀$$

Note that the final expression in Example 1 is a function of x alone. In general, an expression of the form

$$\int_a^x f(t)\, dt$$

represents a function of x—the variable t does not enter into the final result. We now turn to the major result in this section.

5.10.1 THEOREM
The Second Fundamental Theorem of Calculus

Let f be continuous on an open interval I, and let a be any point in I. If F is defined by

$$F(x) = \int_a^x f(t)\, dt \tag{3}$$

then $F'(x) = f(x)$ at each point x in the interval I.

Proof. From the definition of derivative,

$$F'(x) = \lim_{h \to 0} \frac{F(x + h) - F(x)}{h}$$

$$= \lim_{h \to 0} \frac{1}{h}\left[\int_a^{x+h} f(t)\, dt - \int_a^x f(t)\, dt\right]$$

$$= \lim_{h \to 0} \frac{1}{h}\left[\int_a^{x+h} f(t)\, dt + \int_x^a f(t)\, dt\right]$$

$$= \lim_{h \to 0} \frac{1}{h}\int_x^{x+h} f(t)\, dt \qquad\qquad \text{[Theorem 5.9.2]}$$

Applying the Mean-Value Theorem for Integrals (5.9.4) to the last expression, we obtain

$$F'(x) = \lim_{h \to 0} \frac{1}{h}[f(t^*) \cdot h] = \lim_{h \to 0} f(t^*) \tag{4}$$

where t^* is some number between x and $x + h$. Because t^* is between x and $x + h$, it follows that $t^* \to x$ as $h \to 0$. Thus, $f(t^*) \to f(x)$ as $h \to 0$, since f is assumed continuous at x. Therefore, from (4) $F'(x) = f(x)$ ▮

The Second Fundamental Theorem of Calculus ensures that if f is continuous on an open interval, then f has an antiderivative on that interval, namely, $\int_a^x f(t)\, dt$. However, because we can add an arbitrary constant to an antiderivative of f and still have an antiderivative of f, it follows that a function that is continuous on an open interval has infinitely many antiderivatives on the interval. Note, however, that (3) singles out that antiderivative of f whose value at $x = a$ is zero because

$$F(a) = \int_a^a f(t)\, dt = 0$$

▶ **Example 2**   The function $f(x) = x^2$ is continuous on $(-\infty, +\infty)$. Thus, by the Second Fundamental Theorem of Calculus and the remarks above, the function

$$F(x) = \int_a^x f(t)\, dt = \int_2^x t^2\, dt$$

must be that antiderivative of $f(x) = x^2$ on $(-\infty, +\infty)$ with a value of zero at $x = 2$. This is indeed the case since

$$F(x) = \frac{x^3}{3} - \frac{8}{3} \tag{5}$$

(see Example 1) so that

$$F'(x) = x^2 \quad \text{and} \quad F(2) = \frac{8}{3} - \frac{8}{3} = 0 \qquad ◀$$

▶ **Example 3**   Because $f(x) = 1/x$ is continuous on the interval $(0, +\infty)$,

$$F(x) = \int_1^x \frac{1}{t}\, dt$$

is the antiderivative of $1/x$ on $(0, +\infty)$ whose value at $x = 1$ is zero. That is,

$$F'(x) = \frac{1}{x} \quad \text{and} \quad F(1) = 0 \qquad ◀$$

▶ **Example 4**   Find

$$\frac{d}{dx} \int_1^x \frac{\sin t}{t}\, dt$$

Solution.   If we let

$$F(x) = \int_1^x \frac{\sin t}{t}\, dt$$

then it follows from the Second Fundamental Theorem of Calculus that

$$\frac{d}{dx} \int_1^x \frac{\sin t}{t}\, dt = \frac{d}{dx}(F(x)) = F'(x) = \frac{\sin x}{x} \qquad ◀$$

As a result of the Second Fundamental Theorem of Calculus, we now know that continuous functions have antiderivatives. However, there are two possible impediments that can prevent us from applying formula (1), even if f is continuous:

1. The function f might be sufficiently complicated that we are not clever enough to figure out a formula for F in terms of familiar functions.

2. The antiderivative F may not have a formula in terms of familiar functions.

To illustrate the second situation, consider the function $f(x) = 1/x$. Because this function is continuous on $(0, +\infty)$, we know that it has an antiderivative on this interval. In fact, we showed in Example 3 that the function

$$F(x) = \int_1^x \frac{1}{t} \, dt \tag{6}$$

is such an antiderivative. Although this formula is interesting from a theoretical viewpoint, it would be desirable to have a more elementary formula that uses familiar functions and does not involve an integral. It can be proved, however, that no antiderivative of $f(x) = 1/x$ can be expressed by an elementary formula involving finitely many polynomials, rational functions, or trigonometric functions. In short, our repertoire of basic functions is too limited, at present, to produce an antiderivative of $f(x) = 1/x$ which is simpler than (6). In Chapter 7 we will use the Second Fundamental Theorem of Calculus to expand our repertoire of basic functions.

We conclude this section with a word about computing values of functions given by integral formulas. If, for example, we are interested in finding $F(2)$ using formula (6), then we must evaluate

$$F(2) = \int_1^2 \frac{1}{t} \, dt \tag{7}$$

Because the only antiderivative we know for $1/x$ uses an integral formula, the same is true for $1/t$. Thus, we cannot "carry out" the integration in (7), and the only available procedure in this case is to approximate the value of the integral. This can be done using Riemann sums or other numerical techniques such as those that will be developed in Section 9.9.

▶ Exercise Set 5.10

1. Define $F(x)$ by

$$F(x) = \int_1^x (t^3 + 1) \, dt$$

(a) Use the Second Fundamental Theorem of Calculus to find $F'(x)$.

(b) Check the result in (a) by first integrating and then differentiating.

In Exercises 2–5, use the Second Fundamental Theorem of Calculus to find the derivative.

2. $\dfrac{d}{dx} \displaystyle\int_0^x \dfrac{dt}{1 + \sqrt{t}}$.

3. $\dfrac{d}{dx} \displaystyle\int_1^x \sin(\sqrt{t}) \, dt$.

4. $\dfrac{d}{dx} \displaystyle\int_0^x \dfrac{t}{\cos t} \, dt$.

5. $\dfrac{d}{dx} \displaystyle\int_0^x |t| \, dt$.

In Exercises 6–9, express the antiderivatives as integrals.

6. The antiderivative of $1/(1 + x^2)$ on the interval $(-\infty, +\infty)$ whose value at $x = 1$ is 0.

7. The antiderivative of $1/(x - 1)$ on the interval $(1, +\infty)$ whose value at $x = 2$ is 0.

8. The antiderivative of $1/(x - 1)$ on the interval $(-\infty, 1)$ whose value at $x = -3$ is 0.

9. The antiderivative of $1/(x - 1)$ on the interval $(-\infty, 1)$ whose value at $x = 0$ is 0.

10. (a) Over what open interval does the formula

$$F(x) = \int_1^x \frac{1}{t^2 - 9} \, dt$$

represent an antiderivative of

$$f(x) = \frac{1}{x^2 - 9}?$$

(b) Find a point where the graph of F crosses the x-axis.

11. (a) Over what open interval does the formula

$$F(x) = \int_1^x \frac{dt}{t}$$

represent an antiderivative of $f(x) = 1/x$?

(b) Find a point where the graph of F crosses the x-axis.

12. Use the Second Fundamental Theorem of Calculus and the chain rule to show that

$$\frac{d}{dx} \int_a^{g(x)} f(t) \, dt = f(g(x))g'(x)$$

13. Apply the result in Exercise 12 to perform the following differentiations:

(a) $\dfrac{d}{dx} \displaystyle\int_1^{x^3} \frac{1}{t} \, dt$

(b) $\dfrac{d}{dx} \displaystyle\int_3^{\sin x} \frac{1}{1 + t^2} \, dt.$

14. Use Exercise 12 and Theorem 5.9.2 to show that

$$\frac{d}{dx} \int_{h(x)}^{g(x)} f(t) \, dt = f(g(x))g'(x) - f(h(x))h'(x)$$

15. Use the result in Exercise 14 to perform the following differentiations:

(a) $\dfrac{d}{dx} \displaystyle\int_{x^2}^{x^3} \sin^2 t \, dt$ (b) $\dfrac{d}{dx} \displaystyle\int_{-x}^{x} \frac{1}{1 + t} \, dt.$

16. Prove that the function

$$F(x) = \int_x^{3x} \frac{1}{t} \, dt$$

is constant on the interval $(0, +\infty)$ by using Exercise 14 to find $F'(x)$.

17. Prove: If f is continuous on an open interval I and b is any point in I, then at each point in I

$$\frac{d}{dx} \int_x^b f(t) \, dt = -f(x)$$

18. Prove: If f is continuous on an open interval I and a is any point in I, then

$$F(x) = \int_a^x f(t) \, dt$$

is continuous on I.

▶ SUPPLEMENTARY EXERCISES

In Exercises 1–10, evaluate the integrals and check your results by differentiation.

1. $\displaystyle\int \left[\frac{1}{x^3} + \frac{1}{\sqrt{x}} - 5 \sin x \right] dx.$

2. $\displaystyle\int \frac{2t^4 - t + 2}{t^3} \, dt.$ 3. $\displaystyle\int \frac{(\sqrt{x} + 2)^8}{\sqrt{x}} \, dx.$

4. $\displaystyle\int x^3 \cos (2x^4 - 1) \, dx.$ 5. $\displaystyle\int \frac{x \sin \sqrt{2x^2 - 5}}{\sqrt{2x^2 - 5}} \, dx.$

6. $\displaystyle\int \sqrt{\cos \theta} \sin (2\theta) \, d\theta.$ 7. $\displaystyle\int \sqrt{x}(3 + \sqrt[3]{x^4}) \, dx.$

8. $\displaystyle\int \frac{x^{1/3} \, dx}{x^{8/3} + 2x^{4/3} + 1}.$ 9. $\displaystyle\int \sec^2 (\sin 5t) \cos 5t \, dt.$

10. $\displaystyle\int \frac{\cot^2 x}{\sin^2 x}\, dx.$

11. Evaluate $\int y(y^2 + 2)^2\, dy$ two ways: (a) by multiplying out and integrating term by term; and (b) by using the substitution $u = y^2 + 2$. Show that your answers differ by a constant.

In Exercises 12–17, evaluate the definite integral by making the indicated substitution and changing the x-limits of integration to u-limits.

12. $\displaystyle\int_1^0 \sqrt[5]{1 - 2x}\, dx,\ u = 1 - 2x.$

13. $\displaystyle\int_0^{\pi/2} \sin^4 x \cos x\, dx,\ u = \sin x.$

14. $\displaystyle\int_0^{-3} \frac{x\, dx}{\sqrt{x^2 + 16}},\ u = x^2 + 16.$

15. $\displaystyle\int_2^5 \frac{x - 2}{\sqrt{x - 1}}\, dx,\ u = x - 1.$

16. $\displaystyle\int_{\pi/6}^{\pi/4} \frac{\sin 2x\, dx}{\sqrt{1 - \frac{3}{2}\cos 2x}},\ u = 1 - \frac{3}{2}\cos 2x.$

17. $\displaystyle\int_1^4 \frac{1}{\sqrt{x}} \cos\left(\frac{\pi\sqrt{x}}{2}\right) dx,\ u = \frac{\pi\sqrt{x}}{2}.$

In Exercises 18 and 19, evaluate $\int_{-2}^2 f(x)\, dx$.

18. $f(x) = \begin{cases} x^3 & \text{for } x \geq 0 \\ -x & \text{for } x < 0. \end{cases}$

19. $f(x) = |2x - 1|.$

20. A car, initially at rest 10 ft from a pole, moves in a straight line away from the pole with a constant acceleration of a ft/sec². Express the distance $s(t)$ between the car and the pole in terms of a and t and then find a if the car travels 40 ft in the first 2 sec.

21. A projectile is fired upward from ground level. Find the initial velocity if it hits the ground 8 sec later. How high does it go?

22. A particle moving along a straight line is accelerating at a constant rate of 3 meters/sec². Find the initial velocity if the particle moves 40 meters in the first 4 sec.

23. Evaluate:

(a) $\displaystyle\sum_{i=3}^6 5$

(b) $\displaystyle\sum_{i=n}^{n+3} 2$

(c) $\displaystyle\sum_{i=n}^{n+3} n$

(d) $\displaystyle\sum_{k=1}^3 \left(\frac{k - 1}{k + 3}\right)$

(e) $\displaystyle\sum_{k=2}^4 \frac{6}{k^2}$

(f) $\displaystyle\sum_{n=4}^4 (2n + 1)$

(g) $\displaystyle\sum_{k=0}^4 \sin(k\pi/4)$

(h) $\displaystyle\sum_{k=1}^4 \sin^k(\pi/4).$

24. Express in sigma notation and evaluate:
(a) $3 \cdot 1 + 4 \cdot 2 + 5 \cdot 3 + \cdots + 102 \cdot 100$
(b) $200 + 198 + \cdots + 4 + 2.$

25. Express in sigma notation, first starting with $k = 1$, and then with $k = 2$. Do not evaluate.

(a) $\displaystyle\frac{1}{4} - \frac{4}{9} + \frac{9}{16} - \cdots - \frac{64}{81} + \frac{81}{100}$

(b) $\displaystyle\frac{\pi^2}{1} - \frac{\pi^3}{2} + \frac{\pi^4}{3} - \cdots + \frac{\pi^{12}}{11}.$

In Exercises 26–29, use the partition of $[a, b]$ into n subintervals of equal length, and find a closed form for the sum of the areas of: (a) the inscribed rectangles and (b) the circumscribed rectangles. (c) Use your answer in either (a) or (b) to find the area under the curve $y = f(x)$ over the interval $[a, b]$. (Check your answer by integration.)

26. $f(x) = 6 - 2x;\ a = 1,\ b = 3.$

27. $f(x) = 16 - x^2;\ a = 0,\ b = 4.$

28. $f(x) = x^2 + 2;\ a = 1,\ b = 4.$

29. $f(x) = 6;\ a = -1,\ b = 1.$

30. (a) Prove: If $f(-x) = f(x)$ for all x in $[-a, a]$ and f is integrable on the interval $[-a, a]$, then $\int_{-a}^a f(x)\, dx = 2 \int_0^a f(x)\, dx$.
(b) Give a geometric explanation of the result in part (a).

31. In each part evaluate the definite integral geometrically, that is, by interpreting it as a difference of areas and using geometric reasoning to find the areas.

(a) $\displaystyle\int_0^2 (2 - 4x)\, dx$

(b) $\displaystyle\int_{-3}^0 \sqrt{9 - t^2}\, dt$

(c) $\int_{-4}^{2} \frac{|x|}{x}\, dx$

(d) $\int_{-\pi/3}^{\pi/3} \sin x\, dx$

(e) $\int_{0}^{\pi} \cos x\, dx$

(f) $\int_{-10}^{-5} 6\, dx$

(g) $\int_{-5}^{4} f(x)\, dx$ if $f(x) = \begin{cases} -4, & x \le -3 \\ 2, & -3 < x < 0 \\ 1, & x \ge 0. \end{cases}$

32. Given that

$$\int_{-1}^{0} f(x)\, dx = 3, \quad \int_{-1}^{2} f(x)\, dx = -1,$$

and

$$\int_{-1}^{2} g(x)\, dx = 2,$$

evaluate the following:

(a) $\int_{0}^{-1} 8f(x)\, dx$

(b) $\int_{-1}^{2} f(y)\, dy$

(c) $\int_{2}^{2} g(x)\, dx$

(d) $\int_{0}^{2} f(x)\, dx$

(e) $\int_{-1}^{2} [f(x) - 2g(x)]\, dx.$

33. Given that

$$\int_{1}^{5} P(x)\, dx = -1, \quad \int_{3}^{5} P(x)\, dx = 3$$

and

$$\int_{3}^{5} Q(x)\, dx = 4,$$

evaluate the following:

(a) $\int_{3}^{5} [2P(x) + Q(x)]\, dx$

(b) $\int_{5}^{1} P(t)\, dt$

(c) $\int_{-3}^{-5} Q(-x)\, dx$

(d) $\int_{3}^{1} P(x)\, dx.$

34. Suppose that f is continuous and $x^2 \le f(x) \le 6$ for all x in $[-1, 2]$. Find values of A and B such that $A \le \int_{-1}^{2} f(x)\, dx \le B$.

In Exercises 35–38, find: (a) the average value of $f(x)$ over the indicated interval, and (b) all values of x^* described in the Mean-Value Theorem for Integrals.

35. $f(x) = 3x^2;\ [-2, -1]$.

36. $f(x) = \dfrac{x}{\sqrt{x^2 + 9}};\ [0, 4]$.

37. $f(x) = 2 + |x|;\ [-3, 1]$.

38. $f(x) = \sin^2 x;\ [0, \pi]$
  [*Hint:* $\sin^2 x = \frac{1}{2}(1 - \cos 2x).$]

6 applications of the definite integral

6.1 AREA BETWEEN TWO CURVES

In this section we will show how to obtain the area between two curves. Suppose that f and g are continuous functions on $[a, b]$ and

$$f(x) \geq g(x) \quad \text{for} \quad a \leq x \leq b \tag{1}$$

(This means that the curve $y = f(x)$ does not cross below the curve $y = g(x)$ over $[a, b]$.) If f and g are nonnegative on $[a, b]$, then, as illustrated in Figure 6.1.1, the area A bounded above by $y = f(x)$, below by $y = g(x)$, and on the sides by $x = a$ and $x = b$ can be obtained by subtracting the area under $y = g(x)$ from the area under $y = f(x)$, that is

$$A = [\text{area under } f] - [\text{area under } g]$$
$$= \int_a^b f(x) \, dx - \int_a^b g(x) \, dx = \int_a^b [f(x) - g(x)] \, dx$$

This formula can be extended to the case where g has negative values by translating the curves $y = f(x)$ and $y = g(x)$ upward until both are above

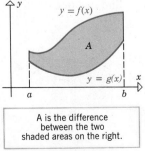

A is the difference between the two shaded areas on the right.

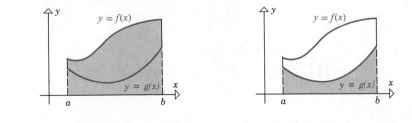

Figure 6.1.1

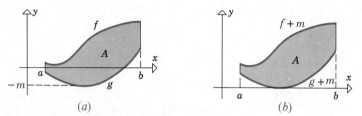

Figure 6.1.2   (a)   (b)

the x-axis. To do this, let $-m$ be the minimum value of $g(x)$ on $[a, b]$ (Figure 6.1.2a). Since $g(x) \geq -m$, it follows that

$$g(x) + m \geq 0$$

so that the functions $g + m$ and $f + m$ are nonnegative on $[a, b]$ (Figure 6.1.2b). It is intuitively clear that the area of a region is unchanged by translation, so the area A between f and g is the same as the area between $f + m$ and $g + m$. Thus,

$$A = [\text{area under } f + m] - [\text{area under } g + m]$$

$$= \int_a^b [f(x) + m] \, dx - \int_a^b [g(x) + m] \, dx = \int_a^b [f(x) - g(x)] \, dx$$

which is the same formula we obtained above. These results suggest the following definition.

6.1.1   DEFINITION   If f and g are continuous functions and if $f(x) \geq g(x)$ for all x in $[a, b]$, then the area of the region bounded above by $y = f(x)$, below by $y = g(x)$, on the left by the line $x = a$, and on the right by the line $x = b$ is

$$A = \int_a^b [f(x) - g(x)] \, dx \qquad (2)$$

When the region is complicated, it may require some careful thought to determine the integrand and limits of integration in (2). To find them, it is often helpful to sketch the region and then draw a vertical line segment through the region at an arbitrary point x connecting the top and bottom boundaries (Figure 6.1.3a). The top endpoint of the line segment will be $f(x)$, the bottom one $g(x)$, and the length of the line segment will be

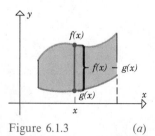

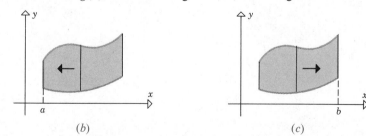

Figure 6.1.3   (a)   (b)   (c)

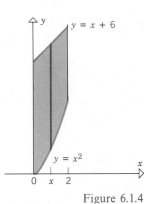

Figure 6.1.4

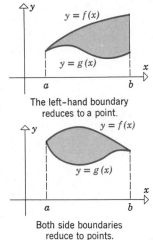

The left-hand boundary
reduces to a point.

Both side boundaries
reduce to points.

Figure 6.1.5

$f(x) - g(x)$, which is the integrand in (2). To determine the limits of integration, imagine moving the line segment left and then right. The leftmost position at which the line segment intersects the region is $x = a$ and the rightmost is $x = b$ (Figures 6.1.3b and 6.1.3c).

▶ **Example 1** Find the area of the region bounded above by $y = x + 6$, bounded below by $y = x^2$, and bounded on the sides by the lines $x = 0$ and $x = 2$.

Solution. The region and a vertical line segment through it are shown in Figure 6.1.4. The line segment extends from $f(x) = x + 6$ on the top to $g(x) = x^2$ on the bottom. If the line segment is moved through the region, its leftmost position will be $x = 0$ and its rightmost will be $x = 2$. Thus, from (2)

$$A = \int_0^2 [(x + 6) - x^2]\, dx$$

$$= \left[\frac{x^2}{2} + 6x - \frac{x^3}{3}\right]_0^2 = \frac{34}{3} - 0 = \frac{34}{3} \qquad ◀$$

Sometimes the upper curve $y = f(x)$ intersects the lower curve $y = g(x)$ at either the left-hand boundary $x = a$, the right-hand boundary $x = b$, or both. When this happens the side of the region reduces to a point, rather than a vertical line segment (Figure 6.1.5).

▶ **Example 2** Find the area of the region enclosed between the curves $y = x^2$ and $y = x + 6$.

Solution. A sketch of the region (Figure 6.1.6) shows that the lower boundary is $y = x^2$ and the upper boundary is $y = x + 6$. At the endpoints of the region, the upper and lower boundaries have the same y-coordinates; thus, to find the endpoints we equate

$$y = x^2 \qquad \text{and} \qquad y = x + 6 \tag{3}$$

This yields

$$x^2 = x + 6$$

or

$$x^2 - x - 6 = 0, \qquad (x + 2)(x - 3) = 0, \qquad x = -2, x = 3$$

Although the y-coordinates of the endpoints are not essential to our solution, they may be obtained from (3) by substituting $x = -2$ and $x = 3$ in either equation. This yields $y = 4$ and $y = 9$, so the upper and lower boundaries intersect at $(-2, 4)$ and $(3, 9)$.

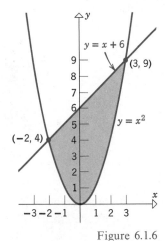

Figure 6.1.6

From (2) with $f(x) = x + 6$, $g(x) = x^2$, $a = -2$, and $b = 3$, we obtain the area

$$A = \int_{-2}^{3} [(x + 6) - x^2] \, dx$$

$$= \left[\frac{x^2}{2} + 6x - \frac{x^3}{3} \right]_{-2}^{3} = \frac{27}{2} - \left(-\frac{22}{3}\right) = \frac{125}{6} \quad \blacktriangleleft$$

▶ **Example 3** Find the area of the region enclosed by the curves $y = \sqrt{x}$, $y = -x + 6$ and $y = 1$.

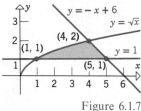

Figure 6.1.7

Solution. The region is shown in Figure 6.1.7. By solving simultaneously, we find that $y = \sqrt{x}$ and $y = -x + 6$ intersect at $(4, 2)$; $y = 1$ and $y = \sqrt{x}$ intersect at $(1, 1)$; and $y = 1$ and $y = -x + 6$ intersect at $(5, 1)$.

The upper boundary of the region consists of two parts: $y = \sqrt{x}$ when $1 \le x \le 4$ and $y = -x + 6$ when $4 \le x \le 5$. Thus, it is necessary to divide the region into parts, R_1 and R_2, as shown in Figure 6.1.8, and find the area of each part separately.

From (2) with $f(x) = \sqrt{x}$, $g(x) = 1$, $a = 1$, and $b = 4$, the area of region R_1 is

$$A_1 = \int_{1}^{4} (\sqrt{x} - 1) \, dx$$

$$= \left[\frac{2}{3} x^{3/2} - x \right]_{1}^{4}$$

$$= \left(\frac{16}{3} - 4 \right) - \left(\frac{2}{3} - 1 \right) = \frac{5}{3}$$

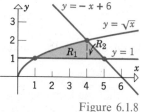

Figure 6.1.8

From (2) with $f(x) = -x + 6$, $g(x) = 1$, $a = 4$, and $b = 5$, the area of region R_2 is

$$A_2 = \int_{4}^{5} [(-x + 6) - 1] \, dx$$

$$= \int_{4}^{5} (5 - x) \, dx = \left[5x - \frac{x^2}{2} \right]_{4}^{5}$$

$$= \left(25 - \frac{25}{2} \right) - (20 - 8) = \frac{1}{2}$$

Thus, the area of the entire region is

$$A = A_1 + A_2 = \frac{5}{3} + \frac{1}{2} = \frac{13}{6} \quad \blacktriangleleft$$

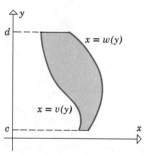

Figure 6.1.9

Sometimes it is necessary to find the area of a region bounded above and below by horizontal lines and bounded on the left and right by the graphs of two functions of y (Figure 6.1.9).

The following definition is similar to Definition 6.1.1, but with the roles of x and y reversed.

6.1.2 DEFINITION If w and v are continuous functions and if $w(y) \geq v(y)$ for all y in $[c, d]$, then the area of the region bounded on the left and right by the curves $x = v(y)$, $x = w(y)$ and above and below by the lines $y = d$, $y = c$, is

$$A = \int_c^d [w(y) - v(y)] \, dy \qquad (4)$$

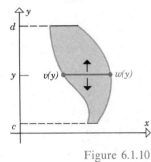

Figure 6.1.10

If the region is complicated, then the integrand and limits of integration in (4) can be obtained by drawing a horizontal line segment through the region from boundary to boundary at an arbitrary point y (Figure 6.1.10). The right-hand endpoint of the line segment will be $w(y)$, the left-hand endpoint $v(y)$, and the length of the line segment will be $w(y) - v(y)$, which is the integrand in (4). To determine the limits of integration, imagine moving the line segment down and then up (Figure 6.1.10). The lowest position where the line segment intersects the region is $y = c$ and the highest is $y = d$.

▶ **Example 4** Find the area of the region in Example 3 using (4).

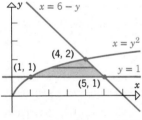

Figure 6.1.11

Solution. Figure 6.1.11 shows the region and a horizontal line extending between the left and right boundaries. (Note that we have rewritten the boundary curves $y = \sqrt{x}$ and $y = -x + 6$ as $x = y^2$ and $x = 6 - y$ so that x is expressed in terms of y.) The left boundary of the region is $x = v(y) = y^2$, the right boundary is $x = w(y) = 6 - y$, and the upper and lower boundaries are $y = 2$ and $y = 1$. Thus, from formula (4) the area of the region is

$$A = \int_1^2 [(6 - y) - y^2] \, dy$$

$$= \left[6y - \frac{y^2}{2} - \frac{y^3}{3} \right]_1^2$$

$$= \left(12 - 2 - \frac{8}{3} \right) - \left(6 - \frac{1}{2} - \frac{1}{3} \right) = \frac{13}{6}$$

which agrees with the answer obtained in Example 3. However, this solution is simpler because it is not necessary to break the region into two parts. ◀

▶ **Example 5** Find the area of the region enclosed by $x = y^2$ and $y = x - 2$.

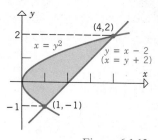

Figure 6.1.12

Solution. The region is shown in Figure 6.1.12. To compute the area by integrating with respect to x [Formula (2)], we would have to split the region into two parts because the equation of the lower boundary changes at $x = 1$. However, if we integrate with respect to y [Formula (4)], no splitting is necessary.

To use (4) the boundary curves must be expressed with x as a function of y. Thus, we rewrite $y = x - 2$ as $x = y + 2$. It now follows from (4) that

$$A = \int_{-1}^{2} [(y + 2) - y^2]\, dy = \frac{y^2}{2} + 2y - \frac{y^3}{3}\Big]_{-1}^{2}$$
$$= \frac{9}{2}$$

◀

▶ Exercise Set 6.1

1. Find the area of the region enclosed by the curves $y = x^2$ and $y = 4x$ by integrating
   (a) with respect to x
   (b) with respect to y.

2. Find the area of the region enclosed by the curves $y^2 = 4x$ and $y = 2x - 4$ by integrating
   (a) with respect to x
   (b) with respect to y.

3. Find the area of the region enclosed by the curves $y^2 = 2x$ and $y = 2x - 2$ by integrating
   (a) with respect to x
   (b) with respect to y.

In Exercises 4–21, sketch the region enclosed by the curves and find its area by any method.

4. $y = x^3$, $y = x$, $x = 0$, $x = 1/2$.

5. $y = x^2$, $y = \sqrt{x}$, $x = 1/4$, $x = 1$.

6. $y = x^3 - 4x$, $y = 0$, $x = 0$, $x = 2$.

7. $y = \cos 2x$, $y = 0$, $x = \pi/4$, $x = \pi/2$.

8. $y = x^3 - 4x^2 + 3x$, $y = 0$, $x = 0$, $x = 3$.

9. $x = y^2 - 4y$, $x = 0$, $y = 0$, $y = 4$.

10. $x = \sin y$, $x = 0$, $y = \pi/4$, $y = 3\pi/4$.

11. $y = \sec^2 x$, $y = 2$, $x = -\pi/4$, $x = \pi/4$.

12. $x^2 = y$, $x = y - 2$.

13. $y = x^2 + 4$, $x + y = 6$.

14. $y = x^3$, $y = -x$, $y = 8$.

15. $y^2 = -x$, $y = x - 6$, $y = -1$, $y = 4$.

16. $y = x$, $y = 4x$, $y = -x + 2$.

17. $y = 2 + |x - 1|$, $y = -\frac{1}{5}x + 7$.

18. $y = x^3 - 4x$, $x = -2$, $x = 2$.

19. $x = y^3 - y$, $x = 0$.

20. $y = x^3 - 2x^2$, $y = 2x^2 - 3x$, $x = 0$, $x = 3$.

21. $y = \sin x$, $y = \cos x$, $x = 0$, $x = 2\pi$.

22. Find a vertical line $x = k$ that divides the area enclosed by $x = \sqrt{y}$, $x = 2$, and $y = 0$ into two equal parts.

23. Find a horizontal line $y = k$ that divides the area between $y = x^2$ and $y = 9$ into two equal parts.

24. Find the area enclosed between the curve $x^{1/2} + y^{1/2} = a^{1/2}$ and the coordinate axes.

25. Suppose that f and g are integrable on $[a, b]$, but neither $f(x) \geq g(x)$ nor $g(x) \geq f(x)$ holds for all x in $[a, b]$. (That is, the curves $y = f(x)$ and $y = g(x)$ are intertwined.)
    (a) What is the geometric significance of the integral
    $$\int_{a}^{b} [f(x) - g(x)]\, dx?$$
    (b) What is the geometric significance of the integral
    $$\int_{a}^{b} |f(x) - g(x)|\, dx?$$

6.2 VOLUMES BY SLICING; DISKS AND WASHERS

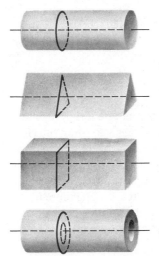

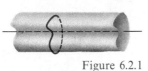

Figure 6.2.1

In this section we will use definite integrals to find volumes of three-dimensional solids.

A right-circular cylinder (Figure 6.2.1, top) can be generated by moving a plane circular disk along a line perpendicular to the disk. In general, we define a *right cylinder* to be any solid that can be generated by moving a plane region along a line or *axis* perpendicular to the region. All of the solids in Figure 6.2.1 are right cylinders. Observe that all cross sections of a right cylinder taken perpendicular to the axis are identical in size and shape.

If a right cylinder is generated by moving a plane region of area A through a distance h (Figure 6.2.2), then the volume V of the cylinder is *defined* to be

$$V = A \cdot h$$

that is, *the volume is the cross-sectional area times the height.*

Volumes of solids that are neither right cylinders nor composed of finitely many right cylinders can be obtained by a technique called "slicing." To illustrate the idea, suppose that a solid S extends along the x-axis and is bounded on the left and right by planes perpendicular to the x-axis at $x = a$ and $x = b$ (Figure 6.2.3). Because S is not assumed to be a right cylinder, its cross sections perpendicular to the x-axis can vary from point to point; we will denote by $A(x)$ the area of the cross section at x (Figure 6.2.3).

Volume $= A \cdot h$

Figure 6.2.2

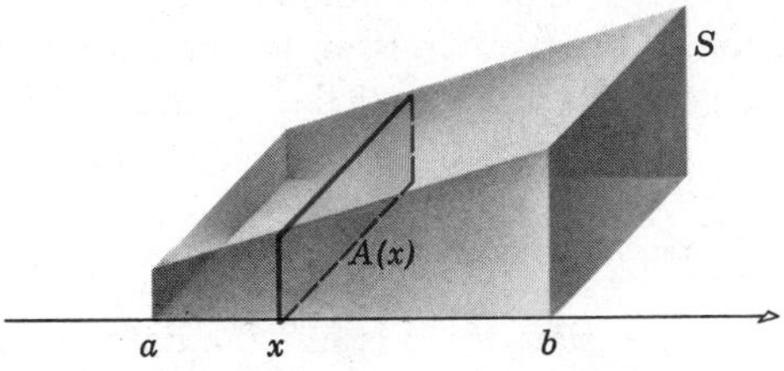

Figure 6.2.3

Let us divide the interval $[a, b]$ into n subintervals with widths

$$\Delta x_1, \Delta x_2, \ldots, \Delta x_n$$

by inserting points

$$x_1, x_2, \ldots, x_{n-1}$$

between a and b, and let us pass a plane perpendicular to the x-axis through each of these points. As illustrated in Figure 6.2.4, these planes cut the solid S into n slices

$$S_1, S_2, \ldots, S_n$$

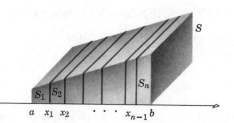

Figure 6.2.4

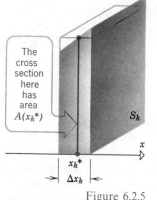

The cross section here has area $A(x_k^*)$

Figure 6.2.5

Consider a typical slice S_k. In general, this slice may not be a right cylinder because its cross section can vary. However, if the slice is very thin, the cross section will not vary much. Therefore, if we choose an arbitrary point x_k^* in the kth subinterval, each cross section of slice S_k will be approximately the same as the cross section at x_k^*, and we can approximate slice S_k by a right cylinder of thickness (height) Δx_k and cross-sectional area $A(x_k^*)$ (Figure 6.2.5).

Thus, the volume V_k of slice S_k is approximately the volume of this cylinder, that is,

$$V_k \approx A(x_k^*) \, \Delta x_k$$

and the volume V of the entire solid is approximately

$$V = V_1 + V_2 + \cdots + V_n \approx \sum_{k=1}^{n} A(x_k^*) \, \Delta x_k \tag{1}$$

If we now increase the number of slices in such a way that $\max \Delta x_k \to 0$, then the slices will become thinner and thinner and our approximations will get better and better. Thus, intuition suggests that approximation (1) will approach the exact value of the volume V as $\max \Delta x_k \to 0$, that is,

$$V = \lim_{\max \Delta x_k \to 0} \sum_{k=1}^{n} A(x_k^*) \, \Delta x_k \tag{2}$$

Since the right side of (2) is just the definite integral

$$\int_a^b A(x) \, dx$$

we are led to the following definition.

6.2.1 DEFINITION
Volumes by Cross Sections Perpendicular to the x-Axis

Let S be a solid bounded by two parallel planes perpendicular to the x-axis at $x = a$ and $x = b$. If, for each x in $[a, b]$, the cross-sectional area of S perpendicular to the x-axis is $A(x)$, then the volume of the solid is

$$V = \int_a^b A(x) \, dx \tag{3}$$

provided $A(x)$ is integrable.

There is a similar result for cross sections perpendicular to the y-axis.

6.2.2   DEFINITION
Volumes by Cross Sections
Perpendicular to the y-Axis

Let S be a solid bounded by two parallel planes perpendicular to the y-axis at $y = c$ and $y = d$. If, for each y in $[c, d]$, the cross-sectional area of S perpendicular to the y-axis is $A(y)$, then the volume of the solid is

$$V = \int_c^d A(y)\, dy \qquad\qquad (4)$$

provided $A(y)$ is integrable.

REMARK.   Although we will not prove it, it can be shown that Definitions 6.2.1 and 6.2.2 are equivalent; that is, when formulas (3) and (4) both apply, they produce the same volume.

▶ **Example 1**   Derive the formula for the volume of a right pyramid whose altitude is h and whose base is a square with sides of length a.

Solution.   As illustrated in Figure 6.2.6a, we introduce a rectangular coordinate system so that the y-axis passes through the apex, and the x-axis passes through the base and is parallel to a side of the base.

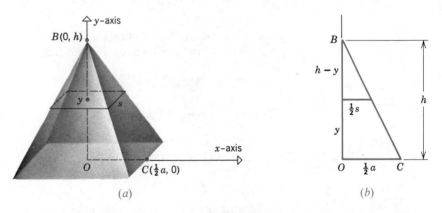

Figure 6.2.6   (a)   (b)

At any point y in the interval $[0, h]$ on the y-axis the cross section perpendicular to the y-axis is a square. If s denotes the length of a side of this square then by similar triangles (Figure 6.2.6b)

$$\frac{\tfrac{1}{2}s}{\tfrac{1}{2}a} = \frac{h - y}{h} \qquad \text{or} \qquad s = \frac{a}{h}(h - y)$$

Thus, the area $A(y)$ of the cross section at y is

$$A(y) = s^2 = \frac{a^2}{h^2}(h - y)^2$$

and by (4) the volume is

$$V = \int_0^h A(y)\, dy = \int_0^h \frac{a^2}{h^2}(h-y)^2\, dy = \frac{a^2}{h^2}\int_0^h (h-y)^2\, dy$$

$$= \frac{a^2}{h^2}\left[-\frac{1}{3}(h-y)^3\right]_0^h$$

$$= \frac{a^2}{h^2}\left[0 + \frac{1}{3}h^3\right] = \frac{1}{3}a^2 h \qquad \blacktriangleleft$$

VOLUMES OF SOLIDS OF REVOLUTION

Let f be nonnegative and continuous on $[a, b]$, and let R be the region bounded above by the graph of f, below by the x-axis, and on the sides by the lines $x = a$ and $x = b$ (Figure 6.2.7a). When this region is revolved

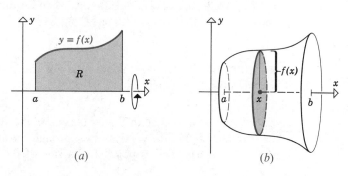

Figure 6.2.7

(a) $\qquad$ (b)

about the x-axis, it generates a solid having circular cross sections (Figure 6.2.7b). Since the cross section at x has radius $f(x)$, the cross-sectional area is

$$A(x) = \pi[f(x)]^2$$

Therefore, from Definition 6.2.1, the volume of the solid is

$$V = \int_a^b \pi[f(x)]^2\, dx \qquad (5)$$

Because the cross sections are circular or disk shaped, the application of this formula is called the **method of disks.**

▶ **Example 2** Find the volume of the solid obtained when the region under the curve $y = \sqrt{x}$ over the interval $[1, 4]$ is revolved about the x-axis (Figure 6.2.8).

Solution. From (5), the volume is

Figure 6.2.8

$$V = \int_a^b \pi[f(x)]^2\, dx = \int_1^4 \pi x\, dx = \frac{\pi x^2}{2}\Big]_1^4 = 8\pi - \frac{\pi}{2} = \frac{15\pi}{2} \qquad \blacktriangleleft$$

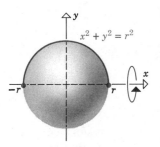

Figure 6.2.9

▶ **Example 3** Derive the formula for the volume of a sphere of radius r.

Solution. As indicated in Figure 6.2.9, a sphere of radius r can be generated by revolving the upper half of the circle

$$x^2 + y^2 = r^2$$

about the x-axis. Since the upper half of this circle is the graph of $y = f(x) = \sqrt{r^2 - x^2}$, it follows from (5) that the volume of the sphere is

$$V = \int_a^b \pi[f(x)]^2 \, dx = \int_{-r}^r \pi(r^2 - x^2) \, dx = \pi\left[r^2 x - \frac{x^3}{3}\right]_{-r}^r = \frac{4}{3}\pi r^3 \quad ◀$$

We will now consider more general solids of revolution. Suppose that f and g are nonnegative continuous functions such that

$$g(x) \le f(x) \qquad \text{for} \qquad a \le x \le b$$

and let R be the region enclosed between the graphs of these functions and the lines $x = a$ and $x = b$ (Figure 6.2.10a). When this region is revolved

Figure 6.2.10

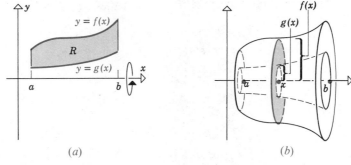

(a)   (b)

about the x-axis, it generates a solid having annular or washer-shaped cross sections (Figure 6.2.10b). Since the cross section at x has inner radius $g(x)$ and outer radius $f(x)$, its area is

$$A(x) = \pi[f(x)]^2 - \pi[g(x)]^2 = \pi([f(x)]^2 - [g(x)]^2)$$

Therefore, from Definition 6.2.1, the volume of the solid is

$$V = \int_a^b \pi([f(x)]^2 - [g(x)]^2) \, dx \tag{6}$$

The application of this formula is called the **method of washers.**

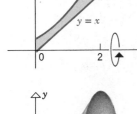

Figure 6.2.11

▶ **Example 4** Find the volume of the solid generated when the region between the graphs of $f(x) = \frac{1}{2} + x^2$ and $g(x) = x$ over the interval $[0, 2]$ is revolved about the x-axis (Figure 6.2.11).

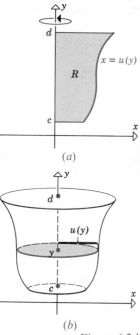

(a)

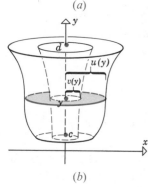

(b)

Figure 6.2.12

Solution. From (6) the volume is

$$V = \int_a^b \pi([f(x)]^2 - [g(x)]^2)\,dx = \int_0^2 \pi([\tfrac{1}{2} + x^2]^2 - x^2)\,dx$$

$$= \int_0^2 \pi\left(\frac{1}{4} + x^4\right) dx = \pi\left[\frac{x}{4} + \frac{x^5}{5}\right]_0^2 = \frac{69\pi}{10} \qquad \blacktriangleleft$$

The methods of disks and washers have analogs for regions revolved about the *y*-axis. If the region of Figure 6.2.12 is revolved about the *y*-axis, the cross sections of the resulting solid taken perpendicular to the *y*-axis are disks, and it follows from Definition 6.2.2 that the volume is

$$V = \int_c^d \pi[u(y)]^2\,dy \qquad (7)$$

(Verify.)

Also, if the region of Figure 6.2.13*a* is revolved about the *y*-axis, the cross sections taken perpendicular to the *y*-axis are washers, and it follows from Definition 6.2.2 that the volume of the solid in Figure 6.2.13*b* is

$$V = \int_c^d \pi([u(y)]^2 - [v(y)]^2)\,dy \qquad (8)$$

(Verify.)

▶ **Example 5** Find the volume of the solid generated when the region enclosed by $y = \sqrt{x}$, $y = 2$, and $x = 0$ is revolved about the *y*-axis (Figure 6.2.14).

Solution. The cross sections taken perpendicular to the *y*-axis are disks, so

(a)

(b)

Figure 6.2.13

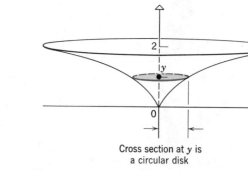

Cross section at *y* is
a circular disk

Figure 6.2.14

we will apply (7). But first we must rewrite $y = \sqrt{x}$ as $x = y^2$. Thus, from (7) with $u(y) = y^2$, the volume is

$$V = \int_c^d \pi[u(y)]^2 \, dy = \int_0^2 \pi y^4 \, dy = \frac{\pi y^5}{5}\bigg]_0^2 = \frac{32\pi}{5} \qquad \blacktriangleleft$$

▶ Exercise Set 6.2

In Exercises 1–12, find the volume of the solid that results when the region enclosed by the given curves is revolved about the x-axis.

1. $y = x^2$, $x = 0$, $x = 2$, $y = 0$.

2. $y = \sec x$, $x = \pi/4$, $x = \pi/3$, $y = 0$.

3. $y = 1 + x^3$, $x = 1$, $x = 2$, $y = 0$.

4. $y = 1/x$, $x = 1$, $x = 4$, $y = 0$.

5. $y = 9 - x^2$, $y = 0$.

6. $y = \sqrt{\cos x}$, $x = \pi/4$, $x = \pi/2$, $y = 0$.

7. $y = x^2$, $y = 4x$.

8. $y = x^2$, $y = 9$.

9. $y = \sin x$, $y = \cos x$, $x = 0$, $x = \pi/4$. [*Hint:* $\cos 2x = \cos^2 x - \sin^2 x$.]

10. $y = x^2 + 1$, $y = x + 3$.

11. $y = \sqrt{x}$, $y = x$.

12. $y = x^2$, $y = x^3$.

In Exercises 13–24, find the volume of the solid that results when the region enclosed by the given curves is revolved about the y-axis.

13. $y = x^3$, $x = 0$, $y = 1$.

14. $x = 1 - y^2$, $x = 0$.

15. $x = \sqrt{1 + y}$, $x = 0$, $y = 3$.

16. $x = \sqrt{\cos y}$, $y = 0$, $y = \pi/2$, $x = 0$.

17. $x = \csc y$, $y = \pi/4$, $y = 3\pi/4$, $x = 0$.

18. $y = 2/x$, $y = 1$, $y = 3$, $x = 0$.

19. $x = \sqrt{9 - y^2}$, $y = 1$, $y = 3$, $x = 0$.

20. $y = x^2 - 1$, $x = 2$, $y = 0$.

21. $y = 1 + x^3$, $x = 1$, $y = 9$.

22. $y = x^2$, $x = y^2$.

23. $x = y^2$, $x = y + 2$.

24. $x = 1 - y^2$, $x = 2 + y^2$, $y = -1$, $y = 1$.

25. Find the volume of the solid that results when the region enclosed by $y = \sqrt{x}$, $y = 0$, and $x = 9$ is revolved about the line $x = 9$.

26. Find the volume of the solid that results when the region in Exercise 25 is revolved about the line $y = 3$.

27. Find the volume of the solid that results when the region enclosed by $x = y^2$ and $x = y$ is revolved about the line $y = -1$.

28. Find the volume of the solid that results when the region in Exercise 27 is revolved about the line $x = -1$.

29. Find the volume of the solid that results when the region above the x-axis and below the curve

$$\frac{x^2}{a^2} + \frac{y^2}{b^2} = 1 \qquad (a > 0, b > 0)$$

is revolved about the x-axis.

30. Let R_1, R_2, R_3, and R_4 be the regions indicated in Figure 6.2.15. Express the following volumes as definite integrals:
    (a) The volume of the solid generated when R_1 is revolved about the x-axis.
    (b) The volume of the solid generated when R_2 is revolved about the x-axis.
    (c) The volume of the solid generated when R_3 is revolved about the y-axis.
    (d) The volume of solid generated when R_4 is revolved about the y-axis.

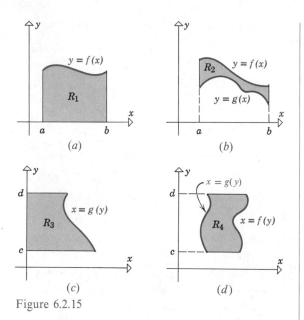

Figure 6.2.15

31. Find the volume of the solid generated when the region enclosed by $y = \sqrt{x + 1}$, $y = \sqrt{2x}$, and $y = 0$ is revolved about the x-axis. [*Hint:* Split the solid into two parts.]

32. Find the volume of the solid generated when the region enclosed by $y = \sqrt{x}$, $y = 6 - x$, and $y = 0$ is revolved about the x-axis. [*Hint:* Split the solid into two parts.]

33. Derive the formula for the volume of a right-circular cone with radius r and height h.

34. A *general cylinder* is any solid that can be generated by translating a plane region along a line passing through but not contained in the region. Derive a formula for the volume of a general cylinder with base area A and height h (Figure 6.2.16).

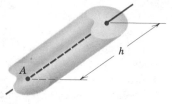

Figure 6.2.16

35. A nose cone for a space reentry vehicle is designed so that a cross section, taken x feet from the tip and perpendicular to the axis of symmetry, is a circle of radius $\frac{1}{4}x^2$ ft. Find the volume of the nose cone given that its length is 20 ft.

36. (a) A vat, shaped like a hemisphere of radius r ft, is filled with a fluid to a depth of h ft. Find the volume of the fluid.
   (b) If fluid enters a hemispherical vat of radius 10 ft at a rate of $\frac{1}{2}$ ft³/min, how fast will the fluid be rising when the depth is 5 ft?

37. The base of a certain solid is the circle $x^2 + y^2 = 9$ and each cross section perpendicular to the x-axis is an equilateral triangle with one side across the base. Find the volume of the solid.

38. A certain solid is 1 ft high and a horizontal cross section taken x ft above the bottom of the solid is an annulus of inner radius x^2 and outer radius $\sqrt{x}$. Find the volume of the solid.

39. The base of a certain solid is the region enclosed by $y = \sin x$, $x = \pi/4$, and $x = 3\pi/4$; and every cross section perpendicular to the x-axis is a square with one side across the base. Find the volume of the solid. [*Hint:* To help with the integration, use the identity $\sin^2 x = \frac{1}{2}(1 - \cos 2x)$.]

40. A hole of radius $r/2$ is drilled through the center of a sphere of radius r. Find the volume of the remaining solid.

41. A wedge is cut from a right-circular cylinder of radius r by two planes, one perpendicular to the axis of the cylinder and the other making an angle θ with the first. Find the volume of the wedge. [*Hint:* Introduce coordinate axes as shown in Figure 6.2.17.]

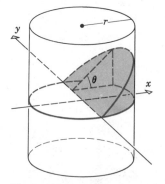

Figure 6.2.17

42. Two right-circular cylinders of radius r have axes that intersect at right angles. Find the volume of the solid common to the two cylinders. [*Hint:* One eighth of the solid is sketched in Figure 6.2.18.]

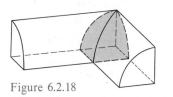

Figure 6.2.18

43. A cocktail glass with a bowl shaped like a hemisphere of diameter 8 cm contains a cherry with a diameter of 2 cm (Figure 6.2.19). If the glass is filled to a depth of h cm, what is the volume of liquid it contains? [*Hint:* First consider the case where the cherry is partially submerged, then the case where it is totally submerged.]

Figure 6.2.19

44. In 1635 Bonaventura Cavalieri, a student of Galileo, stated the following result, called *Cavalieri's Principle: If two solids have the same height, and if the areas of their cross sections taken parallel to and at equal distances from their bases are always equal, then the solids have the same volume.* Prove this result.

6.3 VOLUMES BY CYLINDRICAL SHELLS

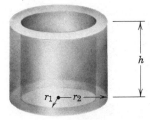

Figure 6.3.1

The methods discussed so far for computing volumes of solids depend on our ability to compute cross-sectional areas of the solid. In this section we will develop an alternate technique for computing volumes that can sometimes be applied when it is inconvenient or difficult to determine the cross-sectional areas or if the integration is too hard. This technique, called the *method of cylindrical shells,* applies only to *solids of revolution,* that is, solids generated by revolving a plane region about a line.

A *cylindrical shell* is a solid enclosed by two concentric right-circular cylinders (Figure 6.3.1). The volume V of a cylindrical shell having inner radius r_1, outer radius r_2, and height h can be written

$$V = [\text{area of cross section}] \cdot [\text{height}]$$
$$= (\pi r_2^2 - \pi r_1^2)h$$
$$= \pi(r_2 + r_1)(r_2 - r_1)h$$
$$= 2\pi \left(\frac{r_2 + r_1}{2} \right) h(r_2 - r_1)$$

Thus,

$$V = 2\pi \cdot [\text{average radius}] \cdot [\text{height}] \cdot [\text{thickness}] \qquad (1)$$

The method of cylindrical shells, which we will now discuss, makes use of this formula.

Let R be a plane region bounded above by a continuous curve $y = f(x)$, bounded below by the x-axis, and bounded on the left and right, respectively, by the lines $x = a$ and $x = b$; let S be the solid generated by revolving the region R about the y-axis (Figure 6.3.2.)

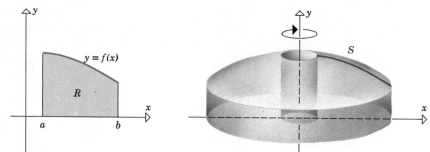

Figure 6.3.2

To find the volume of S, let us divide the interval $[a, b]$ into n subintervals with widths

$$\Delta x_1, \Delta x_2, \ldots, \Delta x_n$$

by inserting points

$$x_1, x_2, \ldots, x_{n-1}$$

between a and b, and let us draw a vertical line through each of these points to divide the region R into n strips $R_1, R_2, \ldots, R_n$ (Figure 6.3.3a). These strips, when revolved about the y-axis, generate solids $S_1, S_2, \ldots, S_n$. As

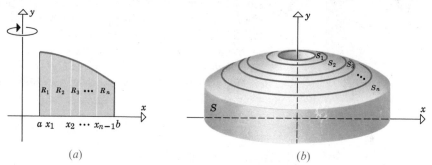

Figure 6.3.3   (a)   (b)

illustrated in Figure 6.3.3b, these solids are nested one inside the other and together form the entire solid S. Thus, the volume of the solid S can be obtained by adding together the volumes of solids $S_1, S_2, \ldots, S_n$:

$$V(S) = V(S_1) + V(S_2) + \cdots + V(S_n) \tag{2}$$

Consider a typical strip R_k and the solid S_k that it generates (Figure 6.3.4).

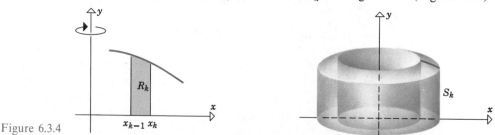

Figure 6.3.4

Although solid S_k resembles a cylindrical shell, it will not, in general, be a cylindrical shell because it can have a curved upper surface. However, if the interval width

$$\Delta x_k = x_k - x_{k-1}$$

is small we can obtain a good approximation to the region R_k by a rectangle of width Δx_k and height $f(x_k^*)$, where

$$x_k^* = \frac{x_k + x_{k-1}}{2}$$

is the midpoint of the interval $[x_{k-1}, x_k]$ (Figure 6.3.5a). This rectangle, when revolved about the y-axis, generates a cylindrical shell, which is a good approximation to the solid S_k (Figure 6.3.5b).

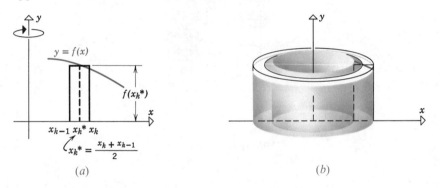

Figure 6.3.5   (a)   (b)

This approximating cylindrical shell has thickness Δx_k, height $f(x_k^*)$, and average radius x_k^*, so that by (1)

$$V(S_k) \approx \text{volume of approximating cylindrical shell} = 2\pi x_k^* f(x_k^*) \, \Delta x_k$$

Thus, from (2), the volume of the entire solid S is approximately

$$V(S) = \sum_{k=1}^{n} V(S_k) \approx \sum_{k=1}^{n} 2\pi x_k^* f(x_k^*) \, \Delta x_k \tag{3}$$

If we now divide $[a, b]$ into more and more subintervals in such a way that $\max \Delta x_k \to 0$, then intuition suggests that our approximations will tend to get better and (3) will approach the exact value of the volume, that is,

$$V(S) = \lim_{\max \Delta x_k \to 0} \sum_{k=1}^{n} 2\pi x_k^* f(x_k^*) \, \Delta x_k \tag{4}$$

Because the right side of (4) is just the definite integral

$$\int_a^b 2\pi x f(x) \, dx$$

we are led to the following definition.

6.3.1 DEFINITION
Volumes by Cylindrical Shells
About the y-Axis

Let R be a plane region bounded above by a continuous curve $y = f(x)$, below by the x-axis, and on the left and right, respectively, by the lines $x = a$ and $x = b$. Then the volume of the solid generated by revolving R about the y-axis is given by

$$V = \int_a^b 2\pi x f(x)\, dx \qquad (5)$$

▶ **Example 1** Use cylindrical shells to find the volume of the solid generated when the region enclosed between $y = \sqrt{x}$, $x = 1$, $x = 4$, and the x-axis is revolved about the y-axis (Figure 6.3.6).

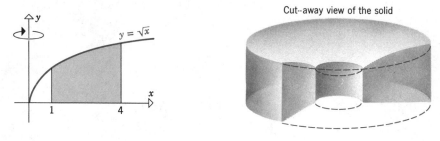

Figure 6.3.6

Solution. Since $f(x) = \sqrt{x}$, $a = 1$, and $b = 4$, Formula (5) yields

$$V = \int_1^4 2\pi x \sqrt{x}\, dx = 2\pi \int_1^4 x^{3/2}\, dx$$

$$= \left[2\pi \cdot \frac{2}{5} x^{5/2} \right]_1^4$$

$$= \frac{4\pi}{5}[32 - 1] = \frac{124\pi}{5} \qquad ◀$$

There is a way of thinking about (5) that is sometimes useful. At each point x in $[a, b]$ the vertical line through x cuts the region R in a line segment that we can view as the vertical "cross section" of R at x (Figure 6.3.7a). When the region R is revolved about the y-axis, the vertical cross section at

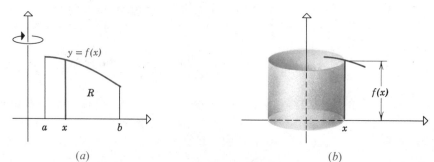

Figure 6.3.7   (a)   (b)

x generates the *surface* of a right-circular cylinder having height $f(x)$ and radius x (Figure 6.3.7*b*). The area of this surface is

$$2\pi x f(x)$$

(see Figure 4.8.1), which is precisely the integrand in (5). Thus, (5) states:

6.3.2 DEFINITION The volume V by cylindrical shells is the integral of the surface area generated by an arbitrary cross section of R taken parallel to the axis about which R is revolved.

Formulation 6.3.2 is helpful for adapting the method of cylindrical shells to problems that are not exactly of the form considered in Definition 6.3.1.

▶ **Example 2** Use cylindrical shells to find the volume of the solid generated when the region R in the first quadrant enclosed between $y = x$ and $y = x^2$ is revolved about the y-axis (Figure 6.3.8).

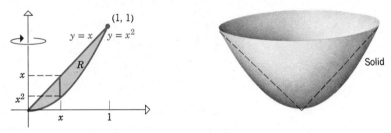

Figure 6.3.8

Solution. At each x in $[0, 1]$ the cross section of R parallel to the y-axis generates a cylindrical surface of height $x - x^2$ and radius x. Since the area of this surface is

$$2\pi x(x - x^2)$$

the volume of the solid is

$$V = \int_0^1 2\pi x(x - x^2)\,dx = 2\pi \int_0^1 (x^2 - x^3)\,dx$$

$$= 2\pi \left[\frac{x^3}{3} - \frac{x^4}{4}\right]_0^1 = 2\pi \left[\frac{1}{3} - \frac{1}{4}\right] = \frac{\pi}{6} \qquad ◀$$

▶ **Example 3** Use cylindrical shells to find the volume of the solid generated when the region R under $y = x^2$ over the interval $[0, 2]$ is revolved about the x-axis (Figure 6.3.9).

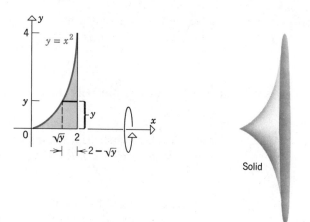

Figure 6.3.9

Solution.   At each y in the interval $0 \leq y \leq 4$, the cross section of R parallel to the x-axis generates a cylindrical surface of height $2 - \sqrt{y}$ and radius y. Since the area of this surface is

$$2\pi y(2 - \sqrt{y})$$

the volume of the solid is

$$V = \int_0^4 2\pi y(2 - \sqrt{y})\, dy = 2\pi \int_0^4 (2y - y^{3/2})\, dy$$

$$= 2\pi \left[y^2 - \frac{2}{5} y^{5/2} \right]_0^4 = \frac{32\pi}{5} \quad \blacktriangleleft$$

The volume in Example 3 can also be obtained by the method of disks [Formula (5) of Section 6.2]. The computations are

$$V = \int_0^2 \pi (x^2)^2\, dx = \int_0^2 \pi x^4\, dx = \frac{\pi x^5}{5} \Big]_0^2 = \frac{32}{5}\pi$$

▶ Exercise Set 6.3

In Exercises 1–6, use cylindrical shells to find the volume of the solid generated when the region enclosed by the given curves is revolved about the y-axis.

1. $y = x^3$, $x = 1$, $y = 0$.

2. $y = \sqrt{x}$, $x = 4$, $x = 9$, $y = 0$.

3. $x^2 + y^3 = 4$, $x = 0$, $x = 4$, $y = 0$.

4. $y = \cos(x^2)$, $x = 0$, $x = \frac{1}{2}\sqrt{\pi}$, $y = 0$.

5. $y = 2x - 1$, $y = -2x + 3$, $x = 2$.

6. $x = y^2$, $y = x^2$.

In Exercises 7–10, use cylindrical shells to find the volume of the solid generated when the region enclosed by the given curves is revolved about the x-axis.

7. $y^2 = x$, $y = 1$, $x = 0$.

8. $x = 2y$, $y = 2$, $y = 3$, $x = 0$.

9. $y = x^2$, $x = 1$, $y = 0$.

10. $xy = 4$, $x + y = 5$.

11. (a)   Use cylindrical shells to find the volume of the solid generated when the region under

the curve $y = x^3 - 3x^2 + 2x$ over $[0, 1]$ is revolved about the y-axis.

(b) For this problem, is the method of cylindrical shells easier or harder than the method of slicing discussed in the last section? Explain.

12. Use cylindrical shells to find the volume of the solid generated when the region enclosed by

$$y = \frac{1}{x^3}, \ x = 1, \ x = 2, \ y = 0$$

is revolved about the line $x = -1$.

13. Use cylindrical shells to find the volume of the solid generated when the region enclosed by

$$y = x^3, \ y = 1, \ x = 0$$

is revolved about the line $y = 1$.

14. Let R_1 and R_2 be regions of the form shown in Figure 6.3.10. Use cylindrical shells to find a formula for the volume of the solid that results when:
(a) region R_1 is revolved about the y-axis
(b) region R_2 is revolved about the x-axis.

15. Use cylindrical shells to find the volume of the cone generated when the triangle with vertices $(0, 0)$, $(0, r)$, $(h, 0)$, where $r > 0$ and $h > 0$ is revolved about the x-axis.

16. The region enclosed between the curve $y^2 = kx$ and the line $x = \frac{1}{4}k$ is revolved about the line $x = \frac{1}{2}k$. Use cylindrical shells to find the volume of the resulting solid. Assume $k > 0$.

17. A round hole of radius a is drilled through the center of a solid sphere of radius r. Use cylindrical shells to find the volume of the portion removed. (Assume $r > a$.)

18. Use cylindrical shells to find the volume of the torus obtained by revolving the circle

$$x^2 + y^2 = a^2$$

about the line $x = b$, where $b > a$. [*Hint:* It may help in the integration to think of an integral as an area.]

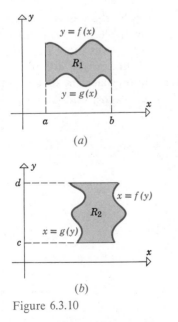

(a)

(b)

Figure 6.3.10

6.4   LENGTH OF A PLANE CURVE

In this section we will use definite integrals to find arc lengths of plane curves. To start, we will consider only curves that are graphs of functions. In a later section, we will extend our results to more general curves.

If f' is continuous on an interval, we shall say that $y = f(x)$ is a **smooth curve** (or f is a **smooth function**) on that interval. We will restrict our discussion of arc length to smooth curves in order to eliminate some complications that would otherwise occur.

6.4.1 PROBLEM   Suppose f is smooth on the interval $[a, b]$. Find the arc length L of the curve $y = f(x)$ over the interval $[a, b]$ (Figure 6.4.1).

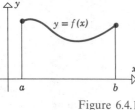

In order to solve Problem 6.4.1 we must first define the term "arc length" precisely. For motivation, consider the graph of a smooth curve $y = f(x)$ over an interval $[a, b]$. As shown in Figure 6.4.2, we divide the interval $[a, b]$ into n subintervals with widths

Figure 6.4.1

$$\Delta x_1, \Delta x_2, \ldots, \Delta x_n$$

by inserting points

$$x_1, x_2, \ldots, x_{n-1}$$

between a and b, and let $P_0, P_1, \ldots, P_n$ be the points on the curve whose x-coordinates are $a, x_1, x_2, \ldots, x_{n-1}, b$. If we now join these points with straight-line segments we obtain a *polygonal path* that we can regard as an approximation to the curve $y = f(x)$. Intuition suggests that the length of the approximating polygonal path will approach the length of the curve if we increase the number of points in such a way that the lengths of the line segments approach zero.

Figure 6.4.2

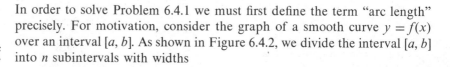

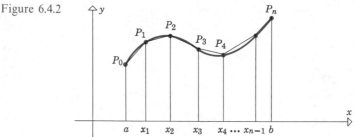

To examine this idea more closely, let us isolate a typical subinterval, say the kth (Figure 6.4.3). As suggested by this figure, the length L_k of the kth line segment in the polygonal path is given by

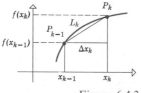

Figure 6.4.3

$$L_k = \sqrt{(\Delta x_k)^2 + [f(x_k) - f(x_{k-1})]^2} \tag{1}$$

By the Mean-Value Theorem (4.10.2) there is a point x_k^* between x_{k-1} and x_k such that

$$\frac{f(x_k) - f(x_{k-1})}{x_k - x_{k-1}} = f'(x_k^*)$$

or

$$f(x_k) - f(x_{k-1}) = f'(x_k^*)\,\Delta x_k$$

Thus, (1) can be rewritten as

$$L_k = \sqrt{1 + [f'(x_k^*)]^2}\,\Delta x_k$$

which means that the length of the *entire* polygonal path is

$$\sum_{k=1}^{n} L_k = \sum_{k=1}^{n} \sqrt{1 + [f'(x_k^*)]^2}\,\Delta x_k$$

If we now increase the number of subintervals in such a way that max $\Delta x_k \to 0$, then the length of the polygonal path will approach the arc length L of the curve $y = f(x)$ over $[a, b]$. Thus,

$$L = \lim_{\max \Delta x_k \to 0} \sum_{k=1}^{n} \sqrt{1 + [f'(x_k^*)]^2}\,\Delta x_k \qquad (2)$$

Since the right side of (2) is just the definite integral

$$\int_a^b \sqrt{1 + [f'(x)]^2}\,dx$$

we are led to the following definition.

6.4.2 DEFINITION If f is a smooth function on $[a, b]$, then the **arc length** L of the curve $y = f(x)$ from $x = a$ to $x = b$ is defined to be

$$L = \int_a^b \sqrt{1 + [f'(x)]^2}\,dx \qquad (3)$$

or equivalently,

$$L = \int_a^b \sqrt{1 + \left(\frac{dy}{dx}\right)^2}\,dx \qquad (3a)$$

Similarly, for a curve expressed in the form $x = g(y)$, where g' is continuous on $[c, d]$, the arc length L from $y = c$ to $y = d$ is given by

$$L = \int_c^d \sqrt{1 + [g'(y)]^2} \, dy \tag{4}$$

or equivalently,

$$L = \int_c^d \sqrt{1 + \left(\frac{dx}{dy}\right)^2} \, dy \tag{4a}$$

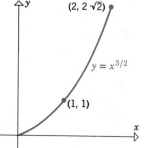

Figure 6.4.4

▶ **Example 1** Find the arc length of the curve $y = x^{3/2}$ from $(1, 1)$ to $(2, 2\sqrt{2})$ (Figure 6.4.4) using:

(a) formula (3)   (b) formula (4).

Solution (a). Since $f(x) = x^{3/2}$,

$$f'(x) = \frac{3}{2}x^{1/2}$$

Thus, from (3), the arc length from $x = 1$ to $x = 2$ is

$$L = \int_1^2 \sqrt{1 + \frac{9}{4}x} \, dx$$

To evaluate this integral we make the u-substitution

$$u = 1 + \frac{9}{4}x, \qquad du = \frac{9}{4} \, dx$$

and change the x-limits $(x = 1, x = 2)$ to the corresponding u-limits $(u = \frac{13}{4}, u = \frac{22}{4})$:

$$L = \frac{4}{9}\int_{13/4}^{22/4} u^{1/2} \, du = \frac{8}{27}u^{3/2}\Big]_{13/4}^{22/4}$$

$$= \frac{8}{27}\left[\left(\frac{22}{4}\right)^{3/2} - \left(\frac{13}{4}\right)^{3/2}\right]$$

$$= \frac{22\sqrt{22} - 13\sqrt{13}}{27}$$

Solution (b). Solving $y = x^{3/2}$ for x in terms of y, we obtain $x = y^{2/3}$. Hence $g(y) = y^{2/3}$ and

$$g'(y) = \frac{2}{3}y^{-1/3}$$

Thus, from (4), the arc length from $y = 1$ to $y = 2\sqrt{2}$ is

$$L = \int_1^{2\sqrt{2}} \sqrt{1 + \frac{4}{9} y^{-2/3}} \, dy$$

$$= \frac{1}{3} \int_1^{2\sqrt{2}} y^{-1/3} \sqrt{9y^{2/3} + 4} \, dy$$

To evaluate this integral we make the u-substitution

$$u = 9y^{2/3} + 4, \qquad du = 6y^{-1/3} \, dy$$

and change the y-limits ($y = 1, y = 2\sqrt{2}$) to the corresponding u-limits ($u = 13, u = 22$). This gives

$$L = \frac{1}{18} \int_{13}^{22} u^{1/2} \, du = \frac{1}{27} u^{3/2} \Big]_{13}^{22}$$

$$= \frac{1}{27} [(22)^{3/2} - (13)^{3/2}]$$

$$= \frac{22\sqrt{22} - 13\sqrt{13}}{27}$$

This result agrees with that in part (a); however, the integration here is more tedious. In problems where there is a choice between using (3) or (4), it is sometimes worthwhile to determine which one leads to the simpler integral. ◀

▶ Exercise Set 6.4

1. Find the arc length of the curve $y = 2x$ from $(1, 2)$ to $(2, 4)$ using
   (a) formula (3)
   (b) formula (4)
   (c) the Theorem of Pythagoras.

2. Find the arc length of the curve $y = mx + b$ from $x = k_1$ to $x = k_2$ ($m \neq 0, k_2 > k_1$) using
   (a) formula (3)
   (b) formula (4)
   (c) the Theorem of Pythagoras.

3. Find the arc length of the curve $y = 3x^{3/2} - 1$ from $x = 0$ to $x = 1$.

4. Find the arc length of the curve $x = \frac{1}{3}(y^2 + 2)^{3/2}$ from $y = 0$ to $y = 1$.

5. Find the arc length of the curve $y = x^{2/3}$ from $x = 1$ to $x = 8$.

6. Find the arc length of the curve $y = \frac{x^4}{16} + \frac{1}{2x^2}$ from $x = 2$ to $x = 3$.

7. Find the arc length of the curve $24xy = y^4 + 48$ from $y = 2$ to $y = 4$.

8. Find the arc length of the curve $x = \frac{1}{8}y^4 + \frac{1}{4}y^{-2}$ from $y = 1$ to $y = 4$.

9. Consider the curve $y = x^{2/3}$.
   (a) Sketch the portion of the curve between $x = -1$ and $x = 8$.
   (b) Explain why (3) cannot be used to find the arc length of the curve sketched in (a).
   (c) Find the arc length of the curve sketched in (a).

10. Find the arc length of the curve $x^{2/3} + y^{2/3} = a^{2/3}$ in the second quadrant from $x = -a$ to $x = -\frac{1}{8}a$ ($a > 0$).

11. Let $y = f(x)$ be a smooth curve and suppose $f'(x) \geq 0$ on the closed interval $[a, b]$.
  (a) Prove: There are numbers m and M such that $m \leq f'(x) \leq M$ for all x in $[a, b]$.
  (b) Prove: The arc length L of $y = f(x)$ over the closed interval $[a, b]$ satisfies the inequalities
  $$(b - a) \sqrt{1 + m^2} \leq L \leq (b - a) \sqrt{1 + M^2}$$

12. Use the result of Exercise 11 to show that the arc length L of $y = \sin x$ over the interval $0 \leq x \leq \frac{\pi}{4}$ satisfies
$$\frac{\pi}{4} \sqrt{\frac{3}{2}} \leq L \leq \frac{\pi}{4} \sqrt{2}$$

6.5 AREA OF A SURFACE OF REVOLUTION

In this section we will apply the definite integral to the problem of finding the area of a surface of revolution. For now we limit the discussion to the following situation:

6.5.1 PROBLEM   Let f be a smooth, nonnegative function on $[a, b]$. Find the area of the surface generated by revolving the portion of the curve $y = f(x)$ between $x = a$ and $x = b$ about the x-axis (Figure 6.5.1).

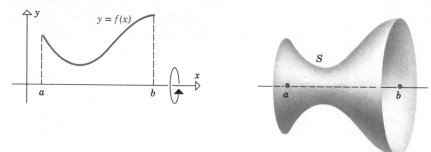

Figure 6.5.1

In order to solve this problem we must first define the term "surface area" precisely. For motivation, let us approximate the curve $y = f(x)$ by a polygonal path of straight line segments connecting the points on the curve that have x-coordinates

$$a, x_1, x_2, \ldots, x_{n-1}, b$$

(Figure 6.5.2a). As usual, let

$$\Delta x_1, \Delta x_2, \ldots, \Delta x_n$$

be the widths of the subintervals determined by these x-coordinates. If these widths are small, then the surface generated by revolving the polygonal path about the x-axis will have approximately the same area as the surface generated by revolving the curve $y = f(x)$ about the x-axis (Figure 6.5.2b). Observe that the surface generated by the polygonal path is made of

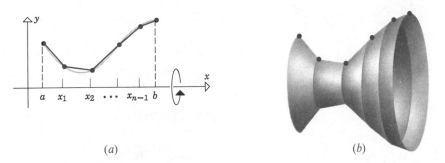

Figure 6.5.2   (a)   (b)

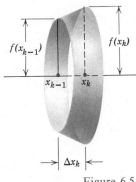

Figure 6.5.3

Figure 6.5.4

parts, each of which is a frustum of a cone. Thus, the lateral area of each of these parts can be obtained from the formula

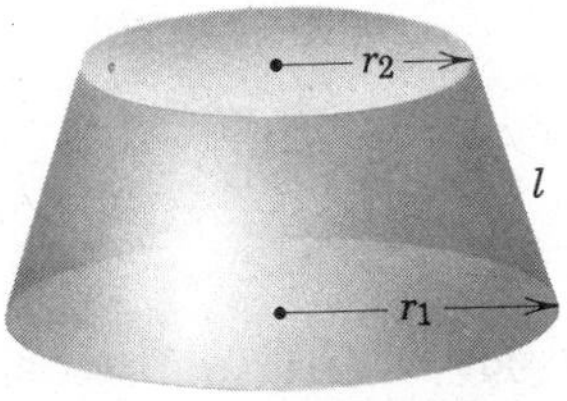

$$S = \pi(r_1 + r_2)l \tag{1}$$

for the lateral area S of a frustum of slant height l and base radii r_1 and r_2 (Figure 6.5.3). A derivation of this formula is discussed in Exercise 15. Intuition suggests that the area of the approximating surface will approach the desired area of the surface if we increase the number of subdivisions in such a way that the lengths of the line segments in the polygonal path approach zero.

To examine this idea more closely, let us isolate a typical section of the approximating surface, say the kth (Figure 6.5.4). Applying (1), we obtain as the lateral area S_k of this kth section

$$S_k = \pi[f(x_{k-1}) + f(x_k)]\sqrt{(\Delta x_k)^2 + [f(x_k) - f(x_{k-1})]^2} \tag{2}$$

By the Mean-Value Theorem (4.10.2) there is a point x_k^* between x_{k-1} and x_k such that

$$\frac{f(x_k) - f(x_{k-1})}{x_k - x_{k-1}} = f'(x_k^*)$$

or

$$f(x_k) - f(x_{k-1}) = f'(x_k^*)\,\Delta x_k$$

Thus, (2) can be rewritten as

$$S_k = \pi[f(x_{k-1}) + f(x_k)]\sqrt{1 + [f'(x_k^*)]^2}\,\Delta x_k \tag{3}$$

Since the arithmetic average of two numbers lies between those numbers, $\frac{1}{2}[f(x_{k-1}) + f(x_k)]$ is between $f(x_{k-1})$ and $f(x_k)$. Since f is continuous on the interval $[x_{k-1}, x_k]$, the Intermediate-Value Theorem (3.7.8) implies that there exists a point x_k^{**} in this interval such that

$$\frac{1}{2}[f(x_{k-1}) + f(x_k)] = f(x_k^{**})$$

Thus, (3) can be rewritten

$$S_k = 2\pi f(x_k^{**})\sqrt{1 + [f'(x_k^*)]^2}\,\Delta x_k$$

so that the area of the *entire* polygonal surface is

$$\sum_{k=1}^{n} S_k = \sum_{k=1}^{n} 2\pi f(x_k^{**})\sqrt{1 + [f'(x_k^*)]^2}\,\Delta x_k$$

If we now increase the number of subintervals in such a way that max $\Delta x_k \to 0$, then the area of the approximating polygonal surface will approach the exact surface area S. Thus,

$$S = \lim_{\max \Delta x_k \to 0} \sum_{k=1}^{n} 2\pi f(x_k^{**})\sqrt{1 + [f'(x_k^*)]^2}\,\Delta x_k \tag{4}$$

If it were the case that $x_k^{**} = x_k^*$, then the right side of (4) would be the definite integral

$$\int_a^b 2\pi f(x)\sqrt{1 + [f'(x)]^2}\,dx \tag{5}$$

However, it is proved in advanced calculus that this is true even if $x_k^* \neq x_k^{**}$ because of the continuity of f and f'. In light of (4) and (5) we are led to the following definition.

6.5.2 DEFINITION Let f be a smooth, nonnegative function on $[a, b]$. Then the **surface area** S generated by revolving the portion of the curve $y = f(x)$ between $x = a$ and $x = b$ about the x-axis is

$$S = \int_a^b 2\pi f(x)\sqrt{1 + [f'(x)]^2}\,dx \tag{6}$$

For a curve expressed in the form $x = g(y)$, where g' is continuous on $[c, d]$, and $g(y) \geq 0$ for $c \leq y \leq d$, the surface area S generated by revolving the portion of the curve from $y = c$ to $y = d$ about the y-axis is given by

$$S = \int_c^d 2\pi g(y)\sqrt{1 + [g'(y)]^2}\,dy \tag{7}$$

▶ **Example 1** Find the surface area generated by revolving the curve

$$y = \sqrt{1 - x^2}, \qquad 0 \leq x \leq \tfrac{1}{2}$$

about the x-axis (Figure 6.5.5).

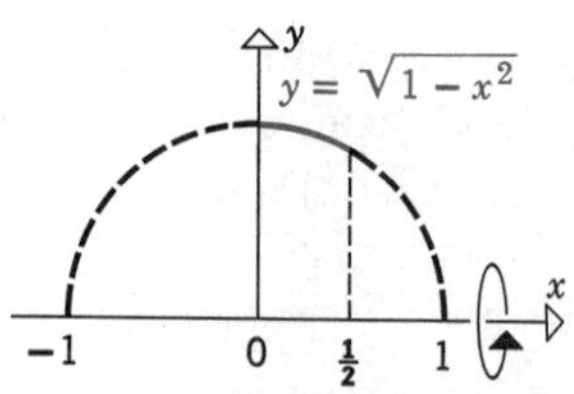

Figure 6.5.5

Solution.   Since $f(x) = \sqrt{1 - x^2}$,

$$f'(x) = -\frac{x}{\sqrt{1 - x^2}}$$

Thus, (6) yields

$$S = \int_0^{1/2} 2\pi \sqrt{1 - x^2} \sqrt{1 + \frac{x^2}{1 - x^2}}\, dx$$

$$= \int_0^{1/2} 2\pi\, dx = 2\pi x \Big]_0^{1/2} = \pi \qquad \blacktriangleleft$$

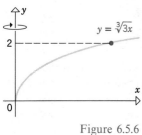

Figure 6.5.6

▶ **Example 2**   Find the surface area generated by revolving the curve

$$y = \sqrt[3]{3x}, \qquad 0 \le y \le 2$$

about the *y*-axis (Figure 6.5.6).

Solution.   We will apply (7) after rewriting $y = \sqrt[3]{3x}$ as $x = g(y) = \frac{1}{3}y^3$. Thus,

$$g'(y) = y^2$$

so that from (7) we obtain

$$S = \int_0^2 2\pi \left(\frac{1}{3} y^3\right) \sqrt{1 + y^4}\, dy$$

$$= \frac{2\pi}{3} \int_0^2 y^3 \sqrt{1 + y^4}\, dy \qquad (8)$$

The *u*-substitution

$$u = 1 + y^4, \qquad du = 4y^3\, dy$$

yields

$$\int y^3 \sqrt{1 + y^4}\, dy = \frac{1}{4} \int \sqrt{u}\, du = \frac{1}{4} \cdot \frac{2}{3} u^{3/2} + C$$

$$= \frac{1}{6}(1 + y^4)^{3/2} + C$$

Thus, from (8),

$$S = \frac{2\pi}{3} \left[\frac{1}{6}(1 + y^4)^{3/2} \right]_0^2 = \frac{\pi}{9}(17^{3/2} - 1) \qquad \blacktriangleleft$$

▶ Exercise Set 6.5

In Exercises 1–6, find the area of the surface generated by revolving the given curve about the x-axis.

1. $y = 7x$, $0 \leq x \leq 1$.

2. $y = \sqrt{x}$, $1 \leq x \leq 4$.

3. $y = \sqrt{4 - x^2}$, $-1 \leq x \leq 1$.

4. $x = \sqrt[3]{y}$, $1 \leq y \leq 8$.

5. $y = \sqrt{x} - \dfrac{1}{3}x^{3/2}$, $1 \leq x \leq 3$.

6. $y = \dfrac{1}{3}x^3 + \dfrac{1}{4}x^{-1}$, $1 \leq x \leq 2$.

In Exercises 7–12, find the area of the surface generated by revolving the given curve about the y-axis.

7. $x = 9y + 1$, $0 \leq y \leq 2$.

8. $x = y^3$, $0 \leq y \leq 1$.

9. $x = \sqrt{9 - y^2}$, $-2 \leq y \leq 2$.

10. $x = 2\sqrt{1 - y}$, $-1 \leq y \leq 0$.

11. $8xy^2 = 2y^6 + 1$, $1 \leq y \leq 2$.

12. $x = |y - 11|$, $0 \leq y \leq 2$.

13. The lateral area S of a right-circular cone with height h and base radius r is $S = \pi r \sqrt{r^2 + h^2}$. Obtain this result using (6).

14. Assume $f(x) \geq 0$ for $a \leq x \leq b$. Derive a formula for the surface area generated when the curve $y = f(x)$, $a \leq x \leq b$ is revolved about the line $y = -k$ ($k > 0$).

15. (*Formula for Surface Area of a Frustum*)
    (a) If a cone of slant height l and base radius r is cut along a lateral edge and laid flat, it becomes a sector of a circle of radius l (Figure 6.5.7).

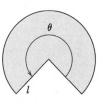

Figure 6.5.7

Use the formula $\frac{1}{2}l^2\theta$ for the area of a sector with radius l and central angle θ (in radians) to show that the lateral surface area of the cone is $\pi r l$.

(b) Use the result in (a) to obtain Formula (1) for the lateral surface area of a frustum.

16. Let $y = f(x)$ be a smooth curve on the interval $[a, b]$ and assume $f(x) \geq 0$ for $a \leq x \leq b$. By the Extreme-Value Theorem (4.6.4), the function f has a maximum value K and a minimum value k on $[a, b]$. Prove: If L is the arc length of the curve $y = f(x)$ between $x = a$ and $x = b$ and if S is the area of the surface that is generated by revolving this curve about the x-axis, then

$$2\pi k L \leq S \leq 2\pi K L$$

17. Let $y = f(x)$ be a smooth curve on $[a, b]$ and assume $f(x) \geq 0$ for $a \leq x \leq b$. Let A be the area under the curve $y = f(x)$ between $x = a$ and $x = b$ and let S be the area of the surface obtained when this section of curve is revolved about the x-axis.

    (a) Prove: $2\pi A \leq S$.

    (b) For what functions f is $2\pi A = S$?

6.6 WORK

If an object moves a distance d along a line while subjected to a *constant* force F applied in the direction of motion, then physicists and engineers define the **work** W done on the object by the force F to be

$$W = F \cdot d$$
$$[\text{work}] = [\text{force}] \cdot [\text{distance}]$$

(1)

If force is measured in pounds and distance in feet, then the unit of work is foot-pounds. In the metric system force is measured in **dynes** (the force required to give a mass of 1 gm an acceleration of 1 cm/sec^2) or **newtons** (the force required to give a mass of 1 kg an acceleration of 1 m/sec^2). In this system common units of work are newton-meters and dyne-centimeters.

▶ **Example 1**   An object moves 5 ft, while subjected to a constant force of 100 lb along its direction of motion. The work done is

$$W = F \cdot d = 100 \cdot 5 = 500 \text{ ft-lb}$$ ◀

Calculus is needed when it is required to calculate the work done by a *variable* force. We shall consider the following problem.

6.6.1 PROBLEM   Suppose an object moves in the positive direction along a coordinate line while subject to a force $F(x)$, in the direction of motion, whose magnitude depends on the coordinate x. Find the work done by the force when the object moves over an interval $[a, b]$.

For example, Figure 6.6.1 shows a block subjected to the force of a compressed spring. As the block moves from a to b the spring expands and the force it applies diminishes. Thus, the force $F(x)$ applied by the spring varies with x.

Figure 6.6.1

Before we can solve Problem 6.6.1 we must define precisely what is meant by the work done by a variable force. To motivate this definition, let us divide the interval $[a, b]$ into n subintervals with widths

$$\Delta x_1, \Delta x_2, \ldots, \Delta x_n$$

by inserting points

$$x_1, x_2, \ldots, x_{n-1}$$

between a and b. If we denote by W_k the work done when the object moves across the kth subinterval, then total work W done when the object moves across the entire interval $[a, b]$ will be

$$W = W_1 + W_2 + \cdots + W_n$$

Let us estimate W_k. If the kth subinterval is short and if $F(x)$ is continuous, then the force will not vary much over this subinterval; it will be almost constant. We can approximate this nearly constant force by $F(x_k^*)$ where x_k^* is any point in the kth subinterval. Thus, from (1), the work done over the kth subinterval is approximately

$$W_k \approx F(x_k^*) \, \Delta x_k$$

and the work done over the entire interval $[a, b]$ is approximately

$$W = \sum_{k=1}^{n} W_k \approx \sum_{k=1}^{n} F(x_k^*) \, \Delta x_k$$

If we now increase the number of subintervals in such a way that $\max \Delta x_k \to 0$, then intuition suggests that our approximations will tend to get better; hence,

$$W = \lim_{\max \Delta x_k \to 0} \sum_{k=1}^{n} F(x_k^*) \, \Delta x_k$$

Since the limit on the right is just the definite integral

$$\int_a^b F(x) \, dx$$

the following definition is suggested.

6.2.2   DEFINITION   If an object moves in the positive direction over the interval $[a, b]$ while subjected to a variable force $F(x)$ in the direction of motion, then the **work** done by the force is

$$W = \int_a^b F(x) \, dx \qquad (2)$$

Hooke's law [Robert Hooke (1635–1703)-English physicist] states that a spring stretched x units beyond its equilibrium position pulls back with a force

$$F(x) = kx$$

where k is a constant (called the ***spring constant***). The value of k depends on such factors as the thickness of the spring, the material used in its composition, and the units of force and distance.

▶ Example 2 A spring whose natural length is 24 in. exerts a force of 5 lb when stretched 10 in. beyond its natural length.

(a) Find the spring constant k.

(b) How much work is required to stretch the spring from its natural length to a length of 42 in.?

Solution (*a*). From Hooke's law,

$$F(x) = kx$$

From the data, $F(x) = 5$ lb when $x = 10$ in. so that

$$5 = k \cdot 10$$

Thus, the spring constant is

$$k = \frac{5}{10} = \frac{1}{2}$$

This means that the force $F(x)$ required to stretch the spring x in. is

$$F(x) = \frac{1}{2}x \tag{3}$$

Solution (*b*). Place the spring along a coordinate line as shown in Figure 6.6.2.

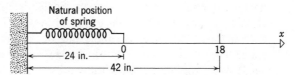

Figure 6.6.2

We want to find the work W required to stretch the spring over the interval from $x = 0$ to $x = 18$. From (2) and (3) the work W required is

$$W = \int_a^b F(x)\,dx = \int_0^{18} \frac{1}{2}x\,dx$$

$$= \frac{x^2}{4}\Bigg]_0^{18} = 81 \text{ in-lb} \qquad \blacktriangleleft$$

▶ Example 3   A cylindrical water tank of radius 10 ft and height 30 ft is half filled with water. How much work is required to pump all the water over the upper rim of the tank?

Solution.   Introduce a coordinate line as shown in Figure 6.6.3. Imagine the water to be divided into n thin layers with thicknesses

$$\Delta x_1, \Delta x_2, \ldots, \Delta x_n$$

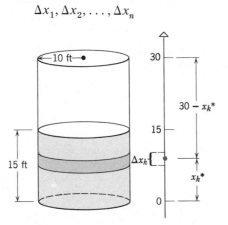

Figure 6.6.3

The force required to move the kth layer equals the weight of the layer, which can be found by multiplying its volume by the density of water (62.4 lb/ft³). Since the kth layer is a cylinder of radius $r = 10$ ft and height Δx_k, the force required to move it is

$$\begin{bmatrix} \text{Force to move} \\ \text{the } k\text{th layer} \end{bmatrix} = (\pi r^2 \, \Delta x_k) \cdot [\text{density of water}]$$

$$= (\pi (10)^2 \, \Delta x_k)(62.4)$$

$$= 6240\pi \, \Delta x_k$$

Because the kth layer has a finite thickness, the upper and lower surfaces are at different distances from the origin. However, if the layer is thin, the difference in these distances is small, and we can reasonably assume that the entire layer is concentrated at a single distance x_k^* from the origin (Figure 6.6.3). With this assumption, the work W_k required to pump the kth layer over the rim will be approximately

$$W_k \approx \underbrace{(30 - x_k^*)}_{\text{distance}} \cdot \underbrace{6240\pi \, \Delta x_k}_{\text{force}}$$

and the work W required to pump all n layers will be approximately

$$W = \sum_{k=1}^{n} W_k \approx \sum_{k=1}^{n} (30 - x_k^*)(6240\pi) \, \Delta x_k$$

To find the *exact* value of the work we take the limit as $\max \Delta x_k \to 0$. This yields

$$W = \lim_{\max \Delta x_k \to 0} \sum_{k=1}^{n} (30 - x_k^*)(6240\pi)\, \Delta x_k$$

$$= \int_0^{15} (30 - x)(6240\pi)\, dx$$

$$= 6240\pi \left(30x - \frac{x^2}{2}\right)\Bigg]_0^{15}$$

$$= 2{,}106{,}000\pi \ \text{ft-lb}$$

◀

▶ Exercise Set 6.6

1. Find the work done when
   (a) a constant force of 30 lb along the x-axis moves an object from $x = -2$ to $x = 5$ in.;
   (b) a variable force of $F(x) = 1/x^2$ lb along the x-axis moves an object from $x = 1$ to $x = 6$ in.

2. A spring whose natural length is 15 in. exerts a force of 45 lb when stretched to a length of 20 in.
   (a) Find the spring constant.
   (b) Find the work done in stretching the spring 3 in. beyond its natural length.
   (c) Find the work done in stretching the spring from a length of 20 in. to a length of 25 in.

3. A spring exerts a force of $\frac{1}{2}$ ton when stretched 5 ft beyond its natural length. How much work is required to stretch the spring 6 ft beyond its natural length?

4. Assume a force of 6 N (newtons) is required to compress a spring from a natural length of 4 m (meters) to a length of $3\frac{1}{2}$ m. Find the work required to compress the spring from its natural length to a length of 2 m. (Hooke's law applies to compression as well as extension.)

5. Assume 10 ft-lb of work is required to stretch a spring 1 ft beyond its natural length. What is the spring constant?

6. A cylindrical tank of radius 5 ft and height 9 ft is two-thirds filled with water. Find the work required to pump all the water over the upper rim.

7. Solve Problem 6 assuming that the tank is two-thirds filled with a liquid of density ρ lb/ft^3.

8. A cone-shaped water reservoir is 20 ft in diameter across the top and 15 ft deep. If the reservoir is filled to a depth of 10 ft, how much work is required to pump all the water to the top of the reservoir?

9. A swimming pool is built in the shape of a rectangular parallelepiped 10 ft deep, 15 ft wide, and 20 ft long.
   (a) If the pool is filled 1 ft below the top, how much work is required to pump all the water into a drain at the top edge of the pool?
   (b) If a one-horsepower motor can do 550 ft-lb of work per second, what size motor is required to empty the pool in one hour?

10. A water tower in the shape of a hemisphere of radius 10 ft is to be filled by pumping water over the top edge from a lake 200 ft below the top. How much work is required to fill the tank? [*Hint:* Solve without calculus.]

11. A 100-ft length of steel chain weighing 15 lb/ft is dangling from a pulley. How much work is required to wind the chain onto the pulley?

12. A rocket weighing 3 tons is filled with 40 tons of liquid fuel. In the initial part of the flight fuel is burned off at a constant rate of 2 tons per 1000 ft of vertical height. How much work is done in lifting the rocket to 3000 ft?

13. (Satellite Problem) The weight of an object is the force exerted on it by the earth's gravity. Thus, if a person weighs 100 lb on the surface of the earth, the earth's gravity is pulling on that person with a force of 100 lb. It is a fundamental law of physics that the force the earth exerts on an object varies inversely as the square of its distance from the earth's center. Thus an object's weight $F(x)$ is related to its distance x from the earth's center by a formula of the form

$$F(x) = k/x^2$$

where k is a constant of proportionality depending on the mass of the object and the units of force and distance.

(a) Assuming that the earth is a sphere of radius 4000 mi, find the constant k in the formula above for a satellite that weighs 6000 lb on the earth's surface.

(b) How much work must be performed to lift this satellite to an orbital position 1000 miles above the earth's surface?

14. (Coulomb's law) It follows from Coulomb's law in physics that two like electrostatic charges repel each other with a force inversely proportional to the square of the distance between them. Suppose that two charges A and B repel with a force of k pounds when they are positioned at points $A(-a, 0)$ and $B(a, 0)$. Find the work W required to move charge A along the x-axis to the origin if charge B remains stationary.

6.7 LIQUID PRESSURE AND FORCE

In this section we use the definite integral to calculate the force exerted by a liquid on a submerged surface.

If a flat surface of area A is submerged horizontally at a depth h in a container of fluid, then the weight of the fluid above exerts a force F on the surface given by

$$F = \rho h A \qquad (1)$$

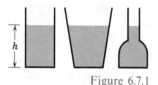

Figure 6.7.1

where ρ is the density of the fluid (weight per unit volume). It is a physical fact that the force F in (1) does not depend on the shape or size of the container. Thus, if the three containers in Figure 6.7.1 have bases of the same area and are filled with a fluid to the same depth, h, then each container will have the same fluid force on its base.

The units in (1) must be compatible. For example, if height is in feet and force in pounds, then the density ρ must be expressed in pounds per cubic foot. The density of water is approximately 62.4 lb/ft³.

Pressure is defined to be force per unit area. Thus, from (1), the pressure p exerted at each point of a flat surface of area A submerged horizontally at a depth h is given by

$$p = \frac{F}{A} = \rho h \qquad (2)$$

▶ **Example 1** If a flat circular plate of radius $r = 2$ ft is submerged horizontally in water so the top surface is at a depth of 3 ft, then the force on the top surface of the plate is

$$F = \rho h A = \rho h(\pi r^2) = (62.4)(3)(4\pi) = 748.8\pi \text{ lb}$$

and the pressure at each point on the plate surface is

$$p = \rho h = (62.4)(3) = 187.2 \text{ lb/ft}^2 \qquad \blacktriangleleft$$

Pascal's* principle in physics states that *fluid pressure is the same in all directions.* Thus, if a flat surface is submerged vertically (or at any angle at all), the pressure at a point of depth h is exactly the same as the pressure at a point of depth h on a horizontal surface (Figure 6.7.2). However, (1) cannot be applied to obtain the total force acting on a flat surface that is not submerged horizontally, because the depth h can vary from point to point on the surface. It is in such problems that calculus comes into play.

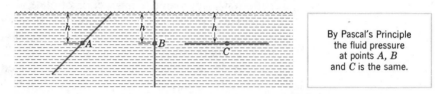

Figure 6.7.2

Suppose we want to compute the total fluid force against a submerged vertical surface. As shown in Figure 6.7.3, let us introduce a vertical x-axis whose positive direction is downward and whose origin is at any convenient point. As indicated in Figure 6.7.3a, let us assume that the submerged portion of the plate extends from $x = a$ to $x = b$ on the x-axis. Also, suppose that a point x on the axis lies $h(x)$ units below the surface and that the cross section of the plate at x has width $w(x)$.

Next, divide the interval $[a, b]$ into n subintervals with lengths

$$\Delta x_1, \Delta x_2, \ldots, \Delta x_n$$

---

* BLAISE PASCAL (1623–1662) French mathematician and scientist. Pascal's mother died when he was three years old and his father, a highly educated magistrate, personally provided the boy's early education. Although Pascal showed an inclination for science and mathematics, his father refused to tutor him in those subjects until he mastered Latin and Greek. Pascal's sister and primary biographer claimed that he independently discovered the first thirty-two propositions of Euclid without ever reading a book on geometry. (However, it is generally agreed that the story is apocryphal.) Nevertheless, the precocious Pascal published a highly respected essay on conic sections by the time he was sixteen years old. Descartes, who read the essay, thought it so brilliant that he could not believe that it was written by such a young man. By age 18 his health began to fail and until his death he was in frequent pain. However, his creativity was unimpaired.

Pascal's contributions to physics include the discovery that air pressure decreases with altitude and the principle of fluid pressure that bears his name. However, the originality of his work is questioned by some historians. Pascal made major contributions to a branch of mathematics called "projective geometry," and he helped to develop probability theory through a series of letters with Fermat.

In 1646, Pascal's health problems resulted in a deep emotional crisis that led him to become increasingly concerned with religious matters. Although born a Catholic, he converted to a religious doctrine called Jansenism, and spent most of his final years writing on religion and philosophy.

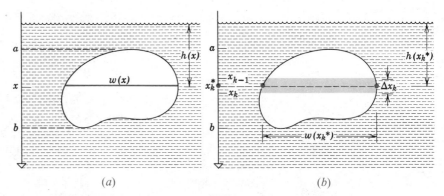

Figure 6.7.3 (a) (b)

and in each subinterval choose an arbitrary point x_k^*. Following a familiar pattern, we approximate the section of plate along the kth subinterval by a rectangle of length $w(x_k^*)$ and width Δx_k^* (Figure 6.7.3b). Because the upper and lower edges of this rectangle are at different depths, (1) cannot be used to calculate the force on this rectangle. However, if Δx_k is small, the difference in depth between the upper and lower edges is small, and we can reasonably assume that the entire rectangle is concentrated at a single depth $h(x_k^*)$ below the surface. With this assumption, (1) can be used to *approximate* the force F_k on the kth rectangle. We obtain

$$F_k \approx \underbrace{\rho h(x_k^*)}_{\text{depth}} \cdot \underbrace{w(x_k^*)\,\Delta x_k}_{\text{area of rectangle}}$$

Thus, the total force F on the plate is approximately

$$F = \sum_{k=1}^{n} F_k \approx \sum_{k=1}^{n} \rho h(x_k^*)w(x_k^*)\,\Delta x_k$$

To find the *exact* value of the force we take the limit as $\max \Delta x_k \to 0$. This yields

$$F = \lim_{\max \Delta x_k \to 0} \sum_{k=1}^{n} \rho h(x_k^*)w(x_k^*)\,\Delta x_k = \int_a^b \rho h(x)w(x)\,dx$$

which suggests the following definition.

6.7.1 DEFINITION   Assume that a plate is immersed vertically in a liquid of density ρ and that the submerged portion extends from $x = a$ to $x = b$ on a vertical x-axis. For $a \le x \le b$, let $w(x)$ be the width of the plate at x and let $h(x)$ be the depth of the point x. Then the total **fluid force** on the plate is

$$F = \int_a^b \rho h(x)w(x)\,dx \tag{3}$$

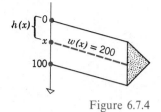

Figure 6.7.4

▶ **Example 2** The face of a dam is a vertical rectangle of height 100 ft and width 200 ft (Figure 6.7.4). Find the total fluid force exerted on the face when the water surface is level with the top of the dam.

Solution. Introduce an x-axis with origin at the water surface as shown in Figure 6.7.4. At a point x on this axis, the width of the dam in feet is $w(x) = 200$ and the depth in feet is $h(x) = x$. Thus, from (3) with $\rho = 62.4$ lb/ft³ (the density of water) we obtain as the total force on the face

$$F = \int_0^{100} (62.4)(x)(200)\, dx$$

$$= 12,480 \int_0^{100} x\, dx$$

$$= 12,480 \left. \frac{x^2}{2} \right]_0^{100}$$

$$= 62,400,000 \text{ lb}$$ ◀

▶ **Example 3** A plate in the form of an isosceles triangle with base 10 ft and altitude 4 ft is submerged vertically in oil as shown in Figure 6.7.5a. Find the fluid force F against a surface of the plate if the oil has density $\rho = 30$ lb/ft³.

Figure 6.7.5   (a)   (b)

Solution. Introduce an x-axis as shown in Figure 6.7.5b. By similar triangles, the width of the plate at a depth of $h(x) = 3 + x$ ft satisfies

$$\frac{w(x)}{10} = \frac{x}{4}$$

so

$$w(x) = \frac{5}{2}x$$

Thus, it follows from (3) that the force on the plate is

$$F = \int_a^b \rho h(x) w(x)\, dx = \int_0^4 (30)(3 + x)\left(\frac{5}{2}x\right) dx$$

$$= 75 \int_0^4 (3x + x^2)\, dx = 75 \left[\frac{3x^2}{2} + \frac{x^3}{3} \right]_0^4 = 3400 \text{ lb}$$ ◀

▶ Exercise Set 6.7

1. A flat square plate with 3-ft sides and negligible thickness is submerged horizontally in a liquid. Find the force and pressure on a surface of the plate if
   (a) the liquid is water and the plate is at a depth of 5 ft;
   (b) the liquid has density 40 lb/ft³ and the plate is at a depth of 10 ft.

In Exercises 2–7, the flat surfaces shown are submerged vertically in water. Find the fluid force against the surface.

2.

3.

4.

5.

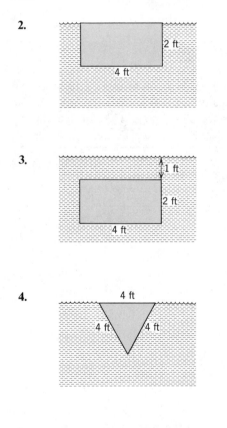

6.

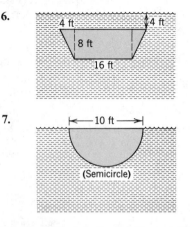

7.

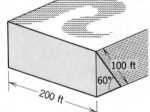

(Semicircle)

8. An oil tank is shaped like a right-circular cylinder of diameter 4 ft. Find the total fluid force against one end when the axis is horizontal and the tank is half filled with oil of density 50 lb/ft³.

9. A square plate of side a ft is dipped in a liquid of density ρ lb/ft³. Find the fluid force on the plate if a vertex is at the surface and a diagonal is perpendicular to the surface.

10. Figure 6.7.6 shows a dam whose face is an inclined rectangle. Find the fluid force on the face when the water is level with the top of this dam.

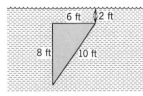

Figure 6.7.6

11. Figure 6.7.7 shows a rectangular swimming pool whose bottom is an inclined plane. Find the fluid force on the bottom when the pool is filled to the top.

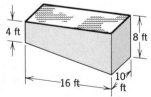

Figure 6.7.7

12. An observation window on a submarine is a square with 2-ft sides. Using ρ_0 for the density of sea water, find the fluid force on the window when the submarine has descended vertically so that the top of the window is at a depth of h feet.

13. (a) Show: If the submarine in Exercise 12 descends vertically at a constant rate, then the fluid force on the window increases at a constant rate.

(b) At what rate is the force on the window increasing if the submarine is descending vertically at 20 ft/min?

14. Find the fluid force on the lateral surface of a right-circular cylinder of base radius r and altitude h when the cylinder is submerged so its top is level with the surface of the water.

► SUPPLEMENTARY EXERCISES

In Exercises 1–3, set up, but do not evaluate, an integral or sum of integrals that gives the area of the region R. (Set up the integral with respect to x or y as directed.)

1. R is the region in the first quadrant enclosed by $y = x^2$, $y = 2 + x$, and $x = 0$.
   (a) Integrate with respect to x.
   (b) Integrate with respect to y.

2. R is enclosed by $x = 4y - y^2$ and $y = \frac{1}{2}x$.
   (a) Integrate with respect to x.
   (b) Integrate with respect to y.

3. R is enclosed by $x = 9$ and $x = y^2$.
   (a) Integrate with respect to x.
   (b) Integrate with respect to y.

In Exercises 4–9, set up, but do not evaluate, an integral or sum of integrals that gives the stated volume. (Set up the integral with respect to x or y as directed.)

4. The volume generated by revolving the region in Exercise 1 about the x-axis.
   (a) Integrate with respect to x.
   (b) Integrate with respect to y.

5. The volume generated by revolving the region in Exercise 1 about the y-axis.
   (a) Integrate with respect to x.
   (b) Integrate with respect to y.

6. The volume generated by revolving the region in Exercise 2 about the x-axis.
   (a) Integrate with respect to x.
   (b) Integrate with respect to y.

7. The volume generated by revolving the region in Exercise 2 about the y-axis.

(a) Integrate with respect to x.
(b) Integrate with respect to y.

8. The volume generated by revolving the region in Exercise 3 about the x-axis.
   (a) Integrate with respect to x.
   (b) Integrate with respect to y.

9. The volume generated by revolving the region in Exercise 3 about the y-axis.
   (a) Integrate with respect to x.
   (b) Integrate with respect to y.

In Exercises 10 and 11, find:
   (a) the area of the region described;
   (b) the volume generated by revolving the region about the indicated line.

10. The region in the first quadrant enclosed by $y = \sin x$, $y = \cos x$, and $x = 0$; revolved about the x-axis.
    [*Hint:* $\cos^2 x - \sin^2 x = \cos 2x$]

11. The region enclosed by the x-axis, the y-axis, and $x = \sqrt{4 - y}$; revolved about the y-axis.

12. Set up a sum of definite integrals that represents the total shaded area between the curves $y = f(x)$ and $y = g(x)$ below.

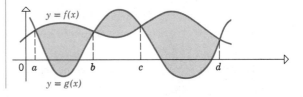

13. Find the *total* area bounded between $y = x^3$ and $y = x$ over the interval $[-1, 2]$. (See previous exercise.)

14. Find the volume of the solid whose base is the region bounded between the curves $y = x$ and $y = x^2$, and whose cross sections perpendicular to the x-axis are squares.

15. Find the volume of the solid whose base is the triangular region with vertices $(0, 0)$, $(a, 0)$, and $(0, b)$, where $a > 0$ and $b > 0$, and whose cross sections perpendicular to the y-axis are semicircles.

In Exercises 16 and 17, find the volume generated by revolving the region described about the axis indicated.

16. The region bounded above by the curve $y = \cos x^2$, on the left by the y-axis, and below by the x-axis; revolved about the y-axis.

17. The region bounded by $y = \sqrt{x}$, $x = 4$, and $y = 0$; revolved about: (a) the line $x = 4$; (b) the line $y = 2$.

18. A football has the shape of the solid generated by revolving the region bounded between the x-axis and the parabola $y = 4R(x^2 - \frac{1}{2}L^2)/L^2$ about the x-axis. Find its volume.

In Exercises 19–22, find the arc length of the indicated curve.

19. $8y^2 = x^2$ between $(0, 0)$ and $(2, 1)$.

20. $y = \frac{1}{3}(x^2 + 2)^{3/2}$, $0 \le x \le 3$.

21. $y = \frac{1}{10}x^5 + \frac{1}{6}x^{-3}$, $1 \le x \le 2$.

22. $y = \frac{1}{3}x^3 + \frac{1}{4x}$, $1 \le x \le 2$.

In Exercises 23–28, find the area of the surface generated by revolving the given curve about the indicated axis.

23. $y = x^3$ between $(1, 1)$ and $(2, 8)$; x-axis.

24. $y^2 = 12x$ between $(0, 0)$ and $(3, 6)$; x-axis.

25. $y = \frac{2}{3}x^{3/2} - \frac{1}{2}x^{1/2}$ between $(0, 0)$ and $(9, 33/2)$; y-axis.

26. The curve in Exercise 25 revolved about the line $x = 9$.

27. $3y = \sqrt{x}(3 - x)$ between $(0, 0)$ and $(3, 0)$; y-axis.

28. $y = \sqrt{2x - x^2}$ between $(\frac{1}{2}, \sqrt{3}/2)$ and $(1, 1)$; x-axis.

29. The natural length of a spring is 6 in. If a force of 2 lb is needed to hold it at a length of 10 in., find the work done in stretching it from 8 in. to 10 in.

30. Find the spring constant if 180 in.-lbs of work are required to stretch a spring 3 in. from its natural length.

31. A 250-lb weight is suspended from a ledge by a uniform 40-ft cable weighing 30 lb. How much work is required to bring the weight up to the ledge?

32. A tank in the shape of a right-circular cone has a 6-ft diameter at the top and a height of 5 ft. It is filled with a liquid of density 64 lb/ft³. How much work can be done by the liquid if it runs out of the bottom of the tank?

33. A vessel has the shape obtained by revolving about the y-axis the part of the parabola $y = 2(x^2 - 4)$ lying below the x-axis. If x and y are in feet, how much work is required to pump all the water in the full vessel to a point 4 ft above its top?

34. Two like magnetic poles repel each other with a force $F = k/x^2$ newtons. Express the work needed to move them along a line from D meters apart to $D/3$ meters apart.

In Exercises 35 and 36, the flat surface shown is submerged vertically in a liquid of density ρ lb/ft³. Find the fluid force against the surface.

35.

36.

(Semicircle)

7 logarithm and exponential functions

7.1 INTRODUCTION

In spite of its seeming simplicity, the integral

$$\int \frac{1}{x}\,dx$$

cannot be evaluated in terms of polynomials, rational functions, or trigonometric functions alone. The same difficulty occurs with such basic integrals as

$$\int \tan x\,dx \quad \text{and} \quad \int \sec x\,dx$$

It is the primary purpose of this chapter to define two new functions, the *natural logarithm function* and *natural exponential function,* which will enable us to evaluate these and other important integrals. We will discover that these new functions have important applications to physics, engineering, biology, and economics.

There are two different ways to define the natural logarithm function, one using properties of exponents and the other using an integral formula. The approach using exponents is attractive because it parallels the definition of logarithms studied in high school. However, when it comes to obtaining properties such as continuity and differentiability, the approach via the integral formula is much less cumbersome.

In Section 7.2 we will review the basic properties of logarithms, explain informally how the natural logarithm can be defined in terms of exponents, and then formally define the natural logarithm as a definite integral. In Section 7.3 we will use the integral definition to develop properties of the natural logarithm, and then in Section 7.4 we will show that the integral and exponent definitions of the natural logarithm are equivalent.

7.2   THE NATURAL LOGARITHM

Recall from algebra that a logarithm is an exponent. More precisely, if b is a positive number, then the symbol

$$\log_b x$$

(read, "the logarithm to the base b of x") represents that power to which b must be raised to produce x. Thus,

$$\log_{10} 100 = 2$$

because 10 must be raised to the second power to produce 100. Similarly,

$$\log_2 8 = 3 \qquad \text{since} \qquad 2^3 = 8$$
$$\log_{10} \frac{1}{1000} = -3 \qquad \text{since} \qquad 10^{-3} = \frac{1}{1000}$$
$$\log_{10} 1 = 0 \qquad \text{since} \qquad 10^0 = 1$$
$$\log_3 81 = 4 \qquad \text{since} \qquad 3^4 = 81$$

In general,

$$y = \log_b x \qquad \text{and} \qquad x = b^y$$

are equivalent statements.*

It is important to note that for every base b, $\log_b x$ is only defined for $x > 0$. This is because $y = \log_b x$ is equivalent to $x = b^y$, and $b^y > 0$ for all real values of y (since b is positive).

The reader should already be familiar with the properties of logarithms listed in the following theorem.

7.2.1   THEOREM (a) $\log_b 1 = 0$

(b) $\log_b ac = \log_b a + \log_b c$

(c) $\log_b \dfrac{a}{c} = \log_b a - \log_b c$

(d) $\log_b a^r = r \log_b a$

(e) $\log_b \dfrac{1}{c} = -\log_b c$

We will prove (a) and (b) and leave the remaining proofs as exercises.

---

*In the expression $x = b^y$, the exponent y is intended to be any real number. However, the meaning of an irrational exponent in expressions such as 2^π and $3^{\sqrt{2}}$ must be carefully defined to make the exponent definition of a logarithm complete and rigorous. It is precisely this difficulty that leads mathematicians to consider the alternate definition of a logarithm that we will introduce shortly.

Proof of (a). Since $b^0 = 1$, it follows that $\log_b 1 = 0$.

Proof of (b). Let

$$x = \log_b a \quad \text{and} \quad y = \log_b c \tag{1}$$

so

$$b^x = a \quad \text{and} \quad b^y = c$$

Therefore,

$$ac = b^x b^y = b^{x+y}$$

or equivalently,

$$\log_b ac = x + y$$

Thus, from (1)

$$\log_b ac = \log_b a + \log_b c \quad\blacksquare$$

The most important logarithms are those with base 10, called *common logarithms,* and those with the base $e \approx 2.71828 \ldots$, called *natural logarithms.* The base e is an irrational number which, as we will prove later, can be expressed as

$$e = \lim_{x \to +\infty} \left(1 + \frac{1}{x}\right)^x \tag{2}$$

On some hand calculators an approximate value of e can be obtained by depressing a single key. However, e can also be approximated from (2) by computing

$$\left(1 + \frac{1}{x}\right)^x$$

for large values of x—the greater the value of x, the more accurate the approximation. Some sample computations, generated with a hand calculator, are shown in Table 7.2.1. The last value in the table is correct to five decimal places.

It has become fairly standard to denote a natural logarithm by the symbol ln (read "ell-en"). Thus,

$$\log_e x \quad \text{and} \quad \ln x$$

both denote the natural logarithm of x. In words, $\ln x$ can be interpreted as that power to which e must be raised to produce x.

Table 7.2.1

x	$1 + \dfrac{1}{x}$	$\left(1 + \dfrac{1}{x}\right)^{x}$
1	2	2.00000
10	1.1	2.59374
100	1.01	2.70481
1000	1.001	2.71692
10,000	1.0001	2.71815
100,000	1.00001	2.71827
1,000,000	1.000001	2.71828

▶ Example 1

$$\ln 1 = 0 \qquad \text{(since } e^0 = 1\text{)}$$
$$\ln e = 1 \qquad \text{(since } e^1 = e\text{)}$$
$$\ln \frac{1}{e} = -1 \qquad \left(\text{since } e^{-1} = \frac{1}{e}\right)$$
$$\ln (e^2) = 2 \qquad\qquad\qquad\qquad ◀$$

For reference, values of the natural logarithm are given in Table 3 of Appendix 3.

We are now ready to give a precise definition of the natural logarithm in terms of an integral.

7.2.2 DEFINITION
The Natural Logarithm Function

The **natural logarithm** function is defined by the formula

$$\ln x = \int_{1}^{x} \frac{1}{t}\, dt, \qquad x > 0 \tag{3}$$

REMARK. We will show in Section 7.4 that $\ln x$, as defined by (3), is equal to $\log_e x$.

Definition 7.2.2 and the Second Fundamental Theorem of Calculus (Theorem 5.10.1) imply that $\ln x$ is an antiderivative of $1/x$ on $(0, +\infty)$; that is,

$$\frac{d}{dx}[\ln x] = \frac{1}{x}, \qquad x > 0 \tag{4}$$

If $u(x) > 0$, and if the function u is differentiable at x, then it follows from the chain rule (Formula (8) in Section 3.5) that

$$\frac{d}{dx}[\ln u] = \frac{1}{u} \cdot \frac{du}{dx} \tag{5}$$

▶ Example 2 Find

$$\frac{d}{dx}[\ln (x^2 + 1)]$$

Solution. From (5) with $u = x^2 + 1$,

$$\frac{d}{dx}[\ln (x^2 + 1)] = \frac{1}{x^2 + 1} \cdot \frac{d}{dx}[x^2 + 1] = \frac{1}{x^2 + 1} \cdot 2x = \frac{2x}{x^2 + 1} \quad ◄$$

▶ Example 3 Find

$$\frac{d}{dx}[\ln |x|]$$

Solution. We consider the cases where $x > 0$ and $x < 0$ separately. [The case where $x = 0$ is excluded, since ln 0 is undefined; see (3).] If $x > 0$, then $|x| = x$, so that

$$\frac{d}{dx}[\ln |x|] = \frac{d}{dx}[\ln x] = \frac{1}{x}$$

If $x < 0$, then $|x| = -x$, so that from (5) with $u = -x$,

$$\frac{d}{dx}[\ln |x|] = \frac{d}{dx}[\ln (-x)] = \frac{1}{(-x)} \cdot \frac{d}{dx}[-x] = \frac{1}{x}$$

Thus,

$$\frac{d}{dx}[\ln |x|] = \frac{1}{x} \tag{6}$$

if $x \neq 0$. ◄

Formula (6) states that the function $\ln |x|$ is an antiderivative of $1/x$ everywhere except at $x = 0$. In contrast, the function $\ln x$ is an antiderivative of $1/x$ only for $x > 0$.

The companion integration formula for (6) is

$$\int \frac{1}{x} dx = \ln |x| + C \tag{7}$$

▶ Example 4 From (6) and the chain rule,

$$\frac{d}{dx}[\ln |\sin x|] = \frac{1}{\sin x} \cdot \frac{d}{dx}[\sin x] = \frac{\cos x}{\sin x} = \cot x \quad ◄$$

▶ **Example 5** Evaluate

$$\int \frac{3x^2}{x^3 + 5} \, dx$$

Solution. Make the substitution

$$u = x^3 + 5 \qquad du = 3x^2 \, dx$$

so that

$$\int \frac{3x^2}{x^3 + 5} \, dx = \int \frac{1}{u} \, du$$
$$= \ln |u| + C \qquad \text{[Formula (7)]}$$
$$= \ln |x^3 + 5| + C \qquad \qquad ◀$$

REMARK. This example illustrates an important point. Any integral of the form

$$\int \frac{g'(x)}{g(x)} \, dx$$

(where the numerator of the integrand is the derivative of the denominator) can be evaluated by the *u*-substitution $u = g(x)$, $du = g'(x) \, dx$, since this substitution yields

$$\int \frac{g'(x)}{g(x)} \, dx = \int \frac{du}{u} = \ln |u| + C = \ln |g(x)| + C$$

▶ **Example 6** Evaluate

$$\int \tan x \, dx$$

Solution.

$$\int \tan x \, dx = \int \frac{\sin x}{\cos x} \, dx$$
$$= -\int \frac{1}{u} \, du \qquad \begin{bmatrix} u = \cos x \\ du = -\sin x \, dx \end{bmatrix}$$
$$= -\ln |u| + C$$
$$= -\ln |\cos x| + C \qquad \qquad ◀$$

▶ Exercise Set 7.2

1. Without using a calculator or tables, find the exact value of the following logarithms.
   (a) $\log_2 16$ (b) $\log_2 (\frac{1}{32})$ (c) $\log_4 4$
   (d) $\log_9 3$ (e) $\log_{10} (.001)$ (f) $\log_{10} (10^4)$
   (g) $\ln (e^3)$ (h) $\ln (\sqrt{e})$.

In Exercises 2–11, solve for x without using a calculator or tables.

2. $\log_{10} (1 + x) = 3$.

3. $\log_{10} (\sqrt{x}) = -1$.

4. $\ln (x^2) = 4$.

5. $\ln (1/x) = -2$.

6. $\log_3 (3^x) = 7$.

7. $\log_5 (5^{2x}) = 8$.

8. $\log_{10} x^2 + \log_{10} x = 30$.

9. $\log_{10} x^{3/2} - \log_{10} \sqrt{x} = 5$.

10. $\ln 4x - 3 \ln (x^2) = \ln 2$.

11. $\ln (1/x) + \ln (2x^3) = \ln 3$.

In Exercises 12–30, find dy/dx.

12. $y = 4 \ln x$.

13. $y = \ln 2x$.

14. $y = \ln (x^3)$.

15. $y = (\ln x)^2$.

16. $y = \ln (\sin x)$.

17. $y = \ln |\tan x|$.

18. $y = \ln (2 + \sqrt{x})$.

19. $y = \ln \left(\dfrac{x}{1 + x^2} \right)$.

20. $y = \ln (\ln x)$.

21. $y = \ln |x^3 - 7x^2 - 3|$.

22. $y = x^3 \ln x$.

23. $y = \sqrt{\ln x}$.

24. $y = \cos (\ln x)$.

25. $y = \sin \left(\dfrac{5}{\ln x} \right)$.

26. $y = x[\ln (x^2 - 2x)]^3$.

27. $y = (x^2 + 1)[\ln (x^2 + 1)]^2$.

28. $y = \dfrac{\ln x}{1 + \ln x}$.

29. $y = \dfrac{x^2}{1 + \ln x}$.

30. $y = \ln \left| \dfrac{1 - \cos \pi x}{1 + \cos \pi x} \right|$.

31. Find dy/dx by implicit differentiation if $y + \ln xy = 1$.

32. Find dy/dx by implicit differentiation if $y = \ln (x \tan y)$.

In Exercises 33–44, evaluate the indefinite integrals.

33. $\displaystyle \int \frac{dx}{2x}$.

34. $\displaystyle \int \frac{5x^4}{x^5 + 1} dx$.

35. $\displaystyle \int \frac{x^2}{x^3 - 4} dx$.

36. $\displaystyle \int \frac{t + 1}{t} dt$.

37. $\displaystyle \int \frac{\sec^2 x}{\tan x} dx$.

38. $\displaystyle \int \cot x \, dx$.

39. $\displaystyle \int \frac{\sin 3\theta}{1 + \cos 3\theta} d\theta$.

40. $\displaystyle \int \frac{dx}{x \ln x}$.

41. $\displaystyle \int \frac{x^3}{x^2 + 1} dx$.

42. $\displaystyle \int \frac{1}{x} \cos (\ln x) \, dx$.

43. $\displaystyle \int \frac{1}{y} (\ln y)^3 \, dy$.

44. $\displaystyle \int \frac{dx}{\sqrt{x}(1 - 2\sqrt{x})}$.

45. Prove:
    (a) $b^{\log_b x} = x$ for $x > 0$
    (b) $\log_b (b^x) = x$ for $-\infty < x < +\infty$.

46. (a) Prove part (c) of Theorem 7.2.1.
    (b) Prove part (d) of Theorem 7.2.1.
    (c) Prove part (e) of Theorem 7.2.1.

47. Prove the following change of base formula for logarithms:

$$\log_b a = \frac{1}{\log_a b} \qquad \text{if } a > 0 \text{ and } b > 0$$

48. The formulas

$$\int \frac{dx}{x} = \ln x + C \quad \text{and} \quad \int \frac{dx}{x} = \ln |x| + C$$

are both correct. What advantage does the second formula have over the first?

49. Magnitudes of earthquakes are measured using the **Richter scale**. On this scale the magnitude R of an earthquake is given by

$$R = \log_{10} \left(\frac{I}{I_0} \right)$$

where I_0 is a fixed standard intensity used for comparison, and I is the intensity of the earthquake being measured.
   (a) Show that if an earthquake measures $R = 3$ on the Richter scale, then its intensity I is 1000 times the standard, that is, $I = 1000I_0$.
   (b) The San Francisco earthquake of 1906 registered $R = 8.2$ on the Richter scale. Express its intensity in terms of the standard intensity.
   (c) How many times more intense is an earthquake measuring $R = 8$ than one measuring $R = 4$?

7.3   PROPERTIES OF THE NATURAL LOGARITHM

In this section we develop basic properties of the logarithm function and obtain the graph of $y = \ln x$.

Our first result shows that the natural logarithm (when defined as an integral) has the basic properties of logarithms listed in Theorem 7.2.1.

7.3.1   THEOREM   *For any positive numbers a and c and any rational number r,*

(a) $\ln 1 = 0$

(b) $\ln ac = \ln a + \ln c$

(c) $\ln \dfrac{a}{c} = \ln a - \ln c$

(d) $\ln a^r = r \ln a$

(e) $\ln \dfrac{1}{c} = -\ln c$

Proof of (a).   From Formula (3) of Section 7.2 with $x = 1$,

$$\ln 1 = \int_1^1 \frac{1}{t}\, dt = 0$$

Proof of (b).   Consider the function $f(x) = \ln ax$. Treating a as constant and differentiating with respect to x, we obtain

$$f'(x) = \frac{d}{dx}[\ln ax] = \frac{1}{ax} \cdot \frac{d}{dx}[ax] = \frac{1}{ax} \cdot a = \frac{1}{x}$$

which shows that $\ln ax$ and $\ln x$ have the same derivative on $(0, +\infty)$. Thus, by Corollary 4.10.4, there is a constant k such that

$$\ln ax - \ln x = k \tag{1}$$

If we let $x = 1$ in this equation and use the fact that $\ln 1 = 0$, we obtain

$$\ln a = k$$

so that (1) can be written

$$\ln ax - \ln x = \ln a$$

In particular, if $x = c$ we obtain

$$\ln ac - \ln c = \ln a$$

or

$$\ln ac = \ln a + \ln c$$

Proofs of (c) and (e). Using part (*a*), we can write

$$0 = \ln 1 = \ln\left(c \cdot \frac{1}{c}\right) = \ln c + \ln \frac{1}{c}$$

so that

$$\ln \frac{1}{c} = -\ln c$$

which proves (*e*). To prove (*c*) we write

$$\ln \frac{a}{c} = \ln\left(a \cdot \frac{1}{c}\right) = \ln a + \ln \frac{1}{c} = \ln a - \ln c$$

Proof of (d). The functions $\ln x^r$ and $r \ln x$ have the same derivative on $(0, +\infty)$, since

$$\frac{d}{dx}[\ln x^r] = \frac{1}{x^r} \cdot \frac{d}{dx}[x^r] = \frac{1}{x^r} \cdot rx^{r-1} = \frac{r}{x}$$

and

$$\frac{d}{dx}[r \ln x] = r\frac{d}{dx}[\ln x] = \frac{r}{x}$$

Thus, there is a constant k such that

$$\ln x^r - r \ln x = k \tag{2}$$

If we let $x = 1$ in this equation and use the fact that $\ln 1 = 0$, we obtain $k = 0$ (verify) so that (2) can be written

$$\ln x^r - r \ln x = 0 \qquad \text{or} \qquad \ln x^r = r \ln x$$

Letting $x = a$ completes the proof. ▨

▶ Example 1   Property (*d*) in the above theorem yields the following useful result:

$$\ln \sqrt[n]{a} = \ln a^{1/n} = \frac{1}{n} \ln a \qquad\qquad ◀$$

The values of $\ln x$ have useful interpretations as areas. If $x > 1$, then

$$\ln x = \int_1^x \frac{1}{t}\,dt$$

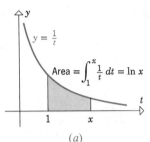

Area $= \int_{1}^{x} \frac{1}{t}\, dt = \ln x$

(a)

Area $= \int_{x}^{1} \frac{1}{t}\, dt = -\ln x$

(b)

Figure 7.3.1

can be viewed as the area under the curve $y = 1/t$ between the points 1 and x on the t-axis (Figure 7.3.1a). On the other hand, if $0 < x < 1$, then by writing

$$\ln x = -\int_{x}^{1} \frac{1}{t}\, dt$$

we can view $\ln x$ as the *negative* of the area under the curve $y = 1/t$ between the points x and 1 on the t-axis (Figure 7.3.1b). Therefore,

$$
\begin{aligned}
\ln x > 0 \quad &\text{if} \quad x > 1 \\
\ln x < 0 \quad &\text{if} \quad 0 < x < 1 \\
\ln x = 0 \quad &\text{if} \quad x = 1
\end{aligned}
$$

Since $\ln x$ is differentiable for $x > 0$, it is continuous on the interval $(0, +\infty)$. The first derivative

$$\frac{d}{dx}[\ln x] = \frac{1}{x}$$

is positive for all x in $(0, +\infty)$, so that $\ln x$ is *increasing* on this interval, and the second derivative

$$\frac{d^2}{dx^2}[\ln x] = \frac{d}{dx}\left[\frac{1}{x}\right] = -\frac{1}{x^2}$$

is negative for all x in $(0, +\infty)$, so that the graph of $\ln x$ is *concave down* on this interval.

To complete the picture of the graph, we must investigate the behavior of $\ln x$ as $x \to 0^+$ and $x \to +\infty$. For this purpose we will need an estimate of $\ln 2$. From the Mean-Value Theorem for integrals (5.9.4) we obtain

$$\ln 2 = \int_{1}^{2} \frac{1}{t}\, dt = (2 - 1) \cdot \frac{1}{t^*} = \frac{1}{t^*} \tag{3}$$

for some t^* satisfying

$$1 \le t^* \le 2 \tag{4}$$

Taking reciprocals in (4) yields

$$\frac{1}{2} \le \frac{1}{t^*} \le 1$$

so that from (3),

$$\frac{1}{2} \le \ln 2 \le 1 \tag{5}$$

We will now use this estimate to deduce the following results.

7.3.2 THEOREM (a) $\lim\limits_{x\to+\infty} \ln x = +\infty$

(b) $\lim\limits_{x\to 0^+} \ln x = -\infty$

Proof of (a). We will show that for any positive number N (no matter how large) the values of $\ln x$ eventually exceed N as $x \to +\infty$. It will then follow that $\lim\limits_{x\to+\infty} \ln x = +\infty$.

Since $\ln x$ is an increasing function, it follows from (5) that for $x > 2^{2N}$ we must have

$$\ln x > \ln 2^{2N} = 2N \ln 2 \geq 2N\left(\frac{1}{2}\right) = N$$

Therefore, $\lim\limits_{x\to+\infty} \ln x = +\infty$.

Proof of (b). If we let

$$v = \frac{1}{x}$$

then $v \to +\infty$ as $x \to 0^+$, so that from part (a) of this theorem and Theorem 7.3.1e we obtain

$$\lim_{x\to 0^+} \ln x = \lim_{x\to 0^+}\left(-\ln\frac{1}{x}\right) = \lim_{v\to+\infty}(-\ln v)$$
$$= -\lim_{v\to+\infty} \ln v = -\infty \quad\blacksquare$$

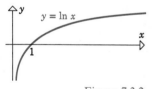

Figure 7.3.2

In light of this theorem and the fact that $y = \ln x$ is increasing and concave down on $(0, +\infty)$, the graph of $\ln x$ is as shown in Figure 7.3.2. The graph crosses the x-axis at $x = 1$ since $\ln 1 = 0$. A more accurate picture can be obtained by plotting some points and sketching some tangent lines. For this purpose we use the approximation,

$$\ln 2 \approx 0.6931$$

which is accurate to four decimal places. (We will show how to obtain this approximation later.) Using this value of $\ln 2$ and applying Theorem 7.3.1 we can estimate the natural logarithm of any power of 2. For example,

$$\ln 8 = \ln 2^3 = 3 \ln 2 \approx 2.0793$$
$$\ln 4 = \ln 2^2 = 2 \ln 2 \approx 1.3862$$
$$\ln \tfrac{1}{4} = \ln 2^{-2} = -2 \ln 2 \approx -1.3862$$
$$\ln \tfrac{1}{8} = \ln 2^{-3} = -3 \ln 2 \approx -2.0793$$
$$\ln \tfrac{1}{2} = -\ln 2 \approx -0.6931$$

In Figure 7.3.3 we have plotted these points and drawn some tangent lines

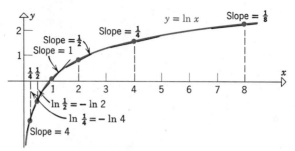

Figure 7.3.3

to obtain a more accurate sketch of the curve $y = \ln x$. The slopes of the tangent lines were obtained from the derivative of $\ln x$.

When possible, the properties of logarithms in Theorem 7.3.1 should be used to convert products, quotients, and exponents into sums, differences, and constant multiples *before* differentiating a function involving natural logarithms.

▶ Example 2

$$\frac{d}{dx}\left[\ln\left(\frac{x^2 \sin x}{\sqrt{1+x}}\right)\right] = \frac{d}{dx}\left[2\ln x + \ln(\sin x) - \frac{1}{2}\ln(1+x)\right]$$

$$= \frac{2}{x} + \frac{\cos x}{\sin x} - \frac{1}{2(1+x)}$$

$$= \frac{2}{x} + \cot x - \frac{1}{2+2x} \qquad \blacktriangleleft$$

We conclude this section by discussing a technique called ***logarithmic differentiation.***

▶ Example 3   The derivative of

$$y = \frac{x^2 \sqrt[3]{7x-14}}{(1+x^2)^4} \tag{6}$$

is messy to calculate directly. However, if we first take ln of both sides and use properties of the natural logarithm, we can write

$$\ln y = 2\ln x + \frac{1}{3}\ln(7x-14) - 4\ln(1+x^2)$$

Differentiating both sides with respect to x yields

$$\frac{1}{y}\frac{dy}{dx} = \frac{2}{x} + \frac{7/3}{7x-14} - \frac{8x}{1+x^2}$$

Thus, on solving for dy/dx and using (6) we obtain

$$\frac{dy}{dx} = \frac{x^2\sqrt[3]{7x - 14}}{(1 + x^2)^4}\left[\frac{2}{x} + \frac{1}{3x - 6} - \frac{8x}{1 + x^2}\right] \qquad \blacktriangleleft$$

Since $\ln y$ is defined only for $y > 0$, results obtained by logarithmic differentiation are valid only at points where this condition is satisfied. The derivative formula obtained in the last example is valid for $x > 2$ since $y > 0$ for such x [see (6)]. As the next example shows, logarithmic differentiation can sometimes be used with functions having negative values by first taking absolute values.

▶ Example 4 The derivative of

$$y = \frac{x\sqrt[3]{x - 5}}{1 + \sin^3 x}$$

can be obtained by writing

$$\ln |y| = \ln\left|\frac{x\sqrt[3]{x - 5}}{1 + \sin^3 x}\right|$$

$$\ln |y| = \ln |x| + \frac{1}{3}\ln |x - 5| - \ln |1 + \sin^3 x|$$

$$\frac{1}{y}\frac{dy}{dx} = \frac{1}{x} + \frac{1}{3(x - 5)} - \frac{3\sin^2 x \cos x}{1 + \sin^3 x}$$

$$\frac{dy}{dx} = \frac{x\sqrt[3]{x - 5}}{1 + \sin^3 x}\left[\frac{1}{x} + \frac{1}{3(x - 5)} - \frac{3\sin^2 x \cos x}{1 + \sin^3 x}\right] \qquad \blacktriangleleft$$

▶ Exercise Set 7.3

1. Let $r = \ln 2$ and $s = \ln 3$. Express the following in terms of r and s:
    (a) $\ln 6$ (b) $\ln 1.5$
    (c) $\ln\frac{1}{2}$ (d) $\ln 9$
    (e) $\ln\sqrt[5]{3}$ (f) $\ln\frac{1}{36}$.

2. Assume that $\ln x_0 = 1$. Solve the following equations for x in terms of x_0:
    (a) $\ln x = -1$ (b) $\ln x = 2$
    (c) $\ln\sqrt[3]{x} = -\frac{1}{2}$.

3. (a) On graph paper, make an accurate sketch of the curve $y = \ln x$ by first plotting the points where $x = \frac{1}{9}, \frac{1}{3}, 1, 3, 9$, then sketching tangent lines at these points, and finally drawing a smooth curve through the points. [Use $\ln 3 \approx 1.1$.]
    (b) Use your sketch to estimate $\ln 2$.
    (c) Use your sketch to estimate x_0 such that $\ln x_0 = 1$.

In Exercises 4–7, draw the graph. Avoid point plotting.

4. $y = \ln |x|$.   5. $y = \ln\frac{1}{x}$.

6. $y = \ln\sqrt{x}$.   7. $y = \ln(x - 1)$.

In Exercises 8–13, use Theorem 7.3.1 to help perform the indicated differentiations and integrations.

8. $\dfrac{d}{dx}\left[\ln\sqrt{\dfrac{x-1}{x+1}}\,\right].$

9. $\dfrac{d}{dx}[\ln(\sqrt{x}\,\sqrt[3]{x+3}\,\sqrt[5]{(3x-2)})].$

10. $\dfrac{d}{dx}\left[\ln\left(\dfrac{\sqrt{x}\,\sqrt[3]{x+1}}{\sin x \sec x}\right)\right].$

11. $\displaystyle\int \dfrac{1}{x}\ln(x^3)\,dx.$

12. $\displaystyle\int \dfrac{dx}{x\ln\sqrt{x}}.$

13. $\displaystyle\int \dfrac{\ln(1/x)}{x}\,dx.$

14. Use an area interpretation of $\ln x$ to obtain (5).

In Exercises 15–18, obtain dy/dx by logarithmic differentiation.

15. $y = x\sqrt[3]{1+x^2}.$

16. $y = \sqrt[5]{\dfrac{x-1}{x+1}}.$

17. $y = \dfrac{(x^2-8)^{1/3}\sqrt{x^3+1}}{x^6-7x+5}.$

18. $y = \dfrac{\sin x \cos x \tan^3 x}{\sqrt{x}}.$

19. (a) Make an appropriate u-substitution to show that

$$\int_a^{ab}\dfrac{1}{t}\,dt = \int_1^b \dfrac{1}{t}\,dt$$

(b) Use this equality to obtain another proof of $\ln(ab) = \ln a + \ln b.$

20. If the interval $[1, x]$ on the t-axis is divided into n equal subintervals and A_1 and A_2 denote, respectively, the total areas of the inscribed and circumscribed rectangles shown in Figure 7.3.4, then

$$A_1 \le \int_1^x \dfrac{1}{t}\,dt \le A_2$$

Use this to solve the following problems.
(a) Find upper and lower estimates for $\ln 2$ using $n = 4$ subintervals.
(b) Find upper and lower estimates for $\ln 2$ using $n = 10$ subintervals.
(c) Compare your answers to the value listed in Appendix 3, Table 3.

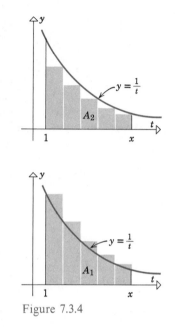

Figure 7.3.4

21. Prove: For all $x > 0$

$$1 - \dfrac{1}{x} \le \ln x \le x - 1$$

[*Hint:* Review the derivation of (5).]

22. Use the result of Exercise 21 to prove that

$$\lim_{x\to 0}\dfrac{\ln(x+1)}{x} = 1$$

23. (a) Use the result of Exercise 21 to prove that $0 < \ln x < x$ when $x > 1.$
(b) Use the result in (a) to prove that

$$0 < \dfrac{\ln x}{x} < \dfrac{2}{\sqrt{x}}$$

when $x > 1.$ [*Hint:* $\ln x = 2\ln\sqrt{x}.$]
(c) Use the result in (b) to prove that

$$\lim_{x\to+\infty}\dfrac{\ln x}{x} = 0$$

(d) Use the result in (c) to prove that

$$\lim_{x\to 0^+} x\ln x = 0$$

[*Remark:* This problem was suggested by an article published by D. S. Greenstein in the *American Mathematical Monthly*, vol. 72 (1965), p. 767.]

24. Sketch the following curves:

(a) $y = \dfrac{\ln x}{x}$

(b) $y = x \ln x$.

[*Remark:* The limits in Exercise 23 may be helpful.]

25. ***Boyle's law*** in physics states that under appropriate conditions the pressure p exerted by a gas is related to its volume v by $pv = c$, where c is a constant depending on the units and various physical factors. Prove: When such a gas increases in volume from v_0 to v_1 the average pressure it exerts is

$$p_{\text{ave}} = \frac{c}{v_1 - v_0} \ln \frac{v_1}{v_0}$$

7.4   THE NUMBER e; THE FUNCTIONS a^x AND e^x

In this section we use the natural logarithm to extend the concept of an exponent. Our objective is to give a definition of a^k in which k can be irrational. This will give meaning to such expressions as

$$2^\pi, \qquad 5^{-\sqrt{2}}, \qquad \text{and} \qquad \sqrt{3}^{\,\sin(\pi/8)}$$

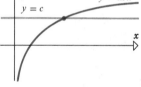

Figure 7.4.1

We also define a new function of fundamental importance, called the *natural exponential function*.

Because $\ln x$ is a continuous function that varies from $-\infty$ to $+\infty$ as x varies over the interval $(0, +\infty)$, it follows that the curve $y = \ln x$ must cross each horizontal line $y = c$ at a point to the right of the y-axis (Figure 7.4.1). Moreover, because $\ln x$ is an increasing function, it cannot cross any horizontal line $y = c$ more than once; otherwise, there would be distinct values of x, say $x_1 < x_2$, such that $\ln x_1 = \ln x_2 = c$, which contradicts the fact that $\ln x$ is increasing. Thus, we have the following result.

7.4.1   THEOREM   *For every real number c there is a unique positive real number x such that $\ln x = c$.*

It follows as a special case that there is a unique positive real number x such that $\ln x = 1$. This number is of special importance and is denoted by e. Thus,

$$\ln e = 1$$

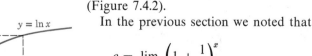

Figure 7.4.2

(Figure 7.4.2).

In the previous section we noted that

$$e = \lim_{x \to +\infty} \left(1 + \frac{1}{x}\right)^x \tag{1}$$

and we used this limit to obtain the approximation

$$e \approx 2.71828 \ldots$$

Later, in Section 10.3, we will show how to derive (1). For reference, approximate values for powers of e appear in Appendix 3, Table 2.

If a is a positive real number and r is a *rational* number, then Theorem 7.3.1d states that

$$\ln a^r = r \ln a$$

This equation gives us a way of defining a^k when k is an *arbitrary* real number.

7.4.2 DEFINITION   If a is any positive real number and k is any real number, then a^k is defined to be that real number whose natural logarithm has value $k \ln a$; that is,

$$\ln a^k = k \ln a \qquad\qquad\qquad (2)$$

▶ Example 1   By definition, 2^π is that real number whose natural logarithm is

$$\ln 2^\pi = \pi \ln 2$$
$$\approx (3.142)(.6931) \qquad \text{(Appendix 3, Table 3)}$$
$$\approx 2.1778$$

From Appendix 3, Table 3, we see that the value of 2^π is between 8.8 and 8.9.   ◀

The following theorem shows that the familiar laws for rational exponents are valid for all real exponents.

7.4.3 THEOREM   *If a and b are positive, then for all real numbers k and l the following laws of exponents hold.*

(a) $a^0 = 1$	(e) $\dfrac{a^k}{a^l} = a^{k-l}$
(b) $a^1 = a$	(f) $a^k \cdot b^k = (ab)^k$
(c) $a^k \cdot a^l = a^{k+l}$	(g) $\dfrac{a^k}{b^k} = \left(\dfrac{a}{b}\right)^k$
(d) $a^{-l} = \dfrac{1}{a^l}$	(h) $(a^k)^l = a^{kl}$

Proof. We will prove (c) and leave the rest as exercises. We can write

$$\ln (a^k \cdot a^l) = \ln a^k + \ln a^l$$
$$= k \ln a + l \ln a$$
$$= (k + l) \ln a$$
$$= \ln (a^{k+l})$$

Thus, $a^k \cdot a^l = a^{k+l}$ since the two sides have the same natural logarithm.   ▮

If a is a positive real constant, then the formula

$$f(x) = a^x \qquad (3)$$

defines a function of x on the interval $-\infty < x < +\infty$. Some examples are

$$2^x, \qquad \pi^x, \qquad \text{and} \qquad e^x$$

The function $f(x) = e^x$, called the ***natural exponential function,*** or usually simply the ***exponential function,*** has special importance because of the following theorem, which states that the natural logarithm and exponential functions undo the effect of one another.

7.4.4   THEOREM   (a) $\ln e^x = x \qquad$ for $-\infty < x < +\infty$
(b) $e^{\ln x} = x \qquad$ for $x > 0$

Proof of (a).

$$\ln e^x = x \ln e = x$$

Proof of (b).   Let $y = e^{\ln x}$. Then

$$\ln y = \ln (e^{\ln x}) = (\ln x)(\ln e) = \ln x$$

Since y and x have the same natural logarithm, it follows that $y = x$ or, equivalently, $e^{\ln x} = x$. ∎

In words, part (b) of this theorem states that $\ln x$ is that power to which e must be raised to produce x. Thus, we have shown that

$$\log_e x = \ln x$$

that is, the exponential and integral definitions of the natural logarithm are equivalent. The following corollary gives a useful computational formula for a^x in terms of the natural logarithm and exponential functions.

7.4.5   COROLLARY   *If a is a positive real number and k is any real number, then*

$$a^k = e^{k \ln a} \qquad (4)$$

Proof.   From (2)

$$\ln a^k = k \ln a$$

so that

$$e^{\ln a^k} = e^{k \ln a}$$

or by Theorem 7.4.4(*b*),

$$a^k = e^{k \ln a} \quad \blacksquare$$

▶ **Example 2** From this corollary and the tables in Appendix 3 we can approximate $3^{\sqrt{2}}$ as follows:

$$3^{\sqrt{2}} = e^{\sqrt{2} \ln 3}$$
$$\approx e^{(1.4)(1.1)} \qquad \text{[Appendix 3, Table 3]}$$
$$\approx e^{1.5}$$
$$\approx 4.5 \qquad \text{[Appendix 3, Table 2]} \qquad \blacktriangleleft$$

Later, we will see that a function of the form $f(x) = a^x \, (a > 0)$ is differentiable everywhere. If we accept this for now, we can obtain the derivative of a^x as follows. Let $y = a^x$ so that $\ln y = \ln a^x = x \ln a$; thus,

$$\frac{d}{dx}[\ln y] = \frac{d}{dx}[x \ln a]$$

Implicit differentiation yields

$$\frac{1}{y}\frac{dy}{dx} = \ln a$$

so

$$\frac{dy}{dx} = y \ln a = a^x \ln a$$

or

$$\frac{d}{dx}[a^x] = a^x \ln a \tag{5}$$

▶ **Example 3**

$$\frac{d}{dx}[\pi^x] = \pi^x \ln \pi$$

Since $\ln e = 1$, it follows as a special case of (5) that

$$\frac{d}{dx}[e^x] = e^x \tag{6}$$

If u is a differentiable function of x, then the chain rule yields the following generalizations of (5) and (6):

$$\frac{d}{dx}[a^u] = a^u \ln a \cdot \frac{du}{dx} \qquad (7)$$

$$\frac{d}{dx}[e^u] = e^u \cdot \frac{du}{dx} \qquad (8)$$

▶ **Example 4**   From (7) with $a = 2$ and $u = \sin x$

$$\frac{d}{dx}[2^{\sin x}] = (2^{\sin x})(\ln 2) \cdot \frac{d}{dx}[\sin x] = (2^{\sin x})(\ln 2)(\cos x)$$

and from (8) with $u = x^3$

$$\frac{d}{dx}[e^{x^3}] = e^{x^3} \cdot \frac{d}{dx}[x^3] = 3x^2 e^{x^3} \qquad ◀$$

The following theorem extends a familiar derivative formula so it applies to arbitrary exponents.

7.4.6   THEOREM   *If x is a positive real number and r is any real exponent, then*

$$\frac{d}{dx}[x^r] = rx^{r-1}$$

Proof.   From Corollary 7.4.5

$$x^r = e^{r \ln x}$$

so that

$$\frac{d}{dx}[x^r] = \frac{d}{dx}[e^{r \ln x}] = e^{r \ln x} \cdot \frac{d}{dx}[r \ln x] = x^r \cdot \frac{r}{x} = rx^{r-1} \quad ▮$$

▶ **Example 5**

$$\frac{d}{dx}[x^\pi] = \pi x^{\pi - 1} \qquad ◀$$

REMARK.   It is important to distinguish between the derivative formulas for a^x (variable exponent) and x^a (constant exponent). Compare, for example, the derivative of x^π in Example 5 and the derivative of π^x in Example 3.

Associated with derivatives (5) and (6) are the companion integration formulas

$$\int a^x \, dx = \frac{a^x}{\ln a} + C \qquad (9)$$

and

$$\int e^x \, dx = e^x + C \qquad\qquad (10)$$

▶ Example 6

$$\int 2^x \, dx = \frac{2^x}{\ln 2} + C \qquad\qquad ◀$$

▶ Example 7   To evaluate

$$\int e^{5x} \, dx$$

let $u = 5x$ so that $du = 5 \, dx$ or $dx = \frac{1}{5} \, du$, which yields

$$\int e^{5x} \, dx = \int e^u \left(\frac{1}{5} \, du\right) = \frac{1}{5} \int e^u \, du = \frac{1}{5} e^u + C$$

$$= \frac{1}{5} e^{5x} + C \qquad\qquad ◀$$

▶ Example 8

$$\int e^{-x} \, dx = -\int e^u \, du \qquad \begin{bmatrix} u = -x \\ du = -dx \end{bmatrix}$$

$$= -e^u + C$$
$$= -e^{-x} + C \qquad\qquad ◀$$

▶ Example 9

$$\int x^2 e^{x^3} \, dx = \int \frac{1}{3} e^u \, du \qquad \begin{bmatrix} u = x^3 \\ du = 3x^2 \, dx \end{bmatrix}$$

$$= \frac{1}{3} e^u + C$$

$$= \frac{1}{3} e^{x^3} + C \qquad\qquad ◀$$

▶ Example 10   Evaluate

$$\int_0^{\ln 3} e^x (1 + e^x)^{1/2} \, dx$$

Solution. Make the u-substitution

$$u = 1 + e^x, \qquad du = e^x \, dx$$

and change the x-limits ($x = 0$, $x = \ln 3$) to the u-limits ($u = 1 + e^0 = 2$, $u = 1 + e^{\ln 3} = 1 + 3 = 4$).

$$\int_0^{\ln 3} e^x (1 + e^x)^{1/2} \, dx = \int_2^4 u^{1/2} \, du = \frac{2}{3} u^{3/2} \bigg]_2^4 = \frac{2}{3} [4^{3/2} - 2^{3/2}]$$

$$= \frac{16}{3} - \frac{4\sqrt{2}}{3} \qquad \blacktriangleleft$$

REMARK. For ease of printing, the exponential function is often denoted by $\exp x$ rather than e^x. For example, in this notation the statements

$$e^{x_1 + x_2} = e^{x_1} e^{x_2}$$

$$e^{\ln x} = x$$

$$\ln (e^x) = x$$

would be written as

$$\exp (x_1 + x_2) = \exp x_1 \cdot \exp x_2$$

$$\exp (\ln x) = x$$

$$\ln (\exp x) = x$$

▶ Exercise Set 7.4

1. Simplify the expression and state the values of x for which your simplification is valid.
   (a) $e^{-\ln x}$ (b) $e^{\ln x^2}$
   (c) $\ln (e^{-x^2})$ (d) $\ln (1/e^x)$
   (e) $\exp (3 \ln x)$ (f) $\ln (xe^x)$
   (g) $\ln (e^{x - \sqrt[3]{x}})$ (h) $e^{x - \ln x}$.

2. Solve for x:
   (a) $\ln (x^2) = 5$ (b) $e^{-4x} = 3$
   (c) $\ln (\ln x) = 0$ (d) $e^x + e^{-x} = 2$.

3. Solve for x:
   (a) $\ln (\sqrt{x}) + \ln (x^{3/2}) = 1$
   (b) $e^{2\pi x} = \sqrt{2}$
   (c) $\ln (\cos x) = 0$, $\quad 0 \le x \le \pi$
   (d) $e^{2x} + 2e^x + 1 = 9$.

4. Where does the graph of $y = e^{2x} - 3e^x - 4$ intersect the x-axis?

5. Use the method of Example 2 to approximate the following. (Round off each computation to two significant digits.)
   (a) $2^{1.7}$ (b) $5^{\sqrt{3}}$ (c) 3^π.

In Exercises 6–21, find dy/dx.

6. $y = e^{7x}$.
7. $y = e^{-5x^2}$.
8. $y = e^{1/x}$.
9. $y = x^3 e^x$.
10. $y = \sin (e^x)$.
11. $y = \dfrac{e^x - e^{-x}}{e^x + e^{-x}}$.
12. $y = \dfrac{e^x}{\ln x}$.
13. $y = e^x \tan x$.
14. $y = \exp (\sqrt{1 + 5x^3})$.
15. $y = e^{(x - e^{3x})}$.
16. $y = \ln (\cos e^x)$.
17. $y = \ln (1 - xe^{-x})$.
18. $y = \sqrt{1 + e^x}$.

19. $y = e^{ax} \cos bx$ (a, b constant).

20. $y = \dfrac{a}{1 + be^{-x}}$ (a, b constant).

21. $y = e^{\ln (x^3+1)}$.

In Exercises 22–26, find $f'(x)$ by Formula (7) and then by logarithmic differentiation.

22. $f(x) = 2^x$.

23. $f(x) = 3^{-x}$.

24. $f(x) = \pi^{\sin x}$.

25. $f(x) = \pi^{x \tan x}$.

26. $f(x) = (\sqrt{2})^{x \ln x}$.

27. (a) Explain why Formula (5) cannot be used to find $(d/dx)[x^x]$.
    (b) Find this derivative by logarithmic differentiation.

In Exercises 28–33, find dy/dx by logarithmic differentiation.

28. $y = x^{\sin x}$.

29. $y = (x^3 - 2x)^{\ln x}$.

30. $y = (x^2 + 3)^{\ln x}$.

31. $y = (\ln x)^{\tan x}$.

32. $y = (1 + x)^{1/x}$.

33. $y = x^{(e^x)}$.

34. Show that $y = e^{3x}$ and $y = e^{-3x}$ both satisfy the equation $y'' - 9y = 0$.

35. Show that for any constants A and B, the function $y = Ae^{2x} + Be^{-4x}$ satisfies the equation $y'' + 2y' - 8y = 0$.

36. Show that for any constants A and k, the function $y = Ae^{kt}$ satisfies the equation $dy/dt = ky$.

37. Let $f(x) = e^{kx}$ and $g(x) = e^{-kx}$. Find
    (a) $f^{(n)}(x)$ (b) $g^{(n)}(x)$.

38. Find dy/dt if $y = e^{-\lambda t}(A \sin \omega t + B \cos \omega t)$, where A, B, λ, and ω are constants.

39. Find $f'(x)$ if
$$f(x) = \frac{1}{\sqrt{2\pi}\sigma} \exp\left[-\frac{1}{2}\left(\frac{x - \mu}{\sigma}\right)^2\right]$$
where μ and σ are constants.

In Exercises 40–67, evaluate the integrals.

40. $\displaystyle\int \frac{dx}{e^x}$.

41. $\displaystyle\int e^{-5x}\, dx$.

42. $\displaystyle\int e^{\tan x} \sec^2 x\, dx$.

43. $\displaystyle\int e^{\sin x} \cos x\, dx$.

44. $\displaystyle\int x^3 e^{x^4}\, dx$.

45. $\displaystyle\int x^2 e^{-2x^3}\, dx$.

46. $\displaystyle\int \frac{e^x + e^{-x}}{e^x - e^{-x}}\, dx$.

47. $\displaystyle\int \frac{e^x}{1 + e^x}\, dx$.

48. $\displaystyle\int \sqrt{e^x}\, dx$.

49. $\displaystyle\int e^{2t} \sqrt{1 + e^{2t}}\, dt$.

50. $\displaystyle\int (x + 3) \exp (x^2 + 6x)\, dx$.

51. $\displaystyle\int \cos x \exp (\sin x)\, dx$.

52. $\displaystyle\int e^x \sin (1 + e^x)\, dx$.

53. $\displaystyle\int e^{-x} \sec^2 (2 - e^{-x})\, dx$.

54. $\displaystyle\int 2^{5x}\, dx$.

55. $\displaystyle\int \pi^{\sin x} \cos x\, dx$.

56. $\displaystyle\int \left[ex^2 + \left(\frac{1}{2}\ln 2\right) \sin x\right] dx$.

57. $\displaystyle\int \left(x \ln 3 - 4\pi e^2 \cos x\right) dx$.

58. $\displaystyle\int e^{2 \ln x}\, dx$.

59. $\displaystyle\int \left[\ln (e^x) + \ln (e^{-x})\right] dx$.

60. $\displaystyle\int \frac{dy}{\sqrt{y}e^{\sqrt{y}}}.$

61. $\displaystyle\int \frac{e^{\sqrt{y}}}{\sqrt{y}}\, dy.$

62. $\displaystyle\int_0^{\ln 2} e^{-3x}\, dx.$ **63.** $\displaystyle\int_0^{\ln 5} e^x(3 - 4e^x)\, dx.$

64. $\displaystyle\int_1^{\sqrt{2}} x4^{-x^2}\, dx.$ **65.** $\displaystyle\int_1^2 (3 - e^x)\, dx.$

66. $\displaystyle\int_0^e \frac{dx}{x + e}.$

67. $\displaystyle\int_{-\ln 3}^{\ln 3} \frac{e^x}{e^x + 4}\, dx.$

68. Show that the equation $e^{1/x} - e^{-1/x} = 0$ has no solution.

69. Prove: If $f'(x) = f(x)$ for all x in $(-\infty, +\infty)$, then $f(x)$ has the form $f(x) = ke^x$ for some constant k. [*Hint:* Let $g(x) = e^{-x} f(x)$ and consider $g'(x)$.]

70. A particle moves along the x-axis so that its x-coordinate at time t is given by $x = ae^{kt} + be^{-kt}$. Show that its acceleration is proportional to x.

71. Find the intersections of the curves $y = 2^x$ and $y = 3^{x+1}$.

72. Find a point on the graph of $y = e^{3x}$ at which the tangent line passes through the origin.

73. Find $f'(x)$ if $f(x) = x^e$.

74. Show that
(a) $y = xe^{-x}$ satisfies the equation
$$xy' = (1 - x)y$$
(b) $y = xe^{-x^2/2}$ satisfies the equation
$$xy' = (1 - x^2)y.$$

75. The equilibrium constant k of a balanced chemical reaction changes with the absolute temperature T according to the law
$$k = k_0 \exp\left(-\frac{q}{2}\frac{T - T_0}{T_0 T}\right)$$
where k_0, q, and T_0 are constants. Find the rate of change of k with respect to T.

76. Find a function $y = f(x)$ such that $e^y - e^{-y} = x$.

77. Evaluate
$$\int \frac{e^{2x}}{e^x + 3}\, dx$$

7.5 ADDITIONAL PROPERTIES OF e^x

In this section we will obtain the graph of $y = e^x$ and derive some additional properties of the exponential function.

The graph of $y = e^x$ is easiest to obtain by first writing this equation in the alternate form

$$\ln y = \ln e^x$$

or

$$x = \ln y$$

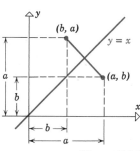

Figure 7.5.1

To obtain the graph of this equation we need some preliminary results. It is easy to prove (Exercise 40) that the points (a, b) and (b, a) are symmetrically positioned about the line $y = x$; that is, the line $y = x$ is the perpendicular bisector of the line segment joining these points (Figure 7.5.1). Thus, interchanging the x- and y-coordinates of a point causes the point to be reflected about the line $y = x$. Consequently, interchanging the x and y variables in an equation causes the graph of that equation to be reflected about this line.

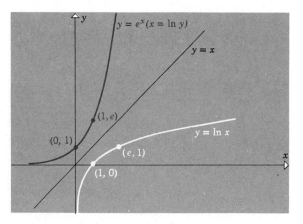

Figure 7.5.2

It follows, therefore, that the graph of the curve $x = \ln y$ (or equivalently $y = e^x$) is the reflection of the curve $y = \ln x$ about the line $y = x$ (Figure 7.5.2).

In Section 8.1 we will prove that $f(x) = e^x$ is everywhere differentiable and consequently everywhere continuous. However, the continuity should be intuitively clear since the graph of $y = e^x$ is the reflection of the continuous curve $y = \ln x$.

The following properties are evident from the graph of e^x (proofs are discussed in the exercises):

$$e^x > 0 \text{ for all } x \tag{1a}$$

$$\lim_{x \to +\infty} e^x = +\infty \tag{1b}$$

$$\lim_{x \to -\infty} e^x = 0 \tag{1c}$$

▶ **Example 1** Sketch the graph of $y = e^{-x}$.

Solution. The effect of replacing x by $-x$ in the equation $y = e^x$ is to reflect the graph of this equation about the y-axis. Thus, the graph of $y = e^{-x}$ is as shown in Figure 7.5.3.

The following limits, which are consequences of (1a), (1b), and (1c), are evident from the graph of e^{-x}:

$$e^{-x} > 0 \qquad \text{for all } x \tag{2a}$$

$$\lim_{x \to +\infty} e^{-x} = \lim_{x \to +\infty} \frac{1}{e^x} = 0 \tag{2b}$$

$$\lim_{x \to -\infty} e^{-x} = \lim_{x \to -\infty} \frac{1}{e^x} = +\infty \tag{2c}$$

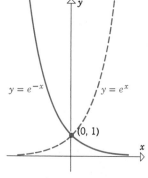

Figure 7.5.3

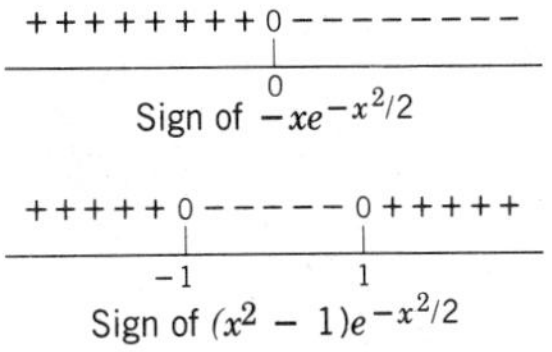

++++++++0----------
——————————————
$\qquad$ 0
Sign of $-xe^{-x^2/2}$

+++++0-----0+++++
——————————————
$\quad$ -1 $\qquad$ 1
Sign of $(x^2-1)e^{-x^2/2}$

Figure 7.5.4

▶ **Example 2** Sketch the graph of $f(x) = e^{-x^2/2}$.

Solution.

$$f'(x) = e^{-x^2/2}\frac{d}{dx}\left[-\frac{x^2}{2}\right] = -xe^{-x^2/2}$$

$$f''(x) = -x\frac{d}{dx}[e^{-x^2/2}] + e^{-x^2/2}\frac{d}{dx}[-x]$$

$$= x^2 e^{-x^2/2} - e^{-x^2/2}$$

$$= (x^2-1)e^{-x^2/2}$$

Since $e^{-x^2/2} > 0$ for all x, the signs of $f'(x)$ and $f''(x)$ are the same as those of $-x$ and $x^2 - 1$, respectively (Figure 7.5.4). The information conveyed by these signs is summarized in Figure 7.5.5a. Since the curve is concave down at $x = 0$, there is a relative maximum at this stationary point. The values of f at this point and at the inflection points, $x = 1$ and $x = -1$, are

$$f(0) = e^0 = 1$$
$$f(1) = e^{-1/2} \approx .61 \qquad \text{(Appendix 3, Table 2)}$$
$$f(-1) = e^{-1/2} \approx .61$$

Since $e^{-x^2/2} > 0$ for all x, it follows that the graph of $y = e^{-x^2/2}$ is entirely above the x-axis. Finally, since $x^2/2 \to +\infty$ as $x \to +\infty$ or $x \to -\infty$, it follows from (2b) that

$$\lim_{x \to +\infty} e^{-x^2/2} = \lim_{x \to -\infty} e^{-x^2/2} = 0$$

Piecing together all of our information yields the graph in Figure 7.5.5b. Note that the graph is symmetric about the y-axis. This could have been anticipated from the fact that the equation $y = e^{-x^2/2}$ is unaltered when x is replaced by $-x$.

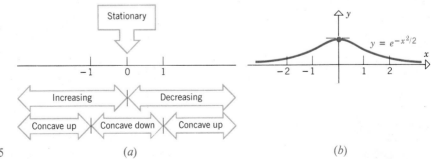

Figure 7.5.5 $\qquad\qquad\qquad\qquad$ (a) $\qquad\qquad\qquad\qquad\qquad\qquad\qquad$ (b)

▶ Exercise Set 7.5

In Exercises 1–6, find the limits.

1. (a) $\lim\limits_{x \to +\infty} 4e^{3x}$ (b) $\lim\limits_{x \to -\infty} 4e^{3x}$.

2. (a) $\lim\limits_{x \to +\infty} 3e^{-2x}$ (b) $\lim\limits_{x \to -\infty} 3e^{-2x}$.

3. (a) $\lim\limits_{x \to +\infty} (e^x + e^{-x})$ (b) $\lim\limits_{x \to -\infty} (e^x + e^{-x})$.

4. (a) $\lim\limits_{x \to +\infty} (e^x - e^{-x})$ (b) $\lim\limits_{x \to -\infty} (e^x - e^{-x})$.

5. (a) $\lim\limits_{x \to +\infty} (1 - e^{-x^2})$ (b) $\lim\limits_{x \to -\infty} (1 - e^{-x^2})$.

6. (a) $\lim\limits_{x \to +\infty} e^{1/x}$ (b) $\lim\limits_{x \to -\infty} e^{1/x}$.

In Exercises 7–12, sketch the graph of f and label the relative extreme points and points of inflection.

7. $f(x) = 4e^{3x}$.

8. $f(x) = 3e^{-2x}$.

9. $f(x) = e^x + e^{-x}$.

10. $f(x) = e^x - e^{-x}$.

11. $f(x) = 1 - e^{-x^2}$.

12. $f(x) = e^{1/x}$.

13. Let $f(x) = e^{|x|}$.
  (a) Is f continuous at $x = 0$?
  (b) Is f differentiable at $x = 0$?
  (c) Sketch the graph of f.

14. In each part determine whether the limit exists. If so, find it.
  (a) $\lim\limits_{x \to +\infty} e^x \cos x$ (b) $\lim\limits_{x \to -\infty} e^x \cos x$.

15. Sketch the graphs of e^x, $-e^x$, and $e^x \cos x$, all in the same figure, and label any points of intersection.

In Exercises 16–19, find the limits.

16. $\lim\limits_{x \to +\infty} \dfrac{e^x + e^{-x}}{e^x - e^{-x}}$.

17. $\lim\limits_{x \to -\infty} \dfrac{e^x + e^{-x}}{e^x - e^{-x}}$.

18. $\lim\limits_{x \to +\infty} \dfrac{2 + e^x}{1 + 3e^x}$.

19. $\lim\limits_{x \to +\infty} e^{(-e^x)}$.

20. One of the fundamental functions of mathematical statistics is

$$f(x) = \frac{1}{\sqrt{2\pi}\sigma} \exp\left[-\frac{1}{2}\left(\frac{x - \mu}{\sigma}\right)^2\right]$$

where μ and σ are constants such that $\sigma > 0$ and $-\infty < \mu < +\infty$.
  (a) Locate the inflection points and relative extreme points.
  (b) Find $\lim\limits_{x \to +\infty} f(x)$ and $\lim\limits_{x \to -\infty} f(x)$.
  (c) Sketch the graph of f.

21. By writing the derivative

$$\frac{d}{dx}[e^x]\Big|_{x=0} = 1$$

as a limit, prove that

$$\lim_{h \to 0} \frac{e^h - 1}{h} = 1$$

In Exercises 22–25, use the method of Exercise 21 to find the limits.

22. $\lim\limits_{x \to 0} \dfrac{e^{2x} - e^x}{x}$.

23. $\lim\limits_{x \to 0} \dfrac{1 - e^{-x}}{x}$.

24. $\lim\limits_{x \to a} \dfrac{e^x - e^a}{x - a}$.

25. $\lim\limits_{x \to +\infty} x(e^{1/x} - 1)$.

In Exercise 40 of this set, it will be shown that

$$\lim_{x \to +\infty} xe^{-x} = 0$$

Use this result, where needed, in Exercises 26–30.

26. (a) Find $\lim\limits_{x \to +\infty} xe^{-2x}$ and $\lim\limits_{x \to -\infty} xe^{-2x}$.
  (b) Sketch the graph of $y = xe^{-2x}$ and label all relative extrema and inflection points.

27. (a) Find $\lim\limits_{x \to +\infty} xe^x$ and $\lim\limits_{x \to -\infty} xe^x$.
  (b) Sketch the graph of $y = xe^x$ and label all relative extrema and inflection points.

28. (a) Find $\lim\limits_{x \to +\infty} x^2 e^{2x}$ and $\lim\limits_{x \to -\infty} x^2 e^{2x}$.
  (b) Sketch the graph of $y = x^2 e^{2x}$ and label all relative extrema and inflection points.

29. (a) Find $\lim\limits_{x \to +\infty} x^2/e^{2x}$ and $\lim\limits_{x \to -\infty} x^2/e^{2x}$.
  (b) Sketch the graph of $y = x^2/e^{2x}$ and label all relative extrema and inflection points.

30. (a) Find $\lim\limits_{x \to +\infty} e^x/x$, $\lim\limits_{x \to -\infty} e^x/x$, $\lim\limits_{x \to 0^+} e^x/x$, and $\lim\limits_{x \to 0^-} e^x/x$.

(b) Sketch the graph of $y = e^x/x$ and label all relative extrema and inflection points.

31. Let c_1, c_2, k_1, and k_2 be positive constants. In each part make a sketch that shows the relative positions of the curves $y = c_1 e^{k_1 x}$ and $y = c_2 e^{k_2 x}$ when:

(a) $c_1 = c_2$ and $k_1 < k_2$

(b) $c_1 < c_2$ and $k_1 = k_2$

(c) $c_1 < c_2$ and $k_2 < k_1$.

32. Repeat Exercise 31 for the curves $y = c_1 e^{-k_1 x}$ and $y = c_2 e^{-k_2 x}$.

33. (a) Sketch the curve $y = 2^x$.

(b) Sketch the curve $y = a^x$ assuming $a > 1$.

(c) Make a sketch that shows the relative positions of the curves $y = a_1^x$ and $y = a_2^x$ if $a_2 > a_1 > 1$.

34. Find the volume of the solid generated when the region bounded by $y = e^x$, $x = 0$, $x = \ln 3$, and $y = 0$ is revolved about the x-axis.

35. Find the area of the region enclosed by the curve $y = e^{-x}$ and the line through the points $(0, 1)$ and $(1, 1/e)$.

36. Prove that the line $y = x$ is the perpendicular bisector of the line segment joining (a, b) and (b, a).

37. Use the Mean-Value Theorem (4.10.2) to prove that

(a) $e^x \geq 1 + x$ if $x \geq 0$

(b) $e^{-x} \geq 1 - x$ if $x \geq 0$.

38. Prove that $\lim\limits_{x \to +\infty} e^x = +\infty$ by showing that for any $N > 0$, there is a point x_0 such that $e^x > N$ whenever $x > x_0$.

39. Prove that $\lim\limits_{x \to -\infty} e^x = 0$ by showing that for any $\epsilon > 0$ there is a point x_0 such that $0 < e^x < \epsilon$ whenever $x < x_0$.

40. It was proved in Exercise 23 of Section 7.3 that $0 < \ln x < x$ for $x > 1$.

(a) Use this result to prove that $\ln x < 2\sqrt{x}$ for $x > 1$.

(b) Use the result in (a) to obtain

$$\frac{1}{e^x} < \frac{x}{e^x} < e^{2\sqrt{x} - x}$$

for $x > 1$.

(c) Use the result in (b) to show that

$$\lim_{x \to +\infty} xe^{-x} = 0$$

(d) By inspection, explain why

$$\lim_{x \to -\infty} xe^{-x} = -\infty.$$

7.6 THE HYPERBOLIC FUNCTIONS

In this section we will study certain combinations of e^x and e^{-x}, called **hyperbolic functions.** These functions have numerous engineering applications and arise naturally in many mathematical problems. It will become evident as we progress that the hyperbolic functions have many properties in common with the trigonometric functions. This similarity is reflected in the names of the hyperbolic functions.

7.6.1 DEFINITION The ***hyperbolic sine*** and ***hyperbolic cosine*** functions, denoted by ***sinh*** and ***cosh,*** respectively, are defined by

$$\sinh x = \frac{e^x - e^{-x}}{2}$$

$$\cosh x = \frac{e^x + e^{-x}}{2}$$

REMARK. sinh rhymes with "cinch" and cosh rhymes with "gosh."

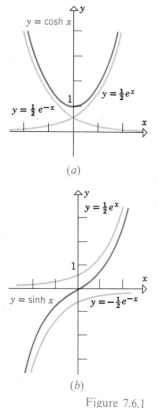

(a)

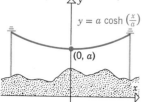

(b)

Figure 7.6.1

The graph of $\cosh x = \frac{1}{2}e^x + \frac{1}{2}e^{-x}$ can be obtained by separately graphing $\frac{1}{2}e^x$ and $\frac{1}{2}e^{-x}$ and then adding the y-coordinates together at each point (Figure 7.6.1a). This graphing technique, called **addition of ordinates**, can also be used to graph $\sinh x = \frac{1}{2}e^x - \frac{1}{2}e^{-x}$ (Figure 7.6.1b).

As an illustration of how hyperbolic functions occur in physical problems, consider a homogeneous flexible cable hanging suspended between two points (for example, an electrical transmission line suspended between two poles). The cable forms a curve called a **catenary** (from the Latin "catena" meaning chain). If a coordinate system is introduced in such a way that the low point of the cable occurs on the y-axis at $(0, a)$ where $a > 0$, then it can be shown using principles of physics that the equation of the curve formed by the cable is

$$y = a \cosh\left(\frac{x}{a}\right),$$

where a depends on the tension and physical properties of the cable (Figure 7.6.2).

The remaining hyperbolic functions, **hyperbolic tangent, hyperbolic cotangent, hyperbolic secant,** and **hyperbolic cosecant** are defined in terms of sinh and cosh as follows.

$$\tanh x = \frac{\sinh x}{\cosh x} = \frac{e^x - e^{-x}}{e^x + e^{-x}}$$

$$\coth x = \frac{\cosh x}{\sinh x} = \frac{e^x + e^{-x}}{e^x - e^{-x}}$$

$$\operatorname{sech} x = \frac{1}{\cosh x} = \frac{2}{e^x + e^{-x}}$$

$$\operatorname{csch} x = \frac{1}{\sinh x} = \frac{2}{e^x - e^{-x}}$$

The graphs of these functions are shown in Figure 7.6.3 (see Exercises 31–34).

The hyperbolic functions satisfy various identities similar to the identities for the trigonometric functions. The most fundamental of these is

$$\cosh^2 x - \sinh^2 x = 1 \tag{1}$$

Figure 7.6.2

which can be proved by writing

$$\cosh^2 x - \sinh^2 x = \left(\frac{e^x + e^{-x}}{2}\right)^2 - \left(\frac{e^x - e^{-x}}{2}\right)^2$$

$$= \frac{1}{4}(e^{2x} + 2e^0 + e^{-2x}) - \frac{1}{4}(e^{2x} - 2e^0 + e^{-2x})$$

$$= 1$$

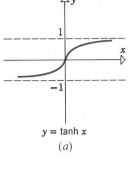

$y = \tanh x$

(a)

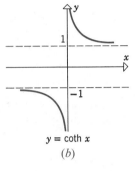

$y = \coth x$

(b)

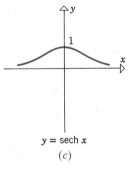

$y = \operatorname{sech} x$

(c)

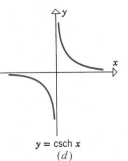

$y = \operatorname{csch} x$

(d)

Figure 7.6.3

If we divide (1) by $\cosh^2 x$ we obtain

$$1 - \tanh^2 x = \operatorname{sech}^2 x \tag{2}$$

and if we divide (1) by $\sinh^2 x$ we obtain

$$\coth^2 x - 1 = \operatorname{csch}^2 x \tag{3}$$

The addition formulas for sinh and cosh are

$$\sinh (x + y) = \sinh x \cosh y + \cosh x \sinh y \tag{4a}$$
$$\cosh (x + y) = \cosh x \cosh y + \sinh x \sinh y \tag{4b}$$

These are easier to prove than the corresponding identities for the trigonometric functions. For example, to prove (4a) we use the relations

$$\cosh x + \sinh x = e^x \tag{5a}$$
$$\cosh x - \sinh x = e^{-x} \tag{5b}$$

which follow immediately from Definition 7.6.1. From (5a) and (5b) it follows that

$$\sinh (x + y) = \frac{e^{(x+y)} - e^{-(x+y)}}{2} = \frac{e^x e^y - e^{-x} e^{-y}}{2}$$

$$= \frac{1}{2}[(\cosh x + \sinh x)(\cosh y + \sinh y)$$
$$- (\cosh x - \sinh x)(\cosh y - \sinh y)]$$

$$= \sinh x \cosh y + \cosh x \sinh y$$

which proves (4a). The proof of (4b) is similar.

If we let $x = y$ in (4a) and (4b) we obtain the following analogs of the trigonometric double angle formulas:

$$\sinh 2x = 2 \sinh x \cosh x \tag{6a}$$
$$\cosh 2x = \cosh^2 x + \sinh^2 x \tag{6b}$$

By using (1) and (6b) the reader should be able to show that

$$\cosh 2x = 2 \sinh^2 x + 1 \tag{7a}$$
$$\cosh 2x = 2 \cosh^2 x - 1 \tag{7b}$$

To obtain subtraction formulas for sinh and cosh we will use the addition formulas and the relations

$$\cosh (-x) = \cosh x \tag{8a}$$
$$\sinh (-x) = -\sinh x \tag{8b}$$

which follow immediately from Definition 7.6.1. If we replace y by $-y$ in (4a) and (4b) and then apply (8a) and (8b) we obtain

$$\sinh (x - y) = \sinh x \cosh y - \cosh x \sinh y \qquad (9a)$$
$$\cosh (x - y) = \cosh x \cosh y - \sinh x \sinh y \qquad (9b)$$

Derivative formulas for $\sinh x$ and $\cosh x$ follow easily from Definition 7.6.1. For example,

$$\frac{d}{dx}[\cosh x] = \frac{d}{dx}\left[\frac{e^x + e^{-x}}{2}\right]$$

$$= \frac{1}{2}\left(\frac{d}{dx}[e^x] + \frac{d}{dx}[e^{-x}]\right)$$

$$= \frac{e^x - e^{-x}}{2} = \sinh x$$

Similarly,

$$\frac{d}{dx}[\sinh x] = \cosh x$$

The derivatives of the remaining hyperbolic functions can be obtained by first expressing them in terms of $\sinh$ and $\cosh$. For example,

$$\frac{d}{dx}[\tanh x] = \frac{d}{dx}\left[\frac{\sinh x}{\cosh x}\right]$$

$$= \frac{\cosh x \dfrac{d}{dx}[\sinh x] - \sinh x \dfrac{d}{dx}[\cosh x]}{\cosh^2 x}$$

$$= \frac{\cosh^2 x - \sinh^2 x}{\cosh^2 x} = \frac{1}{\cosh^2 x} = \text{sech}^2 x$$

Derivations of differentiation formulas for the remaining hyperbolic functions are exercises. For reference we summarize the results.

$$\frac{d}{dx}[\sinh x] = \cosh x$$

$$\frac{d}{dx}[\cosh x] = \sinh x$$

$$\frac{d}{dx}[\tanh x] = \text{sech}^2 x$$

$$\frac{d}{dx}[\coth x] = -\text{csch}^2 x$$

$$\frac{d}{dx}[\text{sech } x] = -\text{sech } x \tanh x$$

$$\frac{d}{dx}[\text{csch } x] = -\text{csch } x \coth x$$

REMARK.   Except for a difference in the pattern of signs, these formulas are similar to the formulas for the derivatives of the trigonometric functions.

These differentiation formulas produce the following companion integration formulas.

$$\int \sinh x \, dx = \cosh x + C$$

$$\int \cosh x \, dx = \sinh x + C$$

$$\int \operatorname{sech}^2 x \, dx = \tanh x + C$$

$$\int \operatorname{csch}^2 x \, dx = -\coth x + C$$

$$\int \operatorname{sech} x \tanh x \, dx = -\operatorname{sech} x + C$$

$$\int \operatorname{csch} x \coth x \, dx = -\operatorname{csch} x + C$$

► Example 1

$$\frac{d}{dx}[\cosh(x^3)] = \sinh(x^3) \cdot \frac{d}{dx}[x^3] = 3x^2 \sinh(x^3)$$

$$\frac{d}{dx}[\ln(\tanh x)] = \frac{1}{\tanh x} \cdot \frac{d}{dx}[\tanh x] = \frac{\operatorname{sech}^2 x}{\tanh x}$$ ◄

► Example 2

$$\int \sinh^5 x \cosh x \, dx = \frac{1}{6} \sinh^6 x + C \qquad \left[\begin{array}{l} u = \sinh x \\ du = \cosh x \, dx \end{array}\right]$$

$$\int \tanh x \, dx = \int \frac{\sinh x}{\cosh x} \, dx$$

$$= \ln|\cosh x| + C \qquad \left[\begin{array}{l} u = \cosh x \\ du = \sinh x \, dx \end{array}\right]$$

$$= \ln(\cosh x) + C$$

We are justified in writing $|\cosh x| = \cosh x$ because $\cosh x$ is positive for all x. In fact, $\cosh x \geq 1$ for all x. This is geometrically clear from Figure 7.6.1a and can be proved using (1) (Exercise 33).   ◄

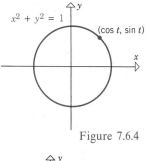

$x^2 + y^2 = 1$

$(\cos t, \sin t)$

Figure 7.6.4

We conclude this section with a brief explanation of why the adjective "hyperbolic" is used for hyperbolic functions.

If t is any real number, then the point $(\cos t, \sin t)$ lies on the circle $x^2 + y^2 = 1$ because

$$\cos^2 t + \sin^2 t = 1$$

(Figure 7.6.4). For this reason sine and cosine are called *circular functions*. Analagously, for any real number t the point $(\cosh t, \sinh t)$ lies on the curve $x^2 - y^2 = 1$ because

$$\cosh^2 t - \sinh^2 t = 1$$

(Figure 7.6.5). As we will see in Section 12.4, this curve is called a hyperbola and accordingly sinh and cosh are called *hyperbolic functions*.

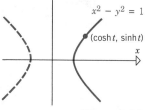

$x^2 - y^2 = 1$

$(\cosh t, \sinh t)$

Figure 7.6.5

▶ Exercise Set 7.6

1. In each part, a value for one of the hyperbolic functions is given at x_0. Find the values of the remaining five hyperbolic functions at x_0.

(a) $\sinh x_0 = -2$

(b) $\cosh x_0 = \dfrac{5}{4}$, $x_0 < 0$

(c) $\tanh x_0 = -\dfrac{4}{5}$

(d) $\coth x_0 = 2$

(e) $\operatorname{sech} x_0 = \dfrac{15}{17}$, $x_0 > 0$

(f) $\operatorname{csch} x_0 = -1$.

In Exercises 2–14, establish the identities.

2. $\cosh(x + y) = \cosh x \cosh y + \sinh x \sinh y$.

3. $\cosh 2x = 2 \sinh^2 x + 1$.

4. $\cosh 2x = 2 \cosh^2 x - 1$.

5. $\cosh(-x) = \cosh x$.

6. $\sinh(-x) = -\sinh x$.

7. $\tanh(x + y) = \dfrac{\tanh x + \tanh y}{1 + \tanh x \tanh y}$.

8. $\tanh(x - y) = \dfrac{\tanh x - \tanh y}{1 - \tanh x \tanh y}$.

9. $\tanh 2x = \dfrac{2 \tanh x}{1 + \tanh^2 x}$.

10. $\cosh \dfrac{1}{2}x = \sqrt{\dfrac{1}{2}(\cosh x + 1)}$.

11. $\sinh \dfrac{1}{2}x = \pm\sqrt{\dfrac{1}{2}(\cosh x - 1)}$.

12. $\sinh x + \sinh y = 2 \sinh\left(\dfrac{x + y}{2}\right) \cosh\left(\dfrac{x - y}{2}\right)$.

13. $\cosh x + \cosh y = 2 \cosh\left(\dfrac{x + y}{2}\right) \cosh\left(\dfrac{x - y}{2}\right)$.

14. $\cosh 3x = 4 \cosh^3 x - 3 \cosh x$.

15. Derive the differentiation formula for

(a) $\sinh x$ (b) $\coth x$

(c) $\operatorname{sech} x$ (d) $\operatorname{csch} x$.

In Exercises 16–21, find dy/dx.

16. $y = \cosh(x^4)$.

17. $y = \sinh(4x - 8)$.

18. $y = \ln(\tanh 2x)$.

19. $y = \coth(\ln x)$.

20. $y = \operatorname{sech}(e^{2x})$.

21. $y = \operatorname{csch}(1/x)$.

In Exercises 22–29, evaluate the integral.

22. $\displaystyle\int \cosh(2x - 3)\, dx$.   **23.** $\displaystyle\int \sinh^6 x \cosh x\, dx$.

24. $\displaystyle\int \operatorname{csch}^2(3x)\, dx$.   **25.** $\displaystyle\int \sqrt{\tanh x}\, \operatorname{sech}^2 x\, dx$.

26. $\int \coth^2 x \, \operatorname{csch}^2 x \, dx.$ **27.** $\int \tanh x \, dx.$

28. $\int \dfrac{e^x - e^{-x}}{e^x + e^{-x}} \, dx.$ **29.** $\int \tanh x \, \operatorname{sech}^3 x \, dx.$

30. Prove the following facts about $\tanh x$:
   (a) $-1 < \tanh x < 1$ for all x
   (b) $\lim\limits_{x \to +\infty} \tanh x = 1$
   (c) $\lim\limits_{x \to -\infty} \tanh x = -1.$

31. Determine the intervals on which $y = \tanh x$ is positive, negative, increasing, decreasing, concave up, and concave down. Then use these results and Exercise 30 to obtain the graph of $y = \tanh x$.

32. Use the results found in Exercises 30 and 31 to help obtain the graph of $y = \coth x$.

33. (a) Prove: $\cosh x \geq 1$ for all x.
   (b) Prove: $0 < \operatorname{sech} x \leq 1$ for all x.
   (c) Obtain the graph of $y = \operatorname{sech} x$.

34. (a) Show that $\lim\limits_{x \to +\infty} \operatorname{csch} x = \lim\limits_{x \to -\infty} \operatorname{csch} x = 0.$
   (b) Show that $\lim\limits_{x \to 0^+} \operatorname{csch} x = +\infty$ and
   $$\lim\limits_{x \to 0^-} \operatorname{csch} x = -\infty$$
   (c) Obtain the graph of $y = \operatorname{csch} x$.

35. Prove: $(\sinh x + \cosh x)^n = \sinh nx + \cosh nx$.

36. (a) It is sometimes said that $\sinh x$ and $\cosh x$ "behave like $\frac{1}{2}e^x$" for x large and positive. Give an informal justification of this statement.
   (b) Make a sketch showing the graphs of $y = \sinh x$, $y = \cosh x$, and $y = \frac{1}{2}e^x$, all in the same figure.

37. Find the arc length of $y = \cosh x$ between $x = 0$ and $x = \ln 2$.

38. Find the arc length of the catenary $y = a \cosh(x/a)$ between $x = 0$ and $x = x_1$ $(x_1 > 0)$.

7.7 FIRST-ORDER DIFFERENTIAL EQUATIONS AND APPLICATIONS

Many problems in science and engineering require the determination of functions satisfying equations containing derivatives. These equations are called *differential equations.* Some examples are

$$\frac{dy}{dx} = 3y \tag{1}$$

$$\frac{d^2y}{dx^2} - 6\frac{dy}{dx} + 8y = 0 \tag{2}$$

$$y' - y = e^{2x} \tag{3}$$

$$\frac{d^3y}{dt^3} - t\frac{dy}{dt} + (t^2 - 1)y = e^t \tag{4}$$

In the first three equations, $y = y(x)$ is an unknown function of x and in the last equation $y = y(t)$ is an unknown function of t. The *order* of a differential equation is the order of the highest derivative that appears in the equation. Thus, (1) and (3) are first-order equations, (2) is second order, and (4) is third order.

A function $y = y(x)$ is a *solution* of a differential equation if the equation is satisfied when $y(x)$ and its derivatives are substituted. For example,

$$y = e^{2x} \tag{5}$$

is a solution of the equation

$$\frac{dy}{dx} - y = e^{2x} \tag{6}$$

since

$$\frac{dy}{dx} - y = 2e^{2x} - e^{2x} = e^{2x}$$

Similarly,

$$y = e^x + e^{2x} \tag{7}$$

is a solution. More generally,

$$y = Ce^x + e^{2x} \tag{8}$$

is a solution for any constant C (verify). This solution is of special importance since it can be proved that *all* solutions of (6) can be obtained by substituting values for the arbitrary constant C. For example, $C = 0$ yields solution (5) and $C = 1$ yields solution (7). A solution of a differential equation from which all other solutions can be derived by substituting values for arbitrary constants is called the **general solution** of the equation. Usually, the general solution of an nth-order equation contains n arbitrary constants. Thus, (8), which is the general solution of the first-order equation (6), contains one arbitrary constant.

Often, solutions of differential equations are expressed as implicitly defined functions. For example,

$$\ln y = xy + C \tag{9}$$

defines a solution of

$$\frac{dy}{dx} = \frac{y^2}{1 - xy} \tag{10}$$

for any value of the constant C, since implicit differentiation of (9) yields

$$\frac{1}{y}\frac{dy}{dx} = x\frac{dy}{dx} + y$$

or

$$\frac{dy}{dx} - xy\frac{dy}{dx} = y^2$$

from which (10) follows.

In this section we will discuss some important classes of first-order differential equations. However, we are only touching the surface of a vast and important field. More complete studies of differential equations are available in numerous textbooks devoted to the subject.

A first-order differential equation is called **separable** if it is expressible in the form

$$\frac{dy}{dx} = \frac{g(x)}{h(y)} \quad .$$

To solve such an equation we rewrite it in the differential form

$$h(y)\, dy = g(x)\, dx \tag{11}$$

and integrate both sides, thereby obtaining the general solution

$$\int h(y)\, dy = \int g(x)\, dx + C$$

where C is an arbitrary constant.

In (11), the x and y variables are "separated" from each other, hence the term *separable* differential equation.

▶ Example 1 Solve the equation

$$\frac{dy}{dx} = \frac{x}{y^2}$$

Solution. Changing to differential form and integrating, yields

$$y^2\, dy = x\, dx$$

$$\int y^2\, dy = \int x\, dx$$

$$\frac{y^3}{3} = \frac{x^2}{2} + C$$

This expresses the solution implicitly. If desired, we can solve explicitly for y to obtain

$$y = \left(\frac{3}{2}x^2 + 3C\right)^{1/3}$$

or

$$y = \left(\frac{3}{2}x^2 + K\right)^{1/3}$$

where $K(= 3C)$ is an arbitrary constant. ◀

▶ **Example 2**   Solve the equation

$$x(y-1)\frac{dy}{dx} = y$$

Solution.   Separating the variables, and integrating yields

$$x(y-1)\,dy = y\,dx$$

$$\frac{y-1}{y}\,dy = \frac{dx}{x}$$

$$\int \frac{y-1}{y}\,dy = \int \frac{dx}{x}$$

$$\int \left(1 - \frac{1}{y}\right) dy = \int \frac{dx}{x}$$

$$y - \ln|y| = \ln|x| + C$$

or, using properties of logarithms,

$$y = \ln|xy| + C \qquad\qquad\qquad ◄$$

When a physical problem leads to a differential equation, there are usually conditions in the problem that determine specific values for the arbitrary constants in the general solution of the equation. For a first-order equation, a condition that specifies the value of the unknown function $y(x)$ at some point $x = x_0$ is called an ***initial condition***. A first-order differential equation together with one initial condition constitutes a ***first-order initial-value problem.***

▶ **Example 3**   Solve the initial-value problem

$$\frac{dy}{dx} = -4xy^2, \qquad y(0) = 1$$

Solution.   We first solve the differential equation:

$$\frac{dy}{y^2} = -4x\,dx$$

$$\int \frac{dy}{y^2} = \int -4x\,dx$$

$$-\frac{1}{y} = -2x^2 + C_1$$

or on multiplying by -1, taking reciprocals, and writing C in place of $-C_1$,

$$y = \frac{1}{2x^2 + C} \qquad\qquad\qquad (12)$$

The initial condition, $y(0) = 1$, requires that $y = 1$ when $x = 0$. Substituting these values in (12) yields $C = 1$. Thus, the solution of the initial-value problem is

$$y = \frac{1}{2x^2 + 1}$$ ◀

Not every first-order differential equation is separable. For example it is impossible to separate the variables in the equation

$$\frac{dy}{dx} + x^2 y = e^x$$

However, this equation can be solved by a different method that we will now consider.

A first-order differential equation is called **linear** if it is expressible in the form

$$\frac{dy}{dx} + p(x)y = q(x) \tag{13}$$

where the functions $p(x)$ and $q(x)$ may be constant. Some examples are

$$\frac{dy}{dx} + x^2 y = e^x \qquad\qquad [p(x) = x^2,\ q(x) = e^x]$$

$$y' - 3e^x y = 0 \qquad\qquad [p(x) = -3e^x,\ q(x) = 0]$$

$$\frac{dy}{dx} + 5y = 2 \qquad\qquad [p(x) = 5,\ q(x) = 2]$$

$$\frac{dy}{dx} + (\sin x)y + x^3 = 0 \qquad [p(x) = \sin x,\ q(x) = -x^3]$$

One procedure for solving (13) is based on the observation that if we define $\rho = \rho(x)$ by

$$\rho = e^{\int p(x)\,dx}$$

then

$$\frac{d\rho}{dx} = e^{\int p(x)\,dx} \cdot \frac{d}{dx}\int p(x)\,dx = \rho p(x)$$

Thus,

$$\frac{d}{dx}(\rho y) = \rho\frac{dy}{dx} + \frac{d\rho}{dx}y = \rho\frac{dy}{dx} + \rho p(x)y \tag{14}$$

If (13) is multiplied through by ρ, it becomes

$$\rho \frac{dy}{dx} + \rho p(x) y = \rho q(x)$$

or from (14),

$$\frac{d}{dx}(\rho y) = \rho q(x)$$

This equation can be solved by integrating both sides to obtain

$$\rho y = \int \rho q(x)\, dx + C$$

or

$$y = \frac{1}{\rho}\left[\int \rho q(x)\, dx + C\right]$$

To summarize, (13) can be solved in three steps:

Step 1. Calculate

$$\rho = e^{\int p(x)\, dx}$$

This is called the ***integrating factor.*** Since any ρ will suffice, we can take the constant of integration to be zero in this step.

Step 2. Multiply both sides of (13) by ρ and express the result as

$$\frac{d}{dx}(\rho y) = \rho q(x)$$

Step 3. Integrate both sides of the equation obtained in Step 2 and then solve for y. Be sure to include a constant of integration in this step.

▶ Example 4   Solve the equation

$$\frac{dy}{dx} - 4xy = x \qquad\qquad (15)$$

Solution.   Since $p(x) = -4x$, the integrating factor is

$$\rho = e^{\int(-4x)\, dx} = e^{-2x^2}$$

If we multiply (15) by ρ we obtain

$$\frac{d}{dx}(e^{-2x^2}y) = xe^{-2x^2}$$

Integrating both sides of this equation yields

$$e^{-2x^2}y = \int xe^{-2x^2}\,dx = -\frac{1}{4}e^{-2x^2} + C$$

and then multiplying both sides by e^{2x^2}, yields

$$y = -\frac{1}{4} + Ce^{2x^2} \qquad \blacktriangleleft$$

▶ **Example 5** Solve the equation

$$x\frac{dy}{dx} - y = x \qquad (x > 0)$$

Solution. To put the equation in form (13), we divide through by x to obtain

$$\frac{dy}{dx} - \frac{1}{x}y = 1 \tag{16}$$

Since $p(x) = -1/x$, the integrating factor is

$$\rho = e^{\int -(1/x)\,dx} = e^{-\ln|x|} = \frac{1}{|x|} = \frac{1}{x}$$

(The absolute value was dropped because of the assumption that $x > 0$.) If we multiply (16) by ρ, we obtain

$$\frac{d}{dx}\left(\frac{1}{x}y\right) = \frac{1}{x}$$

Integrating both sides of this equation yields

$$\frac{1}{x}y = \int \frac{1}{x}\,dx = \ln x + C$$

or

$$y = x\ln x + Cx \qquad \blacktriangleleft$$

We conclude this section with some applications of first-order differential equations.

▶ Example 6 (*Geometry*) Find a curve in the xy-plane that passes through $(0, 3)$ and whose tangent at a point (x, y) has slope $2x/y^2$.

Solution. Since the slope of the tangent is dy/dx, we have

$$\frac{dy}{dx} = \frac{2x}{y^2} \tag{17}$$

and, since the curve passes through $(0, 3)$, we have the initial condition

$$y(0) = 3 \tag{18}$$

Equation (17) is separable and can be written as

$$y^2 \, dy = 2x \, dx$$

so

$$\int y^2 \, dy = \int 2x \, dx$$

or

$$\frac{1}{3} y^3 = x^2 + C$$

From the initial condition, (18), it follows that $C = 9$, and so the curve has the equation

$$\frac{1}{3} y^3 = x^2 + 9$$

or

$$y = (3x^2 + 27)^{1/3} \qquad \blacktriangleleft$$

▶ Example 7 (*Mixing Problems*) At time $t = 0$, a tank contains 4 lb of salt dissolved in 100 gal of water. Suppose that brine containing 2 lb of salt per gallon of water is allowed to enter the tank at a rate of 5 gal/min and that the mixed solution is drained from the tank at the same rate. Find a formula for the amount of salt in the tank after 10 min.

Solution. Let $y(t)$ be the amount of salt (in pounds) at time t. We are interested in finding $y(10)$, the amount of salt at time $t = 10$. We will begin by finding an expression for dy/dt, the rate of change of the amount of salt in the tank at time t. Clearly,

$$\frac{dy}{dt} = \text{rate in} - \text{rate out} \tag{19}$$

where *rate in* is the rate at which salt enters the tank and *rate out* is the rate at which salt leaves the tank. But

$$\text{rate in} = (2 \text{ lb/gal}) \cdot (5 \text{ gal/min}) = 10 \text{ lb/min}$$

At time t, the mixture contains $y(t)$ pounds of salt in 100 gallons of water; thus, the concentration of salt at time t is $y(t)/100$ lb per gal and

$$\text{rate out} = \left(\frac{y(t)}{100} \text{ lb/gal}\right) \cdot (5 \text{ gal/min}) = \frac{y(t)}{20} \text{ lb/min}$$

Therefore, (19) can be written

$$\frac{dy}{dt} = 10 - \frac{y}{20}$$

or

$$\frac{dy}{dt} + \frac{y}{20} = 10 \tag{20}$$

which is a first-order linear differential equation. We also have the initial condition

$$y(0) = 4 \tag{21}$$

since the tank contains 4 lb of salt at time $t = 0$.

Multiplying both sides of (20) by the integrating factor

$$\rho = e^{\int(1/20)\,dt} = e^{t/20}$$

yields

$$\frac{d}{dt}(e^{t/20}y) = 10e^{t/20}$$

so

$$e^{t/20}y = \int 10e^{t/20}\,dt = 200e^{t/20} + C$$

or

$$y(t) = 200 + Ce^{-t/20}$$

From the initial condition, (21), it follows that

$$4 = 200 + C$$

or $C = -196$, so

$$y(t) = 200 - 196e^{-t/20}$$

Thus, after 10 min ($t = 10$), the amount of salt in the tank is

$$y(10) = 200 - 196e^{-0.5} \approx 81.1 \text{ lb}$$ ◀

EXPONENTIAL GROWTH

Many quantities increase or decrease with time in proportion to the amount of the quantity present. Some examples are human population, bacteria in a culture, drug concentration in the bloodstream, radioactivity, and the values of certain kinds of investments. We will show how differential equations can be used to study the growth and decay of such quantities.

7.7.1 DEFINITION A quantity is said to have an ***exponential growth (decay) model*** if at each instant of time its rate of increase (decrease) is proportional to the amount of the quantity present.

Consider a quantity with an exponential growth or decay model and denote by $y(t)$ the amount of the quantity present at time t. We assume that $y(t) > 0$ for all t. Since the rate of change of $y(t)$ is proportional to the amount present, it follows that $y(t)$ satisfies

$$\frac{dy}{dt} = ky \tag{22}$$

where k is a constant of proportionality.

The constant k in (22) is called the ***growth constant*** if $k > 0$ and the ***decay constant*** if $k < 0$. If $k > 0$, then $dy/dt > 0$, so that y is increasing with time (a growth model). If $k < 0$, then $dy/dt < 0$, so that y is decreasing with time (a decay model).

REMARK. In problems of exponential growth and decay, the constant k is often expressed as a percentage. Thus, a growth constant of 3% means $k = 0.03$ and a decay constant of -500% means $k = -5$. We also note that the constant k is often called the ***growth rate***. Strictly speaking, this is not correct; as indicated by (22), the growth rate dy/dt is not k, but rather ky. However, the description of k as a growth rate is so common that we will use this standard terminology in this section.

If we are given a quantity with an exponential growth or decay model, and if we know the amount y_0 present at some initial time $t = 0$, then we can find the amount present at any time t by solving the initial-value problem

$$\frac{dy}{dt} = ky, \qquad y(0) = y_0 \tag{23}$$

The differential equation is both separable and linear and thus can be solved by either of the methods we have studied in this section. We will treat it as a linear equation. Rewriting the differential equation as

$$\frac{dy}{dt} - ky = 0$$

and multiplying through by the integrating factor

$$\rho = e^{\int -k \, dt} = e^{-kt}$$

yields

$$\frac{d}{dt}(e^{-kt}y) = 0$$

After integrating,

$$e^{-kt}y = C$$

or

$$y = Ce^{kt}$$

From the initial condition, $y(0) = y_0$, it follows that $C = y_0$; thus the solution of (23) is

$$y(t) = y_0 e^{kt} \tag{24}$$

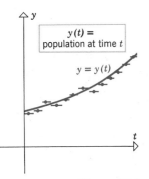

y(t) =
population at time t

$y = y(t)$

Figure 7.7.1

Exponential models have proved useful in studies of population growth. Although populations (e.g., people, bacteria, and flowers) grow in discrete steps, we can apply the results of this section if we are willing to approximate the population graph by a continuous curve $y = y(t)$ (Figure 7.7.1).

▶ **Example 8** *(Population Growth)* According to United Nations data, the world population at the beginning of 1975 was approximately 4 billion and growing at a rate of about 2% per year. Assuming an exponential growth model, estimate the world population at the beginning of the year 2000.

Solution. Let

$t = $ time (in years) elapsed from the beginning of 1975

$y = $ world population in billions

Since the beginning of 1975 corresponds to $t = 0$, it follows from the given data that

$$y_0 = y(0) = 4 \text{ (billion)}$$

Since the growth rate is 2% ($k = 0.02$), it follows from (24) that the world population at time t will be

$$y(t) = y_0 e^{kt} = 4e^{0.02t} \tag{25}$$

Since the beginning of the year 2000 corresponds to $t = 25$ $(2000 - 1975 = 25)$, it follows from (25) that the world population by the year 2000 will be

$$y(25) = 4e^{0.02(25)} = 4e^{0.5}$$

or from Appendix 3, Table 2,

$$y(25) = 4(1.6487) = 6.5948 \text{ (billion)}$$

which is a population of approximately 6.6 billion. ◀

If a quantity has an exponential growth model, then the time required for it to double in size is called the *doubling time*. Similarly, if a quantity has an exponential decay model, then the time required for it to reduce in value by half is called the *halving time*. As it turns out, doubling and halving times depend only on the growth rate and not on the amount present initially. To see why, suppose y has an exponential growth model so that

$$y = y_0 e^{kt} \qquad (k > 0)$$

At any fixed time t_1 let

$$y_1 = y_0 e^{kt_1} \tag{26}$$

be the value of y and let T denote the amount of time required for y to double in size. Thus, at time $t_1 + T$ the value of y will be $2y_1$ so that

$$2y_1 = y_0 e^{k(t_1 + T)} = y_0 e^{kt_1} e^{kT}$$

or, from (26),

$$2y_1 = y_1 e^{kT}$$

Thus,

$$2 = e^{kT} \qquad \text{and} \qquad \ln 2 = kT$$

Therefore, the doubling time T is

$$T = \frac{1}{k} \ln 2 \tag{27}$$

which does not depend on y_0 or t_1. We leave it as an exercise to show that the halving time for a quantity with an exponential decay model $(k < 0)$ is

$$T = -\frac{1}{k} \ln 2 \tag{28}$$

▶ Example 9   It follows from (27) that at the current 2% annual growth rate, the doubling time for the world population is

$$T = \frac{1}{0.02} \ln 2$$

$$= \frac{1}{0.02}(0.6931) \qquad \text{[Appendix 3, Table 3]}$$

$$= 34.655$$

or approximately 35 years. Thus, with a continued 2% annual growth rate the population of 4 billion in 1975 will double to 8 billion by the year 2010 and will double again to 16 billion by 2045.   ◀

Radioactive elements continually undergo a process of disintegration called *radioactive decay.* It is a physical fact that at each instant of time the rate of decay is proportional to the amount of the element present. Consequently, the amount of any radioactive element has an exponential decay model. For radioactive elements, halving time is called *half-life.*

▶ Example 10 (*Radioactive Decay*)   The radioactive element carbon-14 has a half-life of 5750 years. If 100 grams of this element are present initially, how much will be left after 1000 years?

Solution.   From (28) the decay constant is

$$k = -\frac{1}{T} \ln 2$$

$$= -\frac{1}{5750}(0.6931) \qquad \text{[Appendix 3, Table 3]}$$

$$\approx -0.00012$$

Thus, if we take $t = 0$ to be the present time, then $y_0 = y(0) = 100$, so that (24) implies that the amount of carbon-14 after 1000 years will be

$$y(1000) = 100e^{-0.00012(1000)}$$

$$= 100e^{-0.12}$$

$$\approx 100 \, (0.88692)$$

$$= 88.692$$

Thus, about 88.69 grams of carbon-14 will remain.   ◀

► Exercise Set 7.7

In Exercises 1–6, solve the given separable differential equation. Where convenient, express the solution explicitly as a function of x.

1. $\dfrac{dy}{dx} = \dfrac{y}{x}$.

2. $\dfrac{dy}{dx} = \dfrac{x^3}{(1 + x^4)y}$.

3. $\sqrt{1 + x^2}\, y' + x(1 + y) = 0$.

4. $x \tan y - \dfrac{dy}{dx} \sec x = 0$.

5. $e^{-y} \sin x - y' \cos^2 x = 0$.

6. $\dfrac{dy}{dx} = 1 - y + x^2 - yx^2$.

In Exercises 7–12, solve the given first-order linear differential equation. Where convenient, express the solution explicitly as a function of x.

7. $\dfrac{dy}{dx} + 3y = e^{-2x}$.   8. $\dfrac{dy}{dx} - \dfrac{5}{x}y = x$ $(x > 0)$.

9. $y' + y = \cos(e^x)$.   10. $2\dfrac{dy}{dx} + 4y = 1$.

11. $x^2y' + 3xy + 2x^5 = 0$ $(x > 0)$.

12. $\dfrac{dy}{dx} + y - \dfrac{1}{1 + e^x} = 0$.

In Exercises 13–18, solve the initial-value problems.

13. $\dfrac{dy}{dx} - xy = x$, $y(0) = 3$.

14. $2y\dfrac{dy}{dx} = 3x^2(x^3 + 1)^{-1/2}$, $y(2) = 1$.

15. $\dfrac{dy}{dt} + y = 2$, $y(0) = 1$.

16. $y' - xe^y = 2e^y$, $y(0) = 0$.

17. $y^2t\dfrac{dy}{dt} - t + 1 = 0$, $y(1) = 3$.

18. $y' \cosh x + y \sinh x = \cosh^2 x$, $y(0) = \dfrac{1}{4}$.

19. Find a curve in the xy-plane that passes through the point $(1, 2)$ and whose tangent at a point (x, y) has slope $2 - (y/x)$.

20. Find a curve in the xy-plane that passes through the point $(1, 1)$ and whose normal at a point (x, y) has slope $-2y/(3x^2)$.

21. The line normal to a curve at any point (x, y) passes through $(3, 0)$. Given that the curve passes through $(3, 2)$ find its equation.

22. A function f with positive values has the property that its value at every point is twice the slope of the tangent at the point. Given that $f(0) = 2$, find f.

In Exercises 23–40, use a hand-held calculator or the tables in the appendix where needed.

23. At time $t = 0$, a tank contains 25 lb of salt dissolved in 50 gal of water. Then brine containing 4 lb of salt per gallon of water is allowed to enter the tank at a rate of 2 gal/min and the mixed solution is drained from the tank at the same rate.
    (a) How much salt is in the tank at an arbitrary time t?
    (b) How much salt is in the tank after 25 min?

24. A tank initially contains 200 gal of pure water. Then at time $t = 0$ brine containing 5 lb of salt per gallon of water is allowed to enter the tank at a rate of 10 gal/min and the mixed solution is drained from the tank at the same rate.
    (a) How much salt is in the tank at an arbitrary time t?
    (b) How much salt is in the tank after 30 min?

25. A tank with a 1000-gal capacity initially contains 500 gal of brine containing 50 lb of salt. At time $t = 0$, pure water is added at a rate of 20 gal/min and the mixed solution is drained off at a rate of 10 gal/min. How much salt is in the tank when it reaches the point of overflowing?

26. The number of bacteria in a certain culture grows exponentially at a rate of 1% per hour. Assuming that 10,000 bacteria are present initially, find
    (a) the number of bacteria present at any time t;
    (b) the number of bacteria present after 5 hr;
    (c) the time required for the number of bacteria to reach 45,000.

27. Polonium-210 is a radioactive element with a half-life of 140 days. Assume a sample weighs 10 mg initially.

(a) Find a formula for the amount that will remain after t days.

(b) How much will remain after 10 weeks?

28. In a certain chemical reaction a substance decomposes at a rate proportional to the amount present. Tests show that under appropriate conditions 15,000 grams will reduce to 5000 grams in 10 hours.

(a) Find a formula for the amount that will remain from a 15,000 g sample after t hours.

(b) How long will it take for 50% of an initial sample of y_0 grams to decompose?

29. One hundred fruit flies are placed in a breeding container that can support a population of at most 5000 flies. If the population grows exponentially at a rate of 2% per day, how long will it take for the container to reach capacity?

30. In 1960 the American scientist W. F. Libby won the Nobel prize for his discovery of carbon dating, a method for determining the age of certain fossils. Carbon dating is based on the fact that nitrogen is converted to radioactive carbon-14 by cosmic radiation in the upper atmosphere. This radioactive carbon is absorbed by plant and animal tissue through the life processes while the plant or animal lives. However, when the plant or animal dies the absorption process stops and the amount of carbon-14 decreases through radioactive decay. Suppose that tests on a fossil show that 70% of its carbon-14 has decayed. Estimate the age of the fossil, assuming a half-life of 5750 years for carbon-14.

31. Forty percent of a radioactive substance decays in 5 years. Find the half-life of the substance.

32. Assume that if the temperature is constant, then the atmospheric pressure p varies with the altitude h (above sea level) in such a way that

$$\frac{dp}{dh} = kp$$

where k is a constant.

(a) Find a formula for p in terms of k, h, and the atmospheric pressure p_0 at sea level.

(b) Given that p measures 15 lb/in.^{2} at sea level and 12 lb/in.^{2} at 5000 ft above sea level, find the pressure at 10,000 ft (assuming temperature is constant).

33. The town of Grayrock had a population of 10,000 in 1960 and 12,000 in 1970.

(a) Assuming an exponential growth model, estimate the population in 1980.

(b) What is the doubling time for the town's population?

34. Prove: If a quantity A has an exponential growth or decay model and A has values A_1 and A_2 at times t_1 and t_2, respectively, then the growth rate k is

$$k = \frac{1}{t_1 - t_2} \ln\left(\frac{A_1}{A_2}\right)$$

35. Newton's law of cooling states that the rate at which an object cools is proportional to the difference in temperature between the object and the surrounding medium. Show that if C is the constant temperature of a surrounding medium, then the temperature $T(t)$ at time t of a cooling object is given by

$$T(t) = (T_0 - C)e^{kt} + C$$

where T_0 is the temperature of the object at $t = 0$ and k is a negative constant.

36. A liquid with initial temperature 200° is surrounded by a body of air at a constant temperature of 80°. If the liquid cools to 120° in 30 min, what will the temperature be after 1 hr? (Use the result of Exercise 35.)

37. Suppose P dollars is invested at an annual interest rate of $r \times 100\%$. If the accumulated interest is credited to the account at the end of the year, then the interest is said to be *compounded annually;* if it is credited at the end of each six-month period then it is said to be *compounded semi-annually;* and if it is credited at the end of each three-month period, then it is said to be *compounded quarterly.* The more frequently the interest is compounded, the better it is for the investor since more of the interest is itself earning interest.

(a) Show that if interest is compounded n times a year at equally spaced intervals, then the value A of the investment after t years is

$$A = P\left(1 + \frac{r}{n}\right)^{nt}$$

(b) One can imagine interest to be compounded each day, each hour, each minute, and so forth. Carried to the limit one can conceive of interest compounded at each instant of time; this is called *continuous compounding*. Thus, from (a), the value A of P dollars after t years when invested at an annual rate of $r \times 100\%$, compounded continuously, is

$$A = \lim_{n \to +\infty} P\left(1 + \frac{r}{n}\right)^{nt}$$

Use the fact that $\lim_{x \to 0}(1 + x)^{1/x} = e$ to prove that

$$A = Pe^{rt}$$

(c) Use the result in (b) to show that money invested at continuous compound interest increases at a rate proportional to the amount present.

38. (a) If $1000 is invested at 8% per year compounded continuously (Exercise 37) what will the investment be worth after 5 years?

(b) If it is desired that an investment at 8% per year compounded continuously should have a value of $10,000 after 10 years, how much should be invested now?

(c) How long does it take for an investment at 8% per year compounded continuously to double in value?

39. Derive Formula (28) for halving time.

40. Let a quantity have an exponential growth model with growth rate k. How long does it take for the quantity to triple in size?

▶ SUPPLEMENTARY EXERCISES

1. If $r = \ln 2$ and $s = \ln 3$, express the following in terms of r and s:
   (a) $\ln(1/12)$ (b) $\ln(9/\sqrt{8})$
   (c) $\ln(\sqrt[4]{8/3})$.

2. Simplify.
   (a) $e^{2 - \ln x}$ (b) $\exp(\ln x^2 - 2\ln y)$
   (c) $\ln[x^3 \exp(-x^2)]$.

3. Solve for x in terms of $\ln 3$ and $\ln 5$.
   (a) $25^x = 3^{1-x}$ (b) $\sinh x = \frac{1}{4}\cosh x$.

4. Express the following as a rational function of x:
   $3\ln\left(e^{2x}(e^x)^3\right) + 2\exp(\ln 1)$.

5. If $\sinh x = -3/5$, find:
   (a) $\cosh x$ (b) $\tanh x$ (c) $\sinh(2x)$.

6. Use appendix tables to approximate the following quantities to two decimal places.
   (a) 2^e (b) $(\sqrt{2})^\pi$

In Exercises 7–28, find dy/dx. When appropriate, use implicit or logarithmic differentiation.

7. $y = 1/\sqrt{e^x}$.   8. $y = 1/e^{\sqrt{x}}$.

9. $y = x/\ln x$.   10. $y = e^x \ln(1/x)$.

11. $y = x/e^{\ln x}$.   12. $y = \ln\sqrt{x^2 + 2x}$.

13. $y = \ln(10^x/\sin x)$.   14. $y = \cos(e^{-2x})$.

15. $y = e^{\tan x}e^{4\ln x}$.   16. $y = \ln\left|\frac{a + x}{a - x}\right|$.

17. $y = \ln|x + \sqrt{x^2 + a^2}|$.

18. $y = \ln|\tan 3x + \sec 3x|$.

19. $y = [\exp(x^2)]^3$.

20. $y = \ln(x^3/\sqrt{5 + \sin x})$.

21. $y = \sqrt{\ln(\sqrt{x})}$.

22. $y = e^{5x} + (5x)^e$.

23. $y = \pi^x x^\pi$.

24. $y = 4(e^x)^3/\sqrt{\exp{(5x)}}$.

25. $y = \sinh{[\tanh{(5x)}]}$.

26. $x^4 + e^{xy} - y^2 = 20$.

27. $y = e^{3x}(1 + e^{-x})^2$.

28. $y = (\cosh{x})^{x^3}$.

29. Show that $y = e^{ax}\sin{bx}$ satisfies $y'' - 2ay' + (a^2 + b^2)y = 0$ for any real constants a and b.

30. Assuming that u and v are differentiable functions of x, use logarithmic differentiation to derive a formula for $d(u^v)/dx$ in terms of u, v, du/dx, and dv/dx. What formula results when the base u is constant? When the exponent v is constant?

31. Find dy if
(a) $y = e^{-x}$ (b) $y = \ln{(1 + x)}$
(c) $y = 2^{x^2}$.

32. If $y = 3^{2x}5^{7x}$, show that dy/dx is proportional to y.

33. Use the chain rule and the Second Fundamental Theorem of Calculus to find the indicated derivative.

(a) $\dfrac{d}{dx}\left(\displaystyle\int_0^{\ln{x}}\dfrac{dt}{\sqrt{4 + e^t}}\right)$

(b) $\dfrac{d}{dx}\left(\displaystyle\int_1^{e^{5x}}\sqrt{\ln{u} + u}\;du\right)$.

[*Hint:* See Exercise 12, Section 5.10.]

34. Suppose $F(-\pi) = 0$ and that $y = F(x)$ satisfies $dy/dx = (3\sin^2{2x} + 2\cos^2{3x})^{1/2}$. Express $F(x)$ as an integral.

35. If $y = Ce^{kt}$ (C, k constant) and $Y = \ln{y}$, show that the graph of Y versus t is a straight line.

36. Evaluate $\int e^{2x}(4 + e^{2x})\;dx$ two ways: (a) using the substitution $u = 4 + e^{2x}$, and (b) changing the integrand to a sum. Verify that the antiderivatives in (a) and (b) differ by a constant.

In Exercises 37–52, evaluate the indicated integral.

37. $\displaystyle\int\dfrac{e^x}{1 + e^x}\;dx$.

38. $\displaystyle\int\dfrac{1 + e^x}{e^x}\;dx$.

39. $\displaystyle\int x^e\;dx$.

40. $\displaystyle\int\dfrac{x^2}{5 - 2x^3}\;dx$.

41. $\displaystyle\int\dfrac{4x^2 - 3x}{x^3}\;dx$.

42. $\displaystyle\int\dfrac{(\ln{x^2})^2}{x}\;dx$.

43. $\displaystyle\int\dfrac{\sec{x}\tan{x}}{2\sec{x} - 1}\;dx$.

44. $\displaystyle\int\dfrac{e^{5x}}{3 + e^{5x}}\;dx$.

45. $\displaystyle\int(\cos{2x})\exp{(\sin{2x})}\;dx$.

46. $\displaystyle\int\tanh{(3x + 1)}\;dx$.

47. $\displaystyle\int\text{sech}^2{x}\tanh{x}\;dx$.

48. $\displaystyle\int e^{2x}(4 + e^{-3x})\;dx$.

49. $\displaystyle\int_e^{e^2}\dfrac{dx}{x\ln{x}}$.

50. $\displaystyle\int_0^1\dfrac{dx}{\sqrt{e^x}}$.

51. $\displaystyle\int_0^{\pi/4}\dfrac{2\tan{x}}{\cos^2{x}}\;dx$.

52. $\displaystyle\int_1^4\dfrac{dx}{\sqrt{x}e^{\sqrt{x}}}$.

53. Show that $\int e^{kx}\;dx = (e^{kx}/k) + C$ for any nonzero constant k.

54. In each part find the indicated limit by interpreting the expression as a derivative and evaluating the derivative.

(a) $\displaystyle\lim_{h\to 0}\dfrac{10^h - 1}{h}$

(b) $\displaystyle\lim_{h\to 0}\dfrac{e^{(3+h)^2} - e^9}{h}$

(c) $\displaystyle\lim_{h\to 0}\dfrac{\ln{(e^2 + h)} - 2}{h}$

(d) $\displaystyle\lim_{x\to 1}\dfrac{2^x - 2}{x - 1}$.

55. Sketch $y = x^3e^{-x}$, showing all relative extrema and inflection points. You may assume that $\lim_{x\to +\infty}x^3e^{-x} = 0$.

56. Use the second derivative test to determine the nature of the critical points of $f(x) = x^2e^{-x}$.

57. Show that if $y = f(x)$ satisfies $y' = e^y + 2y + x$, then every critical point of y is a relative minimum.

58. Sketch the curve $y = e^{-x/2}\sin{2x}$ on $[-\pi/2, 3\pi/2]$. Show all x-intercepts and all points where $y = \pm e^{-x/2}$.

59. Let A denote the area under the curve $y = e^{-2x}$ over the interval $[0, b]$. Express A as a function of b, and then find the value of b for which $A = \frac{1}{4}$. What is the limiting value of A as $b \to +\infty$.

60. Let R be the region bounded by the curve $y = 4(2x - 1)^{-1/2}$, the x-axis, and the lines $x = 1$ and $x = 3$. Find the volume of the solid generated by revolving R about the x-axis.

61. Find the surface area generated when the curve $y = \cosh x$, $0 \le x \le 1$, is revolved about the x-axis.
[*Hint:* Equation (7b) of Section 7.6 will help to evaluate the integral.]

62. Suppose that a crystal dissolves at a rate proportional to the amount *un*dissolved. If 9 grams are undissolved initially and 6 grams remain undissolved after 1 min, how many grams remain undissolved after 3 min?

63. The population of the United States was 205 million in 1970. Assuming an annual population increase of 1.8%, find: (a) the population in the year 2000, (b) the year in which the population will reach one billion.

8
inverse trigonometric and hyperbolic functions

8.1 INVERSE FUNCTIONS

The functions $f(x) = x^{1/3}$ and $g(x) = x^3$ have the property that each cancels out the effect of the other in the sense that

$$f(g(x)) = f(x^3) = (x^3)^{1/3} = x$$
$$g(f(x)) = g(x^{1/3}) = (x^{1/3})^3 = x$$

Similarly, the functions $f(x) = \ln x$ and $g(x) = e^x$ cancel the effect of one another since

$$f(g(x)) = f(e^x) = \ln(e^x) = x$$
$$g(f(x)) = g(\ln x) = e^{\ln x} = x$$

In this section we will study pairs of functions like these and in subsequent sections we will give applications of our results.

8.1.1 DEFINITION   If two functions f and g satisfy the conditions:

$$f(g(x)) = x \text{ for every } x \text{ in the domain of } g$$
$$g(f(x)) = x \text{ for every } x \text{ in the domain of } f$$

then we say that *f is an inverse of g* and *g is an inverse of f.* We also say that *f and g are inverses of one another.*

▶ Example 1   The functions $x^{1/3}$ and x^3 are inverses of one another, as are e^x and $\ln x$.   ◀

445

It can be shown that a function cannot have two different inverses. Thus, if f has an inverse we are entitled to talk about *the* inverse of f. The inverse of f is commonly denoted by f^{-1} (read, "f inverse"); thus,

$$f(f^{-1}(x)) = x \quad \text{and} \quad f^{-1}(f(x)) = x$$

WARNING.   The symbol f^{-1} does not mean $1/f$.

Given a function f, we will be interested in two important questions:

1.   Does f have an inverse?
2.   If so, how do we find it?

To answer the first question, it will be helpful to understand how the graphs of f and f^{-1} are related when f has an inverse.

In Section 7.5 we saw that the graphs of the inverse functions $f(x) = \ln x$ and $f^{-1}(x) = e^x$ are reflections of one another about the line $y = x$ (Figure 7.5.2). This is not accidental; the graphs of any two inverse functions are related in this way. To prove this, we must show that for each point (a, b) on the graph of f, the reflection (b, a) about $y = x$ lies on the graph of f^{-1}, and conversely. We argue as follows: Let (a, b) be a point on the curve $y = f(x)$. Thus,

$$b = f(a)$$
$$f^{-1}(b) = f^{-1}(f(a))$$
$$f^{-1}(b) = a$$

The last equation implies that (b, a) lies on the curve $y = f^{-1}(x)$. Similarly, we can show that if (a, b) lies on the curve $y = f^{-1}(x)$, then (b, a) lies on the curve $y = f(x)$.

▶ Example 2   The graphs of the inverse functions $f(x) = x^{1/3}$ and $f^{-1}(x) = x^3$ are shown in Figure 8.1.1.   ◀

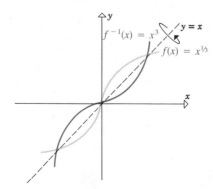

Figure 8.1.1

Let us turn now to the problem of determining conditions that ensure that a function f has an inverse. We begin with a simple example of a function with no inverse, $f(x) = x^2$. To see that no inverse exists, we need only observe that if f had an inverse, then the curve $y = f^{-1}(x)$ would be the reflection about $y = x$ of the curve $y = x^2$ (Figure 8.1.2a). But the reflected curve (Figure 8.1.2b) is not the graph of any function because it does not pass the vertical line test (see Theorem 2.1.4). Thus, $f(x) = x^2$ has no inverse.

Figure 8.1.2   (a)   (b)

Figure 8.1.2 suggests that the reflection about $y = x$ of a curve $y = f(x)$ will fail to pass the vertical line test if there is some *horizontal* line that cuts the curve $y = f(x)$ more than once. Thus, *for a function f to have an inverse, its graph cannot be cut more than once by any horizontal line.* A function f with this property is said to be *one-to-one.*

8.1.2   DEFINITION   A function f is **one-to-one,** written 1–1, if its graph is cut at most once by any horizontal line, or equivalently, if f does not have the same value at two distinct points in its domain.

The following theorem summarizes the preceding discussion. (We will omit the formal proof.)

8.1.3   THEOREM   *If f is one-to-one then f has an inverse, and conversely if f has an inverse then f is one-to-one.*

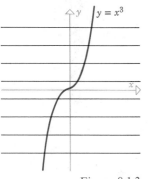

► **Example 3**   The function $f(x) = x^3$ is one-to-one since two different numbers cannot have the same cube, that is, if $x_1 \neq x_2$, then $x_1{}^3 \neq x_2{}^3$. Geometrically, no horizontal line cuts the graph of $f(x) = x^3$ more than once (Figure 8.1.3). ◄

► **Example 4**   The function $f(x) = x^2$ is not one-to-one because there are lines that cut its graph more than once (Figure 8.1.2a). From an algebraic viewpoint, $f(x) = x^2$ is not one-to-one because f can have the same value at two distinct points; for example, $f(-3) = 9$ and $f(3) = 9$. ◄

Figure 8.1.3   ► **Example 5**   Determine whether $f(x) = \sin x$ has an inverse.

Solution. There exist horizontal lines that cut the graph of sin x more than once (Figure 8.1.4), so $f(x) = \sin x$ is not one-to-one and therefore has no inverse.

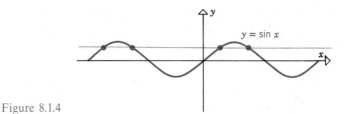

Figure 8.1.4

The following theorem describes a major class of functions with inverses.

8.1.4   THEOREM   *If the domain of f is an interval, and if f is either an increasing function or a decreasing function on that interval, then f has an inverse.*

Proof.   If f is increasing and $x_1 < x_2$ are distinct points in its domain, then $f(x_1) < f(x_2)$ (Definition 4.2.1), so that $f(x_1) \neq f(x_2)$. Thus, f is one-to-one and therefore has an inverse by Theorem 8.1.3. The proof for decreasing functions is similar. ▮

Recall from Theorem 4.2.2 that f is increasing on an open interval if $f'(x) > 0$ on the interval and is decreasing if $f'(x) < 0$. This result makes it easy to apply Theorem 8.1.4 to differentiable functions.

▶ **Example 6**   Show that $f(x) = x^5 + 7x^3 + 4x + 1$ has an inverse.

Solution.   The function f is increasing on $(-\infty, +\infty)$ because

$$f'(x) = 5x^4 + 21x^2 + 4 > 0$$

for all x. Thus, f has an inverse by Theorem 8.1.4.   ◀

We know from Theorem 8.1.3 that the functions with inverses are precisely those that are one-to-one. Suppose that f is such a function; how do we find $f^{-1}(x)$? One possible procedure is to solve the equation

$$f(f^{-1}(x)) = x$$

for $f^{-1}(x)$.

▶ **Example 7**   Find the inverse of $f(x) = 4x - 5$.

Solution.   Replacing x by $f^{-1}(x)$ in the formula for f yields

$$f(f^{-1}(x)) = 4f^{-1}(x) - 5$$

or equivalently,

$$x = 4f^{-1}(x) - 5$$

Solving for $f^{-1}(x)$, we obtain

$$f^{-1}(x) = \frac{1}{4}(x + 5)$$

The graphs of f and f^{-1} are shown in Figure 8.1.5. As expected, they are reflections of one another about $y = x$.

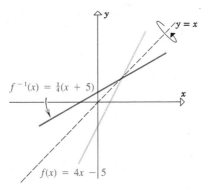

Figure 8.1.5 ◀

▶ **Example 8** Find the inverse of $f(x) = e^{1-x}$.

Solution. From the formula for f we obtain

$$f(f^{-1}(x)) = e^{1-f^{-1}(x)}$$

or equivalently

$$x = e^{1-f^{-1}(x)}$$

Thus,

$$\ln x = 1 - f^{-1}(x)$$

or

$$f^{-1}(x) = 1 - \ln x \qquad ◀$$

▶ **Example 9** In Example 6 we proved that $f(x) = x^5 + 7x^3 + 4x + 1$ has an inverse. However, if we try to find it by the procedure used in the last two examples we obtain

$$f(f^{-1}(x)) = [f^{-1}(x)]^5 + 7[f^{-1}(x)]^3 + 4[f^{-1}(x)] + 1$$

or

$$x = [f^{-1}(x)]^5 + 7[f^{-1}(x)]^3 + 4[f^{-1}(x)] + 1$$

But this equation is too complicated to solve for $f^{-1}(x)$, so we cannot find f^{-1} even though we know it exists. ◄

DERIVATIVES OF INVERSE FUNCTIONS

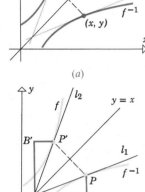

(a)

(b)

Figure 8.1.6

We conclude this section by investigating the relationship between the derivative of a function and the derivative of its inverse. The result we want can be motivated by Figure 8.1.6. In Figure 8.1.6a we show graphs of two inverse functions and tangent lines l_1 and l_2 at the points (x, y) and (y, x). Referring to Figure 8.1.6b, the slopes m_1 and m_2 of tangents l_1 and l_2 are

$$m_1 = \frac{\overline{BP}}{\overline{AB}} \qquad \text{and} \qquad m_2 = \frac{\overline{AB'}}{\overline{B'P'}} \tag{1}$$

But triangles ABP and $AB'P'$ are congruent (verify) so that $\overline{BP} = \overline{B'P'}$ and $\overline{AB} = \overline{AB'}$. From these equalities and (1) it follows that

$$m_1 = \frac{1}{m_2} \tag{2}$$

Since $m_1 = (f^{-1})'(x)$ and $m_2 = f'(y)$, (2) yields

$$(f^{-1})'(x) = \frac{1}{f'(y)} \tag{3}$$

But (x, y) is on the graph of f^{-1}, so $y = f^{-1}(x)$. Consequently, (3) can also be written

$$(f^{-1})'(x) = \frac{1}{f'(f^{-1}(x))}$$

Observe that these formulas break down if $f'(y) = 0$. However, this is to be expected since $f'(y) = 0$ indicates a horizontal tangent for f at y, which reflects about $y = x$ into a vertical tangent for f^{-1} at x.

The following theorem summarizes these informal results.

8.1.5 THEOREM *Suppose that f has an inverse and is differentiable on an open interval I. If $f^{-1}(x)$ is a point in I at which f' is nonzero, then f^{-1} is differentiable at x and*

$$(f^{-1})'(x) = \frac{1}{f'(f^{-1}(x))} \tag{4}$$

REMARK. If we let

$$y = f^{-1}(x) \qquad \text{so that} \qquad x = f(y)$$

then

$$\frac{dy}{dx} = (f^{-1})'(x) \quad \text{and} \quad \frac{dx}{dy} = f'(y) = f'(f^{-1}(x))$$

so that (4) can be written in more compact form as

$$\frac{dy}{dx} = \frac{1}{dx/dy} \tag{5}$$

▶ Example 10   In Section 7.4 we obtained the derivative formula for e^x by *assuming* this function to be differentiable. Using Theorem 8.1.5 we can now prove the differentiability of e^x as follows: Let

$$f(x) = \ln x$$

so that

$$f^{-1}(x) = e^x$$

If x is a point in $(-\infty, +\infty)$, then $f^{-1}(x) = e^x$ is a point in $(0, +\infty)$, since e^x assumes only positive values. Moreover, f' is nonzero on $(0, +\infty)$ since

$$f'(x) = \frac{1}{x} > 0$$

In particular, f' is nonzero at the point $f^{-1}(x) = e^x$. Thus, f^{-1} is differentiable at x by Theorem 8.1.5. Since x is any point in $(-\infty, +\infty)$, we have shown that e^x is differentiable everywhere.   ◀

As a point of interest, the familiar derivative formula for e^x follows from (4) with $f(x) = \ln x$ and $f^{-1}(x) = e^x$, since

$$\frac{d}{dx}[e^x] = (f^{-1})'(x) = \frac{1}{f'(f^{-1}(x))} = \frac{1}{f'(e^x)} = \frac{1}{1/e^x} = e^x$$

The primary purpose of Theorem 8.1.5 is to provide conditions that ensure the differentiability of f^{-1}. Formulas (4) and (5) are of secondary interest because the derivative of f^{-1}, when it exists, can also be obtained by implicit differentiation.

▶ Example 11   In Example 6 we saw that $f(x) = x^5 + 7x^3 + 4x + 1$ has an inverse. Find the derivative of f^{-1} by using Formula (5) and also by implicit differentiation.

Solution.   Let $y = f^{-1}(x)$, so that $f(y) = f(f^{-1}(x)) = x$, which implies that

$$x = y^5 + 7y^3 + 4y + 1 \tag{6}$$

Thus,

$$\frac{dx}{dy} = 5y^4 + 21y^2 + 4$$

so

$$\frac{dy}{dx} = \frac{1}{dx/dy} = \frac{1}{5y^4 + 21y^2 + 4} \tag{7}$$

Since (6) is too complicated to solve for y in terms of x, we must leave (7) in terms of y.

Alternate Solution. Instead of using (5), we can differentiate (6) implicitly with respect to x.
We obtain

$$\frac{d}{dx}[x] = \frac{d}{dx}[y^5 + 7y^3 + 4y + 1]$$

$$1 = 5y^4 \frac{dy}{dx} + 21y^2 \frac{dy}{dx} + 4 \frac{dy}{dx}$$

$$1 = (5y^4 + 21y^2 + 4) \frac{dy}{dx}$$

which yields (7). ◀

▶ Exercise Set 8.1

1. In (a)–(d), determine whether f and g are inverse functions.

   (a) $f(x) = 4x$, $g(x) = \frac{1}{4}x$

   (b) $f(x) = 3x + 1$, $g(x) = 3x - 1$

   (c) $f(x) = \ln(x + 1)$, $g(x) = e^x - 1$

   (d) $f(x) = x^4$, $g(x) = \sqrt[4]{x}$.

In Exercises 2–14, determine whether the function has an inverse.

2. $f(x) = 1 - x$.

3. $f(x) = 3x + 2$.

4. $f(x) = x^2 - 2x + 1$.

5. $f(x) = 2 - x - x^2$.

6. $f(x) = x^3 - 3x + 2$.

7. $f(x) = x^3 - x - 1$.

8. $f(x) = x^3 - 3x^2 + 3x - 1$.

9. $f(x) = x^3 + 3x^2 + 3x + 1$.

10. $f(x) = x^5 + 8x^3 + 2x - 1$.

11. $f(x) = 2x^5 + x^3 + 3x + 2$.

12. $f(x) = x + \frac{1}{x}$, $x > 0$.

13. $f(x) = \sin x$, $-\frac{\pi}{2} < x < \frac{\pi}{2}$.

14. $f(x) = \tan x$, $-\frac{\pi}{2} < x < \frac{\pi}{2}$.

In Exercises 15–25, find $f^{-1}(x)$.

15. $f(x) = x^5$.

16. $f(x) = 6x$.

17. $f(x) = 7x - 6$.

18. $f(x) = \dfrac{x+1}{x-1}$.

19. $f(x) = 3x^3 - 5$.

20. $f(x) = \sqrt[5]{4x+2}$.

21. $f(x) = \sqrt[3]{e^x + 1}$.

22. $f(x) = e^{2x+1}$.

23. $f(x) = e^{1/x}$.

24. $f(x) = 4\ln(x+1)$.

25. $f(x) = 1 - \ln(3x)$.

In Exercises 26–32, use Formula (5) to find the derivative of f^{-1}, and check your work by differentiating implicitly.

26. $f(x) = 2x^3 + 5x + 3$.

27. $f(x) = 5x^3 + x - 7$.

28. $f(x) = e^{3x+4}$.

29. $f(x) = e^{1+x^3}$.

30. $f(x) = x + 2\ln x$.

31. $f(x) = 2x^5 + x^3 + 1$.

32. $f(x) = x^7 + 2x^5 + x^3$.

33. (a) Use Theorem 8.1.4 to show that

$$f(x) = \frac{1}{2}(e^x - e^{-x})$$

has an inverse.
(b) Find $f^{-1}(x)$.

34. (a) Use Theorem 8.1.4 to show that

$$f(x) = \frac{e^x - e^{-x}}{e^x + e^{-x}}$$

has an inverse.
(b) Find $f^{-1}(x)$.

35. Let $f(x) = x^2$, $x > 1$ and $g(x) = \sqrt{x}$.
(a) Show that $f(g(x)) = x$, $x > 1$, and $g(f(x)) = x$, $x > 1$.
(b) Show that f and g are *not* inverses of one another by showing that the graphs of $y = f(x)$ and $y = g(x)$ are not reflections of one another about $y = x$.
(c) Do parts (a) and (b) contradict Definition 8.1.1? Explain.

36. Sometimes the procedure used in Examples 7 and 8 to find $f^{-1}(x)$ produces a correct formula, but

some additional work may be required to find the domain of $f^{-1}(x)$. With this in mind, find $f^{-1}(x)$ if $f(x) = \sqrt{x+3}$.

37. Find $f^{-1}(x)$ if $f(x) = \sqrt{e^x + 1}$. [*Hint:* Read Exercise 36.]

38. Prove that if $d = -a$, then the graph of

$$f(x) = \frac{ax+b}{cx+d}$$

is symmetric about the line $y = x$.

39. (a) Show that $f(x) = (3-x)/(1-x)$ is its own inverse.
(b) What does the result in (a) tell you about the graph of f?

40. Suppose that a line of nonzero slope m intersects the x-axis at $(x_0, 0)$. Find an equation for the reflection of this line about $y = x$.

41. (a) Show that $f(x) = x^3 - 3x^2 + 2x$ is not one-to-one on $(-\infty, +\infty)$.
(b) Find the largest value of k such that f is one-to-one on the interval $(-k, k)$.

42. If f is continuous and one-to-one, do you think that f^{-1} must also be continuous? Justify your conclusion.

43. Let $f(x) = \int_1^x \sqrt[3]{1+t^2}\,dt$.
(a) Without integrating, show that f is one-to-one on the interval $(-\infty, +\infty)$.
(b) Find $\dfrac{d}{dx}[f^{-1}(x)]\Big|_{x=0}$.

44. (a) Prove: If f and g are one-to-one, then so is the composition $f \circ g$.
(b) Prove: If f and g are one-to-one, then

$$(f \circ g)^{-1} = g^{-1} \circ f^{-1}$$

45. Sketch the graph of a function that is one-to-one on $(-\infty, +\infty)$, yet not strictly monotone on $(-\infty, +\infty)$.

46. Prove: A one-to-one function f cannot have two different inverses.

47. Prove: If $a > 0$, then a^x is a differentiable function. [*Hint:* Use Corollary 7.4.5 and the fact that e^x is known to be differentiable.]

8.2 INVERSE TRIGONOMETRIC FUNCTIONS

Because the six basic trigonometric functions are periodic, each of their values is repeated infinitely many times. Thus, the trigonometric functions are not one-to-one, and consequently have no inverses. However, we will show in this section that it is possible to restrict the domains of the trigonometric functions in such a way that the restricted functions do have inverses.

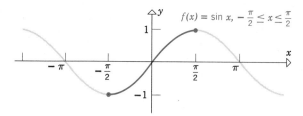

Figure 8.2.1

INVERSE SINE In Figure 8.2.1 we have sketched the graph of

$$f(x) = \sin x, \qquad -\frac{\pi}{2} \le x \le \frac{\pi}{2} \tag{1}$$

Although $\sin x$ is defined on the entire interval $(-\infty, \infty)$, the function $f(x)$ in (1) is defined only on the interval $[-\pi/2, \pi/2]$. Since no horizontal line cuts the graph of $f(x) = \sin x$, $-\pi/2 \le x \le \pi/2$ more than once, this function has an inverse. This inverse is called the **inverse sine function,** and is denoted by

$$\sin^{-1} x$$

The symbol $\sin^{-1} x$ is *never* used to denote $1/\sin x$. If desired, $1/\sin x$ can be rewritten as $(\sin x)^{-1}$ or $\csc x$. In the older literature, $\sin^{-1} x$ is called the **arcsine** function and is denoted by arcsin x. We will not use this terminology or notation.

REMARK. To define $\sin^{-1} x$ we restricted the domain of $\sin x$ to the interval $[-\pi/2, \pi/2]$ to obtain a one-to-one function. There are other ways to restrict the domain of $\sin x$ to obtain one-to-one functions; for example, we might have required that $3\pi/2 \le x \le 5\pi/2$ or $-5\pi/2 \le x \le -3\pi/2$. However, the choice $-\pi/2 \le x \le \pi/2$ is customary.

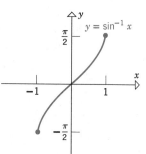

The graph of $\sin^{-1} x$ is obtained by reflecting the graph of $\sin x$, $-\pi/2 \le x \le \pi/2$ about the line $y = x$ (Figure 8.2.2). If we let $y = \sin^{-1} x$, then it is evident from this graph that $-1 \le x \le 1$ and $-\pi/2 \le y \le \pi/2$, so the domain of $\sin^{-1} x$ is the interval $[-1, 1]$ and the range is $[-\pi/2, \pi/2]$.

Because $\sin x$ (restricted) and $\sin^{-1} x$ are inverses of one another, it follows that

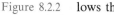

Figure 8.2.2

$$\sin^{-1}(\sin y) = y \quad \text{if} \quad -\frac{\pi}{2} \le y \le \frac{\pi}{2} \tag{2a}$$

$$\sin(\sin^{-1} x) = x \quad \text{if} \quad -1 \le x \le 1 \tag{2b}$$

From these equations we obtain the following important result.

8.2.1 THEOREM *If* $-1 \le x \le 1$ *and* $-\pi/2 \le y \le \pi/2$, *then*

$$y = \sin^{-1} x \quad \text{and} \quad \sin y = x \tag{3}$$

are equivalent statements.

Proof. If $y = \sin^{-1} x$ and $-1 \le x \le 1$, then it follows from (2b) that $\sin y = \sin(\sin^{-1} x) = x$. Conversely, if $\sin y = x$ and $-\pi/2 \le y \le \pi/2$, then $\sin^{-1}(\sin y) = \sin^{-1} x$, or from (2a) $y = \sin^{-1} x$. ▮

REMARK. If we think of y as an angle in radian measure, then it follows from (3) that $y = \sin^{-1} x$ is that angle between $-\pi/2$ and $\pi/2$ whose sine is x.

▶ **Example 1** Find

(a) $\sin^{-1}(\tfrac{1}{2})$ (b) $\sin^{-1}(-1/\sqrt{2})$

Solution (a). Let $y = \sin^{-1}(\tfrac{1}{2})$. From Theorem 8.2.1 this equation is equivalent to

$$\sin y = \tfrac{1}{2}, \quad -\pi/2 \le y \le \pi/2$$

The only value of y satisfying these conditions is $y = \pi/6$, so $\sin^{-1}(\tfrac{1}{2}) = \pi/6$.

Solution (b). Let $y = \sin^{-1}(-1/\sqrt{2})$. This is equivalent to

$$\sin y = -1/\sqrt{2}, \quad -\pi/2 \le y \le \pi/2$$

which is satisfied only by $y = -\pi/4$. Thus, $\sin^{-1}(-1/\sqrt{2}) = -\pi/4$. ◀

In the next section we will encounter functions that are compositions of trigonometric and inverse trigonometric functions. The following example illustrates a technique for simplifying such functions.

▶ **Example 2** Simplify the function $\cos(\sin^{-1} x)$.

Solution. The idea is to express cosine in terms of sine in order to take advantage of the simplification $\sin(\sin^{-1} x) = x$. Thus, we start with the identity

$$\cos^2 \theta = 1 - \sin^2 \theta$$

and substitute $\theta = \sin^{-1} x$ to obtain

$$\cos^2 (\sin^{-1} x) = 1 - \sin^2 (\sin^{-1} x)$$

or on taking square roots

$$|\cos (\sin^{-1} x)| = \sqrt{1 - \sin^2 (\sin^{-1} x)}$$

or

$$|\cos (\sin^{-1} x)| = \sqrt{1 - x^2}$$

Since $-\pi/2 \le \sin^{-1} x \le \pi/2$, it follows that $\cos (\sin^{-1} x)$ is nonnegative; thus, we can drop the absolute value sign and write

$$\cos (\sin^{-1} x) = \sqrt{1 - x^2} \tag{4}$$

◄

Formula (4) can also be obtained geometrically by first constructing a right triangle with an acute angle of $\theta = \sin^{-1} x$ (Figure 8.2.3). For this triangle, $\sin \theta = x$, so we can assume that the side opposite θ has length x and the hypotenuse has length 1. Thus,

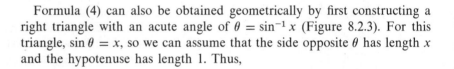

Figure 8.2.3

$$\sin \theta = \frac{\text{opposite side}}{\text{hypotenuse}} = \frac{x}{1} = x$$

By the Theorem of Pythagoras, the side adjacent to θ has length $\sqrt{1 - x^2}$, so

$$\cos \theta = \frac{\text{adjacent side}}{\text{hypotenuse}} = \frac{\sqrt{1 - x^2}}{1} = \sqrt{1 - x^2}$$

or

$$\cos (\sin^{-1} x) = \sqrt{1 - x^2}$$

INVERSE TANGENT   The *inverse tangent* function, $\tan^{-1} x$, is defined to be the inverse of the restricted tangent function,

$$f(x) = \tan x, \qquad -\frac{\pi}{2} < x < \frac{\pi}{2}$$

(Figure 8.2.4*a*). The graph of $\tan^{-1} x$ is obtained by reflecting the graph of f about the line $y = x$ (Figure 8.2.4*b*).

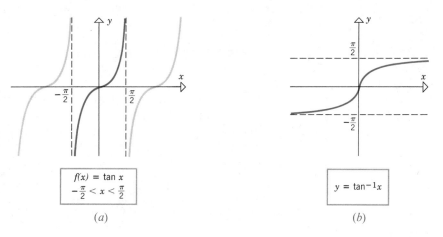

$f(x) = \tan x$
$-\frac{\pi}{2} < x < \frac{\pi}{2}$

$y = \tan^{-1} x$

Figure 8.2.4   (a)   (b)

If we let $y = \tan^{-1} x$, then from Figure 8.2.4*b* we see that $-\infty < x < +\infty$ and $-\pi/2 < y < \pi/2$, so the domain of $\tan^{-1} x$ is $(-\infty, +\infty)$ and the range is $(-\pi/2, \pi/2)$.

Because $\tan x$ (restricted) and $\tan^{-1} x$ are inverses of one another, it follows that

$$\tan^{-1}(\tan y) = y \quad \text{if} \quad -\pi/2 < y < \pi/2$$
$$\tan(\tan^{-1} x) = x \quad \text{if} \quad -\infty < x < +\infty$$

By imitating the proof of Theorem 8.2.1, the reader should be able to prove the following result.

8.2.2 THEOREM   *If $-\infty < x < +\infty$ and $-\pi/2 < y < \pi/2$, then*

$$y = \tan^{-1} x \quad and \quad \tan y = x$$

are equivalent statements.

▶ Example 3   Simplify the function $\sec^2(\tan^{-1} x)$.

Solution. The object is to express secant in terms of tangent to take advantage of the simplification $\tan(\tan^{-1} x) = x$. Thus, if we let $\theta = \tan^{-1} x$ in the identity

$$\sec^2 \theta = 1 + \tan^2 \theta$$

we obtain

$$\sec^2(\tan^{-1} x) = 1 + \tan^2(\tan^{-1} x)$$

or

$$\sec^2(\tan^{-1} x) = 1 + x^2 \tag{5}$$

◀

Figure 8.2.5

Formula (5) can also be obtained from the triangle in Figure 8.2.5. The tangent of the acute angle $\tan^{-1} x$ is x, so we can assume that the side opposite to this angle has length x and the adjacent side length 1. The hypotenuse is determined by the Theorem of Pythagoras. From the triangle we obtain

$$\sec(\tan^{-1} x) = \frac{\text{hypotenuse}}{\text{adjacent side}} = \frac{\sqrt{1 + x^2}}{1} = \sqrt{1 + x^2}$$

or

$$\sec^2(\tan^{-1} x) = 1 + x^2$$

INVERSE SECANT

The ***inverse secant*** function, $\sec^{-1} x$, is defined to be the inverse of the restricted secant function

$$f(x) = \sec x, \qquad 0 \le x < \pi/2 \quad \text{or} \quad \pi \le x < 3\pi/2$$

(Figure 8.2.6a). The graph of $\sec^{-1} x$ is obtained by reflecting the graph of f about the line $y = x$ (Figure 8.2.6b).

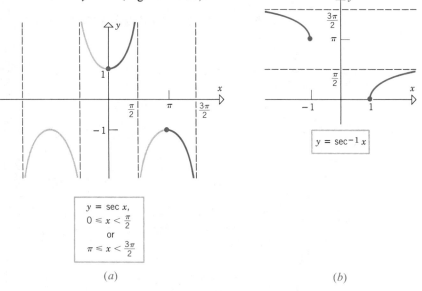

$$y = \sec x,$$
$$0 \le x < \frac{\pi}{2}$$
or
$$\pi \le x < \frac{3\pi}{2}$$

$$y = \sec^{-1} x$$

Figure 8.2.6   (a)   (b)

REMARK. There is no universal agreement among mathematicians about the definition of $\sec^{-1} x$. For example, some writers define $\sec^{-1} x$ by restricting x so $0 \le x < \pi/2$ or $\pi/2 < x \le \pi$. Although there are some advantages to this definition, our definition will produce simpler derivative and integration formulas in the next section.

If we let $y = \sec^{-1} x$, then from Figure 8.2.6b we see that $x \leq -1$ or $x \geq 1$ and $0 \leq y < \pi/2$ or $\pi \leq y < 3\pi/2$. Thus, the domain of $\sec^{-1} x$ is $(-\infty, -1] \cup [1, +\infty)$ and the range is $[0, \pi/2) \cup [\pi, 3\pi/2)$.

Because $\sec x$ (restricted) and $\sec^{-1} x$ are inverses of one another, it follows that

$$\sec^{-1}(\sec y) = y \quad \text{if} \quad 0 \leq y < \pi/2 \quad \text{or} \quad \pi \leq y < 3\pi/2$$
$$\sec(\sec^{-1} x) = x \quad \text{if} \quad x \leq -1 \quad \text{or} \quad x \geq 1$$

Moreover, the following result parallels Theorems 8.2.1 and 8.2.2.

8.2.3 THEOREM *If $x \leq -1$ or $x \geq 1$ and if $0 \leq y < \pi/2$ or $\pi \leq y < 3\pi/2$, then*

$$y = \sec^{-1} x \quad \text{and} \quad \sec y = x$$

are equivalent statements.

INVERSE COSINE, COTANGENT, AND COSECANT

The inverse cosine, cotangent, and cosecant functions are of lessor importance, so we will summarize their properties briefly.

8.2.4 DEFINITION

$y = \cos^{-1} x$	is equivalent to	$x = \cos y$	if $\begin{array}{c} 0 \leq y \leq \pi \\ -1 \leq x \leq 1 \end{array}$
$y = \cot^{-1} x$	is equivalent to	$x = \cot y$	if $\begin{array}{c} 0 < y < \pi \\ -\infty < x < +\infty \end{array}$
$y = \csc^{-1} x$	is equivalent to	$x = \csc y$	if $\begin{cases} 0 < y \leq \pi/2 \\ \text{or} \\ -\pi < y \leq -\pi/2 \\ |x| \geq 1 \end{cases}$

We leave it as an exercise to find the graphs of these three inverse trigonometric functions.

In the definitions of the six inverse trigonometric functions, the restrictions on the domains were selected not only to produce one-to-one functions, but also to ensure that certain "natural" identities hold for the inverse trigonometric functions. For example, if α and β are acute complementary angles, then from basic trigonometry, $\sin \alpha$ and $\cos \beta$ are equal (Figure 8.2.7). Let us write $x = \sin \alpha = \cos \beta$ so that

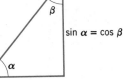
$\sin \alpha = \cos \beta$

Figure 8.2.7

$$\alpha = \sin^{-1} x \quad \text{and} \quad \beta = \cos^{-1} x$$

since $\alpha + \beta = \pi/2$ we obtain the identity

$$\sin^{-1} x + \cos^{-1} x = \frac{\pi}{2} \qquad (6)$$

(Because α and β were assumed to be nonnegative acute angles, this derivation is only valid for $0 \leq x \leq 1$; for a derivation valid for all x in $[-1, 1]$ see Exercise 25.) Similarly, we can obtain the identities

$$\tan^{-1} x + \cot^{-1} x = \frac{\pi}{2} \qquad (7)$$

$$\sec^{-1} x + \csc^{-1} x = \frac{\pi}{2} \qquad (8)$$

The proofs of the following identities, which are evident from Figures 8.2.2, 8.2.4b, and 8.2.6b, are left as exercises.

$$\sin^{-1}(-x) = -\sin^{-1} x \qquad (9)$$

$$\tan^{-1}(-x) = -\tan^{-1} x \qquad (10)$$

$$\sec^{-1}(-x) = \pi + \sec^{-1} x \qquad \text{if } x \geq 1 \qquad (11)$$

▶ Exercise Set 8.2

1. Find the exact value of
   (a) $\sin^{-1}(-1)$ (b) $\cos^{-1}(-1)$
   (c) $\tan^{-1}(-1)$ (d) $\cot^{-1}(1)$
   (e) $\sec^{-1}(1)$ (f) $\csc^{-1}(1)$.

2. Find the exact value of
   (a) $\sin^{-1}(\frac{1}{2}\sqrt{3})$ (b) $\cos^{-1}(\frac{1}{2})$
   (c) $\tan^{-1}(1)$ (d) $\cot^{-1}(-1)$
   (e) $\sec^{-1}(-2)$ (f) $\csc^{-1}(-2)$.

3. Given that $\theta = \sin^{-1}(-\frac{1}{2}\sqrt{3})$, find the exact values of $\cos\theta$, $\tan\theta$, $\cot\theta$, $\sec\theta$, and $\csc\theta$.

4. Given that $\theta = \cos^{-1}(\frac{1}{2})$, find the exact values of $\sin\theta$, $\tan\theta$, $\cot\theta$, $\sec\theta$, and $\csc\theta$.

5. Given that $\theta = \tan^{-1}(\frac{4}{3})$, find the exact values of $\sin\theta$, $\cos\theta$, $\cot\theta$, $\sec\theta$, and $\csc\theta$.

6. Make a table that lists the six inverse trigonometric functions together with their domains and ranges.

7. Find the exact value of
   (a) $\sin^{-1}\left(\sin\frac{\pi}{7}\right)$ (b) $\sin^{-1}(\sin\pi)$
   (c) $\sin^{-1}\left(\sin\frac{5\pi}{7}\right)$.

8. Find the exact value of
   (a) $\cos^{-1}\left(\cos\frac{\pi}{7}\right)$ (b) $\cos^{-1}(\cos\pi)$
   (c) $\cos^{-1}\left(\cos\frac{12\pi}{7}\right)$.

9. For which values of x is it true that
    (a) $\cos^{-1}(\cos x) = x$ (b) $\cos(\cos^{-1}x) = x$
    (c) $\tan^{-1}(\tan x) = x$ (d) $\tan(\tan^{-1}x) = x$
    (e) $\csc^{-1}(\csc x) = x$ (f) $\csc(\csc^{-1}x) = x$.

In Exercises 10–15, find the exact value of the given quantity.

10. $\sec\left[\sin^{-1}\left(-\dfrac{3}{4}\right)\right]$.   **11.** $\sin\left[2\cos^{-1}\left(\dfrac{3}{5}\right)\right]$.

12. $\tan^{-1}\left[\sin\left(-\dfrac{\pi}{2}\right)\right]$.   **13.** $\sin^{-1}\left[\cot\left(\dfrac{\pi}{4}\right)\right]$.

14. $\sin\left[\sin^{-1}\left(\dfrac{2}{3}\right) + \cos^{-1}\left(\dfrac{1}{3}\right)\right]$.

15. $\tan\left[2\sec^{-1}\left(\dfrac{3}{2}\right)\right]$.

16. Prove:

$$\tan^{-1}x + \tan^{-1}y = \tan^{-1}\left(\frac{x+y}{1-xy}\right)$$

provided $-\pi/2 < \tan^{-1}x + \tan^{-1}y < \pi/2$. [*Hint:* Use an identity for $\tan(\alpha + \beta)$.]

17. Use the result in Exercise 16 to show:

    (a) $\tan^{-1}\dfrac{1}{2} + \tan^{-1}\dfrac{1}{3} = \pi/4$

    (b) $2\tan^{-1}\dfrac{1}{3} + \tan^{-1}\dfrac{1}{7} = \pi/4$.

Following Example 2 in this section, we derived the identity $\cos(\sin^{-1}x) = \sqrt{1-x^2}$ by considering a right triangle with an acute angle of $\sin^{-1}x$ (see Figure 8.2.3). In Exercises 18 and 19, complete the identity by constructing an appropriate triangle.

18. (a) $\sin(\cos^{-1}x) = $?
     (b) $\tan(\cos^{-1}x) = $?
     (c) $\csc(\tan^{-1}x) = $?
     (d) $\sin(\tan^{-1}x) = $?

19. (a) $\cos(\tan^{-1}x) = $?
     (b) $\tan(\cot^{-1}x) = $?
     (c) $\sin(\sec^{-1}x) = $?
     (d) $\cot(\csc^{-1}x) = $?

20. Sketch the graphs of $\cos^{-1}x$, $\cot^{-1}x$, and $\csc^{-1}x$.

21. Sketch the graphs of

    (a) $y = \sin^{-1}2x$ (b) $y = \tan^{-1}\dfrac{1}{2}x$.

22. Sketch the graphs of

    (a) $y = \cos^{-1}\dfrac{1}{3}x$ (b) $y = 2\cot^{-1}2x$.

23. Prove:
    (a) $\sin^{-1}(-x) = -\sin^{-1}x$
    (b) $\tan^{-1}(-x) = -\tan^{-1}x$
    (c) $\sec^{-1}(-x) = \pi + \sec^{-1}x$, if $x \geq 1$.

24. Prove: $\cos^{-1}(-x) = \pi - \cos^{-1}x$.

25. In the text we proved that

$$\sin^{-1}x + \cos^{-1}x = \frac{\pi}{2}$$

for $0 \leq x \leq 1$. Use this result together with Exercises 23(a) and 24 to prove this identity for $-1 \leq x \leq 1$.

26. Prove Theorem 8.2.2.

27. Prove Theorem 8.2.3.

28. A camera is positioned x feet from the base of a missile launching pad. If a missile of length a feet is launched vertically, show that when the base of the missile is b feet above the camera lens, the angle θ subtended at the lens by the missile is

$$\theta = \cot^{-1}\frac{x}{a+b} - \cot^{-1}\frac{x}{b}$$

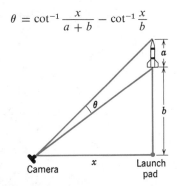

29. In the computer programming language, BASIC, the instruction ATN(X) causes the computer to calculate $\tan^{-1}(X)$. According to the programming manual for a popular microcomputer, this instruction can be used to calculate $\sin^{-1}(X)$ and $\cos^{-1}(X)$ by means of the formulas:

    (a) $\sin^{-1}(X) = \text{ATN}\left(\dfrac{X}{\sqrt{1-X^2}}\right)$

    (b) $\cos^{-1}(X) = -\text{ATN}\left(\dfrac{X}{\sqrt{1-X^2}}\right) + 1.5708$

Derive these results.

8.3 DERIVATIVES AND INTEGRALS INVOLVING INVERSE TRIGONOMETRIC FUNCTIONS

In this section we obtain the derivatives of the inverse trigonometric functions and the related integration formulas.

8.3.1 THEOREM

$$\frac{d}{dx}[\sin^{-1}x] = \frac{1}{\sqrt{1-x^2}} \tag{1}$$

$$\frac{d}{dx}[\cos^{-1}x] = -\frac{1}{\sqrt{1-x^2}} \tag{2}$$

$$\frac{d}{dx}[\tan^{-1}x] = \frac{1}{1+x^2} \tag{3}$$

$$\frac{d}{dx}[\cot^{-1}x] = -\frac{1}{1+x^2} \tag{4}$$

$$\frac{d}{dx}[\sec^{-1}x] = \frac{1}{x\sqrt{x^2-1}} \tag{5}$$

$$\frac{d}{dx}[\csc^{-1}x] = -\frac{1}{x\sqrt{x^2-1}} \tag{6}$$

Proof. To obtain (1), let

$$y = \sin^{-1}x$$

so

$$x = \sin y, \qquad -\frac{\pi}{2} \le y \le \frac{\pi}{2}$$

Thus, from (5) of Section 8.1 and (4) of Section 8.2,

$$\frac{d}{dx}[\sin^{-1}x] = \frac{dy}{dx} = \frac{1}{dx/dy} = \frac{1}{\cos y} = \frac{1}{\cos(\sin^{-1}x)} = \frac{1}{\sqrt{1-x^2}}$$

Formula (3) is obtained similarly; let

$$y = \tan^{-1}x$$

so

$$x = \tan y, \qquad -\frac{\pi}{2} < y < \frac{\pi}{2}$$

Thus, from (5) of Section 8.1 and (5) of Section 8.2,

$$\frac{d}{dx}[\tan^{-1}x] = \frac{dy}{dx} = \frac{1}{dx/dy} = \frac{1}{\sec^2 y} = \frac{1}{\sec^2(\tan^{-1}x)} = \frac{1}{1+x^2}$$

To obtain (5), let

$$y = \sec^{-1} x$$

so

$$x = \sec y, \qquad 0 \le y < \frac{\pi}{2} \quad \text{or} \quad \pi \le y < \frac{3\pi}{2}$$

Thus, from (5) of Section 8.1 and Figure 8.3.1,

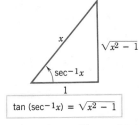

Figure 8.3.1

$$\frac{d}{dx}[\sec^{-1} x] = \frac{dy}{dx} = \frac{1}{dx/dy} = \frac{1}{\sec y \tan y} = \frac{1}{\sec(\sec^{-1} x)\tan(\sec^{-1} x)}$$

$$= \frac{1}{x\sqrt{x^2 - 1}}$$

The remaining formulas can all be obtained from the derivatives of $\sin^{-1}$, $\tan^{-1}$, and $\sec^{-1}$ by using appropriate identities. For example, using identity (6) of Section 8.2 yields

$$\frac{d}{dx}[\cos^{-1} x] = \frac{d}{dx}\left[\frac{\pi}{2} - \sin^{-1} x\right] = -\frac{d}{dx}[\sin^{-1} x] = -\frac{1}{\sqrt{1 - x^2}}$$

Similarly, the derivatives of $\cot^{-1} x$ and $\csc^{-1} x$ follow from identities (7) and (8) of Section 8.2.

▶ **Example 1** Find dy/dx if $y = \sin^{-1}(x^3)$.

Solution. From (1) and the chain rule,

$$\frac{dy}{dx} = \frac{1}{\sqrt{1 - (x^3)^2}}(3x^2) = \frac{3x^2}{\sqrt{1 - x^6}} \qquad\qquad ◀$$

▶ **Example 2** Find dy/dx if $y = \sec^{-1}(e^x)$.

Solution. From (5) and the chain rule,

$$\frac{dy}{dx} = \frac{1}{e^x\sqrt{(e^x)^2 - 1}}(e^x) = \frac{1}{\sqrt{e^{2x} - 1}} \qquad\qquad ◀$$

Differentiation formulas (1)–(6) yield useful integration formulas. Those most commonly needed are:

$$\int \frac{dx}{\sqrt{1 - x^2}} = \sin^{-1} x + C \qquad\qquad (7)$$

$$\int \frac{dx}{1 + x^2} = \tan^{-1} x + C \tag{8}$$

$$\int \frac{dx}{x\sqrt{x^2 - 1}} = \sec^{-1} x + C \tag{9}$$

▶ Example 3   Evaluate

$$\int \frac{dx}{1 + 3x^2}$$

Solution.   Substituting

$$u = \sqrt{3}\,x, \qquad du = \sqrt{3}\,dx$$

yields

$$\int \frac{dx}{1 + 3x^2} = \frac{1}{\sqrt{3}} \int \frac{du}{1 + u^2} = \frac{1}{\sqrt{3}} \tan^{-1} u + C$$

$$= \frac{1}{\sqrt{3}} \tan^{-1}(\sqrt{3}\,x) + C \qquad ◀$$

▶ Example 4   Evaluate

$$\int \frac{e^x}{\sqrt{1 - e^{2x}}}\,dx$$

Solution.   Substituting

$$u = e^x, \qquad du = e^x\,dx$$

yields

$$\int \frac{e^x}{\sqrt{1 - e^{2x}}}\,dx = \int \frac{du}{\sqrt{1 - u^2}} = \sin^{-1} u + C = \sin^{-1}(e^x) + C \qquad ◀$$

▶ Example 5   Evaluate

$$\int \frac{dx}{a^2 + x^2}$$

where $a \neq 0$ is a constant.

Solution.   If we had a 1 in place of a^2, we could use (8). Thus, we look for a u-substitution that will enable us to replace the a^2 with a 1. If we let

$$x = au, \qquad dx = a\,du$$

then

$$\int \frac{dx}{a^2 + x^2} = \int \frac{a\,du}{a^2 + a^2 u^2} = \frac{1}{a}\int \frac{du}{1 + u^2}$$

$$= \frac{1}{a}\tan^{-1} u + C$$

$$= \frac{1}{a}\tan^{-1}\frac{x}{a} + C \qquad \blacktriangleleft$$

The method of Example 5 leads to the following generalizations of (7), (8), and (9).

$$\int \frac{dx}{\sqrt{a^2 - x^2}} = \sin^{-1}\frac{x}{a} + C \qquad (a > 0) \qquad\qquad (10)$$

$$\int \frac{dx}{a^2 + x^2} = \frac{1}{a}\tan^{-1}\frac{x}{a} + C \qquad (a \neq 0) \qquad\qquad (11)$$

$$\int \frac{dx}{x\sqrt{x^2 - a^2}} = \frac{1}{a}\sec^{-1}\frac{x}{a} + C \qquad (a > 0) \qquad\qquad (12)$$

▶ **Example 6** Evaluate

$$\int \frac{dx}{\sqrt{2 - x^2}}$$

Solution. Applying (10) with $a = \sqrt{2}$ yields

$$\int \frac{dx}{\sqrt{2 - x^2}} = \sin^{-1}\frac{x}{\sqrt{2}} + C \qquad \blacktriangleleft$$

▶ **Exercise Set 8.3**

In Exercises 1–9, find dy/dx.

1. (a) $y = \sin^{-1}(\frac{1}{3}x)$ (b) $y = \cos^{-1}(2x + 1)$.

2. (a) $y = \tan^{-1}(x^2)$ (b) $y = \cot^{-1}(\sqrt{x})$.

3. (a) $y = \sec^{-1}(x^7)$ (b) $y = \csc^{-1}(e^x)$.

4. (a) $y = (\tan x)^{-1}$ (b) $y = \dfrac{1}{\tan^{-1} x}$.

5. (a) $y = \sin^{-1}\left(\dfrac{1}{x}\right)$ (b) $y = \cos^{-1}(\cos x)$.

6. (a) $y = \ln(\cos^{-1} x)$ (b) $y = \sqrt{\cot^{-1} x}$.

7. (a) $y = e^x \sec^{-1} x$
   (b) $y = x^2(\sin^{-1} x)^3$.

8. (a) $y = \sin^{-1} x + \cos^{-1} x$
   (b) $y = \sec^{-1} x + \csc^{-1} x$.

9. (a) $y = \tan^{-1}\left(\dfrac{1 - x}{1 + x}\right)$
   (b) $y = (1 + x\csc^{-1} x)^{10}$.

In Exercises 10 and 11, find dy/dx by implicit differentiation.

10. $x^3 + x \tan^{-1} y = e^y$.

11. $\sin^{-1}(xy) = \cos^{-1}(x - y)$.

In Exercises 12–23, evaluate the integral.

12. $\displaystyle\int_0^{1/\sqrt{2}} \frac{dx}{\sqrt{1 - x^2}}$.

13. $\displaystyle\int_{-1}^{1} \frac{dx}{1 + x^2}$.

14. $\displaystyle\int_{\sqrt{2}}^{2} \frac{dx}{x \sqrt{x^2 - 1}}$.

15. $\displaystyle\int_{-\sqrt{2}}^{-2/\sqrt{3}} \frac{dx}{x \sqrt{x^2 - 1}}$.

16. $\displaystyle\int \frac{dx}{\sqrt{1 - 4x^2}}$.

17. $\displaystyle\int \frac{dx}{1 + 16x^2}$.

18. $\displaystyle\int \frac{dx}{x \sqrt{9x^2 - 1}}$.

19. $\displaystyle\int \frac{e^x}{1 + e^{2x}} dx$.

20. $\displaystyle\int \frac{t}{t^4 + 1} dt$.

21. $\displaystyle\int \frac{\sec^2 x \, dx}{\sqrt{1 - \tan^2 x}}$.

22. $\displaystyle\int \frac{\sin \theta}{\cos^2 \theta + 1} d\theta$.

23. $\displaystyle\int \frac{dx}{x \sqrt{1 - (\ln x)^2}}$.

24. Derive integration formulas (10) and (12).

In Exercises 25 and 26, use (10), (11), and (12) to evaluate the integrals.

25. (a) $\displaystyle\int \frac{dx}{\sqrt{9 - x^2}}$
(b) $\displaystyle\int \frac{dx}{5 + x^2}$

(c) $\displaystyle\int \frac{dx}{x \sqrt{x^2 - \pi}}$.

26. (a) $\displaystyle\int \frac{e^x}{4 + e^{2x}} dx$
(b) $\displaystyle\int \frac{dx}{\sqrt{9 - 4x^2}}$

(c) $\displaystyle\int \frac{dy}{y \sqrt{5y^2 - 3}}$.

27. Find the volume of the solid generated when the region bounded by $x = 2$, $x = -2$, $y = 0$, and $y = 1/\sqrt{4 + x^2}$ is revolved about the x-axis.

28. In Exercise 28 of Section 8.2, how far from the launching pad should the camera be positioned to maximize the angle θ subtended at the lens by the missile?

29. Given points $A(2, 1)$ and $B(5, 4)$, find the point P in the interval $[2, 5]$ on the x-axis that maximizes angle APB.

30. An aircraft is flying at an altitude of 4 mi at a speed of 800 mi/hr in a direction away from a tracking station on the ground. How fast is the angle of elevation changing when the aircraft is over a point 10 mi from the station?

31. A lighthouse is located 3 mi off a straight shore. If the light revolves at 2 revolutions per minute, how fast is the beam moving along the coastline at a point 2 mi down the coast?

32. A 25-ft ladder leans against a vertical wall. If the bottom of the ladder slides away from the base of the wall at the rate of 4 ft/sec, how fast is the angle between the ladder and the wall changing when the top of the ladder is 20 ft above the ground?

33. The lower edge of a painting, 10 ft in height, is 2 ft above an observer's eye level. Assuming the best view is obtained when the angle subtended at the observer's eye by the painting is maximum, how far from the wall should the observer stand?

8.4 INVERSE HYPERBOLIC FUNCTIONS

In this section we discuss inverses of hyperbolic functions. We will be interested primarily in the integration formulas that these functions produce.

Because the hyperbolic sine function is increasing on the interval $(-\infty, +\infty)$ (Figure 7.6.1b), it follows that $\sinh x$ is one-to-one on this interval and consequently has an inverse function, $\sinh^{-1} x$. Thus, for all real values of x

$$\sinh^{-1}(\sinh x) = x$$
$$\sinh(\sinh^{-1} x) = x$$

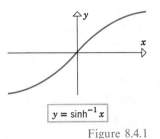

$y = \sinh^{-1} x$

Figure 8.4.1

Moreover,

$$y = \sinh x \qquad \text{and} \qquad x = \sinh^{-1} y$$

are equivalent statements. The graph of $y = \sinh^{-1} x$ can be obtained by reflecting the graph of $y = \sinh x$ about the line $y = x$ (Figure 8.4.1).

Because $\sinh x$ is expressible in terms of e^x, one might suspect that $\sinh^{-1} x$ is expressible in terms of natural logarithms. To see that this is so, we write $y = \sinh^{-1} x$ in the alternate form

$$x = \sinh y = \frac{e^y - e^{-y}}{2}$$

so that

$$e^y - 2x - e^{-y} = 0$$

or, on multiplying through by e^y

$$e^{2y} - 2xe^y - 1 = 0$$

Applying the quadratic formula yields

$$e^y = \frac{2x \pm \sqrt{4x^2 + 4}}{2} = x \pm \sqrt{x^2 + 1}$$

Since $e^y > 0$, the solution involving the minus sign is extraneous and must be discarded. Thus,

$$e^y = x + \sqrt{x^2 + 1}$$

Taking natural logarithms yields

$$y = \ln(x + \sqrt{x^2 + 1})$$

or

$$\sinh^{-1} x = \ln(x + \sqrt{x^2 + 1}) \tag{1}$$

This formula and a table of natural logarithms can be used to evaluate $\sinh^{-1} x$. Moreover, we can use it to obtain the derivative formula for $\sinh^{-1} x$ as follows: from (1),

$$\frac{d}{dx}[\sinh^{-1} x] = \frac{d}{dx}[\ln(x + \sqrt{x^2 + 1})]$$

$$= \frac{1}{x + \sqrt{x^2 + 1}}\left(1 + \frac{x}{\sqrt{x^2 + 1}}\right)$$

$$= \frac{\sqrt{x^2 + 1} + x}{(x + \sqrt{x^2 + 1})(\sqrt{x^2 + 1})}$$

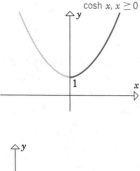

cosh x, $x \geq 0$

$y = \cosh^{-1} x$

(b)

Figure 8.4.2

or, on canceling

$$\frac{d}{dx}[\sinh^{-1} x] = \frac{1}{\sqrt{x^2 + 1}} \qquad (2)$$

Derivatives of the remaining inverse hyperbolic functions are obtained in a similar manner with the exception of $\cosh^{-1}$ and sech^{-1}. As evidenced by Figures 7.6.1a and 7.6.3, $\cosh x$ and $\operatorname{sech} x$ are not one-to-one, so that we must restrict the domains of these functions appropriately before we can obtain inverses. For example, if we let

$$f(x) = \cosh x, \ x \geq 0$$

then f is one-to-one (Figure 8.4.2a) and consequently has an inverse. We call this inverse $\cosh^{-1} x$. The graph of $y = \cosh^{-1} x$ is the reflection of the graph of f about the line $y = x$ (Figure 8.4.2b). The domain of $\cosh^{-1}$ is the interval $[1, +\infty)$, the range is $[0, +\infty)$, and

$$\cosh^{-1}(\cosh x) = x \quad \text{if} \quad x \geq 0$$
$$\cosh(\cosh^{-1} x) = x \quad \text{if} \quad x \geq 1$$

Moreover, if $x \geq 1$ and $y \geq 0$ then the statements

$$y = \cosh^{-1} x \quad \text{and} \quad x = \cosh y$$

are equivalent.

By following the derivations (1) and (2), the reader should be able to show that

$$\cosh^{-1} x = \ln(x + \sqrt{x^2 - 1})$$

and

$$\frac{d}{dx}[\cosh^{-1} x] = \frac{1}{\sqrt{x^2 - 1}}$$

if $x > 1$.

For reference, we summarize the definitions, logarithmic expressions, and derivatives of the inverse hyperbolic functions.

8.4.1 DEFINITION

$y = \sinh^{-1} x$	is equivalent to	$x = \sinh y$	for all x, y
$y = \cosh^{-1} x$	is equivalent to	$x = \cosh y$	if $\begin{array}{l} y \geq 0 \\ x \geq 1 \end{array}$
$y = \tanh^{-1} x$	is equivalent to	$x = \tanh y$	for all x, y
$y = \coth^{-1} x$	is equivalent to	$x = \coth y$	for all x, y
$y = \operatorname{sech}^{-1} x$	is equivalent to	$x = \operatorname{sech} y$	if $\begin{array}{l} y \geq 0 \\ 0 < x \leq 1 \end{array}$
$y = \operatorname{csch}^{-1} x$	is equivalent to	$x = \operatorname{csch} y$	for all x, y

8.4.2 THEOREM

$$\sinh^{-1} x = \ln(x + \sqrt{x^2 + 1}) \qquad -\infty < x < +\infty$$
$$\cosh^{-1} x = \ln(x + \sqrt{x^2 - 1}) \qquad x \geq 1$$
$$\tanh^{-1} x = \frac{1}{2} \ln \frac{1 + x}{1 - x} \qquad -1 < x < 1$$
$$\coth^{-1} x = \frac{1}{2} \ln \frac{x + 1}{x - 1} \qquad |x| > 1$$
$$\text{sech}^{-1} x = \ln \left(\frac{1 + \sqrt{1 - x^2}}{x} \right) \qquad 0 < x \leq 1$$
$$\text{csch}^{-1} x = \ln \left(\frac{1}{x} + \frac{\sqrt{1 + x^2}}{|x|} \right) \qquad x \neq 0$$

8.4.3 THEOREM

$$\frac{d}{dx}[\sinh^{-1} x] = \frac{1}{\sqrt{1 + x^2}}$$
$$\frac{d}{dx}[\cosh^{-1} x] = \frac{1}{\sqrt{x^2 - 1}}$$
$$\frac{d}{dx}[\tanh^{-1} x] = \frac{1}{1 - x^2} \qquad (|x| < 1)$$
$$\frac{d}{dx}[\coth^{-1} x] = \frac{1}{1 - x^2} \qquad (|x| > 1)$$
$$\frac{d}{dx}[\text{sech}^{-1} x] = -\frac{1}{x\sqrt{1 - x^2}}$$
$$\frac{d}{dx}[\text{csch}^{-1} x] = -\frac{1}{|x|\sqrt{1 + x^2}}$$

REMARK. Care must be exercised in differentiating $\tanh^{-1} x$ and $\coth^{-1} x$. In both cases the expression for the derivative is

$$\frac{1}{1 - x^2} \tag{3}$$

However, the domain of $\tanh^{-1} x$ is the interval $-1 < x < 1$ (verify this) so that the derivative formula for $\tanh^{-1} x$ applies only when $|x| < 1$. Similarly, the derivative formula for $\coth^{-1} x$ applies only on the domain of $\coth^{-1} x$, that is, when $|x| > 1$. This distinction becomes important when we integrate (3); for then we must write

$$\int \frac{1}{1 - x^2} \, dx = \begin{cases} \tanh^{-1} x + C & \text{if } |x| < 1 \\ \coth^{-1} x + C & \text{if } |x| > 1 \end{cases}$$

The following integration formulas follow from Theorem 8.4.3.

8.4.4   THEOREM

$$\int \frac{dx}{\sqrt{1+x^2}} = \sinh^{-1} x + C$$

$$\int \frac{dx}{\sqrt{x^2-1}} = \cosh^{-1} x + C$$

$$\int \frac{dx}{1-x^2} = \begin{cases} \tanh^{-1} x + C & \text{if } |x| < 1 \\ \coth^{-1} x + C & \text{if } |x| > 1 \end{cases}$$

$$\int \frac{dx}{x\sqrt{1-x^2}} = -\operatorname{sech}^{-1} |x| + C$$

$$\int \frac{dx}{x\sqrt{1+x^2}} = -\operatorname{csch}^{-1} |x| + C$$

The first three integration formulas follow immediately from the corresponding differentiation formulas. The last two require additional work (see Exercises 25 and 26). By using Theorem 8.4.2 these formulas can also be expressed in terms of the natural logarithm. In particular, we leave it as an exercise to show that the third formula can be written:

$$\int \frac{dx}{1-x^2} = \frac{1}{2} \ln \left| \frac{1+x}{1-x} \right| + C \tag{4}$$

▶ Exercise Set 8.4

1. (a) Prove: $\cosh^{-1} x = \ln (x + \sqrt{x^2 - 1})$, $x \geq 1$.
    (b) Use (a) to obtain the derivative of $\cosh^{-1} x$.

2. (a) Prove: $\tanh^{-1} x = \dfrac{1}{2} \ln \dfrac{1+x}{1-x}$, $-1 < x < 1$.
    (b) Use (a) to obtain the derivative of $\tanh^{-1} x$.

3. (a) Prove: $\operatorname{sech}^{-1} x = \cosh^{-1} \left(\dfrac{1}{x}\right)$, $0 < x \leq 1$

    $\coth^{-1} x = \tanh^{-1} \left(\dfrac{1}{x}\right)$, $|x| > 1$

    $\operatorname{csch}^{-1} x = \sinh^{-1} \left(\dfrac{1}{x}\right)$, $x \neq 0$

    (b) Use (a) and the derivatives of $\sinh^{-1} x$, $\cosh^{-1} x$, and $\tanh^{-1} x$ to find the derivatives of $\operatorname{sech}^{-1} x$, $\coth^{-1} x$, and $\operatorname{csch}^{-1} x$.
    (c) Use (a) and the logarithmic expressions for $\sinh^{-1} x$, $\cosh^{-1} x$, and $\tanh^{-1} x$ to derive the logarithmic expressions for $\operatorname{sech}^{-1} x$, $\coth^{-1} x$, and $\operatorname{csch}^{-1} x$.

4. Without referring to the text, state the domains of the six inverse hyperbolic functions.

In Exercises 5–13, find dy/dx.

5. (a) $y = \sinh^{-1} (\frac{1}{3}x)$
    (b) $y = \cosh^{-1} (2x + 1)$.

6. (a) $y = \tanh^{-1} (x^2)$
    (b) $y = \coth^{-1} (\sqrt{x})$.

7. (a) $y = \operatorname{sech}^{-1} (x^7)$
    (b) $y = \operatorname{csch}^{-1} (e^x)$.

8. (a) $y = (\tanh^{-1} x)^{-1}$
    (b) $y = \dfrac{1}{\tanh^{-1} x}$.

9. (a) $y = \sinh^{-1} \left(\dfrac{1}{x}\right)$
    (b) $y = \cosh^{-1} (\cosh x)$.

10. (a) $y = \ln (\cosh^{-1} x)$
    (b) $y = \sqrt{\coth^{-1} x}$.

11. (a) $y = e^x \operatorname{sech}^{-1} x$
    (b) $y = x^2 (\sinh^{-1} x)^3$.

12. (a) $y = \sinh^{-1} (\tanh x)$
    (b) $y = \cosh^{-1} (\sinh^{-1} x)$.

13. (a) $y = \tanh^{-1}\left(\dfrac{1-x}{1+x}\right)$

   (b) $y = (1 + x \operatorname{csch}^{-1} x)^{10}$.

In Exercises 14–19, evaluate the integral.

14. $\displaystyle\int \frac{dx}{\sqrt{1 + 9x^2}}$.

15. $\displaystyle\int \frac{dx}{\sqrt{x^2 - 2}}$.

16. $\displaystyle\int \frac{dx}{\sqrt{9x^2 - 25}}$.

17. $\displaystyle\int \frac{dx}{\sqrt{1 - e^{2x}}}$.

18. $\displaystyle\int \frac{\sin\theta\,d\theta}{\sqrt{1 + \cos^2\theta}}$.

19. $\displaystyle\int \frac{dx}{x\sqrt{1 + x^6}}$.

In Exercises 20–22, use the natural logarithm table in Appendix 3 to obtain a numerical value for the integral.

20. $\displaystyle\int_0^{1/2} \frac{dx}{1 - x^2}$.

21. $\displaystyle\int_2^3 \frac{dx}{1 - x^2}$.

22. $\displaystyle\int_0^{\sqrt{3}} \frac{dt}{\sqrt{t^2 + 1}}$.

23. Sketch the graphs of

   (a) $\tanh^{-1} x$ (b) $\operatorname{csch}^{-1} x$.

24. Sketch the graphs of

   (a) $\coth^{-1} x$ (b) $\operatorname{sech}^{-1} x$.

25. Show that

$$\frac{d}{dx}[\operatorname{sech}^{-1}|x|] = -\frac{1}{x\sqrt{1 - x^2}}$$

26. Show that

$$\frac{d}{dx}[\operatorname{csch}^{-1}|x|] = -\frac{1}{x\sqrt{1 + x^2}}$$

27. Derive integration formula (4).

28. Let a be a positive constant. Derive an integration formula for:

   (a) $\displaystyle\int \frac{du}{\sqrt{a^2 + u^2}}$ (b) $\displaystyle\int \frac{du}{\sqrt{u^2 - a^2}}$

   (c) $\displaystyle\int \frac{du}{a^2 - u^2}$.

   [*Hint:* See Example 5 of Section 8.3.]

29. Our derivation of the differentiation formula for $\sinh^{-1} x$ used the logarithmic expression for this function. Show that the derivative can also be obtained by the method we used in Section 8.3 to obtain the derivative formula for $\sin^{-1} x$.

30. Find

   (a) $\displaystyle\lim_{x\to+\infty} \sinh^{-1} x$ (b) $\displaystyle\lim_{x\to+\infty} \coth^{-1} x$

   (c) $\displaystyle\lim_{x\to 0^+} \operatorname{csch}^{-1} x$ (d) $\displaystyle\lim_{x\to+\infty}(\cosh^{-1} x - \ln x)$.

► SUPPLEMENTARY EXERCISES

1. In each part determine whether f and g are inverse functions.

   (a) $f(x) = mx$ $g(x) = 1/(mx)$
   (b) $f(x) = 3/(x + 1)$ $g(x) = (3 - x)/x$
   (c) $f(x) = x^3 - 8$ $g(x) = \sqrt[3]{x} + 2$
   (d) $f(x) = x^3 - 1$ $g(x) = \sqrt[3]{x + 1}$
   (e) $f(x) = \sqrt{e^x}$ $g(x) = 2\ln x$.

In Exercises 2–6, find $f^{-1}(x)$ if it exists.

2. $f(x) = 8x^3 - 1$.

3. $f(x) = x^2 - 2x + 1$.

4. $f(x) = x^2 - 2x + 1,\ x \geq 1$.

5. $f(x) = (e^x)^2 + 1$.

6. $f(x) = \exp(x^2) + 1$.

7. Let $f(x) = (ax + b)/(cx + d)$.

   What conditions on a, b, c, d guarantee that f^{-1} exists? Find $f^{-1}(x)$.

8. Show that

$$f(x) = (x + 2)/(x - 1)$$

is its own inverse.

9. Find the largest open interval containing the origin on which f is one-to-one.

   (a) $f(x) = |2x - 5|$ (b) $f(x) = x^2 + 4x$
   (c) $f(x) = \cos(x - 2\pi/3)$.

In Exercises 10–13, find $f^{-1}(x)$, and then use Formula (4) of Section 8.1 to obtain $(f^{-1})'(x)$. Check your work by differentiating $f^{-1}(x)$ directly.

10. $f(x) = x^3 - 8$.

11. $f(x) = 3/(x + 1)$.

12. $f(x) = mx + b\ (m \neq 0)$.   **13.** $f(x) = \sqrt{e^x}$.

14. (a) $\sec^{-1}(-1)$ (b) $\cos^{-1}(-\sqrt{3}/2)$
   (c) $\tan[\sec^{-1}(5/4)]$ (d) $\sec[\sin^{-1}(2/3)]$.

15. (a) $\cos^{-1}(-1/2)$ (b) $\cot^{-1}[\cot(3/4)]$
     (c) $\cos[\sin^{-1}(4/5)]$ (d) $\cos[\sin^{-1}(-4/5)]$.

16. (a) $\tan^{-1}(-1)$ (b) $\csc^{-1}(-2/\sqrt{3})$
     (c) $\cos^{-1}[\cos(-\pi/3)]$ (d) $\sin[-\sec^{-1}(2/\sqrt{3})]$.

17. (a) $\sin^{-1}(1/\sqrt{2})$ (b) $\sin^{-1}[\sin(5\pi/4)]$
     (c) $\tan(\sec^{-1}5)$ (d) $\tan^{-1}[\cot(\pi/6)]$.

18. Use a double angle formula to convert the given expression to an algebraic function of x.
     (a) $\sin(2\csc^{-1}x)$, $|x| \geq 1$
     (b) $\cos(2\sin^{-1}x)$, $|x| \leq 1$
     (c) $\sin(2\tan^{-1}x)$.

19. Express as a rational number:
     (a) $\cos[\cos^{-1}(4/5) + \sin^{-1}(5/13)]$
     (b) $\sin[\sin^{-1}(4/5) + \cos^{-1}(5/13)]$
     (c) $\tan[\tan^{-1}(1/3) + \tan^{-1}(2)]$.

20. If $u = \operatorname{csch}^{-1}(-5/12)$, find $\coth u$, $\sinh u$, $\cosh u$, and $\sinh(2u)$.

21. If $u = \tanh^{-1}(-3/5)$, find $\cosh u$, $\sinh u$, and $\cosh(2u)$.

In Exercises 22 and 23, sketch the graph of f.

22. (a) $f(x) = 3\sin^{-1}(x/2)$
     (b) $f(x) = \cos^{-1}x - \pi/2$.

23. (a) $f(x) = 2\tan^{-1}(-3x)$
     (b) $f(x) = \cos^{-1}x + \sin^{-1}x$.

In Exercises 24–37, find dy/dx, using implicit or logarithmic differentiation where convenient.

24. $y = \sin^{-1}(e^x) + 2\tan^{-1}(3x)$.

25. $y = \dfrac{1}{\sec^{-1}x^2}$.

26. $y = x\sin^{-1}x + \sqrt{1 - x^2}$.

27. $y = \cosh^{-1}(\sec x)$.

28. $\tan^{-1}y = \sin^{-1}x$.

29. $y = x\tanh^{-1}(\ln x)$.

30. $y = \tan^{-1}\left(\dfrac{2x}{1 - x^2}\right)$.

31. $y = \sqrt{\sin^{-1}3x}$.

32. $y = (\sin^{-1}2x)^{-1}$.

33. $y = \exp(\sec^{-1}x)$.

34. $y = (\tan^{-1}x)/\ln x$.

35. $y = \pi^{\sin^{-1}x}$.

36. $y = (\sinh^{-1}x)^\pi$.

37. $y = \tanh^{-1}\left(\dfrac{1}{\coth x}\right)$.

38. Let $f(x) = \tan^{-1}x + \tan^{-1}(1/x)$.
     (a) By considering $f'(x)$, show that $f(x) = C_1$ on $(-\infty, 0)$ and $f(x) = C_2$ on $(0, +\infty)$, where C_1 and C_2 are constants.
     (b) Find C_1 and C_2 in part (a).

39. Show that $y = \tan^{-1}x$ satisfies
$$y'' = -2\sin y\cos^3 y.$$

In Exercises 40–49, evaluate the integral.

40. $\displaystyle\int \frac{dx}{\sqrt{9 - 4x^2}}$.

41. $\displaystyle\int \frac{e^x\,dx}{1 - e^{2x}}$.

42. $\displaystyle\int \frac{\cot x\,dx}{\sqrt{1 - \sin^2 x}}$.

43. $\displaystyle\int \frac{dx}{x\sqrt{(\ln x)^2 - 1}}$.

44. $\displaystyle\int \frac{dx}{e^x\sqrt{1 - e^{-2x}}}$.

45. $\displaystyle\int_{2\sqrt{3}/9}^{2/3} \frac{dx}{3x\sqrt{9x^2 - 1}}$.

46. $\displaystyle\int_0^{\sqrt{2}} \frac{x\,dx}{4 + x^4}$.

47. $\displaystyle\int \frac{dx}{x^{1/2} + x^{3/2}}$.

48. $\displaystyle\int \frac{x^2\,dx}{\sqrt{1 + x^6}}$.

49. $\displaystyle\int_{1/4}^{1/2} \frac{dx}{\sqrt{x}\sqrt{1 - x}}$.

50. The hypotenuse of a right triangle is growing at a rate of a cm/sec and one leg is decreasing at a rate of b cm/sec. How fast is the acute angle between the hypotenuse and the other leg changing at the instant when both legs are 1 cm?

51. Find the value of x at which $f(x) = \tan^{-1}(2x) - \tan^{-1}x$ assumes its maximum value on the interval $[0, +\infty)$.

52. Find the area between the curve $y = 1/(9 + x^2)$, the x-axis, and the lines $x = \pm\sqrt{3}$.

9 techniques of integration

9.1 A BRIEF REVIEW

Although we can now integrate a wide variety of functions, there remain many important kinds of integrals that we cannot yet evaluate. The purpose of this chapter is to develop some additional techniques of integration and also to systematize the procedure of integration. We begin with a review of the basic integration formulas we have encountered thus far.

CONSTANTS, POWERS, EXPONENTIALS

(See Sections 5.2, 7.2, and 7.4)

1. $\displaystyle \int dx = x + C$

2. $\displaystyle \int a\,dx = a \int dx = ax + C$

3. $\displaystyle \int x^r\,dx = \frac{x^{r+1}}{r+1} + C, \quad r \neq -1$

4. $\displaystyle \int \frac{dx}{x} = \ln |x| + C$

5. $\displaystyle \int e^x\,dx = e^x + C$

6. $\displaystyle \int a^x\,dx = \frac{a^x}{\ln a} + C$

TRIGONOMETRIC FUNCTIONS

(See Sections 5.2 and 7.2)

7. $\displaystyle \int \sin x\,dx = -\cos x + C$

8. $\displaystyle \int \cos x\,dx = \sin x + C$

9. $\displaystyle \int \sec^2 x\,dx = \tan x + C$

10. $\displaystyle \int \csc^2 x\,dx = -\cot x + C$

11. $\int \sec x \tan x \, dx = \sec x + C$ **13.** $\int \tan x \, dx = -\ln |\cos x| + C$

12. $\int \csc x \cot x \, dx = -\csc x + C$ **14.** $\int \cot x \, dx = \ln |\sin x| + C$

HYPERBOLIC
FUNCTIONS
(See Section 7.6)

15. $\int \sinh x \, dx = \cosh x + C$ **18.** $\int \operatorname{csch}^2 x \, dx = -\coth x + C$

16. $\int \cosh x \, dx = \sinh x + C$ **19.** $\int \operatorname{sech} x \tanh x \, dx = -\operatorname{sech} x + C$

17. $\int \operatorname{sech}^2 x \, du = \tanh x + C$ **20.** $\int \operatorname{csch} x \coth x \, dx = -\operatorname{csch} x + C$

ALGEBRAIC
FUNCTIONS
(See Sections 8.3 and 8.4)

21. $\displaystyle\int \frac{dx}{\sqrt{1 - x^2}} = \sin^{-1} x + C$

22. $\displaystyle\int \frac{dx}{1 + x^2} = \tan^{-1} x + C$

23. $\displaystyle\int \frac{dx}{x \sqrt{x^2 - 1}} = \sec^{-1} x + C$

24. $\displaystyle\int \frac{dx}{\sqrt{1 + x^2}} = \sinh^{-1} x + C = \ln (x + \sqrt{x^2 + 1}) + C$

25. $\displaystyle\int \frac{dx}{\sqrt{x^2 - 1}} = \cosh^{-1} x + C = \ln (x + \sqrt{x^2 - 1}) + C$

26. $\displaystyle\int \frac{dx}{1 - x^2} = \begin{cases} \tanh^{-1} x + C & \text{if } |x| < 1 \\ \coth^{-1} x + C & \text{if } |x| > 1 \end{cases} = \frac{1}{2} \ln \left| \frac{1 + x}{1 - x} \right| + C$

27. $\displaystyle\int \frac{dx}{x \sqrt{1 - x^2}} = -\operatorname{sech}^{-1} |x| + C = -\ln \left(\frac{1 + \sqrt{1 - x^2}}{x} \right) + C$

28. $\displaystyle\int \frac{dx}{x \sqrt{1 + x^2}} = -\operatorname{csch}^{-1} |x| + C = -\ln \left(\frac{1}{x} + \frac{\sqrt{1 + x^2}}{|x|} \right) + C$

REMARK.   Readers who did not cover Section 8.4 should ignore formulas 24–28. We will study alternative methods for evaluating these integrals in this chapter.

▶ Exercise Set 9.1

With the text closed, complete the following integration formulas.

Constants, Powers, Exponentials

1. $\int dx =$

2. $\int a\,dx =$

3. $\int x^r\,dx =$

4. $\int \dfrac{dx}{x} =$

5. $\int e^x\,dx =$

6. $\int a^x\,dx =$

Trigonometric Functions

7. $\int \sin x\,dx =$

8. $\int \cos x\,dx =$

9. $\int \sec^2 x\,dx =$

10. $\int \csc^2 x\,dx =$

11. $\int \sec x \tan x\,dx =$

12. $\int \csc x \cot x\,dx =$

13. $\int \tan x\,dx =$

14. $\int \cot x\,dx =$

Hyperbolic Functions

15. $\int \sinh x\,dx =$

16. $\int \cosh x\,dx =$

17. $\int \operatorname{sech}^2 x\,dx =$

18. $\int \operatorname{csch}^2 x\,dx =$

19. $\int \operatorname{sech} x \tanh x\,dx =$

20. $\int \operatorname{csch} x \coth x\,dx =$

Algebraic Functions

21. $\int \dfrac{dx}{\sqrt{1 - x^2}} =$

22. $\int \dfrac{dx}{1 + x^2} =$

23. $\int \dfrac{dx}{x\sqrt{x^2 - 1}} =$

24. $\int \dfrac{dx}{\sqrt{1 + x^2}} =$

25. $\int \dfrac{dx}{\sqrt{x^2 - 1}} =$

26. $\int \dfrac{dx}{1 - x^2} =$

27. $\int \dfrac{dx}{x\sqrt{1 - x^2}} =$

28. $\int \dfrac{dx}{x\sqrt{1 + x^2}} =$

9.2 INTEGRATION BY PARTS

In this section, we develop a technique that will help us to evaluate a wide variety of integrals that do not fit any of the basic integration formulas.

If f and g are differentiable functions, then by the rule for differentiating products

$$\frac{d}{dx}[f(x)g(x)] = f(x)g'(x) + g(x)f'(x)$$

Integrating both sides we obtain

$$\int \frac{d}{dx}[f(x)g(x)]\,dx = \int f(x)g'(x)\,dx + \int g(x)f'(x)\,dx$$

or

$$f(x)g(x) + C = \int f(x)g'(x)\,dx + \int g(x)f'(x)\,dx$$

or

$$\int f(x)g'(x)\,dx = f(x)g(x) - \int g(x)f'(x)\,dx + C$$

Since the integral on the right will produce another constant of integration, there is no need to keep the C in this last equation; thus, we obtain

$$\int f(x)g'(x)\,dx = f(x)g(x) - \int f'(x)g(x)\,dx \qquad (1)$$

which is called the formula for *integration by parts.* By using this formula we can sometimes reduce a hard integration problem to an easier one.

In practice, it is usual to rewrite (1) by letting

$$u = f(x) \qquad du = f'(x)\,dx$$
$$v = g(x) \qquad dv = g'(x)\,dx$$

This yields the following formula.

Integration by Parts for Indefinite Integrals

$$\int u\,dv = uv - \int v\,du \qquad (2a)$$

For definite integrals the corresponding formula is

Integration by Parts for Definite Integrals

$$\int_a^b u\,dv = uv\,\Big]_a^b - \int_a^b v\,du \qquad (2b)$$

where a and b are the limits of integration for the variable x.

▶ Example 1 Evaluate

$$\int xe^x\,dx$$

Solution. To apply (2a) we must write the integral in the form

$$\int u \, dv$$

One way to do this is to let

$$u = x \qquad \text{and} \qquad dv = e^x \, dx$$

so that

$$du = dx \qquad \text{and} \qquad v = \int e^x \, dx = e^x$$

Thus, from (2a)

$$\int \underset{u}{\underbrace{x}} \; \underset{dv}{\underbrace{e^x \, dx}} = \underset{u}{\underbrace{x}} \; \underset{v}{\underbrace{e^x}} - \int \underset{v}{\underbrace{e^x}} \; \underset{du}{\underbrace{dx}}$$

or

$$\int xe^x \, dx = xe^x - e^x + C \qquad \qquad \blacktriangleleft$$

REMARK. In the calculation of v from dv above, we omitted the constant of integration and wrote $v = \int e^x \, dx = e^x$. Had we written $v = \int e^x \, dx = e^x + C_1$, the constant C_1 would have eventually canceled out (Exercise 48a). This is always the case in integration by parts (Exercise 48b), so we will usually omit the constant when calculating v from dv.

To use integration by parts successfully, the choice of u and dv must be made so that the new integral is easier than the original. For example, had we decided above to let

$$u = e^x \qquad \qquad dv = x \, dx$$

$$du = e^x \, dx \qquad v = \int x \, dx = \frac{x^2}{2}$$

then we would have obtained

$$\int xe^x \, dx = \int u \, dv = uv - \int v \, du = \frac{x^2}{2}e^x - \frac{1}{2}\int x^2 e^x \, dx$$

For this choice of u and dv the new integral is actually more complicated than the original.

The next example shows that it is sometimes necessary to use integration by parts more than once in the same problem.

▶ Example 2 Evaluate

$$\int x^2 e^{-x}\, dx$$

Solution. Let

$$u = x^2 \qquad dv = e^{-x}\, dx$$

$$du = 2x\, dx \qquad v = \int e^{-x}\, dx = -e^{-x}$$

so that

$$\int x^2 e^{-x}\, dx = \int u\, dv = uv - \int v\, du = -x^2 e^{-x} + 2\int xe^{-x}\, dx \qquad (3)$$

The last integral is similar to the original except that we have replaced x^2 by x. Another integration by parts applied to $\int xe^{-x}\, dx$ will complete the problem. We let

$$u = x \qquad dv = e^{-x}\, dx$$

$$du = dx \qquad v = \int e^{-x}\, dx = -e^{-x}$$

so that

$$\int xe^{-x}\, dx = \int u\, dv = uv - \int v\, du$$

$$= -xe^{-x} + \int e^{-x}\, dx$$

$$= -xe^{-x} - e^{-x} + C_1$$

Substituting in (3) we obtain

$$\int x^2 e^{-x}\, dx = -x^2 e^{-x} + 2(-xe^{-x} - e^{-x} + C_1)$$
$$= -x^2 e^{-x} - 2xe^{-x} - 2e^{-x} + 2C_1$$
$$= -(x^2 + 2x + 2)e^{-x} + C$$

where $C = 2C_1$.

▶ Example 3 Evaluate

$$\int \ln x \, dx$$

Solution. Let

$$u = \ln x \qquad dv = dx$$

$$du = \frac{1}{x} dx \qquad v = \int dx = x$$

so that

$$\int \ln x \, dx = \int u \, dv = uv - \int v \, du$$

$$= x \ln x - \int x \left(\frac{1}{x}\right) dx$$

$$= x \ln x - \int dx = x \ln x - x + C \qquad ◀$$

The next example illustrates how integration by parts can be used to integrate the inverse trigonometric functions.

▶ Example 4 Evaluate

$$\int_0^1 \tan^{-1} x \, dx$$

Solution. Let

$$u = \tan^{-1} x \qquad dv = dx$$

$$du = \frac{1}{1 + x^2} dx \qquad v = \int dx = x$$

Thus,

$$\int_0^1 \tan^{-1} x \, dx = \int_0^1 u \, dv = uv \Big]_0^1 - \int_0^1 v \, du$$

$$= x \tan^{-1} x \Big]_0^1 - \int_0^1 \frac{x}{1 + x^2} dx$$

But

$$\int_0^1 \frac{x}{1 + x^2} dx = \frac{1}{2} \int_0^1 \frac{2x}{1 + x^2} dx = \frac{1}{2} \ln (1 + x^2) \Big]_0^1 = \frac{1}{2} \ln 2$$

so

$$\int_0^1 \tan^{-1} x \, dx = x \tan^{-1} x \Big]_0^1 - \frac{1}{2} \ln 2 = \left(\frac{\pi}{4} - 0\right) - \frac{1}{2} \ln 2$$

$$= \frac{\pi}{4} - \ln \sqrt{2} \qquad \blacktriangleleft$$

▶ Example 5   Evaluate

$$\int e^x \cos x \, dx$$

Solution.   Let

$$u = e^x \qquad dv = \cos x \, dx$$

$$du = e^x \, dx \qquad v = \int \cos x \, dx = \sin x$$

Thus,

$$\int e^x \cos x \, dx = \int u \, dv = uv - \int v \, du$$

$$= e^x \sin x - \int e^x \sin x \, dx \qquad (4)$$

Since the integral $\int e^x \sin x \, dx$ is similar to the original integral $\int e^x \cos x \, dx$, it seems that nothing has been accomplished. However, let us integrate this new integral by parts; we let

$$u = e^x \qquad dv = \sin x \, dx$$

$$du = e^x \, dx \qquad v = \int \sin x \, dx = -\cos x$$

Thus,

$$\int e^x \sin x \, dx = \int u \, dv = uv - \int v \, du$$

$$= -e^x \cos x + \int e^x \cos x \, dx$$

Substituting in (4) yields

$$\int e^x \cos x \, dx = e^x \sin x - \left[-e^x \cos x + \int e^x \cos x \, dx \right]$$

or

$$\int e^x \cos x \, dx = e^x \sin x + e^x \cos x - \int e^x \cos x \, dx$$

which is an equation we can solve for the unknown integral. We obtain

$$2 \int e^x \cos x \, dx = e^x \sin x + e^x \cos x$$

or

$$\int e^x \cos x \, dx = \frac{1}{2} e^x \sin x + \frac{1}{2} e^x \cos x + C \qquad \blacktriangleleft$$

If n is a positive integer, then

$$\int \sin^n x \, dx \qquad \text{and} \qquad \int \cos^n x \, dx$$

can be evaluated by using **reduction formulas.** These are formulas that express the given integral in terms of a similar integral involving a *lower* power. For example, we can obtain a reduction formula for $\int \cos^n x \, dx$ by writing $\cos^n x$ as $\cos^{n-1} x \cdot \cos x$ and letting

$$u = \cos^{n-1} x$$
$$dv = \cos x \, dx$$
$$du = (n-1) \cos^{n-2} x(-\sin x) \, dx = -(n-1) \cos^{n-2} x \sin x \, dx$$
$$v = \int \cos x \, dx = \sin x$$

so that

$$\int \cos^n x \, dx = \int \cos^{n-1} x \cos x \, dx = \int u \, dv$$

$$= uv - \int v \, du$$

$$= \cos^{n-1} x \sin x + (n-1) \int \sin^2 x \cos^{n-2} x \, dx$$

$$= \cos^{n-1} x \sin x + (n-1) \int (1 - \cos^2 x) \cos^{n-2} x \, dx$$

$$= \cos^{n-1} x \sin x + (n-1) \int \cos^{n-2} x \, dx - (n-1) \int \cos^n x \, dx$$

Transposing the last integral on the right to the left side yields

$$n \int \cos^n x \, dx = \cos^{n-1} x \sin x + (n-1) \int \cos^{n-2} x \, dx$$

or

$$\int \cos^n x \, dx = \frac{1}{n} \cos^{n-1} x \sin x + \frac{n-1}{n} \int \cos^{n-2} x \, dx \qquad (5)$$

This reduction formula reduces the exponent by 2. Thus, if we apply it repeatedly, we can eventually express $\int \cos^n x \, dx$ in terms of

$$\int \cos x \, dx = \sin x + C$$

if n is odd, or

$$\int \cos^0 x \, dx = \int dx = x + C$$

if n is even.

▶ Example 6 Evaluate

(a) $\displaystyle\int \cos^3 x \, dx$ (b) $\displaystyle\int \cos^4 x \, dx$

Solution (*a*). From (5) with $n = 3$

$$\int \cos^3 x \, dx = \frac{1}{3} \cos^2 x \sin x + \frac{2}{3} \int \cos x \, dx$$

$$= \frac{1}{3} \cos^2 x \sin x + \frac{2}{3} \sin x + C$$

Solution (*b*). From (5) with $n = 4$

$$\int \cos^4 x \, dx = \frac{1}{4} \cos^3 x \sin x + \frac{3}{4} \int \cos^2 x \, dx$$

and from (5) with $n = 2$

$$\int \cos^2 x \, dx = \frac{1}{2} \cos x \sin x + \frac{1}{2} \int dx = \frac{1}{2} \cos x \sin x + \frac{1}{2} x + C_1$$

so that

$$\int \cos^4 x \, dx = \frac{1}{4} \cos^3 x \sin x + \frac{3}{4} \left(\frac{1}{2} \cos x \sin x + \frac{1}{2} x + C_1 \right)$$

$$= \frac{1}{4} \cos^3 x \sin x + \frac{3}{8} \cos x \sin x + \frac{3}{8} x + C$$

where $C = \frac{3}{4} C_1$ ◀

We leave it as an exercise to derive the following companion formula to (5).

$$\int \sin^n x \, dx = -\frac{1}{n} \sin^{n-1} x \cos x + \frac{n-1}{n} \int \sin^{n-2} x \, dx \qquad (6)$$

▶ **Exercise Set 9.2**

In Exercises 1–26, evaluate the integral.

1. $\int x e^{-x} \, dx.$

2. $\int x e^{3x} \, dx.$

3. $\int \ln (2x + 3) \, dx.$

4. $\int x \ln x \, dx.$

5. $\int x \ln \sqrt{x} \, dx.$

6. $\int \sin^{-1} x \, dx.$

7. $\int \cos^{-1} (2x) \, dx.$

8. $\int x^2 e^{-2x} \, dx.$

9. $\int x^2 e^x \, dx.$

10. $\int x^3 e^{-x} \, dx.$

11. $\int e^x \sin x \, dx.$

12. $\int e^{-3\theta} \sin 3\theta \, d\theta.$

13. $\int \frac{\cos 2\pi x}{e^{2\pi x}} \, dx.$

14. $\int e^{2x} \cos 3x \, dx.$

15. $\int e^{ax} \sin bx \, dx.$

16. $\int x^2 \ln x \, dx.$

17. $\int x^2 \cos x \, dx.$

18. $\int x \tan^{-1} x \, dx.$

19. $\int x \sin (3x + 1) \, dx.$

20. $\int x \sinh x \, dx.$

21. $\int x \sec^2 x \, dx.$

22. $\int \frac{x^3 \, dx}{\sqrt{1 - x^2}}.$

23. $\int \cos (\ln x) \, dx.$

24. $\int \sin (3 \ln x) \, dx$

25. $\int \sin (\ln x) \, dx.$

26. $\int (\ln x)^2 \, dx.$

In Exercises 27–37, evaluate the definite integral.

27. $\int_0^1 x e^{-5x} \, dx.$

28. $\int_0^2 x e^{2x} \, dx.$

29. $\int_1^e x^2 \ln x \, dx.$

30. $\int_{\sqrt{e}}^e \frac{\ln x}{x^2} \, dx.$

31. $\int_{-2}^2 \ln (x + 3) \, dx.$

32. $\int_0^{1/2} \sin^{-1} x \, dx.$

33. $\int_2^4 \sec^{-1} \sqrt{\theta} \, d\theta.$

34. $\int_1^2 x \sec^{-1} x \, dx.$

35. $\int_0^{\pi/2} x \sin 4x \, dx.$

36. $\int_0^\pi (x + x \cos x) \, dx.$

37. $\int_0^1 \frac{x^3}{\sqrt{x^2 + 1}} \, dx.$

38. Solve Exercise 37 without using integration by parts.

39. (a) Find the area of the region enclosed by $y = \ln x$, the line $x = e$, and the x-axis.
   (b) Find the volume of the solid generated when the region in part (a) is revolved about the x-axis.

40. Find the area of the region enclosed by $y = x \sin x$, $y = x$, $x = 0$, and $x = \pi/2$.

41. Find the volume of the solid generated when the region enclosed by $y = \sin x$, $y = 0$, $x = 0$, and $x = \pi$ is revolved about the y-axis.

42. Find the volume of the solid generated when the region enclosed by $y = \sin x \cos x$, $y = 0$, $x = 0$, and $x = \pi/2$ is revolved about the y-axis.

43. Use reduction formula (6) to evaluate

(a) $\displaystyle\int \sin^3 x \, dx$ (b) $\displaystyle\int_0^{\pi/4} \sin^4 x \, dx$.

44. Use reduction formula (5) to evaluate

(a) $\displaystyle\int \cos^5 x \, dx$ (b) $\displaystyle\int_0^{\pi/2} \cos^6 x \, dx$.

45. Derive reduction formula (6).

In Exercises 46 and 47, derive the reduction formula in part (a) and use it to evaluate the integral in part (b).

46. (a) $\displaystyle\int \sec^n x \, dx$

$$= \frac{\sec^{n-2} x \tan x}{n - 1} + \frac{n - 2}{n - 1} \int \sec^{n-2} x \, dx$$

(b) $\displaystyle\int \sec^4 x \, dx$.

47. (a) $\displaystyle\int x^n e^x \, dx = x^n e^x - n \int x^{n-1} e^x \, dx$

(b) $\displaystyle\int x^3 e^x \, dx$.

48. (a) In Example 1, let

$$u = x \qquad dv = e^x \, dx$$

$$du = dx \qquad v = \int e^x \, dx = e^x + C_1$$

and show that the constant C_1 cancels out, thus leading to the same solution we obtained by omitting C_1.

(b) Show that

$$uv - \int v \, du = u(v + C_1) - \int (v + C_1) \, du$$

thereby justifying the omission of the constant of integration when calculating v in integration by parts.

9.3   INTEGRATING POWERS OF SINE AND COSINE

Integrals of the form

$$\int \sin^m x \cos^n x \, dx$$

where m and n are nonnegative integers are evaluated in various ways, depending on the values of m and n. The cases

$$\int \sin^m x \, dx \qquad (n = 0)$$

and

$$\int \cos^n x \, dx \qquad (m = 0)$$

can be treated using reduction formulas (5) and (6) of the previous section.

▶ Example 1   Show that

$$\int \sin^2 x\, dx = \frac{1}{2}x - \frac{1}{4}\sin 2x + C \qquad (1)$$

$$\int \cos^2 x\, dx = \frac{1}{2}x + \frac{1}{4}\sin 2x + C \qquad (2)$$

Solution. Letting $n = 2$ in reduction formulas (5) and (6) of Section 9.2 yields

$$\int \sin^2 x\, dx = -\frac{1}{2}\sin x \cos x + \frac{1}{2}x + C$$

$$= \frac{1}{2}x - \frac{1}{4}\sin 2x + C$$

and

$$\int \cos^2 x\, dx = \frac{1}{2}\cos x \sin x + \frac{1}{2}x + C$$

$$= \frac{1}{2}x + \frac{1}{4}\sin 2x + C$$

Alternate Solution. Apply the identities

$$\sin^2 x = \frac{1}{2}(1 - \cos 2x) \qquad (3a)$$

$$\cos^2 x = \frac{1}{2}(1 + \cos 2x) \qquad (3b)$$

which follow from the double-angle formulas $\cos 2x = 2\cos^2 x - 1$ and $\cos 2x = 1 - 2\sin^2 x$. We obtain

$$\int \sin^2 x\, dx = \frac{1}{2}\int (1 - \cos 2x)\, dx = \frac{1}{2}x - \frac{1}{4}\sin 2x + C$$

$$\int \cos^2 x\, dx = \frac{1}{2}\int (1 + \cos 2x)\, dx = \frac{1}{2}x + \frac{1}{4}\sin 2x + C \qquad \blacktriangleleft$$

► Example 2   The formulas

$$\int \sin^4 x\, dx = \frac{3}{8}x - \frac{1}{4}\sin 2x + \frac{1}{32}\sin 4x + C \qquad (4)$$

$$\int \cos^4 x\, dx = \frac{3}{8}x + \frac{1}{4}\sin 2x + \frac{1}{32}\sin 4x + C \qquad (5)$$

can be obtained from the reduction formulas and basic trigonometric identities. However, a more direct approach is as follows. From (3b)

$$\int \cos^4 x \, dx = \int (\cos^2 x)^2 \, dx = \int \left[\frac{1}{2}(1 + \cos 2x) \right]^2 dx$$

$$= \frac{1}{4} \int (1 + 2 \cos 2x + \cos^2 2x) \, dx$$

To finish, we apply (3b) again and write

$$\cos^2 2x = \frac{1}{2}(1 + \cos 4x) = \frac{1}{2} + \frac{1}{2} \cos 4x$$

which gives

$$\int \cos^4 x \, dx = \frac{1}{4} \int \left(\frac{3}{2} + 2 \cos 2x + \frac{1}{2} \cos 4x \right) dx$$

$$= \frac{3}{8} x + \frac{1}{4} \sin 2x + \frac{1}{32} \sin 4x + C$$

A similar procedure, using (3a), will yield (4). ◀

▶ Example 3   Show that

$$\int \sin^3 x \, dx = -\cos x + \frac{1}{3} \cos^3 x + C \tag{6}$$

$$\int \cos^3 x \, dx = \sin x - \frac{1}{3} \sin^3 x + C \tag{7}$$

Solution.   In the exercises we ask the reader to derive these results from reduction formulas. An alternate approach is as follows.

$$\int \sin^3 x \, dx = \int \sin^2 x \sin x \, dx$$

$$= \int (1 - \cos^2 x) \sin x \, dx$$

$$= \int \sin x \, dx - \int \cos^2 x \sin x \, dx$$

$$= -\cos x + \frac{1}{3} \cos^3 x + C$$

To integrate, let
$u = \cos x,\ du = -\sin x \, dx$

$$\int \cos^3 x \, dx = \int \cos^2 x \cos x \, dx$$

$$= \int (1 - \sin^2 x) \cos x \, dx$$

$$= \int \cos x \, dx - \int \sin^2 x \cos x \, dx \quad \boxed{\begin{array}{l} \text{To integrate, let} \\ u = \sin x, \; du = \cos x \, dx \end{array}}$$

$$= \sin x - \frac{1}{3} \sin^3 x + C \qquad \blacktriangleleft$$

If m and n are both positive integers, then the integral

$$\int \sin^m x \cos^n x \, dx$$

can be evaluated by one of three procedures, depending on whether m and n are odd or even. The procedures are outlined in the following table.

Table 9.3.1

CASE	PROCEDURE	RELEVANT IDENTITIES
n odd	Substitute $u = \sin x$	$\cos^2 x = 1 - \sin^2 x$
m odd	Substitute $u = \cos x$	$\sin^2 x = 1 - \cos^2 x$
m and n even	Use identities to reduce the powers on sin and cos	$\begin{cases} \sin^2 x = \frac{1}{2}(1 - \cos 2x) \\ \cos^2 x = \frac{1}{2}(1 + \cos 2x) \end{cases}$

▶ Example 4 Evaluate

$$\int \sin^4 x \cos^5 x \, dx$$

Solution. Since $n = 5$ is odd, we will follow the first procedure in Table 9.3.1 and make the substitution

$$u = \sin x \qquad du = \cos x \, dx$$

To form the du, we will first split off a factor of $\cos x$.

$$\int \sin^4 x \cos^5 x \, dx = \int \sin^4 x \cos^4 x \cos x \, dx$$

$$= \int \sin^4 x (1 - \sin^2 x)^2 \cos x \, dx$$

$$= \int u^4 (1 - u^2)^2 \, du$$

$$= \int (u^4 - 2u^6 + u^8) \, du$$

$$= \frac{1}{5} u^5 - \frac{2}{7} u^7 + \frac{1}{9} u^9 + C$$

$$= \frac{1}{5} \sin^5 x - \frac{2}{7} \sin^7 x + \frac{1}{9} \sin^9 x + C \qquad \blacktriangleleft$$

▶ **Example 5**   Evaluate

$$\int \sin^3 x \cos^2 x \, dx$$

Solution.   Since $m = 3$ is odd, we will follow the second procedure in Table 9.3.1 and make the substitution

$$u = \cos x \qquad du = -\sin x \, dx$$

To form the *du*, we will split off a factor of $\sin x$.

$$\int \sin^3 x \cos^2 x \, dx = \int \sin^2 x \cos^2 x \sin x \, dx$$

$$= \int (1 - \cos^2 x) \cos^2 x \sin x \, dx$$

$$= -\int (1 - u^2) u^2 \, du$$

$$= \int (u^4 - u^2) \, du$$

$$= \frac{1}{5} u^5 - \frac{1}{3} u^3 + C$$

$$= \frac{1}{5} \cos^5 x - \frac{1}{3} \cos^3 x + C \qquad \blacktriangleleft$$

▶ **Example 6**   Evaluate

$$\int \sin^4 x \cos^4 x \, dx$$

Solution.   Since $m = n = 4$, we will follow the third procedure in Table 9.3.1.

$$\int \sin^4 x \cos^4 x \, dx = \int (\sin^2 x)^2 (\cos^2 x)^2 \, dx$$

$$= \int (\tfrac{1}{2}[1 - \cos 2x])^2 (\tfrac{1}{2}[1 + \cos 2x])^2 \, dx$$

$$= \frac{1}{16} \int (1 - \cos^2 2x)^2 \, dx$$

$$= \frac{1}{16} \int \sin^4 2x \, dx$$

To finish, we will let $u = 2x$, $du = 2 \, dx$ and then use (4).

$$\int \sin^4 x \cos^4 x \, dx = \frac{1}{32} \int \sin^4 u \, du$$

$$= \frac{1}{32} \left(\frac{3}{8} u - \frac{1}{4} \sin 2u + \frac{1}{32} \sin 4u \right) + C$$

$$= \frac{3}{128} x - \frac{1}{128} \sin 4x + \frac{1}{1024} \sin 8x + C \qquad \blacktriangleleft$$

Integrals of the form

$$\int \sin mx \cos nx \, dx \quad \int \sin mx \sin nx \, dx \quad \int \cos mx \cos nx \, dx$$

can be found using the product formulas from trigonometry (see 19a, 19b, and 19c of the Unit I trigonometry review in Appendix 1).

▶ Example 7   Evaluate

$$\int \sin 7x \cos 3x \, dx$$

Solution.   Since

$$\sin 7x \cos 3x = \frac{1}{2} (\sin 4x + \sin 10x)$$

we can write

$$\int \sin 7x \cos 3x \, dx = \frac{1}{2} \int (\sin 4x + \sin 10x) \, dx$$

$$= -\frac{1}{8} \cos 4x - \frac{1}{20} \cos 10x + C \qquad \blacktriangleleft$$

▶ Exercise Set 9.3

In Exercises 1–30, perform the indicated integration.

1. $\int \cos^5 x \sin x \, dx.$

2. $\int \sin^4 3x \cos 3x \, dx.$

3. $\int \sin ax \cos ax \, dx \quad (a \neq 0).$

4. $\int \cos^2 3x \, dx.$

5. $\int \sin^2 5\theta \, d\theta.$

6. $\int \cos^3 at \, dt \quad (a \neq 0).$

7. $\int \cos^4 \left(\frac{x}{4}\right) dx.$

8. $\int \sin^5 x \, dx.$

9. $\int \cos^5 \theta \, d\theta.$

10. $\int \sin^3 x \cos^3 x \, dx.$

11. $\int \sin^2 2t \cos^3 2t \, dt.$

12. $\int \sin^4 x \cos^5 x \, dx.$

13. $\int \cos^4 x \sin^3 x \, dx.$

14. $\int \sin^3 2x \cos^2 2x \, dx.$

15. $\int \sin^5 \theta \cos^4 \theta \, d\theta.$

16. $\int \cos^{1/5} x \sin x \, dx.$

17. $\int \sin^2 x \cos^2 x \, dx.$

18. $\int \sin^2 x \cos^4 x \, dx.$

19. $\int \sin x \cos 2x \, dx.$

20. $\int \sin 3\theta \cos 2\theta \, d\theta.$

21. $\int \sin x \cos \left(\frac{x}{2}\right) dx.$

22. $\int \sin ax \cos bx \, dx \quad (a > 0, \ b > 0, \ a \neq b).$

23. $\int \frac{\sin x}{\cos^8 x} \, dx.$

24. $\int \sqrt{\cos \theta} \sin \theta \, d\theta.$

25. $\int_0^{\pi/4} \cos^3 x \, dx.$

26. $\int_{-\pi}^{\pi} \cos^2 5\theta \, d\theta.$

27. $\int_0^{\pi/3} \sin^4 3x \cos^3 3x \, dx.$

28. $\int_0^{\pi/2} \sin^2 \frac{x}{2} \cos^2 \frac{x}{2} \, dx.$

29. $\int_0^{\pi/6} \sin 2x \cos 4x \, dx.$

30. $\int_0^{2\pi} \sin^2 kx \, dx \quad (k \neq 0).$

31. Let m, n be distinct nonnegative integers. Prove:

(a) $\int_0^{2\pi} \sin mx \cos nx \, dx = 0$

(b) $\int_0^{2\pi} \cos mx \cos nx \, dx = 0$

(c) $\int_0^{2\pi} \sin mx \sin nx \, dx = 0.$

32. The region bounded below by the x-axis and above by the portion of $y = \sin x$ from $x = 0$ to $x = \pi$ is revolved about the x-axis. Find the volume of the resulting solid.

33. Find the volume of the solid that results when the region enclosed by $y = \cos x$, $y = \sin x$, $x = 0$, and $x = \pi/4$ is revolved about the x-axis.

34. (a) Use Formula (6) in Section 9.2 to show that

$$\int_0^{\pi/2} \sin^n x \, dx = \frac{n-1}{n} \int_0^{\pi/2} \sin^{n-2} x \, dx$$

(b) Use this result to derive the **Wallis sine formulas:**

$$\int_0^{\pi/2} \sin^n x \, dx = \frac{\pi}{2} \cdot \frac{1 \cdot 3 \cdot 5 \cdots (n-1)}{2 \cdot 4 \cdot 6 \cdots n} \quad (n \text{ even})$$

$$\int_0^{\pi/2} \sin^n x \, dx = \frac{2 \cdot 4 \cdot 6 \cdots (n-1)}{1 \cdot 3 \cdot 5 \cdots n} \quad \left(\begin{matrix} n \text{ odd} \\ \text{and} \geq 3 \end{matrix}\right).$$

35. Use the Wallis formulas in Exercise 34 to evaluate:

(a) $\int_0^{\pi/2} \sin^3 x \, dx$

(b) $\int_0^{\pi/2} \sin^4 x \, dx$

(c) $\int_0^{\pi/2} \sin^5 x \, dx$

(d) $\int_0^{\pi/2} \sin^6 x \, dx.$

36. Use Formula (5) in Section 9.2 and the method of Exercise 34 to derive the **Wallis cosine formulas:**

$$\int_0^{\pi/2} \cos^n x \, dx = \frac{2 \cdot 4 \cdot 6 \cdots (n-1)}{3 \cdot 5 \cdot 7 \cdots n} \quad \left(\begin{matrix} n \text{ odd} \\ \text{and} \geq 3 \end{matrix}\right)$$

$$\int_0^{\pi/2} \cos^n x \, dx = \frac{\pi}{2} \cdot \frac{1 \cdot 3 \cdot 5 \cdots (n-1)}{2 \cdot 4 \cdot 6 \cdots n} \quad (n \text{ even}).$$

37. Derive (6) and (7) using reduction formulas and appropriate trigonometric identities.

9.4 INTEGRATING POWERS OF SECANT AND TANGENT

In this section we discuss methods for evaluating integrals of the form

$$\int \tan^m x \sec^n x \, dx$$

where m and n are nonnegative integers. We will begin with the integrals

$$\int \tan x \, dx \qquad (m = 1, n = 0)$$

and

$$\int \sec x \, dx \qquad (m = 0, n = 1)$$

The first integral is evaluated by writing

$$\int \tan x \, dx = \int \frac{\sin x}{\cos x} \, dx$$

from which it follows ($u = \cos x$, $du = -\sin x \, dx$) that

$$\int \tan x = -\ln |\cos x| + C \tag{1}$$

or since $-\ln |\cos x| = \ln (1/|\cos x|) = \ln |\sec x|$,

$$\int \tan x = \ln |\sec x| + C \tag{1a}$$

The second integral requires a trick. We write

$$
\begin{aligned}
\int \sec x \, dx &= \int \sec x \left(\frac{\sec x + \tan x}{\sec x + \tan x} \right) dx \\
&= \int \frac{\sec^2 x + \sec x \tan x}{\sec x + \tan x} \, dx \\
&= \int \frac{du}{u} \qquad \begin{bmatrix} u = \sec x + \tan x \\ du = (\sec^2 x + \sec x \tan x) \, dx \end{bmatrix} \\
&= \ln |u| + C
\end{aligned}
$$

from which it follows that

$$\int \sec x \, dx = \ln |\sec x + \tan x| + C \tag{2}$$

Higher powers of secant and tangent can be evaluated using the reduction formulas

$$\int \sec^n x \, dx = \frac{\sec^{n-2} x \tan x}{n-1} + \frac{n-2}{n-1} \int \sec^{n-2} x \, dx \tag{3}$$

$$\int \tan^m x \, dx = \frac{\tan^{m-1} x}{m-1} - \int \tan^{m-2} x \, dx \tag{4}$$

Formula (3) was discussed in Exercise 24a of Section 9.2, and Formula (4) can be obtained from the identity

$$1 + \tan^2 x = \sec^2 x$$

by writing

$$\int \tan^m x \, dx = \int \tan^{m-2} x \tan^2 x \, dx$$

$$= \int \tan^{m-2} x \, (\sec^2 x - 1) \, dx$$

$$= \int \tan^{m-2} x \sec^2 x \, dx - \int \tan^{m-2} x \, dx$$

$$= \frac{\tan^{m-1} x}{m-1} - \int \tan^{m-2} x \, dx \qquad \begin{bmatrix} \text{Let } u = \tan x \\ \text{in the first} \\ \text{integral above} \end{bmatrix}$$

▶ Example 1   Evaluate

$$\int \sec^3 x \, dx$$

Solution.   From (3) with $n = 3$,

$$\int \sec^3 x \, dx = \frac{\sec x \tan x}{2} + \frac{1}{2} \int \sec x \, dx$$

$$= \frac{1}{2} \sec x \tan x + \frac{1}{2} \ln |\sec x + \tan x| + C \qquad \blacktriangleleft$$

▶ Example 2   Evaluate

$$\int \tan^5 x \, dx$$

Solution.   We will use Formula (4) twice.

$$\int \tan^5 x \, dx = \frac{\tan^4 x}{4} - \int \tan^3 x \, dx$$

$$= \frac{\tan^4 x}{4} - \left[\frac{\tan^2 x}{2} - \int \tan x \, dx \right]$$

$$= \frac{1}{4} \tan^4 x - \frac{1}{2} \tan^2 x - \ln |\cos x| + C$$ ◀

If m and n are positive integers, then the integral

$$\int \tan^m x \, \sec^n x \, dx$$

can be evaluated by one of the three procedures in Table 9.4.1.

Table 9.4.1

CASE	PROCEDURE	RELEVANT IDENTITIES
n even	Substitute $u = \tan x$	$\sec^2 x = \tan^2 x + 1$
m odd	Substitute $u = \sec x$	$\tan^2 x = \sec^2 x - 1$
m even and n odd	Reduce to powers of sec x alone	$\tan^2 x = \sec^2 x - 1$

▶ Example 3 Evaluate

$$\int \tan^2 x \, \sec^4 x \, dx$$

Solution. Since $n = 4$ is even, we will follow the first procedure in Table 9.4.1 and make the substitution

$$u = \tan x, \quad du = \sec^2 x \, dx$$

To form the du, we will split off a factor of $\sec^2 x$.

$$\int \tan^2 x \, \sec^4 x \, dx = \int \tan^2 x \, \sec^2 x \, \sec^2 x \, dx$$

$$= \int \tan^2 x \, (\tan^2 x + 1) \sec^2 x \, dx$$

$$= \int u^2 (u^2 + 1) \, du$$

$$= \frac{1}{5} u^5 + \frac{1}{3} u^3 + C = \frac{1}{5} \tan^5 x + \frac{1}{3} \tan^3 x + C$$ ◀

▶ Example 4   Evaluate

$$\int \tan^3 x \sec^3 x \, dx$$

Solution.   Since $m = 3$ is odd, we will follow the second procedure in Table 9.4.1 and make the substitution

$$u = \sec x, \qquad du = \sec x \tan x \, dx$$

To form the du, we will split off the product $\sec x \tan x$.

$$\int \tan^3 x \sec^3 x \, dx = \int \tan^2 x \sec^2 x \, (\sec x \tan x) \, dx$$

$$= \int (\sec^2 x - 1) \sec^2 x \, (\sec x \tan x) \, dx$$

$$= \int (u^2 - 1) u^2 \, du$$

$$= \frac{1}{5} u^5 - \frac{1}{3} u^3 + C = \frac{1}{5} \sec^5 x - \frac{1}{3} \sec^3 x + C \qquad ◀$$

▶ Example 5   Evaluate

$$\int \tan^2 x \sec x \, dx$$

Solution.   Since $m = 2$ and $n = 1$, we will follow the third procedure in Table 9.4.1.

$$\int \tan^2 x \sec x \, dx = \int (\sec^2 x - 1) \sec x \, dx$$

$$= \int \sec^3 x \, dx - \int \sec x \, dx$$

See Example 1 —→

$$= \frac{1}{2} \sec x \tan x + \frac{1}{2} \ln |\sec x + \tan x|$$

$$- \ln |\sec x + \tan x| + C$$

$$= \frac{1}{2} \sec x \tan x - \frac{1}{2} \ln |\sec x + \tan x| + C \qquad ◀$$

▶ Example 6   Instead of using reduction formula (3), the first procedure of Table 9.4.1 can be used to integrate an even power of $\sec x$. For example,

$$\int \sec^6 x \, dx = \int \sec^4 x \sec^2 x \, dx$$

$$= \int (\sec^2 x)^2 \sec^2 x \, dx$$

$$= \int (\tan^2 x + 1)^2 \sec^2 x \, dx$$

$$= \int (u^2 + 1)^2 \, du \qquad (u = \tan x)$$

$$= \int (u^4 + 2u^2 + 1) \, du$$

$$= \frac{1}{5}u^5 + \frac{2}{3}u^3 + u + C$$

$$= \frac{1}{5}\tan^5 x + \frac{2}{3}\tan^3 x + \tan x + C \qquad \blacktriangleleft$$

REMARK. With the aid of the identity

$$1 + \cot^2 x = \csc^2 x$$

some of the techniques in this section can be adapted to treat integrals of the form

$$\int \cot^m x \csc^n x \, dx$$

▶ Exercise Set 9.4

In Exercises 1–30, perform the indicated integration.

1. $\int \sec^2 (3x + 1) \, dx$.

2. $\int \tan 5x \, dx$.

3. $\int \tan^2 x \sec^2 x \, dx$.

4. $\int \tan^5 x \sec^4 x \, dx$.

5. $\int \tan^3 4x \sec^4 4x \, dx$.

6. $\int \tan^4 \theta \sec^4 \theta \, d\theta$.

7. $\int \sec^5 x \tan^3 x \, dx$.

8. $\int \tan^5 \theta \sec \theta \, d\theta$.

9. $\int \tan^4 x \sec x \, dx$.

10. $\int \tan^2 \frac{x}{2} \sec^3 \frac{x}{2} \, dx$.

11. $\int \tan 2t \sec^3 2t \, dt$.

12. $\int \tan x \sec^5 x \, dx$.

13. $\int \sec^4 x \, dx$.

14. $\int \sec^5 x \, dx$.

15. $\int \sec^6 (\pi x) \, dx$.

16. $\int \tan^3 4x \, dx$.

17. $\int \tan^4 x \, dx$.

18. $\int \tan^7 \theta \, d\theta$.

19. $\int x \tan^2 (x^2) \sec^2 (x^2) \, dx$.

20. $\int \tan^2 (1 - 2x) \sec (1 - 2x) \, dx$.

21. $\int \cot^3 x \csc^3 x \, dx$.

22. $\int \cot^2 3t \sec 3t \, dt$.

23. $\int \cot^3 x \, dx$.

24. $\int \csc^4 x \, dx$.

25. $\int \sqrt{\tan x} \sec^4 x \, dx$.

26. $\int \tan x \sec^{3/2} x \, dx$.

27. $\displaystyle\int_0^{\pi/6} \tan^2 2x \, dx.$ **28.** $\displaystyle\int_0^{\pi/6} \sec^3 \theta \tan \theta \, d\theta.$

29. $\displaystyle\int_0^{\pi/2} \tan^5 \frac{x}{2} \, dx.$ **30.** $\displaystyle\int_{\pi/4}^{\pi/2} \csc^3 x \cot x \, dx.$

31. Find the arc length of the curve $y = \ln(\cos x)$ over the interval $[0, \pi/4]$.

32. Find the volume of the solid generated when the region enclosed by $y = \tan x$, $y = 1$, and $x = 0$ is revolved about the x-axis.

33. (a) Show that

$$\int \csc x \, dx = -\ln|\csc x + \cot x| + C$$

(b) Show that the result in (a) can also be written

$$\int \csc x \, dx = \ln|\csc x - \cot x| + C$$

and

$$\int \csc x \, dx = \ln|\tan \tfrac{1}{2}x| + C.$$

34. Rewrite $\sin x + \cos x$ in the form

$$A \sin(x + \phi)$$

and use your result together with Exercise 33 to evaluate

$$\int \frac{dx}{\sin x + \cos x}$$

35. Use the method of Exercise 34 to evaluate

$$\int \frac{dx}{a \sin x + b \cos x} \qquad (a, b \text{ not both zero})$$

9.5   TRIGONOMETRIC SUBSTITUTIONS

Integrals containing expressions of the form

$$\sqrt{a^2 - x^2}, \quad \sqrt{x^2 + a^2}, \quad \text{and} \quad \sqrt{x^2 - a^2}$$

$(a > 0)$ can often be evaluated by making appropriate substitutions involving trigonometric functions. The idea is to choose a substitution that will eliminate the radical. For example, substituting

$$x = a \sin \theta, \qquad -\frac{\pi}{2} \le \theta \le \frac{\pi}{2} \tag{1}$$

in $\sqrt{a^2 - x^2}$ yields

$$\begin{aligned} \sqrt{a^2 - x^2} &= \sqrt{a^2 - a^2 \sin^2 \theta} = \sqrt{a^2(1 - \sin^2 \theta)} \\ &= a\sqrt{\cos^2 \theta} \\ &= a\,|\cos \theta| \\ &= a \cos \theta \qquad \left(\cos \theta \ge 0 \text{ since } -\frac{\pi}{2} \le \theta \le \frac{\pi}{2}\right) \end{aligned}$$

Thus, we have eliminated the radical by this substitution. The purpose of the restriction $-\pi/2 \leq \theta \leq \pi/2$ in (1) is twofold. First it enables us to rewrite (1) as

$$\theta = \sin^{-1}\left(\frac{x}{a}\right)$$

if desired (see Section 8.2), and second this restriction enables us to replace $|\cos\theta|$ by the simpler expression $\cos\theta$ in the resulting calculations.

▶ Example 1   Evaluate

$$\int \frac{dx}{x^2\sqrt{4 - x^2}}$$

Solution.   To eliminate the radical we make the substitution

$$x = 2\sin\theta, \qquad -\frac{\pi}{2} \leq \theta \leq \frac{\pi}{2}$$

so that

$$\frac{dx}{d\theta} = 2\cos\theta \qquad \text{or} \qquad dx = 2\cos\theta\,d\theta$$

This yields

$$\int \frac{dx}{x^2\sqrt{4 - x^2}} = \int \frac{2\cos\theta\,d\theta}{(2\sin\theta)^2\sqrt{4 - 4\sin^2\theta}}$$

$$= \int \frac{2\cos\theta\,d\theta}{(2\sin\theta)^2(2\cos\theta)} = \frac{1}{4}\int \frac{d\theta}{\sin^2\theta}$$

$$= \frac{1}{4}\int \csc^2\theta\,d\theta = -\frac{1}{4}\cot\theta + C$$

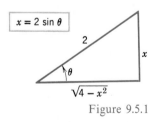

x = 2 sin θ

Figure 9.5.1

To complete the solution we must express $\cot\theta$ in terms of x. This can be done using trigonometric identities (Exercise 25) or more simply by displaying the substitution $x = 2\sin\theta$ ($\sin\theta = x/2$) as in Figure 9.5.1. From the figure we obtain

$$\cot\theta = \frac{\sqrt{4 - x^2}}{x}$$

so that

$$\int \frac{dx}{x^2\sqrt{4 - x^2}} = -\frac{1}{4}\cot\theta + C = -\frac{1}{4}\frac{\sqrt{4 - x^2}}{x} + C \qquad ◀$$

To eliminate the radical in $\sqrt{x^2 + a^2}$ $(a > 0)$, we can make the substitution

$$x = a \tan \theta, \quad -\frac{\pi}{2} < \theta < \frac{\pi}{2} \tag{2}$$

to obtain

$$\sqrt{x^2 + a^2} = \sqrt{a^2 \tan^2 \theta + a^2} = \sqrt{a^2 (1 + \tan^2 \theta)}$$
$$= \sqrt{a^2 \sec^2 \theta}$$
$$= a \, |\sec \theta|$$
$$= a \sec \theta \quad \left(\sec \theta > 0 \text{ since } -\frac{\pi}{2} < \theta < \frac{\pi}{2} \right)$$

As before, the restriction on θ in (2) enables us to write

$$\theta = \tan^{-1} \left(\frac{x}{a} \right)$$

if needed, and also eliminates the absolute value sign in the resulting computation.

▶ Example 2   Evaluate

$$\int \frac{dx}{\sqrt{x^2 + a^2}}$$

Solution.   To eliminate the radical we make the substitution

$$x = a \tan \theta, \quad -\frac{\pi}{2} < \theta < \frac{\pi}{2}$$

so that

$$\frac{dx}{d\theta} = a \sec^2 \theta \quad \text{or} \quad dx = a \sec^2 \theta \, d\theta$$

This yields

$$\int \frac{dx}{\sqrt{x^2 + a^2}} = \int \frac{a \sec^2 d\theta}{\sqrt{a^2 \tan^2 \theta + a^2}} = \int \frac{a \sec^2 \theta \, d\theta}{a \sec \theta}$$
$$= \int \sec \theta \, d\theta = \ln |\sec \theta + \tan \theta| + C$$

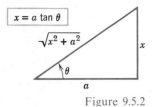

$x = a \tan \theta$

$\sqrt{x^2 + a^2}$

x

θ

a

Figure 9.5.2

To express the solution in terms of x, we use Figure 9.5.2, which yields

$$\int \frac{dx}{\sqrt{x^2 + a^2}} = \ln \left| \frac{\sqrt{x^2 + a^2}}{a} + \frac{x}{a} \right| + C$$

or if we prefer we can rewrite the expression on the right as

$$\ln |\sqrt{x^2 + a^2} + x| - \ln a + C$$

and combine the constant $\ln a$ with the constant of integration to obtain

$$\int \frac{dx}{\sqrt{x^2 + a^2}} = \ln |\sqrt{x^2 + a^2} + x| + C'$$

Moreover, $\sqrt{x^2 + a^2} + x > 0$ for all x, so that we can drop the absolute value sign and write

$$\int \frac{dx}{\sqrt{x^2 + a^2}} = \ln (\sqrt{x^2 + a^2} + x) + C' \tag{3}$$

◄

The integral in the last example can also be evaluated by making the substitution

$$x = au \qquad \text{or} \qquad u = \frac{x}{a}$$

so that

$$dx = a \, du$$

which yields

$$\int \frac{dx}{\sqrt{x^2 + a^2}} = \int \frac{a \, du}{\sqrt{a^2 u^2 + a^2}}$$

$$= \int \frac{du}{\sqrt{u^2 + 1}} = \sinh^{-1} u + C$$

or

$$\int \frac{dx}{\sqrt{x^2 + a^2}} = \sinh^{-1} \left(\frac{x}{a} \right) + C \tag{4}$$

Using the logarithmic expression for $\sinh^{-1}$ in Theorem 8.4.2, the reader can show that (3) and (4) are equivalent.

To eliminate the radical $\sqrt{x^2 - a^2}$, we can make the substitution

$$x = a \sec \theta, \qquad 0 \le \theta < \frac{\pi}{2} \text{ or } \pi \le \theta < \frac{3\pi}{2} \qquad (5)$$

to obtain

$$\sqrt{x^2 - a^2} = \sqrt{a^2 \sec^2 \theta - a^2}$$
$$= \sqrt{a^2 (\sec^2 \theta - 1)} = \sqrt{a^2 \tan^2 \theta}$$
$$= a \, |\tan \theta| = a \tan \theta$$

The removal of the absolute value sign is justified because the restriction on θ in (5) implies that $\tan \theta \ge 0$. This restriction also enables us to write

$$\theta = \sec^{-1}\left(\frac{x}{a}\right)$$

when needed (Definition 8.2.1).

▶ Example 3   Evaluate

$$\int \frac{\sqrt{x^2 - 25}}{x} \, dx$$

Solution.   To eliminate the radical, we make the substitution

$$x = 5 \sec \theta, \qquad 0 \le \theta < \frac{\pi}{2} \text{ or } \pi \le \theta < \frac{3\pi}{2}$$

so that

$$\frac{dx}{d\theta} = 5 \sec \theta \tan \theta \qquad \text{or} \qquad dx = 5 \sec \theta \tan \theta \, d\theta$$

Thus,

$$\int \frac{\sqrt{x^2 - 25}}{x} \, dx = \int \frac{\sqrt{25 \sec^2 \theta - 25}}{5 \sec \theta} (5 \sec \theta \tan \theta) \, d\theta$$

$$= \int \frac{5 \tan \theta}{5 \sec \theta} (5 \sec \theta \tan \theta) \, d\theta$$

$$= 5 \int \tan^2 \theta \, d\theta$$

$$= 5 \int (\sec^2 \theta - 1) \, d\theta$$

$$= 5 \tan \theta - 5\theta + C$$

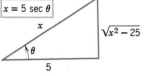

Figure 9.5.3

But $\theta = \sec^{-1}(x/5)$ and from Figure 9.5.3,

$$\tan \theta = \frac{\sqrt{x^2 - 25}}{5}$$

so that

$$\int \frac{\sqrt{x^2 - 25}}{x} \, dx = \sqrt{x^2 - 25} - 5 \sec^{-1}\left(\frac{x}{5}\right) + C \qquad \blacktriangleleft$$

The integral in the next example will arise frequently in later sections.

▶ **Example 4** Evaluate

$$\int_{-a}^{a} \sqrt{a^2 - x^2} \, dx \qquad (a > 0)$$

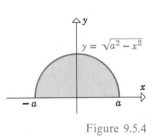

Figure 9.5.4

Solution. This integral can be evaluated by the substitution $x = a \sin \theta$, but the computations are tedious. A better approach is to observe that the integral represents the area of a semicircle of radius a (Figure 9.5.4). Thus,

$$\int_{-a}^{a} \sqrt{a^2 - x^2} \, dx = \frac{1}{2}\pi a^2 \qquad \blacktriangleleft$$

▶ **Exercise Set 9.5**

In Exercises 1–28, perform the indicated integration.

1. $\displaystyle\int \frac{x^2}{\sqrt{9 - x^2}} \, dx.$

2. $\displaystyle\int \frac{dx}{x^2 \sqrt{16 - x^2}}.$

3. $\displaystyle\int \frac{dx}{(4 + x^2)^2}.$

4. $\displaystyle\int \frac{dx}{(x^2 + 1)^{3/2}}.$

5. $\displaystyle\int \frac{\sqrt{x^2 - 9}}{x} \, dx.$

6. $\displaystyle\int \frac{dx}{x^2 \sqrt{x^2 - 16}}.$

7. $\displaystyle\int \frac{x^3}{\sqrt{2 - x^2}} \, dx.$

8. $\displaystyle\int x^3 \sqrt{5 - x^2} \, dx.$

9. $\displaystyle\int \frac{dx}{(3 + x^2)^{3/2}}.$

10. $\displaystyle\int \frac{x^2}{\sqrt{5 + x^2}} \, dx.$

11. $\displaystyle\int \frac{dx}{x^2 \sqrt{4x^2 - 9}}.$

12. $\displaystyle\int \frac{\sqrt{1 + t^2}}{t} \, dt.$

13. $\displaystyle\int \frac{dx}{(1 - x^2)^{3/2}}.$

14. $\displaystyle\int \frac{dx}{x^2 \sqrt{x^2 + 25}}.$

15. $\displaystyle\int \frac{x^2}{1 + x^2} \, dx.$

16. $\displaystyle\int \frac{dx}{1 + 2x^2 + x^4}.$

17. $\displaystyle\int \frac{dx}{x^2 \sqrt{9 - 4x^2}}.$

18. $\displaystyle\int \frac{x^2}{\sqrt{x^2 - 25}} \, dx.$

19. $\displaystyle\int \frac{dx}{(9x^2 - 1)^{3/2}}.$

20. $\displaystyle\int \frac{\cos \theta}{\sqrt{2 - \sin^2 \theta}} \, d\theta.$

21. $\displaystyle\int e^x \sqrt{1 - e^{2x}} \, dx.$

22. $\displaystyle\int_0^{1/3} \frac{dx}{(4 - 9x^2)^2}.$

23. $\displaystyle\int_0^4 x^3 \sqrt{16 - x^2} \, dx.$

24. $\displaystyle\int_{\sqrt{2}}^2 \frac{\sqrt{2x^2 - 4}}{x} \, dx.$

25. $\displaystyle\int_{\sqrt{2}}^2 \frac{dx}{x^2 \sqrt{x^2 - 1}}.$

26. $\displaystyle\int_{-1/\sqrt{2}}^{1/\sqrt{2}} (1 - 2x^2)^{3/2} \, dx.$

27. $\displaystyle\int_1^3 \frac{dx}{x^4 \sqrt{x^2 + 3}}.$

28. $\displaystyle\int_0^3 \frac{x^3}{(3 + x^2)^{5/2}} \, dx.$

29. The integral

$$\int \frac{x}{x^2 + 4}\,dx$$

can be evaluated either by a trigonometric substitution or by the substitution $u = x^2 + 4$. Do it both ways and show that the results are equivalent.

30. By integrating, prove that the area of a circle of radius r is πr^2. [*Hint:* $x^2 + y^2 = r^2$ is the equation of such a circle.]

31. Find the arc length of the curve $y = \ln x$ from $x = 1$ to $x = 2$.

32. Find the arc length of the curve $y = x^2$ from $x = 0$ to $x = 1$.

33. Find the area of the surface generated when the curve in Exercise 32 is revolved about the x-axis.

34. Find the volume of the solid generated when the region enclosed by $x = y(1 - y^2)^{1/4}$, $y = 0$, $y = 1$, and $x = 0$ is revolved about the y-axis.

In cases where the trigonometric substitutions $x = a \sec \theta$ and $x = a \tan \theta$ lead to difficult integrals, it is sometimes possible to use the **hyperbolic substitutions:**

$x = a \sinh u$ for integrals involving $\sqrt{x^2 + a^2}$

$x = a \cosh u$ for integrals involving $\sqrt{x^2 - a^2}$

These substitutions are useful because in each case the hyperbolic identity

$$a^2 \cosh^2 u - a^2 \sinh^2 u = a^2$$

removes the radical.

35. (a) Evaluate

$$\int \frac{dx}{\sqrt{x^2 + 9}}$$

using the hyperbolic substitution suggested above.

(b) Evaluate the integral in (a) by a trigonometric substitution and show that the results in (a) and (b) agree.

36. Follow the directions of Exercise 35 for the integral

$$\int \sqrt{x^2 - 1}\,dx, \qquad x \geq 1$$

37. For the substitution $x = a \sin \theta$, $-\pi/2 \leq \theta \leq \pi/2$, the triangle

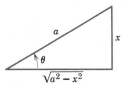

suggests the relationships

(a) $\cos \theta = \dfrac{1}{a} \sqrt{a^2 - x^2}$

(b) $\tan \theta = \dfrac{x}{\sqrt{a^2 - x^2}}$

Use trigonometric identities to prove that these results are correct.

9.6　INTEGRALS INVOLVING $ax^2 + bx + c$

In this section we discuss integrals whose integrands contain a quadratic expression $ax^2 + bx + c$ that cannot be factored into real linear factors. Some cases where factorization is possible are treated in the next section. If $b = 0$, then methods discussed in previous sections will frequently work. For cases where $b \neq 0$, the integral can often be evaluated by first completing the square as follows:

$$ax^2 + bx + c = a\left(x^2 + \frac{b}{a}x\right) + c$$

$$= a\left(x^2 + \frac{b}{a}x + \frac{b^2}{4a^2}\right) + c - \frac{b^2}{4a}$$

$$= a\left(x + \frac{b}{2a}\right)^2 + c - \frac{b^2}{4a}$$

At this point, the substitution

$$u = x + \frac{b}{2a}$$

will reduce the original expression $ax^2 + bx + c$ to the simpler form $au^2 + d$, where $d = c - (b^2/4a)$.

▶ Example 1   Evaluate

$$\int \frac{dx}{x^2 - 2x + 5}$$

Solution. Completing the square yields

$$x^2 - 2x + 5 = (x^2 - 2x + 1) + 5 - 1 = (x - 1)^2 + 4$$

Thus,

$$\int \frac{dx}{x^2 - 2x + 5} = \int \frac{dx}{(x - 1)^2 + 4}$$

$$= \int \frac{du}{u^2 + 4} \qquad \begin{bmatrix} u = x - 1 \\ du = dx \end{bmatrix}$$

$$= \frac{1}{2} \tan^{-1} \frac{u}{2} + C \qquad \begin{bmatrix} \text{Formula (14)} \\ \text{of Section 8.3} \end{bmatrix}$$

$$= \frac{1}{2} \tan^{-1} \left(\frac{x - 1}{2}\right) + C \qquad \blacktriangleleft$$

▶ Example 2   Evaluate

$$\int \frac{dx}{\sqrt{5 - 4x - 2x^2}}$$

Solution. Completing the square yields

$$5 - 4x - 2x^2 = 5 - 2(x^2 + 2x)$$
$$= 5 - 2(x^2 + 2x + 1) + 2$$
$$= 5 - 2(x + 1)^2 + 2 = 7 - 2(x + 1)^2$$

Thus,

$$\int \frac{dx}{\sqrt{5 - 4x - 2x^2}} = \int \frac{dx}{\sqrt{7 - 2(x + 1)^2}}$$

$$= \int \frac{du}{\sqrt{7 - 2u^2}} \qquad \begin{bmatrix} u = x + 1 \\ du = dx \end{bmatrix}$$

$$= \frac{1}{\sqrt{2}} \int \frac{du}{\sqrt{(7/2) - u^2}}$$

$$= \frac{1}{\sqrt{2}} \sin^{-1}\left(\frac{u}{\sqrt{7/2}}\right) + C \qquad \begin{bmatrix} \text{Formula (10),} \\ \text{Section 8.3} \\ \text{with } a = \sqrt{7/2}. \end{bmatrix}$$

$$= \frac{1}{\sqrt{2}} \sin^{-1}\left(\sqrt{\tfrac{2}{7}}u\right) + C$$

$$= \frac{1}{\sqrt{2}} \sin^{-1}\left(\sqrt{\tfrac{2}{7}}(x + 1)\right) + C \qquad \blacktriangleleft$$

▶ **Example 3**   Evaluate

$$\int \frac{x}{x^2 - 4x + 8} dx$$

Solution. Completing the square yields

$$x^2 - 4x + 8 = (x^2 - 4x + 4) + 8 - 4 = (x - 2)^2 + 4$$

Thus, the substitution

$$u = x - 2 \qquad du = dx$$

yields

$$\int \frac{x}{x^2 - 4x + 8} dx = \int \frac{x}{(x - 2)^2 + 4} dx = \int \frac{u + 2}{u^2 + 4} du$$

$$= \int \frac{u}{u^2 + 4} du + 2 \int \frac{du}{u^2 + 4}$$

$$= \frac{1}{2} \int \frac{2u}{u^2 + 4} du + 2 \int \frac{du}{u^2 + 4}$$

$$= \frac{1}{2} \ln (u^2 + 4) + 2\left(\frac{1}{2}\right) \tan^{-1} \frac{u}{2} + C$$

$$= \frac{1}{2} \ln [(x - 2)^2 + 4] + \tan^{-1}\left(\frac{x - 2}{2}\right) + C \qquad \blacktriangleleft$$

▶ Exercise Set 9.6

Evaluate the integrals in Exercises 1–15.

1. $\int \dfrac{dx}{x^2 - 4x + 13}$.

2. $\int \dfrac{dx}{\sqrt{2x - x^2}}$.

3. $\int \dfrac{dx}{\sqrt{8 + 2x - x^2}}$.

4. $\int \dfrac{dx}{16x^2 + 16x + 5}$.

5. $\int \dfrac{dx}{\sqrt{x^2 - 6x + 10}}$.

6. $\int \dfrac{x}{x^2 + 6x + 10}\,dx$.

7. $\int \sqrt{3 - 2x - x^2}\,dx$.

8. $\int \dfrac{e^x}{\sqrt{1 + e^x + e^{2x}}}\,dx$.

9. $\int \dfrac{dx}{2x^2 + 4x + 7}$.

10. $\int \dfrac{\cos \theta}{\sin^2 \theta - 6 \sin \theta + 12}\,d\theta$.

11. $\int \dfrac{2x + 5}{x^2 + 2x + 5}\,dx$.

12. $\int \dfrac{2x + 3}{4x^2 + 4x + 5}\,dx$.

13. $\int \dfrac{x + 3}{\sqrt{x^2 + 2x + 2}}\,dx$.

14. $\int_1^2 \dfrac{dx}{\sqrt{4x - x^2}}$.

15. $\int_0^1 \sqrt{x(4 - x)}\,dx$.

9.7   INTEGRATING RATIONAL FUNCTIONS; PARTIAL FRACTIONS

Recall that a *rational function* is the quotient of two polynomials. Some examples are:

$$\frac{3x}{x^2 + 1}, \quad \frac{6x^2 - x + 2}{(x - 1)(x^2 + 4)}, \quad \frac{\frac{1}{2}x^3 + 2}{x^3 - \frac{1}{3}}, \quad \frac{7x^4 - 8x^2 + 2x + 1}{3x^3 + 2x - 5} \tag{1}$$

In this section we will discuss a method for integrating rational functions.

In algebra we learn to combine two or more fractions into a single fraction; for example,

$$\frac{1}{x} + \frac{3}{x - 1} + \frac{2}{x + 2} = \frac{(x - 1)(x + 2) + 3x(x + 2) + 2x(x - 1)}{x(x - 1)(x + 2)}$$

$$= \frac{6x^2 + 5x - 2}{x^3 + x^2 - 2x} \tag{2}$$

However, the left side of (2) is easier to integrate than the right. Thus, it would be helpful if we knew how to obtain the left side of the equation starting with the right. The procedure for doing this is called ***partial fraction decomposition.***

A rational function

$$\frac{P(x)}{Q(x)} \tag{3}$$

is called *proper* if the degree of $P(x)$ is less than the degree of $Q(x)$; otherwise it is *improper.* Thus, in (1), the first two rational functions are proper and the last two are improper. It can be proved that any proper rational function is expressible as a sum of terms (called *partial fractions*) having the form:

$$\frac{A}{(ax + b)^k} \quad \text{or} \quad \frac{Bx + C}{(ax^2 + bx + c)^k}$$

The exact number of terms of each type depends on how the denominator $Q(x)$ in (3) factors. In theory, a polynomial $Q(x)$ with real coefficients can always be factored into a product of linear and quadratic factors with real coefficients. For example, the polynomial

$$Q(x) = x^3 - 3x^2 + x - 3$$

factors into

$$Q(x) = (x - 3)(x^2 + 1)$$

The quadratic factor $x^2 + 1$ cannot be further decomposed into linear factors without using imaginary numbers $[x^2 + 1 = (x - i)(x + i)]$. Such quadratic factors are said to be *irreducible.*

The first step in the partial fraction decomposition of $P(x)/Q(x)$ is to completely factor the denominator $Q(x)$ into linear and irreducible quadratic factors and then collect all repeated factors so that $Q(x)$ is expressed as a product of *distinct* factors of the form

$$(ax + b)^m \quad \text{and} \quad (ax^2 + bx + c)^m$$

where $ax^2 + bx + c$ is irreducible. Once this is done, the structure of the partial fraction decomposition of $P(x)/Q(x)$ is determined as follows:

Linear Factors
For each factor of the form $(ax + b)^m$, introduce the m terms

$$\frac{A_1}{ax + b} + \frac{A_2}{(ax + b)^2} + \cdots + \frac{A_m}{(ax + b)^m}$$

where $A_1, A_2, \ldots, A_m$ are constants to be determined.

Irreducible Quadratic Factors
For each factor of the form $(ax^2 + bx + c)^m$, introduce m terms

$$\frac{A_1 x + B_1}{ax^2 + bx + c} + \frac{A_2 x + B_2}{(ax^2 + bx + c)^2} + \cdots + \frac{A_m x + B_m}{(ax^2 + bx + c)^m}$$

where $A_1, A_2, \ldots, A_m, B_1, B_2, \ldots, B_m$ are constants to be determined.

▶ Example 1   Evaluate

$$\int \frac{dx}{x^2 + x - 2}$$

Solution.   The integrand can be written

$$\frac{1}{x^2 + x - 2} = \frac{1}{(x - 1)(x + 2)}$$

According to the rule above for linear factors, the factor $x - 1$ introduces one term (since $m = 1$):

$$\frac{A}{x - 1}$$

and the factor $x + 2$ introduces one term:

$$\frac{B}{x + 2}$$

so that the partial fraction decomposition is

$$\frac{1}{(x - 1)(x + 2)} = \frac{A}{x - 1} + \frac{B}{x + 2} \tag{4}$$

where A and B are constants to be determined so that (4) becomes an *identity*. To find these constants we multiply both sides of (4) by $(x - 1)(x + 2)$ to obtain

$$1 = A(x + 2) + B(x - 1) \tag{5}$$

Next we substitute values of x to make the various terms zero. Setting $x = -2$ in (5) yields

$$1 = -3B \quad \text{or} \quad B = -\frac{1}{3}$$

and setting $x = 1$ in (5) yields

$$1 = 3A \quad \text{or} \quad A = \frac{1}{3}$$

Thus, (4) becomes

$$\frac{1}{(x-1)(x+2)} = \frac{1/3}{x-1} + \frac{-1/3}{x+2}$$

and

$$\int \frac{dx}{(x-1)(x+2)} = \frac{1}{3} \int \frac{dx}{x-1} - \frac{1}{3} \int \frac{dx}{x+2}$$

$$= \frac{1}{3} \ln |x-1| - \frac{1}{3} \ln |x+2| + C$$

$$= \frac{1}{3} \ln \left| \frac{x-1}{x+2} \right| + C$$

Alternate solution: The constants A and B in (5) can also be determined by collecting like terms:

$$1 = (A+B)x + (2A - B) \tag{6}$$

and then equating corresponding coefficients on both sides:

$$A + B = 0$$
$$2A - B = 1$$

and then solving these equations simultaneously to obtain $A = \frac{1}{3}$, $B = -\frac{1}{3}$. This method is justified because (6) is an *identity* holding for all x, and two polynomials are equal for all x if and only if their corresponding coefficients are equal (Exercise 52).   ◄

▶ Example 2   Evaluate

$$\int \frac{2x+4}{x^3 - 2x^2} \, dx$$

Solution. The integrand can be written

$$\frac{2x+4}{x^3 - 2x^2} = \frac{2x+4}{x^2(x-2)}$$

Although x^2 is a quadratic factor, it is *not* irreducible since $x^2 = xx$. Thus, by the rule for linear factors, x^2 introduces two terms (since $m = 2$):

$$\frac{A}{x} + \frac{B}{x^2}$$

and the factor $x - 2$ introduces one term (since $m = 1$):

$$\frac{C}{x - 2}$$

so the partial fraction decomposition is

$$\frac{2x + 4}{x^2(x - 2)} = \frac{A}{x} + \frac{B}{x^2} + \frac{C}{x - 2} \tag{7}$$

Multiplying by $x^2(x - 2)$ yields

$$2x + 4 = Ax(x - 2) + B(x - 2) + Cx^2 \tag{8}$$

To determine A, B, and C we will follow the alternate method used in Example 1. Multiplying and collecting like terms in (8) yields

$$2x + 4 = (A + C)x^2 + (-2A + B)x - 2B$$

Equating corresponding coefficients gives

$$
\begin{array}{rl}
A \qquad + C & = 0 \\
-2A + B \qquad & = 2 \\
-2B \qquad & = 4
\end{array}
$$

and solving this system yields

$$A = -2, \qquad B = -2, \qquad C = 2$$

so (7) becomes

$$\frac{2x + 4}{x^2(x - 2)} = \frac{-2}{x} + \frac{-2}{x^2} + \frac{2}{x - 2}$$

Thus,

$$\int \frac{2x + 4}{x^2(x - 2)}\, dx = -2 \int \frac{dx}{x} - 2 \int \frac{dx}{x^2} + 2 \int \frac{dx}{x - 2}$$

$$= -2 \ln |x| + \frac{2}{x} + 2 \ln |x - 2| + C$$

$$= 2 \ln \left| \frac{x - 2}{x} \right| + \frac{2}{x} + C \qquad \blacktriangleleft$$

▶ Example 3   Evaluate

$$\int \frac{x^2 + x - 2}{3x^3 - x^2 + 3x - 1}\, dx$$

Solution. The denominator in the integrand can be factored by grouping:

$$\frac{x^2 + x - 2}{3x^3 - x^2 + 3x - 1} = \frac{x^2 + x - 2}{x^2(3x - 1) + (3x - 1)} = \frac{x^2 + x - 2}{(3x - 1)(x^2 + 1)}$$

By the rule for linear factors, the factor $3x - 1$ introduces one term:

$$\frac{A}{3x - 1}$$

and by the rule for irreducible quadratic factors, the factor $x^2 + 1$ introduces one term:

$$\frac{Bx + C}{x^2 + 1}$$

Thus, the partial fraction decomposition is

$$\frac{x^2 + x - 2}{(3x - 1)(x^2 + 1)} = \frac{A}{3x - 1} + \frac{Bx + C}{x^2 + 1} \tag{9}$$

Multiplying by $(3x - 1)(x^2 + 1)$ yields

$$x^2 + x - 2 = A(x^2 + 1) + (Bx + C)(3x - 1)$$

To determine A, B, and C, we multiply out and collect like terms:

$$x^2 + x - 2 = (A + 3B)x^2 + (-B + 3C)x + (A - C)$$

Equating corresponding coefficients gives

$$
\begin{aligned}
A + 3B \qquad &= \quad 1 \\
- \ B + 3C &= \quad 1 \\
A \qquad - \ C &= -2
\end{aligned}
$$

To solve this system, subtract the third equation from the first to eliminate A. Then use the resulting equation together with the second equation to solve for B and C. Finally, determine A from the first or third equation. This yields (verify):

$$A = -\frac{7}{5}, \qquad B = \frac{4}{5}, \qquad C = \frac{3}{5}$$

Thus, (9) becomes

$$\frac{x^2 + x - 2}{(3x - 1)(x^2 + 1)} = \frac{-\frac{7}{5}}{3x - 1} + \frac{\frac{4}{5}x + \frac{3}{5}}{x^2 + 1}$$

and

$$\int \frac{x^2 + x - 2}{(3x - 1)(x^2 + 1)} dx$$

$$= -\frac{7}{5} \int \frac{dx}{3x - 1} + \frac{4}{5} \int \frac{x}{x^2 + 1} dx + \frac{3}{5} \int \frac{dx}{x^2 + 1}$$

$$= -\frac{7}{15} \ln |3x - 1| + \frac{2}{5} \ln (x^2 + 1) + \frac{3}{5} \tan^{-1} x + C \quad \blacktriangleleft$$

▶ **Example 4** Evaluate

$$\int \frac{3x^4 + 4x^3 + 16x^2 + 20x + 9}{(x + 2)(x^2 + 3)^2} dx$$

Solution. By the rule for linear factors, the factor $x + 2$ introduces one term:

$$\frac{A}{x + 2}$$

and by the rule for irreducible quadratic factors, the factor $(x^2 + 3)^2$ introduces two terms (since $m = 2$):

$$\frac{Bx + C}{x^2 + 3} + \frac{Dx + E}{(x^2 + 3)^2}$$

Thus, the partial fraction decomposition of the integrand is

$$\frac{3x^4 + 4x^3 + 16x^2 + 20x + 9}{(x + 2)(x^2 + 3)^2} = \frac{A}{x + 2} + \frac{Bx + C}{x^2 + 3} + \frac{Dx + E}{(x^2 + 3)^2} \quad (10)$$

Multiplying by $(x + 2)(x^2 + 3)^2$ yields

$$3x^4 + 4x^3 + 16x^2 + 20x + 9$$
$$= A(x^2 + 3)^2 + (Bx + C)(x^2 + 3)(x + 2) + (Dx + E)(x + 2) \quad (11)$$

Multiplying out and collecting terms on the right side of (11) yields

$$(A + B)x^4 + (2B + C)x^3 + (6A + 3B + 2C + D)x^2$$
$$+ (6B + 3C + 2D + E)x + (9A + 6C + 2E)$$

and equating corresponding coefficients with the left side of (11) yields

$$
\begin{aligned}
A + B &= 3 \\
2B + C &= 4 \\
6A + 3B + 2C + D &= 16 \\
6B + 3C + 2D + E &= 20 \\
9A + 6C + 2E &= 9
\end{aligned}
\tag{12}
$$

This system of five equations in five unknowns is tedious to solve. However, we can reduce the work considerably by substituting $x = -2$ in (11) to obtain

$$
49 = 49A \qquad \text{or} \qquad A = 1
$$

Substituting this value in the first equation of (12) gives $B = 2$ and substituting this value of B in the second equation of (11) gives $C = 0$; continuing in this way we are led to the values (verify):

$$
A = 1, \qquad B = 2, \qquad C = 0, \qquad D = 4, \qquad E = 0
$$

Thus, (10) becomes

$$
\frac{3x^4 + 4x^3 + 16x^2 + 20x + 9}{(x + 2)(x^2 + 3)^2} = \frac{1}{x + 2} + \frac{2x}{x^2 + 3} + \frac{4x}{(x^2 + 3)^2}
$$

and

$$
\int \frac{3x^4 + 4x^3 + 16x^2 + 20x + 9}{(x + 2)(x^2 + 3)^2}\, dx
$$

$$
= \int \frac{dx}{x + 2} + \int \frac{2x}{x^2 + 3}\, dx + 4 \int \frac{x}{(x^2 + 3)^2}\, dx
$$

$$
= \ln |x + 2| + \ln (x^2 + 3) - \frac{2}{x^2 + 3} + C
$$

(The third integral on the right was evaluated by the substitution $u = x^2 + 3$.) ◀

As noted in the beginning of this section, partial fraction decomposition only applies to *proper* rational functions. However, the next example shows that improper rational functions can be integrated by first performing a long division, and then working with the remainder term.

▶ Example 5   Evaluate

$$
\int \frac{3x^4 + 3x^3 - 5x^2 + x - 1}{x^2 + x - 2}\, dx
$$

Solution. Since the integrand is an improper rational function, we cannot use a partial fraction decomposition directly. However, if we perform the long division

$$
\begin{array}{r}
3x^2 + 1 \\
x^2 + x - 2 \overline{\big)\, 3x^4 + 3x^3 - 5x^2 + x - 1} \\
3x^4 + 3x^3 - 6x^2 \\
\hline
x^2 + x - 1 \\
x^2 + x - 2 \\
\hline
1
\end{array}
$$

we can write the integrand as the quotient plus the remainder over the divisor, that is

$$\frac{3x^4 + 3x^3 - 5x^2 + x - 1}{x^2 + x - 2} = (3x^2 + 1) + \frac{1}{x^2 + x - 2}$$

Thus,

$$\int \frac{3x^4 + 3x^3 - 5x^2 + x - 1}{x^2 + x - 2}\,dx = \int (3x^2 + 1)\,dx + \int \frac{dx}{x^2 + x - 2}$$

The second integral on the right now involves a proper rational function and can thus be evaluated by a partial fraction decomposition. Using the result of Example 1 we obtain

$$\int \frac{3x^4 + 3x^3 - 5x^2 + x - 1}{x^2 + x - 2}\,dx = x^3 + x + \frac{1}{3}\ln\left|\frac{x-1}{x+2}\right| + C \quad \blacktriangleleft$$

▶ Exercise Set 9.7

In Exercises 1–38, perform the integrations.

1. $\int \dfrac{dx}{x^2 + 3x - 4}$.

2. $\int \dfrac{dx}{x^2 + 8x + 7}$.

3. $\int \dfrac{x}{x^2 - 5x + 6}\,dx$.

4. $\int \dfrac{5x - 4}{x^2 - 4x}\,dx$.

5. $\int \dfrac{11x + 17}{2x^2 + 7x - 4}\,dx$.

6. $\int \dfrac{5x - 5}{3x^2 - 8x - 3}\,dx$.

7. $\int \dfrac{dx}{(x-1)(x+2)(x-3)}$.

8. $\int \dfrac{dx}{x(x^2 - 1)}$.

9. $\int \dfrac{2x^2 - 9x - 9}{x^3 - 9x}\,dx$.

10. $\int \dfrac{2x^2 + 4x - 8}{x^3 - 4x}\,dx$.

11. $\int \dfrac{x^2 + 2}{x + 2}\,dx$.

12. $\int \dfrac{x^2 - 4}{x - 1}\,dx$.

13. $\int \dfrac{3x^2 - 10}{x^2 - 4x + 4}\,dx$.

14. $\int \dfrac{x^2}{x^2 - 3x + 2}\,dx$.

15. $\int \dfrac{x^3}{x^2 - 3x + 2}\,dx$.

16. $\int \dfrac{x^3}{x^2 - x - 6}\,dx$.

17. $\int \dfrac{x^5 + 2x^2 + 1}{x^3 - x}\,dx$.

18. $\int \dfrac{2x^5 - x^3 - 1}{x^3 - 4x}\,dx$.

19. $\int \dfrac{2x^2 + 3}{x(x-1)^2}\,dx$.

20. $\int \dfrac{3x^2 - x + 1}{x^3 - x^2}\,dx$.

21. $\int \dfrac{x^2 + x - 16}{(x+1)(x-3)^2}\,dx$.

22. $\int \dfrac{2x^2 - 2x - 1}{x^3 - x^2}\,dx$.

23. $\int \dfrac{x^2}{(x+2)^3}\,dx.$

24. $\int \dfrac{2x^2 + 3x + 3}{(x+1)^3}\,dx.$

25. $\int \dfrac{2x^2 - 1}{(4x-1)(x^2+1)}\,dx.$

26. $\int \dfrac{dx}{x(x^2 + x + 1)}.$

27. $\int \dfrac{dx}{x^4 - 16}.$

28. $\int \dfrac{dx}{x^3 + x}.$

29. $\int \dfrac{x^3 + 3x^2 + x + 9}{(x^2+1)(x^2+3)}\,dx.$

30. $\int \dfrac{x^3 + x^2 + x + 2}{(x^2+1)(x^2+2)}\,dx.$

31. $\int \dfrac{x^3 - 3x^2 + 2x - 3}{x^2 + 1}\,dx.$

32. $\int \dfrac{x^4 + 6x^3 + 10x^2 + x}{x^2 + 6x + 10}\,dx.$

33. $\int \dfrac{x^2 + 1}{(x^2 + 2x + 3)^2}\,dx.$

34. $\int \dfrac{x^5 + x^4 + 4x^3 + 4x^2 + 4x + 4}{(x^2 + 2)^3}\,dx.$

35. $\int \dfrac{\cos \theta}{\sin^2 \theta + 4 \sin \theta - 5}\,d\theta.$

36. $\int \dfrac{e^t}{e^{2t} - 4}\,dt.$

37. $\int \dfrac{dx}{1 + e^x}.$

38. $\int \dfrac{\sec^2 \theta}{\tan^3 \theta - \tan^2 \theta}\,d\theta.$

39. (a) Find constants a and b such that

$$x^4 + 1 = (x^2 + ax + 1)(x^2 + bx + 1)$$

(b) Use the result in (a) to show that

$$\int_0^1 \dfrac{x}{x^4 + 1}\,dx = \dfrac{\pi}{8}.$$

40. Find the area of the region enclosed by $y = (x-3)/(x^3 + x^2)$, $y = 0$, $x = 1$, and $x = 2$.

41. Find the volume of the solid generated when the region enclosed by $y = x^2/(9 - x^2)$, $y = 0$, $x = 0$, and $x = 2$ is revolved about the x-axis.

In Exercises 42–45, solve the differential equations.

42. $\dfrac{dy}{dx} = y^2 + y.$

43. $\dfrac{dy}{dx} = y^2 - 5y + 6.$

44. $\dfrac{dy}{dt} = t^2 y^2 - 4t^2 y.$

45. $t(t-1)\dfrac{dy}{dt} - (y^2 + y) = 0.$

46. The differential equation

$$\dfrac{dy}{dt} = ay - by^2 \qquad (a > 0, b > 0)$$

which is called the **logistic equation,** first arose in the study of human population growth. By solving the equation, show that its general solution is

$$y = \dfrac{a}{b + Ce^{-at}}$$

where C is an arbitrary constant.

FACTORING POLYNOMIALS—OPTIONAL DISCUSSION AND EXERCISES

The method of partial fractions depends on our ability to carry out the necessary factorization. This is not always easy to do. The following results, usually proved in algebra courses, are helpful to know.

9.7.1 THEOREM *If $p(x)$ is a polynomial and r is a solution of the equation $p(x) = 0$, then $x - r$*
Factor Theorem *is a factor of $p(x)$.*

9.7.2 THEOREM   *Let*

$$p(x) = a_0 x^n + a_1 x^{n-1} + \cdots + a_{n-1} x + a_n$$

be a polynomial with integer coefficients.

(a)   *If r is an integer solution of $p(x) = 0$, then the constant term a_n is an integer multiple of r.*

(b)   *If c/d is a rational solution of $p(x) = 0$, and if c/d is expressed in lowest terms, then the constant term a_n is an integer multiple of c; and the leading coefficient a_0 is an integer multiple of d.*

▶ Example 6   For the equation

$$x^3 + x^2 - 10x + 8 = 0$$

the only possible integer solutions are ± 1, ± 2, ± 4, ± 8.

By substitution, or by using synthetic division, the reader can show that 1, 2, -4 are solutions and the rest are not. It follows that

$$x^3 + x^2 - 10x + 8 = (x - 1)(x - 2)(x + 4)$$ ◀

▶ Example 7   For the equation

$$2x^3 + x^2 + 8x + 4 = 0$$

the only possible numerators for rational solutions are ± 1, ± 2, ± 4, and the only possible denominators are ± 1, ± 2; thus, the only possible rational solutions are

$$\pm 1, \ \pm 2, \ \pm 4, \ \pm \tfrac{1}{2}$$

By substitution, or by using synthetic division, the reader can show that $-\tfrac{1}{2}$ is a solution but the rest are not. It follows by division that

$$2x^3 + x^2 + 8x + 4 = (x + \tfrac{1}{2})(2x^2 + 8) = 2(x + \tfrac{1}{2})(x^2 + 4)$$ ◀

In Exercises 47–50, use Theorems 9.7.1 and 9.7.2.

47. Find all rational solutions, if any, and use your results to factor the polynomial into a product of linear and irreducible quadratic factors.
(a) $x^3 - 6x^2 + 11x - 6 = 0$
(b) $x^3 - 3x^2 + x - 20 = 0$
(c) $x^4 - 5x^3 + 7x^2 - 5x + 6 = 0$.

48. Find all rational solutions, if any, and use your results to factor the polynomial into a product of linear and irreducible quadratic factors.
(a) $8x^3 + 4x^2 - 2x - 1 = 0$
(b) $6x^4 - 7x^3 + 6x^2 - 1 = 0$
(c) $9x^4 - 56x^3 + 57x^2 + 98x - 24 = 0$.

49. Evaluate

$$\int \frac{dx}{x^4 - 3x^3 - 7x^2 + 27x - 18}$$

50. Evaluate

$$\int \frac{dx}{16x^3 - 4x^2 + 4x - 1}$$

51. Use Theorem 9.7.2 to prove:
   (a) $\sqrt{2}$ is irrational

   (b)   If a is a positive integer, then $\sqrt{a}$ is either an integer or is irrational.

52. (a)   Prove: $a_0 x^n + a_1 x^{n-1} + \cdots + a_n \equiv 0$ if and only if $a_0 = a_1 = \cdots = a_n = 0$.
   (b)   Use the result in (a) to prove:
   $$a_0 x^n + a_1 x^{n-1} + \cdots + a_n \equiv$$
   $$b_0 x^n + b_1 x^{n-1} + \cdots + b_n$$
   if and only if $a_0 = b_0, a_1 = b_1, \ldots, a_n = b_n$.

9.8   MISCELLANEOUS SUBSTITUTIONS (OPTIONAL)

In this section we will consider some integrals that do not fit in any of the categories previously studied.

INTEGRALS
INVOLVING
RATIONAL
EXPONENTS

Integrals involving rational powers of x can often be simplified by substituting

$$u = x^{1/n}$$

where n is the least common multiple of the denominators of the exponents. The effect of this substitution is to replace fractional exponents with integer exponents, which are easier to work with.

▶ Example 1   Evaluate

$$\int \frac{\sqrt{x}}{1 + \sqrt[3]{x}} dx$$

Solution.   The integrand involves $x^{1/2}$ and $x^{1/3}$, so we make the substitution

$$u = x^{1/6}$$

or

$$x = u^6 \qquad \text{and} \qquad dx = 6u^5 \, du$$

Thus,

$$\int \frac{\sqrt{x}}{1 + \sqrt[3]{x}} dx = \int \frac{(u^6)^{1/2}}{1 + (u^6)^{1/3}} (6u^5) \, du = 6 \int \frac{u^8}{1 + u^2} \, du$$

By long division

$$\frac{u^8}{1 + u^2} = u^6 - u^4 + u^2 - 1 + \frac{1}{1 + u^2}$$

Thus,

$$\int \frac{\sqrt{x}}{1 + \sqrt[3]{x}} dx = 6 \int \left(u^6 - u^4 + u^2 - 1 + \frac{1}{1 + u^2} \right) du$$

$$= \frac{6}{7} u^7 - \frac{6}{5} u^5 + 2u^3 - 6u + 6 \tan^{-1} u + C$$

$$= \frac{6}{7} x^{7/6} - \frac{6}{5} x^{5/6} + 2x^{1/2} - 6x^{1/6} + 6 \tan^{-1} (x^{1/6}) + C$$

◀

▶ Example 2 Evaluate

$$\int \frac{dx}{2 + 2\sqrt{x}}$$

Solution. The integrand contains $\sqrt{x} = x^{1/2}$, so we make the substitution

$$u = x^{1/2}$$

or

$$x = u^2 \quad \text{and} \quad dx = 2u \, du$$

This yields

$$\int \frac{dx}{2 + 2\sqrt{x}} = \int \frac{2u}{2 + 2u} du = \int \left(1 - \frac{1}{1 + u} \right) du$$

$$= u - \ln |1 + u| + C = \sqrt{x} - \ln |1 + \sqrt{x}| + C \qquad ◀$$

The following example illustrates a variation on the above idea.

▶ Example 3 Evaluate

$$\int \sqrt{1 + e^x} \, dx$$

Solution. The substitution

$$u^2 = 1 + e^x$$

will eliminate the square root. To express dx in terms of du, it is helpful to solve this equation for x and then differentiate. We obtain

$$e^x = u^2 - 1$$
$$x = \ln (u^2 - 1)$$
$$\frac{dx}{du} = \frac{2u}{u^2 - 1}, \qquad dx = \frac{2u}{u^2 - 1} du$$

Thus,

$$\int \sqrt{1 + e^x}\, dx = \int u\left(\frac{2u}{u^2 - 1}\right) du$$

$$= \int \frac{2u^2}{u^2 - 1}\, du$$

$$= \int \left(2 + \frac{2}{u^2 - 1}\right) du \qquad \text{(long division)}$$

$$= 2u + \int \left(\frac{1}{u - 1} - \frac{1}{u + 1}\right) du \qquad \text{(partial fractions)}$$

$$= 2u + \ln|u - 1| - \ln|u + 1| + C$$

$$= 2u + \ln\left|\frac{u - 1}{u + 1}\right| + C$$

$$= 2\sqrt{1 + e^x} + \ln\left|\frac{\sqrt{1 + e^x} - 1}{\sqrt{1 + e^x} + 1}\right| + C \qquad \blacktriangleleft$$

INTEGRALS CONTAINING RATIONAL EXPRESSIONS IN SIN x AND COS x

Functions such as

$$\frac{\sin x + 3\cos^2 x}{\cos x + 4\sin x}, \qquad \frac{\sin x}{1 + \cos x - \cos^2 x}, \qquad \frac{3\sin^5 x}{1 + 4\sin x}$$

are called **rational expressions in sin x and cos x,** since they would become rational functions of x if we were to replace $\sin x$ and $\cos x$ by the letter x. If an integrand is a rational expression in $\sin x$ and $\cos x$, then the substitution

$$u = \tan\left(\frac{x}{2}\right), \qquad -\pi < x < \pi \tag{1}$$

will transform the integrand into a rational function of u (which can then be integrated by methods already discussed.) To see this, observe that

$$\cos\left(\frac{x}{2}\right) = \frac{1}{\sec(x/2)} = \frac{1}{\sqrt{1 + \tan^2(x/2)}} = \frac{1}{\sqrt{1 + u^2}}$$

$$\sin\left(\frac{x}{2}\right) = \tan\left(\frac{x}{2}\right)\cos\left(\frac{x}{2}\right) = u\left(\frac{1}{\sqrt{1 + u^2}}\right)$$

Therefore,

$$\sin x = 2\sin\left(\frac{x}{2}\right)\cos\left(\frac{x}{2}\right) = 2u\left(\frac{1}{\sqrt{1 + u^2}}\right)\left(\frac{1}{\sqrt{1 + u^2}}\right)$$

or

$$\sin x = \frac{2u}{1 + u^2} \tag{2}$$

Also,

$$\cos x = 1 - 2 \sin^2 \left(\frac{x}{2}\right) = 1 - \frac{2u^2}{1 + u^2}$$

or

$$\cos x = \frac{1 - u^2}{1 + u^2} \tag{3}$$

Moreover, from (1)

$$\frac{x}{2} = \tan^{-1} u$$

so that

$$\frac{dx}{du} = \frac{2}{1 + u^2}$$

or

$$dx = \frac{2}{1 + u^2} \, du \tag{4}$$

It follows from (2), (3), and (4) that substitution (1) will yield an integrand that is a rational function of u.

▶ **Example 4** Evaluate

$$\int \frac{dx}{1 + \sin x}$$

Solution. Making substitution (1), it follows from (2) and (4) above that

$$\int \frac{dx}{1 + \sin x} = \int \frac{1}{1 + \left(\dfrac{2u}{1 + u^2}\right)} \left(\frac{2}{1 + u^2}\right) du$$

$$= \int \frac{2}{1 + 2u + u^2} \, du$$

$$= \int \frac{2}{(1 + u)^2} \, du$$

$$= -\frac{2}{1 + u} + C$$

$$= -\frac{2}{1 + \tan\left(\dfrac{x}{2}\right)} + C \qquad \blacktriangleleft$$

REMARK. While substitution (1) is a useful tool, it can lead to cumbersome partial fraction decompositions. Consequently, this method should be used only after looking for simpler methods.

▶ Exercise Set 9.8

In Exercises 1–14, perform the integrations.

1. $\displaystyle\int \frac{dx}{\sqrt{x} + \sqrt[3]{x}}.$

2. $\displaystyle\int \frac{dx}{x - x^{3/5}}.$

3. $\displaystyle\int \frac{dv}{v(1 - v^{1/4})}.$

4. $\displaystyle\int \frac{x^{2/3}}{x + 1} \, dx.$

5. $\displaystyle\int \frac{dt}{t^{1/2} - t^{1/3}}.$

6. $\displaystyle\int \frac{1 + \sqrt{x}}{1 - \sqrt{x}} \, dx.$

7. $\displaystyle\int \frac{x^3}{\sqrt{1 + x^2}} \, dx.$

8. $\displaystyle\int \frac{x}{(x + 3)^{1/5}} \, dx.$

9. $\displaystyle\int \frac{dx}{1 + \sin x + \cos x}.$

10. $\displaystyle\int \frac{dx}{2 + \sin x}.$

11. $\displaystyle\int_{\pi/2}^{\pi} \frac{d\theta}{1 - \cos \theta}.$

12. $\displaystyle\int \frac{dx}{4 \sin x - 3 \cos x}.$

13. $\displaystyle\int \frac{\cos x}{2 - \cos x} \, dx.$

14. $\displaystyle\int \frac{dx}{\sin x + \tan x}.$

15. (a) Use the substitution $u = \tan(x/2)$ to show that

$$\int \sec x \, dx = \ln \left| \frac{1 + \tan \frac{1}{2}x}{1 - \tan \frac{1}{2}x} \right| + C$$

(b) Use the result in (a) to show that

$$\int \sec x \, dx = \ln \left| \tan \left(\frac{\pi}{4} + \frac{x}{2} \right) \right| + C$$

(c) Show that this agrees with (2) of Section 9.4.

16. (a) Use the substitution $u = \tan(x/2)$ to show that

$$\int \csc x \, dx = \frac{1}{2} \ln \left[\frac{1 - \cos x}{1 + \cos x} \right] + C$$

(b) Show that this agrees with Exercise 33a of Section 9.4.

17. Develop a substitution that can be used to integrate rational functions of $\sinh x$ and $\cosh x$ and use your substitution to evaluate

$$\int \frac{dx}{2 \cosh x + \sinh x}$$

without expressing the integrand in terms of e^x and e^{-x}.

9.9 NUMERICAL INTEGRATION; SIMPSON'S RULE

If an antiderivative of f cannot be found, then the integral

$$\int_a^b f(x)\,dx$$

cannot be evaluated by the First Fundamental Theorem of Calculus. In such cases, the value of the integral can be approximated using methods we will study in this section.

Suppose we want to approximate

$$\int_a^b f(x)\,dx$$

where f is continuous on $[a, b]$. Using the definition of a definite integral (Definition 5.7.2), we can write this integral as

$$\int_a^b f(x)\,dx = \lim_{\max \Delta x_k \to 0} \sum_{k=1}^n f(x_k^*)\,\Delta x_k \tag{1}$$

If we divide the interval $[a, b]$ into n subintervals of equal length, then

$$\Delta x_1 = \Delta x_2 = \cdots = \Delta x_n = \frac{b-a}{n}$$

so that we can rewrite (1) as

$$\int_a^b f(x)\,dx = \lim_{n \to +\infty} \sum_{k=1}^n f(x_k^*)\left(\frac{b-a}{n}\right)$$

Thus, we expect the following approximation to be good when n is large:

$$\int_a^b f(x)\,dx \approx \sum_{k=1}^n f(x_k^*)\left(\frac{b-a}{n}\right) = \frac{b-a}{n}\sum_{k=1}^n f(x_k^*) \tag{2}$$

where x_k^* denotes an arbitrary point in the kth subinterval. Expression (2) yields various approximation formulas, depending on how the points $x_1^*, x_2^*, \ldots, x_n^*$ are chosen. For example, let

$$y_0, y_1, \ldots, y_n$$

be the values of f at the endpoints of the subintervals (Figure 9.9.1).

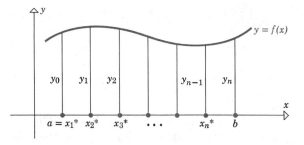

Figure 9.9.1

If we choose $x_1^*, x_2^*, \ldots, x_n^*$ to be the left-hand endpoints of the subintervals, then

$$y_0 = f(x_1^*), \ y_1 = f(x_2^*), \ldots, y_{n-1} = f(x_n^*)$$

so that (2) yields:

Left-hand Endpoint Approximation

$$\int_a^b f(x) \, dx \approx \left(\frac{b-a}{n}\right)[y_0 + y_1 + \cdots + y_{n-1}] \qquad (3)$$

(Figure 9.9.2a). Similarly, if $x_1^*, x_2^*, \ldots, x_n^*$ are the right-hand endpoints of the subintervals, then (2) yields

Right-hand Endpoint Approximation

$$\int_a^b f(x) \, dx \approx \left(\frac{b-a}{n}\right)[y_1 + y_2 + \cdots + y_n] \qquad (4)$$

(Figure 9.9.2b). Averaging the left-hand and right-hand endpoint approximations, we obtain:

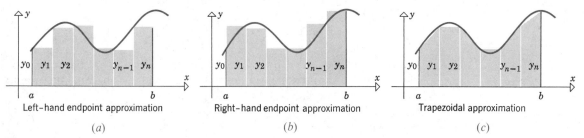

Left–hand endpoint approximation

(a)

Right–hand endpoint approximation

(b)

Trapezoidal approximation

(c)

Figure 9.9.2

Trapezoidal Approximation

$$\int_a^b f(x)\, dx \approx \left(\frac{b-a}{2n}\right)[y_0 + 2y_1 + \cdots + 2y_{n-1} + y_n] \qquad (5)$$

The name *trapezoidal approximation* can be explained by considering the case where $f(x) \geq 0$ on $[a, b]$, so that $\int_a^b f(x)\, dx$ represents the area under $f(x)$ over $[a, b]$. Geometrically, the trapezoidal approximation formula is what results if we approximate this area by the sum of the trapezoidal areas shown in Figure 9.9.2c (Exercise 8). By comparison, the left-hand and right-hand approximation formulas result when we approximate the area using the rectangles shown in Figures 9.9.2a and 9.9.2b.

▶ Example 1 Below we have approximated

$$\int_1^2 \frac{1}{x}\, dx$$

by three methods, left-hand endpoint approximation, right-hand endpoint approximation, and trapezoidal approximation. In each case, we used $n = 10$ subdivisions of the interval $[1, 2]$ so that the subintervals have width

$$\frac{b-a}{n} = \frac{2-1}{10} = \frac{1}{10} = 0.1$$

Left-hand Endpoint Approximation				Right-hand Endpoint Approximation		
	Endpoint				Endpoint	
i	x_i	$y_i = f(x_i) = 1/x_i$		i	x_i	$y_i = f(x_i) = 1/x_i$
0	1.0	1.0000		1	1.1	0.9091
1	1.1	0.9091		2	1.2	0.8333
2	1.2	0.8333		3	1.3	0.7692
3	1.3	0.7692		4	1.4	0.7143
4	1.4	0.7143		5	1.5	0.6667
5	1.5	0.6667		6	1.6	0.6250
6	1.6	0.6250		7	1.7	0.5882
7	1.7	0.5882		8	1.8	0.5556
8	1.8	0.5556		9	1.9	0.5263
9	1.9	0.5263		10	2.0	0.5000
		Total				*Total*
		7.1877				6.6877

From (3)

$$\int_1^2 \frac{1}{x}\,dx \approx (0.1)(7.1877)$$
$$= 0.71877$$

From (4)

$$\int_1^2 \frac{1}{x}\,dx \approx (0.1)(6.6877)$$
$$= 0.66877$$

Trapezoidal Approximation

| | Endpoint | | Multiplier | |
i	x_i	$y_i = f(x_i) = 1/x_i$	w_i	$w_i y_i$
0	1.0	1.0000	1	1.0000
1	1.1	0.9091	2	1.8182
2	1.2	0.8333	2	1.6666
3	1.3	0.7692	2	1.5384
4	1.4	0.7143	2	1.4286
5	1.5	0.6667	2	1.3334
6	1.6	0.6250	2	1.2500
7	1.7	0.5882	2	1.1764
8	1.8	0.5556	2	1.1112
9	1.9	0.5263	2	1.0526
10	2.0	0.5000	1	0.5000
				Total
				13.8754

Since

$$\frac{b - a}{2n} = \frac{2 - 1}{2(10)} = \frac{1}{20} = 0.05$$

we obtain from (5)

$$\int_1^2 \frac{1}{x}\,dx \approx (0.05)(13.8754) = 0.69377 \qquad \blacktriangleleft$$

REMARK.   Because $1/x$ is decreasing on the interval $[1, 2]$, we would expect the left-hand endpoint approximation to be too large and the right-hand endpoint approximation to be too small. This is in fact the case since,

$$\int_1^2 \frac{1}{x}\,dx = \ln 2 \approx 0.6931$$

to four-decimal-place accuracy.

Intuition suggests that we might improve on the endpoint approximations and the trapezoidal approximation by replacing the rectangular and trape-

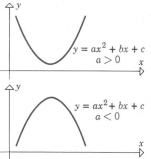

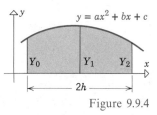

Figure 9.9.3

zoidal strips of Figure 9.9.2 with strips having curved upper boundaries chosen to match the shape of the curve $y = f(x)$. This is the idea behind **Simpson's* rule,** which uses curves of the form

$$y = ax^2 + bx + c$$

to approximate sections of the curve $y = f(x)$. As noted at the end of Section 1.6, these curves are parabolas with axes of symmetry parallel to the y-axis (Figure 9.9.3).

To simplify the description of Simpson's rule we will assume $f(x) \geq 0$ on $[a, b]$ so we can interpret $\int_a^b f(x)\, dx$ as an area. However, the method is valid without this assumption.

The heart of Simpson's rule is the formula

$$A = \frac{h}{3}[Y_0 + 4Y_1 + Y_2] \tag{6}$$

which gives the area under the curve

$$y = ax^2 + bx + c$$

over an arbitrary interval of width $2h$. In this formula Y_0, Y_1, and Y_2 represent the y values at the left-hand endpoint, the midpoint, and the right-hand endpoint of the interval (Figure 9.9.4).

To derive formula (6), denote the midpoint of the interval by m. Since the interval has length $2h$, the left-hand endpoint will be $m - h$ and the right-hand endpoint $m + h$. Thus, the area A under $y = ax^2 + bx + c$ over this interval will be

Figure 9.9.4

---

* THOMAS SIMPSON (1710–1761) English mathematician. Simpson was the son of a weaver. He was trained to follow in his father's footsteps and had little formal education in his early life. His interest in science and mathematics was aroused in 1724, when he witnessed an eclipse of the sun and received two books from a peddler, one on astrology and the other on arithmetic. Simpson quickly absorbed their contents and soon became a successful local fortune teller. His improved financial situation enabled him to give up weaving and marry his landlady, an older woman. Then in 1733 some mysterious "unfortunate incident" forced him to move. He settled in Derby, where he taught in an evening school and worked at weaving during the day. In 1736 he moved to London and published his first mathematical work in a periodical called the *Ladies' Diary* (of which he later became the editor). In 1737 he published a successful calculus textbook that enabled him to give up weaving completely and concentrate on textbook writing and teaching. His fortunes improved further in 1740 when one Robert Heath accused him of plagiarism. The publicity was marvelous, and Simpson proceeded to dash off a succession of best-selling textbooks, *Algebra* (ten editions plus translations), *Geometry* (twelve editions plus translations), *Trigonometry* (five editions plus translations) and numerous others.

It is interesting to note that Simpson did not discover the rule that bears his name. It was a well-known result by Simpson's time.

$$A = \int_{m-h}^{m+h} (ax^2 + bx + c)\, dx$$

$$= \frac{a}{3}x^3 + \frac{b}{2}x^2 + cx \Big]_{m-h}^{m+h}$$

$$= \frac{a}{3}[(m + h)^3 - (m - h)^3] + \frac{b}{2}[(m + h)^2 - (m - h)^2]$$

$$+ c[(m + h) - (m - h)]$$

or on simplifying

$$A = \frac{h}{3}[a(6m^2 + 2h^2) + b(6m) + 6c] \tag{7}$$

But the values of $y = ax^2 + bx + c$ at the left-hand endpoint, the midpoint, and the right-hand endpoint are, respectively,

$$Y_0 = a(m - h)^2 + b(m - h) + c$$
$$Y_1 = am^2 + bm + c$$
$$Y_2 = a(m + h)^2 + b(m + h) + c$$

from which it follows that

$$Y_0 + 4Y_1 + Y_2 = a(6m^2 + 2h^2) + b(6m) + 6c \tag{8}$$

Thus, (6) follows from (7) and (8).

Simpson's rule is obtained by dividing the interval $[a, b]$ into an *even* number of equal subintervals of width h and applying formula (6) to approximate the area under $y = f(x)$ over successive pairs of subintervals. The sum of these approximations then serves as an estimate of $\int_a^b f(x)\, dx$. More precisely, let $[a, b]$ be divided into n subintervals of width h (n even) and let

$$y_0, y_1, \ldots, y_n$$

be the values of $y = f(x)$ at the interval endpoints

$$a = x_0, x_1, \ldots, x_n = b$$

By (6) the area under $y = f(x)$ over the first two subintervals is approximately

$$\frac{h}{3}[y_0 + 4y_1 + y_2]$$

and the area over the second pair of subintervals is approximately

$$\frac{h}{3}[y_2 + 4y_3 + y_4]$$

and the area over the last pair of subintervals is approximately

$$\frac{h}{3}[y_{n-2} + 4y_{n-1} + y_n]$$

Adding all the approximations, collecting terms, and replacing h by $(b-a)/n$ yields:

Simpson's Rule

$$\int_a^b f(x)\, dx \approx \left(\frac{b-a}{3n}\right)[y_0 + 4y_1 + 2y_2 + 4y_3 + 2y_4 + \cdots$$
$$+ 2y_{n-2} + 4y_{n-1} + y_n]$$

▶ Example 2   Below we have approximated

$$\int_1^2 \frac{1}{x}\, dx$$

by Simpson's rule using $n = 10$ subdivisions so that the subintervals have width

$$\frac{b-a}{n} = \frac{2-1}{10} = \frac{1}{10} = 0.1$$

Simpson's Rule

i	Endpoint x_i	$y_i = f(x_i) = 1/x_i$	Multiplier w_i	$w_i y_i$
0	1.0	1.0000	1	1.0000
1	1.1	0.9091	4	3.6364
2	1.2	0.8333	2	1.6666
3	1.3	0.7692	4	3.0768
4	1.4	0.7143	2	1.4286
5	1.5	0.6667	4	2.6668
6	1.6	0.6250	2	1.2500
7	1.7	0.5882	4	2.3528
8	1.8	0.5556	2	1.1112
9	1.9	0.5263	4	2.1052
10	2.0	0.5000	1	0.5000
			Total	20.7944

Since

$$\frac{b-a}{3n} = \frac{2-1}{3(10)} = \frac{1}{30}$$

we obtain from Simpson's rule

$$\int_1^2 \frac{1}{x}\,dx \approx \frac{1}{30}(20.7944) \approx 0.69315$$

To six decimal places, the value of $\ln 2 = \int_1^2 (1/x)\,dx$ is 0.693147 so that with $n = 10$ Simpson's rule is more accurate than the methods of Example 1.

◄

With all the methods studied in this section, there are two sources of error, the *intrinsic* or *truncation error* due to the approximation formula, and the *roundoff* error introduced in the calculations. In general, increasing n reduces the truncation error but increases the roundoff error, since more computations are required for larger n. In practical applications, it is important to know how large n must be taken to ensure a specified degree of accuracy (see Exercise 10). Such problems are studied in a branch of mathematics called **numerical analysis.**

► Exercise Set 9.9

In Exercises 1–6, use the given value of n to approximate the integral by:
(a) the left-hand endpoint rule
(b) the right-hand endpoint rule
(c) the trapezoidal rule
(d) Simpson's rule.

Round off all computations to four decimal places and use the appendix tables of trigonometric functions and exponentials where needed.

1. $\displaystyle\int_0^1 \frac{dx}{x+1}$, $n = 6$. 2. $\displaystyle\int_0^4 x^2\,dx$, $n = 8$.

3. $\displaystyle\int_0^\pi \sin x\,dx$, $n = 4$. 4. $\displaystyle\int_0^1 \frac{dx}{\sqrt{1+x^2}}$, $n = 4$.

5. $\displaystyle\int_0^2 e^{-x^2}\,dx$, $n = 4$. 6. $\displaystyle\int_1^4 \sqrt{1+x^3}\,dx$, $n = 6$.

7. Use the relationship

$$\frac{\pi}{4} = \tan^{-1}(1) = \int_0^1 \frac{dx}{1+x^2}$$

and Simpson's rule with $n = 10$ to estimate π. Round off all computations to four decimal places.

8. Derive the trapezoidal rule by summing the areas of the trapezoids in Figure 9.9.2c.

9. If, in (2), x_k^* is taken as the midpoint of the kth subinterval, we obtain the **midpoint rule** for approximating $\int_a^b f(x)\,dx$. Use the midpoint rule with $n = 10$ to approximate

$$\int_1^2 \frac{1}{x}\,dx$$

(Compare your results to those obtained in Examples 1 and 2.)

10. Let K be any number such that $|f''(x)| \leq K$, and let M be any number such that $|f^{(4)}(x)| \leq M$ for all x in $[a, b]$. It is shown in advanced calculus that the absolute value of the error resulting from Simpson's rule with n subintervals is at most

$$\frac{M(b-a)^5}{180n^4}$$

and the absolute value of the error from the trapezoidal rule is at most

$$\frac{K(b-a)^3}{12n^2}$$

(a) In Examples 1 and 2, $\int_1^2 (1/x)\,dx$ was estimated by the trapezoidal rule and Simpson's rule using $n = 10$. Show that the absolute value of the error for the trapezoidal rule is at most

$$\frac{2}{12 \times 10^2} \approx 0.00167$$

and the absolute value of the error for Simpson's rule is at most

$$\frac{24}{180 \times 10^4} \approx 0.0000133$$

(b) How large should n be taken to ensure that the absolute value of the error in the approximation of $\int_1^2 (1/x)\,dx$ by Simpson's rule is at most 0.0002?

(c) Answer (b) for the trapezoidal rule.

▶ SUPPLEMENTARY EXERCISES

In Exercises 1–64, evaluate the integrals.

1. $\displaystyle\int x \cos 2x \, dx.$

2. $\displaystyle\int x \cos x^2 \, dx.$

3. $\displaystyle\int \tan^3 x \sec x \, dx.$

4. $\displaystyle\int \sin^3 x \cos^2 x \, dx.$

5. $\displaystyle\int \tan^2 3t \sec^2 3t \, dt.$

6. $\displaystyle\int \cot 2x \csc^3 2x \, dx.$

7. $\displaystyle\int \frac{\sin^2 x \, dx}{1 + \cos x}.$

8. $\displaystyle\int \frac{\sin 2x \, dx}{\cos x \,(1 + \cos x)}.$

9. $\displaystyle\int x^2 \cos^2 x \, dx.$

10. $\displaystyle\int \sin^2 2x \cos^2 2x \, dx.$

11. $\displaystyle\int \sec^5 x \sin x \, dx.$

12. $\displaystyle\int \tan^5 2x \, dx.$

13. $\displaystyle\int \sin^4 x \cos^2 x \, dx.$

14. $\displaystyle\int \frac{dx}{\sec^4 x}.$

15. $\displaystyle\int_0^{\pi/4} \sin 5x \sin 3x \, dx.$

16. $\displaystyle\int_{-\pi/10}^{0} \sin 2x \cos 3x \, dx.$

17. $\displaystyle\int_0^1 \sin^2 \pi x \, dx.$

18. $\displaystyle\int_0^{\pi/3} \sin^3 3x \, dx.$

19. $\displaystyle\int_0^{\sqrt{\pi/2}} x \sec^2 (x^2) \, dx.$

20. $\displaystyle\int_0^{\pi/4} \frac{\sec^2 x \, dx}{\sqrt{1 + 3 \tan x}}.$

21. $\displaystyle\int \frac{\sin (\cot^{-1} x) \, dx}{1 + x^2}.$

22. $\displaystyle\int \frac{e^{\tan 3x} \, dx}{\cos^2 3x}.$

23. $\displaystyle\int e^x \sec (e^x) \, dx.$

24. $\displaystyle\int x \sec^2 3x \, dx.$

25. $\displaystyle\int \frac{e^{2x}}{\sqrt{e^{2x} + 1}} \, dx.$

26. $\displaystyle\int \frac{e^x}{\sqrt{e^{2x} + 1}} \, dx.$

27. $\displaystyle\int e^{3x} \sin 2x \, dx.$

28. $\displaystyle\int \ln (a^2 + x^2) \, dx.$

29. $\displaystyle\int_1^2 \sin^{-1} (x/2) \, dx.$

30. $\displaystyle\int (\ln x)^2 \, dx.$

31. $\displaystyle\int (\ln x)/x^2 \, dx.$

32. $\displaystyle\int x^3 e^{-x^2} \, dx.$

33. $\displaystyle\int \frac{x \, dx}{\sqrt{x^2 - 9}}.$

34. $\displaystyle\int_1^2 \frac{\sqrt{4x^2 - 1}}{x} \, dx.$

35. $\displaystyle\int_1^3 \frac{\sqrt{9 - x^2}}{x} \, dx.$

36. $\displaystyle\int_{-\sqrt{8}}^{-\sqrt{8/3}} \frac{ds}{s\sqrt{s^2 + 8}}.$

37. $\displaystyle\int \frac{x^2 \, dx}{\sqrt{2x + 3}}.$

38. $\displaystyle\int \frac{dx}{\sqrt{x^2 + 16}}.$

39. $\displaystyle \int \frac{dt}{\sqrt{3 - 4t - 4t^2}}.$

40. $\displaystyle \int_4^7 \frac{dx}{\sqrt{x^2 - 2x - 8}}.$

41. $\displaystyle \int \frac{dx}{x^2\sqrt{a^2 - x^2}}.$

42. $\displaystyle \int \frac{x^3}{(x^2 + 4)^{1/3}}\, dx.$

43. $\displaystyle \int \sqrt{a^2 - x^2}\, dx.$

44. $\displaystyle \int x\sqrt{a^2 - x^2}\, dx.$

45. $\displaystyle \int \frac{x - 2}{\sqrt{4x - x^2}}\, dx.$

46. $\displaystyle \int_1^3 \frac{dx}{x^2 - 2x + 5}.$

47. $\displaystyle \int \frac{dx}{2x^2 + 3x + 1}.$

48. $\displaystyle \int \frac{dx}{(x^2 + 4)^2}.$

49. $\displaystyle \int \frac{x + 1}{x^3 + x^2 - 6x}\, dx.$

50. $\displaystyle \int \frac{x^3 + 1}{x - 2}\, dx.$

51. $\displaystyle \int \frac{x^2 - 1}{x^3 - 3x}\, dx.$

52. $\displaystyle \int \frac{x - 3}{x^3 - 1}\, dx.$

53. $\displaystyle \int \frac{2x^2 + 5}{x^4 - 1}\, dx.$

54. $\displaystyle \int \frac{x^4 - x^3 - x - 1}{x^3 - x^2}\, dx.$

55. $\displaystyle \int \frac{dx}{(x^2 + 4)(x - 3)}.$

56. $\displaystyle \int \frac{x\, dx}{(x + 1)^3}.$

57. $\displaystyle \int \frac{3x^2 + 12x + 2}{(x^2 + 4)^2}\, dx.$

58. $\displaystyle \int \frac{(4x + 2)\, dx}{x^4 + 2x^3 + x^2}.$

59. $\displaystyle \int \frac{x\, dx}{x^2 + 2x + 5}.$

60. $\displaystyle \int \frac{6x\, dx}{(x^2 + 9)^3}.$

61. $\displaystyle \int \frac{dx}{\sqrt{3 - 2x^2}}.$

62. $\displaystyle \int \frac{1 + t}{\sqrt{t}}\, dt.$

63. $\displaystyle \int \frac{\sqrt{t}\, dt}{1 + t}.$

64. $\displaystyle \int \frac{\sqrt{1 - x^2}}{x^2}\, dx.$

Exercises 65–70 relate to the optional Section 9.8. Evaluate the integrals.

65. $\displaystyle \int_0^1 \frac{x^{2/3}}{1 + x^{1/3}}\, dx.$

66. $\displaystyle \int \frac{dx}{x^{1/2} + x^{1/4}}.$

67. $\displaystyle \int \frac{dx}{1 - \tan x}.$

68. $\displaystyle \int \frac{dx}{3\cos x + 5}.$

69. $\displaystyle \int \frac{dx}{\sin x - \tan x}.$

70. $\displaystyle \int_0^{\pi/3} \frac{dx}{5\sec x - 3}.$

71. Use partial fractions to show that

$$\int \frac{dx}{x^2 - a^2} = \frac{1}{2a} \ln \left| \frac{x - a}{x + a} \right| + C \quad (a \neq 0).$$

72. Find the arc length of: (a) the parabola $y = x^2/2$ from $(0, 0)$ to $(2, 2)$, (b) the curve $y = \ln(\sec x)$ from $(0, 0)$ to $(\pi/4, \frac{1}{2}\ln 2)$.

73. Let R be the region bounded by the curve $y = 1/(4 + x^2)$ and the lines $x = 0$, $y = 0$, and $x = 2$. Find: (a) the area of R, (b) the volume of the solid obtained by revolving R about the x-axis, (c) the volume of the solid obtained by revolving R about the y-axis.

74. Derive the following reduction formulas for $a \neq 0$:

(a) $\displaystyle \int x^n e^{ax}\, dx = \frac{x^n e^{ax}}{a} - \frac{n}{a} \int x^{n-1} e^{ax}\, dx$

(b) $\displaystyle \int x^n \sin ax\, dx =$

$$\frac{-x^n \cos ax}{a} + \frac{n}{a} \int x^{n-1} \cos ax\, dx.$$

$$\int x^n \cos ax\, dx =$$

$$\frac{x^n \sin ax}{a} - \frac{n}{a} \int x^{n-1} \sin ax\, dx$$

(c) $\displaystyle \int \sin^n ax \cos^m ax\, dx = -\frac{\sin^{n-1} ax \cos^{m+1} ax}{a(m + n)}$

$$+ \frac{n - 1}{m + n} \int \sin^{n-2} ax \cos^m ax\, dx$$

$$= \frac{\sin^{n+1} ax \cos^{m-1} ax}{a(m + n)} + \frac{m - 1}{m + n} \int \sin^n ax \cos^{m-2} ax\, dx.$$

75. Use Exercise 74 to evaluate the following integrals.

(a) $\displaystyle \int x^3 e^{2x}\, dx$

(b) $\displaystyle \int_0^{\pi/10} x^2 \sin 5x\, dx$

(c) $\displaystyle \int \sin^2 x \cos^4 x\, dx$

76. Evaluate the following integrals assuming that $a \neq 0$.

(a) $\displaystyle \int x^n \ln ax\, dx \quad (n \neq -1)$

(b) $\displaystyle \int \sec^n ax \tan ax\, dx \quad (n \geq 1).$

77. Find $\int (\sin^3 \theta / \cos^5 \theta)\, d\theta$ two ways: (a) letting $u = \cos \theta$ and (b) expressing the integrand in terms of $\sec \theta$ and $\tan \theta$. Show that your answers differ by a constant.

In Exercises 78–81, approximate the integral using the given value of n and (a) the trapezoidal rule, (b) Simpson's rule. Use a calculator and express the answer to four decimal places.

78. $\displaystyle\int_0^1 \sqrt{x}\, dx, \; n = 4.$

79. $\displaystyle\int_{-4}^2 e^{-x}\, dx, \; n = 6.$

80. $\displaystyle\int_0^4 \sinh x\, dx, \; n = 4.$

81. $\displaystyle\int_4^{5.2} \ln x\, dx, \; n = 6.$

In Exercises 82 and 83, use Simpson's rule with $n = 10$ to approximate the given integral. Use a calculator and express the answer to five decimal places.

82. $\displaystyle\int_0^2 \cos (\sinh x)\, dx.$

83. $\displaystyle\int_1^2 \sin (\ln x)\, dx.$

10

improper integrals;
L'Hôpital's rule

10.1 IMPROPER INTEGRALS

In the definition of $\int_a^b f(x)\,dx$, it is assumed that the interval $[a, b]$ is *finite*. Moreover, in order for the limit

$$\lim_{\max \Delta x_k \to 0} \sum_{k=1}^{n} f(x_k^*)\,\Delta x_k = \int_a^b f(x)\,dx$$

to exist the function f must be *bounded* on the interval $[a, b]$. (See the remark following Theorem 5.7.3.) For example, the integral

$$\int_0^3 \frac{1}{(x-2)^{2/3}}\,dx$$

does not exist, since the integrand approaches $+\infty$ at the point $x = 2$ in the interval of integration.

In this section we will extend the concept of the definite integral to include:

(a) integrals with infinite intervals of integration;
(b) integrals in which the integrand becomes infinite within the interval of integration.

These are called ***improper integrals.***

If f is continuous on the interval $[a, +\infty)$, then we define the improper integral $\int_a^{+\infty} f(x)\,dx$ as the limit:

$$\int_a^{+\infty} f(x)\,dx = \lim_{l \to +\infty} \int_a^l f(x)\,dx \qquad (1)$$

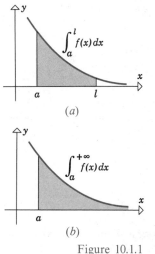

$$\int_a^l f(x)\,dx$$

(a)

$$\int_a^{+\infty} f(x)\,dx$$

(b)

Figure 10.1.1

If this limit exists, the improper integral is said to **converge,** and the value of the limit is the value assigned to the integral. If the limit does not exist, then the improper integral is said to **diverge,** in which case it is not assigned a value.

If f is nonnegative and continuous on $[a, +\infty)$, then (1) has an important geometric interpretation. For each value of $l > a$, the definite integral $\int_a^l f(x)\,dx$ represents the area under the curve $y = f(x)$ over the finite interval $[a, l]$ (Figure 10.1.1a). As we let $l \to +\infty$, this area tends toward the area under $y = f(x)$ over the entire interval $[a, +\infty)$ (Figure 10.1.1b). Thus, $\int_a^{+\infty} f(x)\,dx$ can be regarded as the area under $y = f(x)$ over the interval $[a, +\infty)$.

▶ **Example 1** Find

$$\int_1^{+\infty} \frac{dx}{x^2}$$

Solution. We begin by replacing the infinite upper limit with a finite upper limit l.

$$\int_1^l \frac{dx}{x^2} = -\frac{1}{x}\bigg]_1^l = -\frac{1}{l} - (-1) = 1 - \frac{1}{l}$$

Thus

$$\int_1^{+\infty} \frac{dx}{x^2} = \lim_{l \to +\infty} \int_1^l \frac{dx}{x^2} = \lim_{l \to +\infty} \left(1 - \frac{1}{l}\right) = 1 \qquad ◀$$

▶ **Example 2** Evaluate

$$\int_1^{+\infty} \frac{dx}{x}$$

Solution.

$$\int_1^{+\infty} \frac{dx}{x} = \lim_{l \to +\infty} \int_1^l \frac{dx}{x} = \lim_{l \to +\infty} [\ln |x|]_1^l = \lim_{l \to +\infty} \ln |l| = +\infty$$

Thus, the integral diverges. ◀

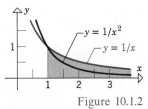

Figure 10.1.2

It is worthwhile to reflect on the results of Examples 1 and 2. Why is it that $\int_1^{+\infty} dx/x^2$ converges while $\int_1^{+\infty} dx/x$ diverges when, in fact, the graphs of $1/x$ and $1/x^2$ look similar over the interval $[1, +\infty)$ (Figure 10.1.2)? The explanation is that $1/x^2$ approaches zero more rapidly than $1/x$ as $x \to +\infty$, so that area accumulates under the curve $y = 1/x$ more rapidly than under the curve $y = 1/x^2$. When we calculate

$$\lim_{l \to +\infty} \int_1^l \frac{dx}{x} \quad \text{and} \quad \lim_{l \to +\infty} \int_1^l \frac{dx}{x^2}$$

the difference is enough that the first limit is infinite while the second is finite.

If f is continuous on the interval $(-\infty, b]$, then we define the improper integral $\int_{-\infty}^b f(x)\, dx$ to be the limit

$$\int_{-\infty}^b f(x)\, dx = \lim_{l \to -\infty} \int_l^b f(x)\, dx$$

As before, the improper integral is said to **converge** if the limit exists and **diverge** if it does not. If f is nonnegative, then the integral represents the area under $y = f(x)$ over the interval $(-\infty, b]$.

▶ **Example 3** Evaluate

$$\int_{-\infty}^0 e^x\, dx$$

Solution. We begin by replacing the infinite lower limit with a finite lower limit l.

$$\int_l^0 e^x\, dx = e^x \Big]_l^0 = e^0 - e^l = 1 - e^l$$

Thus,

$$\int_{-\infty}^0 e^x\, dx = \lim_{l \to -\infty} \int_l^0 e^x\, dx = \lim_{l \to -\infty} (1 - e^l) = 1 \qquad ◀$$

If the two improper integrals $\int_{-\infty}^0 f(x)\, dx$ and $\int_0^{+\infty} f(x)\, dx$ both converge, then we say that $\int_{-\infty}^{+\infty} f(x)\, dx$ **converges** and we define

$$\int_{-\infty}^{+\infty} f(x)\, dx = \int_{-\infty}^0 f(x)\, dx + \int_0^{+\infty} f(x)\, dx \qquad (2)$$

If either integral on the right side of (2) diverges, then we say that $\int_{-\infty}^{+\infty} f(x)\, dx$ **diverges.** If f is nonnegative, then the integral $\int_{-\infty}^{+\infty} f(x)\, dx$ represents the area under $y = f(x)$ over the interval $(-\infty, +\infty)$.

▶ **Example 4** Evaluate

$$\int_{-\infty}^{+\infty} \frac{dx}{1 + x^2}$$

Solution.

$$\int_{0}^{+\infty} \frac{dx}{1 + x^2} = \lim_{l \to +\infty} \int_{0}^{l} \frac{dx}{1 + x^2} = \lim_{l \to +\infty} \left[\tan^{-1} x \right]_{0}^{l} = \lim_{l \to +\infty} (\tan^{-1} l) = \frac{\pi}{2}$$

$$\int_{-\infty}^{0} \frac{dx}{1 + x^2} = \lim_{l \to -\infty} \int_{l}^{0} \frac{dx}{1 + x^2} = \lim_{l \to -\infty} \tan^{-1} x \Big]_{l}^{0} = \lim_{l \to -\infty} (-\tan^{-1} l) =$$

$$-\left(-\frac{\pi}{2}\right) = \frac{\pi}{2}$$

Thus,

$$\int_{-\infty}^{+\infty} \frac{dx}{1 + x^2} = \int_{-\infty}^{0} \frac{dx}{1 + x^2} + \int_{0}^{+\infty} \frac{dx}{1 + x^2} = \frac{\pi}{2} + \frac{\pi}{2} = \pi \qquad ◀$$

REMARK. In (2), the decision to split the integral $\int_{-\infty}^{+\infty} f(x)\, dx$ at $x = 0$ is arbitrary. We can just as well make the split at any other point $x = c$ without affecting the convergence, divergence, or the value of the integral; that is,

$$\int_{-\infty}^{+\infty} f(x)\, dx = \int_{-\infty}^{c} f(x)\, dx + \int_{c}^{+\infty} f(x)\, dx$$

(We omit the proof.)

Next, we consider improper integrals in which the integrand approaches $+\infty$ or $-\infty$ somewhere in the interval of integration. We start with the cases where the integrand becomes infinite at one of the endpoints of integration.

If f is continuous on the interval $[a, b)$, but $f(x) \to +\infty$ or $f(x) \to -\infty$ as x approaches b from the left, then we define the improper integral $\int_{a}^{b} f(x)\, dx$ as the limit:

$$\int_{a}^{b} f(x)\, dx = \lim_{l \to b^-} \int_{a}^{l} f(x)\, dx \qquad (3)$$

To visualize this definition geometrically, consider the case where f is non-negative on the interval $[a, b]$. For each number l satisfying $a \le l < b$, the integral $\int_{a}^{l} f(x)\, dx$ represents the area under $y = f(x)$ over the interval $[a, l]$ (Figure 10.1.3a). As we let l approach b from the left, these areas tend to fill out the entire area under $y = f(x)$ over the interval $[a, b)$ (Figure 10.1.3b).

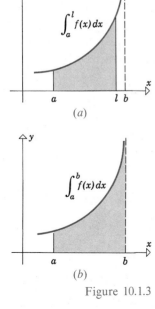

(a)

(b)

Figure 10.1.3

As before, the improper integral $\int_a^b f(x)\, dx$ is said to **converge** or **diverge** depending on whether the limit in (3) does or does not exist.

▶ Example 5   Evaluate

$$\int_0^1 \frac{dx}{\sqrt{1-x}}$$

Solution. The integral is improper because the integrand approaches $+\infty$ as x approaches the upper limit 1 from the left. From (3),

$$\int_0^1 \frac{dx}{\sqrt{1-x}} = \lim_{l\to 1^-} \int_0^l \frac{dx}{\sqrt{1-x}}$$

$$= \lim_{l\to 1^-} \left[-2\sqrt{1-x}\,\right]_0^l$$

$$= \lim_{l\to 1^-} \left[-2\sqrt{1-l} + 2\right] = 2 \qquad ◀$$

If f is continuous on the interval $(a, b]$, but $f(x) \to +\infty$ or $f(x) \to -\infty$ as x approaches a from the right, then we define the improper integral $\int_a^b f(x)\, dx$ as:

$$\int_a^b f(x)\, dx = \lim_{l\to a^+} \int_l^b f(x)\, dx$$

The improper integral **converges** or **diverges** depending on whether the limit does or does not exist.

▶ Example 6   The integral

$$\int_1^2 \frac{dx}{1-x}$$

is improper because the integrand approaches $-\infty$ as x approaches the lower limit 1 from the right. The integral diverges because

$$\int_1^2 \frac{dx}{1-x} = \lim_{l\to 1^+} \int_l^2 \frac{dx}{1-x}$$

$$= \lim_{l\to 1^+} \left[-\ln|1-x|\right]_l^2$$

$$= \lim_{l\to 1^+} \left[-\ln|-1| + \ln|1-l|\right]$$

$$= \lim_{l\to 1^+} \ln|1-l|$$

$$= -\infty \qquad ◀$$

Let f be continuous on the interval $[a, b]$ with the exception that at some point c satisfying $a < c < b$, $f(x)$ becomes infinite (tends to $+\infty$ or $-\infty$) as x approaches c from the left or the right. If the two improper integrals $\int_a^c f(x)\,dx$ and $\int_c^b f(x)\,dx$ both converge, then we say that the improper integral $\int_a^b f(x)\,dx$ **converges,** and we define:

$$\int_a^b f(x)\,dx = \int_a^c f(x)\,dx + \int_c^b f(x)\,dx \qquad (4)$$

If either integral on the right side of (4) diverges, then we say that $\int_a^b f(x)\,dx$ **diverges.**

▶ **Example 7** Evaluate

$$\int_1^4 \frac{dx}{(x-2)^{2/3}}$$

Solution. The integrand approaches $+\infty$ as $x \to 2$, so we use (4) to write

$$\int_1^4 \frac{dx}{(x-2)^{2/3}} = \int_1^2 \frac{dx}{(x-2)^{2/3}} + \int_2^4 \frac{dx}{(x-2)^{2/3}} \qquad (5)$$

But

$$\int_1^2 \frac{dx}{(x-2)^{2/3}} = \lim_{l \to 2^-} \int_1^l \frac{dx}{(x-2)^{2/3}} = \lim_{l \to 2^-} [3(l-2)^{1/3} - 3(1-2)^{1/3}] = 3$$

$$\int_2^4 \frac{dx}{(x-2)^{2/3}} = \lim_{l \to 2^+} \int_l^4 \frac{dx}{(x-2)^{2/3}} = \lim_{l \to 2^+} [3(4-2)^{1/3} - 3(l-2)^{1/3}] = 3\sqrt[3]{2}$$

Thus, from (5)

$$\int_1^4 \frac{dx}{(x-2)^{2/3}} = 3 + 3\sqrt[3]{2} \qquad \blacktriangleleft$$

It is sometimes tempting to apply the Fundamental Theorem of Calculus directly to an improper integral without taking the appropriate limits. To illustrate what can go wrong with this procedure, suppose we ignore the fact that the integral

$$\int_0^2 \frac{dx}{(x-1)^2} \qquad (6)$$

is improper and write

$$\int_0^2 \frac{dx}{(x-1)^2} = -\frac{1}{x-1} \bigg]_0^2 = -1 - (1) = -2$$

This result is clearly nonsense since the integrand is never negative and consequently the integral cannot be negative! To correctly evaluate (6) we should write

$$\int_0^2 \frac{dx}{(x-1)^2} = \int_0^1 \frac{dx}{(x-1)^2} + \int_1^2 \frac{dx}{(x-1)^2} \tag{7}$$

But

$$\int_0^1 \frac{dx}{(x-1)^2} = \lim_{l \to 1^-} \int_0^l \frac{dx}{(x-1)^2} = \lim_{l \to 1^-} \left[-\frac{1}{l-1} - 1 \right] = +\infty$$

so that (6) is a divergent integral.

▶ Exercise Set 10.1

In Exercises 1–20, evaluate the integrals that converge.

1. $\int_0^{+\infty} e^{-x}\, dx.$

2. $\int_1^{+\infty} \frac{dx}{x^3}.$

3. $\int_1^{+\infty} \frac{dx}{\sqrt{x}}.$

4. $\int_{-1}^{+\infty} \frac{x}{1+x^2}\, dx.$

5. $\int_4^{+\infty} \frac{2}{x^2-1}\, dx.$

6. $\int_0^{+\infty} xe^{-x^2}\, dx.$

7. $\int_{-\infty}^0 \frac{dx}{(2x-1)^3}.$

8. $\int_{-\infty}^2 \frac{dx}{x^2+4}.$

9. $\int_{-\infty}^{+\infty} x^3\, dx.$

10. $\int_{-\infty}^{+\infty} \frac{x}{\sqrt{x^2+2}}\, dx.$

11. $\int_{-\infty}^{+\infty} \frac{x}{(x^2+3)^2}\, dx.$

12. $\int_0^8 \frac{dx}{\sqrt[3]{x}}.$

13. $\int_0^{\pi/2} \tan x\, dx.$

14. $\int_0^9 \frac{dx}{\sqrt{9-x}}.$

15. $\int_0^1 \frac{dx}{\sqrt{1-x^2}}.$

16. $\int_{-2}^2 \frac{dx}{x^2}.$

17. $\int_{-1}^8 x^{-1/3}\, dx.$

18. $\int_0^4 \frac{dx}{(x-2)^{2/3}}.$

19. $\int_0^{\pi/2} \frac{\cos\theta}{\sqrt{1-\sin\theta}}\, d\theta.$

20. $\int_1^{+\infty} \frac{dx}{x\sqrt{x^2-1}}.$

21. (a) It is possible for an improper integral to diverge without becoming infinite. Show that $\int_0^{+\infty} \cos x\, dx$ does this.

(b) Evaluate $\int_0^{+\infty} e^{-x} \cos x\, dx.$

22. Show that $\int_1^{+\infty} \frac{dx}{x^p}$ converges if $p > 1$ and diverges if $p \le 1.$

23. Show that $\int_0^1 \frac{dx}{x^p}$ converges if $p < 1$ and diverges if $p \ge 1.$

24. Find the area of the region between the x-axis and the curve $y = 8/(x^2-4)$ for $x \ge 3.$

25. Find the area of the region bounded by $y = 0$, $x = 0$, $x = 4$, and $y = (1-x)^{-2}.$

26. (a) Show that

$$\int_{-1}^0 \frac{dx}{x} = -\infty \quad \text{and} \quad \int_0^1 \frac{dx}{x} = +\infty$$

(b) Evaluate

$$\int_{-1}^1 \frac{dx}{x}.$$

27. Let R be the region to the right of $x = 1$ that is bounded by the x-axis and the curve $y = 1/x$. Show that the solid obtained by revolving R about the x-axis has a finite volume, but an infinite surface area. [It has been suggested that by filling this solid with paint and letting it seep through to the surface one could paint an infinite surface area with a finite amount of paint!]

28. For what values of p does $\int_0^{+\infty} e^{px}\, dx$ converge?

29. In electromagnetic theory, the magnetic potential at a point on the axis of a circular coil is given by

$$u = \frac{2\pi NIr}{k} \int_{a}^{+\infty} \frac{dx}{(r^2 + x^2)^{3/2}}$$

where N, I, r, k, and a are constants. Find u.

30. Sketch the region whose area is $\displaystyle\int_{0}^{+\infty} \frac{dx}{1 + x^2}$, and use your sketch to show that

$$\int_{0}^{+\infty} \frac{dx}{1 + x^2} = \int_{0}^{1} \sqrt{\frac{1 - y}{y}} \, dy$$

31. **(Satellite Problem).** In Exercise 13 of Section 6.6, we determined the work required to lift a 6000-lb satellite to a specified orbital position. The result in part (a) of that problem will be needed here.

(a) Find a definite integral that represents the work required to lift a 6000-lb satellite to a position l miles above the earth's surface.

(b) Find a definite integral that represents the work required to lift a 6000-lb satellite an "infinite distance" above the earth's surface. Evaluate the integral. [The result obtained here is sometimes called the work required to "escape" the earth's gravity.]

10.2 L'HÔPITAL'S RULE (INDETERMINATE FORMS OF TYPE 0/0)

In this section we develop an important new technique for finding limits of functions.

In each of the limits

$$\lim_{x \to 2} \frac{x^2 - 4}{x - 2} \quad \text{and} \quad \lim_{x \to 0} \frac{\sin x}{x} \tag{1}$$

the numerator and denominator both approach zero. It is customary to describe such limits as **indeterminate forms of type 0/0.** The word "indeterminate" conveys the idea that the limit cannot be determined without additional work. In Example 8 of Section 2.5 we evaluated the first limit in (1) by canceling the common factor $x - 2$ from the numerator and denominator, and in Theorem 3.3.4 we resorted to a rather intricate geometric argument to obtain the second limit in (1). Because geometric arguments and the technique of canceling factors apply only to a limited range of problems, it is desirable to have a general method for handling indeterminate forms. This is provided by **L'Hôpital's* rule,** which we now discuss.

---

*GUILLAUME FRANCOIS ANTOINE DE L'HÔPITAL (1661–1704) French mathematician. L'Hôpital, born to parents of the French high nobility, held the title of Marquis de Sainte-Mesme Comte d'Autrement. He showed mathematical talent quite early and at age 15 solved a difficult problem about cycloids posed by Pascal. As a young man he served briefly as a cavalry officer, but resigned because of nearsightedness. In his own time he gained fame as the author of the first textbook ever published on differential calculus, *L'Analyse des Infiniment Petits pour l'Intelligence des Lignes Courbes* (1696). L'Hôpital's rule appeared for the first time in that book. Actually, L'Hôpital's rule and most of the material in the calculus text were due to John Bernoulli who was L'Hôpital's teacher. L'Hôpital dropped his plans for a book on integral calculus when Leibniz informed him that he intended to write such a text.

L'Hôpital was apparently generous and personable and his many contacts with major mathematicians provided the vehicle for disseminating major discoveries in calculus throughout Europe.

10.2.1   THEOREM

L'Hôpital's Rule for Form 0/0

Let lim *stand for one of the limits* $\lim_{x \to a}$, $\lim_{x \to a^+}$, $\lim_{x \to a^-}$, $\lim_{x \to +\infty}$, *or* $\lim_{x \to -\infty}$, *and suppose that* $\lim f(x) = 0$ *and* $\lim g(x) = 0$. *If* $\lim [f'(x)/g'(x)]$ *has a finite value* L, *or if this limit is* $+\infty$ *or* $-\infty$, *then*

$$\lim \frac{f(x)}{g(x)} = \lim \frac{f'(x)}{g'(x)}$$

REMARK.   There are some hypotheses implicit in this theorem. For example, in the case where $x \to a$, the statement $\lim_{x \to a} [f'(x)/g'(x)] = L$ requires that f'/g' be defined in some open interval I containing a (except possibly at a). This implies that f and g are differentiable and $g'(x) \neq 0$ in I (except possibly at a). Similar hypotheses are implicit in the other cases.

In essence, L'Hôpital's rule enables us to replace one limit problem with another that may be simpler. In each of these examples we will follow a three-step process:

Step 1.   Check that $\lim f(x)/g(x)$ is an indeterminate form. If it is not, then L'Hôpital's rule cannot be used.

Step 2.   Differentiate f and g separately.

Step 3.   Find $\lim f'(x)/g'(x)$. If this limit is finite, $+\infty$, or $-\infty$, then it is equal to $\lim f(x)/g(x)$.

▶ Example 1   In Theorem 3.3.4 we proved

$$\lim_{x \to 0} \frac{\sin x}{x} = 1$$

geometrically. We can verify this result using L'Hôpital's rule. Since

$$\lim_{x \to 0} \sin x = 0 \quad \text{and} \quad \lim_{x \to 0} x = 0$$

the limit

$$\lim_{x \to 0} \frac{\sin x}{x}$$

is an indeterminate form of type 0/0. Thus, by L'Hôpital's rule,

$$\lim_{x \to 0} \frac{\sin x}{x} = \lim_{x \to 0} \frac{\dfrac{d}{dx}[\sin x]}{\dfrac{d}{dx}[x]} = \lim_{x \to 0} \frac{\cos x}{1} = 1 \qquad \blacktriangleleft$$

REMARK. To be rigorous, the first equality in this calculation is not justified until the limit on the right is shown to exist. However, for simplicity we will usually arrange the computations as shown when applying L'Hôpital's rule.

▶ **Example 2** Evaluate

$$\lim_{x \to \pi/2} \frac{1 - \sin x}{\cos x}$$

Solution. Since

$$\lim_{x \to \pi/2} (1 - \sin x) = \lim_{x \to \pi/2} \cos x = 0$$

the given limit is an indeterminate form of type $0/0$. Thus, by L'Hôpital's rule

$$\lim_{x \to \pi/2} \frac{1 - \sin x}{\cos x} = \lim_{x \to \pi/2} \frac{\dfrac{d}{dx}[1 - \sin x]}{\dfrac{d}{dx}[\cos x]} = \lim_{x \to \pi/2} \frac{-\cos x}{-\sin x} = \frac{0}{-1} = 0$$

◀

▶ **Example 3** Evaluate

$$\lim_{x \to 0} \frac{e^x - 1}{x^3}$$

Solution. Since

$$\lim_{x \to 0} (e^x - 1) = \lim_{x \to 0} x^3 = 0$$

the given limit is an indeterminate form of type $0/0$. Thus, by L'Hôpital's rule

$$\lim_{x \to 0} \frac{e^x - 1}{x^3} = \lim_{x \to 0} \frac{\dfrac{d}{dx}[e^x - 1]}{\dfrac{d}{dx}[x^3]} = \lim_{x \to 0} \frac{e^x}{3x^2} = +\infty$$

◀

As the following example shows, it is sometimes necessary to apply L'Hôpital's rule more than once in the same problem.

▶ **Example 4** Evaluate

$$\lim_{x \to 0} \frac{1 - \cos x}{x^2}$$

Solution. Since

$$\lim_{x \to 0} (1 - \cos x) = \lim_{x \to 0} x^2 = 0$$

the given limit is an indeterminate form of type 0/0. Thus, by L'Hôpital's rule

$$\lim_{x \to 0} \frac{1 - \cos x}{x^2} = \lim_{x \to 0} \frac{\sin x}{2x}$$

However, the new limit is also an indeterminate form of type 0/0, so we apply L'Hôpital's rule again. This yields

$$\lim_{x \to 0} \frac{1 - \cos x}{x^2} = \lim_{x \to 0} \frac{\sin x}{2x} = \lim_{x \to 0} \frac{\cos x}{2} = \frac{1}{2} \qquad \blacktriangleleft$$

WARNING.   We warn the reader about two possible sources of errors. First, when applying L'Hôpital's rule to $\lim f(x)/g(x)$, the derivatives of $f(x)$ and $g(x)$ are taken separately to yield the new limit, $\lim [f'(x)/g'(x)]$. Do not make the mistake of differentiating $f(x)/g(x)$ according to the quotient rule. Second, make sure that $\lim [f(x)/g(x)]$ is an indeterminate form; otherwise, L'Hôpital's rule does not apply.

▶ Example 5   Evaluate

$$\lim_{x \to 0} \frac{e^x}{x^2}$$

Solution.

$$\lim_{x \to 0} e^x = 1 \qquad \text{and} \qquad \lim_{x \to 0} x^2 = 0$$

so the given problem is not an indeterminate form of type 0/0. By inspection

$$\lim_{x \to 0} \frac{e^x}{x^2} = +\infty \qquad \blacktriangleleft$$

▶ Example 6   Evaluate

$$\lim_{x \to +\infty} \frac{x^{-4/3}}{\sin (1/x)}$$

Solution.   Since

$$\lim_{x \to +\infty} x^{-4/3} = \lim_{x \to +\infty} \sin (1/x) = 0$$

the given limit is an indeterminate form of type $0/0$. Thus, by L'Hôpital's rule

$$\lim_{x \to +\infty} \frac{x^{-4/3}}{\sin(1/x)} = \lim_{x \to +\infty} \frac{-\frac{4}{3}x^{-7/3}}{(-1/x^2)\cos(1/x)} = \lim_{x \to +\infty} \frac{\frac{4}{3}x^{-1/3}}{\cos(1/x)} = \frac{0}{1} = 0 \quad \blacktriangleleft$$

▶ Example 7 Evaluate

$$\lim_{x \to 0^-} \frac{\tan x}{x^2}$$

Solution. Since

$$\lim_{x \to 0^-} \tan x = \lim_{x \to 0^-} x^2 = 0$$

the given limit is an indeterminate form of type $0/0$. Thus, by L'Hôpital's rule

$$\lim_{x \to 0^-} \frac{\tan x}{x^2} = \lim_{x \to 0^-} \frac{\sec^2 x}{2x} = -\infty \quad \blacktriangleleft$$

The proof of L'Hôpital's rule depends on the following result, called the **Extended Mean-Value Theorem** or sometimes the **Cauchy* Mean-Value Theorem.**

---

*AUGUSTIN LOUIS CAUCHY (1789–1857) French mathematician. Cauchy's early education was acquired from his father, a barrister and master of the classics. Cauchy entered L'Ecole Polytechnique in 1805 to study engineering, but because of poor health, was advised to concentrate on mathematics. His major mathematical work began in 1811 with a series of brilliant solutions to some difficult outstanding problems.

In 1814 he wrote a treatise on integrals that was to become the basis for modern complex variable theory; in 1816 there followed a classic paper on wave propagation in liquids that won a prize from the French Academy; and in 1822 he wrote a paper that formed the basis of modern elasticity theory.

Cauchy's mathematical contributions for the next 35 years were brilliant and staggering in quantity, over 700 papers filling 26 modern volumes. Cauchy's work initiated the era of modern analysis. He brought to mathematics standards of precision and rigor undreamed of by Leibniz and Newton.

Cauchy's life was inextricably tied to the political upheavals of the time. A strong partisan of the Bourbons, he left his wife and children in 1830 to follow the Bourbon king Charles X into exile. For his loyalty he was made a baron by the ex-king. Cauchy eventually returned to France, but refused to accept a university position until the government waived its requirement that he take a loyalty oath.

It is difficult to get a clear picture of the man. Devoutly Catholic, he sponsored charitable work for unwed mothers, criminals, and relief for Ireland. Yet other aspects of his life cast him in an unfavorable light. The Norwegian mathematician Abel described him as, "mad, infinitely Catholic, and bigoted." Some writers praise his teaching, yet others say he rambled incoherently and, according to a report of the day, he once devoted an entire lecture to extracting the square root of seventeen to ten decimal places by a method well-known to his students. In any event, Cauchy is undeniably one of the greatest minds in the history of science.

10.2.2   THEOREM   *Let the functions f and g be differentiable on (a, b) and continuous on $[a, b]$. If*
Extended Mean-Value $g'(x) \neq 0$ *for any x in (a, b), then there is at least one point c in (a, b) such that*
Theorem

$$\frac{f'(c)}{g'(c)} = \frac{f(b) - f(a)}{g(b) - g(a)} \tag{2}$$

Proof.   Observe first that $g(b) - g(a) \neq 0$, since otherwise it would follow from the Mean-Value Theorem (4.10.2) that $g'(x) = 0$ at some point x in (a, b), contradicting our hypothesis. For convenience, introduce a new function F defined by

$$F(x) = [f(b) - f(a)]g(x) - [g(b) - g(a)]f(x) \tag{3}$$

It follows from our assumptions about f and g that F is continuous on $[a, b]$ and differentiable on (a, b). Moreover, $F(b) = F(a)$ (verify), so that the Mean-Value Theorem (4.10.2) implies that there is at least one point c in (a, b) where

$$F'(c) = 0$$

Thus, from (3)

$$[f(b) - f(a)]g'(c) - [g(b) - g(a)]f'(c) = 0$$

or

$$\frac{f'(c)}{g'(c)} = \frac{f(b) - f(a)}{g(b) - g(a)}$$

Note that in the special case where $g(x) = x$, formula (2) reduces to

$$f'(c) = \frac{f(b) - f(a)}{b - a}$$

which is precisely the conclusion of the Mean-Value Theorem (4.10.2). Thus, the Extended Mean-Value Theorem is, in fact, an extension of the Mean-Value Theorem.

We will now prove one of the cases of L'Hôpital's rule.

Proof of Theorem 10.2.1 (L'Hôpital's Rule).
We will consider only the case where L is finite and lim is the two-sided limit lim, with a finite.
$\scriptstyle x \to a$

By hypothesis

$$\lim_{x \to a} \frac{f'(x)}{g'(x)} = L \tag{4}$$

As noted in the remark following the statement of Theorem 10.2.1, (4) implies that there is an interval $(a, r]$ extending to the right of a and an interval $[l, a)$ extending to the left of a on which $f'(x)$ and $g'(x)$ are defined and $g'(x) \neq 0$. For convenience, define two new functions F and G by

$$F(x) = \begin{cases} f(x), & x \neq a \\ 0, & x = a \end{cases} \qquad G(x) = \begin{cases} g(x), & x \neq a \\ 0, & x = a \end{cases}$$

Both F and G are continuous on $[l, r]$; they are continuous on the intervals $[l, a)$ and $(a, r]$ because on these intervals $F(x) = f(x)$, $G(x) = g(x)$, and f and g are continuous (since they are differentiable). The continuity of F and G at a follows from the fact that $\lim_{x \to a} f(x)/g(x)$ is an indeterminate form of type $0/0$, so

$$\lim_{x \to a} F(x) = \lim_{x \to a} f(x) = 0$$

$$\lim_{x \to a} G(x) = \lim_{x \to a} g(x) = 0$$

Thus, F and G satisfy the hypotheses of the Extended Mean-Value Theorem (10.2.2) on $[l, a]$ and $[a, r]$. Moreover, the definitions of F and G imply that

$$F'(c) = f'(c) \qquad \text{and} \qquad G'(c) = g'(c) \tag{5}$$

at any point c (different from a) in one of these intervals. If we choose a point $x \neq a$ in one of the intervals $[l, a]$ or $[a, r]$ and apply the Extended Mean-Value Theorem to $[a, x]$ (or $[x, a]$) we conclude that there is a number c between a and x such that

$$\frac{F(x) - F(a)}{G(x) - G(a)} = \frac{F'(c)}{G'(c)} \tag{6}$$

From (5) together with the fact that $F(a) = G(a) = 0$ and $F(x) = f(x)$, $G(x) = g(x)$, we can rewrite (6) as

$$\frac{f(x)}{g(x)} = \frac{f'(c)}{g'(c)}$$

Thus,

$$\lim_{x \to a} \frac{f(x)}{g(x)} = \lim_{x \to a} \frac{f'(c)}{g'(c)} \tag{7}$$

Since c is between a and x, it follows that $c \to a$ as $x \to a$. This fact together with (4) yields

$$\lim_{x \to a} \frac{f'(c)}{g'(c)} = \lim_{c \to a} \frac{f'(c)}{g'(c)} = L$$

Thus, from (7)

$$\lim_{x \to a} \frac{f(x)}{g(x)} = L \quad \blacksquare$$

▶ Exercise Set 10.2

In Exercises 1–20, find the limits.

1. $\lim\limits_{x\to 1} \dfrac{\ln x}{x-1}$.

2. $\lim\limits_{x\to 0} \dfrac{\sin 2x}{\sin 5x}$.

3. $\lim\limits_{x\to 0} \dfrac{e^x-1}{\sin x}$.

4. $\lim\limits_{x\to 3} \dfrac{x-3}{3x^2-13x+12}$.

5. $\lim\limits_{\theta\to 0} \dfrac{\tan\theta}{\theta}$.

6. $\lim\limits_{t\to 0} \dfrac{te^t}{1-e^t}$.

7. $\lim\limits_{x\to 1} \dfrac{\ln x}{\tan \pi x}$.

8. $\lim\limits_{x\to c} \dfrac{x^{1/3}-c^{1/3}}{x-c}$.

9. $\lim\limits_{x\to \pi^+} \dfrac{\sin x}{x-\pi}$.

10. $\lim\limits_{x\to 0^+} \dfrac{\sin x}{x^2}$.

11. $\lim\limits_{x\to \frac{1}{2}\pi^-} \dfrac{\cos x}{\sqrt{\frac{1}{2}\pi - x}}$.

12. $\lim\limits_{x\to 0^+} \dfrac{1-\cos x}{x^3}$.

13. $\lim\limits_{x\to 0} \dfrac{e^x+e^{-x}-2}{1-\cos 2x}$.

14. $\lim\limits_{x\to 0} \dfrac{2\cosh x-2}{1-\cos 2x}$.

15. $\lim\limits_{x\to 0} \dfrac{x-\tan x}{\sin x-x}$.

16. $\lim\limits_{x\to \pi} \dfrac{\sin^2 x}{1+\cos 3x}$.

17. $\lim\limits_{x\to +\infty} \dfrac{\frac{1}{2}\pi - \tan^{-1} x}{\ln\left(1+\dfrac{1}{x^2}\right)}$.

18. $\lim\limits_{x\to 0^+} \dfrac{\tan x}{\tan 2x}$.

19. $\lim\limits_{x\to 0^+} \dfrac{\ln(\cos x)}{\ln(\cos 3x)}$.

20. $\lim\limits_{\theta\to 0} \dfrac{\sin^2\theta - \sin(\theta^2)}{\theta^4}$.

21. (a) Find the error in the following calculation:

$$\lim\limits_{x\to 1} \frac{x^3-x^2+x-1}{x^3-x^2} = \lim\limits_{x\to 1} \frac{3x^2-2x+1}{3x^2-2x}$$
$$= \lim\limits_{x\to 1} \frac{6x-2}{6x-2} = 1$$

(b) Find the correct answer.

22. Find $\lim\limits_{x\to 1} \dfrac{x^4-4x^3+6x^2-4x+1}{x^4-3x^3+3x^2-x}$.

23. Find all values of k and l such that

$$\lim\limits_{x\to 0} \frac{k+\cos lx}{x^2} = -4$$

24. Evaluate

$$\lim\limits_{x\to 0} \frac{(2+x)\ln(1-x)}{(1-e^x)\cos x}$$

25. Evaluate

$$\lim\limits_{x\to +\infty} x\ln\left(\frac{x+1}{x-1}\right)$$

[*Hint:* Rewrite the problem as an indeterminate form of type 0/0.]

26. (a) Explain why L'Hôpital's rule does not apply to the problem

$$\lim\limits_{x\to 0} \frac{x^2\sin(1/x)}{\sin x}$$

(b) Find the limit.

10.3   OTHER INDETERMINATE FORMS
$$\left(\frac{\infty}{\infty},\ 0\cdot\infty,\ 0^0,\ \infty^0,\ 1^\infty,\ \infty-\infty\right)$$

In this section we continue our study of indeterminate forms and introduce another version of L'Hôpital's rule.

We begin by introducing some new notation. In Figure 10.3.1 we have graphed portions of four functions that become "infinite at a." For the first function, $\lim\limits_{x\to a} f(x) = +\infty$ since both one-sided limits are $+\infty$; for the second function, $\lim\limits_{x\to a} f(x) = -\infty$ since both one-sided limits are $-\infty$; while for the

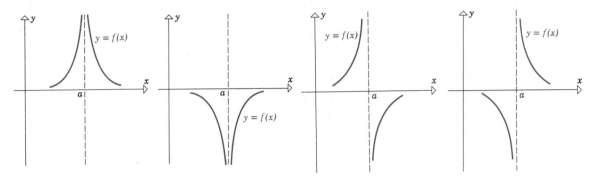

Figure 10.3.1

last two functions, the one-sided limits are infinite, but opposite in sign. We will write

$$\lim_{x\to a} f(x) = \infty$$

to indicate that one of these four possibilities occurs. Also, we will write

$$\lim_{x\to+\infty} f(x) = \infty$$

to mean either $\lim_{x\to+\infty} f(x) = +\infty$ or $\lim_{x\to+\infty} f(x) = -\infty$. Similarly, for the symbol $\lim_{x\to-\infty} f(x) = \infty$.

An **indeterminate form of type** ∞/∞ is a limit $\lim f(x)/g(x)$ in which $\lim f(x) = \infty$ and $\lim g(x) = \infty$. Some examples are

$$\lim_{x\to0^+} \frac{\ln x}{\csc x} \quad \text{(numerator} \to -\infty, \text{denominator} \to +\infty)$$

$$\lim_{x\to+\infty} \frac{x}{e^x} \quad \text{(numerator} \to +\infty, \text{denominator} \to +\infty)$$

$$\lim_{x\to0} \frac{1+\dfrac{1}{x}}{\cot x} \quad \text{(numerator} \to \infty, \text{denominator} \to \infty)$$

The following version of L'Hôpital's rule, usually proved in advanced courses, is applicable to problems like these.

10.3.1 THEOREM

L'Hôpital's Rule for Form $\frac{\infty}{\infty}$

Let $\lim$ *stand for one of the limits* $\lim_{x\to a}, \lim_{x\to a^+}, \lim_{x\to a^-}, \lim_{x\to+\infty}$, *or* $\lim_{x\to-\infty}$, *and suppose that* $\lim f(x) = \infty$ *and* $\lim g(x) = \infty$. *If* $\lim [f'(x)/g'(x)]$ *has a finite value L, or if this limit is* $+\infty$ *or* $-\infty$, *then*

$$\lim \frac{f(x)}{g(x)} = \lim \frac{f'(x)}{g'(x)}$$

REMARK. The remark about hypotheses, which follows Theorem 10.2.1, is also applicable here.

▶ Example 1   Evaluate

$$\lim_{x \to +\infty} \frac{x}{e^x}$$

Solution.

$$\lim_{x \to +\infty} x = \lim_{x \to +\infty} e^x = +\infty$$

so that the given limit is an indeterminate form of type ∞/∞. Thus, by L'Hôpital's rule

$$\lim_{x \to +\infty} \frac{x}{e^x} = \lim_{x \to +\infty} \frac{\frac{d}{dx}[x]}{\frac{d}{dx}[e^x]} = \lim_{x \to +\infty} \frac{1}{e^x} = 0 \qquad \blacktriangleleft$$

▶ Example 2   Evaluate

$$\lim_{x \to 0^+} \frac{\ln x}{\csc x}$$

Solution.

$$\lim_{x \to 0^+} \ln x = -\infty \qquad \text{and} \qquad \lim_{x \to 0^+} \csc x = +\infty$$

so that the given limit is an indeterminate form of type ∞/∞. Thus, by L'Hôpital's rule

$$\lim_{x \to 0^+} \frac{\ln x}{\csc x} = \lim_{x \to 0^+} \frac{1/x}{-\csc x \cot x} \tag{1}$$

This last limit is again an indeterminate form of type ∞/∞. Moreover, any additional applications of L'Hôpital's rule will yield powers of $1/x$ in the numerator and expressions involving $\csc x$ and $\cot x$ in the denominator; thus, repeated application of L'Hôpital's rule simply produces new indeterminate forms. We must try something else. The last limit in (1) can be rewritten

$$\lim_{x\to 0^+}\left(-\frac{\sin x}{x}\tan x\right) = -\lim_{x\to 0^+}\frac{\sin x}{x}\cdot\lim_{x\to 0^+}\tan x = -(1)(0) = 0$$

Thus,

$$\lim_{x\to 0^+}\frac{\ln x}{\csc x} = 0 \qquad\blacktriangleleft$$

A limit, $\lim f(x)g(x)$, is called an **indeterminate form of type** $\mathbf{0\cdot\infty}$ if $\lim f(x) = 0$ and $\lim g(x) = \infty$. Frequently, limit problems of this type can be converted to the form $0/0$ by writing

$$f(x)g(x) = \frac{f(x)}{1/g(x)}$$

or to the form ∞/∞ by writing

$$f(x)g(x) = \frac{g(x)}{1/f(x)}$$

They can then be treated by L'Hôpital's rule.

▶ Example 3 Evaluate

$$\lim_{x\to 0^+} x\ln x$$

Solution. Since

$$\lim_{x\to 0^+} x = 0 \qquad\text{and}\qquad \lim_{x\to 0^+}\ln x = -\infty$$

the given problem is an indeterminate form of type $0\cdot\infty$. We will convert the problem to the form ∞/∞ and apply L'Hôpital's rule:

$$\lim_{x\to 0^+} x\ln x = \lim_{x\to 0^+}\frac{\ln x}{1/x} = \lim_{x\to 0^+}\frac{1/x}{-1/x^2} = \lim_{x\to 0^+}(-x) = 0 \qquad\blacktriangleleft$$

In this example we could have converted the problem to form $0/0$ by writing

$$x\ln x = \frac{x}{1/\ln x}$$

However, this is a less desirable first choice because of the relatively compli-cated derivative of $1/\ln x$.

WARNING. It is tempting to argue that an indeterminate form of type $0 \cdot \infty$ has value 0 since "zero times anything is zero." However, this is fallacious since $0 \cdot \infty$ is not a product of numbers, but rather a statement about limits. The following example gives an indeterminate form of type $0 \cdot \infty$ whose value is not zero.

▶ Example 4   Evaluate

$$\lim_{x \to \pi/4} (1 - \tan x) \sec 2x$$

Solution. The given problem is an indeterminate form of type $0 \cdot \infty$. We convert it to type $0/0$ and apply L'Hôpital's rule as follows:

$$\lim_{x \to \pi/4} (1 - \tan x) \sec 2x = \lim_{x \to \pi/4} \frac{1 - \tan x}{1/\sec 2x}$$

$$= \lim_{x \to \pi/4} \frac{1 - \tan x}{\cos 2x}$$

$$= \lim_{x \to \pi/4} \frac{-\sec^2 x}{-2 \sin 2x}$$

$$= \frac{-2}{-2} = 1 \qquad \qquad \blacktriangleleft$$

Limits of the form

$$\lim_{x \to a} f(x)^{g(x)}$$

give rise to *indeterminate forms of the types* 0^0, ∞^0, and 1^∞. (At this stage the meaning of these symbols should be clear.) All three types are treated by first introducing a dependent variable.

$$y = f(x)^{g(x)}$$

and then calculating

$$\lim_{x \to a} \ln y = \lim_{x \to a} \left(\ln f(x)^{g(x)} \right) = \lim_{x \to a} [g(x) \ln f(x)]$$

Once $\lim_{x \to a} \ln y$ is known it is a simple matter to determine $\lim_{x \to a} y = \lim_{x \to a} f(x)^{g(x)}$ as our examples will show.

▶ Example 5   Evaluate

$$\lim_{x \to 0} (1 + x)^{1/x}$$

Solution. Since

$$\lim_{x\to 0}(1 + x) = 1 \qquad \text{and} \qquad \lim_{x\to 0}\frac{1}{x} = \infty$$

the given limit is an indeterminate form of type 1^∞. As discussed above, we introduce a dependent variable

$$y = (1 + x)^{1/x}$$

and take the natural logarithm of both sides

$$\ln y = \ln (1 + x)^{1/x} = \frac{1}{x}\ln (1 + x) = \frac{\ln (1 + x)}{x}$$

The limit

$$\lim_{x\to 0} \ln y = \lim_{x\to 0} \frac{\ln (1 + x)}{x}$$

is an indeterminate form of type $0/0$, so by L'Hôpital's rule,

$$\lim_{x\to 0} \ln y = \lim_{x\to 0} \frac{\ln (1 + x)}{x} = \lim_{x\to 0} \frac{1/(1 + x)}{1} = 1 \qquad (2)$$

To complete the calculation, we use the known limit, $\lim\limits_{x\to 0} \ln y = 1$, to find the unknown limit, $\lim\limits_{x\to 0} y = \lim\limits_{x\to 0} (1 + x)^{1/x}$. We argue as follows: As $x \to 0$, we have

$$\ln y \to 1 \qquad \text{[from (2)]}$$

so

$$e^{\ln y} \to e^1 \qquad \text{[since } e^x \text{ is a continuous function]}$$

and thus,

$$y \to e$$

Therefore,

$$\lim_{x\to 0}(1 + x)^{1/x} = e \qquad (3)$$

◀

In Section 7.2 [Formula (2)], we stated that e can be expressed as

$$e = \lim_{x \to +\infty} \left(1 + \frac{1}{x}\right)^x \tag{4}$$

This can be derived by letting $x = 1/h$ in (3) to obtain

$$\lim_{h \to +\infty} \left(1 + \frac{1}{h}\right)^h = e$$

which is the same as (4) except for the notation used for the variable.

If a limit problem such as

$$\lim [f(x) - g(x)] \qquad \text{or} \qquad \lim [f(x) + g(x)]$$

leads to one of the expressions

$$(+\infty) - (+\infty), \quad (-\infty) - (-\infty)$$
$$(+\infty) + (-\infty), \quad (-\infty) + (+\infty)$$

then one term tends to make the expression large and the other tends to make it small, resulting in an indeterminate form. These are called **indeterminate forms of type** $\infty - \infty$ Such forms can sometimes be treated by combining the two terms into one and manipulating the result into one of the previous forms.

▶ **Example 6** Evaluate

$$\lim_{x \to 0^+} \left(\frac{1}{x} - \frac{1}{\sin x}\right)$$

Solution. Since

$$\lim_{x \to 0^+} \frac{1}{x} = +\infty \qquad \text{and} \qquad \lim_{x \to 0^+} \frac{1}{\sin x} = +\infty$$

the given limit is an indeterminate form of type $\infty - \infty$. Combining terms yields

$$\lim_{x \to 0^+} \left(\frac{1}{x} - \frac{1}{\sin x}\right) = \lim_{x \to 0^+} \left(\frac{\sin x - x}{x \sin x}\right)$$

which is an indeterminate form of type $0/0$. Applying L'Hôpital's rule twice yields

$$\lim_{x \to 0^+} \left(\frac{\sin x - x}{x \sin x}\right) = \lim_{x \to 0^+} \frac{\cos x - 1}{\sin x + x \cos x}$$

$$= \lim_{x \to 0^+} \frac{-\sin x}{\cos x + \cos x - x \sin x}$$

$$= \frac{0}{2} = 0 \qquad \blacktriangleleft$$

▶ Example 7 Evaluate

$$\lim_{x \to 0^+} (\cot x - \ln x)$$

Solution. Since

$$\lim_{x \to 0^+} \cot x = +\infty \qquad \text{and} \qquad \lim_{x \to 0^+} \ln x = -\infty$$

we are dealing with a problem of the type $(+\infty) - (-\infty)$. This is not an indeterminate form. The first term tends to make the limit large and because of the subtraction, the second term also tends to make the limit large. Thus,

$$\lim_{x \to 0^+} (\cot x - \ln x) = +\infty \qquad \blacktriangleleft$$

▶ Exercise Set 10.3

In Exercises 1–24, find the limits.

1. $\lim\limits_{x \to +\infty} \dfrac{\ln x}{x}$.

2. $\lim\limits_{x \to +\infty} \dfrac{e^{3x}}{x^2}$.

3. $\lim\limits_{x \to 0^+} \dfrac{\cot x}{\ln x}$.

4. $\lim\limits_{x \to 0^+} \dfrac{1 - \ln x}{e^{1/x}}$.

5. $\lim\limits_{x \to +\infty} \dfrac{x \ln x}{x + \ln x}$.

6. $\lim\limits_{x \to +\infty} \dfrac{x^3 - 2x + 1}{4x^3 + 2}$.

7. $\lim\limits_{x \to +\infty} \dfrac{x^{100}}{e^x}$.

8. $\lim\limits_{x \to 0^+} \dfrac{\ln (\sin x)}{\ln (\tan x)}$.

9. $\lim\limits_{x \to +\infty} xe^{-x}$.

10. $\lim\limits_{x \to \pi^-} (x - \pi) \tan \frac{1}{2}x$.

11. $\lim\limits_{x \to +\infty} x \sin \dfrac{\pi}{x}$.

12. $\lim\limits_{x \to 0^+} \tan x \ln x$.

13. $\lim\limits_{x \to 0} (e^x + x)^{1/x}$.

14. $\lim\limits_{x \to 0^+} x^x$.

15. $\lim\limits_{x \to 0^+} x^{\sin x}$.

16. $\lim\limits_{x \to +\infty} (\ln x)^{1/x}$.

17. $\lim\limits_{x \to \frac{1}{2}\pi^-} (\tan x)^{\cos x}$.

18. $\lim\limits_{x \to 0^+} x^{1/\ln x}$.

19. $\lim\limits_{\theta \to 0} \left(\dfrac{1}{1 - \cos \theta} - \dfrac{2}{\sin^2 \theta} \right)$.

20. $\lim\limits_{x \to 0} \left(\dfrac{1}{x} - \dfrac{1}{e^x - 1} \right)$.

21. $\lim\limits_{x \to 0} (\cot x - \csc x)$.

22. $\lim\limits_{x \to +\infty} [\ln x - \ln (1 + x)]$.

23. $\lim\limits_{x \to 0^+} \dfrac{\cot x}{\cot 2x}$.

24. $\lim\limits_{x \to \frac{1}{2}\pi^-} \dfrac{4 \tan x}{1 + \sec x}$.

25. Show that for any positive integer n:

(a) $\lim\limits_{x \to +\infty} \dfrac{x^n}{e^x} = 0$ (b) $\lim\limits_{x \to +\infty} \dfrac{e^x}{x^n} = +\infty$.

26. Show that for any positive integer n:

(a) $\lim\limits_{x \to +\infty} \dfrac{\ln x}{x^n} = 0$ (b) $\lim\limits_{x \to +\infty} \dfrac{x^n}{\ln x} = +\infty$.

27. Limits of the type

$$\frac{0}{\infty}, \quad \frac{\infty}{0}, \quad 0^\infty, \quad \infty \cdot \infty, \quad +\infty + (+\infty),$$

$$+\infty - (-\infty), \quad -\infty + (-\infty), \quad -\infty - (+\infty)$$

are *not* indeterminate forms. Find the following limits by inspection.

(a) $\lim\limits_{x \to 0^+} \dfrac{x}{\ln x}$ (b) $\lim\limits_{x \to +\infty} \dfrac{x^3}{e^{-x}}$

(c) $\lim\limits_{x \to \frac{1}{2}\pi^-} (\cos x)^{\tan x}$

(d) $\lim\limits_{x \to 0^+} (\ln x) \cot x$

(e) $\lim\limits_{x \to \frac{1}{2}\pi^-} \left(\dfrac{1}{\frac{1}{2}\pi - x} + \tan x \right)$

(f) $\lim\limits_{x \to 0^+} \left(\dfrac{1}{x} - \ln x \right)$

(g) $\lim\limits_{x \to -\infty} (x + x^3)$ (h) $\lim\limits_{x \to +\infty} \left(\ln \left(\dfrac{1}{x} \right) - e^x \right)$.

28. Find $\lim\limits_{x \to +\infty} (e^x - x^2)$.

29. Sketch the graph of x^x, $x > 0$.

30. Sketch the graph of $(1/x) \tan x$, $-\pi/2 < x < \pi/2$.

31. Evaluate $\int_0^1 \ln x \, dx$ or else show that the integral diverges.

32. (a)   Show that $\int_1^{+\infty} e^{t^2} \, dt = +\infty$
     [*Hint:* Don't try to carry out the integration. Instead, compare the sizes of $\int_1^l e^t \, dt$ and $\int_1^l e^{t^2} \, dt$.]

   (b)   Use L'Hôpital's rule to evaluate

$$\lim_{x \to +\infty} \frac{1}{x} \int_1^x e^{t^2} \, dt.$$

33. There is a myth that circulates among beginning calculus students which states that all indeterminate forms of types 0^0, ∞^0, and 1^∞ have value 1 because "anything to the zero power is 1" and "1 to any power is 1." The fallacy is that 0^0, ∞^0, and 1^∞ are not powers of numbers, but rather descriptions of limits. The following examples, which were transmitted to me by the late Professor Jack Staib of Drexel University, show that such indeterminate forms can have any positive real value:

(a) $\lim\limits_{x \to 0^+} [x^{(\ln a)/(1+\ln x)}] = 0^0 = a$

(b) $\lim\limits_{x \to +\infty} [x^{(\ln a)/(1+\ln x)}] = \infty^0 = a$

(c) $\lim\limits_{x \to 0} [(x + 1)^{(\ln a)/x}] = 1^\infty = a$.

Prove these results.

► SUPPLEMENTARY EXERCISES

In Exercises 1–14, evaluate the integrals that converge.

1. $\displaystyle\int_{-\infty}^{+\infty} \frac{dx}{x^2 + 4}$.

2. $\displaystyle\int_4^6 \frac{dx}{4 - x}$.

3. $\displaystyle\int_0^1 \frac{x \, dx}{\sqrt{1 - x^2}}$.

4. $\displaystyle\int_0^5 \frac{dx}{\sqrt{25 - x^2}}$.

5. $\displaystyle\int_{-1}^1 \frac{dx}{\sqrt[3]{x^2}}$.

6. $\displaystyle\int_0^{\pi/2} \sec^2 x \, dx$.

7. $\displaystyle\int_0^{+\infty} xe^{-x^2} \, dx$.

8. $\displaystyle\int_{-\infty}^0 xe^x \, dx$.

9. $\displaystyle\int_0^{\pi/2} \tan x \, dx$.

10. $\displaystyle\int_0^{+\infty} \frac{dx}{x^5}$.

11. $\displaystyle\int_e^{+\infty} \frac{dx}{x(\ln x)^2}$.

12. $\displaystyle\int_0^1 \sqrt{x} \ln x \, dx$.

13. $\displaystyle\int_0^{+\infty} \frac{dx}{x^2 + 2x + 2}$.

14. $\displaystyle\int_0^4 \frac{e^{-\sqrt{x}}}{\sqrt{x}} \, dx$.

15. For what values of n does $\displaystyle\int_0^1 x^n \ln x \, dx$ converge? What is its value?

In Exercises 16–31, find the limit.

16. $\lim\limits_{x \to 1} \dfrac{\ln x}{x - 1}$.

17. $\lim\limits_{x \to 0} \dfrac{xe^{3x} - x}{1 - \cos 2x}$.

18. $\lim\limits_{x \to +\infty} \dfrac{\ln (\ln x)}{\sqrt{x}}$.

19. $\lim\limits_{x \to 0^+} x^2 e^{1/x}$.

20. $\lim\limits_{x \to +\infty} (\sqrt{x^2 + x} - x)$.

21. $\lim\limits_{x \to 0^-} x^2 e^{1/x}$.

22. $\lim\limits_{x \to 0^-} (1 - x)^{2/x}$.

23. $\lim\limits_{\theta \to 0} (\csc \theta - \cot \theta)$.

24. $\lim\limits_{x \to 0} \dfrac{x - \tan^{-1} x}{x^4}$.

25. $\lim\limits_{x \to 2} \dfrac{x - 1 - e^{x-2}}{1 - \cos 2\pi x}$.

26. $\lim\limits_{x \to 0} \dfrac{9^x - 3^x}{x}$.

27. $\lim\limits_{x \to 0} \dfrac{\displaystyle\int_0^x \sin t^2 \, dt}{\sin x^2}$.

28. $\lim\limits_{x \to +\infty} x^{1/x}$.

29. $\lim\limits_{x \to +\infty} \dfrac{(\ln x)^3}{x}$.

30. $\lim\limits_{x \to +\infty} \left(\dfrac{x}{x - 3} \right)^x$.

31. $\lim\limits_{x \to 0^+} (1 + x)^{\ln x}$.

32. Find the area of the region in the first quadrant between the curves $y = e^{-x}$ and $y = (e^{-x})^2$.

In Exercises 33 and 34, find:
(a) the area of the region R;
(b) the volume obtained by revolving R about the x-axis.

33. The region R bounded between the x-axis and the curve $y = x^{-2/3}$, $x \geq 8$.

34. The region R bounded between the x-axis and the curve $y = x^{-1/3}$, $0 \leq x \leq 1$.

35. Find the total arc length of the graph of
$f(x) = \sqrt{x - x^2} - \sin^{-1} \sqrt{x}$. [*Hint:* First find the domain of f.]

11 infinite series

11.1 SEQUENCES

In everyday language, we use the term "sequence" to suggest a succession of objects or events given in a specified order. *Informally* speaking, the term "sequence" in mathematics is used to describe an unending succession of numbers. Some possibilities are

$$1, 2, 3, 4, \ldots$$
$$2, 4, 6, 8, \ldots$$
$$1, \tfrac{1}{2}, \tfrac{1}{3}, \tfrac{1}{4}, \ldots$$
$$1, -1, \ 1, \ -1, \ldots$$

In each case, the dots are used to suggest that the sequence continues indefinitely, following the obvious pattern. The numbers in a sequence are called the **terms** of the sequence. The terms may be described according to the positions they occupy. Thus, a sequence has a *first term,* a *second term,* a *third term,* and so forth. Because a sequence continues indefinitely, there is no last term.

   The most common way to specify a sequence is to give a formula for the terms. To illustrate the idea, we have listed in the table below the terms in the sequence $2, 4, 6, 8, \ldots$ together with their term numbers:

term number	1	2	3	4	...
term	2	4	6	8	...

There is a clear relationship here; each term is twice its term number. Thus, for each positive integer n the nth term in the sequence $2, 4, 6, 8, \ldots$ is given by the formula $2n$. This is denoted by writing

$$2, 4, 6, 8, \ldots, 2n, \ldots$$

or more compactly in **bracket notation** as

$$\{2n\}_{n=1}^{+\infty}$$

From the bracket notation, the terms in the sequence can be generated by successively substituting the integer values $n = 1, 2, 3, \ldots$ into the formula $2n$.

▶ **Example 1** List the first five terms of the sequence $\{2^n\}_{n=1}^{+\infty}$.

Solution. Substituting $n = 1, 2, 3, 4, 5$ into the formula 2^n yields

$$2^1, 2^2, 2^3, 2^4, 2^5, \ldots$$

or, equivalently,

$$2, 4, 8, 16, 32, \ldots \qquad ◀$$

▶ **Example 2** Express the following sequences in bracket notation.

(a) $\dfrac{1}{2}, \dfrac{2}{3}, \dfrac{3}{4}, \dfrac{4}{5}, \ldots$

(b) $\dfrac{1}{2}, \dfrac{1}{4}, \dfrac{1}{8}, \dfrac{1}{16}, \ldots$

(c) $1, -1, 1, -1, \ldots$

(d) $\dfrac{1}{2}, -\dfrac{2}{3}, \dfrac{3}{4}, -\dfrac{4}{5}, \ldots$

(e) $1, 3, 5, 7, \ldots$

Solution.
(a) Begin by comparing terms and term numbers:

term number	1	2	3	4	...
term	$\dfrac{1}{2}$	$\dfrac{2}{3}$	$\dfrac{3}{4}$	$\dfrac{4}{5}$	...

In each term, the numerator is the same as the term number, and the denominator is one greater than the term number. Thus, the nth term is $n/(n + 1)$ and the sequence may be written

$$\left\{ \frac{n}{n + 1} \right\}_{n=1}^{+\infty}$$

(b) Observe that the sequence can be rewritten as

$$\frac{1}{2}, \frac{1}{2^2}, \frac{1}{2^3}, \frac{1}{2^4}, \dots$$

and construct a table comparing terms and term numbers:

term number	1	2	3	4	...
term	$\frac{1}{2}$	$\frac{1}{2^2}$	$\frac{1}{2^3}$	$\frac{1}{2^4}$	...

From the table we see that the nth term is $1/2^n$, so the sequence can be written as

$$\left\{\frac{1}{2^n}\right\}_{n=1}^{+\infty}$$

(c) Observe first that $(-1)^r$ is either 1 or -1 according to whether r is an even integer or an odd integer. In the sequence $1, -1, 1, -1, \dots$ the odd-numbered terms are 1's and the even-numbered terms are -1's. Thus, a formula for the nth term can be obtained by raising -1 to a power that will be even when n is odd and odd when n is even. This is accomplished by the formula $(-1)^{n+1}$, so that the sequence can be written as

$$\{(-1)^{n+1}\}_{n=1}^{+\infty}$$

(d) Combining the results in parts (a) and (c), we may write this sequence as

$$\left\{(-1)^{n+1}\frac{n}{n+1}\right\}_{n=1}^{+\infty}$$

(e) Begin by comparing terms and term numbers:

term number	1	2	3	4	...
term	1	3	5	7	...

From the table we see that each term is one less than twice the term number. Thus, the nth term is $2n - 1$, and the sequence can be written as

$$\{2n - 1\}_{n=1}^{+\infty} \qquad \blacktriangleleft$$

Frequently we shall want to write down a sequence without specifying the numerical values of the terms. We do this by writing

$$a_1, a_2, \dots, a_n, \dots$$

or in bracket notation

$$\{a_n\}_{n=1}^{+\infty}$$

or sometimes simply

$$\{a_n\}$$

(There is nothing special about the letter a; any other letter may be used.)

In the beginning of this section we stated informally that a sequence is an unending succession of numbers. However, this is not a satisfactory mathematical definition since the word "succession" is itself an undefined term. The time has come to formally define the term "sequence." When we write a sequence such as

$$2, 4, 6, 8, \ldots, 2n, \ldots$$

in bracket notation

$$\{2n\}_{n=1}^{+\infty} \tag{1}$$

we are specifying a rule that tells us how to associate a numerical value, namely $2n$, with each positive integer n. Stated another way, $\{2n\}_{n=1}^{+\infty}$ may be regarded as a formula for a function whose independent variable n ranges over the positive integers. Indeed, we could rewrite (1) in functional notation as

$$f(n) = 2n, \qquad n = 1, 2, 3, \ldots$$

From this point of view, the notation

$$2, 4, 6, 8, \ldots, 2n, \ldots$$

represents a listing of the function values

$$f(1), f(2), f(3), \ldots, f(n), \ldots$$

This suggests the following definition.

11.1.1 DEFINITION A *sequence* (or *infinite sequence*) is a function whose domain is the set of positive integers.

Because sequences are functions, we may inquire about the graph of a sequence. For example, the graph of the sequence

$$\left\{\frac{1}{n}\right\}_{n=1}^{+\infty}$$

is the graph of the equation

$$y = \frac{1}{n}, \qquad n = 1, 2, 3, \ldots$$

Because the right side of this equation is defined only for positive integer values of n, the graph consists of a succession of isolated points (Figure 11.1.1a). This is in marked distinction to the graph of

$$y = \frac{1}{x}, \qquad x \geq 1$$

which is a continuous curve (Figure 11.1.1b).

Figure 11.1.1

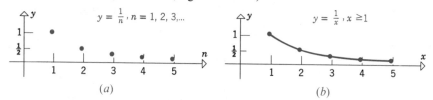

(a) (b)

In Figure 11.1.2 we have sketched the graphs of four sequences:

$$\{n + 1\}_{n=1}^{+\infty}, \qquad \{(-1)^{n+1}\}_{n=1}^{+\infty}, \qquad \left\{\frac{n}{n + 1}\right\}_{n=1}^{+\infty}, \qquad \left\{1 + \left(-\frac{1}{2}\right)^{n}\right\}_{n=1}^{+\infty}$$

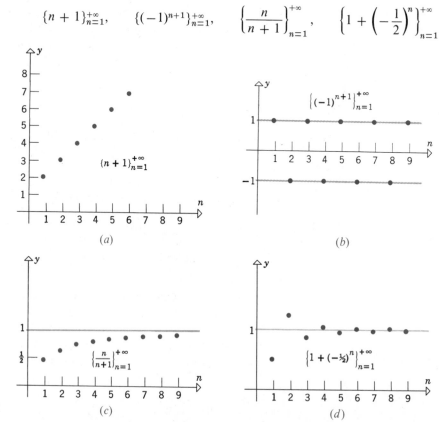

Figure 11.1.2 (c) (d)

Each of these sequences behaves differently as n gets larger and larger. In the sequence $\{n + 1\}$, the terms grow without bound; in the sequence $\{(-1)^{n+1}\}$ the terms oscillate between 1 and -1; in the sequence $\left\{\dfrac{n}{n + 1}\right\}$ the terms increase toward a "limit" of 1; and finally, in the sequence $\{1 + (-\frac{1}{2})^n\}$ the terms also tend toward a "limit" of 1, but do so in an oscillatory fashion.

Let us try to make the term "limit" more precise. To say that a sequence $\{a_n\}_{n=1}^{+\infty}$ approaches a limit L as n gets large is intended to mean that eventually the terms in the sequence become arbitrarily close to the number L. Thus, if we choose *any* positive number ϵ, the terms in the sequence will eventually be within ϵ units of L. Geometrically, this means that if we sketch the lines $y = L + \epsilon$ and $y = L - \epsilon$, the terms in the sequence will eventually be trapped within the band between these lines, and thus be within ϵ units of L (Figure 11.1.3).

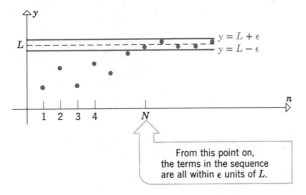

Figure 11.1.3

From this point on, the terms in the sequence are all within ϵ units of L.

The following definition expresses this idea precisely.

11.1.2 DEFINITION A sequence $\{a_n\}_{n=1}^{+\infty}$ is said to have the ***limit*** L if given any $\epsilon > 0$ there is a positive integer N such that $|a_n - L| < \epsilon$ when $n \geq N$.

If a sequence $\{a_n\}_{n=1}^{+\infty}$ has a limit L, we say that the sequence ***converges*** to L and write

$$\lim_{n \to +\infty} a_n = L$$

A sequence that does not have a finite limit is said to ***diverge***.

▶ **Example 3** Figure 11.1.2 suggests that the sequences $\{n + 1\}_{n=1}^{+\infty}$ and $\{(-1)^{n+1}\}_{n=1}^{+\infty}$ diverge, while the sequences $\left\{\dfrac{n}{n + 1}\right\}_{n=1}^{+\infty}$ and $\left\{1 + \left(-\dfrac{1}{2}\right)^n\right\}_{n=1}^{+\infty}$ converge to 1; that is,

$$\lim_{n \to +\infty} \frac{n}{n+1} = 1 \quad \text{and} \quad \lim_{n \to +\infty} \left(1 + \left(-\frac{1}{2}\right)^n\right) = 1 \qquad \blacktriangleleft$$

Many familiar properties of limits apply to limits of sequences.

11.1.3   THEOREM   *Suppose that the sequences $\{a_n\}$ and $\{b_n\}$ converge to limits L_1 and L_2, respectively, and c is a constant. Then*

(a) $\lim\limits_{n \to +\infty} c = c$

(b) $\lim\limits_{n \to +\infty} ca_n = c \lim\limits_{n \to +\infty} a_n = cL_1$

(c) $\lim\limits_{n \to +\infty} (a_n + b_n) = \lim\limits_{n \to +\infty} a_n + \lim\limits_{n \to +\infty} b_n = L_1 + L_2$

(d) $\lim\limits_{n \to +\infty} (a_n - b_n) = \lim\limits_{n \to +\infty} a_n - \lim\limits_{n \to +\infty} b_n = L_1 - L_2$

(e) $\lim\limits_{n \to +\infty} (a_n b_n) = \lim\limits_{n \to +\infty} a_n \cdot \lim\limits_{n \to +\infty} b_n = L_1 L_2$

(f) $\lim\limits_{n \to +\infty} \left(\dfrac{a_n}{b_n}\right) = \dfrac{\lim\limits_{n \to +\infty} a_n}{\lim\limits_{n \to +\infty} b_n} = \dfrac{L_1}{L_2} \quad (\text{if } L_2 \neq 0)$

(We omit the proof.)

▶ **Example 4**   In each part, determine whether the given sequence converges or diverges. If it converges, find the limit.

(a) $\left\{\dfrac{n}{2n+1}\right\}_{n=1}^{+\infty}$ (b) $\left\{(-1)^{n+1} \dfrac{n}{2n+1}\right\}_{n=1}^{+\infty}$

(c) $\left\{(-1)^{n+1} \dfrac{1}{n}\right\}_{n=1}^{+\infty}$ (d) $\{8 - 2n\}_{n=1}^{+\infty}$ (e) $\left\{\dfrac{n}{e^n}\right\}_{n=1}^{+\infty}$

Solution.
(a)   Dividing numerator and denominator by n yields

$$\lim_{n \to +\infty} \frac{n}{2n+1} = \lim_{n \to +\infty} \frac{1}{\left(2 + \dfrac{1}{n}\right)} = \frac{\lim\limits_{n \to +\infty} 1}{\lim\limits_{n \to +\infty} \left(2 + \dfrac{1}{n}\right)} = \frac{\lim\limits_{n \to +\infty} 1}{\lim\limits_{n \to +\infty} 2 + \lim\limits_{n \to +\infty} \dfrac{1}{n}}$$

$$= \frac{1}{2 + 0} = \frac{1}{2}$$

Thus, $\left\{\dfrac{n}{2n+1}\right\}_{n=1}^{+\infty}$ converges to $\dfrac{1}{2}$.

(b)   From part (a),

$$\lim_{n \to +\infty} \frac{n}{2n+1} = \frac{1}{2}$$

Thus, since $(-1)^{n+1}$ oscillates between $+1$ and -1, the product $(-1)^{n+1} \dfrac{n}{2n+1}$ oscillates between positive and negative values, with the odd-numbered terms approaching $\frac{1}{2}$ and the even-numbered terms approaching $-\frac{1}{2}$. Therefore, the sequence $\left\{(-1)^{n+1} \dfrac{n}{2n+1}\right\}_{n=1}^{+\infty}$ approaches no limit—it diverges.

(c) Since $\lim\limits_{n \to +\infty} 1/n = 0$, the product $(-1)^{n+1}(1/n)$ oscillates between positive and negative values, with the odd-numbered terms approaching 0 through positive values and the even-numbered terms approaching 0 through negative values. Thus,

$$\lim_{n \to +\infty} (-1)^{n+1} \frac{1}{n} = 0$$

so that the sequence converges to 0.

(d) $\lim\limits_{n \to +\infty} (8 - 2n) = -\infty$ so the sequence $\{8 - 2n\}_{n=1}^{+\infty}$ diverges.

(e) We want to find $\lim\limits_{n \to +\infty} n/e^n$, which is an indeterminate form of type ∞/∞.

Unfortunately, we cannot apply L'Hôpital's rule directly since e^n and n are not differentiable functions (n assumes only integer values). However, we can apply L'Hôpital's rule to the related problem $\lim\limits_{x \to +\infty} x/e^x$ to obtain

$$\lim_{x \to +\infty} \frac{x}{e^x} = \lim_{x \to +\infty} \frac{1}{e^x} = 0$$

We conclude from this that $\lim\limits_{n \to +\infty} n/e^n = 0$ since the values of n/e^n and x/e^x are the same when x is a positive integer. ◀

▶ Example 5 Show

$$\lim_{n \to +\infty} \sqrt[n]{n} = 1$$

Solution. With the aid of L'Hôpital's rule, $\lim\limits_{n \to +\infty} \dfrac{1}{n} \ln n = 0$. Thus,

$$\lim_{n \to +\infty} \sqrt[n]{n} = \lim_{n \to +\infty} n^{1/n} = \lim_{n \to +\infty} e^{(1/n) \ln n} = e^0 = 1 \qquad ◀$$

▶ Exercise Set 11.1

In Exercises 1–14, show the first five terms of the sequence, determine whether the sequence converges, and if so find the limit. (When writing out the terms of the sequence, you need not find numerical values; leave the terms in the first form you obtain.)

1. $\left\{\dfrac{n}{n+2}\right\}_{n=1}^{+\infty}$.

2. $\left\{\dfrac{n^2}{2n+1}\right\}_{n=1}^{+\infty}$.

3. $\{2\}_{n=1}^{+\infty}$.

4. $\left\{\ln\left(\dfrac{1}{n}\right)\right\}_{n=1}^{+\infty}$.

5. $\left\{\dfrac{\ln n}{n}\right\}_{n=1}^{+\infty}$.

6. $\left\{n\sin\dfrac{\pi}{n}\right\}_{n=1}^{+\infty}$.

7. $\{1+(-1)^n\}_{n=1}^{+\infty}$.

8. $\left\{\dfrac{(-1)^{n+1}}{n^2}\right\}_{n=1}^{+\infty}$.

9. $\left\{(-1)^n\dfrac{2n^3}{n^3+1}\right\}_{n=1}^{+\infty}$.

10. $\left\{\dfrac{n}{2^n}\right\}_{n=1}^{+\infty}$.

11. $\left\{\dfrac{(n+1)(n+2)}{2n^2}\right\}_{n=1}^{+\infty}$.

12. $\left\{\dfrac{\pi^n}{4^n}\right\}_{n=1}^{+\infty}$.

13. $\{n^{1/n}\}_{n=1}^{+\infty}$.

14. $\left\{\left(1-\dfrac{2}{n}\right)^n\right\}_{n=1}^{+\infty}$.

In Exercises 15–22, express the sequence in the notation $\{a_n\}_{n=1}^{+\infty}$, determine whether the sequence converges, and if so find its limit.

15. $\dfrac{1}{2},\dfrac{3}{4},\dfrac{5}{6},\dfrac{7}{8},\ldots$

16. $0,\dfrac{1}{2^2},\dfrac{2}{3^2},\dfrac{3}{4^2},\ldots$

17. $\dfrac{1}{3},\dfrac{1}{9},\dfrac{1}{27},\dfrac{1}{81},\ldots$

18. $-1,2,-3,4,-5,\ldots$

19. $\left(1-\dfrac{1}{2}\right),\left(\dfrac{1}{2}-\dfrac{1}{3}\right),\left(\dfrac{1}{3}-\dfrac{1}{4}\right),\left(\dfrac{1}{4}-\dfrac{1}{5}\right),\ldots$

20. $3,\dfrac{3}{2},\dfrac{3}{2^2},\dfrac{3}{2^3},\ldots$

21. $(\sqrt{2}-\sqrt{3}),(\sqrt{3}-\sqrt{4}),(\sqrt{4}-\sqrt{5}),\ldots$

22. $\dfrac{1}{3^5},-\dfrac{1}{3^6},\dfrac{1}{3^7},-\dfrac{1}{3^8},\ldots$

23. (a) Let $\{a_n\}$ be a sequence for which $a_1 = 3$ and $a_n = 2a_{n-1}$ when $n \geq 2$. Find the first eight terms.

(b) Let $\{a_n\}$ be a sequence for which $a_1 = 1$, $a_2 = 1$, and $a_n = a_{n-1} + a_{n-2}$ when $n \geq 3$. Find the first eight terms.

24. The nth term a_n of the sequence $1, 2, 1, 4, 1, 6, \ldots$ is best written in the form

$$a_n = \begin{cases} 1, & \text{if } n \text{ is odd} \\ n, & \text{if } n \text{ is even} \end{cases}$$

since it would be tedious to find one formula applicable to all terms. By considering even and odd terms separately, find a formula for the nth term of the given sequence.

(a) $1,\dfrac{1}{2^2},3,\dfrac{1}{2^4},5,\dfrac{1}{2^6},\ldots$

(b) $1,\dfrac{1}{3},\dfrac{1}{3},\dfrac{1}{5},\dfrac{1}{5},\dfrac{1}{7},\dfrac{1}{7},\dfrac{1}{9},\dfrac{1}{9},\ldots$

25. Consider the sequence $\{a_n\}_{n=1}^{+\infty}$ where

$$a_n = \dfrac{1}{n^2} + \dfrac{2}{n^2} + \cdots + \dfrac{n}{n^2}$$

(a) Write out the first four terms of the sequence.

(b) Find the limit of the sequence.
   [*Hint:* Sum up the terms in the formula for a_n.]

26. Follow the directions in Exercise 25 with

$$a_n = \dfrac{1^2}{n^3} + \dfrac{2^2}{n^3} + \cdots + \dfrac{n^2}{n^3}$$

27. If we accept the fact that the sequence $\{1/n\}_{n=1}^{+\infty}$ converges to the limit $L = 0$, then according to Definition 11.1.2, for every $\epsilon > 0$, there exists an integer N such that $|a_n - L| = |(1/n) - 0| < \epsilon$ when $n \geq N$. In each part, find the smallest possible value of N for the given value of ϵ.

(a) $\epsilon = 0.5$ (b) $\epsilon = 0.1$ (c) $\epsilon = 0.001$.

28. If we accept the fact that the sequence $\left\{\dfrac{n}{n+1}\right\}_{n=1}^{+\infty}$ converges to the limit $L = 1$, then according to Definition 11.1.2, for every $\epsilon > 0$, there exists an integer N such that $|a_n - L| = \left|\dfrac{n}{n+1} - 1\right| < \epsilon$ when $n \geq N$. In each part, find the smallest value of N for the given value of ϵ.

(a) $\epsilon = 0.25$ (b) $\epsilon = 0.1$ (c) $\epsilon = 0.001$.

29. Prove:

(a)   The sequence $\left\{\dfrac{1}{n}\right\}_{n=1}^{+\infty}$ converges to 0.

(b)   The sequence $\left\{\dfrac{n}{n+1}\right\}_{n=1}^{+\infty}$ converges to 1.

30. Consider the sequence $\{a_n\}_{n=1}^{+\infty}$ whose nth term is

$$a_n = \sum_{k=0}^{n-1} \frac{1}{1+\dfrac{k}{n}} \cdot \frac{1}{n}$$

Show that $\lim\limits_{n \to +\infty} a_n = \ln 2$. [*Hint:* Interpret $\lim\limits_{n \to +\infty} a_n$ as a definite integral.]

31. (a)   Show that a polygon with n equal sides inscribed in a circle of radius r, has perimeter
$$p_n = 2rn \sin (\pi/n).$$

(b)   By finding the limit of the sequence $\{p_n\}_{n=1}^{+\infty}$, show that the perimeters approach the circumference of the circle as n increases.

32. Find $\lim\limits_{n \to +\infty} r^n$, where r is a real number. [*Hint:* Consider the cases $|r| < 1$, $|r| > 1$, $r = 1$, and $r = -1$ separately.]

11.2 MONOTONE SEQUENCES

Sometimes the critical information about a sequence is whether it converges or not, with the limit being of little or no importance. In this section we will discuss results that are used to study convergence of sequences.

We begin with some terminology.

11.2.1   DEFINITION   A sequence $\{a_n\}$ is called

increasing if $a_1 < a_2 < a_3 < \cdots < a_n < \cdots$

nondecreasing if $a_1 \le a_2 \le a_3 \le \cdots \le a_n \le \cdots$

decreasing if $a_1 > a_2 > a_3 > \cdots > a_n > \cdots$

nonincreasing if $a_1 \ge a_2 \ge a_3 \ge \cdots \ge a_n \ge \cdots$

A sequence that is either nondecreasing or nonincreasing is called **monotone**, and a sequence that is increasing or decreasing is called **strictly monotone**. Observe that a strictly monotone sequence is monotone, but not conversely. (Why?)

▶ Example 1

$\dfrac{1}{2}, \dfrac{2}{3}, \dfrac{3}{4}, \ldots, \dfrac{n}{n+1}, \ldots$ is increasing

$1, \dfrac{1}{2}, \dfrac{1}{3}, \ldots, \dfrac{1}{n}, \ldots$ is decreasing

$1, 1, 2, 2, 3, 3, \ldots$ is nondecreasing

$1, 1, \dfrac{1}{2}, \dfrac{1}{2}, \dfrac{1}{3}, \dfrac{1}{3}, \ldots$ is nonincreasing

All four of these sequences are monotone, but the sequence

$$1, -\frac{1}{2}, \frac{1}{3}, -\frac{1}{4}, \ldots, (-1)^{n+1}\frac{1}{n}, \ldots$$

is not. The first and second sequences are strictly monotone.   ◄

In order for a sequence to be increasing, *all* pairs of successive terms, a_n and a_{n+1}, must satisfy $a_n < a_{n+1}$, or equivalently, $a_n - a_{n+1} < 0$. More generally, monotone sequences can be classified as follows:

Difference Between Successive Terms	Classification
$a_n - a_{n+1} < 0$	Increasing
$a_n - a_{n+1} > 0$	Decreasing
$a_n - a_{n+1} \leq 0$	Nondecreasing
$a_n - a_{n+1} \geq 0$	Nonincreasing

Frequently, one can *guess* whether a sequence is increasing, decreasing, nondecreasing, or nonincreasing after writing out some of the initial terms. However, to be certain that the guess is correct, a precise mathematical proof is needed. The following example illustrates a method for doing this.

▶ Example 2   Show that

$$\frac{1}{2}, \frac{2}{3}, \frac{3}{4}, \ldots, \frac{n}{n+1}, \ldots$$

is an increasing sequence.

Solution.   It is intuitively clear that the sequence is increasing. To prove that this is so, let

$$a_n = \frac{n}{n+1}$$

We can obtain a_{n+1} by replacing n by $n+1$ in this formula. This yields

$$a_{n+1} = \frac{n+1}{(n+1)+1} = \frac{n+1}{n+2}$$

Thus, for $n \geq 1$

$$a_n - a_{n+1} = \frac{n}{n+1} - \frac{n+1}{n+2} = \frac{n^2 + 2n - n^2 - 2n - 1}{(n+1)(n+2)}$$

$$= -\frac{1}{(n+1)(n+2)} < 0$$

This proves that the sequence is increasing. ◄

If a_n and a_{n+1} are any successive terms in an increasing sequence, then $a_n < a_{n+1}$. If the terms in the sequence are all positive, then we can divide both sides of this inequality by a_n to obtain $1 < a_{n+1}/a_n$ or equivalently $a_{n+1}/a_n > 1$. More generally, monotone sequences with *positive* terms can be classified as follows:

Ratio of Successive Terms	Classification
$a_{n+1}/a_n > 1$	Increasing
$a_{n+1}/a_n < 1$	Decreasing
$a_{n+1}/a_n \geq 1$	Nondecreasing
$a_{n+1}/a_n \leq 1$	Nonincreasing

► Example 3 Show that the sequence in Example 2 is increasing by examining the ratio of successive terms.

Solution. As shown in the solution of Example 2,

$$a_n = \frac{n}{n+1} \quad \text{and} \quad a_{n+1} = \frac{n+1}{n+2}$$

Thus,

$$\frac{a_{n+1}}{a_n} = \frac{(n+1)/(n+2)}{n/(n+1)} = \frac{n+1}{n+2} \cdot \frac{n+1}{n} = \frac{n^2 + 2n + 1}{n^2 + 2n} \tag{1}$$

Since the numerator in (1) exceeds the denominator, the ratio exceeds 1, that is, $a_{n+1}/a_n > 1$ for $n \geq 1$. This proves that the sequence is increasing.

◄

In our subsequent work we will encounter sequences involving factorials. The reader will recall that if n is a positive integer, then $n!$ (n factorial) is the product of the first n positive integers, that is, $n! = 1 \cdot 2 \cdot 3 \cdots n$. Furthermore, it is agreed that $0! = 1$.

▶ Example 4   Show that the sequence

$$\frac{e}{2!}, \frac{e^2}{3!}, \frac{e^3}{4!}, \ldots, \frac{e^n}{(n+1)!}, \ldots$$

is decreasing.

Solution.   We will examine the ratio of successive terms. Since

$$a_n = \frac{e^n}{(n+1)!}$$

it follows on replacing n by $n + 1$ that

$$a_{n+1} = \frac{e^{n+1}}{[(n+1)+1]!} = \frac{e^{n+1}}{(n+2)!}$$

Thus,

$$\frac{a_{n+1}}{a_n} = \frac{e^{n+1}/(n+2)!}{e^n/(n+1)!} = \frac{e^{n+1}}{e^n} \cdot \frac{(n+1)!}{(n+2)!} = \frac{e}{n+2}$$

For $n \geq 1$, it follows that $n + 2 \geq 3 > e(\approx 2.718\ldots)$, so

$$\frac{a_{n+1}}{a_n} = \frac{e}{n+2} < 1$$

for $n \geq 1$. This proves that the sequence is decreasing.   ◀

The following example illustrates still a third technique for determining whether a sequence is increasing or decreasing.

▶ Example 5   In Examples 2 and 3 we proved that the sequence

$$\frac{1}{2}, \frac{2}{3}, \frac{3}{4}, \ldots, \frac{n}{n+1}, \ldots$$

is increasing by considering the difference and ratio of successive terms. Alternatively, we can proceed as follows. Let

$$f(x) = \frac{x}{x+1}$$

so the nth term in the given sequence is $a_n = f(n)$. The function f is increasing for $x \geq 1$ since

$$f'(x) = \frac{(x+1)(1) - x(1)}{(x+1)^2} = \frac{1}{(x+1)^2} > 0$$

Thus,

$$a_n = f(n) < f(n + 1) = a_{n+1}$$

which proves that the given sequence is increasing.   ◄

In general, if $f(n) = a_n$ is the nth term of a sequence, and if f is differentiable for $x \geq 1$, then we have the following results:

Derivative of f for $x \geq 1$	Classification of the Sequence with nth Term $a_n = f(n)$
$f'(x) > 0$	Increasing
$f'(x) < 0$	Decreasing
$f'(x) \geq 0$	Nondecreasing
$f'(x) \leq 0$	Nonincreasing

We omit the proof.

The following two theorems, whose optional proofs are discussed at the end of this section, show that a monotone sequence either converges or it becomes infinite—divergence by oscillation cannot occur.

11.2.2   THEOREM   *For a nondecreasing sequence $a_1 \leq a_2 \leq a_3 \leq \cdots \leq a_n \leq \cdots$ there are two possibilities:*
(a)   There is a constant M such that $a_n \leq M$ for all n, in which case the sequence converges to a limit L satisfying $L \leq M$.
(b)   No such constant exists, in which case $\lim_{n \to +\infty} a_n = +\infty$.

11.2.3   THEOREM   *For a nonincreasing sequence $a_1 \geq a_2 \geq a_3 \geq \cdots \geq a_n \geq \cdots$ there are two possibilities:*
(a)   There is a constant M such that $a_n \geq M$ for all n, in which case the sequence converges to a limit L satisfying $L \geq M$.
(b)   No such constant exists, in which case $\lim_{n \to +\infty} a_n = -\infty$.

It should be noted that these results do not give a method for obtaining limits; they tell us only whether a limit exists. To prove these theorems we need a preliminary result that takes us to the very foundations of the real number system. In this text we have not been concerned with a logical development of the real numbers; our approach has been to accept the familiar properties of real numbers and to work with them. Indeed, we have not even attempted to define the term "real number." However, by the late nineteenth century, the study of limits and functions in calculus necessitated a precise axiomatic formulation of the real numbers in much the same way that Euclidean geometry is developed from axioms. While we will not attempt to pursue this development, we will have need for the following axiom about real numbers.

11.2.4 AXIOM
The Completeness Axiom

If S is a nonempty set of real numbers, and if there is some real number that is greater than or equal to every number in S, then there is a smallest real number that is greater than or equal to every number in S.

For example, let S be the set of numbers in the interval $(1, 3)$. It is true that there exists a number u greater than or equal to every number in S; some examples are $u = 10$, $u = 100$, and $u = 3.2$. The smallest number u that is greater than or equal to every number in S is $u = 3$.

There is an alternate phrasing of the completeness axiom that is useful to know. Let us call a number u an ***upper bound*** for a set S if u is greater than or equal to every number in S; and if S has a smallest upper bound, call it the ***least upper bound*** of S. Using this terminology the Completeness Axiom states:

11.2.5 AXIOM
The Completeness Axiom
(Alternate Form)

If a nonempty set S of real numbers has an upper bound, then S has a least upper bound.

▶ Example 6 As shown in Examples 2 and 3, the sequence $\left\{ \dfrac{n}{n+1} \right\}_{n=1}^{+\infty}$ is nondecreasing. Since

$$a_n = \frac{n}{n+1} < 1, \qquad n = 1, 2, \ldots$$

the terms in the sequence have $M = 1$ as an upper bound. By Theorem 11.2.2 the sequence must converge to a limit $L \leq M$. This is indeed the case since

$$\lim_{n \to +\infty} \frac{n}{n+1} = \lim_{n \to +\infty} \frac{1}{1 + \dfrac{1}{n}} = 1 \qquad\qquad ◀$$

Because the limit of a sequence $\{a_n\}$ describes the behavior of the terms as n gets *large*, one can alter or even delete a *finite* number of terms in a sequence without affecting either the convergence or the value of the limit. That is, the original and the modified sequence will both converge or both diverge, and in the case of convergence both will have the same limit. We omit the proof.

▶ Example 7   Show that the sequence

$$\left\{ \frac{5^n}{n!} \right\}_{n=1}^{+\infty}$$

converges.

Solution.   It is tedious to determine convergence directly from the limit

$$\lim_{n \to +\infty} \frac{5^n}{n!}$$

Thus, we will proceed indirectly. If we let

$$a_n = \frac{5^n}{n!}$$

then

$$a_{n+1} = \frac{5^{n+1}}{(n + 1)!}$$

so that

$$\frac{a_{n+1}}{a_n} = \frac{5^{n+1}/(n + 1)!}{5^n/n!} = \frac{5^{n+1}}{5^n} \cdot \frac{n!}{(n + 1)!} = \frac{5}{n + 1}$$

For $n = 1$, 2, and 3, the value of a_{n+1}/a_n is greater than 1 so $a_{n+1} > a_n$. Thus,

$$a_1 < a_2 < a_3 < a_4$$

For $n = 4$ the value of a_{n+1}/a_n is 1, so

$$a_4 = a_5$$

For $n \geq 5$ the value of a_{n+1}/a_n is less than 1, so

$$a_5 > a_6 > a_7 > a_8 > \cdots$$

Thus, if we discard the first four terms of the given sequence (which won't affect convergence) the resulting sequence will be decreasing. Moreover, each term in the sequence is positive, so that by Theorem 11.2.3, the sequence converges to some limit that is ≥ 0.   ◀

OPTIONAL

Proof of Theorem 11.2.2.

(a)   Assume there exists a number M such that $a_n \leq M$ for $n = 1, 2, \ldots$
Then M is an upper bound for the set of terms in the sequence. By the

Completeness Axiom there is a least upper bound for the terms, call it L. Now let ϵ be any positive number. Since L is the least upper bound for the terms, $L - \epsilon$ is not an upper bound for the terms, which means that there is at least one term a_N such that

$$a_N > L - \epsilon$$

Moreover, since $\{a_n\}$ is a nondecreasing sequence, we must have

$$a_n \geq a_N > L - \epsilon \qquad (2)$$

when $n \geq N$. But a_n cannot exceed L since L is an upper bound for the terms. This observation together with (2) tells us that $L \geq a_n > L - \epsilon$ for $n \geq N$, so that all terms from the Nth on are within ϵ units of L. This is exactly the requirement to have

$$\lim_{n \to +\infty} a_n = L$$

Observe that $L \leq M$ since M is an upper bound for the terms and L is the least upper bound. This proves (a).

(b) If there is no number M such that $a_n \leq M$ for $n = 1, 2, \ldots$, then no matter how large we choose M, there is a term a_N such that

$$a_N > M$$

and, since the sequence is nondecreasing,

$$a_n \geq a_N > M$$

when $n \geq N$. Thus, the terms in the sequence become arbitrarily large as n increases. That is,

$$\lim_{n \to +\infty} a_n = +\infty \quad \blacksquare$$

The proof of Theorem 11.2.3 will be omitted since it is similar to 11.2.2.

▶ Exercise Set 11.2

In Exercises 1–6, determine whether the given sequence $\{a_n\}$ is monotone by examining $a_n - a_{n+1}$. If so, classify it as increasing, decreasing, nonincreasing, or nondecreasing.

1. $\left\{\dfrac{1}{n}\right\}_{n=1}^{+\infty}$.

2. $\left\{1 - \dfrac{1}{n}\right\}_{n=1}^{+\infty}$.

3. $\left\{\dfrac{n}{2n + 1}\right\}_{n=1}^{+\infty}$.

4. $\left\{\dfrac{n}{4n - 1}\right\}_{n=1}^{+\infty}$.

5. $\{n - 2^n\}_{n=1}^{+\infty}$.

6. $\{n - n^2\}_{n=1}^{+\infty}$.

In Exercises 7–12, determine whether the given sequence $\{a_n\}$ is monotone by examining a_{n+1}/a_n. If so, classify it as increasing, decreasing, nonincreasing, or nondecreasing.

7. $\left\{\dfrac{n}{2n + 1}\right\}_{n=1}^{+\infty}$.

8. $\left\{\dfrac{2^n}{n!}\right\}_{n=1}^{+\infty}$.

9. $\left\{\dfrac{n}{2^n}\right\}_{n=1}^{+\infty}$.

10. $\left\{\dfrac{2^n}{1 + 2^n}\right\}_{n=1}^{+\infty}$.

11. $\left\{\dfrac{10^n}{2^{n^2}}\right\}_{n=1}^{+\infty}$.

12. $\left\{\dfrac{n^n}{n!}\right\}_{n=1}^{+\infty}$.

In Exercises 13–18, use differentiation to show that the sequence is strictly monotone and classify it as increasing or decreasing.

13. $\left\{\dfrac{n}{2n+1}\right\}_{n=1}^{+\infty}$.

14. $\left\{3-\dfrac{1}{n}\right\}_{n=1}^{+\infty}$.

15. $\left\{\dfrac{1}{n+\ln n}\right\}_{n=1}^{+\infty}$.

16. $\{ne^{-2n}\}_{n=1}^{+\infty}$.

17. $\left\{\dfrac{\ln(n+2)}{n+2}\right\}_{n=1}^{+\infty}$.

18. $\{\tan^{-1} n\}_{n=1}^{+\infty}$.

In Exercises 19–24, show that the sequence is monotone and apply Theorem 11.2.2 or 11.2.3 to determine whether it converges.

19. $\left\{\dfrac{n}{5^n}\right\}_{n=1}^{+\infty}$.

20. $\left\{\dfrac{2^n}{(n+1)!}\right\}_{n=1}^{+\infty}$.

21. $\left\{n-\dfrac{1}{n}\right\}_{n=1}^{+\infty}$.

22. $\left\{\cos\dfrac{\pi}{2n}\right\}_{n=1}^{+\infty}$.

23. $\left\{2+\dfrac{1}{n}\right\}_{n=1}^{+\infty}$.

24. $\left\{\dfrac{4n-1}{5n+2}\right\}_{n=1}^{+\infty}$.

25. Find the limit (if it exists) of the sequence.

(a) $1,-1,1,\dfrac{1}{2},\dfrac{1}{3},\dfrac{1}{4},\dfrac{1}{5},\ldots$

(b) $-\dfrac{1}{2},0,0,0,1,2,3,4,\ldots$

26. (a) Is $\left\{\dfrac{100^n}{n!}\right\}_{n=1}^{+\infty}$ a monotone sequence?

(b) Is it a convergent sequence? Justify your answer.

27. Show that $\left\{\dfrac{3^n}{1+3^{2n}}\right\}_{n=1}^{+\infty}$ is a decreasing sequence.

28. Show that $\left\{\dfrac{1\cdot3\cdot5\cdots(2n-1)}{n!}\right\}_{n=1}^{+\infty}$ is an increasing sequence.

29. (a) Show that if $\{a_n\}_{n=1}^{+\infty}$ is a nonincreasing sequence, then $\{-a_n\}_{n=1}^{+\infty}$ is a nondecreasing sequence.

(b) Use part (a) and Theorem 11.2.2 to help prove Theorem 11.2.3.

11.3 INFINITE SERIES

The purpose of this section is to discuss sums

$$u_1 + u_2 + u_3 + \cdots + u_k + \cdots$$

that contain infinitely many terms. The most familiar examples of such sums occur in the decimal representation of real numbers. For example, when we write $\frac{1}{3}$ in the decimal form

$$\frac{1}{3} = 0.3333\ldots$$

we mean

$$\frac{1}{3} = 0.3 + 0.03 + 0.003 + 0.0003 + \cdots$$

$$= \frac{3}{10} + \frac{3}{10^2} + \frac{3}{10^3} + \frac{3}{10^4} + \cdots$$

Our first objective will be to define what is meant by a sum with infinitely many terms. Since it is impossible to "add up" infinitely many numbers, we will deal with infinite sums by means of a limiting process involving sequences.

11.3.1   DEFINITION   An *infinite series* is an expression of the form

$$u_1 + u_2 + u_3 + \cdots + u_k + \cdots$$

or in sigma notation

$$\sum_{k=1}^{\infty} u_k$$

The numbers $u_1, u_2, u_3, \ldots$ are called the *terms* of the series.

Informally speaking, the expression $\sum_{k=1}^{\infty} u_k$ directs us to obtain the "sum" of the terms $u_1, u_2, u_3, \ldots$ To carry out this summation process we proceed as follows: Let s_n denote the sum of the first n terms of the series. Thus,

$$s_1 = u_1$$
$$s_2 = u_1 + u_2$$
$$s_3 = u_1 + u_2 + u_3$$
$$\vdots$$
$$s_n = u_1 + u_2 + u_3 + \cdots + u_n = \sum_{k=1}^{n} u_k$$

The number s_n is called the *nth partial sum* of the series and the sequence $\{s_n\}_{n=1}^{+\infty}$ is called the *sequence of partial sums.*

▶ Example 1   For the infinite series

$$\frac{3}{10} + \frac{3}{10^2} + \frac{3}{10^3} + \frac{3}{10^4} + \cdots$$

the partial sums are

$$s_1 = \frac{3}{10}$$

$$s_2 = \frac{3}{10} + \frac{3}{10^2} = \frac{33}{100}$$

$$s_3 = \frac{3}{10} + \frac{3}{10^2} + \frac{3}{10^3} = \frac{333}{1000}$$

$$s_4 = \frac{3}{10} + \frac{3}{10^2} + \frac{3}{10^3} + \frac{3}{10^4} = \frac{3333}{10000}$$
$$\vdots$$

◀

As n increases, the partial sum $s_n = u_1 + u_2 + \cdots + u_n$ includes more and more terms of the series. Thus, if s_n tends toward a limit as $n \to +\infty$, it is reasonable to view this limit as the sum of *all* the terms in the series. This suggests the following definition.

11.3.2 DEFINITION Let $\{s_n\}$ be the sequence of partial sums of the series $\displaystyle\sum_{k=1}^{\infty} u_k$. If the sequence $\{s_n\}$ converges to a limit S, then the series is said to **converge** and S is called the **sum** of the series. We denote this by writing

$$S = \sum_{k=1}^{\infty} u_k$$

If the sequence of partial sums of a series diverges, then the series is said to **diverge.** A divergent series has no sum.

▶ **Example 2** If Definition 11.3.2 is to be reasonable, it should be the case that

$$\frac{1}{3} = \frac{3}{10} + \frac{3}{10^2} + \frac{3}{10^3} + \cdots + \frac{3}{10^k} + \cdots$$

Let us verify that this is indeed the case. The nth partial sum is

$$s_n = \frac{3}{10} + \frac{3}{10^2} + \cdots + \frac{3}{10^n} \tag{1}$$

The problem of calculating $\displaystyle\lim_{n \to +\infty} s_n$ is complicated by the fact that the number of terms in (1) changes with n. For purposes of calculation, it is desirable to rewrite (1) in closed form (see Section 5.5). To do this, we multiply both sides of (1) by $\frac{1}{10}$ to obtain

$$\frac{1}{10} s_n = \frac{3}{10^2} + \frac{3}{10^3} + \cdots + \frac{3}{10^n} + \frac{3}{10^{n+1}} \tag{2}$$

and then subtract (2) from (1) to obtain

$$s_n - \frac{1}{10} s_n = \frac{3}{10} - \frac{3}{10^{n+1}}$$

or

$$\frac{9}{10} s_n = \frac{3}{10}\left(1 - \frac{1}{10^n}\right)$$

or

$$s_n = \frac{1}{3}\left(1 - \frac{1}{10^n}\right) \tag{3}$$

Since $1/10^n \to 0$ as $n \to +\infty$, it follows from (3) that $S = \lim\limits_{n \to +\infty} s_n = \frac{1}{3}$. Thus,

$$\frac{1}{3} = \frac{3}{10} + \frac{3}{10^2} + \frac{3}{10^3} + \cdots + \frac{3}{10^n} + \cdots$$

◀

▶ Example 3   Determine whether the series

$$1 - 1 + 1 - 1 + 1 - 1 + \cdots$$

converges or diverges. If it converges, find the sum.

Solution.   The partial sums are $s_1 = 1, s_2 = 1 - 1 = 0, s_3 = 1 - 1 + 1 = 1,$ $s_4 = 1 - 1 + 1 - 1 = 0$, and so forth. Thus, the sequence of partial sums is

$$1, 0, 1, 0, 1, 0, \ldots$$

Since this is a divergent sequence, the given series diverges and consequently has no sum.   ◀

The series in Examples 2 and 3 are examples of **geometric series.** A geometric series is one of the form

$$a + ar + ar^2 + ar^3 + \cdots + ar^{k-1} + \cdots \qquad\qquad (a \neq 0)$$

where each term is obtained by multiplying the previous one by a constant r. The multiplier r is called the **ratio** for the series. Some examples of geometric series are:

$$1 + 2 + 4 + 8 + \cdots + 2^{k-1} + \cdots \qquad (a = 1, r = 2)$$

$$3 + \frac{3}{10} + \frac{3}{10^2} + \frac{3}{10^3} + \cdots + \frac{3}{10^{k-1}} + \cdots \qquad \left(a = 3, r = \frac{1}{10}\right)$$

$$\frac{1}{2} - \frac{1}{4} + \frac{1}{8} - \frac{1}{16} + \cdots + (-1)^{k+1}\frac{1}{2^k} + \cdots \qquad \left(a = \frac{1}{2}, r = -\frac{1}{2}\right)$$

$$1 + 1 + 1 + \cdots + 1 + \cdots \qquad (a = 1, r = 1)$$

$$1 - 1 + 1 - 1 + \cdots + (-1)^{k+1} + \cdots \qquad (a = 1, r = -1)$$

The following theorem is the fundamental result on convergence of geometric series.

11.3.3 THEOREM *A geometric series*

$$a + ar + ar^2 + \cdots + ar^{k-1} + \cdots \qquad (a \neq 0)$$

converges if $|r| < 1$ *and diverges if* $|r| \geq 1$. *If the series converges the sum is*

$$\frac{a}{1-r} = a + ar + ar^2 + \cdots + ar^{k-1} + \cdots$$

Proof. Let us treat the case $|r| = 1$ first. If $r = 1$, then the series is

$$a + a + a + \cdots + a + \cdots$$

so that the *n*th partial sum is $s_n = na$ and $\lim\limits_{n \to +\infty} s_n = \lim\limits_{n \to +\infty} na = \pm\infty$ (the sign depending on whether a is positive or negative). This proves divergence. If $r = -1$, the series is

$$a - a + a - a + \cdots$$

so the sequence of partial sums is

$$a, 0, a, 0, a, 0, \ldots$$

which diverges.

Now let us consider the case where $|r| \neq 1$. The *n*th partial sum of the series is

$$s_n = a + ar + ar^2 + \cdots + ar^{n-1} \qquad (4)$$

Multiplying both sides of (4) by r yields

$$rs_n = ar + ar^2 + \cdots + ar^{n-1} + ar^n \qquad (5)$$

and subtracting (5) from (4) gives

$$s_n - rs_n = a - ar^n$$

or

$$(1 - r)s_n = a - ar^n \qquad (6)$$

Since $r \neq 1$ in the case we are considering, this can be rewritten as

$$s_n = \frac{a - ar^n}{1 - r} = \frac{a}{1 - r} - \frac{ar^n}{1 - r} \qquad (7)$$

If $|r| < 1$, then $\lim\limits_{n \to +\infty} r^n = 0$, so that $\{s_n\}$ converges. From (7)

$$\lim_{n \to +\infty} s_n = \frac{a}{1 - r}$$

If $|r| > 1$, then either $r > 1$ or $r < -1$. In the case $r > 1$, $\lim\limits_{n \to +\infty} r^n = +\infty$, and in the case $r < -1$, r^n oscillates between positive and negative values that grow in magnitude, so $\{s_n\}$ diverges. ▮

► Example 4   The series

$$5 + \frac{5}{4} + \frac{5}{4^2} + \cdots + \frac{5}{4^{k-1}} + \cdots$$

is a geometric series with $a = 5$ and $r = \frac{1}{4}$. Since $|r| = \frac{1}{4} < 1$, the series converges and the sum is

$$\frac{a}{1 - r} = \frac{5}{1 - \frac{1}{4}} = \frac{20}{3}$$ ◄

► Example 5   Find the rational number represented by the repeating decimal

$$0.784784784\ldots$$

Solution.   We can write

$$0.784784784\ldots = 0.784 + 0.000784 + 0.000000784 + \cdots$$

so the given decimal is the sum of a geometric series with $a = 0.784$ and $r = 0.001$. Thus,

$$0.784784784\ldots = \frac{a}{1 - r} = \frac{0.784}{1 - 0.001} = \frac{0.784}{0.999} = \frac{784}{999}$$ ◄

► Example 6   Determine whether the series

$$\sum_{k=1}^{\infty} \frac{1}{k(k + 1)} = \frac{1}{1 \cdot 2} + \frac{1}{2 \cdot 3} + \frac{1}{3 \cdot 4} + \frac{1}{4 \cdot 5} + \cdots$$

converges or diverges. If it converges, find the sum.

Solution.   The nth partial sum of the series is

$$s_n = \sum_{k=1}^{n} \frac{1}{k(k + 1)} = \frac{1}{1 \cdot 2} + \frac{1}{2 \cdot 3} + \frac{1}{3 \cdot 4} + \cdots + \frac{1}{n(n + 1)}$$

To calculate $\lim\limits_{n \to +\infty} s_n$ we will rewrite s_n in closed form. This may be accomplished by using the method of partial fractions to obtain (verify):

$$\frac{1}{k(k + 1)} = \frac{1}{k} - \frac{1}{k + 1}$$

from which it follows that

$$
\begin{aligned}
s_n &= \sum_{k=1}^{n} \left(\frac{1}{k} - \frac{1}{k + 1} \right) \\
&= \left(1 - \frac{1}{2}\right) + \left(\frac{1}{2} - \frac{1}{3}\right) + \left(\frac{1}{3} - \frac{1}{4}\right) + \cdots + \left(\frac{1}{n} - \frac{1}{n + 1}\right) \\
&= 1 + \left(-\frac{1}{2} + \frac{1}{2}\right) + \left(-\frac{1}{3} + \frac{1}{3}\right) + \cdots + \left(-\frac{1}{n} + \frac{1}{n}\right) - \frac{1}{n + 1} \quad (8)
\end{aligned}
$$

The sum in (8) is an example of a **telescoping sum,** which means that each term cancels part of the next term, thereby collapsing the sum (like a folding telescope) into only two terms. After the cancellation, (8) can be written as

$$s_n = 1 - \frac{1}{n + 1}$$

so

$$\lim_{n \to +\infty} s_n = \lim_{n \to +\infty} \left(1 - \frac{1}{n + 1}\right) = 1$$

and therefore

$$1 = \sum_{k=1}^{\infty} \frac{1}{k(k + 1)} \qquad \blacktriangleleft$$

▶ **Example 7**   One of the most famous and important of all diverging series is the **harmonic series,**

$$\sum_{k=1}^{\infty} \frac{1}{k} = 1 + \frac{1}{2} + \frac{1}{3} + \frac{1}{4} + \frac{1}{5} + \cdots$$

which arises in connection with the overtones produced by a vibrating musical string. It is not immediately evident that this series diverges. However, the divergence will become apparent when we examine the partial sums in detail. Because the terms in the series are all positive, the partial sums

$$s_1 = 1, \; s_2 = 1 + \frac{1}{2}, \; s_3 = 1 + \frac{1}{2} + \frac{1}{3}, \; s_4 = 1 + \frac{1}{2} + \frac{1}{3} + \frac{1}{4}, \ldots$$

form an increasing sequence

$$s_1 < s_2 < s_3 < \cdots < s_n < \cdots$$

Thus, by Theorem 11.2.2 we can prove divergence by demonstrating that there is no constant M that is greater than or equal to *every* partial sum. To this end, we will consider some selected partial sums, namely $s_2, s_4, s_8, s_{16}, s_{32}, \ldots$. Note that the subscripts are successive powers of 2, so that these are the partial sums of the form s_{2^n}. These partial sums satisfy the inequalities

$$s_2 = 1 + \frac{1}{2} > \frac{1}{2} + \frac{1}{2} = \frac{2}{2}$$

$$s_4 = s_2 + \frac{1}{3} + \frac{1}{4} > s_2 + \left(\frac{1}{4} + \frac{1}{4}\right) = s_2 + \frac{1}{2} > \frac{3}{2}$$

$$s_8 = s_4 + \frac{1}{5} + \frac{1}{6} + \frac{1}{7} + \frac{1}{8} > s_4 + \left(\frac{1}{8} + \frac{1}{8} + \frac{1}{8} + \frac{1}{8}\right) = s_4 + \frac{1}{2} > \frac{4}{2}$$

$$s_{16} = s_8 + \frac{1}{9} + \frac{1}{10} + \frac{1}{11} + \frac{1}{12} + \frac{1}{13} + \frac{1}{14} + \frac{1}{15} + \frac{1}{16}$$

$$> s_8 + \left(\frac{1}{16} + \frac{1}{16} + \frac{1}{16} + \frac{1}{16} + \frac{1}{16} + \frac{1}{16} + \frac{1}{16} + \frac{1}{16}\right) = s_8 + \frac{1}{2} > \frac{5}{2}$$

$$\vdots$$

$$s_{2^n} > \frac{n+1}{2}$$

Now if M is any constant, we can certainly find a positive integer n such that $(n+1)/2 > M$. But for this n

$$s_{2^n} > \frac{n+1}{2} > M$$

so that no constant M is greater than or equal to *every* partial sum of the harmonic series. This proves divergence. ◄

▶ Exercise Set 11.3

1. In each part, find the first four partial sums; find a closed form for the nth partial sum; determine whether the series converges, and if so give the sum.

(a) $\displaystyle\sum_{k=1}^{\infty} \frac{2}{5^{k-1}}$ (b) $\displaystyle\sum_{k=1}^{\infty} \frac{1}{(k+1)(k+2)}$

(c) $\displaystyle\sum_{k=1}^{\infty} \frac{2^{k-1}}{4}$.

In Exercises 2–16 determine whether the series converges or diverges. If it converges, find the sum.

2. $\displaystyle\sum_{k=1}^{\infty} \frac{1}{5^k}$.

3. $\displaystyle\sum_{k=1}^{\infty} \left(-\frac{3}{4}\right)^{k-1}$.

4. $\displaystyle\sum_{k=1}^{\infty} \left(\frac{2}{3}\right)^{k+2}$.

5. $\displaystyle\sum_{k=1}^{\infty} (-1)^{k-1} \frac{7}{6^{k-1}}$.

6. $\displaystyle\sum_{k=1}^{\infty} 4^{k-1}$.

7. $\displaystyle\sum_{k=1}^{\infty} \left(-\frac{3}{2}\right)^{k+1}$.

8. $\displaystyle\sum_{k=1}^{\infty} \left(\frac{1}{k+3} - \frac{1}{k+4} \right).$

9. $\displaystyle\sum_{k=1}^{\infty} \frac{1}{(k+2)(k+3)}.$ **10.** $\displaystyle\sum_{k=1}^{\infty} \left(\frac{1}{2^k} - \frac{1}{2^{k+1}} \right).$

11. $\displaystyle\sum_{k=1}^{\infty} \frac{1}{9k^2 + 3k - 2}.$ **12.** $\displaystyle\sum_{k=2}^{\infty} \frac{1}{k^2 - 1}.$

13. $\displaystyle\sum_{k=1}^{\infty} \frac{4^{k+2}}{7^{k-1}}.$ **14.** $\displaystyle\sum_{k=1}^{\infty} \left(\frac{e}{\pi} \right)^{k-1}.$

15. $\displaystyle\sum_{k=1}^{\infty} \left(-\frac{1}{2} \right)^k.$ **16.** $\displaystyle\sum_{k=3}^{\infty} \frac{5}{k-2}.$

In Exercises 17–22, express the repeating decimal as a fraction.

17. 0.4444...   **18.** 0.9999...

19. 5.373737...   **20.** 0.159159159...

21. 0.782178217821...   **22.** 0.451141414...

23. Find a closed form for the nth partial sum of the series

$$\ln \frac{1}{2} + \ln \frac{2}{3} + \ln \frac{3}{4} + \cdots + \ln \frac{n}{n+1} + \cdots$$

and determine whether the series converges.

24. A ball is dropped from a height of 10 meters. Each time it strikes the ground it bounces vertically to a height that is three-fourths of the previous height. Find the total distance the ball will travel if it is allowed to bounce indefinitely.

25. Show: $\displaystyle\sum_{k=2}^{\infty} \ln (1 - 1/k^2) = -\ln 2.$

26. Show: $\displaystyle\sum_{k=1}^{\infty} \frac{\sqrt{k+1} - \sqrt{k}}{\sqrt{k^2 + k}} = 1.$

27. Use geometric series to show:

(a) $\displaystyle\sum_{k=0}^{\infty} (-1)^k x^k = \frac{1}{1+x}$ if $-1 < x < 1$

(b) $\displaystyle\sum_{k=0}^{\infty} (x-3)^k = \frac{1}{4-x}$ if $2 < x < 4$

(c) $\displaystyle\sum_{k=0}^{\infty} (-1)^k x^{2k} = \frac{1}{1+x^2}$ if $-1 < x < 1.$

28. Prove the following decimal equality assuming $a_n \neq 9$:
$$0.a_1 a_2 \dots a_n 9999 \dots = 0.a_1 a_2 \dots (a_n + 1)0000 \dots$$

29. Let a_1 be any real number and define $a_{n+1} = \frac{1}{2}(a_n + 1)$ for $n = 1, 2, 3, \dots$ Show that the sequence $\{a_n\}_{n=1}^{+\infty}$ converges and find its limit. [*Hint:* Express a_n in terms of a_1.]

11.4   CONVERGENCE; THE INTEGRAL TEST

In the previous section we found sums of series and investigated convergence by first writing the nth partial sum s_n in closed form and then examining the limit $\lim_{n \to +\infty} s_n$. Unfortunately, it is relatively rare that the nth partial sum of a series can be written in closed form; for most series, convergence or divergence is determined by using convergence tests, some of which we will introduce in this section. Once it is established that a series converges, the sum of the series can always be approximated to any degree of accuracy by a partial sum with sufficiently many terms.

Our first theorem states that the terms of an infinite series must tend toward zero if the series is to converge.

11.4.1   THEOREM   *If the series Σu_k converges, then $\displaystyle\lim_{k \to +\infty} u_k = 0.$*

Proof. The term u_k can be written

$$u_k = s_k - s_{k-1} \tag{1}$$

where s_k is the sum of the first k terms and s_{k-1} is the sum of the first $k - 1$ terms. If S denotes the sum of the series, then $\lim\limits_{k \to +\infty} s_k = S$, and since $(k - 1) \to +\infty$ as $k \to +\infty$, we also have $\lim\limits_{k \to +\infty} s_{k-1} = S$. Thus, from (1)

$$\lim_{k \to +\infty} u_k = \lim_{k \to +\infty} (s_k - s_{k-1}) = S - S = 0 \quad \blacksquare$$

The following result is just an alternate phrasing of the above theorem and needs no additional proof.

11.4.2   THEOREM
The Divergence Test

If $\lim\limits_{k \to +\infty} u_k \neq 0$, then the series Σu_k diverges.

▶ **Example 1**   The series

$$\sum_{k=1}^{\infty} \frac{k}{k+1} = \frac{1}{2} + \frac{2}{3} + \frac{3}{4} + \cdots + \frac{k}{k+1} + \cdots$$

diverges since

$$\lim_{k \to +\infty} \frac{k}{k+1} = \lim_{k \to +\infty} \frac{1}{1 + 1/k} = 1 \neq 0 \qquad ◀$$

WARNING.   The converse of Theorem 11.4.1 is false. To prove that a series converges it does not suffice to show that $\lim\limits_{k \to +\infty} u_k = 0$, since this property may hold for divergent as well as convergent series. For example, the kth term of the divergent harmonic series $1 + 1/2 + 1/3 + \cdots + 1/k + \cdots$ tends to zero as $k \to +\infty$, and the kth term of the convergent geometric series $1/2 + 1/2^2 + \cdots + 1/2^k + \cdots$ tends to zero as $k \to +\infty$.

For brevity, the proof of the following result is left for the exercises.

11.4.3   THEOREM   (*a*)   *If Σu_k and Σv_k are convergent series, then $\Sigma(u_k + v_k)$ and $\Sigma(u_k - v_k)$ are convergent series and the sums of these series are related by*

$$\sum_{k=1}^{\infty} (u_k + v_k) = \sum_{k=1}^{\infty} u_k + \sum_{k=1}^{\infty} v_k$$

$$\sum_{k=1}^{\infty} (u_k - v_k) = \sum_{k=1}^{\infty} u_k - \sum_{k=1}^{\infty} v_k$$

(b) If c is a nonzero constant, then the series Σu_k and $\Sigma c u_k$ both converge or both diverge. In the case of convergence, the sums are related by

$$\sum_{k=1}^{\infty} c u_k = c \sum_{k=1}^{\infty} u_k$$

(c) Convergence or divergence is unaffected by deleting a finite number of terms from the beginning of a series; that is, for any positive integer K, the series

$$\sum_{k=1}^{\infty} u_k = u_1 + u_2 + u_3 + \cdots$$

and

$$\sum_{k=K}^{\infty} u_k = u_K + u_{K+1} + u_{K+2} + \cdots$$

both converge or both diverge.

REMARK. Do not read too much into part (c) of this theorem. Although the convergence is not affected when a finite number of terms is deleted from the beginning of a convergent series, the *sum* of the series is changed by the removal of these terms.

▶ Example 2   Find the sum of the series

$$\sum_{k=1}^{\infty} \left(\frac{3}{4^k} - \frac{2}{5^{k-1}} \right)$$

Solution.   The series

$$\sum_{k=1}^{\infty} \frac{3}{4^k} = \frac{3}{4} + \frac{3}{4^2} + \frac{3}{4^3} + \cdots$$

is a convergent geometric series ($a = \frac{3}{4}, r = \frac{1}{4}$), and the series

$$\sum_{k=1}^{\infty} \frac{2}{5^{k-1}} = 2 + \frac{2}{5} + \frac{2}{5^2} + \frac{2}{5^3} + \cdots$$

is also a convergent geometric series ($a = 2, r = \frac{1}{5}$). Thus, from Theorems 11.4.3(a) and 11.3.3 the given series converges and

$$\sum_{k=1}^{\infty}\left(\frac{3}{4^k}-\frac{2}{5^{k-1}}\right)=\sum_{k=1}^{\infty}\frac{3}{4^k}-\sum_{k=1}^{\infty}\frac{2}{5^{k-1}}$$

$$=\frac{\frac{3}{4}}{1-\frac{1}{4}}-\frac{2}{1-\frac{1}{5}}$$

$$=1-\frac{5}{2}=-\frac{3}{2}\qquad\blacktriangleleft$$

▶ Example 3   The series

$$\sum_{k=1}^{\infty}\frac{5}{k}=5+\frac{5}{2}+\frac{5}{3}+\cdots+\frac{5}{k}+\cdots$$

diverges by part (*b*) of Theorem 11.4.3, since

$$\sum_{k=1}^{\infty}\frac{5}{k}=\sum_{k=1}^{\infty}5\left(\frac{1}{k}\right)$$

so each term is a constant times the corresponding term of the divergent harmonic series.   ◀

▶ Example 4   The series

$$\sum_{k=10}^{\infty}\frac{1}{k}=\frac{1}{10}+\frac{1}{11}+\frac{1}{12}+\cdots$$

diverges by part (*c*) of Theorem 11.4.3, since this series results by deleting the first nine terms from the divergent harmonic series.   ◀

If an infinite series $u_1+u_2+u_3+\cdots+u_k+\cdots$ has *positive terms*, then the partial sums $s_1=u_1$, $s_2=u_1+u_2$, $s_3=u_1+u_2+u_3,\ldots$ form an increasing sequence, that is

$$s_1<s_2<s_3<\cdots<s_n<\cdots$$

If there is a finite constant M such that $s_n\le M$ for all n, then according to Theorem 11.2.2, the sequence of partial sums will converge to a limit S, satisfying $S\le M$. If no such constant exists, then $\lim_{n\to+\infty}s_n=+\infty$. This yields the following theorem.

11.4.4   THEOREM   *If Σu_k is a series with positive terms, and if there is a constant M such that*

$$s_n = u_1 + u_2 + \cdots + u_n \leq M$$

for every n, then the series converges and the sum S satisfies $S \leq M$. If no such M exists then the series diverges.

If we have a series with positive terms, say

$$\sum_{k=1}^{\infty} \frac{1}{k^2}$$

and if we form the improper integral

$$\int_1^{+\infty} \frac{1}{x^2}\, dx$$

whose integrand is obtained by replacing the summation index k by x, then there is a relationship between convergence of the series and convergence of the improper integral.

11.4.5   THEOREM
The Integral Test

Let Σu_k be a series with positive terms, and let $f(x)$ be the function that results when k is replaced by x in the formula for u_k. If f is decreasing and continuous for $x \geq 1$, then

$$\sum_{k=1}^{\infty} u_k \quad and \quad \int_1^{+\infty} f(x)\, dx$$

both converge or both diverge.

We will defer the proof to the end of the section and proceed with some examples.

▶ Example 5   Determine whether

$$\sum_{k=1}^{\infty} \frac{1}{k^2}$$

converges or diverges.

Solution.   If we replace k by x in the formula for u_k, we obtain the function

$$f(x) = \frac{1}{x^2}$$

which satisfies the hypotheses of the integral test. (Verify.) Since

$$\int_1^{+\infty} \frac{1}{x^2}\,dx = \lim_{l\to+\infty} \int_1^l \frac{dx}{x^2} = \lim_{l\to+\infty}\left[-\frac{1}{x}\right]_1^l = \lim_{l\to+\infty}\left[1-\frac{1}{l}\right] = 1$$

the integral converges and consequently the series converges.   ◄

REMARK.   In the above example, do *not* erroneously conclude that

$\displaystyle\sum_{k=1}^{\infty} \frac{1}{k^2} = 1$ from the fact that $\displaystyle\int_1^{+\infty} \frac{1}{x^2}\,dx = 1$. (To see that this is false, write

out the series: $1 + 1/2^2 + 1/3^2 + \cdots$; the sum obviously exceeds 1.)

► Example 6   The integral test provides another way to demonstrate divergence of the harmonic series $\displaystyle\sum_{k=1}^{\infty} \frac{1}{k}$. If we replace k by x in the formula

for u_k we obtain the function $f(x) = 1/x$, which satisfies the hypotheses of the integral test. (Verify.) Since

$$\int_1^{+\infty} \frac{1}{x}\,dx = \lim_{l\to+\infty}\int_1^l \frac{1}{x}\,dx = \lim_{l\to+\infty}[\ln l - \ln 1] = +\infty$$

the integral and the series diverge.   ◄

► Example 7   Determine whether the series

$$\frac{1}{e} + \frac{2}{e^4} + \frac{3}{e^9} + \cdots + \frac{k}{e^{k^2}} + \cdots$$

converges or diverges.

Solution.   If we replace k by x in the formula for u_k, we obtain the function

$$f(x) = \frac{x}{e^{x^2}} = xe^{-x^2}$$

For $x \geq 1$, this function has positive values and is continuous. Moreover, for $x \geq 1$ the derivative

$$f'(x) = e^{-x^2} - 2x^2 e^{-x^2} = e^{-x^2}(1 - 2x^2)$$

is negative, so that f is decreasing for $x \geq 1$. Thus, the hypotheses of the integral test are met. But

$$\int_1^{+\infty} xe^{-x^2}\,dx = \lim_{l\to+\infty}\int_1^l xe^{-x^2}\,dx = \lim_{l\to+\infty}\left[-\frac{1}{2}e^{-x^2}\right]_1^l$$

$$= \left(-\frac{1}{2}\right)\lim_{l\to+\infty}\left[e^{-l^2} - e^{-1}\right] = \frac{1}{2e}$$

Thus, the improper integral and the series converge. ◄

The harmonic series and the series in Example 5 are special cases of a class of series called *p-series* or *hyperharmonic series.* A *p*-series is an infinite series of the form

$$\sum_{k=1}^{\infty} \frac{1}{k^p} = 1 + \frac{1}{2^p} + \frac{1}{3^p} + \cdots + \frac{1}{k^p} + \cdots$$

where $p > 0$. Examples of *p*-series are

$$\sum_{k=1}^{\infty} \frac{1}{k} = 1 + \frac{1}{2} + \frac{1}{3} + \cdots + \frac{1}{k} + \cdots \qquad (p = 1)$$

$$\sum_{k=1}^{\infty} \frac{1}{k^2} = 1 + \frac{1}{2^2} + \frac{1}{3^2} + \cdots + \frac{1}{k^2} + \cdots \qquad (p = 2)$$

$$\sum_{k=1}^{\infty} \frac{1}{\sqrt{k}} = 1 + \frac{1}{\sqrt{2}} + \frac{1}{\sqrt{3}} + \cdots + \frac{1}{\sqrt{k}} + \cdots \qquad (p = \tfrac{1}{2})$$

The following theorem tells when a *p*-series converges.

11.4.6 THEOREM
Convergence of p-Series

$$\sum_{k=1}^{\infty} \frac{1}{k^p} = 1 + \frac{1}{2^p} + \frac{1}{3^p} + \cdots + \frac{1}{k^p} + \cdots$$

converges if $p > 1$ and diverges if $0 < p \leq 1$.

Proof. To establish this result when $p \neq 1$, we will use the integral test.

$$\int_1^{+\infty} \frac{1}{x^p} \, dx = \lim_{l \to +\infty} \int_1^l x^{-p} \, dx$$

$$= \lim_{l \to +\infty} \frac{x^{1-p}}{1-p} \Big]_1^l$$

$$= \lim_{l \to +\infty} \left[\frac{l^{1-p}}{1-p} - \frac{1}{1-p} \right]$$

For $p > 1$, $1 - p < 0$ and $l^{1-p} \to 0$ as $l \to +\infty$, so the integral and the series converge. For $0 < p < 1$, $1 - p > 0$ and $l^{1-p} \to +\infty$ as $l \to +\infty$, so the integral and the series diverge. The case $p = 1$ is the harmonic series, which was previously shown to diverge. ▮

► Example 8

$$1 + \frac{1}{\sqrt[3]{2}} + \frac{1}{\sqrt[3]{3}} + \cdots + \frac{1}{\sqrt[3]{k}} + \cdots$$

diverges since it is a p-series with $p = \frac{1}{3} < 1$.   ◄

We conclude this section by proving Theorem 11.4.5 (the integral test).

Proof.   Let $f(x)$ satisfy the hypotheses of the theorem. Since

$$f(1) = u_1, \ f(2) = u_2, \dots, f(n) = u_n, \dots$$

the rectangles shown in Figures 11.4.1a and 11.4.1b have areas $u_1, u_2, \dots, u_n$ as indicated.

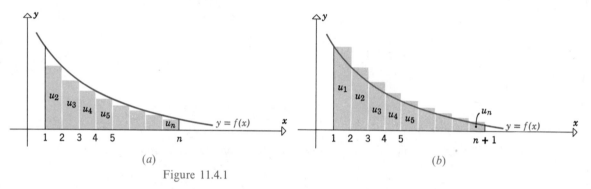

(a) (b)

Figure 11.4.1

In Figure 11.4.1a, the total area of the rectangles is less than the area under the curve from $x = 1$ to $x = n$, so

$$u_2 + u_3 + \cdots + u_n < \int_1^n f(x)\,dx$$

Therefore,

$$u_1 + u_2 + u_3 + \cdots + u_n < u_1 + \int_1^n f(x)\,dx \tag{2}$$

In Figure 11.4.1b, the total area of the rectangles is greater than the area under the curve from $x = 1$ to $x = n + 1$, so that

$$\int_1^{n+1} f(x)\,dx < u_1 + u_2 + \cdots + u_n \tag{3}$$

If we let $s_n = u_1 + u_2 + \cdots + u_n$ be the nth partial sum of the series, then (2) and (3) yield

$$\int_1^{n+1} f(x)\,dx < s_n < u_1 + \int_1^n f(x)\,dx \tag{4}$$

If the integral $\int_1^{+\infty} f(x)\,dx$ converges to a finite value L, then from the right-hand inequality in (4)

$$s_n < u_1 + \int_1^n f(x)\,dx < u_1 + \int_1^{+\infty} f(x)\,dx = u_1 + L$$

Thus, each partial sum is less than the finite constant $u_1 + L$, and the series converges by Theorem 11.4.4. On the other hand, if the integral $\int_1^{+\infty} f(x)\,dx$ diverges, then

$$\lim_{n \to +\infty} \int_1^{n+1} f(x)\,dx = +\infty$$

so that from the left-hand inequality in (4), $\lim\limits_{n \to +\infty} s_n = +\infty$. This means that the series also diverges. ▨

REMARK. If the summation index in a series Σu_k does not begin with $k = 1$, a variation of the integral test may still apply. It can be shown that

$$\sum_{k=K}^{\infty} u_k \qquad \text{and} \qquad \int_K^{+\infty} f(x)\,dx$$

both converge or both diverge provided the hypotheses of Theorem 11.4.5 hold for $x \geq K$.

▶ Exercise Set 11.4

In Exercises 1–4, use Theorem 11.4.3 to find the sum of the series

1. $\displaystyle\sum_{k=1}^{\infty}\left[\frac{1}{2^k} + \frac{1}{4^k}\right].$

2. $\displaystyle\sum_{k=1}^{\infty}\left[\frac{1}{5^k} - \frac{1}{k(k+1)}\right].$

3. $\displaystyle\sum_{k=2}^{\infty}\left[\frac{1}{k^2 - 1} - \frac{7}{10^{k-1}}\right].$

4. $\displaystyle\sum_{k=1}^{\infty}\left[\frac{7}{3^k} + \frac{6}{(k+3)(k+4)}\right].$

5. In each part, determine whether the given p-series converges or diverges.

(a) $\displaystyle\sum_{k=1}^{\infty} \frac{1}{k^3}$

(b) $\displaystyle\sum_{k=1}^{\infty} \frac{1}{\sqrt{k}}$

(c) $\displaystyle\sum_{k=1}^{\infty} k^{-1}$

(d) $\displaystyle\sum_{k=1}^{\infty} k^{-2/3}$

(e) $\displaystyle\sum_{k=1}^{\infty} k^{-4/3}$

(f) $\displaystyle\sum_{k=1}^{\infty} \frac{1}{\sqrt[4]{k}}$

(g) $\displaystyle\sum_{k=1}^{\infty} \frac{1}{\sqrt[3]{k^5}}$

(h) $\displaystyle\sum_{k=1}^{\infty} \frac{1}{k^{\pi}}.$

In Exercises 6–8, use the divergence test to show that the series diverge.

6. (a) $\displaystyle\sum_{k=1}^{\infty} \frac{k+1}{k+2}$

(b) $\displaystyle\sum_{k=1}^{\infty} \ln k.$

7. (a) $\displaystyle\sum_{k=1}^{\infty} \frac{k^2 + k + 3}{2k^2 + 1}$

(b) $\displaystyle\sum_{k=1}^{\infty}\left(1 + \frac{1}{k}\right)^k.$

8. (a) $\displaystyle\sum_{k=1}^{\infty} \cos k\pi$ (b) $\displaystyle\sum_{k=1}^{\infty} \frac{e^k}{k}$.

In Exercises 9–30, determine whether the series converges or diverges.

9. $\displaystyle\sum_{k=1}^{\infty} \frac{1}{k+6}$.

10. $\displaystyle\sum_{k=1}^{\infty} \frac{3}{5k}$.

11. $\displaystyle\sum_{k=1}^{\infty} \frac{1}{5k+2}$.

12. $\displaystyle\sum_{k=1}^{\infty} \frac{k}{1+k^2}$.

13. $\displaystyle\sum_{k=1}^{\infty} \frac{1}{1+9k^2}$.

14. $\displaystyle\sum_{k=1}^{\infty} \frac{1}{(4+2k)^{3/2}}$.

15. $\displaystyle\sum_{k=1}^{\infty} \frac{1}{\sqrt{k+5}}$.

16. $\displaystyle\sum_{k=1}^{\infty} \frac{1}{\sqrt[k]{e}}$.

17. $\displaystyle\sum_{k=1}^{\infty} \frac{1}{\sqrt[3]{2k-1}}$.

18. $\displaystyle\sum_{k=3}^{\infty} \frac{\ln k}{k}$.

19. $\displaystyle\sum_{k=1}^{\infty} \frac{k}{\ln(k+1)}$.

20. $\displaystyle\sum_{k=1}^{\infty} ke^{-k^2}$.

21. $\displaystyle\sum_{k=1}^{\infty} \frac{1}{(k+1)[\ln(k+1)]^2}$.

22. $\displaystyle\sum_{k=1}^{\infty} \frac{k^2+1}{k^2+3}$.

23. $\displaystyle\sum_{k=1}^{\infty} \left(1+\frac{1}{k}\right)^k$.

24. $\displaystyle\sum_{k=1}^{\infty} \frac{1}{\sqrt{k^2+1}}$.

25. $\displaystyle\sum_{k=1}^{\infty} \frac{\tan^{-1} k}{1+k^2}$.

26. $\displaystyle\sum_{k=1}^{\infty} \operatorname{sech}^2 k$.

27. $\displaystyle\sum_{k=5}^{\infty} 7k^{-p} \quad (p>1)$.

28. $\displaystyle\sum_{k=1}^{\infty} 7(k+5)^{-p} \quad (p \le 1)$.

29. $\displaystyle\sum_{k=1}^{\infty} k^2 \sin^2\left(\frac{1}{k}\right)$.

30. $\displaystyle\sum_{k=1}^{\infty} k^2 e^{-k^3}$.

31. Prove: $\displaystyle\sum_{k=2}^{\infty} \frac{1}{k(\ln k)^p}$ converges if $p>1$ and diverges if $p \le 1$.

32. Prove: $\displaystyle\sum_{k=3}^{\infty} \frac{1}{k(\ln k)[\ln(\ln k)]^p}$ converges if $p>1$ and diverges if $p \le 1$.

33. Prove: If Σu_k converges and Σv_k diverges, then $\Sigma(u_k + v_k)$ diverges and $\Sigma(u_k - v_k)$ diverges. [*Hint:* Assume $\Sigma(u_k + v_k)$ converges and use Theorem 11.4.3 to obtain a contradiction. Similarly, for $\Sigma(u_k - v_k)$.]

34. Find examples to show that $\Sigma(u_k + v_k)$ and $\Sigma(u_k - v_k)$ may converge or may diverge if Σu_k and Σv_k both diverge.

35. With the help of Exercise 33, determine whether the given series converges or diverges.

(a) $\displaystyle\sum_{k=1}^{\infty} \left[\left(\frac{2}{3}\right)^{k-1} + \frac{1}{k}\right]$

(b) $\displaystyle\sum_{k=1}^{\infty} \left[\frac{k^2}{1+k^2} + \frac{1}{k(k+1)}\right]$

(c) $\displaystyle\sum_{k=1}^{\infty} \left[\frac{1}{3k+2} + \frac{1}{k^{3/2}}\right]$

(d) $\displaystyle\sum_{k=2}^{\infty} \left[\frac{1}{k(\ln k)^2} - \frac{1}{k^2}\right]$.

11.5 ADDITIONAL CONVERGENCE TESTS

In this section we will develop some additional convergence tests for series with *positive* terms.

Our first result is not only a practical test for convergence, but is also a theoretical tool that will be used to develop other convergence tests.

11.5.1 THEOREM

The Comparison Test

> *Let Σa_k and Σb_k be series with positive terms and suppose*
>
> $$a_1 \le b_1, \, a_2 \le b_2, \, a_3 \le b_3, \ldots, a_k \le b_k, \ldots$$
>
> (a) *If the "bigger series" Σb_k converges, then the "smaller series" Σa_k also converges.*
> (b) *On the other hand, if the "smaller series" Σa_k diverges, then the "bigger series" Σb_k also diverges.*

Proof of (a). Suppose that the series Σb_k converges and its sum is B. Then for all n

$$b_1 + b_2 + \cdots + b_n < \sum_{k=1}^{\infty} b_k = B$$

From our hypothesis it follows that

$$a_1 + a_2 + \cdots + a_n \le b_1 + b_2 + \cdots + b_n$$

so that

$$a_1 + a_2 + \cdots + a_n < B$$

Thus, each partial sum of the series Σa_k is less than B, so that Σa_k converges by Theorem 11.4.4.

Proof of (b). This part is really just an alternate phrasing of part (a). If Σa_k diverges, then Σb_k must diverge since convergence of Σb_k would imply convergence of Σa_k, contrary to hypothesis. ▮

Since the comparison test requires a little ingenuity to use, we will wait until the next section before applying it. For now, we will use the comparison test to develop some other tests that are easier to apply.

11.5.2 THEOREM

The Ratio Test

> *Let Σu_k be a series with positive terms and suppose*
>
> $$\lim_{k \to +\infty} \frac{u_{k+1}}{u_k} = \rho$$
>
> (a) *If $\rho < 1$, the series converges.*
> (b) *If $\rho > 1$ or $\rho = +\infty$, the series diverges.*
> (c) *If $\rho = 1$, the series may converge or diverge, so that another test must be tried.*

Proof of (a). Assume $\rho < 1$, and let $r = \frac{1}{2}(1 + \rho)$. Thus, $\rho < r < 1$, since r is the midpoint between 1 and ρ. It follows that the number

$$\epsilon = r - \rho \tag{1}$$

is positive. Since

$$\rho = \lim_{k \to +\infty} \frac{u_{k+1}}{u_k}$$

it follows that for k sufficiently large, say $k \geq K$, the ratios u_{k+1}/u_k are within ϵ units of ρ. Thus, we will have

$$\frac{u_{k+1}}{u_k} < \rho + \epsilon \qquad \text{when} \quad k \geq K$$

or on substituting (1)

$$\frac{u_{k+1}}{u_k} < r \qquad \text{when} \quad k \geq K$$

that is,

$$u_{k+1} < ru_k \qquad \text{when} \quad k \geq K$$

This yields the inequalities

$$\begin{aligned}
u_{K+1} &< ru_K \\
u_{K+2} &< ru_{K+1} < r^2 u_K \\
u_{K+3} &< ru_{K+2} < r^3 u_K \\
u_{K+4} &< ru_{K+3} < r^4 u_K \\
&\;\;\vdots
\end{aligned} \tag{2}$$

But $|r| < 1$ (why?) so that

$$ru_K + r^2 u_K + r^3 u_K + \cdots$$

is a convergent geometric series. From the inequalities in (2) and the comparison test it follows that

$$u_{K+1} + u_{K+2} + u_{K+3} + \cdots$$

must also be a convergent series. Thus, $u_1 + u_2 + u_3 + \cdots + u_k + \cdots$ converges by Theorem 11.4.3c.

Proof of (b). Assume $\rho > 1$. Thus,

$$\epsilon = \rho - 1 \tag{3}$$

is a positive number. Since

$$\rho = \lim_{k \to +\infty} \frac{u_{k+1}}{u_k}$$

it follows that for k sufficiently large, say $k \geq K$, the ratio u_{k+1}/u_k is within ϵ units of ρ. Thus,

$$\frac{u_{k+1}}{u_k} > \rho - \epsilon \qquad \text{when} \quad k \geq K$$

or on substituting (3)

$$\frac{u_{k+1}}{u_k} > 1 \qquad \text{when} \quad k \geq K$$

that is,

$$u_{k+1} > u_k \qquad \text{when} \quad k \geq K$$

This yields the inequalities

$$\begin{aligned}
u_{K+1} &> u_K \\
u_{K+2} &> u_{K+1} > u_K \\
u_{K+3} &> u_{K+2} > u_K \\
u_{K+4} &> u_{K+3} > u_K \\
&\;\;\vdots
\end{aligned} \tag{4}$$

Since $u_K > 0$, it follows from (4) that $\lim_{k \to +\infty} u_k \neq 0$, so $u_1 + u_2 + \cdots + u_k + \cdots$ diverges by Theorem 11.4.2. The proof in the case where $\rho = +\infty$ is omitted.

Proof of (c). The series

$$\sum_{k=1}^{\infty} \frac{1}{k} \qquad \text{and} \qquad \sum_{k=1}^{\infty} \frac{1}{k^2}$$

both have $\rho = 1$ (verify). Since the first is the divergent harmonic series and the second is a convergent p-series, the ratio test does not distinguish between convergence and divergence when $\rho = 1$.

▶ Example 1   The series

$$\sum_{k=1}^{\infty} \frac{1}{k!}$$

converges by the ratio test since

$$\rho = \lim_{k \to +\infty} \frac{u_{k+1}}{u_k} = \lim_{k \to +\infty} \frac{1/(k + 1)!}{1/k!} = \lim_{k \to +\infty} \frac{k!}{(k + 1)!} = \lim_{k \to +\infty} \frac{1}{k + 1} = 0$$

so that $\rho < 1$.   ◀

▶ Example 2   The series

The series

$$\sum_{k=1}^{\infty} \frac{k}{2^k}$$

converges by the ratio test since

$$\rho = \lim_{k \to +\infty} \frac{u_{k+1}}{u_k} = \lim_{k \to +\infty} \frac{k + 1}{2^{k+1}} \cdot \frac{2^k}{k} = \frac{1}{2} \lim_{k \to +\infty} \frac{k + 1}{k} = \frac{1}{2}$$

so that $\rho < 1$.   ◀

▶ Example 3   The series

$$\sum_{k=1}^{\infty} \frac{k^k}{k!}$$

diverges by the ratio test since

$$\rho = \lim_{k \to +\infty} \frac{u_{k+1}}{u_k} = \lim_{k \to +\infty} \frac{(k + 1)^{k+1}}{(k + 1)!} \cdot \frac{k!}{k^k}$$

$$= \lim_{k \to +\infty} \frac{(k + 1)^k}{k^k}$$

$$= \lim_{k \to +\infty} \left(1 + \frac{1}{k}\right)^k = e \qquad \begin{bmatrix} \text{See Section 10.3} \\ \text{Formula (4).} \end{bmatrix}$$

Since $\rho = e > 1$, the series diverges.   ◀

▶ Example 4 Determine whether the series

$$1 + \frac{1}{3} + \frac{1}{5} + \frac{1}{7} + \cdots + \frac{1}{2k-1} + \cdots$$

converges or diverges.

Solution. The ratio test is of no help since

$$\rho = \lim_{k \to +\infty} \frac{u_{k+1}}{u_k} = \lim_{k \to +\infty} \frac{1}{2(k+1)-1} \cdot \frac{2k-1}{1}$$

$$= \lim_{k \to +\infty} \frac{2k-1}{2k+1} = 1$$

However, the integral test proves that the series diverges since

$$\int_1^{+\infty} \frac{dx}{2x-1} = \lim_{l \to +\infty} \int_1^l \frac{dx}{2x-1} = \lim_{l \to +\infty} \frac{1}{2} \ln (2x-1) \Big]_1^l = +\infty \qquad ◀$$

▶ Example 5 The series

$$\frac{2!}{4} + \frac{4!}{4^2} + \frac{6!}{4^3} + \cdots + \frac{(2k)!}{4^k} + \cdots$$

diverges since

$$\rho = \lim_{k \to +\infty} \frac{u_{k+1}}{u_k} = \lim_{k \to +\infty} \frac{[2(k+1)]!}{4^{k+1}} \cdot \frac{4^k}{(2k)!}$$

$$= \lim_{k \to +\infty} \left(\frac{(2k+2)!}{(2k)!} \cdot \frac{1}{4} \right)$$

$$= \frac{1}{4} \lim_{k \to +\infty} (2k+1)(2k+2) = +\infty \qquad ◀$$

Sometimes the following result is easier to apply than the ratio test.

11.5.3 THEOREM

The Root Test

Let Σu_k be a series with positive terms and suppose

$$\rho = \lim_{k \to +\infty} \sqrt[k]{u_k} = \lim_{k \to +\infty} (u_k)^{1/k}$$

(a) *If $\rho < 1$, the series converges.*
(b) *If $\rho > 1$ or $\rho = +\infty$, the series diverges.*
(c) *If $\rho = 1$, the series may converge or diverge, so that another test must be tried.*

Since the proof of the root test is similar to the proof of the ratio test, we will omit it.

▶ Example 6   The series

$$\sum_{k=1}^{\infty} \left(\frac{4k - 5}{2k + 1} \right)^k$$

diverges by the root test since

$$\rho = \lim_{k \to +\infty} (u_k)^{1/k} = \lim_{k \to +\infty} \frac{4k - 5}{2k + 1} = 2 > 1 \qquad \blacktriangleleft$$

▶ Example 7   The series

$$\sum_{k=1}^{\infty} \frac{1}{(\ln (k + 1))^k}$$

converges by the root test, since

$$\lim_{k \to +\infty} (u_k)^{1/k} = \lim_{k \to +\infty} \frac{1}{\ln (k + 1)} = 0 < 1 \qquad \blacktriangleleft$$

We conclude this section with a remark about notation. Until now we have written most of our infinite series in the form

$$\sum_{k=1}^{\infty} u_k \qquad (5)$$

with the summation index beginning at 1. If the summation index begins at some other integer, it is always possible to rewrite the series in form (5). Thus, for example, the series

$$\sum_{k=0}^{\infty} \frac{2^k}{k!} = 1 + 2 + \frac{2^2}{2!} + \frac{2^3}{3!} + \cdots \qquad (6)$$

can be written as

$$\sum_{k=1}^{\infty} \frac{2^{k-1}}{(k - 1)!} = 1 + 2 + \frac{2^2}{2!} + \frac{2^2}{3!} + \cdots \qquad (7)$$

However, for purposes of applying the convergence tests, it is not necessary that the series have form (5). For example, we can apply the ratio test to (6) without converting to the more complicated form (7). Doing so yields

$$\rho = \lim_{k \to +\infty} \frac{u_{k+1}}{u_k} = \lim_{k \to +\infty} \frac{2^{k+1}}{(k+1)!} \cdot \frac{k!}{2^k} = \lim_{k \to +\infty} \frac{2}{k+1} = 0$$

which shows that the series converges since $\rho < 1$.

▶ Exercise Set 11.5

In Exercises 1–6, apply the ratio test. According to the test, does the series converge, does the series diverge, or are the results inconclusive?

1. $\displaystyle\sum_{k=1}^{\infty} \frac{3^k}{k!}$.

2. $\displaystyle\sum_{k=1}^{\infty} \frac{4^k}{k^2}$.

3. $\displaystyle\sum_{k=2}^{\infty} \frac{1}{5k}$.

4. $\displaystyle\sum_{k=1}^{\infty} k\left(\frac{1}{2}\right)^k$.

5. $\displaystyle\sum_{k=1}^{\infty} \frac{k!}{k^3}$.

6. $\displaystyle\sum_{k=1}^{\infty} \frac{k}{k^2+1}$.

In Exercises 7–10, apply the root test. According to the test, does the series converge, does the series diverge, or are the results inconclusive?

7. $\displaystyle\sum_{k=1}^{\infty} \left(\frac{3k+2}{2k-1}\right)^k$.

8. $\displaystyle\sum_{k=1}^{\infty} \left(\frac{k}{100}\right)^k$.

9. $\displaystyle\sum_{k=1}^{\infty} \frac{k}{5^k}$.

10. $\displaystyle\sum_{k=1}^{\infty} (1 + e^{-k})^k$.

In Exercises 11–32, use any appropriate test to determine whether the series converges.

11. $\displaystyle\sum_{k=1}^{\infty} \frac{2^k}{k^3}$.

12. $\displaystyle\sum_{k=1}^{\infty} \frac{1}{k^2}$.

13. $\displaystyle\sum_{k=0}^{\infty} \frac{7^k}{k!}$.

14. $\displaystyle\sum_{k=1}^{\infty} \frac{1}{2k+1}$.

15. $\displaystyle\sum_{k=1}^{\infty} \frac{k^2}{5^k}$.

16. $\displaystyle\sum_{k=1}^{\infty} \frac{k!\,10^k}{3^k}$.

17. $\displaystyle\sum_{k=1}^{\infty} k^{50} e^{-k}$.

18. $\displaystyle\sum_{k=1}^{\infty} \frac{k^2}{k^3+1}$.

19. $\displaystyle\sum_{k=1}^{\infty} k\left(\frac{2}{3}\right)^k$.

20. $\displaystyle\sum_{k=1}^{\infty} k^k$.

21. $\displaystyle\sum_{k=2}^{\infty} \frac{1}{k \ln k}$.

22. $\displaystyle\sum_{k=1}^{\infty} \frac{2^k}{k^3+1}$.

23. $\displaystyle\sum_{k=1}^{\infty} \left(\frac{4}{7k-1}\right)^k$.

24. $\displaystyle\sum_{k=1}^{\infty} \frac{(k!)^2 2^k}{(2k+2)!}$.

25. $\displaystyle\sum_{k=0}^{\infty} \frac{(k!)^2}{(2k)!}$.

26. $\displaystyle\sum_{k=1}^{\infty} \frac{1}{k^2+25}$.

27. $\displaystyle\sum_{k=1}^{\infty} \frac{1}{1+\sqrt{k}}$.

28. $\displaystyle\sum_{k=1}^{\infty} \frac{k^k}{k!}$.

29. $\displaystyle\sum_{k=1}^{\infty} \frac{\ln k}{e^k}$.

30. $\displaystyle\sum_{k=1}^{\infty} \frac{k!}{e^{k^2}}$.

31. $\displaystyle\sum_{k=0}^{\infty} \frac{(k+4)!}{4!\,k!\,4^k}$.

32. $\displaystyle\sum_{k=1}^{\infty} \left(\frac{k}{k+1}\right)^{k^2}$.

33. Determine whether the following series converges:

$$1 + \frac{1 \cdot 2}{1 \cdot 3} + \frac{1 \cdot 2 \cdot 3}{1 \cdot 3 \cdot 5} + \frac{1 \cdot 2 \cdot 3 \cdot 4}{1 \cdot 3 \cdot 5 \cdot 7} + \cdots$$

34. For which positive values of α does $\displaystyle\sum_{k=1}^{\infty} \frac{\alpha^k}{k^\alpha}$ converge?

35. (a) Show $\displaystyle\lim_{k \to +\infty} (\ln k)^{1/k} = 1$ [*Hint:* Let $y = (\ln x)^{1/x}$ and find $\displaystyle\lim_{x \to +\infty} \ln y$.]

   (b) Use the result in (a) and the root test to show that $\displaystyle\sum_{k=1}^{\infty} \frac{\ln k}{3^k}$ converges.

   (c) Show that the series converges using the ratio test.

36. Prove: $\displaystyle\lim_{k \to +\infty} k!/k^k = 0$ [*Hint:* Exploit Theorem 11.4.1.]

37. Prove: $\displaystyle\lim_{k \to +\infty} \frac{a^k}{k!} = 0$ for every real number a. [See Hint to Exercise 36.]

11.6 APPLYING THE COMPARISON TEST

The comparison test requires some skill and experience to apply. One must first guess whether the given series Σu_k converges or diverges. If the guess is divergence, then one tries to prove it by producing a divergent series whose terms are "smaller" than the corresponding terms of Σu_k, while if the guess is convergence one tries to prove it by producing a convergent series whose terms are "bigger" than the corresponding terms of Σu_k. In this section we will illustrate these ideas through examples and also introduce a useful variation of the comparison test. Before starting, we remind the reader that the comparison test applies only to series with positive terms.

As noted above, the first step in applying the comparison test is to guess whether the series converges or diverges. To aid in this guessing process we have formulated some principles that sometimes *suggest* whether a series is likely to converge or diverge. We have called these "informal principles" because they are not intended as formal theorems. In fact, we will not guarantee that they *always* work. However, they work often enough to be useful as a starting point for the comparison test.

11.6.1 INFORMAL *Constant terms in the denominator of u_k can usually be deleted without affect-*
PRINCIPLE *ing the convergence or divergence of the series.*

▶ **Example 1** Use the above principle to help guess whether the following series converge or diverge.

(a) $\displaystyle\sum_{k=1}^{\infty} \frac{1}{2^k + 1}$ (b) $\displaystyle\sum_{k=5}^{\infty} \frac{1}{\sqrt{k} - 2}$ (c) $\displaystyle\sum_{k=1}^{\infty} \frac{1}{(k + \frac{1}{2})^3}$

Solution.
(a) Deleting the constant 1 suggests that

$$\sum_{k=1}^{\infty} \frac{1}{2^k + 1} \quad \text{behaves like} \quad \sum_{k=1}^{\infty} \frac{1}{2^k}$$

The modified series is a convergent geometric series, so the given series is likely to converge.

(b) Deleting the -2 suggests that

$$\sum_{k=5}^{\infty} \frac{1}{\sqrt{k} - 2} \quad \text{behaves like} \quad \sum_{k=5}^{\infty} \frac{1}{\sqrt{k}}$$

The modified series is a portion of a divergent *p*-series ($p = \frac{1}{2}$), so the given series is likely to diverge.

(c)   Deleting the $\frac{1}{2}$ suggests that

$$\sum_{k=1}^{\infty} \frac{1}{(k+\frac{1}{2})^3} \qquad \text{behaves like} \qquad \sum_{k=1}^{\infty} \frac{1}{k^3}$$

The modified series is a convergent p-series ($p = 3$), so the given series is likely to converge.   ◀

11.6.2   INFORMAL
PRINCIPLE

If a polynomial in k appears as a factor in the numerator or denominator of u_k, all but the highest power of k in the polynomial may usually be deleted without affecting the convergence or divergence behavior of the series.

▶ **Example 2**   Use the above principle to help guess whether the following series converge or diverge.

(a) $\displaystyle\sum_{k=1}^{\infty} \frac{1}{\sqrt{k^3 + 2k}}$ (b) $\displaystyle\sum_{k=1}^{\infty} \frac{6k^4 - 2k^3 + 1}{k^5 + k^2 - 2k}$

Solution.

(a)   Deleting the term $2k$ suggests that

$$\sum_{k=1}^{\infty} \frac{1}{\sqrt{k^3 + 2k}} \qquad \text{behaves like} \qquad \sum_{k=1}^{\infty} \frac{1}{\sqrt{k^3}} = \sum_{k=1}^{\infty} \frac{1}{k^{3/2}}$$

Since the modified series is a convergent p-series ($p = \frac{3}{2}$), the given series is likely to converge.

(b)   Deleting all but the highest powers of k in the numerator and also in the denominator suggests that

$$\sum_{k=1}^{\infty} \frac{6k^4 - 2k^3 + 1}{k^5 + k^2 - 2k} \qquad \text{behaves like} \qquad \sum_{k=1}^{\infty} \frac{6k^4}{k^5} = 6\sum_{k=1}^{\infty} \frac{1}{k}$$

Since the modified series is a constant times the divergent harmonic series, the given series is likely to diverge.   ◀

Once it is decided whether a series is likely to converge or diverge, the second step in applying the comparison test is to produce a series with which the given series may be compared to substantiate the guess. Let us consider the case of convergence first. To prove Σa_k converges by the comparison test, we must find a *convergent* series Σb_k such that

$$a_k \leq b_k$$

for all k. Frequently, b_k is derived from the formula for a_k by either increasing the numerator of a_k, or decreasing the denominator of a_k, or both.

▶ **Example 3**   Use the comparison test to determine whether

$$\sum_{k=1}^{\infty} \frac{1}{2k^2 + k}$$

converges or diverges.

Solution.   Using Principle 11.6.2, the series behaves like the series

$$\sum_{k=1}^{\infty} \frac{1}{2k^2} = \frac{1}{2} \sum_{k=1}^{\infty} \frac{1}{k^2}$$

which is a constant times a convergent p-series. Thus, the given series is likely to converge. To prove the convergence, observe that when we discard the k from the denominator of $1/(2k^2 + k)$, the denominator decreases and the ratio increases, so that

$$\frac{1}{2k^2 + k} < \frac{1}{2k^2}$$

for $k = 1, 2, \ldots$. Since

$$\sum_{k=1}^{\infty} \frac{1}{2k^2} = \frac{1}{2} \sum_{k=1}^{\infty} \frac{1}{k^2}$$

converges, so does $\displaystyle\sum_{k=1}^{\infty} \frac{1}{2k^2 + k}$ by the comparison test.   ◀

▶ **Example 4**   Use the comparison test to determine whether

$$\sum_{k=1}^{\infty} \frac{1}{2k^2 - k}$$

converges or diverges.

Solution.   Using Principle 11.6.2, the series behaves like the convergent series

$$\sum_{k=1}^{\infty} \frac{1}{2k^2} = \frac{1}{2} \sum_{k=1}^{\infty} \frac{1}{k^2}$$

Thus, the given series is likely to converge. However, if we discard the k from the denominator of $1/(2k^2 - k)$, the denominator increases and the ratio decreases, so that

$$\frac{1}{2k^2 - k} > \frac{1}{2k^2}$$

Unfortunately, this inequality is in the wrong direction to prove convergence of the given series. A different approach is needed; we must do something to decrease the denominator, not increase it. We accomplish this by replacing k by k^2 to obtain

$$\frac{1}{2k^2 - k} \leq \frac{1}{2k^2 - k^2} = \frac{1}{k^2}$$

Since $\sum\limits_{k=1}^{\infty} \dfrac{1}{k^2}$ is a convergent p-series, the given series converges by the comparison test.   ◄

To prove that a series Σa_k diverges by the comparison test, we must produce a divergent series Σb_k of positive terms such that $a_k \geq b_k$ for all k.

► **Example 5** Use the comparison test to determine whether

$$\sum_{k=1}^{\infty} \frac{1}{k - \frac{1}{4}}$$

converges or diverges.

Solution. Using Principle 11.6.1, the series behaves like the divergent harmonic series

$$\sum_{k=1}^{\infty} \frac{1}{k}$$

Thus, the given series is likely to diverge. Since

$$\frac{1}{k - \frac{1}{4}} > \frac{1}{k} \qquad \text{for } k = 1, 2, \ldots .$$

and since $\sum\limits_{k=1}^{\infty} \dfrac{1}{k}$ diverges, the given series diverges by the comparison test.   ◄

▶ **Example 6**   Use the comparison test to determine whether

$$\sum_{k=1}^{\infty} \frac{1}{\sqrt{k} + 5}$$

converges or diverges.

Solution.   Using Principle 11.6.1, the series behaves like the divergent *p*-series

$$\sum_{k=1}^{\infty} \frac{1}{\sqrt{k}}$$

Thus, the given series is likely to diverge. For $k \geq 25$ we have

$$\frac{1}{\sqrt{k} + 5} \geq \frac{1}{\sqrt{k} + \sqrt{k}} = \frac{1}{2\sqrt{k}}$$

and since

$$\sum_{k=25}^{\infty} \frac{1}{2\sqrt{k}}$$

diverges (why?), the series

$$\sum_{k=25}^{\infty} \frac{1}{\sqrt{k} + 5}$$

diverges by the comparison test; consequently, the given series diverges by Theorem 11.4.3(*c*).   ◀

The following result is often easier to apply than the comparison test.

11.6.3   THEOREM

The Limit Comparison Test

Let Σa_k and Σb_k be series with positive terms and suppose

$$\rho = \lim_{k \to +\infty} \frac{a_k}{b_k}$$

If ρ is finite and $\rho > 0$, then the series both converge or both diverge.

Proof.   The proof will be complete if we can show that Σb_k converges when Σa_k converges and conversely. To this end, let

$$\epsilon = \frac{\rho}{2} \tag{1}$$

so that $\epsilon > 0$ because $\rho > 0$ by hypothesis. From the assumption that

$$\rho = \lim_{k \to +\infty} \frac{a_k}{b_k}$$

it follows that for k sufficiently large, say $k \geq K$, the ratio a_k/b_k will be within ϵ units of ρ. Thus,

$$\rho - \epsilon < \frac{a_k}{b_k} < \rho + \epsilon \qquad \text{when} \qquad k \geq K$$

or on substituting (1)

$$\frac{1}{2}\rho < \frac{a_k}{b_k} < \frac{3}{2}\rho \qquad \text{when} \quad k \geq K$$

or

$$\frac{1}{2}\rho b_k < a_k < \frac{3}{2}\rho b_k \qquad \text{when} \quad k \geq K \tag{2}$$

If $\sum_{k=1}^{\infty} a_k$ converges, then $\sum_{k=K}^{\infty} a_k$ converges by Theorem 11.4.3(c). It follows from the comparison test and the left-hand inequality in (2), that $\sum_{k=K}^{\infty} \frac{1}{2}\rho b_k$ converges. Thus, $\sum_{k=K}^{\infty} b_k$ converges (Theorem 11.4.3(b)) and $\sum_{k=1}^{\infty} b_k$ converges (Theorem 11.4.3(c)).

Conversely, if $\sum_{k=1}^{\infty} b_k$ converges, then $\sum_{k=K}^{\infty} \frac{3}{2}\rho b_k$ converges (Theorem 11.4.3(b),(c)), so that $\sum_{k=K}^{\infty} a_k$ converges by the comparison test and the right-hand inequality in (2). Thus, $\sum_{k=1}^{\infty} a_k$ converges. ∎

▶ Example 7 Use the limit comparison test to determine whether

$$\sum_{k=1}^{\infty} \frac{3k^3 - 2k^2 + 4}{k^5 - k^3 + 2}$$

converges or diverges.

Solution. From Principle 11.6.2, the series behaves like

$$\sum_{k=1}^{\infty} \frac{3k^3}{k^5} = \sum_{k=1}^{\infty} \frac{3}{k^2} \tag{3}$$

which converges since it is a constant times a convergent *p*-series. Thus, the given series is likely to converge. To substantiate this, we apply the limit comparison test to series (3) and the given series. We obtain

$$\rho = \lim_{k \to +\infty} \frac{\dfrac{3k^3 - 2k^2 + 4}{k^5 - k^3 + 2}}{\dfrac{3}{k^2}} = \lim_{k \to +\infty} \frac{3k^5 - 2k^4 + 4k^2}{3k^5 - 3k^3 + 6} = 1$$

Since $\rho > 0$, the given series converges because series (3) converges. ◄

Unlike the comparison test, the limit comparison test does not require any tricky manipulation of inequalities. However, the test only applies when $0 < \rho < +\infty$. We note, though, that if $\rho = 0$ or $\rho = +\infty$, conclusions about convergence or divergence may be drawn in certain cases (see Exercise 42).

► **Exercise Set 11.6**

In Exercises 1–6, prove that the series converges by the comparison test.

1. $\displaystyle\sum_{k=1}^{\infty} \frac{1}{3^k + 5}$.

2. $\displaystyle\sum_{k=1}^{\infty} \frac{2}{k^4 + k}$.

3. $\displaystyle\sum_{k=1}^{\infty} \frac{1}{5k^2 - k}$.

4. $\displaystyle\sum_{k=1}^{\infty} \frac{k}{8k^3 + 2k^2 - 1}$.

5. $\displaystyle\sum_{k=1}^{\infty} \frac{2^k - 1}{3^k + 2k}$.

6. $\displaystyle\sum_{k=1}^{\infty} \frac{5 \sin^2 k}{k!}$.

In Exercises 7–12, prove that the series diverges by the comparison test.

7. $\displaystyle\sum_{k=1}^{\infty} \frac{3}{k - \frac{1}{4}}$.

8. $\displaystyle\sum_{k=1}^{\infty} \frac{1}{\sqrt{k} + 8}$.

9. $\displaystyle\sum_{k=1}^{\infty} \frac{9}{\sqrt{k} + 1}$.

10. $\displaystyle\sum_{k=2}^{\infty} \frac{k + 1}{k^2 - k}$.

11. $\displaystyle\sum_{k=1}^{\infty} \frac{k^{4/3}}{8k^2 + 5k + 1}$.

12. $\displaystyle\sum_{k=1}^{\infty} \frac{k^{-1/2}}{2 + \sin^2 k}$.

In Exercises 13–18, use the limit comparison test to determine whether the series converges or diverges.

13. $\displaystyle\sum_{k=1}^{\infty} \frac{4k^2 - 2k + 6}{8k^7 + k - 8}$.

14. $\displaystyle\sum_{k=1}^{\infty} \frac{1}{9k + 6}$.

15. $\displaystyle\sum_{k=1}^{\infty} \frac{5}{3^k + 1}$.

16. $\displaystyle\sum_{k=1}^{\infty} \frac{k(k + 3)}{(k + 1)(k + 2)(k + 5)}$.

17. $\displaystyle\sum_{k=1}^{\infty} \frac{1}{\sqrt[3]{8k^2 - 3k}}$.

18. $\displaystyle\sum_{k=1}^{\infty} \frac{1}{(2k + 3)^{17}}$.

In Exercises 19–33, use any method to determine whether the series converges or diverges. In some cases, you may have to use tests from earlier sections.

19. $\displaystyle\sum_{k=1}^{\infty} \frac{1}{k^3 + 2k + 1}$.

20. $\displaystyle\sum_{k=1}^{\infty} \frac{1}{(3 + k)^{2/5}}$.

21. $\displaystyle\sum_{k=1}^{\infty}\frac{1}{9k-2}$.

22. $\displaystyle\sum_{k=1}^{\infty}\frac{\ln k}{k}$.

23. $\displaystyle\sum_{k=1}^{\infty}\frac{\sqrt{k}}{k^3+1}$.

24. $\displaystyle\sum_{k=1}^{\infty}\frac{4}{2+k3^k}$.

25. $\displaystyle\sum_{k=1}^{\infty}\frac{1}{\sqrt{k(k+1)}}$.

26. $\displaystyle\sum_{k=1}^{\infty}\frac{2+(-1)^k}{5^k}$.

27. $\displaystyle\sum_{k=1}^{\infty}\frac{2+\sqrt{k}}{(k+1)^3-1}$.

28. $\displaystyle\sum_{k=1}^{\infty}\frac{4+|\cos k|}{k^3}$.

29. $\displaystyle\sum_{k=1}^{\infty}\frac{1}{4+2^{-k}}$.

30. $\displaystyle\sum_{k=1}^{\infty}\frac{\sqrt{k}\ln k}{k^3+1}$.

31. $\displaystyle\sum_{k=1}^{\infty}\frac{\tan^{-1}k}{k^2}$.

32. $\displaystyle\sum_{k=1}^{\infty}\frac{5^k+k}{k!+3}$.

33. $\displaystyle\sum_{k=1}^{\infty}\frac{\ln k}{k\sqrt{k}}$.

34. Use the limit comparison test to show that

$$\sum_{k=1}^{\infty}\sin\left(\frac{\pi}{k}\right) \text{ diverges. } [\textit{Hint: } \text{Compare with the se-}$$

ries $\displaystyle\sum_{k=1}^{\infty}\frac{\pi}{k}$.]

35. Use the comparison test to determine whether

$$\sum_{k=1}^{\infty}\frac{\ln k}{k^2} \text{ converges or diverges. } [\textit{Hint: } \ln x < 2\sqrt{x}$$

for $x>1$ by Exercise 23 of Section 7.3.]

36. Determine whether $\displaystyle\sum_{k=2}^{\infty}\frac{1}{(\ln k)^2}$ converges or di-

verges. [*Hint:* See Hint to Exercise 35.]

37. Let a, b, and p be positive constants. For which

values of p does the series $\displaystyle\sum_{k=1}^{\infty}\frac{1}{(a+bk)^p}$ con-

verge?

38. (a) Show that $k^k \geq k!$ and use this result to

prove that the series $\displaystyle\sum_{k=1}^{\infty}k^{-k}$ converges by the

comparison test.

   (b) Prove convergence using the root test.

39. Use the limit comparison test to investigate con-

vergence of $\displaystyle\sum_{k=1}^{\infty}\frac{(k+1)^2}{(k+2)!}$.

40. Use the limit comparison test to investigate con-

vergence of the series $1+\frac{1}{3}+\frac{1}{5}+\frac{1}{7}+\cdots$

41. Prove that $\displaystyle\sum_{k=1}^{\infty}\frac{1}{k!}$ converges by comparison with a

suitable geometric series.

42. Let Σa_k and Σb_k be series with positive terms.
Prove:

   (a) If $\lim\limits_{k\to+\infty}(a_k/b_k)=0$ and Σb_k converges, then

Σa_k converges.

   (b) If $\lim\limits_{k\to+\infty}(a_k/b_k)=+\infty$ and Σb_k diverges, then

Σa_k diverges.

11.7 ALTERNATING SERIES; CONDITIONAL CONVERGENCE

So far our emphasis has been on series with positive terms. In this section we discuss series containing negative terms.

Of special importance are series whose terms are alternately positive and negative. These are called **alternating series.** Such series have one of two possible forms:

$$a_1 - a_2 + a_3 - a_4 + \cdots + (-1)^{k+1}a_k + \cdots$$

or

$$-a_1 + a_2 - a_3 + a_4 - \cdots + (-1)^k a_k + \cdots$$

where the a_k's are all positive.

The following theorem is the key result on convergence of alternating series.

11.7.1   THEOREM

Alternating Series Test

An alternating series

$$\sum_{k=1}^{\infty} (-1)^{k+1} a_k \quad \text{or} \quad \sum_{k=1}^{\infty} (-1)^k a_k$$

converges if the following two conditions are satisfied:
(a) $a_1 \geq a_2 \geq a_3 \geq \cdots \geq a_k \geq \cdots$
(b) $\lim_{k \to +\infty} a_k = 0$

To prove this result, we will need the following fact about sequences:

If the even-numbered terms of a sequence tend toward a limit L, and if the odd-numbered terms of the sequence tend toward the same limit L, then the entire sequence tends toward the limit L.

This result should be intuitively obvious; we omit the proof.

Proof of 11.7.1.   We will consider the case

$$a_1 - a_2 + a_3 - a_4 + \cdots + (-1)^{k+1} a_k + \cdots$$

The other case is left as an exercise. Consider the partial sums

$$s_2 = a_1 - a_2$$
$$s_4 = (a_1 - a_2) + (a_3 - a_4)$$
$$s_6 = (a_1 - a_2) + (a_3 - a_4) + (a_5 - a_6)$$
$$s_8 = (a_1 - a_2) + (a_3 - a_4) + (a_5 - a_6) + (a_7 - a_8)$$
$$\vdots$$

By hypothesis (*a*), each of the differences appearing in the parentheses is nonnegative so that

$$s_2 \leq s_4 \leq s_6 \leq s_8 \leq \cdots$$

Moreover, the terms in this sequence are all less than or equal to a_1 since we can write

$$s_2 = a_1 - a_2$$
$$s_4 = a_1 - (a_2 - a_3) - a_4$$

$$s_6 = a_1 - (a_2 - a_3) - (a_4 - a_5) - a_6$$
$$s_8 = a_1 - (a_2 - a_3) - (a_4 - a_5) - (a_6 - a_7) - a_8$$
$$\vdots$$

Thus, the sequence

$$s_2, s_4, s_6, s_8, \ldots s_{2n}, \ldots$$

converges to some limit S, by Theorem 11.2.2. That is,

$$\lim_{n \to +\infty} s_{2n} = S \tag{1}$$

We will show next that the sequence

$$s_1, s_3, s_5, \ldots, s_{2n-1}, \ldots$$

also has limit S. Since the $(2n)$-th term in the given alternating series is $-a_{2n}$, it follows that $s_{2n} - s_{2n-1} = -a_{2n}$, which can be written as

$$s_{2n-1} = s_{2n} + a_{2n} \tag{2}$$

But $2n \to +\infty$ as $n \to +\infty$ so, $\lim_{n \to +\infty} a_{2n} = 0$ by hypothesis (b). Thus, from (1) and (2)

$$\lim_{n \to +\infty} s_{2n-1} = \lim_{n \to +\infty} s_{2n} + \lim_{n \to +\infty} a_{2n} = S + 0 = S \tag{3}$$

From (1) and (3), the sequence $s_1, s_2, s_3, \ldots, s_n, \ldots$ converges to S, so the series converges. ∎

▶ Example 1   The series

$$1 - \frac{1}{2} + \frac{1}{3} - \frac{1}{4} + \cdots + (-1)^{k+1} \frac{1}{k} + \cdots$$

is called the **alternating harmonic series.** Since

$$a_k = \frac{1}{k} > \frac{1}{k+1} = a_{k+1}$$

and

$$\lim_{k \to +\infty} a_k = \lim_{k \to +\infty} \frac{1}{k} = 0$$

this series converges by the alternating series test.   ◀

▶ Example 2   Determine whether the alternating series

$$\sum_{k=1}^{\infty} (-1)^{k+1} \frac{k+3}{k(k+1)}$$

converges or diverges.

Solution.   Requirement (*b*) of the alternating series test is satisfied since

$$\lim_{k \to +\infty} a_k = \lim_{k \to +\infty} \frac{k+3}{k(k+1)} = \lim_{k \to +\infty} \frac{\dfrac{1}{k} + \dfrac{3}{k^2}}{1 + \dfrac{1}{k}} = 0$$

To see if requirement (*a*) is met, we must determine whether the sequence

$$\{a_k\}_{k=1}^{+\infty} = \left\{ \frac{k+3}{k(k+1)} \right\}_{k=1}^{+\infty}$$

is nonincreasing. Since

$$\frac{a_{k+1}}{a_k} = \frac{k+4}{(k+1)(k+2)} \cdot \frac{k(k+1)}{k+3} = \frac{k^2 + 4k}{k^2 + 5k + 6}$$

$$= \frac{k^2 + 4k}{(k^2 + 4k) + (k+6)} < 1$$

we have $a_k > a_{k+1}$, so the series converges by the alternating series test.   ◀

REMARK.   If an alternating series violates condition (*b*) of the alternating series test, then the series must diverge by the divergence test (11.4.2). However, if condition (*b*) is satisfied, but (*a*) is not, the series may either converge or diverge.*

Figure 11.7.1 provides some insight into the way in which an alternating series

$$a_1 - a_2 + a_3 - a_4 + \cdots + (-1)^{k+1} a_k + \cdots$$

converges to its sum S when the hypotheses of the alternating test are satisfied. In the figure we have plotted the successive partial sums on the *x*-axis. Because

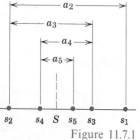

Figure 11.7.1

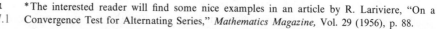

*The interested reader will find some nice examples in an article by R. Lariviere, "On a Convergence Test for Alternating Series," *Mathematics Magazine*, Vol. 29 (1956), p. 88.

$$a_1 \geq a_2 \geq a_3 \geq a_4 \geq \cdots$$

and

$$\lim_{k \to +\infty} a_k = 0$$

the successive partial sums oscillate in smaller and smaller steps, closing in on the sum S. It is of interest to note that the even-numbered partial sums are less than or equal to S and the odd-numbered partial sums are greater than or equal to S. Thus, the sum S falls between any two successive partial sums; that is, for any positive integer n

$$s_n \leq S \leq s_{n+1} \qquad \text{or} \qquad s_{n+1} \leq S \leq s_n$$

depending on whether n is even or odd. In either case,

$$|S - s_n| \leq |s_{n+1} - s_n| \tag{4}$$

But $s_{n+1} - s_n = \pm a_{n+1}$ (the sign depending on whether n is even or odd) so (4) yields

$$|S - s_n| \leq a_{n+1} \tag{5}$$

Since $|S - s_n|$ represents the magnitude of error that results when we approximate the sum of the entire series by the sum of the first n terms, (5) tells us that this error is less than or equal to the magnitude of the $(n + 1)$-st term in the series. The reader can check that (5) also holds for alternating series of the form

$$-a_1 + a_2 - a_3 + a_4 - \cdots + (-1)^k a_k + \cdots$$

In summary, we have the following result.

11.7.2   THEOREM   *If an alternating series satisfies the hypotheses of the alternating series test, and if the sum S of the series is approximated by the nth partial sum s_n, then the absolute value of the error is less than or equal to a_{n+1}.*

▶ **Example 3**   As shown in Example 1, the alternating harmonic series

$$1 - \frac{1}{2} + \frac{1}{3} - \frac{1}{4} + \cdots + (-1)^{k+1} \frac{1}{k} + \cdots$$

satisfies the hypotheses of the alternating series test. If we approximate the sum of the series by

$$1 - \frac{1}{2} + \frac{1}{3} - \frac{1}{4} = \frac{7}{12}$$

then the absolute value of the error is at most $\frac{1}{5} = 0.2$, and if we approximate the sum by

$$1 - \frac{1}{2} + \frac{1}{3} - \frac{1}{4} + \frac{1}{5} - \frac{1}{6} + \frac{1}{7} = \frac{319}{420}$$

then the absolute value of the error is at most $\frac{1}{8} = 0.125$.   ◀

The series

$$1 - \frac{1}{2} - \frac{1}{2^2} + \frac{1}{2^3} + \frac{1}{2^4} - \frac{1}{2^5} - \frac{1}{2^6} + \cdots$$

does not fit in any of the categories studied so far—it has mixed signs, but is not alternating. We will now develop some convergence tests that can be applied to such series.

11.7.3   DEFINITION   A series

$$\sum_{k=1}^{\infty} u_k = u_1 + u_2 + \cdots + u_k + \cdots$$

is said to **converge absolutely** if the series of absolute values

$$\sum_{k=1}^{\infty} |u_k| = |u_1| + |u_2| + \cdots + |u_k| + \cdots$$

converges.

▶ **Example 4**   The series

$$1 - \frac{1}{2} - \frac{1}{2^2} + \frac{1}{2^3} + \frac{1}{2^4} - \frac{1}{2^5} - \frac{1}{2^6} + \cdots$$

converges absolutely since the series of absolute values

$$1 + \frac{1}{2} + \frac{1}{2^2} + \frac{1}{2^3} + \frac{1}{2^4} + \frac{1}{2^5} + \frac{1}{2^6} + \cdots$$

is a convergent geometric series. On the other hand, the alternating harmonic series

$$1 - \frac{1}{2} + \frac{1}{3} - \frac{1}{4} + \frac{1}{5} - \cdots$$

does not converge absolutely since the series of absolute values

$$1 + \frac{1}{2} + \frac{1}{3} + \frac{1}{4} + \frac{1}{5} + \cdots$$

diverges.   ◀

Absolute convergence is of importance because of the following theorem.

11.7.4 THEOREM   *If the series*

$$\sum_{k=1}^{\infty} |u_k| = |u_1| + |u_2| + \cdots + |u_k| + \cdots$$

converges, then so does the series

$$\sum_{k=1}^{\infty} u_k = u_1 + u_2 + \cdots + u_k + \cdots$$

In other words, if a series converges absolutely, then it converges.

Proof. Our proof is based on a trick. We will show that the series

$$\sum_{k=1}^{\infty} (u_k + |u_k|) \tag{6}$$

converges. Since $\Sigma|u_k|$ is assumed to converge, it will then follow from Theorem 11.4.3(*a*) that Σu_k converges, since

$$\sum_{k=1}^{\infty} u_k = \sum_{k=1}^{\infty} [(u_k + |u_k|) - |u_k|]$$

For all k, the value of $u_k + |u_k|$ is either 0 or $2|u_k|$, depending on whether u_k is negative or not. Thus, for all values of k

$$0 \leq u_k + |u_k| \leq 2|u_k| \tag{7}$$

But $\Sigma 2|u_k|$ is a convergent series since it is a constant times the convergent series $\Sigma |u_k|$. Thus, from (7), series (6) converges by the comparison test. ∎

▶ Example 5   In Example 4 we showed that

$$1 - \frac{1}{2} - \frac{1}{2^2} + \frac{1}{2^3} + \frac{1}{2^4} - \frac{1}{2^5} - \frac{1}{2^6} + \cdots$$

converges absolutely. It follows from Theorem 11.7.4 that the series converges. ◀

▶ **Example 6** Show that the series

$$\sum_{k=1}^{\infty} \frac{\cos k}{k^2}$$

converges.

Solution. Since $|\cos k| \leq 1$ for all k

$$\left| \frac{\cos k}{k^2} \right| \leq \frac{1}{k^2}$$

Thus,

$$\sum_{k=1}^{\infty} \left| \frac{\cos k}{k^2} \right|$$

converges by the comparison test, and consequently

$$\sum_{k=1}^{\infty} \frac{\cos k}{k^2}$$

converges. ◀

If $\Sigma |u_k|$ *diverges,* no conclusion can be drawn about the convergence or divergence of Σu_k. For example, consider the two series

$$1 - \frac{1}{2} + \frac{1}{3} - \frac{1}{4} + \cdots + (-1)^{k+1} \frac{1}{k} + \cdots \tag{8}$$

$$-1 - \frac{1}{2} - \frac{1}{3} - \frac{1}{4} - \cdots - \frac{1}{k} - \cdots \tag{9}$$

Series (8), the alternating harmonic series, converges; while series (9), being a constant times the harmonic series, diverges. Yet in each case the series of absolute values is

$$1 + \frac{1}{2} + \frac{1}{3} + \cdots + \frac{1}{k} + \cdots$$

which diverges. A series such as (8), which is convergent, but not absolutely convergent, is called ***conditionally convergent.***

The following version of the ratio test is useful for investigating absolute convergence.

11.7.5 THEOREM

The Ratio Test for Absolute Convergence

Let Σu_k be a series with nonzero terms and suppose

$$\lim_{k \to +\infty} \frac{|u_{k+1}|}{|u_k|} = \rho$$

(a) *If $\rho < 1$, the series Σu_k converges absolutely.*
(b) *If $\rho > 1$ or if $\rho = +\infty$, then the series Σu_k diverges.*
(c) *If $\rho = 1$, no conclusion about convergence can be drawn.*

The proof is discussed in the exercises.

▶ Example 7 The series

$$\sum_{k=1}^{\infty} (-1)^k \frac{2^k}{k!}$$

converges absolutely since

$$\rho = \lim_{k \to +\infty} \frac{|u_{k+1}|}{|u_k|} = \lim_{k \to +\infty} \frac{2^{k+1}}{(k+1)!} \cdot \frac{k!}{2^k} = \lim_{k \to +\infty} \frac{2}{k+1} = 0 < 1 \qquad ◀$$

▶ Example 8 We proved earlier (Theorem 11.3.3) that a geometric series

$$a + ar + ar^2 + \cdots + ar^{k-1} + \cdots$$

converges if $|r| < 1$ and diverges if $|r| \geq 1$. However, a stronger statement can be made—the series converges *absolutely* if $|r| < 1$. This follows from Theorem 11.7.5 since

$$\rho = \lim_{k \to +\infty} \frac{|u_{k+1}|}{|u_k|} = \lim_{k \to +\infty} \frac{|ar^k|}{|ar^{k-1}|} = \lim_{k \to +\infty} |r| = |r|$$

so that $\rho < 1$ if $|r| < 1$. ◀

The following review is included as a ready reference to convergence tests.

Review of Convergence Tests

NAME	STATEMENT	COMMENTS
Divergence test (11.4.2)	If $\lim_{k \to +\infty} u_k \neq 0$, then $\sum u_k$ diverges.	If $\lim_{k \to +\infty} u_k = 0$, $\sum u_k$ may or may not converge.
Integral test (11.4.5)	Let $\sum u_k$ be a series with positive terms and let $f(x)$ be the function that results when k is replaced by x in the formula for u_k. If f is decreasing and continuous for $x \geq 1$, then $$\sum_{k=1}^{\infty} u_k \quad \text{and} \quad \int_1^{+\infty} f(x)\,dx$$ both converge or both diverge.	Use this test when $f(x)$ is easy to integrate. This test only applies to series with positive terms.
Comparison test (11.5.1)	Let $\sum a_k$ and $\sum b_k$ be series with positive terms such that $$a_1 \leq b_1, a_2 \leq b_2, \ldots, a_k \leq b_k, \ldots$$ If $\sum b_k$ converges, then $\sum a_k$ converges; and if $\sum a_k$ diverges, then $\sum b_k$ diverges.	Use this test as a last resort. Other tests are often easier to apply. This test only applies to series with positive terms.
Ratio test (11.5.2)	Let $\sum u_k$ be a series with positive terms and suppose $$\lim_{k \to +\infty} \frac{u_{k+1}}{u_k} = \rho$$ (a) Series converges if $\rho < 1$. (b) Series diverges if $\rho > 1$ or $\rho = +\infty$. (c) No conclusion if $\rho = 1$.	Try this test when u_k involves factorials or kth powers.

Review of Convergence Tests (Continued)

NAME	STATEMENT	COMMENTS
Root test (11.5.3)	Let $\sum u_k$ be a series with positive terms such that $$\rho = \lim_{k \to +\infty} \sqrt[k]{u_k}$$ (a) Series converges if $\rho < 1$. (b) Series diverges if $\rho > 1$ or $\rho = +\infty$. (c) No conclusion if $\rho = 1$.	Try this test when u_k involves kth powers.
Limit comparison test (11.6.3)	Let $\sum a_k$ and $\sum b_k$ be series with positive terms such that $$\rho = \lim_{k \to +\infty} \frac{a_k}{b_k}$$ If $0 < \rho < +\infty$, then both series converge or both diverge.	This is easier to apply than the comparison test, but still requires some skill in choosing the series $\sum b_k$ for comparison.
Alternating series test (11.7.1)	The series $$a_1 - a_2 + a_3 - a_4 + \cdots$$ and $$-a_1 + a_2 - a_3 + a_4 - \cdots$$ converge if (a) $a_1 \geq a_2 \geq a_3 \geq \cdots$ (b) $\lim_{k \to +\infty} a_k = 0$	This test applies only to alternating series.
Ratio test for absolute convergence (11.7.5)	Let $\sum u_k$ be a series with nonzero terms such that $$\rho = \lim_{k \to +\infty} \frac{|u_{k+1}|}{|u_k|}$$ (a) Series converges absolutely if $\rho < 1$. (b) Series diverges if $\rho > 1$ or $\rho = +\infty$. (c) No conclusion if $\rho = 1$.	The series need not have positive terms and need not be alternating to use this test.

▶ Exercise Set 11.7

In Exercises 1–6, use the alternating series test to determine whether the series converges or diverges.

1. $\displaystyle\sum_{k=1}^{\infty} \frac{(-1)^{k+1}}{2k+1}$.

2. $\displaystyle\sum_{k=1}^{\infty} (-1)^{k+1} \frac{k}{3^k}$.

3. $\displaystyle\sum_{k=1}^{\infty} (-1)^{k+1} \frac{k+1}{3k+1}$.

4. $\displaystyle\sum_{k=1}^{\infty} (-1)^{k+1} \frac{k+4}{k^2+k}$.

5. $\displaystyle\sum_{k=1}^{\infty} (-1)^{k+1} e^{-k}$.

6. $\displaystyle\sum_{k=3}^{\infty} (-1)^k \frac{\ln k}{k}$.

In Exercises 7–12, use the ratio test for absolute convergence to determine whether the series converges absolutely or diverges.

7. $\displaystyle\sum_{k=1}^{\infty} \left(-\frac{3}{5}\right)^k$.

8. $\displaystyle\sum_{k=1}^{\infty} (-1)^{k+1} \frac{2^k}{k!}$.

9. $\displaystyle\sum_{k=1}^{\infty} (-1)^{k+1} \frac{3^k}{k^2}$.

10. $\displaystyle\sum_{k=1}^{\infty} (-1)^k \left(\frac{k}{5^k}\right)$.

11. $\displaystyle\sum_{k=1}^{\infty} (-1)^k \left(\frac{k^3}{e^k}\right)$.

12. $\displaystyle\sum_{k=1}^{\infty} (-1)^{k+1} \frac{k^k}{k!}$.

In Exercises 13–30, classify the series as: absolutely convergent, conditionally convergent, or divergent.

13. $\displaystyle\sum_{k=1}^{\infty} \frac{(-1)^{k+1}}{3k}$.

14. $\displaystyle\sum_{k=1}^{\infty} \frac{(-1)^{k+1}}{k^{4/3}}$.

15. $\displaystyle\sum_{k=1}^{\infty} \frac{(-4)^k}{k^2}$.

16. $\displaystyle\sum_{k=1}^{\infty} \frac{(-1)^{k+1}}{k!}$.

17. $\displaystyle\sum_{k=1}^{\infty} \frac{\cos k\pi}{k}$.

18. $\displaystyle\sum_{k=3}^{\infty} \frac{(-1)^k \ln k}{k}$.

19. $\displaystyle\sum_{k=1}^{\infty} (-1)^{k+1} \left(\frac{k+2}{3k-1}\right)^k$.

20. $\displaystyle\sum_{k=1}^{\infty} \frac{(-1)^{k+1}}{k^2+1}$.

21. $\displaystyle\sum_{k=1}^{\infty} (-1)^{k+1} \frac{k+2}{k(k+3)}$.

22. $\displaystyle\sum_{k=1}^{\infty} \frac{(-1)^{k+1}k^2}{k^3+1}$.

23. $\displaystyle\sum_{k=1}^{\infty} \sin\frac{k\pi}{2}$.

24. $\displaystyle\sum_{k=1}^{\infty} \frac{\sin k}{k^3}$.

25. $\displaystyle\sum_{k=2}^{\infty} \frac{(-1)^k}{k \ln k}$.

26. $\displaystyle\sum_{k=1}^{\infty} \frac{(-1)^k}{\sqrt{k(k+1)}}$.

27. $\displaystyle\sum_{k=2}^{\infty} \left(-\frac{1}{\ln k}\right)^k$.

28. $\displaystyle\sum_{k=1}^{\infty} \frac{(-1)^{k+1}}{\sqrt{k+1} + \sqrt{k}}$.

29. $\displaystyle\sum_{k=2}^{\infty} \frac{(-1)^k(k^2+1)}{k^3+2}$.

30. $\displaystyle\sum_{k=1}^{\infty} \frac{k \cos k\pi}{k^2+1}$.

In Exercises 31–34, the given series satisfies the hypotheses of the alternating series test. For the stated value of n, estimate the error that results if the sum of the series is approximated by the nth partial sum.

31. $\displaystyle\sum_{k=1}^{\infty} \frac{(-1)^{k+1}}{k}$; $n = 7$.

32. $\displaystyle\sum_{k=1}^{\infty} \frac{(-1)^{k+1}}{k!}$; $n = 5$.

33. $\displaystyle\sum_{k=1}^{\infty} \frac{(-1)^{k+1}}{\sqrt{k}}$; $n = 99$.

34. $\displaystyle\sum_{k=1}^{\infty} \frac{(-1)^{k+1}}{(k+1)\ln(k+1)}$; $n = 3$.

In Exercises 35–38, the given series satisfies the hypotheses of the alternating series test. Find the smallest value of n for which the nth partial sum approximates the sum of the series to the stated accuracy.

35. $\displaystyle\sum_{k=1}^{\infty} \frac{(-1)^{k+1}}{k}$; $|\text{error}| < 0.0001$.

36. $\displaystyle\sum_{k=1}^{\infty} \frac{(-1)^{k+1}}{k!}$; $|\text{error}| < 0.00001$.

37. $\displaystyle\sum_{k=1}^{\infty} \frac{(-1)^{k+1}}{\sqrt{k}}$; $|\text{error}| < 0.005$.

38. $\displaystyle\sum_{k=1}^{\infty} \frac{(-1)^{k+1}}{(k+1)\ln(k+1)}$; $|\text{error}| < 0.1$.

39. Prove: If Σa_k converges absolutely, then Σa_k^2 converges.

40. Show that the converse of the result in Exercise 39 is false by finding a series for which Σa_k^2 converges, but $\Sigma |a_k|$ diverges.

41. Prove Theorem 11.7.1 for series of the form

$$-a_1 + a_2 - a_3 + a_4 - \cdots + (-1)^k a_k + \cdots$$

42. Prove Theorem 11.7.5. [*Hint:* Theorem 11.7.4 will help in part (*a*). For part (*b*), it may help to review the proof of Theorem 11.5.2.]

11.8 POWER SERIES

In previous sections we studied series with constant terms. In this section we will consider series whose terms involve variables. Such series are of fundamental importance in many branches of mathematics and the physical sciences.

If $c_0, c_1, c_2, \ldots$ are constants and x is a variable, then a series of the form

$$\sum_{k=0}^{\infty} c_k x^k = c_0 + c_1 x + c_2 x^2 + \cdots + c_k x^k + \cdots$$

is called a **power series** in x. Some examples are

$$\sum_{k=0}^{\infty} x^k = 1 + x + x^2 + x^3 + \cdots$$

$$\sum_{k=0}^{\infty} \frac{x^k}{k!} = 1 + x + \frac{x^2}{2!} + \frac{x^3}{3!} + \cdots$$

$$\sum_{k=0}^{\infty} (-1)^k \frac{x^{k+1}}{k+1} = x - \frac{x^2}{2} + \frac{x^3}{3} - \frac{x^4}{4} + \cdots$$

$$\sum_{k=0}^{\infty} (-1)^k \frac{x^{2k}}{(2k)!} = 1 - \frac{x^2}{2!} + \frac{x^4}{4!} - \frac{x^6}{6!} + \cdots$$

$$\sum_{k=0}^{\infty} (-1)^k \frac{x^{2k+1}}{(2k+1)!} = x - \frac{x^3}{3!} + \frac{x^5}{5!} - \frac{x^7}{7!} + \cdots$$

If a numerical value is substituted for x in a power series $\Sigma c_k x^k$, then we obtain a series of constants that may either converge or diverge. This leads to the following basic problem.

A Fundamental Problem. For what values of x does a given power series, $\Sigma c_k x^k$, converge?

The following theorem is the fundamental result on convergence of power series. The proof can be found in most advanced calculus texts.

11.8.1   THEOREM   *For any power series in x, exactly one of the following is true:*

(a)   *The series converges only for $x = 0$.*

(b)   *The series converges absolutely for all x.*

(c)   *The series converges absolutely for all x in some finite open interval $(-R, R)$, and diverges if $x < -R$ or $x > R$ (Figure 11.8.1). At the points $x = R$ and $x = -R$ the series may converge absolutely, converge conditionally, or diverge, depending on the particular series.*

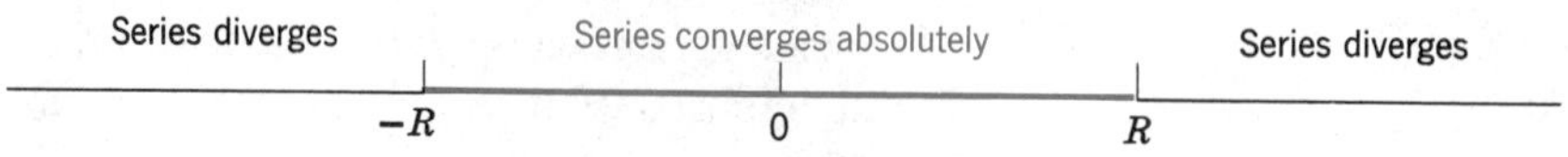

Series diverges	Series converges absolutely	Series diverges
$-R$	0	R

Figure 11.8.1

In case (c), where the power series converges absolutely for $|x| < R$ and diverges for $|x| > R$, we call R the **radius of convergence.** In case (a), where the series converges only for $x = 0$, we define the radius of convergence to be $R = 0$; and in case (b), where the series converges absolutely for all x, we define the radius of convergence to be $R = +\infty$. The set of all values of x for which a power series converges is called the **interval of convergence.**

▶ Example 1   Find the interval of convergence and radius of convergence of the power series

$$\sum_{k=0}^{\infty} x^k = 1 + x + x^2 + \cdots + x^k + \cdots$$

Solution.   For every x, the given series is a geometric series with ratio $r = x$. Thus, by Example 8 of Section 11.7, the series converges absolutely if $-1 < x < 1$ and diverges if $|x| \geq 1$. Therefore, the interval of convergence is $(-1, 1)$ and the radius of convergence is $R = 1$.   ◀

▶ Example 2   Find the interval of convergence and radius of convergence of

$$\sum_{k=0}^{\infty} \frac{x^k}{k!}$$

Solution.   We will apply the ratio test for absolute convergence (11.7.5). For every real number x,

$$\rho = \lim_{k \to +\infty} \left| \frac{u_{k+1}}{u_k} \right| = \lim_{k \to +\infty} \left| \frac{x^{k+1}}{(k+1)!} \cdot \frac{k!}{x^k} \right| = \lim_{k \to +\infty} \left| \frac{x}{k+1} \right| = 0$$

Since $\rho < 1$ for all x, the series converges absolutely for all x. Thus, the interval of convergence is $(-\infty, +\infty)$ and the radius of convergence is $R = +\infty$.   ◀

REMARK. There is a useful byproduct of Example 2: Since

$$\sum_{k=0}^{\infty} \frac{x^k}{k!}$$

converges for all x, Theorem 11.4.1 implies that for all values of x

$$\lim_{k \to +\infty} \frac{x^k}{k!} = 0 \qquad (1)$$

We will need this result later.

▶ Example 3   Find the interval of convergence and radius of convergence of

$$\sum_{k=0}^{\infty} k!x^k$$

Solution. If $x = 0$, the series has only one nonzero term and therefore converges. If $x \neq 0$, the ratio test yields

$$\rho = \lim_{k \to +\infty} \left| \frac{u_{k+1}}{u_k} \right| = \lim_{k \to +\infty} \left| \frac{(k+1)!x^{k+1}}{k!x^k} \right| = \lim_{k \to +\infty} |(k+1)x| = +\infty$$

Therefore, the series converges if $x = 0$, but diverges for all other x. Consequently, the interval of convergence is the single point $x = 0$ and the radius of convergence is $R = 0$. ◀

▶ Example 4   Find the interval of convergence and radius of convergence of

$$\sum_{k=0}^{\infty} \frac{(-1)^k x^k}{3^k(k+1)}$$

Solution. Since $|(-1)^k| = |(-1)^{k+1}| = 1$, we obtain

$$\rho = \lim_{k \to +\infty} \left| \frac{u_{k+1}}{u_k} \right| = \lim_{k \to +\infty} \left| \frac{x^{k+1}}{3^{k+1}(k+2)} \cdot \frac{3^k(k+1)}{x^k} \right|$$

$$= \lim_{k \to +\infty} \left[\frac{|x|}{3} \cdot \left(\frac{k+1}{k+2} \right) \right]$$

$$= \frac{|x|}{3} \lim_{k \to +\infty} \left(\frac{1 + 1/k}{1 + 2/k} \right) = \frac{|x|}{3}$$

The ratio test for absolute convergence implies that the series converges absolutely if

$$|x| < 3$$

and diverges if

$$|x| > 3$$

The ratio test fails when

$$|x| = 3$$

so the cases $x = -3$ and $x = 3$ need separate analyses. Substituting $x = -3$ in the given series yields

$$\sum_{k=0}^{\infty} \frac{(-1)^k(-3)^k}{3^k(k+1)} = \sum_{k=0}^{\infty} \frac{(-1)^k(-1)^k 3^k}{3^k(k+1)} = \sum_{k=0}^{\infty} \frac{1}{k+1}$$

which is the divergent harmonic series $1 + \frac{1}{2} + \frac{1}{3} + \frac{1}{4} + \cdots$. Substituting $x = 3$ in the given series yields

$$\sum_{k=0}^{\infty} \frac{(-1)^k 3^k}{3^k(k+1)} = \sum_{k=0}^{\infty} \frac{(-1)^k}{k+1}$$

which is the conditionally convergent alternating harmonic series $1 - \frac{1}{2} + \frac{1}{3} - \frac{1}{4} + \cdots$. Thus, the interval of convergence for the given series is $(-3, 3]$ and the radius of convergence is $R = 3$. ◀

▶ **Example 5** Find the interval and radius of convergence of the series

$$\sum_{k=0}^{\infty} (-1)^k \frac{x^{2k}}{(2k)!}$$

Solution. Since $|(-1)^k| = |(-1)^{k+1}| = 1$, we have

$$\rho = \lim_{k \to +\infty} \left| \frac{u_{k+1}}{u_k} \right| = \lim_{k \to +\infty} \left| \frac{x^{2(k+1)}}{[2(k+1)]!} \cdot \frac{(2k)!}{x^{2k}} \right| = \lim_{k \to +\infty} \left| \frac{x^{2k+2}}{(2k+2)!} \frac{(2k)!}{x^{2k}} \right|$$

$$= \lim_{k \to +\infty} \left| \frac{x^2}{(2k+2)(2k+1)} \right| = x^2 \lim_{k \to +\infty} \frac{1}{(2k+2)(2k+1)}$$

$$= x^2 \cdot 0 = 0$$

Thus, $\rho < 1$ for all x, which means that the interval of convergence is $(-\infty, +\infty)$ and the radius of convergence is $R = +\infty$. ◀

In addition to power series in x, we shall be interested in series of the form

$$\sum_{k=0}^{\infty} c_k(x-a)^k = c_0 + c_1(x-a) + c_2(x-a)^2 + \cdots + c_k(x-a)^k + \cdots$$

where $c_0, c_1, c_2, \ldots$ and a are constants. Such a series is called a ***power series in*** $x - a$. Some examples are

$$\sum_{k=0}^{\infty} \frac{(x-1)^k}{k+1} = 1 + \frac{(x-1)}{2} + \frac{(x-1)^2}{3} + \frac{(x-1)^3}{4} + \cdots$$

$$\sum_{k=0}^{\infty} \frac{(-1)^k(x+3)^k}{k!} = 1 - (x+3) + \frac{(x+3)^2}{2!} - \frac{(x+3)^3}{3!} + \cdots$$

The first is a power series in $x - 1$ ($a = 1$) and the second a power series in $x + 3$ ($a = -3$).

The convergence properties of a power series $\Sigma c_k(x - a)^k$ may be obtained from Theorem 11.8.1 by substituting $X = x - a$, to obtain

$$\sum_{k=0}^{\infty} a_k X^k$$

which is a power series in X. There are three possibilities for this series; it converges only when $X = 0$, or equivalently only when $x = a$; it converges absolutely for all values of X, or equivalently for all values of x; or finally, it converges absolutely for all X satisfying

$$-R < X < R \tag{2}$$

and diverges when $X < -R$ or $X > R$. But (2) can be written

$$-R < x - a < R$$

or

$$a - R < x < a + R$$

Thus, we are led to the following result.

11.8.2   THEOREM   *For a power series $\Sigma c_k(x - a)^k$, exactly one of the following is true:*
(a)   The series converges only for $x = a$.
(b)   The series converges absolutely for all x.
(c)   The series converges absolutely for all x in some finite open interval $(a - R, a + R)$ and diverges if $x < a - R$ or $x > a + R$ (Figure 11.8.2). At the points $x = a - R$ and $x = a + R$, the series may converge absolutely, converge conditionally, or diverge, depending on the particular series.

Figure 11.8.2

Series diverges	Series converges absolutely	Series diverges
$a - R$	a	$a + R$

In cases (a), (b), and (c) of Theorem 11.8.2 the series is said to have **radius of convergence,** 0, $+\infty$, and R, respectively. The set of all values of x for which the series converges is called the **interval of convergence.**

▶ Example 6   Find the interval of convergence and radius of convergence of the series

$$\sum_{k=1}^{\infty} \frac{(x-5)^k}{k^2}$$

Solution.   We apply the ratio test for absolute convergence.

$$\rho = \lim_{k \to +\infty} \left| \frac{u_{k+1}}{u_k} \right| = \lim_{k \to +\infty} \left| \frac{(x-5)^{k+1}}{(k+1)^2} \cdot \frac{k^2}{(x-5)^k} \right|$$

$$= \lim_{k \to +\infty} \left[|x-5| \left(\frac{k}{k+1} \right)^2 \right]$$

$$= |x-5| \lim_{k \to +\infty} \left(\frac{1}{1+1/k} \right)^2 = |x-5|$$

Thus, the series converges absolutely if $|x-5| < 1$, or $-1 < x - 5 < 1$, or $4 < x < 6$. The series diverges if $x < 4$ or $x > 6$.

To determine the convergence behavior at the endpoints $x = 4$ and $x = 6$, we substitute these values in the given series. If $x = 6$ the series becomes

$$\sum_{k=1}^{\infty} \frac{1^k}{k^2} = \sum_{k=1}^{\infty} \frac{1}{k^2} = 1 + \frac{1}{2^2} + \frac{1}{3^2} + \frac{1}{4^2} + \cdots$$

which is a convergent p-series $(p = 2)$. If $x = 4$ the series becomes

$$\sum_{k=1}^{\infty} \frac{(-1)^k}{k^2} = -1 + \frac{1}{2^2} - \frac{1}{3^2} + \frac{1}{4^2} - \cdots$$

Since this series converges absolutely, the interval of convergence for the given series is [4, 6]. The radius of convergence is $R = 1$ (Figure 11.8.3).   ◀

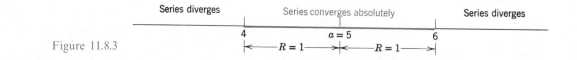

Figure 11.8.3

Series diverges	Series converges absolutely	Series diverges
4	$a = 5$	6
	$\longleftarrow R=1 \longrightarrow \longleftarrow R=1 \longrightarrow$	

▶ Exercise Set 11.8

In Exercises 1–24, find the radius of convergence and the interval of convergence.

1. $\displaystyle\sum_{k=0}^{\infty} \frac{x^k}{k+1}$.

2. $\displaystyle\sum_{k=0}^{\infty} 3^k x^k$.

3. $\displaystyle\sum_{k=0}^{\infty} \frac{(-1)^k x^k}{k!}$.

4. $\displaystyle\sum_{k=0}^{\infty} \frac{k!}{2^k} x^k$.

5. $\displaystyle\sum_{k=1}^{\infty} \frac{5^k}{k^2} x^k$.

6. $\displaystyle\sum_{k=2}^{\infty} \frac{x^k}{\ln k}$.

7. $\displaystyle\sum_{k=1}^{\infty} \frac{x^k}{k(k+1)}$.

8. $\displaystyle\sum_{k=0}^{\infty} \frac{(-2)^k x^{k+1}}{k+1}$.

9. $\displaystyle\sum_{k=1}^{\infty} (-1)^{k-1} \frac{x^k}{\sqrt{k}}$.

10. $\displaystyle\sum_{k=0}^{\infty} \frac{(-1)^k x^{2k}}{(2k)!}$.

11. $\displaystyle\sum_{k=0}^{\infty} (-1)^k \frac{x^{2k+1}}{(2k+1)!}$.

12. $\displaystyle\sum_{k=1}^{\infty} (-1)^k \frac{x^{3k}}{k^{3/2}}$.

13. $\displaystyle\sum_{k=0}^{\infty} \frac{3^k}{k!} x^k$.

14. $\displaystyle\sum_{k=2}^{\infty} (-1)^{k+1} \frac{x^k}{k(\ln k)^2}$.

15. $\displaystyle\sum_{k=0}^{\infty} \frac{x^k}{1+k^2}$.

16. $\displaystyle\sum_{k=0}^{\infty} \frac{(x-3)^k}{2^k}$.

17. $\displaystyle\sum_{k=1}^{\infty} (-1)^{k+1} \frac{(x+1)^k}{k}$.

18. $\displaystyle\sum_{k=0}^{\infty} (-1)^k \frac{(x-4)^k}{(k+1)^2}$.

19. $\displaystyle\sum_{k=0}^{\infty} \left(\frac{3}{4}\right)^k (x+5)^k$.

20. $\displaystyle\sum_{k=1}^{\infty} \frac{(2k+1)!}{k^3} (x-2)^k$.

21. $\displaystyle\sum_{k=1}^{\infty} (-1)^k \frac{(x+1)^{2k+1}}{k^2+4}$.

22. $\displaystyle\sum_{k=1}^{\infty} \frac{(\ln k)(x-3)^k}{k}$.

23. $\displaystyle\sum_{k=0}^{\infty} \frac{\pi^k (x-1)^{2k}}{(2k+1)!}$.

24. $\displaystyle\sum_{k=0}^{\infty} \frac{(2x-3)^k}{4^{2k}}$.

25. Use the root test to find the interval of convergence of $\displaystyle\sum_{k=2}^{\infty} \frac{x^k}{(\ln k)^k}$.

26. Find the radius of convergence of
$$\sum_{k=1}^{\infty} (-1)^k \frac{1 \cdot 2 \cdot 3 \cdots k}{1 \cdot 3 \cdot 5 \cdots (2k-1)} x^{2k+1}$$

27. Find the interval of convergence of
$$\sum_{k=0}^{\infty} \frac{(x-a)^k}{b^k}$$
where $b > 0$.

28. Find the radius of convergence of the power series $\displaystyle\sum_{k=0}^{\infty} \frac{(pk)!}{(k!)^p} x^k$, where p is a positive integer.

29. Find the radius of convergence of the power series $\displaystyle\sum_{k=0}^{\infty} \frac{(k+p)!}{k!(k+q)!} x^k$, where p and q are positive integers.

30. Prove: If $\displaystyle\lim_{k \to +\infty} |c_k|^{1/k} = L$, where $L \neq 0$, then the power series $\displaystyle\sum_{k=0}^{\infty} c_k x^k$ has radius of convergence $1/L$.

31. Prove: If the power series $\displaystyle\sum_{k=0}^{\infty} c_k x^k$ has radius of convergence R, then the series $\displaystyle\sum_{k=0}^{\infty} c_k x^{2k}$ has radius of convergence $\sqrt{R}$.

32. Prove: If the interval of convergence of $\displaystyle\sum_{k=0}^{\infty} c_k (x-a)^k$ is $(a-R, a+R]$, then the series converges conditionally at $a+R$.

11.9 TAYLOR AND MACLAURIN SERIES

One of the early applications of calculus was the calculation of values for functions such as $\sin x$, $\ln x$, and e^x. The basic idea is to approximate the given function by a polynomial in such a way that the resulting error is within some specified tolerance. In this section we will study the approximation of functions by polynomials, and introduce an important class of power series.

Suppose we are interested in approximating a function f by a polynomial

$$p(x) = c_0 + c_1 x + \cdots + c_n x^n \tag{1}$$

over an interval centered at $x = 0$. Because $p(x)$ has $n + 1$ coefficients, it seems reasonable that we will be able to impose $n + 1$ conditions on this polynomial. We will assume that the first n derivatives of f exist at $x = 0$, and we will choose these $n + 1$ conditions to be:

$$f(0) = p(0), \ f'(0) = p'(0), \ f''(0) = p''(0), \ldots, f^{(n)}(0) = p^{(n)}(0) \tag{2}$$

These conditions require that the value of $p(x)$ and its first n derivatives match the value of $f(x)$ and its first n derivatives at $x = 0$. By forcing this high degree of "match" at $x = 0$, it is reasonable to hope that $f(x)$ and $p(x)$ will remain close over some interval (possibly small) centered at $x = 0$. Since

$$p(x) = c_0 + c_1 x + c_2 x^2 + c_3 x^3 + \cdots + c_n x^n$$
$$p'(x) = c_1 + 2c_2 x + 3c_3 x^2 + \cdots + nc_n x^{n-1}$$
$$p''(x) = 2c_2 + 3 \cdot 2c_3 x + \cdots + n(n-1)c_n x^{n-2}$$
$$p'''(x) = 3 \cdot 2c_3 + \cdots + n(n-1)(n-2)c_n x^{n-3}$$
$$\vdots$$
$$p^{(n)}(x) = n(n-1)(n-2) \cdots (1)c_n$$

we obtain on substituting $x = 0$

$$p(0) = c_0$$
$$p'(0) = c_1$$
$$p''(0) = 2c_2 = 2!c_2$$
$$p'''(0) = 3 \cdot 2c_3 = 3!c_3$$
$$\vdots$$
$$p^{(n)}(0) = n(n-1)(n-2) \cdots (1)c_n = n!c_n$$

Thus, from (2),

$$f(0) = c_0$$
$$f'(0) = c_1$$
$$f''(0) = 2!c_2$$
$$f'''(0) = 3!c_3$$
$$\vdots$$
$$f^{(n)}(0) = n!c_n$$

so

$$c_0 = f(0), \; c_1 = f'(0), \; c_2 = \frac{f''(0)}{2!}, \; c_3 = \frac{f'''(0)}{3!}, \ldots, c_n = \frac{f^{(n)}(0)}{n!}$$

Substituting these values in (1) yields a polynomial, called the nth *Maclaurin* polynomial for f.*

11.9.1   DEFINITION   If f can be differentiated n times at 0, then we define the nth ***Maclaurin polynomial*** for f to be

$$p_n(x) = f(0) + f'(0)x + \frac{f''(0)}{2!}x^2 + \frac{f'''(0)}{3!}x^3 + \cdots + \frac{f^{(n)}(0)}{n!}x^n \quad (3)$$

This polynomial has the property that its value and the values of its first n derivatives match the value of $f(x)$ and its first n derivatives when $x = 0$.

▶ **Example 1**   Find the Maclaurin polynomials p_0, p_1, p_2, p_3, and p_n for e^x.

Solution.   Let $f(x) = e^x$. Thus,

$$f'(x) = f''(x) = f'''(x) = \cdots = f^{(n)}(x) = e^x$$

---

* COLIN MACLAURIN (1698–1746) Scottish mathematician. Maclaurin's father, a minister, died when the boy was only six months old, and his mother when he was nine years old. He was then raised by an uncle who was also a minister.

Maclaurin entered Glasgow University as a divinity student, but transferred to mathematics after one year. He received his Master's degree at age 17 and, in spite of his youth, began teaching at Marischal College in Aberdeen, Scotland.

Maclaurin met Isaac Newton during a visit to London in 1719 and from that time on he became Newton's disciple. During that era, some of Newton's analytic methods were bitterly attacked by major mathematicians and much of Newton's important mathematical work resulted from his efforts to defend Newton's ideas geometrically. Maclaurin's work, *A Treatise of Fluxions* (1742), was the first systematic formulation of Newton's methods. The treatise was so carefully done that it was a standard of mathematical rigor in calculus until the work of Cauchy in 1821.

Maclaurin was an outstanding experimentalist. He devised numerous ingenious mechanical devices, made important astronomical observations, performed actuarial computations for insurance societies, and helped to improve maps of the islands around Scotland.

and

$$f(0) = f'(0) = f''(0) = f'''(0) = \cdots = f^{(n)}(0) = e^0 = 1$$

Therefore,

$$p_0(x) = f(0) = 1$$

$$p_1(x) = f(0) + f'(0)x = 1 + x$$

$$p_2(x) = f(0) + f'(0)x + \frac{f''(0)}{2!}x^2 = 1 + x + \frac{x^2}{2!} = 1 + x + \frac{1}{2}x^2$$

$$p_3(x) = f(0) + f'(0)x + \frac{f''(0)}{2!}x^2 + \frac{f'''(0)}{3!}x^3$$

$$= 1 + x + \frac{x^2}{2!} + \frac{x^3}{3!} = 1 + x + \frac{1}{2}x^2 + \frac{1}{6}x^3$$

$$p_n(x) = f(0) + f'(0)x + \frac{f''(0)}{2!}x^2 + \cdots + \frac{f^{(n)}(0)}{n!}x^n$$

$$= 1 + x + \frac{x^2}{2!} + \cdots + \frac{x^n}{n!} \qquad \blacktriangleleft$$

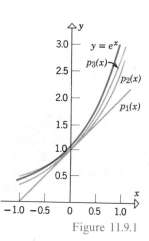

Figure 11.9.1

In Figure 11.9.1 we have sketched the graphs of e^x and its Maclaurin polynomials of degree 1, 2, and 3. Note that the graphs of e^x and $p_3(x)$ are virtually indistinguishable over the interval from -0.5 to $+0.5$. (In the next section we will investigate in detail the accuracy of Maclaurin polynomial approximations.)

▶ **Example 2** Find the nth Maclaurin polynomial for $\ln(x + 1)$.

Solution. Let $f(x) = \ln(x + 1)$ and arrange the computations as follows:

$$f(x) = \ln(x + 1) \qquad\qquad f(0) = \ln 1 = 0$$

$$f'(x) = \frac{1}{x + 1} \qquad\qquad f'(0) = 1$$

$$f''(x) = -\frac{1}{(x + 1)^2} \qquad\qquad f''(0) = -1$$

$$f'''(x) = \frac{2}{(x + 1)^3} \qquad\qquad f'''(0) = 2$$

$$f^{(4)}(x) = -\frac{3 \cdot 2}{(x + 1)^4} \qquad\qquad f^{(4)}(0) = -3!$$

$$f^{(5)}(x) = \frac{4 \cdot 3 \cdot 2}{(x + 1)^5} \qquad\qquad f^{(5)}(0) = 4!$$

$$\vdots \qquad\qquad\qquad\qquad \vdots$$

$$f^{(n)}(x) = (-1)^{n+1}\frac{(n - 1)!}{(x + 1)^n} \qquad\qquad f^{(n)}(0) = (-1)^{n+1}(n - 1)!$$

Substituting these values in (3) yields

$$p_n(x) = x - \frac{x^2}{2} + \frac{x^3}{3} - \cdots + (-1)^{n+1}\frac{x^n}{n}$$

◄

► Example 3 In the Maclaurin polynomials for $\sin x$, only the odd powers of x appear explicitly. To see this, let $f(x) = \sin x$, thus,

$$f(x) = \sin x \qquad f(0) = 0$$
$$f'(x) = \cos x \qquad f'(0) = 1$$
$$f''(x) = -\sin x \qquad f''(0) = 0$$
$$f'''(x) = -\cos x \qquad f'''(0) = -1$$

Since $f^{(4)}(x) = \sin x = f(x)$, the pattern $0, 1, 0, -1$ will repeat over and over as we evaluate successive derivatives at 0. Therefore, the successive Maclaurin polynomials for $\sin x$ are

$$p_1(x) = 0 + x = x$$

$$p_2(x) = 0 + x + 0 = x$$

$$p_3(x) = 0 + x + 0 - \frac{x^3}{3!} = x - \frac{x^3}{3!}$$

$$p_4(x) = 0 + x + 0 - \frac{x^3}{3!} + 0 = x - \frac{x^3}{3!}$$

$$p_5(x) = 0 + x + 0 - \frac{x^3}{3!} + 0 + \frac{x^5}{5!} = x - \frac{x^3}{3!} + \frac{x^5}{5!}$$

$$p_6(x) = 0 + x + 0 - \frac{x^3}{3!} + 0 + \frac{x^5}{5!} + 0 = x - \frac{x^3}{3!} + \frac{x^5}{5!}$$

$$p_7(x) = 0 + x + 0 - \frac{x^3}{3!} + 0 + \frac{x^5}{5!} + 0 - \frac{x^7}{7!} = x - \frac{x^3}{3!} + \frac{x^5}{5!} - \frac{x^7}{7!}$$

$$\vdots$$

In general, the Maclaurin polynomials for $\sin x$ are

$$p_{2n+1}(x) = p_{2n+2}(x) = x - \frac{x^3}{3!} + \frac{x^5}{5!} - \frac{x^7}{7!} + \cdots + (-1)^n\frac{x^{2n+1}}{(2n+1)!}$$

$(n = 0, 1, 2, \ldots)$.

◄

► Example 4 In the Maclaurin polynomials for $\cos x$, only the even powers of x appear explicitly. Using computations similar to those in Example 3, the reader should be able to show that the successive Maclaurin polynomials for $\cos x$ are

$$p_0(x) = p_1(x) = 1$$

$$p_2(x) = p_3(x) = 1 - \frac{x^2}{2!}$$

$$p_4(x) = p_5(x) = 1 - \frac{x^2}{2!} + \frac{x^4}{4!}$$

$$p_6(x) = p_7(x) = 1 - \frac{x^2}{2!} + \frac{x^4}{4!} - \frac{x^6}{6!}$$

$$p_8(x) = p_9(x) = 1 - \frac{x^2}{2!} + \frac{x^4}{4!} - \frac{x^6}{6!} + \frac{x^8}{8!}$$

$$\vdots$$

In general, the Maclaurin polynomials for $\cos x$ are

$$p_{2n}(x) = p_{2n+1}(x) = 1 - \frac{x^2}{2!} + \frac{x^4}{4!} - \cdots + (-1)^n \frac{x^{2n}}{(2n)!}$$

$(n = 0, 1, 2, \ldots)$.   ◄

If we are interested in a polynomial approximation to $f(x)$ on an interval centered at $x = a$, then the idea is to choose the polynomial $p(x)$ so that the values of $p(x)$ and its first n derivatives match the values of $f(x)$ and its first n derivatives at $x = a$. The computations are simplest if the approximating polynomial is expressed in the form

$$p(x) = c_0 + c_1(x - a) + c_2(x - a)^2 + \cdots + c_n(x - a)^n \qquad (4)$$

We leave it as an exercise for the reader to calculate the first n derivatives of $p(x)$ and show

$$p(a) = c_0, \; p'(a) = c_1, \; p''(a) = 2!c_2, \; p'''(a) = 3!c_3, \ldots, p^{(n)}(a) = n!c_n$$

Thus, if we want the values of $p(x)$ and its first n derivatives to match the values of $f(x)$ and its first n derivatives at $x = a$, we must have

$$c_0 = f(a), \; c_1 = f'(a), \; c_2 = \frac{f''(a)}{2!}, \; c_3 = \frac{f'''(a)}{3!}, \ldots, c_n = \frac{f^{(n)}(a)}{n!}$$

Substituting these values in (4) we obtain a polynomial called the *nth Taylor** *polynomial about $x = a$ for f.* (See p. 629 for biography.)

11.9.2   DEFINITION   If f can be differentiated n times at a, then we define the **nth Taylor polynomial for f about $x = a$** to be

$$p_n(x) = f(a) + f'(a)(x - a) + \frac{f''(a)}{2!}(x - a)^2$$

$$+ \frac{f'''(a)}{3!}(x - a)^3 + \cdots + \frac{f^{(n)}(a)}{n!}(x - a)^n \qquad (5)$$

Observe that the Taylor polynomials include the Maclaurin polynomials as a special case (let $a = 0$).

▶ **Example 5** Find the Taylor polynomials $p_1(x)$, $p_2(x)$, and $p_3(x)$ for $\sin x$ about $x = \pi/3$.

Solution. Let $f(x) = \sin x$. Thus,

$$f(x) = \sin x \qquad f(\pi/3) = \sin \frac{\pi}{3} = \frac{\sqrt{3}}{2}$$

$$f'(x) = \cos x \qquad f'(\pi/3) = \cos \frac{\pi}{3} = \frac{1}{2}$$

$$f''(x) = -\sin x \qquad f''(\pi/3) = -\sin \frac{\pi}{3} = -\frac{\sqrt{3}}{2}$$

$$f'''(x) = -\cos x \qquad f'''(\pi/3) = -\cos \frac{\pi}{3} = -\frac{1}{2}$$

Substituting in (5) with $a = \pi/3$ yields

$$p_1(x) = f(\pi/3) + f'(\pi/3)\left(x - \frac{\pi}{3}\right) = \frac{\sqrt{3}}{2} + \frac{1}{2}\left(x - \frac{\pi}{3}\right)$$

$$p_2(x) = f(\pi/3) + f'(\pi/3)\left(x - \frac{\pi}{3}\right) + \frac{f''(\pi/3)}{2!}\left(x - \frac{\pi}{3}\right)^2$$

$$= \frac{\sqrt{3}}{2} + \frac{1}{2}\left(x - \frac{\pi}{3}\right) - \frac{\sqrt{3}}{2 \cdot 2!}\left(x - \frac{\pi}{3}\right)^2$$

---

*BROOK TAYLOR (1685–1731) English mathematician. Taylor was born of well-to-do parents. Musicians and artists were entertained frequently in the Taylor home, which undoubtedly had a lasting influence on young Brook. In later years, Taylor published a definitive work on the mathematical theory of perspective and obtained major mathematical results about the vibrations of strings. There also exists an unpublished work, *On Musick,* that was intended to be part of a joint paper with Isaac Newton.

Taylor's life was scarred with unhappiness, illness, and tragedy. Because his first wife was not rich enough to suit his father, the two men argued bitterly and parted ways. Subsequently, his wife died in childbirth. Then, after he remarried, his second wife also died in childbirth, though his daughter survived.

Taylor's most productive period was from 1714 to 1719, during which time he wrote on a wide range of subjects—magnetism, capillary action, thermometers, perspective, and calculus. In his final years, Taylor devoted his writing efforts to religion and philosophy. According to Taylor, the results that bear his name were motivated by coffeehouse conversations about works of Newton on planetary motion and works of Halley ("Halley's comet") on roots of polynomials.

Taylor's writing style was so terse and hard to understand that he never received credit for many of his innovations.

$$p_3(x) = f(\pi/3) + f'(\pi/3)\left(x - \frac{\pi}{3}\right) + \frac{f''(\pi/3)}{2!}\left(x - \frac{\pi}{3}\right)^2 + \frac{f'''(\pi/3)}{3!}\left(x - \frac{\pi}{3}\right)^3$$

$$= \frac{\sqrt{3}}{2} + \frac{1}{2}\left(x - \frac{\pi}{3}\right) - \frac{\sqrt{3}}{2 \cdot 2!}\left(x - \frac{\pi}{3}\right)^2 - \frac{1}{2 \cdot 3!}\left(x - \frac{\pi}{3}\right)^3 \qquad \blacktriangleleft$$

Frequently, it is convenient to express the defining formula for the Taylor polynomial in sigma notation. To do this, we use the notation $f^{(k)}(a)$ to denote the kth derivative of f at $x = a$, and we make the added convention that $f^{(0)}(a)$ denotes $f(a)$. This enables us to write

$$\sum_{k=0}^{n} \frac{f^{(k)}(a)}{k!}(x - a)^k = f(a) + f'(a)(x - a)$$

$$+ \frac{f''(a)}{2!}(x - a)^2 + \cdots + \frac{f^{(n)}(a)}{n!}(x - a)^n$$

In particular, the nth Maclaurin polynomial for $f(x)$ may be written

$$\sum_{k=0}^{n} \frac{f^{(k)}(0)}{k!}x^k = f(0) + f'(0)x + \frac{f''(0)}{2!}x^2 + \cdots + \frac{f^{(n)}(0)}{n!}x^n$$

Because the value of f and its first n derivatives match the value of the Taylor polynomial and its first n derivatives at $x = a$, we might hope that as n increases the Taylor polynomials for f about $x = a$ will become better and better approximations to $f(x)$—at least in some interval centered at $x = a$. This raises the problem of finding those values of x for which the Taylor polynomials converge to $f(x)$ as $n \to +\infty$. In other words, for which values of x is it true that

$$f(x) = \lim_{n \to +\infty} \sum_{k=0}^{n} \frac{f^{(k)}(a)}{k!}(x - a)^k = \sum_{k=0}^{\infty} \frac{f^{(k)}(a)}{k!}(x - a)^k \qquad (6)$$

The power series in (6) is called the *Taylor series* about $x = a$ for the function f.

11.9.3 DEFINITION If f has derivatives of all orders at a, then we define the *Taylor Series for f about x = a* to be

$$\sum_{k=0}^{\infty} \frac{f^{(k)}(a)}{k!}(x - a)^k = f(a) + f'(a)(x - a)$$

$$+ \frac{f''(a)}{2!}(x - a)^2 + \cdots + \frac{f^{(k)}(a)}{k!}(x - a)^k + \cdots \qquad (7)$$

11.9.4 DEFINITION   If f has derivatives of all orders at 0, then we define the ***Maclaurin series for*** f to be

$$\sum_{k=0}^{\infty} \frac{f^{(k)}(0)}{k!} x^k = f(0) + f'(0)x + \frac{f''(0)}{2!}x^2 + \cdots + \frac{f^{(k)}(0)}{k!}x^k + \cdots \quad (8)$$

Observe that the Maclaurin series for f is just the Taylor series for f about $a = 0$.

▶ Example 6   In Example 1, we found the nth Maclaurin polynomial for e^x to be

$$\sum_{k=0}^{n} \frac{x^k}{k!} = 1 + x + \frac{x^2}{2!} + \cdots + \frac{x^n}{n!}$$

Thus, the Maclaurin series for e^x is

$$\sum_{k=0}^{\infty} \frac{x^k}{k!} = 1 + x + \frac{x^2}{2!} + \frac{x^3}{3!} + \cdots + \frac{x^n}{n!} + \cdots$$

Similarly, from Example 3 the Maclaurin series for $\sin x$ is

$$\sum_{k=0}^{\infty} (-1)^k \frac{x^{2k+1}}{(2k+1)!} = x - \frac{x^3}{3!} + \frac{x^5}{5!} - \frac{x^7}{7!} + \cdots$$

and from Example 4 the Maclaurin series for $\cos x$ is

$$\sum_{k=0}^{\infty} (-1)^k \frac{x^{2k}}{(2k)!} = 1 - \frac{x^2}{2!} + \frac{x^4}{4!} - \frac{x^6}{6!} + \cdots \quad \blacktriangleleft$$

▶ Example 7   Find the Taylor series about $x = 1$ for $1/x$.

Solution.   Let $f(x) = 1/x$ so that

$$f(x) = \frac{1}{x} \qquad\qquad f(1) = 1$$

$$f'(x) = -\frac{1}{x^2} \qquad\qquad f'(1) = -1$$

$$f''(x) = \frac{2}{x^3} \qquad\qquad f''(1) = 2!$$

$$f'''(x) = -\frac{3 \cdot 2}{x^4} \qquad\qquad f'''(1) = -3!$$

$$f^{(4)}(x) = \frac{4 \cdot 3 \cdot 2}{x^5} \qquad\qquad f^{(4)}(1) = 4!$$

$$\vdots \qquad\qquad\qquad \vdots$$

$$f^{(k)}(x) = (-1)^k \frac{k!}{x^{k+1}} \qquad f^{(k)}(1) = (-1)^k k!$$

$$\vdots \qquad\qquad\qquad \vdots$$

Thus, substituting in (7) with $a = 1$ yields

$$\sum_{k=0}^{\infty} \frac{(-1)^k k!}{k!}(x-1)^k = \sum_{k=0}^{\infty} (-1)^k (x-1)^k$$

$$= 1 - (x-1) + (x-1)^2 - (x-1)^3 + \cdots \blacktriangleleft$$

Convergence of Taylor and Maclaurin series will be considered in the next section.

▶ Exercise Set 11.9

In Exercises 1–12, find the fourth Maclaurin polynomial ($n = 4$) for the given function.

1. e^{-2x}.

2. $\dfrac{1}{1+x}$.

3. $\sin 2x$.

4. $e^x \cos x$.

5. $\tan x$.

6. $x^3 - x^2 + 2x + 1$.

7. xe^x.

8. $\tan^{-1} x$.

9. $\sec x$.

10. $\sqrt{1+x}$.

11. $\ln(3 + 2x)$.

12. $\sinh x$.

In Exercises 13–22, find the third Taylor polynomial ($n = 3$) about $x = a$ for the given function.

13. e^x; $a = 1$.

14. $\ln x$; $a = 1$.

15. $\sqrt{x}$; $a = 4$.

16. $x^4 + x - 3$; $a = -2$.

17. $\cos x$; $a = \dfrac{\pi}{4}$.

18. $\tan x$; $a = \dfrac{\pi}{3}$.

19. $\sin \pi x$; $a = -\dfrac{1}{3}$.

20. $\csc x$; $a = \dfrac{\pi}{2}$.

21. $\tan^{-1} x$; $a = 1$.

22. $\cosh x$; $a = \ln 2$.

In Exercises 23–31, find the Maclaurin series for the given function. Express your answer in sigma notation.

23. e^{-x}.

24. e^{ax}.

25. $\dfrac{1}{1+x}$.

26. xe^x.

27. $\ln(1+x)$.

28. $\sin \pi x$.

29. $\cos\left(\dfrac{x}{2}\right)$.

30. $\sinh x$.

31. $\cosh x$.

In Exercises 32–39, find the Taylor series about $x = a$ for the given function. Express your answer in sigma notation.

32. $\dfrac{1}{x}$; $a = 3$.

33. $\dfrac{1}{x}$; $a = -1$.

34. e^x; $a = 2$.

35. $\ln x$; $a = 1$.

36. $\cos x$; $a = \dfrac{\pi}{2}$.

37. $\sin \pi x$; $a = \dfrac{1}{2}$.

38. $\dfrac{1}{x+2}$; $a = 3$.

39. $\sinh x$; $a = \ln 4$.

40. Prove: The value of $p_n(x) = \displaystyle\sum_{k=0}^{n} \frac{f^{(k)}(a)}{k!}(x-a)^k$ and its first n derivatives match the value of $f(x)$ and its first n derivatives at $x = a$.

11.10 TAYLOR FORMULA WITH REMAINDER; CONVERGENCE OF TAYLOR SERIES

In this section we will analyze the error that results when a function f is approximated by a Taylor or Maclaurin polynomial. This work will enable us to investigate convergence properties of Taylor and Maclaurin series.

If we approximate a function f by its nth Taylor polynomial p_n, then the error at a point x is represented by the difference $f(x) - p_n(x)$. This difference is commonly called the **nth remainder** and is denoted by

$$R_n(x) = f(x) - p_n(x)$$

The following theorem gives an important explicit formula for this remainder.

11.10.1 THEOREM

Taylor's Theorem

Suppose that a function f can be differentiated $n + 1$ times at each point in an interval containing the point a, and let

$$p_n(x) = f(a) + f'(a)(x - a) + \frac{f''(a)}{2!}(x - a)^2 + \cdots + \frac{f^{(n)}(a)}{n!}(x - a)^n$$

be the nth Taylor polynomial about $x = a$ for f. Then for each x in the interval, there is at least one point c between a and x such that

$$R_n(x) = f(x) - p_n(x) = \frac{f^{(n+1)}(c)}{(n + 1)!}(x - a)^{n+1} \tag{1}$$

In this theorem, the statement that c is "between" a and x means that c is in the interval (a, x) if $a < x$, or in (x, a) if $x < a$, or $c = a = x$ if $a = x$.

Rewriting (1) as $f(x) = p_n(x) + R_n(x)$ yields the following result, called **Taylor's formula with remainder.**

$$f(x) = f(a) + f'(a)(x - a) + \frac{f''(a)}{2!}(x - a)^2 + \cdots$$

$$+ \frac{f^{(n)}(a)}{n!}(x - a)^n + \frac{f^{(n+1)}(c)}{(n + 1)!}(x - a)^{n+1} \tag{2}$$

This formula expresses $f(x)$ as the sum of its nth Taylor polynomial about $x = a$ plus a remainder or error term,

$$R_n(x) = \frac{f^{(n+1)}(c)}{(n + 1)!}(x - a)^{n+1} \tag{3}$$

in which c is some unspecified number between a and x. The value of c depends on a, x, and n. Historically, (3) was not due to Taylor, but rather to

Joseph Louis Lagrange.* For this reason, (3) is frequently called **Lagrange's form of the remainder.** Other formulas for the remainder may be found in any advanced calculus text. The proof of Taylor's Theorem will be deferred until the end of the section.

For Maclaurin polynomials $(a = 0)$, (2) and (3) become:

$$f(x) = f(0) + f'(0)x + \frac{f''(0)}{2!}x^2 + \cdots + \frac{f^{(n)}(0)}{n!}x^n + \frac{f^{(n+1)}(c)}{(n+1)!}x^{n+1} \qquad (2a)$$

$$R_n(x) = \frac{f^{(n+1)}(c)}{(n+1)!}x^{n+1} \qquad \text{where } c \text{ is between 0 and } x. \qquad (3a)$$

▶ Example 1   From Example 1 of the previous section, the nth Maclaurin polynomial for $f(x) = e^x$ is

$$1 + x + \frac{x^2}{2!} + \cdots + \frac{x^n}{n!}$$

Since $f^{(n+1)}(x) = e^x$, it follows that

$$f^{(n+1)}(c) = e^c$$

Thus, from Taylor's formula with remainder (2a)

---

*JOSEPH LOUIS LAGRANGE (1736–1813) French–Italian mathematician and astronomer. Lagrange, the son of a public official, was born in Turin, Italy. (Baptismal records list his name as Giuseppe Lodovico Lagrangia.) Although his father wanted him to be a lawyer, Lagrange was attracted to mathematics and astronomy after reading a memoir by the astronomer Halley. At age 16 he began to study mathematics on his own and by age 19 was appointed to a professorship at the Royal Artillery School in Turin. The following year Lagrange sent Euler solutions to some famous problems, using new methods that eventually blossomed into a branch of mathematics called calculus of variations. These methods and Lagrange's applications of them to problems in celestial mechanics were so monumental that by age 25 he was regarded by many of his contemporaries as the greatest living mathematician.

In 1776, on the recommendation of Euler, he was chosen to succeed Euler as the director of the Berlin Academy. During his stay in Berlin, Lagrange distinguished himself not only in celestial mechanics, but also in algebraic equations and the theory of numbers. After twenty years in Berlin, he moved to Paris at the invitation of Louis XVI. He was given apartments in the Louvre and treated with great honor, even during the revolution.

Napoleon was a great admirer of Lagrange and showered him with honors—count, senator, and Legion of Honor. The years Lagrange spent in Paris were devoted primarily to didactic treatises summarizing his mathematical conceptions. One of Lagrange's most famous works is a memoir, *Mécanique Analytique* in which he reduced the theory of mechanics to a few general formulas from which all other necessary equations could be derived.

It is an interesting historical fact that Lagrange's father speculated unsuccessfully in several financial ventures, so his family was forced to live quite modestly. Lagrange himself stated that if his family had money, he would not have made mathematics his vocation.

In spite of his fame, Lagrange was always a shy and modest man. On his death, he was buried with honor in the Pantheon.

$$e^x = 1 + x + \frac{x^2}{2!} + \cdots + \frac{x^n}{n!} + \frac{e^c}{(n+1)!} x^{n+1} \tag{4}$$

where c is between 0 and x. Note that (4) is valid for all real values of x since the hypotheses of Taylor's Theorem are satisfied on the interval $(-\infty, +\infty)$ (verify). ◄

▶ **Example 2** In Example 5 of the previous section, we found the Taylor polynomial of degree 3 about $x = \pi/3$ for $f(x) = \sin x$ to be

$$\frac{\sqrt{3}}{2} + \frac{1}{2}\left(x - \frac{\pi}{3}\right) - \frac{\sqrt{3}}{4}\left(x - \frac{\pi}{3}\right)^2 - \frac{1}{12}\left(x - \frac{\pi}{3}\right)^3$$

Since $f^{(4)}(x) = \sin x$ (verify) it follows that $f^{(4)}(c) = \sin c$. Thus, from Taylor's formula with remainder (2) in the case $a = \pi/3$ and $n = 3$ we obtain

$$\sin x = \frac{\sqrt{3}}{2} + \frac{1}{2}\left(x - \frac{\pi}{3}\right) - \frac{\sqrt{3}}{4}\left(x - \frac{\pi}{3}\right)^2 - \frac{1}{12}\left(x - \frac{\pi}{3}\right)^3$$

$$+ \frac{\sin c}{4!}\left(x - \frac{\pi}{3}\right)^4$$

where c is some number between $\pi/3$ and x. ◄

In the previous section we raised the problem of finding those values of x for which the Taylor series for f about $x = a$ converges to $f(x)$, that is

$$f(x) = f(a) + f'(a)(x - a) + \frac{f''(a)}{2!}(x - a)^2$$

$$+ \cdots + \frac{f^{(k)}(a)}{k!}(x - a)^k + \cdots \tag{5}$$

To solve this problem, let us write (5) in sigma notation:

$$f(x) = \sum_{k=0}^{\infty} \frac{f^{(k)}(a)}{k!}(x - a)^k$$

This equality is equivalent to

$$f(x) = \lim_{n \to +\infty} \sum_{k=0}^{n} \frac{f^{(k)}(a)}{k!}(x - a)^k$$

or

$$\lim_{n \to +\infty}\left[f(x) - \sum_{k=0}^{n} \frac{f^{(k)}(a)}{k!}(x - a)^k\right] = 0 \tag{6}$$

But the bracketed expression in (6) is $R_n(x)$, since it is the difference between $f(x)$ and its nth Taylor polynomial about $x = a$. Thus, (6) can be written

$$\lim_{n \to +\infty} R_n(x) = 0$$

which leads us to the following result.

11.10.2 THEOREM   *The equality*

$$f(x) = \sum_{k=0}^{\infty} \frac{f^{(k)}(a)}{k!}(x - a)^k$$

holds if and only if $\lim\limits_{n \to +\infty} R_n(x) = 0.$

To paraphrase this theorem, the *Taylor series for f converges to f(x) at precisely those points where the remainder approaches zero.*

▶ **Example 3**   Show that the Maclaurin series for e^x converges to e^x for all x.

Solution.   We want to show

$$e^x = 1 + x + \frac{x^2}{2!} + \frac{x^3}{3!} + \cdots + \frac{x^n}{n!} + \cdots \tag{7}$$

holds for all x. From (4),

$$e^x = 1 + x + \frac{x^2}{2!} + \cdots + \frac{x^n}{n!} + R_n(x)$$

where

$$R_n(x) = \frac{e^c}{(n + 1)!} x^{n+1}$$

and c is between 0 and x. Thus, we must show that for all x

$$\lim_{n \to +\infty} R_n(x) = \lim_{n \to +\infty} \frac{e^c}{(n + 1)!} x^{n+1} = 0 \tag{8}$$

Our proof of this result will hinge on the limit

$$\lim_{n \to +\infty} \frac{x^{n+1}}{(n + 1)!} = 0 \tag{9}$$

which follows from Formula (1) of Section 11.8. We will consider three cases. If $x > 0$, then

$$0 < c < x$$

from which it follows that

$$0 < e^c < e^x$$

and consequently

$$0 < \frac{e^c}{(n+1)!} x^{n+1} < \frac{e^x}{(n+1)!} x^{n+1} \tag{10}$$

From (9) we obtain

$$\lim_{n \to +\infty} \frac{e^x}{(n+1)!} x^{n+1} = e^x \lim_{n \to +\infty} \frac{x^{n+1}}{(n+1)!} = e^x \cdot 0 = 0$$

Thus, from (10) and the Pinching Theorem (3.3.1), result (8) follows.

In the case $x < 0$, we have $c < 0$ since c is between 0 and x. Thus, $0 < e^c < 1$ and consequently

$$0 < e^c \left| \frac{x^{n+1}}{(n+1)!} \right| < \left| \frac{x^{n+1}}{(n+1)!} \right|$$

or

$$0 < \left| \frac{e^c}{(n+1)!} x^{n+1} \right| < \left| \frac{x^{n+1}}{(n+1)!} \right|$$

or

$$0 < |R_n(x)| < \left| \frac{x^{n+1}}{(n+1)!} \right|$$

From (9) and the Pinching Theorem it follows that $\lim_{n \to +\infty} |R_n(x)| = 0$, and so $\lim_{n \to +\infty} R_n(x) = 0$.

Convergence in the case $x = 0$ is obvious since (7) reduces to

$$e^0 = 1 + 0 + 0 + 0 + \cdots \qquad \blacktriangleleft$$

▶ Example 4 Show that the Maclaurin series for $\sin x$ converges to $\sin x$ for all x.

Solution. We are trying to show that

$$\sin x = x - \frac{x^3}{3!} + \frac{x^5}{5!} - \frac{x^7}{7!} + \cdots$$

holds for all x. Let $f(x) = \sin x$, so that for all x either

$$f^{(n+1)}(x) = \pm \cos x \qquad \text{or} \qquad f^{(n+1)}(x) = \pm \sin x$$

In all of these cases $|f^{(n+1)}(x)| \leq 1$, so that for all possible values of c

$$|f^{(n+1)}(c)| \leq 1$$

Hence,

$$0 \leq |R_n(x)| = \left| \frac{f^{(n+1)}(c)}{(n+1)!} x^{n+1} \right| \leq \frac{|x|^{n+1}}{(n+1)!} \tag{11}$$

From (9) with $|x|$ replacing x,

$$\lim_{n \to +\infty} \frac{|x|^{n+1}}{(n+1)!} = 0$$

so that from (11) and the Pinching Theorem

$$\lim_{n \to +\infty} |R_n(x)| = 0$$

and consequently

$$\lim_{n \to +\infty} R_n(x) = 0$$

for all x.   ◀

▶ **Example 5**   Find the Taylor series for $\sin x$ about $x = \pi/2$ and show that the series converges to $\sin x$ for all x.

Solution.   Let $f(x) = \sin x$. Thus,

$$f(x) = \sin x \qquad f\left(\frac{\pi}{2}\right) = \sin \frac{\pi}{2} = 1$$

$$f'(x) = \cos x \qquad f'\left(\frac{\pi}{2}\right) = \cos \frac{\pi}{2} = 0$$

$$f''(x) = -\sin x \qquad f''\left(\frac{\pi}{2}\right) = -\sin \frac{\pi}{2} = -1$$

$$f'''(x) = -\cos x \qquad f'''\left(\frac{\pi}{2}\right) = -\cos \frac{\pi}{2} = 0$$

Since $f^{(4)}(x) = \sin x$, the pattern $1, 0, -1, 0$ will repeat over and over as we evaluate successive derivatives at $\pi/2$. Thus, the Taylor series representation about $x = \pi/2$ for $\sin x$ is

$$\sin x = 1 - \frac{1}{2!}\left(x - \frac{\pi}{2}\right)^2 + \frac{1}{4!}\left(x - \frac{\pi}{2}\right)^4 - \frac{1}{6!}\left(x - \frac{\pi}{2}\right)^6 + \cdots \quad (12)$$

which we are trying to show is valid for all x. As in Example 4, $|f^{(n+1)}(c)| \le 1$ so that

$$0 \le |R_n(x)| = \left|\frac{f^{(n+1)}(c)}{(n+1)!}\left(x - \frac{\pi}{2}\right)^{n+1}\right| \le \frac{\left|x - \frac{\pi}{2}\right|^{n+1}}{(n+1)!}$$

From (9) with $|x - \pi/2|$ replacing x it follows that

$$\lim_{n \to +\infty} \frac{\left|x - \frac{\pi}{2}\right|^{n+1}}{(n+1)!} = 0$$

so that by the same argument given in Example 4, $\lim_{n \to +\infty} R_n(x) = 0$ for all x.

This shows that (12) is valid for all x. ◄

REMARK. It can be shown that the Taylor series for e^x, $\sin x$, and $\cos x$ about any point $x = a$ converge to these functions for all x.

For reference, we have listed in Table 11.10.1 the Maclaurin series for a number of important functions, and we have indicated the interval over which the series converges to the function.

Sometimes Maclaurin series can be obtained by substituting in other Maclaurin series.

► Example 6 Using the Maclaurin series

$$e^x = 1 + x + \frac{x^2}{2!} + \frac{x^3}{3!} + \frac{x^4}{4!} + \cdots \qquad -\infty < x < +\infty$$

we can derive the Maclaurin series for e^{-x} by substituting $-x$ for x to obtain

$$e^{-x} = 1 + (-x) + \frac{(-x)^2}{2!} + \frac{(-x)^3}{3!} + \frac{(-x)^4}{4!} + \cdots$$
$$-\infty < -x < +\infty$$

or

$$e^{-x} = 1 - x + \frac{x^2}{2!} - \frac{x^3}{3!} + \frac{x^4}{4!} - \cdots \qquad -\infty < x < +\infty$$

Table 11.10.1

MACLAURIN SERIES	INTERVAL OF VALIDITY
$e^x = \sum_{k=0}^{\infty} \dfrac{x^k}{k!} = 1 + x + \dfrac{x^2}{2!} + \dfrac{x^3}{3!} + \dfrac{x^4}{4!} + \cdots$	$-\infty < x < +\infty$
$\sin x = \sum_{k=0}^{\infty} (-1)^k \dfrac{x^{2k+1}}{(2k+1)!} = x - \dfrac{x^3}{3!} + \dfrac{x^5}{5!} - \dfrac{x^7}{7!} + \cdots$	$-\infty < x < +\infty$
$\cos x = \sum_{k=0}^{\infty} (-1)^k \dfrac{x^{2k}}{(2k)!} = 1 - \dfrac{x^2}{2!} + \dfrac{x^4}{4!} - \dfrac{x^6}{6!} + \cdots$	$-\infty < x < +\infty$
$\ln(1+x) = \sum_{k=0}^{\infty} (-1)^k \dfrac{x^{k+1}}{k+1} = x - \dfrac{x^2}{2} + \dfrac{x^3}{3} - \dfrac{x^4}{4} + \cdots$	$-1 < x \leq 1$
$\tan^{-1} x = \sum_{k=0}^{\infty} (-1)^k \dfrac{x^{2k+1}}{2k+1} = x - \dfrac{x^3}{3} + \dfrac{x^5}{5} - \dfrac{x^7}{7} + \cdots$	$-1 \leq x \leq 1$
$\dfrac{1}{1-x} = \sum_{k=0}^{\infty} x^k = 1 + x + x^2 + x^3 + \cdots$	$-1 < x < 1$
$\sinh x = \sum_{k=0}^{\infty} \dfrac{x^{2k+1}}{(2k+1)!} = x + \dfrac{x^3}{3!} + \dfrac{x^5}{5!} + \dfrac{x^7}{7!} + \cdots$	$-\infty < x < +\infty$
$\cosh x = \sum_{k=0}^{\infty} \dfrac{x^{2k}}{(2k)!} = 1 + \dfrac{x^2}{2!} + \dfrac{x^4}{4!} + \dfrac{x^6}{6!} + \cdots$	$-\infty < x < +\infty$

From the Maclaurin series for e^x and e^{-x} we can obtain the Maclaurin series for $\cosh x$ by writing

$$\cosh x = \frac{1}{2}(e^x + e^{-x}) = \frac{1}{2}\left(\left[1 + x + \frac{x^2}{2!} + \frac{x^3}{3!} + \frac{x^4}{4!} + \cdots\right] + \left[1 - x + \frac{x^2}{2!} - \frac{x^3}{3!} + \frac{x^4}{4!} + \cdots\right]\right)$$

or

$$\cosh x = 1 + \frac{x^2}{2!} + \frac{x^4}{4!} + \cdots \qquad -\infty < x < +\infty \qquad \blacktriangleleft$$

We could have also derived this Maclaurin series directly. However, sometimes indirect methods, such as those in this example, are useful when it is messy to calculate the higher derivatives required for the direct calculation of a Maclaurin series. There is, however, a loose thread in the logic of this example. We have produced a power series in x that converges to $\cosh x$ for all x. But isn't it conceivable that we have produced a power series in x *different* from the Maclaurin series? In Section 11.12 we will show that if a power series in $x - a$ converges to $f(x)$ on some interval containing a, then the series must be the Taylor series for f about $x = a$. Thus, we are assured the series obtained for $\cosh x$ is, in fact, the Maclaurin series.

▶ **Example 7** Using the Maclaurin series

$$\frac{1}{1 - x} = 1 + x + x^2 + x^3 + \cdots \qquad -1 < x < 1$$

we can derive the Maclaurin series for $1/(1 - 2x^2)$ by substituting $2x^2$ for x to obtain

$$\frac{1}{1 - 2x^2} = 1 + (2x^2) + (2x^2)^2 + (2x^2)^3 + \cdots \qquad -1 < 2x^2 < 1$$

or

$$\frac{1}{1 - 2x^2} = 1 + 2x^2 + 4x^4 + 8x^6 + \cdots = \sum_{k=0}^{\infty} 2^k x^{2k}, \qquad -1 < 2x^2 < 1$$

Since $2x^2 \geq 0$ for all x, the convergence condition $-1 < 2x^2 < 1$ can be written in the equivalent form $0 \leq 2x^2 < 1$ or $0 \leq x^2 < 1/2$ or $-1/\sqrt{2} < x < 1/\sqrt{2}$. ◀

OPTIONAL

We conclude this section with a proof of Taylor's Theorem.

Proof of Theorem 11.10.1. By hypothesis, f can be differentiated $n + 1$ times at each point in an interval containing the point a. Choose any point b in this interval. To be specific, we will assume $b > a$. (The cases $b < a$ and $b = a$ are left to the reader.) Let $p_n(x)$ be the nth Taylor polynomial for $f(x)$ about $x = a$ and define

$$h(x) = f(x) - p_n(x) \tag{13}$$
$$g(x) = (x - a)^{n+1} \tag{14}$$

Because $f(x)$ and $p_n(x)$ have the same value and the same first n derivatives at $x = a$, it follows that

$$h(a) = h'(a) = h''(a) = \cdots = h^{(n)}(a) = 0 \tag{15}$$

Also, we leave it for the reader to show that

$$g(a) = g'(a) = g''(a) = \cdots = g^{(n)}(a) = 0 \tag{16}$$

and that $g(x)$ and its first n derivatives are nonzero when $x \neq a$.

It is straightforward to check that h and g satisfy the hypotheses of the Extended Mean-Value Theorem (10.2.2) on the interval $[a, b]$, so that there is a point c_1 with $a < c_1 < b$ such that

$$\frac{h(b) - h(a)}{g(b) - g(a)} = \frac{h'(c_1)}{g'(c_1)} \tag{17}$$

or, from (15) and (16),

$$\frac{h(b)}{g(b)} = \frac{h'(c_1)}{g'(c_1)} \tag{18}$$

If we now apply the Extended Mean-Value Theorem to h' and g' over the interval $[a, c_1]$, we may deduce that there is a point c_2 with $a < c_2 < c_1 < b$ such that

$$\frac{h'(c_1) - h'(a)}{g'(c_1) - g'(a)} = \frac{h''(c_2)}{g''(c_2)}$$

or, from (15) and (16),

$$\frac{h'(c_1)}{g'(c_1)} = \frac{h''(c_2)}{h''(c_2)}$$

which, when combined with (18), yields

$$\frac{h(b)}{g(b)} = \frac{h''(c_2)}{g''(c_2)}$$

It should now be clear that if we continue in this way, applying the Extended Mean-Value Theorem to the successive derivatives of h and g, we will eventually obtain a relationship of the form

$$\frac{h(b)}{g(b)} = \frac{h^{(n+1)}(c_{n+1})}{g^{(n+1)}(c_{n+1})} \tag{19}$$

where $a < c_{n+1} < b$. However, $p_n(x)$ is a polynomial of degree n, so that its $(n + 1)$-st derivative is zero. Thus, from (13)

$$h^{(n+1)}(c_{n+1}) = f^{(n+1)}(c_{n+1}) \tag{20}$$

Also, from (14), the $(n + 1)$-st derivative of $g(x)$ is the constant $(n + 1)!$, so that

$$g^{(n+1)}(c_{n+1}) = (n+1)! \tag{21}$$

Substituting (20) and (21) in (19) yields

$$\frac{h(b)}{g(b)} = \frac{f^{(n+1)}(c_{n+1})}{(n+1)!}$$

Letting $c = c_{n+1}$, and using (13) and (14), it follows that

$$f(b) - p_n(b) = \frac{f^{(n+1)}(c)}{(n+1)!}(b-a)^{n+1}$$

But this is precisely (1) in Taylor's Theorem, with the exception that the variable here is b rather than x. Thus, to finish we need only replace b by x. ▮

▶ Exercise Set 11.10

For the functions in Exercises 1–14, find Lagrange's form of the remainder for the given values of a and n.

1. e^{2x}; $a = 0$; $n = 5$. 　 2. $\cos x$; $a = 0$; $n = 8$.

3. $\dfrac{1}{x+1}$; $a = 0$; $n = 4$.

4. $\tan x$; $a = 0$; $n = 2$.

5. xe^x; $a = 0$; $n = 3$.

6. $\ln(1+x)$; $a = 0$; $n = 5$.

7. $\tan^{-1} x$; $a = 0$; $n = 2$.

8. $\sinh x$; $a = 0$; $n = 6$.

9. $\sqrt{x}$; $a = 4$; $n = 3$.

10. $\dfrac{1}{x}$; $a = 1$; $n = 5$.

11. $\sin x$; $a = \dfrac{\pi}{6}$; $n = 4$.

12. $\cos \pi x$; $a = \dfrac{1}{2}$; $n = 2$.

13. $\dfrac{1}{(1+x)^2}$; $a = -2$; $n = 5$.

14. $\csc x$; $a = \dfrac{\pi}{2}$; $n = 1$.

In Exercises 15–18, find Lagrange's form of the remainder $R_n(x)$ when $a = 0$.

15. $f(x) = \dfrac{1}{1-x}$.

16. $f(x) = e^{-x}$.

17. $f(x) = e^{2x}$.

18. $f(x) = \ln(1+x)$.

19. Prove: The Maclaurin series for $\cos x$ converges to $\cos x$ for all x.

20. Prove: The Taylor series for $\sin x$ about $x = \pi/4$ converges to $\sin x$ for all x.

21. Prove: The Taylor series for e^x about $x = 1$ converges to e^x for all x.

22. Prove: The Maclaurin series for $\ln(1+x)$ converges to $\ln(1+x)$ if $0 \le x < 1$.

23. Prove: The Taylor series for e^x about any point $x = a$ converges to e^x for all x.

24. Prove: The Taylor series for $\sin x$ about any point $x = a$ converges to $\sin x$ for all x.

25. Prove: The Taylor series for $\cos x$ about any point $x = a$ converges to $\cos x$ for all x.

In Exercises 26–38, use the known Maclaurin series for e^x, $\sin x$, $\cos x$, and $1/(1-x)$ to find the Maclaurin series for the given function. In each case, specify the interval on which your computations are valid.

26. xe^x. 　 27. e^{-2x}.

28. e^{x^2}. 　 29. $\dfrac{1}{1+x}$.

30. $\dfrac{1}{1-4x^2}$.

31. $\dfrac{x^2}{1+3x}$.

32. $\sin^2 x$. $\left[\textit{Hint: } \sin^2 x = \dfrac{1}{2}(1 - \cos 2x).\right]$

33. $\cos^2 x$. $\left[\textit{Hint: } \cos^2 x = \dfrac{1}{2}(1 + \cos 2x).\right]$

34. $\sinh x$.

35. $\sin(2x)$.

36. $\cos(2x)$.

37. $\cos(x^2)$.

38. $\sin(x^2)$.

39. Use the Maclaurin series for $1/(1-x)$ to express $1/x$ in powers of $x-1$.

40. The purpose of this exercise is to show that the Taylor series of a function f may possibly converge to a value different from $f(x)$ for certain x. Let

$$f(x) = \begin{cases} e^{-1/x^2}, & x \neq 0 \\ 0, & x = 0 \end{cases}$$

(a) Use the definition of a derivative to show that $f'(0) = 0$.

(b) With some difficulty it can be shown that $f^{(n)}(0) = 0$ for $n \geq 2$. Accepting this fact, show that the Maclaurin series of f converges for all x, but converges to $f(x)$ only at the point $x = 0$.

11.11 COMPUTATIONS USING TAYLOR SERIES

In this section we show how Taylor and Maclaurin series can be used to obtain approximate values for trigonometric functions and logarithms.

In the previous section we showed that for all x

$$e^x = 1 + x + \frac{x^2}{2!} + \frac{x^3}{3!} + \frac{x^4}{4!} + \cdots$$

In particular, if we let $x = 1$, we obtain the following expression for e as the sum of an infinite series

$$e = 1 + 1 + \frac{1}{2!} + \frac{1}{3!} + \frac{1}{4!} + \cdots$$

Thus, we can approximate e to any degree of accuracy using an appropriate partial sum

$$e \approx 1 + 1 + \frac{1}{2!} + \frac{1}{3!} + \cdots + \frac{1}{n!} \tag{1}$$

The value of n in this formula is at our disposal, the larger we choose n the more accurate the approximation. For practical applications one would be interested in approximating e to a specified degree of accuracy, in which case it would be important to determine how large n should be chosen to attain the desired accuracy. For example, how large should n be taken in (1) to ensure an error of at most 0.00005? We will now show how Lagrange's remainder formula can be used to answer such questions.

According to Taylor's theorem, the absolute value of the error that results when $f(x)$ is approximated by its nth Taylor polynomial $p_n(x)$ about $x = a$ is

$$|R_n(x)| = |f(x) - p_n(x)| = \left| \frac{f^{(n+1)}(c)}{(n+1)!}(x-a)^{n+1} \right| \qquad (2)$$

In this formula, c is an unknown number between a and x, so that the value of $f^{(n+1)}(c)$ usually cannot be determined. However, it is frequently possible to determine an upper bound on the size of $|f^{(n+1)}(c)|$; that is, we can find a constant M such that $|f^{(n+1)}(c)| \leq M$. For such an M, it follows from (2) that

$$|R_n(x)| = |f(x) - p_n(x)| \leq \frac{M}{(n+1)!}|x-a|^{n+1} \qquad (3)$$

which gives an upper bound on the magnitude of the error $R_n(x)$. The examples to follow illustrate the usefulness of this result; but first let us introduce some terminology.

An approximation is said to be **accurate to n decimal places** if the magnitude of the error is less than 0.5×10^{-n}. For example,

DESCRIPTION	MAGNITUDE OF THE ERROR IS LESS THAN
1 decimal place accuracy	$0.05 \quad = 0.5 \times 10^{-1}$
2 decimal place accuracy	$0.005 \quad = 0.5 \times 10^{-2}$
3 decimal place accuracy	$0.0005 = 0.5 \times 10^{-3}$
$\vdots$	$\vdots$

APPROXIMATING e ▶ **Example 1**   Use (1) to approximate e to four decimal place accuracy.

Solution. From Taylor's formula with remainder (see Example 1, Section 11.10)

$$e^x = 1 + x + \frac{x^2}{2!} + \cdots + \frac{x^n}{n!} + \frac{e^c}{(n+1)!}x^{n+1}$$

where c is between 0 and x. Thus, in the case $x = 1$ we obtain

$$e = 1 + 1 + \frac{1}{2!} + \cdots + \frac{1}{n!} + \frac{e^c}{(n+1)!}$$

where c is between 0 and 1. This tells us that the magnitude of the error in approximation (1) is

$$|R_n| = \left| \frac{e^c}{(n+1)!} \right| = \frac{e^c}{(n+1)!} \qquad (4)$$

where $0 < c < 1$. Since

$$c < 1$$

it follows that

$$e^c < e^1 = e$$

so that from (4)

$$|R_n| < \frac{e}{(n+1)!} \tag{5}$$

This inequality provides an upper bound on the magnitude of the error R_n. Unfortunately, this inequality is not very useful since the right side involves the quantity e, which we are trying to estimate. However, if we use the fact that $e < 3$, then we can replace (5) with the more useful result

$$|R_n| < \frac{3}{(n+1)!}$$

It follows that if we choose n so that

$$|R_n| < \frac{3}{(n+1)!} < 0.5 \times 10^{-4} = 0.00005 \tag{6}$$

then approximation (1) will be accurate to four decimal places. An appropriate value for n may be found by trial and error. For example, using a hand-held calculator or computer one can evaluate $3/(n+1)!$ for $n = 0, 1, 2, \ldots$ until a value of n satisfying (6) is obtained. We leave it for the reader to show that $n = 8$ is the first positive integer satisfying (6). Thus, to four decimal place accuracy

$$e \approx 1 + 1 + \frac{1}{2!} + \frac{1}{3!} + \frac{1}{4!} + \frac{1}{5!} + \frac{1}{6!} + \frac{1}{7!} + \frac{1}{8!} \approx 2.7183 \qquad ◀$$

APPROXIMATING TRIGONOMETRIC FUNCTIONS

▶ **Example 2** Use the Maclaurin series for $\sin x$ to approximate $\sin 3°$ to five decimal place accuracy.

Solution. In the Maclaurin series

$$\sin x = x - \frac{x^3}{3!} + \frac{x^5}{5!} - \frac{x^7}{7!} + \cdots \tag{7}$$

the angle x is assumed to be in radians (because the differentiation formulas for the trigonometric functions were derived with this assumption). Since $3° = \pi/60$ radians, it follows from (7) that

$$\sin 3° = \sin \frac{\pi}{60} = \left(\frac{\pi}{60}\right) - \frac{\left(\frac{\pi}{60}\right)^3}{3!} + \frac{\left(\frac{\pi}{60}\right)^5}{5!} - \frac{\left(\frac{\pi}{60}\right)^7}{7!} + \cdots \qquad (8)$$

We must now decide how many terms in this series must be kept in order to obtain five decimal place accuracy. We will consider two possible approaches, one using Lagrange's remainder formula, and the other exploiting the fact that (8) satisfies the hypotheses of the alternating series test.

If we let $f(x) = \sin x$, then the magnitude of the error that results when $\sin x$ is approximated by its nth Maclaurin polynomial is

$$|R_n| = \left| \frac{f^{(n+1)}(c)}{(n+1)!} x^{n+1} \right|$$

where c is between 0 and x. Since $f^{(n+1)}(c)$ is either $\pm \sin c$ or $\pm \cos c$, it follows that $|f^{(n+1)}(c)| \leq 1$, so

$$|R_n| \leq \frac{|x|^{n+1}}{(n+1)!}$$

In particular, if $x = \pi/60$, then

$$|R_n| \leq \frac{\left(\frac{\pi}{60}\right)^{n+1}}{(n+1)!}$$

Thus, for five decimal place accuracy, we must choose n so that

$$\frac{\left(\frac{\pi}{60}\right)^{n+1}}{(n+1)!} < 0.5 \times 10^{-5} = 0.000005$$

By trial and error with the help of a hand-held calculator, the reader can check that $n = 3$ is the smallest n that works. Thus, in (8) we need only keep terms up to the third power for five decimal place accuracy, that is,

$$\sin 3° \approx \left(\frac{\pi}{60}\right) - \frac{\left(\frac{\pi}{60}\right)^3}{3!} \qquad (9)$$
$$\approx 0.05234$$

An alternate approach to determining n uses the fact that (8) satisfies the hypotheses of the alternating series test, Theorem 11.7.1. (Verify.) Thus, by Theorem 11.7.2, if we use only those terms up to and including

$$\pm \frac{\left(\dfrac{\pi}{60}\right)^m}{m!} \qquad [m \text{ is an odd positive integer.}]$$

then the magnitude of the error will be at most

$$\frac{\left(\dfrac{\pi}{60}\right)^{m+2}}{(m+2)!}$$

Thus, for five decimal place accuracy, we look for the first positive odd integer m such that

$$\frac{\left(\dfrac{\pi}{60}\right)^{m+2}}{(m+2)!} < 0.5 \times 10^{-5} = 0.000005$$

By trial and error, $m = 3$ is the first such integer. Thus, to five decimal place accuracy

$$\sin 3° \approx \left(\frac{\pi}{60}\right) - \frac{\left(\dfrac{\pi}{60}\right)^3}{3!} \approx 0.05234 \tag{10}$$

which is the same as our earlier result. ◄

REMARK. It should be noted that there are two types of errors that result when computing with series. The first, called **truncation error,** is the error that results when the entire series is approximated by a partial sum. The second kind of error, called **round-off error,** results when decimal approximations are used. For example, (9) involves a truncation error of at most 0.5×10^{-5} (five decimal place accuracy). However, to obtain the numerical value (10) it was necessary to approximate π, thereby introducing a round-off error. If one seeks n decimal place accuracy in the final result, then it is common procedure to use $n + 1$ decimal places in each decimal approximation and then round off to n decimal places at the end of all computations. Although this procedure may occasionally not produce n decimal place accuracy, it works often enough that we will use it in this text. Better procedures are considered in a branch of mathematics called **numerical analysis.**

► Example 3 Approximate $\sin 92°$ to five decimal place accuracy.

Solution. We could proceed, as in the previous example, using the Maclaurin series

$$\sin x = x - \frac{x^3}{3!} + \frac{x^5}{5!} - \frac{x^7}{7!} + \cdots \qquad (11)$$

with $x = 92° = \frac{23}{45}\pi$ (radians). However, a better procedure is to work with the Taylor series about $x = \pi/2$ for $\sin x$ (Example 5, Section 11.10):

$$\sin x = 1 - \frac{1}{2!}\left(x - \frac{\pi}{2}\right)^2 + \frac{1}{4!}\left(x - \frac{\pi}{2}\right)^4 - \frac{1}{6!}\left(x - \frac{\pi}{2}\right)^6 + \cdots \quad (12)$$

The advantage of this series over the Maclaurin series is its faster convergence at $x = 92° = \frac{23}{45}\pi$. In general, the rate of convergence of a Taylor series about $x = a$ decreases as one moves farther away from the point a. That is, the farther x becomes from a, the more terms that are required to achieve a given degree of accuracy. Therefore, for the purpose of approximating $\sin 92° = \sin \frac{23}{45}\pi$, series (12) is preferable to series (11) since $a = \pi/2$ is closer to $\frac{23}{45}\pi$ than $a = 0$.

Substituting $x = \frac{23}{45}\pi$ in (12) yields

$$\sin 92° = \sin \frac{23}{45}\pi = 1 - \frac{1}{2!}\left(\frac{\pi}{90}\right)^2 + \frac{1}{4!}\left(\frac{\pi}{90}\right)^4 - \frac{1}{6!}\left(\frac{\pi}{90}\right)^6 + \cdots \quad (13)$$

If we let $f(x) = \sin x$, then the magnitude of the error in approximating $\sin x$ by its nth Taylor polynomial about $a = \pi/2$ is

$$|R_n| = \left| \frac{f^{(n+1)}(c)}{(n+1)!}\left(x - \frac{\pi}{2}\right)^{n+1} \right|$$

where c is between $\pi/2$ and x. As in the previous example, $|f^{(n+1)}(c)| \leq 1$, so that

$$|R_n| \leq \frac{\left|x - \frac{\pi}{2}\right|^{n+1}}{(n+1)!}$$

In particular, if $x = \frac{23}{45}\pi$, then

$$|R_n| \leq \frac{\left(\frac{\pi}{90}\right)^{n+1}}{(n+1)!}$$

For five decimal place accuracy, we must choose n so that

$$\frac{\left(\frac{\pi}{90}\right)^{n+1}}{(n+1)!} < 0.5 \times 10^{-5} = 0.000005$$

By trial and error, the smallest such integer is $n = 3$. Thus, in (13) we need only keep terms up to the third power for five decimal place accuracy, that is,

$$\sin 92° \approx 1 + 0 \cdot \left(\frac{\pi}{90}\right) - \frac{1}{2!}\left(\frac{\pi}{90}\right)^2 + 0 \cdot \left(\frac{\pi}{90}\right)^3$$

$$= 1 - \frac{1}{2!}\left(\frac{\pi}{90}\right)^2$$

$$\approx 0.99939$$

As in Example 2, we could also have obtained this result by exploiting the fact that (13) is an alternating series.   ◄

APPROXIMATING   The Maclaurin series
LOGARITHMS

$$\ln(1 + x) = x - \frac{x^2}{2} + \frac{x^3}{3} - \frac{x^4}{4} + \cdots \qquad -1 < x \leq 1 \qquad (14)$$

is the starting point for the approximation of natural logarithms. Unfortunately, the usefulness of this series is limited because of its slow convergence and the restriction $-1 < x \leq 1$. However, if we replace x by $-x$ in this series, we obtain

$$\ln(1 - x) = -x - \frac{x^2}{2} - \frac{x^3}{3} - \frac{x^4}{4} - \cdots \qquad -1 \leq x < 1 \qquad (15)$$

and on subtracting (15) from (14) we obtain

$$\ln \frac{1 + x}{1 - x} = 2\left(x + \frac{x^3}{3} + \frac{x^5}{5} + \frac{x^7}{7} + \cdots\right) \qquad -1 < x < 1 \qquad (16)$$

Series (16), first obtained by James Gregory* in 1668, can be used to compute the natural logarithm of any positive number y by letting

$$x = \frac{y - 1}{y + 1} \tag{17}$$

---

*JAMES GREGORY (1638–1675) Scottish mathematician and astronomer. Gregory, the son of a minister, was famous in his time as the inventor of the Gregorian reflecting telescope, so named in his honor. Although he is not generally ranked with the great mathematicians, much of his work relating to calculus was studied by Leibniz and Newton and undoubtedly influenced some of their discoveries. There is a manuscript, discovered posthumously, which shows that Gregory had anticipated Taylor series well before Taylor.

so that

$$y = \frac{1 + x}{1 - x}$$

For example, to compute ln 2 we let $y = 2$ in (17), which yields $x = 1/3$. Substituting this value in (16) gives

$$\ln 2 = 2\left[(1/3) + \frac{(1/3)^3}{3} + \frac{(1/3)^5}{5} + \frac{(1/3)^7}{7} + \frac{(1/3)^9}{9} + \cdots\right]$$

$$= 2\left[\frac{1}{3} + \frac{1}{81} + \frac{1}{1{,}215} + \frac{1}{15{,}309} + \frac{1}{177{,}147} + \cdots\right]$$

$$\approx 2[0.33333 + 0.01235 + 0.00082 + 0.00007 + 0.00001 + \cdots]$$

Adding the five terms shown and then rounding to four decimal places at the end yields

$$\ln 2 \approx 2[0.34657] = 0.69314 \approx 0.6931$$

It is of interest to note that in the case $x = 1$, series (14) yields

$$\ln 2 = 1 - \frac{1}{2} + \frac{1}{3} - \frac{1}{4} + \frac{1}{5} - \cdots$$

This result is noteworthy because it is the first time we have been able to obtain the sum of the alternating harmonic series. However, this series converges too slowly to be of any computational value.

APPROXIMATING π If we let $x = 1$ in the Maclaurin series

$$\tan^{-1} x = x - \frac{x^3}{3} + \frac{x^5}{5} - \frac{x^7}{7} + \cdots \qquad -1 \le x \le 1 \qquad (18)$$

we obtain

$$\frac{\pi}{4} = \tan^{-1} 1 = 1 - \frac{1}{3} + \frac{1}{5} - \frac{1}{7} + \cdots$$

or

$$\pi = 4\left[1 - \frac{1}{3} + \frac{1}{5} - \frac{1}{7} + \cdots\right]$$

This famous series, obtained by Leibniz in 1674, converges too slowly to be of computational importance. A more practical procedure for approximating π uses the identity

$$\frac{\pi}{4} = \tan^{-1}\frac{1}{2} + \tan^{-1}\frac{1}{3} \tag{19}$$

By using this identity and series (18) to approximate $\tan^{-1}\frac{1}{2}$ and $\tan^{-1}\frac{1}{3}$, the value of π can be effectively approximated to any degree of accuracy.

▶ Exercise Set 11.11

In this exercise set, a hand-held calculator will be useful.

1. How large should n be taken in (1) to approximate e to:
   (a) five decimal place accuracy?
   (b) ten decimal place accuracy?

In Exercises 2–8, apply Lagrange's form of the remainder.

2. Use $x = -1$ in the Maclaurin series for e^x to approximate $1/e$ to three decimal place accuracy.

3. Use $x = \frac{1}{2}$ in the Maclaurin series for e^x to approximate $\sqrt{e}$ to four decimal place accuracy.

4. Use the Maclaurin series for $\sin x$ to approximate $\sin 4°$ to five decimal place accuracy.

5. Use the Maclaurin series for $\cos x$ to approximate $\cos(\pi/20)$ to four decimal place accuracy.

6. Use an appropriate Taylor series for $\sin x$ to approximate $\sin 85°$ to four decimal place accuracy.

7. Use an appropriate Taylor series for $\cos x$ to approximate $\cos 58°$ to four decimal place accuracy.

8. Use an appropriate Taylor series for $\sin x$ to approximate $\sin 35°$ to four decimal place accuracy.

9. Use series (16) to approximate $\ln 1.25$ to three decimal place accuracy.

10. Use series (16) to approximate $\ln 3$ to four decimal place accuracy.

11. Use the Maclaurin series for $\tan^{-1} x$ to approximate $\tan^{-1} 0.1$ to three decimal place accuracy. [*Hint:* Use the fact that (18) is an alternating series.]

12. Use the Maclaurin series for $\sinh x$ to approximate $\sinh 0.5$ to three decimal place accuracy.

13. Use the Maclaurin series for $\cosh x$ to approximate $\cosh 0.1$ to four decimal place accuracy.

14. Use an appropriate Taylor series for $\sqrt[3]{x}$ to approximate $\sqrt[3]{28}$ to three decimal place accuracy.

15. For what range of x values can $\sin x$ be approximated by $x - x^3/3!$ with three decimal place accuracy?

16. Estimate the maximum error in the approximation $\cos x \approx 1 - x^2/2! + x^4/4!$ if $-0.2 \le x \le 0.2$.

17. For what range of x values can e^x be approximated by $1 + x + x^2/2!$ if the allowable error is at most 0.0005?

18. Estimate the maximum error in the approximation $\ln(1 + x) \approx x$ if $|x| < 0.01$.

19. (a) How many terms in the alternating harmonic series are needed to approximate $\ln 2$ to six decimal place accuracy?
    (b) How many terms in series (16) with $x = \frac{1}{3}$ are needed to approximate $\ln 2$ to six decimal place accuracy?

20. Prove identity (19).

21. Approximate $\tan^{-1}\frac{1}{2}$ and $\tan^{-1}\frac{1}{3}$ to three decimal place accuracy and then use identity (19) to approximate π.

22. The curve $y = \cos x$ is to be approximated by the parabola $x^2 = 2(1 - y)$ over the interval $[-0.2, 0.2]$. Estimate the maximum difference in the y-coordinates of the curves over this interval.

11.12   DIFFERENTIATION AND INTEGRATION OF POWER SERIES

If a function f is expressed as a power series

$$f(x) = \sum_{k=0}^{\infty} c_k(x - a)^k$$

for all x in some interval, then we say that the series **represents** $f(x)$ on that interval. Thus, for example, the geometric series $1 + x + x^2 + x^3 + \cdots$ represents the function $1/(1 - x)$ on the interval $(-1, 1)$ since

$$\frac{1}{1 - x} = 1 + x + x^2 + x^3 + \cdots \qquad -1 < x < 1$$

In this section we show how a power series representation of a function f can be used to obtain power series representations of $f'(x)$ and $\int f(x)\, dx$. Some applications of the results are also discussed.

The following theorem tells us that a power series representation of $f'(x)$ or $\int f(x)\, dx$ can be obtained by differentiating or integrating a power series representation of $f(x)$ term by term. We omit the proofs.

11.12.1   THEOREM   *If the series $\displaystyle\sum_{k=0}^{\infty} c_k(x - a)^k$ has a nonzero radius of convergence R, and if for each x in the interval $(a - R, a + R)$ we have $f(x) = \displaystyle\sum_{k=0}^{\infty} c_k(x - a)^k$, then:*

(a)   *The series $\displaystyle\sum_{k=0}^{\infty} \frac{d}{dx}[c_k(x - a)^k] = \sum_{k=1}^{\infty} kc_k(x - a)^{k-1}$ has radius of convergence R, and for all x in the interval $(a - R, a + R)$*

$$f'(x) = \sum_{k=0}^{\infty} \frac{d}{dx}[c_k(x - a)^k]$$

(b)   *The series $\displaystyle\sum_{k=0}^{\infty} \left[\int c_k(x - a)^k\, dx\right] = \sum_{k=0}^{\infty} \frac{c_k}{k + 1}(x - a)^{k+1}$ has radius of convergence R, and for all x in the interval $(a - R, a + R)$*

$$\int f(x)\, dx = \sum_{k=0}^{\infty} \left[\int c_k(x - a)^k\, dx\right] + C$$

(c)   *For all α and β in the interval $(a - R, a + R)$, the series*

$$\sum_{k=0}^{\infty} \left[\int_{\alpha}^{\beta} c_k(x - a)^k\, dx\right]$$

converges absolutely and

$$\int_{\alpha}^{\beta} f(x)\, dx = \sum_{k=0}^{\infty} \left[\int_{\alpha}^{\beta} c_k (x - a)^k\, dx \right]$$

REMARK. Note that in part (*b*) a separate constant of integration is not introduced for each term in the series $\sum_{k=0}^{\infty} \left[\int c_k (x - a)^k\, dx \right]$; rather a single constant C is added to the entire series.

▶ Example 1 To illustrate Theorem 11.12.1, we will obtain the familiar results

$$\frac{d}{dx}[\sin x] = \cos x \qquad \text{and} \qquad \int \cos x\, dx = \sin x + C$$

using power series. Since

$$\sin x = x - \frac{x^3}{3!} + \frac{x^5}{5!} - \frac{x^7}{7!} + \cdots \qquad -\infty < x < +\infty$$

$$\cos x = 1 - \frac{x^2}{2!} + \frac{x^4}{4!} - \frac{x^6}{6!} + \cdots \qquad -\infty < x < +\infty$$

it follows from part (*a*) of Theorem 11.12.1 that

$$\frac{d}{dx}[\sin x] = \frac{d}{dx}\left[x - \frac{x^3}{3!} + \frac{x^5}{5!} - \frac{x^7}{7!} + \cdots \right]$$

$$= 1 - 3\frac{x^2}{3!} + 5\frac{x^4}{5!} - 7\frac{x^6}{7!} + \cdots$$

$$= 1 - \frac{x^2}{2!} + \frac{x^4}{4!} - \frac{x^6}{6!} + \cdots$$

$$= \cos x$$

Also, from part (*b*) of Theorem 11.12.1

$$\int \cos x\, dx = \int \left[1 - \frac{x^2}{2!} + \frac{x^4}{4!} - \frac{x^6}{6!} + \cdots \right] dx$$

$$= \left[x - \frac{x^3}{3(2!)} + \frac{x^5}{5(4!)} - \frac{x^7}{7(6!)} + \cdots \right] + C$$

$$= \left[x - \frac{x^3}{3!} + \frac{x^5}{5!} - \frac{x^7}{7!} + \cdots \right] + C$$

$$= \sin x + C$$

◀

▶ Example 2   The integral

$$\int_0^1 e^{-x^2}\, dx$$

cannot be evaluated directly because there is no elementary antiderivative of e^{-x^2}. However, it is possible to approximate the integral by some numerical technique such as Simpson's rule. Still, another possibility is to represent e^{-x^2} by its Maclaurin series and then integrate term by term in accordance with part (c) of Theorem 11.12.1. This produces a series that converges to the integral.

The simplest way to obtain the Maclaurin series for e^{-x^2} is to replace x by $-x^2$ in the Maclaurin series

$$e^x = 1 + x + \frac{x^2}{2!} + \frac{x^3}{3!} + \frac{x^4}{4!} + \cdots$$

to obtain

$$e^{-x^2} = 1 - x^2 + \frac{x^4}{2!} - \frac{x^6}{3!} + \frac{x^8}{4!} - \cdots$$

Therefore,

$$\int_0^1 e^{-x^2}\, dx = \int_0^1 \left[1 - x^2 + \frac{x^4}{2!} - \frac{x^6}{3!} + \frac{x^8}{4!} - \cdots \right] dx$$

$$= \left[x - \frac{x^3}{3} + \frac{x^5}{5(2!)} - \frac{x^7}{7(3!)} + \frac{x^9}{9(4!)} - \cdots \right]_0^1$$

$$= 1 - \frac{1}{3} + \frac{1}{5 \cdot 2!} - \frac{1}{7 \cdot 3!} + \frac{1}{9 \cdot 4!} - \cdots \qquad (1)$$

Thus, we have found a series that converges to the value of the integral $\int_0^1 e^{-x^2}\, dx$. Using the first three terms in this series we obtain the approximation

$$\int_0^1 e^{-x^2}\, dx \approx 1 - \frac{1}{3} + \frac{1}{10} = \frac{23}{30} = 0.766\ldots$$

Since series (1) satisfies the hypotheses of the alternating series test, it follows from Theorem 11.7.2 that the magnitude of the error in this approximation is at most $1/(7 \cdot 3!) = 1/42 \approx 0.024$. Greater accuracy can be obtained by using more terms in the series   ◀

The following example illustrates how Theorem 11.12.1 can sometimes be used to obtain new Taylor series from known Taylor series.

▶ Example 3　Derive the Maclaurin series for $\tan^{-1} x$.

Solution.　We could calculate this Maclaurin series directly. However, we can also exploit Theorem 11.12.1 by first writing

$$\int \frac{1}{1 + x^2}\, dx = \tan^{-1} x + C$$

and then integrating the Maclaurin series for $1/(1 + x^2)$ term by term. Since

$$\frac{1}{1 - x} = 1 + x + x^2 + x^3 + x^4 + \cdots$$

it follows on replacing x by $-x^2$ that

$$\frac{1}{1 + x^2} = 1 - x^2 + x^4 - x^6 + x^8 - \cdots \tag{2}$$

Thus,

$$\tan^{-1} x + C = \int \frac{1}{1 + x^2}\, dx = \int [1 - x^2 + x^4 - x^6 + x^8 - \cdots]\, dx$$

or

$$\tan^{-1} x = \left[x - \frac{x^3}{3} + \frac{x^5}{5} - \frac{x^7}{7} + \frac{x^9}{9} - \cdots \right] - C$$

The constant of integration may be evaluated by substituting $x = 0$ and using the condition $\tan^{-1} 0 = 0$. This gives $C = 0$, so that

$$\tan^{-1} x = x - \frac{x^3}{3} + \frac{x^5}{5} - \frac{x^7}{7} + \frac{x^9}{9} - \cdots \tag{3}$$

◀

Although it is conceivable at this point that series (3) is not the Maclaurin series for $\tan^{-1} x$, but rather some other power series converging to $\tan^{-1} x$, the following theorem tells us that this is not the case.

11.12.2　THEOREM　*If*

$$f(x) = c_0 + c_1(x - a) + c_2(x - a)^2 + \cdots + c_n(x - a)^n + \cdots$$

for all x in some open interval containing a, then the series is the Taylor series for f about a.

Proof.　By repeated application of Theorem 11.12.1(a) we obtain

$$f(x) = c_0 + c_1(x - a) + c_2(x - a)^2 + c_3(x - a)^3 + c_4(x - a)^4 + \cdots$$
$$f'(x) = c_1 + 2c_2(x - a) + 3c_3(x - a)^2 + 4c_4(x - a)^3 + \cdots$$

$$f''(x) = 2!c_2 + (3 \cdot 2)c_3(x - a) + (4 \cdot 3)c_4(x - a)^2 + \cdots$$
$$f'''(x) = 3!c_3 + (4 \cdot 3 \cdot 2)c_4(x - a) + \cdots$$
$$\vdots$$

On substituting $x = a$, all the powers of $x - a$ drop out leaving

$$f(a) = c_0 \qquad\qquad c_0 = f(a)$$
$$f'(a) = c_1 \qquad\qquad c_1 = f'(a)$$
$$f''(a) = 2!c_2 \quad \text{or} \quad c_2 = \frac{f''(a)}{2!}$$
$$f'''(a) = 3!c_3 \qquad\qquad c_3 = \frac{f'''(a)}{3!}$$
$$\vdots$$

which shows that the coefficients $c_0, c_1, c_2, c_3, \ldots$ are precisely the coefficients in the Taylor series about a for $f(x)$.

This theorem tells us that no matter how we arrive at a power series in $x - a$ converging to $f(x)$, be it by substitution, by integration, by differentiation, or by algebraic manipulation, the resulting series will be the Taylor series about a for $f(x)$.

REMARK. Parts (a) and (b) of Theorem 11.12.1 say nothing about the behavior of the differentiated and integrated series at the endpoints $a - R$ and $a + R$. Indeed, by differentiating termwise, convergence may be lost at one or both endpoints; and by integrating termwise, convergence may be gained at one or both endpoints. As an illustration, in Example 3 we derived the series (3) for $\tan^{-1} x$ by integrating series (2) for $1/(1 + x^2)$. We leave it as an exercise to show that the series for $\tan^{-1} x$ converges to $\tan^{-1} x$ on the interval $[-1, 1]$, while the series for $1/(1 + x^2)$ converges to $1/(1 + x^2)$ only on $(-1, 1)$. Thus, convergence was gained at both endpoints by integrating.

We conclude this section with some tricks for obtaining Taylor series that would be messy to obtain directly. Our objective is to make the reader aware of these techniques; we will not attempt to justify the results.

▶ Example 4 Find the first three terms in the Maclaurin series for $e^{-x^2} \tan^{-1} x$.

Solution. Using the series for e^{-x^2} and $\tan^{-1} x$ obtained in Examples 2 and 3

$$e^{-x^2} \tan^{-1} x = \left(1 - x^2 + \frac{x^4}{2} - \cdots\right)\left(x - \frac{x^3}{3} + \frac{x^5}{5} + \cdots\right)$$

We now multiply out following a familiar format from elementary algebra

$$1 - x^2 + \frac{x^4}{2} - \cdots$$

$$\times \quad x - \frac{x^3}{3} + \frac{x^5}{5} + \cdots$$

$$x - x^3 + \frac{x^5}{2} - \cdots$$

$$-\frac{x^3}{3} + \frac{x^5}{3} - \frac{x^7}{6} + \cdots$$

$$\frac{x^5}{5} - \frac{x^7}{5} + \cdots$$

$$x - \tfrac{4}{3}x^3 + \tfrac{31}{30}x^5 - \cdots$$

Thus,

$$e^{-x^2}\tan^{-1}x = x - \frac{4}{3}x^3 + \frac{31}{30}x^5 - \cdots$$

If desired, more terms in the series can be obtained by including more terms in the factors.   ◄

▶ **Example 5**   Find the first three terms in the Maclaurin series for $\tan x$.

Solution.   Instead of computing the series directly, we write

$$\tan x = \frac{\sin x}{\cos x} = \frac{x - \dfrac{x^3}{3!} + \dfrac{x^5}{5!} - \cdots}{1 - \dfrac{x^2}{2!} + \dfrac{x^4}{4!} - \cdots}$$

and follow the familiar "long division" process to obtain

$$x + \frac{x^3}{3} + \frac{2x^5}{15} + \cdots$$

$$1 - \frac{x^2}{2} + \frac{x^4}{24} - \cdots \overline{\smash{\big)}\, x - \frac{x^3}{6} + \frac{x^5}{120} - \cdots}$$

$$x - \frac{x^3}{2} + \frac{x^5}{24} - \cdots$$

$$\frac{x^3}{3} - \frac{x^5}{30} + \cdots$$

$$\frac{x^3}{3} - \frac{x^5}{6} + \cdots$$

$$\frac{2x^5}{15} + \cdots$$

Thus,

$$\tan x = x + \frac{x^3}{3} + \frac{2x^5}{15} + \cdots$$ ◀

BINOMIAL SERIES If m is a real number, then the Maclaurin series for $(1 + x)^m$ is called the *binomial series;* it is given by (verify)

$$1 + mx + \frac{m(m-1)}{2!}x^2 + \frac{m(m-1)(m-2)}{3!}x^3 + \cdots$$

It is proved in advanced calculus that the binomial series converges to $(1 + x)^m$ if $|x| < 1$. Thus, for such values of x

$$(1 + x)^m = 1 + mx + \frac{m(m-1)}{2!}x^2 + \cdots$$
$$+ \frac{m(m-1)(m-2) \cdots (m-k+1)}{k!}x^k + \cdots \qquad (4)$$

or in sigma notation

$$(1 + x)^m = 1 + \sum_{k=1}^{\infty} \frac{m(m-1) \cdots (m-k+1)}{k!}x^k \qquad \text{if } |x| < 1 \quad (4a)$$

REMARK. If m is a positive integer, then $f(x) = (1 + x)^m$ is a polynomial of degree m so that

$$f^{(m+1)}(0) = f^{(m+2)}(0) = f^{(m+3)}(0) = \cdots = 0$$

and the binomial series reduces to the familiar binomial expansion

$$(1 + x)^m = 1 + mx + \frac{m(m-1)}{2!}x^2 + \frac{m(m-1)(m-2)}{3!}x^3 + \cdots + x^m$$

▶ Example 6 Express $1/\sqrt{1 + x}$ as a binomial series.

Solution. Substituting $m = -\frac{1}{2}$ in (4) yields

$$\frac{1}{\sqrt{1 + x}} = 1 - \frac{1}{2}x + \frac{(-\frac{1}{2})(-\frac{1}{2} - 1)}{2!}x^2 + \frac{(-\frac{1}{2})(-\frac{1}{2} - 1)(-\frac{1}{2} - 2)}{3!}x^3$$
$$+ \cdots + \frac{(-\frac{1}{2})(-\frac{3}{2})(-\frac{5}{2}) \cdots (-\frac{1}{2} - k + 1)}{k!}x^k + \cdots$$

$$= 1 - \frac{1}{2}x + \frac{1 \cdot 3}{2^2 \cdot 2!}x^2 - \frac{1 \cdot 3 \cdot 5}{2^3 \cdot 3!}x^3 + \cdots$$

$$+ (-1)^k \frac{1 \cdot 3 \cdot 5 \cdots (2k-1)}{2^k k!} x^k + \cdots \quad \blacktriangleleft$$

▶ Exercise Set 11.12

In Exercises 1–4, obtain the stated results by differentiating or integrating Maclaurin series term by term.

1. (a) $\dfrac{d}{dx}[e^x] = e^x$ (b) $\displaystyle\int e^x \, dx = e^x + C.$

2. (a) $\dfrac{d}{dx}[\cos x] = -\sin x$

   (b) $\displaystyle\int \sin x \, dx = -\cos x + C.$

3. (a) $\dfrac{d}{dx}[\sinh x] = \cosh x$

   (b) $\displaystyle\int \sinh x \, dx = \cosh x + C.$

4. (a) $\dfrac{d}{dx}[\ln(1+x)] = \dfrac{1}{1+x}$

   (b) $\displaystyle\int \frac{1}{1+x} \, dx = \ln(1+x) + C.$

5. Derive the Maclaurin series for $1/(1+x)^2$ by differentiating an appropriate Maclaurin series term by term.

6. By differentiating an appropriate series, show that

$$\sum_{k=1}^{\infty} k x^k = \frac{x}{(1-x)^2}.$$

7. By integrating an appropriate series, show that

$$\sum_{k=1}^{\infty} \frac{x^k}{k} = \ln\left(\frac{1}{1-x}\right).$$

In Exercises 8–11, use a binomial series to find the first four nonzero terms in the Maclaurin series for the function and give the radius of convergence.

8. $\sqrt{1+x}.$ **9.** $(1-x)^{1/3}.$

10. $(1+x^2)^{-1/4}.$ **11.** $\dfrac{x}{\sqrt[3]{1-2x}}.$

In Exercises 12–19, use series to approximate the value of the integral to three decimal place accuracy.

12. $\displaystyle\int_0^1 \sin x^2 \, dx.$ **13.** $\displaystyle\int_0^1 \cos \sqrt{x} \, dx.$

14. $\displaystyle\int_0^{0.1} \frac{\sin x}{x} \, dx.$ **15.** $\displaystyle\int_0^{1/2} \frac{dx}{1+x^4}.$

16. $\displaystyle\int_0^{1/2} \tan^{-1} 2x^2 \, dx.$ **17.** $\displaystyle\int_0^{0.1} e^{-x^3} \, dx.$

18. $\displaystyle\int_0^{0.2} \sqrt[3]{1+x^4} \, dx.$ **19.** $\displaystyle\int_0^{1/2} \frac{dx}{\sqrt[4]{x^2+1}}.$

In Exercises 20–27, use any method to find the first four nonzero terms in the Maclaurin series of the given function.

20. $x^4 e^x.$ **21.** $e^{-x^2} \cos x.$

22. $\dfrac{x^2}{1+x^4}.$ **23.** $\dfrac{\sin x}{e^x}.$

24. $\tanh x.$ **25.** $\sin^2 x.$

26. $\dfrac{\ln(1+x)}{1-x}.$ **27.** $x^2 e^{4x} \sqrt{1+x}.$

28. Obtain the familiar result, $\lim\limits_{x \to 0} (\sin x)/x = 1,$ by finding a power series for $(\sin x)/x,$ and taking the limit term by term.

29. Use the method of Exercise 28 to find the limits:

   (a) $\displaystyle\lim_{x \to 0} \frac{1 - \cos x}{\sin x}$

   (b) $\displaystyle\lim_{x \to 0} \frac{\ln \sqrt{1+x} - \sin 2x}{x}.$

30. (a) Use the relationship

$$\int \frac{1}{\sqrt{1-x^2}} \, dx = \sin^{-1} x + C$$

to find the first four nonzero terms in the Maclaurin series for $\sin^{-1} x.$

(b)   Express the series in sigma notation.
(c)   What is the radius of convergence?

31. (a)   Use the relationship

$$\int \frac{1}{\sqrt{1 + x^2}}\, dx = \sinh^{-1} x + C$$

to find the first four nonzero terms in the Maclaurin series for $\sinh^{-1} x$.

(b)   Express the series in sigma notation.
(c)   What is the radius of convergence?

32.   Estimate the maximum error in the approximation $\sqrt{1 + x} \approx 1 + (x/2)$ if $|x| < 0.001$.

33.   Derive (4). (Do not try to prove convergence.)

34.   Prove: If the power series $\sum\limits_{k=0}^{\infty} a_k x^k$ and $\sum\limits_{k=0}^{\infty} b_k x^k$ have the same sum on an interval $(-r, r)$, then $a_k = b_k$ for all values of k.

35.   If m is any real number, and k is a nonnegative integer, then we define the **binomial coefficients** $\dbinom{m}{k}$ by the formulas $\dbinom{m}{0} = 1$ and

$$\binom{m}{k} = \frac{m(m - 1)(m - 2) \cdots (m - k + 1)}{k!}$$

for $k \geq 1$.

Express Formula (4a) in terms of binomial coefficients.

In Exercises 36–41, use the formula obtained in Exercise 35 to express the Maclaurin series for the given function in sigma notation. State the radius of convergence of the series.

36. $(1 + x)^{1/3}$.   **37.** $(1 - x)^{1/3}$.

38. $(1 - x^2)^{2/3}$.   **39.** $(1 + x^2)^{2/3}$.

40. $\dfrac{1}{\sqrt{1 + x}}$.   **41.** $\dfrac{x}{\sqrt{1 + x}}$.

▶ SUPPLEMENTARY EXERCISES

In Exercises 1–6, find $L = \lim\limits_{n \to +\infty} a_n$ if it exists.

1. $a_n = (-1)^n / e^n$.   **2.** $a_n = e^{1/n}$.

3. $a_n = \dfrac{1}{\sqrt{n}} - \dfrac{1}{\sqrt{n + 1}}$.   **4.** $a_n = \sin (\pi n)$.

5. $a_n = \sin\left(\dfrac{(2n - 1)\pi}{2}\right)$.   **6.** $a_n = \dfrac{n + 1}{n(n + 2)}$.

7.   Which of the sequences $\{a_n\}_{n=1}^{+\infty}$ in Exercises 1–6 are (a) decreasing? (b) nondecreasing? (c) alternating?

8.   Suppose $f(x)$ satisfies $f'(x) > 0$ and $f(x) \leq 1 - e^{-x}$ for all $x \geq 1$. What can you conclude about the convergence of $\{a_n\}$ if $a_n = f(n)$, $n = 1, 2, \ldots$?

9.   Use your knowledge of geometric series and p-series to determine all values of q for which the following series converge.

(a) $\sum\limits_{k=0}^{\infty} \pi^k / q^{2k}$ (b) $\sum\limits_{k=1}^{\infty} (1/k^q)^3$

(c) $\sum\limits_{k=2}^{\infty} 1/(\ln q^k)$ (d) $\sum\limits_{k=2}^{\infty} 1/(\ln q)^k$.

10. (a)   Use a suitable test to find all values of q for which $\sum\limits_{k=2}^{\infty} 1/[k\,(\ln k)^q]$ converges.

(b)   Why can't you use the integral test for the series $\sum\limits_{k=1}^{\infty} (2 + \cos k)/k^2$? Test for convergence using a test that does apply.

11.   Express 1.3636… as (a) an infinite series in sigma-notation, (b) a ratio of integers.

12.   In parts (a)–(d), use the comparison test to determine whether the series converges.

(a) $\sum\limits_{k=1}^{\infty} \dfrac{2k - 1}{3k^2 - k}$ (b) $\sum\limits_{k=1}^{\infty} \dfrac{2k + 1}{3k^2 + k}$

(c) $\sum\limits_{k=1}^{\infty} \dfrac{2k - 1}{3k^3 - k^2}$ (d) $\sum\limits_{k=1}^{\infty} \dfrac{2k + 1}{3k^3 + k^2}$.

13.   Find the sum of the series (if it converges).

(a) $\sum\limits_{k=1}^{\infty} \dfrac{2^k + 3^k}{6^{k+1}}$ (b) $\sum\limits_{k=2}^{\infty} \ln\left(1 + \dfrac{1}{k}\right)$

(c) $\sum\limits_{k=1}^{\infty} [k^{-1/2} - (k + 1)^{-1/2}]$.

In Exercises 14–21, determine whether the series converges or diverges. You may use the following limits without proof:

$$\lim_{k \to +\infty} (1 + 1/k)^k = e, \quad \lim_{k \to +\infty} \sqrt[k]{k} = 1,$$

$$\lim_{k \to +\infty} \sqrt[k]{a} = 1$$

14. $\displaystyle\sum_{k=0}^{\infty} e^{-k}.$

15. $\displaystyle\sum_{k=1}^{\infty} ke^{-k^2}.$

16. $\displaystyle\sum_{k=1}^{\infty} \frac{k}{k^2 + 2k + 7}.$

17. $\displaystyle\sum_{k=1}^{\infty} \frac{\sqrt{k}}{k^2 + 7}.$

18. $\displaystyle\sum_{k=1}^{\infty} \left(\frac{k}{k+1}\right)^k.$

19. $\displaystyle\sum_{k=0}^{\infty} \frac{3^k k!}{(2k)!}.$

20. $\displaystyle\sum_{k=0}^{\infty} \frac{k^6 3^k}{(k+1)!}.$

21. $\displaystyle\sum_{k=1}^{\infty} \left(\frac{5k}{2k+1}\right)^{3k}.$

In Exercises 22–25, determine whether the given series is absolutely convergent, conditionally convergent, or divergent.

22. $\displaystyle\sum_{k=1}^{\infty} (-1)^k / e^{1/k}.$

23. $\displaystyle\sum_{k=0}^{\infty} (-2)^k / (3^k + 1).$

24. $\displaystyle\sum_{k=0}^{\infty} (-1)^k / (2k + 1).$

25. $\displaystyle\sum_{k=0}^{\infty} (-1)^k 3^k / 2^{k+1}.$

26. Find the smallest value of n for which the nth partial sum approximates the sum of the series to the stated accuracy:

(a) $\displaystyle\sum_{k=1}^{\infty} \frac{(-1)^k}{k^2 + 1}; \; |\text{error}| < 0.0001.$

(b) $\displaystyle\sum_{k=1}^{\infty} \frac{(-1)^k}{5^k + 1}; \; |\text{error}| < 0.00005.$

In Exercises 27–32, determine the radius of convergence and the interval of convergence of the given power series.

27. $\displaystyle\sum_{k=1}^{\infty} \frac{(x-1)^k}{k\sqrt{k}}.$

28. $\displaystyle\sum_{k=1}^{\infty} \frac{(2x)^k}{3k}.$

29. $\displaystyle\sum_{k=1}^{\infty} \frac{(1-x)^{2k}}{4^k k}.$

30. $\displaystyle\sum_{k=1}^{\infty} \frac{k^2(x+2)^k}{(k+1)!}.$

31. $\displaystyle\sum_{k=1}^{\infty} \frac{k!(x-1)^k}{5^k}.$

32. $\displaystyle\sum_{k=1}^{\infty} \frac{(2k)!x^k}{(2k+1)!}.$

In Exercises 33–35, find:
(a) the nth Taylor polynomial for f about $x = a$ (for the stated values of n and a);
(b) Lagrange's form of $R_n(x)$ (for the stated values of n and a);
(c) an upper bound on the absolute value of the error if $f(x)$ is approximated over the given interval by the Taylor polynomial obtained in part (a).

33. $f(x) = \ln(x-1); \; a = 2; \; n = 3; \; [\frac{3}{2}, 2].$

34. $f(x) = e^{x/2}; \; a = 0; \; n = 4; \; [-1, 0].$

35. $f(x) = \sqrt{x}; \; a = 1; \; n = 2; \; [\frac{4}{9}, 1].$

36. (a) Use the identity $a - x = a(1 - x/a)$ to find the Maclaurin series for $1/(a - x)$ from the geometric series. What is its radius of convergence?
(b) Find the Maclaurin series and radius of convergence of $1/(3 + x)$.
(c) Find the Maclaurin series and radius of convergence of $2x/(4 + x^2)$.
(d) Use partial fractions to find the Maclaurin series and radius of convergence of
$$\frac{1}{(1-x)(2-x)}.$$

37. Use the known Maclaurin series for $\ln(1 + x)$ to find the Maclaurin series and radius of convergence of $\ln(a + x)$ $(a > 0)$.

38. Use the identity $x = a + (x - a)$ and the known Maclaurin series for e^x, $\sin x$, $\cos x$, and $1/(1 - x)$ to find the Taylor series about $x = a$ for: (a) e^x (b) $\sin x$ and (c) $1/x$.

39. Use the series of Example 6 in Section 11.12 to find the Maclaurin series and radius of convergence of $1/\sqrt{9 + x}$.

In Exercises 40–44, use any method to find the first three nonzero terms of the Maclaurin series.

40. $e^{\tan x}.$

41. $\sec x.$

42. $(\sin x)/(e^x - x).$

43. $\sqrt{\cos x}.$

44. $e^x \ln(1 - x).$

45. Find power series for

(a) $\dfrac{1 - \cos x}{x^2}$ \qquad (b) $\dfrac{\ln(1 + x)}{x}$

(c) Use your answers to (a) and (b) to find
$$\lim_{x \to 0} \frac{1 - \cos x}{x^2} \quad \text{and} \quad \lim_{x \to 0} \frac{\ln(1 + x)}{x}.$$

46. How many decimal places of accuracy can be guaranteed if we approximate $\cos x$ by $1 - x^2/2$ for $-0.1 < x < 0.1$?

47. For what values of x can $\sin x$ be replaced by $x - x^3/6 + x^5/120$ with an assured accuracy of 6×10^{-4}?

In Exercises 48–50, approximate the indicated quantity to three decimal place accuracy.

48. $\cos(10°)$.

49. $\displaystyle\int_0^1 \frac{(1 - e^{-t/2})}{t}\, dt.$

50. $\displaystyle\int_0^1 \frac{\sin x}{\sqrt{x}}\, dx.$

51. Show that $y = \displaystyle\sum_{n=0}^{\infty} k^n x^n / n!$ satisfies $y' - ky = 0$ for any fixed k.

12 topics in analytic geometry

12.1 INTRODUCTION TO THE CONIC SECTIONS

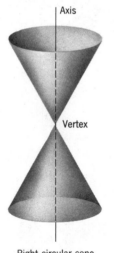

Axis

Vertex

Right circular cone

Figure 12.1.1

The surface shown in Figure 12.1.1 is called a ***double right circular cone*** or sometimes simply a ***cone.*** It is the three-dimensional surface generated by a line that revolves about a fixed ***axis*** in such a way that the line always passes through a fixed point on the axis, called the ***vertex,*** and always makes the same angle with the axis. The cone consists of two parts, or ***nappes,*** that intersect at the vertex.

The curves that can be obtained by intersecting a cone with a plane are called ***conics*** or ***conic sections,*** the most important of which are circles, ellipses, parabolas, and hyperbolas. A ***circle*** is obtained by intersecting a cone with a plane that is perpendicular to the axis and does not contain the vertex (Figure 12.1.2*a*). If the cutting plane is tilted slightly and intersects only one nappe, the resulting intersection is an ***ellipse*** (Figure 12.1.2*b*). If the cutting plane is tilted still further so that it is parallel to a line on the surface of the cone, but intersects only one nappe, the resulting intersection is a ***parabola*** (Figure 12.1.2*c*). Finally, if the plane intersects both nappes, but does not contain the vertex, the resulting intersection is a ***hyperbola*** (Figure 12.1.2*d*). By choosing the cutting plane to pass through the vertex, it is possible to obtain a point or a pair of lines for the intersection. These are called ***degenerate conic sections.***

According to the Alexandrian geographer and astronomer Eratosthenes, the conic sections were first discovered by Menaechmus, a geometer and astronomer in Plato's academy. While it is not known for certain what motivated the discovery of the conic sections, it is commonly believed that they resulted from the study of construction problems. Menaechmus, for example, used them to solve the problem of "doubling the cube," that is, constructing a cube whose volume is twice that of a given cube. Another theory suggests that the conic sections may have originated as a result of work on sundials. With the advent of analytic geometry and calculus, conic

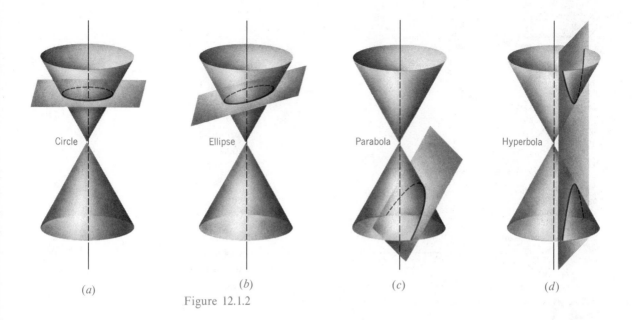

Circle (a)

Ellipse (b)

Parabola (c)

Hyperbola (d)

Figure 12.1.2

sections gained importance in the physical sciences. In 1609 Johannes Kepler published a book known as *Astronomia Nova* in which he presented his landmark discovery that the path of each planet about the sun is an ellipse. Galileo and Newton showed that objects subject to gravitational forces can also move along paths that are parabolas or hyperbolas.

Today, properties of conic sections are used in the construction of telescopes, radar antennas, and navigational systems, and in the determination of satellite orbits. In this chapter we will develop the basic geometric properties and equations of conic sections. For simplicity, our working definitions of the conic sections will be based on their geometric properties rather than their interpretation as intersections of a plane with a cone. This will enable us to keep our work in a two-dimensional setting.

12.2 THE PARABOLA; TRANSLATION OF COORDINATE AXES

12.2.1 DEFINITION   A *parabola* is the set of all points in the plane that are equidistant from a given line and a given point not on the line.

The given line is called the *directrix* of the parabola, and the given point the *focus* (Figure 12.2.1). A parabola is symmetric about the line that passes through the focus at right angles to the directrix. This line of symmetry, called the *axis* of the parabola, meets the parabola at a point called the *vertex.*

The equation of a parabola is simplest if the coordinate axes are positioned so that the vertex is at the origin and the axis of symmetry is along the x-axis or y-axis. The four possible such orientations are shown in Table 12.2.1. In

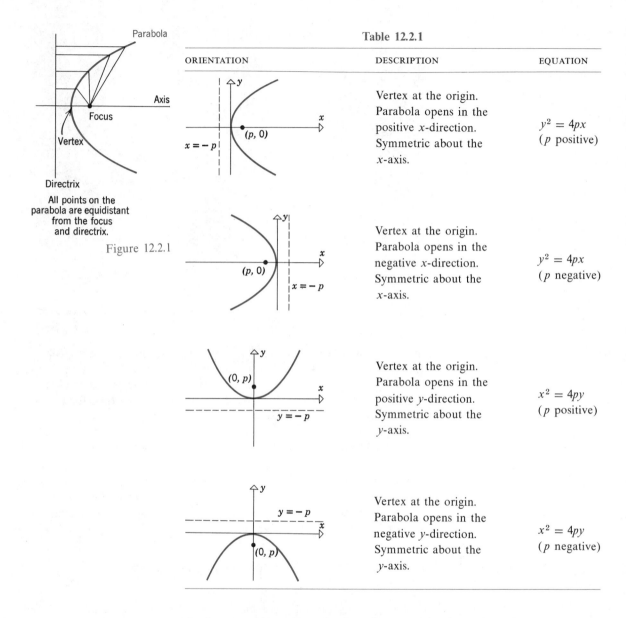

Figure 12.2.1

All points on the parabola are equidistant from the focus and directrix.

Table 12.2.1

ORIENTATION	DESCRIPTION	EQUATION
	Vertex at the origin. Parabola opens in the positive x-direction. Symmetric about the x-axis.	$y^2 = 4px$ (p positive)
	Vertex at the origin. Parabola opens in the negative x-direction. Symmetric about the x-axis.	$y^2 = 4px$ (p negative)
	Vertex at the origin. Parabola opens in the positive y-direction. Symmetric about the y-axis.	$x^2 = 4py$ (p positive)
	Vertex at the origin. Parabola opens in the negative y-direction. Symmetric about the y-axis.	$x^2 = 4py$ (p negative)

the first two orientations, the focus is assumed to be at the point $(p, 0)$, where p is positive in one case and negative in the other. Since the vertex is equidistant from focus and directrix, it follows that the directrix has equation $x = -p$ in these cases. In the last two orientations, the focus is assumed to be at $(0, p)$; and the directrix has equation $y = -p$, where again p may be positive or negative.

To illustrate how the equations in Table 12.2.1 are obtained, let us begin with the first entry in the table—the parabola with focus at $(p, 0)$, directrix $x = -p$, and p positive.

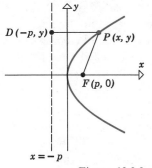

$D(-p, y)$

$P(x, y)$

$F(p, 0)$

$x = -p$

Figure 12.2.2

Let $P(x, y)$ be any point on the parabola. Since P is equidistant from focus and directrix, the distances PF and PD in Figure 12.2.2 are equal; that is,

$$PF = PD \tag{1}$$

where $D(-p, y)$ is the foot of the perpendicular from P to the directrix. From the distance formula, the distances PF and PD are

$$PF = \sqrt{(x-p)^2 + y^2} \quad \text{and} \quad PD = \sqrt{(x+p)^2} \tag{2}$$

Substituting in (1) and squaring yields

$$(x-p)^2 + y^2 = (x+p)^2 \tag{3}$$

and after simplifying

$$y^2 = 4px \tag{4}$$

Conversely, any point $P(x, y)$ satisfying (4) also satisfies (3) (reverse the steps in the simplification); thus, from (2), $PF = PD$, which shows that P is equidistant from focus and directrix. Therefore, each point satisfying (4) lies on the parabola.

The remaining equations in Table 12.2.1 have similar derivations; they are left as exercises.

▶ **Example 1** Find the focus and directrix of the parabola with equation $y^2 = -8x$.

Solution. The equation is of the form $y^2 = 4px$ with $4p = -8$ or $p = -2$. Since y occurs only to an even power, the parabola is symmetric about the x-axis. Thus, the focus is on the x-axis at the point $(-2, 0)$ and the directrix has equation $x = 2$ (Figure 12.2.3). ◀

▶ **Example 2** Find an equation for the parabola that is symmetric about the y-axis, has its vertex at the origin, and passes through the point $(5, 2)$.

Solution. Since the parabola is symmetric about the y-axis and has vertex at the origin, the equation is of the form

$$x^2 = 4py \tag{5}$$

Since the parabola passes through $(5, 2)$ we must have $5^2 = 4p(2)$ or $4p = \frac{25}{2}$. Therefore, (5) becomes

$$x^2 = \frac{25}{2}y \qquad\qquad ◀$$

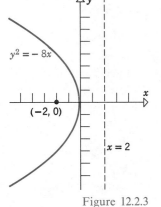

$y^2 = -8x$

$(-2, 0)$

$x = 2$

Figure 12.2.3

If a parabola has its axis of symmetry parallel to one of the coordinate axes, then the location of the focus and directrix may be found by introducing an auxiliary coordinate system with origin at the vertex and axes parallel to the original axes. Before we can discuss the details, we need some preliminary results.

TRANSLATION OF AXES

In Figure 12.2.4a we have translated the axes of an xy-coordinate system to obtain a new $x'y'$-coordinate system whose origin O' is at the point $(x, y) = (h, k)$. As a result, a point P in the plane will have both (x, y)-coordinates and (x', y')-coordinates. As suggested by Figure 12.2.4b, these coordinates are related by

$$x' = x - h, \qquad y' = y - k \tag{6}$$

or, equivalently,

$$x = x' + h, \qquad y = y' + k \tag{7}$$

These are called the **translation equations.**

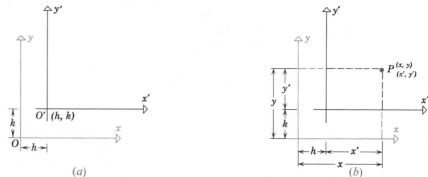

Figure 12.2.4   (a)   (b)

As an illustration, if the new origin is at $(h, k) = (4, -1)$ and the xy-coordinates of a point P are $(2, 5)$, then the $x'y'$-coordinates of P are

$$x' = x - h = 2 - 4 = -2 \quad \text{and} \quad y' = y - k = 5 - (-1) = 6$$

Let us now consider a parabola with vertex V at (h, k) and axis of symmetry parallel to the y-axis. If, as in Figure 12.2.5, we translate the axes so that the vertex V is at the origin of an $x'y'$-coordinate system, then in $x'y'$-coordinates the equation of the parabola will be

$$(x')^2 = 4py'$$

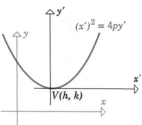

Figure 12.2.5

and from the translation equations (6), the corresponding equation in xy-coordinates will be:

> **Parabola with Vertex (h, k) and Axis Parallel to y-axis**
>
> $$(x - h)^2 = 4p(y - k) \qquad (8)$$

Thus, in xy-coordinates, an equation of form (8) represents a parabola with vertex at (h, k) and axis of symmetry parallel to the y-axis. The parabola opens in the positive or negative y-direction according to whether p is positive or negative. Similarly,

> **Parabola with Vertex (h, k) and Axis Parallel to x-axis**
>
> $$(y - k)^2 = 4p(x - h) \qquad (9)$$

represents a parabola with vertex at (h, k) and axis of symmetry parallel to the x-axis. The parabola opens in the positive or negative x-direction according to whether p is positive or negative.

▶ Example 3 Sketch the parabola

$$(y - 3)^2 = 8(x + 4)$$

and locate the focus and directrix.

Solution. From (9), the equation represents a parabola with vertex at $(h, k) = (-4, 3)$ and axis of symmetry parallel to the x-axis. Moreover, $4p = 8$ or $p = 2$. Since p is positive, the parabola opens in the positive x-direction. This places the focus 2 units to the right of the vertex or at the point $(-2, 3)$. The directrix is 2 units to the left of the vertex (and parallel to the y-axis), so its equation is $x = -6$. The parabola is sketched in Figure 12.2.6. ◀

Sometimes (8) and (9) appear in expanded form, in which case some algebraic manipulations are required to identify the parabola.

▶ Example 4 Show that the curve

$$y = 6x^2 - 12x + 8$$

is a parabola.

Solution. Because the equation involves x to the second power and y to the first power, and because the expanded form of (8) has the same property, we

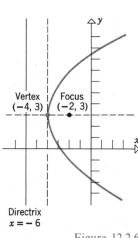

Vertex
$(-4, 3)$

Focus
$(-2, 3)$

Directrix
$x = -6$

Figure 12.2.6

will try to rewrite the equation in form (8). To do this, we first divide by the coefficient of x^2 and collect all the x-terms on one side:

$$\frac{1}{6}y - \frac{8}{6} = x^2 - 2x$$

Next, we complete the square on the x-terms by adding 1 to both sides:

$$\frac{1}{6}y - \frac{2}{6} = (x - 1)^2$$

Finally, we factor out the coefficient of the y-term to obtain

$$(x - 1)^2 = \frac{1}{6}(y - 2)$$

which has form (8) with

$$h = 1, \qquad k = 2, \qquad 4p = \frac{1}{6}, \qquad p = \frac{1}{24}$$

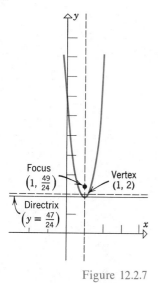

Focus
$\left(1, \frac{49}{24}\right)$

Vertex
$(1, 2)$

Directrix
$\left(y = \frac{47}{24}\right)$

Figure 12.2.7

Thus, the curve is a parabola with vertex at $(1, 2)$ and axis of symmetry parallel to the y-axis. Since $p = \frac{1}{24}$ is positive, the parabola opens in the positive y-direction, the focus is $\frac{1}{24}$ unit above the vertex, and the directrix $\frac{1}{24}$ unit below the vertex. Thus, the focus is at $(1, \frac{49}{24})$ and the directrix has equation $y = \frac{47}{24}$ (Figure 12.2.7). ◀

The procedure used in Example 4 can be used to prove the following general result.

12.2.2 THEOREM *The graph of*

$$y = Ax^2 + Bx + C \qquad (A \neq 0)$$

is a parabola with axis of symmetry parallel to the y-axis; the parabola opens in the posiive y-direction if $A > 0$ and in the negative y-direction if $A < 0$. The graph of

$$x = Ay^2 + By + C \qquad (A \neq 0)$$

is a parabola with axis of symmetry parallel to the x-axis; the parabola opens in the positive x-direction if $A > 0$ and in the negative x-direction if $A < 0$.

▶ **Example 5** Find an equation for the parabola with vertex $(1, 2)$ and focus $(4, 2)$.

Solution. Since the focus and vertex are on a horizontal line, and since the focus is to the right of the vertex, the parabola opens to the right and its equation has the form

$$(y - k)^2 = 4p(x - h) \qquad (p \text{ positive})$$

Since the vertex and focus are three units apart we have $p = 3$, and since the vertex is at $(h, k) = (1, 2)$ we obtain

$$(y - 2)^2 = 12(x - 1) \qquad \blacktriangleleft$$

REFLECTION
PROPERTIES OF
PARABOLAS

Parabolas have important applications in the design of telescopes, radar antennas, and lighting systems. This is due to the following property of parabolas (Exercise 42).

12.2.3 THEOREM
A Geometric Property of Parabolas

The tangent line at a point P on a parabola makes equal angles with the line through P parallel to the axis of symmetry and the line through P and the focus (Figure 12.2.8).

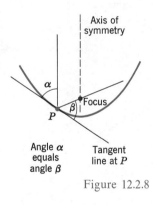

Angle α equals angle β

Figure 12.2.8

It is a principle of physics that when light is reflected from a point P on a surface, the angle of incidence equals the angle of reflection; that is, the angle between the incoming ray and the tangent line at P equals the angle between the outgoing ray and the tangent line at P. Therefore, if the reflecting surface has parabolic cross sections with a common focus, then it follows from Theorem 12.2.3 that all light rays entering parallel to the axis will be reflected through the focus (Figure 12.2.9a). In reflecting telescopes, this principle is used to reflect the (approximately) parallel rays of light from the stars or planets off a parabolic mirror to an eyepiece at the focus of the parabola. Conversely, if a light source is located at the focus of a parabolic reflector, it follows from Theorem 12.2.3 that the reflected rays will form a beam parallel to the axis (Figure 12.2.9b). The parabolic reflectors in flashlights and automobile headlights utilize this principle. The optical principles just discussed also apply to radar signals, which explains the parabolic shape of many radar antennas.

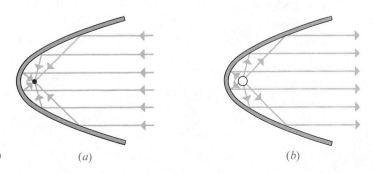

Figure 12.2.9   (*a*)   (*b*)

▶ Exercise Set 12.2

In Exercises 1–16, sketch the parabola. Show the focus, vertex, and directrix.

1. $y^2 = 6x$.
2. $y^2 = -10x$.
3. $x^2 = -9y$.
4. $x^2 = 4y$.
5. $5y^2 = 12x$.
6. $x^2 - 40y = 0$.
7. $(y - 3)^2 = 6(x - 2)$.
8. $(y + 1)^2 = -7(x - 4)$.
9. $(x + 2)^2 = -(y + 2)$.
10. $(x - \frac{1}{2})^2 = 2(y - 1)$.
11. $x^2 - 4x + 2y = 1$.
12. $y^2 - 6y - 2x + 1 = 0$.
13. $x = y^2 - 4y + 2$.
14. $y = 4x^2 + 8x + 5$.
15. $-y^2 + 2y + x = 0$.
16. $y = 1 - 4x - x^2$.

In Exercises 17–31, find an equation for the parabola satisfying the given conditions.

17. Vertex $(0, 0)$; focus $(3, 0)$.
18. Vertex $(0, 0)$; focus $(0, -4)$.
19. Vertex $(0, 0)$; directrix $x = 7$.
20. Vertex $(0, 0)$; directrix $y = \frac{1}{2}$.
21. Vertex $(0, 0)$; symmetric about the x-axis; passes through $(2, 2)$.
22. Vertex $(0, 0)$; symmetric about the y-axis; passes through $(-1, 3)$.
23. Focus $(0, -3)$; directrix $y = 3$.
24. Focus $(6, 0)$; directrix $x = -6$.
25. Axis $y = 0$; passes through $(3, 2)$ and $(2, -3)$.
26. Axis $x = 0$; passes through $(2, -1)$ and $(-4, 5)$.
27. Focus $(3, 0)$; directrix $x = 0$.
28. Vertex $(4, -5)$; focus $(1, -5)$.
29. Vertex $(1, 1)$; directrix $y = -2$.
30. Focus $(-1, 4)$; directrix $x = 5$.
31. Vertex $(5, -3)$; axis parallel to the y-axis; passes through $(9, 5)$.

32. Use the definition of a parabola (12.2.1) to find the equation of the parabola with focus $(2, 1)$ and directrix $x + y + 1 = 0$. [*Hint:* Use the result of Exercise 29, Section 1.5.]

33. (a) Find an equation for the parabola with axis parallel to the y-axis and passing through $(0, 3)$, $(2, 0)$, and $(3, 2)$.
    (b) Find an equation for the parabola with axis parallel to the x-axis and passing through the points in (a).

34. Prove: The line tangent to the parabola $x^2 = 4py$ at (x_0, y_0) is

$$y = \frac{x_0}{2p}x - y_0.$$

35. Find the vertex, focus, and directrix of the parabola $y = Ax^2 + Bx + C$ $(A \neq 0)$.

36. (a) Find an equation for the following parabolic arch with base b and height h.
    (b) Find the area under the arch.

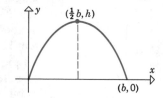

37. Prove: The vertex is the closest point on a parabola to the focus.

38. A comet moves in a parabolic orbit with the sun at its focus. When the comet is 40 million miles from the sun, the line from the sun to the comet makes an angle of $60°$ with the axis of the parabola. How close will the comet come to the sun? [*Hint:* See Exercise 37.]

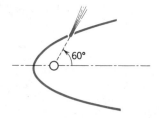

39. How far from the vertex should a light source be placed on the axis of the following parabolic reflector to produce a beam of parallel rays?

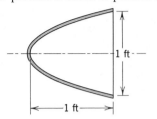

40. Derive the equation $x^2 = 4py$ (p positive) in Table 12.2.1.

41. Derive the equation $y^2 = 4px$ (p negative) in Table 12.2.1.

42. Prove Theorem 12.2.3. [*Hint:* Choose coordinate axes so the parabola has the equation $x^2 = 4py$. Show that the tangent line at $P(x_0, y_0)$ intersects the y-axis at $Q(0, -y_0)$ and that the triangle whose vertices are at P, Q, and the focus is isosceles.]

12.3 THE ELLIPSE

12.3.1 DEFINITION An *ellipse* is the set of all points in the plane, the sum of whose distances from two fixed points is a given positive constant.

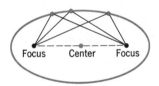

For all points on the ellipse, the sum of the distances to the foci is the same

Figure 12.3.1

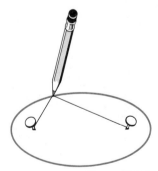

Figure 12.3.2

The two fixed points are called the *foci* (plural of "focus"), and the midpoint of the line segment joining the foci is called the *center* of the ellipse (Figure 12.3.1).

To help visualize Definition 12.3.1, imagine that two ends of a string are tacked to the foci and a pencil traces a curve as it is held tight against the string (Figure 12.3.2). The resulting curve will be an ellipse since the sum of the distances to the foci is a constant, namely the total length of the string. Note that if the foci coincide, the ellipse reduces to a circle.

The line segment through the foci and across the ellipse is called the *major axis* (Figure 12.3.3), while the line segment across the ellipse, through the center, and perpendicular to the major axis is called the *minor axis.* It is traditional in the study of ellipses to denote the length of the major axis by $2a$, the length of the minor axis by $2b$, and the distance between the foci by $2c$ (Figure 12.3.4). The numbers a and b are called the *semiaxes* of the ellipse.

There is a basic relationship between the numbers a, b, and c that can be obtained by considering a point P at the end of the major axis and a point Q

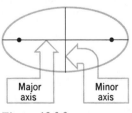

Figure 12.3.3

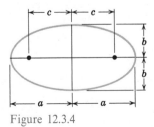

Figure 12.3.4

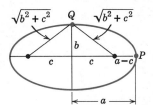

Figure 12.3.5

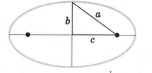

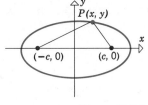

Figure 12.3.6

Figure 12.3.7

at the end of the minor axis. Because P and Q both lie on the ellipse, the sum of the distances from each of them to the foci will be the same. If we express this fact as an equation we obtain (see Figure 12.3.5):

$$2\sqrt{b^2 + c^2} = (a - c) + (a + c)$$

or

$$a = \sqrt{b^2 + c^2} \tag{1}$$

Relationship (1) shows that the distance from a focus to an end of the minor axis is a (Figure 12.3.6), which implies that for *all* points on the ellipse the sum of the distances to the foci is $2a$. (Why?)

It also follows from (1) that $a \geq b$ with the equality holding only when $c = 0$. Geometrically, this means that the major axis of an ellipse is at least as large as the minor axis, and that the two axes have equal length only when the foci coincide, in which case the ellipse is a circle.

The equation of an ellipse is simplest if the coordinate axes are positioned so the center of the ellipse is at the origin and the foci are on the x-axis or y-axis. The two possible such orientations are shown in Table 12.3.1.

To illustrate how the equations in Table 12.3.1 are derived, let us consider the first entry in the table. Let $P(x, y)$ be any point on the ellipse. Since the sum of the distances from P to the foci is $2a$ it follows (Figure 12.3.7) that

$$\sqrt{(x + c)^2 + y^2} + \sqrt{(x - c)^2 + y^2} = 2a$$

Transposing the second radical to the right side of the equation and squaring yields

$$(x + c)^2 + y^2 = 4a^2 - 4a\sqrt{(x - c)^2 + y^2} + (x - c)^2 + y^2$$

and, on simplifying,

$$\sqrt{(x - c)^2 + y^2} = a - \frac{c}{a}x$$

Squaring again and simplifying yields

$$\frac{x^2}{a^2} + \frac{y^2}{a^2 - c^2} = 1$$

which, by virtue of (1), may be written

$$\frac{x^2}{a^2} + \frac{y^2}{b^2} = 1 \tag{2}$$

Table 12.3.1

ORIENTATION	DESCRIPTION	EQUATION
	Foci and major axis on the x-axis. Minor axis on the y-axis. Center at the origin.	$\dfrac{x^2}{a^2} + \dfrac{y^2}{b^2} = 1$
	Foci and major axis on the y-axis. Minor axis on the x-axis. Center at the origin.	$\dfrac{x^2}{b^2} + \dfrac{y^2}{a^2} = 1$

Conversely, it can be shown that any point whose coordinates satisfy (2), has $2a$ as the sum of its distances from the foci, so that such a point is on the ellipse.

The second equation in Table 12.3.1 has a similar derivation.

▶ **Example 1** Does the ellipse

$$\frac{x^2}{9} + \frac{y^2}{4} = 1$$

have its major axis on the x-axis or the y-axis?

Solution. Let us locate the intersections of the ellipse with the coordinate axes. Setting $y = 0$ yields $x^2 = 9$, or $x = \pm 3$, so the ellipse intersects the x-axis at $(-3, 0)$ and $(3, 0)$. Setting $x = 0$ yields $y^2 = 4$, or $y = \pm 2$, so the ellipse intersects the y-axis at $(0, -2)$ and $(0, 2)$. Plotting the points of intersection and making a rough sketch shows that the major axis is along the x-axis (Figure 12.3.8).

Alternate Solution. Since it is always true that $a^2 \geq b^2$, it follows that $a^2 = 9$ and $b^2 = 4$, so that the given equation is of the form

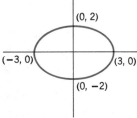

Figure 12.3.8

$$\frac{x^2}{a^2} + \frac{y^2}{b^2} = 1$$

which is an ellipse with major axis along the x-axis (Table 12.3.1). ◀

▶ Example 2 Show that the graph of the equation

$$2x^2 + y^2 = 4$$

is an ellipse, and locate the foci.

Solution. Dividing through by 4 yields

$$\frac{x^2}{2} + \frac{y^2}{4} = 1 \tag{3}$$

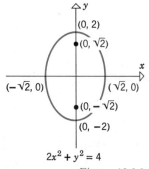

(0, 2)

•(0, √2)

(−√2, 0)

(√2, 0)

•(0, −√2)

(0, −2)

$2x^2 + y^2 = 4$

Figure 12.3.9

which is the equation of an ellipse having $a^2 = 4$ ($a = 2$) and $b^2 = 2$ ($b = \sqrt{2}$). Since the a^2 is under the y^2 in (3), the major axis is along the y-axis. From relationship (1)

$$c = \sqrt{a^2 - b^2} = \sqrt{4 - 2} = \sqrt{2}$$

so the foci are $(0, \sqrt{2})$ and $(0, -\sqrt{2})$. The ellipse is shown in Figure 12.3.9.
◀

▶ Example 3 Find an equation for the ellipse with foci $(0, \pm 2)$ and major axis with endpoints $(0, \pm 4)$.

Solution. From Table 12.3.1, the equation has the form

$$\frac{x^2}{b^2} + \frac{y^2}{a^2} = 1$$

and from the given information, $a = 4$ and $c = 2$. It follows from (1) that

$$b^2 = a^2 - c^2 = 16 - 4 = 12$$

so the equation of the ellipse is

$$\frac{x^2}{12} + \frac{y^2}{16} = 1$$
◀

If the axes of an ellipse are parallel to the coordinate axes but the center is not at the origin, then the equation of the ellipse may be determined by

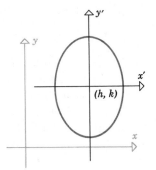

Figure 12.3.10

translation of axes. Specifically, if the center of the ellipse is at the point (h, k), and if we translate the xy-axes so the origin of the $x'y'$-system is at the center (h, k), then the equation of the ellipse in the $x'y'$-system will be

$$\frac{(x')^2}{a^2} + \frac{(y')^2}{b^2} = 1 \quad \text{or} \quad \frac{(x')^2}{b^2} + \frac{(y')^2}{a^2} = 1$$

depending on the location of the major and minor axes (Figure 12.3.10). Thus, from the translation equations $x' = x - h$, $y' = y - k$, the equation of the ellipse in the xy-system will be one of the following:

Ellipse with Center (h, k) and Major Axis Parallel to x-axis

$$\frac{(x - h)^2}{a^2} + \frac{(y - k)^2}{b^2} = 1 \qquad (4)$$

Ellipse with Center (h, k) and Major Axis Parallel to y-axis

$$\frac{(x - h)^2}{b^2} + \frac{(y - k)^2}{a^2} = 1 \qquad (5)$$

Sometimes (4) and (5) appear in expanded form, in which case it is necessary to complete the squares before the ellipse can be analyzed.

▶ Example 4 Show that the curve

$$16x^2 + 9y^2 - 64x - 54y + 1 = 0$$

is an ellipse. Sketch the ellipse and show the location of the foci.

Solution. Our objective is to rewrite the equation in one of the forms, (4) or (5). To do this, group the x-terms and y-terms and take the constant to the right side:

$$(16x^2 - 64x) + (9y^2 - 54y) = -1$$

Next, factor out the coefficients of x^2 and y^2 and complete the squares:

$$16(x^2 - 4x + 4) + 9(y^2 - 6y + 9) = -1 + 64 + 81$$

or

$$16(x - 2)^2 + 9(y - 3)^2 = 144$$

Finally, divide through by 144 to introduce a 1 on the right side:

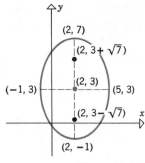

Figure 12.3.11

$$\frac{(x-2)^2}{9} + \frac{(y-3)^2}{16} = 1$$

This is an equation of form (5), with $h = 2, k = 3, a^2 = 16$, and $b^2 = 9$. Thus, the given equation is an ellipse with center $(2, 3)$ and major axis parallel to the y-axis. Since $a = 4$, the major axis extends 4 units above and 4 units below the center, so its endpoints are $(2, 7)$ and $(2, -1)$ (Figure 12.3.11). Since $b = 3$, the minor axis extends 3 units to the left and 3 units to the right of the center, so its endpoints are $(-1, 3)$ and $(5, 3)$. Since

$$c = \sqrt{a^2 - b^2} = \sqrt{16 - 9} = \sqrt{7}$$

the foci lie $\sqrt{7}$ units above and below the center, placing them at the points $(2, 3 + \sqrt{7})$ and $(2, 3 - \sqrt{7})$. ◀

REFLECTION PROPERTIES OF ELLIPSES

In the exercises the reader is asked to prove the following result.

12.3.2 THEOREM
A Geometric Property of Ellipses

A line tangent to an ellipse at a point P makes equal angles with the lines through P and the foci (Figure 12.3.12).

It follows from Theorem 12.3.2 that a ray of light emanating from one focus of an ellipse will be reflected through the other focus. This property of ellipses is used in "whispering galleries." Such rooms have ceilings whose cross sections are elliptical in shape with common foci. As a result, if a person standing at one focus whispers, the sound waves are reflected by the ceiling to the other focus, making it possible for a person at that focus to hear the whispered sound.

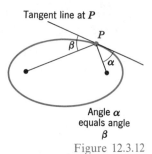

Tangent line at P

Angle α equals angle β

Figure 12.3.12

▶ **Exercise Set 12.3**

In Exercises 1–14, sketch the ellipse. Label the foci and the ends of the major and minor axes.

1. $\dfrac{x^2}{16} + \dfrac{y^2}{9} = 1$.

2. $\dfrac{x^2}{4} + \dfrac{y^2}{25} = 1$.

3. $9x^2 + y^2 = 9$.

4. $4x^2 + 9y^2 = 36$.

5. $x^2 + 3y^2 = 2$.

6. $16x^2 + 4y^2 = 1$.

7. $9(x - 1)^2 + 16(y - 3)^2 = 144$.

8. $(x + 3)^2 + 4(y - 5)^2 = 16$.

9. $3(x + 2)^2 + 4(y + 1)^2 = 12$.

10. $\frac{1}{4}x^2 + \frac{1}{9}(y + 2)^2 - 1 = 0$.

11. $x^2 + 9y^2 + 2x - 18y + 1 = 0$.

12. $9x^2 + 4y^2 + 18x - 24y + 9 = 0$.

13. $4x^2 + y^2 + 8x - 10y = -13$.

14. $5x^2 + 9y^2 - 20x + 54y = -56$.

In Exercises 15–26, find an equation for the ellipse satisfying the given conditions.

15. Ends of major axis $(\pm 3, 0)$; ends of minor axis $(0, \pm 2)$.

16. Ends of major axis $(0, \pm\sqrt{5})$; ends of minor axis $(\pm 1, 0)$.

17. Length of major axis 26; foci $(\pm 5, 0)$.

18. Length of minor axis 16; foci $(0, \pm 6)$.

19. Foci $(\pm 1, 0)$; $b = \sqrt{2}$.

20. Foci $(\pm 3, 0)$; $a = 4$.

21. $c = 2\sqrt{3}$; $a = 4$; center at the origin; foci on a coordinate axis (two answers).

22. $b = 3$; $c = 4$; center at the origin; foci on a coordinate axis (two answers).

23. Ends of major axis $(\pm 6, 0)$; passes through $(2, 3)$.

24. Center at $(0, 0)$; major and minor axes along the coordinate axes; passes through $(3, 2)$ and $(1, 6)$.

25. Foci $(1, 2)$ and $(1, 4)$; minor axis of length 2.

26. Foci $(2, 1)$ and $(2, -3)$; major axis of length 6.

27. Find the equation of the ellipse traced by a point that moves so the sum of its distances to $(4, 1)$ and $(4, 5)$ is 12.

28. Find the equation of the ellipse traced by a point that moves so the sum of its distances to $(0, 0)$ and $(1, 1)$ is 4.

In Exercises 29–32, find all intersections of the given curves and make a sketch of the curves that shows the points of intersection.

29. $x^2 + 4y^2 = 40$ and $x + 2y = 8$.

30. $y^2 = 2x$ and $x^2 + 2y^2 = 12$.

31. $x^2 + 9y^2 = 36$ and $x^2 + y^2 = 20$.

32. $16x^2 + 9y^2 = 36$ and $5x^2 + 18y^2 = 45$.

33. Find two values of k such that the line $x + 2y = k$ is tangent to the ellipse $x^2 + 4y^2 = 8$. Find the points of tangency.

34. Prove: The line tangent to the ellipse $x^2/a^2 + y^2/b^2 = 1$ at (x_0, y_0) has equation $xx_0/a^2 + yy_0/b^2 = 1$.

35. Find the area enclosed by the ellipse $b^2x^2 + a^2y^2 = a^2b^2$.

36. Find the volume of the solid generated when the region enclosed by the ellipse $b^2x^2 + a^2y^2 = a^2b^2$ is

    (a) revolved about the x-axis;

    (b) revolved about the y-axis.

37. Derive the equation $x^2/b^2 + y^2/a^2 = 1$ in Table 12.3.1.

38. Prove Theorem 12.3.2. [*Hint:* Introduce coordinate axes so the ellipse has the equation $x^2/a^2 + y^2/b^2 = 1$, and use the result of Exercise 17 of Section 1.4.]

12.4 THE HYPERBOLA

12.4.1 DEFINITION   A *hyperbola* is the set of all points in the plane, the difference of whose distances from two fixed points is a given positive constant.

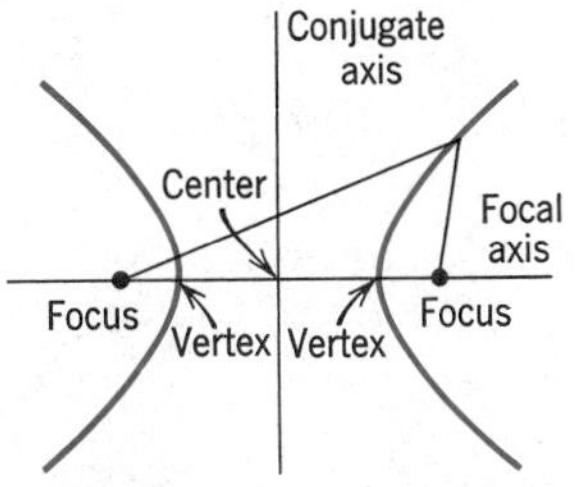

Figure 12.4.1

In this definition the "difference" of the distances is understood to mean the distance to the farther point minus the distance to the closer point.

The two fixed points are called the *foci,* and the midpoint of the line segment joining the foci is called the *center* of the hyperbola (Figure 12.4.1). The line through the foci is called the *focal axis* (or *transverse axis*) and the line through the center and perpendicular to the focal axis is called the *conjugate axis.* The hyperbola intersects the focal axis at two points, called *vertices.* The two separate parts of a hyperbola are called the **branches.**

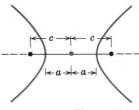

Figure 12.4.2

It is traditional in the study of hyperbolas to denote the distance between the vertices by $2a$, the distance between the foci by $2c$ (Figure 12.4.2), and to define the quantity b as

$$b = \sqrt{c^2 - a^2} \tag{1}$$

(See Figure 12.4.3.)

If V is one vertex of a hyperbola, then, as illustrated in Figure 12.4.4, the distance from V to the farther focus minus the distance from V to the closer focus is

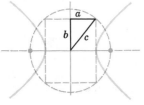

Figure 12.4.3

$$[(c - a) + 2a] - (c - a) = 2a$$

Thus, for *all* points on a hyperbola, the distance to the farther focus minus the distance to the closer focus is $2a$. (Why?)

The equation of a hyperbola is simplest if the coordinate axes are positioned so the center of the hyperbola is at the origin and the foci are on the x-axis or y-axis. The two possible such orientations are shown in Table 12.4.1.

To illustrate how the equations in Table 12.4.1 are derived, let us consider the first entry in the table. Let $P(x, y)$ be any point on the hyperbola, so that the distance from P to the farther focus minus the distance from P to the closer focus is $2a$. Depending on which focus is farther from P, this condition leads either to the equation

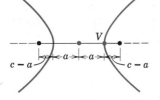

Figure 12.4.4

$$\sqrt{(x + c)^2 + y^2} - \sqrt{(x - c)^2 + y^2} = 2a$$

or to

$$\sqrt{(x - c)^2 + y^2} - \sqrt{(x + c)^2 + y^2} = 2a$$

(See Figure 12.4.5.) In either case,

$$\left| \sqrt{(x + c)^2 + y^2} - \sqrt{(x - c)^2 + y^2} \right| = 2a \tag{2}$$

Squaring both sides of (2) yields

$$(x + c)^2 + y^2 - 2\sqrt{[(x + c)^2 + y^2][(x - c)^2 + y^2]} + (x - c)^2 + y^2 = 4a^2$$

Simplifying, then isolating the radical, then squaring again to remove the radical, ultimately yields (verify)

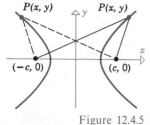

Figure 12.4.5

$$\frac{x^2}{a^2} - \frac{y^2}{c^2 - a^2} = 1$$

Table 12.4.1

ORIENTATION	DESCRIPTION	EQUATION
$(a, 0)$, $(-c, 0)$, $(c, 0)$, $(-a, 0)$	Foci on the x-axis. Conjugate axis on the y-axis. Center at the origin.	$\dfrac{x^2}{a^2} - \dfrac{y^2}{b^2} = 1$
$(0, c)$, $(0, a)$, $(0, -a)$, $(0, -c)$	Foci on the y-axis. Conjugate axis on the x-axis. Center at the origin.	$\dfrac{y^2}{a^2} - \dfrac{x^2}{b^2} = 1$

which, by virtue of (1), may be written

$$\frac{x^2}{a^2} - \frac{y^2}{b^2} = 1 \tag{3}$$

Conversely, it can be shown that for any point P whose coordinates satisfy (3), the distance from P to the farther focus minus the distance from P to the closer focus is $2a$, so that such a point must lie on the hyperbola.

The derivation of the second equation in Table 12.4.1 is similar.

▶ Example 1 Does the hyperbola

$$\frac{x^2}{4} - \frac{y^2}{9} = 1$$

have its focal axis on the x-axis or the y-axis?

Solution. Unlike the case of the ellipse, the orientation of a hyperbola is not determined by examining the relative sizes of a^2 and b^2, but rather by noting where the minus sign occurs in the equation. From Table 12.4.1 we see that the focal axis is on the x-axis when the minus precedes the y^2-term, and it is on the y-axis when the minus precedes the x^2-term. In the given equation the minus precedes the y^2-term so the focal axis is on the x-axis.

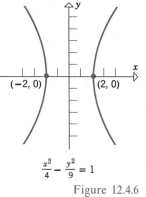

$$\frac{x^2}{4} - \frac{y^2}{9} = 1$$

Figure 12.4.6

Alternately, the orientation of a hyperbola can be obtained by locating its vertices. If we let $x = 0$ in the given equation, we obtain

$$-\frac{y^2}{9} = 1 \quad \text{or} \quad y^2 = -9$$

This equation has no real solutions, so the hyperbola does not intersect the y-axis. This tells us that the focal axis is along the x-axis. The same conclusion can be reached by setting $y = 0$ in the given equation to obtain

$$\frac{x^2}{4} = 1 \quad \text{or} \quad x = \pm 2$$

This tells us that the vertices are at the points $(-2, 0)$ and $(2, 0)$, so again we conclude that the focal axis is on the x-axis (Figure 12.4.6). ◀

The following theorem states that hyperbolas have asymptotes (see Figure 12.4.7).

12.4.2 THEOREM
Asymptotes of Hyperbolas

(a) *The hyperbola*

$$\frac{x^2}{a^2} - \frac{y^2}{b^2} = 1$$

has asymptotes

$$y = \frac{b}{a}x \quad \text{and} \quad y = -\frac{b}{a}x$$

(b) *The hyperbola*

$$\frac{y^2}{a^2} - \frac{x^2}{b^2} = 1$$

has asymptotes

$$y = \frac{a}{b}x \quad \text{and} \quad y = -\frac{a}{b}x$$

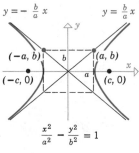

$$\frac{y^2}{a^2} - \frac{x^2}{b^2} = 1$$

Figure 12.4.7

Proof. We will prove part (a); the proof of (b) is similar. If we write the equation $x^2/a^2 - y^2/b^2 = 1$ in the form

$$y^2 = \frac{b^2}{a^2}(x^2 - a^2)$$

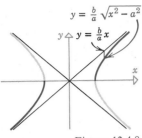

$y = \frac{b}{a}\sqrt{x^2 - a^2}$

$y = \frac{b}{a}x$

Figure 12.4.8

then, in the first quadrant, the vertical distance between the line $y = (b/a)x$ and the hyperbola may be written (Figure 12.4.8).

$$\frac{b}{a}x - \frac{b}{a}\sqrt{x^2 - a^2}$$

But this distance tends to zero as $x \to +\infty$ since

$$\lim_{x \to +\infty}\left(\frac{b}{a}x - \frac{b}{a}\sqrt{x^2 - a^2}\right) = \lim_{x \to +\infty}\frac{b}{a}(x - \sqrt{x^2 - a^2})$$

$$= \lim_{x \to +\infty}\frac{b}{a}\frac{(x - \sqrt{x^2 - a^2})(x + \sqrt{x^2 - a^2})}{x + \sqrt{x^2 - a^2}}$$

$$= \lim_{x \to +\infty}\frac{ab}{x + \sqrt{x^2 - a^2}}$$

$$= 0$$

The analysis in the remaining quadrants is similar. ▪

REMARKS. There is a trick that can be used to avoid memorizing the equations of the asymptotes of a hyperbola. They can be obtained, when needed, by substituting 0 for the 1 on the right side of the hyperbola equation, and then solving for y in terms of x. For example, for the hyperbola

$$\frac{x^2}{a^2} - \frac{y^2}{b^2} = 1$$

we would write

$$\frac{x^2}{a^2} - \frac{y^2}{b^2} = 0 \quad \text{or} \quad y^2 = \frac{b^2}{a^2}x^2 \quad \text{or} \quad y = \pm\frac{b}{a}x$$

which are the equations for the asymptotes. As indicated in Figure 12.4.7, the asymptotes of a hyperbola are along the diagonals of a box extending a units on each side of the origin along the focal axis, and b units on each side of the origin along the conjugate axis.

▶ Example 2 Find the foci, vertices, and asymptotes of the hyperbola

$$\frac{y^2}{16} - \frac{x^2}{9} = 1$$

and sketch the graph.

Solution. Since the minus precedes the x^2-term, this hyperbola has its foci on the y-axis. Also, $a^2 = 16$ and $b^2 = 9$. Since $a = 4$, the vertices are at

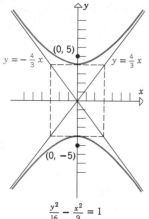

$$\frac{y^2}{16} - \frac{x^2}{9} = 1$$

Figure 12.4.9

(0, −4) and (0, 4). Moreover, from (1), $c = \sqrt{a^2 + b^2} = \sqrt{25} = 5$, so the foci are at (0, −5) and (0, 5).

To obtain the asymptotes, we substitute 0 for 1 in the given equation, which yields

$$\frac{y^2}{16} - \frac{x^2}{9} = 0 \qquad \text{or} \qquad y = \pm \frac{4}{3}x$$

With the aid of a box extending $a = 4$ units above and below the origin in the y-direction and $b = 3$ units to the left and right of the origin in the x-direction, we obtain the sketch in Figure 12.4.9. ◀

▶ **Example 3** Find the equation of the hyperbola with vertices $(0, \pm 8)$ and asymptotes $y = \pm \frac{4}{3}x$.

Solution. Since the vertices are on the y-axis, the equation of the hyperbola has the form $y^2/a^2 - x^2/b^2 = 1$ and the asymptotes are

$$y = \pm \frac{a}{b}x$$

From the location of the vertices we have $a = 8$, so that the given equations of the asymptotes yield

$$y = \pm \frac{a}{b}x = \pm \frac{8}{b}x = \pm \frac{4}{3}x$$

from which it follows that $b = 6$. Thus, the hyperbola has the equation

$$\frac{y^2}{64} - \frac{x^2}{36} = 1 \qquad\qquad ◀$$

If the center of a hyperbola is at the point (h, k) and its focal and conjugate axes are parallel to the coordinate axes, then its equation has one of the forms:

Hyperbola with Center (h, k) and Focal Axis Parallel to x-axis.

$$\frac{(x - h)^2}{a^2} - \frac{(y - k)^2}{b^2} = 1 \qquad\qquad (4)$$

Hyperbola with Center (h, k) and Focal Axis Parallel to y-axis.

$$\frac{(y - k)^2}{a^2} - \frac{(x - h)^2}{b^2} = 1 \qquad\qquad (5)$$

The derivations are similar to those for the parabola and ellipse and will be omitted. As before, the equations of the asymptotes can be obtained by replacing the 1 by a 0 on the right side of (4) or (5), and then solving for y in terms of x.

▶ Example 4 Show that the curve

$$x^2 - y^2 - 4x + 8y - 21 = 0$$

is a hyperbola. Sketch the hyperbola and show the foci, vertices, and asymptotes.

Solution. Our objective is to rewrite the equation in form (4) or (5) by completing the squares. To do this, we first group the x-terms and y-terms and take the constant to the right side:

$$(x^2 - 4x) - (y^2 - 8y) = 21$$

Next, complete the squares:

$$(x^2 - 4x + 4) - (y^2 - 8y + 16) = 21 + 4 - 16$$

or

$$(x - 2)^2 - (y - 4)^2 = 9$$

Finally, divide by 9 to introduce a 1 on the right side:

$$\frac{(x - 2)^2}{9} - \frac{(y - 4)^2}{9} = 1 \tag{6}$$

This is an equation of form (4) with $h = 2$, $k = 4$, $a^2 = 9$, and $b^2 = 9$. Thus, the equation represents a hyperbola with center $(2, 4)$ and focal axis parallel to the x-axis. Since $a = 3$, the vertices are located 3 units to the left and 3 units to the right of the center, or at the points $(-1, 4)$ and $(5, 4)$. From (1), $c = \sqrt{a^2 + b^2} = \sqrt{9 + 9} = 3\sqrt{2}$, so that the foci are located $3\sqrt{2}$ units to the left and right of the center or at the points $(2 - 3\sqrt{2}, 4)$ and $(2 + 3\sqrt{2}, 4)$.

The equations of the asymptotes may be found by substituting 0 for 1 in (6) to obtain

$$\frac{(x - 2)^2}{9} - \frac{(y - 4)^2}{9} = 0 \qquad \text{or} \qquad y - 4 = \pm(x - 2)$$

which yields the asymptotes

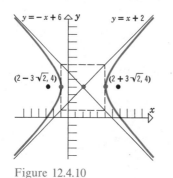

Figure 12.4.10

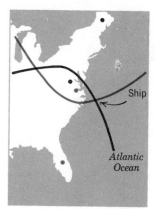

Figure 12.4.11

$$y = x + 2 \qquad \text{and} \qquad y = -x + 6$$

With the aid of a box extending $a = 3$ units left and right of the center and $b = 3$ units above and below the center we obtain the sketch in Figure 12.4.10. ◄

A hyperbola, like that in the last example, is called **equilateral** if $a = b$. For such hyperbolas, the asymptotes are perpendicular. (Verify.)

HYPERBOLIC NAVIGATION

By measuring the difference in reception times of synchronized radio signals from two widely spaced transmitters, a ship's electronic equipment can determine the difference $2a$ in its distances from the two transmitters. This information places the ship somewhere on the hyperbola whose foci are at the transmitters and whose points have $2a$ as their difference in distances from the foci. By using two pairs of transmitters, the position of a ship can be determined as the intersection of two hyperbolas (Figure 12.4.11).

REFLECTION PROPERTIES OF HYPERBOLAS

In the exercises, the reader is asked to prove the following result.

12.4.3 THEOREM
A Geometric Property of Hyperbolas

A line tangent to a hyperbola at a point P makes equal angles with the lines through P and the foci (Figure 12.4.12).

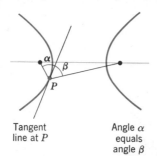

Figure 12.4.12

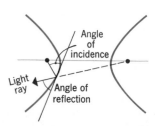

Figure 12.4.13

It follows from Theorem 12.4.3 that a ray of light emanating from one focus of a hyperbola will be reflected back along the line from the opposite focus (Figure 12.4.13). Light reflection properties of hyperbolas are used to advantage in the design of high-quality telescopes.

▶ Exercise Set 12.4

In Exercises 1–16, sketch the hyperbola. Find the coordinates of the vertices and foci, and find equations for the asymptotes.

1. $\dfrac{x^2}{16} - \dfrac{y^2}{4} = 1$.

2. $\dfrac{y^2}{9} - \dfrac{x^2}{25} = 1$.

3. $9y^2 - 4x^2 = 36$.

4. $16x^2 - 25y^2 = 400$.

5. $8x^2 - y^2 = 8$.

6. $3x^2 - y^2 = -9$.

7. $x^2 - y^2 = 1$.

8. $4y^2 - 4x^2 = 1$.

9. $\dfrac{(x-2)^2}{9} - \dfrac{(y-4)^2}{4} = 1$.

10. $\dfrac{(y+4)^2}{3} - \dfrac{(x-2)^2}{5} = 1$.

11. $(y+3)^2 - 9(x+2)^2 = 36$.

12. $16(x+1)^2 - 8(y-3)^2 = 16$.

13. $x^2 - 4y^2 + 2x + 8y - 7 = 0$.

14. $4x^2 - 9y^2 + 16x + 54y - 29 = 0$.

15. $16x^2 - y^2 - 32x - 6y = 57$.

16. $4y^2 - x^2 + 40y - 4x = -60$.

In Exercises 17–30, find an equation for the hyperbola satisfying the given conditions.

17. Vertices $(\pm 2, 0)$; foci $(\pm 3, 0)$.

18. Vertices $(0, \pm 3)$; foci $(0, \pm 5)$.

19. Vertices $(\pm 1, 0)$; asymptotes $y = \pm 2x$.

20. Vertices $(0, \pm 3)$; asymptotes $y = \pm x$.

21. Asymptotes $y = \pm \frac{3}{2}x$; $b = 4$.

22. Foci $(0, \pm 5)$; asymptotes $y = \pm 2x$.

23. Passes through $(5, 9)$; asymptotes $y = \pm x$.

24. Asymptotes $y = \pm \frac{3}{4}x$; $c = 5$.

25. Vertices $(\pm 2, 0)$; passes through $(4, 2)$.

26. Foci $(\pm 3, 0)$; asymptotes $y = \pm 2x$.

27. Vertices $(4, -3)$ and $(0, -3)$; foci 6 units apart.

28. Foci $(1, 8)$ and $(1, -12)$; vertices 4 units apart.

29. Vertices $(2, 4)$ and $(10, 4)$; foci 10 units apart.

30. Asymptotes $y = 2x + 1$ and $y = -2x + 3$; passes through the origin.

31. Find the equation of the hyperbola traced by a point that moves so the difference between its distances to $(0, 0)$ and $(1, 1)$ is 1.

32. Find the equation of the hyperbola traced by a point that moves so the difference between its distances to $(4, -3)$ and $(-2, 5)$ is 6.

In Exercises 33–36, find all intersections of the given curves, and make a sketch of the curves that shows the points of intersection.

33. $x^2 - 4y^2 = 36$ and $x - 2y - 20 = 0$.

34. $y^2 - 8x^2 = 5$ and $y - 2x^2 = 0$.

35. $3x^2 - 7y^2 = 5$ and $9y^2 - 2x^2 = 1$.

36. $x^2 - y^2 = 1$ and $x^2 + y^2 = 7$.

37. A point moves so that the product of its distances from the lines $y = mx$ and $y = -mx$ is a constant k^2. Show that the point moves along a hyperbola having these lines as asymptotes.

38. Prove: The line tangent to the hyperbola $x^2/a^2 - y^2/b^2 = 1$ at (x_0, y_0) has the equation $xx_0/a^2 - yy_0/b^2 = 1$.

39. A line tangent to the hyperbola $4x^2 - y^2 = 36$ intersects the y-axis at the point $(0, 4)$. Find the point(s) of tangency.

40. Let R be the region enclosed between the hyperbola $b^2x^2 - a^2y^2 = a^2b^2$ and the line $x = c$ through the focus $(c, 0)$. Find the volume of the solid generated by revolving R about
(a) the x-axis   (b) the y-axis.

41. Derive the equation $y^2/a^2 - x^2/b^2 = 1$ in Table 12.4.1.

42. Prove Theorem 12.4.3. [*Hint:* Introduce coordinate axes so the hyperbola has the equation $x^2/a^2 - y^2/b^2 = 1$, and use the result of Exercise 17 of Section 1.4.]

12.5 ROTATION OF AXES; SECOND DEGREE EQUATIONS

In previous sections we obtained equations of conic sections with axes parallel to the coordinate axes. In this section we study the equations of conics that are "tilted" relative to the coordinate axes. This will lead us to investigate rotations of coordinate axes.

Each of the equations

$$(y - k)^2 = 4p(x - h) \qquad (x - h)^2 = 4p(y - k)$$

$$\frac{(x - h)^2}{a^2} + \frac{(y - k)^2}{b^2} = 1 \qquad \frac{(x - h)^2}{b^2} + \frac{(y - k)^2}{a^2} = 1$$

$$\frac{(x - h)^2}{a^2} - \frac{(y - k)^2}{b^2} = 1 \qquad \frac{(y - k)^2}{a^2} - \frac{(x - h)^2}{b^2} = 1$$

represents a conic section with axis or axes parallel to the coordinate axes. By squaring out the quadratic terms and simplifying, each of these equations can be rewritten in the form

$$Ax^2 + Cy^2 + Dx + Ey + F = 0 \qquad (1)$$

For example, $(y - k)^2 = 4p(x - h)$ can be written as

$$y^2 - 4px - 2ky + (k^2 + 4ph) = 0$$

which is of form (1) with

$$A = 0, \quad C = 1, \quad D = -4p, \quad E = -2k, \quad F = k^2 + 4ph$$

If a conic section is tilted so that its axes are not parallel to the x- and y-axis, then its equation will contain a "cross-product" term Bxy and the equation will have the form

$$Ax^2 + Bxy + Cy^2 + Dx + Ey + F = 0 \qquad (2)$$

An equation of form (2) in which A, B, and C are not all zero is called a **second degree equation** or **quadratic equation** in x and y.

As an illustration, consider the ellipse with foci $F_1(1, 2)$ and $F_2(-1, -2)$ and such that the sum of the distances from each point $P(x, y)$ on the ellipse to the foci is 6 units. Expressing this condition as an equation, we obtain (Figure 12.5.1):

$$\sqrt{(x - 1)^2 + (y - 2)^2} + \sqrt{(x + 1)^2 + (y + 2)^2} = 6$$

Squaring both sides, then isolating the remaining radical, then squaring again, ultimately yields

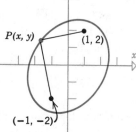

Figure 12.5.1

$$8x^2 - 4xy + 5y^2 = 36$$

as the equation of the ellipse. This is an equation of form (2) with $A = 8$, $B = -4$, $C = 5$, $D = 0$, $E = 0$, $F = -36$.

To study conics that are tilted relative to the coordinate axes, it is frequently helpful to rotate the coordinate axes, so that the rotated coordinate axes are parallel to the axes of the conic. Before we can discuss the details, we need to develop some ideas about rotation of coordinate axes.

ROTATION OF AXES   In Figure 12.5.2a the axes of an xy-coordinate system have been rotated about the origin through an angle θ to produce a new $x'y'$-coordinate system.

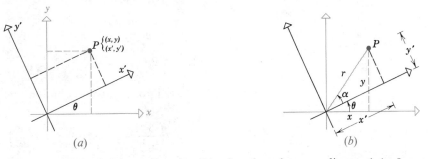

Figure 12.5.2   (a)   (b)

As shown in the figure, each point P in the plane has coordinates (x', y') as well as coordinates (x, y). To see how the two are related, let r be the distance from the common origin to the point P, and let α be the angle shown in Figure 12.5.2b. It follows that

$$x = r \cos(\theta + \alpha), \qquad y = r \sin(\theta + \alpha) \tag{3}$$

and

$$x' = r \cos \alpha, \qquad y' = r \sin \alpha \tag{4}$$

Using familiar trigonometric identities, the relationships in (3) can be written

$$x = r \cos \theta \cos \alpha - r \sin \theta \sin \alpha$$
$$y = r \sin \theta \cos \alpha + r \cos \theta \sin \alpha$$

and on substituting (4) in these equations we obtain the following relationships:

The Rotation Equations

$$x = x' \cos \theta - y' \sin \theta$$
$$y = x' \sin \theta + y' \cos \theta \tag{5}$$

▶ Example 1   Suppose the axes of an xy-coordinate system are rotated through an angle of $\theta = 45°$ to obtain an $x'y'$-coordinate system. Find the equation of the curve

$$x^2 - xy + y^2 - 6 = 0$$

in $x'y'$-coordinates.

Solution.   Substituting the values $\sin \theta = \sin 45° = 1/\sqrt{2}$ and $\cos \theta = \cos 45° = 1/\sqrt{2}$ in (5) yields the rotation equations

$$x = \frac{x'}{\sqrt{2}} - \frac{y'}{\sqrt{2}}$$

$$y = \frac{x'}{\sqrt{2}} + \frac{y'}{\sqrt{2}}$$

Substituting these into the given equation yields

$$\left(\frac{x'}{\sqrt{2}} - \frac{y'}{\sqrt{2}}\right)^2 - \left(\frac{x'}{\sqrt{2}} - \frac{y'}{\sqrt{2}}\right)\left(\frac{x'}{\sqrt{2}} + \frac{y'}{\sqrt{2}}\right) + \left(\frac{x'}{\sqrt{2}} + \frac{y'}{\sqrt{2}}\right)^2 - 6 = 0$$

or

$$\frac{x'^2 - 2x'y' + y'^2 - x'^2 + y'^2 + x'^2 + 2x'y' + y'^2}{2} = 6$$

or

$$\frac{x'^2}{12} + \frac{y'^2}{4} = 1$$

which is the equation of an ellipse (Figure 12.5.3).   ◀

If the rotation equations (5) are solved for x' and y' in terms of x and y, one obtains (Exercise 14):

$$x' = x \cos \theta + y \sin \theta$$

$$y' = -x \sin \theta + y \cos \theta \qquad\qquad (6)$$

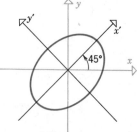

Figure 12.5.3

▶ Example 2   Find the new coordinates of the point $(2, 4)$ if the coordinate axes are rotated through an angle of $\theta = 30°$.

Solution.   From (6) with $x = 2$, $y = 4$, $\cos \theta = \cos 30° = \sqrt{3}/2$, and $\sin \theta = \sin 30° = 1/2$, we obtain

$$x' = 2(\sqrt{3}/2) + 4(1/2) = \sqrt{3} + 2$$
$$y' = -2(1/2) + 4(\sqrt{3}/2) = -1 + 2\sqrt{3}$$

Thus, the new coordinates are $(\sqrt{3} + 2, -1 + 2\sqrt{3})$. ◄

In Example 1, we were able to identify the curve $x^2 - xy + y^2 - 6 = 0$ as an ellipse because the rotation of axes eliminated the xy-term, thereby reducing the equation to a familiar form. This occurred because the new $x'y'$-axes were aligned with the axes of the ellipse. The following theorem tells how to determine an appropriate rotation of axes to eliminate the cross-product term of a second-degree equation in x and y.

12.5.1 THEOREM *If the equation*

$$Ax^2 + Bxy + Cy^2 + Dx + Ey + F = 0 \tag{7}$$

is such that $B \neq 0$, and if an $x'y'$-coordinate system is obtained by rotating the xy-axes through an angle θ satisfying

$$\cot 2\theta = \frac{A - C}{B} \tag{8}$$

then, in $x'y'$-coordinates, equation (7) will have the form

$$A'x'^2 + C'y'^2 + D'x' + E'y' + F' = 0$$

Proof. Substituting (5) into (7) and simplifying yields,

$$A'x'^2 + B'x'y' + C'y'^2 + D'x' + E'y' + F' = 0$$

where

$$
\begin{aligned}
A' &= A\cos^2\theta + B\cos\theta\sin\theta + C\sin^2\theta \\
B' &= B(\cos^2\theta - \sin^2\theta) + 2(C - A)\sin\theta\cos\theta \\
C' &= A\sin^2\theta - B\sin\theta\cos\theta + C\cos^2\theta \\
D' &= D\cos\theta + E\sin\theta \\
E' &= -D\sin\theta + E\cos\theta \\
F' &= F
\end{aligned}
\tag{9}
$$

(Verify.) To complete the proof we must show that $B' = 0$ if

$$\cot 2\theta = \frac{A - C}{B}$$

or equivalently

$$\frac{\cos 2\theta}{\sin 2\theta} = \frac{A - C}{B} \qquad (10)$$

However, by using the trigonometric double angle formulas, we can rewrite B' in the form

$$B' = B\cos 2\theta - (A - C)\sin 2\theta$$

Thus, $B' = 0$ if θ satisfies (10). ∎

REMARK. It is always possible to satisfy (8) with an angle θ in the range $0 < \theta < \pi/2$. We will always use such a value of θ.

▶ Example 3 Identify and sketch the curve $xy = 1$.

Solution. As a first step, we will rotate the coordinate axes to eliminate the cross-product term. Comparing the given equation to (7), we have

$$A = 0, \qquad B = 1, \qquad C = 0$$

Thus, the desired angle of rotation must satisfy

$$\cot 2\theta = \frac{A - C}{B} = \frac{0 - 0}{1} = 0$$

This condition can be met by taking $2\theta = \pi/2$ or $\theta = \pi/4 = 45°$. Substituting $\cos\theta = \cos 45° = 1/\sqrt{2}$ and $\sin\theta = \sin 45° = 1/\sqrt{2}$ in the rotation equations (5) yields

$$x = \frac{x'}{\sqrt{2}} - \frac{y'}{\sqrt{2}}$$
$$y = \frac{x'}{\sqrt{2}} + \frac{y'}{\sqrt{2}} \qquad (11)$$

and substituting these in the equation $xy = 1$ yields

$$\left(\frac{x'}{\sqrt{2}} - \frac{y'}{\sqrt{2}}\right)\left(\frac{x'}{\sqrt{2}} + \frac{y'}{\sqrt{2}}\right) = 1$$

or on simplifying

$$\frac{x'^2}{2} - \frac{y'^2}{2} = 1 \qquad (12)$$

This is an equation of the form

$$\frac{x'^2}{a^2} - \frac{y'^2}{b^2} = 1$$

with $a = b = \sqrt{2}$. With the exception that the variables are x' and y' rather than x and y, this is the familiar equation of a hyperbola (see Table 12.4.1). This hyperbola has its focal axis on the x'-axis, its conjugate axis on the y'-axis. Moreover, $c = \sqrt{a^2 + b^2} = 2$, so that the vertices, foci, and asymptotes are:

vertices: $\qquad (x', y') = (\pm a, 0) = (\pm \sqrt{2}, 0)$

foci: $\qquad (x', y') = (\pm c, 0) = (\pm 2, 0)$

asymptotes: $\quad y' = \pm \dfrac{b}{a} x' \qquad$ or $\qquad y' = \pm x'$

The vertices in xy-coordinates may be obtained by substituting $x' = \sqrt{2}$, $y' = 0$ and $x' = -\sqrt{2}$, $y' = 0$ in (11), and the foci may be obtained by substituting $x' = 2$, $y' = 0$ and $x' = -2$, $y' = 0$. The results are

vertices: $\quad (x, y) = (1, 1) \qquad$ and $\qquad (x, y) = (-1, -1)$

foci: $\qquad (x, y) = (\sqrt{2}, \sqrt{2}) \qquad$ and $\qquad (x, y) = (-\sqrt{2}, -\sqrt{2})$

To obtain the xy-equations of the asymptotes, it is desirable to use form (6) of the rotation equations. Substituting $\cos \theta = \cos 45° = 1/\sqrt{2}$ and $\sin \theta = \sin 45° = 1/\sqrt{2}$ in (6) yields

$$x' = \frac{1}{\sqrt{2}} x + \frac{1}{\sqrt{2}} y$$
$$y' = -\frac{1}{\sqrt{2}} x + \frac{1}{\sqrt{2}} y$$

(13)

Substituting (13) in the asymptote equations $y' = x'$ and $y' = -x'$ and simplifying shows that the asymptotes are

$$x = 0 \qquad \text{and} \qquad y = 0.$$

The curve $xy = 1$ is sketched in Figure 12.5.4. ◀

In problems where it is inconvenient to solve

$$\cot 2\theta = \frac{A - C}{B}$$

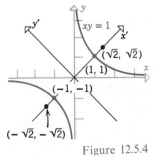

Figure 12.5.4

for θ, the values of $\sin\theta$ and $\cos\theta$ needed for the rotation equations may be obtained by first calculating $\cos 2\theta$ and then computing $\sin\theta$ and $\cos\theta$ from the identities

$$\sin\theta = \sqrt{\frac{1 - \cos 2\theta}{2}} \qquad \cos\theta = \sqrt{\frac{1 + \cos 2\theta}{2}}$$

▶ Example 4   Identify and sketch the curve

$$153x^2 - 192xy + 97y^2 - 30x - 40y - 200 = 0$$

Solution.   We have $A = 153$, $B = -192$, and $C = 97$ so that

$$\cot 2\theta = \frac{A - C}{B} = -\frac{56}{192} = -\frac{7}{24}$$

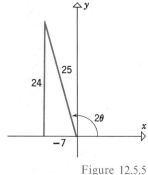

Figure 12.5.5

Since θ is to be chosen in the range $0 < \theta < \pi/2$, this relationship is represented by the triangle in Figure 12.5.5. From that triangle we obtain

$$\cos 2\theta = -\frac{7}{25}$$

which implies that

$$\cos\theta = \sqrt{\frac{1 + \cos 2\theta}{2}} = \sqrt{\frac{1 - 7/25}{2}} = \frac{3}{5}$$

$$\sin\theta = \sqrt{\frac{1 - \cos 2\theta}{2}} = \sqrt{\frac{1 + 7/25}{2}} = \frac{4}{5}$$

Substituting these values in (5) yields the rotation equations

$$x = \frac{3}{5}x' - \frac{4}{5}y'$$

$$y = \frac{4}{5}x' + \frac{3}{5}y'$$

Substituting these expressions in the given equation yields

$$\frac{153}{25}(3x' - 4y')^2 - \frac{192}{25}(3x' - 4y')(4x' + 3y') + \frac{97}{25}(4x' + 3y')^2$$

$$-\frac{30}{5}(3x' - 4y') - \frac{40}{5}(4x' + 3y') - 200 = 0$$

which simplifies to

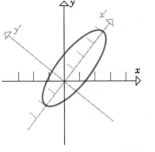

Figure 12.5.6

$$25x'^2 + 225y'^2 - 50x' - 200 = 0$$

or

$$x'^2 + 9y'^2 - 2x' - 8 = 0$$

Completing the square yields

$$\frac{(x' - 1)^2}{9} + y'^2 = 1$$

This is an ellipse whose center is $(1, 0)$ in $x'y'$-coordinates (Figure 12.5.6).

◀

OPTIONAL (THE DISCRIMINANT)

It is possible to describe the graph of a second-degree equation without rotating coordinate axes.

12.5.2 THEOREM   *Consider a second-degree equation*

$$Ax^2 + Bxy + Cy^2 + Dx + Ey + F = 0 \tag{14}$$

(a) *If $B^2 - 4AC < 0$, the equation represents an ellipse, a circle, a point, or else has no graph.*
(b) *If $B^2 - 4AC > 0$, the equation represents a hyperbola or a pair of intersecting lines.*
(c) *If $B^2 - 4AC = 0$, the equation represents a parabola, a line, a pair of parallel lines, or else has no graph.*

The quantity $B^2 - 4AC$ in this theorem is called the **discriminant** of the quadratic equation. To see why Theorem 12.5.2 is true, we need a fact about the discriminant. It can be shown (Exercise 17) that if the coordinate axes are rotated through any angle θ, and if

$$A'x'^2 + B'x'y' + C'y'^2 + D'x' + E'y' + F' = 0 \tag{15}$$

is the equation resulting from (14) after rotation, then

$$B^2 - 4AC = B'^2 - 4A'C' \tag{16}$$

In other words, the discriminant of a quadratic equation is not altered by rotating the coordinate axes; for this reason the discriminant is said to be **invariant** under a rotation of coordinate axes. In particular, if we choose the angle of rotation to eliminate the cross-product term, then (15) becomes

$$A'x'^2 + C'y'^2 + D'x' + E'y' + F' = 0 \tag{17}$$

and since $B' = 0$, (16) tells us that

$$B^2 - 4AC = -4A'C' \tag{18}$$

Part (a) of Theorem 12.5.2 can now be proved as follows. If $B^2 - 4AC < 0$, then from (18), $A'C' > 0$, so that (17) may be divided through by $A'C'$ and written in the form

$$\frac{1}{C'}\left(x'^2 + \frac{D'}{A'}x'\right) + \frac{1}{A'}\left(y'^2 + \frac{D'}{C'}y'\right) = -\frac{F'}{A'C'}$$

By completing the squares, this equation may be rewritten in the form

$$\frac{1}{C'}(x' - h)^2 + \frac{1}{A'}(y' - k)^2 = K \tag{19}$$

But A' and C' have the same sign (since $A'C' > 0$), so that by multiplying (19) through by -1, if necessary, we can assume A' and C' to be positive. Thus, (19) can be written

$$\frac{(x' - h)^2}{(\sqrt{C'})^2} + \frac{(y' - k)^2}{(\sqrt{A'})^2} = K \tag{20}$$

If $K < 0$, this equation has no graph (the left side is nonnegative for all x' and y'); if $K = 0$ the equation is satisfied only by $x' = h$ and $y' = k$, so the graph is the single point (h, k); finally, if $K > 0$, we can rewrite (20) in the form

$$\frac{(x' - h)^2}{(\sqrt{C'}\sqrt{K})^2} + \frac{(y' - k)^2}{(\sqrt{A'}\sqrt{K})^2} = 1$$

which is an ellipse or a circle. The proofs of parts (b) and (c) require a similar kind of analysis.

▶ Example 5   Use the discriminant to identify the graph of

$$8x^2 - 3xy + 5y^2 - 7x + 6 = 0$$

Solution.   We have

$$B^2 - 4AC = (-3)^2 - 4(8)(5) = -151$$

Since the discriminant is negative, the equation represents an ellipse, a point, or else has no graph. (Why can't the graph be a circle?)   ◀

In cases where a quadratic equation represents a point, a pair of parallel lines, a pair of intersecting lines or has no graph, we say that equation represents a **degenerate conic section.** Thus, if we allow for possible degeneracy, it follows from Theorem 12.5.2 that *every quadratic equation has a conic section as its graph.*

▶ Exercise Set 12.5

1. Let an $x'y'$-coordinate system be obtained by rotating an xy-coordinate system through an angle of $\theta = 60°$.
   (a) Find the $x'y'$-coordinates of the point whose xy-coordinates are $(-2, 6)$.
   (b) Find the equation of the curve $\sqrt{3}xy + y^2 = 6$ in $x'y'$-coordinates.
   (c) Sketch the curve in (b), showing both xy-axes and $x'y'$-axes.

2. Let an $x'y'$-coordinate system be obtained by rotating an xy-coordinate system through an angle of $\theta = 30°$.
   (a) Find the $x'y'$-coordinates of the point whose xy-coordinates are $(1, -\sqrt{3})$.
   (b) Find the equation of the curve $2x^2 + 2\sqrt{3}xy = 3$ in $x'y'$-coordinates.
   (c) Sketch the curve in (b), showing both xy-axes and $x'y'$-axes.

In Exercises 3–12, rotate the coordinate axes to remove the xy-term. Then name the conic and sketch its graph.

3. $xy = -9$.
4. $x^2 - xy + y^2 - 2 = 0$.
5. $x^2 + 4xy - 2y^2 - 6 = 0$.
6. $31x^2 + 10\sqrt{3}xy + 21y^2 - 144 = 0$.
7. $x^2 + 2\sqrt{3}xy + 3y^2 + 2\sqrt{3}x - 2y = 0$.
8. $34x^2 - 24xy + 41y^2 - 25 = 0$.
9. $9x^2 - 24xy + 16y^2 - 80x - 60y + 100 = 0$.
10. $5x^2 - 6xy + 5y^2 - 8\sqrt{2}x + 8\sqrt{2}y = 8$.
11. $52x^2 - 72xy + 73y^2 + 40x + 30y - 75 = 0$.
12. $6x^2 + 24xy - y^2 - 12x + 26y + 11 = 0$.

13. Let an $x'y'$-coordinate system be obtained by rotating an xy-coordinate system through an angle θ. Prove: For every choice of θ, the equation $x^2 + y^2 = r^2$ becomes $x'^2 + y'^2 = r^2$. Give a geometric explanation of this result.

14. Solve the rotation equations (5) for x' and y' in terms of x and y.

15. Let an $x'y'$-coordinate system be obtained by rotating an xy-coordinate system through an angle of $45°$. Use (6) to find the equation in xy-coordinates of the curve $3x'^2 + y'^2 = 6$.

16. Derive the expression for B' in (9).

17. Use (9) to prove that $B^2 - 4AC = B'^2 - 4A'C'$ for all values of θ.

18. Use (9) to prove that $A + C = A' + C'$ for all values of θ.

19. Prove: If $A = C$ in (7), then the cross-product term can be eliminated by rotating through $45°$.

20. Prove: If $B \neq 0$, then the graph of $x^2 + Bxy + F = 0$ is a hyperbola if $F \neq 0$ and two intersecting lines if $F = 0$.

In Exercises 21–25, use the discriminant to identify the graph of the given equation.

21. $x^2 - xy + y^2 - 2 = 0$.
22. $x^2 + 4xy - 2y^2 - 6 = 0$.
23. $x^2 + 2\sqrt{3}xy + 3y^2 + 2\sqrt{3}x - 2y = 0$.
24. $6x^2 + 24xy - y^2 - 12x + 26y + 11 = 0$.
25. $34x^2 - 24xy + 41y^2 - 25 = 0$.

26. Each of the following represents a degenerate conic section. Where possible, sketch the graph.
   (a) $x^2 - y^2 = 0$
   (b) $x^2 + 3y^2 + 7 = 0$
   (c) $8x^2 + 7y^2 = 0$
   (d) $x^2 - 2xy + y^2 = 0$
   (e) $9x^2 + 12xy + 4y^2 - 36 = 0$
   (f) $x^2 + y^2 - 2x - 4y = -5$.

27. Prove parts (b) and (c) of Theorem 12.5.2.

▶ SUPPLEMENTARY EXERCISES

In Exercises 1–8, identify the curve as a parabola, ellipse, or hyperbola, and give the following information:
Parabola: the coordinates of the vertex and focus, and the equation of the directrix.
Ellipse: the coordinates of the center and foci, and the lengths of the major and minor axes.
Hyperbola: the coordinates of the center, foci, and vertices, and equations of the asymptotes.

1. $y^2 + 12x - 6y + 33 = 0$.

2. $x^2 - 4y^2 = -1$.

3. $9x^2 + 4y^2 + 36x - 8y + 4 = 0$.

4. $6x + 8y - x^2 - 4y^2 = 12$.

5. $x^2 - 9y^2 - 4x + 18y - 14 = 0$.

6. $4y = x^2 + 2x - 7$.

7. $3x + 2y^2 - 4y - 7 = 0$.

8. $4x^2 = y^2 - 4y$.

In Exercises 9–16, find an equation for the curve described.

9. The parabola with vertex at $(1, 3)$ and directrix $x = -3$.

10. The ellipse with major axis of length 12 and foci at $(2, 7)$ and $(2, -1)$.

11. The hyperbola with foci $(0, \pm 5)$ and vertices 6 units apart.

12. The parabola with axis $y = 1$, vertex $(2, 1)$, and passing through $(3, -1)$.

13. The ellipse with foci $(\pm 3, 0)$ and such that the distances from the foci to $P(x, y)$ on the ellipse add up to 10 units.

14. The hyperbola with vertices $(0, \pm 2)$ and asymptotes $y = \pm 3x$.

15. The curve C with the property that the distance between the point $(3, 4)$ and any point $P(x, y)$ on C is equal to the distance between P and the line $y = 2$.

16. The hyperbola with vertices $(-3, 2)$ and $(1, 2)$ and perpendicular asymptotes.

In Exercises 17–20, sketch the curve whose equation is given in the stated exercise.

17. Exercise 1.   18. Exercise 2.

19. Exercise 3.   20. Exercise 6.

In Exercises 21–26, find the rotation angle θ needed to remove the xy-term; then name the conic and give its equation in $x'y'$-coordinates after the xy-term is removed.

21. $3x^2 - 2xy + 3y^2 = 4$.

22. $7x^2 - 8xy + y^2 = 9$.

23. $11x^2 + 10\sqrt{3}xy + y^2 = 4$.

24. $x^2 + 4xy + 4y^2 - 2\sqrt{5}x + \sqrt{5}y = 0$.

25. $16x^2 - 24xy + 9y^2 - 60x - 80y + 100 = 0$.

26. $73x^2 - 72xy + 52y^2 - 100 = 0$.

In Exercises 27–29, sketch the graph of the equation in the stated exercise, showing the xy- and $x'y'$-axes.

27. Exercise 21.   28. Exercise 23.

29. Exercise 24.

30. For the curve $xy + x - y = 1$,
    (a) find the rotation angle necessary to remove the xy-term;
    (b) find the equation in $x'y'$-coordinates obtained by rotating the xy-coordinate axes through an angle of $\theta = \tan^{-1}(3/4)$ radians and verify the identities $A' + C' = A + C$ and $B'^2 - 4A'C' = B^2 - 4AC$.

13 polar coordinates and parametric equations

13.1 POLAR COORDINATES

Up to now, we have specified the location of points in the plane by means of coordinates relative to a rectangular coordinate system consisting of two perpendicular coordinate axes. In this section, we introduce *polar coordinate systems.* This new kind of coordinate system will be easier to work with in many problems.

To form a polar coordinate system in a plane, we pick a fixed point O, called the *origin* or *pole;* and using the origin as an endpoint, we construct a ray, called the *polar axis.* After selecting a unit of measurement, we may associate with any point P in the plane a pair of *polar coordinates*

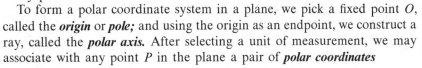

$$(r, \theta)$$

where r is the distance from P to the origin, and θ measures the angle from the polar axis to the line segment OP (Figure 13.1.1). The number r is called the *radial distance* of P and θ is called a *polar angle* of P. In Figure 13.1.2, the points $(6, 45°)$, $(3, 225°)$, $(5, 120°)$, and $(4, 330°)$ are plotted in polar coordinate systems.

Figure 13.1.1

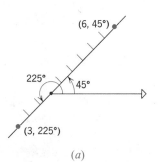

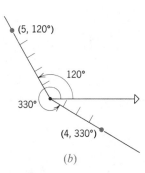

Figure 13.1.2   (*a*)   (*b*)

The polar coordinates of a point are not unique. For example, the polar coordinates

$$(1, 315°), \qquad (1, -45°), \qquad \text{and} \qquad (1, 675°)$$

all represent the same point (Figure 13.1.3). In general, if a point P has polar coordinates (r, θ), then for any integer $n = 0, 1, 2, \ldots$

$$(r, \theta + n \cdot 360°) \qquad \text{and} \qquad (r, \theta - n \cdot 360°)$$

are also polar coordinates of P.

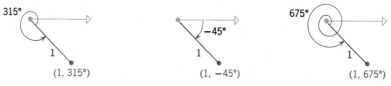

(1, 315°)   (1, −45°)   (1, 675°)

All three pairs of polar coordinates
represent the same point.

Figure 13.1.3

In the case where P is the origin, the line segment OP in Figure 13.1.1 reduces to a point, since $r = 0$. Because there is no clearly defined polar angle in this case, we will agree that an arbitrary polar angle θ may be used. Thus, for arbitrary θ, the point $(0, \theta)$ is the origin.

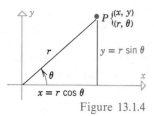

Figure 13.1.4

Frequently, it is helpful to use both polar and rectangular coordinates in the same problem. To do this, we let the positive x-axis of the rectangular coordinate system serve as the polar axis for the polar coordinate system. When this is done, each point P has both polar coordinates (r, θ) and rectangular coordinates (x, y). As suggested by Figure 13.1.4, these coordinates are related by the equations

$$\begin{aligned} x &= r \cos \theta \\ y &= r \sin \theta \end{aligned} \tag{1}$$

and

$$r^2 = x^2 + y^2 \tag{2a}$$

$$\tan \theta = \frac{y}{x} \tag{2b}$$

▶ **Example 1** Find the rectangular coordinates of the point P whose polar coordinates are $(6, 135°)$.

Solution. Substituting the polar coordinates $r = 6$ and $\theta = 135°$ in (1) yields

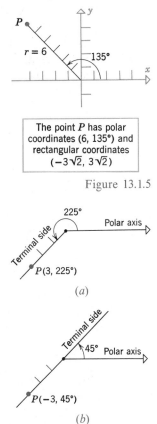

The point P has polar coordinates (6, 135°) and rectangular coordinates $(-3\sqrt{2}, 3\sqrt{2})$

Figure 13.1.5

225°

Terminal side

Polar axis

$P(3, 225°)$

(a)

Terminal side

45° Polar axis

$P(-3, 45°)$

(b)

Figure 13.1.6

$$x = 6 \cos 135° = 6\left(-\frac{\sqrt{2}}{2}\right) = -3\sqrt{2}$$

$$y = 6 \sin 135° = 6\left(\frac{\sqrt{2}}{2}\right) = 3\sqrt{2}$$

Thus, the rectangular coordinates of P are $(-3\sqrt{2}, 3\sqrt{2})$ (Figure 13.1.5). ◄

When we start graphing curves in polar coordinates, it will be desirable to allow negative values for r. This will require a special definition. For motivation, consider the point P with polar coordinates $(3, 225°)$. We can reach this point by rotating the polar axis 225° and then moving *forward* from the origin 3 units along the terminal side of the angle (Figure 13.1.6a). On the other hand, we can also reach the point P by rotating the polar axis 45° and then moving *backward* 3 units from the origin along the *extension* of the terminal side of the angle (Figure 13.1.6b). This suggests that the point $(3, 225°)$ might also be denoted by $(-3, 45°)$, with the minus sign serving to indicate that the point is on the extension of the angle's terminal side rather than on the terminal side itself.

Since the terminal side of the angle $\theta + 180°$ is the extension of the terminal side of the angle θ, we will define

$$(-r, \theta) \quad \text{and} \quad (r, \theta + 180°) \tag{3}$$

to be polar coordinates for the same point. With $r = 3$ and $\theta = 45°$ in (3) it follows that $(-3, 45°)$ and $(3, 225°)$ represent the same point.

In the exercises we ask the reader to show that relationships (1) and (2) remain valid when r is negative.

REMARK. For many purposes it does not matter whether polar angles are represented in degrees or radians. However, later on, when we are concerned with differentiation problems, it will be essential to measure polar angles in radians, since the derivative formulas for the trigonometric functions were derived under this assumption. From here on, we will use radian measure.

► Example 2 Find polar coordinates of the point P whose rectangular coordinates are $(-2, 2\sqrt{3})$.

Solution. We will find polar coordinates (r, θ) of P such that $r > 0$ and $0 \leq \theta < 2\pi$. From (2a),

$$r^2 = x^2 + y^2 = (-2)^2 + (2\sqrt{3})^2 = 4 + 12 = 16$$

so $r = 4$. From (2b),

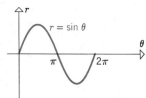

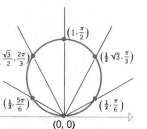

Figure 13.1.7

Figure 13.1.8

$$\tan \theta = \frac{y}{x} = \frac{2\sqrt{3}}{-2} = -\sqrt{3}$$

From this and the fact that $(-2, 2\sqrt{3})$ lies in the second quadrant, it follows that $\theta = 2\pi/3$. Thus, $(4, 2\pi/3)$ are polar coordinates of P. All other polar coordinates of P have the form

$$\left(4, \frac{2\pi}{3} + 2n\pi\right) \quad \text{or} \quad \left(-4, \frac{5\pi}{3} + 2n\pi\right) \text{ where } n \text{ is an integer.} \quad \blacktriangleleft$$

Polar coordinates provide a new way of graphing certain equations. As an example, consider the equation

$$r = \sin \theta \tag{4}$$

By substituting values for θ at increments of $\pi/6$ ($=30°$), and calculating r we can construct the following table:

Table 13.1.1

θ (radians)	0	$\dfrac{\pi}{6}$	$\dfrac{\pi}{3}$	$\dfrac{\pi}{2}$	$\dfrac{2\pi}{3}$	$\dfrac{5\pi}{6}$	π	$\dfrac{7\pi}{6}$	$\dfrac{4\pi}{3}$	$\dfrac{3\pi}{2}$	$\dfrac{5\pi}{3}$	$\dfrac{11\pi}{6}$	2π
$r = \sin \theta$	0	$\dfrac{1}{2}$	$\dfrac{\sqrt{3}}{2}$	1	$\dfrac{\sqrt{3}}{2}$	$\dfrac{1}{2}$	0	$-\dfrac{1}{2}$	$-\dfrac{\sqrt{3}}{2}$	-1	$-\dfrac{\sqrt{3}}{2}$	$-\dfrac{1}{2}$	0

The pairs of values listed in this table may be plotted in one of two ways. One possibility is to view each pair (r, θ) as *rectangular* coordinates of a point and plot each point in the $r\theta$-plane. This yields points on the familiar sine curve (Figure 13.1.7). Alternatively, we can view each pair (r, θ) as *polar* coordinates of a point and plot each point in a polar coordinate system (Figure 13.1.8). Note that there are 13 pairs listed in Table 13.1.1, but only 6 points plotted in Figure 13.1.8. This is because the pairs from $\theta = \pi$ on yield duplicates of the preceding points. For example, $(-1/2, 7\pi/6)$ and $(1/2, \pi/6)$ represent the same point.

The points in Figure 13.1.8 appear to lie on a circle. That this is indeed the case may be seen by expressing (4) in terms of x and y. To do this we first multiply (4) through by r to obtain

$$r^2 = r \sin \theta$$

which may be rewritten using (1) and (2) as

$$x^2 + y^2 = y \quad \text{or} \quad x^2 + y^2 - y = 0$$

or on completing the square

$$x^2 + \left(y - \frac{1}{2}\right)^2 = \frac{1}{4}$$

This is a circle of radius $\frac{1}{2}$ centered at the point $(0, \frac{1}{2})$ in the xy-plane.

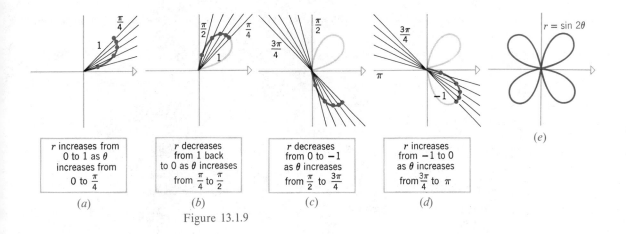

r increases from
0 to 1 as θ
increases from
0 to $\frac{\pi}{4}$

(a)

r decreases
from 1 back
to 0 as θ increases
from $\frac{\pi}{4}$ to $\frac{\pi}{2}$

(b)

r decreases
from 0 to -1
as θ increases
from $\frac{\pi}{2}$ to $\frac{3\pi}{4}$

(c)

r increases
from -1 to 0
as θ increases
from $\frac{3\pi}{4}$ to π

(d)

Figure 13.1.9

▶ **Example 3** Sketch the curve $r = \sin 2\theta$ in polar coordinates.

Solution. A rough sketch of the graph may be obtained without plotting points by observing how the value of r changes with θ.

If θ increases from 0 to $\pi/4$, then 2θ increases from 0 to $\pi/2$ and $r = \sin 2\theta$ increases from 0 to 1 (Figure 13.1.9a).

If θ increases from $\pi/4$ to $\pi/2$, then 2θ increases from $\pi/2$ to π and $r = \sin 2\theta$ decreases from 1 back to 0. Thus, the curve completes a loop (Figure 13.1.9b).

If θ increases from $\pi/2$ to $3\pi/4$, then 2θ increases from π to $3\pi/2$ and r decreases from 0 to -1. Because r is negative, the resulting portion of the graph is in the fourth quadrant although the values of θ are in the second quadrant (Figure 13.1.9c).

If θ increases from $3\pi/4$ to π, then 2θ increases from $3\pi/2$ to 2π and $r = \sin 2\theta$ increases from -1 to 0, and the curve completes a loop in the fourth quadrant (Figure 13.1.9d).

We leave it for the reader to continue the analysis for θ in the range π to 2π and show that the remaining portion of the curve consists of two loops in the second and third quadrants (Figure 13.1.9e). No new points are generated by allowing θ to vary past 2π or through negative values (verify) so that Figure 13.1.9e is the complete graph. ◀

▶ **Exercise Set 13.1**

1. Plot the following points in polar coordinates.
   (a) $(3, \pi/4)$ (b) $(5, 2\pi/3)$
   (c) $(1, \pi/2)$ (d) $(4, 7\pi/6)$
   (e) $(2, 4\pi/3)$ (f) $(0, \pi)$.

2. Plot the following points in polar coordinates.
   (a) $(2, -\pi/3)$ (b) $(3/2, -7\pi/4)$
   (c) $(-3, 3\pi/2)$ (d) $(-5, -\pi/6)$
   (e) $(-6, -\pi)$ (f) $(-1, 9\pi/4)$.

3. Find rectangular coordinates of the points whose polar coordinates are given.

(a) $(6, \pi/6)$ (b) $(7, 2\pi/3)$
(c) $(8, 9\pi/4)$ (d) $(5, 0)$
(e) $(7, 17\pi/6)$ (f) $(0, \pi)$.

4. Find rectangular coordinates of the points whose polar coordinates are given.

(a) $(-8, \pi/4)$ (b) $(7, -\pi/4)$
(c) $(-6, -5\pi/6)$ (d) $(0, -\pi)$
(e) $(-2, -3\pi/2)$ (f) $(-5, 0)$.

5. The following points are given in rectangular coordinates. Express the points in polar coordinates with $r \geq 0$ and $0 \leq \theta < 2\pi$.

(a) $(-5, 0)$ (b) $(2\sqrt{3}, -2)$
(c) $(0, -2)$ (d) $(-8, -8)$
(e) $(-3, 3\sqrt{3})$ (f) $(1, 1)$.

6. Express the points in Exercise 5 in polar coordinates with $r \geq 0$ and $-\pi < \theta \leq \pi$.

7. Express the points in Exercise 5 in polar coordinates with $r \leq 0$ and $0 \leq \theta < 2\pi$.

In Exercises 8–13, identify the curve by transforming to rectangular coordinates.

8. $r = 3$.   **9.** $r \sin \theta = 4$.

10. $r = 2 \sin \theta$.   **11.** $r = \dfrac{6}{2 - \cos \theta}$.

12. $r = 5 \sec \theta$.   **13.** $r + 4 \cos \theta = 0$.

In Exercises 14–19, express the equations in polar coordinates.

14. $x^2 + y^2 = 9$.

15. $x = 7$.

16. $4xy = 9$.

17. $(x^2 + y^2)^2 = 16(x^2 - y^2)$.

18. $x^2 - y^2 = 4$.

19. $x^2 = 9y$.

In Exercises 20–23, use the method of Example 3 to sketch the curve in polar coordinates.

20. $r = \cos 2\theta$.   **21.** $r = 2(1 + \sin \theta)$.

22. $r = 4 \cos 3\theta$.   **23.** $r = 1 - \cos \theta$.

24. Prove that the distance between the points with polar coordinates (r_1, θ_1) and (r_2, θ_2) is

$$d = \sqrt{r_1^2 + r_2^2 - 2r_1 r_2 \cos (\theta_1 - \theta_2)}$$

25. Prove: In polar coordinates, the equation $r = a \sin \theta + b \cos \theta$ represents a circle.

26. Prove: The area of the triangle whose vertices have polar coordinates $(0, 0)$, (r_1, θ_1), and (r_2, θ_2) is $A = \frac{1}{2} r_1 r_2 \sin (\theta_2 - \theta_1)$. [Assume $0 \leq \theta_1 < \theta_2 \leq \pi$ and r_1 and r_2 are positive.]

27. Prove: Equations (1) and (2) hold if r is negative.

28. Prove that $r r_1 \sin (\theta - \theta_1) + r r_2 \sin (\theta_2 - \theta) + r_1 r_2 \sin (\theta_1 - \theta_2) = 0$ is the polar equation of the line through the points (r_1, θ_1) and (r_2, θ_2).

13.2 GRAPHS IN POLAR COORDINATES

In this section we will graph some curves defined in terms of polar coordinates. Since we will be working with both rectangular xy-coordinates and polar coordinates, we assume throughout that the positive x-axis coincides with the polar axis.

LINES IN POLAR
COORDINATES

A line perpendicular to the x-axis and passing through the point with xy-coordinates $(a, 0)$ has the equation

$$x = a$$

To express this equation in polar coordinates we substitute $x = r \cos \theta$ [see (1) in the previous section]; this yields

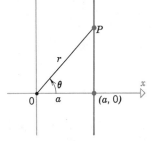

Figure 13.2.1

$$r \cos \theta = a \tag{1}$$

This result makes sense geometrically since each point $P(r, \theta)$ on this line will yield the value a for $r \cos \theta$ (Figure 13.2.1).

A line parallel to the x-axis that meets the y-axis at the point with xy-coordinates $(0, b)$ has the equation

$$y = b$$

Substituting $y = r \sin \theta$ yields

$$r \sin \theta = b \tag{2}$$

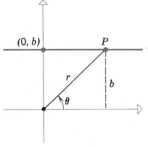

Figure 13.2.2

as the polar equation of this line. This makes sense geometrically since each point $P(r, \theta)$ on this line will yield the value b for $r \sin \theta$ (Figure 13.2.2).

For any constant θ_0, the equation

$$\theta = \theta_0 \tag{3}$$

is satisfied by the coordinates of all points of the form $P(r, \theta_0)$, regardless of the value of r. Thus, the equation represents the line through the origin making an angle of θ_0 (radians) with the polar axis (Figure 13.2.3).

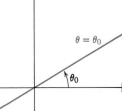

Figure 13.2.3

▶ Example 1 Sketch the graphs of the following equations in polar coordinates.

(a) $r \cos \theta = 3$ (b) $r \sin \theta = -2$ (c) $\theta = \dfrac{3\pi}{4}$

Solution. From (1), (2), and (3), each graph is a line; they are shown in Figure 13.2.4. ◀

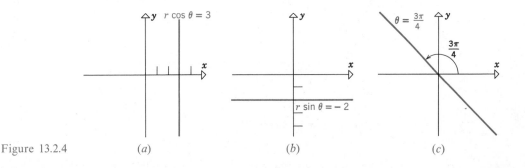

Figure 13.2.4   (a)   (b)   (c)

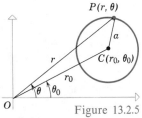

Figure 13.2.5

CIRCLES IN POLAR COORDINATES

By substituting $x = r \cos \theta$ and $y = r \sin \theta$ in the equation $Ax + By + C = 0$, we obtain the **general polar form of a line,**

$$r(A \cos \theta + B \sin \theta) + C = 0$$

Let us try to find the polar equation of a circle whose radius is a and whose center has polar coordinates (r_0, θ_0) (Figure 13.2.5). If we let $P(r, \theta)$ be an arbitrary point on the circle, and if we apply the law of cosines to the triangle OCP in Figure 13.2.5 we obtain

$$r^2 - 2rr_0 \cos (\theta - \theta_0) + r_0{}^2 = a^2 \tag{4}$$

This general equation is rarely used since the corresponding equation in rectangular coordinates is easier to work with. However, we will now consider some special cases of (4) that arise frequently.

A circle of radius a, centered at the origin, has an especially simple polar equation. If we let $r_0 = 0$ in (4), we obtain

$$r^2 = a^2$$

or since $a \geq 0$

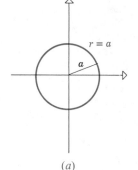

(a)

$$r = a \tag{5}$$

This equation makes sense geometrically since the circle of radius a, centered at the origin, consists of all points $P(r, \theta)$ for which $r = a$, regardless of the value of θ (Figure 13.2.6a).

If a circle of radius a has its center on the x-axis and passes through the origin, then the polar coordinates of the center are either

$$(a, 0) \quad \text{or} \quad (a, \pi)$$

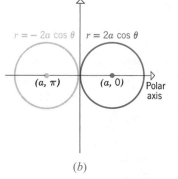

(b)

depending on whether the center is to the right or left of the origin (Figure 13.2.6b). For the circle to the right of the origin we have $r_0 = a$ and $\theta_0 = 0$, so (4) reduces to

$$r^2 - 2ra \cos \theta = 0$$

or

$$r = 2a \cos \theta \tag{6a}$$

For the circle to the left of the origin, $r_0 = a$ and $\theta_0 = \pi$, so (4) reduces to

Figure 13.2.6

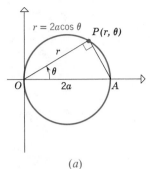

(a)

(b)

Figure 13.2.7

$$r = -2a \cos \theta \qquad (6b)$$

because $\cos (\theta - \pi) = -\cos \theta$.

If a circle of radius a passes through the origin and has its center on the y-axis, then the polar coordinates of the center are $(a, \pi/2)$ or $(a, -\pi/2)$, and (4) reduces to

$$r = 2a \sin \theta \qquad (7a)$$

or

$$r = -2a \sin \theta \qquad (7b)$$

depending on whether the center is above or below the origin (Figure 13.2.6c).

Equations (6a) and (7a) are interpreted geometrically in Figure 13.2.7a and 13.2.7b. In each figure part, OPA is a right triangle (why?) so that for each point $P(r, \theta)$ on the circle, we have $r = 2a \cos \theta$ or $r = 2a \sin \theta$, depending on the orientation of the circle.

▶ **Example 2** Sketch the graphs of the following equations in polar coordinates

(a) $r = 4 \cos \theta$ (b) $r = -5 \sin \theta$ (c) $r = 3$

Solution. These equations are of forms (6a), (7b), and (5), respectively, and therefore represent circles. The graphs are shown in Figure 13.2.8.   ◀

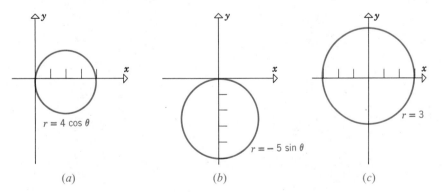

Figure 13.2.8   (a)   (b)   (c)

The points with polar coordinates (r, θ) and $(r, -\theta)$ are symmetric about the x-axis (Figure 13.2.9a). Thus, if a polar equation is unchanged on replacing θ by $-\theta$, then its graph is symmetric about the x-axis. The following tests for symmetry will be useful.

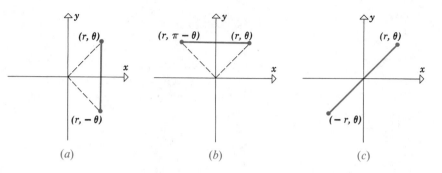

Figure 13.2.9   (a)   (b)   (c)

13.2.1   THEOREM   (a)   *A curve in polar coordinates is symmetric about the x-axis if replacing θ*
Symmetry Tests   *by −θ in its equation produces an equivalent equation (Figure 13.2.9a).*
                      (b)   *A curve in polar coordinates is symmetric about the y-axis if replacing θ*
                            by π − θ in its equation produces an equivalent equation (Figure 13.2.9b).
                      (c)   *A curve in polar coordinates is symmetric about the origin if replacing r*
                            by −r in its equation produces an equivalent equation (Figure 13.2.9c).

CARDIOIDS AND   Equations of the form
LIMAÇONS

$$r = a + b \sin \theta \qquad r = a - b \sin \theta \qquad\qquad (8a\text{–}8b)$$
$$r = a + b \cos \theta \qquad r = a - b \cos \theta \qquad\qquad (8c\text{–}8d)$$

produce polar curves called **limaçons** (from the Latin word "limax," for a slug, a snail-like creature.) There are four possible shapes for a limaçon* that can be determined from the ratio a/b when $a > 0$ and $b > 0$ (Figure 13.2.10).

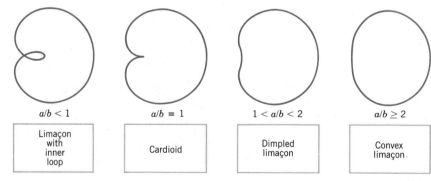

$a/b < 1$	$a/b = 1$	$1 < a/b < 2$	$a/b \geq 2$
Limaçon with inner loop	Cardioid	Dimpled limaçon	Convex limaçon.

Figure 13.2.10

*For a detailed discussion of limaçons and their shapes, the reader is referred to the article by Jane T. Grossman and Michael P. Grossman, "Dimple or No Dimple," *Two Year College Mathematics Journal*, Vol. 13, No. 1 (1982), pp. 52–55.

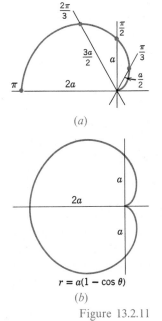

(a)

$r = a(1 - \cos \theta)$

(b)

Figure 13.2.11

The orientation of the limaçon relative to the polar axis depends on whether $\sin \theta$ or $\cos \theta$ appears in the equation and whether the $+$ or $-$ occurs. Because of the heart-shaped appearance of the curve in the case $a = b$, limaçons of this type are called **cardioids**.

▶ **Example 3** Assuming a to be a positive constant, sketch the curve $r = a(1 - \cos \theta)$ in polar coordinates.

Solution. This is an equation of form (8d) with $a = b$ and, therefore, represents a cardioid. Since $\cos(-\theta) = \cos \theta$ the equation is unaltered when θ is replaced by $-\theta$; thus, the cardioid is symmetric about the x-axis. This being the case, we can obtain the entire curve by first sketching the portion of the cardioid above the x-axis and then reflecting this portion about the x-axis.

As θ varies from 0 to π, $\cos \theta$ decreases steadily from 1 to -1, and $1 - \cos \theta$ increases steadily from 0 to 2. Thus, as θ varies from 0 to π, the value of $r = a(1 - \cos \theta)$ will increase steadily from an initial value of $r = 0$ to a final value of $r = 2a$. Using this information, and by plotting the points in the following table, we obtain the graph in Figure 13.2.11a.

Table 13.2.1

θ	0	$\dfrac{\pi}{3}$	$\dfrac{\pi}{2}$	$\dfrac{2\pi}{3}$	π
$r = a(1 - \cos \theta)$	0	$\dfrac{a}{2}$	a	$\dfrac{3a}{2}$	$2a$

On reflecting the curve in Figure 13.2.11a about the x-axis, we obtain the entire cardioid (Figure 13.2.11b). ◀

REMARK. We have sketched the cardioid in Figure 13.2.11 with the line $\theta = 0$ tangent to the curve at the origin. While it is not evident from our discussion that this is the case, we will prove in Section 13.5 that if a polar curve passes through the origin when $\theta = \theta_0$, then the line $\theta = \theta_0$ is tangent to the curve at the origin.

LEMNISCATES Equations of the form

$$r^2 = a^2 \cos 2\theta \qquad r^2 = -a^2 \cos 2\theta \qquad \text{(9a–9b)}$$
$$r^2 = a^2 \sin 2\theta \qquad r^2 = -a^2 \sin 2\theta \qquad \text{(9c–9d)}$$

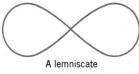

A lemniscate

Figure 13.2.12

represent propeller-shaped curves, called **lemniscates** (Figure 13.2.12). The lemniscates are centered at the origin, but the orientation relative to the polar axis depends on the sign preceding the a^2 and whether $\sin 2\theta$ or $\cos 2\theta$ appears in the equation.

▶ Example 4   Sketch the curve

$$r^2 = 4 \cos 2\theta$$

in polar coordinates.

Solution.   The equation is of type (9a) with $a = 2$, and thus represents a lemniscate. We leave it for the reader to apply the symmetry tests in Theorem 13.2.1 and show that the curve is symmetric about the x-axis and the y-axis. Therefore, we can obtain the entire curve by first sketching the portion of the lemniscate in the range $0 \leq \theta \leq \pi/2$ and then reflecting this portion about the x- and y-axes.

For θ in the range $0 \leq \theta \leq \pi/4$, the quantity $\cos 2\theta$ is nonnegative, so that for each such θ, there are two values of r satisfying $r^2 = 4 \cos 2\theta$, namely $r = 2\sqrt{\cos 2\theta}$ and $r = -2\sqrt{\cos 2\theta}$. As θ varies from 0 to $\pi/4$, the value of $\cos 2\theta$ decreases steadily from 1 to 0, so that $r = 2\sqrt{\cos 2\theta}$ decreases steadily from 2 to 0 and $r = -2\sqrt{\cos 2\theta}$ increases steadily from -2 to 0. With this information and the points plotted from the following table, we obtain the sketch in Figure 13.2.13*a*.

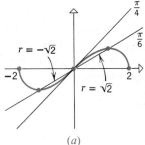

(a)

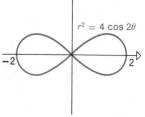

(b)

Figure 13.2.13

Table 13.2.2

θ	0	$\dfrac{\pi}{6}$	$\dfrac{\pi}{4}$
$r = \pm 2\sqrt{\cos 2\theta}$	± 2	$\pm \sqrt{2}$	0

(Note that the curve passes through the origin when $\theta = \pi/4$, so that the line $\theta = \pi/4$ is tangent to the curve at the origin by the remark following Example 3.)

For θ in the range $\pi/4 < \theta < \pi/2$, the quantity $4 \cos 2\theta$ is negative, so there are no real values of r satisfying $r^2 = 4 \cos 2\theta$. Thus, there are no points on the graph for such θ. The entire lemniscate is obtained by reflecting the curve in Figure 13.2.13*a* about the x-axis and then about the y-axis (Figure 13.2.13*b*). (The reflection about the y-axis adds no new points, however.)   ◀

SPIRALS   A curve that "winds around the origin" infinitely many times in such a way that r increases (or decreases) steadily as θ increases is called a *spiral.* The most common example is the *spiral of Archimedes,* which has an equation of the form

$$r = a\theta \qquad (\theta \geq 0) \tag{10a}$$

or

$$r = a\theta \qquad (\theta \leq 0) \tag{10b}$$

In these equations, θ is in radians and a is positive.

▶ Example 5   Sketch the curve

$$r = \theta \qquad (\theta \geq 0)$$

in polar coordinates.

Solution. This is an equation of form (10a) with $a = 1$ and, thus, represents an Archimedean spiral. Since $r = 0$ when $\theta = 0$, the origin is on the curve and the polar axis is tangent to the spiral.

A reasonably accurate sketch may be obtained by plotting the intersections of the spiral with the x- and y-axes and noting that r increases steadily as θ increases. The intersections with the x-axis occur when

$$\theta = 0, \pi, 2\pi, 3\pi, \ldots$$

at which points r has the values

$$r = 0, \pi, 2\pi, 3\pi, \ldots$$

and the intersections with the y-axis occur when

$$\theta = \frac{\pi}{2}, \frac{3\pi}{2}, \frac{5\pi}{2}, \frac{7\pi}{2}, \ldots$$

at which points, r has the values

$$r = \frac{\pi}{2}, \frac{3\pi}{2}, \frac{5\pi}{2}, \frac{7\pi}{2}, \ldots$$

The graph is shown in Figure 13.2.14.   ◀

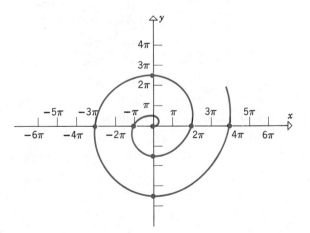

Figure 13.2.14

Starting from the origin, the spiral in Figure 13.2.14 loops counterclockwise around the origin. This is true of all Archimedean spirals of form (10a); Archimedean spirals of form (10b) loop clockwise around the origin.

ROSE CURVES   Equations of the form

$$r = a \sin n\theta \qquad \text{(11a)}$$
$$r = a \cos n\theta \qquad \text{(11b)}$$

represent flower-shaped curves called **roses.** The rose has n equally spaced petals or loops if n is odd and $2n$ equally spaced petals if n is even (Figure 13.2.15). The orientation of the rose relative to the polar axis depends on the sign of the constant a and whether $\sin \theta$ or $\cos \theta$ appears in the equation. (A rose curve occurred in Example 3 of the previous section.)

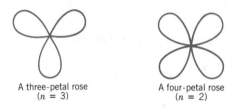

Figure 13.2.15   A three-petal rose   A four-petal rose
                   $(n = 3)$ $(n = 2)$

If $n = 1$ in (11a) or (11b) then we obtain the equation of a circle, which can be regarded as a one-petal rose.

▶ Exercise Set 13.2

In Exercises 1–36, sketch the curve in polar coordinates and give its name.

1. $\theta = \pi/6$.
2. $\theta = -3\pi/4$.
3. $r = 5$.
4. $r = 4 \sin \theta$.
5. $r = -6 \cos \theta$.
6. $r = -3 \sin \theta$.
7. $2r = \cos \theta$.
8. $r = 1 + \sin \theta$.
9. $r = 3(1 - \sin \theta)$.
10. $r - 2 = 2 \cos \theta$.
11. $r = 4 - 4 \cos \theta$.
12. $r = -5 + 5 \sin \theta$.
13. $r = -1 - \cos \theta$.
14. $r = 1 + 2 \sin \theta$.
15. $r = 1 - 2 \cos \theta$.
16. $r = 4 + 3 \cos \theta$.
17. $r = 3 + 2 \sin \theta$.
18. $r = 3 - \cos \theta$.
19. $r = 2 + \sin \theta$.
20. $r - 5 = 3 \sin \theta$.
21. $r = 3 + 4 \cos \theta$.
22. $r = -3 - 4 \sin \theta$.
23. $r = 5 - 2 \cos \theta$.
24. $r^2 = 9 \cos 2\theta$.
25. $r^2 = -9 \cos 2\theta$.
26. $r^2 = \sin 2\theta$.
27. $r^2 = -16 \sin 2\theta$.
28. $r = 4\theta \; (\theta \geq 0)$.
29. $r = 4\theta \; (\theta \leq 0)$.
30. $r = 4\theta$.
31. $r = \cos 2\theta$.
32. $r = 3 \sin 2\theta$.
33. $r = \sin 3\theta$.
34. $r = 2 \cos 3\theta$.
35. $r = 9 \sin 4\theta$.
36. $r = \cos 5\theta$.

In Exercises 37–44, sketch the curve in polar coordinates.

37. $r = 4 \cos \theta + 4 \sin \theta$.
38. $r = e^{\theta}$ (logarithmic spiral).
39. $r = 4 \tan \theta$ (kappa curve).
40. $r = 2 + 2 \sec \theta$ (conchoid of Nichomedes).
41. $r = \sin (\theta/2)$.
42. $r^2 = \theta$ (parabolic spiral).
43. $r = 2 \sin \theta \tan \theta$ (cissoid).
44. $r\theta = 1 \; (\theta > 0)$ (hyperbolic spiral).

13.3   AREA IN POLAR COORDINATES

In this section we will be concerned with the following version of the area problem.

13.3.1   PROBLEM
The Area Problem in
Polar Coordinates

Find the area of the region R between a polar curve $r = f(\theta)$ and two lines, $\theta = \alpha$ and $\theta = \beta$ (Figure 13.3.1).

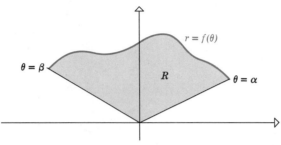

Figure 13.3.1

We will assume that $\alpha < \beta$ and $f(\theta)$ is continuous and nonnegative for $\alpha \le \theta \le \beta$. As a first step toward finding the area of the region R, let $\theta_1, \theta_2, \ldots, \theta_{n-1}$ be any numbers such that

$$\alpha < \theta_1 < \theta_2 < \cdots < \theta_{n-1} < \beta$$

and construct the lines

$$\theta = \theta_1, \ \theta = \theta_2, \ldots, \theta = \theta_{n-1}$$

These lines divide the angle from α to β into n subangles, and they divide the region R into n subregions (Figure 13.3.2). As indicated in this figure, we will denote the subangles by

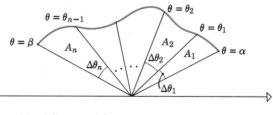

Figure 13.3.2

$$\Delta\theta_1, \Delta\theta_2, \ldots, \Delta\theta_n$$

and the areas of the subregions by

$$A_1, A_2, \ldots, A_n$$

Then the area A of the entire region may be written

$$A = A_1 + A_2 + \cdots + A_n = \sum_{k=1}^{n} A_k$$

Let us now concentrate on the areas $A_1, A_2, \ldots, A_n$ of the subregions. If $\Delta\theta_k$ is not too large, we can approximate the area A_k by the area of a *sector* having central angle $\Delta\theta_k$ and radius $f(\theta_k^*)$, where θ_k^* is an arbitrary angle terminating in the kth subregion (Figure 13.3.3).

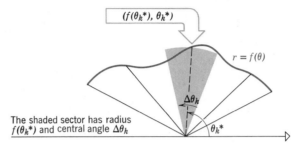

The shaded sector has radius $f(\theta_k^*)$ and central angle $\Delta\theta_k$

Figure 13.3.3

Thus, from the area formula for a sector [Formula (13), Section 3.3],

$$A_k \approx \frac{1}{2}[f(\theta_k^*)]^2 \, \Delta\theta_k$$

and

$$A = \sum_{k=1}^{n} A_k \approx \sum_{k=1}^{n} \frac{1}{2}[f(\theta_k^*)]^2 \, \Delta\theta_k \tag{1}$$

If we now increase n in such a way that $\max \Delta\theta_k \to 0$, then the sectors will become better and better approximations to the subregions and (1) will approach the exact value of the area A; that is,

$$A = \lim_{\max \Delta\theta_k \to 0} \sum_{k=1}^{n} \frac{1}{2}[f(\theta_k^*)]^2 \, \Delta\theta_k \tag{2}$$

Since the limit on the right side of (2) is the definite integral

$$\int_\alpha^\beta \frac{1}{2}[f(\theta)]^2 \, d\theta$$

we are led to the following definition:

13.3.2 DEFINITION If $f(\theta)$ is continuous and nonnegative for $\alpha \leq \theta \leq \beta$, then the area A enclosed by the polar curve $r = f(\theta)$ and the lines $\theta = \alpha$ and $\theta = \beta$ is

$$A = \int_\alpha^\beta \frac{1}{2}[f(\theta)]^2 \, d\theta$$

or equivalently

$$A = \int_{\alpha}^{\beta} \frac{1}{2} r^2 \, d\theta \qquad (3)$$

The hardest part of applying (3) is determining the limits of integration. This can be done as follows:

Step 1. Sketch the region R whose area is to be determined.
Step 2. Draw an arbitrary "radial line" from the origin to the boundary curve $r = f(\theta)$.
Step 3. Ask, "Over what interval of values must θ vary in order for the radial line to sweep out the region R?"
Step 4. Your answer in Step 3 will determine the lower and upper limits of integration.

▶ **Example 1** Find the area of the region in the first quadrant within the cardioid $r = 1 - \cos \theta$.

Solution. The region and a typical radial line are shown in Figure 13.3.4. For the radial line to sweep out the region, θ must vary from 0 to $\pi/2$. Thus, from (3) with $\alpha = 0$ and $\beta = \pi/2$,

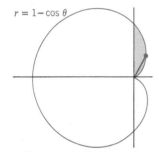

$r = 1 - \cos \theta$

$$A = \int_0^{\pi/2} \frac{1}{2} r^2 \, d\theta = \frac{1}{2} \int_0^{\pi/2} (1 - \cos \theta)^2 \, d\theta$$

$$= \frac{1}{2} \int_0^{\pi/2} (1 - 2 \cos \theta + \cos^2 \theta) \, d\theta$$

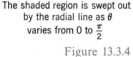

The shaded region is swept out by the radial line as θ varies from 0 to $\frac{\pi}{2}$

Figure 13.3.4

With the help of the identity $\cos^2 \theta = \frac{1}{2}(1 + \cos 2\theta)$, this may be rewritten

$$A = \frac{1}{2} \int_0^{\pi/2} \left(\frac{3}{2} - 2 \cos \theta + \frac{1}{2} \cos 2\theta \right) d\theta$$

$$= \frac{1}{2} \left[\frac{3}{2}\theta - 2 \sin \theta + \frac{1}{4} \sin 2\theta \right]_0^{\pi/2}$$

$$= \frac{3}{8}\pi - 1 \qquad \blacktriangleleft$$

▶ **Example 2** Find the entire area within the cardioid of Example 1.

Solution. For the radial line to sweep out the entire cardioid, θ must vary from 0 to 2π. Thus, from (3) with $\alpha = 0$ and $\beta = 2\pi$

$$A = \int_0^{2\pi} \frac{1}{2} r^2 \, d\theta = \frac{1}{2} \int_0^{2\pi} (1 - \cos \theta)^2 \, d\theta$$

If we proceed as in Example 1, this reduces to

$$A = \frac{1}{2} \int_0^{2\pi} \left(\frac{3}{2} - 2 \cos \theta + \frac{1}{2} \cos 2\theta \right) d\theta = \frac{3\pi}{2}$$

Alternate Solution. Since the cardioid is symmetric about the x-axis, we can calculate the portion of area above the x-axis and double the result. In the portion of the cardioid above the x-axis, θ ranges from 0 to π, so that

$$A = 2 \int_0^\pi \frac{1}{2} r^2 \, d\theta = \int_0^\pi (1 - \cos \theta)^2 \, d\theta = \frac{3\pi}{2} \quad \blacktriangleleft$$

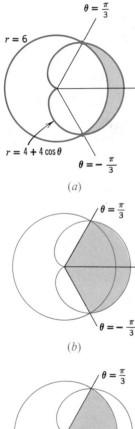

(a)

(b)

▶ **Example 3** Find the area of the region that is inside the cardioid $r = 4 + 4 \cos \theta$ and outside the circle $r = 6$.

Solution. To sketch the region, we need to know where the circle and cardioid intersect. To find these points, we equate the given expressions for r. This yields

$$4 + 4 \cos \theta = 6$$

from which we obtain $\cos \theta = \frac{1}{2}$ or

$$\theta = \frac{\pi}{3} \quad \text{and} \quad \theta = -\frac{\pi}{3}$$

The region is sketched in Figure 13.3.5.
The desired area can be obtained by subtracting the shaded areas in parts (b) and (c) of Figure 13.3.5. Thus,

$$A = \int_{-\pi/3}^{\pi/3} \frac{1}{2} (4 + 4 \cos \theta)^2 \, d\theta - \int_{-\pi/3}^{\pi/3} \frac{1}{2} (6)^2 \, d\theta$$

$$= \int_{-\pi/3}^{\pi/3} \frac{1}{2} [(4 + 4 \cos \theta)^2 - 36] \, d\theta$$

$$= \int_{-\pi/3}^{\pi/3} (16 \cos \theta + 8 \cos^2 \theta - 10) \, d\theta$$

$$= [16 \sin \theta + (4\theta + 2 \sin 2\theta) - 10\theta] \Big]_{-\pi/3}^{\pi/3}$$

$$= 18 \sqrt{3} - 4\pi$$

(4)

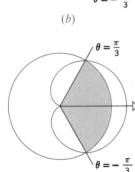

(c)

Figure 13.3.5

Alternate Solution. Using symmetry, we may calculate the portion of area above the x-axis and double it to obtain the entire area A. This yields

$$A = 2 \int_0^{\pi/3} \frac{1}{2} [(4 + 4 \cos \theta)^2 - 36] \, d\theta$$

rather than (4). However, the end result will be the same. ◀

WARNING. In the last example we determined the intersections of the curves $r = 4 + 4 \cos \theta$ and $r = 6$ by equating the right-hand sides and solving for θ. However, it may not be possible to find *all* the intersections of two polar curves, $r = f_1(\theta)$ and $r = f_2(\theta)$, by solving the equation $f_1(\theta) = f_2(\theta)$ for θ. This is because every point has infinitely many pairs of polar coordinates. It is possible, therefore, that a point of intersection may have no single pair of polar coordinates that satisfies both equations. For example, the cardioids

$$r = 1 - \cos \theta \quad \text{and} \quad r = 1 + \cos \theta \tag{5}$$

intersect at three points, the origin, the point $(1, \pi/2)$, and the point $(1, 3\pi/2)$ (Figure 13.3.6).

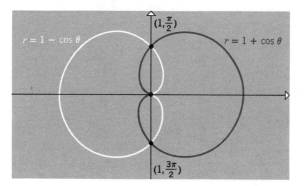

Figure 13.3.6

Equating the right-hand sides in (5) yields $1 - \cos \theta = 1 + \cos \theta$ or $\cos \theta = 0$, so

$$\theta = \frac{\pi}{2} + k\pi, \quad k = 0, \pm 1, \pm 2, \ldots$$

Substituting any of these values in (5) yields $r = 1$, so that we have found only two distinct points of intersection, $(1, \pi/2)$ and $(1, 3\pi/2)$; the origin has been missed.

When calculating intersections in polar coordinates, it is a good idea to make a sketch of the curves to determine how many intersections there should be.

► Exercise Set 13.3

In Exercises 1–17, find the area of the region described.

1. The region in the first quadrant enclosed by the first loop of the spiral $r = \theta$ ($\theta \geq 0$), and the lines $\theta = \pi/6$ and $\theta = \pi/3$.

2. The region in the first quadrant within the cardioid $r = 1 + \sin \theta$.

3. The region enclosed by the cardioid $r = 2 + 2 \cos \theta$.

4. The region outside the cardioid $r = 2 - 2 \cos \theta$ and inside the circle $r = 4$.

5. The region inside the circle $r = 5 \sin \theta$ and outside the limaçon $r = 2 + \sin \theta$.

6. The region enclosed by the inner loop of the limaçon $r = 1 + 2 \cos \theta$.

7. The region inside the cardioid $r = 2 + 2 \cos \theta$ and outside the circle $r = 3$.

8. The region inside the circle $r = 2a \sin \theta$.

9. The region enclosed by the curve $r^2 = \sin 2\theta$.

10. The region common to the circles $r = 4 \cos \theta$ and $r = 4 \sin \theta$.

11. The region enclosed by the rose $r = 4 \cos 3\theta$.

12. The region inside the rose $r = 2a \cos 2\theta$ and outside the circle $r = a\sqrt{2}$.

13. The region common to the circle $r = 3 \cos \theta$ and the cardioid $r = 1 + \cos \theta$.

14. The region between the loops of the limaçon $r = \frac{1}{2} + \cos \theta$.

15. The region inside the circle $r = 10$ and to the right of the line $r \cos \theta = 6$.

16. The region inside the cardioid $r = a(1 + \sin \theta)$ and outside the circle $r = a \sin \theta$.

17. The region enclosed by $r = a \sec^2 \frac{1}{2}\theta$ and the rays whose polar angles are $\theta = 0$ and $\theta = \frac{1}{2}\pi$.

18. (a) Find the error: The area inside the lemniscate $r^2 = a^2 \cos 2\theta$ is

$$A = \int_0^{2\pi} \frac{1}{2} r^2 \, d\theta = \int_0^{2\pi} \frac{1}{2} a^2 \cos 2\theta \, d\theta$$

$$= \frac{1}{4} a^2 \sin 2\theta \bigg]_0^{2\pi} = 0$$

(b) Find the correct area.

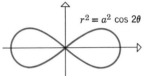

19. Find the area inside the lemniscate $r^2 = 4 \cos 2\theta$ and outside the circle $r = \sqrt{2}$.

20. A radial line is drawn from the origin to the spiral $r = a\theta$ ($a > 0$ and $\theta \geq 0$). Find the area swept out during the second revolution of the radial line that was not swept out during the first revolution.

13.4 PARAMETRIC EQUATIONS

In order to study a curve using calculus, we need some way to represent the curve mathematically. Often, the representation is an equation $y = f(x)$ whose graph is the given curve. However, graphs of functions are specialized curves in the sense that no vertical line can cut such a curve more than once. In order to represent curves that are not graphs of functions, different methods are required. In this section we discuss one of these methods.

Curves that are not graphs of functions can often be specified by using a pair of equations

$$x = x(t)$$
$$y = y(t)$$

to express the coordinates of a point (x, y) on the curve as functions of an auxiliary variable t. These are called *parametric equations* for the curve, and the variable t is called a *parameter.*

Parametric equations arise naturally if one imagines a plane curve C to be traced by a moving point. If we use the parameter t to denote time, then the parametric equations $x = x(t), y = y(t)$ specify how the x- and y-coordinates of the moving point vary with time.

▶ **Example 1** A particle moves in the plane so its x- and y-coordinates vary with time according to the equations

$$x = \tfrac{1}{2}t^3 - 6t$$
$$y = \tfrac{1}{2}t^2$$

Sketch the path of the particle over the time interval $0 \le t \le 4$.

Solution. The path of the particle, shown in Figure 13.4.1, was obtained by calculating the x- and y-coordinates of the particle for $t = 0, 1, 2, 3, 4$ (Table 13.4.1), plotting the points (x, y), and connecting successive points with a smooth curve. The arrows on the curve serve to indicate the direction of motion. ◀

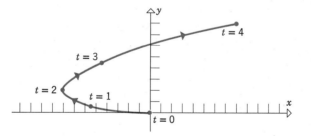

Figure 13.4.1

Table 13.4.1

t	0	1	2	3	4
$x = \tfrac{1}{2}t^3 - 6t$	0	$-\tfrac{11}{2}$	-8	$-\tfrac{9}{2}$	8
$y = \tfrac{1}{2}t^2$	0	$\tfrac{1}{2}$	2	$\tfrac{9}{2}$	8
(x, y)	$(0, 0)$	$(-\tfrac{11}{2}, \tfrac{1}{2})$	$(-8, 2)$	$(-\tfrac{9}{2}, \tfrac{9}{2})$	$(8, 8)$

Although time is a common parameter, it is not the only possibility. Other parameters are illustrated in the following examples.

▶ **Example 2** Let C be a circle of radius a centered at the origin. If $P(x, y)$ is any point on the circle and θ is the angle measured counterclockwise from the positive x-axis to the line segment OP (Figure 13.4.2a), then

$$x = a \cos \theta \quad \text{and} \quad y = a \sin \theta$$

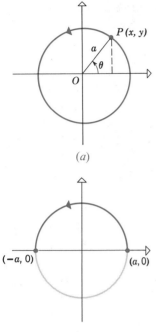

(a)

$$x = a \cos \theta$$
$$y = a \sin \theta, \; 0 \le \theta \le \pi$$

(b)

Figure 13.4.2

As θ varies from 0 to 2π, the point $P(x, y)$ makes one complete revolution. Thus,

$$x = a \cos \theta$$
$$y = a \sin \theta \quad, \; 0 \le \theta \le 2\pi$$

are parametric equations for the circle.

By changing the interval over which θ is allowed to vary, we can obtain different portions of the circle. For example,

$$x = a \cos \theta$$
$$y = a \sin \theta \quad, \; 0 \le \theta \le \pi$$

represents just the upper half of the circle (Figure 13.4.2b).   ◄

A curve that is represented parametrically is traced in a certain direction as the parameter increases. This is called the ***direction of increasing parameter*** or sometimes the ***orientation*** of the curve. For example, the circle

$$x = a \cos \theta$$
$$y = a \sin \theta \quad, \; 0 \le \theta \le 2\pi$$

is traced counterclockwise as θ increases from 0 to 2π, so this is the direction of increasing parameter. Phrased another way, the circle is oriented counterclockwise. We have indicated this in Figure 13.4.2a with an arrow on the circle.

Sometimes a curve given parametrically can be recognized by eliminating the parameter.

▶ Example 3   Sketch the curve

$$x = 2t - 3$$
$$y = 6t - 7$$

Solution.   Solving each equation for t, we obtain

$$t = \frac{x + 3}{2}$$

$$t = \frac{y + 7}{6}$$

Thus,

$$\frac{x + 3}{2} = \frac{y + 7}{6}$$

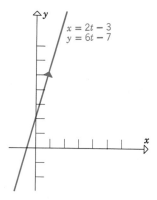

Figure 13.4.3

or

$$y = 3x + 2$$

which is the straight line shown in Figure 13.4.3. The direction of increasing parameter, shown in the figure, was obtained from the original parametric equations by observing that x and y both increase as t increases. ◀

REMARK. In the last example, there were no conditions imposed on the parameter t. When no conditions are stated explicitly, it is understood that the parameter varies over the interval $(-\infty, +\infty)$.

▶ Example 4   Sketch the curve

$$\begin{array}{l} x = a \cos t \\ y = b \sin t \end{array}, \ 0 \le t \le 2\pi$$

where $a > 0$ and $b > 0$.

Solution. We can eliminate t by writing

$$\frac{x}{a} = \cos t \quad \text{and} \quad \frac{y}{b} = \sin t$$

from which it follows that

$$\frac{x^2}{a^2} + \frac{y^2}{b^2} = 1$$

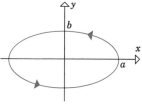

Figure 13.4.4

This is the equation of the ellipse shown in Figure 13.4.4. We leave it for the reader to show that the ellipse has counterclockwise orientation. ◀

▶ Example 5   Let C be the parabola

$$y = x^2$$

and let $m = dy/dx$ be the slope of the tangent to C at the point (x, y). Find parametric equations for C using m as a parameter.

Solution. Our objective is to express the coordinates (x, y) of an arbitrary point on the parabola in terms of the parameter m. But

$$\frac{dy}{dx} = 2x$$

so

$$x = \frac{1}{2}\frac{dy}{dx} = \frac{1}{2}m$$

and

$$y = x^2 = \left(\frac{1}{2}m\right)^2 = \frac{1}{4}m^2$$

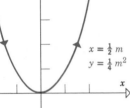

which leads to the parametric equations

$$x = \frac{1}{2}m$$

$$y = \frac{1}{4}m^2$$

$x = \frac{1}{2}m$
$y = \frac{1}{4}m^2$

Figure 13.4.5

The direction of increasing parameter, indicated in Figure 13.4.5, is obtained by noting that x increases as the slope m increases, since $x = \frac{1}{2}m$. ◀

▶ **Example 6** If $r = f(\theta)$ is a polar curve, we can obtain parametric equations for the curve in terms of θ by substituting $r = f(\theta)$ in the relationships

$$x = r\cos\theta$$
$$y = r\sin\theta$$

This yields

$$x = f(\theta)\cos\theta$$
$$y = f(\theta)\sin\theta$$

For example, the cardioid $r = 1 - \cos\theta$ can be represented by the parametric equations

$$x = (1 - \cos\theta)\cos\theta$$
$$y = (1 - \cos\theta)\sin\theta$$

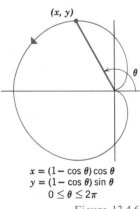

(x, y)

If we impose the added restriction, $0 \leq \theta \leq 2\pi$, then the point (x, y) will traverse the cardioid exactly once as θ varies over $[0, 2\pi]$ (Figure 13.4.6). ◀

$x = (1 - \cos\theta)\cos\theta$
$y = (1 - \cos\theta)\sin\theta$
$0 \leq \theta \leq 2\pi$

Figure 13.4.6

If the parameter is eliminated from a pair of parametric equations, the graph of the resulting equation in x and y may possibly extend beyond the graph of the original parametric equations. For example, consider the curve with parametric equations

$$x = \cosh t$$
$$y = \sinh t \tag{1}$$

We may eliminate the parameter by using the identity

$$\cosh^2 t - \sinh^2 t = 1$$

(Formula (1) of Section 7.6) to obtain

$$x^2 - y^2 = 1 \tag{2}$$

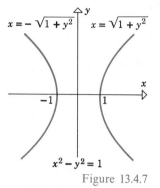

$x = -\sqrt{1 + y^2}$ $x = \sqrt{1 + y^2}$

$x^2 - y^2 = 1$

Figure 13.4.7

which is the equation of the hyperbola shown in Figure 13.4.7. However, the original parametric equations represent only the right branch of this hyperbola since $x = \cosh t$ is never less than 1. To exclude the unwanted left branch we would have to replace (2) by

$$x^2 - y^2 = 1, \qquad x \geq 1$$

or more simply

$$x = \sqrt{1 + y^2}$$

To summarize, whenever the parameter is eliminated from parametric equations, it is important to check that the graph of the resulting equation does not include too much. If it does, added conditions must be imposed.

While parametric equations are of special importance for curves that are not graphs of functions, curves of the form $y = f(x)$ or $x = g(y)$ may also be represented parametrically. This can always be done by introducing a parameter t that is equal to the independent variable.

▶ Example 7   The curve $y = 4x^2 - 1$ can be represented parametrically as

$$x = t$$
$$y = 4t^2 - 1$$

and the curve $x = 5y^3 - y$ can be represented parametrically as

$$x = 5t^3 - t$$
$$y = t \qquad\qquad ◀$$

Sometimes it is difficult to eliminate the parameter from a pair of parametric equations and express y directly in terms of x. Nevertheless, by using the chain rule, it is possible to find the derivative of y with respect to x without eliminating the parameter. We start with the relationship

$$\frac{dy}{dt} = \frac{dy}{dx}\frac{dx}{dt}$$

Provided $dx/dt \neq 0$, this can be rewritten as

$$\frac{dy}{dx} = \frac{dy/dt}{dx/dt} \tag{3}$$

▶ Example 8   Given the curve

$$x = t^2$$
$$y = t^3$$

find dy/dx and d^2y/dx^2 at the point $(1, 1)$ without eliminating the parameter t.

Solution

$$\frac{dy}{dx} = \frac{dy/dt}{dx/dt} = \frac{3t^2}{2t} = \frac{3}{2}t \tag{4a}$$

To find the second derivative, denote dy/dx by y' and use (3) to write

$$\frac{d^2y}{dx^2} = \frac{dy'}{dx} = \frac{dy'/dt}{dx/dt} = \frac{3/2}{2t} = \frac{3}{4t} \tag{4b}$$

Note that these calculations express dy/dx and d^2y/dx^2 in terms of the parameter t. Since the point $(1, 1)$ on the curve corresponds to $t = 1$, it follows from (4a) and (4b) that

$$\left.\frac{dy}{dx}\right|_{t=1} = \frac{3}{2}(1) = \frac{3}{2}$$

and

$$\left.\frac{d^2y}{dx^2}\right|_{t=1} = \frac{3}{4}$$

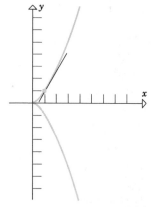

The curve and the tangent at $(1, 1)$ are shown in Figure 13.4.8. Note that the tangent line has slope $\frac{3}{2}$ at $(1, 1)$ and the curve is concave up at $(1, 1)$, in keeping with our calculations.   ◀

REMARK.   Formula (3) does not apply if $dx/dt = 0$. However, at points where $dx/dt = 0$ and $dy/dt \neq 0$ there is generally a **vertical tangent** line. Points (x, y) where both $dx/dt = 0$ and $dy/dt = 0$ are called **singular points.** The behavior of a curve at a singular point must be determined on a case-by-case basis.

Figure 13.4.8

In Section 6.4 we obtained the following formula for the arc length L of a curve $y = f(x)$ from $x = a$ to $x = b$

$$L = \int_a^b \sqrt{1 + [f'(x)]^2}\, dx = \int_a^b \sqrt{1 + \left(\frac{dy}{dx}\right)^2}\, dx \tag{5}$$

This formula is a special case of the following general result about arc length of parametric curves

13.4.1 THEOREM
Arc Length of Parametric Curves

If $x'(t)$ and $y'(t)$ are continuous functions for $a \leq t \leq b$, then the curve

$$\begin{matrix} x = x(t) \\ y = y(t) \end{matrix}, \quad a \leq t \leq b$$

has arc length L given by

$$L = \int_a^b \sqrt{[x'(t)]^2 + [y'(t)]^2}\, dt \tag{6}$$

or equivalently

$$L = \int_a^b \sqrt{\left(\frac{dx}{dt}\right)^2 + \left(\frac{dy}{dt}\right)^2}\, dt \tag{7}$$

The derivation of these formulas is similar to the derivation of (5) and will be omitted. Observe that (5) follows from (7) by expressing the curve $y = f(x)$ in the parametric form

$$\begin{matrix} x = t \\ y = f(t) \end{matrix}$$

in which case

$$\frac{dx}{dt} = 1 \quad \text{and} \quad \frac{dy}{dt} = f'(t) = f'(x) = \frac{dy}{dx}$$

▶ **Example 9** Use (7) to find the arc length of the circle

$$\begin{matrix} x = a\cos t \\ y = a\sin t \end{matrix}, \quad 0 \leq t \leq 2\pi \quad (a > 0)$$

Solution.

$$L = \int_0^{2\pi} \sqrt{\left(\frac{dx}{dt}\right)^2 + \left(\frac{dy}{dt}\right)^2}$$

$$= \int_0^{2\pi} \sqrt{(-a \sin t)^2 + (a \cos t)^2} \, dt$$

$$= \int_0^{2\pi} a \, dt \quad = at \Big]_0^{2\pi} \quad = 2\pi a \qquad \blacktriangleleft$$

OPTIONAL

If a wheel rolls along a straight line without slipping, then a point on the rim of the wheel traces a curve called a *cycloid* (Figure 13.4.9).

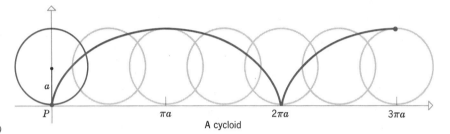

Figure 13.4.9

A cycloid

The cycloid is of special interest because it provides the solution of two famous mathematical problems: the **brachistochrone problem** (from Greek words meaning "shortest time") and the **tautochrone problem** (from Greek words meaning "equal time"). In June of 1696, Johann Bernoulli* posed the brachistochrone problem in the form of a challenge to other mathematicians. The problem was to determine the shape of a wire down which a bead might slide from a point P to another point Q, not directly below, in the *shortest time*. At first, one might conjecture that the wire should form a straight line, since that shape yields the shortest distance from P to Q. However, the correct answer is half of one arch of an inverted cycloid (Figure 13.4.10). In essence, this shape allows the bead to fall rapidly at first, building up sufficient initial

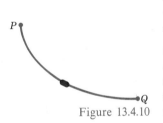

Figure 13.4.10

*BERNOULLI. An amazing Swiss family that included several generations of outstanding mathematicians and scientists. Nikolaus Bernoulli (1623–1708), a druggist, fled from Antwerp to escape religious persecution and ultimately settled in Basel, Switzerland. There he had three sons, Jakob I (also called Jacques or James), Nikolaus, and Johann I (also called Jean or John). (The roman numerals are used to distinguish family members with identical names.)

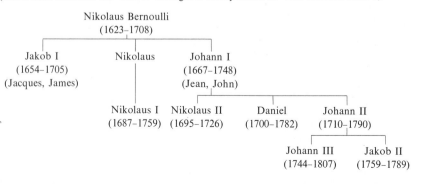

speed to reach Q in the shortest time, even though the path does not provide the shortest distance from P to Q. This problem was solved by Newton, Leibniz, L'Hôpital, Johann Bernoulli, and his older brother Jakob Bernoulli. It was formulated and incorrectly solved years earlier by Galileo, who gave the arc of a circle as the answer.

In the tautochrone problem the object is to find the shape of a wire from P to Q such that two beads started above Q at *any* points on the wire reach Q in the same amount of time. Again, the answer is half of one arch of a cycloid.

We conclude this section by deriving parametric equations for a cycloid.

▶ **Example 10** Let P be a fixed point on the rim of a wheel of radius a. Suppose that the wheel rolls along the positive x-axis of a rectangular coordinate system, and that P is initially at the origin. When the wheel rolls a

---

Following Newton and Leibniz, the Bernoulli brothers, Jakob I and Johann I, are considered by some to be the two most important founders of calculus. Jakob I was self-taught in mathematics. His father wanted him to study for the ministry, but he turned to mathematics and in 1686 became a professor at the University of Basel. When he started working in mathematics, he knew nothing of Newton's and Leibniz' work. He eventually became familiar with Newton's results, but because so little of Leibniz' work was published, Jakob duplicated many of Leibniz' results.

Jakob's younger brother Johann I was urged to enter into business by his father. Instead, he turned to medicine and studied mathematics under the guidance of his older brother. He eventually became a mathematics professor at Gröningen in Holland and then, when Jakob died in 1705, Johann succeeded him as mathematics professor at Basel. Throughout their lives, Jakob I and Johann I had a mutual passion for criticizing each other's work, which frequently erupted into ugly confrontations. Leibniz tried to mediate the disputes, but Jakob, who resented Leibniz' superior intellect, accused him of siding with Johann and thus Leibniz became entangled in the arguments. The brothers often worked on common problems that they posed as challenges to one another. Johann, interested in gaining fame, often used unscrupulous means to make himself appear the originator of his brother's results; Jakob occasionally retaliated. Thus, it is often difficult to determine who deserves credit for many results. However, both men made major contributions to the development of calculus. In addition to his work on calculus, Jakob helped establish fundamental principles in probability, including the Law of Large Numbers, which is a cornerstone of modern probability theory. Johann was Euler's teacher at Basel and L'Hôpital's tutor in France.

Among the other members of the Bernoulli family, Daniel, son of Johann I, is the most famous. He was a professor of mathematics at St. Petersburg Academy in Russia and subsequently a professor of anatomy and then physics at Basel. He did work in calculus and probability, but is best known for his work in physics. A basic law of fluid flow, called Bernoulli's principle, is named in his honor. He won the annual prize of the French Academy 10 times for work on vibrating strings, tides of the sea, and kinetic theory of gases.

Johann II succeeded his father as professor of mathematics at Basel. His research was on the theory of heat and sound. Nikolaus I was a mathematician and law scholar who worked on probability and series. On the recommendation of Leibniz, he was appointed professor of mathematics at Padua and then went to Basel as a professor of logic and then law. Nikolaus II was professor of jurisprudence in Switzerland and then professor of mathematics at St. Petersburg Academy. Johann III was a professor of mathematics and astronomy in Berlin and Jakob II succeeded his uncle Daniel as professor of mathematics at St. Petersburg Academy in Russia. Truly an incredible family!

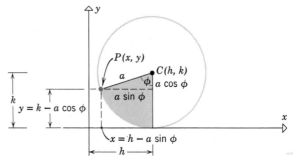

Figure 13.4.11

given distance, the radial line from the center C to the point P rotates through an angle ϕ (Figure 13.4.11), which we will use as our parameter. (It is customary to regard ϕ as a positive angle, even though it is generated by a clockwise rotation.) Our objective is to express the coordinates of $P(x, y)$ in terms of the angle ϕ. Figure 13.4.11 suggests that the coordinates of $P(x, y)$ and the coordinates of the wheel's center $C(h, k)$ are related by

$$
\begin{aligned}
x &= h - a \sin \phi \\
y &= k - a \cos \phi
\end{aligned}
\tag{8}
$$

Since the height of the wheel's center is the radius of the wheel, we have $k = a$; and since the distance h moved by the center is the same as the circular arc length subtended by ϕ (why?), we have $h = a\phi$. Therefore, (8) yields

$$
\begin{aligned}
x &= a\phi - a \sin \phi \\
y &= a - a \cos \phi
\end{aligned}
\quad , \ 0 \le \phi < +\infty
$$

These are the parametric equations of the cycloid. ◄

► Exercise Set 13.4

1. (a) Sketch the curve $x = t$, $y = t^2$ by plotting the points corresponding to $t = -3, -2, -1, 0, 1, 2, 3$.
   (b) Show that the curve is a parabola by eliminating the parameter t.

2. Sketch the curve $x = t^3 - 4t$, $y = 2t$, $-3 \le t \le 3$ by plotting some appropriate points.

In Exercises 3–15, sketch the curve by eliminating the parameter t, and label the direction of increasing t.

3. $x = \cos t$, $y = \sin t$, $0 \le t \le 2\pi$.

4. $x = 1 + \cos t$, $y = 3 - \sin t$, $0 \le t \le 2\pi$.

5. $x = 3t - 4$, $y = 6t + 2$.

6. $x = t - 3$, $y = 3t - 7$, $0 \le t \le 3$.

7. $x = 2 \cos t$, $y = 5 \sin t$, $0 \le t \le 2\pi$.

8. $x = \sqrt{t}$, $y = 2t + 4$.

9. $x = 3 + 2 \cos t$, $y = 2 + 4 \sin t$, $0 \le t \le 2\pi$.

10. $x = 2 \cosh t$, $y = 4 \sinh t$.

11. $x = 4 \sin 2\pi t$, $y = 4 \cos 2\pi t$, $0 \le t \le 1$.

12. $x = \sec t$, $y = \tan t$, $\pi \le t < \dfrac{3\pi}{2}$.

13. $x = \cos 2t$, $y = \sin t$, $-\dfrac{\pi}{2} \le t \le \dfrac{\pi}{2}$.

14. $x = 4t + 3$, $y = 16t^2 - 9$.

15. $x = t^2$, $y = 2 \ln t$, $t \geq 1$.

In Exercises 16–22, find dy/dx at the point corresponding to the given value of the parameter without eliminating the parameter.

16. $x = t^2 + 4$, $y = 8t$; $t = 2$.

17. $x = \cos t$, $y = \sin t$; $t = 3\pi/4$.

18. $x = t + 5$, $y = 5t - 7$; $t = 1$.

19. $x = \sqrt{t}$, $y = 2t + 4$; $t = 9$.

20. $x = \sec \theta$, $y = \tan \theta$; $\theta = \pi/3$.

21. $x = 4 \cos 2\pi s$, $y = 3 \sin 2\pi s$; $s = -1/4$.

22. $x = \sinh t$, $y = \cosh t$; $t = 0$.

In Exercises 23–26, find d^2y/dx^2 at the point corresponding to the given value of the parameter without eliminating the parameter.

23. $x = \frac{1}{2}t^2$, $y = \frac{1}{3}t^3$; $t = 2$.

24. $x = \cos \phi$, $y = \sin \phi$; $\phi = \pi/4$.

25. $x = \sqrt{t}$, $y = 2t + 4$; $t = 1$.

26. $x = \sec t$, $y = \tan t$; $t = \pi/3$.

In Exercises 27–35, find the arc length of the curve.

27. $x = 4t + 3$, $y = 3t - 2$, $0 \leq t \leq 2$.

28. $x = \cos^3 t$, $y = \sin^3 t$, $0 \leq t \leq \pi/2$.

29. $x = \frac{1}{3}t^3$, $y = \frac{1}{2}t^2$, $0 \leq t \leq 1$.

30. $x = \frac{1}{3}t^3$, $y = \frac{1}{2}t^2$, $-1 \leq t \leq 0$.

31. $x = \cos 2t$, $y = \sin 2t$, $0 \leq t \leq \pi/2$.

32. $x = e^t (\sin t + \cos t)$, $y = e^t (\cos t - \sin t)$
    $1 \leq t \leq 4$.

33. $x = (1 + t)^2$, $y = (1 + t)^3$, $0 \leq t \leq 1$.

34. $x = e^t \cos t$, $y = e^t \sin t$, $0 \leq t \leq \pi/2$.

35. One arch of the cycloid
    $x = a(t - \sin t)$, $y = a(1 - \cos t)$.

36. Find parametric equations for the rose
    $r = 2 \cos 2\theta$ using θ as the parameter.

37. Find parametric equations for the limaçon
    $r = 2 + 3 \sin \theta$ using θ as the parameter.

38. Find the equation of the tangent line to the curve
    $x = 2t + 4$, $y = 8t^2 - 2t + 4$ at the point where
    $t = 1$.

39. Find the equation of the tangent line to the curve
    $x = e^t$, $y = e^{-t}$ at the point where $t = 2$.

40. Find all values of t at which the curve $x = 2 \cos t$,
    $y = 4 \sin t$ has a tangent that is
    (a) horizontal   (b) vertical.

41. Find all values of t at which the curve
    $x = 2t^3 - 15t^2 + 24t + 7$, $y = t^2 + t + 1$ has a
    tangent that is
    (a) horizontal   (b) vertical.

42. Show that the curve $x = t^3 - 4t$, $y = t^2$ intersects
    itself at the point $(0, 4)$, and find equations for two
    tangent lines to the curve at the point of intersection.

43. Show that the curve $x = t^2 - 3t + 5$, $y = t^3 + t^2 - 10t + 9$ intersects itself at the point $(3, 1)$,
    and find equations for two tangent lines to the
    curve at the point of intersection.

44. A point traces the circle $x^2 + y^2 = 25$ so that
    $dx/dt = 8$ when the point reaches $(4, 3)$. Find
    dy/dt there.

45. Describe the curve whose parametric equations are
    $x = a \cos t + h$, $y = b \sin t + k$, $0 \leq t \leq 2\pi$.

46. A *hypocycloid* is a curve traced by a point P on the
    circumference of a circle that rolls inside a larger
    fixed circle. Suppose that the fixed circle has radius
    a, the rolling circle has radius b, and the fixed
    circle is centered at the origin. Let ϕ be the angle
    shown in the following figure, and assume that the
    point P is at $(a, 0)$ when $\phi = 0$. Show that the
    hypocycloid generated is given by the parametric
    equations

    $$x = (a - b) \cos \phi + b \cos \left(\frac{a - b}{b} \phi \right)$$

    $$y = (a - b) \sin \phi - b \sin \left(\frac{a - b}{b} \phi \right)$$

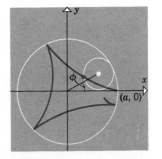

47. If $b = \frac{1}{4}a$ in Exercise 46, then the resulting curve is called a four-cusped hypocycloid.

(a) Sketch this curve.

(b) Show that the curve is given by the parametric equations $x = a \cos^3 \phi$, $y = a \sin^3 \phi$.

(c) Show that curve is given by the equation $x^{2/3} + y^{2/3} = a^{2/3}$ in rectangular coordinates.

If $x'(t)$ and $y'(t)$ are continuous functions for $a \leq t \leq b$, then it can be shown that the area of the surface generated by revolving the curve

$$x = x(t) \atop y = y(t), \quad a \leq t \leq b$$

about the x-axis is

$$S = \int_a^b 2\pi y(t) \sqrt{[x'(t)]^2 + [y'(t)]^2} \, dt$$

and the area of the surface generated by revolving the curve about the y-axis is

$$S = \int_a^b 2\pi x(t) \sqrt{[x'(t)]^2 + [y'(t)]^2} \, dt$$

(The derivations are similar to those used to obtain Formulas (6) and (7) in Section 6.5.) In Exercises 48–53, use these formulas.

48. Find the area of the surface generated by revolving $x = t^2$, $y = 2t$, $0 \leq t \leq 4$ about the x-axis.

49. Find the area of the surface generated by revolving $x = e^t \cos t$, $y = e^t \sin t$, $0 \leq t \leq \pi/2$ about the x-axis.

50. Find the area of the surface generated by revolving $x = \cos^2 t$, $y = \sin^2 t$, $0 \leq t \leq \pi/2$ about the y-axis.

51. Find the area of the surface generated by revolving $x = t$, $y = 2t^2$, $0 \leq t \leq 1$ about the y-axis.

52. By revolving the semicircle $x = r \cos t$, $y = r \sin t$, $0 \leq t \leq \pi$ about the x-axis, show that the surface area of a sphere of radius r is $4\pi r^2$.

53. The equations

$$x = a\phi - a \sin \phi, \quad y = a - a \cos \phi, \quad 0 \leq \phi \leq 2\pi$$

represent one arch of a cycloid. Show that the surface area generated by revolving this curve about the x-axis is $S = 64\pi a^2/3$.

13.5 TANGENT LINES AND ARC LENGTH IN POLAR COORDINATES (OPTIONAL)

In this section we will use our results on parametric equations to derive formulas for arc length and slopes of tangent lines to polar curves.

Let

$$r = f(\theta)$$

be a curve in polar coordinates, for which f is a differentiable function of θ. We can obtain parametric equations for this curve in terms of θ by substituting $r = f(\theta)$ in the relationships

$$x = r \cos \theta$$
$$y = r \sin \theta$$

This yields the parametric equations

$$x = f(\theta) \cos \theta$$
$$y = f(\theta) \sin \theta \tag{1}$$

If we assume the existence of a tangent line to the curve $r = f(\theta)$ at the point $P(r, \theta)$, then the slope of this tangent line is

$$m = \tan \phi = \frac{dy}{dx}$$

where ϕ is the angle of inclination. From (1) we have

$$\frac{dx}{d\theta} = -f(\theta) \sin \theta + f'(\theta) \cos \theta$$

$$\frac{dy}{d\theta} = f(\theta) \cos \theta + f'(\theta) \sin \theta$$

(2)

Thus, if $dx/d\theta \neq 0$, it follows from Equation (3) of Section 13.4 that

$$m = \tan \phi = \frac{dy}{dx} = \frac{dy/d\theta}{dx/d\theta}$$

(3)

or

$$m = \tan \phi = \frac{f(\theta) \cos \theta + f'(\theta) \sin \theta}{-f(\theta) \sin \theta + f'(\theta) \cos \theta} = \frac{r \cos \theta + \sin \theta \dfrac{dr}{d\theta}}{-r \sin \theta + \cos \theta \dfrac{dr}{d\theta}}$$

(4)

which is a formula for the slope of the tangent line to $r = f(\theta)$ at $P(r, \theta)$.

Formula (4) was derived by assuming that $dx/d\theta \neq 0$. If $dx/d\theta = 0$ and $dy/d\theta \neq 0$, we will agree that the curve has a **vertical tangent.** This is reasonable since $\tan \phi$ becomes infinite in this case [see (3)], indicating that $\phi = \pi/2$. Points where both $dx/d\theta = 0$ and $dy/d\theta = 0$ are called **singular points.** No general statement can be made about the behavior of a polar curve at a singular point. Each case requires its own analysis.

If $\cos \theta \neq 0$, we may divide the numerator and denominator of (4) by $\cos \theta$ to obtain the following alternative formula for the slope of a tangent line.

$$m = \tan \phi = \frac{r + \tan \theta \dfrac{dr}{d\theta}}{-r \tan \theta + \dfrac{dr}{d\theta}}$$

(5)

▶ Example 1   Find the slope of the tangent line to the circle

$$r = 4 \cos \theta$$

(6)

at the point where $\theta = \pi/4$.

Solution. Differentiating (6) yields

$$\frac{dr}{d\theta} = -4 \sin \theta \tag{7}$$

If $\theta = \pi/4$ it follows from (6) and (7) that

$$r = 4 \cos \frac{\pi}{4} = 2\sqrt{2} \quad \text{and} \quad \frac{dr}{d\theta} = -4 \sin \frac{\pi}{4} = -2\sqrt{2}$$

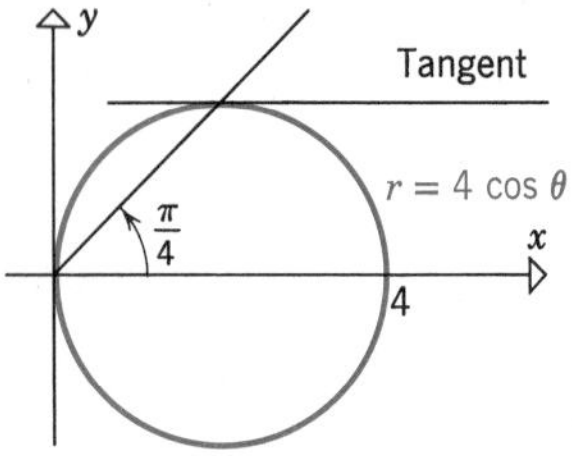

Substituting these values and $\tan \theta = \tan \frac{\pi}{4} = 1$ in (5) yields

$$\tan \phi = \frac{2\sqrt{2} + (1)(-2\sqrt{2})}{-(2\sqrt{2})(1) + (-2\sqrt{2})} = 0$$

Thus, the circle has a horizontal tangent line when $\theta = \pi/4$ (Figure 13.5.1). ◄

Figure 13.5.1

▶ **Example 2**   At what points does the cardioid $r = 1 - \cos \theta$ have a vertical tangent line?

Solution. A vertical tangent line will occur at points where $dx/d\theta = 0$ and $dy/d\theta \neq 0$. The cardioid is given parametrically by the equations

$$\begin{aligned} x &= r \cos \theta = (1 - \cos \theta) \cos \theta \\ y &= r \sin \theta = (1 - \cos \theta) \sin \theta \end{aligned}, \quad 0 \le \theta \le 2\pi$$

Differentiating and then simplifying, we obtain (verify)

$$\frac{dx}{d\theta} = \sin \theta (2 \cos \theta - 1) \tag{8}$$

$$\frac{dy}{d\theta} = (1 - \cos \theta)(1 + 2 \cos \theta) \tag{9}$$

From (8), $dx/d\theta = 0$ if $\sin \theta = 0$ or $\cos \theta = \frac{1}{2}$; this occurs where $\theta = 0$, π, $\pi/3$, $5\pi/3$, or 2π. But from (9), $dy/d\theta \neq 0$ if $\theta = \pi$, $\pi/3$, and $5\pi/3$, so vertical tangents occur at these points. Since $dy/d\theta = 0$ if $\theta = 0$ or $\theta = 2\pi$, these are singular points. Although we will not prove it, the cardioid has a horizontal tangent at the origin (Figure 13.5.2). ◄

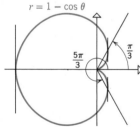

$r = 1 - \cos \theta$

$\frac{5\pi}{3}$ $\frac{\pi}{3}$

Figure 13.5.2

Another way to find the tangent line to a polar curve at a point $P(r, \theta)$ is to find the angle ψ between the **radial line** OP and the tangent line (Figure 13.5.3). By definition, ψ is measured counterclockwise from the radial line OP to the tangent line, and is selected so

$$0 \le \psi < \pi$$

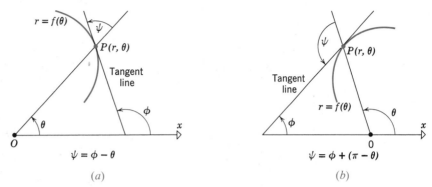

Figure 13.5.3   (a)   (b)

If we choose the polar angle θ in the range $0 \leq \theta < 2\pi$, there is a simple algebraic relationship between θ, ϕ, and ψ that depends on the relative sizes of ϕ and θ. If $\phi \geq \theta$, then

$$\psi = \phi - \theta \tag{10}$$

(Figure 13.5.3a), and if $\phi < \theta$, then the relationship is

$$\psi = \phi + (\pi - \theta) = (\phi - \theta) + \pi \tag{11}$$

(Figure 13.5.3b). In either case, we may write

$$\tan \psi = \tan (\phi - \theta) \tag{12}$$

from which it follows that

$$\tan \psi = \frac{\tan \phi - \tan \theta}{1 + \tan \phi \tan \theta} \tag{13}$$

Substituting (5) into (13) yields

$$\tan \psi = \frac{\left[\left(r + \tan \theta \dfrac{dr}{d\theta}\right)\Big/\left(-r \tan \theta + \dfrac{dr}{d\theta}\right)\right] - \tan \theta}{1 + \left[\left(r + \tan \theta \dfrac{dr}{d\theta}\right)\Big/\left(-r \tan \theta + \dfrac{dr}{d\theta}\right)\right] \tan \theta}$$

$$= \frac{r + \tan \theta \dfrac{dr}{d\theta} + r \tan^2 \theta - \tan \theta \dfrac{dr}{d\theta}}{-r \tan \theta + \dfrac{dr}{d\theta} + r \tan \theta + \tan^2 \theta \dfrac{dr}{d\theta}}$$

$$= \frac{r(1 + \tan^2 \theta)}{(1 + \tan^2 \theta) \dfrac{dr}{d\theta}}$$

or

$$\tan \psi = \frac{r}{dr/d\theta} \tag{14}$$

▶ **Example 3**   For the cardioid $r = 1 - \cos \theta$, find the angle ψ between the tangent line and radial line at the points where the cardioid crosses the y-axis.

Solution.   For the given cardioid we have,

$$\tan \psi = \frac{r}{dr/d\theta} = \frac{1 - \cos \theta}{\sin \theta}$$

so that from the trigonometric identity

$$\tan \frac{\theta}{2} = \frac{1 - \cos \theta}{\sin \theta}$$

(derive) it follows that

$$\tan \psi = \tan \frac{\theta}{2}$$

If $0 \leq \theta < 2\pi$, this implies that

$$\psi = \frac{\theta}{2}$$

since $0 \leq \psi < \pi$. In particular, where the cardioid crosses the y-axis we have $\theta = \pi/2$ and $\theta = 3\pi/2$, so the angles from the radial lines to the tangent lines at these points are $\psi = \pi/4$ and $\psi = 3\pi/4$, respectively (Figure 13.5.4).

◀

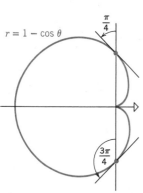

$r = 1 - \cos \theta$

Figure 13.5.4

In Section 13.2 we remarked informally that if a polar curve $r = f(\theta)$ passes through the origin where $\theta = \theta_0$, then the line $\theta = \theta_0$ is tangent to the curve at the origin. To see that this is so, we need only observe that in order for the curve $r = f(\theta)$ to pass through the origin when $\theta = \theta_0$, we must have $r = 0$ for this value of θ. Substituting $r = 0$ in (14) yields

$$\tan \psi = 0$$

(provided $dr/d\theta \neq 0$), which tells us that the angle ψ between the radial line $\theta = \theta_0$ and the tangent line is zero. Thus, the tangent line coincides with the line $\theta = \theta_0$.

The arc length formula for a parametric curve can be used to obtain a formula for the arc length of a polar curve.

13.5.1 THEOREM
Arc Length of a Polar Curve

If a curve has the polar equation $r = f(\theta)$, where $f'(\theta)$ is continuous for $a \le \theta \le b$, then its arc length L from $\theta = a$ to $\theta = b$ is

$$L = \int_a^b \sqrt{[f(\theta)]^2 + [f'(\theta)]^2}\, d\theta \tag{15}$$

or equivalently

$$L = \int_a^b \sqrt{r^2 + \left(\frac{dr}{d\theta}\right)^2}\, d\theta \tag{16}$$

Proof. It follows from (2) that (verify):

$$\left(\frac{dx}{d\theta}\right)^2 + \left(\frac{dy}{d\theta}\right)^2 = [f(\theta)]^2 + [f'(\theta)]^2$$

Thus, from (7) of Theorem 13.4.1 with θ in place of t, we obtain (15). ∎

▶ **Example 4** Find the arc length of the spiral $r = e^\theta$ between $\theta = 0$ and $\theta = 1$.

Solution.

$$L = \int_a^b \sqrt{r^2 + \left(\frac{dr}{d\theta}\right)^2}\, d\theta = \int_0^1 \sqrt{(e^\theta)^2 + (e^\theta)^2}\, d\theta$$

$$= \int_0^1 \sqrt{2}\, e^\theta\, d\theta$$

$$= \sqrt{2}\, e^\theta \Big]_0^1$$

$$= \sqrt{2}(e - 1) \qquad ◀$$

▶ **Example 5** Find the total arc length of the cardioid $r = 1 + \cos\theta$.

Solution. The cardioid is traced out once as θ varies from $\theta = 0$ to $\theta = 2\pi$. Thus,

$$L = \int_a^b \sqrt{r^2 + \left(\frac{dr}{d\theta}\right)^2}\, d\theta = \int_0^{2\pi} \sqrt{(1 + \cos\theta)^2 + (-\sin\theta)^2}\, d\theta$$

$$= \sqrt{2} \int_0^{2\pi} \sqrt{1 + \cos\theta}\, d\theta$$

From the identity $1 + \cos\theta = 2\cos^2\tfrac{1}{2}\theta$, we obtain

$$L = 2\int_0^{2\pi} \sqrt{\cos^2\tfrac{1}{2}\theta}\, d\theta = 2\int_0^{2\pi} |\cos\tfrac{1}{2}\theta|\, d\theta \qquad (17)$$

Since

$$\cos\tfrac{1}{2}\theta \geq 0 \qquad \text{when} \qquad 0 \leq \theta \leq \pi$$

and

$$\cos\tfrac{1}{2}\theta \leq 0 \qquad \text{when} \qquad \pi \leq \theta \leq 2\pi$$

we may rewrite (17) as

$$L = 2\left[\int_0^{\pi}\cos\tfrac{1}{2}\theta\, d\theta - \int_\pi^{2\pi}\cos\tfrac{1}{2}\theta\, d\theta\right]$$

$$= 4\sin\tfrac{1}{2}\theta\,\Big]_0^{\pi} - 4\sin\tfrac{1}{2}\theta\,\Big]_\pi^{2\pi}$$

$$= 8$$

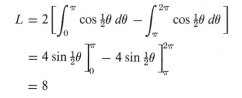

$r = 1 + \cos\theta$

Figure 13.5.5

Alternate Solution. Since the cardioid is symmetric about the x-axis (Figure 13.5.5), the entire arc length is twice the arc length from $\theta = 0$ to $\theta = \pi$. Thus, proceeding as above, we obtain

$$L = 2\sqrt{2}\int_0^{\pi}\sqrt{1 + \cos\theta}\, d\theta = 4\int_0^{\pi}\cos\tfrac{1}{2}\theta\, d\theta = 8 \qquad \blacktriangleleft$$

▶ Exercise Set 13.5

In Exercises 1–6, find the slope of the tangent to the curve at the point with the given value of θ.

1. $r = 2\cos\theta$; $\theta = \pi/3$.

2. $r = 1 + \sin\theta$; $\theta = \pi/4$.

3. $r = 1/\theta$; $\theta = 2$.

4. $r = a\sec 2\theta$; $\theta = \pi/6$.

5. $r = \cos 3\theta$; $\theta = 3\pi/4$.

6. $r = 4 - 3\sin\theta$; $\theta = \pi$.

In Exercises 7–12, find $\tan\psi$ at the point on the curve with the given value of θ.

7. $r = 3(1 - \cos\theta)$; $\theta = \pi/2$.

8. $r = 2\theta$; $\theta = 1$.

9. $r = 5\sin\theta$; $\theta = \pi$.

10. $r = \tan\theta$; $\theta = 3\pi/4$.

11. $r = 4\sin^2\theta$; $\theta = 5\pi/6$.

12. $r = \sin\theta/(1 + \cos\theta)$; $\theta = \pi/3$.

In Exercises 13–19, find the arc length of the curve.

13. $r = e^{3\theta}$ from $\theta = 0$ to $\theta = 2$.

14. The entire circle $r = a$.

15. The entire circle $r = 2a\cos\theta$.

16. $r = \sin^2(\theta/2)$ from $\theta = 0$ to $\theta = \pi$.

17. $r = a\theta^2$ from $\theta = 0$ to $\theta = \pi$.

18. $r = \sin^3(\theta/3)$ from $\theta = 0$ to $\theta = \pi/2$.

19. The entire cardioid $r = a(1 - \cos\theta)$. [*Hint:* See Example 5.]

20. Find all points on the cardioid $r = a(1 + \cos\theta)$ where the tangent is
(a) horizontal   (b) vertical.

21. Find all points on the limaçon $r = 1 - 2\sin\theta$ where the tangent is horizontal.

22. Prove: If the polar curves $r = f_1(\theta)$ and $r = f_2(\theta)$ intersect at a point P, and if β is the smallest nonnegative angle between the tangent lines at P, then

$$\tan\beta = \left| \frac{\tan\psi_2 - \tan\psi_1}{1 + \tan\psi_1 \tan\psi_2} \right|$$

where ψ_1 and ψ_2 are evaluated at P.

23. Show that the curves $r = \sin 2\theta$ and $r = \cos\theta$ intersect at $(\sqrt{3}/2, \pi/6)$ and use the result of Exercise 22 to find the smallest nonnegative angle

between their tangent lines at the point of intersection.

24. Find all points of intersection of the curves $r = \cos\theta$ and $r = 1 - \cos\theta$, and use the result of Exercise 22 to find the smallest nonnegative angle between the tangent lines at each point of intersection.

25. Prove that ψ is the same at each point of the curve $r = e^{a\theta}$.

26. Prove: At all points of intersection of the cardioids $r = a(1 + \cos\theta)$ and $r = b(1 - \cos\theta)$ (excluding the origin), the tangent lines are perpendicular.

▶ SUPPLEMENTARY EXERCISES

Find the rectangular coordinates of the points with the given polar coordinates.

1. (a) $(-2, 4\pi/3)$ (b) $(2, -\pi/2)$
   (c) $(0, -\pi)$ (d) $(-\sqrt{2}, -\pi/4)$
   (e) $(3, \pi)$ (f) $(1, \tan^{-1}(-\tfrac{4}{3}))$.

2. In parts (a)–(c), points are given in rectangular coordinates. Express them in polar coordinates in three ways:
   (i)   with $r \geq 0$ and $0 \leq \theta < 2\pi$;
   (ii) with $r \geq 0$ and $-\pi < \theta \leq \pi$;
   (iii) with $r \leq 0$ and $0 \leq \theta < 2\pi$.
   (a) $(-\sqrt{3}, -1)$ (b) $(-3, 0)$
   (c) $(1, -1)$.

3. Sketch the region in polar coordinates determined by the given inequalities.
   (a) $1 \leq r \leq 2$, $\cos\theta \leq 0$
   (b) $-1 \leq r \leq 1$, $\pi/4 \leq \theta \leq \pi/2$.

In Exercises 4–11, identify the curve by transforming to rectangular coordinates.

4. $r = 2/(1 - \cos\theta)$.   **5.** $r^2 \sin(2\theta) = 1$.

6. $r = \pi/2$.   **7.** $r = -4\csc\theta$.

8. $r = 6/(3 - \sin\theta)$.   **9.** $\theta = \pi/3$.

10. $r = 2\sin\theta + 3\cos\theta$.   **11.** $r = 0$.

In Exercises 12–15, express the given equation in polar coordinates.

12. $x^2 + y^2 = kx$.   **13.** $x = -3$.

14. $y^2 = 4x$.   **15.** $y = 3x$.

In Exercises 16–23, sketch the curve in polar coordinates.

16. $r = -4\sin 3\theta$.   **17.** $r = -1 - 2\cos\theta$.

18. $r = 5\cos\theta$.   **19.** $r = 4 - \sin\theta$.

20. $r = 3(\cos\theta - 1)$.   **21.** $r = \theta/\pi$ $(\theta \geq 0)$.

22. $r = \sqrt{2}\cos(\theta/2)$.   **23.** $r = e^{-\theta/\pi}$ $(\theta \geq 0)$.

In Exercises 24–26, sketch the curves in the same polar coordinate system, and find all points of intersection.

24. $r = 3\cos\theta$, $r = 1 + \cos\theta$.

25. $r = a\cos(2\theta)$, $r = a/2$ $(a > 0)$.

26. $r = 2\sin\theta$, $r = 2 + 2\cos\theta$.

In Exercises 27–29, set up, but *do not evaluate*, definite integrals for the stated area and arc length.

27. (a) The area inside both the circle and cardioid in Exercise 24.
   (b) The arc length of that part of the cardioid outside the circle in Exercise 24.

28. (a) The area inside the rose and outside the circle in Exercise 25.
   (b) The arc length of that part of the rose lying inside the circle in Exercise 25.

29. (a) The area inside the circle and outside the cardioid in Exercise 26.
   (b) The arc length of that portion of the circle lying inside the cardioid in Exercise 26.

In Exercises 30–33, find the area of the region described.

30. One leaf of the rose $r = a \sin 3\theta$.

31. The region outside the circle $r = a$ and inside the lemniscate $r^2 = 2a^2 \cos 2\theta$.

32. The region in part (a) of Exercise 28.

33. The region in part (a) of Exercise 29.

In Exercises 34–37:
(a) Sketch the curve and label the direction of increasing parameter.
(b) Use the parametric equations to find dy/dx, d^2y/dx^2, and the equation of the tangent line at the point on the curve corresponding to the parameter value t_0 (or θ_0).

34. $x = 3 - t^2$, $y = 2 + t$, $0 \le t \le 3$; $t_0 = 1$.

35. $x = 1 + 3 \cos \theta$, $y = -1 + 2 \sin \theta$, $0 \le \theta \le \pi$; $\theta_0 = \pi/2$.

36. $x = 2 \tan \theta$, $y = \sec \theta$, $-\pi/2 < \theta < \pi/2$; $\theta_0 = \pi/3$.

37. $x = 1/t$, $y = \ln t$, $1 \le t \le e$; $t_0 = 2$.

In Exercises 38–43, find the arc length of the curve described.

38. $x = 2t^3$, $y = 3t^2$, $-4 \le t \le 4$.

39. $x = \ln \cos 2t$, $y = 2t$, $0 \le t \le \pi/6$.

40. $x = 3 \cos t - 1$, $y = 3 \sin t + 4$, $0 \le t \le \pi$.

41. $x = 3t^2$, $y = t^3 - 3t$, $0 \le t \le 1$.

42. $x = 1 - \cos t$, $y = t - \sin t$, $-\pi \le t \le \pi$.

43. $r = e^\theta$, $0 \le \theta \le 2\pi$.

In Exercises 44–46, $x(t)$ and $y(t)$ describe the motion of a particle. Find the coordinates of the particle when the instantaneous direction of motion is: (a) horizontal (b) vertical.

44. $x = -2t^2$, $y = t^3 - 3t + 5$.

45. $x = 1 - 2 \sin t$, $y = t + 2 \cos t$, $0 \le t \le \pi$.

46. $x = \ln t$, $y = t^2 - 4t$, $t > 0$.

47. At what instant does the trajectory described in Exercise 46 have a point of inflection?

48. Find a set of parametric equations for
   (a) the line $y = 2x + 3$
   (b) the ellipse $4(x - 2)^2 + y^2 = 4$.

Exercises 49–50 refer to the optional Section 13.5.

49. For the cardioid $r = 2(1 + \cos \theta)$, find the slope of the tangent line and the angle ψ when $\theta = \pi/2$.

50. For the circle $r = 4 \sin \theta$, show that $\psi = \theta$ if $0 \le \theta < \pi$ and find a formula for the inclination angle ϕ as a function of θ for $0 \le \theta < \pi/2$.

14 vectors in the plane

14.1 VECTORS

Many physical quantities like area, length, mass, and temperature are completely described once the magnitude of the quantity is given. Such quantities are called *scalars*. Other physical quantities, called ***vectors,*** are not completely determined until both a magnitude and a direction are specified. For example, wind movement is usually described by giving the speed and the direction, say 20 mph northeast. The wind speed and wind direction together form a vector quantity called the wind *velocity*. Other examples of vectors are force and displacement. In this section we will develop the basic mathematical properties of vectors.

Vectors can be represented geometrically as directed line segments or arrows in two- or three-dimensional space; the direction of the arrow specifies the direction of the vector and the length of the arrow describes its magnitude. The tail of the arrow is called the ***initial point*** of the vector, and the tip of the arrow the ***terminal point.*** We shall denote vectors by lowercase boldface type like $\mathbf{a}, \mathbf{k}, \mathbf{v}, \mathbf{w}$, and $\mathbf{x}$. When discussing vectors, we shall refer to numbers as ***scalars.*** Scalars will be denoted by ordinary lowercase type like a, k, v, w, and x.

If, as in Figure 14.1.1a, the initial point of a vector $\mathbf{v}$ is A and the terminal point is B, we write

$$\mathbf{v} = \overrightarrow{AB}$$

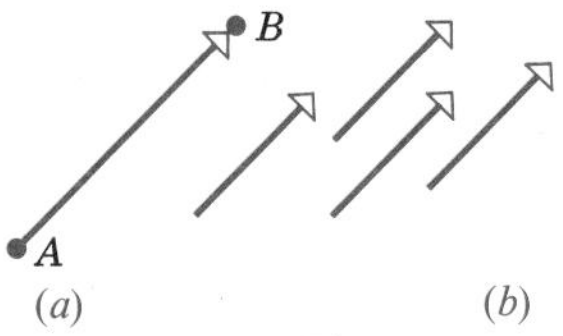

(a)

(b)

Figure 14.1.1

Vectors having the same length and same direction, such as those in Figure 14.1.1b, are called ***equivalent.*** Since we want a vector to be determined solely by its length and direction, equivalent vectors are regarded as **equal** even though they may be located in different positions. If **v** and **w** are equivalent we write

$$\mathbf{v} = \mathbf{w}$$

14.1.1   DEFINITION   If **v** and **w** are any two vectors, then the sum **v** + **w** is the vector determined as follows. Position the vector **w** so that its initial point coincides with the terminal point of **v**. The vector **v** + **w** is represented by the arrow from the initial point of **v** to the terminal point of **w** (Figure 14.1.2*a*).

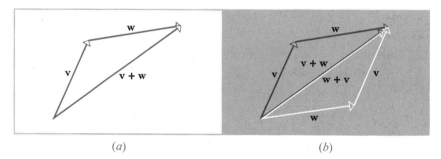

Figure 14.1.2   (*a*)   (*b*)

In Figure 14.1.2*b* we have constructed two sums, **v** + **w** (blue arrows) and **w** + **v** (white arrows). It is evident that

$$\mathbf{v} + \mathbf{w} = \mathbf{w} + \mathbf{v}$$

and that the sum coincides with the diagonal of the parallelogram determined by **v** and **w** when these vectors are located so they have the same initial point.

The vector of length zero is called the **zero vector** and is denoted by **0**. We define

$$\mathbf{0} + \mathbf{v} = \mathbf{v} + \mathbf{0} = \mathbf{v}$$

Since there is no natural direction for the zero vector, we shall agree that it can be assigned any direction that is convenient for the problem at hand.

If **v** is any nonzero vector, then −**v**, the **negative** of **v**, is defined to be the vector having the same magnitude as **v**, but oppositely directed (Figure 14.1.3). This vector has the property

$$\mathbf{v} + (-\mathbf{v}) = \mathbf{0}$$

Figure 14.1.3   (Why?) In addition, we define −**0** = **0**.

14.1.2 DEFINITION If **v** and **w** are any two vectors, then ***subtraction*** of **w** from **v** is defined by

$$\mathbf{v} - \mathbf{w} = \mathbf{v} + (-\mathbf{w})$$

(See Figure 14.1.4a).

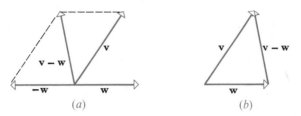

Figure 14.1.4   (a)   (b)

To obtain **v** − **w** without constructing −**w**, position **v** and **w** so their initial points coincide; the vector from the terminal point of **w** to the terminal point of **v** is then the vector **v** − **w** (Figure 14.1.4b).

14.1.3 DEFINITION If **v** is a nonzero vector and k is a nonzero real number (scalar), then the ***product*** $k\mathbf{v}$ is defined to be the vector whose length is $|k|$ times the length of **v** and whose direction is the same as that of **v** if $k > 0$ and opposite to that of **v** if $k < 0$. We define $k\mathbf{v} = \mathbf{0}$ if $k = 0$ or $\mathbf{v} = \mathbf{0}$.

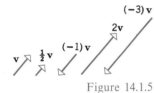

Figure 14.1.5

Figure 14.1.5 illustrates the relation between a vector **v** and the vectors $\frac{1}{2}\mathbf{v}$, $(-1)\mathbf{v}$, $2\mathbf{v}$, and $(-3)\mathbf{v}$. Note that the vector $(-1)\mathbf{v}$ has the same length as **v** but is oppositely directed. Thus, $(-1)\mathbf{v}$ is just the negative of **v**; that is,

$$(-1)\mathbf{v} = -\mathbf{v}$$

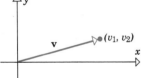

Figure 14.1.6

Problems involving vectors can often be simplified by introducing a rectangular coordinate system. For the moment we shall restrict the discussion to vectors in 2-space (the plane). Let **v** be any vector in the plane with its initial point at the origin of a rectangular coordinate system (Figure 14.1.6). The coordinates (v_1, v_2) of the terminal point of **v** are called the ***components*** of **v**, and we write

$$\mathbf{v} = \langle v_1, v_2 \rangle$$

If equivalent vectors, **v** and **w**, are located so their initial points fall at the origin, then it is obvious that their terminal points must coincide (since the vectors have the same length and direction). The vectors thus have the same components. Conversely, vectors with the same components are equivalent, since they have the same length and same direction. In summary, the vectors

$$\mathbf{v} = \langle v_1, v_2 \rangle \qquad \text{and} \qquad \mathbf{w} = \langle w_1, w_2 \rangle$$

are equivalent if and only if

$$v_1 = w_1 \quad \text{and} \quad v_2 = w_2$$

The operations of vector addition and multiplication by scalars are easy to carry out in terms of components. As illustrated in Figure 14.1.7, if

$$\mathbf{v} = \langle v_1, v_2 \rangle \quad \text{and} \quad \mathbf{w} = \langle w_1, w_2 \rangle$$

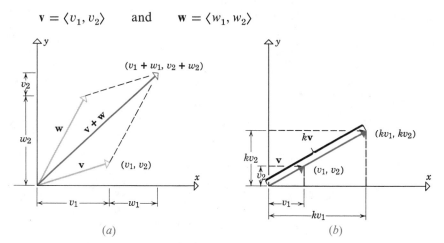

Figure 14.1.7   (a)   (b)

and k is any scalar, then

$$\mathbf{v} + \mathbf{w} = \langle v_1 + w_1, v_2 + w_2 \rangle \tag{1a}$$
$$k\mathbf{v} = \langle kv_1, kv_2 \rangle \tag{1b}$$

Since $\mathbf{v} - \mathbf{w} = \mathbf{v} + (-1)\mathbf{w}$, it follows from formulas (1a) and (1b) that

$$\mathbf{v} - \mathbf{w} = \langle v_1 - w_1, v_2 - w_2 \rangle$$

(Verify.)

▶ Example 1 If

$$\mathbf{v} = \langle 1, -2 \rangle \quad \text{and} \quad \mathbf{w} = \langle 7, 6 \rangle$$

then

$$\mathbf{v} + \mathbf{w} = \langle 1, -2 \rangle + \langle 7, 6 \rangle = \langle 1 + 7, -2 + 6 \rangle = \langle 8, 4 \rangle$$
$$4\mathbf{v} = 4\langle 1, -2 \rangle = \langle 4(1), 4(-2) \rangle = \langle 4, -8 \rangle$$
$$-\mathbf{v} = (-1)\mathbf{v} = (-1)\langle 1, -2 \rangle = \langle -1, 2 \rangle$$
$$\mathbf{v} - \mathbf{w} = \langle 1, -2 \rangle - \langle 7, 6 \rangle = \langle 1 - 7, -2 - 6 \rangle = \langle -6, -8 \rangle \quad ◀$$

Sometimes a vector is positioned so that its initial point is not at the origin. If the coordinates of the initial and terminal points are known, then the components of the vector can be obtained using the following theorem.

14.1.4 THEOREM *If the vector $\overrightarrow{P_1P_2}$ has initial point $P_1(x_1, y_1)$ and terminal point $P_2(x_2, y_2)$, then*

$$\overrightarrow{P_1P_2} = \langle x_2 - x_1, y_2 - y_1 \rangle$$

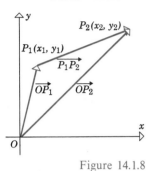

Figure 14.1.8

Proof. The vector $\overrightarrow{P_1P_2}$ is the difference of vectors $\overrightarrow{OP_2}$ and $\overrightarrow{OP_1}$ (Figure 14.1.8). Thus,

$$\overrightarrow{P_1P_2} = \overrightarrow{OP_2} - \overrightarrow{OP_1} = \langle x_2, y_2 \rangle - \langle x_1, y_1 \rangle = \langle x_2 - x_1, y_2 - y_1 \rangle \quad \blacksquare$$

To paraphrase this theorem loosely, *the components of any vector are the coordinates of its terminal point minus the coordinates of its initial point.*

▶ **Example 2** Find the components of the vector with initial point $P_1(1, 3)$ and terminal point $P_2(4, -2)$.

Solution.

$$\overrightarrow{P_1P_2} = \langle 4 - 1, -2 - 3 \rangle = \langle 3, -5 \rangle \qquad ◀$$

The following theorem shows that many of the familiar rules of ordinary arithmetic also hold for vector arithmetic.

14.1.5 THEOREM *For any vectors $\mathbf{u}$, $\mathbf{v}$, and $\mathbf{w}$ and any scalars k and l, the following relationships hold:*

(a) $\mathbf{u} + \mathbf{v} = \mathbf{v} + \mathbf{u}$
(b) $(\mathbf{u} + \mathbf{v}) + \mathbf{w} = \mathbf{u} + (\mathbf{v} + \mathbf{w})$
(c) $\mathbf{u} + \mathbf{0} = \mathbf{0} + \mathbf{u} = \mathbf{u}$
(d) $\mathbf{u} + (-\mathbf{u}) = \mathbf{0}$
(e) $k(l\mathbf{u}) = (kl)\mathbf{u}$
(f) $k(\mathbf{u} + \mathbf{v}) = k\mathbf{u} + k\mathbf{v}$
(g) $(k + l)\mathbf{u} = k\mathbf{u} + l\mathbf{u}$
(h) $1\mathbf{u} = \mathbf{u}$

Before discussing the proof, we note that we have developed two approaches to vectors: *geometric*, in which vectors are represented by arrows or directed line segments, and *analytic*, in which vectors are represented by pairs of numbers called components. As a consequence, the results in this theorem can be established either geometrically or analytically. To illustrate, we shall prove part (b) both ways. The remaining proofs are left as exercises.

Proof of (b) (analytic). Let $\mathbf{u} = \langle u_1, u_2 \rangle$, $\mathbf{v} = \langle v_1, v_2 \rangle$, and $\mathbf{w} = \langle w_1, w_2 \rangle$. Then

$$
\begin{aligned}
(\mathbf{u} + \mathbf{v}) + \mathbf{w} &= (\langle u_1, u_2 \rangle + \langle v_1, v_2 \rangle) + \langle w_1, w_2 \rangle \\
&= \langle u_1 + v_1, u_2 + v_2 \rangle + \langle w_1, w_2 \rangle \\
&= \langle (u_1 + v_1) + w_1, (u_2 + v_2) + w_2 \rangle \\
&= \langle u_1 + (v_1 + w_1), u_2 + (v_2 + w_2) \rangle \\
&= \langle u_1, u_2 \rangle + \langle v_1 + w_1, v_2 + w_2 \rangle \\
&= \mathbf{u} + (\mathbf{v} + \mathbf{w})
\end{aligned}
$$

Proof of (b) (geometric). Let $\mathbf{u}, \mathbf{v},$ and $\mathbf{w}$ be represented by $\overrightarrow{PQ}, \overrightarrow{QR},$ and $\overrightarrow{RS}$ as shown in Figure 14.1.9. Then

$$
\mathbf{v} + \mathbf{w} = \overrightarrow{QS} \quad \text{and} \quad \mathbf{u} + (\mathbf{v} + \mathbf{w}) = \overrightarrow{PS}
$$
$$
\mathbf{u} + \mathbf{v} = \overrightarrow{PR} \quad \text{and} \quad (\mathbf{u} + \mathbf{v}) + \mathbf{w} = \overrightarrow{PS}
$$

Therefore,

$$
(\mathbf{u} + \mathbf{v}) + \mathbf{w} = \mathbf{u} + (\mathbf{v} + \mathbf{w})
$$

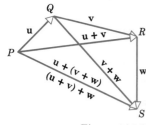

Figure 14.1.9

REMARK. In light of part (*b*) of this theorem, the symbol $\mathbf{u} + \mathbf{v} + \mathbf{w}$ is unambiguous since the same result is obtained no matter where parentheses are inserted. Moreover, if the vectors $\mathbf{u}, \mathbf{v},$ and $\mathbf{w}$ are placed "tip to tail," then the sum $\mathbf{u} + \mathbf{v} + \mathbf{w}$ is the vector from the initial point of $\mathbf{u}$ to the terminal point of $\mathbf{w}$ (Figure 14.1.9).

The *length* of a vector $\mathbf{v}$ is also called the *norm* of $\mathbf{v}$ and is denoted by $\|\mathbf{v}\|$. Motivated by the Theorem of Pythagoras, the norm of a vector $\mathbf{v} = \langle v_1, v_2 \rangle$ is defined as

$$
\|\mathbf{v}\| = \sqrt{v_1{}^2 + v_2{}^2} \tag{2}
$$

(See Figure 14.1.10.)

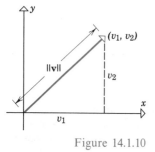

Figure 14.1.10

▶ Example 3 The norm of the vector $\mathbf{v} = \langle -2, 3 \rangle$ is

$$
\|\mathbf{v}\| = \sqrt{(-2)^2 + 3^2} = \sqrt{13} \qquad \blacktriangleleft
$$

Recall from Definition 14.1.3 that the length of $k\mathbf{v}$ is $|k|$ times the length of $\mathbf{v}$. Expressed as an equation, this statement says that

$$
\|k\mathbf{v}\| = |k|\,\|\mathbf{v}\| \tag{3}
$$

It follows from (3) that if a nonzero vector **v** is multiplied by $1/\|\mathbf{v}\|$ (the reciprocal of its length), then the result is a vector of length 1 in the same direction as **v**. (Why?) This process of multiplying **v** by $1/\|\mathbf{v}\|$ to obtain a vector of length 1 is called *normalizing* **v**.

▶ **Example 4** A vector of length 1 in the same direction as

$$\mathbf{v} = \langle 3, 4 \rangle$$

is

$$\frac{1}{\|\mathbf{v}\|}\mathbf{v} = \frac{1}{\sqrt{3^2 + 4^2}}\langle 3, 4 \rangle = \langle \tfrac{3}{5}, \tfrac{4}{5} \rangle \qquad ◀$$

A vector of length 1 is called a *unit vector*. Of special importance are the unit vectors

$$\mathbf{i} = \langle 1, 0 \rangle \qquad \text{and} \qquad \mathbf{j} = \langle 0, 1 \rangle$$

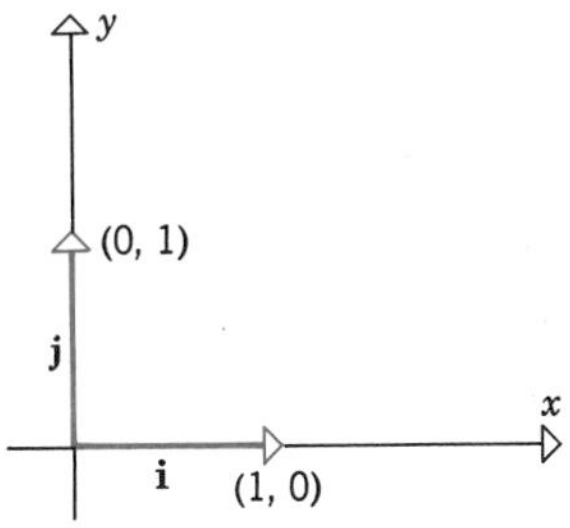

Figure 14.1.11

which run along the positive x-axis and positive y-axis of a rectangular coordinate system (Figure 14.1.11). Every vector $\mathbf{v} = \langle v_1, v_2 \rangle$ is expressible in terms of **i** and **j**, since we can write

$$\mathbf{v} = \langle v_1, v_2 \rangle = \langle v_1, 0 \rangle + \langle 0, v_2 \rangle = v_1\langle 1, 0 \rangle + v_2\langle 0, 1 \rangle = v_1\mathbf{i} + v_2\mathbf{j}$$

For example,

$$\langle 2, 3 \rangle = 2\mathbf{i} + 3\mathbf{j}$$
$$\langle 5, -4 \rangle = 5\mathbf{i} + (-4)\mathbf{j} = 5\mathbf{i} - 4\mathbf{j}$$
$$\langle 0, 2 \rangle = 0\mathbf{i} + 2\mathbf{j} = 2\mathbf{j}$$
$$\langle -4, 0 \rangle = -4\mathbf{i} + 0\mathbf{j} = -4\mathbf{i}$$
$$\langle 0, 0 \rangle = 0\mathbf{i} + 0\mathbf{j} = \mathbf{0}$$

▶ **Example 5**
$$(3\mathbf{i} + 2\mathbf{j}) + (4\mathbf{i} + \mathbf{j}) = 7\mathbf{i} + 3\mathbf{j}$$
$$5(6\mathbf{i} - 2\mathbf{j}) = 30\mathbf{i} - 10\mathbf{j}$$
$$\|2\mathbf{i} - 3\mathbf{j}\| = \sqrt{2^2 + (-3)^2} = \sqrt{13} \qquad ◀$$

▶ **Example 6** If **u** is a unit vector and if ϕ is the angle from the positive x-axis to **u**, then

$$\mathbf{u} = \langle \cos\phi, \sin\phi \rangle = (\cos\phi)\mathbf{i} + (\sin\phi)\mathbf{j}$$

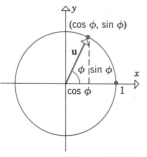

Figure 14.1.12

(Figure 14.1.12); and if $\mathbf{v}$ is an arbitrary vector making an angle ϕ with the positive x-axis, then $\mathbf{v}$ has the same direction as $\mathbf{u}$, but is $\|\mathbf{v}\|$ times as long. Thus,

$$\mathbf{v} = \|\mathbf{v}\| \langle \cos \phi, \sin \phi \rangle$$

or

$$\mathbf{v} = \|\mathbf{v}\| \cos \phi \mathbf{i} + \|\mathbf{v}\| \sin \phi \mathbf{j}$$

For example, the vector of length 2 making an angle of $\pi/4$ with the positive x-axis is

$$\mathbf{v} = 2 \cos \frac{\pi}{4}\mathbf{i} + 2 \sin \frac{\pi}{4}\mathbf{j} = \sqrt{2}\mathbf{i} + \sqrt{2}\mathbf{j} \qquad \blacktriangleleft$$

▶ Exercise Set 14.1

In Exercises 1 and 2, sketch the vectors with the initial point at the origin.

1. (a) $\langle 2, 5 \rangle$
    (b) $\langle -5, -4 \rangle$
    (c) $\langle 2, 0 \rangle$.

2. (a) $\langle -3, 7 \rangle$
    (b) $\langle 6, -2 \rangle$
    (c) $\langle 0, -8 \rangle$.

In Exercises 3 and 4, sketch the vectors with the initial point at the origin.

3. (a) $-5\mathbf{i} + 3\mathbf{j}$
    (b) $3\mathbf{i} - 2\mathbf{j}$
    (c) $-6\mathbf{j}$.

4. (a) $4\mathbf{i} + 2\mathbf{j}$
    (b) $-2\mathbf{i} - \mathbf{j}$
    (c) $4\mathbf{i}$.

In Exercises 5 and 6, find the components of the vector $\overrightarrow{P_1 P_2}$.

5. (a) $P_1(3, 5)$, $P_2(2, 8)$
    (b) $P_1(7, -2)$, $P_2(0, 0)$
    (c) $P_1(-6, -2)$, $P_2(-4, -1)$
    (d) $P_1(0, 0)$, $P_2(-8, 7)$.

6. (a) $P_1(1, 3)$, $P_2(4, 1)$
    (b) $P_1(6, -4)$, $P_2(0, 0)$
    (c) $P_1(-8, -1)$, $P_2(-3, -2)$
    (d) $P_1(0, 0)$, $P_2(-3, -5)$.

7. Find the terminal point of $\mathbf{v} = 3\mathbf{i} - 2\mathbf{j}$ if the initial point is $(1, -2)$.

8. Find the terminal point of $\mathbf{v} = \langle 7, 6 \rangle$ if the initial point is $(2, -1)$.

9. Find the initial point of $\mathbf{v} = \langle -2, 4 \rangle$ if the terminal point is $(2, 0)$.

10. Let $\mathbf{u} = \langle 1, 3 \rangle$, $\mathbf{v} = \langle 2, 1 \rangle$, and $\mathbf{w} = \langle 4, -1 \rangle$. Find:
    (a) $\mathbf{u} - \mathbf{w}$
    (b) $7\mathbf{v} + 3\mathbf{w}$
    (c) $-\mathbf{w} + \mathbf{v}$
    (d) $3(\mathbf{u} - 7\mathbf{v})$
    (e) $-3\mathbf{v} - 8\mathbf{w}$
    (f) $2\mathbf{v} - (\mathbf{u} + \mathbf{w})$.

11. Let $\mathbf{u} = 2\mathbf{i} + 3\mathbf{j}$, $\mathbf{v} = 4\mathbf{i}$, $\mathbf{w} = -\mathbf{i} - 2\mathbf{j}$. Find:
    (a) $\mathbf{w} - \mathbf{v}$
    (b) $6\mathbf{u} + 4\mathbf{w}$
    (c) $-\mathbf{v} - 2\mathbf{w}$
    (d) $4(3\mathbf{u} + \mathbf{v})$
    (e) $-8(\mathbf{v} + \mathbf{w}) + 2\mathbf{u}$
    (f) $3\mathbf{w} - (\mathbf{v} - \mathbf{w})$.

In Exercises 12 and 13, compute the norm of $\mathbf{v}$.

12. (a) $\mathbf{v} = \langle 3, 4 \rangle$
    (b) $\mathbf{v} = -\mathbf{i} + 7\mathbf{j}$
    (c) $\mathbf{v} = -3\mathbf{j}$.

13. (a) $\mathbf{v} = \langle 1, -1 \rangle$
    (b) $\mathbf{v} = \langle 2, 0 \rangle$
    (c) $\mathbf{v} = \sqrt{2}\mathbf{i} - \sqrt{7}\mathbf{j}$.

14. Let $\mathbf{u} = \langle 1, -3 \rangle$, $\mathbf{v} = \langle 1, 1 \rangle$, and $\mathbf{w} = \langle 2, -4 \rangle$. Find:
    (a) $\|\mathbf{u} + \mathbf{v}\|$
    (b) $\|\mathbf{u}\| + \|\mathbf{v}\|$
    (c) $\|-2\mathbf{u}\| + 2\|\mathbf{v}\|$
    (d) $\|3\mathbf{u} - 5\mathbf{v} + \mathbf{w}\|$.

15. Let $\mathbf{u} = \langle 2, -5 \rangle$, $\mathbf{v} = \langle 2, 0 \rangle$, and $\mathbf{w} = \langle 3, 4 \rangle$. Find:
    (a) $\|\mathbf{v} + \mathbf{w}\|$
    (b) $\|\mathbf{v}\| + \|\mathbf{w}\|$
    (c) $\|-3\mathbf{u}\| + 4\|\mathbf{v}\|$
    (d) $\|\mathbf{u} - \mathbf{v} - \mathbf{w}\|$
    (e) $\frac{1}{\|\mathbf{w}\|}\mathbf{w}$
    (f) $\left\|\frac{1}{\|\mathbf{w}\|}\mathbf{w}\right\|$.

16. Let $\mathbf{u} = \langle -1, 1 \rangle$, $\mathbf{v} = \langle 0, 1 \rangle$, and $\mathbf{w} = \langle 3, 4 \rangle$. Find the vector $\mathbf{x}$ that satisfies $\mathbf{u} - 2\mathbf{x} = \mathbf{x} - \mathbf{w} + 3\mathbf{v}$.

17. Let $\mathbf{u} = \langle 1, 3 \rangle$, $\mathbf{v} = \langle 2, 1 \rangle$, $\mathbf{w} = \langle 4, -1 \rangle$. Find the vector $\mathbf{x}$ that satisfies $2\mathbf{u} - \mathbf{v} + \mathbf{x} = 7\mathbf{x} + \mathbf{w}$.

18. Let $\mathbf{u} = 2\mathbf{i} - \mathbf{j}$ and $\mathbf{v} = 4\mathbf{i} + 2\mathbf{j}$. Find scalars c_1 and c_2 such that $c_1\mathbf{u} + c_2\mathbf{v} = -4\mathbf{j}$.

19. Let $\mathbf{u} = \langle 1, -3 \rangle$ and $\mathbf{v} = \langle -2, 6 \rangle$. Show that there do not exist scalars c_1 and c_2 such that $c_1\mathbf{u} + c_2\mathbf{v} = \langle 3, 5 \rangle$.

20. Let $\mathbf{v} = 4\mathbf{i} - 3\mathbf{j}$. Find all scalars k such that $\|k\mathbf{v}\| = 3$.

21. Verify parts (b), (e), (f), and (g) of Theorem 14.1.5 for $\mathbf{u} = \langle 1, -3 \rangle$, $\mathbf{v} = \langle 6, 6 \rangle$, $\mathbf{w} = \langle -8, 1 \rangle$, $k = 3$, and $l = 6$.

22. Find a vector of length 1 having the same direction as $-\mathbf{i} + 4\mathbf{j}$.

23. Find a vector of length 1 oppositely directed to $3\mathbf{i} - 4\mathbf{j}$.

24. Let $\mathbf{p} = \langle x, y \rangle$. Describe the set of points (x, y) for which $\|\mathbf{p}\| = 1$.

25. Let $\mathbf{p}_0 = \langle x_0, y_0 \rangle$ and $\mathbf{p} = \langle x, y \rangle$. Describe the set of all points (x, y) for which $\|\mathbf{p} - \mathbf{p}_0\| = 1$.

26. Find two unit vectors parallel to the line $y = 3x + 2$.

27. (a) Find two unit vectors parallel to the line $x + y = 4$.
(b) Find two unit vectors perpendicular to the line in (a).

28. Let P be the point $(2, 3)$ and Q the point $(7, -4)$. Use vectors to find the point on the line segment joining P and Q that is $\frac{3}{4}$ of the way from P to Q.

29. For the points P and Q in Exercise 28, use vectors to find the point on the line segment joining P and Q that is $\frac{3}{4}$ of the way from Q to P.

30. Find a unit vector making an angle of $135°$ with the x-axis.

31. Find:
(a) a unit vector making an angle of $\pi/3$ with the positive x-axis;
(b) a vector of length 4 making an angle of $3\pi/4$ with the positive x-axis.

32. Use vectors to find the length of the diagonal of the parallelogram determined by $\mathbf{i} + \mathbf{j}$ and $\mathbf{i} - 2\mathbf{j}$.

33. Use vectors to find the fourth vertex of a parallelogram, three of whose vertices are $(0, 0)$, $(1, 3)$, and $(2, 4)$. [*Note:* There is more than one answer.]

34. Prove: $\|\mathbf{u} + \mathbf{v}\| \leq \|\mathbf{u}\| + \|\mathbf{v}\|$ geometrically.

35. Prove parts (a), (c), and (e) of Theorem 14.1.5 analytically.

36. Prove parts (d), (g), and (h) of Theorem 14.1.5 analytically.

37. Prove part (f) of Theorem 14.1.5 geometrically.

38. Use vectors to prove that the midpoints of the sides of a quadrilateral are the vertices of a parallelogram.

39. Use vectors to prove that the line segment joining the midpoints of two sides of a triangle is parallel to the third side and half as long.

40. Prove: The line passing through (x_0, y_0) and parallel to the nonzero vector $a\mathbf{i} + b\mathbf{j}$ has parametric equations

$$x = x_0 + at$$
$$y = y_0 + bt$$

14.2   VECTOR CALCULUS IN TWO DIMENSIONS

In this section we show how vectors can be used to study curves in the plane. Our work here will have important applications to the study of motion along a plane curve.

Let

$$x = x(t)$$
$$y = y(t)$$

be a pair of parametric equations and for each value of t construct the vector $\mathbf{r}(t)$ having $x(t)$ and $y(t)$ as components; that is,

$$\mathbf{r}(t) = x\mathbf{i} + y\mathbf{j} = x(t)\mathbf{i} + y(t)\mathbf{j}$$

As the parameter t varies, the values of $x(t)$ and $y(t)$ vary and consequently the vector $\mathbf{r}(t)$ also varies with t. We call $\mathbf{r}(t)$ a *vector-valued function* of t.

▶ Example 1 If

$$x = t^2$$
$$y = 3t$$

then

$$\mathbf{r}(t) = x\mathbf{i} + y\mathbf{j} = t^2\mathbf{i} + 3t\mathbf{j}$$

and

$$\mathbf{r}(1) = \mathbf{i} + 3\mathbf{j}$$
$$\mathbf{r}(-2) = 4\mathbf{i} - 6\mathbf{j}$$
$$\mathbf{r}(0) = 0\mathbf{i} + 0\mathbf{j} = \mathbf{0}$$

◀

Vector-valued functions provide a new way to think about curves in the plane. Let

$$\mathbf{r}(t) = x(t)\mathbf{i} + y(t)\mathbf{j}$$

be constructed with its initial point at the origin and its terminal point at $(x(t), y(t))$ (Figure 14.2.1). As the parameter t varies, the tip of the vector $\mathbf{r}(t)$ traces out the curve C whose parametric equations are

$$x = x(t)$$
$$y = y(t)$$

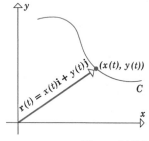

Figure 14.2.1

We call C the **graph** of $\mathbf{r}(t)$, and we call $\mathbf{r}(t)$ the **radius vector** or **position vector** for the curve C. Thus, we may think of the curve C either as the graph of the *two* parametric equations $x = x(t)$, $y = y(t)$ or as the graph of the *single* vector-valued function, $\mathbf{r}(t) = x(t)\mathbf{i} + y(t)\mathbf{j}$.

▶ Example 2 Sketch the graph of the vector-valued function

$$\mathbf{r}(t) = (\cos t)\mathbf{i} + (\sin t)\mathbf{j}, \qquad 0 \le t \le 2\pi$$

Solution. The graph of $\mathbf{r}(t)$ is the same as the graph of the parametric equations

$$x = \cos t$$
$$y = \sin t, \qquad 0 < t \le 2\pi$$

which is a circle of radius 1 centered at the origin. The direction of increasing t is counterclockwise (Figure 14.2.2). ◀

The familiar notions of limit, derivative, and integral may be generalized to vector-valued functions.

14.2.1 DEFINITION   Let lim stand for one of the limits $\lim_{t\to a}$, $\lim_{t\to a^+}$, $\lim_{t\to a^-}$, $\lim_{t\to +\infty}$, or $\lim_{t\to -\infty}$, and let

$$\mathbf{r}(t) = x(t)\mathbf{i} + y(t)\mathbf{j}$$

We define $\lim \mathbf{r}(t)$ by

$$\lim \mathbf{r}(t) = (\lim x(t))\mathbf{i} + (\lim y(t))\mathbf{j}$$

provided $\lim x(t)$ and $\lim y(t)$ exist.

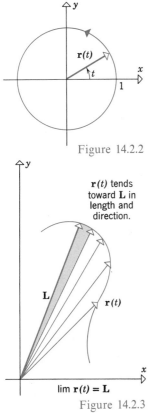

$\mathbf{r}(t) = (\cos t)\mathbf{i} + (\sin t)\mathbf{j}$

Figure 14.2.2

To paraphrase this definition, *the limit of a vector-valued function is computed by taking the limit of each component.* Geometrically, if

$$\lim_{t\to a} \mathbf{r}(t) = \mathbf{L}$$

then the vector $\mathbf{r}(t)$ approaches $\mathbf{L}$ in both length and direction as $t \to a$ (see Figure 14.2.3).

▶ Example 3   Let $\mathbf{r}(t) = t^2\mathbf{i} + 3t\mathbf{j}$. Find $\lim_{t\to 2} \mathbf{r}(t)$.

Solution.

$$\lim_{t\to 2} \mathbf{r}(t) = \lim_{t\to 2} (t^2\mathbf{i} + 3t\mathbf{j}) = (\lim_{t\to 2} t^2)\mathbf{i} + (\lim_{t\to 2} 3t)\mathbf{j} = 4\mathbf{i} + 6\mathbf{j} \quad ◀$$

The definition of a derivative for vector-valued functions is analogous to the definition for real-valued functions.

$\mathbf{r}(t)$ tends toward $\mathbf{L}$ in length and direction.

$\mathbf{L}$

$\mathbf{r}(t)$

$\lim \mathbf{r}(t) = \mathbf{L}$

Figure 14.2.3

14.2.2 DEFINITION   The **derivative** $\mathbf{r}'(t)$ of a vector-valued function $\mathbf{r}(t)$ is defined by

$$\mathbf{r}'(t) = \lim_{\Delta t\to 0} \frac{\mathbf{r}(t + \Delta t) - \mathbf{r}(t)}{\Delta t}$$

provided the limit exists.

To interpret this definition geometrically, let C be the graph of $\mathbf{r}(t)$, and for a fixed value of the parameter t construct the position vectors $\mathbf{r}(t)$ and $\mathbf{r}(t + \Delta t)$. If $\Delta t > 0$, then the tip of the vector $\mathbf{r}(t + \Delta t)$ is in the direction of

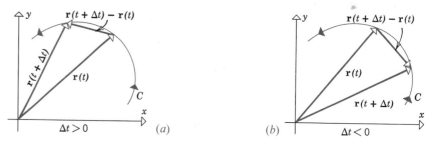

Figure 14.2.4

increasing parameter from the tip of $\mathbf{r}(t)$, while if $\Delta t < 0$ it is the other way around (Figure 14.2.4). In either case, the difference $\mathbf{r}(t + \Delta t) - \mathbf{r}(t)$ coincides with the secant line through the tips of the vectors $\mathbf{r}(t)$ and $\mathbf{r}(t + \Delta t)$. Moreover, since Δt is a scalar, the vector

$$\frac{1}{\Delta t}[\mathbf{r}(t + \Delta t) - \mathbf{r}(t)] = \frac{\mathbf{r}(t + \Delta t) - \mathbf{r}(t)}{\Delta t} \tag{1}$$

also coincides with this secant line. If $\Delta t < 0$, then vector (1) is oppositely directed to $\mathbf{r}(t + \Delta t) - \mathbf{r}(t)$, and if $\Delta t > 0$ these vectors have the same direction. In both cases, vector (1) points in the direction of increasing parameter. As $\Delta t \to 0$, the secant lines through the tips of $\mathbf{r}(t + \Delta t)$ and $\mathbf{r}(t)$ tend toward the tangent line to C at the tip of $\mathbf{r}(t)$. Thus, the vector

$$\mathbf{r}'(t) = \lim_{\Delta t \to 0} \frac{\mathbf{r}(t + \Delta t) - \mathbf{r}(t)}{\Delta t}$$

is tangent to the curve C at the tip of $\mathbf{r}(t)$ and points in the direction of increasing parameter (Figure 14.2.5).

Figure 14.2.5

The derivative $\mathbf{r}'(t)$ is also denoted by

$$\frac{d}{dt}[\mathbf{r}(t)] \text{ or } \frac{d\mathbf{r}}{dt},$$

and if a dependent variable $\mathbf{u} = \mathbf{r}(t)$ is introduced, then the derivative may be denoted by $d\mathbf{u}/dt$.

The following theorem tells us that the derivative of a vector-valued function can be computed by differentiating each component.

14.2.3 THEOREM *If $\mathbf{r}(t) = x(t)\mathbf{i} + y(t)\mathbf{j}$ and if $x(t)$ and $y(t)$ are differentiable, then*

$$\mathbf{r}'(t) = x'(t)\mathbf{i} + y'(t)\mathbf{j}$$

Proof. From Definition 14.2.2

$$\mathbf{r}'(t) = \lim_{\Delta t \to 0} \frac{\mathbf{r}(t + \Delta t) - \mathbf{r}(t)}{\Delta t}$$

$$= \lim_{\Delta t \to 0} \frac{[x(t + \Delta t)\mathbf{i} + y(t + \Delta t)\mathbf{j}] - [x(t)\mathbf{i} + y(t)\mathbf{j}]}{\Delta t}$$

$$= \lim_{\Delta t \to 0} \frac{[x(t + \Delta t) - x(t)]}{\Delta t}\mathbf{i} + \lim_{\Delta t \to 0} \frac{[y(t + \Delta t) - y(t)]}{\Delta t}\mathbf{j}$$

$$= x'(t)\mathbf{i} + y'(t)\mathbf{j} \quad \blacksquare$$

▶ **Example 4** Let $\mathbf{r}(t) = t^2\mathbf{i} + t^3\mathbf{j}$. Find $\mathbf{r}'(t)$ and $\mathbf{r}'(1)$.

Solution.

$$\mathbf{r}'(t) = \frac{d}{dt}[t^2]\mathbf{i} + \frac{d}{dt}[t^3]\mathbf{j} = 2t\mathbf{i} + 3t^2\mathbf{j}$$

Substituting $t = 1$ yields

$$\mathbf{r}'(1) = 2\mathbf{i} + 3\mathbf{j} \qquad\qquad \blacktriangleleft$$

Since derivatives of vector-valued functions are computed by components, most basic theorems on differentiating real-valued functions carry over to vector-valued functions. Specifically, let $\mathbf{r}(t)$, $\mathbf{r}_1(t)$, and $\mathbf{r}_2(t)$ be vector-valued functions, $f(t)$ a real-valued function, k a scalar, and $\mathbf{c}$ a fixed (constant) vector. Then the following rules of differentiation hold.

14.2.4 THEOREM
Rules of Differentiation

$$\frac{d}{dt}[\mathbf{c}] = \mathbf{0}$$

$$\frac{d}{dt}[k\mathbf{r}(t)] = k\frac{d}{dt}[\mathbf{r}(t)]$$

$$\frac{d}{dt}[\mathbf{r}_1(t) + \mathbf{r}_2(t)] = \frac{d}{dt}[\mathbf{r}_1(t)] + \frac{d}{dt}[\mathbf{r}_2(t)]$$

$$\frac{d}{dt}[\mathbf{r}_1(t) - \mathbf{r}_2(t)] = \frac{d}{dt}[\mathbf{r}_1(t)] - \frac{d}{dt}[\mathbf{r}_2(t)]$$

$$\frac{d}{dt}[f(t)\mathbf{r}(t)] = f(t)\frac{d}{dt}[\mathbf{r}(t)] + \mathbf{r}(t)\frac{d}{dt}[f(t)]$$

In addition, if the parameter t is a function of another parameter w, then the following form of the chain rule holds.

14.2.5 THEOREM
The Chain Rule

If $\mathbf{u} = \mathbf{r}(t)$ is a differentiable function of t, and t is a differentiable function of w, then

$$\frac{d\mathbf{u}}{dw} = \frac{d\mathbf{u}}{dt}\frac{dt}{dw}$$

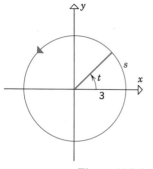

Figure 14.2.6

▶ Example 5 The graph of

$$\mathbf{u} = (3\cos t)\mathbf{i} + (3\sin t)\mathbf{j}, \qquad 0 \le t \le 2\pi$$

is a circle of radius 3 centered at the origin and oriented counterclockwise. For each value of t, let s be the arc length on the circle subtended by the angle t (Figure 14.2.6), so that

$$s = 3t \qquad \text{or} \qquad t = \frac{s}{3}$$

Thus,

$$\mathbf{u} = 3\cos\left(\frac{s}{3}\right)\mathbf{i} + 3\sin\left(\frac{s}{3}\right)\mathbf{j}, \qquad 0 \le s \le 6\pi \tag{2}$$

is a representation of the circle with arc length s as the parameter. From (2) the derivative of $\mathbf{u}$ with respect to s is

$$\frac{d\mathbf{u}}{ds} = \left[-\sin\left(\frac{s}{3}\right)\right]\mathbf{i} + \left[\cos\left(\frac{s}{3}\right)\right]\mathbf{j}$$

This same result can be obtained from the chain rule and the given formula for $\mathbf{u}$ in terms of t by writing

$$\frac{d\mathbf{u}}{ds} = \frac{d\mathbf{u}}{dt}\frac{dt}{ds} = [(-3\sin t)\mathbf{i} + (3\cos t)\mathbf{j}]\left(\frac{1}{3}\right)$$

$$= (-\sin t)\mathbf{i} + (\cos t)\mathbf{j}$$

$$= \left[-\sin\left(\frac{s}{3}\right)\right]\mathbf{i} + \left[\cos\left(\frac{s}{3}\right)\right]\mathbf{j} \qquad ◀$$

Since derivatives of vector-valued functions are calculated by differentiating components, it is natural to define integrals of vector-valued functions in terms of components.

14.2.6 DEFINITION   If $\mathbf{r}(t) = x(t)\mathbf{i} + y(t)\mathbf{j}$, then we define

$$\int \mathbf{r}(t)\, dt = \left(\int x(t)\, dt\right)\mathbf{i} + \left(\int y(t)\, dt\right)\mathbf{j}$$

$$\int_a^b \mathbf{r}(t)\, dt = \left(\int_a^b x(t)\, dt\right)\mathbf{i} + \left(\int_a^b y(t)\, dt\right)\mathbf{j}$$

Observe that $\int \mathbf{r}(t)\, dt$ is, in fact, an antiderivative of $\mathbf{r}(t)$ since

$$\frac{d}{dt}\left[\int \mathbf{r}(t)\,dt\right] = \frac{d}{dt}\left[\int x(t)\,dt\right]\mathbf{i} + \frac{d}{dt}\left[\int y(t)\,dt\right]\mathbf{j}$$
$$= x(t)\mathbf{i} + y(t)\mathbf{j} = \mathbf{r}(t)$$

▶ **Example 6**   Evaluate the indefinite integral of $\mathbf{r}(t) = 2t\mathbf{i} + 3t^2\mathbf{j}$

Solution.

$$\int \mathbf{r}(t)\,dt = \int (2t\mathbf{i} + 3t^2\mathbf{j})\,dt$$
$$= \left(\int 2t\,dt\right)\mathbf{i} + \left(\int 3t^2\,dt\right)\mathbf{j}$$
$$= (t^2 + C_1)\mathbf{i} + (t^3 + C_2)\mathbf{j}$$
$$= (t^2\mathbf{i} + t^3\mathbf{j}) + C_1\mathbf{i} + C_2\mathbf{j}$$
$$= t^2\mathbf{i} + t^3\mathbf{j} + \mathbf{C}$$

where $\mathbf{C} = C_1\mathbf{i} + C_2\mathbf{j}$ is an arbitrary vector constant of integration.   ◀

If $\mathbf{R}(t)$ is an antiderivative of $\mathbf{r}(t)$ in the sense that $\mathbf{R}'(t) = \mathbf{r}(t)$, then it can be shown that

$$\int_a^b \mathbf{r}(t)\,dt = \mathbf{R}(t)\,\Big]_{t=a}^{t=b} = \mathbf{R}(b) - \mathbf{R}(a)$$

and

$$\int \mathbf{r}(t)\,dt = \mathbf{R}(t) + \mathbf{C}$$

where $\mathbf{C}$ is an arbitrary vector constant of integration. We leave the proofs as exercises.

▶ Exercise Set 14.2

In Exercises 1–6, sketch the graph of $\mathbf{r}(t)$ and show the direction of increasing t.

1. $\mathbf{r}(t) = 2\mathbf{i} + t\mathbf{j}$.

2. $\mathbf{r}(t) = \langle 3t - 4, 6t + 2\rangle$.

3. $\mathbf{r}(t) = (1 + \cos t)\mathbf{i} + (3 - \sin t)\mathbf{j}$, $0 \le t \le 2\pi$.

4. $\mathbf{r}(t) = \langle 2\cos t, 5\sin t\rangle$, $0 \le t \le 2\pi$.

5. $\mathbf{r}(t) = (\cosh t)\mathbf{i} + (\sinh t)\mathbf{j}$.

6. $\mathbf{r}(t) = \sqrt{t}\,\mathbf{i} + (2t + 4)\mathbf{j}$.

In Exercises 7–10, calculate the limit.

7. $\lim\limits_{t \to 3} (t^2\mathbf{i} + 2t\mathbf{j})$.

8. $\lim\limits_{t \to \pi/4} \langle \cos t, \sin t\rangle$.

9. $\lim\limits_{t \to 0^+} \left(\sqrt{t}\,\mathbf{i} + \dfrac{\sin t}{t}\mathbf{j}\right)$.

10. $\lim\limits_{t \to +\infty} \left\langle \dfrac{t^2 + 1}{3t^2 + 2}, \dfrac{1}{t}\right\rangle$.

In Exercises 11–14, find $d\mathbf{r}/dt$.

11. $\mathbf{r}(t) = (4 + 5t)\mathbf{i} + (t - t^2)\mathbf{j}$.

12. $\mathbf{r}(t) = 4\mathbf{i} - (\cos t)\mathbf{j}$.

13. $\mathbf{r}(t) = \langle 1/t, \tan t \rangle$.

14. $\mathbf{r}(t) = \langle \tan^{-1} t, t^2 e^{2t} \rangle$.

In Exercises 15–18, find $\mathbf{r}'(t_0)$; then sketch the graph of $\mathbf{r}(t)$ and the tangent vector $\mathbf{r}'(t_0)$.

15. $\mathbf{r}(t) = \langle t, t^2 \rangle$; $t_0 = 2$.

16. $\mathbf{r}(t) = (\cos t)\mathbf{i} + (\sin t)\mathbf{j}$; $t_0 = 3\pi/4$.

17. $\mathbf{r}(t) = \langle e^{-t}, e^{2t} \rangle$; $t_0 = \ln 2$.

18. $\mathbf{r}(t) = (\cos 2t)\mathbf{i} - (4 \sin t)\mathbf{j}$; $t_0 = \pi$.

In Exercises 19–26, evaluate the integral.

19. $\displaystyle \int (3\mathbf{i} + 4t\mathbf{j}) \, dt$.

20. $\displaystyle \int [(\cos t)\mathbf{i} + (\sin t)\mathbf{j}] \, dt$

21. $\displaystyle \int_0^{\pi/3} \langle \cos 3t, -\sin 3t \rangle \, dt$.

22. $\displaystyle \int_0^1 (t^2 \mathbf{i} + t^3 \mathbf{j}) \, dt$.

23. $\displaystyle \int_1^9 (t^{1/2}\mathbf{i} + t^{-1/2}\mathbf{j}) \, dt$.

24. $\displaystyle \int [(t \sin t)\mathbf{i} + \mathbf{j}] \, dt$.

25. $\displaystyle \int \langle te^t, \ln t \rangle \, dt$.

26. $\displaystyle \int_0^2 \|t\mathbf{i} + t^2\mathbf{j}\| \, dt$.

In Exercises 27–29, calculate $d\mathbf{u}/dw$ by the chain rule, and then check your result by expressing $\mathbf{u}$ in terms of w and differentiating.

27. $\mathbf{u} = t\mathbf{i} + t^2\mathbf{j}$; $t = 4w + 1$.

28. $\mathbf{u} = \langle 3 \cos t, 3 \sin t \rangle$; $t = \pi w$.

29. $\mathbf{u} = e^t \mathbf{i} + 4e^{-t}\mathbf{j}$; $t = w^2$.

30. Find $\mathbf{r}(t)$ given that $\mathbf{r}'(t) = t^2\mathbf{i} + 2t\mathbf{j}$ and $\mathbf{r}(0) = \mathbf{i} + \mathbf{j}$.

31. Find $\mathbf{r}(t)$ given that $\mathbf{r}'(t) = (\cos t)\mathbf{i} + (\sin t)\mathbf{j}$ and $\mathbf{r}(0) = \mathbf{i} - \mathbf{j}$.

32. Find $\mathbf{r}(t)$ given that $\mathbf{r}''(t) = \mathbf{i} + e^t\mathbf{j}$, $\mathbf{r}(0) = 2\mathbf{i}$, and $\mathbf{r}'(0) = \mathbf{j}$.

33. Find $\mathbf{r}(t)$ given that $\mathbf{r}''(t) = 12t^2\mathbf{i} - 2\mathbf{j}$, $\mathbf{r}(0) = 2\mathbf{i} - 4\mathbf{j}$, and $\mathbf{r}'(0) = \mathbf{0}$.

34. Prove:

(a) $\displaystyle \int_a^b [\mathbf{r}_1(t) + \mathbf{r}_2(t)] \, dt = \int_a^b \mathbf{r}_1(t) \, dt + \int_a^b \mathbf{r}_2(t) \, dt$

(b) $\displaystyle \int_a^b k\mathbf{r}(t) \, dt = k \int_a^b \mathbf{r}(t) \, dt$ (k a scalar constant).

35. Prove: If $\mathbf{R}'(t) = \mathbf{r}(t)$, then

(a) $\displaystyle \int \mathbf{r}(t) \, dt = \mathbf{R}(t) + \mathbf{C}$, where $\mathbf{C}$ is an arbitrary vector constant

(b) $\displaystyle \int_a^b \mathbf{r}(t) \, dt = \mathbf{R}(b) - \mathbf{R}(a)$.

36. If $\mathbf{r}(t) = x(t)\mathbf{i} + y(t)\mathbf{j}$, is it true that

$$\frac{d}{dt} \|\mathbf{r}(t)\| = \left\| \frac{d\mathbf{r}}{dt} \right\| ?$$

Justify your answer.

37. Use Theorem 14.2.4 to derive the formula

$$\frac{d}{dt}\left[\frac{\mathbf{r}(t)}{\|\mathbf{r}(t)\|} \right] = \frac{\mathbf{r}'(t)}{\|\mathbf{r}(t)\|} - \left[\frac{d}{dt}\|\mathbf{r}(t)\| \right] \frac{\mathbf{r}(t)}{\|\mathbf{r}(t)\|^2}$$

38. Prove Theorem 14.2.4.

39. Prove Theorem 14.2.5.

14.3 UNIT TANGENT AND NORMAL VECTORS; ARC LENGTH AS A PARAMETER

Vectors of length one that are tangent or normal to a curve play an important role in vector calculus. In this section we will study properties of such vectors, and investigate the use of arc length as a parameter in parametric equations.

14.3.1 DEFINITION The **unit tangent vector** $\mathbf{T} = \mathbf{T}(t)$ to the graph C of a vector-valued function $\mathbf{r}(t)$ is defined by

$$\mathbf{T}(t) = \frac{\mathbf{r}'(t)}{\|\mathbf{r}'(t)\|}$$

if $\mathbf{r}'(t) \neq \mathbf{0}$.

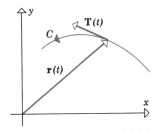

Figure 14.3.1

Because $\mathbf{r}'(t)$ is tangent to the curve C and points in the direction of increasing t, it follows that $\mathbf{T}(t)$ is a vector of length one, tangent to C and pointing in the direction of increasing t (Figure 14.3.1). Observe that $\mathbf{T}(t)$ is defined only at those points where $\mathbf{r}'(t) \neq \mathbf{0}$.

▶ **Example 1**   Find the unit tangent vector to the graph of $\mathbf{r}(t) = t^2\mathbf{i} + t^3\mathbf{j}$ at the point where $t = 2$.

Solution.   Since

$$\mathbf{r}'(t) = 2t\mathbf{i} + 3t^2\mathbf{j}$$

we obtain

$$\mathbf{T}(t) = \frac{\mathbf{r}'(t)}{\|\mathbf{r}'(t)\|} = \frac{2t\mathbf{i} + 3t^2\mathbf{j}}{\|2t\mathbf{i} + 3t^2\mathbf{j}\|} = \frac{2t\mathbf{i} + 3t^2\mathbf{j}}{\sqrt{4t^2 + 9t^4}}$$

Thus,

$$\mathbf{T}(2) = \frac{4\mathbf{i} + 12\mathbf{j}}{\sqrt{160}} = \frac{4\mathbf{i} + 12\mathbf{j}}{4\sqrt{10}} = \frac{1}{\sqrt{10}}\mathbf{i} + \frac{3}{\sqrt{10}}\mathbf{j}$$

The graph of $\mathbf{r}(t)$ and the vector $\mathbf{T}(2)$ are shown in Figure 14.3.2.   ◀

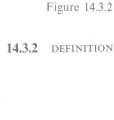

Figure 14.3.2

Let C be the graph of $\mathbf{r}(t) = x(t)\mathbf{i} + y(t)\mathbf{j}$. We say that C is a **smooth curve** or that $\mathbf{r}(t)$ is a **smooth function** if $x'(t)$ and $y'(t)$ are continuous and $\mathbf{r}'(t) = x'(t)\mathbf{i} + y'(t)\mathbf{j} \neq \mathbf{0}$ for any value of t. Smooth curves are sometimes described as having "continuously turning tangents." This is because the tangent vector $\mathbf{T}(t)$ for a smooth curve varies "continuously" without abrupt changes in direction as t increases.

The function $\mathbf{r}(t) = t^2\mathbf{i} + t^3\mathbf{j}$ in Example 1 is not smooth since $\mathbf{r}'(t) = 2t\mathbf{i} + 3t^2\mathbf{j}$ is zero when $t = 0$. Note the behavior of the graph at the origin, which is the point where $t = 0$ (Figure 14.3.2).

14.3.2 DEFINITION   The **unit normal vector** $\mathbf{N} = \mathbf{N}(t)$ to the graph C of a vector-valued function $\mathbf{r}(t)$ is the vector obtained by rotating $\mathbf{T} = \mathbf{T}(t)$ counterclockwise $90°$ (Figure 14.3.3).

Figure 14.3.3   *(a)*   *(b)*

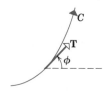

Figure 14.3.4

If $\mathbf{T}$ is a unit tangent vector to a curve C, then we will use ϕ to denote the counterclockwise angle from the direction of the positive x-axis to $\mathbf{T}$ (Figure 14.3.4). Since $\mathbf{T}$ has length 1, we may write (see Example 6 of Section 14.1)

$$\mathbf{T} = (\cos \phi)\mathbf{i} + (\sin \phi)\mathbf{j} \tag{1}$$

and since $\mathbf{N}$ has length 1 and is $90°$ counterclockwise from $\mathbf{T}$, we may write

$$\mathbf{N} = \cos (\phi + \pi/2)\mathbf{i} + \sin (\phi + \pi/2)\mathbf{j}$$

Applying the identities

$$\cos (\phi + \pi/2) = -\sin \phi \quad \text{and} \quad \sin (\phi + \pi/2) = \cos \phi$$

yields

$$\mathbf{N} = (-\sin \phi)\mathbf{i} + (\cos \phi)\mathbf{j} \tag{2}$$

Thus, if $\mathbf{T}$ is given by

$$\mathbf{T} = u_1\mathbf{i} + u_2\mathbf{j}$$

then $u_1 = \cos \phi$ and $u_2 = \sin \phi$, so that

$$\mathbf{N} = -u_2\mathbf{i} + u_1\mathbf{j}$$

▶ **Example 2** Find the unit normal vector to the graph of $\mathbf{r}(t) = t^2\mathbf{i} + t^3\mathbf{j}$ at the point where $t = 2$.

Solution. In Example 1 we found the unit tangent vector at $t = 2$ to be

$$\mathbf{T} = u_1\mathbf{i} + u_2\mathbf{j} = \frac{1}{\sqrt{10}}\mathbf{i} + \frac{3}{\sqrt{10}}\mathbf{j}$$

Thus, the unit normal vector at $t = 2$ is

$$\mathbf{N} = -u_2\mathbf{i} + u_1\mathbf{j} = -\frac{3}{\sqrt{10}}\mathbf{i} + \frac{1}{\sqrt{10}}\mathbf{j} \qquad \blacktriangleleft$$

Formulas (1) and (2) reveal the following fundamental relationship between $\mathbf{T}$ and $\mathbf{N}$.

$$\frac{d\mathbf{T}}{d\phi} = \mathbf{N} \tag{3}$$

In vector calculus it is often convenient to have a curve C given parametrically with arc length as the parameter. This may be done as follows:

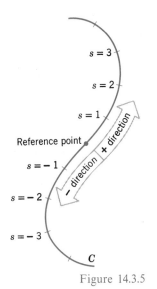

Figure 14.3.5

Step 1. Select any point on the curve C to serve as a *reference point.*

Step 2. Starting from the reference point, choose one direction along the curve to be the *positive direction* and the other to be the *negative direction.*

Step 3. If $P(x, y)$ is a point on the curve, let s be the "signed" arc length along C from the reference point to P, where s is positive if P is in the positive direction from the reference point, and s is negative if P is in the negative direction. Figure 14.3.5 illustrates this idea.

By this procedure, a unique point (x, y) on the curve is determined when a value for s is given. For example, $s = 2$ determines the point that is 2 units along the curve in the positive direction from the reference point, and $s = -\frac{3}{2}$ determines the point that is $\frac{3}{2}$ units along the curve in the negative direction from the reference point.

Let us now treat s as a variable. As the value of s changes, the corresponding point (x, y) moves along C. Thus, x and y are functions of s, say $x = x(s)$ and $y = y(s)$, and the curve C is expressed parametrically by the equations

$$x = x(s)$$
$$y = y(s)$$

having s as the parameter.

REMARK. When defining the parameter s, the choice of positive and negative directions is arbitrary. However, it may be that the curve C is already specified in terms of some other parameter t, in which case we will agree always to take the direction of increasing t as the positive direction for the parameter s. By so doing, s will increase as t increases and vice versa.

14.3.3 THEOREM *Let a curve C be given parametrically by*

$$x = x(t)$$
$$y = y(t)$$

where $x'(t)$ and $y'(t)$ are continuous functions. If an arc length parameter s is introduced with its reference point at $\left(x(t_0), y(t_0)\right)$, then the parameters s and t have the relationships

$$s = \int_{t_0}^{t} \sqrt{\left(\frac{dx}{du}\right)^2 + \left(\frac{dy}{du}\right)^2}\, du \qquad (4)$$

$$\frac{ds}{dt} = \sqrt{\left(\frac{dx}{dt}\right)^2 + \left(\frac{dy}{dt}\right)^2} \qquad (5)$$

Proof. If $t > t_0$, then from Theorem 13.4.1 (with u as the variable of integration rather than t) it follows that

$$\int_{t_0}^{t} \sqrt{\left(\frac{dx}{du}\right)^2 + \left(\frac{dy}{du}\right)^2} \, du \qquad (6)$$

represents the arc length of that portion of the curve C that lies between $(x(t_0), y(t_0))$ and $(x(t), y(t))$. If $t < t_0$, then (6) is the negative of this arc length. In either case, integral (6) represents the "signed" arc length s between these points, which proves (4).

Relationship (5) follows from (4) and the second fundamental theorem of calculus (7.2.1),

$$\frac{ds}{dt} = \frac{d}{dt}\left[\int_{t_0}^{t} \sqrt{\left(\frac{dx}{du}\right)^2 + \left(\frac{dy}{du}\right)^2} \, du\right] = \sqrt{\left(\frac{dx}{dt}\right)^2 + \left(\frac{dy}{dt}\right)^2} \quad \blacksquare$$

REMARK.   Formula (5) reveals two facts worth noting. First, ds/dt does not depend on t_0; that is, the value of ds/dt is independent of where the reference point for the parameter s is located. This is to be expected since changing the position of the reference point shifts each value of s by a constant (the arc length between the reference points), and this constant drops out when we differentiate. The second fact to be noted from (5) is that $ds/dt \geq 0$ for all t. This is also to be expected since s increases with t by the remark preceding Theorem 14.3.3.

▶ Example 3   Find parametric equations for the line

$$x = 2t + 1$$
$$y = 3t - 2$$

using arc length s as a parameter, where the reference point for s is the point on the line where $t = 0$.

Solution.   In (4) of Theorem 14.3.3, u was used as the variable of integration because t was needed as a limit of integration. To apply (4), we first rewrite the given parametric equations with u in place of t; this gives

$$x = 2u + 1$$
$$y = 3u - 2$$

Thus,

$$\frac{dx}{du} = 2, \qquad \frac{dy}{du} = 3$$

so that (4) yields

$$s = \int_{t_0}^{t} \sqrt{\left(\frac{dx}{du}\right)^2 + \left(\frac{dy}{du}\right)^2}\, du = \int_0^t \sqrt{13}\, du = \sqrt{13}\, u \Big]_{u=0}^{u=t} = \sqrt{13}\, t$$

Therefore,

$$t = \frac{1}{\sqrt{13}} s$$

Substituting in the given parametric equations yields

$$x = 2\left(\frac{1}{\sqrt{13}} s\right) + 1 = \frac{2}{\sqrt{13}} s + 1$$

$$y = 3\left(\frac{1}{\sqrt{13}} s\right) - 2 = \frac{3}{\sqrt{13}} s - 2$$

◄

► **Example 4** Find parametric equations for the circle

$$\begin{array}{l} x = a \cos t \\ y = a \sin t \end{array}, \quad 0 \leq t \leq 2\pi$$

using arc length s as a parameter, with the reference point for s being $(a, 0)$.

Solution. Proceeding as in Example 3, we first replace t by u in the given equations so that

$$x = a \cos u, \qquad y = a \sin u$$

and

$$\frac{dx}{du} = -a \sin u, \qquad \frac{dy}{du} = a \cos u$$

Since the reference point $(a, 0)$ corresponds to $t = 0$, we obtain from (4),

$$s = \int_{t_0}^{t} \sqrt{\left(\frac{dx}{du}\right)^2 + \left(\frac{dy}{du}\right)^2}\, du = \int_0^t \sqrt{(-a \sin u)^2 + (a \cos u)^2}\, du$$

$$= \int_0^t a\, du = au \Big]_{u=0}^{u=t} = at$$

Solving for t in terms of s yields

$$t = \frac{s}{a}$$

Substituting this in the given parametric equations and using the fact that $s = at$ ranges from 0 to $2\pi a$ as t ranges from 0 to 2π, we obtain

$$x = a \cos\left(\frac{s}{a}\right)$$
$$y = a \sin\left(\frac{s}{a}\right) \quad, \quad 0 \le s \le 2\pi a$$

◀

Theorem 14.3.3 leads to some useful results about vector-valued functions.

14.3.4   THEOREM   *If* $\mathbf{r}(t) = x(t)\mathbf{i} + y(t)\mathbf{j}$ *is a vector-valued function with* $x'(t)$ *and* $y'(t)$ *continuous, and if s is an arc length parameter for the curve* $\mathbf{r}(t)$, *then*

(a) $\|\mathbf{r}'(t)\| = \dfrac{ds}{dt}$

(b) $\dfrac{d\mathbf{r}}{ds} = \mathbf{T}$

Proof of (a).   Let $x = x(t)$ and $y = y(t)$, so that by (5)

$$\|\mathbf{r}'(t)\| = \|x'(t)\mathbf{i} + y'(t)\mathbf{j}\| = \sqrt{\left(\frac{dx}{dt}\right)^2 + \left(\frac{dy}{dt}\right)^2} = \frac{ds}{dt}$$

Proof of (b).   From Definition 14.3.1, the chain rule, and part (a),

$$\mathbf{T} = \frac{1}{\|\mathbf{r}'(t)\|}\mathbf{r}'(t)$$

$$= \frac{1}{ds/dt}\frac{d\mathbf{r}}{dt} = \frac{1}{ds/dt}\frac{d\mathbf{r}}{ds}\frac{ds}{dt} = \frac{d\mathbf{r}}{ds} \quad ▮$$

▶ **Example 5**   In Example 4 we obtained the parametric equations

$$x = a \cos (s/a)$$
$$y = a \sin (s/a) \quad, \quad 0 \le s \le 2\pi a$$

for a circle of radius a centered at the origin. In vector notation, this circle is the graph of

$$\mathbf{r}(s) = a \cos\left(\frac{s}{a}\right)\mathbf{i} + a \sin\left(\frac{s}{a}\right)\mathbf{j}, \qquad 0 \le s \le 2\pi a$$

so that for each value of s, the unit tangent vector to the circle is

$$\mathbf{T} = \mathbf{T}(s) = \frac{d\mathbf{r}}{ds} = -\sin\left(\frac{s}{a}\right)\mathbf{i} + \cos\left(\frac{s}{a}\right)\mathbf{j}$$ ◀

▶ Exercise Set 14.3

In Exercises 1–8, obtain the unit tangent and unit normal vectors to the curve for the given value of t. Then sketch the vectors and a portion of the curve containing the point of tangency.

1. $\mathbf{r}(t) = (5\cos t)\mathbf{i} + (5\sin t)\mathbf{j}$; $t = \pi/3$.

2. $\mathbf{r}(t) = 2t\mathbf{i} + 4t^2\mathbf{j}$; $t = 1$.

3. $\mathbf{r}(t) = (4 + 5t)\mathbf{i} + (1 - 2t)\mathbf{j}$; $t = 2$.

4. $\mathbf{r}(t) = e^t\mathbf{i} + e^{-t}\mathbf{j}$; $t = 0$.

5. $\mathbf{r}(t) = \frac{1}{2}t^2\mathbf{i} + \frac{1}{3}t^3\mathbf{j}$; $t = 1$.

6. $\mathbf{r}(t) = (\ln t)\mathbf{i} + t\mathbf{j}$; $t = e$.

7. $\mathbf{r}(t) = (4\cos t)\mathbf{i} + (9\sin t)\mathbf{j}$; $t = \pi/4$.

8. $\mathbf{r}(t) = (\ln \sin t)\mathbf{i} + (\ln \cos t)\mathbf{j}$; $t = \pi/6$.

In Exercises 9–15, find parametric equations for the curve using arc length s as a parameter. Use the point on the curve where $t = 0$ as the reference point.

9. $\mathbf{r}(t) = (3t - 2)\mathbf{i} + (4t + 3)\mathbf{j}$.

10. $\mathbf{r}(t) = (3\cos 2t)\mathbf{i} + (3\sin 2t)\mathbf{j}$; $0 \le t \le \pi$.

11. $\mathbf{r}(t) = (3 + \cos t)\mathbf{i} + (2 + \sin t)\mathbf{j}$; $0 \le t \le 2\pi$.

12. $\mathbf{r}(t) = (\cos^3 t)\mathbf{i} + (\sin^3 t)\mathbf{j}$; $0 \le t \le \pi/2$.

13. $\mathbf{r}(t) = \frac{1}{3}t^3\mathbf{i} + \frac{1}{2}t^2\mathbf{j}$; $t \ge 0$.

14. $\mathbf{r}(t) = (1 + t)^2\mathbf{i} + (1 + t)^3\mathbf{j}$; $0 \le t \le 1$.

15. $\mathbf{r}(t) = (e^t \cos t)\mathbf{i} + (e^t \sin t)\mathbf{j}$; $0 \le t \le \pi/2$.

16. Let C be the curve given parametrically in terms of arc length by

$$x = 2\cos\left(\frac{s}{2}\right)$$
$$y = 2\sin\left(\frac{s}{2}\right), \quad 0 \le s \le 4\pi$$

(a) Find $\mathbf{T} = \mathbf{T}(s)$ and $\mathbf{N} = \mathbf{N}(s)$.
(b) Sketch the curve and the vectors $\mathbf{T}(\pi/2)$ and $\mathbf{N}(\pi/2)$.

17. Let C be the curve given parametrically in terms of arc length by

$$x = \frac{3}{5}s + 1$$
$$y = \frac{4}{5}s - 2, \quad 0 \le s \le 10$$

(a) Find $\mathbf{T} = \mathbf{T}(s)$ and $\mathbf{N} = \mathbf{N}(s)$.
(b) Sketch the curve and the vectors $\mathbf{T}(5)$ and $\mathbf{N}(5)$.

18. Find parametric equations for the cycloid

$$x = at - a\sin t$$
$$y = a - a\cos t, \quad 0 \le t \le 2\pi$$

using arc length as the parameter. Take $(0, 0)$ as the reference point.

19. Find the cosine of the smallest positive angle between the vectors $\mathbf{r}(1)$ and $\mathbf{T}(1)$ for the curve $\mathbf{r}(t) = \frac{1}{3}t^3\mathbf{i} + \frac{1}{2}t^2\mathbf{j}$.

20. Prove: If C is given parametrically by the equations $x = x(s)$, $y = y(s)$, where s is an arc length parameter, then

$$\mathbf{T} = \frac{dx}{ds}\mathbf{i} + \frac{dy}{ds}\mathbf{j}, \quad \mathbf{N} = -\frac{dy}{ds}\mathbf{i} + \frac{dx}{ds}\mathbf{j}$$

14.4   CURVATURE

In this section we will obtain a numerical measure of how sharply a curve bends. Our results will have applications in geometry and in the study of motion along a curved path.

Let C be a curve given parametrically in terms of an arc length parameter s. As a point moves along the curve C, the unit tangent vector $\mathbf{T} = \mathbf{T}(s)$ at the point will have constant length one, but will vary in direction. Let $\phi = \phi(s)$ denote the angle measured counterclockwise from the direction of the positive x-axis to $\mathbf{T}$ (Figure 14.4.1). As suggested by Figure 14.4.1, the angle ϕ will vary more rapidly with s for a curve bending sharply, than for a curve bending gradually. This suggests that the rate at which ϕ varies with s can be used as a numerical measure of the "curvature" of C.

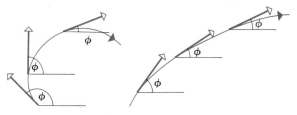

Figure 14.4.1

14.4.1   DEFINITION   Let C be a smooth curve given parametrically in terms of an arc length parameter s, and let $\phi = \phi(s)$ be the counterclockwise angle from the direction of the positive x-axis to the unit tangent vector $\mathbf{T} = \mathbf{T}(s)$. Then the *curvature* at a point on C is denoted by $\kappa = \kappa(s)$ ($\kappa = $ Greek "kappa") and is defined by

$$\kappa = \frac{d\phi}{ds}$$

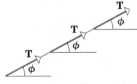

Figure 14.4.2

▶ **Example 1**   For a straight line, the angle $\phi = \phi(s)$ is constant (Figure 14.4.2) so that

$$\kappa = \frac{d\phi}{ds} = 0$$

Thus, a straight line has zero curvature at each point.   ◀

Before proceeding to more examples, it will be helpful to obtain some additional results about curvature.

The sign of $\kappa = d\phi/ds$ depends on whether ϕ increases or decreases as s increases. As suggested by Figure 14.4.3, ϕ increases with s if the concave side of the curve C is on the left as we traverse the curve in the positive direction of s, and ϕ decreases with s if the concave side of the curve is on the right. This yields the following result.

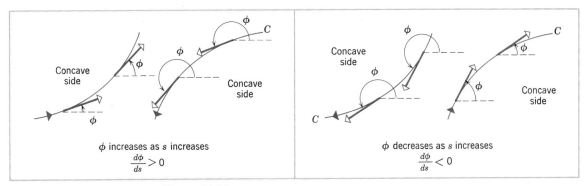

ϕ increases as s increases
$$\frac{d\phi}{ds} > 0$$

ϕ decreases as s increases
$$\frac{d\phi}{ds} < 0$$

Figure 14.4.3

14.4.2 THEOREM
The Sign of the Curvature κ

The curvature $\kappa = d\phi/ds$ is positive if the concave side of the curve is on the left as the curve is traversed in the positive direction of s, and κ is negative if the concave side is on the right.

The curvature κ enters naturally into the following basic derivative formulas:

$$\frac{d\mathbf{T}}{ds} = \kappa \mathbf{N} \tag{1}$$

$$\frac{d\mathbf{T}}{dt} = \kappa \frac{ds}{dt} \mathbf{N} \tag{2}$$

These formulas are obtained from the chain rule as follows:

$$\frac{d\mathbf{T}}{ds} = \frac{d\mathbf{T}}{d\phi}\frac{d\phi}{ds} = \mathbf{N}\kappa = \kappa\mathbf{N}$$

$$\frac{d\mathbf{T}}{dt} = \frac{d\mathbf{T}}{ds}\frac{ds}{dt} = \kappa\mathbf{N}\frac{ds}{dt} = \kappa\frac{ds}{dt}\mathbf{N}$$

REMARK. The reader should be able to deduce from Figure 14.4.3 that $\kappa\mathbf{N}$ always points toward the concave side of C.

Relationships (1) and (2) lead to some useful formulas for calculating the absolute value of the curvature κ. Since $\|\mathbf{N}\| = 1$, it follows from (1) and (2) and the fact that $ds/dt \geq 0$ that

$$\left\|\frac{d\mathbf{T}}{ds}\right\| = \|\kappa\mathbf{N}\| = |\kappa|\,\|\mathbf{N}\| = |\kappa|$$

$$\left\|\frac{d\mathbf{T}}{dt}\right\| = \left\|\kappa\frac{ds}{dt}\mathbf{N}\right\| = |\kappa|\left|\frac{ds}{dt}\right|\|\mathbf{N}\| = |\kappa|\frac{ds}{dt}$$

Thus,

$$|\kappa| = \left\| \frac{d\mathbf{T}}{ds} \right\| \tag{3}$$

$$|\kappa| = \frac{\left\| \dfrac{d\mathbf{T}}{dt} \right\|}{ds/dt} \tag{4}$$

These formulas, in combination with Theorem 14.4.2, can sometimes be used to find the curvature κ.

▶ Example 2 Find the curvature of the circle

$$\begin{array}{l} x = a \cos t \\ y = a \sin t \end{array}, \quad 0 \le t \le 2\pi$$

at each point.

Solution. In Examples 4 and 5 of the previous section, we obtained the parametric equations

$$\begin{array}{l} x = a \cos \left(\dfrac{s}{a} \right) \\ y = a \sin \left(\dfrac{s}{a} \right) \end{array}, \quad 0 \le s \le 2\pi a$$

for the given circle, and we showed that

$$\mathbf{T} = \mathbf{T}(s) = -\sin \left(\frac{s}{a} \right) \mathbf{i} + \cos \left(\frac{s}{a} \right) \mathbf{j}$$

Thus, from formula (3)

$$\begin{aligned} |\kappa| = \left\| \frac{d\mathbf{T}}{ds} \right\| &= \left\| -\frac{1}{a} \cos \left(\frac{s}{a} \right) \mathbf{i} - \frac{1}{a} \sin \left(\frac{s}{a} \right) \mathbf{j} \right\| \\ &= \sqrt{\frac{1}{a^2} \cos^2 \left(\frac{s}{a} \right) + \frac{1}{a^2} \sin^2 \left(\frac{s}{a} \right)} \\ &= \frac{1}{a} \end{aligned}$$

As the circle is traversed in the positive direction for s (counterclockwise) the concave side is on the left, so that $\kappa > 0$. Thus, the circle has constant curvature

$$\kappa = |\kappa| = \frac{1}{a}$$

In words, the curvature is the reciprocal of the radius. We leave it for the reader to obtain this same result by working with formula (4) instead of (3). ◀

The following theorem provides a useful formula for calculating curvature directly from the parametric equations for a curve.

14.4.3 THEOREM *If a smooth curve C has parametric equations $x = x(t)$, $y = y(t)$, then the curvature $\kappa = \kappa(t)$ of C is given by*

$$\kappa = \frac{x'y'' - y'x''}{[x'^2 + y'^2]^{3/2}} \tag{5}$$

Proof. From Theorem 14.3.3

$$\kappa = \frac{d\phi}{ds} = \frac{d\phi/dt}{ds/dt} = \frac{d\phi/dt}{\sqrt{\left(\dfrac{dx}{dt}\right)^2 + \left(\dfrac{dy}{dt}\right)^2}} = \frac{d\phi/dt}{[x'^2 + y'^2]^{1/2}} \tag{6}$$

At each point of the curve, the unit tangent vector $\mathbf{T} = \mathbf{T}(t)$ lies along the tangent line to the curve so that

$$\tan \phi = \text{slope of tangent line} = \frac{dy}{dx}$$

(Figure 14.4.4). Thus,

$$\tan \phi = \frac{dy}{dx} = \frac{dy/dt}{dx/dt} = \frac{y'}{x'} \tag{7}$$

Differentiating the left and right sides of (7) implicitly with respect to t, we obtain

$$\sec^2 \phi \, \frac{d\phi}{dt} = \frac{x'y'' - y'x''}{x'^2}$$

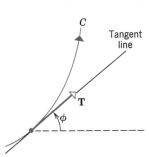

Figure 14.4.4   so that

$$\frac{d\phi}{dt} = \frac{x'y'' - y'x''}{(\sec^2 \phi)(x'^2)} \tag{8}$$

But from (7)

$$\sec^2 \phi = 1 + \tan^2 \phi = 1 + \frac{y'^2}{x'^2}$$

so that (8) may be written

$$\frac{d\phi}{dt} = \frac{x'y'' - y'x''}{x'^2 + y'^2}$$

Substituting this expression in (6) yields (5).   ▮

▶ **Example 3**   The parametric equations

$$\begin{aligned} x &= 2\cos t \\ y &= 3\sin t \end{aligned}, \quad 0 \le t \le 2\pi$$

represent the ellipse

$$\frac{x^2}{4} + \frac{y^2}{9} = 1$$

with counterclockwise orientation (Figure 14.4.5a). Find the curvature at the point $(2, 0)$ and at the point $(0, 3)$.

Solution.   From the given equations,

$$\begin{aligned} x' &= -2\sin t, & x'' &= -2\cos t \\ y' &= 3\cos t, & y'' &= -3\sin t \end{aligned}$$

Thus, from (5),

$$\begin{aligned} \kappa = \kappa(t) &= \frac{x'y'' - y'x''}{[x'^2 + y'^2]^{3/2}} \\ &= \frac{(-2\sin t)(-3\sin t) - (3\cos t)(-2\cos t)}{[(-2\sin t)^2 + (3\cos t)^2]^{3/2}} \\ &= \frac{6\sin^2 t + 6\cos^2 t}{[4\sin^2 t + 9\cos^2 t]^{3/2}} \end{aligned}$$

Therefore,

$$\kappa = \kappa(t) = \frac{6}{[4\sin^2 t + 9\cos^2 t]^{3/2}} \tag{9}$$

The point $(2, 0)$ on the ellipse corresponds to $t = 0$. Substituting $t = 0$ in (9) yields

$$\kappa = \kappa(0) = \frac{6}{9^{3/2}} = \frac{6}{27} = \frac{2}{9}$$

Similarly, the curvature at the point $(0, 3)$ $(t = \pi/2)$ is

$$\kappa = \kappa\left(\frac{\pi}{2}\right) = \frac{6}{4^{3/2}} = \frac{3}{4}$$

Note that the curvature at $(2, 0)$ is smaller than the curvature at $(0, 3)$ as it should be (Figure 14.4.5a). ◀

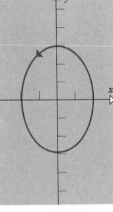

$x = 2 \cos t$
$y = 3 \sin t$

(a)

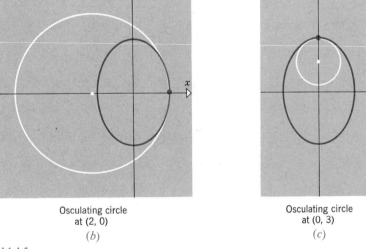

Osculating circle
at $(2, 0)$

(b)

Osculating circle
at $(0, 3)$

(c)

Figure 14.4.5

If a curve C has nonzero curvature κ at a point P, then the circle of radius

$$\rho = \frac{1}{|\kappa|}$$

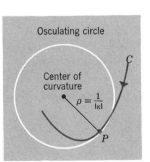

Osculating circle

Center of
curvature

$\rho = \frac{1}{|\kappa|}$

C

P

Figure 14.4.6

sharing a common tangent with C at P, and centered on the concave side of the curve at P is called the *circle of curvature* or *osculating circle* at P. The radius ρ is called the *radius of curvature* at P, and the center of the circle is called the *center of curvature* at P (Figure 14.4.6).

From Example 2, the osculating circle at P has curvature of magnitude $1/\rho = |\kappa|$. Thus, the osculating circle and the curve C not only touch at P but they have curvatures of identical magnitude. In this sense, the osculating circle is the circle that best approximates the curve C near P.

▶ **Example 4**   In Example 3 we found the curvature of the ellipse

$$x = 2 \cos t \atop y = 3 \sin t \quad, \quad 0 \le t \le 2\pi$$

to be 2/9 at (2, 0) and 3/4 at (0, 3). Thus, the radius of curvature is 9/2 at (2, 0) and 4/3 at (0, 3). The osculating circles at these points are shown in Figures 14.4.5b and 14.4.5c.   ◀

The curvature formula in Theorem 14.4.3 presupposes that the curve C is given parametrically. In the event that C is given by an equation of the form $y = f(x)$, we can introduce the parameter $t = x$ and write the curve parametrically as

$$x = t \atop y = f(t)$$

Thus,

$$x' = \frac{dx}{dt} = 1, \qquad x'' = \frac{d^2x}{dt^2} = 0$$

$$y' = \frac{dy}{dt} = \frac{dy}{dx}, \qquad y'' = \frac{d^2y}{dt^2} = \frac{d^2y}{dx^2}$$

so (5) becomes

$$\kappa = \frac{\dfrac{d^2y}{dx^2}}{\left[1 + \left(\dfrac{dy}{dx}\right)^2\right]^{3/2}} \tag{10}$$

▶ **Example 5**   Find the curvature of the parabola $y = x^2$ at the origin.

Solution.   Differentiating $y = x^2$ we obtain

$$\frac{dy}{dx} = 2x, \qquad \frac{d^2y}{dx^2} = 2$$

so that

$$\kappa = \kappa(x) = \frac{2}{[1 + 4x^2]^{3/2}}$$

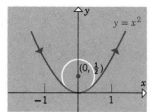

At the origin, $x = 0$ so that the curvature at this point is

$$\kappa = \kappa(0) = 2$$

The parabola and the osculating circle at the origin are shown in Figure 14.4.7. ◀

Figure 14.4.7   REMARK.   Because (10) was derived using $t = x$ as a parameter, it is implicit in this formula that the positive direction for the curve C is the direction of increasing x. Thus, in Example 5, a positive value for κ at the origin was obtained because the concave side of the parabola is on the left as the parabola is traced in the direction of increasing x.

▶ Exercise Set 14.4

In Exercises 1–6, find the curvature at the indicated point.

1. $x = t^2,\ y = t^3;\ t = 1/2.$

2. $x = 4\cos t,\ y = \sin t;\ t = \pi/2.$

3. $x = e^{3t},\ y = e^{-t};\ t = 0.$

4. $x = 1 - t^3,\ y = t - t^2;\ t = 1.$

5. $x = t - \sin t,\ y = 1 - \cos t;\ t = \pi.$

6. $x = 2abt,\ y = b^2 t^2;\ t = 1\ (a > 0,\ b > 0).$

In Exercises 7–12, find the curvature at the indicated point.

7. $y = \sin x;\ x = \pi/2.$ 8. $y = x^3/3;\ x = 0.$

9. $y = 1/x;\ x = 1.$ 10. $y = e^{-x};\ x = 1.$

11. $y = \tan x;\ x = \pi/4.$ 12. $y^2 - 4x^2 = 9;\ (2, 5).$

In Exercises 13–19, sketch the curve, calculate the radius of curvature at the indicated point, and sketch the osculating circle.

13. $y = \sin x;\ x = \pi/2.$ 14. $x = t,\ y = t^2;\ t = 1.$

15. $y = \frac{1}{2}x^2;\ x = -1.$ 16. $x = t,\ y = \sqrt{t};\ t = 2.$

17. $y = \ln x;\ x = 1.$

18. $x = 2t,\ y = 4/t;\ t = 1.$

19. $x = t - \sin t,\ y = 1 - \cos t;\ t = \pi.$

20. At what point(s) does $y = e^x$ have maximum curvature?

21. At what point(s) does $4x^2 + 9y^2 = 36$ have minimum radius of curvature?

22. Show that the curvature of a polar curve $r = f(\theta)$ is

$$\kappa = \frac{r^2 + 2\left(\dfrac{dr}{d\theta}\right)^2 - r\dfrac{d^2 r}{d\theta^2}}{\left[r^2 + \left(\dfrac{dr}{d\theta}\right)^2\right]^{3/2}}$$

[*Hint:*   Use   the   relationships $x = r\cos\theta$, $y = r\sin\theta$.]

In Exercises 23–26, use the result of Exercise 22 to calculate the curvature at the indicated point.

23. $r = 2\sin\theta;\ \theta = \pi/6.$

24. $r = \theta;\ \theta = 1.$

25. $r = a\,(1 + \cos\theta);\ \theta = \pi/2.$

26. $r = e^{2\theta};\ \theta = 1.$

27. Find the radius of curvature of the parabola $y^2 = 4px$ at $(0, 0)$.

28. Find the radius of curvature of $y = \cos x$ at $x = 0$ and $x = \pi$. Sketch the osculating circles at those points.

29. Find the radius of curvature of the ellipse $x = 2\cos t,\ y = \sin t,\ 0 \le t \le 2\pi$ at $t = 0$ and $t = \pi/2$. Sketch the osculating circles at those points.

30. Consider the curve $y = x^4 - 2x^2$.
     (a) Find the radius of curvature at each relative extremum.
     (b) Sketch the curve and show the osculating circles at the relative extrema.

14.5 MOTION IN A PLANE

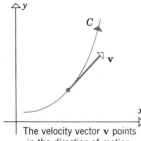

The velocity vector **v** points in the direction of motion and ‖**v**‖ is equal to the speed of the particle

Figure 14.5.1

In Section 5.4 we studied the motion of a particle along a line. In this section we will study the motion of a particle along a curved path in a plane.

As a particle moves along a plane curve C, the direction of motion and the speed of the particle may vary with time. At each instant, the direction of motion is considered to be tangent to the curve C and the speed of the particle is taken to be the instantaneous rate at which the arc length traveled by the particle changes with time. The speed and direction of motion together form a vector quantity called the "velocity" of the particle. This velocity can be represented geometrically by an arrow that points in the direction of motion and has magnitude equal to the speed of the particle (Figure 14.5.1).

The notions of velocity and speed are defined precisely as follows:

14.5.1 DEFINITION If a particle moves along a plane curve C so that the position vector at time t is $\mathbf{r}(t) = x(t)\mathbf{i} + y(t)\mathbf{j}$, then the *velocity* of the particle at time t is

$$\mathbf{v}(t) = \frac{d\mathbf{r}}{dt} = x'(t)\mathbf{i} + y'(t)\mathbf{j} \tag{1}$$

and the *speed* is

$$\|\mathbf{v}(t)\| = \sqrt{[x'(t)]^2 + [y'(t)]^2} \tag{2}$$

This definition agrees with our initial informal discussion, since $d\mathbf{r}/dt$ is tangent to C and $\|d\mathbf{r}/dt\| = ds/dt$ (Theorem 14.3.4). Thus, $\mathbf{v}(t)$ points in the direction of motion and the speed, $\|\mathbf{v}(t)\| = \|d\mathbf{r}/dt\|$, is the rate at which arc length traveled changes with time. For reference, we state the formula

$$\text{speed} = \|\mathbf{v}(t)\| = \frac{ds}{dt} \tag{3}$$

For rectilinear motion, acceleration was defined as the derivative of velocity. We make the same definition for plane motion.

14.5.2 DEFINITION If a particle moves along a plane curve C so that the velocity at time t is $\mathbf{v}(t)$, then the *acceleration* at time t is

$$\mathbf{a}(t) = \frac{d\mathbf{v}}{dt} \tag{4}$$

Since $\mathbf{v}(t) = d\mathbf{r}/dt$, the acceleration vector can be expressed as the second derivative of the position vector

$$\mathbf{a}(t) = \frac{d^2\mathbf{r}}{dt^2} = x''(t)\mathbf{i} + y''(t)\mathbf{j} \tag{5}$$

▶ Example 1   A particle moves around a circle so that its x and y coordinates at time t are

$$x = 2 \cos t, y = 2 \sin t$$

Find the velocity, acceleration, and speed when $t = \pi/4$. Sketch the path of the particle and the velocity and acceleration vectors at $t = \pi/4$.

Solution.   At time t, the position, velocity, acceleration, and speed are

$$\mathbf{r}(t) = 2 \cos t\,\mathbf{i} + 2 \sin t\,\mathbf{j}$$

$$\mathbf{v}(t) = \frac{d\mathbf{r}}{dt} = -2 \sin t\,\mathbf{i} + 2 \cos t\,\mathbf{j}$$

$$\mathbf{a}(t) = \frac{d\mathbf{v}}{dt} = -2 \cos t\,\mathbf{i} - 2 \sin t\,\mathbf{j}$$

$$\|\mathbf{v}(t)\| = \sqrt{(-2 \sin t)^2 + (2 \cos t)^2} = 2$$

Thus, at $t = \pi/4$, the velocity, acceleration, and speed are

$$\mathbf{v}(\pi/4) = -2 \sin \frac{\pi}{4}\mathbf{i} + 2 \cos \frac{\pi}{4}\mathbf{j} = -\sqrt{2}\,\mathbf{i} + \sqrt{2}\,\mathbf{j}$$

$$\mathbf{a}(\pi/4) = -2 \cos \frac{\pi}{4}\mathbf{i} - 2 \sin \frac{\pi}{4}\mathbf{j} = -\sqrt{2}\,\mathbf{i} - \sqrt{2}\,\mathbf{j}$$

$$\|\mathbf{v}(\pi/4)\| = 2$$

Note that the particle moves around the circle with constant speed since $\|\mathbf{v}(t)\| = 2$ does not depend on t. The path of the particle and the velocity and acceleration vectors at $t = \pi/4$ are shown in Figure 14.5.2.   ◀

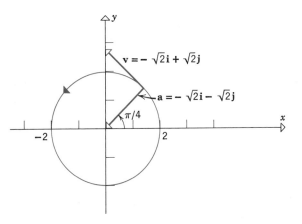

Figure 14.5.2

Since $\mathbf{a}(t)$ is the derivative of $\mathbf{v}(t)$ and $\mathbf{v}(t)$ is the derivative of $\mathbf{r}(t)$, it follows that

$$\mathbf{v}(t) = \int \mathbf{a}(t)\, dt \tag{6}$$

and

$$\mathbf{r}(t) = \int \mathbf{v}(t)\, dt \tag{7}$$

Thus, if the acceleration $\mathbf{a}(t)$ is known, it is possible to determine the velocity $\mathbf{v}(t)$ and the position $\mathbf{r}(t)$ by integration, provided sufficient information is available to determine the constants of integration.

▶ Example 2 For a particle moving in a plane, find the position $\mathbf{r}(t)$ and the velocity $\mathbf{v}(t)$ given that the acceleration is $\mathbf{a}(t) = 6t\mathbf{i} + 12t^2\mathbf{j}$ and the position and velocity at time $t = 0$ are $\mathbf{r}(0) = \mathbf{i} + \mathbf{j}$ and $\mathbf{v}(0) = \mathbf{0}$.

Solution. From (6)

$$\mathbf{v}(t) = \int (6t\mathbf{i} + 12t^2\mathbf{j})\, dt = 3t^2\mathbf{i} + 4t^3\mathbf{j} + \mathbf{c}_1 \tag{8}$$

where $\mathbf{c}_1$ is a vector constant of integration. Since $\mathbf{v}(0) = \mathbf{0}$, it follows on substituting $t = 0$ in (8) that $\mathbf{c}_1 = \mathbf{0}$. Thus,

$$\mathbf{v}(t) = 3t^2\mathbf{i} + 4t^3\mathbf{j}$$

From (7)

$$\mathbf{r}(t) = \int (3t^2\mathbf{i} + 4t^3\mathbf{j})\, dt = t^3\mathbf{i} + t^4\mathbf{j} + \mathbf{c}_2 \tag{9}$$

Since $\mathbf{r}(0) = \mathbf{i} + \mathbf{j}$, it follows on substituting $t = 0$ in (9) that $\mathbf{c}_2 = \mathbf{i} + \mathbf{j}$. Thus,

$$\mathbf{r}(t) = (t^3 + 1)\mathbf{i} + (t^4 + 1)\mathbf{j} \qquad ◀$$

It is a fact of physics that an object of mass m, when subjected to a force $\mathbf{F}$, undergoes an acceleration $\mathbf{a}$ satisfying

$$\mathbf{F} = m\mathbf{a} \tag{10}$$

This fundamental result is called **Newton's Second Law of Motion.** We will show how this principle, together with vector calculus, can be used to study the motion of an object moving near the surface of the earth in a vertical plane.

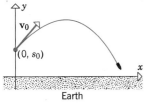

Figure 14.5.3

Consider an xy-coordinate system whose origin is on the surface of the earth and whose positive y-axis points vertically up. From an initial point s_0 units up on the y-axis, let a projectile of mass m be fired, thrown, or released at time $t = 0$ with an initial velocity vector of v_0 (Figure 14.5.3). We will obtain formulas for the position and velocity of the projectile as functions of t, where t is the amount of time elapsed (in seconds) from the initial firing or release.

We shall assume that once the projectile is fired or released, the only force acting on it is the downward force of the earth's gravity. For a projectile of mass m, it is shown in physics that the downward force of gravity is

$$\mathbf{F} = -mg\mathbf{j} \tag{11}$$

where g is a constant approximately equal to 32 ft/sec^2 if distance is measured in feet and 9.8 m/sec^2 if distance is measured in meters. By Newton's Second Law of Motion (10), this force produces an acceleration $\mathbf{a} = \mathbf{a}(t)$ satisfying

$$\mathbf{F} = m\mathbf{a}$$

or from (11)

$$-mg\mathbf{j} = m\mathbf{a} \tag{12}$$

or on canceling m

$$\mathbf{a} = -g\mathbf{j} \tag{13}$$

From (6) and (13) it follows that

$$\mathbf{v}(t) = \int -g\mathbf{j}\, dt$$

or since g is constant

$$\mathbf{v}(t) = -gt\mathbf{j} + \mathbf{c}_1 \tag{14}$$

where $\mathbf{c}_1$ is a vector constant of integration. Since the initial velocity at $t = 0$ is v_0, it follows on substituting $t = 0$ in (14) that

$$\mathbf{v}_0 = \mathbf{c}_1$$

so that (14) may be written

$$\mathbf{v}(t) = -gt\mathbf{j} + \mathbf{v}_0 \tag{15}$$

This formula specifies the velocity of the object at any time t.

The position function $\mathbf{r}(t)$ can be obtained using (7) and (15). These yield

$$\mathbf{r}(t) = \int (-gt\mathbf{j} + \mathbf{v}_0)\, dt$$

or since $\mathbf{v}_0$ (the initial velocity) is constant

$$\mathbf{r}(t) = -\frac{1}{2}gt^2\mathbf{j} + \mathbf{v}_0 t + \mathbf{c}_2 \tag{16}$$

Since the object is initially s_0 units up on the y-axis, the position vector at $t = 0$ is $\mathbf{r}(0) = s_0\mathbf{j}$. It follows on substituting $t = 0$ in (16) that

$$s_0\mathbf{j} = \mathbf{c}_2$$

so that (16) may be written

$$\mathbf{r}(t) = \left(-\frac{1}{2}gt^2 + s_0\right)\mathbf{j} + \mathbf{v}_0 t \tag{17}$$

This formula specifies the position of the object at any time t.

REMARK.   Because the mass m cancels in (12), the mass of the projectile does not enter into the final formulas for velocity and position. Physically, this means that the mass has no influence on the path or the velocity of the projectile—these are completely determined by the initial position and velocity. This explains the famous observation of Galileo, that two objects of different mass, released from the same height, will reach the ground at the same time.

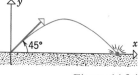

Figure 14.5.4

▶ Example 3   A shell, fired from a cannon, has a muzzle speed (the speed as it leaves the barrel) of 800 ft/sec. The barrel makes an angle of 45° with the horizontal and, for simplicity, the barrel opening is assumed to be at ground level.

(a)   Find parametric equations for the shell's trajectory relative to the coordinate system in Figure 14.5.4.

(b)   How high does the shell rise?

(c)   How far does the shell travel horizontally? ˙

(d)   What is the speed of the shell at its point of impact with the ground?

Solution (a).   From the given data, the initial velocity vector $\mathbf{v}_0$ has a magnitude of 800 and makes an angle of 45° with the positive x-axis. Thus,

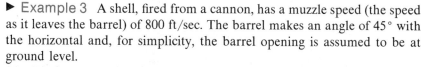

$$\mathbf{v}_0 = (800\cos 45°)\mathbf{i} + (800\sin 45°)\mathbf{j} = 400\sqrt{2}\,\mathbf{i} + 400\sqrt{2}\,\mathbf{j}$$

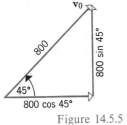

Figure 14.5.5

(Figure 14.5.5). From the given data, the initial y-coordinate of the shell is $s_0 = 0$. Substituting s_0 and $\mathbf{v}_0$ in (17) and taking $g = 32$ (since distance is measured in feet), we obtain the position vector of the shell after t seconds:

$$\mathbf{r}(t) = -16t^2\mathbf{j} + (400\sqrt{2}\,\mathbf{i} + 400\sqrt{2}\,\mathbf{j})t$$

or

$$\mathbf{r}(t) = (400\sqrt{2})t\mathbf{i} + (400\sqrt{2}\,t - 16t^2)\mathbf{j} \tag{18}$$

which yields the parametric equations

$$
\begin{aligned}
x &= 400\sqrt{2}\,t \\
y &= 400\sqrt{2}\,t - 16t^2
\end{aligned}
,\ t \geq 0
\qquad
\begin{aligned}
&\text{(19a)} \\
&\text{(19b)}
\end{aligned}
$$

(The reader should be able to verify that the trajectory is a parabola.)

Solution (b). The shell is rising when $dy/dt > 0$ (y is increasing with t) and is falling when $dy/dt < 0$ (y is decreasing with t). Consequently, the peak occurs when $dy/dt = 0$, since this is the transition between rise and fall. Thus, from (19b) the peak occurs when

$$400\sqrt{2} - 32t = 0$$

or

$$t = \frac{25\sqrt{2}}{2}$$

From (19b), the value of y at this time is

$$y = 5000 \ \text{(feet)}$$

Solution (c). The shell will hit the ground when $y = 0$. From (19b), this occurs when

$$400\sqrt{2}\,t - 16t^2 = 0$$

or

$$t(400\sqrt{2} - 16t) = 0$$

The solution $t = 0$ corresponds to the initial position of the shell and the solution $t = 25\sqrt{2}$ is the time of impact. Substituting the latter value in (19a) yields

$$x = 20{,}000 \ \text{ft}$$

as the horizontal distance traveled by the shell.

Solution (d).   The velocity at time t is obtained by differentiating (18):

$$\mathbf{v}(t) = 400\sqrt{2}\,\mathbf{i} + (400\sqrt{2} - 32t)\mathbf{j}$$

From (c), impact occurs when $t = 25\sqrt{2}$, so that the velocity vector at this point is

$$\mathbf{v}(25\sqrt{2}) = 400\sqrt{2}\,\mathbf{i} + [400\sqrt{2} - 32(25\sqrt{2})]\mathbf{j} = 400\sqrt{2}\,\mathbf{i} - 400\sqrt{2}\,\mathbf{j}$$

Thus, the speed at impact is

$$\|\mathbf{v}(25\sqrt{2})\| = \sqrt{(400\sqrt{2})^2 + (-400\sqrt{2})^2} = 800 \text{ ft/sec} \qquad \blacktriangleleft$$

NORMAL AND TANGENTIAL COMPONENTS OF ACCELERATION

If a particle moves along a plane curve C so that its position vector is

$$\mathbf{r}(t) = x(t)\mathbf{i} + y(t)\mathbf{j}$$

then the velocity and acceleration vectors are

$$\mathbf{v}(t) = x'(t)\mathbf{i} + y'(t)\mathbf{j}$$

and

$$\mathbf{a}(t) = x''(t)\mathbf{i} + y''(t)\mathbf{j}$$

These formulas express $\mathbf{v}(t)$ and $\mathbf{a}(t)$ in terms of components parallel to the coordinate axes. For many purposes, however, it is desirable to resolve the velocity and acceleration vectors into components parallel to the unit tangent and unit normal vectors to the curve C.

14.5.3   THEOREM   *For a particle moving along a smooth plane curve C, the velocity and acceleration vectors can be written as*

$$\mathbf{v} = \frac{ds}{dt}\mathbf{T} \tag{20}$$

$$\mathbf{a} = \frac{d^2s}{dt^2}\mathbf{T} + \kappa\left(\frac{ds}{dt}\right)^2\mathbf{N} \tag{21}$$

where s is an arc length parameter for the curve, and $\mathbf{T}, \mathbf{N}$, *and* κ *denote the unit tangent vector, unit normal vector, and curvature (Figure 14.5.6).*

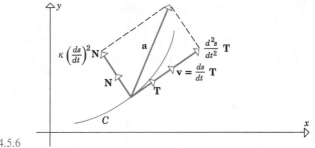

Figure 14.5.6

Proof. By Theorem 14.3.4 and the chain rule

$$\mathbf{v} = \frac{d\mathbf{r}}{dt} = \frac{d\mathbf{r}}{ds}\frac{ds}{dt} = \mathbf{T}\frac{ds}{dt} = \frac{ds}{dt}\mathbf{T} \tag{22}$$

which proves (20). To obtain (21) we differentiate (22), using the product rule of Theorem 14.2.4:

$$\mathbf{a} = \frac{d\mathbf{v}}{dt} = \frac{ds}{dt}\frac{d\mathbf{T}}{dt} + \mathbf{T}\frac{d^2s}{dt^2}$$

$$= \frac{ds}{dt}\frac{d\mathbf{T}}{ds}\frac{ds}{dt} + \mathbf{T}\frac{d^2s}{dt^2}$$

$$= \frac{ds}{dt}(\kappa\mathbf{N})\frac{ds}{dt} + \mathbf{T}\frac{d^2s}{dt^2} \qquad \begin{bmatrix} \text{Formula (1),} \\ \text{Section 14.4} \end{bmatrix}$$

Thus,

$$\mathbf{a} = \frac{d^2s}{dt^2}\mathbf{T} + \kappa\left(\frac{ds}{dt}\right)^2\mathbf{N} \quad \blacksquare$$

Formula (20) is in keeping with the fact that the velocity vector is tangent to *C* and has magnitude equal to the speed *ds/dt* of the particle. Formula (21) should make sense from your experience as an automobile passenger. If a car increases its speed rapidly, the effect is to press the passenger back against the seat. This occurs because the rapid increase in speed results in a large value for

$$\frac{d}{dt}\left(\frac{ds}{dt}\right) = \frac{d^2s}{dt^2}$$

which produces a large tangential component of acceleration in formula (21). On the other hand, if the car rounds a turn in the road, the passenger is thrown to the side. The greater the curvature in the road or the greater the speed of the car, the greater the force with which the passenger is thrown.

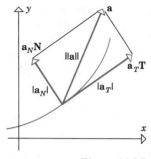

Figure 14.5.7

This occurs because ds/dt (speed) and κ (curvature) both enter as factors in the normal component of acceleration in (21).

The coefficients of $\mathbf{T}$ and $\mathbf{N}$ in (21) are called the *scalar tangential* and *scalar normal components of acceleration;* they are denoted by

$$a_T = \frac{d^2s}{dt^2} \qquad \text{and} \qquad a_N = \kappa \left(\frac{ds}{dt}\right)^2 \tag{23}$$

so that

$$\mathbf{a} = a_T\mathbf{T} + a_N\mathbf{N}$$

It follows from the Theorem of Pythagoras (Figure 14.5.7) that

$$\|\mathbf{a}\|^2 = a_T{}^2 + a_N{}^2 \tag{24}$$

It is a consequence of the remark following (1) and (2) of Section 14.4 that the vector $a_N\mathbf{N}$ points toward the concave side of the path of motion.

▶ Example 4 A particle moves so that its position vector at time t is $\mathbf{r}(t) = t\mathbf{i} + \frac{1}{2}t^2\mathbf{j}$. Find the magnitudes of the tangential and normal scalar components of acceleration at time $t = 1$.

Solution. Since

$$\mathbf{v} = \frac{d\mathbf{r}}{dt} = \mathbf{i} + t\mathbf{j} \tag{25}$$

the speed of the particle at time t is

$$\frac{ds}{dt} = \|\mathbf{v}\| = \sqrt{1 + t^2}$$

from which it follows that

$$|a_T| = \left|\frac{d^2s}{dt^2}\right| = \left|\frac{t}{\sqrt{1 + t^2}}\right| \tag{26}$$

Rather than compute a_N from (23), which requires the calculation of κ, we will use (24). From (25)

$$\mathbf{a} = \frac{d\mathbf{v}}{dt} = \mathbf{j}$$

so that

$$\|\mathbf{a}\| = 1$$

Thus,

$$a_N{}^2 = \|\mathbf{a}\|^2 - a_T{}^2 = 1 - \frac{t^2}{1 + t^2} = \frac{1}{1 + t^2}$$

so that

$$|a_N| = \frac{1}{\sqrt{1 + t^2}} \tag{27}$$

From (26) and (27) the values of $|a_T|$ and $|a_N|$ at time $t = 1$ are

$$|a_T| = \frac{1}{\sqrt{2}} \quad \text{and} \quad |a_N| = \frac{1}{\sqrt{2}} \qquad \blacktriangleleft$$

▶ Exercise Set 14.5

In Exercises 1–6, $\mathbf{r}(t)$ is the position vector of a particle moving in the plane. Find the velocity, acceleration, and speed at an arbitrary time t; then sketch the path of the particle together with the velocity and acceleration vectors at the indicated time t.

1. $\mathbf{r}(t) = 3 \cos t\,\mathbf{i} + 3 \sin t\,\mathbf{j}$; $t = \pi/3$.

2. $\mathbf{r}(t) = t\mathbf{i} + t^2\mathbf{j}$; $t = 2$.

3. $\mathbf{r}(t) = e^t\mathbf{i} + e^{-t}\mathbf{j}$; $t = 0$.

4. $\mathbf{r}(t) = (2 + 4t)\mathbf{i} + (1 - t)\mathbf{j}$; $t = 1$.

5. $\mathbf{r}(t) = \cosh t\,\mathbf{i} + \sinh t\,\mathbf{j}$; $t = \ln 2$.

6. $\mathbf{r}(t) = 4 \cos t\,\mathbf{i} + 9 \sin t\,\mathbf{j}$; $t = \pi/2$.

In Exercises 7–10, use the given information to find the position and velocity vectors of the particle.

7. $\mathbf{a}(t) = -32\mathbf{j}$; $\mathbf{v}(0) = \mathbf{0}$; $\mathbf{r}(0) = \mathbf{0}$.

8. $\mathbf{a}(t) = t\mathbf{j}$; $\mathbf{v}(0) = \mathbf{i} + \mathbf{j}$; $\mathbf{r}(0) = \mathbf{0}$.

9. $\mathbf{a}(t) = -\cos t\,\mathbf{i} - \sin t\,\mathbf{j}$; $\mathbf{v}(0) = \mathbf{i}$; $\mathbf{r}(0) = \mathbf{j}$.

10. $\mathbf{a}(t) = \mathbf{i} + e^{-t}\mathbf{j}$; $\mathbf{v}(0) = 2\mathbf{i} + \mathbf{j}$; $\mathbf{r}(0) = \mathbf{i} - \mathbf{j}$.

11. A shell is fired from ground level with a muzzle speed of 320 ft/sec and elevation angle of 60°. Find:
    (a) parametric equations for the shell's trajectory;

(b) the maximum height reached by the shell;
(c) the horizontal distance traveled by the shell;
(d) the speed of the shell at impact.

12. Solve Exercise 11 assuming that the muzzle speed is 980 m/sec and the elevation angle is 45°.

13. A rock is thrown downward from the top of a building, 168 ft high, at an angle of 60° with the horizontal. How far from the base of the building will the rock land if its initial speed is 80 ft/sec?

14. Solve Exercise 13 assuming that the rock is thrown horizontally at a speed of 80 ft/sec.

15. A shell is to be fired from ground level at an elevation angle of 30°. What should the muzzle speed be in order for the maximum height of the shell to be 2500 ft?

16. A shell, fired from ground level at an elevation angle of 45°, hits the ground 24,500 m away. Calculate the muzzle speed of the shell.

17. Find two elevation angles that will enable a shell, fired from ground level, with a muzzle speed of 800 ft/sec, to hit a ground-level target 10,000 ft away.

18. A ball rolls off a table 4 ft high while moving at a constant speed of 5 ft/sec.
 (a) How long does it take for the ball to hit the floor after it leaves the table?
 (b) At what speed does the ball hit the floor?
 (c) If a ball were dropped from table height (initial speed 0) at the same time the rolling ball leaves the table, which ball would hit the ground first?

In Exercises 19–24, find the magnitudes of the tangential and normal components of acceleration at the stated time t.

19. $\mathbf{r}(t) = 2\cos t\,\mathbf{i} + 2\sin t\,\mathbf{j}; \ t = \pi/3.$

20. $\mathbf{r}(t) = t\,\mathbf{i} + t^2\,\mathbf{j}; \ t = 1.$

21. $\mathbf{r}(t) = e^{-t}\,\mathbf{i} + e^t\,\mathbf{j}; \ t = 0.$

22. $\mathbf{r}(t) = \cos t^2\,\mathbf{i} + \sin t^2\,\mathbf{j}; \ t = \sqrt{\pi}/2.$

23. $\mathbf{r}(t) = (t^3 - 2t)\,\mathbf{i} + (t^2 - 4)\,\mathbf{j}; \ t = 1.$

24. $\mathbf{r}(t) = e^t\cos t\,\mathbf{i} + e^t\sin t\,\mathbf{j}; \ t = \pi/4.$

25. Prove: If a particle moves with constant speed along a plane curve, then the velocity and acceleration vectors are perpendicular at each point of the curve.

26. Prove: If a particle moves in the plane so that its acceleration vector is always $\mathbf{0}$, then the path of the particle is a line.

27. Prove: If a particle moves along the curve $y = f(x)$ so x increases with time, then the acceleration vector is tangent to the curve at a point of inflection.

28. A shell is fired from ground level at an elevation angle of α and with a muzzle speed of v_0.
 (a) Find parametric equations for the shell's trajectory.
 (b) Show that the maximum height reached by the shell is
$$\text{maximum height} = \frac{(v_0\sin\alpha)^2}{2g}$$
 (c) With v_0 kept fixed, what elevation angle will enable the shell to travel the largest possible horizontal distance?

29. The nuclear accelerator at the Enrico Fermi Laboratory is circular with a radius of 1 kilometer. Find the magnitude of the normal component of acceleration of a proton moving around the accelerator with a constant speed of 3×10^5 kilometers per second.

▶ SUPPLEMENTARY EXERCISES

1. Find $\overrightarrow{P_1P_2}$ and $\|\overrightarrow{P_1P_2}\|$.
 (a) $P_1(2, 3), P_2(5, -1)$ (b) $P_1(2, -1), P_2(1, 3)$.

In Exercises 2–7, find all vectors satisfying the given conditions.

2. A vector of length 1, that is perpendicular to the line $x + y = -1$.

3. The vector oppositely directed to $3\mathbf{i} - 4\mathbf{j}$, and having the same length.

4. The vector obtained by rotating $\mathbf{i}$ counterclockwise through an angle θ.

5. The vector with initial point $(1, 2)$ and a terminal point that is 3/5 of the way from $(1, 2)$ to $(5, 5)$.

6. A vector of length 2 that is parallel to the tangent to the curve $y = x^2$ at $(-1, 1)$.

7. The vector of length 12 that makes an angle of $120°$ with the x-axis.

8. Solve for c_1 and c_2 given that
$c_1\langle -2, 5\rangle + 3c_2\langle 1, 3\rangle = \langle -6, -51\rangle$

9. Solve for $\mathbf{u}$ if $3\mathbf{u} - (\mathbf{i} + \mathbf{j}) = \mathbf{i} + \mathbf{u}$.

10. Solve for $\mathbf{u}$ and $\mathbf{v}$ if $3\mathbf{u} - 4\mathbf{v} = 3\mathbf{v} - 2\mathbf{u} = \langle 1, 2\rangle$.

11. Two forces $\mathbf{F}_1 = 2\mathbf{i} - \mathbf{j}$ and $\mathbf{F}_2 = -3\mathbf{i} - 4\mathbf{j}$ are applied at a point. What force $\mathbf{F}_3$ must be applied at the point to cancel the effect of $\mathbf{F}_1$ and $\mathbf{F}_2$?

12. Given the points $P(3, 4)$, $Q(1, 1)$, and $R(5, 2)$, use vector methods to find the coordinates of the fourth vertex of the parallelogram whose adjacent sides are $\overrightarrow{PQ}$ and $\overrightarrow{QR}$.

In Exercises 13–15,
(a) find $\mathbf{v} = d\mathbf{r}/dt$ and $\mathbf{a} = d^2\mathbf{r}/dt^2$;
(b) sketch the graph of $\mathbf{r}(t)$, showing the direction of increasing t, and find the vectors $\mathbf{r}'(t)$ and $\mathbf{r}''(t)$ at the points corresponding to $t = t_0$ and $t = t_1$.

13. $\mathbf{r}(t) = \sqrt{t + 4}\,\mathbf{i} + 2t\mathbf{j}$; $t_0 = -3$, $t_1 = 0$.

14. $\mathbf{r}(t) = \langle 2 + \cosh t, 1 - 2\sinh t \rangle$; $t_0 = 0$, $t_1 = \ln 2$.

15. $\mathbf{r}(t) = \langle 2t^3 - 1, t^3 + 1 \rangle$; $t_0 = 0$, $t_1 = -\frac{1}{2}$.

16. Find the limits.

(a) $\lim\limits_{t \to e} \langle t + \ln t^2, \ln t + t^2 \rangle$

(b) $\lim\limits_{t \to \pi/6} (\cos 2t\mathbf{i} - 3t\mathbf{j})$.

17. Evaluate the integrals.

(a) $\displaystyle\int (k\mathbf{i} + m\mathbf{j})\,dt$

(b) $\displaystyle\int_0^{\ln 3} \langle e^{2t}, 2e^t \rangle\,dt$

(c) $\displaystyle\int_0^2 \|\cos t\mathbf{i} + \sin t\mathbf{j}\|\,dt$

(d) $\displaystyle\int \frac{d}{dt}[\sqrt{t^2 + 3}\,\mathbf{i} + \ln(\sin t)\mathbf{j}]\,dt$.

In Exercises 18–20, find: (a) ds/dt (b) parametric equations for the curve with arc length s as a parameter assuming that the point corresponding to t_0 is the reference point.

18. $\mathbf{r}(t) = (3e^t + 2)\mathbf{i} + (e^t - 1)\mathbf{j}$; $t_0 = 0$.

19. $\mathbf{r}(t) = \left\langle \dfrac{t^2 + 1}{t}, \ln t^2 \right\rangle$, where $t > 0$; $t_0 = 1$.

20. $\mathbf{r}(t) = \langle t^3, t^2 \rangle$, where $t \geq 0$; $t_0 = 0$.

For the curves given in Exercises 21–23, find:
(a) the unit tangent vector $\mathbf{T}$ at P_0;
(b) the unit normal vector $\mathbf{N}$ at P_0;
(c) the curvature κ at P_0.

21. $\mathbf{r}(t) = (t^2 + 1)\mathbf{i} + (1/t)\mathbf{j}$; $P_0(2, 1)$.

22. $y = \ln x$; $P_0(1, 0)$.

23. $x = (y - 1)^2$; $P_0(0, 1)$.

In Exercises 24–27, find the curvature κ of the given curve at P_0.

24. $xy^2 = 1$; $P_0(1, 1)$.

25. $\mathbf{r}(t) = (t + t^3)\mathbf{i} + (t + t^2)\mathbf{j}$; $P_0(2, 2)$.

26. $y = a\cosh(x/a)$; $P_0(a, a\cosh 1)$.

27. $e^x = \sec y$; $P_0(0, 0)$.

28. Find the smallest radius of curvature and the point at which it occurs.
(a) $y = e^x$ (b) $\mathbf{r}(t) = \langle e^{2t}, e^{-2t} \rangle$.

29. Find the equation of the osculating circle for the parabola $y = (x - 1)^2$ at the point $(1, 0)$. Verify that y' and y'' for the parabola are the same as y' and y'' for the osculating circle at $(1, 0)$.

In Exercises 30 and 31, calculate $d\mathbf{u}/dw$ by the chain rule, and check the result by first expressing $\mathbf{u}$ in terms of w.

30. $\mathbf{u} = \langle \sin t, 2\cos 2t \rangle$; $t = e^{w/2}$.

31. $\mathbf{u} = \langle e^t - 1, 2e^{2t} \rangle$; $t = \ln w$.

In Exercises 32 and 33, find the magnitudes of the tangential and normal components of acceleration.

32. $\mathbf{r}(t) = \langle \cosh 2t, \sinh 2t \rangle$, $t \geq 0$.

33. $\mathbf{r}(t) = \langle \sin t - t\cos t, \cos t + t\sin t \rangle$, $t \geq 0$.

For the motion described in Exercises 34 and 35:
(a) find $\mathbf{v}$, $\mathbf{a}$ and ds/dt at P_0;
(b) find $\mathbf{T}$, $\mathbf{N}$, and κ at P_0;
(c) find a_T and a_N at P_0;
(d) describe the trajectory;
(e) find the center of the osculating circle at P_0.

34. $\mathbf{r}(t) = (1 - t^2)\mathbf{i} + 2t\mathbf{j}$; $P_0(0, 2)$.

35. $\mathbf{r}(t) = \langle e^{-t}, e^t \rangle$; $P_0(1, 1)$.

36. At $t = 0$, a particle of mass m is located at the point $(-2/m, 0)$ and has a velocity of $(2\mathbf{i} - 3\mathbf{j})/m$. Find the position function $\mathbf{r}(t)$ if the particle is acted upon by a force $\mathbf{F} = \langle 2\cos t, 3\sin t \rangle$ for $t \geq 0$.

37. The force acting on a particle of unit mass $(m = 1)$ is $\mathbf{F} = (\sin t)\mathbf{i} + (4e^{2t})\mathbf{j}$. If the particle starts at the origin with an initial velocity $\mathbf{v}_0 = \mathbf{i} + 2\mathbf{j}$, find the position function $\mathbf{r}(t)$ at any $t \geq 0$.

38. A curve in a railroad track has the shape of the parabola $x = y^2/100$. If a train is loaded so that its scalar normal component of acceleration cannot exceed 25 units/sec^2, what is its maximum possible speed as it rounds the curve at $(0, 0)$?

39. A particle moves along the parabola $y = 2x - x^2$ with a constant velocity of 4 ft/sec in the x-direction. Find the magnitudes of the scalar tangential and normal components of acceleration at the points (a) $(1, 1)$ and (b) $(0, 0)$.

40. If a particle moves along the curve $y = 2x^2$ with constant speed $ds/dt = 10$, what are a_T and a_N at $P(x, 2x^2)$?

41. A child twirls a weight at the end of a 2-meter string at a rate of 1 revolution/second. Find a_T and a_N for the motion of the weight.

For the motion described in Exercises 42–43, find (a) ds/dt and (b) the distance traveled over the interval described.

42. $\mathbf{r}(t) = \langle 2 \sinh t, \sinh^2 t \rangle$, $0 \le t \le 1$.

43. $\mathbf{r}(t) = e^t \langle \sin 2t, \cos 2t \rangle$, $0 \le t \le \ln 3$.

15 three-dimensional space

15.1 RECTANGULAR COORDINATES IN 3-SPACE; SPHERES; CYLINDRICAL SURFACES

Just as points in a plane can be placed in one-to-one correspondence with pairs of real numbers by using two perpendicular coordinate lines, so points in three-dimensional space can be placed in one-to-one correspondence with triples of real numbers by using three mutually perpendicular coordinate lines. To obtain this correspondence, we choose the coordinate lines so they intersect at their origins, and we call these lines the *x-axis,* the *y-axis,* and the *z-axis* (Figure 15.1.1*a*).

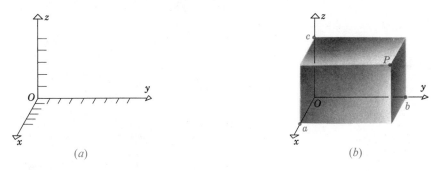

Figure 15.1.1   (*a*)   (*b*)

The three coordinate axes form a three-dimensional *rectangular* or *Cartesian coordinate system* and the point of intersection of the coordinate axes is called the *origin* of the coordinate system.

Each pair of coordinate axes determines a plane called a *coordinate plane.* These are referred to as the *xy-plane,* the *xz-plane,* and the *yz-plane.* To each point *P* in 3-space we assign a triple of numbers (*a, b, c*), called the *coordi-*

nates of P, by passing three planes through P parallel to the coordinate planes, and letting a, b, and c be the coordinates of the intersections of these planes, with the x, y, and z axes, respectively (Figure 15.1.1b). The notation $P(a, b, c)$ will sometimes be used to denote a point P with coordinates (a, b, c).

In Figure 15.1.2 we have constructed the points whose coordinates are $(4, 5, 6)$ and $(-3, 2, -4)$.

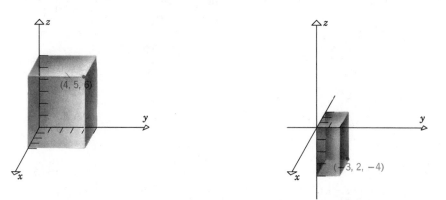

Figure 15.1.2

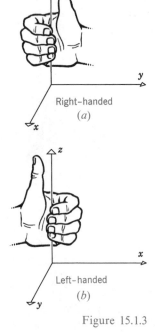

Right-handed
(*a*)

Left-handed
(*b*)

Figure 15.1.3

In this text we will call three-dimensional space *3-space,* two-dimensional space (a plane) *2-space,* and one-dimensional space (a line) *1-space.*

Rectangular coordinate systems in 3-space fall into two categories: *left-handed* and *right-handed.* A right-handed system has the property that when the fingers of the right hand are cupped so they curve from the positive x-axis toward the positive y-axis, the thumb points (roughly) in the direction of the positive z-axis (Figure 15.1.3a). A system that is not right-handed is called left-handed (Figure 15.1.3b). We shall use only right-handed coordinate systems.

Just as the coordinate axes in a two-dimensional coordinate system divide 2-space into four quadrants, so the coordinate planes of a three-dimensional coordinate system divide 3-space into eight parts, called *octants* (count them). Those points having three positive coordinates form the *first octant;* the remaining octants have no standard numbering.

The reader should be able to visualize the following results about three dimensional rectangular coordinate systems:

REGION	DESCRIPTION
xy-plane	consists of all points of the form $(x, y, 0)$
xz-plane	consists of all points of the form $(x, 0, z)$
yz-plane	consists of all points of the form $(0, y, z)$
x-axis	consists of all points of the form $(x, 0, 0)$
y-axis	consists of all points of the form $(0, y, 0)$
z-axis	consists of all points of the form $(0, 0, z)$

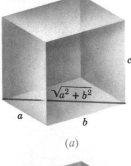

To obtain a formula for the distance between two points in 3-space, we first consider a box (rectangular parallelepiped) with dimensions a, b, and c (Figure 15.1.4a). By the Theorem of Pythagoras, the length of the diagonal of the base is $\sqrt{a^2 + b^2}$. Moreover, the diagonal of the box is the hypotenuse of a right triangle having a diagonal of the base for one side and a vertical edge for the other side (Figure 15.1.4b). Thus, by the Theorem of Pythagoras, the length of the diagonal of the box is the square root of $[\sqrt{a^2 + b^2}]^2 + c^2$ or more simply

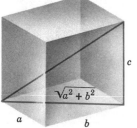

$$\begin{bmatrix} \text{the length of a} \\ \text{diagonal of a} \\ \text{box} \end{bmatrix} = \sqrt{a^2 + b^2 + c^2} \tag{1}$$

Suppose now that $P_1(x_1, y_1, z_1)$ and $P_2(x_2, y_2, z_2)$ are points in 3-space. As shown in Figure 15.1.5, these points are on diagonally opposite corners of a box whose dimensions are

$$|x_2 - x_1|, \qquad |y_2 - y_1|, \qquad \text{and} \qquad |z_2 - z_1|$$

Thus, from (1) the distance d between P_1 and P_2 is

$$d = \sqrt{|x_2 - x_1|^2 + |y_2 - y_1|^2 + |z_2 - z_1|^2}$$

or equivalently

$$d = \sqrt{(x_2 - x_1)^2 + (y_2 - y_1)^2 + (z_2 - z_1)^2} \tag{2}$$

(a)

(b)

Figure 15.1.4

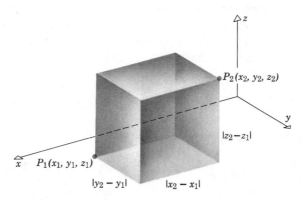

Figure 15.1.5

▶ Example 1   The distance between the points $(4, -1, 3)$ and $(2, 3, -1)$ is

$$d = \sqrt{(4 - 2)^2 + (-1 - 3)^2 + (3 + 1)^2} = \sqrt{36} = 6 \qquad ◀$$

In 2-space the midpoint of the line segment joining $P_1(x_1, y_1)$ and $P_2(x_2, y_2)$ is

$$\left(\frac{x_1 + x_2}{2}, \frac{y_1 + y_2}{2} \right)$$

In 3-space the formula is similar:

$$\begin{bmatrix} \textit{The midpoint of the} \\ \textit{line segment joining} \\ P_1(x_1, y_1, z_1) \textit{ and } P_2(x_2, y_2, z_2) \end{bmatrix} = \left(\frac{x_1 + x_2}{2}, \frac{y_1 + y_2}{2}, \frac{z_1 + z_2}{2} \right)$$

(We omit the proof.)

▶ Example 2   The midpoint of the line segment joining the points $(-1, 3, -8)$ and $(3, 1, 0)$ is

$$\left(\frac{-1 + 3}{2}, \frac{3 + 1}{2}, \frac{-8 + 0}{2} \right) = (1, 2, -4) \qquad \blacktriangleleft$$

SPHERES   In 2-space, the graph of an equation relating the variables x and y is the set of all points (x, y) whose coordinates satisfy the equation. Usually, such graphs are curves. Similarly, in 3-space the graph of an equation relating x, y, and z is the set of all points (x, y, z) whose coordinates satisfy the equation. Usually, such graphs are surfaces in 3-space. For example, in 3-space, a sphere with center (x_0, y_0, z_0) and radius r consists of all points (x, y, z) at a distance of r units from (x_0, y_0, z_0) (Figure 15.1.6). Thus, from (2)

$$\sqrt{(x - x_0)^2 + (y - y_0)^2 + (z - z_0)^2} = r$$

or equivalently

Equation of a Sphere

$$(x - x_0)^2 + (y - y_0)^2 + (z - z_0)^2 = r^2 \qquad (3)$$

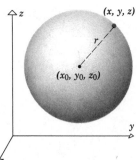

Figure 15.1.6

This equation represents the sphere with center (x_0, y_0, z_0) and radius r; it is called the **standard equation** of a sphere. Some examples are:

EQUATION	GRAPH
$(x - 3)^2 + (y - 2)^2 + (z - 1)^2 = 9$	Sphere with center $(3, 2, 1)$ and radius 3
$(x + 1)^2 + y^2 + (z + 4)^2 = 5$	Sphere with center $(-1, 0, -4)$ and radius $\sqrt{5}$
$x^2 + y^2 + z^2 = 1$	Sphere with center $(0, 0, 0)$ and radius 1

▶ **Example 3** Describe the graph of

$$x^2 + y^2 + z^2 - 2x - 4y + 8z + 17 = 0$$

Solution. We can put the equation in the form of (3) by completing the squares:

$$(x^2 - 2x) + (y^2 - 4y) + (z^2 + 8z) = -17$$
$$(x^2 - 2x + 1) + (y^2 - 4y + 4) + (z^2 + 8z + 16) = -17 + 21$$
$$(x - 1)^2 + (y - 2)^2 + (z + 4)^2 = 4$$

Thus, the graph is a sphere of radius 2 centered at $(1, 2, -4)$.   ◀

By squaring out and collecting terms in (3), we can rewrite the equation of a sphere in the form:

$$x^2 + y^2 + z^2 + Gx + Hy + Iz + J = 0 \qquad (4)$$

However, it is not true that every equation of form (4) has a sphere as its graph; for if we begin with an equation of form (4) and complete the squares as in Example 2, we will obtain an equation of the form:

$$(x - x_0)^2 + (y - y_0)^2 + (z - z_0)^2 = k$$

where k is a constant. If $k > 0$, then the equation represents a sphere of radius $\sqrt{k}$. However, if $k = 0$ the graph is the single point (x_0, y_0, z_0), and if $k < 0$ the equation is not satisfied by any real values of x, y, and z; thus, there is no graph. To summarize:

15.1.1 THEOREM *An equation of the form*

$$x^2 + y^2 + z^2 + Gx + Hy + Iz + J = 0$$

represents a sphere or a point, or else has no graph.

CYLINDRICAL
SURFACES

Now that we have introduced graphs in 3-space, an equation in x and y such as

$$x + y = 1 \tag{5}$$

becomes ambiguous. Should we graph this equation in 2-space (Figure 15.1.7*a*) or should we take the position that (5) is really the equation

$$x + y + 0z = 1 \tag{6}$$

in which case it should be graphed in 3-space? If we take the latter position, then the graph is the plane shown in Figure 15.1.7*b* since (x, y, z) will satisfy (6) for *arbitrary z* as long as x and y satisfy (5).

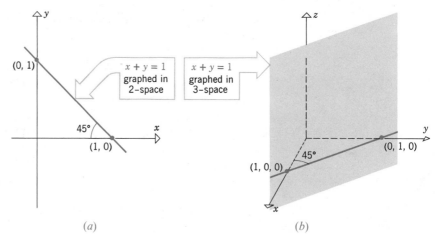

Figure 15.1.7   (a)   (b)

Actually, it is correct to graph (5) either in 2-space or 3-space. However, the appropriate choice will usually be clear from the context in which the equation arises.

A three-dimensional surface generated by a line traversing a plane curve and moving parallel to a fixed line is sometimes called a ***cylindrical surface;*** the moving line is called the ***generator*** of the surface.

In general, if an equation in x and y has a curve C for its graph in the xy-plane, then the same equation graphed in 3-space produces the cylindrical surface that is traced out as a generator parallel to the z-axis traverses the curve C in the xy-plane. Similarly, an equation in x and z only represents a cylindrical surface in 3-space with generator parallel to the y-axis; and an equation in y and z alone represents a cylindrical surface in 3-space with generator parallel to the x-axis. In brief:

An equation containing only two of the three variables x, y, and z represents a cylindrical surface in 3-space. The generator of the surface is parallel to the axis corresponding to the missing variable.

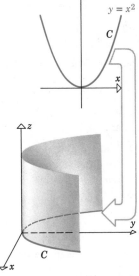

▶ Example 4 Sketch the surface $y = x^2$ in 3-space.

Solution. (Figure 15.1.8.)   ◀

▶ Example 5 Sketch the graph of $x^2 + z^2 = 1$ in 3-space.

Solution. Since y does not appear in this equation, the graph is a cylindrical surface with generator parallel to the y-axis. It is helpful to begin with a sketch of the equation in 2-space. In the xz-plane the curve $x^2 + z^2 = 1$ is a circle (Figure 15.1.9a). Thus, in 3-space this equation represents a right-circular cylinder parallel to the y-axis (15.1.9b).

Figure 15.1.8

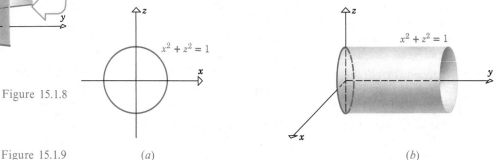

Figure 15.1.9   (a)   (b)

▶ Example 6 Sketch the graph of $z = \sin y$.

Solution. (Figure 15.1.10).

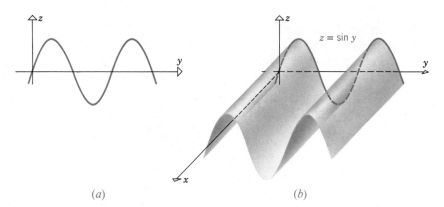

Figure 15.1.10   (a)   (b)

▶ Exercise Set 15.1

1. Plot the points P and Q in a right-handed coordinate system. Then, find the distance between them and the midpoint of the line segment joining them.
   (a) $P(0, 0, 0)$; $Q(2, 1, 3)$
   (b) $P(5, 2, 3)$; $Q(4, 1, 6)$
   (c) $P(-2, -1, 3)$; $Q(3, 0, 5)$
   (d) $P(-1, -1, -3)$; $Q(4, 3, -2)$.

2. A cube of side 4 has its geometric center at the origin and its faces parallel to the coordinate planes. Sketch the cube and give the coordinates of the vertices.

3. A rectangular parallelepiped has its faces parallel to the coordinate planes and has $(4, 2, -2)$ and $(-6, 1, 1)$ as endpoints of a diagonal. Sketch the parallelepiped and give the coordinates of the vertices.

4. Show that $(4, 5, 2)$, $(1, 7, 3)$, and $(2, 4, 5)$ are vertices of an equilateral triangle.

5. (a) Show that $(2, 1, 6)$, $(4, 7, 9)$, and $(8, 5, -6)$ are the vertices of a right triangle.
   (b) Which vertex is at the $90°$ angle?
   (c) Find the area of the triangle.

6. Find the distance from the point $(-5, 2, -3)$ to the:
   (a) xy-plane   (b) xz-plane
   (c) yz-plane   (d) x-axis
   (e) y-axis   (f) z-axis.

7. Show that the distance from a point (x_0, y_0, z_0) to the z-axis is $\sqrt{x_0{}^2 + y_0{}^2}$, and find the distances from the point to the x- and y-axes.

In Exercises 8–11, find an equation for the sphere with center C and radius r.

8. $C(0, 0, 0)$; $r = 8$.

9. $C(-2, 4, -1)$; $r = 6$.

10. $C(5, -2, 4)$; $r = \sqrt{7}$.

11. $C(0, 1, 0)$; $r = 3$.

12. In each part find an equation for the sphere with center $(-3, 5, -4)$ and satisfying the given condition:
    (a) tangent to the xy-plane

    (b) tangent to the xz-plane
    (c) tangent to the yz-plane.

13. In each part, find an equation for the sphere with center $(2, -1, -3)$ and satisfying the given condition.
    (a) tangent to the xy-plane
    (b) tangent to the xz-plane
    (c) tangent to the yz-plane.

In Exercises 14–19, find the standard equation of the sphere satisfying the given conditions.

14. Center $(1, 0, -1)$; diameter $= 8$.

15. A diameter has endpoints $(-1, 2, 1)$ and $(0, 2, 3)$.

16. Center $(-1, 3, 2)$ and passing through the origin.

17. Center $(3, -2, 4)$ and passing through $(7, 2, 1)$.

18. Center $(-3, 5, -4)$; tangent to the sphere of radius 1 centered at the origin (two answers).

19. Center $(0, 0, 0)$; tangent to the sphere of radius 1 centered at $(3, -2, 4)$ (two answers).

In Exercises 20–25, describe the surface whose equation is given.

20. $x^2 + y^2 + z^2 - 2x - 6y - 8z + 1 = 0$.

21. $x^2 + y^2 + z^2 + 10x + 4y + 2z - 19 = 0$.

22. $x^2 + y^2 + z^2 - y = 0$.

23. $2x^2 + 2y^2 + 2z^2 - 2x - 3y + 5z - 2 = 0$.

24. $x^2 + y^2 + z^2 + 2x - 2y + 2z + 3 = 0$.

25. $x^2 + y^2 + z^2 - 3x + 4y - 8z + 25 = 0$.

In Exercises 26–29, sketch the surface whose equation is given.

26. (a) $y = x$ (b) $y = z$
    (c) $x = z$.

27. (a) $x^2 + y^2 = 25$ (b) $y^2 + z^2 = 25$
    (c) $x^2 + z^2 = 25$.

28. (a) $y = x^2$ (b) $z = x^2$
    (c) $y = z^2$.

29. (a) $y = e^x$ (b) $x = \ln z$
    (c) $yz = 1$.

30. Find an equation for the right-circular cylinder of radius a shown.

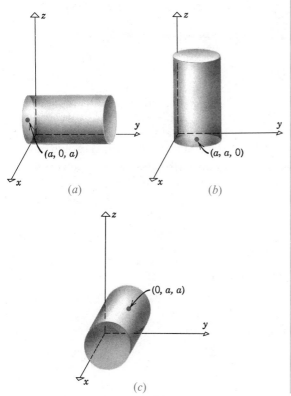

    (a)   (b)

          (c)

31. Show that for all values of θ and ϕ, the point $(a \sin \phi \cos \theta,\ a \sin \phi \sin \theta,\ a \cos \phi)$ lies on the sphere $x^2 + y^2 + z^2 = a^2$.

32. Consider the equation

$$x^2 + y^2 + z^2 + Gx + Hy + Iz + J = 0$$

and let $K = G^2 + H^2 + I^2 - 4J$.

(a) Prove the equation represents a sphere, a point, or no graph, according to whether $K > 0$, $K = 0$, or $K < 0$.

(b) In the case where $K > 0$, find the center and radius of the sphere.

15.2 VECTORS AND LINES IN 3-SPACE

In Section 14.1 we defined a vector to be a directed line segment or arrow in two-dimensional space or three-dimensional space. The geometric definitions of equality of vectors, addition, subtraction, and scalar multiplication given in that section apply equally well to vectors in 2-space and 3-space. However, where a vector $\mathbf{v}$ in 2-space is represented in a rectangular coordinate system by a pair of components, $\mathbf{v} = \langle v_1, v_2 \rangle$, we will see that it requires three components to specify a vector in 3-space.

Let $\mathbf{v}$ be a vector in 3-space positioned so its initial point is at the origin of a rectangular coordinate system (Figure 15.2.1). The coordinates (v_1, v_2, v_3) of the terminal point of $\mathbf{v}$ are called the **components** of $\mathbf{v}$ and we write

$$\mathbf{v} = \langle v_1, v_2, v_3 \rangle$$

The **zero vector** in 3-space is

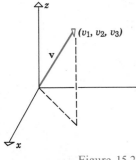

Figure 15.2.1

$$\mathbf{0} = \langle 0, 0, 0 \rangle$$

If $\mathbf{v} = \langle v_1, v_2, v_3 \rangle$ and $\mathbf{w} = \langle w_1, w_2, w_3 \rangle$, then

$\mathbf{v} = \mathbf{w}$ if and only if $v_1 = w_1$, $v_2 = w_2$, and $v_3 = w_3$

$\mathbf{v} + \mathbf{w} = \langle v_1 + w_1, v_2 + w_2, v_3 + w_3 \rangle$

$\mathbf{v} - \mathbf{w} = \langle v_1 - w_1, v_2 - w_2, v_3 - w_3 \rangle$

$k\mathbf{v} = \langle kv_1, kv_2, kv_3 \rangle$

Moreover, the results listed in Theorem 14.1.5 remain valid for vectors in 3-space. The proofs of these results are analogous to the corresponding proofs in 2-space, and will be omitted. In addition, the reader should have no trouble modifying the proof of Theorem 14.1.4 to show that the vector $\overrightarrow{P_1P_2}$ with initial point $P_1(x_1, y_1, z_1)$ and terminal point $P_2(x_2, y_2, z_2)$ is

$$\overrightarrow{P_1P_2} = \langle x_2 - x_1, y_2 - y_1, z_2 - z_1 \rangle \tag{1}$$

(See Figure 15.2.2.)

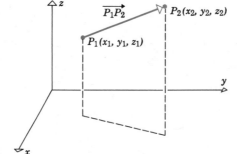

Figure 15.2.2

▶ **Example 1** Formula (1) can be used to show that the midpoint M of the line segment joining the points $P_1(x_1, y_1, z_1)$ and $P_2(x_2, y_2, z_2)$ has coordinates $M(\bar{x}, \bar{y}, \bar{z})$ given by

$$\bar{x} = \frac{x_1 + x_2}{2}, \qquad \bar{y} = \frac{y_1 + y_2}{2}, \qquad \bar{z} = \frac{z_1 + z_2}{2} \tag{2}$$

We need only observe that

$$\overrightarrow{P_1M} = \tfrac{1}{2}\overrightarrow{P_1P_2}$$

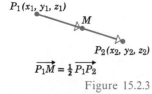

$\overrightarrow{P_1M} = \tfrac{1}{2}\overrightarrow{P_1P_2}$

Figure 15.2.3

(Figure 15.2.3), so that

$$\langle \bar{x} - x_1, \bar{y} - y_1, \bar{z} - z_1 \rangle = \tfrac{1}{2}\langle x_2 - x_1, y_2 - y_1, z_2 - z_1 \rangle$$

or, on equating components,

$$\bar{x} - x_1 = \tfrac{1}{2}(x_2 - x_1), \qquad \bar{y} - y_1 = \tfrac{1}{2}(y_2 - y_1), \qquad \bar{z} - z_1 = \tfrac{1}{2}(z_2 - z_1)$$

from which (2) follows. ◀

The **length** or **norm** of a vector $\mathbf{v} = \langle v_1, v_2, v_3 \rangle$, denoted by $\|\mathbf{v}\|$, is the distance from the origin to the point (v_1, v_2, v_3) (Figure 15.2.1). Thus,

$$\|\mathbf{v}\| = \sqrt{v_1^2 + v_2^2 + v_3^2}$$

In 3-space, as in 2-space, we have the basic relationship

$$\|k\mathbf{v}\| = |k|\,\|\mathbf{v}\|$$

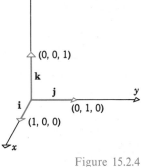

Figure 15.2.4

since multiplying $\mathbf{v}$ by the scalar k multiplies the length of $\mathbf{v}$ by a factor of $|k|$. Of special importance in 3-space are the unit vectors

$$\mathbf{i} = \langle 1, 0, 0 \rangle, \qquad \mathbf{j} = \langle 0, 1, 0 \rangle, \qquad \mathbf{k} = \langle 0, 0, 1 \rangle$$

which run along the positive x-axis, y-axis, and z-axis (Figure 15.2.4). Every vector $\mathbf{v} = \langle v_1, v_2, v_3 \rangle$ in 3-space is expressible in terms of $\mathbf{i}, \mathbf{j}$, and $\mathbf{k}$ since we can write

$$\begin{aligned}\mathbf{v} = \langle v_1, v_2, v_3 \rangle &= v_1 \langle 1, 0, 0 \rangle + v_2 \langle 0, 1, 0 \rangle + v_3 \langle 0, 0, 1 \rangle \\ &= v_1 \mathbf{i} + v_2 \mathbf{j} + v_3 \mathbf{k}\end{aligned}$$

▶ Example 2

$$\langle 2, -3, 4 \rangle = 2\mathbf{i} - 3\mathbf{j} + 4\mathbf{k}$$
$$\langle 0, 3, 0 \rangle = 3\mathbf{j}$$
$$\langle 4, 1, 2 \rangle + 2\langle 6, -1, 3 \rangle = \langle 4, 1, 2 \rangle + \langle 12, -2, 6 \rangle = \langle 16, -1, 8 \rangle$$
$$(3\mathbf{i} + 2\mathbf{j} - \mathbf{k}) - (4\mathbf{i} - \mathbf{j} + 2\mathbf{k}) = -\mathbf{i} + 3\mathbf{j} - 3\mathbf{k}$$
$$2(\mathbf{i} + \mathbf{j} - \mathbf{k}) + 4(\mathbf{i} - \mathbf{j}) = 6\mathbf{i} - 2\mathbf{j} - 2\mathbf{k}$$
$$\|\langle 1, 2, -3 \rangle\| = \|\mathbf{i} + 2\mathbf{j} - 3\mathbf{k}\| = \sqrt{1^2 + 2^2 + (-3)^2} = \sqrt{14} \quad ◀$$

Just as a curve in 2-space may be represented by a pair of parametric equations

$$x = x(t)$$
$$y = y(t)$$

so a curve C in 3-space can be represented by three parametric equations

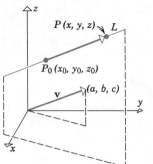

Figure 15.2.5

$$x = x(t)$$
$$y = y(t)$$
$$z = z(t)$$

As the parameter t varies, the point $(x, y, z) = \big(x(t), y(t), z(t)\big)$ traces out the curve C. As an illustration, we will derive parametric equations for a line in three-dimensional space.

Suppose L is the line in 3-space passing through the point $P_0(x_0, y_0, z_0)$ and parallel to the nonzero vector $\mathbf{v} = a\mathbf{i} + b\mathbf{j} + c\mathbf{k}$. It is clear (Figure 15.2.5) that L consists precisely of those points $P(x, y, z)$ for which the vector $\overrightarrow{P_0P}$ is parallel to $\mathbf{v}$. In other words, the point $P(x, y, z)$ is on L if and only if $\overrightarrow{P_0P}$ is a scalar multiple of $\mathbf{v}$, say

$$\overrightarrow{P_0P} = t\mathbf{v}$$

In terms of components, this equation can be written

$$\langle x - x_0, y - y_0, z - z_0 \rangle = \langle ta, tb, tc \rangle$$

from which it follows that

$$x - x_0 = ta, \qquad y - y_0 = tb, \qquad z - z_0 = tc$$

This yields the following result:

15.2.1   THEOREM
Parametric Equations for a Line in 3-space

The line passing through the point (x_0, y_0, z_0) and parallel to the nonzero vector $\mathbf{v} = \langle a, b, c \rangle = a\mathbf{i} + b\mathbf{j} + c\mathbf{k}$ has parametric equations

$$x = x_0 + at$$
$$y = y_0 + bt$$
$$z = z_0 + ct$$

▶ **Example 3**   The line passing through the point $(1, 2, -3)$ and parallel to the vector $\mathbf{v} = 4\mathbf{i} + 5\mathbf{j} - 7\mathbf{k}$ has parametric equations

$$x = 1 + 4t$$
$$y = 2 + 5t$$
$$z = -3 - 7t$$

▶ **Example 4**   (a)   Find parametric equations for the line L passing through the points $P_1(2, 4, -1)$ and $P_2(5, 0, 7)$.
(b)   Where does the line intersect the xy-plane?

Solution (a). Since the vector $\overrightarrow{P_1P_2} = \langle 3, -4, 8 \rangle$ is parallel to L and $P_1(2, 4, -1)$ lies on L, the line L is given by

$$x = 2 + 3t$$
$$y = 4 - 4t$$
$$z = -1 + 8t$$

Solution (b). The line intersects the xy-plane at the point where $z = -1 + 8t = 0$; that is, when $t = \frac{1}{8}$. Substituting this value of t in the parametric equations for L yields the point of intersection

$$(x, y, z) = \left(\frac{19}{8}, \frac{7}{2}, 0 \right)$$
◄

▶ **Example 5** Find parametric equations for the line segment joining the points $P_1(2, 4, -1)$ and $P_2(5, 0, 7)$.

Solution. From Example 4, the line through P_1 and P_2 has parametric equations $x = 2 + 3t$, $y = 4 - 4t$, $z = -1 + 8t$. With these equations, the point P_1 corresponds to $t = 0$ and P_2 to $t = 1$. Thus, the line segment from P_1 to P_2 has equations

$$x = 2 + 3t$$
$$y = 4 - 4t, \qquad 0 \le t \le 1$$
$$z = -1 + 8t$$
◄

▶ **Example 6** Let L_1 and L_2 be the lines

$$L_1: \begin{array}{l} x = 1 + 4t \\ y = 5 - 4t \\ z = -1 + 5t \end{array} \qquad L_2: \begin{array}{l} x = 2 + 8t \\ y = 4 - 3t \\ z = 5 + t \end{array}$$

(a) Are the lines parallel?
(b) Do the lines intersect?

Solution (a). The line L_1 is parallel to the vector $4\mathbf{i} - 4\mathbf{j} + 5\mathbf{k}$, and the line L_2 is parallel to the vector $8\mathbf{i} - 3\mathbf{j} + \mathbf{k}$. These vectors are not parallel since neither is a scalar multiple of the other. Thus, the lines are not parallel.

Solution (b). In order for the lines to intersect at some point (x_0, y_0, z_0) these coordinates would have to satisfy the equations of both L_1 and L_2. In other words, there would have to exist values t_1 and t_2 for the parameters such that

$$x_0 = 1 + 4t_1 \qquad\qquad x_0 = 2 + 8t_2$$
$$y_0 = 5 - 4t_1 \quad \text{and} \quad y_0 = 4 - 3t_2$$
$$z_0 = -1 + 5t_1 \qquad\qquad z_0 = 5 + t_2$$

This leads to three conditions on t_1 and t_2,

$$1 + 4t_1 = 2 + 8t_2$$
$$5 - 4t_1 = 4 - 3t_2 \qquad\qquad (3)$$
$$-1 + 5t_1 = 5 + t_2$$

We will try to solve these equations for t_1 and t_2. If we obtain a solution, then the lines intersect; if we find that there is no solution, then the lines do not intersect, since the three conditions cannot be satisfied.

The first two equations in (3) may be solved by adding them together to obtain

$$6 = 6 + 5t_2$$

or $t_2 = 0$. Substituting $t_2 = 0$ in the first equation yields

$$1 + 4t_1 = 2$$

or $t_1 = \frac{1}{4}$. However, the values $t_1 = \frac{1}{4}$, $t_2 = 0$ do not satisfy the third equation in (3), so that there is no simultaneous solution to the three equations. Thus, the lines do not intersect. ◀

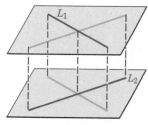

Parallel planes containing skew lines L_1 and L_2 may be determined by translating each line until it intersects the other.

Figure 15.2.6

Two lines in 3-space that are not parallel and do not intersect (such as those in Example 6) are called **skew** lines. As illustrated in Figure 15.2.6, any two skew lines lie in parallel planes.

▶ Exercise Set 15.2

1. Find the components of the vector $\overrightarrow{P_1 P_2}$.
   (a) $P_1(5, -2, 1)$, $P_2(2, 4, 2)$
   (b) $P_1(-1, 3, 5)$, $P_2(0, 0, 0)$
   (c) $P_1(0, 0, 0)$, $P_2(-1, 6, 1)$
   (d) $P_1(4, 1, -3)$, $P_2(9, 1, -3)$.

2. Find a vector with the same direction as $v = \langle 7, 0, -6 \rangle$ but with twice the length of v.

3. Find a vector, oppositely directed to $v = -3i + 4j + k$, but with twice the length of v.

4. Let $u = \langle 2, -1, 3 \rangle$, $v = \langle 4, 0, -2 \rangle$, $w = \langle 1, 1, 3 \rangle$.
   Find:
   (a) $u - w$ (b) $7v + 3w$
   (c) $-w + v$ (d) $3(u - 7v)$
   (e) $-3v - 8w$ (f) $2v - (u + w)$.

5. Compute the norm of v.
   (a) $v = i + j + k$ (b) $v = -3i + 2j + k$
   (c) $v = \langle -1, 2, 4 \rangle$ (d) $v = \langle 0, -3, 0 \rangle$·

6. Let $\mathbf{u} = \mathbf{i} - 3\mathbf{j} + 2\mathbf{k}$, $\mathbf{v} = \mathbf{i} + \mathbf{j}$, and $\mathbf{w} = 2\mathbf{i} + 2\mathbf{j} - 4\mathbf{k}$. Find:
   (a) $\|\mathbf{u} + \mathbf{v}\|$ (b) $\|\mathbf{u}\| + \|\mathbf{v}\|$
   (c) $\|-2\mathbf{u}\| + 2\|\mathbf{v}\|$ (d) $\|3\mathbf{u} - 5\mathbf{v} + \mathbf{w}\|$
   (e) $\dfrac{1}{\|\mathbf{w}\|}\mathbf{w}$ (f) $\left\|\dfrac{1}{\|\mathbf{w}\|}\mathbf{w}\right\|$

In Exercises 7–10, find parametric equations for the line through P_1 and P_2.

7. $P_1(5, -2, 1)$, $P_2(2, 4, 2)$.
8. $P_1(-1, 3, 5)$, $P_2(-1, 3, 2)$.
9. $P_1(0, 0, 0)$, $P_2(-1, 6, 1)$.
10. $P_1(4, 0, 7)$, $P_2(-1, -1, 2)$.

In Exercises 11–14, find parametric equations for the line segment joining P_1 and P_2.

11. $P_1(5, -2, 1)$, $P_2(2, 4, 2)$.
12. $P_1(-1, 3, 5)$, $P_2(-1, 3, 2)$.
13. $P_1(0, 0, 0)$, $P_2(-1, 6, 1)$.
14. $P_1(4, 0, 7)$, $P_2(-1, -1, 2)$.

In Exercises 15–19, find parametric equations for the line.

15. The line through $(-1, 2, 4)$ and parallel to $3\mathbf{i} - 4\mathbf{j} + \mathbf{k}$.

16. The line through $(2, -1, 5)$ and parallel to $\langle -1, 2, 7 \rangle$.

17. The line through $(-2, 0, 5)$ and parallel to the line $x = 1 + 2t$, $y = 4 - t$, $z = 6 + 2t$.

18. The line through the origin and parallel to the line $x = t$, $y = -1 + t$, $z = 2$.

19. The line through $(3, 7, 0)$ and parallel to the x-axis.

20. Where does the line $x = -1 + 2t$, $y = 3 + t$, $z = 4 - t$ intersect
   (a) the xy-plane
   (b) the xz-plane
   (c) the yz-plane?

21. Where does the line $x = -2$, $y = 4 + 2t$, $z = -3 + t$ intersect
   (a) the xy-plane
   (b) the xz-plane
   (c) the yz-plane?

22. Find parametric equations for the line through (x_0, y_0, z_0) and (x_1, y_1, z_1).

23. Find parametric equations for the line through (x_1, y_1, z_1) and parallel to the line $x = x_0 + at$, $y = y_0 + bt$, $z = z_0 + ct$.

24. Prove: If a, b, and c are nonzero, then each point on the line $x = x_0 + at$, $y = y_0 + bt$, $z = z_0 + ct$ satisfies

$$\frac{x - x_0}{a} = \frac{y - y_0}{b} = \frac{z - z_0}{c}$$

and conversely, each point (x, y, z) satisfying these equations lies on the line. (These are called the *symmetric equations* of the line.)

25. Show that the lines

$$\begin{array}{lll} x = 2 + t & & x = 2 + t \\ y = 2 + 3t & \text{and} & y = 3 + 4t \\ z = 3 + t & & z = 4 + 2t \end{array}$$

intersect and find the point of intersection.

26. Show that the lines

$$\begin{array}{lll} x + 1 = 4t & & x + 13 = 12t \\ y - 3 = t & \text{and} & y - 1 = 6t \\ z - 1 = 0 & & z - 2 = 3t \end{array}$$

intersect and find the point of intersection.

27. Show that the lines

$$\begin{array}{lll} x = 1 + 7t & & x = 4 - t \\ y = 3 + t & \text{and} & y = 6 \\ z = 5 - 3t & & z = 7 + 2t \end{array}$$

are skew.

28. Show that the lines

$$\begin{array}{lll} x = 2 + 8t & & x = 3 + 8t \\ y = 6 - 8t & \text{and} & y = 5 - 3t \\ z = 10t & & z = 6 + t \end{array}$$

are skew.

29. Determine whether P_1, P_2 and P_3 lie on the same line.
   (a) $P_1(6, 9, 7)$, $P_2(9, 2, 0)$, $P_3(0, -5, -3)$
   (b) $P_1(1, 0, 1)$, $P_2(3, -4, -3)$, $P_3(4, -6, -5)$.

30. Find k_1 and k_2 so that the point $(k_1, 1, k_2)$ lies on the line passing through $(0, 2, 3)$ and $(2, 7, 5)$.

31. Find the point on the line segment joining $P_1(1, 4, -3)$ and $P_2(1, 5, -1)$ that is $\frac{2}{3}$ of the way from P_1 to P_2.

32. In each part, determine whether the lines are parallel.

(a) $x = 3 - 2t$
$y = 4 + t$ and $x = 5 - 4t$
$z = 6 - t$ $y = -2 + 2t$
                   $z = 7 - 2t$

(b) $x = 5 + 3t$
$y = 4 - 2t$ and $x = -1 + 9t$
$z = -2 + 3t$ $y = 5 - 6t$
                    $z = 3 + 8t.$

15.3   DOT PRODUCT; PROJECTIONS

In this section we introduce a kind of multiplication of vectors in 2-space and 3-space. We discuss the arithmetic properties of this multiplication and give some applications.

Let **u** and **v** be two nonzero vectors in 2-space or 3-space, and assume these vectors have been positioned so their initial points coincide. By the *angle between **u** and **v***, we shall mean the angle θ determined by **u** and **v** that satisfies $0 \le \theta \le \pi$ (Figure 15.3.1).

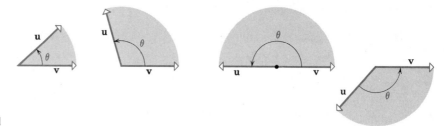

Figure 15.3.1

15.3.1   DEFINITION   If **u** and **v** are vectors in 2-space or 3-space and θ is the angle between **u** and **v**, then the *dot product* or *Euclidean inner product* **u · v** is defined by

$$\mathbf{u} \cdot \mathbf{v} = \begin{cases} \|\mathbf{u}\|\,\|\mathbf{v}\| \cos \theta, & \text{if } \mathbf{u} \ne \mathbf{0} \text{ and } \mathbf{v} \ne \mathbf{0} \\ 0 & , \quad \text{if } \mathbf{u} = \mathbf{0} \text{ or } \mathbf{v} = \mathbf{0} \end{cases}$$

► **Example 1**   As shown in Figure 15.3.2, the angle between vectors

$$\mathbf{u} = \langle 0, 2 \rangle \quad \text{and} \quad \mathbf{v} = \langle 1, 1 \rangle$$

is 45°. Thus,

$$\mathbf{u} \cdot \mathbf{v} = \|\mathbf{u}\|\,\|\mathbf{v}\| \cos \theta = \sqrt{0^2 + 2^2}\sqrt{1^2 + 1^2} \cos 45° = (2)(\sqrt{2})\frac{1}{\sqrt{2}} = 2$$

◄

Figure 15.3.2

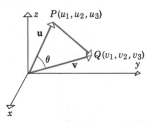

Figure 15.3.3

For purposes of computation, it is desirable to have a formula that expresses the dot product of two vectors in terms of the components of the vectors. We will derive such a formula for vectors in 3-space; the derivation for vectors in 2-space is similar.

Let $\mathbf{u} = \langle u_1, u_2, u_3 \rangle$ and $\mathbf{v} = \langle v_1, v_2, v_3 \rangle$ be two nonzero vectors. If, as in Figure 15.3.3, θ is the angle between $\mathbf{u}$ and $\mathbf{v}$, then the law of cosines yields

$$\|\overrightarrow{PQ}\|^2 = \|\mathbf{u}\|^2 + \|\mathbf{v}\|^2 - 2\|\mathbf{u}\|\,\|\mathbf{v}\| \cos \theta \tag{1}$$

Since $\overrightarrow{PQ} = \mathbf{v} - \mathbf{u}$, we can rewrite (1) as

$$\|\mathbf{u}\|\,\|\mathbf{v}\| \cos \theta = \tfrac{1}{2}(\|\mathbf{u}\|^2 + \|\mathbf{v}\|^2 - \|\mathbf{v} - \mathbf{u}\|^2)$$

or

$$\mathbf{u} \cdot \mathbf{v} = \tfrac{1}{2}(\|\mathbf{u}\|^2 + \|\mathbf{v}\|^2 - \|\mathbf{v} - \mathbf{u}\|^2)$$

Substituting

$$\|\mathbf{u}\|^2 = u_1{}^2 + u_2{}^2 + u_3{}^2, \qquad \|\mathbf{v}\|^2 = v_1{}^2 + v_2{}^2 + v_3{}^2$$

and

$$\|\mathbf{v} - \mathbf{u}\|^2 = (v_1 - u_1)^2 + (v_2 - u_2)^2 + (v_3 - u_3)^2$$

we obtain, after simplifying,

$$\mathbf{u} \cdot \mathbf{v} = u_1 v_1 + u_2 v_2 + u_3 v_3 \tag{2a}$$

If $\mathbf{u} = \langle u_1, u_2 \rangle$ and $\mathbf{v} = \langle v_1, v_2 \rangle$ are two vectors in 2-space, then the formula corresponding to (2a) is

$$\mathbf{u} \cdot \mathbf{v} = u_1 v_1 + u_2 v_2 \tag{2b}$$

If $\mathbf{u}$ and $\mathbf{v}$ are nonzero vectors, the formula in Definition 15.3.1 can be written as

$$\cos \theta = \frac{\mathbf{u} \cdot \mathbf{v}}{\|\mathbf{u}\|\,\|\mathbf{v}\|} \tag{3}$$

▶ Example 2 Consider the vectors

$$\mathbf{u} = 2\mathbf{i} - \mathbf{j} + \mathbf{k} \qquad \text{and} \qquad \mathbf{v} = \mathbf{i} + \mathbf{j} + 2\mathbf{k}$$

Find $\mathbf{u} \cdot \mathbf{v}$ and determine the angle θ between $\mathbf{u}$ and $\mathbf{v}$.

Solution.

$$\mathbf{u} \cdot \mathbf{v} = u_1 v_1 + u_2 v_2 + u_3 v_3 = (2)(1) + (-1)(1) + (1)(2) = 3$$

For the given vectors, $\|\mathbf{u}\| = \|\mathbf{v}\| = \sqrt{6}$, so that

$$\cos \theta = \frac{3}{\sqrt{6}\sqrt{6}} = \frac{1}{2}$$

Thus, $\theta = 60°$. ◀

▶ Example 3 Find the angle between a diagonal of a cube and one of its edges.

Solution. Let k be the length of an edge and let us introduce a coordinate system as shown in Figure 15.3.4.

If we let $\mathbf{u}_1 = \langle k, 0, 0 \rangle$, $\mathbf{u}_2 = \langle 0, k, 0 \rangle$, and $\mathbf{u}_3 = \langle 0, 0, k \rangle$, then the vector

$$\mathbf{d} = \langle k, k, k \rangle = \mathbf{u}_1 + \mathbf{u}_2 + \mathbf{u}_3$$

is a diagonal of the cube. The angle θ between $\mathbf{d}$ and the edge $\mathbf{u}_1$ satisfies

$$\cos \theta = \frac{\mathbf{u}_1 \cdot \mathbf{d}}{\|\mathbf{u}_1\| \|\mathbf{d}\|} = \frac{k^2}{(k)(\sqrt{3k^2})} = \frac{1}{\sqrt{3}}$$

Thus,

$$\theta = \cos^{-1} \frac{1}{\sqrt{3}} \approx 54°44'$$ ◀

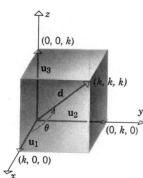

Figure 15.3.4

The sign of the dot product provides useful information about the angle between two vectors.

15.3.2 THEOREM *If $\mathbf{u}$ and $\mathbf{v}$ are nonzero vectors in 2-space or 3-space, and if θ is the angle between them, then*

θ is acute	if and only if	$\mathbf{u} \cdot \mathbf{v} > 0$
θ is obtuse	if and only if	$\mathbf{u} \cdot \mathbf{v} < 0$
$\theta = \pi/2$	if and only if	$\mathbf{u} \cdot \mathbf{v} = 0$

Proof. Since $\mathbf{u}$ and $\mathbf{v}$ are nonzero vectors, $\|\mathbf{u}\| > 0$ and $\|\mathbf{v}\| > 0$. Thus,

$$\mathbf{u} \cdot \mathbf{v} = \|\mathbf{u}\| \|\mathbf{v}\| \cos \theta$$

is positive, negative, or zero according to whether $\cos \theta$ is positive, negative, or zero. Since $0 \leq \theta \leq \pi$, it follows that θ is acute if and only if

$\cos \theta > 0$; θ is obtuse if and only if $\cos \theta < 0$, and $\theta = \pi/2$ if and only if $\cos \theta = 0$. ∎

▶ Example 4 If $\mathbf{u} = \mathbf{i} - 2\mathbf{j} + 3\mathbf{k}$, $\mathbf{v} = -3\mathbf{i} + 4\mathbf{j} + 2\mathbf{k}$, and $\mathbf{w} = 3\mathbf{i} + 6\mathbf{j} + 3\mathbf{k}$, then

$$\mathbf{u} \cdot \mathbf{v} = (1)(-3) + (-2)(4) + (3)(2) = -5$$
$$\mathbf{v} \cdot \mathbf{w} = (-3)(3) + (4)(6) + (2)(3) = 21$$
$$\mathbf{u} \cdot \mathbf{w} = (1)(3) + (-2)(6) + (3)(3) = 0$$

Therefore $\mathbf{u}$ and $\mathbf{v}$ make an obtuse angle, $\mathbf{v}$ and $\mathbf{w}$ make an acute angle, and $\mathbf{u}$ and $\mathbf{w}$ are perpendicular. ◀

Perpendicular vectors are also called *orthogonal* vectors. In light of Theorem 15.3.2, two nonzero vectors are orthogonal if and only if their dot product is zero. If we agree to consider $\mathbf{u}$ and $\mathbf{v}$ to be perpendicular when either or both of these vectors is $\mathbf{0}$, then we can state without exception that two vectors $\mathbf{u}$ and $\mathbf{v}$ are orthogonal (perpendicular) if and only if $\mathbf{u} \cdot \mathbf{v} = 0$.

▶ Example 5 Show that in 2-space the vector $a\mathbf{i} + b\mathbf{j}$ is perpendicular to the line $ax + by + c = 0$.

Solution. Let $P_1(x_1, y_1)$ and $P_2(x_2, y_2)$ be distinct points on the line, so that

$$ax_1 + by_1 + c = 0$$
$$ax_2 + by_2 + c = 0$$
(4)

Since the vector $\overrightarrow{P_1 P_2} = (x_2 - x_1)\mathbf{i} + (y_2 - y_1)\mathbf{j}$ runs along the line, we need only show that $a\mathbf{i} + b\mathbf{j}$ and $\overrightarrow{P_1 P_2}$ are perpendicular. But on subtracting the equations in (4), we obtain

$$a(x_2 - x_1) + b(y_2 - y_1) = 0$$

which can be expressed in the form

$$(a\mathbf{i} + b\mathbf{j}) \cdot [(x_2 - x_1)\mathbf{i} + (y_2 - y_1)\mathbf{j}] = 0$$

or

$$(a\mathbf{i} + b\mathbf{j}) \cdot \overrightarrow{P_1 P_2} = 0$$

so that $a\mathbf{i} + b\mathbf{j}$ and $\overrightarrow{P_1 P_2}$ are perpendicular. ◀

Of special interest are the angles α, β, and γ that a vector $\mathbf{u}$ in 3-space makes with the vectors $\mathbf{i}$, $\mathbf{j}$, and $\mathbf{k}$ (Figure 15.3.5). These are called the

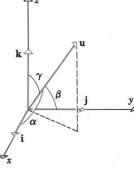

Figure 15.3.5

direction angles of **u**. The numbers $\cos \alpha$, $\cos \beta$, and $\cos \gamma$ are called the *direction cosines* of **u**. Formulas for the direction cosines follow easily from (3).

15.3.3   THEOREM   *The direction cosines of a nonzero vector* $\mathbf{u} = u_1\mathbf{i} + u_2\mathbf{j} + u_3\mathbf{k}$ *in 3-space are*

$$\cos \alpha = \frac{u_1}{\|\mathbf{u}\|}, \qquad \cos \beta = \frac{u_2}{\|\mathbf{u}\|}, \qquad \cos \gamma = \frac{u_3}{\|\mathbf{u}\|}$$

Proof.   Since $\mathbf{u} \cdot \mathbf{i} = (u_1)(1) + (u_2)(0) + (u_3)(0) = u_1$, it follows that

$$\cos \alpha = \frac{\mathbf{u} \cdot \mathbf{i}}{\|\mathbf{u}\| \, \|\mathbf{i}\|} = \frac{u_1}{\|\mathbf{u}\|}$$

Similarly for $\cos \beta$ and $\cos \gamma$.   ∎

The direction cosines of a vector $\mathbf{u} = u_1\mathbf{i} + u_2\mathbf{j} + u_3\mathbf{k}$ can be obtained by simply reading off the components of the unit vector $\mathbf{u}/\|\mathbf{u}\|$ since

$$\frac{\mathbf{u}}{\|\mathbf{u}\|} = \frac{u_1}{\|\mathbf{u}\|}\mathbf{i} + \frac{u_2}{\|\mathbf{u}\|}\mathbf{j} + \frac{u_3}{\|\mathbf{u}\|}\mathbf{k} = (\cos \alpha)\mathbf{i} + (\cos \beta)\mathbf{j} + (\cos \gamma)\mathbf{k}$$

▶ **Example 6**   Find the direction cosines of the vector $\mathbf{u} = 2\mathbf{i} - 4\mathbf{j} + 4\mathbf{k}$, and estimate the direction angles to the nearest degree.

Solution. $\|\mathbf{u}\| = \sqrt{4 + 16 + 16} = 6$, so that $\mathbf{u}/\|\mathbf{u}\| = \frac{1}{3}\mathbf{i} - \frac{2}{3}\mathbf{j} + \frac{2}{3}\mathbf{k}$. Thus,

$$\cos \alpha = \frac{1}{3}, \qquad \cos \beta = -\frac{2}{3}, \qquad \cos \gamma = \frac{2}{3}$$

With the help of trigonometric tables or a hand-held calculator that can compute inverse trigonometric functions, we obtain

$$\alpha = \cos^{-1}\left(\frac{1}{3}\right) \approx 71°, \qquad \beta = \cos^{-1}\left(-\frac{2}{3}\right) \approx 132°,$$

$$\gamma = \cos^{-1}\left(\frac{2}{3}\right) \approx 48° \qquad\qquad\qquad ◀$$

The following arithmetic properties of the dot product are useful in calculations involving vectors.

15.3.4   THEOREM   *If* **u**, **v**, *and* **w** *are vectors in 2- or 3-space and k is a scalar, then*
    (a) $\mathbf{u} \cdot \mathbf{v} = \mathbf{v} \cdot \mathbf{u}$
    (b) $\mathbf{u} \cdot (\mathbf{v} + \mathbf{w}) = \mathbf{u} \cdot \mathbf{v} + \mathbf{u} \cdot \mathbf{w}$
    (c) $k(\mathbf{u} \cdot \mathbf{v}) = (k\mathbf{u}) \cdot \mathbf{v} = \mathbf{u} \cdot (k\mathbf{v})$
    (d) $\mathbf{v} \cdot \mathbf{v} = \|\mathbf{v}\|^2$

Proof. We shall prove parts (*c*) and (*d*) for vectors in 3-space and omit the remaining proofs.

(*c*)   Let $\mathbf{u} = \langle u_1, u_2, u_3 \rangle$ and $\mathbf{v} = \langle v_1, v_2, v_3 \rangle$; then

$$\begin{aligned} k(\mathbf{u} \cdot \mathbf{v}) &= k(u_1 v_1 + u_2 v_2 + u_3 v_3) \\ &= (k u_1) v_1 + (k u_2) v_2 + (k u_3) v_3 \\ &= (k\mathbf{u}) \cdot \mathbf{v} \end{aligned}$$

Similarly, $k(\mathbf{u} \cdot \mathbf{v}) = \mathbf{u} \cdot (k\mathbf{v})$.

(*d*) $\mathbf{v} \cdot \mathbf{v} = v_1 v_1 + v_2 v_2 + v_3 v_3 = v_1{}^2 + v_2{}^2 + v_3{}^2 = \|\mathbf{v}\|^2.$ ∎

Part (*d*) of the last theorem is sometimes expressed in the following alternate form.

$$\|\mathbf{v}\| = \sqrt{\mathbf{v} \cdot \mathbf{v}} \tag{5}$$

In many applications it is of interest to "decompose" a vector **u** into a sum of two terms, one parallel to a specified nonzero vector **a** and the other perpendicular to **a**. If **u** and **a** are positioned so their initial points coincide at a point Q, we may decompose the vector **u** as follows (Figure 15.3.6): Drop a perpendicular from the tip of **u** to the line through **a**, and construct the vector $\mathbf{w}_1$ from Q to the foot of this perpendicular. Next form the difference

$$\mathbf{w}_2 = \mathbf{u} - \mathbf{w}_1$$

As indicated in Figure 15.3.6, the vector $\mathbf{w}_1$ is parallel to **a**, the vector $\mathbf{w}_2$ is perpendicular to **a**, and

$$\mathbf{w}_1 + \mathbf{w}_2 = \mathbf{w}_1 + (\mathbf{u} - \mathbf{w}_1) = \mathbf{u}$$

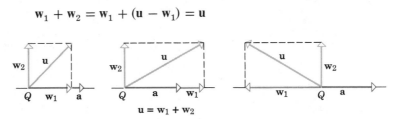

Figure 15.3.6

$\mathbf{u} = \mathbf{w}_1 + \mathbf{w}_2$

The vector $\mathbf{w}_1$ is called the ***orthogonal projection of* u *on* a** or sometimes the ***vector component of* u *along* a.** It is denoted by

$$\text{proj}_{\mathbf{a}}\,\mathbf{u} \tag{6}$$

The vector $\mathbf{w}_2$ is called the ***vector component of* u *orthogonal to* a.** Since $\mathbf{w}_2 = \mathbf{u} - \mathbf{w}_1$, this vector may be written in notation (6) as

$$\mathbf{w}_2 = \mathbf{u} - \text{proj}_{\mathbf{a}}\,\mathbf{u}$$

The following theorem gives formulas for calculating the vectors $\text{proj}_a \mathbf{u}$ and $\mathbf{u} - \text{proj}_a \mathbf{u}$.

15.3.5 THEOREM *If* $\mathbf{u}$ *and* $\mathbf{a}$ *are vectors in 2-space or 3-space and if* $\mathbf{a} \neq \mathbf{0}$, *then*

$$\text{proj}_a \mathbf{u} = \frac{\mathbf{u} \cdot \mathbf{a}}{\|\mathbf{a}\|^2}\mathbf{a} \qquad \text{(vector component of } \mathbf{u} \text{ along } \mathbf{a}\text{)}$$

$$\mathbf{u} - \text{proj}_a \mathbf{u} = \mathbf{u} - \frac{\mathbf{u} \cdot \mathbf{a}}{\|\mathbf{a}\|^2}\mathbf{a} \qquad \text{(vector component of } \mathbf{u} \text{ orthogonal to } \mathbf{a}\text{)}$$

Proof. Let $\mathbf{w}_1 = \text{proj}_a \mathbf{u}$ and $\mathbf{w}_2 = \mathbf{u} - \text{proj}_a \mathbf{u}$. Since $\mathbf{w}_1$ is parallel to $\mathbf{a}$, it must be a scalar multiple of $\mathbf{a}$, so it can be written in the form $\mathbf{w}_1 = k\mathbf{a}$. Thus,

$$\mathbf{u} = \mathbf{w}_1 + \mathbf{w}_2 = k\mathbf{a} + \mathbf{w}_2 \tag{7}$$

Taking the dot product of both sides of (7) with $\mathbf{a}$ and using Theorem 15.3.4 yields

$$\mathbf{u} \cdot \mathbf{a} = (k\mathbf{a} + \mathbf{w}_2) \cdot \mathbf{a} = k\|\mathbf{a}\|^2 + \mathbf{w}_2 \cdot \mathbf{a} \tag{8}$$

But $\mathbf{w}_2 \cdot \mathbf{a} = 0$ since $\mathbf{w}_2$ is perpendicular to $\mathbf{a}$; thus, (8) yields

$$k = \frac{\mathbf{u} \cdot \mathbf{a}}{\|\mathbf{a}\|^2}$$

Since $\text{proj}_a \mathbf{u} = \mathbf{w}_1 = k\mathbf{a}$, we obtain

$$\text{proj}_a \mathbf{u} = \frac{\mathbf{u} \cdot \mathbf{a}}{\|\mathbf{a}\|^2}\mathbf{a} \quad \blacksquare$$

▶ **Example 7** Let $\mathbf{u} = 2\mathbf{i} - \mathbf{j} + 3\mathbf{k}$ and $\mathbf{a} = 4\mathbf{i} - \mathbf{j} + 2\mathbf{k}$. Find the vector component of $\mathbf{u}$ along $\mathbf{a}$ and the vector component of $\mathbf{u}$ orthogonal to $\mathbf{a}$.

Solution.

$$\mathbf{u} \cdot \mathbf{a} = (2)(4) + (-1)(-1) + (3)(2) = 15$$
$$\|\mathbf{a}\|^2 = 4^2 + (-1)^2 + 2^2 = 21$$

Thus, the vector component of $\mathbf{u}$ along $\mathbf{a}$ is

$$\text{proj}_a \mathbf{u} = \frac{\mathbf{u} \cdot \mathbf{a}}{\|\mathbf{a}\|^2}\mathbf{a} = \frac{15}{21}(4\mathbf{i} - \mathbf{j} + 2\mathbf{k}) = \frac{20}{7}\mathbf{i} - \frac{5}{7}\mathbf{j} + \frac{10}{7}\mathbf{k}$$

and the vector component of $\mathbf{u}$ orthogonal to $\mathbf{a}$ is

$$u - \text{proj}_a \, u = (2i - j + 3k) - \left(\frac{20}{7}i - \frac{5}{7}j + \frac{10}{7}k\right)$$

$$= -\frac{6}{7}i - \frac{2}{7}j + \frac{11}{7}k$$

As a check, the reader may wish to verify that the vectors $u - \text{proj}_a \, u$ and a are perpendicular by showing that their dot product is zero. ◄

A formula for the length of the vector component of u along a may be obtained by writing

$$\|\text{proj}_a \, u\| = \left\|\frac{u \cdot a}{\|a\|^2}a\right\|$$

$$= \left|\frac{u \cdot a}{\|a\|^2}\right| \|a\| \quad \left[\text{Since } \frac{u \cdot a}{\|a\|^2} \text{ is a scalar.}\right]$$

$$= \frac{|u \cdot a|}{\|a\|^2} \|a\| \quad [\text{Since } \|a\|^2 > 0.]$$

which yields

$$\|\text{proj}_a \, u\| = \frac{|u \cdot a|}{\|a\|} \tag{9}$$

If θ denotes the angle between u and a, then $u \cdot a = \|u\| \|a\| \cos\theta$, so that (9) can also be written as

$$\|\text{proj}_a \, u\| = \|u\| \, |\cos\theta| \tag{10}$$

(Verify.) A geometric interpretation of this result is given in Figure 15.3.7.
As an example we will use vector methods to derive a formula for the distance from a point in the plane to a line.

Figure 15.3.7

► Example 8 Find a formula for the distance D between the point $P_0(x_0, y_0)$ and the line $ax + by + c = 0$.

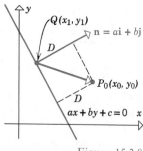

Figure 15.3.8

Solution. Let $Q(x_1, y_1)$ be any point on the line and position the vector

$$\mathbf{n} = a\mathbf{i} + b\mathbf{j}$$

so that its initial point is at Q.

By virtue of Example 5, the vector $\mathbf{n}$ is perpendicular to the line (Figure 15.3.8). As indicated in the figure, the distance D is equal to the length of the orthogonal projection of $\overrightarrow{QP_0}$ on $\mathbf{n}$; thus, from (9),

$$D = \|\text{proj}_{\mathbf{n}} \overrightarrow{QP_0}\| = \frac{|\overrightarrow{QP_0} \cdot \mathbf{n}|}{\|\mathbf{n}\|}$$

But

$$\overrightarrow{QP_0} = \langle x_0 - x_1, y_0 - y_1 \rangle$$
$$\overrightarrow{QP_0} \cdot \mathbf{n} = a(x_0 - x_1) + b(y_0 - y_1)$$
$$\|\mathbf{n}\| = \sqrt{a^2 + b^2}$$

so that

$$D = \frac{|a(x_0 - x_1) + b(y_0 - y_1)|}{\sqrt{a^2 + b^2}} \tag{11}$$

Since the point $Q(x_1, y_1)$ lies on the line, its coordinates satisfy the equation of the line, so

$$ax_1 + by_1 + c = 0$$

or

$$c = -ax_1 - by_1$$

Substituting this expression in (11) yields the formula

$$D = \frac{|ax_0 + by_0 + c|}{\sqrt{a^2 + b^2}} \tag{12}$$

◀

As an illustration, the distance D from the point $(1, -2)$ to the line $3x + 4y - 6 = 0$ is

$$D = \frac{|(3)(1) + 4(-2) - 6|}{\sqrt{3^2 + 4^2}} = \frac{|-11|}{\sqrt{25}} = \frac{11}{5}$$

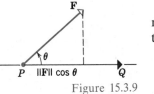

Figure 15.3.9

Recall from Section 6.6 that the work W done by a constant force $\mathbf{F}$ of magnitude $\|\mathbf{F}\|$ acting in the direction of motion on a particle moving from P to Q on a line is

$$W = (\text{force}) \cdot (\text{distance}) = \|\mathbf{F}\| \|\overrightarrow{PQ}\|$$

If a force $\mathbf{F}$ is constant, but makes an angle θ with the direction of motion (Figure 15.3.9), then we define the work done by $\mathbf{F}$ to be

$$W = \mathbf{F} \cdot \overrightarrow{PQ} = (\|\mathbf{F}\| \cos \theta) \|\overrightarrow{PQ}\| \tag{13}$$

The quantity $\|\mathbf{F}\| \cos \theta$ is the "component" of force in the direction of motion and $\|\overrightarrow{PQ}\|$ is the distance traveled by the particle.

▶ Example 9 A wagon is pulled horizontally by exerting a force of 10 lb on the handle at an angle of 60° with the horizontal. How much work is done in moving the wagon 50 ft?

Solution. Introduce an xy-coordinate system so the wagon moves from $P(0, 0)$ to $Q(50, 0)$ along the x-axis (Figure 15.3.10). In this coordinate system

$$\overrightarrow{PQ} = 50\mathbf{i}$$

and

$$\mathbf{F} = (10 \cos 60°)\mathbf{i} + (10 \sin 60°)\mathbf{j} = 5\mathbf{i} + 5\sqrt{3}\mathbf{j}$$

so that the work done is

$$W = \mathbf{F} \cdot \overrightarrow{PQ} = (5\mathbf{i} + 5\sqrt{3}\mathbf{j}) \cdot (50\mathbf{i}) = 250 \text{ (foot-pounds)} \qquad ◀$$

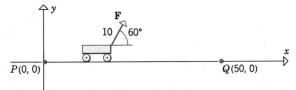

Figure 15.3.10

▶ Exercise Set 15.3

1. Find $\mathbf{u} \cdot \mathbf{v}$.
   (a) $\mathbf{u} = \mathbf{i} + 2\mathbf{j}, \mathbf{v} = 6\mathbf{i} - 8\mathbf{j}$
   (b) $\mathbf{u} = \langle -7, -3 \rangle, \mathbf{v} = \langle 0, 1 \rangle$
   (c) $\mathbf{u} = \mathbf{i} - 3\mathbf{j} + 7\mathbf{k}, \mathbf{v} = 8\mathbf{i} - 2\mathbf{j} - 2\mathbf{k}$
   (d) $\mathbf{u} = \langle -3, 1, 2 \rangle, \mathbf{v} = \langle 4, 2, -5 \rangle$.

2. In each part of Exercise 1, find the cosine of the angle θ between $\mathbf{u}$ and $\mathbf{v}$.

3. Determine whether $\mathbf{u}$ and $\mathbf{v}$ make an acute angle, an obtuse angle, or are orthogonal.
   (a) $\mathbf{u} = 7\mathbf{i} + 3\mathbf{j} + 5\mathbf{k}, \mathbf{v} = -8\mathbf{i} + 4\mathbf{j} + 2\mathbf{k}$
   (b) $\mathbf{u} = 6\mathbf{i} + \mathbf{j} + 3\mathbf{k}, \mathbf{v} = 4\mathbf{i} - 6\mathbf{k}$
   (c) $\mathbf{u} = \langle 1, 1, 1 \rangle, \mathbf{v} = \langle -1, 0, 0 \rangle$
   (d) $\mathbf{u} = \langle 4, 1, 6 \rangle, \mathbf{v} = \langle -3, 0, 2 \rangle$.

4. Find the orthogonal projection of $\mathbf{u}$ on $\mathbf{a}$.
   (a) $\mathbf{u} = 2\mathbf{i} + \mathbf{j}, \mathbf{a} = -3\mathbf{i} + 2\mathbf{j}$
   (b) $\mathbf{u} = \langle 2, 6 \rangle, \mathbf{a} = \langle -9, 3 \rangle$
   (c) $\mathbf{u} = -7\mathbf{i} + \mathbf{j} + 3\mathbf{k}, \mathbf{a} = 5\mathbf{i} + \mathbf{k}$
   (d) $\mathbf{u} = \langle 0, 0, 1 \rangle, \mathbf{a} = \langle 8, 3, 4 \rangle$.

5. In each part of Exercise 4, find the vector component of $\mathbf{u}$ orthogonal to $\mathbf{a}$.

6. Find $\|\text{proj}_{\mathbf{a}} \mathbf{u}\|$
   (a) $\mathbf{u} = 2\mathbf{i} - \mathbf{j}, \mathbf{a} = 3\mathbf{i} + 4\mathbf{j}$
   (b) $\mathbf{u} = \langle 4, 5 \rangle, \mathbf{a} = \langle 1, -2 \rangle$
   (c) $\mathbf{u} = 2\mathbf{i} - \mathbf{j} + 3\mathbf{k}, \mathbf{a} = \mathbf{i} + 2\mathbf{j} + 2\mathbf{k}$
   (d) $\mathbf{u} = \langle 4, -1, 7 \rangle, \mathbf{a} = \langle 2, 3, -6 \rangle$.

7. Verify Theorem 15.3.4 for $\mathbf{u} = 6\mathbf{i} - \mathbf{j} + 2\mathbf{k}, \mathbf{v} = 2\mathbf{i} + 7\mathbf{j} + 4\mathbf{k}$, and $k = -5$.

8. Find two vectors in 2-space of norm 1 that are orthogonal to $3\mathbf{i} - 2\mathbf{j}$.

9. Let $\mathbf{u} = \langle 1, 2 \rangle$, $\mathbf{v} = \langle 4, -2 \rangle$, and $\mathbf{w} = \langle 6, 0 \rangle$. Find:
   (a) $\mathbf{u} \cdot (7\mathbf{v} + \mathbf{w})$ (b) $\|(\mathbf{u} \cdot \mathbf{w})\mathbf{w}\|$
   (c) $\|\mathbf{u}\|(\mathbf{v} \cdot \mathbf{w})$ (d) $(\|\mathbf{u}\|\mathbf{v}) \cdot \mathbf{w}$.

10. Explain why each of the following expressions makes no sense.
   (a) $\mathbf{u} \cdot (\mathbf{v} \cdot \mathbf{w})$ (b) $(\mathbf{u} \cdot \mathbf{v}) + \mathbf{w}$
   (c) $\|\mathbf{u} \cdot \mathbf{v}\|$ (d) $k \cdot (\mathbf{u} + \mathbf{v})$.

11. Use vectors to find the cosines of the interior angles of the triangle with vertices $(-1, 0), (2, -1)$, and $(1, 4)$.

12. Find two unit vectors that make an angle of $45°$ with $4\mathbf{i} + 3\mathbf{j}$.

13. Show that $A(2, -1, 1)$, $B(3, 2, -1)$, and $C(7, 0, -2)$ are vertices of a right triangle. At which vertex is the right angle?

14. Suppose $\mathbf{a} \cdot \mathbf{b} = \mathbf{a} \cdot \mathbf{c}$ and $\mathbf{a} \neq \mathbf{0}$. Does it follow that $\mathbf{b} = \mathbf{c}$? Explain.

15. Let $\mathbf{a} = k\mathbf{i} + \mathbf{j}$ and $\mathbf{b} = 4\mathbf{i} + 3\mathbf{j}$. Find k so that
   (a) $\mathbf{a}$ and $\mathbf{b}$ are orthogonal
   (b) the angle between $\mathbf{a}$ and $\mathbf{b}$ is $\pi/4$
   (c) the angle between $\mathbf{a}$ and $\mathbf{b}$ is $\pi/6$
   (d) $\mathbf{a}$ and $\mathbf{b}$ are parallel.

16. Find the direction cosines of $\mathbf{u}$ and estimate the direction angles to the nearest degree.
   (a) $\mathbf{u} = \mathbf{i} + \mathbf{j} - \mathbf{k}$ (b) $\mathbf{u} = 2\mathbf{i} - 2\mathbf{j} + \mathbf{k}$
   (c) $\mathbf{u} = 3\mathbf{i} - 2\mathbf{j} - 6\mathbf{k}$ (d) $\mathbf{u} = 3\mathbf{i} - 4\mathbf{k}$.

17. Prove: The direction cosines of a vector satisfy the equation $\cos^2 \alpha + \cos^2 \beta + \cos^2 \gamma = 1$.

18. Prove: Two nonzero vectors $\mathbf{u}_1$ and $\mathbf{u}_2$ are perpendicular if and only if their direction cosines satisfy

$$\cos \alpha_1 \cos \alpha_2 + \cos \beta_1 \cos \beta_2 + \cos \gamma_1 \cos \gamma_2 = 0$$

19. Use Formula (12) to calculate the distance between the point and the line.
   (a) $3x + 4y + 7 = 0; (1, -2)$
   (b) $y = -2x + 1; (-3, 5)$
   (c) $2x + y = 8; (2, 6)$.

20. A boat travels 100 meters due north while the wind exerts a force of 50 newtons toward the northeast. How much work does the wind do?

21. Find the work done by a force $\mathbf{F} = -3\mathbf{j}$ (pounds) applied to a point that moves on a line from $(1, 3)$ to $(4, 7)$. Assume distance is measured in feet.

22. Prove: $\|\mathbf{u} + \mathbf{v}\|^2 + \|\mathbf{u} - \mathbf{v}\|^2 = 2\|\mathbf{u}\|^2 + 2\|\mathbf{v}\|^2$.

23. Prove: $\mathbf{u} \cdot \mathbf{v} = \frac{1}{4}\|\mathbf{u} + \mathbf{v}\|^2 - \frac{1}{4}\|\mathbf{u} - \mathbf{v}\|^2$.

24. Find the angle between the diagonal of a cube and one of its faces.

25. Prove: If $\mathbf{v}$ is orthogonal to $\mathbf{w}_1$ and $\mathbf{w}_2$, then $\mathbf{v}$ is orthogonal to $k_1\mathbf{w}_1 + k_2\mathbf{w}_2$ for all scalars k_1 and k_2.

26. Let $\mathbf{u}$ and $\mathbf{v}$ be nonzero vectors, and let $k = \|\mathbf{u}\|$ and $l = \|\mathbf{v}\|$. Prove that

$$\mathbf{w} = l\mathbf{u} + k\mathbf{v}$$

bisects the angle between $\mathbf{u}$ and $\mathbf{v}$.

15.4 CROSS PRODUCT

In many applications of vectors to problems in geometry, physics, and engineering, it is of interest to construct a vector in 3-space that is perpendicular to two given vectors. In this section we introduce a type of vector multiplication that facilitates this construction.

We begin with some notation. If a, b, c, and d are real numbers, then the symbol

$$\begin{vmatrix} a & b \\ c & d \end{vmatrix}$$

called a **2 × 2 (two-by-two) determinant,** denotes the number

$$\begin{vmatrix} a & b \\ c & d \end{vmatrix} = ad - bc$$

For example,

$$\begin{vmatrix} 3 & -2 \\ 4 & 5 \end{vmatrix} = (3)(5) - (-2)(4) = 15 + 8 = 23$$

A **3 × 3 determinant**

$$\begin{vmatrix} a_1 & a_2 & a_3 \\ b_1 & b_2 & b_3 \\ c_1 & c_2 & c_3 \end{vmatrix}$$

is defined in terms of 2 × 2 determinants by the formula

$$\begin{vmatrix} a_1 & a_2 & a_3 \\ b_1 & b_2 & b_3 \\ c_1 & c_2 & c_3 \end{vmatrix} = a_1 \begin{vmatrix} b_2 & b_3 \\ c_2 & c_3 \end{vmatrix} - a_2 \begin{vmatrix} b_1 & b_3 \\ c_1 & c_3 \end{vmatrix} + a_3 \begin{vmatrix} b_1 & b_2 \\ c_1 & c_2 \end{vmatrix}$$

The right side of this formula is easily remembered by noting that a_1, a_2, and a_3 are the entries in the first "row" of the left side, and the 2 × 2 determinants on the right side arise by deleting the first row and an appropriate column from the left side. The pattern is as follows:

$$
\begin{vmatrix} a_1 & a_2 & a_3 \\ b_1 & b_2 & b_3 \\ c_1 & c_2 & c_3 \end{vmatrix} = a_1 \begin{vmatrix} a_1 & a_2 & a_3 \\ b_1 & b_2 & b_3 \\ c_1 & c_2 & c_3 \end{vmatrix} - a_2 \begin{vmatrix} a_1 & a_2 & a_3 \\ b_1 & b_2 & b_3 \\ c_1 & c_2 & c_3 \end{vmatrix} + a_3 \begin{vmatrix} a_1 & a_2 & a_3 \\ b_1 & b_2 & b_3 \\ c_1 & c_2 & c_3 \end{vmatrix}
$$

For example,

$$
\begin{vmatrix} 3 & -2 & -5 \\ 1 & 4 & -4 \\ 0 & 3 & 2 \end{vmatrix} = 3 \begin{vmatrix} 4 & -4 \\ 3 & 2 \end{vmatrix} - (-2) \begin{vmatrix} 1 & -4 \\ 0 & 2 \end{vmatrix} + (-5) \begin{vmatrix} 1 & 4 \\ 0 & 3 \end{vmatrix}
$$

$$
= 3(20) + 2(2) - 5(3)
$$

$$
= 49
$$

15.4.1   DEFINITION   If $\mathbf{u} = \langle u_1, u_2, u_3 \rangle$ and $\mathbf{v} = \langle v_1, v_2, v_3 \rangle$ are vectors in 3-space, then the *cross product* $\mathbf{u} \times \mathbf{v}$ is the vector defined by

$$
\mathbf{u} \times \mathbf{v} = \begin{vmatrix} u_2 & u_3 \\ v_2 & v_3 \end{vmatrix} \mathbf{i} - \begin{vmatrix} u_1 & u_3 \\ v_1 & v_3 \end{vmatrix} \mathbf{j} + \begin{vmatrix} u_1 & u_2 \\ v_1 & v_2 \end{vmatrix} \mathbf{k} \tag{1}
$$

Formula (1) can be remembered by writing it in the form,

$$
\mathbf{u} \times \mathbf{v} = \begin{vmatrix} \mathbf{i} & \mathbf{j} & \mathbf{k} \\ u_1 & u_2 & u_3 \\ v_1 & v_2 & v_3 \end{vmatrix} \tag{2}
$$

However, this is just a mnemonic device since the entries in a determinant must be numbers, not vectors.

▶ Example 1   Find $\mathbf{u} \times \mathbf{v}$, where $\mathbf{u} = \langle 1, 2, -2 \rangle$ and $\mathbf{v} = \langle 3, 0, 1 \rangle$.

Solution.

$$
\mathbf{u} \times \mathbf{v} = \begin{vmatrix} \mathbf{i} & \mathbf{j} & \mathbf{k} \\ 1 & 2 & -2 \\ 3 & 0 & 1 \end{vmatrix}
$$

$$
= \begin{vmatrix} 2 & -2 \\ 0 & 1 \end{vmatrix} \mathbf{i} - \begin{vmatrix} 1 & -2 \\ 3 & 1 \end{vmatrix} \mathbf{j} + \begin{vmatrix} 1 & 2 \\ 3 & 0 \end{vmatrix} \mathbf{k} = 2\mathbf{i} - 7\mathbf{j} - 6\mathbf{k} \quad ◀
$$

Observe that the cross product of two vectors is another vector, whereas the dot product of two vectors is a scalar.

The following theorem gives an important relationship between dot product and cross product and also shows that $\mathbf{u} \times \mathbf{v}$ is orthogonal to both $\mathbf{u}$ and $\mathbf{v}$.

15.4.2 THEOREM *If* $\mathbf{u}$ *and* $\mathbf{v}$ *are vectors in 3-space, then:*

(a) $\mathbf{u} \cdot (\mathbf{u} \times \mathbf{v}) = 0$ ($\mathbf{u} \times \mathbf{v}$ *is orthogonal to* $\mathbf{u}$)

(b) $\mathbf{v} \cdot (\mathbf{u} \times \mathbf{v}) = 0$ ($\mathbf{u} \times \mathbf{v}$ *is orthogonal to* $\mathbf{v}$)

(c) $\|\mathbf{u} \times \mathbf{v}\|^2 = \|\mathbf{u}\|^2 \|\mathbf{v}\|^2 - (\mathbf{u} \cdot \mathbf{v})^2$ (*Lagrange's identity*)

Proof. Let $\mathbf{u} = \langle u_1, u_2, u_3 \rangle$ and $\mathbf{v} = \langle v_1, v_2, v_3 \rangle$.

(a) By definition

$$\mathbf{u} \times \mathbf{v} = \begin{vmatrix} u_2 & u_3 \\ v_2 & v_3 \end{vmatrix} \mathbf{i} - \begin{vmatrix} u_1 & u_3 \\ v_1 & v_3 \end{vmatrix} \mathbf{j} + \begin{vmatrix} u_1 & u_2 \\ v_1 & v_2 \end{vmatrix} \mathbf{k}$$

which may be rewritten

$$\mathbf{u} \times \mathbf{v} = \langle u_2 v_3 - u_3 v_2, \ u_3 v_1 - u_1 v_3, \ u_1 v_2 - u_2 v_1 \rangle \tag{3}$$

so that

$$\mathbf{u} \cdot (\mathbf{u} \times \mathbf{v}) = u_1(u_2 v_3 - u_3 v_2) + u_2(u_3 v_1 - u_1 v_3)$$
$$+ \ u_3(u_1 v_2 - u_2 v_1) = 0$$

(b) Similar to (a).

(c) From (3) we obtain

$$\|\mathbf{u} \times \mathbf{v}\|^2 = (u_2 v_3 - u_3 v_2)^2 + (u_3 v_1 - u_1 v_3)^2 + (u_1 v_2 - u_2 v_1)^2 \tag{4}$$

Moreover,

$$\|\mathbf{u}\|^2 \|\mathbf{v}\|^2 - (\mathbf{u} \cdot \mathbf{v})^2 = (u_1^2 + u_2^2 + u_3^2)(v_1^2 + v_2^2 + v_3^2)$$
$$- (u_1 v_1 + u_2 v_2 + u_3 v_3)^2 \tag{5}$$

Lagrange's identity can be established by "multiplying out" the right sides of (4) and (5) and verifying their equality.

▶ **Example 2** Let $\mathbf{u} = \langle 1, 2, -2 \rangle$ and $\mathbf{v} = \langle 3, 0, 1 \rangle$.

In Example 1 we showed that

$$\mathbf{u} \times \mathbf{v} = \langle 2, -7, -6 \rangle$$

Thus,

$$\mathbf{u} \cdot (\mathbf{u} \times \mathbf{v}) = (1)(2) + (2)(-7) + (-2)(-6) = 0$$

and

$$\mathbf{v} \cdot (\mathbf{u} \times \mathbf{v}) = (3)(2) + (0)(-7) + (1)(-6) = 0$$

so that $\mathbf{u} \times \mathbf{v}$ is orthogonal to both $\mathbf{u}$ and $\mathbf{v}$ as guaranteed by Theorem 15.4.2.

◀

If $\mathbf{u}$ and $\mathbf{v}$ are nonzero vectors in 3-space, then the length of $\mathbf{u} \times \mathbf{v}$ has a useful geometric interpretation. Lagrange's identity, given in Theorem 15.4.2, states that

$$\|\mathbf{u} \times \mathbf{v}\|^2 = \|\mathbf{u}\|^2 \|\mathbf{v}\|^2 - (\mathbf{u} \cdot \mathbf{v})^2 \tag{6}$$

If θ denotes the angle between $\mathbf{u}$ and $\mathbf{v}$, then $\mathbf{u} \cdot \mathbf{v} = \|\mathbf{u}\| \|\mathbf{v}\| \cos \theta$, so (6) can be rewritten as

$$\begin{aligned}
\|\mathbf{u} \times \mathbf{v}\|^2 &= \|\mathbf{u}\|^2 \|\mathbf{v}\|^2 - \|\mathbf{u}\|^2 \|\mathbf{v}\|^2 \cos^2 \theta \\
&= \|\mathbf{u}\|^2 \|\mathbf{v}\|^2 (1 - \cos^2 \theta) \\
&= \|\mathbf{u}\|^2 \|\mathbf{v}\|^2 \sin^2 \theta
\end{aligned}$$

Since $0 \leq \theta \leq \pi$, it follows that $\sin \theta \geq 0$, so

$$\|\mathbf{u} \times \mathbf{v}\| = \|\mathbf{u}\| \|\mathbf{v}\| \sin \theta \tag{7}$$

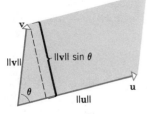

Figure 15.4.1

But $\|\mathbf{v}\| \sin \theta$ is the altitude of the parallelogram determined by $\mathbf{u}$ and $\mathbf{v}$ (Figure 15.4.1). Thus, from (7), the area A of this parallelogram is given by

$$A = (\text{base})(\text{altitude}) = \|\mathbf{u}\| \|\mathbf{v}\| \sin \theta = \|\mathbf{u} \times \mathbf{v}\|$$

In other words, the length of $\mathbf{u} \times \mathbf{v}$ is numerically equal to the area of the parallelogram determined by $\mathbf{u}$ and $\mathbf{v}$.

▶ Example 3 Find the area of the triangle determined by the points $P_1(2, 2, 0)$, $P_2(-1, 0, 2)$, and $P_3(0, 4, 3)$.

Solution. The area A of the triangle is half the area of the parallelogram determined by the vectors $\overrightarrow{P_1P_2}$ and $\overrightarrow{P_1P_3}$ (Figure 15.4.2). But $\overrightarrow{P_1P_2} = \langle -3, -2, 2 \rangle$ and $\overrightarrow{P_1P_3} = \langle -2, 2, 3 \rangle$ so that

$$\overrightarrow{P_1P_2} \times \overrightarrow{P_1P_3} = \langle -10, 5, -10 \rangle$$

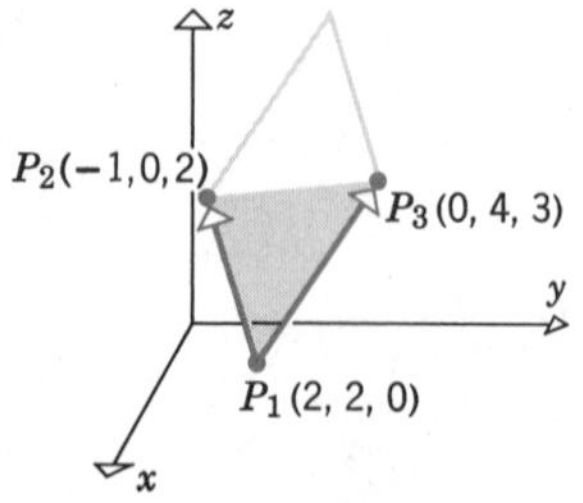

(verify) and consequently

$$A = \frac{1}{2} \|\overrightarrow{P_1P_2} \times \overrightarrow{P_1P_3}\| = \frac{15}{2}$$

◀

Figure 15.4.2

It follows from (7) that $\mathbf{u} \times \mathbf{v} = \mathbf{0}$ if and only if

$$\mathbf{u} = \mathbf{0} \quad \text{or} \quad \mathbf{v} = \mathbf{0} \quad \text{or} \quad \sin \theta = 0$$

In all three cases the vectors $\mathbf{u}$ and $\mathbf{v}$ are parallel. For the first two cases this is true because $\mathbf{0}$ is parallel to every vector, and in the third case, $\sin \theta = 0$ implies that the angle θ between $\mathbf{u}$ and $\mathbf{v}$ is $\theta = 0$ or $\theta = \pi$. In summary, we have the following result.

15.4.3 THEOREM *If $\mathbf{u}$ and $\mathbf{v}$ are vectors in 3-space, then $\mathbf{u} \times \mathbf{v} = \mathbf{0}$ if and only if $\mathbf{u}$ and $\mathbf{v}$ are parallel vectors.*

The main arithmetic properties of the cross product are listed in the next theorem.

15.4.4 THEOREM *If $\mathbf{u}$, $\mathbf{v}$, and $\mathbf{w}$ are any vectors in 3-space and k is any scalar, then:*
(a) $\mathbf{u} \times \mathbf{v} = -(\mathbf{v} \times \mathbf{u})$
(b) $\mathbf{u} \times (\mathbf{v} + \mathbf{w}) = (\mathbf{u} \times \mathbf{v}) + (\mathbf{u} \times \mathbf{w})$
(c) $(\mathbf{u} + \mathbf{v}) \times \mathbf{w} = (\mathbf{u} \times \mathbf{w}) + (\mathbf{v} \times \mathbf{w})$
(d) $k(\mathbf{u} \times \mathbf{v}) = (k\mathbf{u}) \times \mathbf{v} = \mathbf{u} \times (k\mathbf{v})$
(e) $\mathbf{u} \times \mathbf{0} = \mathbf{0} \times \mathbf{u} = \mathbf{0}$
(f) $\mathbf{u} \times \mathbf{u} = \mathbf{0}$

We will prove (a) and leave the remaining proofs as exercises.

Proof (a). First note that interchanging the rows of a 2×2 determinant changes the sign of the determinant, since

$$\begin{vmatrix} c & d \\ a & b \end{vmatrix} = bc - ad = -(ad - bc) = - \begin{vmatrix} a & b \\ c & d \end{vmatrix}$$

It follows that interchanging $\mathbf{u}$ and $\mathbf{v}$ in (1) interchanges the rows of the three determinants on the right side of (1) and thereby changes the sign of each component in the cross product. Thus, $\mathbf{u} \times \mathbf{v} = -(\mathbf{v} \times \mathbf{u})$. ▊

Cross products of the unit vectors $\mathbf{i}$, $\mathbf{j}$, and $\mathbf{k}$ are of special interest. We obtain, for example

$$\mathbf{i} \times \mathbf{j} = \begin{vmatrix} \mathbf{i} & \mathbf{j} & \mathbf{k} \\ 1 & 0 & 0 \\ 0 & 1 & 0 \end{vmatrix} = \begin{vmatrix} 0 & 0 \\ 1 & 0 \end{vmatrix} \mathbf{i} - \begin{vmatrix} 1 & 0 \\ 0 & 0 \end{vmatrix} \mathbf{j} + \begin{vmatrix} 1 & 0 \\ 0 & 1 \end{vmatrix} \mathbf{k} = \mathbf{k}$$

The reader should have no trouble obtaining the following list of cross products:

$$\mathbf{i} \times \mathbf{j} = \mathbf{k} \qquad \mathbf{j} \times \mathbf{k} = \mathbf{i} \qquad \mathbf{k} \times \mathbf{i} = \mathbf{j}$$
$$\mathbf{j} \times \mathbf{i} = -\mathbf{k} \qquad \mathbf{k} \times \mathbf{j} = -\mathbf{i} \qquad \mathbf{i} \times \mathbf{k} = -\mathbf{j}$$
$$\mathbf{i} \times \mathbf{i} = 0 \qquad \mathbf{j} \times \mathbf{j} = 0 \qquad \mathbf{k} \times \mathbf{k} = 0$$

The following diagram is helpful for remembering these results.

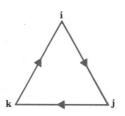

In this diagram, the cross product of two consecutive vectors going clockwise is the next vector around, and the cross product of two consecutive vectors going counterclockwise is the negative of the next vector around.

WARNING. It is *not* true in general that $\mathbf{u} \times (\mathbf{v} \times \mathbf{w}) = (\mathbf{u} \times \mathbf{v}) \times \mathbf{w}$. For example,

$$\mathbf{i} \times (\mathbf{j} \times \mathbf{j}) = \mathbf{i} \times 0 = 0$$

and

$$(\mathbf{i} \times \mathbf{j}) \times \mathbf{j} = \mathbf{k} \times \mathbf{j} = -\mathbf{i}$$

so that

$$\mathbf{i} \times (\mathbf{j} \times \mathbf{j}) \neq (\mathbf{i} \times \mathbf{j}) \times \mathbf{j}$$

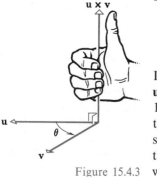

Figure 15.4.3

We know from Theorem 15.4.2 that $\mathbf{u} \times \mathbf{v}$ is orthogonal to both $\mathbf{u}$ and $\mathbf{v}$. It can be shown that if $\mathbf{u}$ and $\mathbf{v}$ are nonzero vectors, then the direction of $\mathbf{u} \times \mathbf{v}$ can be determined using the following "right-hand rule"* (Figure 15.4.3). Let θ be the angle between $\mathbf{u}$ and $\mathbf{v}$, and suppose $\mathbf{u}$ is rotated through the angle θ until it coincides with $\mathbf{v}$. If the fingers of the right hand are cupped so they point in the direction of rotation, then the thumb indicates (roughly) the direction of $\mathbf{u} \times \mathbf{v}$. The reader may find it instructive to practice this rule with the products

---

*Recall that we agreed to consider only right-handed coordinate systems in this text. Had we used left-handed systems instead, a "left-hand rule" would apply here.

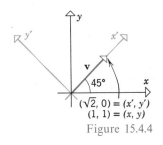

Figure 15.4.4

$$\mathbf{i} \times \mathbf{j} = \mathbf{k} \qquad \mathbf{j} \times \mathbf{k} = \mathbf{i} \qquad \mathbf{k} \times \mathbf{i} = \mathbf{j}$$

Initially, we defined a vector to be a directed line segment or arrow in 2-space or 3-space; coordinate systems and components were introduced later in order to simplify computations with vectors. Thus, a vector has a "mathematical existence" regardless of whether a coordinate system has been introduced. Furthermore, the components of a vector are not determined by the vector alone; they depend as well on the coordinate system chosen. For example, in Figure 15.4.4 we have indicated a fixed-plane vector $\mathbf{v}$ and two different coordinate systems. In the xy-coordinate system the components of $\mathbf{v}$ are $\langle 1, 1 \rangle$, and in the $x'y'$-system they are $\langle \sqrt{2}, 0 \rangle$.

This raises an important question about our definition of cross product. Since we defined the cross product $\mathbf{u} \times \mathbf{v}$ in terms of the components of $\mathbf{u}$ and $\mathbf{v}$, and since these components depend on the coordinate system chosen, it seems possible that two *fixed* vectors $\mathbf{u}$ and $\mathbf{v}$ might yield different vectors for $\mathbf{u} \times \mathbf{v}$ in different coordinate systems. Fortunately, this is not the case. To see this, we need only recall that:

(i) $\mathbf{u} \times \mathbf{v}$ is perpendicular to both $\mathbf{u}$ and $\mathbf{v}$.

(ii)   The orientation of $\mathbf{u} \times \mathbf{v}$ is determined by the right-hand rule.

(iii) $\|\mathbf{u} \times \mathbf{v}\| = \|\mathbf{u}\| \, \|\mathbf{v}\| \sin \theta$.

These three properties completely determine the vector $\mathbf{u} \times \mathbf{v}$. Properties (i) and (ii) determine the direction, and property (iii) determines the length. Since these properties depend only on the lengths and relative positions of $\mathbf{u}$ and $\mathbf{v}$ and not on the particular right-handed coordinate system being used, the vector $\mathbf{u} \times \mathbf{v}$ will remain unchanged if a different right-handed coordinate system is introduced. Thus, we say that the definition of $\mathbf{u} \times \mathbf{v}$ is ***coordinate free***. This result is of importance to physicists and engineers who often work with many coordinate systems in the same problem.

If $\mathbf{a} = \langle a_1, a_2, a_3 \rangle$, $\mathbf{b} = \langle b_1, b_2, b_3 \rangle$, and $\mathbf{c} = \langle c_1, c_2, c_3 \rangle$ are vectors in 3-space, then the number

$$\mathbf{a} \cdot (\mathbf{b} \times \mathbf{c})$$

is called the ***triple scalar product*** of $\mathbf{a}$, $\mathbf{b}$, and $\mathbf{c}$. The triple scalar product may be conveniently calculated from the formula

$$\mathbf{a} \cdot (\mathbf{b} \times \mathbf{c}) = \begin{vmatrix} a_1 & a_2 & a_3 \\ b_1 & b_2 & b_3 \\ c_1 & c_2 & c_3 \end{vmatrix} \tag{8}$$

The validity of this formula may be seen by writing

$$\mathbf{a} \cdot (\mathbf{b} \times \mathbf{c}) = \mathbf{a} \cdot \left(\begin{vmatrix} b_2 & b_3 \\ c_2 & c_3 \end{vmatrix} \mathbf{i} - \begin{vmatrix} b_1 & b_3 \\ c_1 & c_3 \end{vmatrix} \mathbf{j} + \begin{vmatrix} b_1 & b_2 \\ c_1 & c_2 \end{vmatrix} \mathbf{k} \right)$$

$$= \begin{vmatrix} b_2 & b_3 \\ c_2 & c_3 \end{vmatrix} a_1 - \begin{vmatrix} b_1 & b_3 \\ c_1 & c_3 \end{vmatrix} a_2 + \begin{vmatrix} b_1 & b_2 \\ c_1 & c_2 \end{vmatrix} a_3$$

$$= \begin{vmatrix} a_1 & a_2 & a_3 \\ b_1 & b_2 & b_3 \\ c_1 & c_2 & c_3 \end{vmatrix}$$

▶ **Example 4**   Calculate the triple scalar product $\mathbf{a} \cdot (\mathbf{b} \times \mathbf{c})$ of the vectors

$$\mathbf{a} = 3\mathbf{i} - 2\mathbf{j} - 5\mathbf{k}, \qquad \mathbf{b} = \mathbf{i} + 4\mathbf{j} - 4\mathbf{k}, \qquad \mathbf{c} = 3\mathbf{j} + 2\mathbf{k}$$

Solution.

$$\mathbf{a} \cdot (\mathbf{b} \times \mathbf{c}) = \begin{vmatrix} 3 & -2 & -5 \\ 1 & 4 & -4 \\ 0 & 3 & 2 \end{vmatrix}$$

$$= 3 \begin{vmatrix} 4 & -4 \\ 3 & 2 \end{vmatrix} - (-2) \begin{vmatrix} 1 & -4 \\ 0 & 2 \end{vmatrix} + (-5) \begin{vmatrix} 1 & 4 \\ 0 & 3 \end{vmatrix}$$

$$= 60 + 4 - 15 = 49 \qquad \blacktriangleleft$$

REMARK.   The symbol $(\mathbf{a} \cdot \mathbf{b}) \times \mathbf{c}$ makes no sense since we cannot form the cross product of a scalar and a vector. Thus, no ambiguity arises if we write $\mathbf{a} \cdot \mathbf{b} \times \mathbf{c}$ rather than $\mathbf{a} \cdot (\mathbf{b} \times \mathbf{c})$. However, for clarity we will usually keep the parentheses.

From the following calculations, we see that interchanging the rows of a 2×2 determinant changes the sign of its value, and that converting the rows into columns has no effect on the value.

$$\begin{vmatrix} c & d \\ a & b \end{vmatrix} = bc - ad = -(ad - bc) = - \begin{vmatrix} a & b \\ c & d \end{vmatrix}$$

$$\begin{vmatrix} a & c \\ b & d \end{vmatrix} = ad - bc = \begin{vmatrix} a & b \\ c & d \end{vmatrix}$$

The following analogous result holds for 3×3 determinants. We omit the proof.

15.4.5   THEOREM   (a)   *Interchanging two rows of a 3×3 determinant changes the sign of its value.*

(b) $\begin{vmatrix} a_1 & a_2 & a_3 \\ b_1 & b_2 & b_3 \\ c_1 & c_2 & c_3 \end{vmatrix} = \begin{vmatrix} a_1 & b_1 & c_1 \\ a_2 & b_2 & c_2 \\ a_3 & b_3 & c_3 \end{vmatrix}$

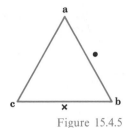

Figure 15.4.5

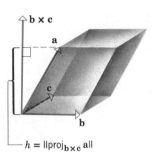

$h = \|\text{proj}_{\mathbf{b} \times \mathbf{c}}\, \mathbf{a}\|$

Figure 15.4.6

It follows from part (*a*) of this theorem that

$$\mathbf{a} \cdot (\mathbf{b} \times \mathbf{c}) = \mathbf{c} \cdot (\mathbf{a} \times \mathbf{b}) = \mathbf{b} \cdot (\mathbf{c} \times \mathbf{a}) \tag{9}$$

since the 3×3 determinants that represent these products can be obtained from one another by *two* row interchanges. (Verify.) These relationships may be remembered by moving the vectors **a**, **b**, and **c** clockwise around the triangle in Figure 15.4.5.

The triple scalar product $\mathbf{a} \cdot (\mathbf{b} \times \mathbf{c})$ has a useful geometric interpretation. If we assume, for the moment, that the vectors **a**, **b**, **c** do not all lie in the same plane, then the three vectors form adjacent sides of a parallelepiped (Figure 15.4.6) when they are positioned with a common initial point. If the parallelogram determined by **b** and **c** is regarded as the base of the parallelepiped, then the area of the base is $\|\mathbf{b} \times \mathbf{c}\|$, and the height h is the length of the orthogonal projection of **a** on $\mathbf{b} \times \mathbf{c}$ (Figure 15.4.6). Therefore, by (9) of Section 15.3.1

$$h = \|\text{proj}_{\mathbf{b} \times \mathbf{c}}\, \mathbf{a}\| = \frac{|\mathbf{a} \cdot (\mathbf{b} \times \mathbf{c})|}{\|\mathbf{b} \times \mathbf{c}\|}$$

It follows that the volume V of the parallelepiped is

$$V = (\text{area of base}) \cdot \text{height} = \|\mathbf{b} \times \mathbf{c}\| \frac{|\mathbf{a} \cdot (\mathbf{b} \times \mathbf{c})|}{\|\mathbf{b} \times \mathbf{c}\|}$$

or, on simplifying,

$$V = \begin{bmatrix} \text{volume of parallelepiped with} \\ \text{adjacent sides } \mathbf{a},\ \mathbf{b},\ \text{and } \mathbf{c} \end{bmatrix} = |\mathbf{a} \cdot (\mathbf{b} \times \mathbf{c})| \tag{10}$$

REMARK. It follows from this formula that

$$\mathbf{a} \cdot (\mathbf{b} \times \mathbf{c}) = \pm V$$

where the $+$ or $-$ results depending on whether **a** makes an acute or obtuse angle with $\mathbf{b} \times \mathbf{c}$ (Theorem 15.3.2).

If the vectors **a**, **b**, and **c** do not lie in the same plane, then they determine a parallelepiped of positive volume. Thus, from (10), $\mathbf{a} \cdot (\mathbf{b} \times \mathbf{c}) \neq 0$ if **a**, **b**, and **c** do not lie in the same plane. It follows, therefore, that if $\mathbf{a} \cdot (\mathbf{b} \times \mathbf{c}) = 0$, then **a**, **b**, and **c** lie in the same plane. Conversely, it can be shown that if **a**, **b**, and **c** lie in the same plane, then $\mathbf{a} \cdot (\mathbf{b} \times \mathbf{c}) = 0$. In summary, we have the following result.

15.4.6 THEOREM   *If the vectors* $\mathbf{a} = \langle a_1, a_2, a_3 \rangle$, $\mathbf{b} = \langle b_1, b_2, b_3 \rangle$, *and* $\mathbf{c} = \langle c_1, c_2, c_3 \rangle$ *have the same initial point, then they lie in the same plane if and only if*

$$\mathbf{a} \cdot (\mathbf{b} \times \mathbf{c}) = \begin{vmatrix} a_1 & a_2 & a_3 \\ b_1 & b_2 & b_3 \\ c_1 & c_2 & c_3 \end{vmatrix} = 0$$

▶ Exercise Set 15.4

In Exercises 1–4, find $\mathbf{a} \times \mathbf{b}$.

1. $\mathbf{a} = \langle 1, 2, -3 \rangle$, $\mathbf{b} = \langle -4, 1, 2 \rangle$.

2. $\mathbf{a} = 3\mathbf{i} + 2\mathbf{j} - \mathbf{k}$, $\mathbf{b} = -\mathbf{i} - 3\mathbf{j} + \mathbf{k}$.

3. $\mathbf{a} = \langle 0, 1, -2 \rangle$, $\mathbf{b} = \langle 3, 0, -4 \rangle$.

4. $\mathbf{a} = 4\mathbf{i} + \mathbf{k}$, $\mathbf{b} = 2\mathbf{i} - \mathbf{j}$.

5. Let $\mathbf{u} = \langle 2, -1, 3 \rangle$, $\mathbf{v} = \langle 0, 1, 7 \rangle$, and $\mathbf{w} = \langle 1, 4, 5 \rangle$. Find:
(a) $\mathbf{u} \times (\mathbf{v} \times \mathbf{w})$
(b) $(\mathbf{u} \times \mathbf{v}) \times \mathbf{w}$
(c) $\mathbf{u} \times (\mathbf{v} - 2\mathbf{w})$
(d) $(\mathbf{u} \times \mathbf{v}) - 2\mathbf{w}$
(e) $(\mathbf{u} \times \mathbf{v}) \times (\mathbf{v} \times \mathbf{w})$
(f) $(\mathbf{v} \times \mathbf{w}) \times (\mathbf{u} \times \mathbf{v})$.

6. Find a vector orthogonal to both $\mathbf{u}$ and $\mathbf{v}$.
(a) $\mathbf{u} = -7\mathbf{i} + 3\mathbf{j} + \mathbf{k}$, $\mathbf{v} = 2\mathbf{i} + 4\mathbf{k}$
(b) $\mathbf{u} = \langle -1, -1, -1 \rangle$, $\mathbf{v} = \langle 2, 0, 2 \rangle$.

7. Verify Theorem 15.4.2 for the vectors $\mathbf{u} = \mathbf{i} - 5\mathbf{j} + 6\mathbf{k}$ and $\mathbf{v} = 2\mathbf{i} + \mathbf{j} + 2\mathbf{k}$.

8. Verify Theorem 15.4.4 for $\mathbf{u} = \langle 2, 0, -1 \rangle$, $\mathbf{v} = \langle 6, 7, 4 \rangle$, $\mathbf{w} = \langle 1, 1, 1 \rangle$, and $k = -3$.

9. Find the area of the triangle having vertices P, Q, and R.
(a) $P(1, 5, -2)$, $Q(0, 0, 0)$, $R(3, 5, 1)$
(b) $P(2, 0, -3)$, $Q(1, 4, 5)$, $R(7, 2, 9)$.

10. Use the cross product to find the sine of the angle between the vectors $\mathbf{a} = 2\mathbf{i} + 3\mathbf{j} - 6\mathbf{k}$ and $\mathbf{b} = 2\mathbf{i} + 3\mathbf{j} + 6\mathbf{k}$

11. What is wrong with the expression $\mathbf{u} \times \mathbf{v} \times \mathbf{w}$?

In Exercises 12–15, find $\mathbf{a} \cdot (\mathbf{b} \times \mathbf{c})$.

12. $\mathbf{a} = \langle 1, -2, 2 \rangle$, $\mathbf{b} = \langle 0, 3, 2 \rangle$, $\mathbf{c} = \langle -4, 1, -3 \rangle$.

13. $\mathbf{a} = 2\mathbf{i} - 3\mathbf{j} + \mathbf{k}$, $\mathbf{b} = 4\mathbf{i} + \mathbf{j} - 3\mathbf{k}$, $\mathbf{c} = \mathbf{j} + 5\mathbf{k}$.

14. $\mathbf{a} = \langle 2, 1, 0 \rangle$, $\mathbf{b} = \langle 1, -3, 1 \rangle$, $\mathbf{c} = \langle 4, 0, 1 \rangle$.

15. $\mathbf{a} = \mathbf{i}$, $\mathbf{b} = \mathbf{i} + \mathbf{j}$, $\mathbf{c} = \mathbf{i} + \mathbf{j} + \mathbf{k}$.

16. Suppose $\mathbf{u} \cdot (\mathbf{v} \times \mathbf{w}) = 3$. Find
(a) $\mathbf{u} \cdot (\mathbf{w} \times \mathbf{v})$
(b) $(\mathbf{v} \times \mathbf{w}) \cdot \mathbf{u}$
(c) $\mathbf{w} \cdot (\mathbf{u} \times \mathbf{v})$
(d) $\mathbf{v} \cdot (\mathbf{u} \times \mathbf{w})$
(e) $(\mathbf{u} \times \mathbf{w}) \cdot \mathbf{v}$
(f) $\mathbf{v} \cdot (\mathbf{w} \times \mathbf{w})$.

17. Find the volume of the parallelepiped with sides $\mathbf{a}$, $\mathbf{b}$, and $\mathbf{c}$.
(a) $\mathbf{a} = \langle 2, -6, 2 \rangle$, $\mathbf{b} = \langle 0, 4, -2 \rangle$, $\mathbf{c} = \langle 2, 2, -4 \rangle$
(b) $\mathbf{a} = 3\mathbf{i} + \mathbf{j} + 2\mathbf{k}$, $\mathbf{b} = 4\mathbf{i} + 5\mathbf{j} + \mathbf{k}$, $\mathbf{c} = \mathbf{i} + 2\mathbf{j} + 4\mathbf{k}$.

18. Determine whether $\mathbf{u}$, $\mathbf{v}$, and $\mathbf{w}$ lie in the same plane.
(a) $\mathbf{u} = \langle 1, -2, 1 \rangle$, $\mathbf{v} = \langle 3, 0, -2 \rangle$, $\mathbf{w} = \langle 5, -4, 0 \rangle$
(b) $\mathbf{u} = 5\mathbf{i} - 2\mathbf{j} + \mathbf{k}$, $\mathbf{v} = 4\mathbf{i} - \mathbf{j} + \mathbf{k}$, $\mathbf{w} = \mathbf{i} - \mathbf{j}$
(c) $\mathbf{u} = \langle 4, -8, 1 \rangle$, $\mathbf{v} = \langle 2, 1, -2 \rangle$, $\mathbf{w} = \langle 3, -4, 12 \rangle$.

19. Consider the parallelepiped with sides

$$\mathbf{a} = 3\mathbf{i} + 2\mathbf{j} + \mathbf{k}$$
$$\mathbf{b} = \mathbf{i} + \mathbf{j} + 2\mathbf{k}$$
$$\mathbf{c} = \mathbf{i} + 3\mathbf{j} + 3\mathbf{k}$$

(a) Find the volume.
(b) Find the area of the face determined by $\mathbf{a}$ and $\mathbf{c}$.
(c) Find the angle between $\mathbf{a}$ and the plane containing the face determined by $\mathbf{b}$ and $\mathbf{c}$.

20. Find a vector $\mathbf{n}$ perpendicular to the plane determined by the points $A(0, -2, 1)$, $B(1, -1, -2)$, and $C(-1, 1, 0)$.

21. Prove:
(a) $(\mathbf{u} + k\mathbf{v}) \times \mathbf{v} = \mathbf{u} \times \mathbf{v}$
(b) $\mathbf{u} \cdot \mathbf{v} \times \mathbf{z} = \mathbf{u} \times \mathbf{v} \cdot \mathbf{z}$.

22. Let **u**, **v**, and **w** be vectors in 3-space with the same initial point and such that no two are collinear. Prove:

(a) **u** × (**v** × **w**) lies in the plane determined by **v** and **w**;

(b) (**u** × **v**) × **w** lies in the plane determined by **u** and **v**.

23. Prove parts (b) and (c) of Theorem 15.4.4.

24. Prove parts (d), (e), and (f) of Theorem 15.4.4.

25. Prove that **x** × (**y** × **z**) = (**x** · **z**)**y** − (**x** · **y**)**z**. [*Hint:* First prove the result for the case **z** = **i**, then for **z** = **j**, and then for **z** = **k**. Finally, prove it for an arbitrary vector **z** = z_1**i** + z_2**j** + z_3**k**.]

26. Let **a** = **i** + 3**j** − **k**, **b** = **i** + **j** + 2**k**, and **c** = 3**i** − **j** + 2**k**. Calculate **a** × (**b** × **c**) using Exercise 25 and then check your result by calculating directly.

27. Prove: If **a**, **b**, **c**, and **d** lie in the same plane, then

$$(\mathbf{a} \times \mathbf{b}) \times (\mathbf{c} \times \mathbf{d}) = \mathbf{0}.$$

28. It is a theorem of solid geometry that the volume of a tetrahedron is $\frac{1}{3}$(area of base) · (height). Use this result to prove that the volume of a tetrahedron whose sides are the vectors **a**, **b**, and **c** is $\frac{1}{6}|\mathbf{a} \cdot (\mathbf{b} \times \mathbf{c})|$.

29. Use the result of Exercise 28 to find the volume of the tetrahedron with vertices P, Q, R, and S.

(a) $P(-1, 2, 0)$, $Q(2, 1, -3)$, $R(1, 0, 1)$, $S(3, -2, 3)$

(b) $P(0, 0, 0)$, $Q(1, 2, -1)$, $R(3, 4, 0)$, $S(-1, -3, 4)$.

15.5 CURVES IN 3-SPACE

In Chapter 14 we studied two-dimensional vector-valued functions, and motion along a curved path in the plane. Since many of those two-dimensional results extend with little modification to three dimensions, we will discuss those ideas with a minimum of proof.

A curve C in 3-space given by the parametric equations

$$x = x(t)$$
$$y = y(t)$$
$$z = z(t)$$

can also be represented by a single *vector-valued function,*

$$\mathbf{r}(t) = x(t)\mathbf{i} + y(t)\mathbf{j} + z(t)\mathbf{k}$$

called the *position vector* or *position function* for the curve (Figure 15.5.1). As in two dimensions, limits of vector-valued functions are defined componentwise:

$$\lim \mathbf{r}(t) = \lim [x(t)\mathbf{i} + y(t)\mathbf{j} + z(t)\mathbf{k}]$$
$$= (\lim x(t))\mathbf{i} + (\lim y(t))\mathbf{j} + (\lim z(t))\mathbf{k}$$

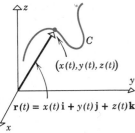

Figure 15.5.1 The derivative of a three-dimensional vector-valued function is defined by

$$\mathbf{r}'(t) = \lim_{\Delta t \to 0} \frac{\mathbf{r}(t + \Delta t) - \mathbf{r}(t)}{\Delta t}$$

and it follows, as in two dimensions, that derivatives can be calculated componentwise:

$$\mathbf{r}'(t) = \frac{d}{dt}[x(t)\mathbf{i} + y(t)\mathbf{j} + z(t)\mathbf{k}] = x'(t)\mathbf{i} + y'(t)\mathbf{j} + z'(t)\mathbf{k}$$

The rules of differentiation listed in Theorem 14.2.4 and the chain rule, Theorem 14.2.5, hold for three-dimensional vector-valued functions. In addition, we have the following rules for differentiating dot products and cross products of three-dimensional vector-valued functions:

$$\frac{d}{dt}[\mathbf{r}_1(t) \cdot \mathbf{r}_2(t)] = \mathbf{r}_1(t) \cdot \frac{d\mathbf{r}_2}{dt} + \frac{d\mathbf{r}_1}{dt} \cdot \mathbf{r}_2(t) \tag{1}$$

$$\frac{d}{dt}[\mathbf{r}_1(t) \times \mathbf{r}_2(t)] = \mathbf{r}_1(t) \times \frac{d\mathbf{r}_2}{dt} + \frac{d\mathbf{r}_1}{dt} \times \mathbf{r}_2(t) \tag{2}$$

The proofs of (1) and (2) are left as exercises.

REMARK.   In (1), the order of the factors in each term on the right does not matter, but in (2) it does.

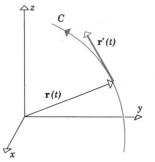

Figure 15.5.2

If $\mathbf{r}(t)$ is the position vector for a curve C in 3-space, and if $\mathbf{r}'(t) \neq \mathbf{0}$, then the vector $\mathbf{r}'(t)$ is *tangent* to the curve C at the tip of $\mathbf{r}(t)$ and points in the direction of increasing parameter (Figure 15.5.2), just as in two dimensions. We will call C a **smooth curve** or $\mathbf{r}(t)$ a **smooth function** if $x'(t), y'(t)$, and $z'(t)$ are continuous and $\mathbf{r}'(t) = x'(t)\mathbf{i} + y'(t)\mathbf{j} + z'(t)\mathbf{k} \neq \mathbf{0}$ for any value of t.

As in two dimensions, integrals of vector-valued functions are defined in terms of components:

$$\int \mathbf{r}(t)\, dt = \int [x(t)\mathbf{i} + y(t)\mathbf{j} + z(t)\mathbf{k}]\, dt$$

$$= \left(\int x(t)\, dt\right)\mathbf{i} + \left(\int y(t)\, dt\right)\mathbf{j} + \left(\int z(t)\, dt\right)\mathbf{k}$$

$$\int_a^b \mathbf{r}(t)\, dt = \int_a^b [x(t)\mathbf{i} + y(t)\mathbf{j} + z(t)\mathbf{k}]\, dt$$

$$= \left(\int_a^b x(t)\, dt\right)\mathbf{i} + \left(\int_a^b y(t)\, dt\right)\mathbf{j} + \left(\int_a^b z(t)\, dt\right)\mathbf{k}$$

If a curve C in 3-space has parametric equations

$$x = x(t)$$
$$y = y(t)$$
$$z = z(t)$$

where $x'(t)$, $y'(t)$, and $z'(t)$ are continuous for $a \leq t \leq b$, then the portion of the curve from $t = a$ to $t = b$ has arc length

$$L = \int_a^b \sqrt{\left(\frac{dx}{dt}\right)^2 + \left(\frac{dy}{dt}\right)^2 + \left(\frac{dz}{dt}\right)^2} \, dt \qquad (3)$$

This formula generalizes the arc-length formula for two-dimensional curves [Formula (7), Section 13.4]. If the three parametric equations $x = x(t)$, $y = y(t)$, $z = z(t)$ for C are replaced by the single position vector $\mathbf{r}(t) = x(t)\mathbf{i} + y(t)\mathbf{j} + z(t)\mathbf{k}$, then

$$\mathbf{r}'(t) = \left(\frac{dx}{dt}\right)\mathbf{i} + \left(\frac{dy}{dt}\right)\mathbf{j} + \left(\frac{dz}{dt}\right)\mathbf{k}$$

and

$$\|\mathbf{r}'(t)\| = \sqrt{\left(\frac{dx}{dt}\right)^2 + \left(\frac{dy}{dt}\right)^2 + \left(\frac{dz}{dt}\right)^2}$$

so that (3) can be written in the alternate form,

$$L = \int_a^b \|\mathbf{r}'(t)\| \, dt \qquad (4)$$

To illustrate these ideas, let us consider an example. If a and c are positive constants, then the equations

$$x = a \cos t$$
$$y = a \sin t$$
$$z = ct$$

represent a curve called a ***circular helix*** (Figure 15.5.3). This curve winds like a corkscrew around a right circular cylinder of radius a centered on the z-axis. To visualize this, observe that $z = ct$ increases with t so that a point (x, y, z) on the helix moves upward as t increases. However, as t increases the point (x, y, z) also moves in a path directly over the circle

$$x = a \cos t$$
$$y = a \sin t$$

in the xy-plane. The combination of these upward and circular motions produces the resulting corkscrew curve.

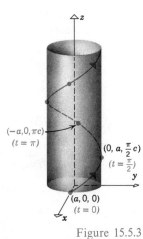

Figure 15.5.3

▶ Example 1   Find the arc length of that portion of the circular helix

$$x = \cos t$$
$$y = \sin t$$
$$z = t$$

from $t = 0$ to $t = \pi$.

Solution.   From (3) the arc length is

$$L = \int_0^{\pi} \sqrt{\left(\frac{dx}{dt}\right)^2 + \left(\frac{dy}{dt}\right)^2 + \left(\frac{dz}{dt}\right)^2}\, dt$$

$$= \int_0^{\pi} \sqrt{(-\sin t)^2 + (\cos t)^2 + 1}\, dt$$

$$= \int_0^{\pi} \sqrt{2}\, dt \ = \ \sqrt{2}\,\pi \qquad ◀$$

It is often convenient to represent a curve C in 3-space by parametric equations with arc length as a parameter. The procedure in 3-space is the same as in 2-space. From an arbitrary reference point on the curve, we choose positive and negative directions along the curve and then associate with each point $P(x, y, z)$ on the curve the "signed" arc length s from the reference point to P. In this way, x, y, and z become functions of s

$$x = x(s)$$
$$y = y(s)$$
$$z = z(s)$$

which are parametric equations for C in terms of s (Figure 15.5.4). If the curve C is already specified in terms of a parameter t, we take the direction of increasing t as the positive direction for s. As a result, s increases as t increases and conversely; thus,

$$\frac{ds}{dt} \ge 0$$

We leave it as an exercise for the reader to state and prove the three-dimensional version of Theorem 14.3.3 showing that

$$\frac{ds}{dt} = \sqrt{\left(\frac{dx}{dt}\right)^2 + \left(\frac{dy}{dt}\right)^2 + \left(\frac{dz}{dt}\right)^2} \qquad (5)$$

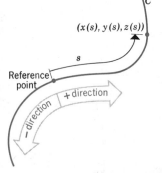

Figure 15.5.4   or, equivalently,

$$\frac{ds}{dt} = \|\mathbf{r}'(t)\| \tag{6}$$

where $\mathbf{r}(t) = x(t)\mathbf{i} + y(t)\mathbf{j} + z(t)\mathbf{k}$ is the position vector for C.

The definitions of velocity, speed, and acceleration of particles moving in 3-space are the same as in 2-space. If a particle moves along a curve C so that its position vector at time t is

$$\mathbf{r}(t) = x(t)\mathbf{i} + y(t)\mathbf{j} + z(t)\mathbf{k}$$

then the **velocity, speed,** and **acceleration** are defined by

$$\mathbf{v} = \frac{d\mathbf{r}}{dt} = \frac{dx}{dt}\mathbf{i} + \frac{dy}{dt}\mathbf{j} + \frac{dz}{dt}\mathbf{k}$$

$$\text{speed} = \|\mathbf{v}\| = \|\mathbf{r}'(t)\| = \frac{ds}{dt} = \sqrt{\left(\frac{dx}{dt}\right)^2 + \left(\frac{dy}{dt}\right)^2 + \left(\frac{dz}{dt}\right)^2}$$

$$\mathbf{a} = \frac{d\mathbf{v}}{dt} = \frac{d^2\mathbf{r}}{dt^2} = \frac{d^2x}{dt^2}\mathbf{i} + \frac{d^2y}{dt^2}\mathbf{j} + \frac{d^2z}{dt^2}\mathbf{k}$$

At each instant, the velocity vector is tangent to the curve C (Figure 15.5.2) and points in the direction of motion. Moreover, the speed of the particle is the instantaneous rate ds/dt at which arc length s measured from a fixed reference point changes with time [(5) and (6)].

▶ Example 2 A particle moves along a curve so that its coordinates at time t are

$$x = t, \qquad y = \frac{1}{2}t^2, \qquad z = \frac{1}{3}t^3$$

Find the velocity, speed, and acceleration at time $t = 1$.

Solution. The position vector of the particle is

$$\mathbf{r}(t) = t\mathbf{i} + \frac{1}{2}t^2\mathbf{j} + \frac{1}{3}t^3\mathbf{k}$$

so that

$$\mathbf{v} = \frac{d\mathbf{r}}{dt} = \mathbf{i} + t\mathbf{j} + t^2\mathbf{k}$$

$$\text{speed} = \frac{ds}{dt} = \|\mathbf{v}\| = \sqrt{1 + t^2 + t^4}$$

$$\mathbf{a} = \frac{d\mathbf{v}}{dt} = \mathbf{j} + 2t\mathbf{k}$$

Thus, at time $t = 1$

$$\mathbf{v} = \mathbf{i} + \mathbf{j} + \mathbf{k}, \qquad \frac{ds}{dt} = \sqrt{3}, \qquad \mathbf{a} = \mathbf{j} + 2\mathbf{k} \qquad \blacktriangleleft$$

The *unit tangent vector* $\mathbf{T} = \mathbf{T}(t)$ to a curve C in 3-space is defined exactly as in 2-space:

$$\mathbf{T}(t) = \frac{\mathbf{r}'(t)}{\|\mathbf{r}'(t)\|}$$

where $\mathbf{r}(t)$ is the position vector for C. For each value of t, the vector $\mathbf{T}(t)$ is tangent to C, has length 1, and points in the direction of increasing t. The unit tangent vector is defined only at points where $\mathbf{r}'(t) \neq \mathbf{0}$.

In two dimensions, we defined the unit normal vector $\mathbf{N}$ to be the vector resulting from a 90° counterclockwise rotation of $\mathbf{T}$. We then showed in Section 14.4 that

$$\frac{d\mathbf{T}}{ds} = \kappa\mathbf{N} \tag{7}$$

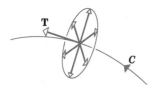

T

C

There are infinitely
many unit vectors
perpendicular to **T**

Figure 15.5.5

where $\kappa = d\phi/ds$ is the curvature, and ϕ is the counterclockwise angle from the direction of the positive x-axis to $\mathbf{T}$. In three dimensions the problem of defining $\mathbf{N}$ is complicated by the fact that there are infinitely many unit vectors perpendicular to $\mathbf{T}$ (Figure 15.5.5). Which one should be selected for $\mathbf{N}$? A similar complication arises in the definition of curvature, because a three-dimensional curve can "bend" in infinitely many different directions. Our objective will be to define the curvature and select the unit normal vector for three-dimensional curves in such a way that (7) remains valid. This can be done as follows.

15.5.1   DEFINITION   At each point of a smooth curve C in 3-space, the *unit normal vector* $\mathbf{N}$ and the *curvature* κ are defined by

$$\mathbf{N} = \frac{d\mathbf{T}/ds}{\|d\mathbf{T}/ds\|} \qquad \text{and} \qquad \kappa = \|d\mathbf{T}/ds\|$$

The vector **N** is sometimes called the ***principal unit normal.***

With these definitions it follows that the basic formula

$$\frac{d\mathbf{T}}{ds} = \kappa \mathbf{N}$$

applies in three dimensions as well as two dimensions. It is clear from the definition that **N** has length 1. The following theorem justifies the adjective "normal."

15.5.2 THEOREM *At each point of a smooth curve C in 3-space, the unit normal vector* **N** *is perpendicular to the unit tangent vector* **T.**

Proof. Since

$$\frac{1}{\|d\mathbf{T}/ds\|}$$

is a positive scalar, the vector $d\mathbf{T}/ds$ and the vector

$$\mathbf{N} = \frac{d\mathbf{T}/ds}{\|d\mathbf{T}/ds\|}$$

have the same direction. Thus, it suffices to prove that $d\mathbf{T}/ds$ and **T** are perpendicular. But $\mathbf{T} = \mathbf{T}(s)$ has a constant length 1, so that

$$1 = \|\mathbf{T}\|^2 = \mathbf{T} \cdot \mathbf{T}$$

Differentiating both sides of this equation with respect to s and using (1) yields

$$0 = \mathbf{T} \cdot \frac{d\mathbf{T}}{ds} + \frac{d\mathbf{T}}{ds} \cdot \mathbf{T} = 2\left(\frac{d\mathbf{T}}{ds} \cdot \mathbf{T}\right)$$

Thus,

$$\frac{d\mathbf{T}}{ds} \cdot \mathbf{T} = 0$$

which proves that $d\mathbf{T}/ds$ and **T** are perpendicular. ∎

REMARKS. For a curve in 2-space the curvature $\kappa = d\phi/ds$ may be positive or negative. However, for a curve in 3-space, the curvature

$$\kappa = \left\|\frac{d\mathbf{T}}{ds}\right\| \tag{8}$$

is always nonnegative. Moreover, (8) applies only in 3-space; in 2-space the analogous relationship is [see (3), Section 14.4],

$$|\kappa| = \left\| \frac{d\mathbf{T}}{ds} \right\|$$

To use the formulas for $\mathbf{N}$ and κ given in Definition 15.5.1, the curve C must be expressed with arc length s as the parameter. The following result is useful for calculating $\mathbf{N}$ and κ when some other parameter is used.

15.5.3 THEOREM *If $\mathbf{r}(t)$ is the position vector of a smooth curve C in 3-space, then*

$$\mathbf{N} = \frac{d\mathbf{T}/dt}{\|d\mathbf{T}/dt\|} \tag{9}$$

$$\kappa = \frac{\left\| \dfrac{d\mathbf{r}}{dt} \times \dfrac{d^2\mathbf{r}}{dt^2} \right\|}{\left\| \dfrac{d\mathbf{r}}{dt} \right\|^3} \tag{10}$$

Proof. Using the chain rule and the fact that s increases with t, we obtain

$$\frac{\dfrac{d\mathbf{T}}{dt}}{\left\| \dfrac{d\mathbf{T}}{dt} \right\|} = \frac{\dfrac{d\mathbf{T}}{ds}\dfrac{ds}{dt}}{\left\| \dfrac{d\mathbf{T}}{ds}\dfrac{ds}{dt} \right\|} = \frac{\dfrac{d\mathbf{T}}{ds}\dfrac{ds}{dt}}{\left| \dfrac{ds}{dt} \right| \left\| \dfrac{d\mathbf{T}}{ds} \right\|} = \frac{\dfrac{d\mathbf{T}}{ds}\dfrac{ds}{dt}}{\dfrac{ds}{dt} \left\| \dfrac{d\mathbf{T}}{ds} \right\|} = \frac{\dfrac{d\mathbf{T}}{ds}}{\left\| \dfrac{d\mathbf{T}}{ds} \right\|} = \mathbf{N}$$

which establishes (9).

By using the definition of the unit tangent vector and (6), we may write

$$\mathbf{T} = \frac{d\mathbf{r}/dt}{\|d\mathbf{r}/dt\|} = \frac{d\mathbf{r}/dt}{ds/dt}$$

so that

$$\frac{d\mathbf{r}}{dt} = \frac{ds}{dt}\mathbf{T} \tag{11}$$

and

$$\frac{d^2\mathbf{r}}{dt^2} = \frac{d^2s}{dt^2}\mathbf{T} + \frac{ds}{dt}\frac{d\mathbf{T}}{dt} \tag{12}$$

But

$$\frac{d\mathbf{T}}{dt} = \frac{d\mathbf{T}}{ds}\frac{ds}{dt} = \kappa\mathbf{N}\frac{ds}{dt}$$

so (12) may be written

$$\frac{d^2\mathbf{r}}{dt^2} = \frac{d^2s}{dt^2}\mathbf{T} + \kappa \left(\frac{ds}{dt}\right)^2 \mathbf{N} \tag{13}$$

Therefore, from (11)

$$\frac{d\mathbf{r}}{dt} \times \frac{d^2\mathbf{r}}{dt^2} = \left(\frac{ds}{dt}\right)\left(\frac{d^2s}{dt^2}\right)(\mathbf{T} \times \mathbf{T}) + \kappa \left(\frac{ds}{dt}\right)^3 (\mathbf{T} \times \mathbf{N})$$

or, since $\mathbf{T} \times \mathbf{T} = \mathbf{0}$,

$$\frac{d\mathbf{r}}{dt} \times \frac{d^2\mathbf{r}}{dt^2} = \kappa \left(\frac{ds}{dt}\right)^3 (\mathbf{T} \times \mathbf{N})$$

Since $\mathbf{T}$ and $\mathbf{N}$ are unit vectors, so is $\mathbf{T} \times \mathbf{N}$; thus,

$$\left\| \frac{d\mathbf{r}}{dt} \times \frac{d^2\mathbf{r}}{dt^2} \right\| = \kappa \left(\frac{ds}{dt}\right)^3$$

But $ds/dt = \|d\mathbf{r}/dt\|$, so (10) follows. ▊

▶ **Example 3** Find $\mathbf{T}$, $\mathbf{N}$, and κ for the circular helix

$$\begin{aligned} x &= a \cos t \\ y &= a \sin t \quad (a > 0) \\ z &= ct \end{aligned}$$

Solution. The position vector for the helix is

$$\mathbf{r} = (a \cos t)\mathbf{i} + (a \sin t)\mathbf{j} + (ct)\mathbf{k}$$

Thus,

$$\frac{d\mathbf{r}}{dt} = (-a \sin t)\mathbf{i} + (a \cos t)\mathbf{j} + c\mathbf{k}$$

$$\left\| \frac{d\mathbf{r}}{dt} \right\| = \sqrt{(-a \sin t)^2 + (a \cos t)^2 + c^2} = \sqrt{a^2 + c^2}$$

$$\mathbf{T} = \frac{d\mathbf{r}/dt}{\|d\mathbf{r}/dt\|} = -\frac{a \sin t}{\sqrt{a^2 + c^2}}\mathbf{i} + \frac{a \cos t}{\sqrt{a^2 + c^2}}\mathbf{j} + \frac{c}{\sqrt{a^2 + c^2}}\mathbf{k}$$

Next, we will find $\mathbf{N}$.

$$\frac{d\mathbf{T}}{dt} = -\frac{a\cos t}{\sqrt{a^2 + c^2}}\mathbf{i} - \frac{a\sin t}{\sqrt{a^2 + c^2}}\mathbf{j}$$

$$\left\|\frac{d\mathbf{T}}{dt}\right\| = \sqrt{\left(-\frac{a\cos t}{\sqrt{a^2 + c^2}}\right)^2 + \left(-\frac{a\sin t}{\sqrt{a^2 + c^2}}\right)^2} = \sqrt{\frac{a^2}{a^2 + c^2}}$$

$$= \frac{a}{\sqrt{a^2 + c^2}}$$

$$\mathbf{N} = \frac{d\mathbf{T}/dt}{\|d\mathbf{T}/dt\|} = (-\cos t)\mathbf{i} - (\sin t)\mathbf{j}$$

Finally, we will find κ.

$$\frac{d^2\mathbf{r}}{dt^2} = (-a\cos t)\mathbf{i} - (a\sin t)\mathbf{j}$$

$$\frac{d\mathbf{r}}{dt} \times \frac{d^2\mathbf{r}}{dt^2} = \begin{vmatrix} \mathbf{i} & \mathbf{j} & \mathbf{k} \\ -a\sin t & a\cos t & c \\ -a\cos t & -a\sin t & 0 \end{vmatrix}$$

$$= (ac\sin t)\mathbf{i} - (ac\cos t)\mathbf{j} + a^2\mathbf{k}$$

$$\left\|\frac{d\mathbf{r}}{dt} \times \frac{d^2\mathbf{r}}{dt^2}\right\| = \sqrt{(ac\sin t)^2 + (-ac\cos t)^2 + a^4}$$

$$= \sqrt{a^2c^2 + a^4} = a\sqrt{a^2 + c^2}$$

Therefore,

$$\kappa = \frac{\left\|\dfrac{d\mathbf{r}}{dt} \times \dfrac{d^2\mathbf{r}}{dt^2}\right\|}{\left\|\dfrac{d\mathbf{r}}{dt}\right\|^3} = \frac{a\sqrt{a^2 + c^2}}{(\sqrt{a^2 + c^2})^3} = \frac{a}{a^2 + c^2}$$

Note that κ does not depend on t, which means that the helix has constant curvature. ◀

REMARK. If the parameter t represents time, then (13) can be written

$$\mathbf{a} = \frac{d^2s}{dt^2}\mathbf{T} + \kappa\left(\frac{ds}{dt}\right)^2\mathbf{N} \tag{14}$$

This result, which expresses the acceleration vector $\mathbf{a}$ in terms of normal and tangential components, is identical to the corresponding result in two dimensions. It follows from (14) that the acceleration vector $\mathbf{a}$ of a particle moving in three dimensions is always in the plane determined by $\mathbf{T}$ and $\mathbf{N}$. (Why?)

▶ Exercise Set 15.5

In Exercises 1–6, sketch the graph of the curve.

1. $\mathbf{r}(t) = t\mathbf{i} + t\mathbf{j} + t\mathbf{k}$.

2. $\mathbf{r}(t) = (1 + 3t)\mathbf{i} + (-1 + t)\mathbf{j} + 2t\mathbf{k}$.

3. $\mathbf{r}(t) = (2 \cos t)\mathbf{i} + (2 \sin t)\mathbf{j} + t\mathbf{k}$.

4. $x = 9 \cos t, \ y = 4 \sin t, \ z = t$.

5. $x = t, \ y = t^2, \ z = 2$.

6. $x = t, \ y = t, \ z = \sin t; \ 0 \leq t \leq 2\pi$.

In Exercises 7–14, find the unit tangent vector **T**, the unit normal vector **N**, and the curvature κ at the indicated point.

7. $\mathbf{r}(t) = 4 \cos t\mathbf{i} + 4 \sin t\mathbf{j} + t\mathbf{k}; \ t = \pi/2$.

8. $x = e^t, \ y = e^{-t}, \ z = t; \ t = 0$.

9. $\mathbf{r}(t) = t\mathbf{i} + \frac{1}{2}t^2\mathbf{j} + \frac{1}{3}t^3\mathbf{k}; \ t = 0$.

10. $x = \sin t, \ y = \cos t, \ z = \frac{1}{2}t^2; \ t = 0$.

11. $\mathbf{r}(t) = 3 \cos t\mathbf{i} + 4 \sin t\mathbf{j} + t\mathbf{k}; \ t = \pi/2$.

12. $x = e^t \cos t, \ y = e^t \sin t, \ z = e^t; \ t = 0$.

13. $\mathbf{r}(t) = \mathbf{i} + t\mathbf{j} + t^2\mathbf{k}; \ t = 1$.

14. $x = \cosh t, \ y = \sinh t, \ z = t; \ t = \ln 2$.

In Exercises 15–20, find the velocity, speed, and acceleration at the given time t of a particle moving along the given curve.

15. $\mathbf{r}(t) = t\mathbf{i} + \frac{1}{2}t^2\mathbf{j} + \frac{1}{3}t^3\mathbf{k}; \ t = 1$.

16. $x = 1 + 3t, \ y = 2 - 4t, \ z = 7 + t; \ t = 2$.

17. $\mathbf{r}(t) = 2 \cos t\mathbf{i} + 2 \sin t\mathbf{j} + t\mathbf{k}; \ t = \pi/4$.

18. $\mathbf{r}(t) = 3t\mathbf{i} + 2t^2\mathbf{j} + \ln t\mathbf{k}; \ t = 1$.

19. $\mathbf{r}(t) = e^t \sin t\mathbf{i} + e^t \cos t\mathbf{j} + t\mathbf{k}; \ t = \pi/2$.

20. $\mathbf{r}(t) = 2t\mathbf{i} + t^2\mathbf{j} + \ln t\mathbf{k}; \ t = 2$.

In Exercises 21–26, find the arc length of the curve.

21. $\mathbf{r}(t) = (4 + 3t)\mathbf{i} + (2 - 2t)\mathbf{j} + (5 + t)\mathbf{k};$
    $3 \leq t \leq 4$.

22. $\mathbf{r}(t) = 3 \cos t\mathbf{i} + 3 \sin t\mathbf{j} + t\mathbf{k}; \ 0 \leq t \leq 2\pi$.

23. $\mathbf{r}(t) = t^3\mathbf{i} + t\mathbf{j} + \frac{1}{2}\sqrt{6}t^2\mathbf{k}; \ 1 \leq t \leq 3$.

24. $x = \cos^3 t, \ y = \sin^3 t, \ z = 2; \ 0 \leq t \leq \pi/2$.

25. $\mathbf{r}(t) = \langle e^t, e^{-t}, \sqrt{2}t \rangle; \ 0 \leq t \leq 1$.

26. $x = \frac{1}{2}t, \ y = \frac{1}{3}(1 - t)^{3/2}, \ z = \frac{1}{3}(1 + t)^{3/2};$
    $-1 \leq t \leq 1$.

In Exercises 27–30, perform the integration.

27. $\int \left[t^2\mathbf{i} - 2t\mathbf{j} + \frac{1}{t}\mathbf{k} \right] dt$.

28. $\int \langle e^{-t}, e^t, 3t^2 \rangle \, dt$.

29. $\int_0^1 (e^{2t}\mathbf{i} + e^{-t}\mathbf{j} + t\mathbf{k}) \, dt$.

30. $\int_{-3}^3 \langle (3 - t)^{3/2}, (3 + t)^{3/2}, 1 \rangle \, dt$.

31. Calculate $(d/dt) [\mathbf{r}_1(t) \cdot \mathbf{r}_2(t)]$ two ways; first using formula (1), and then differentiating $\mathbf{r}_1(t) \cdot \mathbf{r}_2(t)$ directly.
    (a) $\mathbf{r}_1(t) = 2t\mathbf{i} + 3t^2\mathbf{j} + t^3\mathbf{k}, \ \mathbf{r}_2(t) = t^4\mathbf{k}$
    (b) $\mathbf{r}_1(t) = 3 \sec t\mathbf{i} - t\mathbf{j} + \ln t\mathbf{k},$
        $\mathbf{r}_2(t) = 4t\mathbf{i} - \sin t\mathbf{k}$.

32. Calculate $\dfrac{d}{dt}[\mathbf{r}_1(t) \times \mathbf{r}_2(t)]$ two ways; first using formula (2), and then by differentiating $\mathbf{r}_1(t) \times \mathbf{r}_2(t)$ directly.
    (a) $\mathbf{r}_1(t) = 2t\mathbf{i} + 3t^2\mathbf{j} + t^3\mathbf{k}, \ \mathbf{r}_2(t) = t^4\mathbf{k}$
    (b) $\mathbf{r}_1(t) = 3 \sec t\mathbf{i} - t\mathbf{j} + \ln t\mathbf{k},$
        $\mathbf{r}_2(t) = 4t\mathbf{i} - \sin t\mathbf{k}$.

33. Let $\mathbf{r}(t) = \ln t\mathbf{i} + 2t\mathbf{j} + t^2\mathbf{k}$.
    Find:
    (a) ds/dt (b) $\|\mathbf{r}'(t)\|$
    (c) $\int_1^3 \|\mathbf{r}'(t)\| \, dt$.

34. Let $x = \cos t, \ y = \sin t, \ z = t^{3/2}$.
    Find
    $$\int_0^2 \left(\frac{ds}{dt} \right) dt$$

35. Prove that the angle between the unit tangent vector **T** and the positive z-axis is the same at each point of the helix $x = a \cos t, \ y = a \sin t, \ z = bt$.

36. Prove that a straight line in 3-space has zero curvature at each point.

37. Prove: $(d/dt) [\mathbf{r}(t) \times \mathbf{r}'(t)] = \mathbf{r}(t) \times \mathbf{r}''(t)$.

38. Let $\mathbf{r}(t) = (a \cos t)\mathbf{i} + (a \sin t)\mathbf{j}$ $(a > 0)$.

    (a) Show that $\mathbf{r}(t) \cdot \mathbf{r}'(t) = 0$.

    (b) Give a geometric explanation of the result in part (a).

39. Prove that $\|\mathbf{r}(t)\|$ is constant if and only if $\mathbf{r}(t) \cdot \mathbf{r}'(t) = 0$.

40. Prove: If the speed of a particle is constant, then the acceleration and velocity vectors are orthogonal.

41. Prove: If the acceleration of a particle is zero for all t, then the particle moves on a straight line.

42. Suppose that a particle in 3-space undergoes a constant acceleration of $-g\mathbf{k}$, and assume that at the initial time, $t = 0$, the position vector is $\mathbf{r}_0$ and the velocity vector is $\mathbf{v}_0$. By integration, show that the position of the particle at any time t is $\mathbf{r}(t) = -\frac{1}{2}gt^2\mathbf{k} + t\mathbf{v}_0 + \mathbf{r}_0$.

43. Use the result of Exercise 42 to find the position, velocity, and speed of a particle having the given acceleration, initial position, and initial velocity.

    (a) $\mathbf{a}(t) = -32\mathbf{k}$; $\mathbf{v}_0 = \mathbf{0}$; $\mathbf{r}_0 = \mathbf{0}$

    (b) $\mathbf{a}(t) = -32\mathbf{k}$; $\mathbf{v}_0 = \mathbf{0}$; $\mathbf{r}_0 = \mathbf{i} + 2\mathbf{j} + \mathbf{k}$

    (c) $\mathbf{a}(t) = -32\mathbf{k}$; $\mathbf{v}_0 = \mathbf{i} + \mathbf{k}$; $\mathbf{r}_0 = 3\mathbf{i} - \mathbf{j} + 4\mathbf{k}$.

44. Let $\mathbf{u} = \mathbf{u}(t)$, $\mathbf{v} = \mathbf{v}(t)$, and $\mathbf{w} = \mathbf{w}(t)$ be differentiable vector-valued functions. Use formulas (1) and (2) to show that

$$\frac{d}{dt}[\mathbf{u} \cdot (\mathbf{v} \times \mathbf{w})] = \frac{d\mathbf{u}}{dt} \cdot [\mathbf{v} \times \mathbf{w}] + \mathbf{u} \cdot \left[\frac{d\mathbf{v}}{dt} \times \mathbf{w}\right]$$
$$+ \mathbf{u} \cdot \left[\mathbf{v} \times \frac{d\mathbf{w}}{dt}\right]$$

45. Let u_1, u_2, u_3, v_1, v_2, v_3, w_1, w_2, and w_3 be differentiable functions of t. Use Exercise 44 to show that:

$$\frac{d}{dt}\begin{vmatrix} u_1 & u_2 & u_3 \\ v_1 & v_2 & v_3 \\ w_1 & w_2 & w_3 \end{vmatrix}$$

$$= \begin{vmatrix} u_1' & u_2' & u_3' \\ v_1 & v_2 & v_3 \\ w_1 & w_2 & w_3 \end{vmatrix} + \begin{vmatrix} u_1 & u_2 & u_3 \\ v_1' & v_2' & v_3' \\ w_1 & w_2 & w_3 \end{vmatrix} + \begin{vmatrix} u_1 & u_2 & u_3 \\ v_1 & v_2 & v_3 \\ w_1' & w_2' & w_3' \end{vmatrix}$$

46. The **binormal vector** $\mathbf{B}$ at a point on the graph of $\mathbf{r}(t)$ is defined by

$$\mathbf{B} = \mathbf{T} \times \mathbf{N}$$

    (a) Show that $\mathbf{B}$ has length 1 and that $\mathbf{T}$, $\mathbf{N}$, and $\mathbf{B}$ are mutually perpendicular.

    (b) Find the binormal at a general point on the helix $x = a \cos t$, $y = a \sin t$, $z = bt$.

47. Derive formulas (1) and (2).

15.6 PLANES IN 3-SPACE

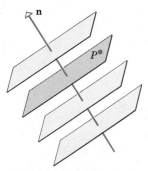

The shaded plane is uniquely determined by the point P and the vector $\mathbf{n}$ perpendicular to the plane

Figure 15.6.1

In this section we will use vectors to derive equations of planes in 3-space, and we will use these equations to solve some basic geometric problems.

A plane in 3-space is uniquely determined by specifying a point in the plane and a vector perpendicular to the plane (Figure 15.6.1). A vector perpendicular to a plane is called a **normal** to the plane.

Suppose we want the equation of the plane passing through the point $P_0(x_0, y_0, z_0)$ and perpendicular to the nonzero vector $\mathbf{n} = \langle a, b, c \rangle$. It is evident from Figure 15.6.2 that the plane consists precisely of those points $P(x, y, z)$ for which the vector $\overrightarrow{P_0P}$ is perpendicular to $\mathbf{n}$; or phrased as an equation

$$\mathbf{n} \cdot \overrightarrow{P_0P} = 0 \tag{1}$$

Since $\overrightarrow{P_0P} = \langle x - x_0, y - y_0, z - z_0 \rangle$, (1) can be rewritten as

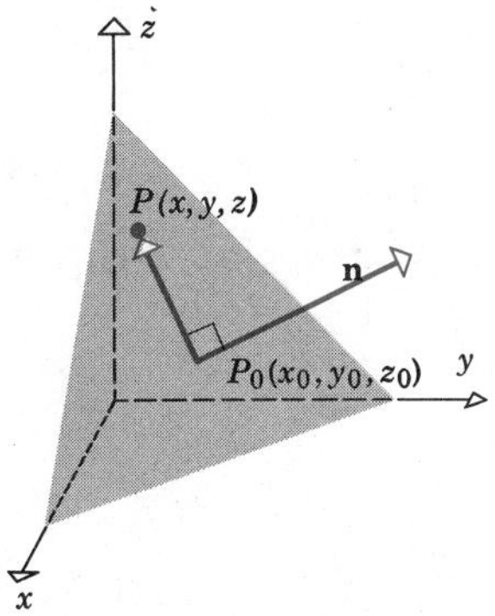

Figure 15.6.2

$$a(x - x_0) + b(y - y_0) + c(z - z_0) = 0 \qquad (2)$$

We will call this the ***point-normal form*** of the equation of a plane.

▶ Example 1   Find an equation of the plane passing through the point $(3, -1, 7)$ and perpendicular to the vector $\mathbf{n} = \langle 4, 2, -5 \rangle$.

Solution.   From (2), a point-normal form of the equation is

$$4(x - 3) + 2(y + 1) - 5(z - 7) = 0 \qquad ◀$$

By multiplying out and collecting terms, (2) can be rewritten in the form

$$ax + by + cz + d = 0 \qquad (3)$$

where a, b, c, and d are constants, and a, b, and c are not all zero. To illustrate, the equation in Example 1 can be rewritten as

$$4x + 2y - 5z + 25 = 0$$

As our next theorem shows, every equation of form (3) represents a plane in 3-space.

15.6.1   THEOREM   *If a, b, c, and d are constants and a, b, and c are not all zero, then the graph of the equation*

$$ax + by + cz + d = 0$$

is a plane having the vector $\mathbf{n} = \langle a, b, c \rangle$ as a normal.

Proof.   By hypothesis, a, b, and c are not all zero. Assume, for the moment, that $a \neq 0$. The equation $ax + by + cz + d = 0$ can be rewritten as $a[x + (d/a)] + by + cz = 0$. But this is a point-normal form of the plane passing through the point $(-d/a, 0, 0)$ and having $\mathbf{n} = (a, b, c)$ as a normal.

If $a = 0$, then either $b \neq 0$ or $c \neq 0$. A straightforward modification of the above argument will handle these other cases.   ▮

The equation in Theorem 15.6.1 is called the ***general form*** of the equation of a plane.

▶ Example 2   Determine whether the planes

$$3x - 4y + 5z = 0$$

and

$$-6x + 8y - 10z - 4 = 0$$

are parallel.

Solution. It is clear geometrically that two planes are parallel if and only if their normals are parallel vectors. A normal to the first plane is

$$\mathbf{n}_1 = \langle 3, -4, 5 \rangle$$

and a normal to the second plane is

$$\mathbf{n}_2 = \langle -6, 8, -10 \rangle$$

Since $\mathbf{n}_2$ is a scalar multiple of $\mathbf{n}_1$, the normals are parallel. Thus, the planes are also parallel. ◄

A plane cannot be specified by giving a point on it and *one* vector parallel to it, since there are infinitely many such planes (Figure 15.6.3*a*). However, as shown in Figure 15.6.3*b*, a plane is uniquely determined by giving a point in the plane and *two* nonparallel vectors that are parallel to the plane. A plane is also uniquely determined by specifying three noncollinear points in the plane (Figure 15.6.3*c*).

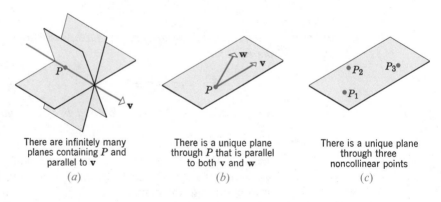

There are infinitely many planes containing P and parallel to $\mathbf{v}$	There is a unique plane through P that is parallel to both $\mathbf{v}$ and $\mathbf{w}$	There is a unique plane through three noncollinear points
(*a*)	(*b*)	(*c*)

Figure 15.6.3

► **Example 3**   Find an equation of the plane through the points $P_1(1, 2, -1)$, $P_2(2, 3, 1)$, and $P_3(3, -1, 2)$.

Solution. Since the points P_1, P_2, and P_3 lie in the plane, the vectors $\overrightarrow{P_1P_2} = \langle 1, 1, 2 \rangle$ and $\overrightarrow{P_1P_3} = \langle 2, -3, 3 \rangle$ are parallel to the plane. Therefore,

$$\overrightarrow{P_1P_2} \times \overrightarrow{P_1P_3} = \begin{vmatrix} \mathbf{i} & \mathbf{j} & \mathbf{k} \\ 1 & 1 & 2 \\ 2 & -3 & 3 \end{vmatrix} = 9\mathbf{i} + \mathbf{j} - 5\mathbf{k}$$

is normal to the plane, since it is perpendicular to both $\overrightarrow{P_1P_2}$ and $\overrightarrow{P_1P_3}$. By using this normal and the point $P_1(1, 2, -1)$ in the plane, we obtain the point-normal form

$$9(x - 1) + (y - 2) - 5(z + 1) = 0$$

which may be rewritten as

$$9x + y - 5z - 16 = 0 \qquad \blacktriangleleft$$

▶ Example 4 Determine whether the line

$$x = 3 + 8t$$
$$y = 4 + 5t$$
$$z = -3 - t$$

is parallel to the plane $x - 3y + 5z = 12$.

Solution. The vector $\mathbf{v} = \langle 8, 5, -1 \rangle$ is parallel to the line and the vector $\mathbf{n} = \langle 1, -3, 5 \rangle$ is normal to the plane. In order for the line and plane to be parallel, the vectors $\mathbf{v}$ and $\mathbf{n}$ must be perpendicular. But this is not so, since the dot product

$$\mathbf{v} \cdot \mathbf{n} = (8)(1) + (5)(-3) + (-1)(5) = -12$$

is nonzero. Thus, the line and plane are not parallel. $\qquad \blacktriangleleft$

▶ Example 5 Find the intersection of the line and plane in Example 4.

Solution. If we let (x_0, y_0, z_0) be the point of intersection, then the coordinates of this point satisfy the equation of the plane and the parametric equations of the line. Thus,

$$x_0 - 3y_0 + 5z_0 = 12 \qquad (4)$$

and for some value of t, say $t = t_0$,

$$x_0 = 3 + 8t_0$$
$$y_0 = 4 + 5t_0 \qquad (5)$$
$$z_0 = -3 - t_0$$

Substituting (5) in (4) yields

$$(3 + 8t_0) - 3(4 + 5t_0) + 5(-3 - t_0) = 12$$

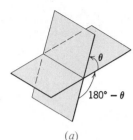

(a)

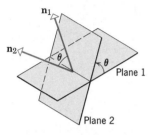

In this figure the angle between n_1 and n_2 is θ. If the direction of one of the normals is reversed, then the angle between the normals would be $180° - \theta$

(b)

Figure 15.6.4

Solving for t_0 yields

$$t_0 = -3$$

and on substituting this value in (5), we obtain

$$(x_0, y_0, z_0) = (-21, -11, 0) \qquad \blacktriangleleft$$

Two intersecting planes determine two angles of intersection, an acute angle θ ($0 \le \theta \le 90°$) and its supplement $180° - \theta$ (Figure 15.6.4a). If n_1 and n_2 are normals to the planes, then the angle between n_1 and n_2 is θ or $180° - \theta$, depending on the directions of the normals (Figure 15.6.4b). Thus, the angles of intersection of two planes may be determined from the normals.

▶ **Example 6** Find the acute angle of intersection of the planes $2x - 4y + 4z = 7$ and $6x + 2y - 3z = 2$.

Solution. From the given equations, we obtain the normals $n_1 = \langle 2, -4, 4 \rangle$ and $n_2 = \langle 6, 2, -3 \rangle$. Since the dot product $n_1 \cdot n_2 = -8$ is negative, the normals make an obtuse angle. To obtain normals making an acute angle, we can reverse the direction of either n_1 or n_2. To be specific let us reverse the direction of n_1. Thus, the acute angle θ between the planes will be the angle between $-n_1 = \langle -2, 4, -4 \rangle$ and $n_2 = \langle 6, 2, -3 \rangle$. From (3) of Section 15.3 we obtain

$$\cos \theta = \frac{(-n_1) \cdot n_2}{\|-n_1\| \, \|n_2\|} = \frac{8}{\sqrt{36}\sqrt{49}} = \frac{4}{21}$$

With the aid of trigonometric tables or a hand-held calculator that can compute inverse trigonometric functions we obtain

$$\theta = \cos^{-1}\left(\frac{4}{21}\right) \approx 79° \qquad \blacktriangleleft$$

We conclude this section by discussing three basic "distance problems" in 3-space:

(a) Find the distance between a point and a plane.
(b) Find the distance between two parallel planes.
(c) Find the distance between two skew lines.

The three problems are related. If we can find the distance between a point and a plane, then we can find the distance between parallel planes by computing the distance between one of the planes and an arbitrary point P_0 in the other (Figure 15.6.5a). Moreover, we can find the distance between two skew lines by computing the distance between parallel planes containing them (Figure 15.6.5b).

Figure 15.6.5   (a)   (b)

15.6.2 THEOREM *The distance D between a point $P_0(x_0, y_0, z_0)$ and the plane $ax + by + cz + d = 0$ is*

$$D = \frac{|ax_0 + by_0 + cz_0 + d|}{\sqrt{a^2 + b^2 + c^2}} \tag{6}$$

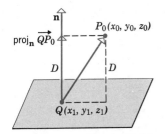

Figure 15.6.6

Proof. Let $Q(x_1, y_1, z_1)$ be any point in the plane, and position the normal $\mathbf{n} = \langle a, b, c \rangle$ so that its initial point is at Q. As illustrated in Figure 15.6.6, the distance D is equal to the length of the orthogonal projection of QP_0 on $\mathbf{n}$. Thus, from (9) of Section 15.3,

$$D = \|\operatorname{proj}_{\mathbf{n}} \overrightarrow{QP_0}\| = \frac{|\overrightarrow{QP_0} \cdot \mathbf{n}|}{\|\mathbf{n}\|}$$

But

$$\overrightarrow{QP_0} = \langle x_0 - x_1, y_0 - y_1, z_0 - z_1 \rangle$$
$$\overrightarrow{QP_0} \cdot \mathbf{n} = a(x_0 - x_1) + b(y_0 - y_1) + c(z_0 - z_1)$$
$$\|\mathbf{n}\| = \sqrt{a^2 + b^2 + c^2}$$

Thus,

$$D = \frac{|a(x_0 - x_1) + b(y_0 - y_1) + c(z_0 - z_1)|}{\sqrt{a^2 + b^2 + c^2}} \tag{7}$$

Since the point $Q(x_1, y_1, z_1)$ lies in the plane, its coordinates satisfy the equation of the plane, so that

$$ax_1 + by_1 + cz_1 + d = 0$$

or

$$d = -ax_1 - by_1 - cz_1$$

Substituting this expression in (7) yields (6). ▮

REMARK. Note the similarity between (6) and the formula for the distance between a point and a line in 2-space [(12) of Section 15.3].

▶ **Example 7** Find the distance D between the point $(1, -4, -3)$ and the plane $2x - 3y + 6z = -1$.

Solution. To apply (6), we first rewrite the equation of the plane in the form

$$2x - 3y + 6z + 1 = 0$$

Then

$$D = \frac{|(2)(1) + (-3)(-4) + 6(-3) + 1|}{\sqrt{2^2 + (-3)^2 + 6^2}} = \frac{|-3|}{7} = \frac{3}{7} \qquad \blacktriangleleft$$

▶ **Example 8** The planes

$$x + 2y - 2z = 3 \quad \text{and} \quad 2x + 4y - 4z = 7$$

are parallel since their normals, $\langle 1, 2, -2 \rangle$ and $\langle 2, 4, -4 \rangle$, are parallel vectors. Find the distance between these planes.

Solution. To find the distance D between the planes, we may select an arbitrary point in one of the planes and compute its distance to the other plane. By setting $y = z = 0$ in the equation $x + 2y - 2z = 3$, we obtain the point $P_0(3, 0, 0)$ in this plane. From (6), the distance from P_0 to the plane $2x + 4y - 4z = 7$ is

$$D = \frac{|(2)(3) + 4(0) + (-4)(0) - 7|}{\sqrt{2^2 + 4^2 + (-4)^2}} = \frac{1}{6} \qquad \blacktriangleleft$$

▶ **Example 9** It was shown in Example 6 of Section 15.2 that the lines

$$L_1: \begin{aligned} x &= 1 + 4t \\ y &= 5 - 4t \\ z &= -1 + 5t \end{aligned} \qquad L_2: \begin{aligned} x &= 2 + 8t \\ y &= 4 - 3t \\ z &= 5 + t \end{aligned}$$

are skew. Find the distance between them.

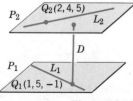

Figure 15.6.7

Solution. Let P_1 and P_2 denote parallel planes containing L_1 and L_2, respectively (Figure 15.6.7). To find the distance D between L_1 and L_2, we will calculate the distance from a point in P_1 to the plane P_2. Setting $t = 0$ in the equations of L_1 yields the point $Q_1(1, 5, -1)$ in plane P_1. Next we will obtain an equation for plane P_2.

The vector $\mathbf{u}_1 = \langle 4, -4, 5 \rangle$ is parallel to line L_1, and therefore also parallel to plane P_1. Similarly, $\mathbf{u}_2 = \langle 8, -3, 1 \rangle$ is parallel to L_2 and P_2.

Therefore, the cross product

$$\mathbf{n} = \mathbf{u}_1 \times \mathbf{u}_2 = \begin{vmatrix} \mathbf{i} & \mathbf{j} & \mathbf{k} \\ 4 & -4 & 5 \\ 8 & -3 & 1 \end{vmatrix} = 11\mathbf{i} + 36\mathbf{j} + 20\mathbf{k}$$

is normal to both P_1 and P_2. Using this normal and the point $Q_2(2, 4, 5)$ found by setting $t = 0$ in the equations of L_2, we obtain an equation of P_2:

$$11(x - 2) + 36(y - 4) + 20(z - 5) = 0$$

or

$$11x + 36y + 20z - 266 = 0$$

The distance between $Q_1(1, 5, -1)$ and this plane is

$$D = \frac{|(11)(1) + (36)(5) + (20)(-1) - 266|}{\sqrt{11^2 + 36^2 + 20^2}} = \frac{95}{\sqrt{1817}}$$

This is also the distance between L_1 and L_2. ◀

▶ Exercise Set 15.6

In Exercises 1–4, find an equation of the plane passing through P and have $\mathbf{n}$ as a normal.

1. $P(2, 6, 1)$; $\mathbf{n} = \langle 1, 4, 2 \rangle$.

2. $P(-1, -1, 2)$; $\mathbf{n} = \langle -1, 7, 6 \rangle$.

3. $P(1, 0, 0)$; $\mathbf{n} = \langle 0, 0, 1 \rangle$.

4. $P(0, 0, 0)$; $\mathbf{n} = \langle 2, -3, -4 \rangle$.

5. Find an equation of the plane passing through the given points.
   (a) $(-2, 1, 1)$, $(0, 2, 3)$, and $(1, 0, -1)$
   (b) $(3, 2, 1)$, $(2, 1, -1)$, and $(-1, 3, 2)$.

6. Determine whether the planes are parallel.
   (a) $3x - 2y + z = 4$ and $6x - 4y + 3z = 7$
   (b) $2x - 8y - 6z - 2 = 0$ and
       $-x + 4y + 3z - 5 = 0$
   (c) $y = 4x - 2z + 3$ and $x = \frac{1}{4}y + \frac{1}{2}z$.

7. Determine whether the line and plane are parallel.
   (a) $x = 4 + 2t$, $y = -t$, $z = -1 - 4t$;
       $3x + 2y + z - 7 = 0$
   (b) $x = t$, $y = 2t$, $z = 3t$; $x - y + 2z = 5$.

8. Determine whether the planes are perpendicular.
   (a) $x - y + 3z - 2 = 0$, $2x + z = 1$
   (b) $3x - 2y + z = 1$, $4x + 5y - 2z = 4$.

9. Determine whether the line and plane are perpendicular.
   (a) $x = -1 + 2t$, $y = 4 + t$, $z = 1 - t$;
       $4x + 2y - 2z = 7$
   (b) $x = 3 - t$, $y = 2 + t$, $z = 1 - 3t$;
       $2x + 2y - 5 = 0$.

10. Find the point of intersection of the line and plane.
   (a) $x = t$, $y = t$, $z = t$; $3x - 2y + z - 5 = 0$
   (b) $x = 1 + t$, $y = -1 + 3t$, $z = 2 + 4t$;
       $x - y + 4z = 7$
   (c) $x = 2 - t$, $y = 3 + t$, $z = t$;
       $2x + y + z = 1$.

11. Find the acute angle of intersection of the planes (to the nearest degree).
   (a) $x = 0$ and $2x - y + z - 4 = 0$
   (b) $x + 2y - 2z = 5$ and $6x - 3y + 2z = 8$.

12. Find an equation of the plane through $(-1, 4, -3)$ and perpendicular to the line $x - 2 = t, y + 3 = 2t, z = -t$.

13. Find an equation of
   (a) the xy-plane   (b) the xz-plane
   (c) the yz-plane.

14. Find an equation of the plane that contains the point (x_0, y_0, z_0) and is
   (a) parallel to the xy-plane
   (b) parallel to the yz-plane
   (c) parallel to the xz-plane.

15. Find an equation of the plane through the origin that is parallel to the plane $4x - 2y + 7z + 12 = 0$.

16. Find an equation of the plane containing the line $x = -2 + 3t, y = 4 + 2t, z = 3 - t$ and perpendicular to the plane $x - 2y + z = 5$.

17. Find an equation of the plane through $(-1, 4, 2)$ and containing the line of intersection of the planes $4x - y + z - 2 = 0$ and $2x + y - 2z - 3 = 0$.

18. Show that the points $(1, 0, -1), (0, 2, 3), (-2, 1, 1)$, and $(4, 2, 3)$ lie in the same plane.

19. Find parametric equations of the line through $(5, 0, -2)$ that is parallel to the planes $x - 4y + 2z = 0$ and $2x + 3y - z + 1 = 0$.

20. Find an equation of the plane through $(-1, 2, -5)$ and perpendicular to the planes $2x - y + z = 1$ and $x + y - 2z = 3$.

21. Find an equation of the plane through $(1, 2, -1)$ and perpendicular to the line of intersection of the planes $2x + y + z = 2$ and $x + 2y + z = 3$.

22. Find a plane through the points $P_1(-2, 1, 4)$, $P_2(1, 0, 3)$ and perpendicular to the plane $4x - y + 3z = 2$.

23. Show that the lines

$$x = -2 + t \qquad x = 3 - t$$
$$y = 3 + 2t \quad \text{and} \quad y = 4 - 2t$$
$$z = 4 - t \qquad z = t$$

are parallel and find an equation of the plane they determine.

24. Find an equation of the plane that contains the point $(2, 0, 3)$ and the line $x = -1 + t, y = t, z = -4 + 2t$.

25. Find an equation of the plane, each of whose points is equidistant from $(2, -1, 1)$ and $(3, 1, 5)$.

26. Find an equation of the plane containing the line $x = 3t, y = 1 + t, z = 2t$ and parallel to the intersection of the planes $2x - y + z = 0$ and $y + z + 1 = 0$.

27. Show that the line $x = 0, y = t, z = t$
   (a) lies in the plane $6x + 4y - 4z = 0$
   (b) is parallel to and below the plane $5x - 3y + 3z = 1$
   (c) is parallel to and above the plane $6x + 2y - 2z = 3$.

28. Show that the lines

$$x + 1 = 4t \qquad\qquad x + 13 = 12t$$
$$y - 3 = t \quad \text{and} \quad y - 1 = 6t$$
$$z - 1 = 0 \qquad\qquad z - 2 = 3t$$

intersect and find an equation of the plane they determine.

29. Find parametric equations of the line of intersection of the planes
   (a) $-2x + 3y + 7z + 2 = 0$ and
        $x + 2y - 3z + 5 = 0$
   (b) $3x - 5y + 2z = 0$ and $z = 0$.

30. Show that the plane whose intercepts with the coordinate axes are $x = a$, $y = b$, and $z = c$ has equation

$$\frac{x}{a} + \frac{y}{b} + \frac{z}{c} = 1$$

provided a, b, and c are nonzero.

In Exercises 31–33, find the distance between the point and the plane.

31. $(1, -2, 3); 2x - 2y + z = 4$.

32. $(0, 1, 5); 3x + 6y - 2z - 5 = 0$.

33. $(7, 2, -1); 20x - 4y - 5z = 0$.

In Exercises 34–36, find the distance between the given parallel planes.

34. $2x - 3y + 4z = 7, 4x - 6y + 8z = 3$.

35. $-2x + y + z = 0, 6x - 3y - 3z - 5 = 0$.

36. $x + y + z = 1, x + y + z = -1$.

In Exercises 37–39, find the distance between the given skew lines.

37. $x = 1 + 7t$, $y = 3 + t$, $z = 5 - 3t$; $x = 4 - t$, $y = 6$, $z = 7 + 2t$.

38. $x = 3 - t$, $y = 4 + 4t$, $z = 1 + 2t$; $x = t$, $y = 3$, $z = 2t$.

39. $x = 2 + 4t$, $y = 6 - 4t$, $z = 5t$; $x = 3 + 8t$, $y = 5 - 3t$, $z = 6 + t$.

40. Show that the line $x = -1 + t$, $y = 3 + 2t$, $z = -t$ and the plane $2x - 2y - 2z + 3 = 0$ are parallel, and find the distance between them.

41. Prove: The planes $a_1 x + b_1 y + c_1 z = d_1$ and $a_2 x + b_2 y + c_2 z = d_2$ are perpendicular if and only if $a_1 a_2 + b_1 b_2 + c_1 c_2 = 0$.

15.7 CRAMER'S RULE (OPTIONAL)

In this section we show how determinants can be used to solve systems of two linear equations in two unknowns and three linear equations in three unknowns.

Recall that a ***solution*** of a system of two linear equations in two unknowns

$$a_1 x + b_1 y = k_1$$
$$a_2 x + b_2 y = k_2 \tag{1}$$

is a value for x and a value for y that satisfy *both* equations. The graphs of these equations are lines, which we will denote by L_1 and L_2. Since a point (x, y) lies on a line if and only if the numbers x and y satisfy the equation of the line, the solutions of the system will correspond to points of intersection of L_1 and L_2. There are three possibilities (Figure 15.7.1).

(a) The lines L_1 and L_2 are parallel and distinct, in which case there is no intersection and, consequently, no solution to the system.

(b) The lines L_1 and L_2 intersect at only one point, in which case the system has exactly one solution.

(c) The lines L_1 and L_2 coincide, in which case there are infinitely many points of intersection, and consequently infinitely many solutions to the system.

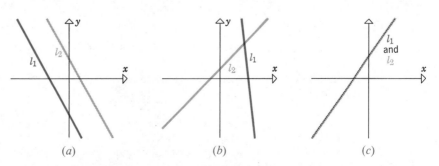

Figure 15.7.1   (a)   (b)   (c)

▶ **Example 1**   If we multiply the second equation of the system

$$x + y = 4$$
$$2x + 2y = 6$$

by $\frac{1}{2}$, it becomes evident that there is no solution since the two equations in the resulting system

$$x + y = 4$$
$$x + y = 3$$

contradict each other. Geometrically, the lines $x + y = 4$ and $2x + 2y = 6$ are distinct and parallel.   ◀

▶ **Example 2**   Since the second equation of the system

$$x + y = 4$$
$$2x + 2y = 8$$

is a multiple of the first equation, it is evident that any solution of the first equation will satisfy the second equation automatically. But the first equation, $x + y = 4$, has infinitely many solutions since we may assign x an arbitrary value and determine y from the relationship $y = 4 - x$. (Some possibilities are $x = 0$, $y = 4$; $x = 1$, $y = 3$; $x = -1$, $y = 5$.) Thus, the system has infinitely many solutions. Geometrically, the lines $x + y = 4$ and $2x + 2y = 8$ coincide.   ◀

▶ **Example 3**   If we add the equations of the system

$$x + y = 3$$
$$2x - y = 6$$

we obtain $3x = 9$ or $x = 3$, and if we substitute this value in the first equation we obtain $y = 0$, so that this system has the unique solution $x = 3$, $y = 0$. Geometrically, the lines $x + y = 3$ and $2x - y = 6$ intersect only at the point $(3, 0)$.   ◀

A *solution* of a system of three linear equations in three unknowns

$$a_1x + b_1y + c_1z = k_1$$
$$a_2x + b_2y + c_2z = k_2 \tag{2}$$
$$a_3x + b_3y + c_3z = k_3$$

consists of values of x, y, and z that satisfy all three equations. The graphs of these equations are planes in 3-space, and the solutions of this system correspond to points where all three planes intersect. It follows that system (2), like system (1), has zero, one, or infinitely many solutions, depending on the relative orientation of the planes (Figure 15.7.2).

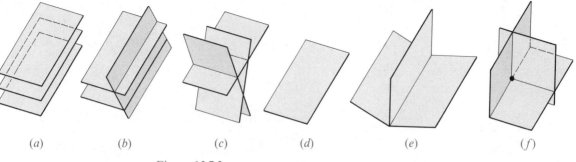

(a) (b) (c) (d) (e) (f)

Figure 15.7.2

If a system of two linear equations in two unknowns or three linear equations in three unknowns has a unique solution, the solution can be expressed as a ratio of determinants by using the following theorems, which are cases of a general result called *Cramer's* Rule*.

---

*GABRIEL CRAMER (1704–1752) Swiss mathematician. Although Cramer does not rank with the great mathematicians of his time, his contributions as a disseminator of mathematical ideas have earned him a well-deserved place in the history of mathematics. The son of a physician, Cramer was born and educated in Geneva, Switzerland. At age 20 he competed for, but failed to secure, the chair of philosophy at the Académie de Calvin at Geneva. However, the awarding magistrates were sufficiently impressed with Cramer and a fellow competitor to create a new chair of mathematics for both men to share. Alternately, each assumed the full responsibility and salary associated with the chair for two or three years while the other traveled. During his travels Cramer met many of the great mathematicians and scientists of his day, the Bernoullis, Euler, Halley, D'Alembert, and others. Many of these contacts and friendships led to extensive correspondence in which information about new mathematical discoveries was transmitted. Eventually, Cramer became sole possessor of the mathematics chair and the chair of philosophy as well.

Cramer's mathematical work was primarily in geometry and probability; he had relatively little knowledge of calculus and did not use it to any great extent in his work. In 1730 he finished second to Johann I Bernoulli in a competition for a prize offered by the Paris Academy to explain properties of planetary orbits.

Cramer's most widely known work, *Introduction á l'analyse des lignes courbes algébriques* (1750) was a study and classification of algebraic curves; Cramer's rule appeared in the appendix. Although the rule bears his name, variations of the basic idea were formulated earlier by Leibniz (and even earlier by Chinese mathematicians). However, Cramer's superior notation helped clarify and popularize the technique.

Perhaps Cramer's most important contributions stemmed from his work as an editor of the mathematical creations of others. He edited and published the works of Jacob I Bernoulli and Leibniz.

Overwork combined with a fall from a carriage eventually led to his death in 1752. Cramer was apparently a good-natured and pleasant person, though he never married. His interests were broad. He wrote on philosophy of law and government and the history of mathematics. He served in public office, participated in artillery and fortifications activities for the government, instructed workers on techniques of cathedral repair, and undertook excavations of cathedral archives. Cramer received numerous honors for his activities.

15.7.1 THEOREM   *If the system of equations*

Cramer's Rule for
Two Unknowns

$$a_1 x + b_1 y = k_1$$
$$a_2 x + b_2 y = k_2$$

is such that

$$\begin{vmatrix} a_1 & b_1 \\ a_2 & b_2 \end{vmatrix} \neq 0$$

then the system has a unique solution. This solution is

$$x = \frac{\begin{vmatrix} k_1 & b_1 \\ k_2 & b_2 \end{vmatrix}}{\begin{vmatrix} a_1 & b_1 \\ a_2 & b_2 \end{vmatrix}}, \qquad y = \frac{\begin{vmatrix} a_1 & k_1 \\ a_2 & k_2 \end{vmatrix}}{\begin{vmatrix} a_1 & b_1 \\ a_2 & b_2 \end{vmatrix}}$$

15.7.2 THEOREM   *If the system of equations*

Cramer's Rule for
Three Unknowns

$$a_1 x + b_1 y + c_1 z = k_1$$
$$a_2 x + b_2 y + c_2 z = k_2$$
$$a_3 x + b_3 y + c_3 z = k_3$$

is such that

$$\begin{vmatrix} a_1 & b_1 & c_1 \\ a_2 & b_2 & c_2 \\ a_3 & b_3 & c_3 \end{vmatrix} \neq 0$$

then the system has a unique solution. This solution is

$$x = \frac{\begin{vmatrix} k_1 & b_1 & c_1 \\ k_2 & b_2 & c_2 \\ k_3 & b_3 & c_3 \end{vmatrix}}{\begin{vmatrix} a_1 & b_1 & c_1 \\ a_2 & b_2 & c_2 \\ a_3 & b_3 & c_3 \end{vmatrix}}, y = \frac{\begin{vmatrix} a_1 & k_1 & c_1 \\ a_2 & k_2 & c_2 \\ a_3 & k_3 & c_3 \end{vmatrix}}{\begin{vmatrix} a_1 & b_1 & c_1 \\ a_2 & b_2 & c_2 \\ a_3 & b_3 & c_3 \end{vmatrix}}, z = \frac{\begin{vmatrix} a_1 & b_1 & k_1 \\ a_2 & b_2 & k_2 \\ a_3 & b_3 & k_3 \end{vmatrix}}{\begin{vmatrix} a_1 & b_1 & c_1 \\ a_2 & b_2 & c_2 \\ a_3 & b_3 & c_3 \end{vmatrix}}$$

REMARK.   There is a pattern to the formulas in these theorems. In each formula the determinant in the denominator is formed from the coefficients of the unknowns; and the determinant in the numerator differs from the determinant in the denominator in that the coefficients of the unknown being calculated are replaced by the k's. It is assumed in these theorems that the system is written so that like unknowns are aligned vertically and the constants appear by themselves on the right side of each equation.

We will prove Cramer's rule for the case of three unknowns. However, let us first look at some examples.

▶ Example 4 Use Cramer's rule to solve

$$5x - 2y = -1$$
$$2x + 3y = 3$$

Solution.

$$x = \frac{\begin{vmatrix} -1 & -2 \\ 3 & 3 \end{vmatrix}}{\begin{vmatrix} 5 & -2 \\ 2 & 3 \end{vmatrix}} = \frac{3}{19}; \qquad y = \frac{\begin{vmatrix} 5 & -1 \\ 2 & 3 \end{vmatrix}}{\begin{vmatrix} 5 & -2 \\ 2 & 3 \end{vmatrix}} = \frac{17}{19} \qquad \blacktriangleleft$$

▶ Example 5 Use Cramer's rule to solve

$$x \qquad + 2z = 6$$
$$-3x + 4y + 6z = 30$$
$$- x - 2y + 3z = 8$$

Solution.

$$x = \frac{\begin{vmatrix} 6 & 0 & 2 \\ 30 & 4 & 6 \\ 8 & -2 & 3 \end{vmatrix}}{\begin{vmatrix} 1 & 0 & 2 \\ -3 & 4 & 6 \\ -1 & -2 & 3 \end{vmatrix}} = \frac{-10}{11}; \quad y = \frac{\begin{vmatrix} 1 & 6 & 2 \\ -3 & 30 & 6 \\ -1 & 8 & 3 \end{vmatrix}}{\begin{vmatrix} 1 & 0 & 2 \\ -3 & 4 & 6 \\ -1 & -2 & 3 \end{vmatrix}} = \frac{18}{11};$$

$$z = \frac{\begin{vmatrix} 1 & 0 & 6 \\ -3 & 4 & 30 \\ -1 & -2 & 8 \end{vmatrix}}{\begin{vmatrix} 1 & 0 & 2 \\ -3 & 4 & 6 \\ -1 & -2 & 3 \end{vmatrix}} = \frac{38}{11} \qquad \blacktriangleleft$$

Proof of Cramer's Rule for Three Unknowns. Since two vectors are equal if and only if their corresponding components are equal, the system

$$a_1 x + b_1 y + c_1 z = k_1$$
$$a_2 x + b_2 y + c_2 z = k_2 \qquad (3)$$
$$a_3 x + b_3 y + c_3 z = k_3$$

is equivalent to the single vector equation

$$\langle a_1 x + b_1 y + c_1 z, a_2 x + b_2 y + c_2 z, a_3 x + b_3 y + c_3 z \rangle = \langle k_1, k_2, k_3 \rangle$$

or, equivalently,

$$x \langle a_1, a_2, a_3 \rangle + y \langle b_1, b_2, b_3 \rangle + z \langle c_1, c_2, c_3 \rangle = \langle k_1, k_2, k_3 \rangle \qquad (4)$$

If we define

$$\mathbf{a} = \langle a_1, a_2, a_3 \rangle, \quad \mathbf{b} = \langle b_1, b_2, b_3 \rangle, \quad \mathbf{c} = \langle c_1, c_2, c_3 \rangle, \quad \mathbf{k} = \langle k_1, k_2, k_3 \rangle$$

then (4) can be written

$$x\mathbf{a} + y\mathbf{b} + z\mathbf{c} = \mathbf{k} \qquad (5)$$

Therefore, solving system (3) is equivalent to solving the vector equation (5) for the unknown scalars x, y, and z. If we take the dot product of both sides of (5) with $\mathbf{b} \times \mathbf{c}$, we obtain

$$(x\mathbf{a} + y\mathbf{b} + z\mathbf{c}) \cdot (\mathbf{b} \times \mathbf{c}) = \mathbf{k} \cdot (\mathbf{b} \times \mathbf{c})$$

or

$$x(\mathbf{a} \cdot \mathbf{b} \times \mathbf{c}) + y(\mathbf{b} \cdot \mathbf{b} \times \mathbf{c}) + z(\mathbf{c} \cdot \mathbf{b} \times \mathbf{c}) = \mathbf{k} \cdot \mathbf{b} \times \mathbf{c} \qquad (6)$$

By Theorem 15.4.2,

$$\mathbf{b} \cdot \mathbf{b} \times \mathbf{c} = 0 \qquad \text{and} \qquad \mathbf{c} \cdot \mathbf{b} \times \mathbf{c} = 0$$

so that (6) reduces to

$$x(\mathbf{a} \cdot \mathbf{b} \times \mathbf{c}) = \mathbf{k} \cdot \mathbf{b} \times \mathbf{c}$$

Expressing these triple scalar products as determinants yields

$$x \begin{vmatrix} a_1 & a_2 & a_3 \\ b_1 & b_2 & b_3 \\ c_1 & c_2 & c_3 \end{vmatrix} = \begin{vmatrix} k_1 & k_2 & k_3 \\ b_1 & b_2 & b_3 \\ c_1 & c_2 & c_3 \end{vmatrix}$$

Solving for x and applying Theorem 15.4.5b yields

$$x = \frac{\begin{vmatrix} k_1 & b_1 & c_1 \\ k_2 & b_2 & c_2 \\ k_3 & b_3 & c_3 \end{vmatrix}}{\begin{vmatrix} a_1 & b_1 & c_1 \\ a_2 & b_2 & c_2 \\ a_3 & b_3 & c_3 \end{vmatrix}}$$

The formulas for y and z have similar derivations.

As a matter of logic, we have shown that if there exists a solution of system (3), then the solution is given by the formulas in Cramer's rule. However, it is logically possible that the system has no solution. This may be ruled out by substituting the values of x, y, and z given by Cramer's rule into (3) and verifying that the equations are satisfied. The details are tedious and will be omitted. ▮

▶ Exercise Set 15.7

In Exercises 1–8, solve the system using Cramer's rule.

1. $3x - 4y = -5$
$2x + y = 4.$

2. $-x + 3y = 8$
$2x + 5y = 7.$

3. $2x_1 - 5x_2 = -2$
$4x_1 + 6x_2 = 1.$

4. $3a + 2b = 4$
$-a + b = 7.$

5. $x + 2y + z = 3$
$2x + y - z = 0$
$x - y + z = 6.$

6. $x - 3y + z = 4$
$2x - y = -2$
$4x - 3z = 0.$

7. $x_1 + x_2 - 2x_3 = 1$
$2x_1 - x_2 + x_3 = 2$
$x_1 - 2x_2 - 4x_3 = -4.$

8. $r + s + t = 2$
$r - s - 2t = 0$
$-r + 2s + t = 4.$

9. Use Cramer's rule to solve the rotation equations

$$x = x' \cos \theta - y' \sin \theta$$
$$y = x' \sin \theta + y' \cos \theta$$

for x' and y' in terms of x and y.

10. Solve the following system of equations for the unknown angles α, β, and γ, where $0 \le \alpha \le 2\pi$, $0 \le \beta \le 2\pi$, and $0 \le \gamma < \pi$

$$2 \sin \alpha - \cos \beta + 3 \tan \gamma = 3$$
$$4 \sin \alpha + 2 \cos \beta - 2 \tan \gamma = 2$$
$$6 \sin \alpha - 3 \cos \beta + \tan \gamma = 9.$$

[*Hint:* First solve for $\sin \alpha$, $\cos \beta$, and $\tan \gamma$.]

15.8 QUADRIC SURFACES

Figure 15.8.1

In 2-space the general shape of a curve can be obtained by plotting points. However, for surfaces in 3-space, point plotting is not generally helpful since too many points are needed to obtain even a crude picture of the surface. It is better to build up the shape of the surface by using curves of intersection with some well-chosen planes. The curve of intersection of a plane and a surface is called the *trace* of the surface in the plane (Figure 15.8.1). In Figure 15.8.2, we get an excellent picture of the surface $z = x^3 - 3xy^2$ from the traces shown. (This surface is called a "monkey saddle" because a monkey sitting astride the x-axis has a place for its two feet and tail.)

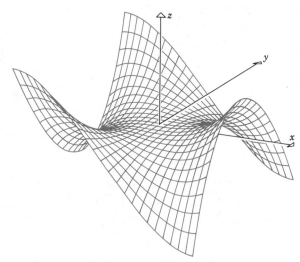

Figure 15.8.2

Earlier in the text, we saw that the graph in two dimensions of a second-degree equation in x and y,

$$Ax^2 + Bxy + Cy^2 + Dx + Ey + F = 0$$

is a conic section. In three dimensions, the graph of a second-degree equation in x, y, and z,

$$Ax^2 + By^2 + Cz^2 + Dxy + Exz + Fyz + Gx + Hy + Iz + J = 0$$

is called a *quadric surface* or a *quadric.*

Figure 15.8.3 shows the equations and graphs of six important quadric surfaces. In the equations, a, b, and c are positive constants. The stated equations apply only when the quadric surfaces are in the positions shown. If

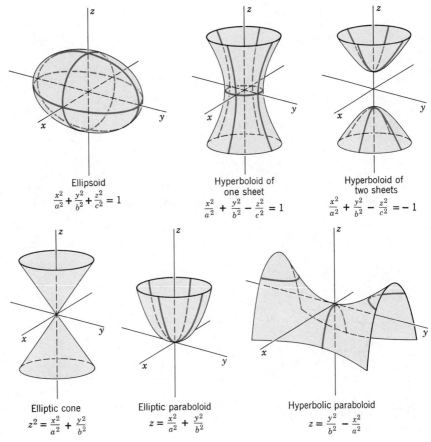

Ellipsoid
$$\frac{x^2}{a^2} + \frac{y^2}{b^2} + \frac{z^2}{c^2} = 1$$

Hyperboloid of
one sheet
$$\frac{x^2}{a^2} + \frac{y^2}{b^2} - \frac{z^2}{c^2} = 1$$

Hyperboloid of
two sheets
$$\frac{x^2}{a^2} + \frac{y^2}{b^2} - \frac{z^2}{c^2} = -1$$

Elliptic cone
$$z^2 = \frac{x^2}{a^2} + \frac{y^2}{b^2}$$

Elliptic paraboloid
$$z = \frac{x^2}{a^2} + \frac{y^2}{b^2}$$

Hyperbolic paraboloid
$$z = \frac{y^2}{b^2} - \frac{x^2}{a^2}$$

Figure 15.8.3

the quadric surfaces are rotated or translated from these positions, then the equations change. We will consider this point in more detail later.

The traces of the quadric surfaces in planes parallel to the coordinate planes are described in Table 15.8.1.

To study the traces of a quadric surface, we will need equations of planes parallel to the coordinate planes. The plane parallel to the xy-plane and passing through $(0, 0, k)$ (Figure 15.8.4, page 850) consists of all points (x, y, z) with $z = k$. Thus, its equation is

$$z = k$$

Similarly, $x = k$ represents a plane parallel to the yz-plane and passing through $(k, 0, 0)$, while $y = k$ represents a plane parallel to the xz-plane and passing through $(0, k, 0)$. The equation of the trace of a surface in the plane $z = k$ is obtained by substituting $z = k$ in the equation of the surface. Similarly, for traces in the planes $x = k$ and $y = k$.

Table 15.8.1

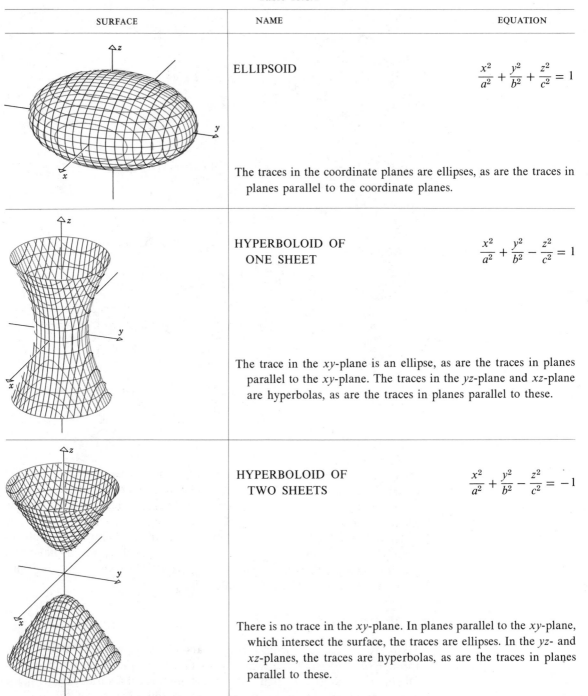

SURFACE	NAME	EQUATION
	ELLIPSOID	$\dfrac{x^2}{a^2} + \dfrac{y^2}{b^2} + \dfrac{z^2}{c^2} = 1$

The traces in the coordinate planes are ellipses, as are the traces in planes parallel to the coordinate planes.

| | HYPERBOLOID OF ONE SHEET | $\dfrac{x^2}{a^2} + \dfrac{y^2}{b^2} - \dfrac{z^2}{c^2} = 1$ |

The trace in the xy-plane is an ellipse, as are the traces in planes parallel to the xy-plane. The traces in the yz-plane and xz-plane are hyperbolas, as are the traces in planes parallel to these.

| | HYPERBOLOID OF TWO SHEETS | $\dfrac{x^2}{a^2} + \dfrac{y^2}{b^2} - \dfrac{z^2}{c^2} = -1$ |

There is no trace in the xy-plane. In planes parallel to the xy-plane, which intersect the surface, the traces are ellipses. In the yz- and xz-planes, the traces are hyperbolas, as are the traces in planes parallel to these.

SURFACE	NAME	EQUATION
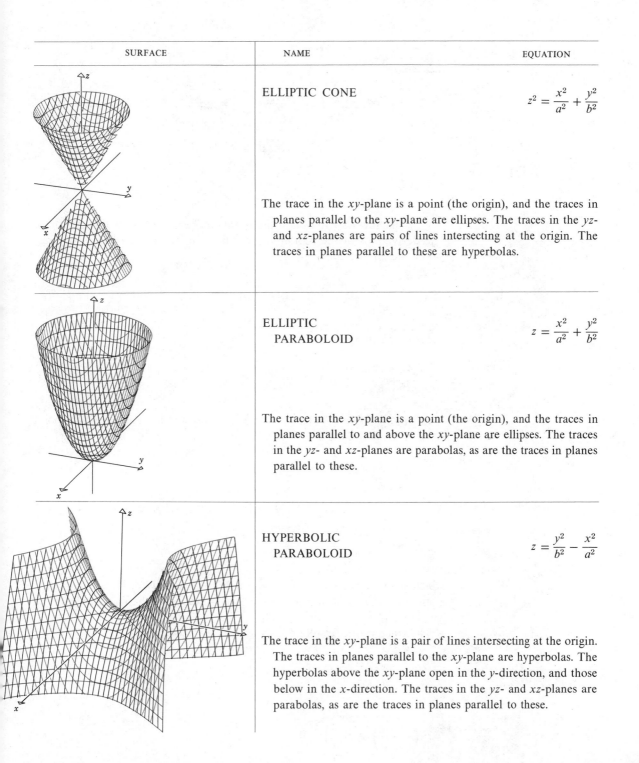	ELLIPTIC CONE	$z^2 = \dfrac{x^2}{a^2} + \dfrac{y^2}{b^2}$
	The trace in the xy-plane is a point (the origin), and the traces in planes parallel to the xy-plane are ellipses. The traces in the yz- and xz-planes are pairs of lines intersecting at the origin. The traces in planes parallel to these are hyperbolas.	
	ELLIPTIC PARABOLOID	$z = \dfrac{x^2}{a^2} + \dfrac{y^2}{b^2}$
	The trace in the xy-plane is a point (the origin), and the traces in planes parallel to and above the xy-plane are ellipses. The traces in the yz- and xz-planes are parabolas, as are the traces in planes parallel to these.	
	HYPERBOLIC PARABOLOID	$z = \dfrac{y^2}{b^2} - \dfrac{x^2}{a^2}$
	The trace in the xy-plane is a pair of lines intersecting at the origin. The traces in planes parallel to the xy-plane are hyperbolas. The hyperbolas above the xy-plane open in the y-direction, and those below in the x-direction. The traces in the yz- and xz-planes are parabolas, as are the traces in planes parallel to these.	

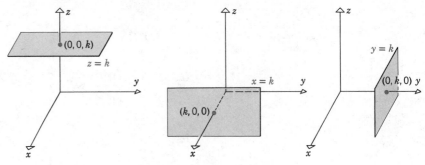

Figure 15.8.4

We will show how to obtain the results in Table 15.8.1 for elliptic cones. The derivations of the remaining results in the table are left as exercises.
For the elliptic cone

$$z^2 = \frac{x^2}{a^2} + \frac{y^2}{b^2} \qquad (a > 0, b > 0)$$

the trace in the xy-plane is

$$\frac{x^2}{a^2} + \frac{y^2}{b^2} = 0 \qquad (z = 0)$$

which implies that $x = 0$, $y = 0$, $z = 0$. Thus, the trace is the single point $(0, 0, 0)$. In the plane $z = k$ parallel to the xy-plane, the trace is

$$\frac{x^2}{a^2} + \frac{y^2}{b^2} = k^2$$

or

$$\frac{x^2}{(ak)^2} + \frac{y^2}{(bk)^2} = 1$$

which is an ellipse. As $|k|$ increases, so do $(ak)^2$ and $(bk)^2$. Thus, the dimensions of these ellipses increase as the planes containing them recede from the xy-plane.
The trace in the xz-plane is

$$z^2 = \frac{x^2}{a^2} \qquad (y = 0)$$

or

$$\left(z - \frac{x}{a}\right)\left(z + \frac{x}{a}\right) = 0$$

Thus, the trace is the pair of lines

$$z = \frac{x}{a} \quad \text{and} \quad z = -\frac{x}{a}$$

which intersect at the origin. Similarly, the trace in the yz-plane is the pair of intersecting lines

$$z = \frac{y}{b} \quad \text{and} \quad z = -\frac{y}{b} \quad (x = 0)$$

The trace in the plane $y = k$ parallel to the xz-plane is

$$z^2 = \frac{x^2}{a^2} + \frac{k^2}{b^2}$$

or

$$z^2 - \frac{x^2}{a^2} = \frac{k^2}{b^2}$$

or

$$\frac{z^2}{k^2/b^2} - \frac{x^2}{a^2 k^2/b^2} = 1$$

which is a hyperbola in the plane $y = k$ opening along a line parallel to the z-axis. Similarly, the trace in the plane $x = k$ is

$$z^2 = \frac{k^2}{a^2} + \frac{y^2}{b^2}$$

or

$$z^2 - \frac{y^2}{b^2} = \frac{k^2}{a^2}$$

or

$$\frac{z^2}{k^2/a^2} - \frac{y^2}{b^2 k^2/a^2} = 1$$

which is a hyperbola in the plane $x = k$ opening along a line parallel to the z-axis.

We will now discuss the problem of sketching quadric surfaces.

ELLIPSOIDS A fairly accurate graph of the ellipsoid

$$\frac{x^2}{a^2} + \frac{y^2}{b^2} + \frac{z^2}{c^2} = 1 \quad (a > 0, b > 0, c > 0) \tag{1}$$

can be obtained by first plotting the intercepts with the coordinate axes. The x-intercepts occur where $y = z = 0$. Substituting these values in (1) yields $x^2/a^2 = 1$ or $x = \pm a$. Thus, the x-intercepts are

$$(a, 0, 0) \qquad \text{and} \qquad (-a, 0, 0)$$

The y-intercepts (obtained by setting $x = z = 0$) are $(0, \pm b, 0)$, and the z-intercepts (obtained by setting $x = y = 0$) are $(0, 0, \pm c)$. The ellipsoid is sketched in Figure 15.8.5.

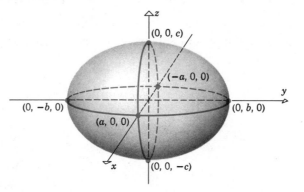

Figure 15.8.5

Observe that if $a = b = c$, then the ellipsoid represented by (1) reduces to a sphere of radius a centered at the origin.

HYPERBOLOIDS OF ONE SHEET

A fairly accurate graph of the hyperboloid of one sheet

$$\frac{x^2}{a^2} + \frac{y^2}{b^2} - \frac{z^2}{c^2} = 1 \qquad (a > 0, b > 0, c > 0) \qquad (2)$$

can be obtained by first sketching its traces in the coordinate planes. The trace in the xy-plane is the ellipse,

$$\frac{x^2}{a^2} + \frac{y^2}{b^2} = 1 \qquad (z = 0)$$

which intersects the x-axis at $(\pm a, 0, 0)$ and the y-axis at $(0, \pm b, 0)$ (Figure 15.8.6). The trace in the yz-plane is the hyperbola

$$\frac{y^2}{b^2} - \frac{z^2}{c^2} = 1 \qquad (x = 0)$$

which intersects the y-axis at $(0, \pm b, 0)$ and has asymptotes

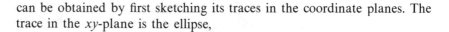

$$y = \pm \frac{b}{c} z$$

Figure 15.8.6

The trace in the xz-plane is the hyperbola

$$\frac{x^2}{a^2} - \frac{z^2}{c^2} = 1 \qquad (y = 0)$$

which intersects the x-axis at $(\pm a, 0, 0)$ and has asymptotes

$$x = \pm \frac{a}{c} z$$

HYPERBOLOIDS OF
TWO SHEETS

A fairly accurate graph of the hyperboloid of two sheets,

$$\frac{x^2}{a^2} + \frac{y^2}{b^2} - \frac{z^2}{c^2} = -1 \qquad (a > 0, b > 0, c > 0) \tag{3}$$

can be obtained by plotting the intercepts with the z-axis and sketching the traces in the coordinate planes.

The z-intercepts occur where $x = y = 0$. Substituting these values in (3) yields

$$-\frac{z^2}{c^2} = -1$$

or $z = \pm c$. Thus, the z-intercepts are $(0, 0, \pm c)$ (Figure 15.8.7). The trace in the yz-plane is the hyperbola

$$\frac{z^2}{c^2} - \frac{y^2}{b^2} = 1 \qquad (x = 0)$$

which intersects the z-axis at $(0, 0, \pm c)$ and has asymptotes

$$z = \pm \frac{c}{b} y$$

The trace in the xz-plane is the hyperbola,

$$\frac{z^2}{c^2} - \frac{x^2}{a^2} = 1 \qquad (y = 0)$$

which intersects the z-axis at the points $(0, 0, \pm c)$ and has asymptotes

$$z = \pm \frac{c}{a} x$$

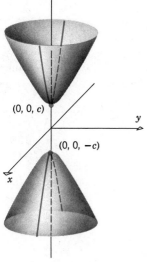

Figure 15.8.7

ELLIPTIC CONES   A fairly accurate graph of the elliptic cone

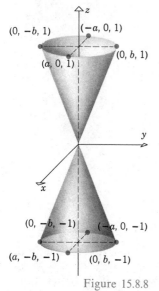

$$z^2 = \frac{x^2}{a^2} + \frac{y^2}{b^2} \tag{4}$$

can be obtained by first sketching the traces in the planes $z = 1$ and $z = -1$ (Figure 15.8.8). In both planes, the traces are ellipses with the equation

$$\frac{x^2}{a^2} + \frac{y^2}{b^2} = 1$$

If $a = b$, then all traces of the cone in planes parallel to the xy-plane are circles, so the surface is called a *circular cone.*

Figure 15.8.8

ELLIPTIC
PARABOLOIDS

A fairly accurate graph of the elliptic paraboloid,

$$z = \frac{x^2}{a^2} + \frac{y^2}{b^2} \qquad (a > 0, b > 0) \tag{5a}$$

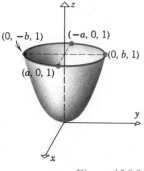

can be obtained by first sketching its trace in the plane $z = 1$ (Figure 15.8.9). This trace is the ellipse with the equation

$$\frac{x^2}{a^2} + \frac{y^2}{b^2} = 1$$

If $a = b$, then all traces of the paraboloid in planes parallel to the xy-plane are circles, so the surface is called a *circular paraboloid.*

Later, we will encounter equations of the form

Figure 15.8.9

$$z = -\left(\frac{x^2}{a^2} + \frac{y^2}{b^2}\right) \tag{5b}$$

An equation of this type represents an elliptic paraboloid opening in the negative z-direction.

HYPERBOLIC
PARABOLOIDS

The hyperbolic paraboloid

$$z = \frac{y^2}{b^2} - \frac{x^2}{a^2} \qquad (a > 0, b > 0) \tag{6}$$

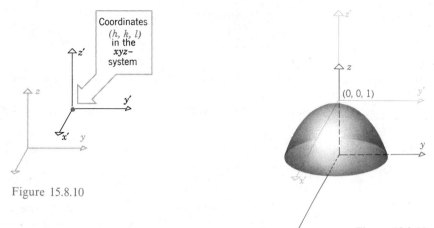

Figure 15.8.10

Figure 15.8.11

is a difficult surface to draw accurately. Fortunately, a rough sketch, simply showing the position of the surface, suffices for most purposes.

TRANSLATION OF AXES IN 3-SPACE

Let an $x'y'z'$-coordinate system be obtained by translating an xyz-coordinate system so the $x'y'z'$-origin is at the point whose xyz-coordinates are $(x, y, z) = (h, k, l)$ (Figure 15.8.10). It can be shown that the $x'y'z'$-coordinates and xyz-coordinates of a point P are related by

$$x' = x - h, \qquad y' = y - k, \qquad z' = z - l \tag{7}$$

[Compare this to (6) in Section 12.2.]

▶ Example 1   Sketch the surface $z = 1 - x^2 - y^2$.

Solution.   Rewrite the equation in the form

$$z - 1 = -(x^2 + y^2) \tag{8}$$

If we translate the coordinate axes so the new origin is at the point $(h, k, l) = (0, 0, 1)$, then the translation equations (7) are

$$x' = x, \qquad y' = y, \qquad z' = z - 1$$

so that in $x'y'z'$-coordinates (8) becomes

$$z' = -(x'^2 + y'^2)$$

which is of form (5b) with $a = b = 1$. Thus, the surface is a circular paraboloid, opening down, with vertex $(0, 0, 1)$ in xyz-coordinates (Figure 15.8.11). ◀

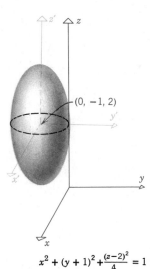

$$x^2 + (y + 1)^2 + \frac{(z-2)^2}{4} = 1$$

Figure 15.8.12

▶ Example 2   Sketch the surface

$$4x^2 + 4y^2 + z^2 + 8y - 4z = -4$$

Solution.   Completing the squares yields

$$4x^2 + 4(y + 1)^2 + (z - 2)^2 = -4 + 4 + 4$$

or

$$x^2 + (y + 1)^2 + \frac{(z - 2)^2}{4} = 1 \tag{9}$$

If we translate the coordinate axes so the new origin is at the point $(h, k, l) = (0, -1, 2)$, then the translation equations (7) are

$$x' = x, \qquad y' = y + 1, \qquad z' = z - 2$$

so that in $x'y'z'$-coordinates (9) becomes

$$x'^2 + y'^2 + \frac{z'^2}{4} = 1$$

which is of form (1) with $a = 1$, $b = 1$, and $c = 2$. Thus, the surface is the ellipsoid shown in Figure 15.8.12.   ◀

▶ Exercise Set 15.8

In Exercises 1–12, name and sketch the quadric surfaces.

1. $x^2 + \dfrac{y^2}{4} + \dfrac{z^2}{9} = 1.$ **2.** $\dfrac{z^2}{4} - \dfrac{y^2}{9} - \dfrac{x^2}{4} = 1.$

3. $\dfrac{x^2}{4} + \dfrac{y^2}{9} - \dfrac{z^2}{16} = 1.$ **4.** $z = x^2 + y^2.$

5. $z^2 = \dfrac{x^2}{4} + \dfrac{y^2}{9}.$ **6.** $z - y^2 + x^2 = 0.$

7. $16z + x^2 + 4y^2 = 0.$ **8.** $36 - x^2 - 4y^2 = 9z^2.$

9. $4x^2 + y^2 - z^2 = 16.$ **10.** $9z^2 - 4y^2 - x^2 = 36.$

11. $9x^2 - 4y^2 = 36z.$ **12.** $2x^2 - 8z + y^2 = 0.$

13. The following equations represent quadric surfaces with orientations different from those in Figure 15.8.3. Name and sketch the surface.

(a) $\dfrac{z^2}{c^2} - \dfrac{y^2}{b^2} + \dfrac{x^2}{a^2} = 1$ (b) $\dfrac{x^2}{a^2} - \dfrac{y^2}{b^2} - \dfrac{z^2}{c^2} = 1$

(c) $x = \dfrac{y^2}{b^2} + \dfrac{z^2}{c^2}$ (d) $x^2 = \dfrac{y^2}{b^2} + \dfrac{z^2}{c^2}$

(e) $y = \dfrac{z^2}{c^2} - \dfrac{x^2}{a^2}$ (f) $y = -\left(\dfrac{x^2}{a^2} + \dfrac{z^2}{c^2}\right).$

In Exercises 14–17, sketch the graph of the equation.

14. $z = \sqrt{1 - x^2 - y^2}.$ **15.** $z = \sqrt{x^2 + y^2}.$

16. $z = \sqrt{1 + x^2 + y^2}.$ **17.** $z = \sqrt{x^2 + y^2 - 1}.$

In Exercises 18–23, name and sketch the surface.

18. $\dfrac{(x - 1)^2}{4} + \dfrac{(y - 2)^2}{9} + \dfrac{(z - 4)^2}{16} = 1.$

19. $z = (x + 2)^2 + (y - 3)^2 - 9.$

20. $4x^2 - y^2 + 16(z - 2)^2 = 100.$

21. $9x^2 + y^2 + 4z^2 - 18x + 2y + 16z = 10.$

22. $z^2 = 4x^2 + y^2 + 8x - 2y + 4z.$

23. $z = 4 - x^2 - y^2 - 2y.$

24. Obtain the results in Table 15.8.1 for the ellipsoid, $x^2/a^2 + y^2/b^2 + z^2/c^2 = 1$.

25. Obtain the results in Table 15.8.1 for the hyperboloid of one sheet, $x^2/a^2 + y^2/b^2 - z^2/c^2 = 1$.

26. Obtain the results in Table 15.8.1 for the hyperboloid of two sheets, $x^2/a^2 + y^2/b^2 - z^2/c^2 = -1$.

27. Obtain the results in Table 15.8.1 for the elliptic paraboloid $z = x^2/a^2 + y^2/b^2$.

28. Obtain the results in Table 15.8.1 for the hyperbolic paraboloid $z = y^2/b^2 - x^2/a^2$.

29. For the elliptic paraboloid
$$z = \frac{x^2}{9} + \frac{y^2}{4}$$

find

(a) the focus and vertex of the (parabolic) trace in the plane $x = k$;

(b) the foci and the endpoints of the major and minor axes of the (elliptic) trace in the plane $z = k$.

30. Use the method of slicing to find the volume of the ellipsoid
$$\frac{x^2}{a^2} + \frac{y^2}{b^2} + \frac{z^2}{c^2} = 1$$

[*Hint:* The area of the ellipse $x^2/a^2 + y^2/b^2 = 1$ is πab.]

15.9 SPHERICAL AND CYLINDRICAL COORDINATES

In this section we discuss two new ways of describing the position of a point in three dimensions.

The **cylindrical coordinates** (r, θ, z) of a point $P(x, y, z)$ are obtained by replacing x and y by polar coordinates and keeping z the same (Figure 15.9.1). For simplicity, we will require that

$$r \geq 0 \quad \text{and} \quad 0 \leq \theta < 2\pi$$

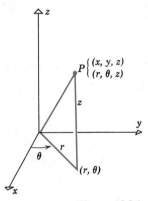

From the familiar relationship between rectangular and polar coordinates in the xy-plane (Section 13.1), it follows that the rectangular and cylindrical coordinates of a point in 3-space are related by

$$x = r \cos \theta, \quad y = r \sin \theta, \quad z = z \tag{1}$$

Figure 15.9.1

or expressed another way

$$r = \sqrt{x^2 + y^2}, \quad \tan \theta = \frac{y}{x}, \quad z = z \tag{2}$$

▶ **Example 1** The point whose cylindrical coordinates are $(r, \theta, z) = (4, \pi/3, -3)$ has rectangular coordinates

$$x = 4 \cos \frac{\pi}{3} = 2, \quad y = 4 \sin \frac{\pi}{3} = 2\sqrt{3}, \quad z = -3 \quad ◀$$

▶ **Example 2**   The equation in cylindrical coordinates of the surface $z = x^2 + y^2 - 2x + y$ is

$$z = r^2 - 2r \cos \theta + r \sin \theta$$ ◀

▶ **Example 3**   The equation in rectangular coordinates of the surface

$$r = 4 \cos \theta$$

can be obtained by multiplying both sides by r to get $r^2 = 4r \cos \theta$, from which it follows that

$$x^2 + y^2 = 4x$$

or equivalently

$$(x - 2)^2 + y^2 = 4$$

This is a right-circular cylinder parallel to the z-axis.   ◀

In cylindrical coordinates, surfaces of the forms

$$r = r_0, \qquad \theta = \theta_0, \qquad \text{and} \qquad z = z_0$$

(r_0, θ_0, and z_0 constants) are of special interest. These surfaces are shown in Figure 15.9.2.

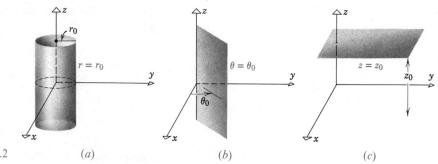

Figure 15.9.2   (a)   (b)   (c)

The surface $r = r_0$ is a right-circular cylinder of radius r_0 centered on the z-axis. At each point (r, θ, z) on this cylinder, r has the value r_0, but θ and z are unrestricted except for our blanket assumption that $0 \le \theta < 2\pi$.

The surface $\theta = \theta_0$ is a half plane attached along the z-axis and making an angle θ_0 with the positive x-axis. At each point (r, θ, z) on this surface, θ has

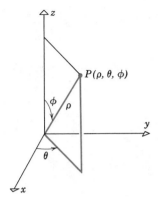

Figure 15.9.3

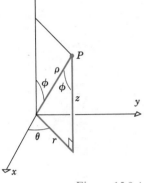

Figure 15.9.4

the value θ_0, but r and z are unrestricted except for our blanket assumption that $r \geq 0$.

The surface $z = z_0$ is a horizontal plane. At each point (r, θ, z) on this plane, z has the value z_0, but r and θ are unrestricted except for our blanket assumptions.

The **spherical coordinates** (ρ, θ, ϕ) of a point P in 3-space are illustrated in Figure 15.9.3. The coordinate ρ is the distance from P to the origin, the coordinate θ is the same as in polar coordinates, and the coordinate ϕ is the angle from the positive z-axis to the line segment joining P and the origin. We will require that

$$\rho \geq 0, \qquad 0 \leq \theta < 2\pi, \qquad 0 \leq \phi \leq \pi$$

The following relationships between spherical and cylindrical coordinates of a point P are immediate from Figure 15.9.4:

$$r = \rho \sin \phi, \qquad \theta = \theta, \qquad z = \rho \cos \phi \qquad (3)$$

Combining these with the equations $x = r \cos \theta$ and $y = r \sin \theta$, we obtain the following relationships between the spherical and rectangular coordinates of P:

$$x = \rho \sin \phi \cos \theta, \qquad y = \rho \sin \phi \sin \theta, \qquad z = \rho \cos \phi \qquad (4)$$

Moreover, since ρ is the distance between the origin and P, we have

$$\rho = \sqrt{x^2 + y^2 + z^2} \qquad (5)$$

▶ Example 4 The point with spherical coordinates $(\rho, \theta, \phi) = (4, \pi/3, \pi/4)$ has rectangular coordinates

$$x = \rho \sin \phi \cos \theta = 4 \sin \frac{\pi}{4} \cos \frac{\pi}{3} = \sqrt{2}$$

$$y = \rho \sin \phi \sin \theta = 4 \sin \frac{\pi}{4} \sin \frac{\pi}{3} = \sqrt{6}$$

$$z = \rho \cos \phi = 4 \cos \frac{\pi}{4} = 2\sqrt{2} \qquad ◀$$

▶ Example 5 Find the equation of the paraboloid $z = x^2 + y^2$ in spherical coordinates

Solution. Substituting (4) in this equation yields

$$\rho \cos \phi = \rho^2 \sin^2 \phi \cos^2 \theta + \rho^2 \sin^2 \phi \sin^2 \theta$$
$$\rho \cos \phi = \rho^2 \sin^2 \phi \, (\cos^2 \theta + \sin^2 \theta)$$
$$\rho \cos \phi = \rho^2 \sin^2 \phi$$

which simplifies to $\rho \sin^2 \phi = \cos \phi$ ◀

In spherical coordinates, surfaces of the forms

$$\rho = \rho_0, \qquad \theta = \theta_0, \qquad \phi = \phi_0$$

(ρ_0, θ_0, ϕ_0 constants) are of special interest. These surfaces are shown in Figure 15.9.5.

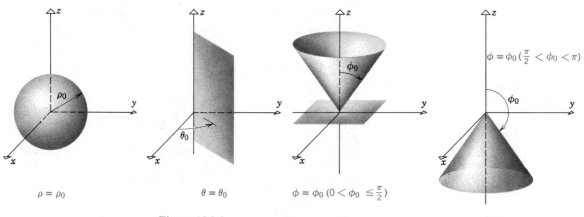

$$\rho = \rho_0 \qquad\qquad \theta = \theta_0 \qquad\qquad \phi = \phi_0 \; (0 < \phi_0 \leq \tfrac{\pi}{2})$$

Figure 15.9.5

The surface $\rho = \rho_0$ consists of all points whose distance ρ from the origin is ρ_0. Assuming ρ_0 to be nonnegative, this is a sphere of radius ρ_0 centered at the origin.

As in cylindrical coordinates, the surface $\theta = \theta_0$ is a half plane attached along the z-axis, making an angle of θ_0 with the positive x-axis.

The surface $\phi = \phi_0$ consists of all points from which a line segment to the origin makes an angle of ϕ_0 with the positive z-axis. Depending on whether $0 < \phi_0 < \pi/2$, or $\pi/2 < \phi_0 < \pi$, or $\phi_0 = \pi/2$ this will be a cone opening up, a cone opening down, or the xy-plane.

SPHERICAL
COORDINATES IN
NAVIGATION

Spherical coordinates are related to longitude and latitude coordinates used in navigation. Let us construct a right-hand rectangular coordinate system with origin at the earth's center, positive z-axis passing through the north pole, and positive x-axis passing through the prime meridian (Figure 15.9.6). If we assume the earth to be a perfect sphere of radius $\rho = 4000$ miles, then each point on the earth has spherical coordinates of the form (4000, θ, ϕ) where ϕ and θ determine the latitude and longitude of the point. It is usual to specify longitudes in degrees east or west of the prime meridian and latitudes in degrees north or south of the equator. However, it is a simple matter to determine ϕ and θ from such data.

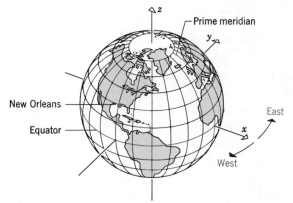

Figure 15.9.6

▶ Example 6 The city of New Orleans is located at 90° West longitude and 30° North latitude. Find its spherical and rectangular coordinates relative to the coordinate axes of Figure 15.9.6. (Take miles as the unit of distance.)

Solution. A longitude of 90° West corresponds to $\theta = 360° - 90° = 270°$ or $\theta = 3\pi/2$ radians; and a latitude of 30° North corresponds to $\phi = 90° - 30° = 60°$ or $\phi = \pi/3$ radians. Thus, the spherical coordinates of New Orleans are $(4000, 3\pi/2, \pi/3)$.

From (4), the rectangular coordinates of New Orleans are

$$x = 4000 \sin \frac{\pi}{3} \cos \frac{3\pi}{2} = 4000 \frac{\sqrt{3}}{2}(0) = 0 \text{ miles}$$

$$y = 4000 \sin \frac{\pi}{3} \sin \frac{3\pi}{2} = 4000 \frac{\sqrt{3}}{2}(-1) = -2000 \sqrt{3} \text{ miles}$$

$$z = 4000 \cos \frac{\pi}{3} = 4000 \left(\frac{1}{2}\right) = 2000 \text{ miles} \qquad ◀$$

▶ Exercise Set 15.9

1. Convert from rectangular to cylindrical coordinates.
   (a) $(4\sqrt{3}, 4, -4)$ (b) $(-5, 5, 6)$
   (c) $(0, 2, 0)$ (d) $(4, -4\sqrt{3}, 6)$
   (e) $(\sqrt{2}, -\sqrt{2}, 1)$ (f) $(0, 0, 1)$.

2. Convert from cylindrical to rectangular coordinates.
   (a) $(4, \pi/6, 3)$ (b) $(8, 3\pi/4, -2)$
   (c) $(5, 0, 4)$ (d) $(7, \pi, -9)$
   (e) $(6, 5\pi/3, 7)$ (f) $(1, \pi/2, 0)$.

3. Convert from rectangular to spherical coordinates.
 (a) $(1, \sqrt{3}, -2)$ (b) $(1, -1, \sqrt{2})$
 (c) $(0, 3\sqrt{3}, 3)$ (d) $(-5\sqrt{3}, 5, 0)$
 (e) $(4, 4, 4\sqrt{6})$ (f) $(1, -\sqrt{3}, -2)$.

4. Convert from spherical to rectangular coordinates.
 (a) $(5, \pi/6, \pi/4)$ (b) $(7, 0, \pi/2)$
 (c) $(1, \pi, 0)$ (d) $(2, 3\pi/2, \pi/2)$
 (e) $(1, 2\pi/3, 3\pi/4)$ (f) $(3, 7\pi/4, 5\pi/6)$.

5. Convert from cylindrical to spherical coordinates.
 (a) $(\sqrt{3}, \pi/6, 3)$ (b) $(1, \pi/4, -1)$
 (c) $(2, 3\pi/4, 0)$ (d) $(6, 1, -2\sqrt{3})$
 (e) $(4, 5\pi/6, 4)$ (f) $(2, 0, -2)$.

6. Convert from spherical to cylindrical coordinates.
 (a) $(5, \pi/4, 2\pi/3)$ (b) $(1, 7\pi/6, \pi)$
 (c) $(3, 0, 0)$ (d) $(4, \pi/6, \pi/2)$
 (e) $(5, \pi/2, 0)$ (f) $(6, 0, 3\pi/4)$.

In Exercises 7–14, an equation is given in cylindrical coordinates. Express the equation in rectangular coordinates and sketch the graph.

7. $r = 3$. **8.** $\theta = \pi/4$.

9. $z = r^2$. **10.** $z = r\cos\theta$.

11. $r = 4\sin\theta$. **12.** $r = 2\sec\theta$.

13. $r^2 + z^2 = 1$. **14.** $r^2\cos 2\theta = z$.

In Exercises 15–22, an equation is given in spherical coordinates. Express the equation in rectangular coordinates and sketch the graph.

15. $\rho = 3$. **16.** $\theta = \pi/3$.

17. $\phi = \pi/4$. **18.** $\rho = 2\sec\phi$.

19. $\rho = 4\cos\phi$. **20.** $\rho\sin\phi = 1$.

21. $\rho\sin\phi = 2\cos\theta$. **22.** $\rho - 2\sin\phi\cos\theta = 0$.

In Exercises 23–28, express the equation in: (a) cylindrical coordinates, (b) spherical coordinates.

23. $x^2 + y^2 + z^2 = 9$. **24.** $z^2 = x^2 - y^2$.

25. $2x + 3y + 4z = 1$. **26.** $x^2 + y^2 - z^2 = 1$.

27. $x^2 = 16 - z^2$. **28.** $x^2 + y^2 + z^2 = 2z$.

29. Leningrad, Russia, is located at 30° East longitude and 60° North latitude. Find its spherical and rectangular coordinates relative to the coordinate axes of Figure 15.9.6. Take miles as the unit of distance and assume the earth to be a sphere of radius 4000 miles.

30. Sketch the surface whose equation in cylindrical coordinates is $z = \sin\theta$ for $0 \le \theta \le \pi/2$.

31. Sketch the surface whose equation in spherical coordinates is $\rho = a(1 - \cos\phi)$. [*Hint.* The surface is shaped like a familiar fruit.]

▶ SUPPLEMENTARY EXERCISES

In Exercises 1 and 2, find
(a) $\|\mathbf{a}\|$ (b) $\mathbf{a}\cdot\mathbf{b}$
(c) $\mathbf{a}\times\mathbf{b}$ (d) $\mathbf{b}\times\mathbf{a}$
(e) the area of the triangle with sides $\mathbf{a}$ and $\mathbf{b}$
(f) $3\mathbf{a} - 2\mathbf{b}$.

1. $\mathbf{a} = \langle 1, 2, -1\rangle$, $\mathbf{b} = \langle 2, -1, 3\rangle$.

2. $\mathbf{a} = \langle 1, -2, 2\rangle$, $\mathbf{b} = \langle 3, 4, -5\rangle$.

In Exercises 3 and 4, find:
(a) $\|\text{proj}_{\mathbf{b}}\,\mathbf{a}\|$ (b) $\|\text{proj}_{\mathbf{a}}\,\mathbf{b}\|$
(c) the angle between $\mathbf{a}$ and $\mathbf{b}$
(d) the direction cosines of $\mathbf{a}$.

3. $\mathbf{a} = 3\mathbf{i} - 4\mathbf{j}$, $\mathbf{b} = 2\mathbf{i} + 2\mathbf{j} - \mathbf{k}$.

4. $\mathbf{a} = -\mathbf{j}$, $\mathbf{b} = \mathbf{i} + \mathbf{j}$.

5. Verify the identity $\mathbf{a}\times(\mathbf{b}\times\mathbf{c}) = (\mathbf{a}\cdot\mathbf{c})\mathbf{b} - (\mathbf{a}\cdot\mathbf{b})\mathbf{c}$ for $\mathbf{a} = \mathbf{i} + \mathbf{j}$, $\mathbf{b} = 2\mathbf{i} - \mathbf{k}$, $\mathbf{c} = \mathbf{j} - \mathbf{k}$.

6. Find the vector with length 5 and direction angles $\alpha = 60°$, $\beta = 120°$, $\gamma = 135°$.

7. Find the vector with length 3 and direction cosines $-1/\sqrt{2}$, 0, and $1/\sqrt{2}$.

8. For the points $P(6, 5, 7)$ and $Q(7, 3, 9)$ find:
 (a) the midpoint of the line segment PQ
 (b) the length and direction cosines of $\vec{PQ}$.

9. If $\mathbf{u} = \mathbf{i} + 2\mathbf{j} - 3\mathbf{k}$ and $\mathbf{v} = \mathbf{i} + \mathbf{j} + 2\mathbf{k}$, find:
 (a) the vector component of $\mathbf{u}$ along $\mathbf{v}$,
 (b) the vector component of $\mathbf{u}$ orthogonal to $\mathbf{v}$.

10. Consider the points $O(0, 0, 0)$, $A(0, a, a)$, and $B(-3, 4, 2)$. Find all nonzero values of a that make $\overrightarrow{OA}$ orthogonal to $\overrightarrow{AB}$.

11. (a) Under what conditions are $\mathbf{u} + \mathbf{v}$ and $\mathbf{u} - \mathbf{v}$ orthogonal?
    (b) Prove: $\|\mathbf{a}\|^2\|\mathbf{b}\|^2 = (\mathbf{a} \cdot \mathbf{b})^2 + \|\mathbf{a} \times \mathbf{b}\|^2$.

12. If $M(3, -1, 5)$ is the midpoint of the line segment PQ and if the coordinates of P are $(1, 2, 3)$, find the coordinates of Q.

13. Find all possible vectors of length 1 orthogonal to both $\mathbf{a} = \langle 3, -2, 1 \rangle$ and $\mathbf{b} = \langle -2, 1, -3 \rangle$.

14. Find the distance from the point $P(2, 3, 4)$ to the plane containing the points $A(0, 0, 1)$, $B(1, 0, 0)$, and $C(0, 2, 0)$.

In Exercises 15–18, find an equation for the plane that satisfies the given conditions.

15. The plane through $A(1, 2, 3)$ and $B(2, 4, 2)$ that is parallel to $\mathbf{v} = \langle -3, -1, -2 \rangle$.

16. The plane through $P_0(-1, 2, 3)$ that is perpendicular to the planes $2x - 3y + 5 = 0$ and $3x - y - 4z + 6 = 0$.

17. The plane that passes through $P(1, 1, 1)$, $Q(2, 3, 0)$, and $R(2, 1, 2)$.

18. The plane with intercepts $x = 2$, $y = -3$, $z = 10$.

19. Let L be the line through $P(1, 2, 8)$ that is parallel to $\mathbf{v} = \langle 3, -1, -4 \rangle$.
    (a) For what values of k and l will the point $Q(k, 3, l)$ be on L?
    (b) If L' has parametric equations $x = -8 - 3t$, $y = 5 + t$, $z = 0$, show that L' intersects L and find the point of intersection.
    (c) Find the point at which L intersects the plane through $R(-4, 0, 3)$ having a normal vector $\langle 3, -2, 6 \rangle$.

20. Consider the lines L_1 and L_2 with symmetric equations,

$$L_1: \frac{x - 1}{2} = \frac{y + \frac{3}{2}}{1} = \frac{z + 1}{2}$$

$$L_2: \frac{4 - x}{1} = \frac{3 - y}{2} = \frac{4 + z}{2}$$

(see Exercise 24, Section 15.2).
    (a) Are L_1 and L_2 parallel? Perpendicular?
    (b) Find parametric equations for L_1 and L_2.
    (c) Do L_1 and L_2 intersect? If so, where?

21. Find parametric equations for the line through P_1 and P_2.
    (a) $P_1(1, -1, 2)$, $P_2(3, 2, -1)$
    (b) $P_1(1, -3, 4)$, $P_2(1, 2, -3)$.

22. For the points $A(1, -1, 2)$, $B(2, -3, 0)$, $C(-1, -2, 0)$, and $D(2, 1, -1)$ find:
    (a) $\overrightarrow{AB} \times \overrightarrow{AC}$ (b) the area of triangle ABC
    (c) the volume of the parallelepiped determined by the vectors $\overrightarrow{AB}$, $\overrightarrow{AC}$, $\overrightarrow{AD}$
    (d) the distance from D to the plane containing A, B, and C.

23. (a) Find parametric equations for the intersection of the planes $2x + y - z = 3$ and $x + 2y + z = 3$.
    (b) Find the acute angle between the two planes.

In Exercises 24–26, describe the region satisfying the given conditions:

24. (a) $x^2 + 9y^2 + 4z^2 > 36$
    (b) $x^2 + y^2 + z^2 - 6x + 2y - 6 < 0$.

25. (a) $z > 4x^2 + 9y^2$
    (b) $x^2 + 4y^2 + z^2 = 0$.

26. (a) $y^2 + 4z^2 = 4$, $0 \le x \le 2$
    (b) $9x^2 + 4y^2 + 36x - 8y = -60$.

In Exercises 27–31, identify the quadric surface whose equation is given.

27. $100x^2 + 225y^2 - 36z^2 = 0$.

28. $x^2 - z^2 + y = 0$.

29. $400x^2 + 25y^2 + 16z^2 = 400$.

30. $4x^2 - y^2 + 4z^2 = 4$.

31. $-16x^2 - 100y^2 + 25z^2 = 400$.

32. Identify the surface by completing the squares:
    (a) $x^2 + 4y^2 - z^2 - 6x + 8y + 4z = 0$
    (b) $x^2 + y^2 + z^2 + 6x - 4y + 12z = 0$.

33. Find the work done by a constant force $\mathbf{F} = 3\mathbf{i} - 4\mathbf{j} + \mathbf{k}$ (pounds) acting on a particle that moves along the line segment from $P(5, 7, 0)$ to $Q(6, 6, 6)$ (units in feet).

34. Two forces $F_1 = i - 3j + k$ and $F_2 = i + 2j + 2k$ (pounds) act on a particle as it moves in a straight line from $P(-1, -2, 3)$ to $Q(0, 2, 0)$ (units in feet). How much work is done?

In Exercises 35 and 36, sketch the graph of the curve, showing the direction of increasing t.

35. $r(t) = \langle t, t^2 + 1, 1 \rangle$.

36. $x = t, y = t, z = 2 \cos (\pi t/2), 0 \leq t \leq 2$.

37. Find the velocity, speed, acceleration, unit tangent vector, unit normal vector, and curvature when $t = 0$ for the motion given by $x = a \sin t$, $y = a \cos t$, $z = a \ln (\cos t)$ $(a > 0)$.

38. The position vector of a particle is
$$r(t) = \sin (2t)i + \cos (2t)j + 2e^t k.$$
(a) Find the velocity, acceleration, and speed as functions of t.
(b) Find the tangential and normal components of acceleration, and the curvature when $t = 0$.

In Exercises 39 and 40, find the arc length of the curve.

39. $x = 2t, y = 4 \sin 3t, z = 4 \cos 3t, 0 \leq t \leq 2\pi$.

40. $r(t) = \langle e^{-t}, \sqrt{2}t, e^t \rangle, 0 \leq t \leq \ln 2$.

41. Consider the curve whose position vector is
$$r(t) = \langle e^{-t}, e^{2t}, t^3 + 1 \rangle$$
Find parametric equations for the tangent line to the curve at the point where $t = 0$.

42. (a) Prove: If the speed is constant, then the velocity and acceleration vectors of a moving particle are orthogonal.
(b) Prove: If the tangential component of acceleration of a moving particle is zero at every point, then dr/dt has constant length.
(c) Prove: The curvature of a line in 3-space is zero at every point.

43. If $u = \langle 2t, 3, -t^2 \rangle$ and $v = \langle 0, t^2, t \rangle$, find:
(a) $\int_0^3 u \, dt$ (b) $d(u \times v)/dt$.

44. If $r(t) = \langle \cos (\pi e^t), \sin (\pi e^t), \pi t \rangle$, find the angle between the acceleration a and the velocity v when $t = 0$.

45. Use Cramer's Rule to solve
(a) $3x - y + z = 10$ (b) $2x + y = 0$
$\quad x + 2y + 3z = 11$ $\qquad 6x - 2y = 5.$
$\quad 2x - 2y - z = 2$

46. Convert from rectangular coordinates to: (i) cylindrical coordinates (ii) spherical coordinates.
(a) $(2, 2, 2\sqrt{6})$ (b) $(1, \sqrt{3}, 0)$.

47. Convert $(\sqrt{2}, \pi/4, 1)$ from cylindrical coordinates to
(a) rectangular coordinates
(b) spherical coordinates.

48. Express the equation in terms of: (i) cylindrical coordinates, (ii) spherical coordinates:
(a) $x^2 + y^2 = z$ (b) $x^2 - y^2 - z^2 = 0$.

49. Express the equation in terms of rectangular coordinates:
(a) $z = r^2 \cos 2\theta$ (b) $\rho^2 \sin \phi \cos \phi \cos \theta = 1$.

50. Sketch the set of points defined by the given conditions.
(a) $0 \leq \theta \leq \pi/2, 0 \leq r \leq \cos \theta, 0 \leq z \leq 2$ (cylindrical coordinates)
(b) $0 \leq \theta \leq \pi/2, 0 \leq \phi \leq \pi/4, 0 \leq \rho \leq 2 \sec \phi$ (spherical coordinates)
(c) $r = 2 \sin \theta, 0 \leq z \leq 2$ (cylindrical coordinates)
(d) $\rho = 2 \cos \phi$ (spherical coordinates).

16

partial derivatives

16.1 FUNCTIONS OF TWO VARIABLES

There are many familiar formulas in which a given variable depends on two or more other variables. For example, the area A of a triangle depends on the base length b and height h by the formula

$$A = \tfrac{1}{2}bh$$

We say that A is a *function* of the two variables b and h. Similarly, the volume V of a rectangular box depends on the length l, the width w, and the height h by the formula

$$V = lwh$$

We say that V is a function of the three variables l, w, and h.

The terminology and notation for functions of two or more variables is similar to that used for functions of one variable. For example, the expression

$$z = f(x, y)$$

means that z is a function of x and y in the sense that a unique value of the *dependent variable z* is determined by specifying values for the *independent variables x* and *y*. Similarly,

$$w = f(x, y, z)$$

means that the dependent variable w is uniquely determined by specifying values for the independent variables x, y, and z. For the time being, we will concentrate on functions of two variables. In a later section we will take up functions of three or more variables.

The functional relationship

$$z = f(x, y) \tag{1}$$

has a useful geometric interpretation. When values of the independent variables x and y are specified, a point (x, y) in the xy-plane is determined. Thus, the dependent variable z in (1) may be viewed as a numerical value associated with the point (x, y). This suggests the following definition.

16.1.1　DEFINITION　A *function f of two real variables,* x and y, is a rule that assigns a unique real number $f(x, y)$ to each point (x, y) in some set D of the xy-plane.

The set D in this definition is called the **domain** of the function; it is the set of points in the xy-plane at which the function is defined. If $f(x, y)$ is specified by a formula and the domain of f is not stated explicitly, then it is understood that the domain consists of all points at which the formula makes sense and yields a real number for the value of the function.

▶ Example 1　Let

$$f(x, y) = 3x^2 \sqrt{y} - 1$$

Find $f(1, 4)$, $f(0, 9)$, $f(t^2, t)$, $f(ab, 9b)$, and the domain of f.

Solution.　By substitution

$$\begin{aligned}
f(1, 4) &= 3(1)^2 \sqrt{4} - 1 = 5, \\
f(0, 9) &= 3(0)^2 \sqrt{9} - 1 = -1 \\
f(t^2, t) &= 3(t^2)^2 \sqrt{t} - 1 = 3t^4 \sqrt{t} - 1 \\
f(ab, 9b) &= 3(ab)^2 \sqrt{9b} - 1 = 9a^2 b^2 \sqrt{b} - 1
\end{aligned}$$

Because of the $\sqrt{y}$, we must have $y \geq 0$ to avoid imaginary values for $f(x, y)$. Thus, the domain of f consists of all points in the xy-plane that are on or above the x-axis. (See Figure 16.1.1.)　◀

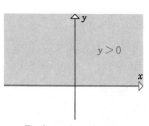

The heavy boundary line is included in the domain.

Figure 16.1.1

▶ Example 2　Sketch the domain of the function $f(x, y) = \ln(x^2 - y)$.

Solution.　$\ln(x^2 - y)$ is defined only when $0 < x^2 - y$ or $y < x^2$. To sketch this region we use the fact that the curve $y = x^2$ separates the region where $y < x^2$ from the region where $y > x^2$. To determine the region where $y < x^2$ holds, we can select an arbitrary "test point" off the boundary $y = x^2$ and determine whether $y < x^2$ or $y > x^2$ at the test point. For example, if we choose the test point $(x, y) = (0, 1)$, then $x^2 = 0, y = 1$, so that this point lies in the region where $y > x^2$. Thus, the region where $y < x^2$ is the one that does *not* contain the test point (Figure 16.1.2). ◄

16.1.2 DEFINITION   We define the *graph* of a function $f(x, y)$ to be the graph of the equation $z = f(x, y)$.

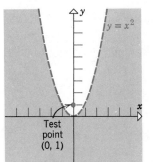

The dotted boundary does not belong to the domain.

Figure 16.1.2

▶ **Example 3** Describe the graph of the function $f(x, y) = 1 - x - \frac{1}{2}y$.

Solution. By definition, the graph of the given function is the graph of the equation

$$z = 1 - x - \tfrac{1}{2}y$$

or equivalently

$$x + \tfrac{1}{2}y + z = 1$$

which is a plane. A triangular portion of the plane can be sketched by plotting the intersections with the coordinate axes and joining them with line segments (Figure 16.1.3). ◄

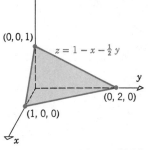

Figure 16.1.3

▶ **Example 4** Sketch the graph of the function $f(x, y) = \sqrt{1 - x^2 - y^2}$.

Solution. By definition, the graph of the given function is the graph of the equation

$$z = \sqrt{1 - x^2 - y^2} \tag{2}$$

After squaring both sides, this can be rewritten as

$$x^2 + y^2 + z^2 = 1$$

which represents a sphere of radius 1, centered at the origin. Since (2) imposes the added condition that $z \geq 0$, the graph is just the upper hemisphere (Figure 16.1.4). ◄

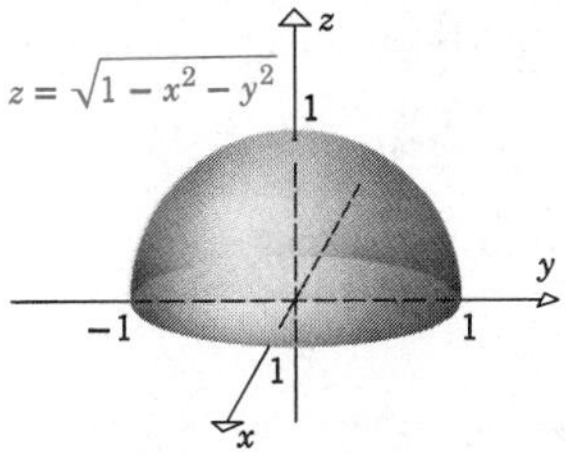

▶ **Example 5** Sketch the graph of the function $f(x, y) = -\sqrt{x^2 + y^2}$.

Solution. The graph of the given function is the graph of the equation

$$z = -\sqrt{x^2 + y^2} \tag{3}$$

Figure 16.1.4   After squaring, we obtain

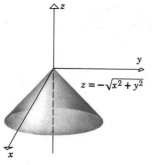

$$z^2 = x^2 + y^2$$

which is the equation of a right circular cone [(4) of Section 15.8]. Since (3) imposes the condition that $z \leq 0$, the graph is just the lower nappe of the cone (Figure 16.1.5). ◀

$z = -\sqrt{x^2 + y^2}$

Figure 16.1.5

▶ Example 6   Sketch the graph of the function $f(x, y) = x^2 + \frac{1}{4}y^2$.

The graph of f is the graph of the equation

$$z = x^2 + \frac{1}{4}y^2$$

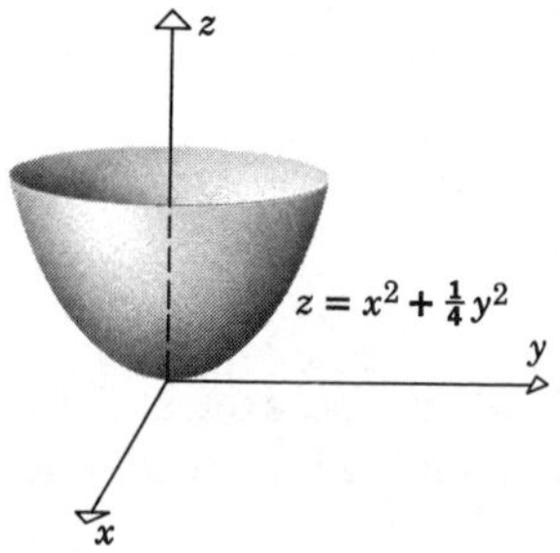

As discussed in Section 15.8 [Equation (5a)], this is an elliptic paraboloid (Figure 16.1.6). ◀

$z = x^2 + \frac{1}{4}y^2$

Figure 16.1.6

We are all familiar with topographic (or contour) maps in which a three-dimensional landscape, such as a mountain range, is represented by two-dimensional contour lines or curves of constant elevation. Consider, for example, the model hill and its contour map shown in Figure 16.1.7. The contour map is constructed by passing planes of constant elevation through the hill, projecting the resulting contours onto a flat surface, and labeling the contours with their elevations. In Figure 16.1.7, note how the two gullies appear as indentations in the contour lines and how the curves are close together on the contour map where the hill has a steep slope and become more widely spaced where the slope is gradual.

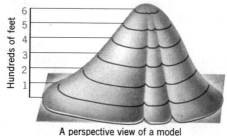

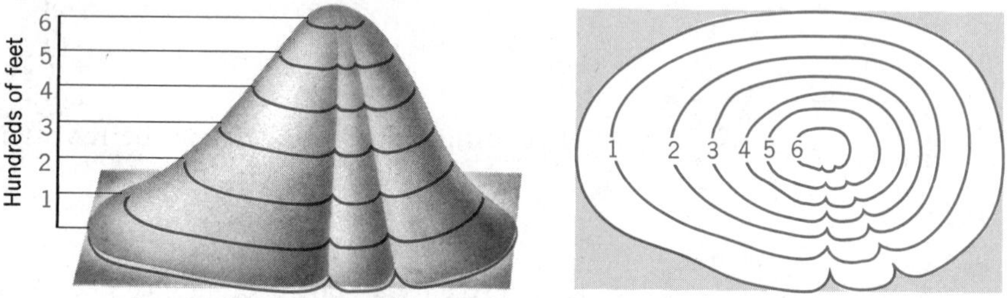

Figure 16.1.7

A perspective view of a model hill with two gullies.

A contour map of the model hill.

Contour maps are useful for studying functions of two variables. If the surface

$$z = f(x, y) \tag{4}$$

is cut by the horizontal plane

$$z = k \tag{5}$$

and if the resulting curve is projected onto the xy-plane, then we obtain what is called the **level curve of height k** for the function. It follows from (4) and (5) that this level curve has the equation

$$f(x, y) = k$$

(See Figure 16.1.8.)

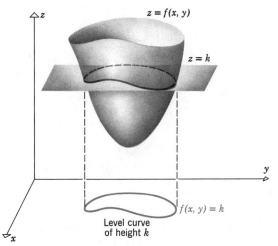

Figure 16.1.8

▶ **Example 7** The function $f(x, y) = x^2 + \frac{1}{4}y^2$ has the paraboloid $z = x^2 + \frac{1}{4}y^2$ for its graph (Figure 16.1.9). The level curves have equations of the form

$$x^2 + \tfrac{1}{4}y^2 = k \tag{6}$$

For $k > 0$ these are ellipses; for $k = 0$, it is the single point $(0, 0)$; and for $k < 0$ there are no level curves, since (6) is not satisfied by any real values of x and y. Some sample level curves are shown in Figure 16.1.9. ◀

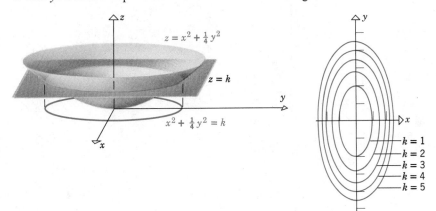

Figure 16.1.9

▶ Example 8   The function $f(x, y) = 2 - x - y$ has the plane $z = 2 - x - y$ as its graph. The level curves have equations of the form $2 - x - y = k$, or $y = -x + (2 - k)$. These form a family of parallel lines of slope -1 (Figure 16.1.10). ◀

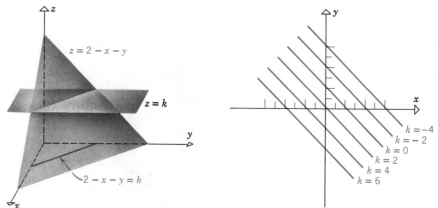

Figure 16.1.10

▶ Example 9   The function $f(x, y) = y^2 - x^2$ has the hyperbolic paraboloid (saddle curve) $z = y^2 - x^2$ as its graph. The level curves have equations of the form $y^2 - x^2 = k$. For $k > 0$ these curves are hyperbolas opening along lines parallel to the y-axis; for $k < 0$ they are hyperbolas opening along lines parallel to the x-axis; and for $k = 0$ the level curve consists of the intersecting lines $y + x = 0$ and $y - x = 0$ (Figure 16.1.11). ◀

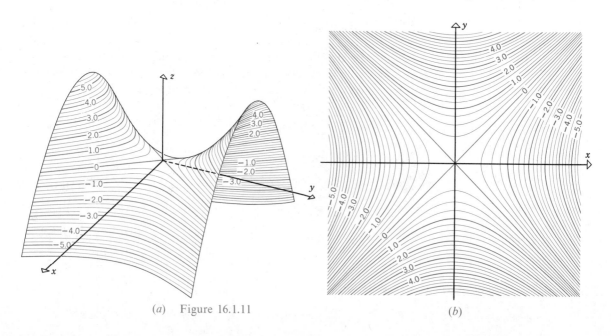

(a)   Figure 16.1.11   (b)

COMPUTER GRAPHICS (OPTIONAL) In recent years, computer technology has been used to generate graphic representations of mathematical surfaces in three dimensions. The purpose of this brief section is to illustrate various forms of such computer graphics. It is not our objective to study this topic in detail, but simply to acquaint the reader with this extremely valuable tool.

REPRESENTATIONS OF SOLIDS Computers can be used to generate various kinds of solids (Figure 16.1.12).

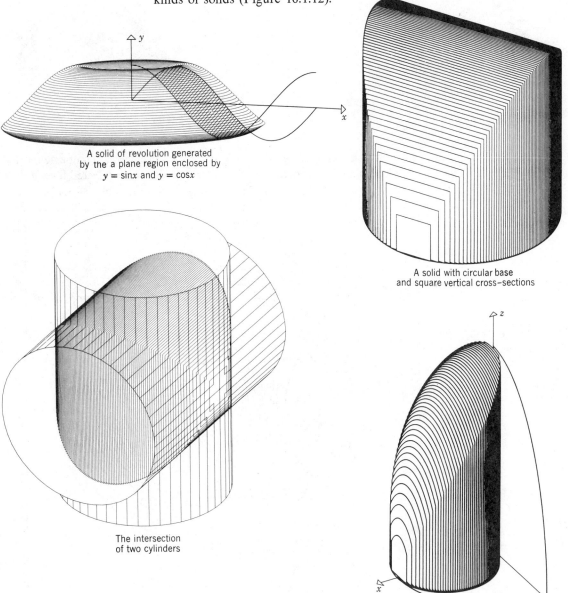

A solid of revolution generated by the a plane region enclosed by $y = \sin x$ and $y = \cos x$

A solid with circular base and square vertical cross–sections

The intersection of two cylinders

The intersection of a cylinder and a paraboloid

Figure 16.1.12

MESH PERSPECTIVES A portion of a surface $z = f(x, y)$ can be viewed in perspective and delineated by a rectangular mesh over a rectangular portion of the xy-plane. By altering program parameters, it is possible to change the section of surface shown, the point of view, and the apparent distance from the observer to the origin (Figure 16.1.13).

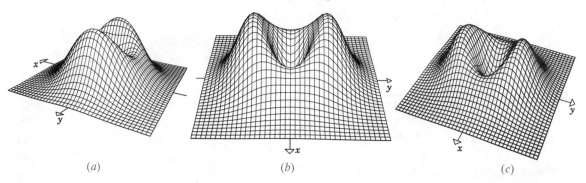
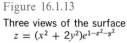

(a) (b) (c)

Figure 16.1.13

Three views of the surface
$z = (x^2 + 2y^2)e^{1-x^2-y^2}$

2D CONTOUR MAPS A two-dimensional plot of level curves can be generated with labels and numerical data. Some programs permit selected level curves to be drawn more heavily for emphasis or clarity (Figure 16.1.14).

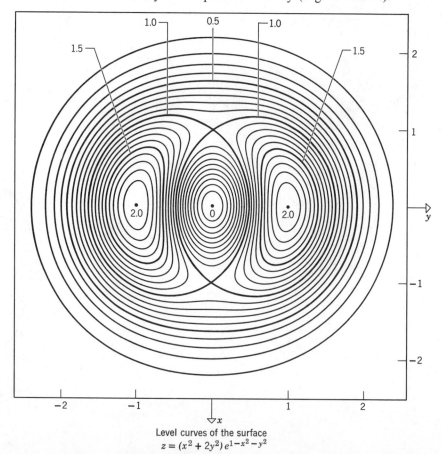

Figure 16.1.14

Level curves of the surface
$z = (x^2 + 2y^2)e^{1-x^2-y^2}$

3D CONTOUR MAPS A perspective view of a two-dimensional contour map can be generated to relate the map to the corresponding portion of the surface (Figure 16.1.15).

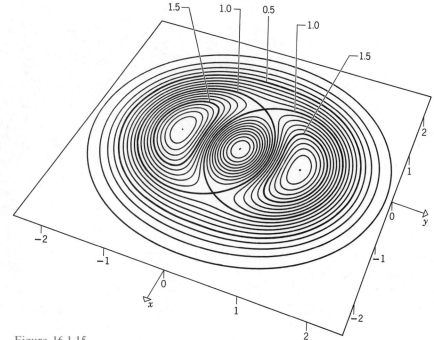

Figure 16.1.15

POLAR MESH PERSPECTIVES A surface $z = f(r, \theta)$ in cylindrical coordinates can be delineated by a mesh corresponding to constant values of r and θ (Figure 16.1.16).

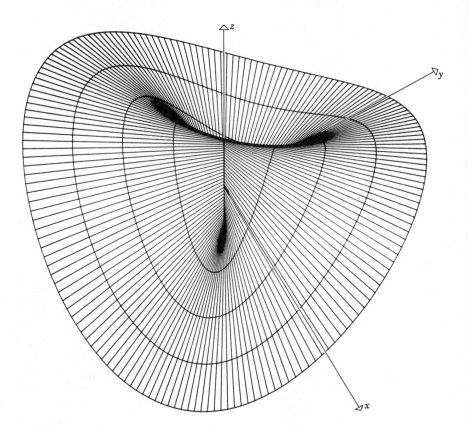

Figure 16.1.16

► Exercise Set 16.1

1. Let $f(x, y) = x^2 y + 1$. Find
   (a) $f(2, 1)$ (b) $f(1, 2)$ (c) $f(0, 0)$
   (d) $f(1, -3)$ (e) $f(3a, a)$ (f) $f(ab, a - b)$.

2. Let $f(x, y) = x + \sqrt[3]{xy}$. Find
   (a) $f(t, t^2)$ (b) $f(x, x^2)$ (c) $f(2y^2, 4y)$.

3. Let $f(x, y) = xy + 3$. Find
   (a) $f(x + y, x - y)$ (b) $f(xy, 3x^2 y^3)$.

4. Let $g(x) = x \sin x$. Find
   (a) $g\left(\dfrac{x}{y}\right)$ (b) $g(xy)$ (c) $g(x - y)$.

5. Find $F(g(x), h(y))$ if $F(x, y) = xe^{xy}$, $g(x) = x^3$, and $h(y) = 3y + 1$.

6. Find $g(u(x, y), v(x, y))$ if $g(x, y) = y \sin(x^2 y)$, $u(x, y) = x^2 y^3$, and $v(x, y) = \pi xy$.

7. Let $f(x, y) = x + 3x^2 y^2$, $x(t) = t^2$, and $y(t) = t^3$. Find
   (a) $f(x(t), y(t))$ (b) $f(x(0), y(0))$
   (c) $f(x(2), y(2))$.

In Exercises 8–13, sketch the domain of f. Use solid lines for portions of the boundary included in the domain and dashed lines for portions not included.

8. $f(x, y) = xy\sqrt{y - 1}$.

9. $f(x, y) = \ln(1 - x^2 - y^2)$.

10. $f(x, y) = \sqrt{x^2 + y^2 - 4}$.

11. $f(x, y) = \dfrac{1}{x - y^2}$.

12. $f(x, y) = \ln xy$.

13. $f(x, y) = \sqrt{\dfrac{x^2 + y^2}{x^2 - y^2}}.$

In Exercises 14–25, sketch the graph of f.

14. $f(x, y) = 3$.

15. $f(x, y) = 4 - 2x - 4y$.

16. $f(x, y) = \sqrt{9 - x^2 - y^2}$.

17. $f(x, y) = \sqrt{x^2 + y^2}$.

18. $f(x, y) = x^2 + y^2$.

19. $f(x, y) = x^2 - y^2$.

20. $f(x, y) = 4 - x^2 - y^2$.

21. $f(x, y) = -\sqrt{1 - x^2/4 - y^2/9}$.

22. $f(x, y) = \sqrt{x^2 + y^2 - 1}$.

23. $f(x, y) = \sqrt{x^2 + y^2 + 1}$.

24. $f(x, y) = x^2$.

25. $f(x, y) = y + 1$.

In Exercises 26–33, sketch the level curve $z = k$ for the specified values of k.

26. $z = 3x + y$; $k = -2, -1, 0, 1, 2$.

27. $z = x^2 + y^2$; $k = 0, 1, 2, 3, 4$.

28. $z = y/x$; $k = -2, -1, 0, 1, 2$.

29. $z = x^2 + y$; $k = -2, -1, 0, 1, 2$.

30. $z = x^2 + 9y^2$; $k = 0, 1, 2, 3, 4$.

31. $z = x^2 - y^2$; $k = -2, -1, 0, 1, 2$.

32. $z = y \csc x$; $k = -2, -1, 0, 1, 2$.

33. $z = \sqrt{\dfrac{x + y}{x - y}}$; $k = 0, 1, 2, 3, 4$.

34. Let $f(x, y) = yx^2 + 1$. Find the equation of the level curve that passes through the point

    (a) $(1, 2)$ (b) $(-2, 4)$ (c) $(0, 0)$.

35. Let $f(x, y) = x^2 - 2x^3 + 3xy$. Find an equation of the level curve that passes through the point

    (a) $(-1, 1)$ (b) $(0, 0)$ (c) $(2, -1)$.

36. Let $f(x, y) = ye^x$. Find an equation for the level curve that passes through the point

    (a) $(\ln 2, 1)$ (b) $(0, 3)$ (c) $(1, -2)$.

37. If $V(x, y)$ is the voltage or potential at a point (x, y) in the xy-plane, then the level curves of V are called *equipotential curves*. Along such a curve, the voltage remains constant. Given that

$$V(x, y) = \frac{8}{\sqrt{16 + x^2 + y^2}}$$

sketch the equipotential curves at which $V = 2.0$, $V = 1.0$, and $V = 0.5$.

38. If $T(x, y)$ is the temperature at a point (x, y) on a thin metal plate in the xy-plane, then the level curves of T are called *isothermal curves*. All points on such a curve are at the same temperature. Suppose that a plate occupies the first quadrant and $T(x, y) = xy$.

    (a) Sketch the isothermal curves on which

$$T = 1, \quad T = 2, \quad T = 3$$

    (b) An ant, initially at $(1, 4)$, wants to walk on the plate so the temperature along its path remains constant. What path should the ant take?

16.2 PARTIAL DERIVATIVES

In this section we will study derivatives associated with functions of two variables.

Let f be a function of x and y. If we hold y constant, say $y = y_0$, and view x as a variable, then $f(x, y_0)$ is a function of x alone. If this function is differentiable at $x = x_0$, then the value of this derivative is denoted by

$$f_x(x_0, y_0) \tag{1}$$

and is called the **partial derivative of f with respect to x** at the point (x_0, y_0). Similarly, if we hold x constant, say $x = x_0$, then $f(x_0, y)$ is a function of y alone. If this function is differentiable at $y = y_0$, then the value of this derivative is denoted by

$$f_y(x_0, y_0) \tag{2}$$

and is called the **partial derivative of f with respect to y** at (x_0, y_0).

The values of $f_x(x_0, y_0)$ and $f_y(x_0, y_0)$ are usually obtained by finding expressions for $f_x(x, y)$ and $f_y(x, y)$ at a general point (x, y) and then substituting $x = x_0$ and $y = y_0$ in these expressions. To obtain $f_x(x, y)$ we differentiate $f(x, y)$ with respect to x, *treating y as a constant;* and to obtain $f_y(x, y)$ we differentiate $f(x, y)$ with respect to y, *treating x as a constant.*

▶ **Example 1** Find $f_x(1, 2)$ and $f_y(1, 2)$ if

$$f(x, y) = 2x^3y^2 + 2y + 4x$$

Solution. Treating y as a constant and differentiating with respect to x, we obtain

$$f_x(x, y) = 6x^2y^2 + 4$$

Treating x as a constant and differentiating with respect to y, we obtain

$$f_y(x, y) = 4x^3y + 2$$

Substituting $x = 1$ and $y = 2$ in these partial-derivative formulas yields

$$f_x(1, 2) = 6(1)^2(2)^2 + 4 = 28$$
$$f_y(1, 2) = 4(1)^3(2) + 2 = 10 \qquad ◀$$

The partial derivatives f_x and f_y are also denoted by the symbols*

---

*The symbol ∂ is called a partial derivative sign. It is not a letter of any alphabet, but rather an invented mathematical symbol.

$$\frac{\partial f}{\partial x} \quad \text{and} \quad \frac{\partial f}{\partial y}$$

and if a dependent variable $z = f(x, y)$ is introduced, then the symbols

$$\frac{\partial z}{\partial x} \quad \text{and} \quad \frac{\partial z}{\partial y}$$

may be used. Some typical notations for the partial derivatives at a point (x_0, y_0) are

$$\frac{\partial f}{\partial x}\bigg|_{x=x_0,\, y=y_0} \qquad \frac{\partial z}{\partial y}\bigg|_{(x_0,\, y_0)} \qquad \frac{\partial f}{\partial x}\bigg|_{(x_0,\, y_0)} \qquad \frac{\partial f}{\partial x}(x_0, y_0)$$

▶ **Example 2** Find $\partial z/\partial x$ and $\partial z/\partial y$ if

$$z = x^4 \sin (xy^3)$$

Solution.

$$\frac{\partial z}{\partial x} = \frac{\partial}{\partial x}[x^4 \sin (xy^3)]$$

$$= x^4 \frac{\partial}{\partial x}[\sin (xy^3)] + \sin (xy^3) \cdot \frac{\partial}{\partial x}(x^4)$$

$$= x^4 \cos (xy^3) \cdot y^3 + \sin (xy^3) \cdot 4x^3$$

$$= x^4 y^3 \cos (xy^3) + 4x^3 \sin (xy^3)$$

$$\frac{\partial z}{\partial y} = \frac{\partial}{\partial y}[x^4 \sin (xy^3)]$$

$$= x^4 \frac{\partial}{\partial y}[\sin (xy^3)] + \sin (xy^3) \cdot \frac{\partial}{\partial y}(x^4)$$

$$= x^4 \cos (xy^3) \cdot 3xy^2 + \sin (xy^3) \cdot 0$$

$$= 3x^5 y^2 \cos (xy^3) \qquad \blacktriangleleft$$

The partial derivatives of $f(x, y)$ have a useful geometric interpretation. Let P be a point on the surface

$$z = f(x, y)$$

If y is held constant, say $y = y_0$, and x is allowed to vary, then the point P moves along the curve C_1 that is the intersection of the surface with the vertical plane $y = y_0$ (Figure 16.2.1a). Thus, the partial derivative $f_x(x_0, y_0)$ can be interpreted as the slope (change in z per unit change in x) of the tangent line to the curve C_1 at the point (x_0, y_0). Similarly, if x is held constant, say $x = x_0$, and y is allowed to vary, then the point P moves along the curve C_2 that is the intersection of the surface with the vertical plane $x = x_0$. Thus, the partial derivative $f_y(x_0, y_0)$ can be interpreted as the slope of the tangent line (change in z per unit change in y) to the curve C_2 at the point (x_0, y_0) (Figure 16.2.1b).

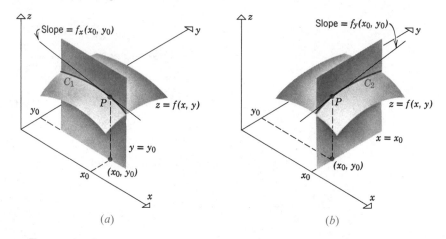

Figure 16.2.1   (a)   (b)

▶ Example 3   A point Q moves along the intersection of the sphere $x^2 + y^2 + z^2 = 1$ with the plane $x = \frac{2}{3}$. At what rate is z changing with y when the point is at $P(\frac{2}{3}, \frac{1}{3}, \frac{2}{3})$?

Solution.   Since the z-coordinate of the point $P(\frac{2}{3}, \frac{1}{3}, \frac{2}{3})$ is positive, this point lies on the upper hemisphere

$$z = \sqrt{1 - x^2 - y^2} \tag{3}$$

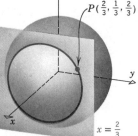

Figure 16.2.2

(Figure 16.2.2). Since $x = \frac{2}{3}$, $y = \frac{1}{3}$ at P, the rate at which z changes with y at this point (as Q moves along the circle of intersection) is

$$\frac{\partial z}{\partial y}\bigg|_{x = \frac{2}{3}, y = \frac{1}{3}}$$

From (3) we obtain

$$\frac{\partial z}{\partial y} = \frac{\partial}{\partial y}[(1 - x^2 - y^2)^{1/2}] = \frac{1}{2}(1 - x^2 - y^2)^{-1/2}(-2y)$$

$$= -\frac{y}{\sqrt{1 - x^2 - y^2}}$$

so that

$$\frac{\partial z}{\partial y}\bigg|_{x = \frac{2}{3}, y = \frac{1}{3}} = -\frac{\frac{1}{3}}{\sqrt{1 - (\frac{2}{3})^2 - (\frac{1}{3})^2}} = -\frac{1}{2}$$

Alternate Solution. Instead of solving $x^2 + y^2 + z^2 = 1$ explicitly for z as a function of x and y, we can obtain $\partial z / \partial y$ by implicit differentiation. Differentiating both sides of $x^2 + y^2 + z^2 = 1$ with respect to y and treating z as a function of x and y yields

$$\frac{\partial}{\partial y}[x^2 + y^2 + z^2] = \frac{\partial}{\partial y}[1]$$

$$2y + 2z\frac{\partial z}{\partial y} = 0$$

$$\frac{\partial z}{\partial y} = -\frac{y}{z}$$

Substituting the y- and z-coordinates of the point $(\frac{2}{3}, \frac{1}{3}, \frac{2}{3})$ yields

$$\frac{\partial z}{\partial y} = -\frac{\frac{1}{3}}{\frac{2}{3}} = -\frac{1}{2}$$

which agrees with our previous calculation. ◀

▶ **Example 4** According to the ideal gas law of physics, the pressure P exerted by a confined gas is related to its volume V and its temperature T by

$$P = k\frac{T}{V}$$

where k is a constant depending on the amount of gas present and the units of measurement. Show that if the temperature of such a gas remains constant, then the pressure and volume satisfy the condition

$$\frac{\partial P}{\partial V} = -\frac{P}{V} \tag{4}$$

Solution. Treating T as a constant we obtain

$$\frac{\partial P}{\partial V} = \frac{\partial}{\partial V}\left(k\frac{T}{V}\right) = (kT)\left(-\frac{1}{V^2}\right) = -k\frac{T}{V} \cdot \frac{1}{V} = -\frac{P}{V} \qquad ◀$$

▶ **Example 5** The diagonal D of a rectangle is given by $D = \sqrt{x^2 + y^2}$, where x and y are the lengths of the sides.

(a) Find a formula for the instantaneous rate of change of D with respect to x if x varies while y remains constant.

(b) Suppose $y = 4$ in. Find the rate of change of D with respect to x at the instant when $x = 3$ in.

Solution (a).

$$\frac{\partial D}{\partial x} = \frac{1}{2}(x^2 + y^2)^{-1/2}(2x) = \frac{x}{\sqrt{x^2 + y^2}}$$

Solution (b).

$$\frac{\partial D}{\partial x}\bigg|_{x=3, \, y=4} = \frac{3}{\sqrt{3^2 + 4^2}} = \frac{3}{5}$$

Thus, the diagonal is increasing at a rate of $\frac{3}{5}$ inches in D per inch increase in x. ◀

Since the partial derivatives $\partial f/\partial x$ and $\partial f/\partial y$ are functions of x and y, each can have partial derivatives. This gives rise to four possible **second-order** partial derivatives of f, which are defined by

$$\frac{\partial^2 f}{\partial x^2} = \frac{\partial}{\partial x}\left(\frac{\partial f}{\partial x}\right), \qquad \frac{\partial^2 f}{\partial y^2} = \frac{\partial}{\partial y}\left(\frac{\partial f}{\partial y}\right)$$

$$\frac{\partial^2 f}{\partial x \, \partial y} = \frac{\partial}{\partial x}\left(\frac{\partial f}{\partial y}\right), \qquad \frac{\partial^2 f}{\partial y \, \partial x} = \frac{\partial}{\partial y}\left(\frac{\partial f}{\partial x}\right)$$

Henceforth, we will call $\partial f/\partial x$ and $\partial f/\partial y$ the **first-order** partial derivatives of f.

▶ **Example 6** Find the second-order partial derivatives of $f(x, y) = x^2 y^3 + x^4 y$.

Solution. We have

$$\frac{\partial f}{\partial x} = 2xy^3 + 4x^3 y$$

$$\frac{\partial f}{\partial y} = 3x^2 y^2 + x^4$$

so that

$$\frac{\partial^2 f}{\partial x^2} = \frac{\partial}{\partial x}\left(\frac{\partial f}{\partial x}\right) = \frac{\partial}{\partial x}(2xy^3 + 4x^3 y) = 2y^3 + 12x^2 y$$

$$\frac{\partial^2 f}{\partial y^2} = \frac{\partial}{\partial y}\left(\frac{\partial f}{\partial y}\right) = \frac{\partial}{\partial y}(3x^2 y^2 + x^4) = 6x^2 y$$

$$\frac{\partial^2 f}{\partial x \, \partial y} = \frac{\partial}{\partial x}\left(\frac{\partial f}{\partial y}\right) = \frac{\partial}{\partial x}(3x^2 y^2 + x^4) = 6xy^2 + 4x^3$$

$$\frac{\partial^2 f}{\partial y \, \partial x} = \frac{\partial}{\partial y}\left(\frac{\partial f}{\partial x}\right) = \frac{\partial}{\partial y}(2xy^3 + 4x^3 y) = 6xy^2 + 4x^3$$

◀

REMARK. The derivatives

$$\frac{\partial^2 f}{\partial y\,\partial x} \quad \text{and} \quad \frac{\partial^2 f}{\partial x\,\partial y}$$

are called the ***mixed second-order partial derivatives*** or ***mixed second-partials.*** For most functions that arise in applications these mixed partial derivatives are equal (as in the last example). In the next section we will state precise conditions under which equality holds.

By successively differentiating, we can obtain third-order partial derivatives and higher. Some possibilities are

$$\frac{\partial^3 f}{\partial x^3} = \frac{\partial}{\partial x}\left(\frac{\partial^2 f}{\partial x^2}\right), \qquad \frac{\partial^3 f}{\partial y^2\,\partial x} = \frac{\partial}{\partial y}\left(\frac{\partial^2 f}{\partial y\,\partial x}\right)$$

$$\frac{\partial^3 f}{\partial y\,\partial x^2} = \frac{\partial}{\partial y}\left(\frac{\partial^2 f}{\partial x^2}\right), \qquad \frac{\partial^4 f}{\partial y^2\,\partial x^2} = \frac{\partial}{\partial y}\left(\frac{\partial^3 f}{\partial y\,\partial x^2}\right)$$

Higher order partial derivatives can be denoted more compactly with subscript notation. For example,

$$\frac{\partial^2 f}{\partial y\,\partial x} = \frac{\partial}{\partial y}\left(\frac{\partial f}{\partial x}\right) = \frac{\partial}{\partial y}(f_x) = (f_x)_y$$

It is usual to drop the parentheses and write simply

$$\frac{\partial^2 f}{\partial y\,\partial x} = f_{xy}$$

Note that in "∂" notation the sequence of differentiations is obtained by reading from right to left, but in the subscript notation it is left to right. Some other examples are

$$f_{xx} = \frac{\partial^2 f}{\partial x^2}, \quad f_{yyx} = \frac{\partial^3 f}{\partial x\,\partial y^2}, \quad f_{xxyy} = \frac{\partial^4 f}{\partial y^2\,\partial x^2}$$

▶ **Example 7** Let $f(x, y) = y^2 e^x + y$. Find f_{xyy}.

Solution.

$$f_{xyy} = \frac{\partial^3 f}{\partial y^2\,\partial x} = \frac{\partial^2}{\partial y^2}\left(\frac{\partial f}{\partial x}\right) = \frac{\partial^2}{\partial y^2}(y^2 e^x) = \frac{\partial}{\partial y}(2y e^x) = 2e^x \qquad ◀$$

▶ Exercise Set 16.2

In Exercises 1–6, find $\partial z/\partial x$ and $\partial z/\partial y$.

1. $z = 3x^3y^2$.

2. $z = 4x^2 - 2y + 7x^4y^5$.

3. $z = 4e^{x^2y^3}$.

4. $z = \cos(x^5y^4)$.

5. $z = x^3 \ln(1 + xy^{-3/5})$. **6.** $z = e^{xy} \sin 4y^2$.

In Exercises 7–12, find $f_x(x, y)$ and $f_y(x, y)$

7. $f(x, y) = \sqrt{3x^5y - 7x^3y}$.

8. $f(x, y) = \dfrac{x + y}{x - y}$.

9. $f(x, y) = y^{-3/2} \tan^{-1}(x/y)$.

10. $f(x, y) = x^3e^{-y} + y^3 \sec \sqrt{x}$.

11. $f(x, y) = (y^2 \tan x)^{-4/3}$.

12. $f(x, y) = \cosh(\sqrt{x}) \sinh^2(xy^2)$.

13. Given $f(x, y) = 9 - x^2 - 7y^3$, find
(a) $f_x(3, 1)$ (b) $f_y(3, 1)$.

14. Given $f(x, y) = x^2ye^{xy}$, find
(a) $\dfrac{\partial f}{\partial x}\Big|_{(1,1)}$ (b) $\dfrac{\partial f}{\partial y}\Big|_{(1,1)}$.

15. Given $z = \sqrt{x^2 + 4y^2}$, find
(a) $\dfrac{\partial z}{\partial x}\Big|_{(1,2)}$ (b) $\dfrac{\partial z}{\partial y}\Big|_{(1,2)}$.

16. Given $w = x^2 \cos xy$, find
(a) $\dfrac{\partial w}{\partial x}(\tfrac{1}{2}, \pi)$ (b) $\dfrac{\partial w}{\partial y}(\tfrac{1}{2}, \pi)$.

In Exercises 17–20, calculate $\partial z/\partial x$ and $\partial z/\partial y$ using implicit differentiation. Leave your answers in terms of x, y, and z.

17. $(x^2 + y^2 + z^2)^{3/2} = 1$.

18. $\ln(2x^2 + y - z^3) = x$.

19. $x^2 + z \sin xyz = 0$.

20. $e^{xy} \sinh z - z^2x + 1 = 0$.

In Exercises 21–26, find f_{xx}, f_{yy}, f_{xy}, and f_{yx}.

21. $f(x, y) = 4x^2 - 8xy^4 + 7y^5 - 3$.

22. $f(x, y) = \sqrt{x^2 + y^2}$.   **23.** $f(x, y) = e^x \cos y$.

24. $f(x, y) = e^{x - y^z}$.   **25.** $f(x, y) = \ln(4x - 5y)$.

26. $f(x, y) = (x^2 - y^2)/(x^2 + y^2)$.

27. Given $f(x, y) = x^3y^5 - 2x^2y + x$, find
(a) f_{xxy} (b) f_{yxy} (c) f_{yyy}.

28. Given $z = (2x - y)^5$, find
(a) $\dfrac{\partial^3 z}{\partial y \, \partial x \, \partial y}$ (b) $\dfrac{\partial^3 z}{\partial x^2 \, \partial y}$ (c) $\dfrac{\partial^4 z}{\partial x^2 \, \partial y^2}$.

29. Given $f(x, y) = y^3e^{-5x}$, find
(a) $f_{xyy}(0, 1)$ (b) $f_{xxx}(0, 1)$ (c) $f_{yyxx}(0, 1)$.

30. Given $w = e^y \cos x$, find
(a) $\dfrac{\partial^3 w}{\partial y^2 \, \partial x}\Big|_{(\pi/4, 0)}$ (b) $\dfrac{\partial^3 w}{\partial x^2 \, \partial y}\Big|_{(\pi/4, 0)}$.

31. Express the following derivatives in "∂" notation.
(a) f_{xxx} (b) f_{xyy}
(c) f_{yyxx} (d) f_{xyyy}.

32. Express the following derivatives in "subscript" notation.
(a) $\dfrac{\partial^3 f}{\partial y^2 \, \partial x}$ (b) $\dfrac{\partial^4 f}{\partial x^4}$
(c) $\dfrac{\partial^4 f}{\partial y^2 \, \partial x^2}$ (d) $\dfrac{\partial^5 f}{\partial x^2 \, \partial y^3}$.

33. Show that the following functions satisfy
$$\frac{\partial^2 f}{\partial x^2} + \frac{\partial^2 f}{\partial y^2} = 0$$
(This is called **Laplace's equation.**)
(a) $f(x, y) = e^x \sin y + e^y \cos x$
(b) $f(x, y) = \ln(x^2 + y^2)$
(c) $f(x, y) = \tan^{-1} \dfrac{2xy}{x^2 - y^2}$.

34. Show that $u(x, y)$ and $v(x, y)$ satisfy
$$\frac{\partial u}{\partial x} = \frac{\partial v}{\partial y} \quad \text{and} \quad \frac{\partial u}{\partial y} = -\frac{\partial v}{\partial x}$$
(These are called the **Cauchy-Riemann equations.**)
(a) $u = x^2 - y^2$, $v = 2xy$
(b) $u = e^x \cos y$, $v = e^x \sin y$
(c) $u = \ln(x^2 + y^2)$, $v = 2 \tan^{-1}(y/x)$.

35. A point moves along the intersection of the paraboloid $z = x^2 + 3y^2$ and the plane $x = 2$. At what rate is z changing with y when the point is at $(2, 1, 7)$?

36. A point moves along the intersection of the surface $z = \sqrt{29 - x^2 - y^2}$ and the plane $y = 3$. At what rate is z changing with x when the point is at $(4, 3, 2)$?

37. Find the slope of the tangent line at $(-1, 1, 5)$ to the curve of intersection of the surface $z = x^2 + 4y^2$ and
(a) the plane $x = -1$
(b) the plane $y = 1$.

38. Find the slope of the tangent line at (2, 1, 2) to the curve of intersection of the surface $x^2 + y^2 + z^2 = 9$ and
(a) the plane $x = 2$ (b) the plane $y = 1$.

39. The volume V of a right-circular cylinder is given by $V = \pi r^2 h$, where r is the radius and h is the height.
(a) Find a formula for the instantaneous rate of change of V with respect to r if h remains constant.
(b) Find a formula for the instantaneous rate of change of V with respect to h if r remains constant.
(c) Suppose h has a constant value of 4 in., but r varies. Find the rate of change of V with respect to r at the instant when $r = 6$ in.
(d) Suppose r has a constant value of 8 in., but h varies. Find the instantaneous rate of change of V with respect to h at the instant when $h = 10$ in.

40. The volume V of a right-circular cone is given by

$$V = \frac{\pi}{24} d^2 \sqrt{4s^2 - d^2}$$

where s is the slant height and d is the diameter of the base.
(a) Find a formula for the instantaneous rate of change of V with respect to s if d remains constant.
(b) Find a formula for the instantaneous rate of change of V with respect to d if s remains constant.
(c) Suppose d has a constant value of 16 cm, but s varies. Find the rate of change of V with respect to s at the instant when $s = 10$ cm.
(d) Suppose s has a constant value of 10 cm, but d varies. Find the rate of change of V with respect to d at the instant when $d = 16$ cm.

41. According to the ideal gas law (Example 4), the pressure, temperature, and volume of a gas are related by $P = kT/V$. Suppose that for a certain gas, $k = 10$.
(a) Find the instantaneous rate of change of pressure (lb/in.²) with respect to temperature if the temperature is 80°F and the volume remains fixed at 50 in.³.

(b) Find the instantaneous rate of change of volume with respect to pressure if the volume is 50 in.³ and the temperature remains fixed at 80°F.

42. Find parametric equations for the tangent line at (1, 3, 3) to the curve of intersection of the surface $z = x^2 y$ and
(a) the plane $x = 1$
(b) the plane $y = 3$.

43. Suppose

$$\sin (x + z) + \sin (x - y) = 1$$

Use implicit differentiation to find $\partial z/\partial x$, $\partial z/\partial y$, and $\partial^2 z/\partial x \partial y$ in terms of x, y, and z.

44. The volume of a right-circular cone of radius r and height h is $V = \frac{1}{3}\pi r^2 h$. Show that if the height remains constant while the radius changes, then the volume satisfies

$$\frac{\partial V}{\partial r} = \frac{2V}{r}$$

45. The temperature at a point (x, y) on a metal plate in the xy-plane is $T(x, y) = x^3 + 2y^2 + x$ degrees. Assume that distance is measured in centimeters and find the rate at which temperature changes with distance if we start at the point (1, 2) and move
(a) to the right and parallel to the x-axis
(b) upward and parallel to the y-axis.

46. When two resistors having resistances R_1 ohms and R_2 ohms are connected in parallel, their combined resistance R in ohms is
$$R = R_1 R_2/(R_1 + R_2).$$
Show that

$$\frac{\partial^2 R}{\partial R_1{}^2} \frac{\partial^2 R}{\partial R_2{}^2} = \frac{4R^2}{(R_1 + R_2)^4}$$

47. Prove: If $u(x, y)$ and $v(x, y)$ each have equal mixed second partials, and if u and v satisfy the Cauchy-Riemann equations (Exercise 34), then u and v both satisfy Laplace's equation (Exercise 33).

48. Recall that for a function of one variable the derivative $f'(x)$ can be expressed as the limit

$$f'(x) = \lim_{h \to 0} \frac{f(x + h) - f(x)}{h}$$

Express $f_x(x, y)$ and $f_y(x, y)$ as limits.

16.3   LIMITS, CONTINUITY, AND DIFFERENTIABILITY

The purpose of this section is to introduce notions of limit, continuity, and differentiability for functions of two variables. We will not go into great detail; our objective is to develop the basic ideas accurately, and to obtain results needed in later sections. A more extensive study of these topics is usually given in advanced calculus.

Let us begin with some terminology. If D is a set of points in the xy-plane, then a point (x_0, y_0) is called an ***interior point*** of D if there is *some* circular disk with positive radius, centered at (x_0, y_0), and containing only points in D (Figure 16.3.1). A point (x_0, y_0) is called a ***boundary point*** of D if *every* circular disk with positive radius and centered at (x_0, y_0) contains both points in D and points not in D (Figure 16.3.1). A boundary point of D may or may not belong to the set D. A set D is called ***open*** if it contains *none* of its boundary points and ***closed*** if it contains *all* of its boundary points. (As an analogy, an open interval on a line contains neither of its endpoints, while a closed interval contains both of its endpoints.)

We will now show how to define limits for functions of two variables. For a function f of one variable, the statement,

$$\lim_{x \to x_0} f(x) = L$$

means that the value of $f(x)$ can be made arbitrarily close to L by making x sufficiently close to (but different from) x_0. Similarly, the statement

$$\lim_{(x,y) \to (x_0, y_0)} f(x, y) = L$$

means that the value of $f(x, y)$ can be made arbitrarily close to L by making the point (x, y) sufficiently close to (but different from) (x_0, y_0). In other words, if we pick any positive number ϵ, we must be able to make $f(x, y)$ lie within ϵ units of L, by restricting (x, y) sufficiently close to (but different from) (x_0, y_0). This idea is captured precisely in the following formal definition (see also Figure 16.3.2).

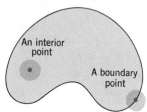

An interior point

A boundary point

Figure 16.3.1

16.3.1   DEFINITION   Let f be a function of two variables, and suppose that f is defined at each point in some circular region centered at (x_0, y_0), except possibly at (x_0, y_0) itself. We will write

$$\lim_{(x,y) \to (x_0, y_0)} f(x, y) = L \tag{1}$$

if given any number $\epsilon > 0$, we can find a number $\delta > 0$ such that $f(x, y)$ satisfies

$$|f(x, y) - L| < \epsilon$$

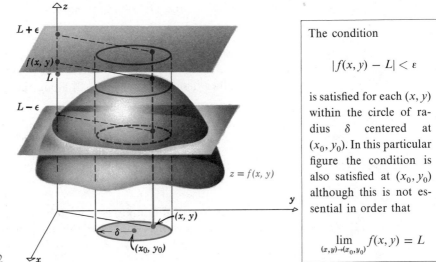

The condition

$$|f(x, y) - L| < \varepsilon$$

is satisfied for each (x, y) within the circle of radius δ centered at (x_0, y_0). In this particular figure the condition is also satisfied at (x_0, y_0) although this is not essential in order that

$$\lim_{(x,y)\to(x_0,y_0)} f(x, y) = L$$

Figure 16.3.2

whenever the distance between (x, y) and (x_0, y_0) satisfies the condition

$$0 < \sqrt{(x - x_0)^2 + (y - y_0)^2} < \delta$$

Equation (1) can also be written

$$f(x, y) \to L \quad \text{as} \quad (x, y) \to (x_0, y_0)$$

Recall that if

$$\lim_{x \to x_0} f(x) = L$$

then $f(x) \to L$ as $x \to x_0$ from the left and the right. Similarly, if

$$\lim_{(x,y)\to(x_0,y_0)} f(x, y) = L$$

then it can be shown that $f(x, y) \to L$ as $(x, y) \to (x_0, y_0)$ along *any* path in the domain of f leading to (x_0, y_0). Thus, if $f(x, y)$ approaches different values along two paths leading to (x_0, y_0), then

$$\lim_{(x,y)\to(x_0,y_0)} f(x, y)$$

does not exist.

▶ Example 1 We will show that

$$\lim_{(x,y)\to(0,0)} \frac{x^2 - y^2}{x^2 + y^2}$$

does not exist. If $(x, y) \to (0, 0)$ along the x-axis ($y = 0$), then

$$\lim_{(x,y)\to(0,0)} \frac{x^2 - y^2}{x^2 + y^2} = \lim_{(x,y)\to(0,0)} \frac{x^2}{x^2} = \lim_{(x,y)\to(0,0)} 1 = 1$$

On the other hand, if $(x, y) \to (0, 0)$ along the y-axis ($x = 0$), then

$$\lim_{(x,y)\to(0,0)} \frac{x^2 - y^2}{x^2 + y^2} = \lim_{(x,y)\to(0,0)} \frac{-y^2}{y^2} = \lim_{(x,y)\to(0,0)} (-1) = -1$$

Since the limiting values are different along the two paths, the limit does not exist. ◄

The definition of continuity for functions of two variables is similar to the definition for functions of one variable (see Definition 3.7.1).

16.3.2 DEFINITION A function f of two variables is called ***continuous*** at the point (x_0, y_0) if the following conditions are satisfied:

1. $f(x_0, y_0)$ is defined.

2. $\displaystyle\lim_{(x,y)\to(x_0,y_0)} f(x, y)$ exists.

3. $\displaystyle\lim_{(x,y)\to(x_0,y_0)} f(x, y) = f(x_0, y_0)$.

The requirement that $f(x_0, y_0)$ be defined eliminates the possibility of a hole in the surface $z = f(x, y)$ above the point (x_0, y_0); the requirement that $\lim_{(x,y)\to(x_0,y_0)} f(x, y)$ exists ensures that $z = f(x, y)$ does not become "infinite" at (x_0, y_0); and the requirement that $\lim_{(x,y)\to(x_0,y_0)} f(x, y) = f(x_0, y_0)$ ensures that the surface does not have a vertical jump or step above the point (x_0, y_0).

A function of two variables is called ***continuous on a region R*** of the xy-plane if it is continuous at each point of R. A function that is continuous on the entire xy-plane is called ***continuous everywhere*** or simply ***continuous.***

Intuitively, we may imagine the graph of a continuous function of two variables to be constructed from a thin sheet of clay that has been hollowed and pinched into peaks and valleys without creating tears or pinholes.

To help identify continuous functions, we will use the following theorem, which we state without proof.

16.3.3 THEOREM *(a) If g and h are continuous functions of one variable, then $f(x, y) = g(x)h(y)$ is a continuous function of x and y.*

(b) If g is a continuous function of one variable and h is a continuous function of two variables, then their composition $f(x, y) = g(h(x, y))$ is a continuous function of x and y.

▶ **Example 2**   Since $g(x) = 3x^2$ and $h(y) = y^5$ are continuous functions, it follows from part (*a*) of the above theorem that $f(x, y) = 3x^2 y^5$ is a continuous function of x and y. In general, any function of the form $f(x, y) = Ax^m y^n$ (*m* and *n* nonnegative integers) is continuous since it is the product of the continuous functions Ax^m and y^n.   ◀

▶ **Example 3**   Since $g(x) = \sin x$ is continuous, and since $h(x, y) = xy^2$ is continuous by Example 2, it follows from part (*b*) of the above theorem that $g(h(x, y)) = g(xy^2) = \sin(xy^2)$ is a continuous function of x and y. By a similar argument, each of the following is continuous:

$$(x^4 y^5)^{1/3}, \qquad e^{xy}, \qquad \cosh(x^3 y)$$ ◀

▶ **Example 4**   By Example 3, e^{xy} is continuous. Thus, $\cos(e^{xy})$ is continuous by part (*b*) of Theorem 16.3.3.   ◀

Just as for functions of one variable, the sum, difference, and product of continuous functions of two variables are also continuous, and the ratio of continuous functions is continuous except where the denominator is zero.

▶ **Example 5**   The following functions are continuous since they are sums, differences, and products of continuous functions:

$$3 - 2x^2 y + 9x^4 y^8, \qquad e^{xy} \cos(xy^2 + 1), \qquad (3 + ye^x)^{17}$$ ◀

▶ **Example 6**   Since the function

$$f(x, y) = \frac{x^3 y^2}{1 - xy}$$

is a ratio of continuous functions, it is continuous except where $1 - xy = 0$. Thus, $f(x, y)$ is continuous everywhere except on the hyperbola $xy = 1$.   ◀

▶ **Example 7**   Evaluate

$$\lim_{(x,y) \to (-1,2)} \frac{xy}{x^2 + y^2}$$

Solution.   Since $f(x, y) = xy/(x^2 + y^2)$ is continuous at $(-1, 2)$ (why?), it follows from part 3 of Definition 16.3.2 that

$$\lim_{(x,y) \to (-1,2)} \frac{xy}{x^2 + y^2} = \frac{(-1)(2)}{(-1)^2 + (2)^2} = -\frac{2}{5}$$ ◀

Recall that a function f of one variable is called differentiable at x_0 if it has a derivative at x_0, or, in other words, if the limit

$$f'(x_0) = \lim_{\Delta x \to 0} \frac{f(x_0 + \Delta x) - f(x_0)}{\Delta x} \tag{2}$$

exists. A function f that is differentiable at a point x_0 enjoys two important properties:

1. $f(x)$ is continuous at x_0.
2. The curve $y = f(x)$ has a nonvertical tangent line at x_0.

Our next objective is to extend the notion of differentiability to functions of two variables in such a way that the natural analogs of these two properties hold. More precisely, when $f(x, y)$ is differentiable at (x_0, y_0), we will want it to be the case that

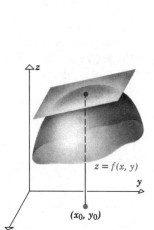

$z = f(x, y)$

(x_0, y_0)

Figure 16.3.3

1. $f(x, y)$ is continuous at (x_0, y_0).
2. The surface $z = f(x, y)$ has a nonvertical tangent plane at (x_0, y_0) (Figure 16.3.3).

A precise definition of a tangent plane will be given later.

It would not be unreasonable to guess that a function f of two variables should be called differentiable at (x_0, y_0) if the two partial derivatives $f_x(x_0, y_0)$ and $f_y(x_0, y_0)$ exist at (x_0, y_0). Unfortunately, this condition is not strong enough to meet our objectives, since there are functions that have partial derivatives at a point, but are not continuous at that point. For example, the function

$$f(x, y) = \begin{cases} -1 & \text{if } x \text{ and } y \text{ are positive} \\ 0 & \text{otherwise} \end{cases}$$

is discontinuous at $(0, 0)$, but has partial derivatives at $(0, 0)$. To be precise, $f_x(0, 0) = 0$ and $f_y(0, 0) = 0$. These facts should be evident from Figure 16.3.4.

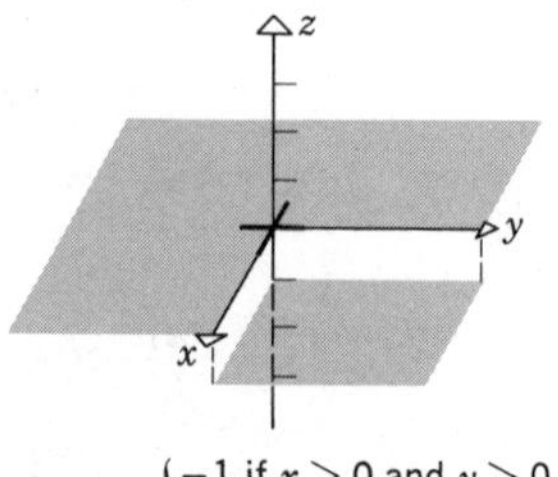

To motivate the appropriate definition of differentiability for functions of two variables, we will reformulate (2) in an alternate form. With the increment notation

$$\Delta f = f(x_0 + \Delta x) - f(x_0)$$

(2) can be rewritten

$$f(x, y) = \begin{cases} -1 \text{ if } x > 0 \text{ and } y > 0 \\ 0 \text{ otherwise} \end{cases}$$

Figure 16.3.4

$$f'(x_0) = \lim_{\Delta x \to 0} \frac{\Delta f}{\Delta x} \tag{3}$$

Let us define ϵ as

$$\epsilon = \frac{\Delta f}{\Delta x} - f'(x_0) \tag{4}$$

so that

$$\Delta f = f'(x_0)\,\Delta x + \epsilon\,\Delta x \tag{5}$$

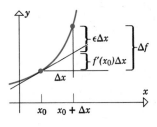

Figure 16.3.5

The quantities in this equation are pictured in Figure 16.3.5. The term Δf represents the change in height that results when a point moves along the graph of f so that the x-coordinate changes from x_0 to $x_0 + \Delta x$; the term $f'(x_0)\,\Delta x$ represents the change in height that results when a point moves along the tangent line at $(x_0, f(x_0))$ so that the x-coordinate changes from x_0 to $x_0 + \Delta x$; finally, the term $\epsilon\,\Delta x$ represents the difference between Δf and $f'(x_0)\,\Delta x$. It is evident from Figure 16.3.5 that $\epsilon\,\Delta x \to 0$ as $\Delta x \to 0$. However, it is also true that $\epsilon \to 0$ as $\Delta x \to 0$. This is not at all evident from Figure 16.3.5, but follows immediately from (3) and (4).

If f is a function of x and y, then the symbol Δf, called the **increment** of f, denotes the change in the value of $f(x, y)$ that results when (x, y) varies from some initial position (x_0, y_0) to some new position $(x_0 + \Delta x, y_0 + \Delta y)$; thus,

$$\Delta f = f(x_0 + \Delta x, y_0 + \Delta y) - f(x_0, y_0)$$

(See Figure 16.3.6.) If a dependent variable $z = f(x, y)$ is used, then we will sometimes write Δz rather than Δf.

Motivated by the form of (5), we now make the following definition of differentiability for functions of two variables.

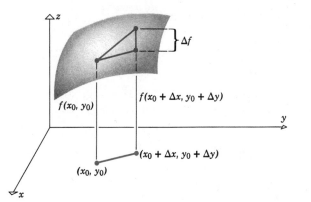

Figure 16.3.6

16.3.4 DEFINITION   A function f of two variables is said to be **differentiable** at (x_0, y_0) if $f_x(x_0, y_0)$ and $f_y(x_0, y_0)$ exist and Δf can be written in the form

$$\Delta f = f_x(x_0, y_0)\,\Delta x + f_y(x_0, y_0)\,\Delta y + \epsilon_1\,\Delta x + \epsilon_2\,\Delta y$$

where ϵ_1 and ϵ_2 are functions of Δx and Δy such that $\epsilon_1 \to 0$ and $\epsilon_2 \to 0$ as $(\Delta x, \Delta y) \to (0, 0)$.

A function is called ***differentiable on a region*** R of the xy-plane if it is differentiable at each point of R. A function that is differentiable on the entire xy-plane is called ***everywhere differentiable*** or simply ***differentiable.***

REMARK.   Before proceeding further, we want to reemphasize an important point. For functions of one variable, the terms "differentiable" and "has a derivative" are synonymous. However, for functions of two variables differentiability is a more stringent requirement than the mere existence of partial derivatives.

Earlier, we set two goals for our definition of differentiability—first, we want a function f that is differentiable at (x_0, y_0) to be continuous at (x_0, y_0), and second we want the graph of f to have a nonvertical tangent plane at (x_0, y_0). The next theorem shows that the continuity criterion is met; the existence of a nonvertical tangent plane will be demonstrated in a later section.

16.3.5   THEOREM   *If f is differentiable at (x_0, y_0), then f is continuous at (x_0, y_0).*

Proof.   In order to prove that

$$\lim_{(x,y)\to(x_0,y_0)} f(x, y) = f(x_0, y_0)$$

we will show that

$$\lim_{(x,y)\to(x_0,y_0)} [f(x, y) - f(x_0, y_0)] = 0 \tag{6}$$

Let $\Delta x = x - x_0$ and $\Delta y = y - y_0$, so that $x = x_0 + \Delta x$, $y = y_0 + \Delta y$. Since $\Delta x \to 0$ as $x \to x_0$ and $\Delta y \to 0$ as $y \to y_0$, (6) can be rewritten in the form

$$\lim_{(\Delta x,\Delta y)\to(0,0)} [f(x_0 + \Delta x, y_0 + \Delta y) - f(x_0, y_0)] = 0$$

or more briefly

$$\lim_{(\Delta x,\Delta y)\to(0,0)} \Delta f = 0 \tag{7}$$

Let us now prove (7). By hypothesis

$$\Delta f = f_x(x_0, y_0)\, \Delta x + f_y(x_0, y_0)\, \Delta y + \epsilon_1\, \Delta x + \epsilon_2\, \Delta y$$

where $\epsilon_1 \to 0$, $\epsilon_2 \to 0$ as $(\Delta x, \Delta y) \to (0, 0)$. Thus,

$$\lim_{(\Delta x, \Delta y) \to (0,0)} \Delta f = \lim_{(\Delta x, \Delta y) \to (0,0)} [f_x(x_0, y_0)\, \Delta x + f_y(x_0, y_0)\, \Delta y + \epsilon_1\, \Delta x + \epsilon_2\, \Delta y] = 0$$

which completes the proof.   ▪

The next theorem gives simple conditions for a function of two variables to be differentiable at a point.

16.3.6   THEOREM   *If f has first-order partial derivatives at each point in some circular region centered at (x_0, y_0), and if these partial derivatives are continuous at (x_0, y_0), then f is differentiable at (x_0, y_0).*

The proof is optional and may be found at the end of this section.

▶ **Example 8**   Show that $f(x, y) = x^3 y$ is a differentiable function.

Solution.   The partial derivatives $f_x = 3x^2 y$ and $f_y = x^3$ are defined and continuous everywhere in the xy-plane. Thus, the hypotheses of Theorem 16.3.6 are satisfied at each point (x_0, y_0) in the xy-plane. Thus, $f(x, y) = x^3 y$ is everywhere differentiable.   ◀

The following theorem gives conditions under which the mixed second partial derivatives of a function are equal. We omit the proof.

16.3.7   THEOREM
Equality of Mixed Partials   *Let f be a function of two variables. If f, f_x, f_y, f_{xy}, and f_{yx} are continuous on an open set, then $f_{xy} = f_{yx}$ at each point of the set.*

In general, the order of differentiation in an n-th order partial derivative can be changed without affecting the final result whenever the function and all its partial derivatives of order n or less are continuous. For example, if f and its partial derivatives of the first, second, and third orders are continuous on an open set, then at each point of that set,

$$f_{xyy} = f_{yxy} = f_{yyx}$$

or in another notation,

$$\frac{\partial^3 f}{\partial y^2\, \partial x} = \frac{\partial^3 f}{\partial y\, \partial x\, \partial y} = \frac{\partial^3 f}{\partial x\, \partial y^2}$$

OPTIONAL

Proof of Theorem 16.3.6.

We must show that

$$\Delta f = f(x_0 + \Delta x, y_0 + \Delta y) - f(x_0, y_0)$$

can be expressed in the form

$$\Delta f = f_x(x_0, y_0)\,\Delta x + f_y(x_0, y_0)\,\Delta y + \epsilon_1\,\Delta x + \epsilon_2\,\Delta y$$

where $\epsilon_1 \to 0, \epsilon_2 \to 0$ as $(\Delta x, \Delta y) \to (0, 0)$. In our discussion, we will restrict Δx and Δy to be sufficiently small that the points

$$B(x_0 + \Delta x, y_0) \qquad \text{and} \qquad C(x_0 + \Delta x, y_0 + \Delta y)$$

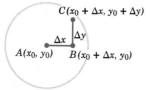

Figure 16.3.7

lie inside a circle where f_x and f_y exist (Figure 16.3.7). The increment Δf represents the change in the value of f as (x, y) varies from the initial point $A(x_0, y_0)$ to the final point $C(x_0 + \Delta x, y_0 + \Delta y)$. Let us resolve Δf into two terms

$$\Delta f = \Delta f_1 + \Delta f_2 \tag{8}$$

where Δf_1 is the change in the value of f as (x, y) varies from A to B and Δf_2 is the change in the value of f as (x, y) varies from B to C. Thus,

$$\Delta f_1 = f(x_0 + \Delta x, y_0) - f(x_0, y_0) \tag{9}$$

$$\Delta f_2 = f(x_0 + \Delta x, y_0 + \Delta y) - f(x_0 + \Delta x, y_0) \tag{10}$$

Along the line segment from A to B, y has a constant value of y_0, so that $f(x, y) = f(x, y_0)$ is a function of x alone. Let us denote this function by

$$f_1(x) = f(x, y_0)$$

Similarly, along the line segment from B to C, x has a constant value of $x = x_0 + \Delta x$, so that $f(x, y) = f(x_0 + \Delta x, y)$ is a function of y alone. Let us denote this function by

$$f_2(y) = f(x_0 + \Delta x, y)$$

With these definitions of f_1 and f_2, (9) and (10) can be written

$$\Delta f_1 = f_1(x_0 + \Delta x) - f_1(x_0) \tag{11}$$

$$\Delta f_2 = f_2(y_0 + \Delta y) - f_2(y_0) \tag{12}$$

The reader should be able to show that f_1 and f_2 satisfy the hypotheses of the Mean-Value Theorem (4.10.2). Accepting this, we can rewrite (11) and (12) in the form

$$\Delta f_1 = f_1'(x^*)\,\Delta x \tag{13}$$

$$\Delta f_2 = f_2'(y^*)\,\Delta y \tag{14}$$

where x^* is between x_0 and $x_0 + \Delta x$ and y^* is between y_0 and $y_0 + \Delta y$. From the definitions of $f_1(x)$ and $f_2(y)$

$$f_1'(x) = \frac{d}{dx}[f_1(x)] = f_x(x, y_0)$$

$$f_2'(y) = \frac{d}{dy}[f_2(y)] = f_y(x_0 + \Delta x, y)$$

so that (13) and (14) can be rewritten as

$$\Delta f_1 = f_x(x^*, y_0)\,\Delta x \tag{15}$$

$$\Delta f_2 = f_y(x_0 + \Delta x, y^*)\,\Delta y \tag{16}$$

If we define

$$\epsilon_1 = f_x(x^*, y_0) - f_x(x_0, y_0) \tag{17}$$

$$\epsilon_2 = f_y(x_0 + \Delta x, y^*) - f_y(x_0, y_0) \tag{18}$$

then (15) and (16) can be rewritten as

$$\Delta f_1 = f_x(x_0, y_0)\,\Delta x + \epsilon_1\,\Delta x$$

$$\Delta f_2 = f_y(x_0, y_0)\,\Delta y + \epsilon_2\,\Delta y$$

and substituting these in (8) yields

$$\Delta f = f_x(x_0, y_0)\,\Delta x + f_y(x_0, y_0)\,\Delta y + \epsilon_1\,\Delta x + \epsilon_2\,\Delta y$$

To finish, we must show that $\epsilon_1 \to 0$, $\epsilon_2 \to 0$ as $(\Delta x, \Delta y) \to (0, 0)$.

Since x^* is between x_0 and $x_0 + \Delta x$. and y^* is between y_0 and $y_0 + \Delta y$, it follows that

$$x^* \to x_0 \quad \text{as} \quad \Delta x \to 0$$

$$y^* \to y_0 \quad \text{as} \quad \Delta y \to 0$$

But this, together with the hypothesis that f_x and f_y are continuous at (x_0, y_0), implies that

$$f_x(x^*, y_0) \to f_x(x_0, y_0)$$

$$f_y(x_0 + \Delta x, y^*) \to f_y(x_0, y_0)$$

as $(\Delta x, \Delta y) \to (0, 0)$. Thus, from (17) and (18) it follows that $\epsilon_1 \to 0$ and $\epsilon_2 \to 0$ as $(\Delta x, \Delta y) \to (0, 0)$. ∎

▶ Exercise Set 16.3

1. Find Δf given that $f(x, y) = x^2y$, $(x_0, y_0) = (1, 3)$, $\Delta x = 0.1$, and $\Delta y = 0.2$.

2. Find Δz given that $z = 3x^2 - 2y$, $(x_0, y_0) = (-2, 4)$, $\Delta x = 0.02$, and $\Delta y = -0.03$.

3. Find the increment of $f(x, y) = x/y$ as (x, y) varies from $(-1, 2)$ to $(3, 1)$.

4. Find the increment of $g(u, v) = 2uv - v^3$ as (u, v) varies from $(0, 1)$ to $(4, -2)$.

In Exercises 5–12, sketch the region where f is continuous.

5. $f(x, y) = y \ln (1 + x)$.

6. $f(x, y) = \sqrt{x - y}$.

7. $f(x, y) = \dfrac{x^2y}{\sqrt{25 - x^2 - y^2}}$.

8. $f(x, y) = \ln (2x - y + 1)$.

9. $f(x, y) = \cos \left(\dfrac{xy}{1 + x^2 + y^2} \right)$.

10. $f(x, y) = e^{(1-xy)}$.

11. $f(x, y) = \sin^{-1} (xy)$.

12. $f(x, y) = \tan^{-1} (y - x)$.

In Exercises 13–26, find the limit, if it exists.

13. $\lim\limits_{(x,y) \to (1,3)} (4xy^2 - x)$.

14. $\lim\limits_{(x,y) \to (1/2, \pi)} (xy^2 \sin xy)$.

15. $\lim\limits_{(x,y) \to (-1,2)} \dfrac{xy^3}{x + y}$.

16. $\lim\limits_{(x,y) \to (1,-3)} e^{2x-y^2}$.

17. $\lim\limits_{(x,y) \to (0,0)} \ln (1 + x^2y^3)$.

18. $\lim\limits_{(x,y) \to (4,-2)} x \sqrt[3]{y^3 + 2x}$.

19. $\lim\limits_{(x,y) \to (0,0)} \dfrac{x - y}{x^2 + y^2}$.

20. $\lim\limits_{(x,y) \to (0,0)} \dfrac{3}{x^2 + 2y^2}$.

21. $\lim\limits_{(x,y) \to (0,0)} \dfrac{\sin(x^2 + y^2)}{x^2 + y^2}$.

22. $\lim\limits_{(x,y) \to (0,0)} \dfrac{1 - \cos(x^2 + y^2)}{x^2 + y^2}$.

23. $\lim\limits_{(x,y) \to (0,0)} \dfrac{x^4 - y^4}{x^2 + y^2}$.

24. $\lim\limits_{(x,y) \to (0,0)} \dfrac{x^4 - 16y^4}{x^2 + 4y^2}$.

25. $\lim\limits_{(x,y) \to (0,0)} \dfrac{xy}{x^2 + y^2}$.

26. $\lim\limits_{(x,y) \to (0,0)} \dfrac{1 - x^2 - y^2}{x^2 + y^2}$.

27. (a) Show that $\dfrac{x^2y}{x^4 + y^2}$ approaches zero as $(x, y) \to (0, 0)$ along any straight line, $y = mx$.

(b) Show that $\lim\limits_{(x,y) \to (0,0)} \dfrac{x^2y}{x^4 + y^2}$ does not exist by letting $(x, y) \to (0, 0)$ along the parabola $y = x^2$.

28. Find
$$\lim\limits_{(x,y) \to (0,1)} \tan^{-1} \left[\dfrac{x^2 + 1}{x^2 + (y - 1)^2} \right].$$

29. Find
$$\lim\limits_{(x,y) \to (0,1)} \tan^{-1} \left[\dfrac{x^2 - 1}{x^2 + (y - 1)^2} \right].$$

In Exercises 30–34, find f_{xy} and f_{yx} and verify their equality.

30. $f(x, y) = 2xy - 3y^2$

31. $f(x, y) = 4x^3y + 3x^2y$.

32. $f(x, y) = \dfrac{x^3}{y}$.

33. $f(x, y) = \sin (x^2 + y^3)$.

34. $f(x, y) = \sqrt{x^2 + y^2 - 1}$.

35. Let f be a function of two variables with continuous third- and fourth-order partial derivatives.
    (a) How many of the third-order partial derivatives can be distinct?
    (b) How many of the fourth order?

36. Express the partial derivatives $f_x(x_0, y_0)$ and $f_y(x_0, y_0)$ as limits. [*Hint:* For a function f of one variable.
$$f'(x_0) = \lim\limits_{h \to 0} \dfrac{f(x_0 + h) - f(x_0)}{h} \Big]$$

37. Let
$$f(x, y) = \begin{cases} \dfrac{xy}{x^2 + y^2} & \text{if} \quad (x, y) \neq (0, 0) \\ 0 & \text{if} \quad (x, y) = (0, 0) \end{cases}$$

Prove: $f_x(0, 0)$ and $f_y(0, 0)$ exist, but f is not continuous at $(0, 0)$. [*Hint:* Show that $f_x(0, 0) = 0$ and $f_y(0, 0) = 0$ by expressing these derivatives as limits (see Exercise 36). To prove that f is not continuous at $(0, 0)$, show that
$$\lim\limits_{(x,y) \to (0,0)} f(x, y)$$
does not exist by letting $(x, y) \to (0, 0)$ along $y = x$ and along $y = -x$.]

16.4   THE CHAIN RULE FOR FUNCTIONS OF TWO VARIABLES

If y is a differentiable function of x and x is a differentiable function of t, then the chain rule for functions of one variable states that

$$\frac{dy}{dt} = \frac{dy}{dx}\frac{dx}{dt}$$

In this section we will derive versions of the chain rule for functions of two variables.

Assume that z is a function of x and y, say

$$z = f(x, y) \tag{1}$$

and suppose that x and y, in turn, are functions of a single variable t, say

$$x = x(t), \qquad y = y(t)$$

On substituting these functions of t in (1), we obtain the relationship

$$z = f\big(x(t), y(t)\big)$$

which expresses z as a function of the single variable t. Thus, we may ask for the derivative dz/dt and we may inquire about its relationship to the derivatives $\partial z/\partial x$, $\partial z/\partial y$, dx/dt, and dy/dt.

16.4.1   THEOREM

Chain Rule

If $x = x(t)$ and $y = y(t)$ are differentiable at t, and if $z = f(x, y)$ is differentiable at the point $\big(x(t), y(t)\big)$, then $z = f\big(x(t), y(t)\big)$ is differentiable at t, and

$$\frac{dz}{dt} = \frac{\partial z}{\partial x}\frac{dx}{dt} + \frac{\partial z}{\partial y}\frac{dy}{dt} \tag{2}$$

Proof.   From the definition of derivative for functions of one variable

$$\frac{dz}{dt} = \lim_{\Delta t \to 0} \frac{\Delta z}{\Delta t} \tag{3}$$

Since $z = f(x, y)$ is differentiable at the point $(x, y) = (x(t), y(t))$, we can express Δz in the form

$$\Delta z = \frac{\partial z}{\partial x}\Delta x + \frac{\partial z}{\partial y}\Delta y + \epsilon_1 \Delta x + \epsilon_2 \Delta y \tag{4}$$

where the partial derivatives are evaluated at $\big(x(t), y(t)\big)$ and $\epsilon_1 \to 0, \epsilon_2 \to 0$ as $(\Delta x, \Delta y) \to (0, 0)$. Thus, from (3) and (4),

$$\frac{dz}{dt} = \lim_{\Delta t \to 0} \frac{\Delta z}{\Delta t} = \lim_{\Delta t \to 0}\left[\frac{\partial z}{\partial x}\frac{\Delta x}{\Delta t} + \frac{\partial z}{\partial y}\frac{\Delta y}{\Delta t} + \epsilon_1 \frac{\Delta x}{\Delta t} + \epsilon_2 \frac{\Delta y}{\Delta t}\right] \tag{5}$$

But

$$\lim_{\Delta t \to 0} \frac{\Delta x}{\Delta t} = \frac{dx}{dt} \quad \text{and} \quad \lim_{\Delta t \to 0} \frac{\Delta y}{\Delta t} = \frac{dy}{dt}$$

Therefore, if we can show that $\epsilon_1 \to 0, \epsilon_2 \to 0$ as $\Delta t \to 0$, then the proof will be complete, since (5) will reduce to (2). But $\Delta x \to 0$ and $\Delta y \to 0$ as $\Delta t \to 0$, since

$$\lim_{\Delta t \to 0} \Delta x = \lim_{\Delta t \to 0} \frac{\Delta x}{\Delta t} \Delta t = \frac{dx}{dt} \cdot 0 = 0$$

and similarly for Δy. Thus, as Δt tends to zero, $(\Delta x, \Delta y) \to (0, 0)$, which implies that $\epsilon_1 \to 0, \epsilon_2 \to 0$. ∎

▶ Example 1   Suppose

$$z = x^2 y, \qquad x = t^2, \qquad y = t^3$$

Use the chain rule to find dz/dt, and check the result by expressing z as a function of t and differentiating directly.

Solution. By the chain rule

$$\frac{dz}{dt} = \frac{\partial z}{\partial x}\frac{dx}{dt} + \frac{\partial z}{\partial y}\frac{dy}{dt}$$
$$= (2xy)(2t) + (x^2)(3t^2)$$
$$= (2t^5)(2t) + (t^4)(3t^2)$$
$$= 7t^6$$

Alternately, we may express z directly as a function of t

$$z = x^2 y = (t^2)^2(t^3) = t^7$$

and then differentiate to obtain $dz/dt = 7t^6$. However, this procedure is not always convenient. ◀

▶ Example 2   Suppose

$$z = \sqrt{xy + y}, \qquad x = \cos\theta, \qquad y = \sin\theta$$

Use the chain rule to find $dz/d\theta$ when $\theta = \pi/2$.

Solution. From the chain rule with θ in place of t,

$$\frac{dz}{d\theta} = \frac{\partial z}{\partial x}\frac{dx}{d\theta} + \frac{\partial z}{\partial y}\frac{dy}{d\theta}$$

we obtain

$$\frac{dz}{d\theta} = \frac{1}{2}(xy + y)^{-1/2}(y)(-\sin\theta) + \frac{1}{2}(xy + y)^{-1/2}(x + 1)(\cos\theta)$$

when $\theta = \pi/2$, we have

$$x = \cos\frac{\pi}{2} = 0, \qquad y = \sin\frac{\pi}{2} = 1$$

Substituting $x = 0$, $y = 1$, $\theta = \pi/2$ in the formula for $dz/d\theta$ yields

$$\left.\frac{dz}{d\theta}\right|_{\theta=\pi/2} = \frac{1}{2}(1)(1)(-1) + \frac{1}{2}(1)(1)(0) = -\frac{1}{2} \qquad \blacktriangleleft$$

REMARK.   There are many variations in derivative notations, each of which gives the chain rule a different look. If $z = f(x, y)$, where x and y are functions of t, then some possibilities are

$$\frac{dz}{dt} = f_x\frac{dx}{dt} + f_y\frac{dy}{dt}$$

$$\frac{df}{dt} = \frac{\partial f}{\partial x}\frac{dx}{dt} + \frac{\partial f}{\partial y}\frac{dy}{dt}$$

$$\frac{df}{dt} = f_x x'(t) + f_y y'(t)$$

The reader may be able to construct other variations as well.

In the special case where $z = F(x, y)$ and y is a differentiable function of x, chain-rule formula (2) yields

$$\frac{dz}{dx} = \frac{\partial F}{\partial x}\frac{dx}{dx} + \frac{\partial F}{\partial y}\frac{dy}{dx} = \frac{\partial F}{\partial x} + \frac{\partial F}{\partial y}\frac{dy}{dx} \qquad (6)$$

This result can be used to find derivatives of functions that are defined implicitly. Suppose that the equation

$$F(x, y) = 0 \qquad (7)$$

defines y implicitly as a differentiable function of x, and we are interested in finding dy/dx. Differentiating both sides of (7) with respect to x and applying (6) yields

$$\frac{\partial F}{\partial x} + \frac{\partial F}{\partial y}\frac{dy}{dx} = 0$$

or

$$\frac{dy}{dx} = -\frac{\partial F/\partial x}{\partial F/\partial y} \qquad (8)$$

provided $\partial F/\partial y \neq 0$.

▶ Example 3 Given that

$$x^3 + y^2 x - 3 = 0$$

find dy/dx using (8) and check the result using implicit differentiation.

Solution. By (8)

$$\frac{dy}{dx} = -\frac{\partial F/\partial x}{\partial F/\partial y} = -\frac{3x^2 + y^2}{2yx}$$

On the other hand, differentiating the given equation implicitly yields

$$3x^2 + y^2 + x\left(2y\frac{dy}{dx}\right) + 0 = 0$$

or

$$\frac{dy}{dx} = -\frac{3x^2 + y^2}{2yx}$$

which agrees with results obtained by (8). ◀

In Theorem 16.4.1 the variables x and y are each functions of a single variable t. We now consider the case where x and y are each functions of two variables. Let

$$z = f(x, y) \qquad (9)$$

and suppose x and y are functions of u and v, say

$$x = x(u, v), \qquad y = y(u, v)$$

On substituting these functions of u and v into (9) we obtain the relationship

$$z = f(x(u, v), y(u, v))$$

which expresses z as a function of the two variables u and v. Thus, we may ask for the partial derivatives $\partial z/\partial u$ and $\partial z/\partial v$; and we may inquire about the relationship between these derivatives and the derivatives $\partial z/\partial x$, $\partial z/\partial y$, $\partial x/\partial u$, $\partial x/\partial v$, $\partial y/\partial u$, and $\partial y/\partial v$.

16.4.2 THEOREM

Chain Rule

If $x = x(u, v)$ *and* $y = y(u, v)$ *have first-order partial derivatives at the point* (u, v) *and if* $z = f(x, y)$ *is differentiable at the point* $(x(u, v), y(u, v))$, *then* $z = f(x(u, v), y(u, v))$ *has first-order partial derivatives at* (u, v), *given by*

$$\frac{\partial z}{\partial u} = \frac{\partial z}{\partial x}\frac{\partial x}{\partial u} + \frac{\partial z}{\partial y}\frac{\partial y}{\partial u}$$

$$\frac{\partial z}{\partial v} = \frac{\partial z}{\partial x}\frac{\partial x}{\partial v} + \frac{\partial z}{\partial y}\frac{\partial y}{\partial v}$$

Proof. If v is held fixed, then $x = x(u, v)$ and $y = y(u, v)$ become functions of u alone. Thus, we are back in the case of Theorem 16.4.1. If we apply that theorem with u in place of t and if we use ∂ rather than d to indicate that the variable v is fixed, we obtain

$$\frac{\partial z}{\partial u} = \frac{\partial z}{\partial x}\frac{\partial x}{\partial u} + \frac{\partial z}{\partial y}\frac{\partial y}{\partial u}$$

The formula for $\partial z/\partial v$ is derived similarly.   ▮

▶ **Example 4** Given that

$$z = e^{xy}, \qquad x = 2u + v, \qquad y = u/v$$

find $\partial z/\partial u$ and $\partial z/\partial v$ using the chain rule.

Solution.

$$\frac{\partial z}{\partial u} = \frac{\partial z}{\partial x}\frac{\partial x}{\partial u} + \frac{\partial z}{\partial y}\frac{\partial y}{\partial u}$$

$$= (ye^{xy})(2) + (xe^{xy})(1/v)$$

$$= \left[2y + \frac{x}{v}\right]e^{xy}$$

$$= \left[\frac{2u}{v} + \frac{2u + v}{v}\right]e^{(2u+v)(u/v)}$$

$$= \left[\frac{4u}{v} + 1\right]e^{(2u+v)(u/v)}$$

$$\frac{\partial z}{\partial v} = \frac{\partial z}{\partial x}\frac{\partial x}{\partial v} + \frac{\partial z}{\partial y}\frac{\partial y}{\partial v}$$

$$= (ye^{xy})(1) + (xe^{xy})\left(-\frac{u}{v^2}\right)$$

$$= \left[y - x \left(\frac{u}{v^2} \right) \right] e^{xy}$$

$$= \left[\frac{u}{v} - (2u + v) \left(\frac{u}{v^2} \right) \right] e^{(2u+v)(u/v)}$$

$$= -\frac{2u^2}{v^2} e^{(2u+v)(u/v)}$$ ◄

The chain rules are useful in related rates problems.

▶ Example 5   At what rate is the area of a rectangle changing if its length is 15 ft and increasing at 3 ft/sec while its width is 6 ft and increasing at 2 ft/sec?

Solution.   Let

x = length of the rectangle in feet
y = width of the rectangle in feet
A = area of the rectangle in square feet
t = time in seconds

We are given that

$$\frac{dx}{dt} = 3 \quad \text{and} \quad \frac{dy}{dt} = 2 \tag{10}$$

at the instant when

$$x = 15, \quad y = 6 \tag{11}$$

We want to find dA/dt at that instant.
From the area formula $A = xy$, we obtain

$$\frac{dA}{dt} = \frac{\partial A}{\partial x} \frac{dx}{dt} + \frac{\partial A}{\partial y} \frac{dy}{dt} = y \frac{dx}{dt} + x \frac{dy}{dt}$$

Substituting (10) and (11) in this equation yields

$$\frac{dA}{dt} = 6(3) + 15(2) = 48$$

Thus, the area is increasing at a rate of 48 ft²/sec at the given instant. ◄

▶ Exercise Set 16.4

In Exercises 1–6, find dz/dt using the chain rule.

1. $z = 3x^2y^3$; $x = t^4$, $y = t^2$.

2. $z = \ln(2x^2 + y)$; $x = \sqrt{t}$, $y = t^{2/3}$.

3. $z = 3\cos x - \sin xy$; $x = 1/t$, $y = 3t$.

4. $z = \sqrt{1 + x - 2xy^4}$; $x = \ln t$, $y = t$.

5. $z = e^{1-xy}$; $x = t^{1/3}$, $y = t^3$.

6. $z = \cosh^2 xy$; $x = t/2$, $y = e^t$.

In Exercises 7–13, find $\partial z/\partial u$ and $\partial z/\partial v$ by the chain rule.

7. $z = 8x^2y - 2x + 3y$; $x = uv$, $y = u - v$.

8. $z = x^2 - y\tan x$; $x = u/v$, $y = u^2v^2$.

9. $z = x/y$; $x = 2\cos u$, $y = 3\sin v$.

10. $z = 3x - 2y$; $x = u + v\ln u$, $y = u^2 - v\ln v$.

11. $z = e^{x^2y}$; $x = \sqrt{uv}$, $y = 1/v$.

12. $z = \cos x \sin y$; $x = u - v$, $y = u^2 + v^2$.

13. $z = \tan^{-1}(x^2 + y^2)$; $x = e^u\sin v$; $y = e^u\cos v$.

14. Let $w = rs/(r^2 + s^2)$; $r = uv$, $s = u - 2v$. Use the chain rule to find $\partial w/\partial u$ and $\partial w/\partial v$.

15. Let $T = x^2y - xy^3 + 2$; $x = r\cos\theta$, $y = r\sin\theta$. Use the chain rule to find $\partial T/\partial r$ and $\partial T/\partial\theta$.

16. Let $R = e^{2s-t^2}$; $s = 3\phi$, $t = \phi^{1/2}$. Use the chain rule to find $dR/d\phi$.

17. Let $t = u/v$; $u = x^2 - y^2$, $v = 4xy^3$. Use the chain rule to find $\partial t/\partial x$ and $\partial t/\partial y$.

18. Use the chain rule to find $\left.\dfrac{dz}{dt}\right|_{t=3}$ if $z = x^2y$; $x = t^2$, $y = t + 7$.

19. Use the chain rule to find $\left.\dfrac{dw}{ds}\right|_{s=1/4}$ if $w = r^2 - r\tan\theta$; $r = \sqrt{s}$, $\theta = \pi s$.

20. Use the chain rule to find

$$\left.\frac{\partial f}{\partial u}\right|_{u=1,\,v=-2} \quad \text{and} \quad \left.\frac{\partial f}{\partial v}\right|_{u=1,\,v=-2}$$

if $f(x, y) = x^2y^2 - x + 2y$; $x = \sqrt{u}$, $y = uv^3$.

21. Use the chain rule to find

$$\left.\frac{\partial z}{\partial r}\right|_{r=2,\,\theta=\pi/6} \quad \text{and} \quad \left.\frac{\partial z}{\partial\theta}\right|_{r=2,\,\theta=\pi/6}$$

if $z = xye^{x/y}$; $x = r\cos\theta$, $y = r\sin\theta$.

In Exercises 22–25, use (8) to find dy/dx and check your result using implicit differentiation.

22. $x^2y^3 + \cos y = 0$.

23. $x^3 - 3xy^2 + y^3 = 5$.

24. $e^{xy} + ye^y = 1$.

25. $x - \sqrt{xy} + 3y = 4$.

26. The portion of a tree usable for lumber may be viewed as a right-circular cylinder. If the height of a tree increases 24 in. per year and the diameter increases 3 in. per year, how fast is the volume of usable lumber increasing when the tree is 240 in. high and the diameter is 30 in.?

27. Two straight roads intersect at right angles. Car A, moving on one of the roads, approaches the intersection at 25 mi/hr and car B, moving on the other road, approaches the intersection at 30 mi/hr. At what rate is the distance between the cars changing when A is 0.3 miles from the intersection and B is 0.4 miles from the intersection?

28. Use the ideal gas law (Example 4, Section 16.2) with $k = 10$ to find the rate at which the temperature of a gas is changing when the volume is 200 in.^{3} and increasing at the rate of 4 in.^{3}/sec, while the pressure is 5 lb/in.^{2} and decreasing at the rate of 1 lb/in.^{2} per sec.

29. Two sides of a triangle have lengths $a = 4$ cm and $b = 3$ cm, but are increasing at the rate of 1 cm/sec. If the area of the triangle remains constant, at what rate is the angle θ between a and b changing when $\theta = \pi/6$?

30. Two sides of a triangle have lengths $a = 5$ cm and $b = 10$ cm, and the included angle is $\theta = \pi/3$. If a is increasing at a rate of 2 cm/sec, b is increasing at a rate of 1 cm/sec, and θ remains constant, at what rate is the third side changing? Is it increasing or decreasing? [*Hint:* Use the law of cosines.]

31. Let $z = f(y + cx) + g(y - cx)$, where $c \neq 0$. Show that

$$\frac{\partial^2 z}{\partial x^2} = c^2\frac{\partial^2 z}{\partial y^2}$$

[*Hint:* Introduce new variables $u = y + cx$ and $v = y - cx$.]

32. Let $z = f(x - y, y - x)$. Show that if f is differentiable then $\partial z/\partial x + \partial z/\partial y = 0$.

33. Let $z = f(x, y)$, $x = r \cos \theta$, $y = r \sin \theta$. Show that

(a) $\dfrac{\partial z}{\partial x} = \dfrac{\partial z}{\partial r} \cos \theta - \dfrac{1}{r} \dfrac{\partial z}{\partial \theta} \sin \theta$

(b) $\dfrac{\partial z}{\partial y} = \dfrac{\partial z}{\partial r} \sin \theta + \dfrac{1}{r} \dfrac{\partial z}{\partial \theta} \cos \theta$

(c) $\left(\dfrac{\partial z}{\partial x}\right)^2 + \left(\dfrac{\partial z}{\partial y}\right)^2 = \left(\dfrac{\partial z}{\partial r}\right)^2 + \dfrac{1}{r^2}\left(\dfrac{\partial z}{\partial \theta}\right)^2$.

34. Let $z = f(a + bt, c + dt)$. Express dz/dt in terms of partial derivatives of z.

35. Let $z = f(r, \theta)$, $r = \sqrt{x^2 + y^2}$, $\theta = \tan^{-1}\dfrac{y}{x}$. Express $\partial z/\partial x$ and $\partial z/\partial y$ in terms of $\partial z/\partial \theta$ and $\partial z/\partial r$.

36. Show that if $u(x, y)$ and $v(x, y)$ satisfy the Cauchy-Riemann equations (Exercise 34, Section 16.2), and if $x = r \cos \theta$ and $y = r \sin \theta$, then

$$\frac{\partial u}{\partial r} = \frac{1}{r}\frac{\partial v}{\partial \theta} \quad \text{and} \quad \frac{\partial v}{\partial r} = -\frac{1}{r}\frac{\partial u}{\partial \theta}$$

37. Prove: If f, f_x, and f_y are continuous on a circular region containing $A(x_0, y_0)$ and $B(x_1, y_1)$, then there is a point (x^*, y^*) on the line segment joining A and B such that

$$f(x_1, y_1) - f(x_0, y_0) \\ = f_x(x^*, y^*)(x_1 - x_0) + f_y(x^*, y^*)(y_1 - y_0)$$

This result is the two-dimensional version of the Mean-Value Theorem. [*Hint:* Express the line segment joining A and B in parametric form and use the Mean-Value Theorem for functions of one variable.]

38. Prove: If $f_x(x, y) = 0$ and $f_y(x, y) = 0$ throughout a circular region, then $f(x, y)$ is constant on that region. [*Hint:* Use the result of Exercise 37.]

16.5   DIRECTIONAL DERIVATIVES; GRADIENT

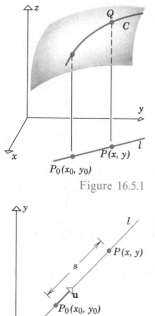

Figure 16.5.1

Figure 16.5.2

The partial derivatives $f_x(x, y)$ and $f_y(x, y)$ represent the rates of change of $f(x, y)$ in directions parallel to the x- and y-axes. In this section we will investigate rates of change of $f(x, y)$ in other directions.

Let $z = f(x, y)$, let (x_0, y_0) be a fixed point in the xy-plane, and let l be a line in the xy-plane passing through the point $P_0(x_0, y_0)$. If a point $P(x, y)$ moves along the line l, there is a companion point Q, directly above it, that moves along the surface $z = f(x, y)$, tracing out some curve C (Figure 16.5.1). If we let s be the distance from P_0 to P along l, then we can inquire about the rate at which the z-coordinate of Q changes with s.

To study this rate of change, let us introduce a *unit* vector

$$\mathbf{u} = \langle u_1, u_2 \rangle$$

with initial point $P_0(x_0, y_0)$ and pointing in the direction of motion of $P(x, y)$ (See Figure 16.5.2). Let s be the (signed) distance along l from $P_0(x_0, y_0)$ to $P(x, y)$, where s is nonnegative if P is in the direction of $\mathbf{u}$ from P_0 and is negative otherwise. The vector $\overrightarrow{P_0P}$ is related to the vector $\mathbf{u}$ by

$$\overrightarrow{P_0P} = s\mathbf{u}$$

or

$$\langle x - x_0, y - y_0 \rangle = \langle su_1, su_2 \rangle$$

(Figure 16.5.2). By equating corresponding components, we get the following parametric equations of l:

$$x = x_0 + su_1$$
$$y = y_0 + su_2 \tag{1}$$

Since the point Q is directly above $P(x, y)$ and lies on the surface, the z-coordinate of Q is

$$z = f(x, y)$$

or, in terms of s,

$$z = f(x_0 + su_1, y_0 + su_2)$$

The rate at which z changes with s can now be calculated using the chain rule. We obtain

$$\frac{dz}{ds} = f_x(x, y)\frac{dx}{ds} + f_y(x, y)\frac{dy}{ds}$$

or from (1)

$$\frac{dz}{ds} = f_x(x, y)u_1 + f_y(x, y)u_2 \tag{2}$$

where x and y in this formula are expressed in terms of s by equations (1). Equation (2) can be used to obtain dz/ds anywhere along C. For example, by setting $s = 0$ in (1) and substituting in (2) we obtain the instantaneous rate of change of z with s at the point $P_0(x_0, y_0)$:

$$\left.\frac{dz}{ds}\right|_{s=0} = f_x(x_0, y_0)u_1 + f_y(x_0, y_0)u_2 \tag{3}$$

This quantity is commonly denoted by $D_{\mathbf{u}}z(x_0, y_0)$ or $D_{\mathbf{u}}f(x_0, y_0)$.

16.5.1 DEFINITION If $f(x, y)$ is differentiable at (x_0, y_0) and if $\mathbf{u} = \langle u_1, u_2 \rangle$ is a unit vector, then the **directional derivative** of f at (x_0, y_0) in the direction of $\mathbf{u}$ is defined by

$$D_{\mathbf{u}}f(x_0, y_0) = f_x(x_0, y_0)u_1 + f_y(x_0, y_0)u_2 \tag{4}$$

The directional derivative $D_{\mathbf{u}}f(x_0, y_0)$ represents the instantaneous rate at which $z = f(x, y)$ changes with distance as (x, y) moves from (x_0, y_0) in the direction of $\mathbf{u}$. Geometrically, $D_{\mathbf{u}}f(x_0, y_0)$ may be interpreted as the slope of

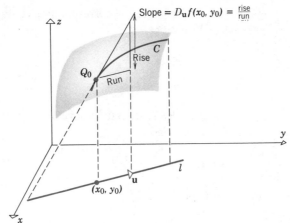

Figure 16.5.3

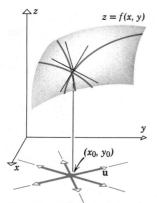

$z = f(x, y)$

The rate at which z changes with distance at (x_0, y_0) depends on the direction of $\mathbf{u}$

Figure 16.5.4

the tangent line to the curve C of Figure 16.5.3 at the point Q_0. It should be kept in mind that there are infinitely many directional derivatives of $z = f(x, y)$ at a point (x_0, y_0), one for each possible choice of the direction vector $\mathbf{u}$ (Figure 16.5.4).

▶ **Example 1**   Find the directional derivative of $f(x, y) = 3x^2y$ at the point $(1, 2)$ in the direction of the vector $\mathbf{a} = 3\mathbf{i} + 4\mathbf{j}$.

Solution.   The partial derivatives of f are

$$f_x(x, y) = 6xy, \qquad f_y(x, y) = 3x^2$$

so that

$$f_x(1, 2) = 12, \qquad f_y(1, 2) = 3$$

Thus, the directional derivative at $(1, 2)$ is

$$D_{\mathbf{u}}f(1, 2) = f_x(1, 2)u_1 + f_y(1, 2)u_2 = 12u_1 + 3u_2 \tag{5}$$

where u_1, u_2 are components of the *unit* vector in the direction of differentiation. We obtain these components by normalizing $\mathbf{a} = 3\mathbf{i} + 4\mathbf{j}$,

$$\mathbf{u} = \frac{\mathbf{a}}{\|\mathbf{a}\|} = \frac{1}{\sqrt{25}}(3\mathbf{i} + 4\mathbf{j}) = \frac{3}{5}\mathbf{i} + \frac{4}{5}\mathbf{j}$$

Thus, from (5)

$$D_{\mathbf{u}}f(1, 2) = 12\left(\frac{3}{5}\right) + 3\left(\frac{4}{5}\right) = \frac{48}{5}$$

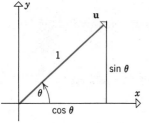

Figure 16.5.5

If $\mathbf{u} = u_1\mathbf{i} + u_2\mathbf{j}$ is a unit vector making an angle θ with the positive x-axis, then

$$u_1 = \cos\theta \quad \text{and} \quad u_2 = \sin\theta$$

(See Figure 16.5.5) so that (4) can be written in the form

$$D_{\mathbf{u}}f(x_0, y_0) = f_x(x_0, y_0)\cos\theta + f_y(x_0, y_0)\sin\theta \qquad (6)$$

▶ **Example 2** Find the directional derivative of e^{xy} at $(-2, 0)$ in the direction of the unit vector $\mathbf{u}$ that makes an angle of $\pi/3$ with the positive x-axis.

Solution. Let $f(x, y) = e^{xy}$, so that

$$f_x(x, y) = ye^{xy}, \qquad f_y(x, y) = xe^{xy}$$
$$f_x(-2, 0) = 0, \qquad f_y(-2, 0) = -2$$

From (6)

$$D_{\mathbf{u}}f(-2, 0) = f_x(-2, 0)\cos\frac{\pi}{3} + f_y(-2, 0)\sin\frac{\pi}{3}$$

$$= 0\left(\frac{1}{2}\right) + (-2)\left(\frac{\sqrt{3}}{2}\right) = -\sqrt{3} \qquad ◀$$

As might be expected, directional derivatives in the directions of the positive coordinate axes reduce to partial derivatives:

$$D_{\mathbf{i}}f(x_0, y_0) = f_x(x_0, y_0)(1) + f_y(x_0, y_0)(0) = f_x(x_0, y_0)$$
$$D_{\mathbf{j}}f(x_0, y_0) = f_x(x_0, y_0)(0) + f_y(x_0, y_0)(1) = f_y(x_0, y_0)$$

Moreover, directional derivatives in the directions of the negative coordinate axes reduce to negatives of partial derivatives:

$$D_{-\mathbf{i}}f(x_0, y_0) = f_x(x_0, y_0)(-1) + f_y(x_0, y_0)(0) = -f_x(x_0, y_0)$$
$$D_{-\mathbf{j}}f(x_0, y_0) = f_x(x_0, y_0)(0) + f_y(x_0, y_0)(-1) = -f_y(x_0, y_0)$$

The directional derivative formula

$$D_{\mathbf{u}}f(x, y) = f_x(x, y)u_1 + f_y(x, y)u_2$$

can be expressed in the form of a dot product by writing

$$D_{\mathbf{u}}f(x, y) = (f_x(x, y)\mathbf{i} + f_y(x, y)\mathbf{j}) \cdot (u_1\mathbf{i} + u_2\mathbf{j}) \qquad (7)$$

The second vector in the dot product is **u**. However, the first vector is something new; it is called the *gradient of f* and is denoted by the symbol ∇f or $\nabla f(x, y)$.*

16.5.2 DEFINITION If f is a function of x and y, then the **gradient of f** is defined by

$$\nabla f(x, y) = f_x(x, y)\mathbf{i} + f_y(x, y)\mathbf{j} \tag{8}$$

With the gradient notation, formula (7) for the directional derivative can be written in the following compact form:

$$D_\mathbf{u} f(x, y) = \nabla f(x, y) \cdot \mathbf{u} \tag{9}$$

In words, the dot product of the gradient of f with a unit vector **u** produces the directional derivative of f in the direction of **u**.

▶ **Example 3** Find the gradient of $f(x, y) = 3x^2 y$ at the point (1, 2) and use it to calculate the directional derivative of f at (1, 2) in the direction of the vector $\mathbf{a} = 3\mathbf{i} + 4\mathbf{j}$.

Solution. From (8)

$$\nabla f(x, y) = f_x(x, y)\mathbf{i} + f_y(x, y)\mathbf{j} = 6xy\mathbf{i} + 3x^2\mathbf{j}$$

so that the gradient of f at (1, 2) is

$$\nabla f(1, 2) = 12\mathbf{i} + 3\mathbf{j}$$

The unit vector in the direction of **a** is

$$\mathbf{u} = \frac{\mathbf{a}}{\|\mathbf{a}\|} = \frac{1}{5}(3\mathbf{i} + 4\mathbf{j}) = \tfrac{3}{5}\mathbf{i} + \tfrac{4}{5}\mathbf{j}$$

Thus, from (9)

$$D_\mathbf{u} f(1, 2) = \nabla f(1, 2) \cdot \mathbf{u} = (12\mathbf{i} + 3\mathbf{j}) \cdot (\tfrac{3}{5}\mathbf{i} + \tfrac{4}{5}\mathbf{j}) = \frac{48}{5}$$

which agrees with the result obtained in Example 1. ◀

The gradient is not merely a notational device to simplify the formula for the directional derivative; the length and direction of the gradient ∇f provide important information about the function f.

---

* The symbol ∇ (read, "del") is an inverted delta. In older books this symbol is sometimes called a "nabla" because of its similarity in form to an ancient Hebrew ten-stringed harp of that name.

16.5.3   THEOREM   *Let f be a function of two variables that is differentiable at (x_0, y_0).*

   (a)   If $\nabla f(x_0, y_0) = \mathbf{0}$, then all directional derivatives of f at (x_0, y_0) are zero.

   (b)   If $\nabla f(x_0, y_0) \neq \mathbf{0}$, then among all possible directional derivatives of f at (x_0, y_0), the derivative in the direction of $\nabla f(x_0, y_0)$ has the largest value. The value of that directional derivative is $\|\nabla f(x_0, y_0)\|$.

   (c)   If $\nabla f(x_0, y_0) \neq \mathbf{0}$, then among all possible directional derivatives of f at (x_0, y_0), the derivative in the direction opposite to that of $\nabla f(x_0, y_0)$ has the smallest value. The value of that directional derivative is $-\|\nabla f(x_0, y_0)\|$.

Proof of (a).   If $\nabla f(x_0, y_0) = \mathbf{0}$, then for all choices of $\mathbf{u}$ we have

$$D_{\mathbf{u}} f(x_0, y_0) = \nabla f(x_0, y_0) \cdot \mathbf{u} = \mathbf{0} \cdot \mathbf{u} = 0$$

Proofs of (b) and (c).   Assume $\nabla f(x_0, y_0) \neq \mathbf{0}$ and let θ be the angle between $\nabla f(x_0, y_0)$ and an arbitrary unit vector $\mathbf{u}$. By the definition of dot product,

$$D_{\mathbf{u}} f(x_0, y_0) = \nabla f(x_0, y_0) \cdot \mathbf{u} = \|\nabla f(x_0, y_0)\| \, \|\mathbf{u}\| \cos \theta$$

or, since $\|\mathbf{u}\| = 1$,

$$D_{\mathbf{u}} f(x_0, y_0) = \|\nabla f(x_0, y_0)\| \cos \theta$$

Thus, the maximum value of $D_{\mathbf{u}} f(x_0, y_0)$ is $\|\nabla f(x_0, y_0)\|$, and this occurs when $\cos \theta = 1$, or when $\mathbf{u}$ has the same direction as $\nabla f(x_0, y_0)$ (since $\theta = 0$). The minimum value of $D_{\mathbf{u}} f(x_0, y_0)$ is $-\|\nabla f(x_0, y_0)\|$, and this occurs when $\cos \theta = -1$ or when $\mathbf{u}$ and $\nabla f(x_0, y_0)$ are oppositely directed (since $\theta = \pi$). ■

▶ **Example 4**   For the function $f(x, y) = x^2 e^y$, find the maximum value of a directional derivative at $(-2, 0)$, and give a unit vector in the direction of the maximum.

Solution.   Since

$$\nabla f(x, y) = f_x(x, y)\mathbf{i} + f_y(x, y)\mathbf{j} = 2xe^y \mathbf{i} + x^2 e^y \mathbf{j}$$

the gradient of f at $(-2, 0)$ is

$$\nabla f(-2, 0) = -4\mathbf{i} + 4\mathbf{j}$$

By Theorem 16.5.3, the maximum value of the directional derivative is

$$\|\nabla f(-2, 0)\| = \sqrt{(-4)^2 + 4^2} = \sqrt{32} = 4\sqrt{2}$$

This maximum occurs in the direction of $\nabla f(-2, 0)$. A unit vector in this direction is

$$\frac{\nabla f(-2, 0)}{\|\nabla f(-2, 0)\|} = \frac{1}{4\sqrt{2}}(-4\mathbf{i} + 4\mathbf{j}) = -\frac{1}{\sqrt{2}}\mathbf{i} + \frac{1}{\sqrt{2}}\mathbf{j} \qquad \blacktriangleleft$$

There is an important geometric relationship between the gradient $\nabla f(x, y)$ and the level curves of $f(x, y)$. If (x_0, y_0) is any point in the domain of f, and if $f(x_0, y_0) = c$, then the equation

$$f(x, y) = c$$

defines a level curve of f that passes through the point (x_0, y_0) (Figure 16.5.6a). Let us recall that the function f has its maximum rate of increase at (x_0, y_0) in the direction of $\nabla f(x_0, y_0)$ and has its maximum rate of decrease in

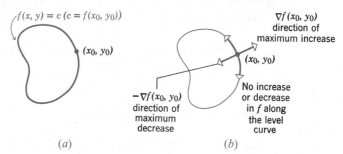

Figure 16.5.6   (a)   (b)

the direction of $-\nabla f(x_0, y_0)$. The level curve through (x_0, y_0) is the "dividing line" between these extremes in the sense that $f(x, y)$ neither increases nor decreases along this curve (Figure 16.5.6b). The following theorem shows that the gradient of f at (x_0, y_0) is perpendicular to the level curve of f through (x_0, y_0). (A vector is **perpendicular** or **normal** to a curve at a point if it is perpendicular to a nonzero tangent vector at that point.)

16.5.4   THEOREM   *If f is differentiable at (x_0, y_0) and $\nabla f(x_0, y_0) \neq \mathbf{0}$, then $\nabla f(x_0, y_0)$ is perpendicular to the level curve of f through (x_0, y_0).*

To see why this result is true, let

$$f(x, y) = c \qquad (10)$$

be the level curve through (x_0, y_0). It is proved in advanced calculus that with the given hypotheses, this level curve can be represented parametrically by equations

$$x = x(t)$$

$$y = y(t)$$

so that the level curve has a nonzero tangent vector at (x_0, y_0). More precisely,

$$x'(t_0)\mathbf{i} + y'(t_0)\mathbf{j} \neq \mathbf{0}$$

where t_0 is the value of the parameter corresponding to (x_0, y_0).

Differentiating both sides of (10) with respect to t, and applying the chain rule yields

$$f_x(x(t), y(t))x'(t) + f_y(x(t), y(t))y'(t) = 0$$

Substituting $t = t_0$, and using the fact that $x(t_0) = x_0$ and $y(t_0) = y_0$, we obtain

$$f_x(x_0, y_0)x'(t_0) + f_y(x_0, y_0)y'(t_0) = 0$$

which can be written as

$$\nabla f(x_0, y_0) \cdot (x'(t_0)\mathbf{i} + y'(t_0)\mathbf{j}) = 0$$

This equation tells us that $\nabla f(x_0, y_0)$ is perpendicular to the tangent vector $x'(t_0)\mathbf{i} + y'(t_0)\mathbf{j}$ and, therefore, is perpendicular to the level curve through (x_0, y_0).

▶ **Example 5** For the function $f(x, y) = x^2 + y^2$, sketch the level curve through the point (3, 4), and draw the gradient vector at this point.

Solution. Since $f(3, 4) = 25$, the level curve through the point (3, 4) has the equation $f(x, y) = 25$, which is the circle

$$x^2 + y^2 = 25$$

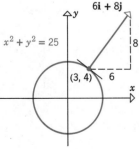

Figure 16.5.7

Since

$$\nabla f(x, y) = f_x(x, y)\mathbf{i} + f_y(x, y)\mathbf{j} = 2x\mathbf{i} + 2y\mathbf{j}$$

the gradient vector at (3, 4) is

$$\nabla f(3, 4) = 6\mathbf{i} + 8\mathbf{j}$$

(Figure 16.5.7). Note that the gradient vector is perpendicular to the circle at (3, 4) as guaranteed by Theorem 16.5.4. ◀

▶ Exercise Set 16.5

In Exercises 1–4, find ∇z.

1. $z = 4x - 8y$.

2. $z = e^{-3y} \cos 4x$.

3. $z = \ln \sqrt{x^2 + y^2}$.

4. $z = e^{-5x} \sec x^2 y$.

In Exercises 5–8, find the gradient of f at the indicated point.

5. $f(x, y) = (x^2 + xy)^3$; $(-1, -1)$.

6. $f(x, y) = (x^2 + y^2)^{-1/2}$; $(3, 4)$.

7. $f(x, y) = y \ln (x + y)$; $(-3, 4)$.

8. $f(x, y) = y^2 \tan^3 x$; $(\pi/4, -3)$.

In Exercises 9–12, find $D_{\mathbf{u}} f$ at P.

9. $f(x, y) = (1 + xy)^{3/2}$; $P(3, 1)$; $\mathbf{u} = \dfrac{1}{\sqrt{2}} \mathbf{i} + \dfrac{1}{\sqrt{2}} \mathbf{j}$.

10. $f(x, y) = e^{2xy}$; $P(4, 0)$; $\mathbf{u} = -\frac{3}{5}\mathbf{i} + \frac{4}{5}\mathbf{j}$.

11. $f(x, y) = \ln (1 + x^2 + y)$; $P(0, 0)$;

$$\mathbf{u} = -\frac{1}{\sqrt{10}}\mathbf{i} - \frac{3}{\sqrt{10}}\mathbf{j}.$$

12. $f(x, y) = \dfrac{cx + dy}{x - y}$; $P(3, 4)$; $\mathbf{u} = \frac{4}{5}\mathbf{i} + \frac{3}{5}\mathbf{j}$.

In Exercises 13–18, find the directional derivative of f at P in the direction of $\mathbf{a}$.

13. $f(x, y) = 4x^3 y^2$; $P(2, 1)$; $\mathbf{a} = 4\mathbf{i} - 3\mathbf{j}$.

14. $f(x, y) = x^2 - 3xy + 4y^3$; $P(-2, 0)$; $\mathbf{a} = \mathbf{i} + 2\mathbf{j}$.

15. $f(x, y) = y^2 \ln x$; $P(1, 4)$; $\mathbf{a} = -3\mathbf{i} + 3\mathbf{j}$.

16. $f(x, y) = e^x \cos y$; $P(0, \pi/4)$; $\mathbf{a} = 5\mathbf{i} - 2\mathbf{j}$.

17. $f(x, y) = \tan^{-1}\left(\dfrac{y}{x}\right)$; $P(-2, 2)$; $\mathbf{a} = -\mathbf{i} - \mathbf{j}$.

18. $f(x, y) = xe^y - ye^x$; $P(0, 0)$; $\mathbf{a} = 5\mathbf{i} - 2\mathbf{j}$.

In Exercises 19–22, find the directional derivative of f at P in the direction of a vector making the angle θ with the positive x-axis.

19. $f(x, y) = \sqrt{xy}$; $P(1, 4)$; $\theta = \pi/3$.

20. $f(x, y) = \dfrac{x - y}{x + y}$; $P(-1, -2)$; $\theta = \pi/2$.

21. $f(x, y) = \tan (2x + y)$; $P(\pi/6, \pi/3)$; $\theta = 7\pi/4$.

22. $f(x, y) = \sinh x \cosh y$; $P(0, 0)$; $\theta = \pi$.

In Exercises 23–26, sketch the level curve of $f(x, y)$ that passes through P and draw the gradient vector at P.

23. $f(x, y) = 4x - 2y + 3$; $P(1, 2)$.

24. $f(x, y) = y/x^2$; $P(-2, 2)$.

25. $f(x, y) = x^2 + 4y^2$; $P(-2, 0)$.

26. $f(x, y) = x^2 - y^2$; $P(2, -1)$.

In Exercises 27–30, find a unit vector in the direction in which f increases most rapidly at P; and find the rate of change in f at P in that direction.

27. $f(x, y) = 4x^3 y^2$; $P(-1, 1)$.

28. $f(x, y) = 3x - \ln y$; $P(2, 4)$.

29. $f(x, y) = \sqrt{x^2 + y^2}$; $P(4, -3)$.

30. $f(x, y) = \dfrac{x}{x + y}$; $P(0, 2)$.

In Exercises 31–34, find a unit vector in the direction in which f decreases most rapidly at P; and find the rate of change of f at P in that direction.

31. $f(x, y) = 20 - x^2 - y^2$; $P(-1, -3)$.

32. $f(x, y) = e^{xy}$; $P(2, 3)$.

33. $f(x, y) = \cos (3x - y)$; $P\left(\dfrac{\pi}{6}, \dfrac{\pi}{4}\right)$.

34. $f(x, y) = \sqrt{\dfrac{x - y}{x + y}}$; $P(3, 1)$.

35. Find the directional derivative of $f(x, y) = \dfrac{x}{x + y}$ at $P(1, 0)$ in the direction to $Q(-1, -1)$.

36. Find the directional derivative of $f(x, y) = e^{-x} \sec y$ at $P(0, \pi/4)$ in the direction of the origin.

37. Find the directional derivative of $f(x, y) = \sqrt{xy} \, e^y$ at $P(1, 1)$ in the direction of the negative y-axis.

38. Let $f(x, y) = \dfrac{y}{x + y}$. Find a unit vector $\mathbf{u}$ for which $D_{\mathbf{u}} f(2, 3) = 0$.

39. Find a unit vector $\mathbf{u}$ that is perpendicular at $P(1, -2)$ to the level curve of $f(x, y) = 4x^2 y$ through P.

40. Find a unit vector **u** that is perpendicular at $P(2, -3)$ to the level curve of $f(x, y) = 3x^2y - xy$ through P.

41. Given that $D_{\mathbf{u}}f(1, 2) = -5$ if $\mathbf{u} = \frac{3}{5}\mathbf{i} - \frac{4}{5}\mathbf{j}$ and $D_{\mathbf{v}}f(1, 2) = 10$ if $\mathbf{v} = \frac{4}{5}\mathbf{i} + \frac{3}{5}\mathbf{j}$, find
(a) $f_x(1, 2)$
(b) $f_y(1, 2)$
(c) the directional derivative of f at $(1, 2)$ in the direction of the origin.

42. Given that $f_x(-5, 1) = -3$ and $f_y(-5, 1) = 2$, find the directional derivative of f at $P(-5, 1)$ in the direction to $Q(-4, 3)$.

43. Given that $\nabla f(4, -5) = 2\mathbf{i} - \mathbf{j}$, find the directional derivative of f at $(4, -5)$ in the direction of $\mathbf{a} = 5\mathbf{i} + 2\mathbf{j}$.

44. The temperature at a point (x, y) on a metal plate in the xy-plane is $T(x, y) = \dfrac{xy}{1 + x^2 + y^2}$ degrees Celsius.
(a) Find the rate of change of temperature at $(1, 1)$ in the direction of $\mathbf{a} = 2\mathbf{i} - \mathbf{j}$.
(b) An ant at $(1, 1)$ wants to walk in the direction in which the temperature drops most rapidly. Find a unit vector in that direction.

45. If the electric potential at a point (x, y) in the xy-plane is $V(x, y)$, then the **electric intensity vector** at (x, y) is $\mathbf{E} = -\nabla V(x, y)$. Suppose $V(x, y) = e^{-2x} \cos 2y$.
(a) Find the electric intensity vector at $(\pi/4, 0)$.
(b) Show that at each point in the plane, the electric potential decreases most rapidly in the direction of **E**.

46. On a certain mountain, the elevation z above a point (x, y) in a horizontal xy-plane at sea level is $z = 2000 - 2x^2 - 4y^2$ ft. The positive x-axis points east, and the positive y-axis north. A climber is at the point $(-20, 5, 1100)$.
(a) If the climber uses a compass reading to walk due west, will he ascend or descend?
(b) If the climber uses a compass reading to walk northeast, will he ascend or descend? At what rate?
(c) In what compass direction should the climber walk to travel a level path?

47. Let $\mathbf{u}_r$ be a unit vector making an angle θ with the positive x-axis, and let $\mathbf{u}_\theta$ be a unit vector $90°$ counterclockwise from $\mathbf{u}_r$. Show that if $z = f(x, y)$, $x = r\cos\theta$, and $y = r\sin\theta$, then

$$\nabla z = \frac{\partial z}{\partial r}\mathbf{u}_r + \frac{1}{r}\frac{\partial z}{\partial \theta}\mathbf{u}_\theta$$

[*Hint:* Use parts (a) and (b) of Exercise 33, Section 16.4.]

48. Prove: If f and g are differentiable, then
(a) $\nabla(f + g) = \nabla f + \nabla g$
(b) $\nabla(cf) = c\nabla f$ (c constant)
(c) $\nabla(fg) = f\nabla g + g\nabla f$
(d) $\nabla\left(\dfrac{f}{g}\right) = \dfrac{g\nabla f - f\nabla g}{g^2}$
(e) $\nabla(f^p) = pf^{p-1}\nabla f$.

49. Prove: If $x = x(t)$ and $y = y(t)$ are differentiable at t, and if $z = f(x, y)$ is differentiable at the point $(x(t), y(t))$, then

$$\frac{dz}{dt} = \nabla z \cdot \mathbf{r}'(t)$$

where $\mathbf{r}(t) = x(t)\mathbf{i} + y(t)\mathbf{j}$.

50. Prove: If f, f_x, and f_y are continuous on a circular region, and if $\nabla f(x, y) = \mathbf{0}$ throughout the region, then $f(x, y)$ is constant on the region. [*Hint:* Exercise 38, Section 16.4.]

16.6 TANGENT PLANES

In this section we discuss tangent planes to surfaces in three-dimensional space. We are concerned with three main questions: What is a tangent plane? When do tangent planes exist? How do we find equations of tangent planes?

We will begin with some terminology. Let $P_0(x_0, y_0, z_0)$ be a point on a curve C in 3-space. If the curve has a unit tangent vector **T** and a unit normal vector **N** at a point P_0, then the line through P_0 parallel to **T** will be called the *tangent line* to C at P_0, and the line through P_0 parallel to **N** will be called the *normal line* to C at P_0 (Figure 16.6.1).

16.6.1 DEFINITION Let $P_0(x_0, y_0, z_0)$ be a point on a surface $z = f(x, y)$, and suppose that f is differentiable at $(x_0 \, y_0)$. Consider the set of curves on the surface that are intersections of the surface S with vertical planes through P_0. If each of these curves has a tangent line at P_0, and if these tangent lines lie in a common plane, then this plane is called the *tangent plane* to the surface at P_0. The line through P_0 perpendicular to the tangent plane is called the *normal line* to the surface at P_0 and a vector normal to the tangent plane at P_0 is said to be *normal* to the surface at P_0 (Figure 16.6.2).

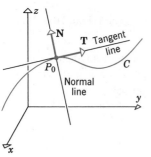

Figure 16.6.1

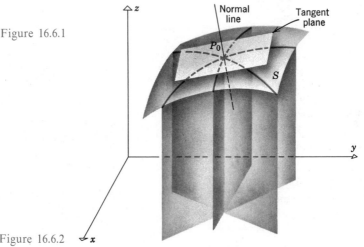

Figure 16.6.2

16.6.2 THEOREM *Let $P_0(x_0, y_0, z_0)$ be any point on the surface $z = f(x, y)$. If $f(x, y)$ is differentiable at (x_0, y_0), then the surface has a nonvertical tangent plane at P_0, and this plane has the equation*

$$f_x(x_0, y_0)(x - x_0) + f_y(x_0, y_0)(y - y_0) - (z - z_0) = 0 \qquad (1)$$

Moreover, the vector

$$\mathbf{n} = \langle f_x(x_0, y_0), f_y(x_0, y_0), -1 \rangle \qquad (2)$$

is normal to the surface at P_0, and the normal line to the surface at P_0 has parametric equations

$$x = x_0 + f_x(x_0, y_0)t$$
$$y = y_0 + f_y(x_0, y_0)t \qquad\qquad (3)$$
$$z = z_0 - t$$

Proof. We must show that all vertical planes through $P_0(x_0, y_0, z_0)$ produce curves on the surface $z = f(x, y)$ whose tangent lines at P_0 lie in single plane.

Let C be the intersection of the surface $z = f(x, y)$ with a vertical plane through $P_0(x_0, y_0, z_0)$. We obtain parametric equations of C as follows. The vertical plane that produces the trace C intersects the xy-plane in a line L passing through the point (x_0, y_0) (Figure 16.6.3). If $\mathbf{u} = u_1\mathbf{i} + u_2\mathbf{j}$ denotes a unit vector along L, then L can be represented parametrically by the equations

$$x = x_0 + su_1$$
$$y = y_0 + su_2$$

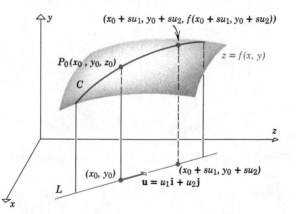

Figure 16.6.3

[See the derivation of (1) in Section 16.5.] Each point (x, y) on L has a companion point (x, y, z) on the trace C, where $z = f(x, y)$. Thus, as s varies over the interval $(-\infty, +\infty)$ the point $(x, y) = (x_0 + su_1, y_0 + su_2)$ varies over L, and the companion point

$$(x, y, z) = \big(x_0 + su_1, y_0 + su_2, f(x_0 + su_1, y_0 + su_2)\big)$$

varies over the trace C (Figure 16.6.3). Thus, C may be represented parametrically by the equations

$$x = x_0 + su_1$$
$$y = y_0 + su_2 \qquad\qquad (4)$$
$$z = f(x_0 + su_1, y_0 + su_2)$$

or equivalently by the single vector-valued function

$$\mathbf{r}(s) = (x_0 + su_1)\mathbf{i} + (y_0 + su_2)\mathbf{j} + f(x_0 + su_1, y_0 + su_2)\mathbf{k}$$

From the chain rule, the derivative of $\mathbf{r}(s)$ with respect to s may be written

$$\mathbf{r}'(s) = u_1\mathbf{i} + u_2\mathbf{j} + [f_x(x_0 + su_1, y_0 + su_2)u_1 \\ + f_y(x_0 + su_1, y_0 + su_2)u_2]\mathbf{k}$$

Geometrically, $\mathbf{r}'(s)$ is tangent to the curve C at the point corresponding to s. In particular, the vector

$$\mathbf{t} = \mathbf{r}'(0) = u_1\mathbf{i} + u_2\mathbf{j} + [f_x(x_0, y_0)u_1 + f_y(x_0, y_0)u_2]\mathbf{k} \qquad (5)$$

is tangent to C at $P_0(x_0, y_0, z_0)$ since the point P_0 corresponds to $s = 0$.

To show that all the vectors of form (5) lie in a single plane, we will choose two of these vectors and show that the rest lie in the plane determined by these two. Setting $u_1 = 0$, $u_2 = 1$ in (5) yields the vector

$$\mathbf{t}_1 = \mathbf{j} + f_y(x_0, y_0)\mathbf{k}$$

and setting $u_1 = 1$, $u_2 = 0$ in (5) yields the vector

$$\mathbf{t}_2 = \mathbf{i} + f_x(x_0, y_0)\mathbf{k}$$

(The vectors $\mathbf{t}_1$ and $\mathbf{t}_2$ are tangent vectors at P_0 to the traces by vertical planes parallel to the y-axis and x-axis, respectively.) Forming a 3×3 determinant from the components of $\mathbf{t}$, $\mathbf{t}_1$, and $\mathbf{t}_2$ we obtain (verify)

$$\begin{vmatrix} u_1 & u_2 & f_x(x_0, y_0)u_1 + f_y(x_0, y_0)u_2 \\ 0 & 1 & f_y(x_0, y_0) \\ 1 & 0 & f_x(x_0, y_0) \end{vmatrix} = 0$$

Thus, $\mathbf{t}$, $\mathbf{t}_1$, and $\mathbf{t}_2$ lie in the same plane by Theorem 15.4.6. This proves the existence of a tangent plane at P_0.

Since the vectors $\mathbf{t}_1$ and $\mathbf{t}_2$ lie in the tangent plane at P_0, their cross product

$$\mathbf{n} = \mathbf{t}_1 \times \mathbf{t}_2 = \begin{vmatrix} \mathbf{i} & \mathbf{j} & \mathbf{k} \\ 0 & 1 & f_y(x_0, y_0) \\ 1 & 0 & f_x(x_0, y_0) \end{vmatrix} = f_x(x_0, y_0)\mathbf{i} + f_y(x_0, y_0)\mathbf{j} - \mathbf{k}$$

is normal to the tangent plane. This normal $\mathbf{n}$ and the point (x_0, y_0, z_0) yield the equation of the tangent plane:

$$f_x(x_0, y_0)(x - x_0) + f_y(x_0, y_0)(y - y_0) - (z - z_0) = 0$$

This tangent plane is not vertical since the normal **n** has a nonzero **k** component.

Finally, since the vector **n** is parallel to the line normal to the surface at P_0, this normal line has parametric equations

$$x = x_0 + f_x(x_0, y_0)t$$
$$y = y_0 + f_y(x_0, y_0)t$$
$$z = z_0 - t$$

▶ **Example 1** Find equations of the tangent plane and normal line to the surface $z = x^2y$ at the point $(2, 1, 4)$.

Solution. Since $f(x, y) = x^2y$, it follows that

$$f_x(x, y) = 2xy \quad \text{and} \quad f_y(x, y) = x^2$$

so that with $x = 2$, $y = 1$,

$$f_x(2, 1) = 4 \quad \text{and} \quad f_y(2, 1) = 4$$

Thus, a vector normal to the surface at $(2, 1, 4)$ is

$$\mathbf{n} = f_x(2, 1)\mathbf{i} + f_y(2, 1)\mathbf{j} - \mathbf{k} = 4\mathbf{i} + 4\mathbf{j} - \mathbf{k}$$

Therefore, the tangent plane has the equation

$$4(x - 2) + 4(y - 1) - (z - 4) = 0$$

or

$$4x + 4y - z = 8$$

and the normal line has equations

$$x = 2 + 4t$$
$$y = 1 + 4t$$
$$z = 4 - t \qquad ◀$$

Recall that if $y = f(x)$ is a function of one variable, then the differential

$$dy = f'(x_0)\, dx$$

represents the change in y along the *tangent line* at (x_0, y_0) produced by a change dx in x. In contrast

$$\Delta y = f(x_0 + \Delta x) - f(x_0)$$

represents the change in y along the *curve* $y = f(x)$ produced by a change Δx in x. Analogously, if $z = f(x, y)$ is a function of two variables, we will define dz to be the change in z along the *tangent plane* at (x_0, y_0, z_0) to the surface $z = f(x, y)$ produced by changes dx and dy in x and y. This is in contrast to

$$\Delta z = f(x_0 + \Delta x, y_0 + \Delta y) - f(x_0, y_0)$$

which represents the change in z *along the surface* produced by changes Δx and Δy in x and y. A comparison of dz and Δz is shown in Figure 16.6.4 in the case where $dx = \Delta x$ and $dy = \Delta y$.

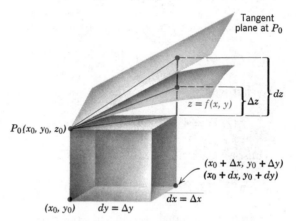

Figure 16.6.4

To derive a formula for dz, let $P_0(x_0, y_0, z_0)$ be a fixed point on the surface $z = f(x, y)$. If we assume f to be differentiable at (x_0, y_0), then the surface has a tangent plane at P_0, given by the equation

$$f_x(x_0, y_0)(x - x_0) + f_y(x_0, y_0)(y - y_0) - (z - z_0) = 0$$

or

$$z = z_0 + f_x(x_0, y_0)(x - x_0) + f_y(x_0, y_0)(y - y_0) \tag{6}$$

From (6) it follows that the tangent plane has height z_0 when $x = x_0, y = y_0$, and it has height

$$z_0 + f_x(x_0, y_0)\, dx + f_y(x_0, y_0)\, dy \tag{7}$$

when $x = x_0 + dx$, $y = y_0 + dy$. Thus, the change dz in the height of the tangent plane as (x, y) varies from (x_0, y_0) to $(x_0 + dx, y_0 + dy)$ is obtained by subtracting z_0 from expression (7). This yields

$$dz = f_x(x_0, y_0) \, dx + f_y(x_0, y_0) \, dy$$

This quantity is called the **total differential** of z at (x_0, y_0). Often we omit the subscripts on x_0 and y_0 and write

$$dz = f_x(x, y) \, dx + f_y(x, y) \, dy \qquad (8)$$

In this formula, dx and dy are usually viewed as variables and x and y fixed numbers. Formula (8) may also be written using df rather than dz.

If $z = f(x, y)$ is differentiable at the point (x, y), then the increment Δz can be written

$$\Delta z = f_x(x, y) \, \Delta x + f_y(x, y) \, \Delta y + \epsilon_1 \, \Delta x + \epsilon_2 \, \Delta y \qquad (9)$$

where $\epsilon_1 \to 0$, $\epsilon_2 \to 0$ as $(\Delta x, \Delta y) \to (0, 0)$. In the case, where $dx = \Delta x$ and $dy = \Delta y$ it follows from (8) and (9) that

$$\Delta z = dz + \epsilon_1 \, \Delta x + \epsilon_2 \, \Delta y$$

Thus, when $\Delta x = dx$ and $\Delta y = dy$ are small, we can approximate Δz by

$$\Delta z \approx dz \qquad (10)$$

Geometrically, this approximation tells us that the change in z along the surface and the change in z along the tangent plane are approximately equal when $\Delta x = dx$ and $\Delta y = dy$ are small (see Figure 16.6.4).

▶ **Example 2** Let $z = 4x^3y^2$. Find dz.

Solution. Since $f(x, y) = 4x^3y^2$,

$$f_x(x, y) = 12x^2y^2 \qquad \text{and} \qquad f_y(x, y) = 8x^3y$$

so

$$dz = 12x^2y^2 \, dx + 8x^3y \, dy \qquad ◀$$

▶ **Example 3** Let $f(x, y) = \sqrt{x^2 + y^2}$. Use a total differential to approximate the change in $f(x, y)$ as (x, y) varies from the point $(3, 4)$ to the point $(3.04, 3.98)$.

Solution. We will approximate Δf (the change in f) by

$$df = f_x(x, y) \, dx + f_y(x, y) \, dy = \frac{x}{\sqrt{x^2 + y^2}} \, dx + \frac{y}{\sqrt{x^2 + y^2}} \, dy$$

Since $x = 3$, $y = 4$, $dx = 0.04$, and $dy = -0.02$, we obtain

$$\Delta f \approx df = \frac{3}{\sqrt{3^2 + 4^2}}(0.04) + \frac{4}{\sqrt{3^2 + 4^2}}(-0.02)$$

$$= \frac{3}{5}(0.04) - \frac{4}{5}(0.02)$$

$$= 0.008$$

The reader may want to use a hand calculator to show that the true value of Δf to five decimal places is

$$\Delta f = \sqrt{(3.04)^2 + (3.98)^2} - \sqrt{3^2 + 4^2} \approx 0.00819 \qquad \blacktriangleleft$$

▶ **Example 4**   The volume V of a right-circular cone of radius r and height h is given by $V = \frac{1}{3}\pi r^2 h$. Suppose the height increases from 10 cm to 10.01 cm, while the radius decreases from 12 cm to 11.95 cm. Use a total differential to approximate the change ΔV in volume.

Solution.   We will approximate ΔV by

$$dV = \frac{\partial V}{\partial r}\, dr + \frac{\partial V}{\partial h}\, dh = \tfrac{2}{3}\pi r h\, dr + \tfrac{1}{3}\pi r^2\, dh$$

With $h = 10$, $dh = 0.01$, $r = 12$, and $dr = -0.05$, we obtain

$$\Delta V \approx dV = \tfrac{2}{3}\pi(12)(10)(-0.05) + \tfrac{1}{3}\pi(12)^2(0.01)$$

or

$$\Delta V \approx -3.52\pi \approx -11.06$$

Thus, the volume decreases by approximately 11.06 cm^3.   ◀

▶ **Example 5**   The radius of a right-circular cylinder is measured with an error of at most 2%, and the height is measured with an error of at most 4%. Estimate the maximum possible percentage error in the calculated volume V.

Solution.   Let r, h, and V be the true radius, height, and volume of the cylinder, and let Δr, Δh, and ΔV be the errors in these quantities. We are given that

$$\left| \frac{\Delta r}{r} \right| \leq 0.02 \qquad \text{and} \qquad \left| \frac{\Delta h}{h} \right| \leq 0.04$$

We want to find the maximum possible value of $|\Delta V / V|$. Since the volume of the cylinder is $V = \pi r^2 h$, it follows from (8) that

$$dV = \frac{\partial V}{\partial r}\, dr + \frac{\partial V}{\partial h}\, dh = 2\pi rh\, dr + \pi r^2\, dh$$

If we choose $dr = \Delta r$ and $dh = \Delta h$, then we may use the approximations

$$\Delta V \approx dV \qquad \text{and} \qquad \frac{\Delta V}{V} \approx \frac{dV}{V}$$

But

$$\frac{dV}{V} = \frac{2\pi rh\, dr + \pi r^2\, dh}{\pi r^2 h} = 2\frac{dr}{r} + \frac{dh}{h}$$

so by the triangle inequality (Theorem 1.2.5)

$$\left|\frac{dV}{V}\right| = \left|2\frac{dr}{r} + \frac{dh}{h}\right| \le 2\left|\frac{dr}{r}\right| + \left|\frac{dh}{h}\right| \le 2(0.02) + (0.04) = 0.08$$

Thus, the maximum percentage error in V is approximately 8%. ◀

▶ Exercise Set 16.6

In Exercises 1–8, find equations for the tangent plane and normal line to the given surface at the point P.

1. $z = 4x^3 y^2 + 2y$; $P(1, -2, 12)$.

2. $z = \frac{1}{2}x^7 y^{-2}$; $P(2, 4, 4)$.

3. $z = xe^{-y}$; $P(1, 0, 1)$.

4. $z = \ln \sqrt{x^2 + y^2}$; $P(-1, 0, 0)$.

5. $z = e^{3y} \sin 3x$; $P(\pi/6, 0, 1)$.

6. $z = x^{1/2} + y^{1/2}$; $P(4, 9, 5)$.

7. $x^2 + y^2 + z^2 = 25$; $P(-3, 0, 4)$.

8. $x^2 y - 4z^2 = -7$; $P(-3, 1, -2)$.

In Exercises 9–12, find dz.

9. $z = 7x - 2y$.   10. $z = 5x^2 y^5 - 2x + 4y + 7$.

11. $z = \tan^{-1} xy$.   12. $z = \sec^2 (x - 3y)$.

In Exercises 13–16, use a total differential to approximate the change in $f(x, y)$ as (x, y) varies from P to Q.

13. $f(x, y) = x^2 + 2xy - 4x$; $P(1, 2)$, $Q(1.01, 2.04)$.

14. $f(x, y) = x^{1/3} y^{1/2}$; $P(8, 9)$, $Q(7.78, 9.03)$.

15. $f(x, y) = \dfrac{x + y}{xy}$; $P(-1, -2)$, $Q(-1.02, -2.04)$.

16. $f(x, y) = \ln \sqrt{1 + xy}$; $P(0, 2)$, $Q(-0.09, 1.98)$

17. Find all points on the surface at which the tangent plane is horizontal.
    (a) $z = x^3 y^2$
    (b) $z = x^2 - xy + y^2 - 2x + 4y$.

18. Find a point on the surface $z = 3x^2 - y^2$ at which the tangent plane is parallel to the plane $6x + 4y - z = 5$.

19. Find a point on the surface $z = 8 - 3x^2 - 2y^2$ at which the tangent plane is perpendicular to the line, $x = 2 - 3t$, $y = 7 + 8t$, $z = 5 - t$.

20. Show that the surfaces $z = \sqrt{x^2 + y^2}$ and $z = \frac{1}{10}(x^2 + y^2) + \frac{5}{2}$ intersect at $(3, 4, 5)$ and have a common tangent plane at that point.

21. Show that the surfaces $z = \sqrt{16 - x^2 - y^2}$ and $z = \sqrt{x^2 + y^2}$ intersect at $(2, 2, 2\sqrt{2})$ and have tangent planes that are perpendicular at that point.

22. (a) Show that every line normal to the cone $z = \sqrt{x^2 + y^2}$ passes through the z-axis.
    (b) Show that every tangent plane intersects the cone in a line passing through the origin.

23. One leg of a right triangle increases from 3 cm to 3.2 cm, while the other leg decreases from 4 cm to 3.96 cm. Use a total differential to approximate the change in the length of the hypotenuse.

24. The volume V of a right-circular cone of radius r and height h is given by $V = \frac{1}{3}\pi r^2 h$. Suppose the height decreases from 20 in. to 19.95 in., while the radius increases from 4 in. to 4.05 in. Use a total differential to approximate the change in volume.

25. The length and width of a rectangle are measured with errors of at most 3% and 5%, respectively. Use differentials to estimate the percentage error in the calculated area.

26. The radius and height of a right-circular cone are measured with errors of at most 1% and 4%, respectively. Use differentials to estimate the percentage error in the calculated volume.

27. The length and width of a rectangle are measured with errors of at most r%. Use differentials to estimate the percentage error in the calculated length of the diagonal.

28. The legs of a right triangle are measured to be 3 cm and 4 cm, with a maximum error of 0.05 cm in each measurement. Use differentials to estimate the maximum possible error in the calculated value of (a) the hypotenuse and (b) the area of the triangle.

29. The total resistance R of two resistances R_1 and R_2, connected in parallel, is

$$R = \frac{R_1 R_2}{R_1 + R_2}$$

Suppose R_1 and R_2 are measured to be 200 ohms and 400 ohms, respectively, with a maximum error of 2% in each. Use differentials to estimate the maximum percentage error in the calculated value of R.

30. According to the ideal gas law, the pressure, temperature, and volume of a confined gas are related by $P = kT/V$ where k is a constant. Use differentials to estimate the percentage change in pressure if the temperature of a gas is increased 3% and the volume is increased 5%.

31. An angle θ of a right triangle is calculated by the formula

$$\theta = \sin^{-1}\frac{a}{c}$$

where a is the length of the side opposite to θ, and c is the length of the hypotenuse. Suppose that the measurements $a = 3$ in. and $c = 5$ in. each have a maximum possible error of 0.01 in. Use differentials to estimate the maximum possible error in the calculated value of θ.

32. An open cylindrical can has an inside radius of 2 cm and an inside height of 5 cm. Use differentials to estimate the volume of metal in the can if it is 0.01 cm thick. [*Hint:* The volume of metal is the difference, ΔV, in the volumes of two cylinders.]

33. (a) Find all points of intersection of the line $x = -1 + t$, $y = 2 + t$, $z = 2t + 7$, and the surface $z = x^2 + y^2$.
   (b) At each point of intersection, find the cosine of the acute angle between the given line and the line normal to the surface.

34. Show that if f is differentiable and $z = xf(x/y)$, then all tangent planes to the graph of this equation pass through a common point.

35. Show that the equation of the plane tangent to the ellipsoid

$$\frac{x^2}{a^2} + \frac{y^2}{b^2} + \frac{z^2}{c^2} = 1$$

at (x_0, y_0, z_0) can be written in the form

$$\frac{x_0 x}{a^2} + \frac{y_0 y}{b^2} + \frac{z_0 z}{c^2} = 1$$

36. Show that the equation of the plane tangent to the paraboloid

$$z = \frac{x^2}{a^2} + \frac{y^2}{b^2}$$

at (x_0, y_0, z_0) can be written in the form

$$z + z_0 = \frac{2x_0 x}{a^2} + \frac{2y_0 y}{b^2}$$

37. Prove: If $z = f(x, y)$ and $z = g(x, y)$ intersect at $P(x_0, y_0, z_0)$ and if f and g are differentiable at (x_0, y_0), then the normal lines at P are perpendicular if and only if

$$f_x(x_0, y_0)g_x(x_0, y_0) + f_y(x_0, y_0)g_y(x_0, y_0) = -1$$

16.7 FUNCTIONS OF THREE VARIABLES

In this section we discuss functions involving three independent variables. Where the results are natural extensions of results for functions of two variables we will omit the proofs.

16.7.1 DEFINITION A **function f of three variables** is a rule that assigns a unique real number $f(x, y, z)$ to each point (x, y, z) in some set D of three-dimensional space.

The set D in this definition is called the **domain** of the function; it is the set of points in 3-space at which the function is defined. If f is given by a formula and no domain is specified, it is understood that the domain consists of all points (x, y, z) at which the formula makes sense and yields a real number for the value of the function.

▶ **Example 1** Let

$$f(x, y, z) = \sqrt{1 - x^2 - y^2 - z^2}$$

Find $f(0, \frac{1}{2}, -\frac{1}{2})$ and the domain of f.

Solution. By substitution,

$$f(0, \tfrac{1}{2}, -\tfrac{1}{2}) = \sqrt{1 - (0)^2 - (\tfrac{1}{2})^2 - (-\tfrac{1}{2})^2} = \sqrt{\tfrac{1}{2}}$$

Because of the square root sign, we must have $0 \le 1 - x^2 - y^2 - z^2$ in order to have a real value for $f(x, y, z)$. Rewriting this inequality in the form

$$x^2 + y^2 + z^2 \le 1$$

we see that the domain of f consists of all points on or within the sphere $x^2 + y^2 + z^2 = 1$. ◀

When it is necessary to introduce a dependent variable for a function $f(x, y, z)$, we will usually use the letter w and write $w = f(x, y, z)$. Because this equation involves four variables, it cannot be graphed in three dimensions—"four dimensions" are needed. However, if k is a constant, then an equation of the form $f(x, y, z) = k$ will, in general, represent a surface in three-dimensional space. (For example, $x^2 + y^2 + z^2 = 1$ represents a sphere.)

16.7.2 DEFINITION If k is a constant, then the graph of the equation $f(x, y, z) = k$ is called a **level surface** of the function $f(x, y, z)$.

Level surfaces for functions of three variables are analogous to level curves for functions of two variables. Geometrically, the function $f(x, y, z)$ has a constant value of k as (x, y, z) varies over the level surface $f(x, y, z) = k$.

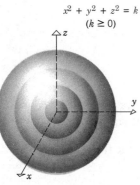

$x^2 + y^2 + z^2 = k$
$(k \geq 0)$

Figure 16.7.1

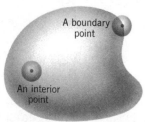

Level surfaces of
$f(x,y,z) = z^2 - x^2 - y^2$

$k > 0$ $k = 0$ $k < 0$

Figure 16.7.2

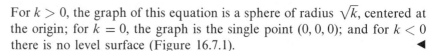

A boundary
point

An interior
point

Figure 16.7.3

REMARK. The term "level surface" can be confusing. A level surface need *not* be level in the sense of being horizontal. It is simply a surface on which all values of f are the same.

▶ **Example 2** Describe the level surfaces of $f(x, y, z) = x^2 + y^2 + z^2$.

Solution. The level surfaces have equations of the form

$$x^2 + y^2 + z^2 = k$$

For $k > 0$, the graph of this equation is a sphere of radius $\sqrt{k}$, centered at the origin; for $k = 0$, the graph is the single point $(0, 0, 0)$; and for $k < 0$ there is no level surface (Figure 16.7.1). ◀

▶ **Example 3** Describe the level surfaces of $f(x, y, z) = z^2 - x^2 - y^2$.

Solution. The level surfaces have equations of the form

$$z^2 - x^2 - y^2 = k$$

As discussed in Section 15.8, this equation represents a cone if $k = 0$, a hyperboloid of two sheets if $k > 0$, and a hyperboloid of one sheet if $k < 0$ (Figure 16.7.2). ◀

If D is a set of points in 3-space, then a point (x_0, y_0, z_0) is called an **interior point** of D if there is some spherical ball with positive radius, centered at (x_0, y_0, z_0), and containing only points in D (Figure 16.7.3). A point (x_0, y_0, z_0) is called a **boundary point** of D if *every* spherical ball with positive radius and centered at (x_0, y_0, z_0) contains both points in D and points not in D (Figure 16.7.3). A boundary point of D may or may not belong to D. A set D is called **open** if it contains *none* of its boundary points and **closed** if it contains *all* of its boundary points.

Notions of limit, continuity, and differentiability for functions of three variables are similar to the corresponding notions for functions of two variables. Thus, the statement

$$\lim_{(x,y,z) \to (x_0,y_0,z_0)} f(x, y, z) = L$$

tells us that the value of $f(x, y, z)$ can be made arbitrarily close to L by making the point (x, y, z) sufficiently close to (but different from) (x_0, y_0, z_0). The formulation of a precise definition is left as an exercise.

The definition of continuity for functions of three variables is similar to Definition 16.3.2 for functions of two variables.

16.7.3 DEFINITION   A function f of three variables is called ***continuous at a point*** (x_0, y_0, z_0) if

1. $f(x_0, y_0, z_0)$ is defined.

2. $\displaystyle\lim_{(x,y,z)\to(x_0,y_0,z_0)} f(x, y, z)$ exists.

3. $\displaystyle\lim_{(x,y,z)\to(x_0,y_0,z_0)} f(x, y, z) = f(x_0, y_0, z_0)$.

If f is continuous at each point of a region R in 3-space, we say that f is ***continuous on R.*** A function that is continuous at every point in 3-space is called ***continuous everywhere*** or simply ***continuous.***

For a function $f(x, y, z)$ of three variables, there are three ***partial derivatives:***

$$f_x(x, y, z), \qquad f_y(x, y, z), \qquad f_z(x, y, z)$$

The partial derivative f_x is calculated by holding y and z constant, and differentiating with respect to x. For f_y, the variables x and z are held constant, and for f_z the variables x and y are held constant. If a dependent variable

$$w = f(x, y, z)$$

is used, then the three partial derivatives of f may be denoted by

$$\frac{\partial w}{\partial x}, \qquad \frac{\partial w}{\partial y}, \qquad \text{and} \qquad \frac{\partial w}{\partial z}$$

▶ **Example 4**   If $f(x, y, z) = x^3 y^2 z^4 + 2xy + z$, then

$$f_x(x, y, z) = 3x^2 y^2 z^4 + 2y$$
$$f_y(x, y, z) = 2x^3 yz^4 + 2x$$
$$f_z(x, y, z) = 4x^3 y^2 z^3 + 1$$
$$f_z(-1, 1, 2) = 4(-1)^3(1)^2(2)^3 + 1 = -31 \qquad ◀$$

The definition of differentiability for functions of three variables is a direct generalization of Definition 16.3.4 for functions of two variables.

16.7.4 DEFINITION   A function f of three variables is said to be ***differentiable*** at (x_0, y_0, z_0) if $f_x(x_0, y_0, z_0), f_y(x_0, y_0, z_0)$, and $f_z(x_0, y_0, z_0)$ exist and

$$\Delta f = f(x_0 + \Delta x, y_0 + \Delta y, z_0 + \Delta z) - f(x_0, y_0, z_0)$$

can be written in the form

$$\Delta f = f_x(x_0, y_0, z_0)\,\Delta x + f_y(x_0, y_0, z_0)\,\Delta y + f_z(x_0, y_0, z_0)\,\Delta z$$
$$+ \epsilon_1\,\Delta x + \epsilon_2\,\Delta y + \epsilon_3\,\Delta z$$

where $\epsilon_1, \epsilon_2,$ and ϵ_3 are functions of $\Delta x, \Delta y,$ and Δz such that $\epsilon_1 \to 0, \epsilon_2 \to 0,$ and $\epsilon_3 \to 0$ as $(\Delta x, \Delta y, \Delta z) \to (0, 0, 0)$.

The meaning of the terms **differentiable on a region R** and **differentiable** (i.e., **everywhere differentiable**) should be clear.

As in the cases of one and two variables, a function $f(x, y, z)$ that is differentiable at a point must also be continuous at that point. Moreover, the following generalization of Theorem 16.3.6 gives simple conditions for $f(x, y, z)$ to be differentiable at a point.

16.7.5   THEOREM   *If f has first-order partial derivatives at each point of some spherical region centered at (x_0, y_0, z_0), and if these partial derivatives are continuous at (x_0, y_0, z_0), then f is differentiable at (x_0, y_0, z_0).*

The reader should have no trouble modifying the proof of Theorem 16.4.1 to obtain the following form of the chain rule.

16.7.6   THEOREM   *If $x = x(t), y = y(t),$ and $z = z(t)$ are differentiable at t and $w = f(x, y, z)$ is*
Chain Rule   *differentiable at the point $(x(t), y(t), z(t))$, then $w = f(x(t), y(t), z(t))$ is differentiable at t, and*

$$\frac{dw}{dt} = \frac{\partial w}{\partial x}\frac{dx}{dt} + \frac{\partial w}{\partial y}\frac{dy}{dt} + \frac{\partial w}{\partial z}\frac{dz}{dt}$$

▶ **Example 5**   Suppose

$$w = x^3 y^2 z, \qquad x = t^2, \qquad y = t^3, \qquad z = t^4$$

Use the chain rule to find dw/dt.

Solution.   By the chain rule,

$$\frac{dw}{dt} = \frac{\partial w}{\partial x}\frac{dx}{dt} + \frac{\partial w}{\partial y}\frac{dy}{dt} + \frac{\partial w}{\partial z}\frac{dz}{dt}$$
$$= (3x^2 y^2 z)(2t) + (2x^3 yz)(3t^2) + (x^3 y^2)(4t^3)$$
$$= (3t^{14})(2t) + (2t^{13})(3t^2) + (t^{12})(4t^3) = 16t^{15} \qquad ◀$$

Other variations of the chain rule for functions of three variables will be considered in the next section.

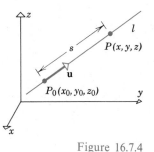

Figure 16.7.4

We will now show how to define directional derivatives for functions of three variables. Suppose $w = f(x, y, z)$, where f is differentiable at the point $P_0(x_0, y_0, z_0)$. Let l be a line passing through $P_0(x_0, y_0, z_0)$ and

$$\mathbf{u} = \langle u_1, u_2, u_3 \rangle$$

a unit vector along l. Let s be the (signed) distance from $P_0(x_0, y_0, z_0)$ to a point $P(x, y, z)$ on l, where s is nonnegative if P is in the direction of $\mathbf{u}$ from P_0 and negative otherwise (Figure 16.7.4).

Since $\overrightarrow{P_0 P} = s\mathbf{u}$, it follows that l is given parametrically by

$$
\begin{aligned}
x &= x_0 + su_1 \\
y &= y_0 + su_2 \\
z &= z_0 + su_3
\end{aligned}
\tag{1}
$$

As the parameter s varies, the point $P(x, y, z)$ moves along l and the value of $w = f(x, y, z)$ changes.

The reader should have no trouble modifying the derivation of (3) of Section 16.5 to show that the instantaneous rate of change of w with s at (x_0, y_0, z_0) (i.e., when $s = 0$) is

$$\left. \frac{dw}{ds} \right|_{s=0} = f_x(x_0, y_0, z_0)u_1 + f_y(x_0, y_0, z_0)u_2 + f_z(x_0, y_0, z_0)u_3$$

This quantity is denoted by $D_{\mathbf{u}} w(x_0, y_0, z_0)$ or $D_{\mathbf{u}} f(x_0, y_0, z_0)$.

16.7.7   DEFINITION   If f is differentiable at (x_0, y_0, z_0), and if $\mathbf{u} = \langle u_1, u_2, u_3 \rangle$ is a unit vector, then the *directional derivative* of f at (x_0, y_0, z_0) in the direction of $\mathbf{u}$ is defined by

$$D_{\mathbf{u}} f(x_0, y_0, z_0) = f_x(x_0, y_0, z_0)u_1 + f_y(x_0, y_0, z_0)u_2 + f_z(x_0, y_0, z_0)u_3 \tag{2}$$

The directional derivative $D_{\mathbf{u}} f(x_0, y_0, z_0)$ represents the instantaneous rate at which $w = f(x, y, z)$ changes with distance as (x, y, z) moves from (x_0, y_0, z_0) in the direction of the unit vector $\mathbf{u}$.

Note the similarity between the formula for $D_{\mathbf{u}} f(x_0, y_0, z_0)$ and the formula for $D_{\mathbf{u}} f(x_0, y_0)$ in Definition 16.5.1. If we now define the *gradient* of f by

$$\nabla f(x, y, z) = f_x(x, y, z)\mathbf{i} + f_y(x, y, z)\mathbf{j} + f_z(x, y, z)\mathbf{k} \tag{3}$$

then the directional derivative at (x, y, z),

$$D_{\mathbf{u}} f(x, y, z) = f_x(x, y, z)u_1 + f_y(x, y, z)u_2 + f_z(x, y, z)u_3$$

can be expressed in the form

$$D_{\mathbf{u}} f(x, y, z) = \nabla f(x, y, z) \cdot \mathbf{u} \tag{4}$$

which parallels (9) of Section 16.5. Using this result, the reader should have no trouble proving the following extension of Theorem 16.5.3.

16.7.8   THEOREM   *Let f be a function of three variables that is differentiable at (x_0, y_0, z_0).*
(a)   *If $\nabla f(x_0, y_0, z_0) = \mathbf{0}$, then all directional derivatives of f at (x_0, y_0, z_0) are zero.*
(b)   *If $\nabla f(x_0, y_0, z_0) \neq \mathbf{0}$, then among all possible directional derivatives of f at (x_0, y_0, z_0), the derivative in the direction of $\nabla f(x_0, y_0, z_0)$ has the largest value. The value of that directional derivative is $\|\nabla f(x_0, y_0, z_0)\|$.*
(c)   *If $\nabla f(x_0, y_0, z_0) \neq \mathbf{0}$, then among all possible directional derivatives of f at (x_0, y_0, z_0), the derivative in the direction opposite to that of $\nabla f(x_0, y_0, z_0)$ has the smallest value. The value of that directional derivative is $-\|\nabla f(x_0, y_0, z_0)\|$.*

▶ **Example 6**   Find the directional derivative of $f(x, y, z) = x^2 y - yz^3 + z$ at the point $P(1, -2, 0)$ in the direction of the vector $\mathbf{a} = 2\mathbf{i} + \mathbf{j} - 2\mathbf{k}$, and find the maximum rate of increase of f at P.

Solution.   Since

$$f_x(x, y, z) = 2xy, \qquad f_y(x, y, z) = x^2 - z^3, \qquad f_z(x, y, z) = -3yz^2 + 1$$

it follows that

$$\nabla f(x, y, z) = 2xy\mathbf{i} + (x^2 - z^3)\mathbf{j} + (-3yz^2 + 1)\mathbf{k}$$
$$\nabla f(1, -2, 0) = -4\mathbf{i} + \mathbf{j} + \mathbf{k}$$

A unit vector in the direction of $\mathbf{a}$ is

$$\mathbf{u} = \frac{\mathbf{a}}{\|\mathbf{a}\|} = \frac{1}{\sqrt{9}} (2\mathbf{i} + \mathbf{j} - 2\mathbf{k}) = \tfrac{2}{3}\mathbf{i} + \tfrac{1}{3}\mathbf{j} - \tfrac{2}{3}\mathbf{k}$$

Therefore

$$D_{\mathbf{u}} f(1, -2, 0) = \nabla f(1, -2, 0) \cdot \mathbf{u} = (-4)(\tfrac{2}{3}) + (1)(\tfrac{1}{3}) + (1)(-\tfrac{2}{3}) = -3$$

The maximum rate of increase of f at P is

$$\|\nabla f(1, -2, 0)\| = \sqrt{(-4)^2 + (1)^2 + (1)^2} = 3\sqrt{2}$$ ◀

If $P_0(x_0, y_0, z_0)$ is a point in the domain of the function f and $f(x_0, y_0, z_0) = c$, then the equation

$$f(x, y, z) = c$$

defines a level surface of f that passes through the point $P_0(x_0, y_0, z_0)$. The following theorem states that the gradient of f at P_0 is perpendicular to this surface at P_0 (Figure 16.7.5). (Recall that for functions of two variables, the gradient at a point is perpendicular to the level *curve* through the point.)

16.7.9 THEOREM *If f is differentiable at (x_0, y_0, z_0) and $\nabla f(x_0, y_0, z_0) \neq \mathbf{0}$, then $\nabla f(x_0, y_0, z_0)$ is perpendicular to the level surface of $f(x, y, z)$ through (x_0, y_0, z_0).*

We omit the proof.

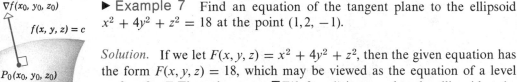

Figure 16.7.5

▶ **Example 7** Find an equation of the tangent plane to the ellipsoid $x^2 + 4y^2 + z^2 = 18$ at the point $(1, 2, -1)$.

Solution. If we let $F(x, y, z) = x^2 + 4y^2 + z^2$, then the given equation has the form $F(x, y, z) = 18$, which may be viewed as the equation of a level surface for F. Thus, the vector $\nabla F(1, 2, -1)$ is normal to the ellipsoid at the point $(1, 2, -1)$. To find this vector we write

$$\nabla F(x, y, z) = \frac{\partial F}{\partial x}\mathbf{i} + \frac{\partial F}{\partial y}\mathbf{j} + \frac{\partial F}{\partial z}\mathbf{k} = 2x\mathbf{i} + 8y\mathbf{j} + 2z\mathbf{k}$$

$$\nabla F(1, 2, -1) = 2\mathbf{i} + 16\mathbf{j} - 2\mathbf{k}$$

Using this normal and the point $(1, 2, -1)$, we obtain as the equation of the tangent plane

$$2(x - 1) + 16(y - 2) - 2(z + 1) = 0$$

or

$$x + 8y - z = 18$$

Alternate Solution. Solving the given equation for z, we obtain $z = \pm\sqrt{18 - x^2 - 4y^2}$. Since the point $(1, 2, -1)$ has a negative z-coordinate, this point lies on the lower half of the ellipsoid, which is given by

$$z = -\sqrt{18 - x^2 - 4y^2}$$

Using this expression and Theorem 16.6.2, we can obtain the equation of the tangent plane as follows:

$$\frac{\partial z}{\partial x} = \frac{x}{\sqrt{18 - x^2 - 4y^2}}, \qquad \frac{\partial z}{\partial y} = \frac{4y}{\sqrt{18 - x^2 - 4y^2}}$$

so that

$$\left.\frac{\partial z}{\partial x}\right|_{x=1,y=2} = 1, \qquad \left.\frac{\partial z}{\partial y}\right|_{x=1,y=2} = 8$$

Thus, from (1) of Theorem 16.6.2 the tangent plane at $(1, 2, -1)$ is given by

$$(1)(x - 1) + 8(y - 2) - (z + 1) = 0$$

or

$$x + 8y - z = 18$$

which agrees with our previous solution. ◄

REMARK. The first method in the last example is useful when it is inconvenient or impossible to express z explicitly as a function of x and y.

If $w = f(x, y, z)$, then we define the **increment** Δw (also written Δf) to be

$$\Delta w = f(x + \Delta x, y + \Delta y, z + \Delta z) - f(x, y, z)$$

and we define the **total differential** dw (also written df) to be

$$dw = f_x(x, y, z)\, dx + f_y(x, y, z)\, dy + f_z(x, y, z)\, dz$$

where $\Delta x, \Delta y, \Delta z, dx, dy$, and dz are all variables representing changes in the values of x, y, and z.

The increment Δw represents the change in the value of $w = f(x, y, z)$ when x, y, and z are changed by amounts Δx, Δy, and Δz, respectively. However, for functions of three variables, the total differential dw has no natural geometric interpretation.

If we let

$$dx = \Delta x, \qquad dy = \Delta y, \qquad dz = \Delta z$$

and if we assume that $w = f(x, y, z)$ is differentiable at (x, y, z), then it follows from Definition 16.7.4 that

$$\Delta w = dw + \epsilon_1 \Delta x + \epsilon_2 \Delta y + \epsilon_3 \Delta z \tag{5}$$

where $\epsilon_1 \to 0, \epsilon_2 \to 0, \epsilon_3 \to 0$ as $(\Delta x, \Delta y, \Delta z) \to (0, 0, 0)$. Thus, when $\Delta x, \Delta y$, and Δz are small, it follows from (5) that

$$\Delta w \approx dw$$

▶ Example 8 The length, width, and height of a rectangular box are each measured with an error of at most 5%. Estimate the maximum possible percentage error that results if these quantities are used to calculate the diagonal of the box.

Solution. Let x, y, z, and D be the true length, width, height, and diagonal of the box, respectively; and let $\Delta x, \Delta y, \Delta z$, and ΔD be the errors in these quantities. We are given that

$$\left|\frac{\Delta x}{x}\right| \le 0.05, \quad \left|\frac{\Delta y}{y}\right| \le 0.05, \quad \left|\frac{\Delta z}{z}\right| \le 0.05$$

We want to estimate $|\Delta D/D|$. Since the diagonal D is related to the length, width, and height by

$$D = \sqrt{x^2 + y^2 + z^2}$$

it follows that

$$dD = \frac{\partial D}{\partial x}\,dx + \frac{\partial D}{\partial y}\,dy + \frac{\partial D}{\partial z}\,dz$$

$$= \frac{x}{\sqrt{x^2 + y^2 + z^2}}\,dx + \frac{y}{\sqrt{x^2 + y^2 + z^2}}\,dy + \frac{z}{\sqrt{x^2 + y^2 + z^2}}\,dz$$

If we choose $\Delta x = dx, \Delta y = dy, \Delta z = dz$, then we can use the approximation $\Delta D/D \approx dD/D$. But,

$$\frac{dD}{D} = \frac{x}{x^2 + y^2 + z^2}\,dx + \frac{y}{x^2 + y^2 + z^2}\,dy + \frac{z}{x^2 + y^2 + z^2}\,dz$$

or

$$\frac{dD}{D} = \frac{x^2}{x^2 + y^2 + z^2}\frac{dx}{x} + \frac{y^2}{x^2 + y^2 + z^2}\frac{dy}{y} + \frac{z^2}{x^2 + y^2 + z^2}\frac{dz}{z}$$

Thus,

$$\left|\frac{dD}{D}\right| = \left|\frac{x^2}{x^2 + y^2 + z^2}\frac{dx}{x} + \frac{y^2}{x^2 + y^2 + z^2}\frac{dy}{y} + \frac{z^2}{x^2 + y^2 + z^2}\frac{dz}{z}\right|$$

$$\le \left|\frac{x^2}{x^2 + y^2 + z^2}\frac{dx}{x}\right| + \left|\frac{y^2}{x^2 + y^2 + z^2}\frac{dy}{y}\right| + \left|\frac{z^2}{x^2 + y^2 + z^2}\frac{dz}{z}\right|$$

$$\le \frac{x^2}{x^2 + y^2 + z^2}(0.05) + \frac{y^2}{x^2 + y^2 + z^2}(0.05) + \frac{z^2}{x^2 + y^2 + z^2}(0.05)$$

$$= 0.05$$

Therefore, the maximum percentage error in D is approximately 5%. ◀

▶ **Exercise Set 16.7**

1. Let $f(x, y, z) = xy^2z^3 + 3$. Find
  (a) $f(2, 1, 2)$ (b) $f(-3, 2, 1)$
  (c) $f(0, 0, 0)$ (d) $f(a, a, a)$
  (e) $f(t, t^2, -t)$ (f) $f(a + b, a - b, b)$.

2. Let $f(x, y, z) = zxy + x$. Find
  (a) $f(x + y, x - y, x^2)$ (b) $f(xy, y/x, xz)$.

3. Find $F(f(x), g(y), h(z))$ if $F(x, y, z) = ye^{xyz}$, $f(x) = x^2$, $g(y) = y + 1$, and $h(z) = z^2$.

4. Find $g(u(x, y, z), v(x, y, z), w(x, y, z))$ if $g(x, y, z) = z \sin xy$, $u(x, y, z) = x^2z^3$, $v(x, y, z) = \pi xyz$, and $w(x, y, z) = xy/z$.

5. Let $f(x, y, z) = x^2y^2z^4$, $x(t) = t^3$, $y(t) = t^2$, and $z(t) = t$. Find
  (a) $f(x(t), y(t), z(t))$ (b) $f(x(0), y(0), z(0))$
  (c) $f(x(2), y(2), z(2))$.

In Exercises 6–9, describe the domain of f.

6. $f(x, y, z) = \sqrt{25 - x^2 - y^2 - z^2}$.

7. $f(x, y, z) = \dfrac{xyz}{x + y + z}$.   **8.** $f(x, y, z) = e^{xyz}$.

9. $f(x, y, z) = z + \ln[1 - x^2 - y^2]$.

In Exercises 10–13, sketch the level surface $f(x, y, z) = k$.

10. $f(x, y, z) = 4x - 2y + z$; $k = 1$.

11. $f(x, y, z) = 4x^2 + y^2 + 4z^2$; $k = 16$.

12. $f(x, y, z) = x^2 + y^2 - z^2$; $k = 0$.

13. $f(x, y, z) = z - x^2 - y^2 + 4$; $k = 7$.

14. Let $f(x, y, z) = xyz + 3$. Find the equation of the level surface that passes through the point
  (a) $(1, 0, 2)$ (b) $(-2, 4, 1)$ (c) $(0, 0, 0)$.

In Exercises 15–19, find $\partial w/\partial x$, $\partial w/\partial y$, and $\partial w/\partial z$.

15. $w = x^2y^4z^3 + xy + z^2 + 1$.

16. $w = ye^z \sin x$.   **17.** $w = \dfrac{x^2 - y^2}{y^2 + z^2}$.

18. $w = y^3e^{2x + 3z}$.   **19.** $w = \sqrt{x^2 + y^2 + z^2}$.

In Exercises 20–24, find f_x, f_y, and f_z.

20. $f(x, y, z) = z \ln(x^2y \cos z)$.

21. $f(x, y, z) = \tan^{-1}\left(\dfrac{1}{xy^2z^3}\right)$.

22. $f(x, y, z) = y^{-3/2} \sec\left(\dfrac{xz}{y}\right)$.

23. $f(x, y, z) = \cosh(\sqrt{z}) \sinh^2(x^2yz)$.

24. $f(x, y, z) = \left(\dfrac{xz}{1 - z^2 - y^2}\right)^{-3/4}$.

25. Let $f(x, y, z) = 5x^2yz^3$. Find
  (a) $f_x(1, -1, 2)$ (b) $f_y(1, -1, 2)$
  (c) $f_z(1, -1, 2)$.

26. Let $f(x, y, z) = y^2e^{xz}$. Find
  (a) $\left.\dfrac{\partial f}{\partial x}\right|_{(1, 1, 1)}$ (b) $\left.\dfrac{\partial f}{\partial y}\right|_{(1, 1, 1)}$ (c) $\left.\dfrac{\partial f}{\partial z}\right|_{(1, 1, 1)}$.

27. Let $w = \sqrt{x^2 + 4y^2 - z^2}$. Find
  (a) $\left.\dfrac{\partial w}{\partial x}\right|_{(2, 1, -1)}$ (b) $\left.\dfrac{\partial w}{\partial y}\right|_{(2, 1, -1)}$
  (c) $\left.\dfrac{\partial w}{\partial z}\right|_{(2, 1, -1)}$.

28. Let $w = x \sin xyz$. Find
  (a) $\dfrac{\partial w}{\partial x}(1, \frac{1}{2}, \pi)$ (b) $\dfrac{\partial w}{\partial y}(1, \frac{1}{2}, \pi)$
  (c) $\dfrac{\partial w}{\partial z}(1, \frac{1}{2}, \pi)$.

In Exercises 29–32, find $\partial w/\partial x$, $\partial w/\partial y$, and $\partial w/\partial z$ using implicit differentiation. Leave your answers in terms of x, y, z, and w.

29. $(x^2 + y^2 + z^2 + w^2)^{3/2} = 4$.

30. $\ln(2x^2 + y - z^3 + 3w) = z$.

31. $w^2 + w \sin xyz = 0$.

32. $e^{xy} \sinh w - z^2w + 1 = 0$.

33. Let $f(x, y, z) = x^3y^5z^7 + xy^2 + y^3z$. Find
  (a) f_{xy} (b) f_{yz} (c) f_{xz} (d) f_{zz}
  (e) f_{zyy} (f) f_{xxy} (g) f_{zyx} (h) f_{xxyz}.

34. Let $w = (4x - 3y + 2z)^5$. Find
  (a) $\dfrac{\partial^2 w}{\partial x \, \partial z}$ (b) $\dfrac{\partial^3 w}{\partial x \, \partial y \, \partial z}$ (c) $\dfrac{\partial^4 w}{\partial z^2 \, \partial y \, \partial x}$.

In Exercises 35–38, find dw/dt using the chain rule.

35. $w = 5x^2y^3z^4$; $x = t^2$, $y = t^3$, $z = t^5$.

36. $w = \ln(3x^2 - 2y + 4z^3)$; $x = t^{1/2}$, $y = t^{2/3}$, $z = t^{-2}$.

37. $w = 5 \cos xy - \sin xz$; $x = 1/t$, $y = t$, $z = t^3$.

38. $w = \sqrt{1 + x - 2yz^4 x}$; $x = \ln t$, $y = t$, $z = 4t$.

39. Use the chain rule to find $\dfrac{dw}{dt}\Big|_{t=1}$ if $w = x^3 y^2 z^4$;

$x = t^2$, $y = t + 2$, $z = 2t^4$.

40. Use the chain rule to find $\dfrac{dw}{dt}\Big|_{t=0}$ if $w = x \sin yz^2$;

$x = \cos t$, $y = t^2$, $z = e^t$.

In Exercises 41–44, find the gradient of f at P, and then use the gradient to calculate $D_{\mathbf{u}} f$ at P.

41. $f(x, y, z) = 4x^5 y^2 z^3$; $P(2, -1, 1)$;

$\mathbf{u} = \dfrac{1}{3}\mathbf{i} + \dfrac{2}{3}\mathbf{j} - \dfrac{2}{3}\mathbf{k}$.

42. $f(x, y, z) = ye^{xz} + z^2$; $P(0, 2, 3)$;

$\mathbf{u} = \dfrac{2}{7}\mathbf{i} - \dfrac{3}{7}\mathbf{j} + \dfrac{6}{7}\mathbf{k}$.

43. $f(x, y, z) = \ln(x^2 + 2y^2 + 3z^2)$; $P(-1, 2, 4)$;

$\mathbf{u} = -\dfrac{3}{13}\mathbf{i} - \dfrac{4}{13}\mathbf{j} - \dfrac{12}{13}\mathbf{k}$.

44. $f(x, y, z) = \sin xyz$; $P(\frac{1}{2}, \frac{1}{3}, \pi)$;

$\mathbf{u} = \dfrac{1}{\sqrt{3}}\mathbf{i} - \dfrac{1}{\sqrt{3}}\mathbf{j} + \dfrac{1}{\sqrt{3}}\mathbf{k}$.

In Exercises 45–48, find the directional derivative of f at P in the direction of $\mathbf{a}$.

45. $f(x, y, z) = x^3 z - yx^2 + z^2$; $P(2, -1, 1)$;

$\mathbf{a} = 3\mathbf{i} - \mathbf{j} + 2\mathbf{k}$.

46. $f(x, y, z) = y - \sqrt{x^2 + z^2}$; $P(-3, 1, 4)$;

$\mathbf{a} = 2\mathbf{i} - 2\mathbf{j} - \mathbf{k}$.

47. $f(x, y, z) = \dfrac{z - x}{z + y}$; $P(1, 0, -3)$;

$\mathbf{a} = -6\mathbf{i} + 3\mathbf{j} - 2\mathbf{k}$.

48. $f(x, y, z) = e^{x+y+3z}$; $P(-2, 2, -1)$;

$\mathbf{a} = 20\mathbf{i} - 4\mathbf{j} + 5\mathbf{k}$.

In Exercises 49–52, find a unit vector in the direction in which f increases most rapidly at P, and find the rate of increase of f in that direction.

49. $f(x, y, z) = x^3 z^2 + y^3 z + z - 1$; $P(1, 1, -1)$.

50. $f(x, y, z) = \sqrt{x - 3y + 4z}$; $P(0, -3, 0)$.

51. $f(x, y, z) = \dfrac{x}{z} + \dfrac{z}{y^2}$; $P(1, 2, -2)$.

52. $f(x, y, z) = \tan^{-1}\left(\dfrac{x}{y + z}\right)$; $P(4, 2, 2)$.

In Exercises 53 and 54, find a unit vector in the direction in which f decreases most rapidly at P, and find the rate of change of f in that direction.

53. $f(x, y, z) = \dfrac{x + z}{z - y}$; $P(5, 7, 6)$.

54. $f(x, y, z) = 4e^{xy} \cos z$; $P(0, 1, \pi/4)$.

55. Find the directional derivative of $f(x, y, z) = \dfrac{y}{x + z}$ at $P(2, 1, -1)$ in the direction from P to $Q(-1, 2, 0)$.

56. Find the directional derivative of $f(x, y, z) = x^3 y^2 z^5 - 2xz + yz + 3x$ at $P(-1, -2, 1)$ in the direction of the negative z-axis.

In Exercises 57–60, find equations for the tangent plane and the normal line to the given surface at the point P.

57. $x^2 + y^2 + z^2 = 49$; $P(-3, 2, -6)$.

58. $xz - yz^3 + yz^2 = 2$; $P(2, -1, 1)$.

59. $\sqrt{\dfrac{z + x}{y - 1}} = z^2$; $P(3, 5, 1)$.

60. $\sin xz - 4 \cos yz = 4$; $P(\pi, \pi, 1)$.

61. Show that every line normal to the sphere $x^2 + y^2 + z^2 = 1$ passes through the origin.

62. Find all points on the ellipsoid $2x^2 + 3y^2 + 4z^2 = 9$ at which the tangent plane is parallel to the plane $x - 2y + 3z = 5$.

63. Find all points on the surface $x^2 + y^2 - z^2 = 1$ at which the normal line is parallel to the line through $P(1, -2, 1)$ and $Q(4, 0, -1)$.

64. The temperature (in degrees Celsius) at a point (x, y, z) in a metal solid is

$$T(x, y, z) = \dfrac{xyz}{1 + x^2 + y^2 + z^2}$$

(a) Find the rate of change of temperature at $(1, 1, 1)$ in the direction of the origin.

(b) Find the direction in which the temperature rises most rapidly at $(1, 1, 1)$. (Express your answer as a unit vector.)

(c) Find the rate at which the temperature rises moving from $(1, 1, 1)$ in the direction obtained in (b).

65. Show that the following functions satisfy *Laplace's equation,*

$$\frac{\partial^2 f}{\partial x^2} + \frac{\partial^2 f}{\partial y^2} + \frac{\partial^2 f}{\partial z^2} = 0$$

(a) $f(x, y, z) = (x^2 + y^2 + z^2)^{-1/2}$
(b) $f(x, y, z) = e^{4x} e^{3y} \cos 5z$.

66. Show that the ellipsoid $2x^2 + 3y^2 + z^2 = 9$ and the sphere

$$x^2 + y^2 + z^2 - 6x - 8y - 8z + 24 = 0$$

have a common tangent plane at the point $(1, 1, 2)$.

In Exercises 67–70, find dw.

67. $w = 8x - 3y + 4z$.

68. $w = 4x^2 y^3 z^7 - 3xy + z + 5$.

69. $w = \tan^{-1}(xyz)$.

70. $w = \sqrt{x} + \sqrt{y} + \sqrt{z}$.

71. Use a total differential to approximate the change in $f(x, y, z) = 2xy^2 z^3$ as (x, y, z) varies from $P(1, -1, 2)$ to $Q(0.99, -1.02, 2.02)$.

72. Use a total differential to approximate the change in $f(x, y, z) = xyz/(x + y + z)$ as (x, y, z) varies from $P(-1, -2, 4)$ to $Q(-1.04, -1.98, 3.97)$.

73. The length, width, and height of a rectangular box are measured to be 3 cm, 4 cm, and 5 cm, respectively, with a maximum error of 0.05 cm in each measurement. Use differentials to approximate the maximum error in the calculated volume.

74. The total resistance R of three resistances R_1, R_2, and R_3, connected in parallel, is given by

$$\frac{1}{R} = \frac{1}{R_1} + \frac{1}{R_2} + \frac{1}{R_3}$$

Suppose that R_1, R_2, and R_3 are measured to be 100 ohms, 200 ohms, and 500 ohms, respectively, with a maximum error of 10% in each. Use differentials to approximate the maximum percentage error in the calculated value of R.

75. The area of a triangle is to be computed from the formula $A = \frac{1}{2} ab \sin \theta$, where a and b are the lengths of two sides and θ is the included angle. Suppose that a, b, and θ are measured to be 40 ft, 50 ft, and $30°$, respectively. Use differentials to approximate the maximum error in the calculated value of A if the maximum errors in a, b, and θ are $\frac{1}{2}$ ft, $\frac{1}{4}$ ft, and $2°$, respectively.

76. The length, width, and height of a rectangular box are measured with errors of at most $r\%$. Use differentials to approximate the percentage error in the computed value of the volume.

77. Use differentials to approximate the percentage error in $w = xy^2 z^3$ if x, y, and z have errors of at most 1%, 2%, and 3%, respectively.

78. Two surfaces are said to be *orthogonal* at a point of intersection if their normal lines are perpendicular at that point. Prove that the surfaces $f(x, y, z) = 0$ and $g(x, y, z) = 0$ are orthogonal at a point of intersection, (x_0, y_0, z_0), if and only if

$$f_x g_x + f_y g_y + f_z g_z = 0$$

at (x_0, y_0, z_0). (Assume that $\nabla f(x_0, y_0, z_0) \neq \mathbf{0}$ and $\nabla g(x_0, y_0, z_0) \neq \mathbf{0}$.)

79. Use the result of Exercise 78 to show that the sphere $x^2 + y^2 + z^2 = a^2$ and the cone $z^2 = x^2 + y^2$ are orthogonal at every point of intersection.

80. Use Definition 16.3.1 as a model to define

$$\lim_{(x, y, z) \to (x_0, y_0, z_0)} f(x, y, z) = L$$

16.8 FUNCTIONS OF n VARIABLES; MORE ON THE CHAIN RULE

In this section we will discuss functions involving more than three variables. Our main objective is to develop forms of the chain rule for such functions.

All the definitions and theorems we have stated for functions of two and three variables can be extended to functions of four or more variables. If

$$w = f(v_1, v_2, \ldots, v_n)$$

is a function of n variables, then we can define n partial derivatives

$$\frac{\partial w}{\partial v_1}, \frac{\partial w}{\partial v_2}, \dots, \frac{\partial w}{\partial v_n}$$

each of which is calculated by holding $n - 1$ of the variables fixed and differentiating with respect to the remaining variable. In order to define directional derivatives for a function of n variables, it is first necessary to define the notion of a vector in "n-dimensional space." This topic is studied in a branch of mathematics called **_linear algebra,_** and will not be considered in this text.

If $w = f(v_1, v_2, \dots, v_n)$, we define the **_increment_** Δw and the **_total differential_** dw by

$$\Delta w = f(v_1 + \Delta v_1, v_2 + \Delta v_2, \dots, v_n + \Delta v_n) - f(v_1, v_2, \dots, v_n) \quad (1)$$

and

$$dw = \frac{\partial w}{\partial v_1} dv_1 + \frac{\partial w}{\partial v_2} dv_2 + \cdots + \frac{\partial w}{\partial v_n} dv_n \quad (2)$$

where $\Delta v_1, \Delta v_2, \dots, \Delta v_n$ and $dv_1, dv_2, \dots, dv_n$ are variables representing changes in the values of $v_1, v_2, \dots, v_n$.

If $v_1, v_2, \dots, v_n$ are functions of a single variable t, then $w = f(v_1, v_2, \dots, v_n)$ is a function of t, and a chain rule formula for dw/dt is

$$\frac{dw}{dt} = \frac{\partial w}{\partial v_1} \frac{dv_1}{dt} + \frac{\partial w}{\partial v_2} \frac{dv_2}{dt} + \cdots + \frac{\partial w}{\partial v_n} \frac{dv_n}{dt} \quad (3)$$

which is a natural extension of the chain rule formulas in Theorems 16.4.1 and 16.7.6. Observe that (3) results if we formally divide both sides of (2) by dt.

Other forms of the chain rule arise, depending on the number of variables involved. For example, in Theorem 16.4.2 we obtained the chain rule formulas

$$\frac{\partial z}{\partial u} = \frac{\partial z}{\partial x} \frac{\partial x}{\partial u} + \frac{\partial z}{\partial y} \frac{\partial y}{\partial u} \quad (4)$$

$$\frac{\partial z}{\partial v} = \frac{\partial z}{\partial x} \frac{\partial x}{\partial v} + \frac{\partial z}{\partial y} \frac{\partial y}{\partial v} \quad (5)$$

for the case where z is a function of two variables $z = f(x, y)$; and x and y in turn are functions of two other variables, $x = x(u, v)$, $y = y(u, v)$.

Formulas (4) and (5) can be represented schematically by a "tree diagram" constructed as follows (Figure 16.8.1). Starting with z at the top of the diagram and moving downward, join each variable by lines (or branches) to those variables that depend *directly* on it. Thus, z is joined to x and y and these in turn are each joined to u and v. Next label each branch with a derivative whose "numerator" contains the variable at the top end of that branch, and whose "denominator" contains the variable at the bottom end of that branch. This completes the "tree." To find the formula for $\partial z / \partial u$ trace all paths through the tree that start at z and end at u. Each such path produces one of the terms in the formula for $\partial z / \partial u$ (Figure 16.8.1a). Similarly, each term in the formula for $\partial z / \partial v$ corresponds to a path starting at z and ending at v (Figure 16.8.1b).

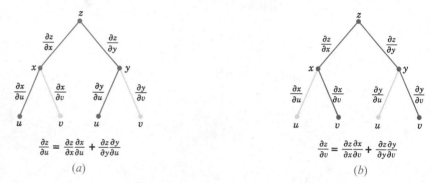

$$\frac{\partial z}{\partial u} = \frac{\partial z}{\partial x}\frac{\partial x}{\partial u} + \frac{\partial z}{\partial y}\frac{\partial y}{\partial u}$$

$$\frac{\partial z}{\partial v} = \frac{\partial z}{\partial x}\frac{\partial x}{\partial v} + \frac{\partial z}{\partial y}\frac{\partial y}{\partial v}$$

Figure 16.8.1   (a)   (b)

The following examples illustrate how tree diagrams can be used to construct other forms of the chain rule.

▶ **Example 1** Suppose

$$w = e^{xyz}, \qquad x = 3r + s, \qquad y = 3r - s, \qquad z = r^2 s$$

Use appropriate forms of the chain rule to find $\partial w / \partial r$ and $\partial w / \partial s$.

Solution. From the tree diagram

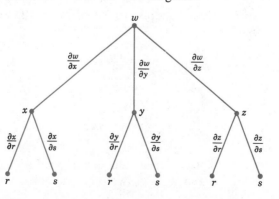

we obtain the formulas

$$\frac{\partial w}{\partial r} = \frac{\partial w}{\partial x}\frac{\partial x}{\partial r} + \frac{\partial w}{\partial y}\frac{\partial y}{\partial r} + \frac{\partial w}{\partial z}\frac{\partial z}{\partial r}$$

$$\frac{\partial w}{\partial s} = \frac{\partial w}{\partial x}\frac{\partial x}{\partial s} + \frac{\partial w}{\partial y}\frac{\partial y}{\partial s} + \frac{\partial w}{\partial z}\frac{\partial z}{\partial s}$$

from which we obtain

$$\frac{\partial w}{\partial r} = yze^{xyz}(3) + xze^{xyz}(3) + xye^{xyz}(2rs) = e^{xyz}(3yz + 3xz + 2xyrs)$$

and

$$\frac{\partial w}{\partial s} = yze^{xyz}(1) + xze^{xyz}(-1) + xye^{xyz}(r^2) = e^{xyz}(yz - xz + xyr^2)$$

If desired, we can express $\partial w/\partial r$ and $\partial w/\partial s$ in terms of r and s alone by replacing x, y, and z by their expressions in terms of r and s. ◀

▶ **Example 2** Suppose $w = x^2 + y^2 - z^2$ and

$$x = \rho \sin \phi \cos \theta, \qquad y = \rho \sin \phi \sin \theta, \qquad z = \rho \cos \phi$$

Use appropriate forms of the chain rule to find $\partial w/\partial \rho$ and $\partial w/\partial \theta$.

Solution. From the tree diagram

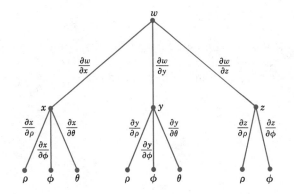

we obtain the formulas

$$\frac{\partial w}{\partial \rho} = \frac{\partial w}{\partial x}\frac{\partial x}{\partial \rho} + \frac{\partial w}{\partial y}\frac{\partial y}{\partial \rho} + \frac{\partial w}{\partial z}\frac{\partial z}{\partial \rho}$$

$$\frac{\partial w}{\partial \theta} = \frac{\partial w}{\partial x}\frac{\partial x}{\partial \theta} + \frac{\partial w}{\partial y}\frac{\partial y}{\partial \theta}$$

from which we obtain

$$\frac{\partial w}{\partial \rho} = (2x) \sin \phi \cos \theta + 2y \sin \phi \sin \theta - 2z \cos \phi$$

$$= 2\rho \sin^2 \phi \cos^2 \theta + 2\rho \sin^2 \phi \sin^2 \theta - 2\rho \cos^2 \phi$$
$$= 2\rho \sin^2 \phi (\cos^2 \theta + \sin^2 \theta) - 2\rho \cos^2 \phi$$
$$= 2\rho (\sin^2 \phi - \cos^2 \phi)$$
$$= -2\rho \cos 2\phi$$

$$\frac{\partial w}{\partial \theta} = (2x)(-\rho \sin \phi \sin \theta) + (2y) \rho \sin \phi \cos \theta$$

$$= -2\rho^2 \sin^2 \phi \sin \theta \cos \theta + 2\rho^2 \sin^2 \phi \sin \theta \cos \theta$$
$$= 0$$

This result is explained by the fact that w does not vary with θ. We may see this directly by expressing w in terms of ρ, ϕ, and θ. If this is done, the expressions involving θ will cancel, leaving w as a function of ρ and ϕ alone. (Verify.) ◀

In many applications of the chain rule, some of the variables in the function $w = f(v_1, v_2, \ldots, v_n)$ are functions of the remaining variables. Tree diagrams are especially helpful in such situations.

▶ Example 3 Suppose

$$w = xy + yz, \qquad y = \sin x, \qquad z = e^x$$

Use an appropriate form of the chain rule to find dw/dx.

Solution. From the tree diagram

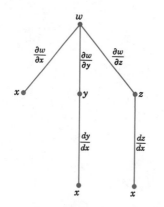

we obtain the formula

$$\frac{dw}{dx} = \frac{\partial w}{\partial x} + \frac{\partial w}{\partial y}\frac{dy}{dx} + \frac{\partial w}{\partial z}\frac{dz}{dx}$$

from which it follows that

$$\frac{dw}{dx} = y + (x + z)\cos x + ye^x$$
$$= \sin x + (x + e^x)\cos x + e^x \sin x$$

This same result can be obtained by first expressing *w* explicitly in terms of *x*,

$$w = x \sin x + e^x \sin x$$

and then differentiating with respect to *x*; however, such direct substitution is not always convenient.   ◄

REMARK.   Unlike the differential *dz*, a partial symbol ∂z has no meaning of its own. For example, if we were to "cancel" partial symbols in the chain rule formula

$$\frac{\partial z}{\partial u} = \frac{\partial z}{\partial x}\frac{\partial x}{\partial u} + \frac{\partial z}{\partial y}\frac{\partial y}{\partial u}$$

we would obtain

$$\frac{\partial z}{\partial u} = \frac{\partial z}{\partial u} + \frac{\partial z}{\partial u}$$

which is nonsense in cases where $\partial z/\partial u \neq 0$.

In each of the expressions

$$z = \sin xy, \qquad z = \frac{xy}{1 + xy}, \qquad z = e^{xy}$$

the independent variables occur only in the combination *xy*, so the substitution $t = xy$ reduces the expression to a function of one variable

$$z = \sin t, \qquad z = \frac{t}{1 + t}, \qquad z = e^t$$

Conversely, if we begin with a function of one variable $z = f(t)$ and substitute $t = xy$, we obtain a function $z = f(xy)$ in which the variables appear only in the combination xy. Functions whose variables occur in fixed combinations arise frequently in applications.

▶ **Example 4** Show that a function of the form $z = f(xy)$ satisfies the equation

$$x \frac{\partial z}{\partial x} - y \frac{\partial z}{\partial y} = 0$$

(assuming the derivatives exist).

Solution. Let $t = xy$, so that $z = f(t)$. From the tree diagram

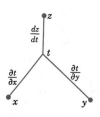

we obtain

$$\frac{\partial z}{\partial x} = \frac{dz}{dt}\frac{\partial t}{\partial x} = y\frac{dz}{dt}$$

$$\frac{\partial z}{\partial y} = \frac{dz}{dt}\frac{\partial t}{\partial y} = x\frac{dz}{dt}$$

Thus,

$$x \frac{\partial z}{\partial x} - y \frac{\partial z}{\partial y} = xy\frac{dz}{dt} - yx\frac{dz}{dt} = 0$$

◀

▶ **Exercise Set 16.8**

1. Let $f(v, w, x, y) = 4v^2w^3x^4y^5$. Find $\partial f/\partial v$, $\partial f/\partial w$, $\partial f/\partial x$, and $\partial f/\partial y$.

2. Let $w = r \cos st + e^u \sin ur$. Find $\partial w/\partial r$, $\partial w/\partial s$, $\partial w/\partial t$, $\partial w/\partial u$.

3. Let $f(v_1, v_2, v_3, v_4) = \dfrac{v_1^2 - v_2^2}{v_3^2 + v_4^2}$. Find $\partial f/\partial v_1$, $\partial f/\partial v_2$, $\partial f/\partial v_3$, $\partial f/\partial v_4$.

4. Let $V = xe^{2x-y} + we^{zw} + yw$. Find V_x, V_y, V_z, and V_w.

5. Let $f(v, w, x, y) = 2v^{1/2}w^4x^{1/2}y^{2/3}$. Find $f_v(1, -2, 4, 8)$, $f_w(1, -2, 4, 8)$, $f_x(1, -2, 4, 8)$, and $f_y(1, -2, 4, 8)$.

6. Let $u(w, x, y, z) = xe^{yw} \sin^2 z$. Find

   (a) $\dfrac{\partial u}{\partial x}(0, 0, 1, \pi)$ (b) $\dfrac{\partial u}{\partial y}(0, 0, 1, \pi)$

   (c) $\dfrac{\partial u}{\partial w}(0, 0, 1, \pi)$ (d) $\dfrac{\partial u}{\partial z}(0, 0, 1, \pi)$

   (e) $\dfrac{\partial^4 u}{\partial x \, \partial y \, \partial w \, \partial z}$ (f) $\dfrac{\partial^4 u}{\partial w \, \partial z \, \partial y^2}$.

In Exercises 7–13, use appropriate forms of the chain rule to find the derivatives.

7. Let $v = 7w^2x^3y^4z^5$, where $w = t^4$, $x = t^3$, $y = t^2$, $z = t$. Find dv/dt.

8. Let $z = u^7$, where $u = 5x^2 - 2y^3$. Find $\partial z/\partial x$ and $\partial z/\partial y$.

9. Let $z = \ln(x^2 + 1)$, where $x = r\cos\theta$. Find $\partial z/\partial r$ and $\partial z/\partial\theta$.

10. Let $u = rs^2\ln t$, $r = x^2$, $s = 4y + 1$, $t = xy^3$. Find $\partial u/\partial x$ and $\partial u/\partial y$.

11. Let $w = 4x^2 + 4y^2 + z^2$, $x = \rho\sin\phi\cos\theta$, $y = \rho\sin\phi\sin\theta$, $z = \rho\cos\phi$. Find $\partial w/\partial\rho$, $\partial w/\partial\phi$, and $\partial w/\partial\theta$.

12. Let $w = 3xy^2z^3$, $y = 3x^2 + 2$, $z = \sqrt{x - 1}$. Find dw/dx.

13. Let $w = \sqrt{x^2 + y^2 + z^2}$, $x = \cos 2y$, $z = \sqrt{y}$. Find dw/dy.

14. The length, width, and height of a rectangular box are increasing at rates of 1 in./sec, 2 in./sec and 3 in./sec, respectively.
    (a) At what rate is the volume increasing when the length is 2 in., the width is 3 in., and the height is 6 in.?
    (b) At what rate is the diagonal increasing at that instant?

15. Angle A of triangle ABC is increasing at a rate of $\pi/60$ radians/sec, side AB is increasing at a rate of 2 cm/sec, and side AC is increasing at a rate of 4 cm/sec. At what rate is the length of BC changing when angle A is $\pi/3$ radians, $AB = 20$ cm and $AC = 10$ cm? Is the length of BC increasing or decreasing? [*Hint:* Use the law of cosines.]

16. The area A of a triangle is given by $A = \frac{1}{2}ab\sin\theta$, where a and b are the lengths of two sides and θ is the angle between these sides. Suppose $a = 5$, $b = 10$, and $\theta = \pi/3$. Find
    (a) the rate at which A changes with a if b and θ are held constant
    (b) the rate at which A changes with θ if a and b are held constant
    (c) the rate at which b changes with a if A and θ are held constant.

17. Suppose $x^2 + 4xz + z^2 - 3yz + 5 = 0$. Find $\partial z/\partial x$ and $\partial z/\partial y$ by implicit differentiation.

18. Suppose $e^{xy}\cos yz - e^{yz}\sin xz + 2 = 0$. Find $\partial z/\partial x$ and $\partial z/\partial y$ by implicit differentiation.

19. Let $f(w, x, y, z) = wz\tan^{-1}\dfrac{x}{y} + 5w$. Show that
$$f_{ww} + f_{xx} + f_{yy} + f_{zz} = 0$$

In the remaining exercises, you may assume that all derivatives mentioned exist.

20. Let f be a function of one variable, and let $z = f(x + 2y)$. Show that
$$2\frac{\partial z}{\partial x} - \frac{\partial z}{\partial y} = 0$$

21. Let f be a function of one variable and let $z = f(x^2 + y^2)$. Show that
$$y\frac{\partial z}{\partial x} - x\frac{\partial z}{\partial y} = 0$$

22. Let f be a function of one variable, and let $w = f(\rho)$, where $\rho = (x^2 + y^2 + z^2)^{1/2}$. Show that
$$\left(\frac{\partial w}{\partial x}\right)^2 + \left(\frac{\partial w}{\partial y}\right)^2 + \left(\frac{\partial w}{\partial z}\right)^2 = \left(\frac{dw}{d\rho}\right)^2$$

23. Let f be a function of three variables and let $w = f(x - y, y - z, z - x)$. Show that
$$\frac{\partial w}{\partial x} + \frac{\partial w}{\partial y} + \frac{\partial w}{\partial z} = 0$$

24. Let $w = f(x, y, z)$, $x = \rho\sin\phi\cos\theta$, $y = \rho\sin\phi\sin\theta$, and $z = \rho\cos\phi$. Express $\partial w/\partial\rho$, $\partial w/\partial\phi$, and $\partial w/\partial\theta$ in terms of $\partial w/\partial x$, $\partial w/\partial y$, and $\partial w/\partial z$.

25. Let $w = \ln(e^r + e^s + e^t + e^u)$. Show that
$$w_{rstu} = -6e^{r+s+t+u-4w}$$
[*Hint:* Take advantage of the relationship $e^w = e^r + e^s + e^t + e^u$.]

26. Suppose w is a function of x_1, x_2, and x_3, and
$$x_1 = a_1y_1 + b_1y_2$$
$$x_2 = a_2y_1 + b_2y_2$$
$$x_3 = a_3y_1 + b_3y_2$$
where the a's and b's are constants. Express $\partial w/\partial y_1$ and $\partial w/\partial y_2$ in terms of $\partial w/\partial x_1$, $\partial w/\partial x_2$, and $\partial w/\partial x_3$.

27. (a) Let w be a function of $x_1, x_2, x_3,$ and x_4, and let each x_i be a function of t. Find a chain rule formula for dw/dt.

   (b) Let w be a function of $x_1, x_2, x_3,$ and x_4, and let each x_i be a function of $v_1, v_2,$ and v_3. Find chain rule formulas for $\partial w/\partial v_1$, $\partial w/\partial v_2$, and $\partial w/\partial v_3$.

28. Let $w = (x_1^2 + x_2^2 + \cdots + x_n^2)^k$, where $n > 2$. For what values of k does

$$\frac{\partial^2 w}{\partial x_1^2} + \frac{\partial^2 w}{\partial x_2^2} + \cdots + \frac{\partial^2 w}{\partial x_n^2} = 0$$

hold?

29. Show that

$$\frac{d}{dx}\left[\int_{a(x)}^{b(x)} f(t)\, dt\right] = f(b(x))b'(x) - f(a(x))a'(x)$$

This result is called **Leibniz' rule.** [*Hint:* Let $u = a(x)$, $v = b(x)$, and

$$F(u, v) = \int_u^v f(t)\, dt$$

Then use a chain rule and Theorem 5.10.1.}

30. Use the result of Exercise 29 to compute the following derivatives without performing the integrations.

   (a) $\dfrac{d}{dx}\displaystyle\int_x^{x^2} e^{t^2}\, dt$ (b) $\dfrac{d}{dx}\displaystyle\int_{\sin x}^{\cos x} (t^3 + 2)^{2/3}\, dt$

   (c) $\dfrac{d}{dx}\displaystyle\int_{3x}^{x^3} \sin^5 t\, dt$ (d) $\dfrac{d}{dx}\displaystyle\int_{e^x}^{e^{2x}} (\ln t)^4\, dt.$

16.9 MAXIMA AND MINIMA OF FUNCTIONS OF TWO VARIABLES

In Chapter 4 we learned how to find maximum and minimum values of a function of one variable. In this section we will develop similar techniques for functions of two variables.

The graphs of many functions of two variables form hills and valleys; the tops of the hills are called *relative maxima,* and the bottoms of the valleys are called *relative minima* (Figure 16.9.1).

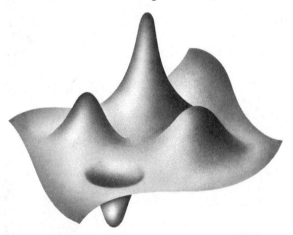

Figure 16.9.1

Geometrically, relative maxima and minima are the high and low points in their immediate vicinity. These ideas are expressed in the following definitions.

16.9.1 DEFINITION A function f of two variables is said to have a ***relative maximum*** at a point (x_0, y_0) if there is a circle centered at (x_0, y_0) such that $f(x_0, y_0) \geq f(x, y)$ for all points (x, y) inside the circle.

16.9.2 DEFINITION A function f of two variables is said to have a ***relative minimum*** at a point (x_0, y_0) if there is a circle centered at (x_0, y_0) such that $f(x_0, y_0) \leq f(x, y)$ for all points (x, y) inside the circle.

16.9.3 DEFINITION A function f of two variables is said to have a ***relative extremum*** at (x_0, y_0) if f has either a relative maximum or a relative minimum at (x_0, y_0).

It is essential to understand the implication of the adjective "relative." If the inequality

$$f(x_0, y_0) \geq f(x, y)$$

holds for *all* points (x, y) in the domain of f, then $f(x_0, y_0)$ is the largest value of f, and we say that f has a ***maximum*** or ***absolute maximum*** at (x_0, y_0). In contrast, for f to have a relative maximum at (x_0, y_0), this inequality need only hold within *some* circle centered at (x_0, y_0). Similarly, f has a ***minimum*** or ***absolute minimum*** at (x_0, y_0) if

$$f(x_0, y_0) \leq f(x, y)$$

holds for all (x, y) in the domain of f.

Let f be a function of two variables whose first partial derivatives exist at (x_0, y_0). If f has a relative maximum at (x_0, y_0), then it is clear geometrically that the traces of the surface $z = f(x, y)$ on the planes $x = x_0$ and $y = y_0$ have horizontal tangent lines at (x_0, y_0) (Figure 16.9.2). Thus,

$$f_x(x_0, y_0) = 0 \quad \text{and} \quad f_y(x_0, y_0) = 0$$

The same result holds if f has a relative minimum at (x_0, y_0).

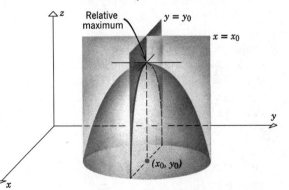

Figure 16.9.2

16.9.4 THEOREM   *If f has a relative extremum at a point (x_0, y_0), and if the first partial derivatives of f exist at this point, then*

$$f_x(x_0, y_0) = 0 \quad and \quad f_y(x_0, y_0) = 0$$

A point (x_0, y_0) at which $f_x(x_0, y_0) = f_y(x_0, y_0) = 0$ is called a **critical point** of f. Thus, the last theorem tells us that the relative extrema of functions having first partial derivatives occur at critical points.

For a function of one variable, the condition $f'(x_0) = 0$ is *not* sufficient to guarantee that f has a relative extremum at x_0. (The graph of f may have an inflection point with a horizontal tangent line at x_0.) Similarly, the conditions $f_x(x_0, y_0) = 0$ and $f_y(x_0, y_0) = 0$ are not sufficient to guarantee that a function f of two variables has a relative extremum at (x_0, y_0). For example, consider the function $f(x, y) = y^2 - x^2$. The graph of f is the hyperbolic paraboloid

$$z = y^2 - x^2$$

shown in Figure 16.9.3.

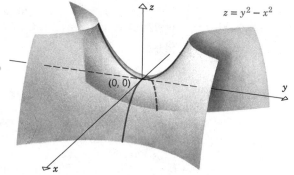

Figure 16.9.3

At the point $(0, 0)$, the trace in the xz-plane and the trace in the yz-plane both have horizontal tangent lines, that is, $f_x(0, 0) = 0$ and $f_y(0, 0) = 0$. However, f does not have a relative extremum at $(0, 0)$. To see this, observe that inside any circle in the xy-plane centered at $(0, 0)$, there exist points where $f(x, y)$ is positive (points on the y-axis) and there exist points where $f(x, y)$ is negative (points on the x-axis). Thus, $f(0, 0) = 0$ is neither the largest nor the smallest value of $f(x, y)$ in the circle.

A critical point at which a function f does not have a relative extremum is called a **saddle point** of f. Thus, the point $(0, 0)$ is a saddle point of the function $f(x, y) = y^2 - x^2$ graphed in Figure 16.9.3. (More advanced books distinguish between various types of saddle points. We will not do this, however.)

▶ **Example 1**   Locate all relative maxima, relative minima, and saddle points of $f(x, y) = 9 - x^2 - y^2$.

Solution. To obtain the critical points, we set the partial derivatives

$$f_x(x, y) = -2x \quad \text{and} \quad f_y(x, y) = -2y$$

equal to zero. This yields $x = 0$ and $y = 0$, so $(0, 0)$ is the only critical point. At this critical point we have $f(0, 0) = 9$, while at all points (x, y) different from $(0, 0)$ we have $f(x, y) < 9$ since $f(x, y) = 9 - x^2 - y^2 = 9 - (x^2 + y^2)$. Thus, f has a relative maximum (and, in fact, an absolute maximum) at $(0, 0)$. ◀

For functions more complicated than that in Example 1, we will need other methods for finding relative extrema and saddle points. The following theorem, which is usually proved in advanced calculus, is analogous to the second derivative test for functions of one variable.

16.9.5 THEOREM

The Second-Partials Test

Let f be a function of two variables with continuous second-order partial derivatives in some circle centered at a critical point (x_0, y_0); and let

$$D = f_{xx}(x_0, y_0)f_{yy}(x_0, y_0) - f_{xy}{}^2(x_0, y_0)$$

(a) If $D > 0$ and $f_{xx}(x_0, y_0) > 0$, then f has a relative minimum at (x_0, y_0).
(b) If $D > 0$ and $f_{xx}(x_0, y_0) < 0$, then f has a relative maximum at (x_0, y_0).
(c) If $D < 0$, then f has a saddle point at (x_0, y_0).
(d) If $D = 0$, then no conclusion can be drawn.

▶ **Example 2** Locate all relative extrema and saddle points of

$$f(x, y) = 3x^2 - 2xy + y^2 - 8y.$$

Solution. Since $f_x(x, y) = 6x - 2y$ and $f_y(x, y) = -2x + 2y - 8$, the critical points of f satisfy the equations

$$6x - 2y = 0$$
$$-2x + 2y - 8 = 0$$

Solving these for x and y yields $x = 2$, $y = 6$ (verify), so $(2, 6)$ is the only critical point. To apply Theorem 16.9.5 we need the second-order partial derivatives

$$f_{xx}(x, y) = 6, \quad f_{yy}(x, y) = 2, \quad f_{xy}(x, y) = -2$$

At the point $(2, 6)$ we have

$$D = f_{xx}(2, 6)f_{yy}(2, 6) - f_{xy}{}^2(2, 6) = (6)(2) - (-2)^2 = 8 > 0$$

and

$$f_{xx}(2, 6) = 6 > 0$$

so that f has a relative minimum at $(2, 6)$ by part (a) of the second-partials test.   ◀

▶ **Example 3**   Locate all relative extrema and saddle points of

$$f(x, y) = 4xy - x^4 - y^4$$

Solution.   Since

$$\begin{aligned} f_x(x, y) &= 4y - 4x^3 \\ f_y(x, y) &= 4x - 4y^3 \end{aligned} \tag{1}$$

the critical points of f have coordinates satisfying the equations

$$\begin{array}{ccc} 4y - 4x^3 = 0 & & y = x^3 \\ & \text{or} & \\ 4x - 4y^3 = 0 & & x = y^3 \end{array} \tag{2}$$

Substituting the top equation in the bottom yields $x = (x^3)^3$ or $x^9 - x = 0$ or $x(x^8 - 1) = 0$, which has solutions $x = 0$, $x = 1$, $x = -1$. Substituting these values in the top equation of (2) we obtain the corresponding y values $y = 0$, $y = 1$, $y = -1$. Thus, the critical points of f are $(0, 0)$, $(1, 1)$, and $(-1, -1)$.

From (1),

$$f_{xx}(x, y) = -12x^2, \qquad f_{yy}(x, y) = -12y^2, \qquad f_{xy}(x, y) = 4$$

which yields the following table:

CRITICAL POINT (x_0, y_0)	$f_{xx}(x_0, y_0)$	$f_{yy}(x_0, y_0)$	$f_{xy}(x_0, y_0)$	$D = f_{xx}f_{yy} - f_{xy}^2$
$(0, 0)$	0	0	4	-16
$(1, 1)$	-12	-12	4	128
$(-1, -1)$	-12	-12	4	128

At the points $(1, 1)$ and $(-1, -1)$, we have $D > 0$ and $f_{xx} < 0$ so that relative maxima occur at these critical points. At $(0, 0)$ there is a saddle point since $D < 0$. The surface and its contour map are shown in Figure 16.9.4.   ◀

▶ **Example 4**   Determine the dimensions of a rectangular box, open at the top, having a volume of 32 ft^3, and requiring the least amount of material for its construction.

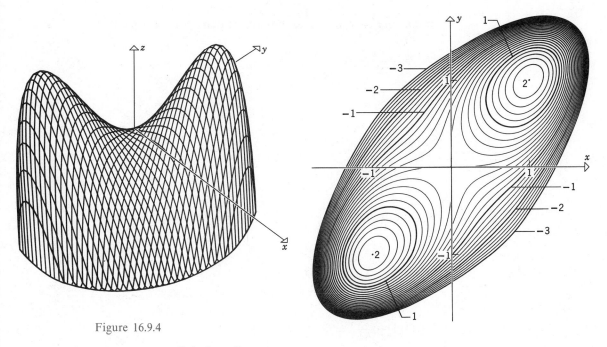

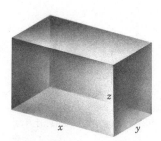

Figure 16.9.4

Solution. Let

x = length of the box (in feet)
y = width of the box (in feet)
z = height of the box (in feet)
S = surface area of the box (in square feet)

We may reasonably assume that the box with least surface area requires the least amount of material, so that our objective is to minimize the surface area

$$S = xy + 2xz + 2yz \qquad (3)$$

(see Figure 16.9.5) subject to the volume requirement

$$xyz = 32 \qquad (4)$$

From (4) we obtain $z = 32/xy$, so (3) can be rewritten as

$$S = xy + \frac{64}{y} + \frac{64}{x} \qquad (5)$$

which expresses S as a function of two variables. Differentiating (5) we obtain,

Two sides have area xz
Two sides have area yz
One side has area xy

Figure 16.9.5

$$\frac{\partial S}{\partial x} = y - \frac{64}{x^2}, \qquad \frac{\partial S}{\partial y} = x - \frac{64}{y^2} \qquad (6)$$

so that the coordinates of the critical points of S satisfy

$$y - \frac{64}{x^2} = 0$$

$$x - \frac{64}{y^2} = 0$$

Solving the first equation for y yields

$$y = \frac{64}{x^2} \tag{7}$$

and substituting this expression in the second equation yields

$$x - \frac{64}{(64/x^2)^2} = 0$$

which can be rewritten as

$$x\left(1 - \frac{x^3}{64}\right) = 0$$

The solutions of this equation are $x = 0$ and $x = 4$. Since we cannot construct a box of width zero, the only solution of significance is $x = 4$. Substituting this value in (7) yields $y = 4$. To see that we have located a relative minimum, we use the second-partials test. From (6),

$$\frac{\partial^2 S}{\partial x^2} = \frac{128}{x^3}, \qquad \frac{\partial^2 S}{\partial y^2} = \frac{128}{y^3}, \qquad \frac{\partial^2 S}{\partial y \, \partial x} = 1$$

Thus, when $x = 4$ and $y = 4$, we have

$$\frac{\partial^2 S}{\partial x^2} = 2, \qquad \frac{\partial^2 S}{\partial y^2} = 2, \qquad \frac{\partial^2 S}{\partial y \, \partial x} = 1$$

and

$$D = \frac{\partial^2 S}{\partial x^2} \frac{\partial^2 S}{\partial y^2} - \left(\frac{\partial^2 S}{\partial y \, \partial x}\right)^2 = (2)(2) - (1)^2 = 3$$

Since $\partial^2 S / \partial x^2 > 0$ and $D > 0$, it follows from the second-partials test that a relative minimum occurs when $x = y = 4$. Substituting these values in (4) yields $z = 2$, so that the box using least material has a height of 2 ft and a square base whose sides are 4 ft long. ◄

REMARK   Strictly speaking, the solution in the last example is incomplete since we have not shown that an *absolute minimum* for S occurs when

$x = y = 4$ and $z = 2$, only a relative minimum. The problem of showing that a relative extremum is also an absolute extremum can be difficult for functions of two or more variables and will not be considered in this text. However, in applied problems it is often obvious from physical or geometric considerations that an absolute extremum has been found.

Definitions of relative extrema can be given for functions of three or more variables. For example, if f is a function of three variables, then f has a **relative maximum** at (x_0, y_0, z_0) if $f(x_0, y_0, z_0) \geq f(x, y, z)$ for all points (x, y, z) within some sphere centered at (x_0, y_0, z_0), and for a **relative minimum** the inequality is reversed. If $f(x, y, z)$ has first partial derivatives, then the relative extrema occur at **critical points,** that is, points where

$$f_x(x_0, y_0, z_0) = f_y(x_0, y_0, z_0) = f_z(x_0, y_0, z_0) = 0$$

The extension of the second-partials test (Theorem 16.9.5) to functions of three or more variables is given in advanced calculus texts.

▶ Exercise Set 16.9

In Exercises 1–20, locate all relative maxima, relative minima, and saddle points

1. $f(x, y) = 3x^2 + 2xy + y^2$.

2. $f(x, y) = x^3 - 3xy - y^3$.

3. $f(x, y) = y^2 + xy + 3y + 2x + 3$.

4. $f(x, y) = x^2 + xy - 2y - 2x + 1$.

5. $f(x, y) = x^2 + xy + y^2 - 3x$.

6. $f(x, y) = xy - x^3 - y^2$.

7. $f(x, y) = x^2 + 2y^2 - x^2y$.

8. $f(x, y) = 2x^2 - 4xy + y^4 + 2$.

9. $f(x, y) = x^2 + y^2 + 2/(xy)$.

10. $f(x, y) = x^3 + y^3 - 3x - 3y$.

11. $f(x, y) = x^2 + y - e^y$.

12. $f(x, y) = xe^y$.

13. $f(x, y) = e^x \sin y$.

14. $f(x, y) = xy + 2/x + 4/y$.

15. $f(x, y) = 2y^2x - yx^2 + 4xy$.

16. $f(x, y) = y \sin x$.

17. $f(x, y) = e^{-(x^2+y^2+2x)}$.

18. $f(x, y) = xy + \dfrac{a^3}{x} + \dfrac{b^3}{y}$ $(a \neq 0, b \neq 0)$.

19. $f(x, y) = \sin x + \sin y$, $0 < x < \pi, 0 < y < \pi$.

20. $f(x, y) = \sin x + \sin y + \sin (x + y)$, $0 < x < \pi/2, 0 < y < \pi/2$.

21. (a) Show that the second-partials test provides no information about the critical points of $f(x, y) = x^4 + y^4$.
    (b) Classify all critical points as relative maxima, relative minima, or saddle points.

22. (a) Show that the second-partials test provides no information about the critical points of $f(x, y) = x^4 - y^4$.
    (b) Classify all critical points as relative maxima, relative minima, or saddle points.

23. Show that $f(x, y) = y^2 - 2xy + x^2$ has an absolute minimum at each point on the line $y = x$.

24. Find three positive numbers whose sum is 48 and such that their product is as large as possible.

25. Find three positive numbers whose sum is 27 and such that the sum of their squares is as small as possible.

26. Find all points on the plane $x + y + z = 5$ in the first octant at which $f(x, y, z) = xy^2z^2$ has a maximum value.

27. Find the points on the surface $x^2 - yz = 5$ that are closest to the origin.

28. Find the dimensions of the rectangular box of maximum volume that can be inscribed in a sphere of radius a.

29. Find the maximum volume of a rectangular box with three faces in the coordinate planes and a vertex in the first octant on the plane $x + y + z = 1$.

30. A manufacturer makes two models of an item, standard and deluxe. It costs \$40 to manufacture the standard model and \$60 for the deluxe. A market research firm estimates that if the standard model is priced at x dollars and the deluxe at y dollars, then the manufacturer will sell $500(y - x)$ of the standard items and $45{,}000 + 500(x - 2y)$ of the deluxe each year. How should the items be priced to maximize the profit?

31. A closed rectangular box with a volume of 16 ft^3 is made from two kinds of materials. The top and bottom are made of material costing 10¢ per square foot and the sides from material costing 5¢ per square foot. Find the dimensions of the box so that the cost of materials is minimized.

32. Use the methods of this section to find the distance between the lines

$$
\begin{array}{ccc}
x = 3t & & x = 2t \\
y = 2t & \text{and} & y = 2t + 3 \\
z = t & & z = 2t
\end{array}
$$

33. Use the methods of this section to find the distance from the point $(-1, 3, 2)$ to the plane $x - 2y + z = 4$.

34. Show that among all parallelograms with perimeter l, a square with sides of length $l/4$ has maximum area. [*Hint:* The area of a parallelogram is $A = ab \sin \alpha$, where a and b are the lengths of two adjacent sides and α is the angle between them.]

35. Determine the dimensions of a rectangular box, open at the top, having volume V, and requiring the least amount of material for its construction.

36. A length of sheet metal 27 in. wide is to be made into a water trough by bending up two sides as shown in the accompanying figure. Find x and ϕ so that the trapezoid-shaped cross section has a maximum area.

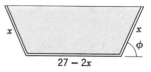

$$27 - 2x$$

16.10 LAGRANGE MULTIPLIERS

In Example 4 of the previous section, we dealt with the problem of minimizing

$$S = xy + 2xz + 2yz$$

subject to the constraint

$$xyz - 32 = 0$$

This is a special case of the following general problem, which we will study in this section:

Three-Variable Extremum Problem with One Constraint

Maximize or minimize the function

$$f(x, y, z)$$

subject to the constraint

$$g(x, y, z) = 0$$

We will also be interested in the two-variable version of this problem.

Two-Variable Extremum Problem with One Constraint

Maximize or minimize the function

$$f(x, y)$$

subject to the constraint

$$g(x, y) = 0$$

One way to attack these problems is to solve the constraint equation for one of the variables in terms of the rest and substitute the result into f. The resulting function of one or two variables can then be maximized or minimized by finding its critical points. (See the solution of Example 4 in the previous section.) However, if the constraint equation is too complicated to solve for one of the variables in terms of the rest, then other techniques must be used. We will discuss one such technique, called *the method of Lagrange multipliers.* Since a rigorous discussion of this topic requires results from advanced calculus, we will not be too finicky about all the technical details; we will emphasize computational techniques.

Let us begin with the two-variable problem of maximizing or minimizing $f(x, y)$ subject to the constraint $g(x, y) = 0$. The graph of $g(x, y) = 0$ is some curve C in the xy-plane. Geometrically, we are concerned with finding the maximum or minimum value of $f(x, y)$ as (x, y) varies over the constraint curve C. If (x_0, y_0) is a point on the constraint curve C, then we will say that $f(x, y)$ has a *constrained relative maximum* at (x_0, y_0) if there is a circle centered at (x_0, y_0) such that

$$f(x_0, y_0) \geq f(x, y) \tag{1}$$

for all points (x, y) on C within the circle (Figure 16.10.1). For a *constrained relative minimum* at (x_0, y_0), the inequality in (1) is reversed. We will say that

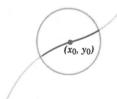

$g(x, y) = 0$

(x_0, y_0)

C

At a constrained relative maximum, the inequality $f(x_0, y_0) \geq f(x, y)$ holds at all points of C within some circle centered at (x_0, y_0)

Figure 16.10.1

f has a **constrained relative extremum** at (x_0, y_0) if f has either a constrained relative maximum or a constrained relative minimum at (x_0, y_0). The following result is the key to finding constrained relative extrema.

16.10.1   THEOREM
The Constrained-Extremum Principle for Two Variables

Let f and g be functions of two variables with continuous first partial derivatives on some open set containing the curve $g(x, y) = 0$, and assume that $\nabla g \neq 0$ at any point on this curve. If f has a constrained relative extremum on the constraint curve $g(x, y) = 0$, then this extremum occurs at a point (x_0, y_0) where the gradient vectors $\nabla f(x_0, y_0)$ and $\nabla g(x_0, y_0)$ are parallel, that is,

$$\nabla f(x_0, y_0) = \lambda \nabla g(x_0, y_0) \tag{2}$$

for some number λ.

The number λ in (2) is called a **Lagrange multiplier.**

To see why the constrained extremum principle is true, let C denote the constraint curve $g(x, y) = 0$, and suppose that a constrained relative extremum occurs at the point (x_0, y_0) on C. It is proved in advanced calculus that under the hypotheses of this theorem, the curve C can be represented parametrically by equations

$$x = x(t)$$
$$y = y(t)$$

so that C has a nonzero tangent vector at (x_0, y_0). More precisely,

$$x'(t_0)\mathbf{i} + y'(t_0)\mathbf{j} \neq 0$$

where t_0 is the value of the parameter corresponding to (x_0, y_0). In terms of the parameter t, the value of f at a point (x, y) on the curve C is $f(x, y) = f(x(t), y(t))$. Since f has a constrained relative extremum at the point (x_0, y_0), and since this point corresponds to $t = t_0$, it follows that the function

$$F(t) = f(x(t), y(t))$$

has a relative extremum at $t = t_0$, that is,

$$F'(t_0) = 0 \tag{3}$$

By the chain rule

$$F'(t) = \frac{d}{dt}[f(x(t), y(t))] = f_x(x, y)\frac{dx}{dt} + f_y(x, y)\frac{dy}{dt}$$

so that condition (3) yields the equation

$$f_x(x_0, y_0)x'(t_0) + f_y(x_0, y_0)y'(t_0) = 0$$

which may be expressed in the form

$$\nabla f(x_0, y_0) \cdot [x'(t_0)\mathbf{i} + y'(t_0)\mathbf{j}] = 0 \qquad (4)$$

Since the vector $x'(t_0)\mathbf{i} + y'(t_0)\mathbf{j}$ is tangent to the curve C at (x_0, y_0), it follows from (4) that $\nabla f(x_0, y_0)$ is perpendicular to C at (x_0, y_0). But $\nabla g(x_0, y_0)$ is also perpendicular to C at (x_0, y_0) since C is the level curve $g(x, y) = 0$ for the function g (see Theorem 16.5.4). Since $\nabla f(x_0, y_0)$ and $\nabla g(x_0, y_0)$ are both perpendicular to C at (x_0, y_0), these vectors are parallel.

▶ Example 1   At what point on the circle $x^2 + y^2 = 1$ is the product xy a maximum?

Solution.   We want to maximize

$$f(x, y) = xy$$

subject to the constraint

$$g(x, y) = x^2 + y^2 - 1 = 0 \qquad (5)$$

At a constrained relative extremum we must have

$$\nabla f = \lambda \nabla g$$

or

$$y\mathbf{i} + x\mathbf{j} = \lambda(2x\mathbf{i} + 2y\mathbf{j})$$

which is equivalent to the pair of equations

$$y = 2x\lambda \qquad \text{and} \qquad x = 2y\lambda \qquad (6)$$

Since the maximum value of xy on the circle $x^2 + y^2 = 1$ is obviously greater than zero, we must have $x \neq 0$ and $y \neq 0$ at a constrained maximum. Thus, the equations in (6) can be rewritten as

$$\lambda = \frac{y}{2x} \qquad \text{and} \qquad \lambda = \frac{x}{2y}$$

from which we obtain

$$\frac{y}{2x} = \frac{x}{2y}$$

or

$$y^2 = x^2 \tag{7}$$

Substituting this in (5) yields

$$2x^2 - 1 = 0$$

or

$$x = \frac{1}{\sqrt{2}} \quad \text{and} \quad x = -\frac{1}{\sqrt{2}}$$

Substituting $x = 1/\sqrt{2}$ in (7) yields $y = \pm 1/\sqrt{2}$ and substituting $x = -1/\sqrt{2}$ in (7) yields $y = \pm 1/\sqrt{2}$, so that there are four candidates for the location of a maximum

$$\left(\frac{1}{\sqrt{2}}, \frac{1}{\sqrt{2}}\right), \left(\frac{1}{\sqrt{2}}, -\frac{1}{\sqrt{2}}\right), \left(-\frac{1}{\sqrt{2}}, \frac{1}{\sqrt{2}}\right), \left(-\frac{1}{\sqrt{2}}, -\frac{1}{\sqrt{2}}\right)$$

At the first and fourth points the function $f(x, y) = xy$ has value $\frac{1}{2}$, while at the second and third points the value is $-\frac{1}{2}$. Thus, the constrained maximum value of $\frac{1}{2}$ occurs at $(1/\sqrt{2}, 1/\sqrt{2})$ and $(-1/\sqrt{2}, -1/\sqrt{2})$. ◄

REMARK.   If c is a constant, then the functions $g(x, y)$ and $g(x, y) - c$ have the same gradient since the constant c drops out when we differentiate. Consequently, it is *not* essential to rewrite a constraint of the form $g(x, y) = c$ as $g(x, y) - c = 0$ in order to apply the constrained extremum principle. Thus, in the last example, we could have kept the constraint in the form $x^2 + y^2 = 1$ and taken $g(x, y) = x^2 + y^2$ rather than $g(x, y) = x^2 + y^2 - 1$.

In Exercise 6 of Section 4.7, it was stated that among all rectangles of perimeter p, a square of side $p/4$ has maximum area. This result can be obtained using Lagrange multipliers.

▶ Example 2   Find the dimensions of a rectangle having perimeter p and maximum area.

Solution.   Let

$$x = \text{length of the rectangle}$$

$$y = \text{width of the rectangle}$$

$$A = \text{area of the rectangle}$$

We want to maximize

$$A = xy$$

subject to the perimeter constraint

$$2x + 2y = p \tag{8}$$

From the constrained extremum principle with $f(x, y) = xy$ and $g(x, y) = 2x + 2y$, we must have $\nabla f = \lambda \nabla g$, or

$$y\mathbf{i} + x\mathbf{j} = \lambda(2\mathbf{i} + 2\mathbf{j})$$

at a constrained relative maximum. This is equivalent to the two equations

$$y = 2\lambda \quad \text{and} \quad x = 2\lambda$$

Eliminating λ from these equations we obtain

$$x = y$$

Using this condition and constraint (8), we obtain $x = p/4$, $y = p/4$. ◀

Lagrange multipliers can also be used in the three-variable problem of maximizing or minimizing $f(x, y, z)$ subject to the constraint $g(x, y, z) = 0$. The graph of $g(x, y, z) = 0$ is some surface S in 3-space. Geometrically we are concerned with finding the maximum or minimum of the function $f(x, y, z)$ as (x, y, z) varies over the surface S. We will say that $f(x, y, z)$ has a *constrained relative maximum* at (x_0, y_0, z_0) if there is a sphere centered at (x_0, y_0, z_0) such that

$$f(x_0, y_0, z_0) \geq f(x, y, z)$$

for all points (x, y, z) on S within the sphere (Figure 16.10.2). The meaning of the terms, *constrained relative minimum* and *constrained relative extremum* should be clear. It can be shown that a constrained relative extremum can only occur at a point (x_0, y_0, z_0) where $\nabla f(x_0, y_0, z_0)$ and $\nabla g(x_0, y_0, z_0)$ are parallel, that is,

$$\nabla f(x_0, y_0, z_0) = \lambda \nabla g(x_0, y_0, z_0)$$

for some number λ.

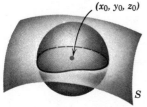

(x_0, y_0, z_0)

S

$g(x, y, z) = 0$

At a constrained relative maximum, the inequality $f(x_0, y_0, z_0) \geq f(x, y, z)$ holds at all points of S inside some sphere centered at (x_0, y_0, z_0)

Figure 16.10.2

▶ **Example 3** Find the points on the sphere $x^2 + y^2 + z^2 = 36$ that are closest to and farthest from $(1, 2, 2)$.

Solution. To avoid radicals, we will find points on the sphere that minimize and maximize the *square* of the distance to $(1, 2, 2)$. Thus, we want to extremize

$$f(x, y, z) = (x - 1)^2 + (y - 2)^2 + (z - 2)^2$$

subject to the constraint

$$x^2 + y^2 + z^2 = 36 \tag{9}$$

Therefore, with $g(x, y, z) = x^2 + y^2 + z^2$, we must have $\nabla f(x, y, z) = \lambda \nabla g(x, y, z)$ at a constrained relative extremum; that is,

$$2(x - 1)\mathbf{i} + 2(y - 2)\mathbf{j} + 2(z - 2)\mathbf{k} = \lambda(2x\mathbf{i} + 2y\mathbf{j} + 2z\mathbf{k})$$

which leads to the equations

$$2(x - 1) = 2x\lambda$$
$$2(y - 2) = 2y\lambda \tag{10}$$
$$2(z - 2) = 2z\lambda$$

It is clear geometrically that the points on the sphere $x^2 + y^2 + z^2 = 36$ that are closest to and farthest from $(1, 2, 2)$ do not lie in the coordinate planes (make a sketch). Thus, we can assume that x, y, and z are nonzero, and we can rewrite (10) as

$$\frac{x - 1}{x} = \lambda$$

$$\frac{y - 2}{y} = \lambda$$

$$\frac{z - 2}{z} = \lambda$$

The first two equations imply that

$$\frac{x - 1}{x} = \frac{y - 2}{y}$$
$$xy - y = xy - 2x$$
$$y = 2x \tag{11}$$

Similarly, the first and third equations imply that

$$z = 2x \tag{12}$$

Substituting (11) and (12) in the constraint equation (9), we obtain

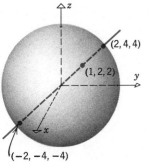

Figure 16.10.3

$$9x^2 = 36$$

or

$$x = \pm 2$$

Substituting these values in (11) and (12) yields two points

$$(2, 4, 4) \quad \text{and} \quad (-2, -4, -4)$$

Since $f(2, 4, 4) = 9$ and $f(-2, -4, -4) = 81$, it follows that $(2, 4, 4)$ is the point on the sphere closest to $(1, 2, 2)$, and $(-2, -4, -4)$ is the point that is farthest (Figure 16.10.3). ◄

Next we will use Lagrange multipliers to solve the problem of Example 4 in the previous section.

▶ Example 4   Use Lagrange multipliers to determine the dimensions of a rectangular box, open at the top, having a volume of 32 ft³, and requiring the least amount of material for its construction.

Solution.   With the notation of Example 4 in Section 16.9, the problem is to minimize the surface area

$$S = xy + 2xz + 2yz$$

subject to the volume constraint

$$xyz = 32 \tag{13}$$

If we let $f(x, y, z) = xy + 2xz + 2yz$ and $g(x, y, z) = xyz$, then we must have $\nabla f = \lambda \nabla g$, that is,

$$(y + 2z)\mathbf{i} + (x + 2z)\mathbf{j} + (2x + 2y)\mathbf{k} = \lambda(yz\mathbf{i} + xz\mathbf{j} + xy\mathbf{k})$$

This condition yields the three equations

$$y + 2z = \lambda yz$$
$$x + 2z = \lambda xz$$
$$2x + 2y = \lambda xy$$

Because of the volume constraint $xyz = 32$, we know that x, y, and z are nonzero, so that these equations can be rewritten

$$\frac{1}{z} + \frac{2}{y} = \lambda$$

$$\frac{1}{z} + \frac{2}{x} = \lambda$$

$$\frac{2}{y} + \frac{2}{x} = \lambda$$

From the first two equations,

$$y = x \tag{14}$$

and from the second and third equations, $z = \frac{1}{2}y$. This and (14) imply that

$$z = \frac{1}{2}x \tag{15}$$

Substituting (14) and (15) in the volume constraint (13) yields

$$\tfrac{1}{2}x^3 = 32$$

This equation, together with (14) and (15), yields

$$x = 4, \qquad y = 4, \qquad z = 2$$

which agrees with the result obtained in Example 4 of the previous section.

◀

Lagrange multipliers can also be used in problems involving two or more constraints. However, we will not pursue this topic here.

▶ Exercise Set 16.10

In Exercises 1–7, use Lagrange multipliers to find the maximum and minimum values of f subject to the given constraint. Also, find the points at which these extreme values occur.

1. $f(x, y) = xy$; $4x^2 + 8y^2 = 16$.

2. $f(x, y) = x^2 - y$; $x^2 + y^2 = 25$.

3. $f(x, y) = 4x^3 + y^2$; $2x^2 + y^2 = 1$.

4. $f(x, y) = x - 3y - 1$; $x^2 + 3y^2 = 16$.

5. $f(x, y, z) = 2x + y - 2z$; $x^2 + y^2 + z^2 = 4$.

6. $f(x, y, z) = 3x + 6y + 2z$; $2x^2 + 4y^2 + z^2 = 70$.

7. $f(x, y, z) = xyz$; $x^2 + y^2 + z^2 = 1$.

In Exercises 8–16, solve using Lagrange multipliers.

8. Find the point on the line $2x - 4y = 3$ that is closest to the origin.

9. Find the point on the line $y = 2x + 3$ that is closest to $(4, 2)$.

10. Find the point on the plane $x + 2y + z = 1$ that is closest to the origin.

11. Find the point on the plane $4x + 3y + z = 2$ that is closest to $(1, -1, 1)$.

12. Find the points on the surface $xy - z^2 = 1$ that are closest to the origin.

13. Find the maximum value of $\sin x \sin y$, where x and y denote the acute angles of a right triangle.

14. Find a vector in 3-space whose length is 5 and whose components have the largest possible sum.

15. Find the points on the circle $x^2 + y^2 = 45$ that are closest to and farthest from $(1, 2)$.

16. The temperature at a point (x, y) on a metal plate is $T(x, y) = 4x^2 - 4xy + y^2$. An ant, walking on the plate, traverses a circle of radius 5 centered at the origin. What are the highest and lowest temperatures encountered by the ant?

In Exercises 17–24, use Lagrange multipliers to solve the indicated problems from Section 16.9.

17. Exercise 25. 18. Exercise 26.

19. Exercise 27. 20. Exercise 28.

21. Exercise 31. 22. Exercise 33.

23. Exercise 34. 24. Exercise 35.

▶ SUPPLEMENTARY EXERCISES

1. Let $f(x, y) = e^x \ln y$. Find:
   (a) $f(0, e)$ (b) $f(\ln y, e^x)$ (c) $f(r + s, rs)$.

2. Sketch the domain of f using solid lines for portions of the boundary included in the domain and dashed lines for portions not included.
   (a) $f(x, y) = \sqrt{x - y}/(2x - y)$
   (b) $f(x, y) = \ln (xy - 1)$
   (c) $f(x, y) = (\sin^{-1} x)/e^y$.

3. Describe the graph of f.
   (a) $f(x, y) = \sqrt{x^2 + 4y^2}$
   (b) $f(x, y) = 1 - x/a - y/b$.

4. Find f_x, f_y, and f_z if $f(x, y, z) = x^2/(y^2 + z^2)$.

5. Find $\partial w/\partial r$ if $w = \ln (xy)/\sin yz$, $x = r + s$, $y = s$, $z = 3r - s$.

6. Find $\partial f/\partial x$, $\partial f/\partial y$, and $\partial f/\partial u$ if $f(x, y, z, u) = (e^{yz}/x) + \ln (u - x)$.

7. Find f_x, f_y, f_{yx}, and f_{yzx} at $(0, \pi/2, 1)$ if $f(x, y, z) = e^{xy} \sin yz$.

8. Find $g_{xy}(0, 3)$ and $g_{yy}(2, 0)$ if $g(x, y) = \sin (xy) + xe^y$.

9. Find $\partial w/\partial \theta|_{r=1, \theta=0}$ if $w = \ln (x^2 + y^2)$, $x = re^\theta$, $y = \tan (r\theta)$.

10. Find $\partial w/\partial r$ if $w = x \cos y + y \sin x$, $x = rs^2$, $y = r + s$.

11. Find $\partial w/\partial s$ if $w = \ln (x^2 + y^2 + 2z)$, $x = r + s$, $y = r - s$, $z = 2rs$.

In Exercises 12–15, verify the assertion.

12. If $w = \tan (x^2 + y^2) + x\sqrt{y}$, then $w_{xy} = w_{yx}$.

13. If $w = \ln (3x - 3y) + \cos (x + y)$, then $\partial^2 w/\partial x^2 = \partial^2 w/\partial y^2$.

14. If $F(x, y, z) = 2z^3 - 3(x^2 + y^2)z$, then $F_{xx} + F_{yy} + F_{zz} = 0$.

15. If $f(x, y, z) = xyz + x^2 + \ln (y/z)$, then $f_{xyzx} = f_{zxxy}$.

16. Find the slope of the tangent line at $(1, -2, -3)$ to the curve of intersection of the surface $z = 5 - 4x^2 - y^2$ with
   (a) the plane $x = 1$ (b) the plane $y = -2$.

17. The pressure in newtons/m^2 of a gas in a cylinder is given by $P = 10T/V$ with T in °C and V in m^3.
   (a) If T is increasing at a rate of 3°C/min with V held fixed at 2.5 m^3, find the rate at which the pressure is changing when $T = 50$°C.
   (b) If T is held fixed at 50°C while V is decreasing at the rate of 3 m^3/min, find the rate at which the pressure is changing when $V = 2.5$ m^3.

In Exercises 18 and 19:
   (a) Find the limit of $f(x, y)$ as $(x, y) \to (0, 0)$ if it exists.
   (b) Determine if f is continuous at $(0, 0)$.

18. $f(x, y) = \dfrac{x^4 - x + y - x^3y}{x - y}$.

19. $f(x, y) = \begin{cases} \dfrac{x^4 - y^4}{x^2 + y^2} & \text{if } (x, y) \neq (0, 0) \\ 0 & \text{if } (x, y) = (0, 0). \end{cases}$

20. Find dw/dt using the chain rule.
(a) $w = \sin xy + y \ln xz + z$, $x = e^t$, $y = t^2$, $z = 1$
(b) $w = \sqrt{xy - e^z}$, $x = \sin t$, $y = 3t$, $z = \cos t$.

21. Use Formula (8) of Section 16.4 to find dy/dx.
(a) $3x^2 - 5xy + \tan xy = 0$
(b) $x \ln y + \sin(x - y) = \pi$.

22. If $F(x, y) = 0$, find a formula for d^2y/dx^2 in terms of partial derivatives of F. [*Hint:* Use Formula (8) of Section 16.4.]

23. The voltage V across a fixed resistance R in series with a variable resistance r is $V = RE/(r + R)$, where E is the source voltage. Express dV/dt in terms of dE/dt and dr/dt.

24. Let $f(x, y, z) = 1/(z - x^2 - 4y^2)$.
(a) Describe the domain of f.
(b) Describe the level surface $f(x, y, z) = 2$.
(c) Find $f(3t, uv, e^{3t})$.

In Exercises 25–29, find:
(a) the gradient of f at P_0
(b) the directional derivative at P_0 in the indicated direction.

25. $f(x, y) = x^2y^5$, $P_0(3, 1)$; from P_0 toward $P_1(4, -3)$.

26. $f(x, y, z) = ye^x \sin z$, $P_0(\ln 2, 2, \pi/4)$; in the direction of $\mathbf{a} = \langle 1, -2, 2 \rangle$.

27. $f(x, y, z) = \ln(xyz)$, $P_0(3, 2, 6)$; $\mathbf{u} = \langle -1, 1, 1 \rangle / \sqrt{3}$.

28. $f(x, y) = x^2y + 2xy^2$, $P_0(1, 2)$; $\mathbf{u}$ makes an angle of $60°$ with the positive x-axis.

29. $f(x, y, z) = xy + yz + zx$, $P_0(1, -1, 2)$; from P_0 toward $P_1(11, 10, 0)$.

30. Let $f(x, y, z) = (x + y)^2 + (y + z)^2 + (z + x)^2$. Find the maximum rate of decrease of f at $P_0(2, -1, 2)$ and the direction in which this rate of decrease occurs.

31. Find all unit vectors $\mathbf{u}$ such that $D_{\mathbf{u}}f = 0$ at P_0.
(a) $f(x, y) = x^3y^3 - xy$, $P_0(1, -1)$
(b) $f(x, y) = xe^y$, $P_0(-2, 0)$.

32. The directional derivative $D_{\mathbf{u}}f$ at (x_0, y_0) is known to be 2 when $\mathbf{u}$ makes an angle of $30°$ with the positive x-axis, and 8 when this angle is $150°$. Find $D_{\mathbf{u}}f(x_0, y_0)$ in the direction of the vector $\sqrt{3}\mathbf{i} + 2\mathbf{j}$.

33. At the point $(1, 2)$, the directional derivative $D_{\mathbf{u}}f$ is $2\sqrt{2}$ toward $P_1(2, 3)$ and -3 toward $P_2(1, 0)$. Find $D_{\mathbf{u}}f(1, 2)$ toward the origin.

In Exercises 34 and 35:
(a) Find a normal vector $\mathbf{N}$ at $P_0(x_0, y_0, f(x_0, y_0))$.
(b) Find an equation for the tangent plane at P_0.

34. $f(x, y) = 4x^2 + y^2 + 1$; $P_0(1, 2, 9)$.

35. $f(x, y) = 2\sqrt{x^2 + y^2}$; $P_0(4, -3, 10)$.

36. Find equations for the tangent plane and normal line to the given surface at P_0.
(a) $z = x^2e^{2y}$; $P_0(1, \ln 2, 4)$
(b) $x^2y^3z^4 + xyz = 2$; $P_0(2, 1, -1)$.

37. Find all points P_0 on the surface $z = 2 - xy$ at which the normal line passes through the origin.

38. Show that for all tangent planes to the surface $x^{2/3} + y^{2/3} + z^{2/3} = 1$, the sum of the squares of the x-, y-, and z-intercepts is 1.

39. Find all points on the paraboloid $z = 9x^2 + 4y^2$ at which the normal line is parallel to the line through the points $P(4, -2, 5)$ and $Q(-2, -6, 4)$.

40. If $w = x^2y - 2xy + y^2x$, find the increment Δw and the differential dw if (x, y) varies from $(1, 0)$ to $(1.1, -0.1)$.

41. Use differentials to estimate the change in the volume $V = \frac{1}{3}x^2h$ of a pyramid with a square base when its height h is increased from 2 to 2.2 m, while its base dimension x is decreased from 1 to 0.9 m. Compare this to ΔV.

42. If $f(x, y, z) = x^2y^4/(1 + z^2)$, use differentials to estimate $f(4.996, 1.003, 1.995)$.

In Exercises 43–45, locate all relative minima, relative maxima, and saddle points.

43. $f(x, y) = x^2 + 3xy + 3y^2 - 6x + 3y$.

44. $f(x, y) = x^2y - 6y^2 - 3x^2$.

45. $f(x, y) = x^3 - 3xy + \frac{1}{2}y^2$.

Solve Exercises 46 and 47 two ways:
(a) Use the constraint to eliminate a variable.
(b) Use Lagrange multipliers.

46. Find all relative extrema of x^2y^2 subject to the constraint $4x^2 + y^2 = 8$.

47. Find the dimensions of the rectangular box of maximum volume that can be inscribed in the ellipsoid $(x/a)^2 + (y/b)^2 + (z/c)^2 = 1$.

In Exercises 48 and 49, use Lagrange multipliers.

48. Find the points on the curve $5x^2 - 6xy + 5y^2 = 8$ whose distance from the origin is (i) minimum (ii) maximum.

49. A current I branches into currents I_1, I_2, and I_3 through resistors R_1, R_2, and R_3 (see figure) in such a way that the total energy to the three resistors is a minimum. If the energy delivered to R_i is $I_i^2 R_i$ ($i = 1, 2, 3$), find the ratios $I_1 : I_2 : I_3$.

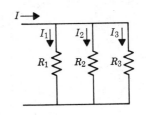

17 multiple integrals

17.1 DOUBLE INTEGRALS

The notion of a definite integral can be extended to functions of two or more variables. In this section we will discuss the *double integral*, which is the extension to functions of two variables.

To set the groundwork, let us review the steps needed to define the definite integral $\int_a^b f(x)\,dx$ of a function of one variable.

Step 1. Divide the interval $[a, b]$ into n subintervals having lengths

$$\Delta x_1, \Delta x_2, \ldots, \Delta x_n$$

Step 2. In each subinterval, choose an arbitrary point, and let these points be denoted by

$$x_1^*, x_2^*, \ldots, x_n^*$$

Step 3. Form the Riemann sum

$$\sum_{k=1}^{n} f(x_k^*)\,\Delta x_k$$

Step 4. Repeat this process with more and more subdivisions, so that the length of each subinterval approaches zero, and n, the number of subintervals, approaches $+\infty$. Define

$$\int_{a}^{b} f(x)\, dx = \lim_{n \to +\infty} \sum_{k=1}^{n} f(x_k^*)\, \Delta x_k$$

Recall that when $f(x)$ is nonnegative on the interval $[a, b]$, the Riemann sum

$$\sum_{k=1}^{n} f(x_k^*)\, \Delta x_k$$

can be interpreted as an approximation by rectangles to the area under the curve $y = f(x)$ over the interval $[a, b]$ (Figure 17.1.1) and the integral $\int_{a}^{b} f(x)\, dx$ can be interpreted as the exact area under the curve.

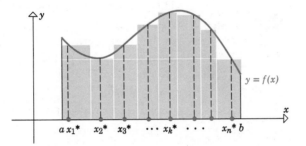

Figure 17.1.1

In the case where $f(x)$ assumes both positive and negative values on $[a, b]$, the integral $\int_{a}^{b} f(x)\, dx$ can be interpreted as a difference of areas, the area above the x-axis between $[a, b]$ and $y = f(x)$ minus the area below the x-axis between $[a, b]$ and $y = f(x)$.

The definition of an integral for a function of two variables is a natural extension of these ideas. Whereas the integration of $f(x)$ takes place over a closed interval on the x-axis, the integration of $f(x, y)$ takes place over a closed region R in the xy-plane. Later we will place more restrictions on the region R, but for now let us just assume that the entire region R can be enclosed within some suitably large rectangle with sides parallel to the coordinate axes. This ensures that R does not extend indefinitely in any direction (Figure 17.1.2).

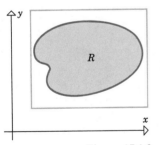

Figure 17.1.2

To define a double integral we proceed as follows:

Step 1. Using lines parallel to the coordinate axes, divide a rectangle enclosing the region R into subrectangles, and exclude from consideration all those subrectangles that contain any points outside of R. This leaves only rectangles that are subsets of R (Figure 17.1.3). Denote the areas of these rectangles by

$$\Delta A_1, \Delta A_2, \ldots, \Delta A_n$$

Step 2. Choose an arbitrary point in each of these subrectangles, and denote them by

$$(x_1^*, y_1^*), (x_2^*, y_2^*), \ldots, (x_n^*, y_n^*)$$

(see Figure 17.1.3).

Step 3. Form the sum

$$\sum_{k=1}^{n} f(x_k^*, y_k^*)\, \Delta A_k$$

This is called a ***Riemann sum***.

Step 4. Repeat this process with more and more subdivisions, so that the length and width of each rectangle approach zero; and n, the number of rectangles, approaches $+\infty$. Define

$$\iint\limits_{R} f(x, y)\, dA = \lim_{n \to +\infty} \sum_{k=1}^{n} f(x_k^*, y_k^*)\, \Delta A_k \tag{1}$$

The symbol $\displaystyle\iint\limits_{R} f(x, y)\, dA$ is called the ***double integral*** of $f(x, y)$ over R.

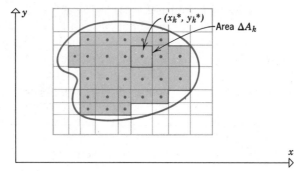

Figure 17.1.3

A precise definition of expression (1) (in terms of ϵ's and δ's) and conditions under which the double integral exists are studied in advanced calculus. However, for our purposes it suffices to say that existence is ensured when f is continuous on R and the region R is not too "complicated."

In the special case where $f(x, y)$ is *nonnegative* on the region R, the double integral may be interpreted as the volume of the solid S bounded above by the surface $z = f(x, y)$ and below by the region R (Figure 17.1.4). To see this, observe that the product

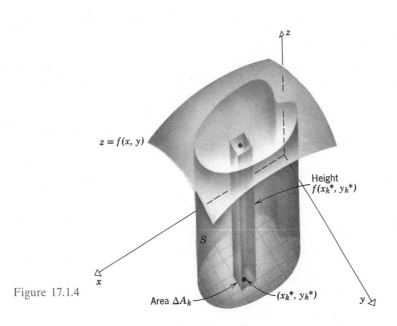

Figure 17.1.4

$$f(x_k^*, y_k^*) \, \Delta A_k$$

represents the volume of a rectangular parallelepiped whose height is $f(x_k^*, y_k^*)$ and whose base has area ΔA_k (Figure 17.1.4). Consequently, the Riemann sum

$$\sum_{k=1}^{n} f(x_k^*, y_k^*) \, \Delta A_k$$

may be interpreted as an approximation by rectangular parallelepipeds to the volume of the solid S. There are two sources of error in this approximation. First, the bases of the parallelepipeds will not, in general, cover the region R exactly and, second, the approximating parallelepipeds have flat tops, whereas the upper surface of the solid may be curved. However, it is intuitively clear that as we use more and more subrectangles of decreasing size, both errors diminish, so (1) gives the exact volume.

In the case where $f(x, y)$ has both positive and negative values on the region R, the double integral of f over R may be interpreted as a difference of volumes, the volume above the xy-plane between $z = f(x, y)$ and R minus the volume below the xy-plane between $z = f(x, y)$ and R. (Explain why.)

Double integrals enjoy many properties analogous to those of single integrals:

$$\iint\limits_{R} cf(x, y)\, dA = c \iint\limits_{R} f(x, y)\, dA \qquad (c \text{ a constant})$$

$$\iint\limits_{R} [f(x, y) + g(x, y)]\, dA = \iint\limits_{R} f(x, y)\, dA + \iint\limits_{R} g(x, y)\, dA$$

$$\iint\limits_{R} [f(x, y) - g(x, y)]\, dA = \iint\limits_{R} f(x, y)\, dA - \iint\limits_{R} g(x, y)\, dA$$

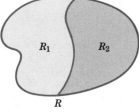

Figure 17.1.5

Moreover, if the region R is subdivided into two regions R_1 and R_2 (Figure 17.1.5) then

$$\iint\limits_{R} f(x, y)\, dA = \iint\limits_{R_1} f(x, y)\, dA + \iint\limits_{R_2} f(x, y)\, dA$$

We omit the proofs.

Except in the simplest cases, it is impractical to obtain the value of a double integral from the limit in (1). However, we will now show how to evaluate double integrals by calculating two successive single integrals. For the remainder of this section, we will limit our discussion to the case where R is a rectangle. In the next section we will consider double integrals over more complicated regions.

The partial derivatives of a function $f(x, y)$ are calculated by holding one of the variables fixed and differentiating with respect to the other variable. Let us consider the inverse of this process, *partial integration.* The symbol

$$\int_{a}^{b} f(x, y)\, dx$$

is a *partial definite integral with respect to x.* It is evaluated by holding y fixed and integrating with respect to x. Similarly, the *partial definite integral with respect to y,*

$$\int_{c}^{d} f(x, y)\, dy$$

is evaluated by holding x fixed and integrating with respect to y.

► Example 1

$$\int_0^1 xy^2\, dx = y^2 \int_0^1 x\, dx = \left. \frac{y^2 x^2}{2} \right|_{x=0}^1 = \frac{y^2}{2}$$

$$\int_0^1 xy^2\, dy = x \int_0^1 y^2\, dy = \left. \frac{xy^3}{3} \right|_{y=0}^1 = \frac{x}{3}$$

◄

As this example shows, an integral of the form $\int_a^b f(x, y)\, dx$ produces a function of y as the result, while an integral of the form $\int_c^d f(x, y)\, dy$ produces a function of x. This being the case, we can consider the following types of calculations:

$$\int_c^d \left[\int_a^b f(x, y)\, dx \right] dy \tag{2a}$$

$$\int_a^b \left[\int_c^d f(x, y)\, dy \right] dx \tag{2b}$$

In (2a), the inside integration, $\int_a^b f(x, y)\, dx$, yields a function of y, which is then integrated over the interval $c \le y \le d$. In (2b), the integration, $\int_c^d f(x, y)\, dy$, yields a function of x, which is then integrated over the interval $a \le x \le b$. Expressions (2a) and (2b) are called **iterated** (or **repeated**) **integrals.** Often the brackets are omitted and these expressions are written

$$\int_c^d \int_a^b f(x, y)\, dx\, dy = \int_c^d \left[\int_a^b f(x, y)\, dx \right] dy \tag{3a}$$

$$\int_a^b \int_c^d f(x, y)\, dy\, dx = \int_a^b \left[\int_c^d f(x, y)\, dy \right] dx \tag{3b}$$

► Example 2   Evaluate

(a) $\displaystyle \int_0^3 \int_1^2 (1 + 8xy)\, dy\, dx$ (b) $\displaystyle \int_1^2 \int_0^3 (1 + 8xy)\, dx\, dy$

Solution (a).

$$\int_0^3 \int_1^2 (1 + 8xy)\, dy\, dx = \int_0^3 \left[\int_1^2 (1 + 8xy)\, dy \right] dx$$

$$= \int_0^3 \left[y + 4xy^2 \right]_{y=1}^2 dx$$

$$= \int_0^3 [(2 + 16x) - (1 + 4x)]\, dx$$

$$= \int_0^3 (1 + 12x)\, dx$$

$$= x + 6x^2 \Big]_0^3 = 57$$

Solution (b).

$$\int_1^2 \int_0^3 (1 + 8xy)\, dx\, dy = \int_1^2 \left[\int_0^3 (1 + 8xy)\, dx \right] dy$$

$$= \int_1^2 \left[x + 4x^2 y \right]_{x=0}^3 dy$$

$$= \int_1^2 (3 + 36y)\, dy$$

$$= 3y + 18y^2 \Big]_1^2$$

$$= 57 \qquad\qquad \blacktriangleleft$$

It is no accident that the two iterated integrals in the last example have the same value; it is a consequence of the following theorem.

17.1.1   THEOREM   *Let R be the rectangle defined by the inequalities*

$$a \le x \le b, \qquad c \le y \le d$$

If f(x, y) is continuous on this rectangle, then

$$\iint\limits_R f(x, y)\, dA = \int_c^d \int_a^b f(x, y)\, dx\, dy = \int_a^b \int_c^d f(x, y)\, dy\, dx$$

This major theorem enables us to evaluate a double integral over a rectangle by calculating an iterated integral. Moreover, the theorem tells us that the order of integration in the iterated integral does not matter. We will not formally prove this result. However, we offer the following geometric argument for the case where $f(x, y)$ is nonnegative on R.

If $f(x, y)$ is nonnegative on R, the double integral

$$\iint\limits_R f(x, y)\, dA$$

represents the volume of the solid S bounded above by the surface $z = f(x, y)$ and below by the rectangle R. However, as discussed in Section 6.2 (Definition 6.2.2), the volume of the solid S is also given by the formula

$$\text{Vol } (S) = \int_c^d A(y) \, dy \tag{4}$$

where $A(y)$ represents the area of the cross section perpendicular to the y-axis taken at the point y (Figure 17.1.6).

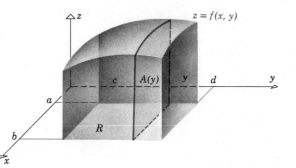

Figure 17.1.6

But consider how we might compute the cross-sectional area $A(y)$. For each *fixed y* in the interval $c \leq y \leq d$, the function $f(x, y)$ is a function of x alone, and $A(y)$ may be viewed as the area under the graph of this function along the interval $a \leq x \leq b$ (see the blue shaded area in Figure 17.1.6). Thus,

$$A(y) = \int_a^b f(x, y) \, dx$$

Substituting this expression in (4) yields

$$\text{Vol } (S) = \int_c^d \left[\int_a^b f(x, y) \, dx \right] dy = \int_c^d \int_a^b f(x, y) \, dx \, dy \tag{5}$$

The volume of the solid S can also be obtained using cross sections perpendicular to the x-axis. By Definition 6.2.1,

$$\text{Vol } (S) = \int_a^b A(x) \, dx \tag{6}$$

where $A(x)$ is the area of the cross section perpendicular to the x-axis taken at the point x (Figure 17.1.7). For each *fixed x* in the interval $a \leq x \leq b$, the function $f(x, y)$ is a function of y alone, so that the area $A(x)$ is given by

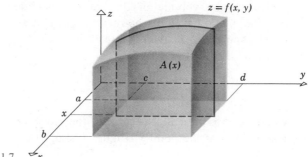

Figure 17.1.7

$$A(x) = \int_c^d f(x, y)\, dy$$

Substituting this expression in (6) yields

$$\text{Vol }(S) = \int_a^b \left[\int_c^d f(x, y)\, dy \right] dx = \int_a^b \int_c^d f(x, y)\, dy\, dx \tag{7}$$

Since the volume of S is also given by the double integral $\iint\limits_R f(x, y)\, dA$, it follows from (5) and (7) that

$$\iint\limits_R f(x, y)\, dA = \int_c^d \int_a^b f(x, y)\, dx\, dy = \int_a^b \int_c^d f(x, y)\, dy\, dx$$

which is what we intended to show.

▶ **Example 3** Evaluate the double integral

$$\iint\limits_R y^2 x\, dA$$

over the rectangle $R = \{(x, y): -3 \le x \le 2,\ 0 \le y \le 1\}$.

Solution. In view of Theorem 17.1.1, the value of the double integral may be obtained from either of the iterated integrals

$$\int_{-3}^2 \int_0^1 y^2 x\, dy\, dx \qquad \text{or} \qquad \int_0^1 \int_{-3}^2 y^2 x\, dx\, dy \tag{8}$$

Using the first of these, we obtain

$$\iint_R y^2x\, dA = \int_{-3}^{2}\int_0^1 y^2x\, dy\, dx$$

$$= \int_{-3}^{2}\left[\frac{1}{3}y^3x\right]_{y=0}^{1} dx$$

$$= \int_{-3}^{2}\frac{1}{3}x\, dx$$

$$= \frac{x^2}{6}\Bigg]_{-3}^{2} = -\frac{5}{6}$$

The reader should check this result by evaluating the second integral in (8). ◀

▶ **Example 4** Use a double integral to find the volume of the solid bounded above by the plane $z = 4 - x - y$ and below by the rectangle $R = \{(x, y): 0 \le x \le 1, 0 \le y \le 2\}$ (Figure 17.1.8).

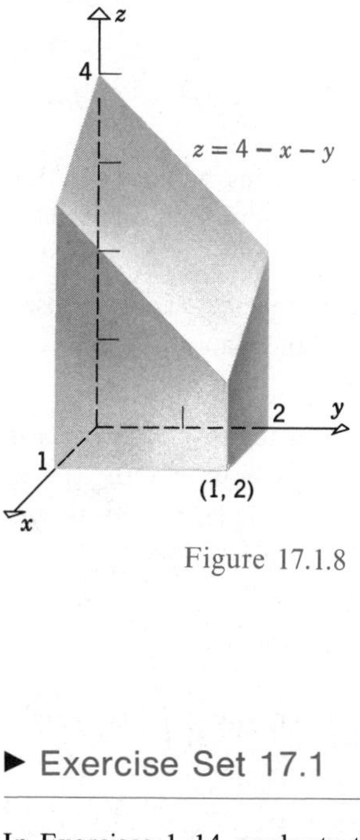

Solution.

$$V = \iint_R (4 - x - y)\, dA = \int_0^2\int_0^1 (4 - x - y)\, dx\, dy$$

$$= \int_0^2\left[4x - \frac{x^2}{2} - xy\right]_{x=0}^{1} dy = \int_0^2\left(\frac{7}{2} - y\right)dy$$

$$= \left[\frac{7}{2}y - \frac{y^2}{2}\right]_0^2 = 5$$

Figure 17.1.8   The volume can also be obtained by first integrating with respect to y and then with respect to x. ◀

$z = 4 - x - y$

$(1, 2)$

▶ Exercise Set 17.1

In Exercises 1–14, evaluate the iterated integrals.

1. $\displaystyle\int_0^1\int_0^2 (x + 3)\, dy\, dx.$

2. $\displaystyle\int_1^3\int_{-1}^1 (2x - 4y)\, dy\, dx.$

5. $\displaystyle\int_0^{\ln 3}\int_0^{\ln 2} e^{x+y}\, dy\, dx.$

6. $\displaystyle\int_0^2\int_0^1 y\sin x\, dy\, dx.$

3. $\displaystyle\int_2^4\int_0^1 x^2y\, dx\, dy.$

4. $\displaystyle\int_{-2}^0\int_{-1}^2 (x^2 + y^2)\, dx\, dy.$

7. $\displaystyle\int_0^3\int_0^1 x(x^2 + y)^{1/2}\, dx\, dy.$

8. $\displaystyle\int_{-1}^{2}\int_{2}^{4}(2x^2y + 3xy^2)\,dx\,dy.$

9. $\displaystyle\int_{-1}^{0}\int_{2}^{5}dx\,dy.$

10. $\displaystyle\int_{4}^{6}\int_{-3}^{7}dy\,dx.$

11. $\displaystyle\int_{0}^{1}\int_{0}^{1}\frac{x}{(xy+1)^2}\,dy\,dx.$

12. $\displaystyle\int_{\pi/2}^{\pi}\int_{1}^{2}x\cos xy\,dy\,dx.$

13. $\displaystyle\int_{0}^{\ln 2}\int_{0}^{1}xy\,e^{y^2x}\,dy\,dx.$

14. $\displaystyle\int_{3}^{4}\int_{1}^{2}\frac{1}{(x+y)^2}\,dy\,dx.$

In Exercises 15–19, evaluate the double integral over the rectangular region R.

15. $\displaystyle\iint_{R}4xy^3\,dA;$

$R = \{(x,y): -1 \le x \le 1, -2 \le y \le 2\}.$

16. $\displaystyle\iint_{R}\frac{xy}{\sqrt{x^2+y^2+1}}\,dA;$

$R = \{(x,y): 0 \le x \le 1, 0 \le y \le 1\}.$

17. $\displaystyle\iint_{R}x\sqrt{1-x^2}\,dA;$

$R = \{(x,y): 0 \le x \le 1, 2 \le y \le 3\}.$

18. $\displaystyle\iint_{R}(x\sin y - y\sin x)\,dA;$

$R = \{(x,y): 0 \le x \le \pi/2, 0 \le y \le \pi/3\}.$

19. $\displaystyle\iint_{R}\cos(x+y)\,dA;$

$R = \{(x,y): -\pi/4 \le x \le \pi/4, 0 \le y \le \pi/4\}.$

In Exercises 20–25, the iterated integral represents the volume of a solid. Make an accurate sketch of the solid. (You do *not* have to find the volume.)

20. $\displaystyle\int_{0}^{5}\int_{1}^{2}4\,dx\,dy.$

21. $\displaystyle\int_{0}^{1}\int_{0}^{1}(2-x-y)\,dy\,dx.$

22. $\displaystyle\int_{2}^{3}\int_{3}^{4}y\,dx\,dy.$

23. $\displaystyle\int_{0}^{3}\int_{0}^{4}\sqrt{25-x^2-y^2}\,dy\,dx.$

24. $\displaystyle\int_{-2}^{2}\int_{-2}^{2}(x^2+y^2)\,dx\,dy.$

25. $\displaystyle\int_{0}^{1}\int_{-1}^{1}\sqrt{4-x^2}\,dy\,dx.$

In Exercises 26–30, use a double integral to find the volume.

26. The volume under the plane $z = 2x + y$ and over the rectangle $R = \{(x,y): 3 \le x \le 5, 1 \le y \le 2\}.$

27. The volume under the surface $z = 3x^3 + 3x^2y$ and over the rectangle $R = \{(x,y): 1 \le x \le 3, 0 \le y \le 2\}.$

28. The volume in the first octant bounded by the coordinate planes, the plane $y = 4$, and the plane $x/3 + z/5 = 1.$

29. The volume of the solid in the first octant enclosed by the surface $z = x^2$ and the planes $x = 2, y = 3, y = 0,$ and $z = 0.$

30. The volume of the solid in the first octant that is enclosed by the planes $x = 0$, $z = 0$, $x = 5$, $z - y = 0$, and $z = -2y + 6$. [*Hint:* Break the solid into two parts.]

31. Suppose that $f(x,y) = g(x)h(y)$ and $R = \{(x,y): a \le x \le b, c \le y \le d\}$. Show that

$$\iint_{R}f(x,y)\,dA = \left[\int_{a}^{b}g(x)\,dx\right]\left[\int_{c}^{d}h(y)\,dy\right]$$

32. Evaluate

$$\iint_{R}x\cos(xy)\cos^2\pi x\,dA$$

where $R = \{(x,y): 0 \le x \le \tfrac{1}{2}, 0 \le y \le \pi\}$. [*Hint:* One order of integration leads to a simpler solution than the other.]

17.2 DOUBLE INTEGRALS OVER NONRECTANGULAR REGIONS

In this section we will show how to evaluate double integrals over regions other than rectangles.

Until now we have considered only iterated integrals with constant limits of integration. However, our work in this section will lead us to iterated integrals of the following types:

$$\int_a^b \int_{g_1(x)}^{g_2(x)} f(x, y) \, dy \, dx = \int_a^b \left[\int_{g_1(x)}^{g_2(x)} f(x, y) \, dy \right] dx \tag{1a}$$

$$\int_c^d \int_{h_1(y)}^{h_2(y)} f(x, y) \, dx \, dy = \int_c^d \left[\int_{h_1(y)}^{h_2(y)} f(x, y) \, dx \right] dy \tag{1b}$$

► **Example 1** Evaluate

$$\int_0^2 \int_{x^2}^{x} y^2 x \, dy \, dx.$$

Solution.

$$\int_0^2 \int_{x^2}^{x} y^2 x \, dy \, dx = \int_0^2 \left[\int_{x^2}^{x} y^2 x \, dy \right] dx$$

$$= \int_0^2 \left[\frac{y^3 x}{3} \right]_{y=x^2}^{x} dx$$

$$= \int_0^2 \left(\frac{x^4}{3} - \frac{x^7}{3} \right) dx$$

$$= \left[\frac{x^5}{15} - \frac{x^8}{24} \right]_0^2$$

$$= \frac{32}{15} - \frac{256}{24} = -\frac{128}{15} \qquad ◄$$

► **Example 2** Evaluate

$$\int_0^{\pi} \int_0^{\cos y} x \sin y \, dx \, dy.$$

Solution.

$$\int_0^{\pi} \int_0^{\cos y} x \sin y \, dx \, dy = \int_0^{\pi} \left[\int_0^{\cos y} x \sin y \, dx \right] dy$$

$$= \int_0^\pi \left[\frac{x^2}{2} \sin y \right]_{x=0}^{\cos y} dy$$

$$= \int_0^\pi \frac{1}{2} \cos^2 y \sin y \, dy$$

$$= \left[-\frac{1}{6} \cos^3 y \right]_0^\pi = \frac{1}{3} \qquad \blacktriangleleft$$

We will be concerned with evaluating double integrals over two types of closed regions, which we will call *type I* and *type II* (Figure 17.2.1).

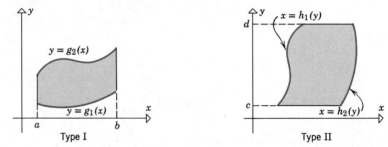

Figure 17.2.1　　　　Type I　　　　　　　　　　Type II

A type I region is bounded on the left and right by vertical lines $x = a$ and $x = b$ and is bounded below and above by smooth curves $y = g_1(x)$ and $y = g_2(x)$, where $g_1(x) \le g_2(x)$ for $a \le x \le b$.

A type II region is bounded below and above by horizontal lines $y = c$ and $y = d$ and is bounded on the left and right by smooth curves $x = h_1(y)$ and $x = h_2(y)$ satisfying $h_1(y) \le h_2(y)$ for $c \le y \le d$.

The following theorem will enable us to evaluate double integrals over type I and type II regions using iterated integrals.

17.2.1　THEOREM　　(*a*)　*If R is a type I region on which f(x, y) is continuous, then*

$$\iint\limits_R f(x, y) \, dA = \int_a^b \int_{g_1(x)}^{g_2(x)} f(x, y) \, dy \, dx \tag{2a}$$

(*b*)　*If R is a type II region on which f(x, y) is continuous, then*

$$\iint\limits_R f(x, y) \, dA = \int_c^d \int_{h_1(y)}^{h_2(y)} f(x, y) \, dx \, dy \tag{2b}$$

Although we will not prove this theorem, the results are easy to visualize geometrically. Suppose $f(x, y)$ is nonnegative on a type I region R, so that the integral

$$\iint\limits_{R} f(x, y)\, dA \tag{3}$$

represents the volume of the solid S bounded above by the surface $z = f(x, y)$ and below by the region R. By the method of cross sections, the volume of S is also given by

$$\text{Vol}\,(S) = \int_{a}^{b} A(x)\, dx \tag{4}$$

where $A(x)$ is the area of the cross section at the fixed point x. As shown in Figure 17.2.2, this cross-sectional area extends from $g_1(x)$ to $g_2(x)$ in the y-direction, so

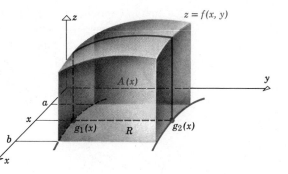

Figure 17.2.2

$$A(x) = \int_{g_1(x)}^{g_2(x)} f(x, y)\, dy$$

Substituting this in (4) we obtain

$$\text{Vol}\,(S) = \int_{a}^{b} \int_{g_1(x)}^{g_2(x)} f(x, y)\, dy\, dx$$

Since the volume of S is also given by (3), we obtain

$$\iint\limits_{R} f(x, y)\, dA = \int_{a}^{b} \int_{g_1(x)}^{g_2(x)} f(x, y)\, dy\, dx$$

Part (b) of Theorem 17.2.1 can be motivated in a similar way by using cross sections parallel to the x-axis.

To apply Theorem 17.2.1, it is usual to start with a two-dimensional sketch of the region R. [It is not necessary to graph $f(x, y)$.] For a type I region, the limits of integration in the formula

$$\iint\limits_{R} f(x, y)\, dA = \int_{a}^{b} \int_{g_1(x)}^{g_2(x)} f(x, y)\, dy\, dx \tag{5}$$

may be obtained as follows:

Step 1. Since x is held fixed for the first integration, we draw a vertical line through the region R at an arbitrary fixed point x (Figure 17.2.3). This line crosses the boundary of R twice. The lower point of intersection is on the curve $y = g_1(x)$ and the higher point is on the curve $y = g_2(x)$. These intersections determine the y-limits of integration in (5).

Step 2. Imagine moving the line drawn in Step 1 first to the left and then to the right (Figure 17.2.3). The leftmost position where the line intersects the region R is $x = a$ and the rightmost position where the line intersects the region R is $x = b$. This yields the limits for the x-integration in (5).

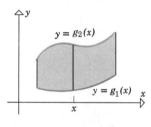 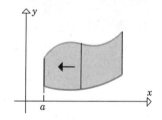 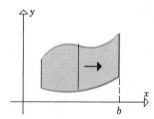

Figure 17.2.3

▶ Example 3 Evaluate

$$\iint\limits_{R} xy\, dA$$

over the region R enclosed between $y = \frac{1}{2}x$, $y = \sqrt{x}$, $x = 2$, and $x = 4$.

Solution. We view R as a type I region. The region R and a vertical line corresponding to a fixed x are shown in Figure 17.2.4. This line meets the region R at the lower boundary $y = \frac{1}{2}x$ and the upper boundary $y = \sqrt{x}$. These are the y-limits of integration. Moving this line first left and then right yields the x-limits of integration, $x = 2$ and $x = 4$. Thus,

$$\iint\limits_{R} xy \, dA = \int_{2}^{4} \int_{x/2}^{\sqrt{x}} xy \, dy \, dx = \int_{2}^{4} \left[\frac{xy^2}{2} \right]_{y=x/2}^{\sqrt{x}} dx = \int_{2}^{4} \left(\frac{x^2}{2} - \frac{x^3}{8} \right) dx$$

$$= \left[\frac{x^3}{6} - \frac{x^4}{32} \right]_{2}^{4} = \left(\frac{64}{6} - \frac{256}{32} \right) - \left(\frac{8}{6} - \frac{16}{32} \right) = \frac{11}{6} \quad \blacktriangleleft$$

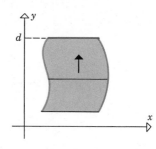

Figure 17.2.4

If R is a type II region, then the limits of integration in the formula

$$\iint\limits_{R} f(x, y) \, dA = \int_{c}^{d} \int_{h_1(y)}^{h_2(y)} f(x, y) \, dx \, dy \tag{6}$$

may be obtained as follows:

Step 1. Since y is held fixed for the first integration, we draw a horizontal line through the region R at a fixed point y (Figure 17.2.5). This line crosses the boundary of R twice. The leftmost point of intersection is on the curve $x = h_1(y)$ and the rightmost point is on the curve $x = h_2(y)$. These intersections determine x-limits of integration in (6).

Step 2. Imagine moving the line drawn in Step 1 first down and then up (Figure 17.2.5). The lowest position where the line intersects the region R is $y = c$, and the highest position where the line intersects the region R is $y = d$. This yields the y-limits of integration in (6).

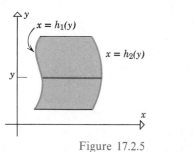

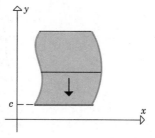

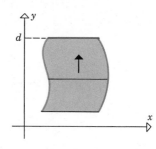

Figure 17.2.5

▶ Example 4 Evaluate

$$\iint\limits_{R} (2x - y^2)\, dA$$

over the triangular region R enclosed between the lines $y = -x + 1$, $y = x + 1$, and $y = 3$.

Solution. We view R as a type II region. The region R and a horizontal line corresponding to a fixed y are shown in Figure 17.2.6.

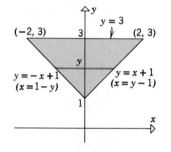

Figure 17.2.6

This line meets the region R at its left-hand boundary $x = 1 - y$ and its right-hand boundary $x = y - 1$. These are the x-limits of integration. Moving this line first down and then up yields the y-limits, $y = 1$ and $y = 3$. Thus,

$$\iint\limits_{R} (2x - y^2)\, dA = \int_{1}^{3} \int_{1-y}^{y-1} (2x - y^2)\, dx\, dy$$

$$= \int_{1}^{3} \left[x^2 - y^2 x \right]_{x=1-y}^{y-1} dy$$

$$= \int_{1}^{3} [(1 - 2y + 2y^2 - y^3) - (1 - 2y + y^3)]\, dy$$

$$= \int_{1}^{3} (2y^2 - 2y^3)\, dy$$

$$= \left[\frac{2y^3}{3} - \frac{y^4}{2} \right]_{1}^{3} = -\frac{68}{3} \qquad ◀$$

REMARK. To integrate over a type II region, the left- and right-hand boundaries must be expressed in the form $x = h_1(y)$ and $x = h_2(y)$. This is why we rewrote the boundary equations $y = -x + 1$ and $y = x + 1$ as $x = 1 - y$ and $x = y - 1$ in the last example.

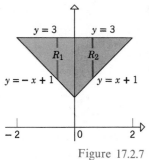

$y = 3$ $y = 3$

R_1 R_2

$y = -x + 1$ $y = x + 1$

-2 0 2

Figure 17.2.7

In Example 4 we could have treated R as a type I region, but with an added complication. Viewed as a type I region, the upper boundary of R is the line $y = 3$ (Figure 17.2.7) and the lower boundary consists of two parts, the line $y = -x + 1$ to the left of the origin and the line $y = x + 1$ to the right of the origin. To carry out the integration it is necessary to decompose the region R into two parts R_1 and R_2, as shown in Figure 17.2.7, and write:

$$\iint_R (2x - y^2)\, dA = \iint_{R_1} (2x - y^2)\, dA + \iint_{R_2} (2x - y^2)\, dA$$

$$= \int_{-2}^{0} \int_{-x+1}^{3} (2x - y^2)\, dy\, dx + \int_{0}^{2} \int_{x+1}^{3} (2x - y^2)\, dy\, dx$$

This will yield the same result that was obtained in Example 4.

Sometimes the evaluation of an iterated integral can be simplified by reversing the order of integration. The next example illustrates how this is done.

▶ Example 5   Since there is no elementary antiderivative of e^{x^2}, the integral

$$\int_{0}^{2} \int_{y/2}^{1} e^{x^2}\, dx\, dy$$

cannot be evaluated by performing the x-integration first. Evaluate this integral by expressing it as an equivalent iterated integral with the order of integration reversed.

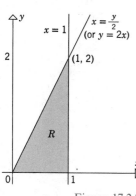

$x = 1$ $x = \dfrac{y}{2}$ (or $y = 2x$)

2 $(1, 2)$

R

0 1

Figure 17.2.8

Solution. For the inside integration, y is fixed and x varies from the line $x = y/2$ to the line $x = 1$ (Figure 17.2.8). For the outside integration, y varies from 0 to 2, so the given iterated integral is equal to a double integral over the triangular region R in Figure 17.2.8.

To reverse the order of integration, we treat R as a type I region, which enables us to write the given integral as,

$$\int_{0}^{2} \int_{y/2}^{1} e^{x^2}\, dx\, dy = \iint_R e^{x^2}\, dA$$

$$= \int_{0}^{1} \int_{0}^{2x} e^{x^2}\, dy\, dx$$

$$= \int_{0}^{1} \left[e^{x^2} y \right]_{y=0}^{2x} dx$$

$$= \int_0^1 2xe^{x^2}\,dx$$

$$= e^{x^2}\Big]_0^1 = e - 1 \qquad \blacktriangleleft$$

The double integral

$$\iint\limits_R 1\,dA = \iint\limits_R dA$$

represents the volume of a solid of constant height 1 above the region R. Numerically, this is the same as the area of the region R. Thus,

$$\text{area of } R = \iint\limits_R dA \qquad\qquad (7)$$

▶ **Example 6**   Use a double integral to find the area of the region R enclosed between the parabola $y = \frac{1}{2}x^2$ and the line $y = 2x$.

Solution.   The region R may be treated equally well as type I (Figure 17.2.9a) or type II (Figure 17.2.9b). Treating R as type I yields,

$$\text{area of } R = \iint\limits_R dA = \int_0^4 \int_{x^2/2}^{2x} dy\,dx$$

$$= \int_0^4 \Big[y\Big]_{y=x^2/2}^{2x} dx$$

$$= \int_0^4 \left(2x - \frac{1}{2}x^2\right) dx = \left[x^2 - \frac{x^3}{6}\right]_0^4 = \frac{16}{3}$$

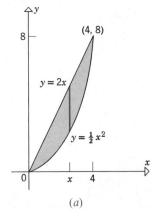

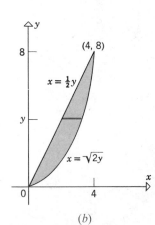

Figure 17.2.9

(a)

(b)

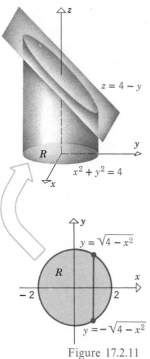

Figure 17.2.10

Treating R as type II yields

$$\text{area of } R = \iint_R dA = \int_0^8 \int_{y/2}^{\sqrt{2y}} dx\, dy$$

$$= \int_0^8 \Big[x \Big]_{x=y/2}^{\sqrt{2y}} dy$$

$$= \int_0^8 \left(\sqrt{2y} - \frac{1}{2}y \right) dy$$

$$= \left[\frac{2\sqrt{2}}{3} y^{3/2} - \frac{y^2}{4} \right]_0^8 = \frac{16}{3} \qquad \blacktriangleleft$$

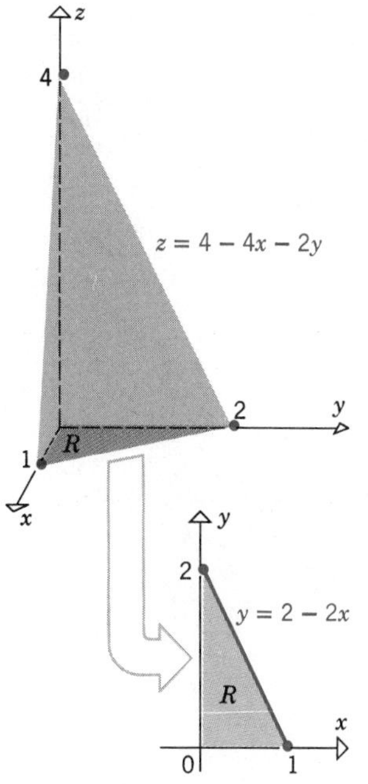

▶ **Example 7** Use a double integral to find the volume of the tetrahedron bounded by the coordinate planes and the plane $z = 4 - 4x - 2y$.

Solution. The tetrahedron in question is bounded above by the plane

$$z = 4 - 4x - 2y \qquad (8)$$

and below by the triangular region R shown in Figure 17.2.10. Thus, the volume is given by

$$V = \iint_R (4 - 4x - 2y)\, dA$$

The region R is bounded by the x-axis, the y-axis, and the line $y = 2 - 2x$ [set $z = 0$ in (8)], so that treating R as a type I region yields,

$$V = \iint_R (4 - 4x - 2y)\, dA = \int_0^1 \int_0^{2-2x} (4 - 4x - 2y)\, dy\, dx$$

$$= \int_0^1 \Big[4y - 4xy - y^2 \Big]_{y=0}^{2-2x} dx = \int_0^1 (4 - 8x + 4x^2)\, dx = \frac{4}{3} \qquad \blacktriangleleft$$

▶ **Example 8** Find the volume of the solid bounded by the cylinder $x^2 + y^2 = 4$ and the planes $y + z = 4$ and $z = 0$.

Figure 17.2.11

Solution. The solid shown in Figure 17.2.11 is bounded above by the plane $z = 4 - y$ and below by the region R within the circle $x^2 + y^2 = 4$. The volume is given by

$$V = \iint_R (4 - y)\, dA$$

Treating R as a type I region we obtain

$$V = \int_{-2}^{2} \int_{-\sqrt{4-x^2}}^{\sqrt{4-x^2}} (4 - y)\, dy\, dx$$

$$= \int_{-2}^{2} \left[4y - \frac{1}{2} y^2 \right]_{y=-\sqrt{4-x^2}}^{\sqrt{4-x^2}} dx$$

$$= \int_{-2}^{2} 8\sqrt{4 - x^2}\, dx$$

$$= 8(2\pi) = 16\pi \qquad \text{(See Example 4 of Section 9.5.)} \qquad \blacktriangleleft$$

▶ Exercise Set 17.2

In Exercises 1–12, evaluate the iterated integral.

1. $\displaystyle\int_0^1 \int_{x^2}^x xy^2\, dy\, dx.$

2. $\displaystyle\int_1^2 \int_y^{3-y} y\, dx\, dy.$

3. $\displaystyle\int_0^3 \int_0^{\sqrt{9-y^2}} y\, dx\, dy.$

4. $\displaystyle\int_{1/4}^1 \int_{x^2}^x \sqrt{\frac{x}{y}}\, dy\, dx.$

5. $\displaystyle\int_{\sqrt{\pi}}^{\sqrt{2\pi}} \int_0^{x^3} \sin\frac{y}{x}\, dy\, dx.$

6. $\displaystyle\int_{-1}^1 \int_{-x^2}^{x^2} (x^2 - y)\, dy\, dx.$

7. $\displaystyle\int_{\pi/2}^{\pi} \int_0^{x^2} \frac{1}{x} \cos\frac{y}{x}\, dy\, dx.$

8. $\displaystyle\int_0^{\pi/2} \int_0^{\sin y} e^x \cos y\, dx\, dy.$

9. $\displaystyle\int_0^a \int_0^{\sqrt{a^2-x^2}} (x + y)\, dy\, dx.$

10. $\displaystyle\int_1^2 \int_0^{y^2} e^{x/y^2}\, dx\, dy.$

11. $\displaystyle\int_0^1 \int_0^x y\sqrt{x^2 - y^2}\, dy\, dx.$

12. $\displaystyle\int_0^1 \int_0^x e^{x^2}\, dy\, dx.$

In Exercises 13–28, evaluate the double integral.

13. $\displaystyle\iint_R 6xy\, dA$; R is the region bounded by $y = 0$, $x = 2$, and $y = x^2$.

14. $\displaystyle\iint_R xy\, dA$; R is the region bounded by the trapezoid with vertices $(1, 3)$, $(5, 3)$, $(2, 1)$, and $(4, 1)$.

15. $\displaystyle\iint_R x \cos xy\, dA$; R is the region enclosed by $x = 1$, $x = 2$, $y = \pi/2$, and $y = 2\pi/x$.

16. $\displaystyle\iint_R (x + y)\, dA$; R is the region enclosed between the curves $y = x^2$ and $y = \sqrt{x}$.

17. $\displaystyle\iint_R x^2\, dA$; R is the region bounded by $y = 16/x$, $y = x$, and $x = 8$.

18. $\displaystyle\iint_R xy^2\, dA$; R is the region enclosed by $y = 1$, $y = 2$, $x = 0$, and $y = x$.

19. $\iint\limits_R x(1 + y^2)^{-1/2} \, dA$; R is the region in the first quadrant enclosed by $y = x^2$, $y = 4$, and $x = 0$. [*Hint:* Choose your order of integration carefully.]

20. $\iint\limits_R x \cos y \, dA$; R is the triangular region bounded by $y = x$, $y = 0$, and $x = \pi$.

21. $\iint\limits_R (3x - 2y) \, dA$; R is the region enclosed by the circle $x^2 + y^2 = 1$.

22. $\iint\limits_R y \, dA$; R is the region in the first quadrant enclosed between the circle $x^2 + y^2 = 25$ and the line $x + y = 5$.

23. $\iint\limits_R \frac{1}{1 + x^2} \, dA$; R is the triangular region with vertices $(0, 0)$, $(1, 1)$, and $(0, 1)$.

24. $\iint\limits_R (x^2 - xy) \, dA$; R is the region enclosed by $y = x$ and $y = 3x - x^2$.

25. $\iint\limits_R xy \, dA$; R is the region enclosed by $y = \sqrt{x}$, $y = 6 - x$, and $y = 0$.

26. $\iint\limits_R x \, dA$; R is the region enclosed by $y = \sin^{-1} x$, $x = 1/\sqrt{2}$, and $y = 0$.

27. $\iint\limits_R (x - 1) \, dA$; R is the region enclosed between $y = x$ and $y = x^3$.

28. $\iint\limits_R x^2 \, dA$; R is the region in the first quadrant enclosed by $xy = 1$, $y = x$, and $y = 2x$.

In Exercises 29–34, use double integration to find the area of the plane region enclosed by the given curves.

29. $x + y = 5$, $x = 0$, and $y = 0$.

30. $y = x^2$ and $y = 4x$.

31. $y = \sin x$ and $y = \cos x$, for $0 \leq x \leq \pi/4$.

32. $y^2 = -x$ and $3y - x = 4$.

33. $y^2 = 9 - x$ and $y^2 = 9 - 9x$.

34. $y = \cosh x$, $y = \sinh x$, $x = 0$, and $x = 1$.

In Exercises 35–48, use double integration to find the volume of each solid.

35. The tetrahedron in the first octant bounded by the coordinate planes and the plane $z = 5 - 2x - y$.

36. The solid bounded by the cylinder $x^2 + y^2 = 9$ and the planes $z = 0$ and $z = 3 - x$.

37. The solid bounded above by the plane $z = x + 2y + 2$, below by the xy-plane, and laterally by $y = 0$ and $y = 1 - x^2$.

38. The solid in the first octant bounded above by the paraboloid $z = x^2 + 3y^2$, below by the plane $z = 0$, and laterally by $y = x^2$ and $y = x$.

39. The solid bounded above by the paraboloid $z = 9x^2 + y^2$, below by the plane $z = 0$, and laterally by the planes $x = 0$, $y = 0$, $x = 3$, and $y = 2$.

40. The solid enclosed by $y^2 = x$, $z = 0$, and $x + z = 1$.

41. The wedge cut from the cylinder $4x^2 + y^2 = 9$ by the planes $z = 0$ and $z = y + 3$.

42. The solid in the first octant bounded above by $z = 9 - x^2$, below by $z = 0$, and laterally by $y^2 = 3x$.

43. The solid in the first octant bounded by the coordinate planes and the planes $x + 2y = 4$ and $x + 8y - 4z = 0$.

44. The solid in the first octant bounded by the surface $z = e^{y-x}$, the plane $x + y = 1$, and the coordinate planes.

45. The solid bounded above by the paraboloid $z = 1 - x^2 - y^2$ and below by the xy-plane. [*Hint:* Use a trigonometric substitution to evaluate the integral.]

46. The solid in the first octant bounded by the paraboloid $z = x^2 + y^2$ and the cylinder $x^2 + y^2 = 4$. [*Hint:* Use a trigonometric substitution to evaluate the integral.]

47. The solid common to the cylinders $x^2 + y^2 = 25$ and $x^2 + z^2 = 25$.

48. The solid bounded above by the paraboloid $z = x^2 + y^2$, laterally by the cylinder $x^2 + (y-1)^2 = 1$, and below by the xy-plane.

In Exercises 49–56, express the integral as an equivalent integral with the order of integration reversed.

49. $\displaystyle\int_0^2 \int_0^{\sqrt{x}} f(x, y) \, dy \, dx.$

50. $\displaystyle\int_0^4 \int_{2y}^8 f(x, y) \, dx \, dy.$

51. $\displaystyle\int_0^2 \int_1^{e^y} f(x, y) \, dx \, dy.$

52. $\displaystyle\int_1^e \int_0^{\ln x} f(x, y) \, dy \, dx.$

53. $\displaystyle\int_{-2}^2 \int_{-\sqrt{1-(x^2/4)}}^{\sqrt{1-(x^2/4)}} f(x, y) \, dy \, dx.$

54. $\displaystyle\int_0^1 \int_{y^2}^{\sqrt{y}} f(x, y) \, dx \, dy.$

55. $\displaystyle\int_0^1 \int_{\sin^{-1}y}^{\pi/2} f(x, y) \, dx \, dy.$

56. $\displaystyle\int_{-3}^1 \int_{x^2+6x}^{4x+3} f(x, y) \, dy \, dx.$

In Exercises 57 and 58, evaluate the integral by first reversing the order of integration.

57. $\displaystyle\int_0^1 \int_{4x}^4 e^{-y^2} \, dy \, dx.$

58. $\displaystyle\int_0^1 \int_{\sin^{-1}y}^{\pi/2} \sec^2(\cos x) \, dx \, dy.$

59. Evaluate $\displaystyle\iint_R \sin(y^3) \, dA$, where R is the region bounded by $y = \sqrt{x}$, $y = 2$, and $x = 0$. [*Hint:* Choose the order of integration carefully.]

60. Evaluate $\displaystyle\iint_R x \, dA$, where R is the region bounded by $x = \ln y$, $x = 0$, and $y = e$ [*Hint:* Choose the order of integration carefully.]

61. In each part evaluate $\displaystyle\iint_R xy^2 \, dA$.

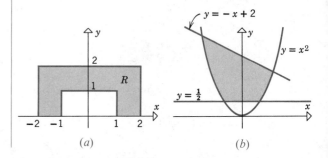

(a)   (b)

17.3 DOUBLE INTEGRALS IN POLAR COORDINATES

In this section we will be concerned with double integrals of functions of the form $f(r, \theta)$, where r and θ denote the polar coordinates of a point in the xy-plane. If we introduce a dependent variable

$$z = f(r, \theta)$$

then this equation represents a three-dimensional surface expressed in cylindrical coordinates (Figure 17.3.1).

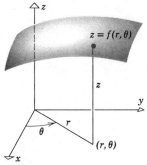

Figure 17.3.1

Our discussion of integration in this section will be limited to closed regions of the form shown in Figure 17.3.2a. The region is enclosed between two rays $\theta = \alpha$ and $\theta = \beta$ ($\beta > \alpha$) and between two smooth polar curves $r = r_1(\theta)$ and $r = r_2(\theta)$, which satisfy $r_1(\theta) \leq r_2(\theta)$ for $\alpha \leq \theta \leq \beta$.

In the special case where $r_1(\theta)$ is identically zero, the graph of $r_1(\theta)$ reduces to a single point, the origin, and the region R has a form like that in Figure 17.3.2b. If, in addition, $\alpha = 0$ and $\beta = 2\pi$, then R has a form like that in Figure 17.3.2c.

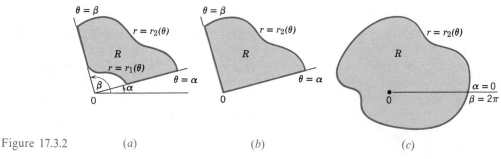

Figure 17.3.2   (a)   (b)   (c)

If $f(r, \theta)$ is defined on a region R having one of the above forms, then we define the *double integral* of $f(r, \theta)$ over R as follows:

Step 1. Cover the region R with a grid of circular arcs centered at the origin and rays emanating from the origin (Figure 17.3.3). The blocks formed by this grid are called ***polar rectangles.*** Exclude from consideration all polar rectangles that contain any points outside of R. This leaves only polar rectangles that are subsets of R. Denote the areas of these polar rectangles by

$$\Delta A_1, \Delta A_2, \ldots, \Delta A_n$$

Step 2. Choose an arbitrary point in each of these polar rectangles, and denote them by

$$(r_1^*, \theta_1^*), (r_2^*, \theta_2^*), \ldots, (r_n^*, \theta_n^*)$$

(Figure 17.3.3).

Step 3. Form the ***Riemann sum***

$$\sum_{k=1}^{n} f(r_k^*, \theta_k^*) \, \Delta A_k$$

Step 4. Repeat this process using more and more subdivisions in such a way that the dimensions of all the polar rectangles tend to zero. Thus n, the number of interior polar rectangles, will tend to $+\infty$. Define

$$\iint\limits_{R} f(r,\theta)\, dA = \lim_{n \to +\infty} \sum_{k=1}^{n} f(r_k^*, \theta_k^*)\, \Delta A_k$$

The symbol $\iint\limits_{R} f(r,\theta)\, dA$ is called the ***polar double integral*** of $f(r,\theta)$ over R.

In the special case where $f(r,\theta)$ is nonnegative on the region R, the polar double integral may be interpreted as the volume of the solid S bounded above by the surface $z = f(r,\theta)$ and below by the region R (Figure 17.3.4). To see this, observe that the product

$$f(r_k^*, \theta_k^*)\, \Delta A_k \tag{1}$$

represents the volume of a cylinder whose base is the kth polar rectangle and whose height is $f(r_k^*, \theta_k^*)$. Thus, (1) is an approximation to the volume of the portion of S above the kth polar rectangle; and the sum

$$\sum_{k=1}^{n} f(r_k^*, \theta_k^*)\, \Delta A_k$$

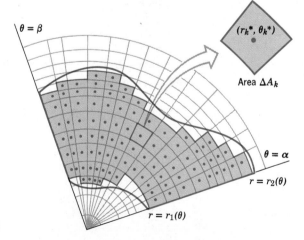

Figure 17.3.3

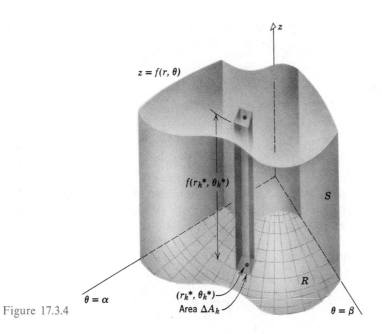

$z = f(r, \theta)$

$f(r_k{}^*, \theta_k{}^*)$

S

$\theta = \alpha$

$(r_k{}^*, \theta_k{}^*)$

Area ΔA_k

$\theta = \beta$

R

Figure 17.3.4

is an approximation to the entire volume of S. There are two sources of error in this approximation. First, the polar rectangles may not fill up the region R exactly and, second, the approximating solids have flat tops, whereas the upper surface of the solid may be curved. However, as we use more and more polar rectangles of decreasing dimensions, it is intuitively clear that the errors in the approximation diminish and the limit in Step 4 above gives the exact volume.

In the case where $f(r, \theta)$ has both positive and negative values on the region R, the double polar integral of $f(r, \theta)$ over R may be interpreted as a difference of volumes, the volume above the xy-plane between $z = f(r, \theta)$ and R minus the volume below the xy-plane between $z = f(r, \theta)$ and R.

The following theorem will enable us to evaluate double polar integrals by using iterated integrals.

17.3.1   THEOREM   *If R is a region of the type shown in Figure 17.3.2 and if $f(r, \theta)$ is continuous on R, then*

$$\iint\limits_{R} f(r, \theta) \, dA = \int_{\alpha}^{\beta} \int_{r_1(\theta)}^{r_2(\theta)} f(r, \theta) \, r \, dr \, d\theta \qquad (2)$$

REMARK.   In (2) note that the dA in the double integral becomes $r \, dr \, d\theta$ in the iterated integral.

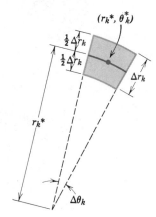

Figure 17.3.5

We will not formally prove this theorem. However, the appearance of the factor r in the iterated integral may be explained by returning to the definition

$$\iint_R f(r, \theta)\, dA = \lim_{n \to +\infty} \sum_{k=1}^{n} f(r_k^*, \theta_k^*)\, \Delta A_k \tag{3}$$

In the Riemann sum $\sum_{k=1}^{n} f(r_k^*, \theta_k^*)\, \Delta A_k$, suppose the arbitrary point (r_k^*, θ_k^*) is chosen at the "center" of the kth polar rectangle, that is, at the point halfway between the bounding circular arcs on the ray that bisects them (Figure 17.3.5). Suppose also that this polar rectangle has a central angle $\Delta \theta_k$ and a "radial thickness" Δr_k. Thus, the inner radius of this polar rectangle is $r_k^* - \frac{1}{2}\Delta r_k$ and the outer radius is $r_k^* + \frac{1}{2}\Delta r_k$. Treating the area ΔA_k of this polar rectangle as the difference in area of two sectors, we obtain

$$\Delta A_k = \frac{1}{2}\left(r_k^* + \frac{1}{2}\Delta r_k\right)^2 \Delta \theta_k - \frac{1}{2}\left(r_k^* - \frac{1}{2}\Delta r_k\right)^2 \Delta \theta_k$$

which simplifies to

$$\Delta A_k = r_k^* \,\Delta r_k \,\Delta \theta_k \tag{4}$$

(Verify.) Substituting this expression in (3) yields

$$\iint_R f(r, \theta)\, dA = \lim_{n \to +\infty} \sum_{k=1}^{n} f(r_k^*, \theta_k^*) r_k^* \,\Delta r_k \,\Delta \theta_k$$

which explains the form of the integrand in (2).

To apply Theorem 17.3.1, we start with a sketch of the region R. From this sketch the limits of integration in the formula,

$$\iint_R f(r, \theta)\, dA = \int_{\alpha}^{\beta} \int_{r_1(\theta)}^{r_2(\theta)} f(r, \theta)\, r\, dr\, d\theta \tag{5}$$

may be obtained as follows:

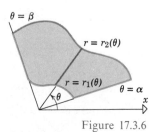

Figure 17.3.6

Step 1. Since θ is held fixed for the first integration, draw a radial line from the origin through the region R at a fixed angle θ (Figure 17.3.6). This line crosses the boundary of R at most twice. The innermost point of intersection is on the curve $r = r_1(\theta)$ and the outermost point is on the curve $r = r_2(\theta)$. These intersections determine the r-limits of integration in (5).

Step 2. Imagine rotating a ray along the positive x-axis one revolution counterclockwise about the origin. The smallest angle at which this ray intersects the region R is $\theta = \alpha$ and the largest angle is $\theta = \beta$. This yields the θ-limits of integration.

▶ Example 1 Evaluate

$$\iint\limits_{R} \sin\theta \, dA$$

where R is the region in the first quadrant that is outside the circle $r = 2$ and inside the cardioid $r = 2(1 + \cos\theta)$.

Solution. The region R is sketched in Figure 17.3.7. Following the two steps outlined above we obtain

$$\iint\limits_{R} \sin\theta \, dA = \int_{0}^{\pi/2} \int_{2}^{2(1+\cos\theta)} (\sin\theta) \, r \, dr \, d\theta$$

$$= \int_{0}^{\pi/2} \frac{1}{2} r^2 \sin\theta \Big]_{r=2}^{2(1+\cos\theta)} d\theta$$

$$= 2 \int_{0}^{\pi/2} [(1 + \cos\theta)^2 \sin\theta - \sin\theta] \, d\theta$$

$$= 2 \left[-\frac{1}{3}(1 + \cos\theta)^3 + \cos\theta \right]_{0}^{\pi/2}$$

$$= 2 \left[-\frac{1}{3} - \left(-\frac{5}{3} \right) \right] = \frac{8}{3} \qquad ◀$$

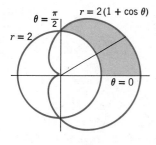

$\theta = \dfrac{\pi}{2}$ $r = 2(1 + \cos\theta)$

$r = 2$

$\theta = 0$

Figure 17.3.7

▶ Example 2 In cylindrical coordinates, the equation $r^2 + z^2 = a^2$ represents a sphere of radius a centered at the origin. (It is the sphere whose equation in rectangular coordinates is $x^2 + y^2 + z^2 = a^2$.) Use a double polar integral to calculate the volume of this sphere.

Solution. The upper hemisphere is given by the (cylindrical) equation

$$z = \sqrt{a^2 - r^2}$$

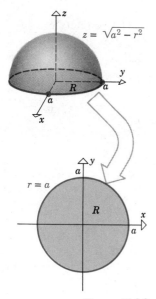

Figure 17.3.8

so that the volume enclosed by the entire sphere is

$$V = 2 \iint\limits_{R} \sqrt{a^2 - r^2} \, dA$$

where R is the circular region shown in Figure 17.3.8. Thus,

$$V = 2 \iint\limits_{R} \sqrt{a^2 - r^2} \, dA$$

$$= \int_0^{2\pi} \int_0^a \sqrt{a^2 - r^2} \,(2r) \, dr \, d\theta$$

$$= \int_0^{2\pi} \left[-\frac{2}{3}(a^2 - r^2)^{3/2} \right]_{r=0}^a d\theta$$

$$= \int_0^{2\pi} \frac{2}{3} a^3 \, d\theta$$

$$= \left[\frac{2}{3} a^3 \theta \right]_0^{2\pi} = \frac{4}{3} \pi a^3 \qquad \blacktriangleleft$$

The double polar integral

$$\iint\limits_{R} 1 \, dA = \iint\limits_{R} dA$$

represents the volume of a solid of constant height 1 above the region R. Numerically, this volume is the same as the area of the region R. Thus,

$$\text{area of } R = \iint\limits_{R} dA \qquad (6)$$

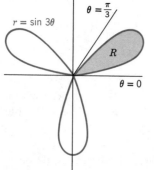

Figure 17.3.9

▶ **Example 3** Use a double polar integral to find the area enclosed by the three-leaved rose $r = \sin 3\theta$.

Solution. The rose is sketched in Figure 17.3.9. We will calculate the area of the petal R in the first quadrant and multiply by three.

$$A = 3 \iint\limits_{R} dA$$

$$= 3 \int_0^{\pi/3} \int_0^{\sin 3\theta} r \, dr \, d\theta$$

$$= \frac{3}{2} \int_0^{\pi/3} \sin^2 3\theta \, d\theta$$

$$= \frac{3}{4} \int_0^{\pi/3} (1 - \cos 6\theta) \, d\theta$$

$$= \left[\frac{3}{4}\theta - \frac{3}{24} \sin 6\theta \right]_0^{\pi/3}$$

$$= \frac{1}{4}\pi \qquad\qquad\qquad\qquad \blacktriangleleft$$

Sometimes a double integral

$$\iint_R f(x, y) \, dA$$

that is hard to evaluate in rectangular coordinates is easy to evaluate when the integrand and the region R are expressed in polar coordinates. This is especially true when the integrand or the boundary of R involve expressions of the form

$$x^2 + y^2 \qquad \text{or} \qquad \sqrt{x^2 + y^2}$$

since these simplify to

$$r^2 \qquad \text{and} \qquad r$$

in polar coordinates.

▶ **Example 4** Use polar coordinates to evaluate

$$\int_{-1}^{1} \int_0^{\sqrt{1-x^2}} (x^2 + y^2)^{3/2} \, dy \, dx$$

Solution. The first step is to express this iterated integral as a double integral in rectangular coordinates. For fixed x, the y-integration runs from the lower boundary $y = 0$ up to the semicircle $y = \sqrt{1 - x^2}$. The fixed x can extend from -1 on the left to $+1$ on the right, so

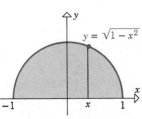

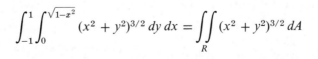

$$\int_{-1}^{1} \int_0^{\sqrt{1-x^2}} (x^2 + y^2)^{3/2} \, dy \, dx = \iint_R (x^2 + y^2)^{3/2} \, dA$$

Figure 17.3.10

where R is the region shown in Figure 17.3.10. Expressing this double integral as a double integral in polar coordinates yields

$$\int_{-1}^{1} \int_{0}^{\sqrt{1-x^2}} (x^2 + y^2)^{3/2} \, dy \, dx = \iint\limits_{R} (r^2)^{3/2} \, dA$$

$$= \int_{0}^{\pi} \int_{0}^{1} (r^3) r \, dr \, d\theta$$

$$= \int_{0}^{\pi} \frac{1}{5} \, d\theta = \frac{\pi}{5} \quad \blacktriangleleft$$

▶ Exercise Set 17.3

In Exercises 1–6, evaluate the iterated integral.

1. $\displaystyle\int_{0}^{\pi/2} \int_{0}^{\sin \theta} r \cos \theta \, dr \, d\theta$. **2.** $\displaystyle\int_{0}^{\pi} \int_{0}^{1+\cos \theta} r \, dr \, d\theta$.

3. $\displaystyle\int_{-\pi/2}^{\pi/2} \int_{0}^{a \sin \theta} r^2 \, dr \, d\theta$. **4.** $\displaystyle\int_{0}^{\pi/3} \int_{0}^{\cos 3\theta} r \, dr \, d\theta$.

5. $\displaystyle\int_{0}^{\pi} \int_{0}^{1-\sin \theta} r^2 \cos \theta \, dr \, d\theta$. **6.** $\displaystyle\int_{0}^{\pi} \int_{0}^{\cos \theta} r^3 \, dr \, d\theta$.

In Exercises 7–12, use a double integral in polar coordinates to find the area of the region described.

7. The region enclosed by the cardioid $r = 1 - \cos \theta$.

8. The region enclosed by the rose $r = \sin 2\theta$.

9. The region in the first quadrant bounded by $r = 1$ and $r = \sin 2\theta$, with $\pi/4 \le \theta \le \pi/2$.

10. The region inside the circle $x^2 + y^2 = 4$ and to the right of the line $x = 1$.

11. The region inside the circle $r = 4 \sin \theta$ and outside the circle $r = 2$.

12. The region inside the circle $r = 1$ and outside the cardioid $r = 1 + \cos \theta$.

In Exercises 13–18, use a double integral in polar coordinates to find the volume of the solid.

13. The solid common to the sphere $x^2 + y^2 + z^2 = 9$ and the cylinder $x^2 + y^2 = 1$.

14. The solid common to the sphere $r^2 + z^2 = 4$ and the cylinder $r = 2 \cos \theta$.

15. The solid bounded above by the cone $z = \sqrt{x^2 + y^2}$, below by the xy-plane, and laterally by the cylinder $x^2 + y^2 = 2y$.

16. The solid bounded above by the surface $z = (x^2 + y^2)^{-1/2}$, below by the xy-plane, and

enclosed between the cylinders $x^2 + y^2 = 1$ and $x^2 + y^2 = 9$.

17. The solid bounded above by the paraboloid $z = 1 - x^2 - y^2$, below by the xy-plane, and laterally by the cylinder $x^2 + y^2 - x = 0$.

18. The solid in the first octant bounded above by the plane $z = r \sin \theta$, below by the xy-plane, and laterally by the plane $x = 0$ and the cylinder $r = 3 \sin \theta$.

In Exercises 19–22, use polar coordinates to evaluate the double integral.

19. $\displaystyle\iint\limits_{R} e^{-(x^2+y^2)} \, dA$, where R is the region enclosed by the circle $x^2 + y^2 = 1$.

20. $\displaystyle\iint\limits_{R} \sqrt{9 - x^2 - y^2} \, dA$, where R is the region in the first quadrant within the circle $x^2 + y^2 = 9$.

21. $\displaystyle\iint\limits_{R} \frac{1}{1 + x^2 + y^2} \, dA$, where R is the sector in the first quadrant bounded by $y = 0$, $y = x$, and $x^2 + y^2 = 4$.

22. $\displaystyle\iint\limits_{R} 2y \, dA$, where R is the region in the first quadrant, bounded above by the circle $(x - 1)^2 + y^2 = 1$ and below by the line $y = x$.

In Exercises 23–28, evaluate the iterated integral by converting to polar coordinates.

23. $\displaystyle\int_{0}^{1} \int_{0}^{\sqrt{1-x^2}} (x^2 + y^2) \, dy \, dx$.

24. $\displaystyle\int_{-2}^{2}\int_{-\sqrt{4-y^2}}^{\sqrt{4-y^2}} e^{-(x^2+y^2)}\, dx\, dy.$

25. $\displaystyle\int_{0}^{2}\int_{0}^{\sqrt{2x-x^2}} \sqrt{x^2+y^2}\, dy\, dx.$

26. $\displaystyle\int_{0}^{1}\int_{0}^{\sqrt{1-y^2}} \cos(x^2+y^2)\, dx\, dy.$

27. $\displaystyle\int_{0}^{a}\int_{0}^{\sqrt{a^2-x^2}} \frac{dy\, dx}{(1+x^2+y^2)^{3/2}}.$

28. $\displaystyle\int_{0}^{1}\int_{y}^{\sqrt{y}} \sqrt{x^2+y^2}\, dx\, dy.$

29. Use polar coordinates to find the volume of the solid bounded above by the ellipsoid $x^2/a^2 + y^2/a^2 + z^2/c^2 = 1$, below by the xy-plane, and laterally by the cylinder $x^2 + y^2 - ay = 0$.

30. Find the area of the region enclosed by the lemniscate $r^2 = 2a^2 \cos 2\theta$. [*Hint:* Find the area in the first quadrant, and use symmetry.]

31. Find the area in the first quadrant inside the circle $r = 4 \sin \theta$ and outside the lemniscate $r^2 = 8 \cos 2\theta$. [*Hint:* Divide the region into two parts.]

32. Show that the shaded area in the following figure is $a^2\phi - \frac{1}{2}a^2 \sin 2\phi$.

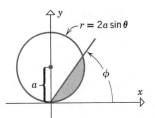

33. The integral

$$\int_{0}^{+\infty} e^{-x^2}\, dx$$

which arises in probability theory, can be evaluated using a trick. Let the value of the integral be I. Thus,

$$I = \int_{0}^{+\infty} e^{-x^2}\, dx = \int_{0}^{+\infty} e^{-y^2}\, dy$$

since the letter used for the variable of integration in a definite integral does not matter.

(a) Show that

$$I^2 = \int_{0}^{+\infty}\int_{0}^{+\infty} e^{-(x^2+y^2)}\, dx\, dy$$

(b) Evaluate the iterated integral in (a) by converting to polar coordinates.

(c) Use the result in (b) to find I.

17.4 SURFACE AREA

In Section 6.5 we showed how to find the surface area of a surface of revolution. In this section we consider the following more general surface-area problem.

17.4.1 PROBLEM Let f be a function defined on a closed region R of the xy-plane. Find the area of that portion of the surface $z = f(x, y)$ whose projection on the xy-plane is the region R (Figure 17.4.1).

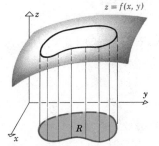

$z = f(x, y)$

Figure 17.4.1

In the special case where the surface is a plane and the region R is rectangular, we have the following result.

17.4.2   THEOREM   *Let R be a closed rectangular region in the xy-plane. If R has sides of length l and w, then the surface area S of that portion of the plane $z = ax + by + c$ that projects onto the region R is given by*

$$S = \sqrt{a^2 + b^2 + 1}\ lw \tag{1}$$

Proof. The portion of the plane that projects onto the region R, is a parallelogram (Figure 17.4.2). Thus, if we can find vectors **a** and **b** forming adjacent sides of this parallelogram, then we can obtain its surface area S from the formula

$$S = \|\mathbf{a} \times \mathbf{b}\| \tag{2}$$

(see Section 15.4).

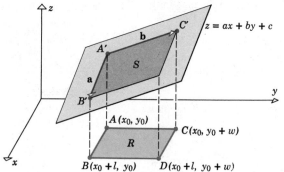

Figure 17.4.2

To find the vectors **a** and **b**, suppose that the four corners of the rectangle R are

$$A(x_0, y_0), \qquad B(x_0 + l, y_0), \qquad C(x_0, y_0 + w), \qquad D(x_0 + l, y_0 + w)$$

(Figure 17.4.2). Then those points in the plane $z = ax + by + c$ that lie above the corners A, B, and C are

$A'(x_0, y_0, ax_0 + by_0 + c)$
$B'(x_0 + l, y_0, a[x_0 + l] + by_0 + c)$
$C'(x_0, y_0 + w, ax_0 + b[y_0 + w] + c)$

Thus, the vectors

$\mathbf{a} = \overrightarrow{A'B'} = l\mathbf{i} + 0\mathbf{j} + al\mathbf{k}$
$\mathbf{b} = \overrightarrow{A'C'} = 0\mathbf{i} + w\mathbf{j} + bw\mathbf{k}$

form adjacent sides of the parallelogram in question. Since

$$\mathbf{a} \times \mathbf{b} = \begin{vmatrix} \mathbf{i} & \mathbf{j} & \mathbf{k} \\ l & 0 & al \\ 0 & w & bw \end{vmatrix} = -alw\mathbf{i} - lbw\mathbf{j} + lw\mathbf{k}$$

it follows from (2) that

$$S = \|\mathbf{a} \times \mathbf{b}\| = \sqrt{(-alw)^2 + (-lbw)^2 + (lw)^2} = \sqrt{a^2 + b^2 + 1}\ lw$$

Theorem 17.4.2 leads to the following more general result.

17.4.3 DEFINITION

Surface Area

If f has continuous first partial derivatives on a closed region R of the xy-plane, then the area S of that portion of the surface $z = f(x, y)$ that projects onto R is

$$S = \iint\limits_{R} \sqrt{\left(\frac{\partial z}{\partial x}\right)^2 + \left(\frac{\partial z}{\partial y}\right)^2 + 1}\ dA \tag{3}$$

Formula (3) can be motivated as follows:

Step 1. Enclose R within a rectangle whose sides are parallel to the x- and y-axes. Using lines parallel to the x- and y-axes, divide this rectangle into subrectangles, and exclude from consideration all those subrectangles that contain any points outside of R. This leaves only rectangles that are subsets of R (Figure 17.4.3a). Let these rectangles be denoted by

$$R_1, R_2, \ldots, R_n$$

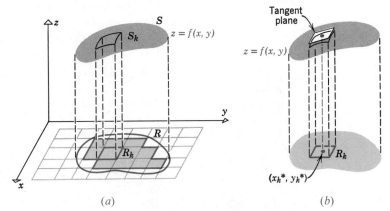

Figure 17.4.3   (a)   (b)

and suppose that the sides of rectangle R_k have lengths

$$\Delta x_k \quad \text{and} \quad \Delta y_k$$

Step 2. When projected up to the surface $z = f(x, y)$, each subrectangle determines a patch of area on the surface (Figure 17.4.3a). If we denote the areas of these patches by

$$S_1, S_2, \ldots, S_n$$

then the entire surface area S may be approximated by adding the areas of these patches:

$$S \approx S_1 + S_2 + \cdots + S_n \tag{4}$$

(This is only an approximation since the subrectangles may not completely fill out the region R.)

Step 3. We will now approximate each of the areas $S_1, S_2, \ldots, S_n$. Let

$$(x_k^*, y_k^*)$$

be an arbitrary point in the kth subrectangle and above this point construct the tangent plane to the surface $z = f(x, y)$ (Figure 17.4.3b). It follows from Theorem 16.6.2, that the equation of this tangent plane can be written in the form

$$z = f_x(x_k^*, y_k^*)x + f_y(x_k^*, y_k^*)y + c$$

where c is an appropriate constant. If the rectangle R_k is small, then we can reasonably approximate the area S_k of the kth patch on the surface by the portion of area on the tangent plane over R_k (Figure 17.4.3b). Thus, by Theorem 17.4.2

$$S_k \approx \sqrt{f_x(x_k^*, y_k^*)^2 + f_y(x_k^*, y_k^*)^2 + 1}\ \Delta x_k\, \Delta y_k$$

Substituting this expression in (4) and writing $\Delta A_k = \Delta x_k\, \Delta y_k$ for the area of the kth subrectangle, we obtain the Riemann sum

$$S \approx \sum_{k=1}^{n} \sqrt{f_x(x_k^*, y_k^*)^2 + f_y(x_k^*, y_k^*)^2 + 1}\ \Delta A_k \tag{5}$$

Step 4. There are two sources of error in approximation (5). First, the rectangles $R_1, R_2, \ldots, R_n$ may not fill up the region R completely and, second, we have approximated patches of area on the surface by areas on tangent planes. However, let us repeat this approximation process using more and more rectangles of decreasing dimensions. It is intuitively clear that both kinds of errors diminish and the exact surface area S is

$$S = \lim_{n \to +\infty} \sum_{k=1}^{n} \sqrt{f_x(x_k^*, y_k^*)^2 + f_y(x_k^*, y_k^*)^2 + 1} \; \Delta A_k$$

or equivalently

$$S = \iint\limits_R \sqrt{f_x(x, y)^2 + f_y(x, y)^2 + 1} \; dA$$

This is the motivation for Definition 17.4.3.

▶ **Example 1**   Find the surface area of the portion of the cylinder $x^2 + z^2 = 4$, above the rectangle $R = \{(x, y): 0 \le x \le 1, 0 \le y \le 4\}$ in the xy-plane.

Solution.   The surface is shown in Figure 17.4.4. The portion of the cylinder, $x^2 + z^2 = 4$, that lies above the xy-plane is given by $z = \sqrt{4 - x^2}$. Thus, from (3)

$$S = \iint\limits_R \sqrt{\left(\frac{\partial z}{\partial x}\right)^2 + \left(\frac{\partial z}{\partial y}\right)^2 + 1} \; dA$$

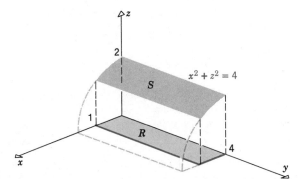

Figure 17.4.4

$$= \iint\limits_{R} \sqrt{\left(-\frac{x}{\sqrt{4-x^2}}\right)^2 + 0 + 1} \, dA$$

$$= \int_0^4 \int_0^1 \frac{2}{\sqrt{4-x^2}} \, dx \, dy$$

$$= 2 \int_0^4 \left[\sin^{-1}\frac{1}{2}x\right]_{x=0}^1 dy \qquad \begin{bmatrix} \text{Formula (13)} \\ \text{Section 8.3} \end{bmatrix}$$

$$= 2 \int_0^4 \frac{\pi}{6} \, dy = \frac{4}{3}\pi \qquad\qquad\qquad \blacktriangleleft$$

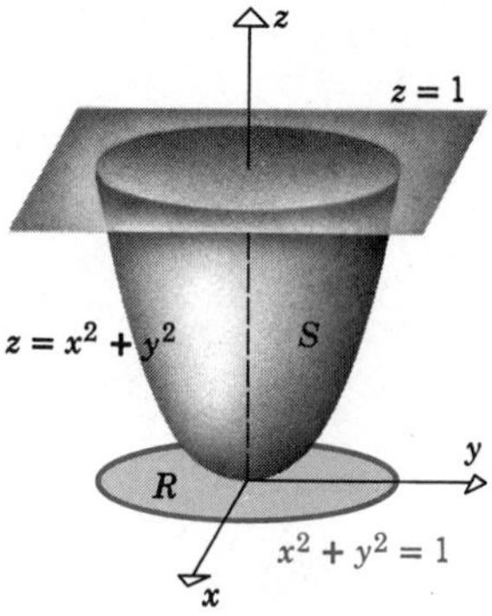

▶ **Example 2** Find the surface area of the portion of the paraboloid $z = x^2 + y^2$ below the plane $z = 1$.

Solution. The surface is shown in Figure 17.4.5. From the equations $z = 1$ and $z = x^2 + y^2$, we see that the plane and the paraboloid intersect in a circle whose projection on the xy-plane has the equation $x^2 + y^2 = 1$. Consequently, the surface whose area we seek projects onto the region R enclosed by this circle. Since the surface has the equation

$$z = x^2 + y^2$$

it follows from (3) that

$$S = \iint\limits_{R} \sqrt{4x^2 + 4y^2 + 1} \, dA$$

This integral is best evaluated in polar coordinates. Replacing $x^2 + y^2$ by r^2 and substituting $r \, dr \, d\theta$ for dA we obtain

$$S = \int_0^{2\pi} \int_0^1 \sqrt{4r^2 + 1} \, r \, dr \, d\theta = \int_0^{2\pi} \left[\frac{1}{12}(4r^2+1)^{3/2}\right]_{r=0}^1 d\theta$$

$$= \int_0^{2\pi} \frac{1}{12}(5\sqrt{5} - 1) \, d\theta = \frac{1}{6}\pi(5\sqrt{5} - 1) \qquad\qquad \blacktriangleleft$$

Figure 17.4.5

▶ **Exercise Set 17.4**

---

1. Find the surface area of the portion of the cylinder $y^2 + z^2 = 9$ above the rectangle $R = \{(x, y): 0 \le x \le 2, -3 \le y \le 3\}$.

2. By integration, find the surface area of the portion of the plane $2x + 2y + z = 8$ in the first octant that is cut off by the three coordinate planes.

3. Find the surface area of the portion of the cone $z^2 = 4x^2 + 4y^2$ that is above the region in the first quadrant bounded by the line $y = x$ and the parabola $y = x^2$.

4. Find the surface area of the portion of the cone $z = \sqrt{x^2 + y^2}$ that lies inside the cylinder

$x^2 + y^2 = 2x$.

5. Find the surface area of the portion of the paraboloid $z = 1 - x^2 - y^2$ that is above the xy-plane.

6. Find the area of the portion of the surface $z = 2x + y^2$ that is above the triangular region with vertices $(0, 0)$, $(0, 1)$, and $(1, 1)$.

7. Find the area of the portion of the surface $z = xy$ that is above the sector in the first quadrant bounded by the lines $y = x/\sqrt{3}$, $y = 0$, and the circle $x^2 + y^2 = 9$.

8. Find the surface area of the portion of the paraboloid $2z = x^2 + y^2$ that is inside the cylinder $x^2 + y^2 = 8$.

9. Find the surface area of the portion of the sphere $x^2 + y^2 + z^2 = 16$ between the planes $z = 1$ and $z = 2$.

10. Find the surface area of the portion of the sphere $x^2 + y^2 + z^2 = 8$ that is cut out by the cone $z = \sqrt{x^2 + y^2}$.

11. Find the total surface area of the portion of the sphere $x^2 + y^2 + z^2 = a^2$ inside the cylinder $x^2 + y^2 = ay$.

12. Use a double integral to derive the formula for the surface area of a sphere of radius a.

13. Find the surface area of the portion of the cylinder $x^2 + z^2 = 16$ that lies inside the cylinder $x^2 + y^2 = 16$. [*Hint:* Find the area in the first octant and use symmetry.]

14. Find the surface area of the portion of the cylinder $x^2 + z^2 = 5x$ that lies inside the sphere $x^2 + y^2 + z^2 = 25$.

15. The portion of the surface

$$z = \frac{h}{a} \sqrt{x^2 + y^2} \qquad (a, h > 0)$$

between the xy-plane and the plane $z = h$ is a right-circular cone of height h and radius a. Use a double integral to show that the lateral surface area of this cone is $S = \pi a \sqrt{a^2 + h^2}$.

17.5 TRIPLE INTEGRALS

Earlier we defined the notion of a double integral for functions of two variables. In this section we will discuss *triple integrals* for functions of three variables.

Whereas a double integral

$$\iint_R f(x, y)\, dA$$

is evaluated over a closed region R in the xy-plane, a triple integral of a function $f(x, y, z)$ is evaluated over a closed three-dimensional region G. We will assume that G can be enclosed within some suitably large box (rectangular parallelepiped) with sides parallel to the coordinate planes (Figure 17.5.1). This ensures that the region does not extend indefinitely in any direction.

To define a triple integral, we proceed as follows:

Step 1. Enclose G in a box with sides parallel to the coordinate planes, and use planes parallel to the coordinate planes to divide this box into subboxes. Exclude

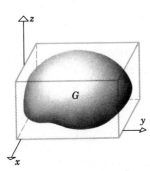

Figure 17.5.1

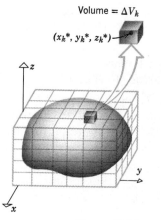

Volume $= \Delta V_k$

$(x_k{}^*, y_k{}^*, z_k{}^*)$

Figure 17.5.2

from consideration all subboxes that contain any points outside of G. This leaves only boxes that are subsets of G (Figure 17.5.2). Denote the volumes of these boxes by

$$\Delta V_1, \ \Delta V_2, \ldots, \Delta V_n$$

Step 2. Choose an arbitrary point in each box, and let these be denoted by

$$(x_1^*, y_1^*, z_1^*), \ (x_2^*, y_2^*, z_2^*), \ldots, (x_n^*, y_n^*, z_n^*)$$

(See Figure 17.5.2.)

Step 3. Form the sum

$$\sum_{k=1}^{n} f(x_k^*, y_k^*, z_k^*) \, \Delta V_k$$

This is called a **_Riemann sum._**

Step 4. Repeat this process with more and more subdivisions, so that the length, width, and height of each box approach zero, and n, the number of boxes, approaches $+\infty$. Define

$$\iiint\limits_{G} f(x, y, z) \, dV = \lim_{n \to +\infty} \sum_{k=1}^{n} f(x_k^*, y_k^*, z_k^*) \, \Delta V_k \qquad (1)$$

The symbol

$$\iiint\limits_{G} f(x, y, z) \, dV$$

is called the **_triple integral_** of $f(x, y, z)$ over the region G. Conditions under which the triple integral exists are studied in advanced calculus. However, for our purposes it suffices to say that existence is ensured when f is continuous on G and the region G is not too "complicated."

Triple integrals have a number of physical interpretations, some of which we will consider in the next section. In the special case where $f(x, y, z) = 1$ the triple integral over G represents the volume of the solid G, that is,

$$\text{volume of } G = \iiint\limits_{G} dV \qquad (2)$$

[Compare this to (7) of Section 17.2.] To obtain (2), let $f(x, y, z) = 1$ in (1). This yields

$$\iiint\limits_{G} dV = \lim_{n \to +\infty} \sum_{k=1}^{n} \Delta V_k$$

As suggested by Figure 17.5.2, the sum $\Sigma \, \Delta V_k$ represents the total volume of the boxes interior to the solid G. As n increases and the dimensions of these boxes approach zero, the boxes tend to fill up the solid G, so that their total volume approaches the volume of G, that is,

$$\iiint\limits_{G} dV = \lim_{n \to +\infty} \sum_{k=1}^{n} \Delta V_k = \text{volume of } G$$

Triple integrals enjoy many properties of single and double integrals:

$$\iiint\limits_{G} cf(x, y, z) \, dV = c \iiint\limits_{G} f(x, y, z) \, dV \qquad (c \text{ a constant})$$

$$\iiint\limits_{G} [f(x, y, z) + g(x, y, z)] \, dV = \iiint\limits_{G} f(x, y, z) \, dV + \iiint\limits_{G} g(x, y, z) \, dV$$

$$\iiint\limits_{G} [f(x, y, z) - g(x, y, z)] \, dV = \iiint\limits_{G} f(x, y, z) \, dV - \iiint\limits_{G} g(x, y, z) \, dV$$

Moreover, if the region G is subdivided into two subregions G_1 and G_2 (Figure 17.5.3), then

$$\iiint\limits_{G} f(x, y, z) \, dV = \iiint\limits_{G_1} f(x, y, z) \, dV + \iiint\limits_{G_2} f(x, y, z) \, dV$$

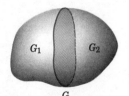

G_1 G_2

G

Figure 17.5.3

We omit the proofs.

Just as a double integral can be evaluated by two single integrations, a triple integral can be evaluated by three single integrations.

17.5.1 THEOREM *Let G be the rectangular box defined by the inequalities*

$$a \le x \le b, \qquad c \le y \le d, \qquad k \le z \le l$$

If f is continuous on the region G, then

$$\iiint\limits_{G} f(x, y, z) \, dV = \int_{a}^{b} \int_{c}^{d} \int_{k}^{l} f(x, y, z) \, dz \, dy \, dx \qquad (3)$$

Moreover, the iterated integral on the right can be replaced with any of the five other iterated integrals that result by altering the order of integration.

We omit the proof.

▶ Example 1 Evaluate the triple integral

$$\iiint\limits_{G} 12xy^2z^3 \, dV$$

over the rectangular box G defined by the inequalities: $-1 \le x \le 2$, $0 \le y \le 3$, $0 \le z \le 2$.

Solution. Of the six possible iterated integrals we might use, we will choose the one in (3). Thus, we will first integrate with respect to z, holding x and y fixed, then with respect to y holding x fixed, and finally with respect to x.

$$\iiint\limits_{G} 12xy^2z^3 \, dV = \int_{-1}^{2}\int_{0}^{3}\int_{0}^{2} 12xy^2z^3 \, dz \, dy \, dx$$

$$= \int_{-1}^{2}\int_{0}^{3} \left[3xy^2z^4 \right]_{z=0}^{2} dy \, dx$$

$$= \int_{-1}^{2}\int_{0}^{3} 48xy^2 \, dy \, dx$$

$$= \int_{-1}^{2} \left[16xy^3 \right]_{y=0}^{3} dx$$

$$= \int_{-1}^{2} 432x \, dx$$

$$= 216x^2 \bigg]_{-1}^{2} = 648$$

We will also be concerned with evaluating triple integrals over solid regions other than rectangular boxes. For simplicity, we will restrict our discussion to solid regions constructed as follows. Let R be a closed region in the xy-plane and let $g_1(x, y)$ and $g_2(x, y)$ be continuous functions satisfying

$$g_1(x, y) \le g_2(x, y)$$

for all (x, y) in R. Geometrically, this condition states that the surface $z = g_2(x, y)$ does not dip below the surface $z = g_1(x, y)$ over R (Figure

17.5.4*a*). We will call $z = g_1(x, y)$ the **lower surface** and $z = g_2(x, y)$ the **upper surface.** Let G be the solid consisting of all points above or below the region R that lie between the upper surface and the lower surface (Figure 17.5.4*b*). A solid G constructed in this way will be called a **simple solid,** and the region R will be called the **projection** of G on the xy-plane.

The following theorem, which we state without proof, will enable us to evaluate triple integrals over simple solids.

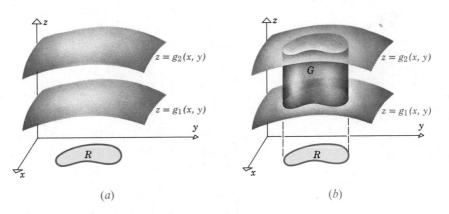

Figure 17.5.4   (*a*)   (*b*)

17.5.2 THEOREM *Let G be a simple solid with upper surface $z = g_2(x, y)$ and lower surface $z = g_1(x, y)$; and let R be the projection of G on the xy-plane. If f(x, y, z) is continuous on G, then*

$$\iiint_G f(x, y, z)\, dV = \iint_R \left[\int_{g_1(x,y)}^{g_2(x,y)} f(x, y, z)\, dz \right] dA \qquad (4)$$

In (4), the first integration is with respect to z, after which a function of x and y remains. This function of x and y is then integrated over the region R in the xy-plane. To apply (4), it is usual to begin with a three-dimensional sketch of the solid G, from which the limits of integration can be obtained as follows:

Step 1. Find an equation $z = g_2(x, y)$ for the upper surface and an equation $z = g_1(x, y)$ for the lower surface of G. The functions $g_1(x, y)$ and $g_2(x, y)$ determine the z-limits of integration.

Step 2. Make a two-dimensional sketch of the projection R of the solid on the xy-plane. From this sketch determine the limits of integration for the double integral over R in (4).

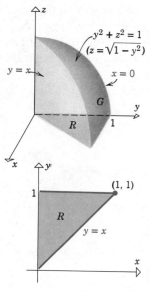

Figure 17.5.5

▶ **Example 2**   Let G be the wedge in the first octant cut from the cylinder $y^2 + z^2 = 1$ by the planes $y = x$ and $x = 0$. Evaluate

$$\iiint_G z \, dV$$

Solution.   The solid G and its projection R on the xy-plane are shown in Figure 17.5.5. The upper surface of the solid is formed by the cylinder and the lower surface by the xy-plane. Since the portion of the cylinder $y^2 + z^2 = 1$ that lies above the xy-plane has the equation $z = \sqrt{1 - y^2}$, and the xy-plane has the equation $z = 0$, it follows from (4) that

$$\iiint_G z \, dV = \iint_R \left[\int_0^{\sqrt{1-y^2}} z \, dz \right] dA \qquad (5)$$

For the double integral over R, the x and y integrations can be performed in either order, since R is both a type I and type II region. We will integrate with respect to x first. With this choice, (5) yields

$$\iiint_G z \, dV = \int_0^1 \int_0^y \int_0^{\sqrt{1-y^2}} z \, dz \, dx \, dy = \int_0^1 \int_0^y \frac{1}{2} z^2 \Big]_{z=0}^{\sqrt{1-y^2}} dx \, dy$$

$$= \int_0^1 \int_0^y \frac{1}{2}(1 - y^2) \, dx \, dy = \frac{1}{2} \int_0^1 (1 - y^2)x \Big]_{x=0}^{y} dy$$

$$= \frac{1}{2} \int_0^1 (y - y^3) \, dy = \frac{1}{2} \left[\frac{1}{2} y^2 - \frac{1}{4} y^4 \right]_0^1 = \frac{1}{8} \qquad ◀$$

▶ **Example 3**   Use a triple integral to find the volume of the solid enclosed between the cylinder $x^2 + y^2 = 9$ and the planes $z = 1$ and $x + z = 5$.

Solution.   The solid G and its projection R on the xy-plane are shown in Figure 17.5.6. The lower surface of the solid is the plane $z = 1$; and the upper surface is the plane $x + z = 5$, or equivalently, $z = 5 - x$. Thus, from (2) and (4)

$$\text{volume of } G = \iiint_G dV = \iint_R \left[\int_1^{5-x} dz \right] dA \qquad (6)$$

For the double integral over R, we will integrate with respect to y first. Thus, (6) yields

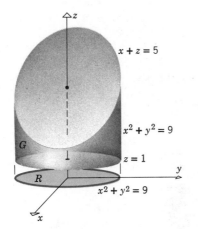

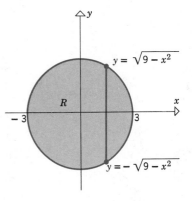

Figure 17.5.6   (a)   (b)

$$\text{volume of } G = \int_{-3}^{3}\int_{-\sqrt{9-x^2}}^{\sqrt{9-x^2}}\int_{1}^{5-x} dz\, dy\, dx$$

$$= \int_{-3}^{3}\int_{-\sqrt{9-x^2}}^{\sqrt{9-x^2}} z\Big]_{z=1}^{5-x} dy\, dx$$

$$= \int_{-3}^{3}\int_{-\sqrt{9-x^2}}^{\sqrt{9-x^2}} (4-x)\, dy\, dx$$

$$= \int_{-3}^{3} (8-2x)\sqrt{9-x^2}\, dx$$

$$= 8\int_{-3}^{3}\sqrt{9-x^2}\, dx - \int_{-3}^{3} 2x\sqrt{9-x^2}\, dx \quad \begin{bmatrix}\text{For the first integral,}\\ \text{see Example 4, Section}\\ 9.5.\end{bmatrix}$$

$$= 8\left(\frac{9}{2}\pi\right) - \int_{-3}^{3} 2x\sqrt{9-x^2}\, dx \quad \begin{bmatrix}\text{Let } u = 9-x^2 \text{ or better, apply}\\ \text{the result in Exercise 30}\\ \text{of Section 5.9.}\end{bmatrix}$$

$$= 8\left(\frac{9}{2}\pi\right) - 0 \; = 36\pi \qquad \blacktriangleleft$$

▶ **Example 4** Find the volume of the solid enclosed by the paraboloids

$$z = 5x^2 + 5y^2 \quad \text{and} \quad z = 6 - 7x^2 - y^2$$

Solution. The solid G and its projection R on the xy-plane are shown in Figure 17.5.7. The projection R is obtained by solving the given equations simultaneously to determine where the paraboloids intersect. We obtain

$$5x^2 + 5y^2 = 6 - 7x^2 - y^2$$

or

$$2x^2 + y^2 = 1 \tag{7}$$

which tells us that the paraboloids intersect in a curve on the elliptic cylinder given by (7).

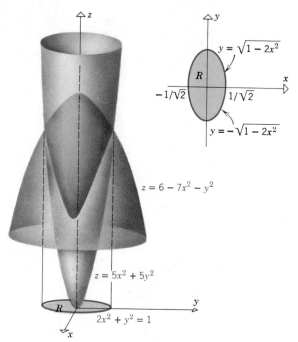

Figure 17.5.7

The projection of this intersection on the xy-plane is an ellipse with this same equation. Therefore,

$$\text{volume of } G = \iiint\limits_{G} dV$$

$$= \iint\limits_{R} \left[\int_{5x^2+5y^2}^{6-7x^2-y^2} dz \right] dA$$

$$\text{volume of } G = \int_{-1/\sqrt{2}}^{1/\sqrt{2}} \int_{-\sqrt{1-2x^2}}^{\sqrt{1-2x^2}} \int_{5x^2+5y^2}^{6-7x^2-y^2} dz \, dy \, dx$$

$$= \int_{-1/\sqrt{2}}^{1/\sqrt{2}} \int_{-\sqrt{1-2x^2}}^{\sqrt{1-2x^2}} (6 - 12x^2 - 6y^2) \, dy \, dx$$

$$= \int_{-1/\sqrt{2}}^{1/\sqrt{2}} \left[6(1 - 2x^2)y - 2y^3 \right]_{y=-\sqrt{1-2x^2}}^{\sqrt{1-2x^2}} dx$$

$$= 8 \int_{-1/\sqrt{2}}^{1/\sqrt{2}} (1 - 2x^2)^{3/2} \, dx \qquad \qquad \left[\text{Let } x = \frac{1}{\sqrt{2}} \sin \theta \right]$$

$$= \frac{8}{\sqrt{2}} \int_{-\pi/2}^{\pi/2} \cos^4 \theta \, d\theta$$

$$= \frac{3\pi}{\sqrt{2}} \qquad \text{[Formula (5), Section 9.3]} \qquad \blacktriangleleft$$

For certain regions, triple integrals are best evaluated by integrating first with respect to x or y rather than z. For example, if the solid G is bounded on the left and right by the surfaces $y = g_1(x, z)$ and $y = g_2(x, z)$ and bounded laterally by a cylinder extending in the y-direction (Figure 17.5.8a), then

$$\iiint\limits_{G} f(x, y, z) \, dV = \iint\limits_{R} \left[\int_{g_1(x,z)}^{g_2(x,z)} f(x, y, z) \, dy \right] dA$$

where R is the projection of the solid G on the xz-plane.

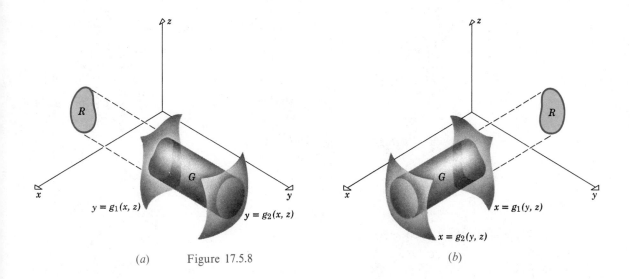

(a)   Figure 17.5.8   (b)

Similarly, if the solid G is bounded in the back and front by the surfaces $x = g_1(y, z)$ and $x = g_2(y, z)$ and bounded laterally by a cylinder extending in the x-direction, then

$$\iiint_G f(x, y, z)\, dV = \iint_R \left[\int_{g_1(y,z)}^{g_2(y,z)} f(x, y, z)\, dx \right] dA$$

where R is the projection of the solid on the yz-plane (Figure 17.5.8b).

▶ Example 5   In Example 2, we evaluated

$$\iiint_G z\, dV$$

over the wedge in Figure 17.5.5 by integrating first with respect to z. Evaluate this integral by integrating first with respect to x.

Solution. The solid is bounded in the back by the plane $x = 0$ and in the front by the plane $x = y$, so

$$\iiint_G z\, dV = \iint_R \left[\int_0^y z\, dx \right] dA$$

where R is the projection of G on the yz-plane (Figure 17.5.9). The integration over R can be performed first with respect to z and then y or vice versa. Performing the z-integration first yields

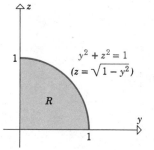

$y^2 + z^2 = 1$
$(z = \sqrt{1 - y^2})$

R

Figure 17.5.9

$$\iiint_G z\, dV = \int_0^1 \int_0^{\sqrt{1-y^2}} \int_0^y z\, dx\, dz\, dy = \int_0^1 \int_0^{\sqrt{1-y^2}} zx \Big]_{x=0}^y dz\, dy$$

$$= \int_0^1 \int_0^{\sqrt{1-y^2}} zy\, dz\, dy = \int_0^1 \frac{1}{2} z^2 y \Big]_{z=0}^{\sqrt{1-y^2}} = \int_0^1 \frac{1}{2}(1 - y^2)y\, dy = \frac{1}{8}$$

which agrees with the result in Example 2.   ◀

▶ Exercise Set 17.5

In Exercises 1–8, evaluate the iterated integral.

1. $\displaystyle\int_{-1}^1 \int_0^2 \int_0^1 (x^2 + y^2 + z^2)\, dx\, dy\, dz.$

2. $\displaystyle\int_{1/3}^{1/2} \int_0^\pi \int_0^1 zx \sin xy\, dz\, dy\, dx.$

3. $\displaystyle\int_0^2 \int_{-1}^{y^2} \int_1^z yz\, dx\, dz\, dy.$

4. $\displaystyle\int_0^{\pi/4} \int_0^1 \int_0^{x^2} x \cos y\, dz\, dx\, dy.$

5. $\displaystyle\int_0^3 \int_0^{\sqrt{9-z^2}} \int_0^x xy\, dy\, dx\, dz.$

6. $\displaystyle\int_1^3 \int_x^{x^2} \int_0^{\ln z} xe^y\, dy\, dz\, dx.$

7. $\displaystyle\int_0^2 \int_0^{\sqrt{4-x^2}} \int_{-5+x^2+y^2}^{3-x^2-y^2} x \, dz \, dy \, dx.$

8. $\displaystyle\int_1^2 \int_z^2 \int_0^{\sqrt{3y}} \frac{y}{x^2+y^2} \, dx \, dy \, dz.$

In Exercises 9–12, evaluate the triple integral.

9. $\displaystyle\iiint_G xy \sin yz \, dV$, where G is the rectangular box defined by the inequalities $0 \le x \le \pi$, $0 \le y \le 1$, $0 \le z \le \pi/6$.

10. $\displaystyle\iiint_G y \, dV$, where G is the solid enclosed by the plane $z = y$, the xy-plane, and the parabolic cylinder $y = 1 - x^2$.

11. $\displaystyle\iiint_G xyz \, dV$, where G is the solid in the first octant bounded by the parabolic cylinder $z = 2 - x^2$ and the planes $z = 0$, $y = x$, and $y = 0$.

12. $\displaystyle\iiint_G \cos(z/y) \, dV$, where G is the solid defined by the inequalities, $\pi/6 \le y \le \pi/2$, $y \le x \le \pi/2$, $0 \le z \le xy$.

In Exercises 13–21, use a triple integral to find the volume of the solid.

13. The solid in the first octant bounded by the coordinate planes and the plane $3x + 6y + 4z = 12$.

14. The solid bounded by the surface $z = \sqrt{y}$ and the planes $x + y = 1$ and $z = 0$.

15. The solid bounded by the surface $y = x^2$ and the planes $y + z = 4$ and $z = 0$.

16. The wedge in the first octant cut from the cylinder $y^2 + z^2 = 1$ by the planes $y = x$ and $x = 0$.

17. The solid enclosed between the elliptic cylinder $x^2 + 9y^2 = 9$ and the planes $z = 0$ and $z = x + 3$.

18. The solid common to the cylinders $x^2 + y^2 = 1$ and $x^2 + z^2 = 1$.

19. The solid bounded by the paraboloid $z = 4x^2 + y^2$ and the parabolic cylinder $z = 4 - 3y^2$.

20. The solid enclosed between the paraboloids $z = 8 - x^2 - y^2$ and $z = 3x^2 + y^2$.

21. The solid enclosed by the sphere $x^2 + y^2 + z^2 = 2a^2$ and the paraboloid $az = x^2 + y^2$ $(a > 0)$. [*Hint:* Polar coordinates will help in this problem.]

22. In each part sketch the solid whose volume is given by the integral.

(a) $\displaystyle\int_0^3 \int_{x^2}^9 \int_0^2 dz \, dy \, dx$

(b) $\displaystyle\int_0^2 \int_0^{2-y} \int_0^{2-x-y} dz \, dx \, dy.$

23. In each part sketch the solid whose volume is given by the integral.

(a) $\displaystyle\int_{-1}^1 \int_{-\sqrt{1-x^2}}^{\sqrt{1-x^2}} \int_0^{y+1} dz \, dy \, dx$

(b) $\displaystyle\int_0^9 \int_0^{y/3} \int_0^{\sqrt{y^2-9x^2}} dz \, dx \, dy.$

24. In each part sketch the solid whose volume is given by the integral.

(a) $\displaystyle\int_0^1 \int_0^{\sqrt{1-x^2}} \int_0^2 dy \, dz \, dx$

(b) $\displaystyle\int_{-2}^2 \int_0^{4-y^2} \int_0^2 dx \, dz \, dy.$

25. Let G be the tetrahedron in the first octant bounded by the coordinate planes and the plane

$$\frac{x}{a} + \frac{y}{b} + \frac{z}{c} = 1 \qquad (a > 0, b > 0, c > 0)$$

(a) List six different iterated integrals that represent the volume of G.

(b) Evaluate any one of the six to show that the volume of G is $\frac{1}{6}abc$.

26. In parts (a)–(c), express the integral as an equivalent integral in which the z-integration is performed first, the y-integration second, and the x-integration last.

(a) $\displaystyle\int_0^3 \int_0^{\sqrt{9-z^2}} \int_0^{\sqrt{9-y^2-z^2}} f(x, y, z)\, dx\, dy\, dz$

(b) $\displaystyle\int_0^4 \int_0^2 \int_0^{x/2} f(x, y, z)\, dy\, dz\, dx$

(c) $\displaystyle\int_0^4 \int_0^{4-y} \int_0^{\sqrt{z}} f(x, y, z)\, dx\, dz\, dy.$

27. Use a triple integral to find the volume of the tetrahedron with vertices $(0, 0, 0)$, $(a, a, 0)$, $(a, 0, 0)$ and $(a, 0, a)$.

28. Let G be the rectangular box defined by the inequalities $a \le x \le b$, $c \le y \le d$, $k \le z \le l$. Show that

$$\iiint\limits_G f(x)g(y)h(z)\, dV$$

$$= \left[\int_a^b f(x)\, dx\right]\left[\int_c^d g(y)\, dy\right]\left[\int_k^l h(z)\, dz\right]$$

29. Use the result of Exercise 28 to evaluate

(a) $\displaystyle\iiint\limits_G xy^2 \sin z\, dV$, where G is the set of points satisfying

$$-1 \le x \le 1,\ 0 \le y \le 1,\ 0 \le z \le \frac{\pi}{2}$$

(b) $\displaystyle\iiint\limits_G e^{2x+y-z}\, dV$, where G is the set of points satisfying

$$0 \le x \le 1,\ 0 \le y \le \ln 3,\ 0 \le z \le \ln 2$$

30. Use a triple integral to find the volume of the ellipsoid

$$\frac{x^2}{a^2} + \frac{y^2}{b^2} + \frac{z^2}{c^2} = 1$$

17.6 CENTROIDS, CENTERS OF GRAVITY, THEOREM OF PAPPUS

Suppose that a physical body is acted on by a gravitational field. Because the body is composed of many particles, each of which is affected by gravity, the action of the gravitational field on the body consists of a large number of forces distributed over the entire body. However, it is a physical fact that a body subjected to gravity behaves as if the entire force of gravity acts at a single point. This point is called the **center of gravity** or **center of mass** of the body. In this section we will show how double and triple integrals can be used to locate centers of gravity.

The thickness of a lamina is negligible

Figure 17.6.1

To begin, let us consider an idealized flat object that is sufficiently thin to be viewed as a two-dimensional solid (Figure 17.6.1). Such a solid is called a **lamina**. A lamina is called **homogeneous** if its composition and structure is uniform throughout, and is called **inhomogeneous** otherwise.

For a *homogeneous* lamina of mass M and area A, we define the **density** δ of the lamina by

$$\delta = \frac{M}{A}$$

Thus, δ measures the mass per unit area.

For an inhomogeneous lamina the composition may vary from point to point. If the lamina is placed in the xy-plane, then the "density" of the

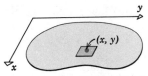

Figure 17.6.2

lamina at a general point (x, y) can be specified by a function $\delta(x, y)$, called the ***density function*** for the lamina. Informally, the density function can be visualized as follows. Construct a small rectangle centered at (x, y) and let ΔM and ΔA be the mass and area of the portion of lamina enclosed by this rectangle (Figure 17.6.2). If the ratio $\Delta M/\Delta A$ tends toward a limiting value as the dimensions of the rectangle tend to zero, then this limit is considered to be the density of the lamina at (x, y). Symbolically,

$$\delta(x, y) = \lim \frac{\Delta M}{\Delta A} \tag{1}$$

From this relationship we obtain the approximation

$$\Delta M \approx \delta(x, y)\,\Delta A \tag{2}$$

which relates the mass and area of a small rectangular portion of lamina centered at (x, y). It is assumed that as the dimensions of the rectangle tend to zero, the error in this approximation also tends to zero.

We can now obtain a formula for the mass of a lamina.

17.6.1 DEFINITION
Mass of a Lamina

If a lamina with density function $\delta(x, y)$ occupies a region R in the xy-plane, then its total mass M is defined to be

$$M = \iint\limits_{R} \delta(x, y)\, dA \tag{3}$$

To motivate this result, let a rectangle enclosing the region R be divided into subrectangles by lines parallel to the coordinate axes, and exclude from consideration all those rectangles that contain any points outside of R. Let

$$\Delta A_1, \Delta A_2, \ldots, \Delta A_n$$

be the areas of the remaining rectangles and let

$$\Delta M_1, \Delta M_2, \ldots, \Delta M_n$$

be the masses of the portions of lamina enclosed by these rectangles. We can approximate the total mass of the lamina by adding the masses of these pieces

$$M = \text{total mass} \approx \sum_{k=1}^{n} \Delta M_k \tag{4}$$

If we now let (x_k^*, y_k^*) be the center of the kth rectangle, then it follows from (2) that

$$\Delta M_k \approx \delta(x_k^*, y_k^*)\,\Delta A_k$$

Substituting this in (4) yields

$$M \approx \sum_{k=1}^{n} \delta(x_k^*, y_k^*)\,\Delta A_k \tag{5}$$

If we now increase n in such a way that the dimensions of the rectangles tend to zero, then the errors in our approximations will diminish so that

$$M = \lim_{n \to +\infty} \sum_{k=1}^{n} \delta(x_k^*, y_k^*)\,\Delta A_k = \iint_R \delta(x, y)\,dA$$

▶ **Example 1**   A triangular lamina with vertices $(0, 0)$, $(0, 1)$, and $(1, 0)$ has density function $\delta(x, y) = xy$. Find its total mass.

Solution.   Referring to (3) and Figure 17.6.3, the mass M of the lamina is

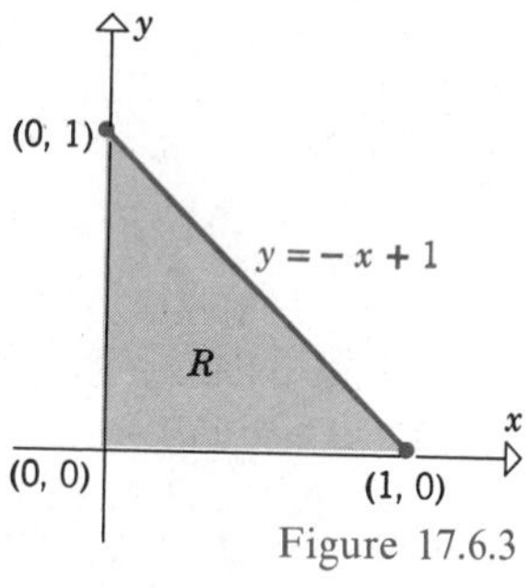

$$M = \iint_R \delta(x, y)\,dA = \iint_R xy\,dA = \int_0^1 \int_0^{-x+1} xy\,dy\,dx$$

$$= \int_0^1 \left[\frac{1}{2}xy^2 \right]_{y=0}^{-x+1} dx = \int_0^1 \left[\frac{1}{2}x^3 - x^2 + \frac{1}{2}x \right] dx$$

$$= \frac{1}{24} \text{ (units of mass)} \qquad \blacktriangleleft$$

(0, 1)

$y = -x + 1$

R

(0, 0)   (1, 0)

Figure 17.6.3

Let us now consider the following problem.

17.6.2   PROBLEM   Suppose a lamina with density function $\delta(x, y)$ occupies a region R in a horizontal xy-plane. Find the coordinates $(\bar{x}, \bar{y})$ of the center of gravity.

To motivate the solution, consider what happens if we try to balance the lamina on a knife edge parallel to the x-axis. Suppose the lamina in Figure 17.6.4 is placed on a knife edge along a line $y = c$ that does not pass through the center of gravity. Because the lamina behaves as if its entire mass is concentrated at the center of gravity $(\bar{x}, \bar{y})$, the lamina will be rotationally unstable and the force of gravity will cause a rotation about $y = c$. Similarly, the lamina will undergo a rotation if placed on a knife edge along $y = d$. However, if the knife edge runs along the line $y = \bar{y}$ through the center of gravity, the lamina will be in perfect balance. Similarly, the lamina will be in perfect balance on a knife edge along the line $x = \bar{x}$ through the center of gravity.

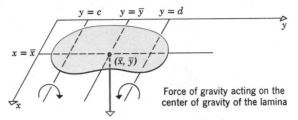

Figure 17.6.4

Force of gravity acting on the center of gravity of the lamina

This discussion suggests that the center of gravity of a lamina can be determined as the intersection of two lines of balance, one parallel to the x-axis and the other parallel to the y-axis. In order to find these lines of balance, we will need some preliminary results about rotations.

Children on a seesaw learn by experience that a lighter child can balance a heavier one by sitting farther from the fulcrum or pivot point. This is because the tendency for a mass to produce rotation is proportional not only to the magnitude of the mass but also the distance between the mass and the fulcrum. To be precise, if a point-mass m is located on a coordinate axis at a point x, then the tendency for that mass to produce a rotation about a point a on the axis is measured by the following quantity, called the **moment of m about a,**

$$\begin{bmatrix} \text{moment of } m \\ \text{about } a \end{bmatrix} = m(x - a)$$

The number $x - a$ is called the **lever arm.** Depending on whether the mass is to the right or left of a, the lever arm is either the distance between x and a or the negative of this distance (Figure 17.6.5). Positive lever arms result in positive moments and clockwise rotations, while negative lever arms result in negative moments and counterclockwise rotations.

is positive
(clockwise rotation)

is negative
(counterclockwise rotation)

Figure 17.6.5

Suppose masses $m_1, m_2, \ldots, m_n$ are located at points $x_1, x_2, \ldots, x_n$ on a coordinate axis and a fulcrum is positioned at the point a (Figure 17.6.6). Depending on whether the sum of the moments about a,

$$\sum_{k=1}^{n} m_k(x_k - a) = m_1(x_1 - a) + m_2(x_2 - a) + \cdots + m_n(x_n - a)$$

is positive, negative, or zero, the axis will rotate clockwise about a, rotate counterclockwise about a, or balance perfectly. In the last case, the system of masses is said to be in **equilibrium.**

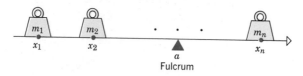

Figure 17.6.6

▶ **Example 2** Where should a fulcrum be placed so that the following system of masses is in equilibrium?

Solution. Introduce an x-axis with origin at the mass m_1, and let a be the unknown coordinate of the fulcrum (Figure 17.6.7). (Our choice of location for the origin is arbitrary. Any convenient location will suffice.)

Figure 17.6.7

We will determine a so that the total moment about a is zero. The sum of the moments about a is

$$\text{total moment} = 20(0 - a) + 10(3 - a) + 40(15 - a) = 630 - 70a$$

For equilibrium, a should be chosen so that

$$630 - 70a = 0$$

or $a = 9$. Thus, the fulcrum should be located 9 feet to the right of mass m_1.

◀

The ideas we have discussed can be extended to masses distributed in two-dimensional space. If a mass m is located at a point (x, y) in a horizontal xy-plane, then we define the **moment of m about the line $x = a$** and the **moment of m about the line $y = c$** by

$$\begin{bmatrix} \text{moment of } m \\ \text{about the} \\ \text{line } x = a \end{bmatrix} = m(x - a) \qquad \binom{\text{mass times}}{\text{lever arm}} \qquad (6)$$

$$\begin{bmatrix} \text{moment of } m \\ \text{about the} \\ \text{line } y = c \end{bmatrix} = m(y - c) \qquad \binom{\text{mass times}}{\text{lever arm}} \qquad (7)$$

(See Figure 17.6.8.)

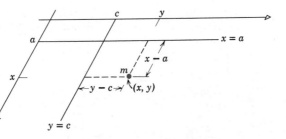

Figure 17.6.8

If a number of masses are distributed in the xy-plane and if we imagine the xy-plane to be a weightless sheet, then this sheet will balance on the line $x = a$ (or $y = c$) if the sum of the moments about that line is zero.

Suppose now that we are dealing not with isolated point masses, but with a lamina in the xy-plane. How should we define the moments about the lines $x = a$ and $y = c$? To motivate the appropriate definitions, assume the lamina occupies the region R in the xy-plane and has density function $\delta(x, y)$. Following a now familiar procedure, let R be partitioned into subregions and let R_k be a typical rectangle interior to R. Let ΔA_k be the area of R_k, let ΔM_k be the mass of that portion of the lamina enclosed by R_k, and let (x_k^*, y_k^*) be the center of the rectangle R_k. From (2), the mass ΔM_k may be approximated by

$$\Delta M_k \approx \delta(x_k^*, y_k^*) \, \Delta A_k$$

Moreover, if the rectangle R_k is small, there should be little error in assuming that the entire mass ΔM_k is concentrated at the center (x_k^*, y_k^*) of the rectangle. Thus, from (6) and (7), the moment of the mass ΔM_k about the line $x = a$ is approximately

$$(x_k^* - a) \, \Delta M_k \approx (x_k^* - a)\delta(x_k^*, y_k^*) \, \Delta A_k$$

and the moment about the line $y = c$ is approximately

$$(y_k^* - c) \, \Delta M_k \approx (y_k^* - c)\delta(x_k^*, y_k^*) \, \Delta A_k$$

Therefore, the total moment about $x = a$ produced by all the rectangular regions interior to R is approximately

$$\sum_{k=1}^{n} (x_k^* - a)\delta(x_k^*, y_k^*)\,\Delta A_k \tag{8}$$

and the total moment about $y = c$ is approximately

$$\sum_{k=1}^{n} (y_k^* - c)\delta(x_k^*, y_k^*)\,\Delta A_k \tag{9}$$

If we assume that the errors in our approximations diminish as the dimensions of the rectangles tend to zero, then (8) and (9) suggest the definitions.

$$\begin{bmatrix} \text{moment of} \\ \text{lamina } R \\ \text{about the} \\ \text{line } x = a \end{bmatrix} = \lim_{n \to +\infty} \sum_{k=1}^{n} (x_k^* - a)\delta(x_k^*, y_k^*)\,\Delta A_k = \iint_{R} (x - a)\delta(x, y)\,dA \tag{10}$$

$$\begin{bmatrix} \text{moment of} \\ \text{lamina } R \\ \text{about the} \\ \text{line } y = c \end{bmatrix} = \lim_{n \to +\infty} \sum_{k=1}^{n} (y_k^* - c)\delta(x_k^*, y_k^*)\,\Delta A_k = \iint_{R} (y - c)\delta(x, y)\,dA \tag{11}$$

Of special importance are the moments about the y-axis ($a = 0$) and about the x-axis ($c = 0$). These moments are denoted by

$$M_y = \begin{bmatrix} \text{moment of lamina } R \\ \text{about the } y\text{-axis} \end{bmatrix} = \iint_{R} x\delta(x, y)\,dA \tag{12}$$

$$M_x = \begin{bmatrix} \text{moment of lamina } R \\ \text{about the } x\text{-axis} \end{bmatrix} = \iint_{R} y\delta(x, y)\,dA \tag{13}$$

We are now in a position to obtain formulas for the center of gravity $(\bar{x}, \bar{y})$ of a lamina. Since a lamina balances on the lines $x = \bar{x}$ and $y = \bar{y}$ through the center of gravity, the moments of the lamina about these lines must be zero. Thus, from (10) and (11)

$$\iint_{R} (x - \bar{x})\delta(x, y)\,dA = 0$$

$$\iint_{R} (y - \bar{y})\delta(x, y)\,dA = 0$$

Since $\bar{x}$ and $\bar{y}$ are constant, these equations can be rewritten

$$\iint\limits_{R} x\delta(x, y)\, dA = \bar{x} \iint\limits_{R} \delta(x, y)\, dA$$

$$\iint\limits_{R} y\delta(x, y)\, dA = \bar{y} \iint\limits_{R} \delta(x, y)\, dA$$

from which we obtain the following formulas:

Center of gravity of a lamina

$$\bar{x} = \frac{\displaystyle\iint\limits_{R} x\delta(x, y)\, dA}{\displaystyle\iint\limits_{R} \delta(x, y)\, dA} \qquad \bar{y} = \frac{\displaystyle\iint\limits_{R} y\delta(x, y)\, dA}{\displaystyle\iint\limits_{R} \delta(x, y)\, dA} \qquad (14a)$$

By virtue of (3), (12), and (13), these formulas can be written as

$$\bar{x} = \frac{\begin{bmatrix} \text{moment about} \\ \text{the } y\text{-axis} \end{bmatrix}}{\begin{bmatrix} \text{mass of the} \\ \text{lamina} \end{bmatrix}} = \frac{M_y}{M} \qquad \bar{y} = \frac{\begin{bmatrix} \text{moment about} \\ \text{the } x\text{-axis} \end{bmatrix}}{\begin{bmatrix} \text{mass of the} \\ \text{lamina} \end{bmatrix}} = \frac{M_x}{M} \qquad (14b)$$

▶ **Example 3**   Find the center of gravity of the triangular lamina with vertices $(0, 0)$, $(0, 1)$, and $(1, 0)$ and density function $\delta(x, y) = xy$.

Solution.   The lamina is shown in Figure 17.6.3. In Example 1 we found the mass of the lamina to be

$$M = \iint\limits_{R} \delta(x, y)\, dA = \iint\limits_{R} xy\, dA = \frac{1}{24}$$

The moment of the lamina about the y-axis is

$$M_y = \iint\limits_{R} x\delta(x, y)\, dA = \iint\limits_{R} x^2 y\, dA = \int_{0}^{1} \int_{0}^{-x+1} x^2 y\, dy\, dx$$

$$= \int_{0}^{1} \left[\frac{1}{2} x^2 y^2\right]_{y=0}^{-x+1} dx = \int_{0}^{1} \left(\frac{1}{2} x^4 - x^3 + \frac{1}{2} x^2\right) dx = \frac{1}{60}$$

and the moment about the x-axis is

$$M_x = \iint\limits_R y\delta(x, y)\, dA = \iint\limits_R xy^2\, dA = \int_0^1 \int_0^{-x+1} xy^2\, dy\, dx$$

$$= \int_0^1 \left[\frac{1}{3}xy^3\right]_{y=0}^{-x+1} dx = \int_0^1 \left(-\frac{1}{3}x^4 + x^3 - x^2 + \frac{1}{3}x\right) dx = \frac{1}{60}$$

From (14b),

$$\bar{x} = \frac{M_y}{M} = \frac{1/60}{1/24} = \frac{2}{5}, \qquad \bar{y} = \frac{M_x}{M} = \frac{1/60}{1/24} = \frac{2}{5}$$

so the center of gravity is $(2/5, 2/5)$. ◀

In the special case of a *homogeneous* lamina the center of gravity is called the **centroid of the lamina** or sometimes the **centroid of the region R.** Because the density function δ is constant for a homogeneous lamina, the factor δ may be moved through the integral signs in (14a) and canceled. Thus, the centroid $(\bar{x}, \bar{y})$ of a region R is given by the formulas

Centroid of a region R

$$\bar{x} = \frac{\displaystyle\iint\limits_R x\, dA}{\displaystyle\iint\limits_R dA} = \frac{1}{\text{area of } R}\iint\limits_R x\, dA \qquad (15a)$$

$$\bar{y} = \frac{\displaystyle\iint\limits_R y\, dA}{\displaystyle\iint\limits_R dA} = \frac{1}{\text{area of } R}\iint\limits_R y\, dA \qquad (15b)$$

▶ Example 4 Find the centroid of the following semicircular region:

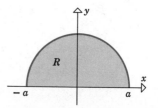

Solution. By symmetry, $\bar{x} = 0$ since the *y*-axis is obviously a line of balance. From (15b),

$$\bar{y} = \frac{\displaystyle\iint_R y \, dA}{\text{area of } R} = \frac{\displaystyle\iint_R y \, dA}{\frac{1}{2}\pi a^2} \tag{16}$$

The integral in (16) is best evaluated in polar coordinates:

$$\iint_R y \, dA = \int_0^\pi \int_0^a (r \sin \theta) r \, dr \, d\theta = \int_0^\pi \left[\frac{1}{3} r^3 \sin \theta \right]_{r=0}^a d\theta$$

$$= \frac{1}{3} a^3 \int_0^\pi \sin \theta \, d\theta = \frac{2}{3} a^3$$

Thus, from (16)

$$\bar{y} = \frac{\frac{2}{3} a^3}{\frac{1}{2} \pi a^2} = \frac{4a}{3\pi}$$

so the centroid is $\left(0, \dfrac{4a}{3\pi} \right)$.   ◄

For three-dimensional solids the formulas for moments, centers of gravity, and centroids are similar to the formulas for laminas. If a solid *G* of mass *M* and volume *V* is homogeneous in composition, we define the ***density*** δ of the solid to be

$$\delta = \frac{M}{V}$$

which measures the mass per unit volume. If the solid *G* is not homogeneous, then the density at a general point (x, y, z) is specified by a ***density function*** $\delta(x, y, z)$ that may be viewed as a limit

$$\delta(x, y, z) = \lim \frac{\Delta M}{\Delta V}$$

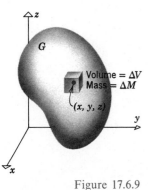

Figure 17.6.9

where ΔM and ΔV represent the mass and volume of a rectangular parallelepiped, centered at (x, y, z), whose dimensions tend to zero (Figure 17.6.9). The following formulas generalize the corresponding results for laminas:

$$\text{mass of } G = \iiint_G \delta(x, y, z) \, dV \tag{17}$$

Center of gravity $(\bar{x}, \bar{y}, \bar{z})$ of a solid G

$$\bar{x} = \frac{\iiint\limits_{G} x\delta(x, y, z)\, dV}{\iiint\limits_{G} \delta(x, y, z)\, dV} = \frac{1}{\text{mass of } G} \iiint\limits_{G} x\delta(x, y, z)\, dV \qquad (18a)$$

$$\bar{y} = \frac{\iiint\limits_{G} y\delta(x, y, z)\, dV}{\iiint\limits_{G} \delta(x, y, z)\, dV} = \frac{1}{\text{mass of } G} \iiint\limits_{G} y\delta(x, y, z)\, dV \qquad (18b)$$

$$\bar{z} = \frac{\iiint\limits_{G} z\delta(x, y, z)\, dV}{\iiint\limits_{G} \delta(x, y, z)\, dV} = \frac{1}{\text{mass of } G} \iiint\limits_{G} z\delta(x, y, z)\, dV \qquad (18c)$$

Centroid $(\bar{x}, \bar{y}, \bar{z})$ of a solid G

$$\bar{x} = \frac{\iiint\limits_{G} x\, dV}{\iiint\limits_{G} dV} = \frac{\iiint\limits_{G} x\, dV}{\text{volume of } G} \qquad (19a)$$

$$\bar{y} = \frac{\iiint\limits_{G} y\, dV}{\iiint\limits_{G} dV} = \frac{\iiint\limits_{G} y\, dV}{\text{volume of } G} \qquad (19b)$$

$$\bar{z} = \frac{\iiint\limits_{G} z\, dV}{\iiint\limits_{G} dV} = \frac{\iiint\limits_{G} z\, dV}{\text{volume of } G} \qquad (19c)$$

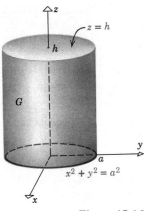

Figure 17.6.9

▶ **Example 5** Find the mass and the center of gravity of the adjacent cylindrical solid, assuming that the density at each point is proportional to the distance between the point and the base of the solid.

Solution. Since the density is proportional to the distance z from the base, the density function has the form $\delta(x, y, z) = kz$ where k is some (unknown) constant of proportionality. From (17) the mass of the solid is

$$M = \iiint_G \delta(x, y, z)\, dV = \int_{-a}^{a} \int_{-\sqrt{a^2-x^2}}^{\sqrt{a^2-x^2}} \int_{0}^{h} kz\, dz\, dy\, dx$$

$$= k \int_{-a}^{a} \int_{-\sqrt{a^2-x^2}}^{\sqrt{a^2-x^2}} \tfrac{1}{2}h^2\, dy\, dx$$

$$= kh^2 \int_{-a}^{a} \sqrt{a^2 - x^2}\, dx$$

$$= \tfrac{1}{2}kh^2\pi a^2 \qquad \begin{bmatrix}\text{See Example 4} \\ \text{of Section 9.5}\end{bmatrix}$$

Without additional information, the constant k cannot be determined. However, as we shall now see, the value of k does not affect the center of gravity.

From (18),

$$\bar{z} = \frac{\displaystyle\iiint_G z\delta(x, y, z)\, dV}{\text{mass of } G} = \frac{\displaystyle\iiint_G z\delta(x, y, z)\, dV}{\tfrac{1}{2}kh^2\pi a^2}$$

But

$$\iiint_G z\delta(x, y, z)\, dV = \int_{-a}^{a} \int_{-\sqrt{a^2-x^2}}^{\sqrt{a^2-x^2}} \int_{0}^{h} z(kz)\, dz\, dy\, dx$$

$$= k \int_{-a}^{a} \int_{-\sqrt{a^2-x^2}}^{\sqrt{a^2-x^2}} \tfrac{1}{3}h^3\, dy\, dx$$

$$= \tfrac{2}{3}kh^3 \int_{-a}^{a} \sqrt{a^2 - x^2}\, dx$$

$$= \tfrac{1}{3}kh^3\pi a^2$$

so that

$$\bar{z} = \frac{\tfrac{1}{3}kh^3\pi a^2}{\tfrac{1}{2}kh^2\pi a^2} = \tfrac{2}{3}h$$

Similar calculations using (18) will yield $\bar{x} = \bar{y} = 0$. However, this is evident by inspection, since it follows from the symmetry of the solid and the form of its density function that the center of gravity is on the z-axis. Thus, the center of gravity is $(0, 0, \frac{2}{3}h)$.   ◄

The next theorem, due to the Greek mathematician Pappus,* gives an interesting and useful relationship between the centroid of a plane region R and the volume of the solid generated when the region is revolved about a line.

17.6.3 THEOREM

Theorem of Pappus

If R is a plane region and L is a line that lies in the plane of R, but does not intersect R, then the volume of the solid formed by revolving R about L is given by

$$volume = (area\ of\ R) \cdot \binom{distance\ traveled}{by\ the\ centroid}$$

Proof. Introduce an xy-coordinate system so that L is along the y-axis and the region R is in the first quadrant (Figure 17.6.10). Let R be partitioned into subregions in the usual way and let R_k be a typical rectangle interior to R. If (x_k^*, y_k^*) is the center of R_k and if the area of R_k is $\Delta A_k = \Delta x_k \Delta y_k$, then the volume generated by R_k as it revolves about L is

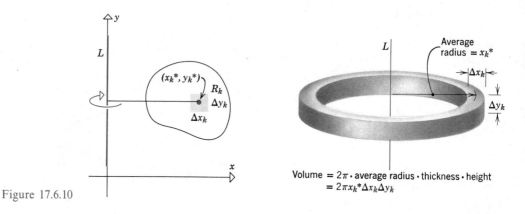

Figure 17.6.10

*PAPPUS OF ALEXANDRIA (4th Century A.D.) Greek mathematician. Pappus lived during the early Christian era when mathematical activity was in a period of decline. His main contributions to mathematics appeared in a series of eight books called *The Collection* (written about 340 A.D.). This work, which survives only partially, contained some original results, but was devoted mostly to statements, refinements, and proofs of results by earlier mathematicians. Pappus' Theorem, stated without proof in Book VII of *The Collection,* was probably known and proved in earlier times. This result is sometimes called Guldin's Theorem in recognition of the Swiss mathematician, Paul Guldin (1577–1643), who rediscovered it independently.

$$2\pi x_k^* \, \Delta x_k \, \Delta y_k = 2\pi x_k^* \, \Delta A_k$$

[See Figure 17.6.10 and (1) of Section 6.3.] Therefore, the total volume of the solid is approximately

$$V \approx \sum_{k=1}^{n} 2\pi x_k^* \, \Delta A_k$$

from which it follows that the exact volume is

$$V = \iint_R 2\pi x \, dA = 2\pi \iint_R x \, dA$$

Thus, from (15a)

$$V = 2\pi \cdot \bar{x} \cdot [\text{area of } R]$$

This completes the proof since $2\pi\bar{x}$ is the distance traveled by the centroid when R is revolved about the y-axis. ▮

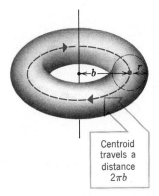

Centroid travels a distance $2\pi b$

Figure 17.6.11

▶ **Example 6** Use Pappus' Theorem to find the volume V of the torus generated by revolving a circular region of radius r about a line at a distance b (greater than r) from the center of the circle (Figure 17.6.11).

Solution. By symmetry, the centroid of a circular region is its center. Thus, the distance traveled by the centroid is $2\pi b$. Since the area of a circle is πr^2, it follows from Pappus' Theorem that the volume of the torus is

$$V = (2\pi b)(\pi r^2) = 2\pi^2 b r^2 \qquad\qquad ◀$$

▶ **Exercise Set 17.6**

1. Where should the fulcrum be placed so the seesaw is in equilibrium?

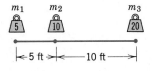

m_1 m_2 m_3
⎡5⎤   ⎡10⎤   ⎡20⎤

|← 5 ft →|←   10 ft   →|

2. A rectangular lamina with density $\delta(x, y) = xy^2$ has vertices $(0, 0), (0, 2), (3, 0), (3, 2)$. Find its mass and center of gravity.

3. A lamina with density $\delta(x, y) = x + y$ is bounded by the x-axis, the line $x = 1$, and the curve $y = \sqrt{x}$. Find its mass and center of gravity.

4. A lamina with density $\delta(x, y) = y$ is bounded by $y = \sin x$, $y = 0$, $x = 0$, and $x = \pi$. Find its mass and center of gravity.

5. A lamina with density $\delta(x, y) = xy$ is in the first quadrant and is bounded by the circle $x^2 + y^2 = a^2$ and the coordinate axes. Find its mass and center of gravity.

6. A lamina with density $\delta(x, y) = x^2 + y^2$ is bounded by the x-axis and the upper half of the circle $x^2 + y^2 = 1$. Find its mass and center of gravity.

In Exercises 7–13, find the centroid of the region.

7. The triangular region bounded by $y = x$, $x = 1$, and the x-axis.

8. The region bounded by $y = x^2$, $x = 1$, and the x-axis.

9. The region enclosed between $y = x$ and $y = 2 - x^2$.

10. The region enclosed by the parabolas $x = 3y^2 - 6y$ and $x = 2y - y^2$.

11. The region above the x-axis and between the circles $x^2 + y^2 = a^2$ and $x^2 + y^2 = b^2$ $(a < b)$.

12. The region enclosed between the y-axis and the right half of the circle $x^2 + y^2 = a^2$.

13. The region enclosed between $y = |x|$ and the line $y = 4$.

14. A rectangular box defined by the inequalities $0 \le x \le 1$, $0 \le y \le 1$, $0 \le z \le 1$ has density $\delta(x, y, z) = 3xyz$. Find its mass and center of gravity.

15. A cube defined by the inequalities $0 \le x \le a$, $0 \le y \le a$, $0 \le z \le a$ has density $\delta(x, y, z) = a - x$. Find its mass and center of gravity.

16. The cylindrical solid enclosed by $x^2 + y^2 = a^2$, $z = 0$, and $z = h$ has density $\delta(x, y, z) = h - z$. Find its mass and center of gravity.

17. The solid enclosed by the surface $z = 1 - y^2$ (where $y \ge 0$) and the planes $z = 0$, $x = -1$, $x = 1$ has density $\delta(x, y, z) = yz$. Find its mass and center of gravity.

18. The solid enclosed by the surface $y = 9 - x^2$ (where $x \ge 0$) and the planes $y = 0$, $z = 0$, and $z = 1$ has density $\delta(x, y, z) = xz$. Find its mass and center of gravity.

In Exercises 19–23, find the centroid of the solid.

19. The tetrahedron in the first octant enclosed by the coordinate planes and the plane $x + y + z = 1$.

20. The solid enclosed by the xy-plane and the hemisphere $z = \sqrt{a^2 - x^2 - y^2}$.

21. The solid bounded by the surface $z = y^2$, and the planes $x = 0$, $x = 1$, and $z = 1$.

22. The solid in the first octant bounded by the surface $z = xy$ and the planes $z = 0$, $x = 2$, and $y = 2$.

23. The solid in the first octant bounded by the sphere $x^2 + y^2 + z^2 = a^2$ and the coordinate planes.

24. Find the center of gravity of the square lamina with vertices $(0, 0)$, $(1, 0)$, $(0, 1)$, and $(1, 1)$ if
    (a) the density is proportional to the square of the distance from the origin;
    (b) the density is proportional to the distance from the y-axis.

25. Find the mass of a circular lamina of radius a whose density at a point is proportional to the distance from the center. (Let k denote the constant of proportionality.)

26. Find the center of gravity of the cube determined by the inequalities $0 \le x \le 1$, $0 \le y \le 1$, $0 \le z \le 1$ if
    (a) the density is proportional to the square of the distance to the origin;
    (b) the density is proportional to the sum of the distances to the faces that lie in the coordinate planes.

27. Show that in polar coordinates the formulas for the centroid $(\bar{x}, \bar{y})$ of a region R are

$$\bar{x} = \frac{1}{\text{area of } R} \iint\limits_{R} r^2 \cos\theta \, dr \, d\theta$$

$$\bar{y} = \frac{1}{\text{area of } R} \iint\limits_{R} r^2 \sin\theta \, dr \, d\theta$$

28. Use the result of Exercise 27 to find the centroid $(\bar{x}, \bar{y})$ of the region enclosed by the cardioid $r = a(1 + \sin\theta)$.

29. Use the result of Exercise 27 to find the centroid $(\bar{x}, \bar{y})$ of the petal of the rose $r = \sin 2\theta$ that lies in the first quadrant.

30. Let R be the rectangle bounded by the lines $x = 0$, $x = 3$, $y = 0$, and $y = 2$. By inspection, find the centroid of R and use it to evaluate

$$\iint\limits_{R} x \, dA \quad \text{and} \quad \iint\limits_{R} y \, dA$$

31. Use the Theorem of Pappus and the fact that the volume of a sphere of radius a is $V = \frac{4}{3}\pi a^3$ to show that the centroid of the lamina bounded by the x-axis and the semicircle $y = \sqrt{a^2 - x^2}$ is $(0, 4a/3\pi)$. (This problem was solved directly in Example 4.)

32. Use the Theorem of Pappus and the result of Exercise 31 to find the volume of the solid generated when the region bounded by the x-axis and the semicircle $y = \sqrt{a^2 - x^2}$ is revolved about
(a) the line $y = -a$
(b) the line $y = x - a$.

33. Use the Theorem of Pappus and the fact that the area of an ellipse with semiaxes a and b is πab to find the volume of the elliptical torus generated by revolving the ellipse

$$\frac{(x - k)^2}{a^2} + \frac{y^2}{b^2} = 1$$

about the y-axis. Assume $k > a$.

34. Use the Theorem of Pappus to find the volume of the solid generated when the region enclosed by $y = x^2$ and $y = 8 - x^2$ is revolved about the x-axis.

35. Use the Theorem of Pappus to find the centroid of the triangular region with vertices $(0, 0)$, $(a, 0)$, and $(0, b)$, where $a > 0$ and $b > 0$. [*Hint:* Revolve the region about the x-axis to obtain $\bar{y}$ and about the y-axis to obtain $\bar{x}$.]

17.7 TRIPLE INTEGRALS IN CYLINDRICAL AND SPHERICAL COORDINATES

In Section 17.3 we saw that some double integrals are easier to evaluate in polar coordinates. Similarly, some triple integrals are easier to evaluate in cylindrical or spherical coordinates. In this section we will study triple integrals in these coordinate systems.

In rectangular coordinates the triple integral of a function f over a solid region G is defined as

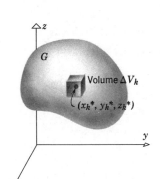

Figure 17.7.1

$$\iiint\limits_{G} f(x, y, z)\, dV = \lim_{n \to +\infty} \sum_{k=1}^{n} f(x_k^*, y_k^*, z_k^*)\, \Delta V_k$$

where ΔV_k denotes the volume of a rectangular parallelepiped interior to G and (x_k^*, y_k^*, z_k^*) is a point in this parallelepiped (Figure 17.7.1). Triple integrals in cylindrical and spherical coordinates are similarly defined, except that the region G is not divided into rectangular parallelepipeds, but into regions more appropriate to these coordinate systems.

In cylindrical coordinates, the simplest equations are of the form

$$r = \text{constant}, \qquad \theta = \text{constant}, \qquad z = \text{constant}$$

As indicated in Figure 15.9.2, the first equation represents a right circular cylinder centered on the z-axis, the second a vertical half plane hinged on the z-axis, and the third a horizontal plane. These surfaces can be paired up to determine solids called *cylindrical wedges* or *cylindrical elements of volume.* To

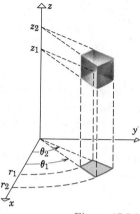

Figure 17.7.2

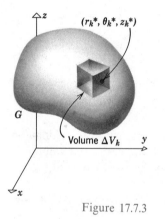

Figure 17.7.3

be precise, a cylindrical wedge is a solid enclosed between six surfaces of the following type:

two cylinders $\quad r = r_1,\ r = r_2 \quad (r_1 < r_2)$

two half planes $\quad \theta = \theta_1,\ \theta = \theta_2 \quad (\theta_1 < \theta_2)$

two planes $\quad z = z_1,\ z = z_2 \quad (z_1 < z_2)$

(Figure 17.7.2). The dimensions $\theta_2 - \theta_1$, $r_2 - r_1$, and $z_2 - z_1$ are called the *central angle*, *thickness*, and *height* of the wedge.

To define the triple integral over G of a function $f(r, \theta, z)$ in cylindrical coordinates, we proceed as follows:

Step 1. Construct a three-dimensional grid consisting of right-circular cylinders centered on the z-axis, half planes hinged on the z-axis, and horizontal planes. The blocks determined by this grid will be cylindrical wedges. Using this grid, divide the region G into subregions and exclude from consideration subregions that contain any portion of the surface of the solid G. This leaves only cylindrical wedges interior to G (Figure 17.7.3). Denote the volumes of these interior cylindrical wedges by

$$\Delta V_1,\ \Delta V_2, \ldots, \Delta V_n$$

Step 2. Choose an arbitrary point in each interior cylindrical wedge, and let these be denoted by

$$(r_1^*, \theta_1^*, z_1^*),\ (r_2^*, \theta_2^*, z_2^*), \ldots, (r_n^*, \theta_n^*, z_n^*)$$

(See Figure 17.7.3).

Step 3. Form the sum

$$\sum_{k=1}^{n} f(r_k^*, \theta_k^*, z_k^*)\, \Delta V_k$$

Step 4. Repeat this process with more and more subdivisions, so that the height, thickness, and central angle of each interior cylindrical wedge approaches zero, and n, the number of such wedges, approaches $+\infty$. Define

$$\iiint\limits_{G} f(r, \theta, z)\, dV = \lim_{n \to +\infty} \sum_{k=1}^{n} f(r_k^*, \theta_k^*, z_k^*)\, \Delta V_k$$

In Section 17.5, we defined simple solids. The following theorem will enable us to evaluate triple integrals in cylindrical coordinates over such solids by repeated integration.

17.7.1 THEOREM *Let G be a simple solid whose upper surface has the equation $z = g_2(r, \theta)$ and whose lower surface has the equation $z = g_1(r, \theta)$ in cylindrical coordinates. If R is the projection of the solid on the xy-plane and if $f(r, \theta, z)$ is continuous on G, then*

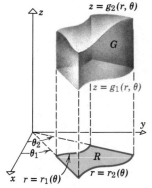

z = g₂(r, θ)

$$\iiint\limits_{G} f(r, \theta, z)\, dV = \iint\limits_{R} \left[\int_{g_1(r, \theta)}^{g_2(r, \theta)} f(r, \theta, z)\, dz \right] dA \qquad (1)$$

where the double integral over R is evaluated in polar coordinates. In particular, if the projection R is as shown in Figure 17.3.2, then (1) can be written

$$\iiint\limits_{G} f(r, \theta, z)\, dV = \int_{\theta_1}^{\theta_2} \int_{r_1(\theta)}^{r_2(\theta)} \int_{g_1(r, \theta)}^{g_2(r, \theta)} f(r, \theta, z)\, r\, dz\, dr\, d\theta \qquad (2)$$

Figure 17.7.4   *The type of solid to which formula (2) applies is illustrated in Figure 17.7.4.*

We will not formally prove Theorem 17.7.1. However, the appearance of the factor r in the integrand of (2) can be explained by returning to the definition

$$\iiint\limits_{G} f(r, \theta, z)\, dV = \lim_{n \to +\infty} \sum_{k=1}^{n} f(r_k^*, \theta_k^*, z_k^*)\, \Delta V_k \qquad (3)$$

In this formula, the volume ΔV_k of the kth cylindrical wedge can be written

$$\Delta V_k = [\text{area of base}] \cdot [\text{height}] \qquad (4)$$

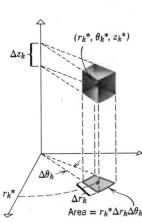

$(r_k^*, \theta_k^*, z_k^*)$

Area = $r_k^* \Delta r_k \Delta \theta_k$

Figure 17.7.5

If we denote the thickness, central angle, and height of this wedge by Δr_k, $\Delta \theta_k$, and Δz_k, and if we choose the arbitrary point $(r_k^*, \theta_k^*, z_k^*)$ to lie above the "center" of the base (Figures 17.3.5 and 17.7.5), then it follows from (4) of Section 17.3 that the base has area $r_k^* \Delta r_k \Delta \theta_k$. Thus, (4) can be written

$$\Delta V_k = r_k^* \Delta r_k \Delta \theta_k \Delta z_k$$

Substituting this expression in (3) yields

$$\iiint\limits_{G} f(r, \theta, z)\, dV = \lim_{n \to +\infty} \sum_{k=1}^{n} f(r_k^*, \theta_k^*, z_k^*) r_k^* \Delta z_k \Delta r_k \Delta \theta_k$$

which explains the form of the integrand in (2).

To apply (1) and (2) it is best to begin with a three-dimensional sketch of the solid G, from which the limits of integration can be obtained as follows:

Step 1. Identify the upper surface $z = g_2(r, \theta)$ and the lower surface $z = g_1(r, \theta)$ of the solid. The functions $g_1(r, \theta)$ and $g_2(r, \theta)$ determine the z-limits of integration. (If the upper and lower surfaces are given in rectangular coordinates, convert them to cylindrical coordinates.)

Step 2. Make a two-dimensional sketch of the projection R of the solid on the xy-plane. From this sketch the r- and θ-limits of integration may be obtained exactly as with double integrals in polar coordinates.

▶ Example 1   Use triple integration in cylindrical coordinates to find the volume and the centroid of the solid G that is bounded above by the hemisphere $z = \sqrt{25 - x^2 - y^2}$, below by the xy-plane, and laterally by the cylinder $x^2 + y^2 = 9$.

Solution. The solid G and its projection R on the xy-plane are shown in Figure 17.7.6. In cylindrical coordinates, the upper surface of G is the hemisphere $z = \sqrt{25 - r^2}$ and the lower surface is the plane $z = 0$. Thus, from (1), the volume of G is

$$V = \iiint_G dV = \iint_R \left[\int_0^{\sqrt{25-r^2}} dz \right] dA$$

For the double integral over R, we use polar coordinates:

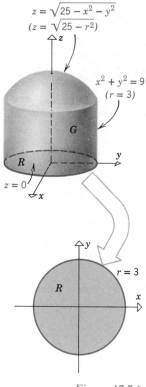

$z = \sqrt{25 - x^2 - y^2}$
$(z = \sqrt{25 - r^2})$

$x^2 + y^2 = 9$
$(r = 3)$

$r = 3$

Figure 17.7.6

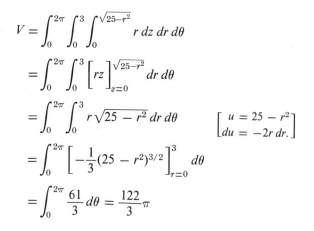

$$V = \int_0^{2\pi} \int_0^3 \int_0^{\sqrt{25-r^2}} r \, dz \, dr \, d\theta$$

$$= \int_0^{2\pi} \int_0^3 \left[rz \right]_{z=0}^{\sqrt{25-r^2}} dr \, d\theta$$

$$= \int_0^{2\pi} \int_0^3 r\sqrt{25 - r^2} \, dr \, d\theta \qquad \left[\begin{array}{l} u = 25 - r^2 \\ du = -2r \, dr. \end{array} \right]$$

$$= \int_0^{2\pi} \left[-\frac{1}{3}(25 - r^2)^{3/2} \right]_{r=0}^{3} d\theta$$

$$= \int_0^{2\pi} \frac{61}{3} \, d\theta = \frac{122}{3}\pi$$

From this result and (19) of Section 17.6,

$$\bar{z} = \frac{1}{V} \iiint\limits_{G} z \, dV = \frac{3}{122\pi} \iiint\limits_{G} z \, dV$$

$$= \frac{3}{122\pi} \iint\limits_{R} \left[\int_0^{\sqrt{25-r^2}} z \, dz \right] dA$$

$$= \frac{3}{122\pi} \int_0^{2\pi} \int_0^3 \int_0^{\sqrt{25-r^2}} zr \, dz \, dr \, d\theta$$

$$= \frac{3}{122\pi} \int_0^{2\pi} \int_0^3 \left[\frac{1}{2} rz^2 \right]_{z=0}^{\sqrt{25-r^2}} dr \, d\theta$$

$$= \frac{3}{244\pi} \int_0^{2\pi} \int_0^3 (25r - r^3) \, dr \, d\theta$$

$$= \frac{3}{244\pi} \int_0^{2\pi} \frac{369}{4} \, d\theta = \frac{1107}{488}$$

By symmetry, the centroid $(\bar{x}, \bar{y}, \bar{z})$ of G lies on the z-axis, so $\bar{x} = \bar{y} = 0$. Thus, the centroid is at the point $(0, 0, 1107/488)$. ◀

Frequently, a triple integral that is hard to evaluate in rectangular coordinates, is easy to evaluate when the integrand and the limits of integration are expressed in cylindrical coordinates. This is especially true when the integrand or the limits of integration involve expressions of the form

$$x^2 + y^2 \quad \text{or} \quad \sqrt{x^2 + y^2}$$

since these simplify to

$$r^2 \quad \text{and} \quad r$$

in cylindrical coordinates.

▶ **Example 2** Use cylindrical coordinates to evaluate

$$\int_{-3}^3 \int_{-\sqrt{9-x^2}}^{\sqrt{9-x^2}} \int_0^{9-x^2-y^2} x^2 \, dz \, dy \, dx$$

Solution. In problems of this type, it is helpful to sketch the region of integration G and its projection R on the xy-plane. From the z-limits of integration, the upper surface of G is the paraboloid $z = 9 - x^2 - y^2$ and the lower surface is the xy-plane, $z = 0$. From the x- and y-limits of integra-

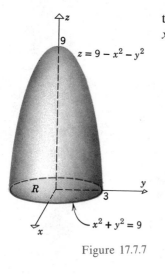

$z = 9 - x^2 - y^2$

R

3

$x^2 + y^2 = 9$

Figure 17.7.7

tion, the projection R is the region in the xy-plane enclosed by the circle $x^2 + y^2 = 9$ (Figure 17.7.7). Thus,

$$\int_{-3}^{3} \int_{-\sqrt{9-x^2}}^{\sqrt{9-x^2}} \int_{0}^{9-x^2-y^2} x^2 \, dz \, dy \, dx = \iiint_{G} x^2 \, dV$$

$$= \iiint_{G} (r \cos \theta)^2 \, dV = \iint_{R} \left[\int_{0}^{9-r^2} r^2 \cos^2 \theta \, dz \right] dA$$

$$= \int_{0}^{2\pi} \int_{0}^{3} \int_{0}^{9-r^2} r^3 \cos^2 \theta \, dz \, dr \, d\theta = \int_{0}^{2\pi} \int_{0}^{3} \left[zr^3 \cos^2 \theta \right]_{z=0}^{9-r^2} dr \, d\theta$$

$$= \int_{0}^{2\pi} \int_{0}^{3} (9r^3 - r^5) \cos^2 \theta \, dr \, d\theta = \int_{0}^{2\pi} \left[\left(\frac{9r^4}{4} - \frac{r^6}{6} \right) \cos^2 \theta \right]_{r=0}^{3} d\theta$$

$$= \frac{243}{4} \int_{0}^{2\pi} \cos^2 \theta \, d\theta = \frac{243}{4} \int_{0}^{2\pi} \frac{1}{2} (1 + \cos 2\theta) \, d\theta = \frac{243\pi}{4} \quad \blacktriangleleft$$

SPHERICAL COORDINATES

In spherical coordinates, the simplest equations are of the form

$$\rho = \text{constant}, \qquad \theta = \text{constant}, \qquad \phi = \text{constant}$$

As indicated in Figure 15.9.5, the first equation represents a sphere centered at the origin, the second a half plane hinged on the z-axis, and the third a right-circular cone with vertex at the origin and centered on the z-axis. By a *spherical wedge* or *spherical element of volume* we mean a solid enclosed between six surfaces of the following form:

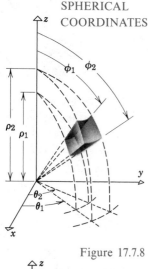

Figure 17.7.8

two spheres	$\rho = \rho_1, \rho = \rho_2$	$(\rho_1 < \rho_2)$
two half planes	$\theta = \theta_1, \theta = \theta_2$	$(\theta_1 < \theta_2)$
two right-circular cones	$\phi = \phi_1, \phi = \phi_2$	$(\phi_1 < \phi_2)$

(Figure 17.7.8). We will refer to the numbers $\rho_2 - \rho_1$, $\theta_2 - \theta_1$, and $\phi_2 - \phi_1$ as the *dimensions* of a spherical wedge.

If G is a solid region in three-dimensional space, then the triple integral over G of a function $f(\rho, \theta, \phi)$ in spherical coordinates is similar in definition to the triple integral in cylindrical coordinates, except that the solid G is partitioned into *spherical* wedges by a three-dimensional grid consisting of spheres centered at the origin, half planes hinged on the z-axis, and right-circular cones with vertices at the origin and opening along the z-axis (Figure 17.7.9). The defining equation of a triple integral in spherical coordinates is

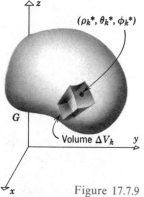

$(\rho_k^*, \theta_k^*, \phi_k^*)$

G

Volume ΔV_k

Figure 17.7.9

$$\iiint_{G} f(\rho, \theta, \phi) \, dV = \lim_{n \to +\infty} \sum_{k=1}^{n} f(\rho_k^*, \theta_k^*, \phi_k^*) \, \Delta V_k \qquad (5)$$

where ΔV_k is the volume of the kth spherical wedge interior to G, $(\rho_k^*, \theta_k^*, \phi_k^*)$ is an arbitrary point in this wedge, and n increases in such a way that the dimensions of each interior spherical wedge tend to zero.

In the exercises we will help the reader to show that if the point $(\rho_k^*, \theta_k^*, \phi_k^*)$ is suitably chosen, then the volume ΔV_k in (5) can be written

$$\Delta V_k = \rho_k^{*2} \sin \phi_k^* \, \Delta \rho_k \, \Delta \phi_k \, \Delta \theta_k$$

where $\Delta \rho_k$, $\Delta \phi_k$, and $\Delta \theta_k$ are the dimensions of the wedge. Substituting this in (5) we obtain

$$\iiint\limits_{G} f(\rho, \theta, \phi) \, dV = \lim_{n \to +\infty} \sum_{k=1}^{n} f(\rho_k^*, \theta_k^*, \phi_k^*) \rho_k^{*2} \sin \phi_k^* \, \Delta \rho_k \, \Delta \phi_k \, \Delta \theta_k$$

which suggests the following formula for evaluating a triple integral in spherical coordinates by repeated integration:

$$\iiint\limits_{G} f(\rho, \theta, \phi) \, dV = \iiint\limits_{\substack{\text{appropriate} \\ \text{limits of} \\ \text{integration}}} f(\rho, \theta, \phi) \rho^2 \sin \phi \, d\rho \, d\phi \, d\theta \qquad (6)$$

REMARK. Note the extra factor of $\rho^2 \sin \phi$ that appears in the integrand of the iterated integral. This is analogous to the extra factor of r that appeared when we integrated in cylindrical coordinates.

In (6) we omitted the limits of integration. Instead, we will give some examples that explain how to obtain such limits. In all our examples we will use the same order of integration: first with respect to ρ, then ϕ, and then θ.

Suppose it is desired to integrate $f(\rho, \theta, \phi)$ over the spherical solid G enclosed by the sphere $\rho = \rho_0$. The basic idea is to choose the limits of integration so that every point of the solid is accounted for in the integration process. Figure 17.7.10 illustrates one way of doing this. Holding θ and ϕ fixed for the first integration, we let ρ vary from 0 to ρ_0. This covers a radial line from the origin to the surface of the sphere. Next, keeping θ fixed, we let ϕ vary from 0 to π so that the radial line sweeps out a fan-shaped region. Finally, we let θ vary from 0 to 2π so that the fan-shaped region makes a complete revolution, thereby sweeping out the entire sphere. Thus, the triple integral of $f(\rho, \theta, \phi)$ over the spherical solid G may be evaluated by writing

$$\iiint\limits_{G} f(\rho, \theta, \phi) \, dV = \int_0^{2\pi} \int_0^{\pi} \int_0^{\rho_0} f(\rho, \theta, \phi) \rho^2 \sin \phi \, d\rho \, d\phi \, d\theta$$

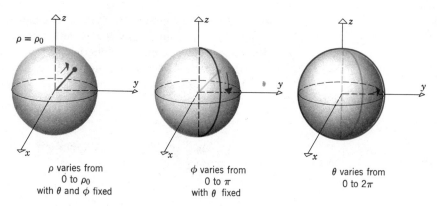

$\rho = \rho_0$

ρ varies from
0 to ρ_0
with θ and ϕ fixed

ϕ varies from
0 to π
with θ fixed

θ varies from
0 to 2π

Figure 17.7.10

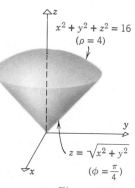

$x^2 + y^2 + z^2 = 16$
$(\rho = 4)$

$z = \sqrt{x^2 + y^2}$
$(\phi = \frac{\pi}{4})$

Figure 17.7.11

Table 17.7.1 (pages 1031–32) suggests how the limits of integration in spherical coordinates can be obtained for some other common solids.

▶ **Example 3** Use spherical coordinates to find the volume and the centroid of the solid G bounded above by the sphere $x^2 + y^2 + z^2 = 16$ and below by the cone $z = \sqrt{x^2 + y^2}$.

Solution. The solid G is sketched in Figure 17.7.11. In spherical coordinates, the equation of the sphere $x^2 + y^2 + z^2 = 16$ is $\rho = 4$ and the equation of the cone $z = \sqrt{x^2 + y^2}$ is

$$\rho \cos \phi = \sqrt{\rho^2 \sin^2 \phi \cos^2 \theta + \rho^2 \sin^2 \phi \sin^2 \theta}$$

which simplifies to

$$\rho \cos \phi = \rho \sin \phi$$

$$\tan \phi = 1$$

$$\phi = \frac{\pi}{4}$$

The volume of G is

$$V = \iiint\limits_{G} dV = \int_0^{2\pi} \int_0^{\pi/4} \int_0^4 \rho^2 \sin \phi \, d\rho \, d\phi \, d\theta$$

$$= \int_0^{2\pi} \int_0^{\pi/4} \left[\frac{\rho^3}{3} \sin \phi \right]_{\rho=0}^4 d\phi \, d\theta$$

$$= \int_0^{2\pi} \int_0^{\pi/4} \frac{64}{3} \sin \phi \, d\phi \, d\theta$$

$$= \frac{64}{3} \int_0^{2\pi} \left[-\cos \phi \right]_{\phi=0}^{\pi/4} d\theta = \frac{64}{3} \int_0^{2\pi} \left(1 - \frac{\sqrt{2}}{2} \right) d\theta$$

By symmetry, the centroid $(\bar{x}, \bar{y}, \bar{z})$ is on the z-axis, so $\bar{x} = \bar{y} = 0$. From (19) of Section 17.6 and the volume calculated above

$$\bar{z} = \frac{1}{V} \iiint\limits_{G} z \, dV = \frac{1}{V} \int_0^{2\pi} \int_0^{\pi/4} \int_0^4 (\rho \cos \phi) \rho^2 \sin \phi \, d\rho \, d\phi \, d\theta$$

$$= \frac{1}{V} \int_0^{2\pi} \int_0^{\pi/4} \left[\frac{\rho^4}{4} \cos \phi \sin \phi \right]_{\rho=0}^4 d\phi \, d\theta$$

$$= \frac{64}{V} \int_0^{2\pi} \int_0^{\pi/4} \sin \phi \cos \phi \, d\phi \, d\theta = \frac{64}{V} \int_0^{2\pi} \left[\frac{1}{2} \sin^2 \phi \right]_{\phi=0}^{\pi/4} d\theta$$

$$= \frac{16}{V} \int_0^{2\pi} d\theta = \frac{32\pi}{V} = \frac{3}{2(2 - \sqrt{2})}$$

With the help of a hand calculator, $\bar{z} \approx 2.56$ (to two decimal places), so the approximate location of the centroid is $(0, 0, 2.56)$. ◄

Table 17.7.1

DETERMINATION OF LIMITS	INTEGRAL
This is the portion of the sphere $\rho = \rho_0$ that lies in the first octant. 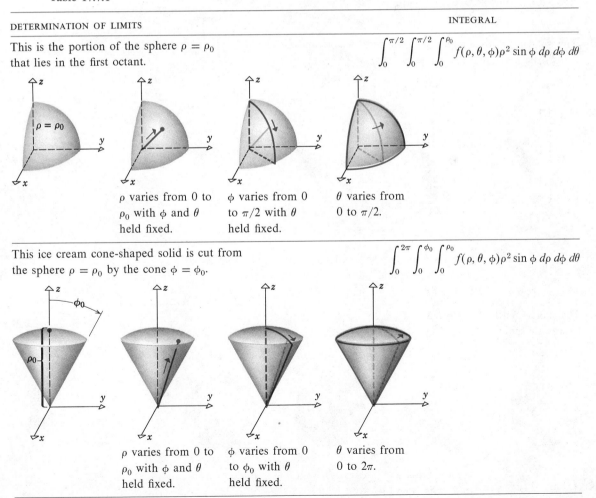 ρ varies from 0 to ρ_0 with ϕ and θ held fixed. ϕ varies from 0 to $\pi/2$ with θ held fixed. θ varies from 0 to $\pi/2$.	$\int_0^{\pi/2} \int_0^{\pi/2} \int_0^{\rho_0} f(\rho, \theta, \phi) \rho^2 \sin \phi \, d\rho \, d\phi \, d\theta$
This ice cream cone-shaped solid is cut from the sphere $\rho = \rho_0$ by the cone $\phi = \phi_0$. ρ varies from 0 to ρ_0 with ϕ and θ held fixed. ϕ varies from 0 to ϕ_0 with θ held fixed. θ varies from 0 to 2π.	$\int_0^{2\pi} \int_0^{\phi_0} \int_0^{\rho_0} f(\rho, \theta, \phi) \rho^2 \sin \phi \, d\rho \, d\phi \, d\theta$

DETERMINATION OF LIMITS	INTEGRAL

This solid is cut from the sphere $\rho = \rho_0$ by two cones $\phi = \phi_1$ and $\phi = \phi_2$.

$$\int_0^{2\pi} \int_{\phi_1}^{\phi_2} \int_0^{\rho_0} f(\rho, \theta, \phi) \rho^2 \sin \phi \, d\rho \, d\phi \, d\theta$$

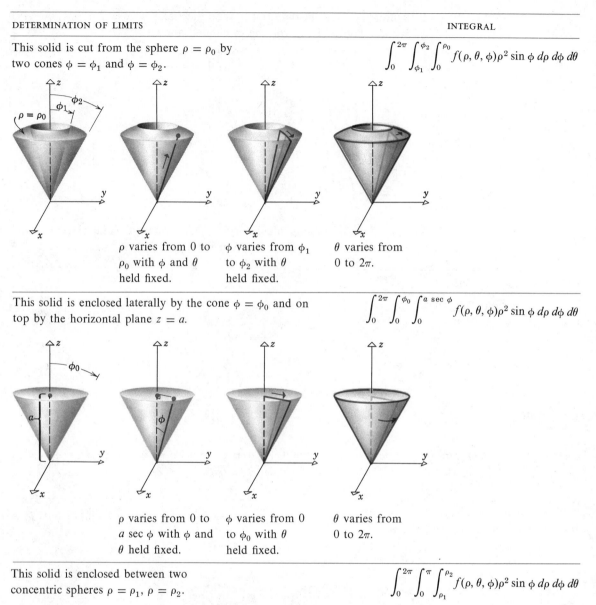

ρ varies from 0 to ρ_0 with ϕ and θ held fixed.

ϕ varies from ϕ_1 to ϕ_2 with θ held fixed.

θ varies from 0 to 2π.

This solid is enclosed laterally by the cone $\phi = \phi_0$ and on top by the horizontal plane $z = a$.

$$\int_0^{2\pi} \int_0^{\phi_0} \int_0^{a \sec \phi} f(\rho, \theta, \phi) \rho^2 \sin \phi \, d\rho \, d\phi \, d\theta$$

ρ varies from 0 to $a \sec \phi$ with ϕ and θ held fixed.

ϕ varies from 0 to ϕ_0 with θ held fixed.

θ varies from 0 to 2π.

This solid is enclosed between two concentric spheres $\rho = \rho_1$, $\rho = \rho_2$.

$$\int_0^{2\pi} \int_0^{\pi} \int_{\rho_1}^{\rho_2} f(\rho, \theta, \phi) \rho^2 \sin \phi \, d\rho \, d\phi \, d\theta$$

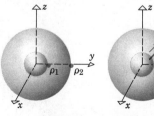
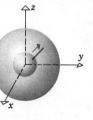
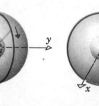

ρ varies from ρ_1 to ρ_2 with ϕ and θ held fixed.

ϕ varies from 0 to π with θ held fixed.

θ varies from 0 to 2π.

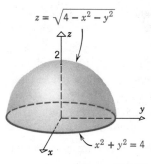

$z = \sqrt{4 - x^2 - y^2}$

$x^2 + y^2 = 4$

Figure 17.7.12

▶ **Example 4**   Use spherical coordinates to evaluate

$$\int_{-2}^{2}\int_{-\sqrt{4-x^2}}^{\sqrt{4-x^2}}\int_{0}^{\sqrt{4-x^2-y^2}} z^2\sqrt{x^2+y^2+z^2}\,dz\,dy\,dx$$

Solution.   In problems like this, it is helpful to begin (when possible) with a sketch of the region G of integration. From the z-limits of integration, the upper surface of G is the hemisphere $z = \sqrt{4 - x^2 - y^2}$, and the lower surface is the xy-plane, $z = 0$. From the x- and y-limits of integration, the projection of the solid G on the xy-plane is the region enclosed by the circle $x^2 + y^2 = 4$. From this information we obtain the sketch of G in Figure 17.7.12. Thus,

$$\int_{-2}^{2}\int_{-\sqrt{4-x^2}}^{\sqrt{4-x^2}}\int_{0}^{\sqrt{4-x^2-y^2}} z^2\sqrt{x^2+y^2+z^2}\,dz\,dy\,dx$$

$$= \iiint_{G} z^2\sqrt{x^2+y^2+z^2}\,dV = \iiint_{G} (\rho^2\cos^2\phi)\rho\,dV$$

$$= \int_{0}^{2\pi}\int_{0}^{\pi/2}\int_{0}^{2} \rho^5\cos^2\phi\sin\phi\,d\rho\,d\phi\,d\theta$$

$$= \int_{0}^{2\pi}\int_{0}^{\pi/2} \frac{32}{3}\cos^2\phi\sin\phi\,d\phi\,d\theta$$

$$= \frac{32}{3}\int_{0}^{2\pi}\left[-\frac{1}{3}\cos^3\phi\right]_{\phi=0}^{\pi/2} d\theta = \frac{32}{9}\int_{0}^{2\pi} d\theta = \frac{64}{9}\pi \quad ◀$$

▶ **Exercise Set 17.7**

In Exercises 1–4, evaluate the iterated integral.

1. $\displaystyle\int_{0}^{2\pi}\int_{0}^{1}\int_{0}^{\sqrt{1-r^2}} zr\,dz\,dr\,d\theta$.

2. $\displaystyle\int_{0}^{\pi/2}\int_{0}^{\cos\theta}\int_{0}^{r^2} r\sin\theta\,dz\,dr\,d\theta$.

3. $\displaystyle\int_{0}^{\pi/2}\int_{0}^{\pi/2}\int_{0}^{1} \rho^3\sin\phi\cos\phi\,d\rho\,d\phi\,d\theta$.

4. $\displaystyle\int_{0}^{2\pi}\int_{0}^{\pi/4}\int_{0}^{a\sec\phi} \rho^2\sin\phi\,d\rho\,d\phi\,d\theta$.

In Exercises 5–9, use cylindrical coordinates to find the volume of the solid.

5. The solid bounded by the paraboloid $z = x^2 + y^2$ and the plane $z = 9$.

6. The solid bounded above and below by the sphere $x^2 + y^2 + z^2 = 9$ and laterally by the cylinder $x^2 + y^2 = 4$.

7. The solid enclosed between the surface $r^2 + z^2 = 20$ and the surface $z = r^2$.

8. The solid bounded by the cone $z = \dfrac{h}{a}r$, and $z = h$.

9. The solid in the first octant below the sphere $x^2 + y^2 + z^2 = 16$ and inside the cylinder $x^2 + y^2 = 4x$.

In Exercises 10–14, use spherical coordinates to find the volume of the solid.

10. The solid bounded above by the sphere $\rho = 4$ and below by the cone $\phi = \pi/3$.

11. The solid in the first octant bounded by the sphere $\rho = 2$, the coordinate planes, and the cones $\phi = \pi/6$ and $\phi = \pi/3$.

12. The solid within the cone $\phi = \pi/4$ and between the spheres $\rho = 1$ and $\rho = 2$.

13. The solid within the sphere $x^2 + y^2 + z^2 = 9$, outside the cone $z = \sqrt{x^2 + y^2}$, and above the xy-plane.

14. The solid bounded by the sphere $x^2 + y^2 + z^2 = 4a^2$ and the planes $z = 0$ and $z = a$.

15. Find the volume of $x^2 + y^2 + z^2 = a^2$ using
    (a) cylindrical coordinates
    (b) spherical coordinates.

In Exercises 16–18, use cylindrical coordinates.

16. Find the mass of the solid bounded by the cone $z = \sqrt{x^2 + y^2}$ and $z = 3$ if the density is $\delta(x, y, z) = 3 - z$.

17. Find the mass of the solid in the first octant bounded above by the paraboloid $z = 4 - x^2 - y^2$, below by the plane $z = 0$, and laterally by the cylinder $x^2 + y^2 = 2x$ and the planes $x = 0$ and $y = 0$ assuming the density to be $\delta(x, y, z) = z$. [Hint: Formula (5) of Section 9.2 and Formula (5) of Section 9.3 will help with the integration.]

18. Find the mass of a right-circular cylinder of radius a and height h if the density is proportional to the distance from the base. (Let k be the constant of proportionality.)

In Exercises 19–21, use spherical coordinates.

19. Find the mass of the solid common to the sphere $x^2 + y^2 + z^2 = 1$ and the cone $z = \sqrt{x^2 + y^2}$ if the density is $\delta(x, y, z) = \sqrt{x^2 + y^2 + z^2}$.

20. Find the mass of the solid enclosed between the spheres $x^2 + y^2 + z^2 = 1$ and $x^2 + y^2 + z^2 = 4$ if the density is $\delta(x, y, z) = (x^2 + y^2 + z^2)^{-1/2}$.

21. Find the mass of a spherical solid of radius a if the density is proportional to the distance from the center. (Let k be the constant of proportionality.)

In Exercises 22–24, use cylindrical coordinates to find the centroid of the solid.

22. The solid bounded by the cone $z = \sqrt{x^2 + y^2}$ and the plane $z = 2$.

23. The solid bounded above by the sphere $x^2 + y^2 + z^2 = 2$ and below by the paraboloid $z = x^2 + y^2$.

24. The solid bounded above by the paraboloid $z = x^2 + y^2$, below by the plane $z = 0$, and laterally by the cylinder $(x - 1)^2 + y^2 = 1$. [Hint: Formula (5) of Section 9.2 and Formula (5) of Section 9.3 will help with the integration.]

In Exercises 25–27, use spherical coordinates to find the centroid.

25. The solid in the first octant bounded by the coordinate planes and the sphere $x^2 + y^2 + z^2 = a^2$.

26. The solid bounded above by the sphere $\rho = 4$ and below by the cone $\phi = \pi/3$.

27. The solid that is enclosed by the hemispheres $y = \sqrt{9 - x^2 - z^2}$, $y = \sqrt{4 - x^2 - z^2}$, and the plane $y = 0$.

In Exercises 28–31, use cylindrical or spherical coordinates to evaluate the integral.

28. $\displaystyle\int_0^a \int_0^{\sqrt{a^2 - x^2}} \int_0^{a^2 - x^2 - y^2} x^2 \, dz \, dy \, dx.$

29. $\displaystyle\int_{-1}^1 \int_0^{\sqrt{1 - x^2}} \int_0^{\sqrt{1 - x^2 - y^2}} e^{-(x^2 + y^2 + z^2)^{3/2}} \, dz \, dy \, dx.$

30. $\displaystyle\int_0^2 \int_0^{\sqrt{4 - y^2}} \int_{\sqrt{x^2 + y^2}}^{\sqrt{8 - x^2 - y^2}} z^2 \, dz \, dx \, dy.$

31. $\displaystyle\int_{-3}^3 \int_{-\sqrt{9 - y^2}}^{\sqrt{9 - y^2}} \int_{-\sqrt{9 - x^2 - y^2}}^{\sqrt{9 - x^2 - y^2}} \sqrt{x^2 + y^2 + z^2} \, dz \, dx \, dy.$

32. Let G be the solid in the first octant bounded by the sphere $x^2 + y^2 + z^2 = 4$ and the coordinate planes. Evaluate

$$\iiint_G xyz \, dV$$

using
(a) rectangular coordinates
(b) cylindrical coordinates
(c) spherical coordinates.

Solve Exercises 33–38 using either cylindrical or spherical coordinates, whichever seems appropriate.

33. Find the center of gravity of the solid hemisphere bounded by $z = \sqrt{a^2 - x^2 - y^2}$ and $z = 0$ if the

density is proportional to the distance from the origin.

34. Find the center of gravity of the solid in the first octant bounded by the cylinder $x^2 + y^2 = a^2$, the coordinate planes, and the plane $z = a$ if the density is $\delta(x, y, z) = xyz$.

35. Find the center of gravity of the solid bounded by the paraboloid $z = 1 - x^2 - y^2$ and the xy-plane if the density is $\delta(x, y, z) = x^2 + y^2 + z^2$.

36. Find the center of gravity of the solid bounded by the cylinder $x^2 + y^2 = 1$, the cone $z = \sqrt{x^2 + y^2}$, and the xy-plane if the density is $\delta(x, y, z) = z$.

37. Find the volume common to the spheres $x^2 + y^2 + z^2 = 9$ and $x^2 + y^2 + (z - 2)^2 = 4$.

38. Suppose the density at a point on a spherical planet is assumed to be

$$\delta = \delta_0 e^{[(\rho/R^3)-1]}$$

where δ_0 is a constant, R is the radius of the planet, and ρ is the distance from the point to the planet's center. Calculate the mass of the planet.

39. In this exercise we will obtain a formula for the volume of the spherical wedge in Figure 17.7.8.

(a) Use a triple integral in cylindrical coordinates to show that the volume of the solid bounded above by a sphere $\rho = \rho_0$, below by a cone $\phi = \phi_0$, and on the sides by $\theta = \theta_1$ and $\theta = \theta_2$ ($\theta_1 < \theta_2$) is

$$V = \frac{1}{3}\rho_0{}^3(1 - \cos\phi_0)(\theta_2 - \theta_1)$$

[*Hint:* In cylindrical coordinates, the sphere has the equation $r^2 + z^2 = \rho_0{}^2$ and the cone has the equation $z = r\cot\phi_0$. For simplicity, consider only the case $0 < \phi_0 < \pi/2$.]

(b) Subtract appropriate volumes and use the result in (a) to deduce that the volume ΔV of the spherical wedge is

$$\Delta V = \frac{\rho_2{}^3 - \rho_1{}^3}{3}(\cos\phi_1 - \cos\phi_2)(\theta_2 - \theta_1)$$

(c) Apply the Mean-Value Theorem to the functions $\cos\phi$ and ρ^3 to deduce that the formula in (b) can be written as

$$\Delta V = \rho^{*2}\sin\phi^* \, \Delta\rho \, \Delta\phi \, \Delta\theta$$

where ρ^* is between ρ_1 and ρ_2, ϕ^* is between ϕ_1 and ϕ_2, and $\Delta\rho = \rho_2 - \rho_1$, $\Delta\phi = \phi_2 - \phi_1$, $\Delta\theta = \theta_2 - \theta_1$.

▶ SUPPLEMENTARY EXERCISES

In Exercises 1–4, evaluate the iterated integrals.

1. $\displaystyle\int_{1/2}^{1}\int_{0}^{2x} \cos(\pi x^2)\, dy\, dx.$

2. $\displaystyle\int_{0}^{2}\int_{-y}^{2y} xe^{y^3}\, dx\, dy.$

3. $\displaystyle\int_{-1}^{0}\int_{0}^{y^2}\int_{xy}^{1} 2y\, dz\, dx\, dy.$

4. $\displaystyle\int_{0}^{1}\int_{0}^{z}\int_{0}^{\sqrt{yz}} x\, dx\, dy\, dz.$

In Exercises 5 and 6, express the iterated integral as an equivalent integral with the order of integration reversed.

5. $\displaystyle\int_{0}^{2}\int_{0}^{x/2} e^x e^y\, dy\, dx.$

6. $\displaystyle\int_{0}^{\pi}\int_{y}^{\pi} \frac{\sin x}{x}\, dx\, dy.$

7. Use a double integral to find the area of the region bounded by $y = 2x^3$, $2x + y = 4$, and the x-axis.

8. Use a double integral to find the area of the region bounded by $x = y^2$ and $x = 4y - y^2$.

9. Sketch the region R whose area is given by the iterated integral.

(a) $\displaystyle\int_{0}^{\pi/2} \int_{\tan(x/2)}^{\sin x} dy\, dx.$

(b) $\displaystyle\int_{\pi/6}^{\pi/2} \int_{a}^{a(1+\cos\theta)} r\, dr\, d\theta \quad (a > 0).$

In Exercises 10–12, evaluate the double integral over R using either rectangular or polar coordinates.

10. $\displaystyle\iint\limits_{R} xy\, dA$; R is the region bounded by $y = \sqrt{x}$, $y = 2 - \sqrt{x}$, and the y-axis.

11. $\displaystyle\iint\limits_{R} x^2 \sin y^2\, dA$; R is the region bounded by $y = x^3$, $y = -x^3$, and $y = 8$.

12. $\displaystyle\iint\limits_{R} (4 - x^2 - y^2)\, dA$; R is the sector in the first quadrant bounded by the circle $x^2 + y^2 = 4$ and the coordinate axes.

In Exercises 13–15, use a double integral in rectangular or polar coordinates to find the volume of the solid.

13. The solid in the first octant bounded by the coordinate planes and the plane $3x + 2y + z = 6$.

14. The solid enclosed by the cylinders $y = 3x + 4$ and $y = x^2$, and such that $0 \le z \le \sqrt{y}$.

15. The solid $G = \{(x, y, z): 2 \le x^2 + y^2 \le 4$ and $0 \le z \le 1/(x^2 + y^2)^2\}$.

16. Convert to polar coordinates and evaluate:

$$\int_{0}^{\sqrt{2}} \int_{x}^{\sqrt{4-x^2}} 4xy\, dy\, dx.$$

17. Convert to rectangular coordinates and evaluate:

$$\int_{0}^{\pi/2} \int_{0}^{2a \sin\theta} r \sin 2\theta\, dr\, d\theta.$$

In Exercises 18–20, find the area of the region using a double integral in polar coordinates.

18. The region outside the circle $r = \sqrt{2}\,a$ and inside the lemniscate $r^2 = 4a^2 \cos 2\theta$.

19. The region enclosed by the rose $r = \cos 3\theta$.

20. The region inside the circle $r = 2\sqrt{3} \sin\theta$ and outside the circle $r = 3$.

In Exercises 21–23, find the area of the surface described.

21. The part of the paraboloid $z = 3x^2 + 3y^2 - 3$ below the xy-plane.

22. The part of the plane $2x + 2y + z = 7$ in the first octant.

23. The part of the cone $z^2 = x^2 + y^2$ between the planes $z = 1$ and $z = 4$.

24. Evaluate $\displaystyle\iiint\limits_{G} x^2 yz\, dV$, where G is the set of points satisfying $0 \le x \le 2$, $-x \le y \le x^2$, $0 \le z \le x + y$.

25. Evaluate $\displaystyle\iiint\limits_{G} \sqrt{x^2 + y^2}\, dV$, where G is the set of points satisfying $x^2 + y^2 \le 16$, $0 \le z \le 4 - y$.

26. If $G = \{(x, y, z): x^2 + y^2 \le z \le 4x\}$, express the volume of G as a triple integral in (a) rectangular coordinates (b) cylindrical coordinates.

27. In each part find an equivalent integral of the form

$$\iiint\limits_{G} (\quad)\, dx\, dz\, dy.$$

(a) $\displaystyle\int_{0}^{1} \int_{0}^{(1-x)/2} \int_{0}^{1-x-2y} z\, dz\, dy\, dx$

(b) $\displaystyle\int_{0}^{2} \int_{x^2}^{4} \int_{0}^{4-y} 3\, dz\, dy\, dx.$

28. (a) Change to cylindrical coordinates and then evaluate:

$$\int_{-2}^{2} \int_{-\sqrt{4-x^2}}^{\sqrt{4-x^2}} \int_{(x^2+y^2)^2}^{16} x^2\, dz\, dy\, dx$$

(b) Change to spherical coordinates and evaluate:

$$\int_0^1 \int_0^{\sqrt{1-x^2}} \int_0^{\sqrt{1-x^2-y^2}} \frac{1}{1 + x^2 + y^2 + z^2}\, dz\, dy\, dx.$$

29. If G is the region bounded above by the sphere $\rho = a$ and below by the cone $\phi = \pi/3$, express

$$\iiint_G (x^2 + y^2)\, dV$$ as an iterated integral in (a) spherical coordinates, (b) cylindrical coordinates, (c) Cartesian coordinates.

In Exercises 30–33, find the volume of G.

30. $G = \{(r, \theta, z): 0 \le r \le 2 \sin \theta,\ 0 \le z \le r \sin \theta \}$.

31. G is the solid enclosed by the "inverted apple" $\rho = a(1 + \cos \phi)$.

32. G is the solid enclosed between the surfaces $x = y^2 + z^2$ and $x = 1 - y^2$.

33. G is the solid bounded below by the upper nappe of the cone $\phi = \pi/6$ and above by the plane $z = a$.

In Exercises 34–36, find the centroid $(\bar{x}, \bar{y})$ of the plane region R.

34. The region R is the upper half of the ellipse $(x/a)^2 + (y/b)^2 = 1$.

35. The region R is enclosed by the cardioid $r = a(1 + \sin \theta)$.

36. The region R is bounded by $y^2 = 4x$ and $y^2 = 8(x - 2)$.

In Exercises 37 and 38, find the center of gravity of the lamina with density δ.

37. The triangular lamina with vertices $(a, 0)$, $(-a, 0)$, and $(0, b)$, where $a > 0$ and $b > 0$; and $\delta(x, y)$ is proportional to the distance from (x, y) to the y-axis.

38. The lamina enclosed by the circle $r = 3 \cos \theta$, but outside the cardioid $r = 1 + \cos \theta$, and with $\delta(r, \theta)$ proportional to the distance from (r, θ) to the x-axis.

In Exercises 39 and 40, find the mass of the solid G if its density is δ.

39. The solid G is the part of the first octant under the plane $x/a + y/b + z/c = 1$, where a, b, c are positive, and $\delta(x, y, z) = kz$.

40. The spherical solid G is bounded by $\rho = a$ and $\delta(x, y, z)$ is twice the distance from (x, y, z) to the origin.

In Exercises 41–43, find the centroid of G.

41. The solid G bounded by $y = x^2, y = 4, z = 0$, and $y + z = 4$.

42. The solid G is the part of the sphere $\rho \le a$ lying within the cone $\phi \le \phi_0$, where $\phi_0 \le \pi/2$.

43. The solid G is bounded by the cone with vertex $(0, 0, h)$, and base $x^2 + y^2 \le R^2$ in the xy-plane.

18 topics in vector calculus

18.1 LINE INTEGRALS

In previous chapters we considered three kinds of integrals in rectangular coordinates: definite integrals over intervals, double integrals over two-dimensional regions, and triple integrals over three-dimensional regions. In this section we will discuss *line integrals*, which are integrals over *curves* in two- or three-dimensional space.

18.1.1 DEFINITION Let C be a smooth plane curve given parametrically by the equations

$$\begin{aligned} x &= x(t) \\ y &= y(t) \end{aligned}, \qquad a \le t \le b$$

and let $f(x, y)$ and $g(x, y)$ be continuous on some open region containing the curve C. Then we define

$$\int_C f(x, y)\, dx = \int_a^b f(x(t), y(t)) x'(t)\, dt \tag{1}$$

$$\int_C g(x, y)\, dy = \int_a^b g(x(t), y(t)) y'(t)\, dt \tag{2}$$

The symbol $\int_C f(x, y)\, dx$ is called the **line integral of f along C with respect to x** and $\int_C g(x, y)\, dy$ is called the **line integral of g along C with respect to y.**

REMARK. In words, the line integrals $\int_C f(x, y)\, dx$ and $\int_C g(x, y)\, dy$ are evaluated by replacing x and y by their expressions in terms of t, writing dx and

dy as $dx = x'(t)\, dt$ and $dy = y'(t)\, dt$, and then integrating in the usual way over the interval $a \le t \le b$.

Frequently, the line integrals in (1) and (2) occur in combination, in which case we dispense with one of the integral signs and write:

$$\int_C f(x, y)\, dx + g(x, y)\, dy = \int_C f(x, y)\, dx + \int_C g(x, y)\, dy \tag{3}$$

The expression $f(x, y)\, dx + g(x, y)\, dy$ is called a **differential form.**

▶ Example 1   Evaluate

$$\int_C 2xy\, dx + (x^2 + y^2)\, dy$$

over the circular arc C given by $x = \cos t,\ y = \sin t,\ 0 \le t \le \pi/2$ (Figure 18.1.1).

Solution.   From (1) and (2)

$$\int_C 2xy\, dx = \int_0^{\pi/2} (2 \cos t \sin t) \left[\frac{d}{dt} (\cos t) \right] dt$$

$$= -2 \int_0^{\pi/2} \sin^2 t \cos t\, dt = -\frac{2}{3} \sin^3 t \Big]_0^{\pi/2} = -\frac{2}{3}$$

$$\int_C (x^2 + y^2)\, dy = \int_0^{\pi/2} (\cos^2 t + \sin^2 t) \left[\frac{d}{dt} (\sin t) \right] dt$$

$$= \int_0^{\pi/2} \cos t\, dt = \sin t \Big]_0^{\pi/2} = 1$$

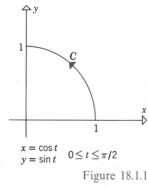

$x = \cos t$
$y = \sin t$ $0 \le t \le \pi/2$

Figure 18.1.1

Thus, from (3)

$$\int_C 2xy\, dx + (x^2 + y^2)\, dy = \int_C 2xy\, dx + \int_C (x^2 + y^2)\, dy$$

$$= -\frac{2}{3} + 1 = \frac{1}{3} \qquad ◀$$

Since the parametric equations of a curve are used to evaluate line integrals over that curve, it seems possible that two different parametrizations of a curve C might produce different values for the same line integral over C. Fortunately, this is not the case.

18.1.2   THEOREM
Independence of
Parametrization

If two different sets of parametric equations represent the same curve C, and if the curve is traced in the same direction in both cases, then the two sets of equations produce the same value for a given line integral over C.

18.1.3 THEOREM
Reversal of Orientation

If two different sets of parametric equations represent the same curve C, but the curve is traced in opposite directions in the two cases, then the two sets of equations produce values for a line integral over C that are negatives of one another.

We omit the proofs. Instead, we will illustrate these facts with some examples.

▶ **Example 2** Evaluate

$$\int_C 2xy\, dx + (x^2 + y^2)\, dy$$

over the curve C given parametrically by

(a) $x = \cos 2t,\ y = \sin 2t$, where $0 \le t \le \dfrac{\pi}{4}$

(b) $x = \cos(t^2),\ y = \sin(t^2)$, where $0 \le t \le \sqrt{\pi/2}$

(c) $x = \cos(-t),\ y = \sin(-t)$, where $\dfrac{3\pi}{2} \le t \le 2\pi$

(d) $x = t,\ y = \sqrt{1 - t^2}$, where $0 \le t \le 1$

Solution. In parts (a) and (b), the curve C is the quarter circle shown in Figure 18.1.1, traced in the counterclockwise direction. Thus, from Example 1 and Theorem 18.1.2,

$$\int_C 2xy\, dx + (x^2 + y^2)\, dy = \frac{1}{3}$$

in both cases. In parts (c) and (d), the curve is the same quarter circle, but traced in the clockwise direction. Thus,

$$\int_C 2xy\, dx + (x^2 + y^2)\, dy = -\frac{1}{3}$$

These results should be verified by direct evaluation, using (1), (2), and (3). ◀

If a curve C is traced in a certain direction, then the same curve traced in the opposite direction is often denoted by the symbol $-C$. Thus, it follows from Theorem 18.1.3 that

$$\int_{-C} f(x, y)\, dx + g(x, y)\, dy = -\int_C f(x, y)\, dx + g(x, y)\, dy \qquad (4)$$

In the definition of a line integral, we required the curve C to be smooth. However, the definition can be extended to curves formed by finitely many smooth curves $C_1, C_2, \ldots, C_n$ joined end to end. Such a curve is called **piecewise smooth** (Figure 18.1.2). We define a line integral over a piecewise smooth curve C to be the sum of the integrals over the pieces:

$$\int_C = \int_{C_1} + \int_{C_2} + \cdots + \int_{C_n}$$

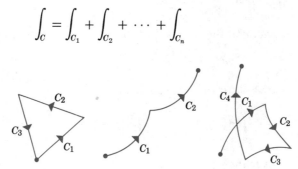

Figure 18.1.2

▶ Example 3   Evaluate

$$\int_C x^2 y \, dx + x \, dy$$

in a counterclockwise direction around the triangular path shown in Figure 18.1.3.

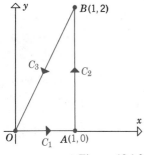

Figure 18.1.3

Solution.   We will integrate over C_1, C_2, and C_3 separately and add the results. For each of the three integrals we must find parametric equations that trace the path of integration in the correct direction.

The horizontal line segment C_1 can be represented as follows using $x = t$ as the parameter:

$$\begin{aligned} x &= t \\ y &= 0 \end{aligned} \quad , \ 0 \le t \le 1$$

The vertical line segment C_2 can be represented as follows using $y = t$ as the parameter:

$$\begin{aligned} x &= 1 \\ y &= t \end{aligned} \quad , \ 0 \le t \le 2$$

To find parametric equations for C_3 that trace C_3 in the direction shown, consider the vector $\overrightarrow{BO} = -\mathbf{i} - 2\mathbf{j}$. The line segment C_3 forms a portion of the line that is parallel to this vector and passes through the point $B(1, 2)$. From the two-dimensional analog of Theorem 15.2.1 (see Exercise 40 of Section 14.1), this line has parametric equations $x = 1 - t$, $y = 2 - 2t$. Since these equations yield the coordinates of B when $t = 0$ and of O when $t = 1$, it follows that a set of parametric equations for C_3 is

$$\begin{aligned} x &= 1 - t \\ y &= 2 - 2t \end{aligned}, \quad 0 \leq t \leq 1$$

We can now carry out the calculations.

$$\int_{C_1} x^2 y \, dx + x \, dy = \int_0^1 (t^2)(0) \frac{d}{dt}[t] \, dt + \int_0^1 (t) \frac{d}{dt}[0] \, dt = 0$$

$$\int_{C_2} x^2 y \, dx + x \, dy = \int_0^2 (1^2)(t) \frac{d}{dt}[1] \, dt + \int_0^2 (1) \frac{d}{dt}[t] \, dt = 0 + 2 = 2$$

$$\int_{C_3} x^2 y \, dx + x \, dy = \int_0^1 (1 - t)^2 (2 - 2t) \frac{d}{dt}[1 - t] \, dt$$

$$+ \int_0^1 (1 - t) \frac{d}{dt}[2 - 2t] \, dt$$

$$= 2 \int_0^1 (t - 1)^3 \, dt + 2 \int_0^1 (t - 1) \, dt$$

$$= -\frac{1}{2} - 1 = -\frac{3}{2}$$

Thus,

$$\int_C x^2 y \, dx + x \, dy = 0 + 2 + \left(-\frac{3}{2}\right) = \frac{1}{2} \qquad \blacktriangleleft$$

REMARK. The procedure used in the last example to find parametric equations for C_3 is worth noting. In general, the line segment between two points P_1 and P_2 traced in the direction from P_1 to P_2 can be represented parametrically by using the vector $\overrightarrow{P_1 P_2}$ and the point P_1 to find parametric equations for the line through P_1 and P_2, and then restricting t to the interval $0 \leq t \leq 1$.

For many applications it is helpful to express the line integral

$$\int_C f(x, y) \, dx + g(x, y) \, dy \tag{5}$$

in vector notation. To do this, we introduce the vector-valued function

$$\mathbf{F}(x, y) = f(x, y)\mathbf{i} + g(x, y)\mathbf{j} \tag{6}$$

and express the curve C: $x = x(t)$, $y = y(t)$ in terms of its position vector

$$\mathbf{r}(t) = x(t)\mathbf{i} + y(t)\mathbf{j}$$

If we formally multiply both sides of the equation

$$\frac{d\mathbf{r}}{dt} = \frac{dx}{dt}\mathbf{i} + \frac{dy}{dt}\mathbf{j}$$

by dt, we are led to the following convenient notation

$$d\mathbf{r} = \frac{d\mathbf{r}}{dt}\,dt = dx\mathbf{i} + dy\mathbf{j} \tag{7}$$

If we now calculate the dot product of (6) and (7), we obtain

$$\mathbf{F}(x, y) \cdot d\mathbf{r} = f(x, y)\,dx + g(x, y)\,dy$$

which is precisely the integrand in (5). This suggests the notation

$$\int_C f(x, y)\,dx + g(x, y)\,dy = \int_C \mathbf{F}(x, y) \cdot d\mathbf{r} \tag{8}$$

Moreover, the equation

$$\int_C f(x, y)\,dx + g(x, y)\,dy = \int_a^b \left[f(x(t), y(t))\,\frac{dx}{dt} + g(x(t), y(t))\,\frac{dy}{dt} \right] dt$$

can be written as

$$\int_C \mathbf{F}(x, y) \cdot d\mathbf{r} = \int_a^b \left[\mathbf{F}(x(t), y(t)) \cdot \frac{d\mathbf{r}}{dt} \right] dt \tag{9}$$

WORK   One of the most important applications of line integrals occurs in the study of work. Recall that if a constant force $\mathbf{F}$ acts on a particle moving along a straight line from a point P to a point Q, then the work done by $\mathbf{F}$ is

$$W = \mathbf{F} \cdot \overrightarrow{PQ} = (\|\mathbf{F}\| \cos \theta)\|\overrightarrow{PQ}\| \tag{10}$$

where $\|\mathbf{F}\| \cos \theta$ is the component of force in the direction of motion and $\|\overrightarrow{PQ}\|$ is the distance traveled by the particle [see (13), Section 15.3].

This elementary notion of work is inadequate for many applications. For example, if a rocket moves through the earth's gravitational field, the path is usually a curve rather than a straight line; and the force of gravity on the rocket does not remain constant, but changes with the position of the rocket relative to the earth's center. Thus, we must consider the more general notion of work done on a particle moving along a curve C while subject to a force $\mathbf{F}(x, y)$ that varies from point to point along the curve (Figure 18.1.4).

Let us assume that C is a smooth curve given parametrically by the equations

$$x = x(t) \atop y = y(t), \quad a \le t \le b$$

and that the force vector at a point (x, y) on the curve is

$$\mathbf{F}(x, y) = f(x, y)\mathbf{i} + g(x, y)\mathbf{j}$$

where $f(x, y)$ and $g(x, y)$ are continuous functions.

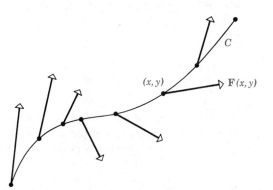

Figure 18.1.4

Let $\mathbf{r}(t) = x(t)\mathbf{i} + y(t)\mathbf{j}$ $(a \le t \le b)$ denote the position vector for C, and let us define the function $W(t)$ by

$$W(t) = \begin{bmatrix} \text{the work done as the particle moves from} \\ \text{the initial point } \mathbf{r}(a) \text{ to the point } \mathbf{r}(t) \end{bmatrix}$$

(See Figure 18.1.5). Thus, the work done as the particle moves from $\mathbf{r}(a)$ to $\mathbf{r}(t + \Delta t)$ is $W(t + \Delta t)$, and the work done as the particle moves from $\mathbf{r}(t)$ to $\mathbf{r}(t + \Delta t)$ is $W(t + \Delta t) - W(t)$.

If Δt is small, then we may reasonably approximate the portion of curve between $\mathbf{r}(t)$ and $\mathbf{r}(t + \Delta t)$ by a straight line. If the force vector $\mathbf{F}$ does not vary much over this small section of curve, we may assume the force to be constant between $\mathbf{r}(t)$ and $\mathbf{r}(t + \Delta t)$ and equal to its value at $\mathbf{r}(t)$. Thus, (10) implies that the work done as the particle moves from $\mathbf{r}(t)$ to $\mathbf{r}(t + \Delta t)$ can be approximated by

$$W(t + \Delta t) - W(t) \approx \mathbf{F}\big(x(t), y(t)\big) \cdot [\mathbf{r}(t + \Delta t) - \mathbf{r}(t)]$$

if Δt is small. If we assume that the error in this approximation diminishes to zero as $\Delta t \to 0$, then it follows on dividing by Δt and taking the limit that

$$\lim_{\Delta t \to 0} \frac{W(t + \Delta t) - W(t)}{\Delta t} = \lim_{\Delta t \to 0} \mathbf{F}\big(x(t), y(t)\big) \cdot \frac{\mathbf{r}(t + \Delta t) - \mathbf{r}(t)}{\Delta t}$$

or

$$W'(t) = \mathbf{F}\big(x(t), y(t)\big) \cdot \mathbf{r}'(t) \tag{11}$$

Figure 18.1.5 — $W(t) = $ Work done in moving the particle from $\mathbf{r}(a)$ to $\mathbf{r}(t)$

Since $W(t)$ represents the total work done as the particle moves from $\mathbf{r}(a)$ to $\mathbf{r}(t)$, it follows that

$$W(a) = 0$$

Thus, the total work W done as the particle moves from $\mathbf{r}(a)$ to $\mathbf{r}(b)$ can be written

$$W = W(b) = W(b) - W(a) = \int_a^b W'(t)\, dt$$

or from (11)

$$W = \int_a^b \mathbf{F}\big(x(t), y(t)\big) \cdot \mathbf{r}'(t)\, dt$$

Since this is precisely the line integral over C of the force $\mathbf{F}$ [see (9)], we are led to the following definition of the work W done by a force $\mathbf{F}(x, y)$ acting on a particle as it moves along a curve C:

$$W = \int_C \mathbf{F}(x, y) \cdot d\mathbf{r} \tag{12}$$

▶ **Example 4** Find the work done if a particle moves from $(-2, 4)$ to $(1, 1)$ along the parabola $y = x^2$, while subject to the force

$$\mathbf{F}(x, y) = x^3 y \mathbf{i} + (x - y)\mathbf{j}$$

Solution. If we use $x = t$ as the parameter, the path of the particle is represented by

$$\begin{aligned} x &= t \\ y &= t^2 \end{aligned}, \quad -2 \le t \le 1$$

or

$$\mathbf{r}(t) = t\mathbf{i} + t^2 \mathbf{j} \quad (-2 \le t \le 1)$$

Since $\mathbf{F}(x, y) = x^3 y \mathbf{i} + (x - y)\mathbf{j}$, it follows that

$$\mathbf{F}\big(x(t), y(t)\big) = (t^3)(t^2)\mathbf{i} + (t - t^2)\mathbf{j} = t^5 \mathbf{i} + (t - t^2)\mathbf{j}$$

Thus, from (12) and (9)

$$W = \int_C \mathbf{F}(x, y) \cdot d\mathbf{r} = \int_a^b \left[\mathbf{F}\big(x(t), y(t)\big) \cdot \frac{d\mathbf{r}}{dt} \right] dt$$

$$= \int_{-2}^{1} [t^5\mathbf{i} + (t - t^2)\mathbf{j}] \cdot [\mathbf{i} + 2t\mathbf{j}] \, dt$$

$$= \int_{-2}^{1} (t^5 + 2t^2 - 2t^3) \, dt$$

$$= \frac{1}{6}t^6 + \frac{2}{3}t^3 - \frac{1}{2}t^4 \Big]_{-2}^{1}$$

$$= 3 \quad \text{(units of work)} \quad \blacktriangleleft$$

LINE INTEGRALS IN THREE DIMENSIONS The notion of a line integral can be extended to three dimensions. If C is a smooth curve in three dimensions given parametrically by

$$x = x(t)$$
$$y = y(t), \quad a \le t \le b$$
$$z = z(t)$$

and if $f(x, y, z)$, $g(x, y, z)$ and $h(x, y, z)$ are continuous on some region containing C, then we define

$$\int_C f(x, y, z) \, dx = \int_a^b f(x(t), y(t), z(t))x'(t) \, dt$$

$$\int_C g(x, y, z) \, dy = \int_a^b g(x(t), y(t), z(t))y'(t) \, dt \qquad (13)$$

$$\int_C h(x, y, z) \, dz = \int_a^b h(x(t), y(t), z(t))z'(t) \, dt$$

$$\int_C f(x, y, z) \, dx + g(x, y, z) \, dy + h(x, y, z) \, dz$$

$$= \int_C f(x, y, z) \, dx + \int_C g(x, y, z) \, dy + \int_C h(x, y, z) \, dz \qquad (14)$$

If we let

$$\mathbf{F}(x, y, z) = f(x, y, z)\mathbf{i} + g(x, y, z)\mathbf{j} + h(x, y, z)\mathbf{k}$$

$$\mathbf{r}(t) = x(t)\mathbf{i} + y(t)\mathbf{j} + z(t)\mathbf{k}$$

$$d\mathbf{r} = \frac{d\mathbf{r}}{dt} dt = dx\mathbf{i} + dy\mathbf{j} + dz\mathbf{k}$$

then (14) can be written

$$\int_C f(x, y, z)\, dx + g(x, y, z)\, dy + h(x, y, z)\, dz = \int_C \mathbf{F}(x, y, z) \cdot d\mathbf{r} \quad (15)$$

Moreover, it follows from (13) that

$$\int_C \mathbf{F}(x, y, z) \cdot d\mathbf{r} = \int_a^b \left[\mathbf{F}(x(t), y(t), z(t)) \cdot \frac{d\mathbf{r}}{dt} \right] dt \quad (16)$$

Finally, if a particle moves along a curve C in three dimensions while subject to a force $\mathbf{F}(x, y, z) = f(x, y, z)\mathbf{i} + g(x, y, z)\mathbf{j} + h(x, y, z)\mathbf{k}$, then we define the work W done by the force on the particle to be

$$W = \int_C \mathbf{F}(x, y, z) \cdot d\mathbf{r} \quad (17)$$

▶ **Example 5**   Evaluate $\int_C \mathbf{F} \cdot d\mathbf{r}$ if $\mathbf{F}(x, y, z) = yz\mathbf{i} + xz\mathbf{j} + xy\mathbf{k}$ and C is the curve $\mathbf{r}(t) = t\mathbf{i} + t^2\mathbf{j} + t^3\mathbf{k}$ $(0 \le t \le 1)$.

Solution.   From (16)

$$\int_C \mathbf{F} \cdot d\mathbf{r} = \int_0^1 \left[\mathbf{F}(x(t), y(t), z(t)) \cdot \frac{d\mathbf{r}}{dt} \right] dt$$

$$= \int_0^1 (t^2 t^3 \mathbf{i} + t t^3 \mathbf{j} + t t^2 \mathbf{k}) \cdot (\mathbf{i} + 2t\mathbf{j} + 3t^2\mathbf{k})\, dt$$

$$= \int_0^1 (t^5 + 2t^5 + 3t^5)\, dt = \int_0^1 6t^5\, dt = 1 \qquad ◀$$

▶ **Exercise Set 18.1**

---

1. Let C be the curve $x = t$, $y = t^2$, $0 \le t \le 1$. Find

(a) $\displaystyle\int_C (2x + y)\, dx$ (b) $\displaystyle\int_C (x^2 - y)\, dy$

(c) $\displaystyle\int_C (2x + y)\, dx + (x^2 - y)\, dy$.

2. Let $\mathbf{F} = (3x + 2y)\mathbf{i} + (2x - y)\mathbf{j}$.   Evaluate $\int_C \mathbf{F} \cdot d\mathbf{r}$, where C is
(a) the line segment from $(0, 0)$ to $(1, 1)$
(b) the parabolic arc $y = x^2$ from $(0, 0)$ to $(1, 1)$
(c) the curve $y = \sin(\pi x/2)$ from $(0, 0)$ to $(1, 1)$
(d) the curve $x = y^3$ from $(0, 0)$ to $(1, 1)$.

In Exercises 3–10, evaluate the line integral.

3. $\displaystyle\int_C y\, dx - x^2\, dy$, where C is the curve $x = t$, $y = \tfrac{1}{2}t^2$, $0 \le t \le 2$.

4. $\displaystyle\int_C e^{y-x}\, dx + xy\, dy$, where C is the curve $x = 2 + t$, $y = 3 - t$, $0 \le t \le 1$.

5. $\displaystyle\int_C (x + 2y)\, dx + (x - y)\, dy$, where C is the curve $x = 2\cos t$, $y = 4\sin t$, $0 \le t \le \pi/4$.

6. $\int_C (x^2 - y^2)\,dx + x\,dy$, where C is the curve $x = t^{2/3}$, $y = t$, $-1 \leq t \leq 1$.

7. $\int_C -y\,dx + x\,dy$ along $y^2 = 3x$ from $(3, 3)$ to $(0, 0)$.

8. $\int_C (y - x)\,dx + x^2y\,dy$ along $y^2 = x^3$ from $(1, -1)$ to $(1, 1)$.

9. $\int_C (x^2 + y^2)\,dx - x\,dy$ along the quarter circle $x^2 + y^2 = 1$ from $(1, 0)$ to $(0, 1)$.

10. $\int_C (y - x)\,dx + xy\,dy$ along the line segment from $(3, 4)$ to $(2, 1)$.

11. Find $\int_C \mathbf{F} \cdot d\mathbf{r}$, where $\mathbf{F}(x, y) = x^2\mathbf{i} + xy\mathbf{j}$ and C is the semicircle $\mathbf{r}(t) = 2 \cos t\mathbf{i} + 2 \sin t\mathbf{j}$, $0 \leq t \leq \pi$.

12. Find $\int_C \mathbf{F} \cdot d\mathbf{r}$, where $\mathbf{F}(x, y) = x^2y\mathbf{i} + 4\mathbf{j}$ and C is the curve $\mathbf{r}(t) = e^t\mathbf{i} + e^{-t}\mathbf{j}$, $0 \leq t \leq 1$.

13. Find $\int_C \mathbf{F} \cdot d\mathbf{r}$, where
$\mathbf{F}(x, y) = (x^2 + y^2)^{-3/2}(x\mathbf{i} + y\mathbf{j})$ and C is the curve $\mathbf{r}(t) = e^t \sin t\mathbf{i} + e^t \cos t\mathbf{j}$, $0 \leq t \leq 1$.

14. Evaluate $\int_C y\,dx - x\,dy$ over each of the following curves:

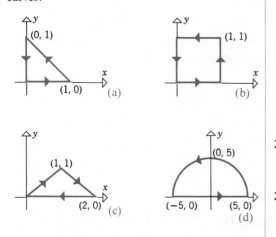

(a)

(b)

(c)

(d)

In Exercises 15–18, evaluate the three-dimensional line integral.

15. $\int_C yz\,dx - xz\,dy + xy\,dz$, where C is the curve $x = e^t$, $y = e^{3t}$, $z = e^{-t}$, $0 \leq t \leq 1$.

16. $\int_C y\,dx + z\,dy - x\,dz$, where C is the line segment from $(1, 1, 1)$ to $(-3, 2, 0)$.

17. $\int_C \mathbf{F} \cdot d\mathbf{r}$, where $\mathbf{F}(x, y, z) = z\mathbf{i} + x\mathbf{j} + y\mathbf{k}$ and C is the curve $\mathbf{r}(t) = \sin t\mathbf{i} + 3 \sin t\mathbf{j} + \sin^2 t\mathbf{k}$, $0 \leq t \leq \pi/2$.

18. $\int_C x^2z\,dx - yx^2\,dy + 3xz\,dz$, where C is the triangular path from $(0, 0, 0)$ to $(1, 1, 0)$ to $(1, 1, 1)$ to $(0, 0, 0)$.

19. Find the work done if a particle moves along the parabolic arc $x = y^2$ from $(0, 0)$ to $(1, 1)$ while subject to the force $\mathbf{F}(x, y) = xy\mathbf{i} + x^2\mathbf{j}$.

20. Find the work done if a particle moves along the curve $x = t$, $y = 1/t$, $1 \leq t \leq 3$ while subject to the force $\mathbf{F}(x, y) = (x^2 + xy)\mathbf{i} + (y - x^2y)\mathbf{j}$.

21. Find the work done by the force

$$\mathbf{F}(x, y) = \frac{1}{x^2 + y^2}\mathbf{i} + \frac{4}{x^2 + y^2}\mathbf{j}$$

acting on a particle that moves along the curve C shown.

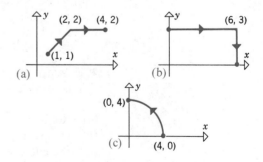

(a)

(b)

(c)

22. Find the work done by a force $\mathbf{F}(x, y, z) = xy\mathbf{i} + yz\mathbf{j} + xz\mathbf{k}$ acting on a particle that moves along the curve $\mathbf{r}(t) = t\mathbf{i} + t^2\mathbf{j} + t^3\mathbf{k}$, $0 \leq t \leq 1$.

23. Find the work that is done by a force $\mathbf{F}(x, y, z) = (x + y)\mathbf{i} + xy\mathbf{j} - z^2\mathbf{k}$ acting on a particle that moves along the line segment from $(0, 0, 0)$ to $(1, 3, 1)$ and then along the line segment from $(1, 3, 1)$ to $(2, -1, 4)$.

24. Evaluate $\displaystyle\int_{-c} \frac{x\,dy - y\,dx}{x^2 + y^2}$, where C is the circle $x^2 + y^2 = a^2$ traversed counterclockwise.

25. A particle moves from the point $(0, 0)$ to the point $(1, 0)$ along the curve $x = t$, $y = \lambda t(1 - t)$ while subject to the force $\mathbf{F}(x, y) = xy\mathbf{i} + (x - y)\mathbf{j}$. For what value of λ is the work done equal to 1?

26. (a) Show that $\displaystyle\int_C f(x, y)\,dx = 0$ if C is a vertical line segment in the xy-plane.

   (b) Show that $\displaystyle\int_C f(x, y)\,dy = 0$ if C is a horizontal line segment in the xy-plane.

27. A particle moves counterclockwise along the upper half of the circle $x^2 + y^2 = a^2$ from $(a, 0)$ to $(-a, 0)$ while subject to the force

$$\mathbf{F}(x, y) = \frac{k\mathbf{r}}{\|\mathbf{r}\|^3}$$

where k is a constant and $\mathbf{r} = x\mathbf{i} + y\mathbf{j}$. Find the work done.

28. A farmer carries a sack of grain weighing 20 lb up a circular helical staircase encircling a silo of radius 25 ft. As the farmer climbs, grain leaks from the sack at a rate of 1 lb per 10 ft of ascent. How much work is performed by the farmer on the sack of grain if he climbs 60 ft in exactly four revolutions?

18.2   LINE INTEGRALS INDEPENDENT OF PATH

In general, the value of a line integral $\int_C \mathbf{F} \cdot d\mathbf{r}$ depends on the curve C. However, we will show in this section that when the integrand satisfies appropriate conditions, the value of the integral depends only on the location of the endpoints of the curve C and not on the shape of the curve joining the endpoints. In such cases the evaluation of the line integral is greatly simplified.

Let us begin with an example.

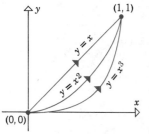

Figure 18.2.1

▶ **Example 1**   Let $\mathbf{F}(x, y) = y\mathbf{i} + x\mathbf{j}$. Evaluate the line integral $\int_C \mathbf{F} \cdot d\mathbf{r} = \int_C y\,dx + x\,dy$ over the following curves (Figure 18.2.1):

(a)   the line segment $y = x$ from $(0, 0)$ to $(1, 1)$
(b)   the parabola $y = x^2$ from $(0, 0)$ to $(1, 1)$
(c)   the cubic $y = x^3$ from $(0, 0)$ to $(1, 1)$

Solution (a).   With $x = t$ as the parameter, the path of integration is given by $\mathbf{r}(t) = t\mathbf{i} + t\mathbf{j}$ $(0 \le t \le 1)$. Since $\mathbf{F}(x, y) = y\mathbf{i} + x\mathbf{j}$, it follows that $\mathbf{F}(x(t), y(t)) = t\mathbf{i} + t\mathbf{j}$. Thus,

$$\int_C \mathbf{F} \cdot d\mathbf{r} = \int_0^1 \left[\mathbf{F}(x(t), y(t)) \cdot \frac{d\mathbf{r}}{dt} \right] dt = \int_0^1 (t\mathbf{i} + t\mathbf{j}) \cdot (\mathbf{i} + \mathbf{j})\,dt$$

$$= \int_0^1 2t\,dt = 1$$

Solution (b). With $x = t$ as the parameter, $\mathbf{r}(t) = t\mathbf{i} + t^2\mathbf{j}$ $(0 \leq t \leq 1)$, and $\mathbf{F}(x(t), y(t)) = t^2\mathbf{i} + t\mathbf{j}$. Thus,

$$\int_C \mathbf{F} \cdot d\mathbf{r} = \int_0^1 \left[\mathbf{F}(x(t), y(t)) \cdot \frac{d\mathbf{r}}{dt}\right] dt = \int_0^1 (t^2\mathbf{i} + t\mathbf{j}) \cdot (\mathbf{i} + 2t\mathbf{j}) \, dt$$

$$= \int_0^1 3t^2 \, dt = 1$$

Solution (c). With $x = t$ as the parameter, $\mathbf{r}(t) = t\mathbf{i} + t^3\mathbf{j}$ $(0 \leq t \leq 1)$, and $\mathbf{F}(x(t), y(t)) = t^3\mathbf{i} + t\mathbf{j}$.

$$\int_C \mathbf{F} \cdot d\mathbf{r} = \int_0^1 \left[\mathbf{F}(x(t), y(t)) \cdot \frac{d\mathbf{r}}{dt}\right] dt = \int_0^1 (t^3\mathbf{i} + t\mathbf{j}) \cdot (\mathbf{i} + 3t^2\mathbf{j}) \, dt$$

$$= \int_0^1 4t^3 \, dt = 1 \qquad \blacktriangleleft$$

In this example we obtained the same value for the line integral even though we integrated over three different paths joining $(0, 0)$ to $(1, 1)$. This is not accidental. The following theorem shows that this is the case because $\mathbf{F}(x, y) = y\mathbf{i} + x\mathbf{j}$ is the gradient of some function ϕ [specifically $\mathbf{F}(x, y) = \nabla\phi$, where $\phi(x, y) = xy$].

18.2.1 THEOREM
The Fundamental Theorem of
Line Integrals

Suppose that $\mathbf{F}(x, y) = f(x, y)\mathbf{i} + g(x, y)\mathbf{j}$, where f and g are continuous in some open region containing the points (x_0, y_0) and (x_1, y_1). If

$$\mathbf{F}(x, y) = \nabla\phi(x, y)$$

at each point of this region, then for any piecewise smooth curve C that starts at (x_0, y_0), ends at (x_1, y_1), and lies entirely in the region,

$$\int_C \mathbf{F}(x, y) \cdot d\mathbf{r} = \phi(x_1, y_1) - \phi(x_0, y_0) \tag{1}$$

Proof. We will give the proof for a smooth curve C. The proof for piecewise smooth curves is obtained by considering each piece separately. The details are left for the exercises.

If C is given parametrically by $x = x(t)$, $y = y(t)$ $(a \leq t \leq b)$, then the initial and final points of the curve C are

$$(x_0, y_0) = (x(a), y(a))$$
$$(x_1, y_1) = (x(b), y(b))$$

Since $\mathbf{F}(x, y) = \nabla\phi$, it follows that

$$\mathbf{F}(x, y) = \frac{\partial\phi}{\partial x}\mathbf{i} + \frac{\partial\phi}{\partial y}\mathbf{j}$$

so

$$\int_C \mathbf{F}(x, y) \cdot d\mathbf{r} = \int_C \frac{\partial \phi}{\partial x} dx + \frac{\partial \phi}{\partial y} dy$$

$$= \int_a^b \left[\frac{\partial \phi}{\partial x} \frac{dx}{dt} + \frac{\partial \phi}{\partial y} \frac{dy}{dt} \right] dt$$

$$= \int_a^b \frac{d}{dt} [\phi(x(t), y(t))] \, dt$$

$$= \phi[x(t), y(t)] \Big]_{t=a}^{b}$$

$$= \phi(x(b), y(b)) - \phi(x(a), y(a))$$

$$= \phi(x_1, y_1) - \phi(x_0, y_0) \quad \blacksquare$$

Since the right side of (1) involves only the values of ϕ at the endpoints (x_0, y_0) and (x_1, y_1), it follows that the integral on the left has the same value for *every* piecewise smooth curve C joining these endpoints. We say that the integral is ***independent of the path C.*** If $\mathbf{F}(x, y)$ is the gradient of a function $\phi(x, y)$ in some open region, then $\mathbf{F}$ is said to be ***conservative*** in that region, and ϕ is called a ***potential function*** for $\mathbf{F}$ in that region. In brief, Theorem 18.2.1 tells us that if the function $\mathbf{F}$ is conservative in a region, and C is a path (curve) in that region, then the line integral of $\mathbf{F} \cdot d\mathbf{r}$ is independent of the path and the value of the integral can be determined from the values of a potential function at the endpoints of the path.

▶ **Example 2**   The function $\mathbf{F}(x, y) = y\mathbf{i} + x\mathbf{j}$ is conservative in the entire xy-plane since it is the gradient of $\phi(x, y) = xy$. Thus, for any piecewise smooth curve C running from $(0, 0)$ to $(1, 1)$, it follows from (1) that

$$\int_C y \, dx + x \, dy = \int_C \mathbf{F} \cdot d\mathbf{r} = \phi(1, 1) - \phi(0, 0) = 1 - 0 = 1$$

which explains the results obtained in Example 1.   ◀

A curve C given by $\mathbf{r}(t) = x(t)\mathbf{i} + y(t)\mathbf{j}$ $(a \leq t \leq b)$ is called ***closed*** if the initial point $\mathbf{r}(a) = \langle x_0, y_0 \rangle$ and the terminal point $\mathbf{r}(b) = \langle x_1, y_1 \rangle$ coincide (Figure 18.2.2). Line integrals over smooth closed curves require no new computational techniques. However, if $\mathbf{F}(x, y) = f(x, y)\mathbf{i} + g(x, y)\mathbf{j}$ is the gradient of $\phi(x, y)$ then it follows from the Fundamental Theorem of Line Integrals (18.2.1) that

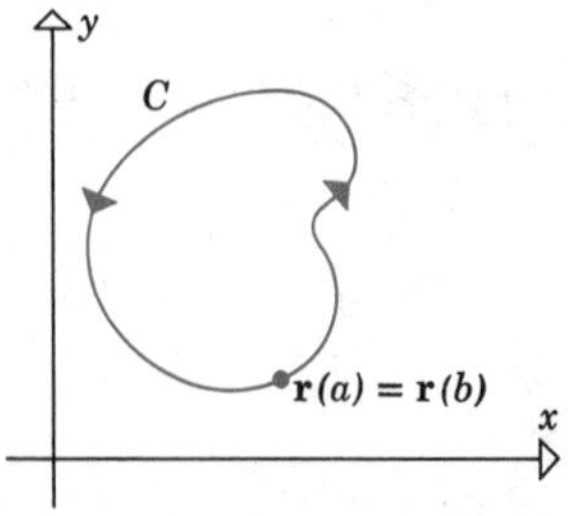

Figure 18.2.2

$$\int_C \mathbf{F} \cdot d\mathbf{r} = \int_C f(x, y) \, dx + g(x, y) \, dy = \phi(x_1, y_1) - \phi(x_0, y_0) = 0$$

Thus, we have the following result.

18.2.2   THEOREM   *Suppose that* $\mathbf{F}(x, y) = f(x, y)\mathbf{i} + g(x, y)\mathbf{j}$, *where f and g are continuous in some open region. If* $\mathbf{F}$ *is conservative in that region, then*

$$\int_C \mathbf{F} \cdot d\mathbf{r} = 0$$

for every piecewise smooth closed curve C in the region.

Theorem 18.2.1 raises two basic questions that must be answered if Formula (1) is to be useful for calculations.

1. How can we tell whether $\mathbf{F}(x, y) = f(x, y)\mathbf{i} + g(x, y)\mathbf{j}$ is the gradient of some function $\phi(x, y)$?

2. If $\mathbf{F}(x, y)$ is the gradient of some function $\phi(x, y)$, how do we find ϕ?

To answer these questions, we will need some terminology. A plane curve $\mathbf{r} = \mathbf{r}(t)$ $(a \leq t \leq b)$ is said to be **simple** if it does not intersect itself anywhere between its endpoints. A simple curve may or may not be closed (Figure 18.2.3).

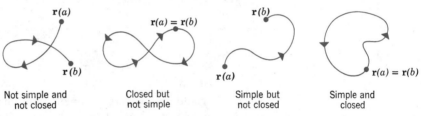

Figure 18.2.3

Not simple and not closed   Closed but not simple   Simple but not closed   Simple and closed

A plane region whose boundary consists of *one* simple closed curve is called **simply connected.** As illustrated in Figure 18.2.4, a simply connected region cannot consist of two or more separated pieces and cannot contain any holes.

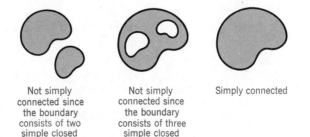

Figure 18.2.4

Not simply connected since the boundary consists of two simple closed curves   Not simply connected since the boundary consists of three simple closed curves   Simply connected

The following theorem will help us to determine whether a function $\mathbf{F}(x, y)$ is conservative.

18.2.3   THEOREM   *Let $\mathbf{F}(x, y) = f(x, y)\mathbf{i} + g(x, y)\mathbf{j}$, where f and g have continuous first partial derivatives in an open simply connected region. Then $\mathbf{F}(x, y)$ is the gradient of some function $\phi(x, y)$ on that region if and only if*

$$\frac{\partial f}{\partial y} = \frac{\partial g}{\partial x} \tag{2}$$

at each point of the region.

Proof. We will prove that if $\mathbf{F}$ is the gradient of some function, then condition (2) holds. The proof of the converse is omitted since it requires results from advanced calculus.

Suppose that $\mathbf{F}$ is the gradient of some function ϕ, that is,

$$\mathbf{F}(x, y) = f(x, y)\mathbf{i} + g(x, y)\mathbf{j} = \frac{\partial \phi}{\partial x}\mathbf{i} + \frac{\partial \phi}{\partial y}\mathbf{j} = \nabla\phi$$

It follows that

$$f(x, y) = \frac{\partial \phi}{\partial x} \quad \text{and} \quad g(x, y) = \frac{\partial \phi}{\partial y}$$

so

$$\frac{\partial f}{\partial y} = \frac{\partial^2 \phi}{\partial y\, \partial x} \quad \text{and} \quad \frac{\partial g}{\partial x} = \frac{\partial^2 \phi}{\partial x\, \partial y}$$

Since $\partial f/\partial y$ and $\partial g/\partial x$ were assumed continuous, it follows from Theorem 16.3.7 that the two mixed partial derivatives of ϕ are equal. Thus, $\partial f/\partial y = \partial g/\partial x$. ∎

WARNING.   In (2), the **i**-component of **F** is differentiated with respect to y and the **j**-component with respect to x. It is easy to get this backwards by mistake.

▶ **Example 3**   Show that $\mathbf{F}(x, y) = 2xy^3\mathbf{i} + (1 + 3x^2y^2)\mathbf{j}$ is the gradient of some function ϕ, and find ϕ.

Solution.   Since $f(x, y) = 2xy^3$ and $g(x, y) = 1 + 3x^2y^2$ we have

$$\frac{\partial f}{\partial y} = 6xy^2 = \frac{\partial g}{\partial x}$$

so that (2) holds. Thus, $\mathbf{F}$ is the gradient of some function ϕ, that is,

$$\mathbf{F}(x, y) = 2xy^3\mathbf{i} + (1 + 3x^2y^2)\mathbf{j} = \frac{\partial \phi}{\partial x}\mathbf{i} + \frac{\partial \phi}{\partial y}\mathbf{j} = \nabla\phi \tag{3}$$

To find ϕ, we begin by equating components in (3):

$$\frac{\partial \phi}{\partial x} = 2xy^3 \tag{4}$$

$$\frac{\partial \phi}{\partial y} = 1 + 3x^2y^2 \tag{5}$$

Integrating (4) with respect to x (and treating y as constant), we obtain

$$\phi = \int 2xy^3 \, dx = x^2y^3 + k(y) \tag{6}$$

where $k(y)$ represents the "constant" of integration. We are justified in treating the constant of integration as a function of y, since y is held constant in the integration process, so

$$\frac{\partial}{\partial x}[k(y)] = 0$$

To find $k(y)$, we differentiate (6) with respect to y:

$$\frac{\partial \phi}{\partial y} = 3x^2y^2 + k'(y)$$

Equating this with (5) yields

$$1 + 3x^2y^2 = 3x^2y^2 + k'(y)$$

Thus,

$$k'(y) = 1 \quad \text{or} \quad k(y) = \int 1 \, dy = y + K$$

where K is a (numerical) constant of integration. Substituting in (6) we obtain

$$\phi = x^2y^3 + y + K$$

The appearance of the arbitrary constant K tells us that ϕ is not unique. The reader may want to verify that $\nabla \phi = \mathbf{F}$.

Alternate Solution. Instead of integrating (4) with respect to x, as above, we can begin by integrating (5) with respect to y (treating x as a constant in the integration process). We obtain

$$\phi = \int (1 + 3x^2y^2) \, dy = y + x^2y^3 + k(x) \tag{7}$$

where $k(x)$ is the "constant" of integration. Differentiating this expression with respect to x yields

$$\frac{\partial \phi}{\partial x} = 2xy^3 + k'(x)$$

and equating this with (4) gives

$$2xy^3 = 2xy^3 + k'(x)$$

so $k'(x) = 0$ and $k(x) = K$, where K is a (numerical) constant of integration. Substituting this in (7) gives

$$\phi = y + x^2y^3 + K$$

which agrees with our previous solution.   ◀

If a curve C runs from the initial point (x_0, y_0) to the terminal point (x_1, y_1), and if the line integral $\int_C \mathbf{F} \cdot d\mathbf{r}$ is independent of the path joining these points, then we write

$$\int_C \mathbf{F} \cdot d\mathbf{r} = \int_{(x_0, y_0)}^{(x_1, y_1)} \mathbf{F} \cdot d\mathbf{r}$$

or, equivalently,

$$\int_C f(x, y)\, dx + g(x, y)\, dy = \int_{(x_0, y_0)}^{(x_1, y_1)} f(x, y)\, dx + g(x, y)\, dy$$

▶ **Example 4**   Evaluate

$$\int_{(1,4)}^{(3,1)} 2xy^3\, dx + (1 + 3x^2y^2)\, dy$$

Solution.   From Example 3, $\mathbf{F}(x, y) = 2xy^3\mathbf{i} + (1 + 3x^2y^2)\mathbf{j}$ is the gradient of $\phi(x, y) = y + x^2y^3 + K$. Thus, from the Fundamental Theorem of Line Integrals (18.2.1),

$$\int_{(1,4)}^{(3,1)} 2xy^3\, dx + (1 + 3x^2y^2)\, dy = \int_{(1,4)}^{(3,1)} \mathbf{F} \cdot d\mathbf{r}$$
$$= \phi(3, 1) - \phi(1, 4)$$
$$= (10 + K) - (68 + K) = -58 \quad ◀$$

REMARK.   Note that the constant K drops out. In future integration problems we will omit K from the computations.

▶ **Example 5**   A particle moves over the semicircle C: $\mathbf{r}(t) = \cos t\,\mathbf{i} + \sin t\,\mathbf{j}$ $(0 \leq t \leq \pi)$ while subject to the force $\mathbf{F}(x, y) = e^y\mathbf{i} + xe^y\mathbf{j}$. Find the work done.

Solution.   The work done is

$$W = \int_C \mathbf{F} \cdot d\mathbf{r} = \int_C e^y \, dx + xe^y \, dy \tag{8}$$

If we try to evaluate this integral directly, we must deal with the complicated integral

$$W = \int_C \mathbf{F} \cdot d\mathbf{r} = \int_0^\pi \left[\mathbf{F}(x(t), y(t)) \cdot \frac{d\mathbf{r}}{dt} \right] dt$$

$$= \int_0^\pi [e^{\sin t}\mathbf{i} + (\cos t)e^{\sin t}\mathbf{j}] \cdot [-\sin t\,\mathbf{i} + \cos t\,\mathbf{j}] \, dt$$

$$= \int_0^\pi [(-\sin t)e^{\sin t} + (\cos^2 t)e^{\sin t}] \, dt$$

To avoid this difficulty, we will try an alternate approach.

The force $\mathbf{F}$ is the gradient of some function ϕ, since

$$\frac{\partial}{\partial y}(e^y) = e^y = \frac{\partial}{\partial x}(xe^y)$$

Thus, the integral in (8) is independent of the path and we are free to replace C by a straight line segment C_1 joining the endpoint $(1, 0)$ to the endpoint $(-1, 0)$ (Figure 18.2.5). This will simplify the computations.

The straight line segment C_1 can be represented parametrically by

$$x = 1 - 2t \atop y = 0 \quad , \quad 0 \le t \le 1$$

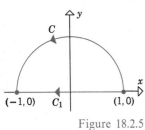

Figure 18.2.5

(See the remark following Example 3 of Section 18.1.) Thus,

$$\mathbf{r}(t) = (1 - 2t)\mathbf{i} + 0\mathbf{j} = (1 - 2t)\mathbf{i} \qquad (0 \le t \le 1)$$

and

$$\frac{d\mathbf{r}}{dt} = -2\mathbf{i}$$

Therefore, the work done is

$$W = \int_{C_1} \mathbf{F} \cdot d\mathbf{r} = \int_{C_1} \left[(e^y\mathbf{i} + xe^y\mathbf{j}) \cdot \frac{d\mathbf{r}}{dt} \right] dt$$

$$= \int_0^1 [e^0\mathbf{i} + (1 - 2t)e^0\mathbf{j}] \cdot (-2\mathbf{i}) \, dt$$

$$= \int_0^1 [\mathbf{i} + (1 - 2t)\mathbf{j}] \cdot (-2\mathbf{i}) \, dt$$

$$= \int_0^1 -2\,dt = -2 \qquad \text{(units of work)}$$

Second Solution. Since the line segment C_1 has initial point $(1, 0)$ and terminal point $(-1, 0)$, the line segment $-C_1$ has initial point $(-1, 0)$ and terminal point $(1, 0)$. Thus, $-C_1$ can be represented parametrically by

$$\begin{aligned} x &= t \\ y &= 0 \end{aligned}, \quad -1 \le t \le 1$$

It follows that

$$\mathbf{r}(t) = t\mathbf{i} + 0\mathbf{j} = t\mathbf{i} \qquad (-1 \le t \le 1)$$

and

$$\frac{d\mathbf{r}}{dt} = \mathbf{i}$$

From (4) of Section 18.1,

$$
\begin{aligned}
W &= \int_{C_1} \mathbf{F} \cdot d\mathbf{r} = -\int_{-C_1} \mathbf{F} \cdot d\mathbf{r} \\
&= -\int_{-C_1} \left[(e^y\mathbf{i} + xe^y\mathbf{j}) \cdot \frac{d\mathbf{r}}{dt} \right] dt \\
&= -\int_{-1}^1 [(e^0\mathbf{i} + te^0\mathbf{j}) \cdot \mathbf{i}]\,dt \\
&= -\int_{-1}^1 [(\mathbf{i} + t\mathbf{j}) \cdot \mathbf{i}]\,dt \\
&= -\int_{-1}^1 dt = -2
\end{aligned}
$$

Third Solution. Since $\mathbf{F}(x, y)$ is the gradient of some function ϕ, we can proceed, as in Example 4, by finding ϕ and applying (1). Since $\nabla\phi = \mathbf{F}(x, y) = e^y\mathbf{i} + xe^y\mathbf{j}$,

$$\frac{\partial \phi}{\partial x} = e^y \quad \text{and} \quad \frac{\partial \phi}{\partial y} = xe^y \qquad (9)$$

Thus,

$$\phi = \int e^y\,dx = xe^y + k(y) \qquad (10)$$

and consequently

$$\frac{\partial \phi}{\partial y} = xe^y + k'(y)$$

Equating this with the expression for $\partial \phi / \partial y$ in (9), we obtain $k'(y) = 0$ or $k(y) = K$. Thus, from (10),

$$\phi = xe^y + K$$

Since the curve C begins at $(1, 0)$ and ends at $(-1, 0)$, it follows that

$$W = \int_C \mathbf{F} \cdot d\mathbf{r} = \int_{(1,0)}^{(-1,0)} e^y \, dx + xe^y \, dy = \phi(-1, 0) - \phi(1, 0)$$

$$= (-1)e^0 - (1)e^0 = -2 \quad \blacktriangleleft$$

▶ **Example 6** A particle moves around the circle C given by $\mathbf{r}(t) = \cos t\mathbf{i} + \sin t\mathbf{j}$ $(0 \le t \le 2\pi)$ while subject to the force $\mathbf{F}(x, y) = e^y\mathbf{i} + xe^y\mathbf{j}$. Find the work done.

Solution. In Example 5 we showed that $\mathbf{F} \cdot d\mathbf{r} = e^y \, dx + xe^y \, dy$ is conservative. Since C is a closed curve, it follows that

$$W = \int_C \mathbf{F} \cdot d\mathbf{r} = \int_C e^y \, dx + xe^y \, dy = 0 \quad \blacktriangleleft$$

▶ **Exercise Set 18.2**

In Exercises 1–6, determine whether $\mathbf{F}$ is conservative. If it is, find a potential function for it.

1. $\mathbf{F}(x, y) = x\mathbf{i} + y\mathbf{j}$.

2. $\mathbf{F}(x, y) = 3y^2\mathbf{i} + 6xy\mathbf{j}$.

3. $\mathbf{F}(x, y) = x^2y\mathbf{i} + 5xy^2\mathbf{j}$.

4. $\mathbf{F}(x, y) = e^x \cos y\mathbf{i} - e^x \sin y\mathbf{j}$.

5. $\mathbf{F}(x, y) = (\cos y + y \cos x)\mathbf{i} + (\sin x - x \sin y)\mathbf{j}$.

6. $\mathbf{F}(x, y) = x \ln y\mathbf{i} + y \ln x\mathbf{j}$.

7. Show that $\displaystyle\int_{(-1,2)}^{(1,3)} y^2 \, dx + 2xy \, dy$ is independent of the path and evaluate the integral by:
    (a) using Theorem 18.2.1
    (b) integrating along the line segment from $(-1, 2)$ to $(1, 3)$.

8. Show that $\displaystyle\int_{(0,1)}^{(\pi,-1)} y \sin x \, dx - \cos x \, dy$ is independent of the path, and evaluate the integral by:

(a) using Theorem 18.2.1
(b) integrating along the line segment from $(0, 1)$ to $(\pi, -1)$.

In Exercises 9–14, show that the integral is independent of the path, and find its value by any method.

9. $\displaystyle\int_{(1,2)}^{(4,0)} 3y \, dx + 3x \, dy$.

10. $\displaystyle\int_{(0,0)}^{(1,\pi/2)} e^x \sin y \, dx + e^x \cos y \, dy$.

11. $\displaystyle\int_{(0,0)}^{(3,2)} 2xe^y \, dx + x^2e^y \, dy$.

12. $\displaystyle\int_{(-1,2)}^{(0,1)} (3x - y + 1) \, dx - (x + 4y + 2) \, dy$.

13. $\displaystyle\int_{(2,-2)}^{(-1,0)} 2xy^3 \, dx + 3y^2x^2 \, dy$.

14. $\displaystyle\int_{(1,1)}^{(3,3)} \left(e^x \ln y - \frac{e^y}{x}\right) dx + \left(\frac{e^x}{y} - e^y \ln x\right) dy$, where x and y are positive.

In Exercises 15–18, find the work done by the conservative force **F** as it acts on a particle moving from P to Q.

15. $\mathbf{F}(x, y) = xy^2\mathbf{i} + x^2y\mathbf{j}$; $P(1, 1)$, $Q(0, 0)$.

16. $\mathbf{F}(x, y) = ye^{xy}\mathbf{i} + xe^{xy}\mathbf{j}$; $P(-1, 1)$, $Q(2, 0)$.

17. $\mathbf{F}(x, y) = \dfrac{y}{x^2 + y^2}\mathbf{i} - \dfrac{x}{x^2 + y^2}\mathbf{j}$; $P(0, 2)$, $Q(3, 3)$.
[Assume $y > 0$.]

18. $\mathbf{F}(x, y) = e^{-y}\cos x\mathbf{i} - e^{-y}\sin x\mathbf{j}$; $P(\pi/2, 1)$, $Q(-\pi/2, 0)$.

19. Let

$$\mathbf{F}(x, y) = (e^y + ye^x)\mathbf{i} + (xe^y + e^x)\mathbf{j}$$

Find the work done by the force **F** if it acts on a particle that moves
(a) from $(a, 0)$ to $(-a, 0)$ along the x-axis
(b) from $(a, 0)$ to $(-a, 0)$ along the upper half of the circle $x^2 + y^2 = a^2$
(c) from $(a, 0)$ to $(-a, 0)$ along the upper half of the ellipse

$$\frac{x^2}{a^2} + \frac{y^2}{b^2} = 1$$

(d) once around the circle $x^2 + y^2 = a^2$.

20. Evaluate

$$\int_C (\sin y \sinh x + \cos y \cosh x)\, dx +$$

$$(\cos y \cosh x - \sin y \sinh x)\, dy$$

where C is the line segment

$$\begin{aligned} x &= t \\ y &= \tfrac{1}{2}\pi(t - 1) \end{aligned}, \quad 1 \le t \le 2.$$

21. Let

$$\mathbf{F}(x, y) = \frac{k}{\|\mathbf{r}\|^3}\mathbf{r}$$

where $\mathbf{r} = x\mathbf{i} + y\mathbf{j}$. Show that **F** is conservative on any region that does not contain the origin.

22. Find a function h for which

$$\mathbf{F}(x, y) = h(x)[x \sin y + y \cos y]\mathbf{i} + h(x)[x \cos y - y \sin y]\mathbf{j}$$

is conservative.

23. (a) Prove: $\mathbf{F}(x, y) = f(x, y)\mathbf{i} + g(x, y)\mathbf{j}$ is conservative in a region if and only if there is a function $\phi(x, y)$ such that

$$d\phi = f(x, y)\, dx + g(x, y)\, dy$$

on the region.
(b) Prove: If $\mathbf{F}(x, y)$ is conservative on a simply connected open region containing (x_0, y_0) and (x_1, y_1), then

$$\int_{(x_0, y_0)}^{(x_1, y_1)} d\phi = \phi(x_1, y_1) - \phi(x_0, y_0)$$

where ϕ is the function in part (a).

24. Prove Theorem 18.2.1 in the case where C is a piecewise smooth curve composed of smooth curves $C_1, C_2, \ldots, C_n$.

18.3 GREEN'S THEOREM

In this section we will discuss a remarkable and beautiful theorem that expresses the double integral over a plane region in terms of a line integral over the boundary.

18.3.1 THEOREM
Green's* Theorem

Let R be a simply connected plane region whose boundary is a simple, closed piecewise smooth curve C traversed counterclockwise. If f(x, y) and g(x, y) have continuous first partial derivatives on some open set containing R, then

$$\int_C f(x, y)\, dx + g(x, y)\, dy = \iint_R \left(\frac{\partial g}{\partial x} - \frac{\partial f}{\partial y}\right) dA \qquad (1)$$

Proof. For simplicity, we will prove the theorem only for regions that are simultaneously type I and type II (see Section 17.2). Such a region is shown in Figure 18.3.1

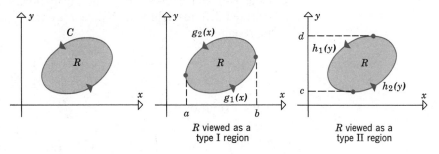

Figure 18.3.1

R viewed as a type I region

R viewed as a type II region

The crux of the proof is to show that

$$\int_C f(x, y) \, dx = -\iint_R \frac{\partial f}{\partial y} \, dA \tag{2a}$$

and

$$\int_C g(x, y) \, dy = \iint_R \frac{\partial g}{\partial x} \, dA \tag{2b}$$

Formula (1) will then follow by adding (2a) and (2b).

To prove (2a), view R as a type I region and let C_1 and C_2 be the lower and upper boundary curves, oriented as in Figure 18.3.2. Then

$$\int_C f(x, y) \, dx = \int_{C_1} f(x, y) \, dx + \int_{C_2} f(x, y) \, dx$$

or, equivalently,

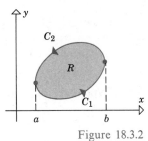

Figure 18.3.2

_____

*GEORGE GREEN (1793–1841), English mathematician and physicist. Green left school at an early age to work in his father's bakery and consequently had little early formal education. When his father opened a mill, the boy used the top room as a study in which he taught himself physics and mathematics from library books. In 1828 Green published his most important work, *An Essay on the Application of Mathematical Analysis to the Theories of Electricity and Magnetism.* Although Green's Theorem appeared in that paper, the result went virtually unnoticed because of the small press run and local distribution. Following the death of his father in 1829, Green was urged by friends to seek a college education. After four years of self-study to close the gaps in his elementary education, Green was admitted to Caius College, Cambridge in 1833. He graduated four years later, but with a disappointing performance on his final examinations—possibly because he was more interested in his own research. After a succession of works on light and sound, he was named to be Perse Fellow at Caius College. Two years later he died. In 1845, four years after his death, his paper of 1828 was republished and the theories developed therein by this obscure, self-taught baker's son helped pave the way to the modern theories of electricity and magnetism.

$$\int_C f(x, y)\, dx = \int_{C_1} f(x, y)\, dx - \int_{-C_2} f(x, y)\, dx \tag{3}$$

(This step will help simplify our calculations since C_1 and $-C_2$ are both oriented left to right.) With $x = t$ as the parameter, C_1 and $-C_2$ are represented by

$$C_1: x = t,\ y = g_1(t) \qquad (a \le t \le b)$$
$$-C_2: x = t,\ y = g_2(t) \qquad (a \le t \le b)$$

Thus, (3) yields

$$\int_C f(x, y)\, dx = \int_a^b f(t, g_1(t)) \left(\frac{dx}{dt}\right) dt - \int_a^b f(t, g_2(t)) \left(\frac{dx}{dt}\right) dt$$

$$= \int_a^b f(t, g_1(t))\, dt - \int_a^b f(t, g_2(t))\, dt$$

$$= -\int_a^b [f(t, g_2(t)) - f(t, g_1(t))]\, dt$$

$$= -\int_a^b \left[f(t, y) \right]_{y=g_1(t)}^{y=g_2(t)} dt$$

$$= -\int_a^b \left[\int_{g_1(t)}^{g_2(t)} \frac{\partial f}{\partial y}\, dy \right] dt$$

$$= -\int_a^b \int_{g_1(x)}^{g_2(x)} \frac{\partial f}{\partial y}\, dy\, dx \qquad \text{[Since } x = t.\text{]}$$

$$= -\iint_R \frac{\partial f}{\partial y}\, dA$$

The proof of (2b) is obtained similarly by treating R as a type II region. The details are omitted.

▶ Example 1   Use Green's Theorem to evaluate

$$\int_C x^2 y\, dx + x\, dy$$

over the triangular path shown in Figure 18.3.3.

Solution.   Since $f(x, y) = x^2 y$ and $g(x, y) = x$, it follows from (1) that

$$\int_C x^2 y\, dx + x\, dy = \iint_R \left[\frac{\partial}{\partial x}(x) - \frac{\partial}{\partial y}(x^2 y) \right] dA$$

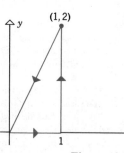

Figure 18.3.3

$$= \int_0^1 \int_0^{2x} (1 - x^2) \, dy \, dx$$

$$= \int_0^1 (2x - 2x^3) \, dx$$

$$= \left[x^2 - \frac{x^4}{2} \right]_0^1 = \frac{1}{2}$$

This agrees with the result obtained in Example 3 of Section 18.1, where we evaluated the line integral directly. Note how much simpler this solution is. ◀

▶ **Example 2** Suppose that a particle travels once around the unit circle in the counterclockwise direction, while subject to the force $\mathbf{F}(x, y) = (e^x - y^3)\mathbf{i} + (\cos y + x^3)\mathbf{j}$. Use Green's Theorem to find the work done by the force.

Solution.

$$W = \int_C (e^x - y^3) \, dx + (\cos y + x^3) \, dy$$

$$= \iint_R \left[\frac{\partial}{\partial x}(\cos y + x^3) - \frac{\partial}{\partial y}(e^x - y^3) \right] dA$$

$$= \iint_R (3x^2 + 3y^2) \, dA$$

$$= 3 \iint_R (x^2 + y^2) \, dA \qquad \left[\begin{array}{l} \text{Use polar coordinates to} \\ \text{evaluate this integral.} \end{array} \right]$$

$$= 3 \int_0^{2\pi} \int_0^1 (r^2) r \, dr \, d\theta$$

$$= \frac{3}{4} \int_0^{2\pi} d\theta = \frac{3\pi}{2} \qquad \blacktriangleleft$$

Green's Theorem leads to some new formulas for area. Letting $f(x, y) = 0$ and $g(x, y) = x$ in (1) yields

$$\int_C x \, dy = \iint_R dA = \text{area of } R \qquad \qquad (4a)$$

Letting $f(x, y) = -y$ and $g(x, y) = 0$ in (1) yields

$$\int_C -y\,dx = \iint_R dA = \text{area of } R \tag{4b}$$

Adding (4a) and (4b) and dividing the result by 2 yields

$$\frac{1}{2}\int_C -y\,dx + x\,dy = \text{area of } R \tag{4c}$$

Formulas (4a), (4b), and (4c) each express the area of R in terms of a line integral taken counterclockwise around the boundary of R. These formulas apply to those regions that satisfy the hypotheses of Green's Theorem. Although (4c) is more complicated than (4a) or (4b), it frequently results in an easier integral to evaluate.

▶ **Example 3**   Use (4c) to find the area enclosed by the ellipse $x^2/a^2 + y^2/b^2 = 1$.

Solution.   The ellipse, with counterclockwise orientation, can be represented parametrically by

$$\begin{aligned} x &= a\cos t \\ y &= b\sin t \end{aligned}, \quad 0 \le t \le 2\pi$$

If we denote this curve by C, then the area enclosed by the ellipse is

$$\begin{aligned} A &= \frac{1}{2}\int_C -y\,dx + x\,dy \\ &= \frac{1}{2}\int_0^{2\pi} [(-b\sin t)(-a\sin t) + (a\cos t)(b\cos t)]\,dt \\ &= \frac{1}{2}ab\int_0^{2\pi} (\sin^2 t + \cos^2 t)\,dt \\ &= \frac{1}{2}ab\int_0^{2\pi} dt = \pi ab \end{aligned}$$

This result can also be obtained from Formula (4a) or (4b).   ◀

REMARK.   Green's Theorem provides another viewpoint about some earlier results. If $\partial f/\partial y = \partial g/\partial x$ throughout a simply connected open region R with boundary C, then it follows from Green's Theorem that

$$\int_C f\,dx + g\,dy = \iint_R \left(\frac{\partial g}{\partial x} - \frac{\partial f}{\partial y}\right) dA = 0$$

However, this is to be expected since the condition $\partial f/\partial y = \partial g/\partial x$ implies that $\mathbf{F}(x, y) = f(x, y)\mathbf{i} + g(x, y)\mathbf{j}$ is conservative in R (Theorem 18.2.3) and we already know that a line integral around a closed curve has value zero in this case (Theorem 18.2.2).

▶ Exercise Set 18.3

In Exercises 1 and 2, evaluate the line integral using Green's Theorem and check the answer by evaluating directly.

1. $\int_C y^2\, dx + x^2\, dy$, where C is the square with vertices $(0, 0), (1, 0), (1, 1)$, and $(0, 1)$ traversed counterclockwise.

2. $\int_C y\, dx + x\, dy$, where C is the unit circle traversed counterclockwise.

In Exercises 3–13, use Green's Theorem to evaluate the integral. In each exercise, assume that the curve C is traversed counterclockwise.

3. $\int_C 3xy\, dx + 2xy\, dy$, where C is the rectangle bounded by $x = -2, x = 4, y = 1$, and $y = 2$.

4. $\int_C (x^2 - y^2)\, dx + x\, dy$, where C is the circle $x^2 + y^2 = 9$.

5. $\int_C x \cos y\, dx - y \sin x\, dy$, where C is the square with vertices $(0, 0), (0, \pi/2), (\pi/2, \pi/2)$ and $(\pi/2, 0)$.

6. $\int_C y \tan^2 x\, dx + \tan x\, dy$, where C is the circle $(x - 1)^2 + (y + 1)^2 = 1$.

7. $\int_C (x^2 - y)\, dx + x\, dy$, where C is the circle $x^2 + y^2 = 4$.

8. $\int_C (e^x + y^2)\, dx + (e^y + x^2)\, dy$, where C is the boundary of the region enclosed between $y = x^2$ and $y = x$.

9. $\int_C \ln(1 + y)\, dx - \dfrac{xy}{1 + y}\, dy$, where C is the triangle with vertices, $(0, 0), (2, 0)$, and $(0, 4)$.

10. $\int_C x^2 y\, dx - y^2 x\, dy$, where C is the boundary of the region in the first quadrant, enclosed between the coordinate axes and the circle $x^2 + y^2 = 16$.

11. $\int_C \tan^{-1} y\, dx - \dfrac{y^2 x}{1 + y^2}\, dy$, where C is the square with vertices $(0, 0), (0, 1), (1, 1)$, and $(1, 0)$.

12. $\int_C \cos x \sin y\, dx + \sin x \cos y\, dy$, where C is the triangle with vertices $(0, 0), (3, 3)$, and $(0, 3)$.

13. $\int_C x^2 y\, dx + (y + xy^2)\, dy$, where C is the boundary of the region enclosed by $y = x^2$ and $x = y^2$.

14. Let C be the boundary of the region enclosed between $y = x^2$ and $y = 2x$. Assuming that C is traversed counterclockwise, evaluate the following integrals by Green's Theorem:

(a) $\int_C (6xy - y^2)\, dx$ (b) $\int_C (6xy - y^2)\, dy$.

15. Find the area of the ellipse in Example 3 using (a) Formula (4a)   (b) Formula (4b).

16. Use a line integral to find the area of the region enclosed by
$$\begin{aligned} x &= a \cos^3 t \\ y &= a \sin^3 t \end{aligned}, \quad 0 \le t \le 2\pi.$$

17. Use a line integral to find the area of the triangle with vertices $(0, 0), (a, 0)$, and $(0, b)$, where a and b are positive.

18. Use a line integral to find the area of the region in the first quadrant enclosed by $y = x, y = 1/x$, and $y = x/9$.

19. A particle, starting at $(5, 0)$, traverses the upper semicircle $x^2 + y^2 = 25$ and returns to its starting point along the x-axis. Use Green's Theorem to find the work done on the particle by a force $\mathbf{F}(x, y) = xy\mathbf{i} + (\frac{1}{2}x^2 + yx)\mathbf{j}$.

20. A particle subject to a force $\mathbf{F}(x, y) = \sqrt{y}\,\mathbf{i} + \sqrt{x}\,\mathbf{j}$ moves counterclockwise around the closed curve formed by $y = 0$, $x = 2$, and $y = x^3/4$. Use Green's Theorem to find the work done.

21. (a) Let R be a plane region with area A whose boundary is a piecewise smooth simple closed curve C. Use Green's Theorem to prove that the centroid $(\bar{x}, \bar{y})$ of R is given by

$$\bar{x} = \frac{1}{2A}\int_C x^2\,dy, \qquad \bar{y} = -\frac{1}{2A}\int_C y^2\,dx$$

  (b) Use the result in (a) to find the centroid of the region enclosed between the x-axis and the upper half of the circle $x^2 + y^2 = a^2$.

22. Evaluate $\displaystyle\int_C y\,dx - x\,dy$, where C is the cardioid $r = a(1 + \cos\theta)$, $0 \le \theta \le 2\pi$.

23. (a) Let C be the line segment from a point (a, b) to a point (c, d). Show that

$$\int_C x\,dy - y\,dx = ad - bc$$

  (b) Use the result in (a) to show that the area A of a triangle with successive vertices (x_1, y_1), (x_2, y_2), and (x_3, y_3) going counterclockwise is

$$A = \tfrac{1}{2}[(x_1 y_2 - x_2 y_1) \\ + (x_2 y_3 - x_3 y_2) + (x_3 y_1 - x_1 y_3)]$$

  (c) Find a formula for the area of a polygon with successive vertices $(x_1, y_1), (x_2, y_2), \ldots,$ (x_n, y_n) going counterclockwise.

24. Find a simple closed curve C that maximizes the value of

$$\int_C \frac{1}{3}y^3\,dx + \left(x - \frac{1}{3}x^3\right)dy$$

25. (a) Let R be the region enclosed by the circle, C: $x = \cos t$, $y = \sin t$, $0 \le t \le 2\pi$. By evaluating directly, show that

$$\int_C -\frac{y}{x^2 + y^2}\,dx + \frac{x}{x^2 + y^2}\,dy = 2\pi$$

  (b) Let f and g be the functions

$$f(x, y) = -\frac{y}{x^2 + y^2}, \quad g(x, y) = \frac{x}{x^2 + y^2}$$

  Show that $\partial g/\partial x = \partial f/\partial y$.

  (c) Find the error in the following argument: By Green's Theorem

$$\int_C -\frac{y}{x^2 + y^2}\,dx + \frac{x}{x^2 + y^2}\,dy$$

$$= \int_C f(x, y)\,dx + g(x, y)\,dy$$

$$= \iint_R \left(\frac{\partial g}{\partial x} - \frac{\partial f}{\partial y}\right)dA = 0$$

[which contradicts the result in (a)].

18.4 INTRODUCTION TO SURFACE INTEGRALS

In previous sections we considered four kinds of integrals—integrals over intervals, double integrals over two-dimensional regions, triple integrals over three-dimensional solids, and line integrals over curves in two- or three-dimensional space. In this section we will discuss *surface integrals,* which are integrals over surfaces in three-dimensional space. Such integrals occur in problems of fluid and heat flow, electricity, magnetism, mass, and center of gravity.

We will motivate the definition of a surface integral by considering a problem about mass. In Section 17.6 we defined a lamina to be an idealized flat object that is sufficiently thin to be viewed as a two-dimensional solid. If

The thickness of a curved lamina is negligible

Figure 18.4.1

Figure 18.4.2

we imagine a flat lamina to be bent into some curved shape, the result is a *curved lamina* (Figure 18.4.1). The density of a curved lamina at a point (x, y, z) can be specified by a function $\delta(x, y, z)$, called the *density function* for the lamina. Informally, the density function can be visualized as follows: Consider a small section of the lamina containing the point (x, y, z), and let ΔM and ΔS be the mass and surface area of this section (Figure 18.4.2). If the ratio $\Delta M/\Delta S$ tends toward a limiting value when the small section of lamina is allowed to shrink down to the point (x, y, z), then this limit is the density of the lamina at (x, y, z). Symbolically,

$$\delta(x, y, z) = \lim \frac{\Delta M}{\Delta S}$$

From this equation we obtain the approximation

$$\Delta M \approx \delta(x, y, z)\, \Delta S \tag{1}$$

which relates the mass and surface area of a small section of lamina containing the point (x, y, z).

We will now use (1) to obtain a formula for the mass of a curved lamina.

18.4.1   DEFINITION   If a curved lamina with density $\delta(x, y, z)$ has the equation $z = f(x, y)$, and if the projection of this lamina on the xy-plane is the region R (Figure 18.4.3), then the mass M of the lamina is defined to be

$$M = \iint_R \delta[x, y, f(x, y)]\sqrt{f_x(x, y)^2 + f_y(x, y)^2 + 1}\; dA \tag{2a}$$

or in an alternate notation

$$M = \iint_R \delta[x, y, f(x, y)]\sqrt{\left(\frac{\partial z}{\partial x}\right)^2 + \left(\frac{\partial z}{\partial y}\right)^2 + 1}\; dA \tag{2b}$$

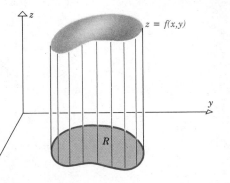

Figure 18.4.3

We will motivate this formula in four steps that closely parallel the four steps in the motivation of Definition 17.4.3 for surface area. (The reader should review the discussion following that definition before continuing.)

Step 1. As shown in Figure 18.4.4, subdivide a rectangle containing R into subrectangles and discard all those subrectangles that contain any points outside of R. This leaves only rectangles $R_1, R_2, \ldots, R_n$ that are subsets of R. We will denote the area of R_k by ΔA_k.

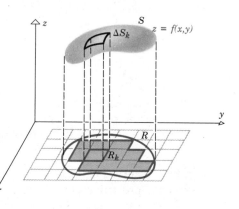

Figure 18.4.4

Step 2. As illustrated in Figure 18.4.4, project each subrectangle onto the lamina, let the surface areas of the resulting sections of lamina be

$$\Delta S_1, \Delta S_2, \ldots, \Delta S_n$$

and let $(x_k^*, y_k^*, z_k^*) = [x_k^*, y_k^*, f(x_k^*, y_k^*)]$ be an arbitrary point in the kth section.

Step 3. As in Step 3 of the surface area discussion, the surface area ΔS_k can be approximated by

$$\Delta S_k \approx \sqrt{f_x(x_k^*, y_k^*)^2 + f_y(x_k^*, y_k^*)^2 + 1} \, \Delta A_k \quad (3)$$

It follows from (1) that the mass ΔM_k of the kth section of lamina is approximately

$$\Delta M_k \approx \delta(x_k^*, y_k^*, z_k^*) \, \Delta S_k = \delta[x_k^*, y_k^*, f(x_k^*, y_k^*)] \, \Delta S_k \quad (4)$$

Substituting (3) into (4) and adding the masses of the individual sections yields the following approximation to the total mass M of the lamina

$$M \approx \sum_{k=1}^{n} \delta[x_k^*, y_k^*, f(x_k^*, y_k^*)] \sqrt{f_x(x_k^*, y_k^*)^2 + f_y(x_k^*, y_k^*)^2 + 1} \, \Delta A_k$$

Step 4. If we repeat the subdivision process using more and more rectangles with dimensions that decrease to zero as $n \to +\infty$, then it is reasonable to assume that the errors in the approximations diminish to zero and the exact mass of the lamina is

$$M = \lim_{n \to +\infty} \sum_{k=1}^{n} \delta[x_k^*, y_k^*, f(x_k^*, y_k^*)] \sqrt{f_x(x_k^*, y_k^*)^2 + f_y(x_k^*, y_k^*)^2 + 1} \, \Delta A_k$$

or, equivalently,

$$M = \iint_R \delta[x, y, f(x, y)] \sqrt{f_x(x, y)^2 + f_y(x, y)^2 + 1} \, dA$$

which agrees with (2a).

▶ **Example 1** A curved lamina is the portion of the paraboloid $z = x^2 + y^2$ below the plane $z = 1$ (Figure 18.4.5) and has constant density $\delta(x, y, z) = \delta_0$. Find the mass of the lamina.

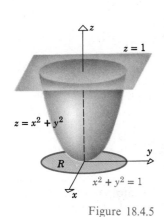

Figure 18.4.5

Solution. Since $z = f(x, y) = x^2 + y^2$, it follows that

$$\frac{\partial z}{\partial x} = 2x \quad \text{and} \quad \frac{\partial z}{\partial y} = 2y$$

Substituting these expressions and $\delta(x, y, z) = \delta(x, y, f(x, y)) = \delta_0$ into (2b) yields

$$M = \iint_R \delta_0 \sqrt{(2x)^2 + (2y)^2 + 1} \, dA = \delta_0 \iint_R \sqrt{4x^2 + 4y^2 + 1} \, dA \quad (5)$$

where R is the circular region enclosed by $x^2 + y^2 = 1$. To evaluate (5) we use polar coordinates:

$$M = \delta_0 \int_0^{2\pi} \int_0^1 \sqrt{4r^2 + 1} \, r \, dr \, d\theta = \frac{\delta_0}{12} \int_0^{2\pi} (4r^2 + 1)^{3/2} \Big]_{r=0}^1 d\theta$$

$$= \frac{\delta_0}{12} \int_0^{2\pi} (5^{3/2} - 1) \, d\theta = \frac{\pi \delta_0}{6} (5\sqrt{5} - 1) \qquad \blacktriangleleft$$

We will now generalize the procedure used to derive formulas (2a) and (2b) to develop the concept of a *surface integral*.

Let σ be a surface with finite surface area and $g(x, y, z)$ a continuous function defined on σ. (We will *not* assume in this discussion that σ can be represented by an equation of the form $z = f(x, y)$. Thus, the surface might be a sphere $x^2 + y^2 + z^2 = a^2$ or perhaps a surface represented by an equation of the form $y = f(x, z)$ or $x = f(y, z)$.)

Subdivide σ into n parts $\sigma_1, \sigma_2, \ldots, \sigma_n$ with surface areas $\Delta S_1, \Delta S_2, \ldots, \Delta S_n$ and form the sum

$$\sum_{k=1}^{n} g(x_k^*, y_k^*, z_k^*)\, \Delta S_k \tag{6}$$

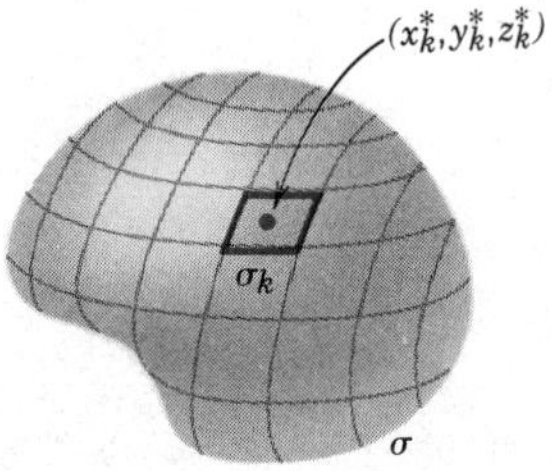

where (x_k^*, y_k^*, z_k^*) is an arbitrary point in the kth part σ_k (Figure 18.4.6).

Now, repeat the subdivision process dividing σ into more and more parts, so that each part shrinks to a point as $n \to +\infty$. If (6) tends toward a limit as $n \to +\infty$, and if this limit does not depend on the way the subdivisions are made or how the points (x_k^*, y_k^*, z_k^*) are chosen, then this limit is called the *surface integral* of $g(x, y, z)$ over σ, and is denoted by

Figure 18.4.6

$$\iint_{\sigma} g(x, y, z)\, dS = \lim_{n \to +\infty} \sum_{k=1}^{n} g(x_k^*, y_k^*, z_k^*)\, \Delta S_k \tag{7}$$

▶ **Example 2**   In the special case where $g(x, y, z) = 1$, the integral in (7) is the surface area of σ, that is

$$S = \text{surface area of } \sigma = \iint_{\sigma} dS \tag{8}$$

To see this, observe that for every subdivision of σ into parts with surface areas $\Delta S_1, \Delta S_2, \ldots, \Delta S_n$, these surface areas add up to the total surface area S, that is

$$S = \sum_{k=1}^{n} \Delta S_k$$

Since the right side has the value S for all n, it follows that

$$S = \lim_{n \to +\infty} \sum_{k=1}^{n} \Delta S_k.$$

Formula (8) now follows from this and (7).   ◀

▶ **Example 3** If σ is a curved lamina with density $\delta(x, y, z)$ and mass M, then

$$M = \iint_\sigma \delta(x, y, z)\, dS \tag{9}$$

To see this, let σ be subdivided into n parts with surface areas $\Delta S_1, \Delta S_2, \ldots, \Delta S_n$ and for $k = 1, 2, \ldots, n$ let (x_k^*, y_k^*, z_k^*) be an arbitrary point in the kth part. It follows from (1) that

$$\delta(x_k^*, y_k^*, z_k^*)\, \Delta S_k$$

is approximately the mass of the kth part, and so the entire mass M can be approximated by

$$M \approx \sum_{k=1}^n \delta(x_k^*, y_k^*, z_k^*)\, \Delta S_k$$

If we assume that the error in this approximation tends to zero as the number of subdivisions tends to infinity and the parts shrink to points, then the exact mass is

$$M = \lim_{n \to +\infty} \sum_{k=1}^n \delta(x_k^*, y_k^*, z_k^*)\, \Delta S_k = \iint_\sigma \delta(x, y, z)\, dS \qquad \blacktriangleleft$$

If σ is a curved lamina with density $\delta(x, y, z)$ and equation $z = f(x, y)$, then (2b) expresses the mass of the lamina as a double integral and (9) as a surface integral. Equating (2b) and (9) yields the following relationship between the surface integral and the double integral:

$$\iint_\sigma \delta(x, y, z)\, dS = \iint_R \delta[x, y, f(x, y)] \sqrt{\left(\frac{\partial z}{\partial x}\right)^2 + \left(\frac{\partial z}{\partial y}\right)^2 + 1}\, dA$$

This is a special case of the following general result that can be used to evaluate many surface integrals.

18.4.2 THEOREM (a) *Let σ be a surface with equation $z = f(x, y)$ and R its projection on the xy-plane. If f has continuous first partial derivatives on R and $g(x, y, z)$ is continuous on σ, then*

$$\iint_\sigma g(x, y, z)\, dS = \iint_R g[x, y, f(x, y)] \sqrt{\left(\frac{\partial z}{\partial x}\right)^2 + \left(\frac{\partial z}{\partial y}\right)^2 + 1}\, dA \tag{10}$$

(b) *Let σ be a surface with equation $y = f(x, z)$ and let R be its projection on the xz-plane. If f has continuous first partial derivatives on R and $g(x, y, z)$ is continuous on σ, then*

$$\iint_\sigma g(x, y, z)\, dS = \iint_R g[x, f(x, z), z] \sqrt{\left(\frac{\partial y}{\partial x}\right)^2 + \left(\frac{\partial y}{\partial z}\right)^2 + 1}\, dA \quad (11)$$

(c) *Let σ be a surface with equation $x = f(y, z)$ and let R be its projection on the yz-plane. If f has continuous first partial derivatives on R and $g(x, y, z)$ is continuous on σ, then*

$$\iint_\sigma g(x, y, z)\, dS = \iint_R g[f(y, z), y, z] \sqrt{\left(\frac{\partial x}{\partial y}\right)^2 + \left(\frac{\partial x}{\partial z}\right)^2 + 1}\, dA \quad (12)$$

We omit the proof.

▶ **Example 4** Evaluate the surface integral

$$\iint_\sigma xz\, dS$$

where σ is the part of the plane $x + y + z = 1$ that lies in the first octant.

Solution. Since the equation of the plane can be written as

$$z = 1 - x - y$$

which is of the form $z = f(x, y)$, we can apply Formula (10) with $z = f(x, y) = 1 - x - y$ and $g(x, y, z) = xz$. Thus,

$$\frac{\partial z}{\partial x} = -1 \quad \text{and} \quad \frac{\partial z}{\partial y} = -1$$

so (10) becomes

$$\iint_\sigma xz\, dS = \iint_R x(1 - x - y)\sqrt{(-1)^2 + (-1)^2 + 1}\, dA \quad (13)$$

where R is the projection of σ on the xy-plane (Figure 18.4.7a). Rewriting the double integral in (13) as an iterated integral yields

$$\iint_\sigma xz\, dS = \sqrt{3} \int_0^1 \int_0^{1-x} (x - x^2 - xy)\, dy\, dx$$

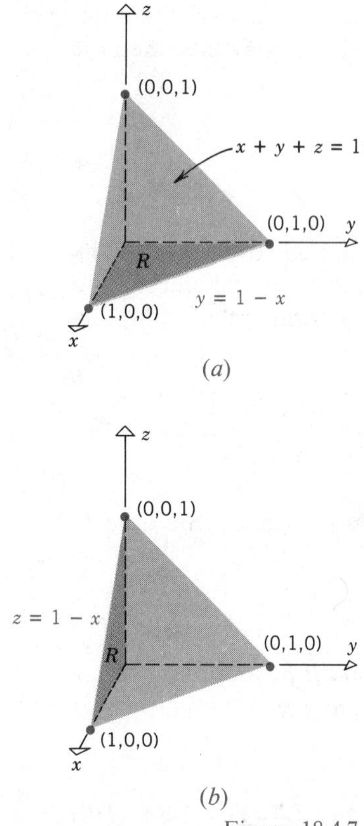

(a)

(b)

Figure 18.4.7

$$= \sqrt{3} \int_0^1 xy - x^2 y - \frac{xy^2}{2} \Big]_{y=0}^{1-x} dx$$

$$= \sqrt{3} \int_0^1 \left(\frac{x}{2} - x^2 + \frac{x^3}{2} \right) dx$$

$$= \sqrt{3} \left[\frac{x^2}{4} - \frac{x^3}{3} + \frac{x^4}{8} \right]_0^1 = \frac{\sqrt{3}}{24}$$

Alternate Solution. Since the equation of the plane can be written as

$$y = 1 - x - z$$

which is of the form $y = f(x, z)$, we can apply Formula (11) with $y = f(x, z) = 1 - x - z$. Thus,

$$\frac{\partial y}{\partial x} = -1 \qquad \text{and} \qquad \frac{\partial y}{\partial z} = -1$$

so (11) yields

$$\iint_\sigma xz \, dS = \iint_R xz \sqrt{(-1)^2 + (-1)^2 + 1} \, dA \tag{14}$$

where R is the projection of σ on the xz-plane (Figure 18.4.7*b*). Rewriting the double integral in (14) as an iterated integral yields

$$\iint_\sigma xz \, dS = \sqrt{3} \int_0^1 \int_0^{1-x} xz \, dz \, dx$$

$$= \sqrt{3} \int_0^1 \frac{xz^2}{2} \Big]_{z=0}^{1-x} dx$$

$$= \frac{\sqrt{3}}{2} \int_0^1 (x - 2x^2 + x^3) \, dx$$

$$= \frac{\sqrt{3}}{2} \left[\frac{x^2}{2} - \frac{2x^3}{3} + \frac{x^4}{4} \right]_0^1 = \frac{\sqrt{3}}{24} \qquad \blacktriangleleft$$

▶ Example 5 Evaluate the surface integral

$$\iint_\sigma y^2 z^2 \, dS$$

where σ is the part of the cone $z = \sqrt{x^2 + y^2}$ between the planes $z = 1$ and $z = 2$ (Figure 18.4.8).

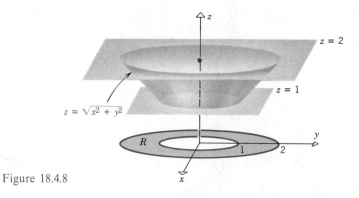

Figure 18.4.8

Solution. We will apply Formula (10) with $z = f(x, y) = \sqrt{x^2 + y^2}$ and $g(x, y, z) = y^2 z^2$. Thus,

$$\frac{\partial z}{\partial x} = \frac{x}{\sqrt{x^2 + y^2}} \quad \text{and} \quad \frac{\partial z}{\partial y} = \frac{y}{\sqrt{x^2 + y^2}}$$

so

$$\sqrt{\left(\frac{\partial z}{\partial x}\right)^2 + \left(\frac{\partial z}{\partial y}\right)^2 + 1} = \sqrt{2}$$

(verify), and (10) yields

$$\iint_\sigma y^2 z^2 \, dS = \iint_R y^2 (\sqrt{x^2 + y^2})^2 \sqrt{2} \, dA = \sqrt{2} \iint_R y^2 (x^2 + y^2) \, dA$$

where R is the annulus enclosed between $x^2 + y^2 = 1$ and $x^2 + y^2 = 4$ (Figure 18.4.8). Using polar coordinates to evaluate this double integral over the annulus R yields

$$\iint_\sigma y^2 z^2 \, dS = \sqrt{2} \int_0^{2\pi} \int_1^2 (r \sin\theta)^2 (r^2) r \, dr \, d\theta$$

$$= \sqrt{2} \int_0^{2\pi} \int_1^2 r^5 \sin^2\theta \, dr \, d\theta$$

$$= \sqrt{2} \int_0^{2\pi} \frac{r^6}{6} \sin^2\theta \Big]_{r=1}^2 \, d\theta$$

$$= \frac{21}{\sqrt{2}} \int_0^{2\pi} \sin^2\theta \, d\theta$$

$$= \frac{21}{\sqrt{2}} \left[\frac{1}{2}\theta - \frac{1}{4}\sin 2\theta\right]_0^{2\pi} = \frac{21\pi}{\sqrt{2}} \quad \begin{bmatrix} \text{See Example 1,} \\ \text{Section 9.3} \end{bmatrix}$$

◄

Sometimes Formulas (10), (11), and (12) cannot be applied because the partial derivatives in these formulas do not exist at every point of the region R. However, it may still be possible to compute surface integrals in such cases by means of a limiting procedure. The next example illustrates this.

▶ **Example 6** Evaluate the surface integral

$$\iint_\sigma dS \tag{15}$$

where σ is the upper hemisphere of radius a given by $z = \sqrt{a^2 - x^2 - y^2}$ (Figure 18.4.9).

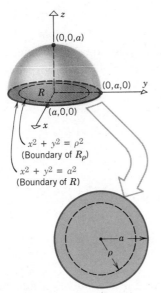

Figure 18.4.9

Solution. The partial derivatives

$$\frac{\partial z}{\partial x} = \frac{-x}{\sqrt{a^2 - x^2 - y^2}} \quad \text{and} \quad \frac{\partial z}{\partial y} = \frac{-y}{\sqrt{a^2 - x^2 - y^2}}$$

do not exist on the boundary of the region R since $x^2 + y^2 = a^2$ there. In order to overcome this problem we evaluate the double integral over a slightly smaller circular region R_ρ of radius ρ, and then let ρ approach a. The computations are as follows:

$$\iint_\sigma dS = \lim_{\rho \to a} \iint_{R_\rho} \sqrt{\left(\frac{\partial z}{\partial x}\right)^2 + \left(\frac{\partial z}{\partial y}\right)^2 + 1}\, dA$$

$$= \lim_{\rho \to a} \iint_{R_\rho} \sqrt{\frac{x^2}{a^2 - x^2 - y^2} + \frac{y^2}{a^2 - x^2 - y^2} + 1}\, dA$$

$$= \lim_{\rho \to a} \iint_{R_\rho} \frac{a}{\sqrt{a^2 - x^2 - y^2}}\, dA$$

$$= \lim_{\rho \to a} \int_0^{2\pi} \int_0^{\rho} \frac{a}{\sqrt{a^2 - r^2}} r\, dr\, d\theta$$

$$= \lim_{\rho \to a} \int_0^{2\pi} \left. -a\sqrt{a^2 - r^2}\, \right]_{r=0}^{\rho} d\theta$$

$$= \lim_{\rho \to a} \int_0^{2\pi} (a^2 - a\sqrt{a^2 - \rho^2})\, d\theta$$

$$= \lim_{\rho \to a} 2\pi(a^2 - a\sqrt{a^2 - \rho^2}) = 2\pi a^2 \qquad \blacktriangleleft$$

Note that (15) is the surface area of the hemisphere σ [see (8)], so we have shown that the surface area of a *sphere* of radius a is $4\pi a^2$.

We conclude this section by noting without proof that the standard properties of integrals also hold for surface integrals. More precisely:

$$\iint_\sigma (f + g)\, dS = \iint_\sigma f\, dS + \iint_\sigma g\, dS$$

$$\iint_\sigma (f - g)\, dS = \iint_\sigma f\, dS - \iint_\sigma g\, dS$$

$$\iint_\sigma kf\, dS = k \iint_\sigma f\, dS \;\;(k\ \text{constant})$$

and, finally, if a surface σ is subdivided into finitely many parts, then a surface integral over σ can be computed by computing it over each part and adding the results. Thus, if σ is subdivided into two parts, σ_1 and σ_2 (Figure 18.4.10), then

Figure 18.4.10

$$\iint_\sigma f\, dS = \iint_{\sigma_1} f\, dS + \iint_{\sigma_2} f\, dS$$

▶ Exercise Set 18.4

In Exercises 1–4, find the mass of the given lamina assuming the density to be a constant δ_0.

1. The lamina that is the portion of the paraboloid $z = 1 - x^2 - y^2$ above the xy-plane.

2. The lamina that is the portion of the plane $2x + 2y + z = 8$ in the first octant.

3. The lamina that is the portion of the cylinder $x^2 + z^2 = 4$ that lies above the rectangle

$R = \{(x, y): 0 \le x \le 1, 0 \le y \le 4\}$ in the xy-plane.

4. The lamina that is the portion of the paraboloid $2z = x^2 + y^2$ inside the cylinder $x^2 + y^2 = 8$.

In Exercises 5–10, evaluate the surface integrals.

5. $\iint_{\sigma} z^2 \, dS$, where σ is the portion of the cone $z = \sqrt{x^2 + y^2}$ between the planes $z = 1$ and $z = 2$.

6. $\iint_{\sigma} xyz \, dS$, where σ is the portion of the plane $x + y + z = 1$ lying in the first octant.

7. $\iint_{\sigma} x^2 y \, dS$, where σ is the portion of the cylinder $x^2 + z^2 = 1$ between the planes $y = 0, y = 1$, and above the xy-plane.

8. $\iint_{\sigma} (x^2 + y^2) z \, dS$, where σ is the portion of the sphere $x^2 + y^2 + z^2 = 4$ above the plane $z = 1$.

9. $\iint_{\sigma} (x + y + z) \, dS$, where σ is the portion of the plane $x + y = 1$ in the first octant between $z = 0$ and $z = 1$.

10. $\iint_{\sigma} (x + y) \, dS$, where σ is the portion of the plane $z = 6 - 2x - 3y$ in the first octant.

11. Evaluate

$$\iint_{\sigma} (x + y + z) \, dS$$

over the surface of the cube defined by the inequalities, $0 \le x \le 1$, $0 \le y \le 1$, $0 \le z \le 1$. [*Hint:* Integrate over each surface separately.]

12. Evaluate

$$\iint_{\sigma} zx^2 \, dS$$

over the portion of the cylinder $x^2 + y^2 = 1$ be-

tween $z = 0$ and $z = 1$. [*Hint:* Divide the surface into two parts and evaluate by projecting on the xz-plane.]

13. Evaluate

$$\iint_{\sigma} \sqrt{x^2 + y^2 + z^2} \, dS$$

over the portion of the cone $z = \sqrt{x^2 + y^2}$ below the plane $z = 1$.

14. Evaluate

$$\iint_{\sigma} (z + 1) \, dS$$

where σ is the upper hemisphere $z = \sqrt{1 - x^2 - y^2}$.

15. Evaluate

$$\iint_{\sigma} (x^2 + y^2) \, dS$$

over the surface of the sphere $x^2 + y^2 + z^2 = a^2$. [*Hint:* Divide the sphere into two parts and integrate over each part separately.]

16. Find the mass of the lamina that is the portion of the cone

$$z = \sqrt{x^2 + y^2}$$

between $z = 1$ and $z = 4$ if the density is

$$\delta(x, y, z) = x^2 z$$

17. Find the mass of the lamina that is the portion of the surface $y^2 = 4 - z$ between the planes $x = 0$, $x = 3$, $y = 0$, and $y = 3$, if the density is $\delta(x, y, z) = y$.

18. Find the mass of the lamina that is the portion of the paraboloid $z = x^2 + y^2$ below the plane $z = 1$ if the density is $\delta(x, y, z) = \sqrt{x^2 + y^2}$.

In Exercises 19 and 20 set up, but do not evaluate, an iterated integral equal to the given surface integral by projecting σ on (a) the xy-plane, (b) the yz-plane, and (c) the xz-plane.

19. $\displaystyle\iint_{\sigma} xyz \, dS$, where σ is the portion of the plane $2x + 3y + 4z = 12$ in the first octant.

20. $\displaystyle\iint_{\sigma} xz \, dS$, where σ is the portion of the sphere $x^2 + y^2 + z^2 = a^2$ in the first octant.

In Exercises 21 and 22, set up, but do not evaluate, two different iterated integrals equal to the given surface integral.

21. $\displaystyle\iint_{\sigma} xyz \, dS$, where σ is the portion of the surface $y^2 = x$ between the planes $z = 0$, $z = 4$, $y = 1$, and $y = 2$.

22. $\displaystyle\iint_{\sigma} x^2 y \, dS$, where σ is the portion of the cylinder $y^2 + z^2 = a^2$ in the first octant between the planes $x = 0$, $x = 9$, $z = y$, and $z = 2y$.

23. Prove: If a curved lamina has constant density δ_0, then its mass is the density times the surface area.

24. Show that the mass of the spherical lamina $x^2 + y^2 + z^2 = a^2$ is $2\pi a^3$ if the density at any point is equal to the distance from the point to the xy-plane.

18.5   SURFACE INTEGRALS OF VECTOR FUNCTIONS

In this section we will study surface integrals with integrands that involve vector functions. Applications of the results we obtain here will be given in later sections.

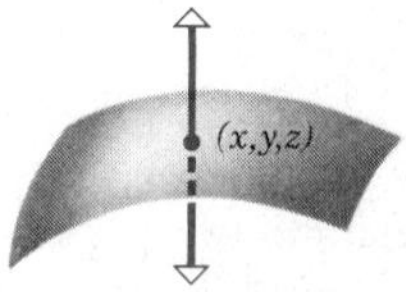

If a surface σ has a nonzero normal vector at a point (x, y, z), then there are two oppositely directed *unit normal* vectors at that point (Figure 18.5.1). These vectors are described by various names, depending on the signs of their components. For example, if the unit normal vector $\mathbf{n}$ has a positive $\mathbf{k}$ component, then $\mathbf{n}$ points roughly in the upward direction and is called an *upward unit normal* (Figure 18.5.2). If $\mathbf{n}$ has a negative $\mathbf{k}$ component, then $\mathbf{n}$ points roughly downward and is called a *downward unit normal*.

Table 18.5.1 explains the terminology used to describe unit normal vectors.

Figure 18.5.1

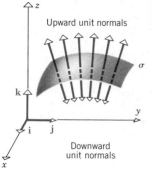

Figure 18.5.2

▶ **Example 1**   If the vector

$$\mathbf{n} = \frac{1}{\sqrt{3}}\mathbf{i} - \frac{1}{\sqrt{3}}\mathbf{j} + \frac{1}{\sqrt{3}}\mathbf{k}$$

is normal to a surface σ, then $\mathbf{n}$ is an upward unit normal (positive $\mathbf{k}$ component), a left unit normal (negative $\mathbf{j}$ component), and a forward unit normal (positive $\mathbf{i}$ component).   ◀

To compute the unit normals to a surface σ given by an equation $z = z(x, y)$, we first rewrite this equation as

$$z - z(x, y) = 0$$

Table **18.5.1**

TERMINOLOGY	MATHEMATICAL MEANING	PICTURE
Upward unit normal	Positive **k** component	
Downward unit normal	Negative **k** component	
Right unit normal	Positive **j** component	
Left unit normal	Negative **j** component	
Forward unit normal	Positive **i** component	
Backward unit normal	Negative **i** component	

which is a level surface for the function

$$G(x, y, z) = z - z(x, y).$$

By Theorem 16.7.9, the gradient

$$\nabla G = -\frac{\partial z}{\partial x}\mathbf{i} - \frac{\partial z}{\partial y}\mathbf{j} + \mathbf{k} \tag{1}$$

is normal to this surface. To obtain a *unit* normal vector, we must normalize (1). Since

$$\|\nabla G\| = \sqrt{\left(\frac{\partial z}{\partial x}\right)^2 + \left(\frac{\partial z}{\partial y}\right)^2 + 1}$$

it follows that

$$\frac{\nabla G}{\|\nabla G\|} = \frac{-\frac{\partial z}{\partial x}\mathbf{i} - \frac{\partial z}{\partial y}\mathbf{j} + \mathbf{k}}{\sqrt{\left(\frac{\partial z}{\partial x}\right)^2 + \left(\frac{\partial z}{\partial y}\right)^2 + 1}} \tag{2}$$

is a unit normal vector to σ at the point (x, y, z). Moreover, because the $\mathbf{k}$ component is positive, (2) is the upward unit normal. To find the downward unit normal, we must multiply this vector by -1. This yields

$$-\frac{\nabla G}{\|\nabla G\|} = \frac{\frac{\partial z}{\partial x}\mathbf{i} + \frac{\partial z}{\partial y}\mathbf{j} - \mathbf{k}}{\sqrt{\left(\frac{\partial z}{\partial x}\right)^2 + \left(\frac{\partial z}{\partial y}\right)^2 + 1}} \tag{3}$$

For surfaces expressed in the form $y = y(x, z)$ or $x = x(y, z)$, the unit normal vectors are obtained by normalizing the gradients of the functions $y - y(x, z)$ and $x - x(y, z)$, respectively. The resulting formulas are listed in Table 18.5.2.

Table 18.5.2

EQUATION OF σ	NORMALS TO σ	
	Upward unit normal	Downward unit normal
$z = z(x, y)$	$\dfrac{-\frac{\partial z}{\partial x}\mathbf{i} - \frac{\partial z}{\partial y}\mathbf{j} + \mathbf{k}}{\sqrt{\left(\frac{\partial z}{\partial x}\right)^2 + \left(\frac{\partial z}{\partial y}\right)^2 + 1}}$	$\dfrac{\frac{\partial z}{\partial x}\mathbf{i} + \frac{\partial z}{\partial y}\mathbf{j} - \mathbf{k}}{\sqrt{\left(\frac{\partial z}{\partial x}\right)^2 + \left(\frac{\partial z}{\partial y}\right)^2 + 1}}$
	Right unit normal	Left unit normal
$y = y(x, z)$	$\dfrac{-\frac{\partial y}{\partial x}\mathbf{i} + \mathbf{j} - \frac{\partial y}{\partial z}\mathbf{k}}{\sqrt{\left(\frac{\partial y}{\partial x}\right)^2 + \left(\frac{\partial y}{\partial z}\right)^2 + 1}}$	$\dfrac{\frac{\partial y}{\partial x}\mathbf{i} - \mathbf{j} + \frac{\partial y}{\partial z}\mathbf{k}}{\sqrt{\left(\frac{\partial y}{\partial x}\right)^2 + \left(\frac{\partial y}{\partial z}\right)^2 + 1}}$
	Forward unit normal	Backward unit normal
$x = x(y, z)$	$\dfrac{\mathbf{i} - \frac{\partial x}{\partial y}\mathbf{j} - \frac{\partial x}{\partial z}\mathbf{k}}{\sqrt{\left(\frac{\partial x}{\partial y}\right)^2 + \left(\frac{\partial x}{\partial z}\right)^2 + 1}}$	$\dfrac{-\mathbf{i} + \frac{\partial x}{\partial y}\mathbf{j} + \frac{\partial x}{\partial z}\mathbf{k}}{\sqrt{\left(\frac{\partial x}{\partial y}\right)^2 + \left(\frac{\partial x}{\partial z}\right)^2 + 1}}$

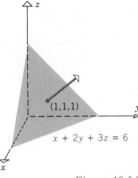

Figure 18.5.3

▶ **Example 2** Find the unit normal vector at the point $(1, 1, 1)$ to $x + 2y + 3z = 6$ shown in Figure 18.5.3.

Solution. Since the vector points above the xy-plane, to the front of the yz-plane, and to the right of the xz-plane, it can be calculated from the formula for an upward normal, a right normal, or a forward normal. For illustrative purposes we will calculate the vector two ways—as an upward normal and as a forward normal.

To use the formula for an upward normal we rewrite the equation of the plane as

$$z = 2 - \frac{1}{3}x - \frac{2}{3}y$$

from which we obtain

$$\frac{\partial z}{\partial x} = -\frac{1}{3} \quad \text{and} \quad \frac{\partial z}{\partial y} = -\frac{2}{3}.$$

Thus, from Table 18.5.2, the upward unit normal is

$$\mathbf{n} = \frac{\frac{1}{3}\mathbf{i} + \frac{2}{3}\mathbf{j} + \mathbf{k}}{\sqrt{\left(-\frac{1}{3}\right)^2 + \left(-\frac{2}{3}\right)^2 + 1}} = \frac{3}{\sqrt{14}}\left(\frac{1}{3}\mathbf{i} + \frac{2}{3}\mathbf{j} + \mathbf{k}\right)$$

$$= \frac{1}{\sqrt{14}}\mathbf{i} + \frac{2}{\sqrt{14}}\mathbf{j} + \frac{3}{\sqrt{14}}\mathbf{k}$$

To use the formula for a forward unit normal we rewrite the equation of the plane as

$$x = 6 - 2y - 3z$$

from which we obtain

$$\frac{\partial x}{\partial y} = -2 \quad \text{and} \quad \frac{\partial x}{\partial z} = -3$$

Thus, from Table 18.5.2, the forward unit normal is

$$\mathbf{n} = \frac{\mathbf{i} + 2\mathbf{j} + 3\mathbf{k}}{\sqrt{(-2)^2 + (-3)^2 + 1}} = \frac{1}{\sqrt{14}}\mathbf{i} + \frac{2}{\sqrt{14}}\mathbf{j} + \frac{3}{\sqrt{14}}\mathbf{k}$$

which agrees with result obtained above. ◀

A surface σ is said to be ***oriented*** if a unit normal vector is constructed at each point of the surface in such a way that the vectors vary continuously (have no abrupt changes in direction) as we traverse any curve on the sur-

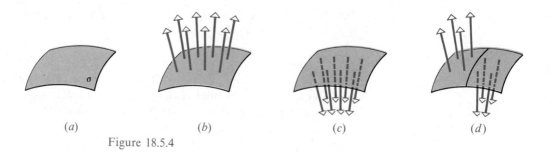

Figure 18.5.4

face. The unit normal vectors are then said to form an **orientation** of the surface. For example, the surface σ in Figure 18.5.4a can be oriented by constructing an upward unit normal at each point (Figure 18.5.4b) or a downward unit normal at each point (Figure 18.5.4c). However, the unit normals in Figure 18.5.4d do not form an orientation because the directions of these vectors change abruptly as we cross the curve drawn on the surface.

Most common surfaces can be oriented by constructing the unit normal vectors appropriately; such surfaces are called **orientable**. (A famous nonorientable surface, called a Möbius strip, is discussed in Exercise 22). It is proved in advanced courses that an orientable surface has only two possible orientations. For example, the surface σ in Figure 18.5.4 can be oriented by upward normals or by downward normals, but any mixture of the two is not an orientation since abrupt direction changes in the normals are introduced.

▶ **Example 3**   The two possible orientations of a sphere are by **inward normals** or **outward normals** (Figure 18.5.5).   ◀

We now turn to the main topic of this section, surface integrals of vector-valued functions.

18.5.1   DEFINITION   If the vector function $\mathbf{F}(x, y, z) = f(x, y, z)\mathbf{i} + g(x, y, z)\mathbf{j} + h(x, y, z)\mathbf{k}$ has continuous components on the oriented surface σ, and if $\mathbf{n} = \mathbf{n}(x, y, z)$ is the unit normal vector of the orientation at (x, y, z), then

$$\iint_\sigma \mathbf{F} \cdot \mathbf{n}\, dS$$

is called the **flux integral of F over σ**, or the **surface integral of F over σ**, or the **surface integral of the normal component of F over σ**.

In Section 18.8, we will see that surface integrals of vector-valued functions have important applications and physical interpretations. However, for now we will concentrate on computing them.

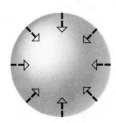

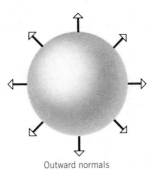

Figure 18.5.5   Outward normals

► **Example 4** Let σ be the portion of the surface $z = 1 - x^2 - y^2$ above the xy-plane oriented by upward normals (Figure 18.5.6), and let $\mathbf{F}(x, y, z) = x\mathbf{i} + y\mathbf{j} + z\mathbf{k}$. Evaluate

$$\iint_\sigma \mathbf{F} \cdot \mathbf{n}\, dS$$

Solution. From the first line in Table 18.5.2, the upward unit normal to the surface

$$z = 1 - x^2 - y^2 \tag{4}$$

is given by

$$\mathbf{n} = \frac{-\dfrac{\partial z}{\partial x}\mathbf{i} - \dfrac{\partial z}{\partial y}\mathbf{j} + \mathbf{k}}{\sqrt{\left(\dfrac{\partial z}{\partial x}\right)^2 + \left(\dfrac{\partial z}{\partial y}\right)^2 + 1}}$$

Thus, from Formula (10) in Theorem 18.4.2

$$\iint_\sigma \mathbf{F} \cdot \mathbf{n}\, dS = \iint_R \mathbf{F} \cdot \left[\frac{-\dfrac{\partial z}{\partial x}\mathbf{i} - \dfrac{\partial z}{\partial y}\mathbf{j} + \mathbf{k}}{\sqrt{\left(\dfrac{\partial z}{\partial x}\right)^2 + \left(\dfrac{\partial z}{\partial y}\right)^2 + 1}}\right] \sqrt{\left(\dfrac{\partial z}{\partial x}\right)^2 + \left(\dfrac{\partial z}{\partial y}\right)^2 + 1}\ dA$$

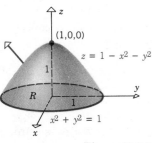

$z = 1 - x^2 - y^2$

$x^2 + y^2 = 1$

Figure 18.5.6

or, on canceling square roots,

$$\iint_\sigma \mathbf{F} \cdot \mathbf{n}\, dS = \iint_R \mathbf{F} \cdot \left(-\frac{\partial z}{\partial x}\mathbf{i} - \frac{\partial z}{\partial y}\mathbf{j} + \mathbf{k}\right) dA \tag{5}$$

where R is the region shown in 18.5.6. From (4) and the given formula for $\mathbf{F}$, it follows that

$$\mathbf{F} \cdot \left(-\frac{\partial z}{\partial x}\mathbf{i} - \frac{\partial z}{\partial y}\mathbf{j} + \mathbf{k} \right) = (x\mathbf{i} + y\mathbf{j} + z\mathbf{k}) \cdot (2x\mathbf{i} + 2y\mathbf{j} + \mathbf{k})$$

$$= 2x^2 + 2y^2 + z$$
$$= 2x^2 + 2y^2 + (1 - x^2 - y^2)$$
$$= x^2 + y^2 + 1$$

Substituting this expression in (5) yields

$$\iint_\sigma \mathbf{F} \cdot \mathbf{n} \, dS = \iint_R (x^2 + y^2 + 1) \, dA$$

$$= \int_0^{2\pi} \int_0^1 (r^2 + 1)r \, dr \, d\theta \quad \begin{bmatrix} \text{Using polar coordinates} \\ \text{to evaluate the integral.} \end{bmatrix}$$

$$= \int_0^{2\pi} \left(\frac{3}{4} \right) d\theta = \frac{3\pi}{2} \qquad \blacktriangleleft$$

The procedure used to derive Formula (5) in the last example can be used to obtain the following general results.

18.5.2   THEOREM   *If the surface σ is given by an equation of the form z = z(x, y), and if R is the projection of the surface on the xy-plane, then*

(a)
$$\iint_\sigma \mathbf{F} \cdot \mathbf{n} \, dS = \iint_R \mathbf{F} \cdot \left(-\frac{\partial z}{\partial x}\mathbf{i} - \frac{\partial z}{\partial y}\mathbf{j} + \mathbf{k} \right) dA \qquad (6)$$

if σ is oriented by upward normals.

(b)
$$\iint_\sigma \mathbf{F} \cdot \mathbf{n} \, dS = \iint_R \mathbf{F} \cdot \left(\frac{\partial z}{\partial x}\mathbf{i} + \frac{\partial z}{\partial y}\mathbf{j} - \mathbf{k} \right) dA \qquad (7)$$

if σ is oriented by downward normals.

Similar formulas that apply to surfaces that are expressed in the form $x = x(y, z)$ and $y = y(x, z)$ are discussed in the exercises.

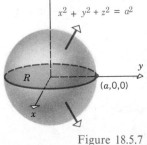

▶ **Example 5**   Let σ be the sphere $x^2 + y^2 + z^2 = a^2$ oriented by outward normals (Figure 18.5.7), and let $\mathbf{F}(x, y, z) = z\mathbf{k}$. Evaluate

$$\iint_\sigma \mathbf{F} \cdot \mathbf{n} \, dS$$

Figure 18.5.7

Solution.   On the upper hemisphere the outward unit normal is upward, while on the lower hemisphere it is downward. Since different formulas for these normals apply on the two hemispheres, it is desirable to write

$$\iint_{\sigma} \mathbf{F} \cdot \mathbf{n}\, dS = \iint_{\sigma_1} \mathbf{F} \cdot \mathbf{n}\, dS + \iint_{\sigma_2} \mathbf{F} \cdot \mathbf{n}\, dS \tag{8}$$

where σ_1 is the upper hemisphere and σ_2 is the lower hemisphere.
   The upper hemisphere σ_1 has the equation

$$z = \sqrt{a^2 - x^2 - y^2} \tag{9}$$

so that (6) yields

$$\iint_{\sigma_1} \mathbf{F} \cdot \mathbf{n}\, dS = \iint_{R} (z\mathbf{k}) \cdot \left(\frac{x}{\sqrt{a^2 - x^2 - y^2}}\mathbf{i} + \frac{y}{\sqrt{a^2 - x^2 - y^2}}\mathbf{j} + \mathbf{k} \right) dA$$

$$= \iint_{R} z\, dA \qquad \left[\begin{array}{c} R \text{ is the region shown} \\ \text{in Figure 18.5.7} \end{array} \right]$$

$$= \iint_{R} \sqrt{a^2 - x^2 - y^2}\, dA$$

$$= \int_{0}^{2\pi} \int_{0}^{a} \sqrt{a^2 - r^2}\, r\, dr\, d\theta$$

$$= \int_{0}^{2\pi} \left. -\frac{1}{3}(a^2 - r^2)^{3/2} \right]_{0}^{a} d\theta$$

$$= \int_{0}^{2\pi} \frac{1}{3}a^3\, d\theta = \frac{2\pi a^3}{3}$$

The lower hemisphere σ_2 has the equation

$$z = -\sqrt{a^2 - x^2 - y^2}$$

so that (7) yields

$$\iint_{\sigma_2} \mathbf{F} \cdot \mathbf{n}\, dS = \iint_{R} (z\mathbf{k}) \cdot \left(\frac{x}{\sqrt{a^2 - x^2 - y^2}}\mathbf{i} + \frac{y}{\sqrt{a^2 - x^2 - y^2}}\mathbf{j} - \mathbf{k} \right) dA$$

$$= \iint_{R} -z\, dA = \iint_{R} \sqrt{a^2 - x^2 - y^2}\, dA \qquad \left[\begin{array}{c} \text{The computations} \\ \text{are identical to} \\ \text{those above.} \end{array} \right]$$

$$= \frac{2\pi a^3}{3}$$

Thus, from (8)

$$\iint_{\sigma} \mathbf{F} \cdot \mathbf{n}\, dS = \frac{2\pi a^3}{3} + \frac{2\pi a^3}{3} = \frac{4\pi a^3}{3} \qquad \blacktriangleleft$$

▶ Exercise Set 18.5

1. Use three different formulas in Table 18.5.2 to calculate the unit normal to $2x + 3y + 4z = 9$ at $(1, 1, 1)$ that has positive components.

2. Use three different formulas in Table 18.5.2 to calculate the unit normal to $x^2 + y^2 + z^2 = 9$ at $(2, 1, -2)$ that points below the xy-plane.

3. In each part, use any appropriate formula to calculate the indicated unit normal.
   (a) The unit normal to $z = x^2 + y^2$ at $(1, 2, 5)$ that points toward the z-axis;
   (b) the unit normal to $z = \sqrt{x^2 + y^2}$ at $(-3, 4, 5)$ that points toward the xz-plane;
   (c) the unit normal to the cylinder $x^2 + z^2 = 25$ at $(3, 2, -4)$ that points away from the xy-plane.

4. In each part, use any appropriate formula to calculate the indicated unit normal.
   (a) The unit normal to the surface $y^2 = x$ at $(1, 1, 2)$ that points toward the xz-plane;
   (b) the unit normal to the hyperbolic paraboloid $y = z^2 - x^2$ at $(1, 3, 2)$ that points toward the yz-plane;
   (c) the unit normal to the cone $x^2 = y^2 + z^2$ at $(\sqrt{2}, -1, -1)$ that points away from the xy-plane.

In Exercises 5–15, evaluate $\iint\limits_{\sigma} \mathbf{F} \cdot \mathbf{n} \, dS$.

5. $\mathbf{F}(x, y, z) = x\mathbf{i} + y\mathbf{j} + 2z\mathbf{k}$; σ is the portion of the surface $z = 1 - x^2 - y^2$ above the xy-plane oriented by upward normals.

6. $\mathbf{F}(x, y, z) = (x + y)\mathbf{i} + (y + z)\mathbf{j} + (z + x)\mathbf{k}$; σ is the portion of the plane $x + y + z = 1$ in the first octant oriented by unit normals with positive components.

7. $\mathbf{F}(x, y, z) = z^2\mathbf{k}$; σ is the upper hemisphere $z = \sqrt{1 - x^2 - y^2}$ oriented by upward unit normals.

8. $\mathbf{F}(x, y, z) = x^2\mathbf{i} + yx\mathbf{j} + zx\mathbf{k}$; σ is the portion of the plane $6x + 3y + 2z = 6$ in the first octant oriented by unit normals with positive components.

9. $\mathbf{F}(x, y, z) = x\mathbf{i} + y\mathbf{j} + z\mathbf{k}$; σ is the upper hemisphere $z = \sqrt{9 - x^2 - y^2}$ oriented by upward unit normals.

10. $\mathbf{F}(x, y, z) = \mathbf{i} + \mathbf{j} + \mathbf{k}$; σ is the portion of the cone $z = \sqrt{x^2 + y^2}$ below the plane $z = 1$, oriented by downward unit normals.

11. $\mathbf{F}(x, y, z) = x\mathbf{i} + y\mathbf{j} + 2z\mathbf{k}$; σ is the portion of the cone $z^2 = x^2 + y^2$ between the planes $z = 1$ and $z = 2$, oriented by upward unit normals.

12. $\mathbf{F}(x, y, z) = y\mathbf{j} + \mathbf{k}$; σ is the portion of the paraboloid $z = x^2 + y^2$ below the plane $z = 4$, oriented by downward unit normals.

13. $\mathbf{F}(x, y, z) = x\mathbf{k}$; σ is the portion of the paraboloid $z = x^2 + y^2$ below the plane $z = y$, oriented by downward unit normals.

14. $\mathbf{F}(x, y, z) = x\mathbf{i} + y\mathbf{j} + z\mathbf{k}$; σ is the portion of the cylinder $z^2 = 1 - x^2$ between the planes $y = 1$ and $y = -2$, oriented by outward unit normals.

15. $\mathbf{F}(x, y, z) = x\mathbf{i} + y\mathbf{j} + z\mathbf{k}$; σ is the sphere $x^2 + y^2 + z^2 = a^2$ oriented by outward unit normals.

16. Let σ be the surface of the cube bounded by the planes $x = \pm 1$, $y = \pm 1$, $z = \pm 1$, oriented by outward unit normals. In each part, evaluate the flux integral of $\mathbf{F}$ over σ.
   (a) $\mathbf{F}(x, y, z) = x\mathbf{i}$
   (b) $\mathbf{F}(x, y, z) = x\mathbf{i} + y\mathbf{j} + z\mathbf{k}$
   (c) $\mathbf{F}(x, y, z) = x^2\mathbf{i} + y^2\mathbf{j} + z^2\mathbf{k}$

17. Show that reversing the orientation of σ reverses the sign of
$$\iint\limits_{\sigma} \mathbf{F} \cdot \mathbf{n} \, dS$$

18. Derive Formulas (6) and (7) in Theorem 18.5.2. [*Hint:* Read the derivation of (5) in Example 4.]

19. Obtain the analogs of Formulas (6) and (7) for:
   (a) surfaces of the form $x = x(y, z)$;
   (b) surfaces of the form $y = y(x, z)$.

20. Evaluate $\iint\limits_{\sigma} \mathbf{F} \cdot \mathbf{n} \, dS$ if σ is the portion of the paraboloid $x = y^2 + z^2$ with $x \leq 1$ and $z \geq 0$ orien-

ted by backward unit normals and $\mathbf{F}(x, y, z) = y\mathbf{i} - z\mathbf{j} + 8\mathbf{k}$. [*Hint:* See Exercise 19(a).]

21. Evaluate $\displaystyle\iint_\sigma \mathbf{F} \cdot \mathbf{n} \, dS$ if σ is the hemisphere $y = \sqrt{1 - x^2 - z^2}$ oriented by right unit normals and $\mathbf{F}(x, y, z) = x\mathbf{i} + y\mathbf{j} + z\mathbf{k}$. [*Hint:* See Exercise 19(b).]

22. The best-known example of a nonorientable surface is the **Möbius strip,** which can be visualized by taking a band of paper, twisting it once, and gluing the ends together (Figures 18.5.8a and 18.5.8b). In Figure 18.5.8c, we have tried to construct unit normal vectors whose directions vary continuously moving counterclockwise around the dotted curve starting and finishing on the vertical line.

   (a) Explain why these vectors are not part of an orientation of the surface.
   (b) Explain why the surface is not orientable.

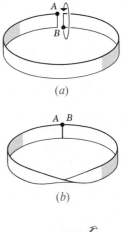

(a)

(b)

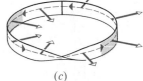

(c)

Figure 18.5.8

18.6 THE DIVERGENCE THEOREM

In two dimensions, Green's Theorem shows how to express a double integral over a plane region in terms of a line integral over the boundary. In three dimensions there is a corresponding result, called the *Divergence Theorem* or *Gauss'* Theorem* (see p. 1088), which expresses a triple integral over a solid in terms of a surface integral over the boundary. In this section we will discuss this important result, and in Section 18.8 we will illustrate some of its applications.

We will begin with some terminology.

18.6.1 DEFINITION If $\mathbf{F}(x, y, z) = f(x, y, z)\mathbf{i} + g(x, y, z)\mathbf{j} + h(x, y, z)\mathbf{k}$, then we define the *divergence of F,* written div **F**, by

$$\operatorname{div} \mathbf{F} = \frac{\partial f}{\partial x} + \frac{\partial g}{\partial y} + \frac{\partial h}{\partial z}$$

▶ Example 1 The divergence of

$$\mathbf{F}(x, y, z) = xy^2z^2\mathbf{i} + (3xz^2 + y^4)\mathbf{j} + x^3y^2z\mathbf{k}$$

is

$$\text{div } \mathbf{F} = \frac{\partial}{\partial x}(xy^2z^2) + \frac{\partial}{\partial y}(3xz^2 + y^4) + \frac{\partial}{\partial z}(x^3y^2z)$$

$$= y^2z^2 + 4y^3 + x^3y^2 \qquad \blacktriangleleft$$

18.6.2   THEOREM
Divergence Theorem
(Gauss' Theorem)

Let G be a solid with surface σ oriented by outward unit normals. If

$$\mathbf{F}(x, y, z) = f(x, y, z)\mathbf{i} + g(x, y, z)\mathbf{j} + h(x, y, z)\mathbf{k}$$

where f, g, and h have continuous first partial derivatives on some open set containing G, then

$$\iint_{\sigma} \mathbf{F} \cdot \mathbf{n} \, dS = \iiint_{G} \text{div } \mathbf{F} \, dV \tag{1}$$

---

[*]KARL FRIEDRICH GAUSS (1777–1855)—German mathematician and scientist. Sometimes called the "prince of mathematicians," Gauss ranks with Newton and Archimedes as one of the three greatest mathematicians who ever lived. His father, a laborer, was an uncouth but honest man who would have liked Gauss to take up a trade such as gardening or bricklaying; but the boy's genius for mathematics was not to be denied. In the entire history of mathematics there may never have been a child so precocious as Gauss—by his own account he worked out the rudiments of arithmetic before he could talk. One day, before he was even three years old, his genius became apparent to his parents in a very dramatic way. His father was preparing the weekly payroll for the laborers under his charge while the boy watched quietly from a corner. At the end of the long and tedious calculation, Gauss informed his father that there was an error in the result and stated the answer, which he had worked out in his head. To the astonishment of his parents, a check of the computations showed Gauss to be correct!

For his elementary education Gauss was enrolled in a squalid school run by a man named Büttner whose main teaching technique was thrashing. Büttner was in the habit of assigning long addition problems which, unknown to his students, were arithmetic progressions that he could sum up using formulas. On the first day that Gauss entered the arithmetic class, the students were asked to sum the numbers from 1 to 100. But no sooner had Büttner stated the problem than Gauss turned over his slate and exclaimed in his peasant dialect, "Ligget se'." (Here it lies.) For nearly an hour Büttner glared at Gauss, who sat with folded hands while his classmates toiled away. When Büttner examined the slates at the end of the period, Gauss's slate contained a single number, 5050—the only correct solution in the class.

To his credit, Büttner recognized the genius of Gauss and with the help of his assistant, John Bartels, had him brought to the attention of Karl Wilhelm Ferdinand, Duke of Brunswick. The shy and awkward boy, who was then fourteen, so captivated the Duke that he subsidized him through preparatory school, college, and the early part of his career.

From 1795 to 1798 Gauss studied mathematics at the University of Göttingen, receiving his degree in absentia from the University of Helmstadt. For his dissertation, he gave the first complete proof of the fundamental theorem of algebra, which states that every polynomial equation has as many solutions as its degree. At age 19 he solved a problem that baffled Euclid, inscribing a regular polygon of seventeen sides in a circle using straightedge and compass; and in 1801, at age 24, he published his first masterpiece, *Disquisitiones Arithmeticae,* considered by many to be one of the most brilliant achievements in mathematics. In that paper Gauss systematized the study of number theory (properties of the integers) and formulated the basic concepts that form the foundation of that subject. (Continued on p. 1089.)

The proof of this theorem for a general solid G is too difficult to present here. However, we can give a proof for the special case where G is a simple solid (see Figure 17.5.4 and the discussion preceding it).

Proof (for simple solids). Let G be a simple solid with upper surface $z = g_2(x, y)$, lower surface $z = g_1(x, y)$, and projection R on the xy-plane. Let σ_1 denote the lower surface, σ_2 the upper surface, and σ_3 the lateral surface (Figure 18.6.1a). If the upper surface and lower surface meet as in Figure 18.6.1b, then there is no lateral surface σ_3. Our proof will allow for both cases shown in those figures.

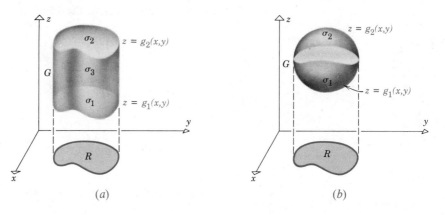

Figure 18.6.1   (a)   (b)

In the same year that the *Disquisitiones* was published, Gauss again applied his phenomenal computational skills in a dramatic way. The astronomer Giuseppi Piazzi had observed the asteroid Ceres for $\frac{1}{40}$ of its orbit, but lost it in the sun. Using only three observations and the "method of least squares" that he had developed in 1795, Gauss computed the orbit with such accuracy that astronomers had no trouble relocating it the following year. This achievement brought him instant recognition as the premier mathematician in Europe, and in 1807 he was made Professor of Astronomy and head of the astronomical observatory at Göttingen.

In the years that followed, Gauss revolutionized mathematics by bringing to it standards of precision and rigor undreamed of by his predecessors. He had a passion for perfection that drove him to polish and rework his papers rather than publish less finished work in greater numbers—his favorite saying was "Pauca, sed matura" (Few, but ripe). As a result, many of his important discoveries were squirreled away in diaries that remained unpublished until years after his death.

Among his myriad achievements, Gauss discovered the Gaussian or "bell-shaped" error curve fundamental in probability, gave the first geometric interpretation of complex numbers and established their fundamental role in mathematics, developed methods of characterizing surfaces intrinsically by means of the curves that they contain, developed the theory of conformal (angle-preserving) maps, and discovered non-Euclidean geometry 30 years before the ideas were published by others. In physics he made major contributions to the theory of lenses and capillary action, and with Wilhelm Weber he did fundamental work in electromagnetism. Gauss invented the heliotrope, bifilar magnetometer, and an electrotelegraph.

Gauss was deeply religious and aristocratic in demeanor. He mastered foreign languages with ease, read extensively, and enjoyed minerology and botany as hobbies. He disliked teaching and was usually cool and discouraging to other mathematicians, possibly because he had already anticipated their work. It has been said that if Gauss had published all of his discoveries, the current state of mathematics would be advanced by 50 years. He was without a doubt the greatest mathematician of the modern era.

We want to show that

$$\iint_{\sigma} \mathbf{F} \cdot \mathbf{n} \, dS = \iiint_{G} \operatorname{div} \mathbf{F} \, dV$$

or equivalently

$$\iint_{\sigma} [f(x, y, z)\mathbf{i} + g(x, y, z)\mathbf{j} + h(x, y, z)\mathbf{k}] \cdot \mathbf{n} \, dS = \iiint_{G} \left(\frac{\partial f}{\partial x} + \frac{\partial g}{\partial y} + \frac{\partial h}{\partial z} \right) dV$$

To prove this, it suffices to prove the following three equalities:

$$\iint_{\sigma} [f(x, y, z)\mathbf{i} \cdot \mathbf{n}] \, dS = \iiint_{G} \frac{\partial f}{\partial x} \, dV \tag{2a}$$

$$\iint_{\sigma} [g(x, y, z)\mathbf{j} \cdot \mathbf{n}] \, dS = \iiint_{G} \frac{\partial g}{\partial y} \, dV \tag{2b}$$

$$\iint_{\sigma} [h(x, y, z)\mathbf{k} \cdot \mathbf{n}] \, dS = \iiint_{G} \frac{\partial h}{\partial z} \, dV \tag{2c}$$

Since the proofs of all three formulas are similar, we shall prove only the third.

It follows from Theorem 17.5.2 that

$$\iiint_{G} \frac{\partial h}{\partial z} \, dV = \iint_{R} \left[\int_{g_1(x,y)}^{g_2(x,y)} \frac{\partial h}{\partial z} \, dz \right] dA = \iint_{R} \left[h(x, y, z) \right]_{z=g_1(x,y)}^{g_2(x,y)} dA$$

so

$$\iiint_{G} \frac{\partial h}{\partial z} \, dV = \iint_{R} [h(x, y, g_2(x, y)) - h(x, y, g_1(x, y))] \, dA \tag{3}$$

We will evaluate the surface integral in (2c) by integrating over each surface of G separately. If there is a lateral surface σ_3, then at each point of this surface $\mathbf{n} \cdot \mathbf{k} = 0$ since $\mathbf{n}$ is horizontal and $\mathbf{k}$ is vertical. Thus,

$$\iint_{\sigma_3} [h(x, y, z)\mathbf{k} \cdot \mathbf{n}] \, dS = 0$$

Therefore, regardless of whether G has a lateral surface or not, we can write

$$\iint_{\sigma} [h(x, y, z)\mathbf{k} \cdot \mathbf{n}] \, dS = \iint_{\sigma_1} [h(x, y, z)\mathbf{k} \cdot \mathbf{n}] \, dS + \iint_{\sigma_2} [h(x, y, z)\mathbf{k} \cdot \mathbf{n}] \, dS \tag{4}$$

On the upper surface σ_2, the outer normal is an upward normal, and on the lower surface σ_1 the outer normal is a downward normal. Thus, Theorem 18.5.2 implies that

$$\iint_{\sigma_1} [h(x, y, z)\mathbf{k} \cdot \mathbf{n}]\, dS = \iint_R \left[h(x, y, g_2(x, y))\mathbf{k} \cdot \left(-\frac{\partial z}{\partial x}\mathbf{i} - \frac{\partial z}{\partial y}\mathbf{j} + \mathbf{k} \right) \right] dA$$

$$= \iint_R [h(x, y, g_2(x, y))]\, dA \qquad (5)$$

and

$$\iint_{\sigma_2} [h(x, y, z)\mathbf{k} \cdot \mathbf{n}]\, dS = \iint_R \left[h(x, y, g_1(x, y))\mathbf{k} \cdot \left(\frac{\partial z}{\partial x}\mathbf{i} + \frac{\partial z}{\partial y}\mathbf{j} - \mathbf{k} \right) \right] dA$$

$$= -\iint_R [h(x, y, g_1(x, y))]\, dA \qquad (6)$$

Substituting (5) and (6) into (4) and combining the terms into a single integral yields:

$$\iint_{\sigma} [h(x, y, z)\mathbf{k} \cdot \mathbf{n}]\, dS = \iint_R [h(x, y, g_2(x, y)) - h(x, y, g_1(x, y))]\, dA \qquad (7)$$

Equation (2c) now follows from (3) and (7). ▌

▶ **Example 2** Let σ be the sphere $x^2 + y^2 + z^2 = a^2$ oriented by outward normals and let $\mathbf{F}(x, y, z) = z\mathbf{k}$. Use the Divergence Theorem to evaluate

$$\iint_{\sigma} \mathbf{F} \cdot \mathbf{n}\, dS$$

Solution.   Let G be the spherical solid enclosed by σ. Since

$$\text{div } \mathbf{F} = \frac{\partial z}{\partial z} = 1$$

it follows from (1) that

$$\iint_{\sigma} \mathbf{F} \cdot \mathbf{n}\, dS = \iiint_G dV = \text{volume of } G = \frac{4\pi a^3}{3} \qquad \blacktriangleleft$$

REMARK. Notice how much simpler this solution is than the direct calculation of $\iint_{\sigma} \mathbf{F} \cdot \mathbf{n}\, dS$ in Example 5 of Section 18.5.

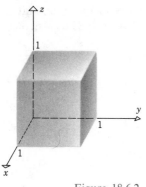

Figure 18.6.2

▶ **Example 3** Let G be the cube in the first octant shown in Figure 18.6.2, and let σ be its surface. Use the Divergence Theorem to compute

$$\iint_{\sigma} \mathbf{F} \cdot \mathbf{n} \, dS$$

where $\mathbf{F}(x, y, z) = 2x\mathbf{i} + 3y\mathbf{j} + z^2\mathbf{k}$, and $\mathbf{n}$ is the outward unit normal.

Solution. Since

$$\operatorname{div} \mathbf{F} = \frac{\partial}{\partial x}(2x) + \frac{\partial}{\partial y}(3y) + \frac{\partial}{\partial z}(z^2) = 5 + 2z$$

it follows from (1) that

$$\iint_{\sigma} \mathbf{F} \cdot \mathbf{n} \, dS = \iiint_{G} (5 + 2z) \, dV = \int_0^1 \int_0^1 \int_0^1 (5 + 2z) \, dz \, dy \, dx$$

$$= \int_0^1 \int_0^1 \left[\frac{1}{4}(5 + 2z)^2 \right]_{z=0}^1 dy \, dx = \int_0^1 \int_0^1 6 \, dy \, dx = 6 \quad ◀$$

▶ **Example 4** Let G be the cylindrical solid bounded by $x^2 + y^2 = 9$, $z = 0$, and $z = 2$ (Figure 18.6.3), and let σ be its surface. Use the Divergence Theorem to evaluate

$$\iint_{\sigma} \mathbf{F} \cdot \mathbf{n} \, dS$$

where $\mathbf{F}(x, y, z) = x^3\mathbf{i} + y^3\mathbf{j} + z^2\mathbf{k}$, and $\mathbf{n}$ is the outward unit normal to σ.

Solution. Since

$$\operatorname{div} \mathbf{F} = \frac{\partial}{\partial x}(x^3) + \frac{\partial}{\partial y}(y^3) + \frac{\partial}{\partial z}(z^2) = 3x^2 + 3y^2 + 2z$$

it follows from (1) that

$$\iint_{\sigma} \mathbf{F} \cdot \mathbf{n} \, dS = \iiint_{G} (3x^2 + 3y^2 + 2z) \, dV$$

We will use cylindrical coordinates to evaluate the triple integral. We obtain

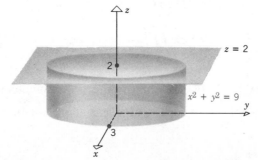

Figure 18.6.3

$$\iint_\sigma \mathbf{F} \cdot \mathbf{n}\, dS = \int_0^{2\pi} \int_0^3 \int_0^2 (3r^2 + 2z)r\, dz\, dr\, d\theta$$

$$= \int_0^{2\pi} \int_0^3 \left[3r^3 z + z^2 r \right]_{z=0}^{2} dr\, d\theta$$

$$= \int_0^{2\pi} \int_0^3 (6r^3 + 4r)\, dr\, d\theta$$

$$= \int_0^{2\pi} \left[\frac{3r^4}{2} + 2r^2 \right]_0^3 d\theta$$

$$= \int_0^{2\pi} \frac{279}{2}\, d\theta = 279\pi \qquad \blacktriangleleft$$

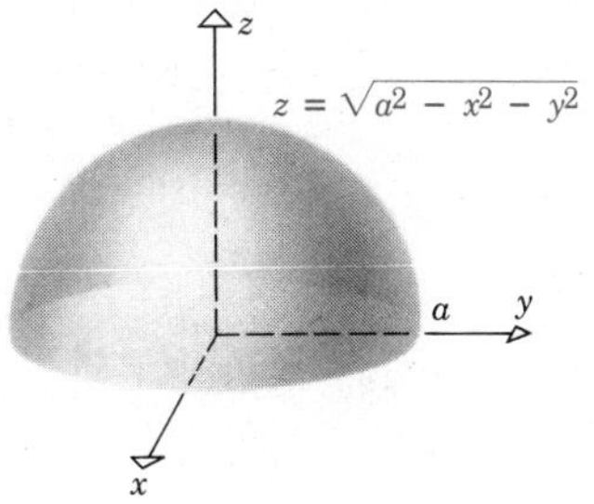

$z = \sqrt{a^2 - x^2 - y^2}$

▶ **Example 5** Let G be the solid bounded above by the hemisphere $z = \sqrt{a^2 - x^2 - y^2}$ and below by the plane $z = 0$ (Figure 18.6.4). Find

$$\iint_\sigma \mathbf{F} \cdot \mathbf{n}\, dS$$

where $\mathbf{F}(x, y, z) = x^3\mathbf{i} + y^3\mathbf{j} + z^3\mathbf{k}$, σ is the surface of G, and $\mathbf{n}$ is the outer unit normal.

Figure 18.6.4   *Solution.* Since

$$\text{div } \mathbf{F} = \frac{\partial}{\partial x}(x^3) + \frac{\partial}{\partial y}(y^3) + \frac{\partial}{\partial z}(z^3) = 3x^2 + 3y^2 + 3z^2$$

it follows from (1) that

$$\iint_\sigma \mathbf{F} \cdot \mathbf{n}\, dS = \iiint_G (3x^2 + 3y^2 + 3z^2)\, dV$$

We will use spherical coordinates to evaluate the triple integral. We obtain

$$\iint_\sigma \mathbf{F} \cdot \mathbf{n}\, dS = \int_0^{2\pi} \int_0^{\pi/2} \int_0^a (3\rho^2)\rho^2 \sin\phi\, d\rho\, d\phi\, d\theta$$

$$= 3 \int_0^{2\pi} \int_0^{\pi/2} \int_0^a \rho^4 \sin\phi\, d\rho\, d\phi\, d\theta$$

$$= 3 \int_0^{2\pi} \int_0^{\pi/2} \left[\frac{\rho^5}{5} \sin\phi \right]_{\rho=0}^{a} d\phi\, d\theta$$

$$= \frac{3a^5}{5} \int_0^{2\pi} \int_0^{\pi/2} \sin\phi\, d\phi\, d\theta$$

$$= \frac{3a^5}{5} \int_0^{2\pi} \left[-\cos\phi \right]_0^{\pi/2} d\theta$$

$$= \frac{3a^5}{5} \int_0^{2\pi} d\theta = \frac{6\pi a^5}{5} \qquad \blacktriangleleft$$

► Exercise Set 18.6

In Exercises 1–5, find the divergence of **F**.

1. $\mathbf{F}(x, y, z) = xz^3\mathbf{i} + 2y^4x^2\mathbf{j} + 5z^2y\mathbf{k}.$

2. $\mathbf{F}(x, y, z) = 7y^3z^2\mathbf{i} - 8x^2z^5\mathbf{j} - 3xy^4\mathbf{k}.$

3. $\mathbf{F}(x, y, z) = e^{xy}\mathbf{i} - \cos y\mathbf{j} + \sin^2 z\mathbf{k}.$

4. $\mathbf{F}(x, y, z) = \dfrac{1}{\sqrt{x^2 + y^2 + z^2}}(x\mathbf{i} + y\mathbf{j} + z\mathbf{k}).$

5. $\mathbf{F}(x, y, z) = \ln x\mathbf{i} + e^{xyz}\mathbf{j} + \tan^{-1}\left(\dfrac{z}{x}\right)\mathbf{k}.$

In Exercises 6–17, use the Divergence Theorem to evaluate $\displaystyle\iint\limits_{\sigma} \mathbf{F} \cdot \mathbf{n}\, dS$, where **n** is the outer unit normal to σ.

6. $\mathbf{F}(x, y, z) = 2x\mathbf{i} + 2y\mathbf{j} + 2z\mathbf{k};$ σ is the sphere $x^2 + y^2 + z^2 = 9.$

7. $\mathbf{F}(x, y, z) = 4x\mathbf{i} - 3y\mathbf{j} + 7z\mathbf{k};$ σ is the surface of the cube bounded by the coordinate planes and the planes $x = 1,\ y = 1,$ and $z = 1.$

8. $\mathbf{F}(x, y, z) = z^3\mathbf{i} - x^3\mathbf{j} + y^3\mathbf{k};$ σ is the sphere $x^2 + y^2 + z^2 = a^2.$

9. $\mathbf{F}(x, y, z) = (x - z)\mathbf{i} + (y - x)\mathbf{j} + (z - y)\mathbf{k};$ σ is the surface of the cylindrical solid bounded by $x^2 + y^2 = a^2,\ z = 0,$ and $z = 1.$

10. $\mathbf{F}(x, y, z) = x\mathbf{i} + y\mathbf{j} + z\mathbf{k};$ σ is the surface of the solid bounded by the paraboloid $z = 1 - x^2 - y^2$ and the xy-plane.

11. $\mathbf{F}(x, y, z) = x^3\mathbf{i} + y^3\mathbf{j} + z^3\mathbf{k};$ σ is the surface of the cylindrical solid bounded by $x^2 + y^2 = 4,$ $z = 0,$ and $z = 3.$

12. $\mathbf{F}(x, y, z) = (x^3 - e^y)\mathbf{i} + (y^3 + \sin z)\mathbf{j} + (z^3 - xy)\mathbf{k};$ σ is the surface of the solid bounded by $z = \sqrt{4 - x^2 - y^2}$ and the xy-plane. [*Hint:* Use spherical coordinates.]

13. $\mathbf{F}(x, y, z) = (x^2 + y)\mathbf{i} + xy\mathbf{j} - (2xz + y)\mathbf{k};$ σ is the surface of the tetrahedron in the first octant bounded by $x + y + z = 1$ and the coordinate planes.

14. $\mathbf{F}(x, y, z) = 2xz\mathbf{i} + yz\mathbf{j} + z^2\mathbf{k};$ σ is the surface of the hemispherical solid bounded by $z = \sqrt{a^2 - x^2 - y^2}$ and the xy-plane.

15. $\mathbf{F}(x, y, z) = x^2\mathbf{i} + y^2\mathbf{j} + z^2\mathbf{k};$ σ is the surface of the conical solid bounded by $z = \sqrt{x^2 + y^2}$ and $z = 1.$

16. $\mathbf{F}(x, y, z) = x^2y\mathbf{i} - xy^2\mathbf{j} + (z + 2)\mathbf{k};$ σ is the surface of the solid bounded above by the plane $z = 2x$ and below by the paraboloid $z = x^2 + y^2.$

17. $\mathbf{F}(x, y, z) = x^3\mathbf{i} + x^2y\mathbf{j} + xy\mathbf{k};$ σ is the surface of the solid bounded by $z = 4 - x^2,\ y + z = 5,$ $z = 0,$ and $y = 0.$

In Exercises 18–20, verify the Divergence Theorem.

18. $\mathbf{F}(x, y, z) = x^2\mathbf{i} + y^2\mathbf{j} + z^2\mathbf{k};$ σ is the surface of the cube bounded by $x = 1, y = 1, z = 1,$ and the coordinate planes.

19. $\mathbf{F}(x, y, z) = x\mathbf{i} + y\mathbf{j} + z\mathbf{k};$ σ is the sphere $x^2 + y^2 + z^2 = a^2.$

20. $\mathbf{F}(x, y, z) = (x + y)\mathbf{i} + (y + z)\mathbf{j} + (z + x)\mathbf{k};$ σ is the surface of the cylindrical solid bounded by $x^2 + y^2 = 9,\ z = 0,$ and $z = 5.$

21. Prove the following properties of divergence:
    (a) $\text{div}\,(\mathbf{F} + \mathbf{G}) = \text{div}\,\mathbf{F} + \text{div}\,\mathbf{G};$
    (b) $\text{div}\,(f\mathbf{F}) = f\,\text{div}\,\mathbf{F} + \nabla f \cdot \mathbf{F},$ where $f(x, y, z)$ is a scalar-valued function and $\mathbf{F}(x, y, z)$ a vector-valued function.

22. Prove that if σ is the surface of a solid G oriented by outward unit normals, then the volume of G is given by
$$\text{vol}(G) = \frac{1}{3}\iint\limits_{\sigma} \mathbf{F} \cdot \mathbf{n}\, dS$$
where $\mathbf{F}(x, y, z) = x\mathbf{i} + y\mathbf{j} + z\mathbf{k}$ and **n** is an outer unit normal to σ.

23. Use the formula in Exercise 22 to find the volume of a right-circular cylinder of radius a and height h.

24. Prove: If $\mathbf{F}(x, y, z) = a\mathbf{i} + b\mathbf{j} + c\mathbf{k},$ where a, b, and c are constant, and σ is the surface of a solid G, then
$$\iint\limits_{\sigma} \mathbf{F} \cdot \mathbf{n}\, dS = 0$$
where **n** is an outer unit normal to σ.

18.7 STOKES' THEOREM

In this section we will discuss another generalization of Green's Theorem to three dimensions, called *Stokes'* Theorem. In two dimensions, Green's Theorem expresses a double integral over a plane region in terms of a line integral over the boundary of the region. In three dimensions, Stokes' Theorem expresses a surface integral in terms of a line integral over the boundary of the surface.

We will begin with some terminology.

18.7.1 DEFINITION If $\mathbf{F}(x, y, z) = f(x, y, z)\mathbf{i} + g(x, y, z)\mathbf{j} + h(x, y, z)\mathbf{k}$, then we define the ***curl of* F** by

$$\text{curl } \mathbf{F} = \left(\frac{\partial h}{\partial y} - \frac{\partial g}{\partial z}\right)\mathbf{i} + \left(\frac{\partial f}{\partial z} - \frac{\partial h}{\partial x}\right)\mathbf{j} + \left(\frac{\partial g}{\partial x} - \frac{\partial f}{\partial y}\right)\mathbf{k} \tag{1}$$

REMARK. This formula can be remembered by writing it in the determinant form

$$\text{curl } \mathbf{F} = \begin{vmatrix} \mathbf{i} & \mathbf{j} & \mathbf{k} \\ \dfrac{\partial}{\partial x} & \dfrac{\partial}{\partial y} & \dfrac{\partial}{\partial z} \\ f & g & h \end{vmatrix} \tag{2}$$

---

*GEORGE GABRIEL STOKES (1819–1903), Irish mathematician and physicist. Born in Skreen, Ireland, Stokes came from a family deeply rooted in the Church of Ireland. His father was a rector, his mother the daughter of a rector, and three of his brothers took holy orders. He received his early education from his father and a local parish clerk. In 1837, he entered Pembroke College and after graduating with top honors accepted a fellowship at the college. In 1847 he was appointed Lucasian professor of mathematics at Cambridge, a position once held by Isaac Newton, but one that had lost its esteem through the years. By virtue of his accomplishments, Stokes ultimately restored the position to the eminence it once held. Unfortunately, the position paid very little and Stokes was forced to teach at the Government School of Mines during the 1850s to supplement his income.

Stokes was one of several outstanding nineteenth century scientists who helped turn the physical sciences in a more empirical direction. He systematically studied hydrodynamics, elasticity of solids, behavior of waves in elastic solids, and diffraction of light. For Stokes, mathematics was a tool for his physical studies. He wrote classic papers on the motion of viscous fluids that laid the foundation for modern hydrodynamics; he elaborated on the wave theory of light; and he wrote papers on gravitational variation that established him as a founder of the modern science of geodesy.

Stokes was honored in his later years with degrees, medals, and memberships in foreign societies. He was knighted in 1889. Throughout his life, Stokes gave generously of his time to learned societies and readily assisted those who sought his help in solving problems. He was deeply religious and vitally concerned with the relationship between science and religion.

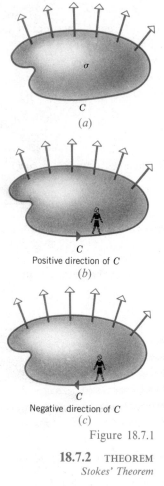

σ

C

(a)

C

Positive direction of C

(b)

C

Negative direction of C

(c)

Figure 18.7.1

The reader should verify that Formula (1) results if we compute this determinant by interpreting a "product" such as $(\partial/\partial x)(f)$ to mean $\partial f/\partial x$. We note, however, that (2) is just a mnemonic device and not a true determinant, since the entries in a determinant must be numbers, not vectors and partial derivative symbols.

▶ **Example 1**   If $\mathbf{F}(x, y, z) = x^2 y\mathbf{i} + 2y^3 z\mathbf{j} + 3z\mathbf{k}$, then

$$\text{curl } \mathbf{F} = \begin{vmatrix} \mathbf{i} & \mathbf{j} & \mathbf{k} \\ \dfrac{\partial}{\partial x} & \dfrac{\partial}{\partial y} & \dfrac{\partial}{\partial z} \\ x^2 y & 2y^3 z & 3z \end{vmatrix}$$

$$= \left[\frac{\partial}{\partial y}(3z) - \frac{\partial}{\partial z}(2y^3 z) \right]\mathbf{i} - \left[\frac{\partial}{\partial x}(3z) - \frac{\partial}{\partial z}(x^2 y) \right]\mathbf{j}$$

$$+ \left[\frac{\partial}{\partial x}(2y^3 z) - \frac{\partial}{\partial y}(x^2 y) \right]\mathbf{k}$$

$$= -2y^3\mathbf{i} - x^2\mathbf{k} \qquad \blacktriangleleft$$

To state Stokes' Theorem we need some terminology about oriented surfaces. If σ is an oriented surface bounded by a curve C (Figure 18.7.1), then the orientation of σ induces an orientation or direction along C as follows. If we imagine a woman walking along the curve C with her head in the (general) direction of the normals that orient σ, then the woman is walking in the **positive direction** of C if the surface σ is on the woman's left and in the **negative direction** if σ is on the right (Figure 18.7.1*b* and 18.7.1*c*).

18.7.2   THEOREM   *Let σ be an oriented surface, bounded by a curve C. If the components of*
Stokes' Theorem $\mathbf{F}(x, y, z) = f(x, y, z)\mathbf{i} + g(x, y, z)\mathbf{j} + h(x, y, z)\mathbf{k}$ *have continuous first partial derivatives on some open set containing σ, then*

$$\int_C \mathbf{F} \cdot d\mathbf{r} = \iint_\sigma (\text{curl } \mathbf{F}) \cdot \mathbf{n} \, dS \qquad (3)$$

where the line integral is taken in the positive direction of C.

The proof of Stokes' Theorem is too difficult for a first course and will be omitted.

If s is an arc length parameter for the curve C, and if the positive direction of C is in the direction of increasing s, then

$$\frac{d\mathbf{r}}{ds} = \mathbf{T}$$

where **T** is the unit tangent vector to C that points in the positive direction of C. (The proof is similar to that in Theorem 14.3.4.) This suggests the notation

$$d\mathbf{r} = \mathbf{T} \, ds$$

which enables us to write (3) as

$$\int_C \mathbf{F} \cdot \mathbf{T} \, ds = \iint_\sigma (\text{curl } \mathbf{F}) \cdot \mathbf{n} \, dS \qquad (4)$$

▶ Example 2   Verify Stokes' Theorem if σ is the portion of the paraboloid $z = 4 - x^2 - y^2$ for which $z \geq 0$ and $\mathbf{F}(x, y, z) = 2z\mathbf{i} + 3x\mathbf{j} + 5y\mathbf{k}$.

Solution. If σ is oriented by outward normals, then the positive direction for the boundary curve C is as shown in Figure 18.7.2. Since the curve C is a circle in the xy-plane with radius 2 and traversed counterclockwise looking down the z-axis, it is represented (with the proper direction) by the parametric equations

$$\begin{aligned} x &= 2 \cos t \\ y &= 2 \sin t \,, \quad 0 \leq t \leq 2\pi \\ z &= 0 \end{aligned} \qquad (5)$$

or equivalently by the position function

$$\mathbf{r}(t) = 2 \cos t\mathbf{i} + 2 \sin t\mathbf{j} + 0 = 2 \cos t\mathbf{i} + 2 \sin t\mathbf{j}$$

where $0 \leq t \leq 2\pi$. Since $\mathbf{F}(x, y, z) = 2z\mathbf{i} + 3x\mathbf{j} + 5y\mathbf{k}$, it follows from (5) that

$$\begin{aligned} \mathbf{F}(x(t), y(t), z(t)) &= 2(0)\mathbf{i} + 3(2 \cos t)\mathbf{j} + 5(2 \sin t)\mathbf{k} \\ &= 6 \cos t\mathbf{j} + 10 \sin t\mathbf{k} \end{aligned}$$

Thus,

$$\begin{aligned} \int_C \mathbf{F} \cdot d\mathbf{r} &= \int_0^{2\pi} \left[\mathbf{F}(x(t), y(t), z(t)) \cdot \frac{d\mathbf{r}}{dt} \right] dt \\ &= \int_0^{2\pi} (6 \cos t\mathbf{j} + 10 \sin t\mathbf{k}) \cdot (-2 \sin t\mathbf{i} + 2 \cos t\mathbf{j}) \, dt \\ &= \int_0^{2\pi} 12 \cos^2 t \, dt = 12 \left[\frac{1}{2}t + \frac{1}{4} \sin 2t \right]_0^{2\pi} = 12\pi \end{aligned}$$

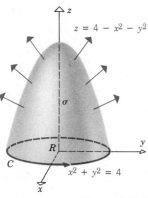

$z = 4 - x^2 - y^2$

σ

R

C

$x^2 + y^2 = 4$

Figure 18.7.2

On the other hand,

$$\text{curl } \mathbf{F} = \begin{vmatrix} \mathbf{i} & \mathbf{j} & \mathbf{k} \\ \dfrac{\partial}{\partial x} & \dfrac{\partial}{\partial y} & \dfrac{\partial}{\partial z} \\ 2z & 3x & 5y \end{vmatrix} = 5\mathbf{i} + 2\mathbf{j} + 3\mathbf{k}$$

Since the outward normals are upward normals, and the paraboloid is given by $z = z(x, y) = 4 - x^2 - y^2$, it follows from Formula (6) of Section 18.5 (with curl $\mathbf{F}$ replacing $\mathbf{F}$) that

$$\iint_{\sigma} (\text{curl } \mathbf{F}) \cdot \mathbf{n} \, dS = \iint_{R} (\text{curl } \mathbf{F}) \cdot \left(-\frac{\partial z}{\partial x}\mathbf{i} - \frac{\partial z}{\partial y}\mathbf{j} + \mathbf{k} \right) dA$$

$$= \iint_{R} (5\mathbf{i} + 2\mathbf{j} + 3\mathbf{k}) \cdot (2x\mathbf{i} + 2y\mathbf{j} + \mathbf{k}) \, dA$$

$$= \iint_{R} (10x + 4y + 3) \, dA$$

$$= \int_{0}^{2\pi} \int_{0}^{2} (10r \cos \theta + 4r \sin \theta + 3) r \, dr \, d\theta$$

$$= \int_{0}^{2\pi} \left[\frac{10r^3}{3} \cos \theta + \frac{4r^3}{3} \sin \theta + \frac{3r^2}{2} \right]_{r=0}^{2} d\theta$$

$$= \int_{0}^{2\pi} \left(\frac{80}{3} \cos \theta + \frac{32}{3} \sin \theta + 6 \right) d\theta$$

$$= \left[\frac{80}{3} \sin \theta - \frac{32}{3} \cos \theta + 6\theta \right]_{0}^{2\pi} = 12\pi$$

which is the same as the value of the line integral.   ◄

REMARK.   Had we oriented σ by inward normals in this example, then the surface integral and the line integral would have had value -12π. (Why?)

► **Example 3**   Let C be the rectangle in the plane $z = y$ oriented as in Figure 18.7.3, and let $\mathbf{F}(x, y, z) = x^2\mathbf{i} + 4xy^3\mathbf{j} + y^2x\mathbf{k}$. Find

$$\int_{C} \mathbf{F} \cdot d\mathbf{r}$$

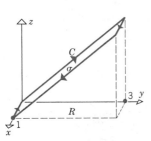

Figure 18.7.3

Solution.   To evaluate the integral directly would require four separate integrations, one over each side of the rectangle. Instead, we will apply Stokes' Theorem, so we need only evaluate a single surface integral over the rectangular surface σ bounded by C. For the positive direction of C to be as shown in Figure 18.7.3, the surface σ must be oriented by downward normals. Since the surface σ has equation $z = y$ and

$$\text{curl } \mathbf{F} = \begin{vmatrix} \mathbf{i} & \mathbf{j} & \mathbf{k} \\ \dfrac{\partial}{\partial x} & \dfrac{\partial}{\partial y} & \dfrac{\partial}{\partial z} \\ x^2 & 4xy^3 & y^2x \end{vmatrix} = 2yx\mathbf{i} - y^2\mathbf{j} + 4y^3\mathbf{k}$$

it follows from Formula (7) of Theorem 18.5.1 that

$$\iint_{\sigma} (\text{curl } \mathbf{F}) \cdot \mathbf{n} \, dS = \iint_{R} (\text{curl } \mathbf{F}) \cdot \left(\frac{\partial z}{\partial x} \mathbf{i} + \frac{\partial z}{\partial y} \mathbf{j} - \mathbf{k} \right) dA$$

$$= \iint_{R} (2yx\mathbf{i} - y^2\mathbf{j} + 4y^3\mathbf{k}) \cdot (0\mathbf{i} + \mathbf{j} - \mathbf{k}) \, dA$$

$$= \int_{0}^{1} \int_{0}^{3} (-y^2 - 4y^3) \, dy \, dx$$

$$= -\int_{0}^{1} \left[\frac{y^3}{3} + y^4 \right]_{y=0}^{3} dx$$

$$= -\int_{0}^{1} 90 \, dx = -90 \qquad \blacktriangleleft$$

► Exercise Set 18.7

In Exercises 1–5, find the curl of **F**.

1. $\mathbf{F}(x, y, z) = x\mathbf{i} + y\mathbf{j} + z\mathbf{k}$.

2. $\mathbf{F}(x, y, z) = xz^3\mathbf{i} + 2y^4x^2\mathbf{j} + 5z^2y\mathbf{k}$.

3. $\mathbf{F}(x, y, z) = e^{xy}\mathbf{i} - \cos y\mathbf{j} + \sin^2 z\mathbf{k}$.

4. $\mathbf{F}(x, y, z) = 7y^3z^2\mathbf{i} - 8x^2z^5\mathbf{j} - 3xy^4\mathbf{k}$.

5. $\mathbf{F}(x, y, z) = \ln x\mathbf{i} + e^{xyz}\mathbf{j} + \tan^{-1}\left(\dfrac{z}{x}\right)\mathbf{k}$.

In Exercises 6–12, use Stokes' Theorem to evaluate $\displaystyle\int_{C} \mathbf{F} \cdot d\mathbf{r}$.

6. $\mathbf{F}(x, y, z) = xz\mathbf{i} + 3x^2y^2\mathbf{j} + yx\mathbf{k}$; C is the rectangle in the plane $z = y$ shown in Figure 18.7.3.

7. $\mathbf{F}(x, y, z) = 3z\mathbf{i} + 4x\mathbf{j} + 2y\mathbf{k}$; C is the boundary of the paraboloid shown in Figure 18.7.2.

8. $\mathbf{F}(x, y, z) = -3y^2\mathbf{i} + 4z\mathbf{j} + 6x\mathbf{k}$; C is the triangle in the plane $z = \frac{1}{2}y$ with vertices $(2, 0, 0)$, $(0, 2, 1)$, and $(0, 0, 0)$ with a counterclockwise orientation looking down the positive z-axis.

9. $\mathbf{F}(x, y, z) = xy\mathbf{i} + x^2\mathbf{j} + z^2\mathbf{k}$; C is the intersection of the paraboloid $z = x^2 + y^2$ and the plane $z = y$ with a counterclockwise orientation looking down the positive z-axis.

10. $\mathbf{F}(x, y, z) = xy\mathbf{i} + yz\mathbf{j} + zx\mathbf{k}$; C is the triangle in the plane $x + y + z = 1$ with vertices $(1, 0, 0)$, $(0, 1, 0)$, and $(0, 0, 1)$ with a counterclockwise orientation looking from the first octant toward the origin.

11. $\mathbf{F}(x, y, z) = (x - y)\mathbf{i} + (y - z)\mathbf{j} + (z - x)\mathbf{k}$; C is the circle $x^2 + y^2 = a^2$ in the xy-plane with counterclockwise orientation looking down the positive z-axis.

12. $\mathbf{F}(x, y, z) = (z + \sin x)\mathbf{i} + (x + y^2)\mathbf{j} + (y + e^z)\mathbf{k}$; C is the intersection of the sphere $x^2 + y^2 + z^2 = 1$ and the cone $z = \sqrt{x^2 + y^2}$ with counterclockwise orientation looking down the positive z-axis.

In Exercises 13–16, verify Stokes' Theorem by computing the line and surface integrals in (3) and showing that they are equal.

13. $\mathbf{F}(x, y, z) = (x - y)\mathbf{i} + (y - z)\mathbf{j} + (z - x)\mathbf{k}$; σ is the portion of the plane $x + y + z = 1$ in the first octant.

14. $\mathbf{F}(x, y, z) = x^2\mathbf{i} + y^2\mathbf{j} + z^2\mathbf{k}$; σ is the portion of the cone $z = \sqrt{x^2 + y^2}$ below the plane $z = 1$.

15. $\mathbf{F}(x, y, z) = x\mathbf{i} + y\mathbf{j} + z\mathbf{k}$; σ is the upper hemisphere $z = \sqrt{a^2 - x^2 - y^2}$.

16. $\mathbf{F}(x, y, z) = (z - y)\mathbf{i} + (z + x)\mathbf{j} - (x + y)\mathbf{k}$; σ is the portion of the paraboloid $z = 9 - x^2 - y^2$ above the xy-plane.

17. Use Stokes' Theorem to evaluate

$$\int_C z^2 \, dx + 2x \, dy - y^3 \, dz$$

over the circle $x^2 + y^2 = 1$ in the xy-plane traversed counterclockwise looking down the positive z-axis. [*Hint:* Find a surface σ whose boundary is C.]

18. Use Stokes' Theorem to evaluate

$$\int_C y^2 \, dx + z^2 \, dy + x^2 \, dz$$

over the triangle with vertices $P(a, 0, 0)$, $Q(0, a, 0)$, $R(0, 0, a)$ traversed counterclockwise looking toward the origin from the first octant.

19. Prove: If the components of $\mathbf{F}$ and their first- and second-order partial derivatives are continuous then

$$\mathrm{div}\,(\mathrm{curl}\,\mathbf{F}) = 0$$

20. Use Exercise 19 and the Divergence Theorem to prove that

$$\iint_\sigma (\mathrm{curl}\,\mathbf{F}) \cdot \mathbf{n} \, dS = 0$$

if σ is a sphere, $\mathbf{n}$ is the unit outer normal to σ, and the components of $\mathbf{F}$ satisfy the conditions stated in Exercise 19.

21. Prove: If f and the components of $\mathbf{F}$ are continuous and have continuous second-order partial derivatives, then
   (a) $\mathrm{curl}\,(\nabla f) = \mathbf{0}$;
   (b) $\mathrm{curl}\,(\nabla f + \mathrm{curl}\,\mathbf{F}) = \mathrm{curl}\,(\mathrm{curl}\,\mathbf{F})$. [*Hint:* You may use the property $\mathrm{curl}\,(\mathbf{F}_1 + \mathbf{F}_2) = \mathrm{curl}\,\mathbf{F}_1 + \mathrm{curl}\,\mathbf{F}_2$ without proof.]

22. (a) Use Stokes' Theorem to prove: If σ_1 and σ_2 are oriented surfaces with the same boundary C, and if C has the same positive orientation relative to each surface, then

$$\iint_{\sigma_1} (\mathrm{curl}\,\mathbf{F}) \cdot \mathbf{n} \, dS = \iint_{\sigma_2} (\mathrm{curl}\,\mathbf{F}) \cdot \mathbf{n} \, dS$$

   (b) Use the result in (a) to evaluate

$$\iint_\sigma (\mathrm{curl}\,\mathbf{F}) \cdot \mathbf{n} \, dS$$

   where $\mathbf{F}(x, y, z) = x^3\mathbf{i} + y\mathbf{j} + z^2 e^{xy}\mathbf{k}$ and the surface σ is the upper hemisphere $z = \sqrt{a^2 - x^2 - y^2}$ oriented by upward unit normals. [*Hint:* Find a circular disk in the xy-plane that shares a common boundary with the hemisphere.]

23. Use Stokes' Theorem to prove the result in Exercise 20. [*Hint:* See Exercise 22(a) and note that the upper and lower hemispheres share a common boundary.]

18.8 APPLICATIONS

In this section we will discuss applications of Stokes' Theorem and the Divergence Theorem to problems of fluid flow. These theorems also have applications to problems in electricity, magnetism, and mechanics that we will not discuss here.

Consider a body of fluid flowing in a region of three-dimensional space, and suppose that at each point in the region the velocity of the fluid is given by

$$\mathbf{F}(x, y, z) = f(x, y, z)\mathbf{i} + g(x, y, z)\mathbf{j} + h(x, y, z)\mathbf{k}$$

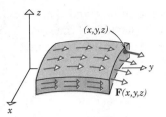

Figure 18.8.1

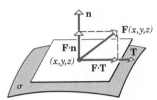

Figure 18.8.2

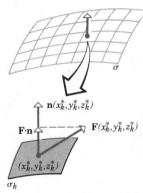

Figure 18.8.3

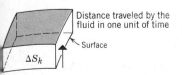

Distance traveled by the
fluid in one unit of time

Surface

ΔS_k

Figure 18.8.4

The function $\mathbf{F}(x, y, z)$ is called the *flow field* for the fluid motion (Figure 18.8.1).

For simplicity, we will make two assumptions:

1. The velocity of the fluid at each point of the region does not vary with time. Such a flow is said to be *steady* over the region.

2. The density (mass per unit volume) of the fluid is constant throughout the region. Such a fluid is called *incompressible* over the region. To keep our formulas simple, we will assume this density to be 1.

If σ is an oriented surface within such a three-dimensional body of fluid, then at each point (x, y, z) on the surface the velocity vector $\mathbf{F}(x, y, z)$ can be resolved into two components, a component $\mathbf{F} \cdot \mathbf{T}$ tangent to the surface, and a component $\mathbf{F} \cdot \mathbf{n}$ along the unit normal vector $\mathbf{n}$ in the orientation of σ (Figure 18.8.2). The component $\mathbf{F} \cdot \mathbf{T}$ is the component of flow *along* the surface, and $\mathbf{F} \cdot \mathbf{n}$ is the component *across* the surface. Fluid crosses σ in the direction of orientation where $\mathbf{F} \cdot \mathbf{n}$ is positive and opposite to that direction where $\mathbf{F} \cdot \mathbf{n}$ is negative. (Why?) In a specified time interval, the volume of fluid crossing σ in the direction of orientation minus the volume crossing in the opposite direction is called the *net volume* of fluid crossing σ during the time interval and the surface integral

$$\iint_{\sigma} \mathbf{F} \cdot \mathbf{n} \, dS$$

is called the *flux of F across σ*; it represents the net volume of fluid that crosses the oriented surface σ in one unit of time. To see this, subdivide σ into n parts $\sigma_1, \sigma_2, \ldots, \sigma_n$ with areas

$$\Delta S_1, \Delta S_2, \ldots, \Delta S_n$$

If the parts are small and the flow is not too erratic, it is reasonable to assume that the velocity does not vary much on each part. Thus, if (x_k^*, y_k^*, z_k^*) is any point in the kth part, we can assume that $\mathbf{F}(x, y, z)$ is constant and equal to $\mathbf{F}(x_k^*, y_k^*, z_k^*)$ throughout the part, and that the component of flow across the surface σ_k is

$$\mathbf{F}(x_k^*, y_k^*, z_k^*) \cdot \mathbf{n}(x_k^*, y_k^*, z_k^*) \tag{1}$$

where $\mathbf{n}(x_k^*, y_k^*, z_k^*)$ is the unit normal at (x_k^*, y_k^*, z_k^*) in the orientation of σ (Figure 18.8.3).

If we observe the flow across σ_k for one unit of time, then the section of fluid initially on the surface will move to some new position and the volume of fluid that crosses σ_k will form a solid (Figure 18.8.4) with a volume that can be approximated by

$$\text{volume} \approx \begin{bmatrix} \text{distance traveled by} \\ \text{the fluid in one unit} \\ \text{of time} \end{bmatrix} \Delta S_k \tag{2}$$

But (1) is the component of fluid velocity across the surface and therefore represents either the distance traveled by the fluid in one unit of time or the negative of that distance, depending on whether $\mathbf{F} \cdot \mathbf{n}$ is positive or negative. Thus, from (2)

$$\mathbf{F}(x_k^*, y_k^*, z_k^*) \cdot \mathbf{n}(x_k^*, y_k^*, z_k^*)\, \Delta S_k$$

is either the approximate volume of fluid crossing σ_k in one unit of time or the negative of that approximate volume. It is positive if the fluid is crossing in the same direction as the orienting vector $\mathbf{n}(x_k^*, y_k^*, z_k^*)$ and negative if it is crossing in the opposite direction. Thus,

$$\sum_{k=1}^{n} \mathbf{F}(x_k^*, y_k^*, z_k^*) \cdot \mathbf{n}(x_k^*, y_k^*, z_k^*)\, \Delta S_k \tag{3}$$

is approximately the volume of fluid crossing σ in the direction of orientation minus the volume crossing in the opposite direction.

If we now repeat the subdivision process, dividing σ into more and more parts, in such a way that each part shrinks to a point, then we can assume that the errors in our approximations diminish to zero, and

$$\lim_{n \to +\infty} \sum_{k=1}^{n} \mathbf{F}(x_k^*, y_k^*, z_k^*) \cdot \mathbf{n}(x_k^*, y_k^*, z_k^*)\, \Delta S_k = \iint_{\sigma} \mathbf{F}(x, y, z) \cdot \mathbf{n}(x, y, z)\, dS$$

represents the *exact* net volume of fluid crossing σ in one unit of time.

In summary, we have the following result:

$$\begin{bmatrix} \text{flux of } \mathbf{F} \\ \text{across } \sigma \end{bmatrix} = \begin{bmatrix} \text{net volume of fluid crossing} \\ \sigma \text{ in the direction of orien-} \\ \text{tation in one unit of time.} \end{bmatrix} = \iint_{\sigma} \mathbf{F} \cdot \mathbf{n}\, dS \tag{4}$$

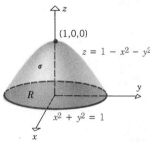

Figure 18.8.5

▶ **Example 1** Let σ be the portion of the surface $z = 1 - x^2 - y^2$ above the xy-plane oriented by upward normals (Figure 18.8.5), and suppose the flow field is

$$\mathbf{F}(x, y, z) = x\mathbf{i} + y\mathbf{j} + z\mathbf{k}$$

Find the net volume of fluid that crosses σ in the direction of orientation in one unit of time.

Solution. From (4), the net volume of fluid crossing σ in one unit of time is

$$\iint_{\sigma} \mathbf{F} \cdot \mathbf{n}\, dS$$

We showed in Example 4 of Section 18.5 that

$$\iint_\sigma \mathbf{F} \cdot \mathbf{n} \, dS = \iint_\sigma (x^2 + y^2 + 1) \, dA = \frac{3\pi}{2}$$

so there is a net volume of $3\pi/2$ cubic units of fluid crossing σ in one unit of time. ◀

We now have the background to give a physical interpretation of the divergence of a flow field

$$\mathbf{F}(x, y, z) = f(x, y, z)\mathbf{i} + g(x, y, z)\mathbf{j} + h(x, y, z)\mathbf{k}$$

If G is a *small* spherical region centered at (x, y, z) and div $\mathbf{F}$ is continuous on G, then div $\mathbf{F}$ will not vary much from its value at the center (x, y, z); and we can assume that div $\mathbf{F}$ is constant over G. If σ is the surface of G oriented by outward normals, then the Divergence Theorem (18.6.2) yields

$$\iint_\sigma \mathbf{F} \cdot \mathbf{n} \, dS = \iiint_G \text{div } \mathbf{F} \, dV \approx \text{div } \mathbf{F}(x, y, z) \iiint_G dV$$

where div $\mathbf{F}(x, y, z)$ is the divergence at the center of the sphere. Therefore,

$$\iint_\sigma \mathbf{F} \cdot \mathbf{n} \, dS \approx \text{div } \mathbf{F}(x, y, z) \, [\text{volume } (G)] \tag{5}$$

or

$$\text{div } \mathbf{F}(x, y, z) \approx \frac{1}{\text{vol } (G)} \iint_\sigma \mathbf{F} \cdot \mathbf{n} \, dS = \frac{\text{flux of } \mathbf{F} \text{ across } \sigma}{\text{vol } (G)} \tag{6}$$

If we assume that the errors in our approximations diminish to zero over smaller and smaller spheres, then it follows from (6) that

$$\text{div } \mathbf{F}(x, y, z) = \lim_{\text{vol } (G) \to 0} \frac{\text{flux of } \mathbf{F} \text{ across } \sigma}{\text{vol } (G)}$$

The quantity

$$\frac{\text{flux of } \mathbf{F} \text{ across } \sigma}{\text{vol } (G)}$$

is called the **flux density of F over the sphere G,** and its limit as vol $(G) \to 0$ is called the **flux density of F** at the point (x, y, z). Thus, we have the following physical description of the divergence:

$$\text{div } \mathbf{F}(x, y, z) = \text{flux density of } \mathbf{F} \text{ at } (x, y, z) \tag{7}$$

The flux density at (x, y, z) is a measure of the rate at which fluid is approaching or departing from the point (x, y, z). To see why, consider a sphere G centered at (x, y, z) with its surface σ oriented by outward unit normals. If the sphere is small, then the volume of fluid approaching or departing from (x, y, z) in one unit of time is approximately the volume of fluid crossing the surface σ in that time. But (4) and (5) imply that the net volume of fluid crossing σ in one unit of time is directly proportional to the divergence of $\mathbf{F}$ at (x, y, z).

We leave it for the reader to show that the net fluid flow is away from (x, y, z) if div $\mathbf{F}(x, y, z) > 0$, and toward (x, y, z) if div $\mathbf{F}(x, y, z) < 0$. Points where div $\mathbf{F}(x, y, z) > 0$ are called **sources,** and points where div $\mathbf{F}(x, y, z) < 0$ are called **sinks.** At a source, fluid is entering the flow, and at a sink it is leaving the flow. If div $\mathbf{F}(x, y, z) = 0$ throughout the flow, then there are no sources or sinks.

▶ **Example 2**   The fluid with flow field $\mathbf{F}(x, y, z) = yz\mathbf{i} + xz\mathbf{j} + xy\mathbf{k}$ has no sources or sinks since

$$\text{div } \mathbf{F}(x, y, z) = \frac{\partial}{\partial x}(yz) + \frac{\partial}{\partial y}(xz) + \frac{\partial}{\partial z}(xy) = 0 \qquad ◀$$

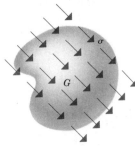

Figure 18.8.6

We are now in a position to give an important physical interpretation of the Divergence Theorem. Let σ be an oriented surface that encloses a finite three-dimensional region G inside a moving body of fluid (Figure 18.8.6). If $\mathbf{F}(x, y, z)$ is the flow field, then it follows from the Divergence Theorem (Formula (1), Section 18.6) that

$$\iint_{\sigma} \mathbf{F} \cdot \mathbf{n} \, dS = \iiint_{G} \text{div } \mathbf{F} \, dV$$

But from (4) and (7) this equation states that

$$\begin{bmatrix} \text{flux of } \mathbf{F} \\ \text{across } \sigma \end{bmatrix} = \iiint_{G} [\text{flux density}] \, dV$$

In other words, the Divergence Theorem asserts that *the net volume of fluid that crosses σ in one unit of time is obtained by integrating the flux density over the region enclosed by σ.*

We conclude this section by obtaining a physical interpretation of the curl of a flow field

$$\mathbf{F}(x, y, z) = f(x, y, z)\mathbf{i} + g(x, y, z)\mathbf{j} + h(x, y, z)\mathbf{k}$$

Let σ_a be a *small* oriented disk-shaped region of radius a, centered at (x, y, z), and having boundary C_a (Figure 18.8.7a). If the components of

(a)

(b)

Figure 18.8.7

curl **F** are continuous on σ_a, then curl **F** will not vary much from its "value" at the center (x, y, z), and we can assume curl **F** to be approximately constant on σ_a. Since σ_a is a flat surface, the unit normal vectors that orient σ_a are all the same as the unit normal $\mathbf{n} = \mathbf{n}(x, y, z)$ at the center (x, y, z) (Figure 18.8.7b). Thus, curl $\mathbf{F} \cdot \mathbf{n}$ is also constant on σ_a, so Equation (4) of Section 18.7 and Stokes' Theorem (18.7.2) yield

$$\int_{C_a} \mathbf{F} \cdot \mathbf{T} \, ds = \iint_{\sigma_a} (\text{curl } \mathbf{F}) \cdot \mathbf{n} \, dS \approx \text{curl } \mathbf{F}(x, y, z) \cdot \mathbf{n} \iint_{\sigma_a} dS$$

where the line integral is taken in the positive direction of C_a, and **T** is the unit tangent vector to C_a pointing in the positive direction. Therefore,

$$\int_{C_a} \mathbf{F} \cdot \mathbf{T} \, ds \approx [\text{curl } \mathbf{F}(x, y, z) \cdot \mathbf{n}] \, [\text{surface area of } \sigma_a]$$

or

$$\int_{C_a} \mathbf{F} \cdot \mathbf{T} \, ds \approx \pi a^2 [\text{curl } \mathbf{F}(x, y, z) \cdot \mathbf{n}]$$

or

$$\text{curl } \mathbf{F}(x, y, z) \cdot \mathbf{n} \approx \frac{1}{\pi a^2} \int_{C_a} \mathbf{F} \cdot \mathbf{T} \, ds \tag{8}$$

If we assume that the errors in our approximations diminish to zero over smaller and smaller disks, then it follows from (8) that

$$\text{curl } \mathbf{F}(x, y, z) \cdot \mathbf{n} = \lim_{a \to 0} \frac{1}{\pi a^2} \int_{C_a} \mathbf{F} \cdot \mathbf{T} \, ds \tag{9}$$

The integral in (9) has a physical interpretation that we will now explain. At each point of C_a, the flow field **F** has a component $\mathbf{F} \cdot \mathbf{u}$ along the outward normal to C_a and a component $\mathbf{F} \cdot \mathbf{T}$ tangent to C_a (Figure 18.8.8). When **F** is closely aligned with **T**, the magnitude of $\mathbf{F} \cdot \mathbf{T}$ is large compared to the magnitude of $\mathbf{F} \cdot \mathbf{u}$, so the fluid tends to move along the circle C_a rather than pass through it. Thus, physicists call the integral

$$\int_{C_a} \mathbf{F} \cdot \mathbf{T} \, ds$$

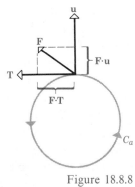

Figure 18.8.8

the **circulation of F around C_a** and use it as a measure of the tendency for fluid to circulate in the positive direction of C_a.

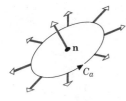

(a)

If the flow is normal to the circle at each point of C_a, then $\mathbf{F} \cdot \mathbf{T} = 0$ so

$$\int_{C_a} \mathbf{F} \cdot \mathbf{T} \, ds = 0$$

and there is no tendency for the fluid to circulate around C_a (Figure 18.8.9a). However, the more closely $\mathbf{F}$ is aligned with $\mathbf{T}$, the larger the value of $\mathbf{F} \cdot \mathbf{T}$ and the larger the value of the circulation. The maximum circulation around C_a occurs when $\mathbf{F}$ is aligned with $\mathbf{T}$ at each point of C_a (Figure 18.8.9b).

The quantity

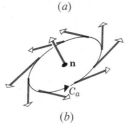

(b)

Figure 18.8.9

$$\frac{1}{\pi a^2} \int_{C_a} \mathbf{F} \cdot \mathbf{T} \, ds = \frac{\text{circulation of } \mathbf{F} \text{ around } C_a}{\text{area of the disk } \sigma_a}$$

is called the **specific circulation** of $\mathbf{F}$ over the disk σ_a and its limit as $a \to 0$ is called the **rotation of $\mathbf{F}$ about $\mathbf{n}$** at (x, y, z). Thus, from (9),

Curl $\mathbf{F}(x,y,z)$

$$\text{curl } \mathbf{F}(x, y, z) \cdot \mathbf{n} = \text{rotation of } \mathbf{F} \text{ about } \mathbf{n} \text{ at } (x, y, z)$$

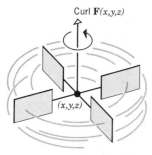

(x,y,z)

Figure 18.8.10

Since curl $\mathbf{F} \cdot \mathbf{n}$ has its maximum value when $\mathbf{n}$ is in the same direction as curl $\mathbf{F}(x, y, z)$, it follows that the rotation of $\mathbf{F}$ at (x, y, z) has its maximum value about this $\mathbf{n}$. Thus, in the vicinity of the point (x, y, z) the maximum circulation occurs around small circles in the plane normal to curl $\mathbf{F}$. Physically, if a paddle wheel is immersed in the fluid so the pivot point is at (x, y, z), then the paddles will turn most quickly when the spindle is aligned with curl $\mathbf{F}(x, y, z)$ (Figure 18.8.10).

► Exercise Set 18.8

In Exercises 1–4, a flow field $\mathbf{F}(x, y, z)$ is given. Find the net volume of fluid that crosses the surface σ in the direction of orientation in one unit of time.

1. $\mathbf{F}(x, y, z) = x\mathbf{i} + y\mathbf{j} + 2z\mathbf{k}$; σ is the portion of the surface $z = 1 - x^2 - y^2$ oriented by upward unit normals.

2. $\mathbf{F}(x, y, z) = x^2\mathbf{i} + yx\mathbf{j} + zx\mathbf{k}$; σ is the portion of the plane $6x + 3y + 2z = 6$ in the first octant oriented by upward unit normals.

3. $\mathbf{F}(x, y, z) = (x + y)\mathbf{i} + (y + z)\mathbf{j} + (z + x)\mathbf{k}$; σ is the upper hemisphere $z = \sqrt{9 - x^2 - y^2}$ oriented by upward unit normals.

4. $\mathbf{F}(x, y, z) = x\mathbf{i} + y\mathbf{j} + 2z\mathbf{k}$; σ is the portion of the cone $z^2 = x^2 + y^2$ between the planes $z = 1$ and $z = 2$, oriented by upward unit normals.

In Exercises 5–8, a flow field $\mathbf{F}(x, y, z)$ is given. Use the Divergence Theorem to find the net volume of fluid that crosses the surface σ in the direction of orientation in one unit of time.

5. $\mathbf{F}(x, y, z) = x\mathbf{i} + y\mathbf{j} + z\mathbf{k}$; σ is the sphere $x^2 + y^2 + z^2 = a^2$ oriented by outward unit normals.

6. $\mathbf{F}(x, y, z) = xy\mathbf{i} + yz\mathbf{j} + xz\mathbf{k}$; σ is the cube bounded by the coordinate planes and the planes $x = 2, y = 2, z = 2$, and oriented by outward unit normals.

7. $F(x, y, z) = (x^3 - y)i + (y^3 + z)j + (z^3 - 2xy)k$; σ is the surface of the cylinder (including top and bottom) bounded by $x^2 + y^2 = 4$, $z = 0$, and $z = 3$, oriented by outward unit normals.

8. $F(x, y, z) = x^2i + xyj - 2xzk$; σ is the tetrahedron bounded by the coordinate planes and the plane $x + y + z = 1$, oriented by outward unit normals.

In Exercises 9–12, determine whether the flow field $F(x, y, z)$ is free of sources and sinks. If it is not, find the location of all sources and sinks.

9. $F(x, y, z) = (y + z)i - xz^3j + (x^2 \sin y)k$.

10. $F(x, y, z) = xyi - xyj + y^2k$.

11. $F(x, y, z) = x^3i + y^3j + z^3k$.

12. $F(x, y, z) = (x^3 - x)i + (y^3 - y)j + (z^3 - z)k$.

13. For the flow field $F(x, y, z) = (x - z)i + (y - x)j + (z - xy)k$:
    (a) use Stokes' Theorem to find the circulation around the triangle with vertices $A(1, 0, 0)$, $B(0, 2, 0)$, and $C(0, 0, 1)$ oriented counterclockwise looking from the origin toward the first octant;
    (b) find the rotation of F about the vector k at the origin;
    (c) find the unit vector n such that the rotation of F about n at the origin is maximum.

▶ SUPPLEMENTARY EXERCISES

In Exercises 1–6, evaluate the line integral by any method.

1. $\int_C 2y \, dx + 3 \, dy$ along $y = \sin x$ from the point $(0, 0)$ to the point $(\pi, 0)$.

2. $\int_C x^5 \, dy$ along the curve C given by $x = 1/t$, $y = 4t^2$, $1 \leq t \leq 2$.

3. $\int_C \langle 2e^y, -x \rangle \cdot dr$ along $y = \ln x$ from the point $(1, 0)$ to the point $(3, \ln 3)$.

4. $\int_C (yi + zj + xk) \cdot dr$ along the curve C given by $r = \langle t^2 - 2t, -2t, t - 2 \rangle$, $0 \leq t \leq 2$.

5. $\int_C z \, dx + y \, dy - x \, dz$ along the line segment from $(0, 0, 0)$ to $(1, 2, 3)$.

6. $\int_C x \sin xy \, dx - y \sin xy \, dy$ along the line segment from $(0, 0)$ to $(1, \pi)$.

In Exercises 7–9, determine whether F is conservative. If it is, find a potential function for it.

7. $F(x, y) = y \sin xyi - x \cos xyj$.

8. $F(x, y) = 2x(\ln y - 1)i + \left(\dfrac{x^2}{y} - 3y^2 \right)j$.

9. $F(x, y) = \left(3x^2 - \dfrac{y^2}{x^2} \right)i + \left(\dfrac{2y}{x} + 4y \right)j$.

10. Let $F(x, y) = yi - 2xj$. Find functions $h(x)$ and $g(y)$ such that $h(x)F(x, y)$ and $g(y)F(x, y)$ are each conservative.

In Exercises 11–14, determine whether the integral is independent of the path. If it is, find its value by any method.

11. $\int_{(1, \pi/4)}^{(2, \pi/4)} (\cos 2y - 3x^2y^2) \, dx$
    $\qquad + (\cos 2y - 2x \sin 2y - 2x^3y) \, dy$.

12. $\int_{(0,0)}^{(2,-2)} x^2y^4 \, dx + y^2x^4 \, dy$.

13. $\int_{(1, 2)}^{(2, 1)} \dfrac{y \, dx - x \, dy}{y^2}$.

14. $\int_{(0, 1)}^{(1, 0)} (ye^{xy} - 1) \, dx + (xe^{xy}) \, dy$.

In Exercises 15–18, evaluate the line integral counterclockwise over C using Green's Theorem.

15. $\int_C 2xy \, dx + xy^2 \, dy$ along the triangle with vertices $(0, 0)$, $(2, 0)$, $(2, 4)$.

16. $\int_C 3y \, dx - 4x \, dy$ along the boundary of the semicircular region $\{(x, y): x^2 + y^2 \leq 2 \text{ and } 0 \leq x\}$.

17. $\int_C -2y\,dx + (x - 4)\,dy$; C runs from $(0, 2)$ to $(0, -2)$ on the y-axis, then to $(0, 2)$ along the curve $x = 4 - y^2$.

18. $\int_C 5xy\,dx + x^3\,dy$ along the curve $y = x^2$ from $(0, 0)$ to $(2, 4)$, then back to $(0, 0)$ along $y = 2x$.

19. Use Green's Theorem to evaluate $\int_C y^3\,dx - x^3\,dy$ counterclockwise around the boundary of the region $R = \{(x, y): 1 \le x^2 + y^2 \le 4 \text{ and } 0 \le y\}$.

20. Find the area of the circle $r = 2\cos\theta$ using Formula (4c) of Section 18.3.

In Exercises 21–23, use any method to find the work done by the force **F** as it acts on a particle moving along C.

21. $\mathbf{F} = (2x + 3y)\mathbf{i} + (xy)\mathbf{j}$ along $y = x^2$ from $(0, 0)$ to $(1, 1)$.

22. $\mathbf{F} = 3x^2y^3\mathbf{i} + 3x^3y^2\mathbf{j}$ clockwise once around the cardioid $r = 3 + 3\cos\theta$.

23. $\mathbf{F} = z\mathbf{i} + y\mathbf{j} - x\mathbf{k}$ on the line segment from $(0, 0, 0)$ to $(1, 2, 3)$.

24. Assuming that $\phi(x, y)$ has continuous first- and second-order partial derivatives, show that
 (a) $\text{curl}(\nabla\phi) = \mathbf{0}$ (b) $\text{div}(\nabla\phi) = \phi_{xx} + \phi_{yy}$.
The centroid of a curved lamina σ is defined by

$$\bar{x} = \frac{\displaystyle\iint_\sigma x\,dS}{\text{area of }\sigma}, \quad \bar{y} = \frac{\displaystyle\iint_\sigma y\,dS}{\text{area of }\sigma}, \quad \bar{z} = \frac{\displaystyle\iint_\sigma z\,dS}{\text{area of }\sigma}$$

In Exercises 25 and 26, find the centroid of the lamina.

25. The lamina that is the portion of the paraboloid $z = \frac{1}{2}(x^2 + y^2)$ below the plane $z = 4$.

26. The lamina that is the portion of the sphere $x^2 + y^2 + z^2 = 4$ above the plane $z = 1$.

27. Let σ be the portion of the sphere $x^2 + y^2 + z^2 = 1$ that lies in the first octant, **n** an outward unit normal (with respect to the sphere), and $D_\mathbf{n}f$ the directional derivative of $f(x, y, z) = \ln\sqrt{x^2 + y^2 + z^2}$. Evaluate the surface integral

$$\iint_\sigma D_\mathbf{n}f\,dS$$

28. Let σ be the sphere $x^2 + y^2 + z^2 = 1$, **n** an inward unit normal, and $D_\mathbf{n}f$ the directional derivative of $f(x, y, z) = x^2 + y^2 + z^2$. Evaluate the surface integral

$$\iint_\sigma D_\mathbf{n}f\,dS$$

29. Let G be a solid with surface σ oriented by outward unit normals, suppose that ϕ has continuous first and second partial derivatives in some open set containing G, and let $D_\mathbf{n}\phi$ be the directional derivative of ϕ. Show that

$$\iint_\sigma D_\mathbf{n}\phi\,dS = \iiint_G \left[\frac{\partial^2\phi}{\partial x^2} + \frac{\partial^2\phi}{\partial y^2} + \frac{\partial^2\phi}{\partial z^2}\right]dV$$

30. Show that Green's Theorem is a special case of Stokes' Theorem by applying Stokes' Theorem to $\mathbf{F}(x, y) = f(x, y)\mathbf{i} + g(x, y)\mathbf{j}$.

appendix 1 trigonometry review

UNIT I

The purpose of this section and the next is to review the fundamentals of trigonometry.

We shall be concerned with angles in the plane that are generated by rotating a ray (or half-line) about its endpoint. The starting position of the ray is called the ***initial side*** of the angle and the final position of the ray is called the ***terminal side.*** The initial and terminal sides meet at a point called the ***vertex*** of the angle (Figure A.1).

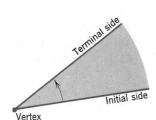

Figure A.1

An angle is considered ***positive*** if it is generated by a counterclockwise rotation and ***negative*** if it is generated by a clockwise rotation. In a rectangular coordinate system, an angle is said to be in ***standard position*** if its vertex is at the origin and its initial side is along the positive *x*-axis (Figure A.2).

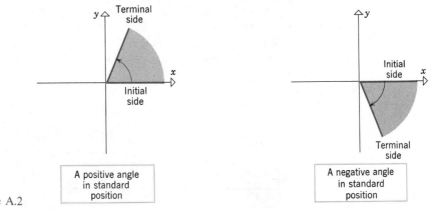

A positive angle in standard position

A negative angle in standard position

Figure A.2

A1

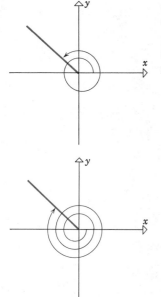

Figure A.3

As shown in Figure A.3, an angle may be generated by making more than one complete revolution.

The **sine, cosine, tangent, cosecant, secant,** *and* **cotangent** of a positive acute angle θ can be defined as ratios of the sides of a right triangle. Using the notation from Figure A.4, these definitions take the form:

$$\sin \theta = \frac{\text{side opposite } \theta}{\text{hypotenuse}} = \frac{y}{r} \tag{1}$$

$$\cos \theta = \frac{\text{side adjacent to } \theta}{\text{hypotenuse}} = \frac{x}{r} \tag{2}$$

$$\tan \theta = \frac{\text{side opposite } \theta}{\text{side adjacent to } \theta} = \frac{y}{x} \tag{3}$$

$$\csc \theta = \frac{\text{hypotenuse}}{\text{side oppposite } \theta} = \frac{r}{y} \tag{4}$$

$$\sec \theta = \frac{\text{hypotenuse}}{\text{side adjacent to } \theta} = \frac{r}{x} \tag{5}$$

$$\cot \theta = \frac{\text{side adjacent to } \theta}{\text{side opposite to } \theta} = \frac{x}{y} \tag{6}$$

We will call sin, cos, tan, csc, sec, and cot the **trigonometric functions.** Because similar triangles have proportional sides, the values of the trigonometric functions depend only on the size of θ and not on the particular right triangle used to compute the ratios.

Figure A.4

▶ **Example 1**   Recall from high school geometry that the two legs of a 45°–45°–90° triangle are of equal size. This fact and the Theorem of Pythagoras yield Figure A.5a. Recall also that the hypotenuse of a 30°–60°–90° triangle is twice the size of the shorter leg and the shorter leg is opposite the 30° angle. These facts and the Theorem of Pythagoras yield Figure A.5b.

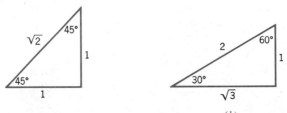

Figure A.5

(b)

From Figure A.5 we obtain

$$\sin 45° = \frac{1}{\sqrt{2}} \qquad \cos 45° = \frac{1}{\sqrt{2}} \qquad \tan 45° = 1$$

$$\csc 45° = \sqrt{2} \qquad \sec 45° = \sqrt{2} \qquad \cot 45° = 1$$

$$\sin 30° = \frac{1}{2} \qquad \cos 30° = \frac{\sqrt{3}}{2} \qquad \tan 30° = \frac{1}{\sqrt{3}}$$

$$\csc 30° = 2 \qquad \sec 30° = \frac{2}{\sqrt{3}} \qquad \cot 30° = \sqrt{3}$$

$$\sin 60° = \frac{\sqrt{3}}{2} \qquad \cos 60° = \frac{1}{2} \qquad \tan 60° = \sqrt{3}$$

$$\csc 60° = \frac{2}{\sqrt{3}} \qquad \sec 60° = 2 \qquad \cot 60° = \frac{1}{\sqrt{3}} \qquad \blacktriangleleft$$

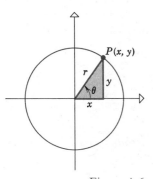

Since a right triangle cannot have an angle larger than 90°, formulas (1)–(6) are not applicable if θ is obtuse. To obtain definitions of the trigonometric functions that apply to all angles, we take the following approach: Given an angle θ, introduce a coordinate system so that the angle is in standard position. Then construct a circle of arbitrary radius r centered at the origin and let $P(x, y)$ be the point where the terminal side of the angle intersects the circle (Figure A.6). We make the following definition:

Figure A.6

DEFINITION **1**

$$\sin \theta = \frac{y}{r}, \qquad \cos \theta = \frac{x}{r}, \qquad \tan \theta = \frac{y}{x},$$

$$\csc \theta = \frac{r}{y}, \qquad \sec \theta = \frac{r}{x}, \qquad \cot \theta = \frac{x}{y}.$$

The formulas in Definition 1 agree with formulas (1)–(6). However, Definition 1 applies to all angles—positive, negative, acute, or obtuse. Note that $\tan \theta$ and $\sec \theta$ are undefined if the terminal side of θ is on the y-axis (since $x = 0$), while $\csc \theta$ and $\cot \theta$ are undefined if the terminal side of θ is on the x-axis (since $y = 0$). Moreover,

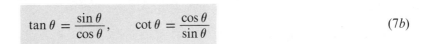

$$\csc \theta = \frac{1}{\sin \theta}, \qquad \sec \theta = \frac{1}{\cos \theta}, \qquad \cot \theta = \frac{1}{\tan \theta} \qquad (7a)$$

$$\tan \theta = \frac{\sin \theta}{\cos \theta}, \qquad \cot \theta = \frac{\cos \theta}{\sin \theta} \qquad (7b)$$

(verify.)

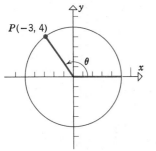

Figure A.7

▶ Example 2   In Figure A.7,

$$x = -3, \qquad y = 4, \qquad \text{and} \qquad r = 5$$

so that for the angle θ shown,

$$\sin \theta = \frac{4}{5} \qquad \cos \theta = -\frac{3}{5} \qquad \tan \theta = -\frac{4}{3}$$

$$\csc \theta = \frac{5}{4} \qquad \sec \theta = -\frac{5}{3} \qquad \cot \theta = -\frac{3}{4}$$

◀

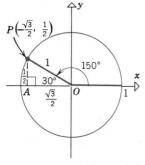

(cos θ, sin θ)

θ

1

Figure A.8

Since the value of a trigonometric function is determined by the point where the terminal side of the angle intersects a circle of radius r, it follows that the value of a trigonometric function will be the same for any two angles with the same terminal side (like those in Figure A.3.)

Using properties of similar triangles, it can be shown that the values of the trigonometric functions in Definition 1 are the same for every value r. In particular, for the *unit circle* (the circle of radius $r = 1$ centered at the origin) the formulas for $\sin \theta$ and $\cos \theta$ in Definition 1 become

$$x = \cos \theta, \qquad y = \sin \theta$$

Thus, we have the following geometric interpretation of $\sin \theta$ and $\cos \theta$ (Figure A.8):

THEOREM 2

Geometric Interpretation of $\sin \theta$ and $\cos \theta$

If an angle θ is placed in standard position, then its terminal side intersects the unit circle at the point $(\cos \theta, \sin \theta)$

▶ **Example 3** Evaluate the trigonometric functions of $\theta = 150°$.

Solution. Construct a unit circle and place the angle $\theta = 150°$ in standard position (Figure A.9). Since angle AOP is $30°$ and $\triangle OAP$ is a $30°$–$60°$–$90°$ triangle, the leg AP has length $\frac{1}{2}$ (half the hypotenuse) and the leg OA has length $\sqrt{3}/2$ by the Theorem of Pythagoras. Thus, the coordinates of P are $(-\sqrt{3}/2, 1/2)$ and Theorem 2 yields

$$\sin 150° = \frac{1}{2}, \qquad \cos 150° = -\frac{\sqrt{3}}{2}$$

Moreover, from Formulas (7a) and (7b)

$$\tan 150° = \frac{\sin 150°}{\cos 150°} = \frac{1/2}{-\sqrt{3}/2} = -\frac{1}{\sqrt{3}}$$

$$\cot 150° = \frac{1}{\tan 150°} = -\sqrt{3}$$

$$\sec 150° = \frac{1}{\cos 150°} = -\frac{2}{\sqrt{3}}$$

$$\csc 150° = \frac{1}{\sin 150°} = 2$$

◀

$P\left(-\frac{\sqrt{3}}{2}, \frac{1}{2}\right)$

$\frac{1}{2}$ 1

$30°$ $150°$

A $\frac{\sqrt{3}}{2}$ O 1

Figure A.9

The following useful theorem is suggested by Figure A.10. We omit the proof.

THEOREM 3   *If an angle θ is generated by a ray with endpoint (x_0, y_0) and initial side extending in the positive x-direction, then the intersection of the terminal side with the circle of radius r centered at (x_0, y_0) is*

$$(x_0 + r \cos \theta, y_0 + r \sin \theta)$$

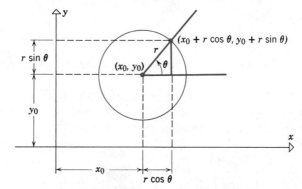

Figure A.10

RADIAN MEASURE

In calculus, angles are measured in *radians* rather than degrees, minutes, and seconds because it simplifies many important formulas. Before we can define radian measure, we need some preliminary ideas.

Let Q be an arbitrary point on a circle of radius r and imagine that a particle, initially at Q, moves around the circle and stops at a point P (Figure A.11). During its motion the point will travel a distance d along the circle. To distinguish between clockwise and counterclockwise motions we introduce the **signed distance** or **signed arc length** s traveled by the point; it is defined by:

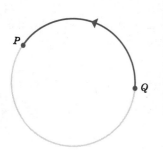

Figure A.11

$$s = d \qquad \text{if the motion is counterclockwise,}$$
$$s = -d \qquad \text{if the motion is clockwise,}$$
$$s = 0 \qquad \text{if there is no motion}$$

For example, if $s = 3$, the point has traveled 3 units in the counterclockwise direction from Q; and, if $s = -\pi$, the point has traveled π units in the clockwise direction from Q.

DEFINITION 4   To define the radian measure of an angle, we place the vertex of the angle at the center of a circle of arbitrary radius r. As a ray sweeps out the angle, the intersection of the ray with the circle moves some signed distance s along the

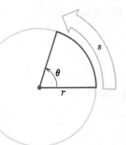

Figure A.12

circle (Figure A.12). We define

$$\theta = \frac{s}{r} \qquad (8)$$

to be the **radian measure** of the angle.

REMARK. We leave it as an exercise to show that the radian measure of an angle depends only on the size of the angle, and not on the radius r selected for the circle.

Figure A.13

▶ **Example 4** Express the angle $90°$ in radian measure.

Solution. An angle of $90°$ placed with its vertex at the center of a circle of radius r intercepts one fourth of the circumference (Figure A.13). Since the full circumference is $2\pi r$, we obtain

$$s = \tfrac{1}{4}(2\pi r) = \tfrac{1}{2}\pi r$$

Thus, the radian measure of the angle is

$$\theta = \frac{s}{r} = \frac{\tfrac{1}{2}\pi r}{r} = \frac{\pi}{2} \qquad \text{(radians)} \qquad ◀$$

▶ **Example 5** From the fact that $90°$ corresponds to $\pi/2$ radians, the reader should be able to obtain the relationships in the following table:

Degrees	30°	45°	60°	90°	120°	135°	150°	180°	270°	360°
Radians	$\frac{\pi}{6}$	$\frac{\pi}{4}$	$\frac{\pi}{3}$	$\frac{\pi}{2}$	$\frac{2\pi}{3}$	$\frac{3\pi}{4}$	$\frac{5\pi}{6}$	π	$\frac{3\pi}{2}$	2π

Some angles and their radian measure are shown in the following figure. ◀

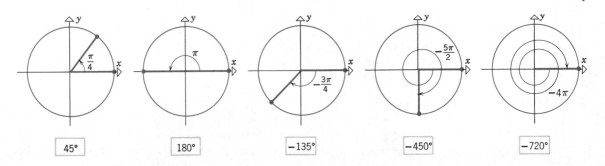

The following table will be useful in the next example.

$$360° = 2\pi \text{ radians}$$

$$180° = \pi \text{ radians} \approx 3.14159 \ldots \text{ radians}$$

$$\left(\frac{180}{\pi}\right)° = 1 \text{ radian} \approx 57° \; 17' \; 44.8''$$

$$1° = \frac{\pi}{180} \text{ radian} \approx 0.01745 \text{ radian}$$

▶ Example 6

    (a)   Express 146° in radians.

    (b)   Express 3 radians in degrees.

Solution (a). Since

$$1° = \frac{\pi}{180} \text{ radians}$$

it follows that

$$146° = \frac{\pi}{180} \cdot 146 \text{ radians} = \frac{73\pi}{90} \text{ radians}$$

Solution (b). Since

$$1 \text{ radian} = \left(\frac{180}{\pi}\right)°$$

it follows that

$$3 \text{ radians} = \left(3 \cdot \frac{180}{\pi}\right)° = \left(\frac{540}{\pi}\right)° \approx 171.9° \qquad ◀$$

REMARK. By convention, an angle written without any units is understood to be measured in radians. Thus, the statement $\theta = 3.2$ means $\theta = 3.2$ radians, *not* $\theta = 3.2°$.

▶ Example 7   Evaluate the trigonometric functions at $\theta = 5\pi/6$.

Solution. Since $5\pi/6 = 150°$, this problem is equivalent to that of Example 3. From that example we obtain

$$\sin\frac{5\pi}{6} = \frac{1}{2}, \qquad \cos\frac{5\pi}{6} = -\frac{\sqrt{3}}{2}, \qquad \tan\frac{5\pi}{6} = -\frac{1}{\sqrt{3}}$$

$$\csc\frac{5\pi}{6} = 2, \qquad \sec\frac{5\pi}{6} = -\frac{2}{\sqrt{3}}, \qquad \cot\frac{5\pi}{6} = -\sqrt{3} \qquad \blacktriangleleft$$

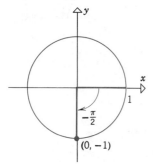

Figure A.14

▶ **Example 8**   Evaluate the trigonometric functions at $\theta = -\pi/2$.

Solution.   Construct a unit circle and place the angle $\theta = -\pi/2 = -90°$ in standard position (Figure A.14). The terminal side of θ intersects the circle at $P(0, -1)$ so that from Theorem 2 we obtain

$$\sin\left(-\frac{\pi}{2}\right) = -1, \quad \cos\left(-\frac{\pi}{2}\right) = 0$$

and from Formulas (7a) and (7b)

$$\tan\left(-\frac{\pi}{2}\right) = \frac{\sin\left(-\dfrac{\pi}{2}\right)}{\cos\left(-\dfrac{\pi}{2}\right)} = \frac{-1}{0} = \text{undefined}$$

$$\cot\left(-\frac{\pi}{2}\right) = \frac{\cos\left(-\dfrac{\pi}{2}\right)}{\sin\left(-\dfrac{\pi}{2}\right)} = \frac{0}{-1} = 0$$

$$\sec\left(-\frac{\pi}{2}\right) = \frac{1}{\cos\left(-\dfrac{\pi}{2}\right)} = \frac{1}{0} = \text{undefined}$$

$$\csc\left(-\frac{\pi}{2}\right) = \frac{1}{\sin\left(-\dfrac{\pi}{2}\right)} = \frac{1}{-1} = -1 \qquad \blacktriangleleft$$

It follows from Theorem 2 that $\cos\theta$ is positive if θ terminates in the first or fourth quadrant, and is negative if θ terminates in the second or third quadrant. (To visualize this, look at Figure A.8.) Similarly, $\sin\theta$ is positive if θ terminates in the first or second quadrant, and is negative if θ terminates in the third or fourth quadrant. Moreover, since $\tan\theta = \sin\theta/\cos\theta$, it follows that $\tan\theta$ is positive where $\sin\theta$ and $\cos\theta$ have the same sign (first and third quadrants) and negative where they have opposite signs (second and fourth quadrants).

The signs of $\sin\theta$, $\cos\theta$, and $\tan\theta$ can be remembered from the adjacent diagram in which each quadrant is labeled with the trigonometric functions that are positive there.

It follows from Formulas (7a) and (7b) that $\csc\theta$ has the same sign as $\sin\theta$, $\cot\theta$ the same sign as $\tan\theta$, and $\sec\theta$ the same sign as $\cos\theta$.

▶ Example 9   Find θ if $\sin\theta = \frac{1}{2}$.

Solution.   We will begin by looking for positive angles that satisfy the equation.

Because $\sin\theta$ is positive, the angle θ must terminate in the first or second quadrant. If it terminates in the first quadrant, then the hypotenuse of $\triangle OAP$ in Figure A.15a is double the leg AP, so

$$\theta = 30° = \frac{\pi}{6} \text{ radians}$$

If θ terminates in the second quadrant, then the hypotenuse of $\triangle OAP$ is double the leg AP, so $\angle AOP = 30°$, which implies that

$$\theta = 180° - 30° = 150° = \frac{5\pi}{6} \text{ radians}$$

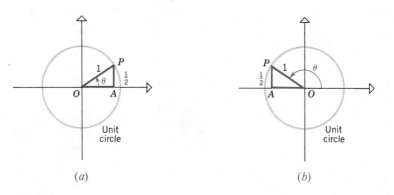

Figure A.15

(a) (b)

Now that we have found these two solutions, all other solutions are obtained by adding or subtracting multiples of $360°$ (2π radians) to them. Thus, the entire set of solutions is given by the formulas

$$\theta = 30° \pm n \cdot 360° \qquad n = 0, 1, 2, \ldots$$

and

$$\theta = 150° \pm n \cdot 360° \qquad n = 0, 1, 2, \ldots$$

or in radian measure

$$\theta = \frac{\pi}{6} \pm n \cdot 2\pi \qquad n = 0, 1, 2, \ldots$$

and

$$\theta = \frac{5\pi}{6} \pm n \cdot 2\pi \qquad n = 0, 1, 2, \ldots \qquad ◀$$

Table 1 lists the values of the trigonometric functions for some frequently used angles. A more extensive table is given in Appendix 3. Dashes denote undefined quantities.

Table 1

	$\theta = 0$	$\dfrac{\pi}{6}$	$\dfrac{\pi}{4}$	$\dfrac{\pi}{3}$	$\dfrac{\pi}{2}$	$\dfrac{2\pi}{3}$	$\dfrac{3\pi}{4}$	$\dfrac{5\pi}{6}$	π	$\dfrac{3\pi}{2}$	2π
$\sin\theta$	0	$1/2$	$1/\sqrt{2}$	$\sqrt{3}/2$	1	$\sqrt{3}/2$	$1/\sqrt{2}$	$1/2$	0	-1	0
$\cos\theta$	1	$\sqrt{3}/2$	$1/\sqrt{2}$	$1/2$	0	$-1/2$	$-1/\sqrt{2}$	$-\sqrt{3}/2$	-1	0	1
$\tan\theta$	0	$1/\sqrt{3}$	1	$\sqrt{3}$	—	$-\sqrt{3}$	-1	$-1/\sqrt{3}$	0	—	0
$\csc\theta$	—	2	$\sqrt{2}$	$2/\sqrt{3}$	1	$2/\sqrt{3}$	$\sqrt{2}$	2	—	-1	—
$\sec\theta$	1	$2/\sqrt{3}$	$\sqrt{2}$	2	—	-2	$-\sqrt{2}$	$-2/\sqrt{3}$	-1	—	1
$\cot\theta$	—	$\sqrt{3}$	1	$1/\sqrt{3}$	0	$-1/\sqrt{3}$	-1	$-\sqrt{3}$	—	0	—

TRIGONOMETRIC IDENTITIES

Earlier, we obtained the relationships

$$\csc\theta = \frac{1}{\sin\theta}, \qquad \sec\theta = \frac{1}{\cos\theta}, \qquad \cot\theta = \frac{1}{\tan\theta}$$

and

$$\tan\theta = \frac{\sin\theta}{\cos\theta}, \qquad \cot\theta = \frac{\cos\theta}{\sin\theta}$$

These results are examples of *identities,* which means that the equalities hold for all values of θ where both sides are defined. One of the most important identities in trigonometry can be derived by applying the Theorem of Pythagoras to the triangle in Figure A.6 to obtain

$$x^2 + y^2 = r^2$$

Dividing both sides by r^2 and using the definitions of $\sin\theta$ and $\cos\theta$ (Definition 1), we obtain the following fundamental result

$$\sin^2\theta + \cos^2\theta = 1 \tag{9}$$

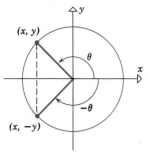

Figure A.16

Figure A.16 shows two angles, θ and $-\theta$, having equal magnitude but opposite sign. By symmetry, the terminal sides of these angles intersect a circle centered at the origin at points whose x-coordinates are equal and whose y-coordinates differ only in sign. Thus,

$$\sin(-\theta) = \frac{-y}{r} = -\frac{y}{r} = -\sin\theta$$

$$\cos(-\theta) = \frac{x}{r} = \cos\theta$$

In summary, we have the identities

$$\sin(-\theta) = -\sin\theta \qquad (10a)$$
$$\cos(-\theta) = \cos\theta \qquad (10b)$$

As previously noted, a trigonometric function has the same value for two angles with the same terminal side. In radian measure, the angles $\theta, \theta + 2\pi$, and $\theta - 2\pi$ all have the same terminal side, so

$$\sin\theta = \sin(\theta + 2\pi) = \sin(\theta - 2\pi)$$
$$\cos\theta = \cos(\theta + 2\pi) = \cos(\theta - 2\pi) \qquad (11)$$

More generally, any multiple of 2π can be added to or subtracted from an angle θ in radians without changing the terminal side. Thus,

$$\sin\theta = \sin(\theta \pm 2n\pi), \qquad n = 0, 1, 2, \ldots$$
$$\cos\theta = \cos(\theta \pm 2n\pi), \qquad n = 0, 1, 2, \ldots \qquad (12)$$

The following result, called the **law of cosines,** generalizes the Theorem of Pythagoras:

THEOREM 5   *If the sides of a triangle have lengths a, b, and c, and if θ is the angle between*
Law of Cosines   *the sides with lengths a and b, then*

$$c^2 = a^2 + b^2 - 2ab\cos\theta$$

Proof. Introduce a coordinate system so that θ is in standard position and the side of length a falls along the positive x-axis (Figure A.17).

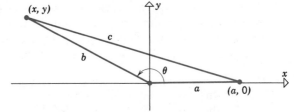

Figure A.17

As shown in Figure A.17, the side of length a extends from the origin to $(a, 0)$ and the side of length b extends from the origin to some point (x, y). From the definition of $\sin \theta$ and $\cos \theta$ we have

$$\sin \theta = \frac{y}{b}, \qquad \cos \theta = \frac{x}{b}$$

so

$$y = b \sin \theta, \qquad x = b \cos \theta \tag{13}$$

From the distance formula in Theorem 1.3.1 we obtain

$$c^2 = (x - a)^2 + (y - 0)^2$$

so that, from (13),

$$
\begin{aligned}
c^2 &= (b \cos \theta - a)^2 + b^2 \sin^2 \theta \\
&= a^2 + b^2 (\cos^2 \theta + \sin^2 \theta) - 2ab \cos \theta \\
&= a^2 + b^2 - 2ab \cos \theta
\end{aligned}
$$

which completes the proof. ▮

Using the law of cosines we will now be able to establish the following identities, called the **addition formulas** for sine and cosine.

$$\sin (\alpha - \beta) = \sin \alpha \cos \beta - \cos \alpha \sin \beta \tag{14a}$$
$$\cos (\alpha - \beta) = \cos \alpha \cos \beta + \sin \alpha \sin \beta \tag{14b}$$

$$\sin (\alpha + \beta) = \sin \alpha \cos \beta + \cos \alpha \sin \beta \tag{15a}$$
$$\cos (\alpha + \beta) = \cos \alpha \cos \beta - \sin \alpha \sin \beta \tag{15b}$$

Identities (15a) and (15b) can be obtained by substituting $-\beta$ for β in (14a) and (14b) and using the identities

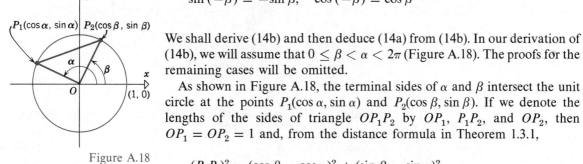

Figure A.18

$$\sin(-\beta) = -\sin\beta, \quad \cos(-\beta) = \cos\beta$$

We shall derive (14b) and then deduce (14a) from (14b). In our derivation of (14b), we will assume that $0 \le \beta < \alpha < 2\pi$ (Figure A.18). The proofs for the remaining cases will be omitted.

As shown in Figure A.18, the terminal sides of α and β intersect the unit circle at the points $P_1(\cos\alpha, \sin\alpha)$ and $P_2(\cos\beta, \sin\beta)$. If we denote the lengths of the sides of triangle OP_1P_2 by OP_1, P_1P_2, and OP_2, then $OP_1 = OP_2 = 1$ and, from the distance formula in Theorem 1.3.1,

$$
\begin{aligned}
(P_1P_2)^2 &= (\cos\beta - \cos\alpha)^2 + (\sin\beta - \sin\alpha)^2 \\
&= (\sin^2\alpha + \cos^2\alpha) + (\sin^2\beta + \cos^2\beta) \\
&\quad - 2(\cos\alpha\cos\beta + \sin\alpha\sin\beta) \\
&= 2 - 2(\cos\alpha\cos\beta + \sin\alpha\sin\beta)
\end{aligned}
$$

But angle $P_2OP_1 = \alpha - \beta$, so that the law of cosines yields

$$
\begin{aligned}
(P_1P_2)^2 &= (OP_1)^2 + (OP_2)^2 - 2(OP_1)(OP_2)\cos(\alpha - \beta) \\
&= 2 - 2\cos(\alpha - \beta).
\end{aligned}
$$

Equating the two expressions for $(P_1P_2)^2$ and simplifying, we obtain

$$\cos(\alpha - \beta) = \cos\alpha\cos\beta + \sin\alpha\sin\beta$$

which completes the derivation of (14b).

To derive identity (14a), we will need the identities

$$\cos\left(\frac{\pi}{2} - \alpha\right) = \sin\alpha \tag{16a}$$

$$\sin\left(\frac{\pi}{2} - \alpha\right) = \cos\alpha \tag{16b}$$

which state that *the sine of an angle is the cosine of its complement* and *the cosine of an angle is the sine of its complement*. The derivations of (16a) and (16b) are left as exercises.

We can now derive (14a) as follows:

$$
\begin{aligned}
\sin(\alpha - \beta) &= \cos\left[\frac{\pi}{2} - (\alpha - \beta)\right] \\
&= \cos\left[\left(\frac{\pi}{2} - \alpha\right) + \beta\right] \\
&= \cos\left[\left(\frac{\pi}{2} - \alpha\right) - (-\beta)\right]
\end{aligned}
$$

$$= \cos\left(\frac{\pi}{2} - \alpha\right)\cos\left(-\beta\right) + \sin\left(\frac{\pi}{2} - \alpha\right)\sin\left(-\beta\right)$$

$$= \cos\left(\frac{\pi}{2} - \alpha\right)\cos\beta - \sin\left(\frac{\pi}{2} - \alpha\right)\sin\beta$$

$$= \sin\alpha\cos\beta - \cos\alpha\sin\beta$$

In the special case where $\alpha = \beta$, (15a) and (15b) reduce to the identities

$$\sin 2\alpha = 2\sin\alpha\cos\alpha \tag{17a}$$
$$\cos 2\alpha = \cos^2\alpha - \sin^2\alpha \tag{17b}$$

These are called the **double-angle formulas.** By using the identity $\sin^2\alpha + \cos^2\alpha = 1$, (17b) can be rewritten in the alternate forms

$$\cos 2\alpha = 2\cos^2\alpha - 1 \tag{17c}$$
$$\cos 2\alpha = 1 - 2\sin^2\alpha \tag{17d}$$

If we replace α by $\alpha/2$ in (17c) and (17d) and use some algebra, we obtain the **half-angle formulas.**

$$\cos^2\frac{\alpha}{2} = \frac{1 + \cos\alpha}{2} \tag{18a}$$

$$\sin^2\frac{\alpha}{2} = \frac{1 - \cos\alpha}{2} \tag{18b}$$

Identities (14a) and (15a) and some algebra yield the first of the following **product formulas.**

$$\sin\alpha\cos\beta = \frac{1}{2}[\sin(\alpha - \beta) + \sin(\alpha + \beta)] \tag{19a}$$

$$\sin\alpha\sin\beta = \frac{1}{2}[\cos(\alpha - \beta) - \cos(\alpha + \beta)] \tag{19b}$$

$$\cos\alpha\cos\beta = \frac{1}{2}[\cos(\alpha - \beta) + \cos(\alpha + \beta)] \tag{19c}$$

The derivations of (19b) and (19c) are left as exercises.

We can deduce some important identities involving tangent, cotangent, secant, and cosecant from (8) and (9). The first of the identities

$$\tan^2\theta + 1 = \sec^2\theta \tag{20a}$$
$$1 + \cot^2\theta = \csc^2\theta \tag{20b}$$

can be obtained by dividing

$$\sin^2 \theta + \cos^2 \theta = 1$$

by $\cos^2 \theta$ and the second by dividing by $\sin^2 \theta$.

In the exercises we have asked the reader to deduce the first of the identities

$$\tan(\alpha + \beta) = \frac{\tan \alpha + \tan \beta}{1 - \tan \alpha \tan \beta} \qquad (21a)$$

$$\tan(\alpha - \beta) = \frac{\tan \alpha - \tan \beta}{1 + \tan \alpha \tan \beta} \qquad (21b)$$

from the relationship $\tan(\alpha + \beta) = \sin(\alpha + \beta)/\cos(\alpha + \beta)$. Identity (21b) is obtained from (21a) by substituting $-\beta$ for β and using the relationship

$$\tan(-\beta) = \frac{\sin(-\beta)}{\cos(-\beta)} = \frac{-\sin \beta}{\cos \beta} = -\tan \beta \qquad (22)$$

If we let $\alpha = \beta$ in (21a), we obtain the double-angle formula:

$$\tan 2\alpha = \frac{2 \tan \alpha}{1 - \tan^2 \alpha} \qquad (23)$$

If we let $\beta = \pi$ in (21a) and (21b) and use the fact that $\tan \pi = 0$, we obtain the identities

$$\tan(\alpha + \pi) = \tan \alpha \qquad (24a)$$
$$\tan(\alpha - \pi) = \tan \alpha \qquad (24b)$$

The trigonometric identities obtained in this section are the ones most commonly used. In later sections other identities will be derived as they are needed.

▶ Trigonometry Exercises I

In Exercises 1 and 2, express the angles in radians.

1. (a) 270°   (b) 390°   (c) 20°
   (d) 138°   (e) $\left(\dfrac{\pi}{10}\right)^{\circ}$.

2. (a) 150°   (b) 420°   (c) 15°
   (d) 117°   (e) 165°.

In Exercises 3 and 4, express the angles in degrees.

3. (a) $\dfrac{\pi}{15}$ (b) $\dfrac{3\pi}{2}$ (c) 4.5   (d) $\dfrac{8\pi}{5}$
   (e) 5π.

4. (a) $\dfrac{\pi}{10}$ (b) 3.2   (c) $\dfrac{\pi}{6}$ (d) $\dfrac{2\pi}{5}$
   (e) 2.

5. Sketch the following angles in standard position.
   (a) $\frac{\pi}{3}$　(b) $-\frac{5\pi}{6}$　(c) $\frac{9\pi}{4}$　(d) $-\pi$
   (e) 4π.

In Exercises 6 and 7, evaluate $\sin\theta$, $\cos\theta$, $\tan\theta$, $\csc\theta$, $\sec\theta$, $\cot\theta$ for the given values of θ.

6. (a) $\theta = \frac{\pi}{3}$　(b) $\theta = -\frac{5\pi}{6}$　(c) $\theta = \frac{9\pi}{4}$
   (d) $\theta = -\pi$.　(e) $\theta = 4\pi$.

7. (a) $\theta = \frac{\pi}{6}$　(b) $\theta = -\frac{7\pi}{3}$　(c) $\theta = -\frac{5\pi}{4}$
   (d) $\theta = -3\pi$　(e) $\theta = \pi$.

8. Find all values of θ (in radians) such that
   (a) $\sin\theta = 1$　(b) $\cos\theta = 1$　(c) $\tan\theta = 1$
   (d) $\csc\theta = 1$　(e) $\sec\theta = 1$　(f) $\cot\theta = 1$.

9. Find all values of θ (in radians) such that
   (a) $\sin\theta = 0$　(b) $\cos\theta = 0$　(c) $\tan\theta = 0$
   (d) $\csc\theta$ is undefined
   (e) $\sec\theta$ is undefined
   (f) $\cot\theta$ is undefined.

In Exercises 10–25, find all values of θ (in radians) that satisfy the given equation. Do not use any tables.

10. $\cos\theta = -\frac{1}{\sqrt{2}}$.
11. $\sin\theta = -\frac{1}{\sqrt{2}}$.
12. $\tan\theta = -1$.
13. $\cos\theta = \frac{1}{2}$.
14. $\sin\theta = -\frac{1}{2}$.
15. $\tan\theta = \sqrt{3}$.
16. $\tan\theta = 1/\sqrt{3}$.
17. $\sin\theta = -\frac{\sqrt{3}}{2}$.
18. $\sin\theta = -1$.
19. $\cos\theta = -1$.
20. $\cot\theta = 1$.
21. $\cot\theta = \sqrt{3}$.
22. $\sec\theta = -2$.
23. $\csc\theta = -2$.
24. $\csc\theta = \frac{2}{\sqrt{3}}$.
25. $\sec\theta = \frac{2}{\sqrt{3}}$.

26. Find $\sin\theta$, $\cos\theta$, $\tan\theta$, $\csc\theta$, $\sec\theta$, $\cot\theta$ for the angles sketched in Figures A.19a and A.19b.

27. Use identities and Table 1 to find the values of
    (a) $\cos 15°$　(b) $\sin 22.5°$　(c) $\sin 37.5°$.

28. Use identities and Table 1 to find the values of
    (a) $\sin 75°$　(b) $\tan 75°$.

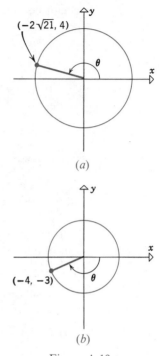

(a)

(b)

Figure A.19

29. (a) Let θ be an acute angle such that $\sin\theta = a/3$. Express the remaining trigonometric functions in terms of a by using the triangle in Figure A.20.

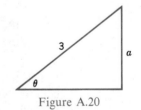

Figure A.20

(b) Let θ be an acute angle such that $\tan\theta = a/5$. Construct an appropriate triangle, and use it to express the remaining trigonometric functions in terms of a.

(c) Let θ be an acute angle such that $\sec\theta = a$. Construct an appropriate triangle, and use it to express the remaining trigonometric functions in terms of a.

30. How could you use a ruler and protractor to estimate $\sin 17°$ and $\cos 17°$?

31. Let ABC be a triangle whose angles at A and B are $30°$ and $45°$. If the side opposite the angle B has length 9, find the lengths of the remaining sides and the size of the angle C.

Exercises 32–34 refer to an arbitrary triangle ABC in which side a is opposite angle A, side b is opposite angle B, and side c is opposite angle C.

32. Prove: The area of a triangle ABC can be written

$$\text{area} = \frac{1}{2}bc \sin A$$

Find two other similar formulas for the area.

33. Prove the **law of sines:** In any triangle, the ratios of the sides to the sines of the opposite angles are equal. That is,

$$\frac{a}{\sin A} = \frac{b}{\sin B} = \frac{c}{\sin C}$$

34. (a) Use Exercises 16 and 17 to prove that the area of a triangle ABC can be written

$$\text{area} = \frac{b^2 \sin A \sin C}{2 \sin B}$$

(b) Find two other similar formulas for the area.

35. Prove: If θ is positive and measured in radians, then the area of a sector with angle θ and radius r is

$$\text{area} = \frac{1}{2}r^2\theta$$

36. Derive the identities

$$\cos\left(\frac{\pi}{2} - \alpha\right) = \sin\alpha$$

$$\sin\left(\frac{\pi}{2} - \alpha\right) = \cos\alpha$$

37. From the identities in Exercise 36, obtain the identities

$$\cos\left(\frac{\pi}{2} + \alpha\right) = -\sin\alpha$$

$$\sin\left(\frac{\pi}{2} + \alpha\right) = \cos\alpha$$

38. Derive identities (18a) and (18b).

39. Derive identities (19b) and (19c).

40. Derive identity (21).

41. Express

$$\sin\left(\frac{3\pi}{2} - \theta\right) \quad\text{and}\quad \cos\left(\frac{3\pi}{2} + \theta\right)$$

in terms of $\sin\theta$ and $\cos\theta$.

In Exercises 42–53, verify the given identities.

42. $\tan\dfrac{\theta}{2} = \dfrac{1 - \cos\theta}{\sin\theta}$.

43. $\tan\dfrac{\theta}{2} = \dfrac{\sin\theta}{1 + \cos\theta}$.

44. $2\csc 2\theta = \sec\theta\csc\theta$.

45. $\tan\theta + \cot\theta = 2\csc 2\theta$.

46. $\sin 3\theta + \sin\theta = 2\sin 2\theta\cos\theta$.

47. $\sin 3\theta - \sin\theta = 2\cos 2\theta\sin\theta$.

48. $\dfrac{\cos\theta\sec\theta}{1 + \tan^2\theta} = \cos^2\theta$.

49. $\dfrac{\cos\theta\tan\theta + \sin\theta}{\tan\theta} = 2\cos\theta$.

50. $\dfrac{\sin 2\theta}{\sin\theta} - \dfrac{\cos 2\theta}{\cos\theta} = \sec\theta$.

51. $\dfrac{\sin\theta + \cos 2\theta - 1}{\cos\theta - \sin 2\theta} = \tan\theta$.

52. $\cos\left(\dfrac{\pi}{3} + \theta\right) + \cos\left(\dfrac{\pi}{3} - \theta\right) = \cos\theta$.

53. $\sin\left(\dfrac{3\pi}{2} - \theta\right) + \sin\left(\dfrac{3\pi}{2} + \theta\right) = -2\cos\theta$.

54. Beginning with identity $\sin^2\theta + \cos^2\theta = 1$, obtain the identity

$$1 + \cot^2\theta = \csc^2\theta$$

55. Express $\sin 3\theta$ and $\cos 3\theta$ in terms of $\sin\theta$ and $\cos\theta$.

56. (a) Express $3\sin\alpha + 5\cos\alpha$ in the form $C\sin(\alpha + \phi)$.

(b) Show that a sum of the form $A\sin\alpha + B\cos\alpha$ can be rewritten in the form $C\sin(\alpha + \phi)$.

57. Show that the length of the diagonal of the parallelogram in Figure A.21 is

$$d = \sqrt{a^2 + b^2 + 2ab\cos\theta}$$

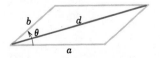

Figure A.21

58. Prove that the values of the trigonometric functions, as defined in Definition 1, do not depend on the value of r.

59. Prove that the ratio s/r in Formula (8) depends only on the angle and not on the radius r selected for the circle. [*Hint:* Use the result from plane geometry that states that the arc subtended on a circle by a central angle is proportional to the angle.]

UNIT II

Recall that an angle written without any units is understood to be measured in radians. Thus, the expression

$$\sin 3$$

means, the sine of 3 radians. More generally, for any real number x, the expression

$$\sin x \tag{1}$$

means, the sine of x radians (that is, the sine of an angle whose radian measure is x).

REMARK.   The use of the letter x in expression (1) rather than θ, α, β, or some other symbol is dangerous but virtually unavoidable in many problems. The danger is that x had a different meaning in the definition of the trigonometric functions (Definition 1 of Unit I). While it is obviously not correct to give a symbol two different meanings in the *same* problem, there is nothing wrong with using a symbol differently in *different* problems as long as we clearly understand how it is being used.

The expression $\sin x$ is defined for every real number x so that

$$f(x) = \sin x$$

defines a function f with domain $(-\infty, +\infty)$. By definition, the graph of $\sin x$ is the graph of the equation

$$y = \sin x$$

The general shape of this graph can be obtained geometrically by constructing an angle of x radians in standard position and recalling from the previous section that the terminal side of this angle intersects the unit circle at the point

$$(\cos x, \sin x)$$

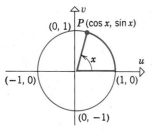

Figure A.22

This is illustrated in Figure A.22, where we have used u and v to label the coordinate axes so we can use x for the angle. Thus, by observing the ordinate of the point P in Figure A.22, we can see how the value of $\sin x$ varies with x. When $x = 0$, P is the point $(1, 0)$, so that $\sin x = 0$. As x increases from 0 to $\pi/2$, P moves counterclockwise along the circle from $(1, 0)$ to the point $(0, 1)$; thus, $\sin x$ increases from 0 to 1 (Figure A.23a). As x increases from $\pi/2$ to π, P moves counterclockwise along the circle from $(0, 1)$ to $(-1, 0)$; thus, $\sin x$ decreases from 1 to 0 (Figure A.23b). As x increases from π to $3\pi/2$, P moves counterclockwise along the circle from $(-1, 0)$ to $(0, -1)$; thus, $\sin x$ decreases from 0 to -1 (Figure A.23c). As x increases from $3\pi/2$ to 2π, P completes one full revolution by moving counterclockwise from $(0, -1)$ to $(1, 0)$; thus, $\sin x$ increases from -1 to 0 (Figure A.23d). As x increases past 2π, the point P begins a second revolution around the circle so that the values of $\sin x$ begin to repeat. In fact, the graph of

$$y = \sin x$$

will repeat itself every 2π units, since P makes a complete revolution whenever x changes by 2π. For negative values of x, the graph of $\sin x$ can be obtained by observing the ordinate of P during a *clockwise* rotation around the unit circle. Alternatively we can use the identity

$$\sin(-x) = -\sin x$$

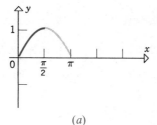

(a)

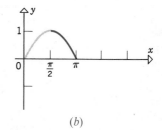

(b)

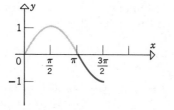

Figure A.23   (c)

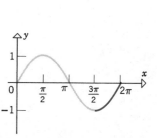

(d)

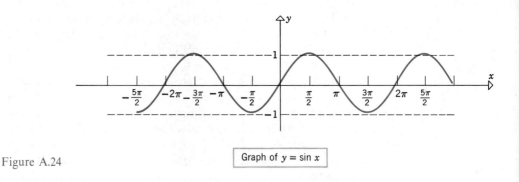

Figure A.24

Graph of $y = \sin x$

to observe that the values of the sine function at x and $-x$ differ only in sign. Either way we are led to the graph in Figure A.24.

The fact that the graph of $y = \sin x$ repeats itself every 2π units could have been anticipated from the identities

$$\sin(x + 2\pi) = \sin x, \qquad \sin(x - 2\pi) = \sin x$$

It is this *periodic* or repetitive nature of trigonometric functions that makes them useful for studying repetitive physical phenomena such as vibrations and wave motion.

By observing the abscissa of the point P in Figure A.22, we can see how the value of $\cos x$ varies with x. An analysis similar to the one we used to graph $\sin x$ yields the graph of $y = \cos x$ shown in Figure A.25.

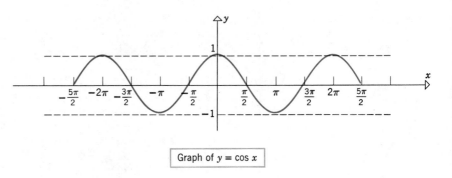

Figure A.25

Graph of $y = \cos x$

Observe that the graph of $\cos x$ is symmetric about the y-axis. This could have been anticipated from the identity

$$\cos(-x) = \cos x$$

which shows that for all x the cosine function assigns the same value to $-x$ and x.

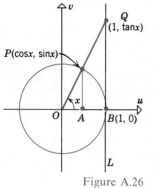

Figure A.26

The graph of

$$y = \tan x$$

can also be obtained geometrically. To begin, let us assume that

$$0 \le x < \frac{\pi}{2}$$

If we construct the angle x in standard position and let Q be the intersection of its terminal side with the vertical line L through $(1, 0)$ (Figure A.26), then the coordinates of Q are

$$(1, \tan x)$$

To see this, observe that triangles OAP and OBQ are similar, so that

$$\frac{BQ}{OB} = \frac{AP}{OA} \tag{2}$$

But $OB = 1$, $OA = \cos x$, and $AP = \sin x$, so that from (2)

$$BQ = \frac{\sin x}{\cos x} = \tan x$$

Thus, as shown in Figure A.26, the ordinate of Q is $\tan x$.

As x increases from 0 toward $\pi/2$, the point Q travels from $(1, 0)$ up the vertical line L getting higher and higher without any bound. Thus, as x increases from 0 toward $\pi/2$, the value of $\tan x$ increases without bound from an initial value of zero (Figure A.27a).

From the identity

$$\tan(-x) = -\tan x$$

and Figure A.27a, we can obtain the portion of the graph shown in Figure A.27b.

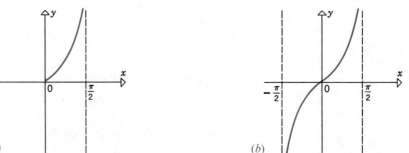

Figure A.27   (a)   (b)

Finally, from the identities

$$\tan (x + \pi) = \tan x \quad \text{and} \quad \tan (x - \pi) = \tan x$$

we see that the graph of $y = \tan x$ repeats itself every π units. This fact together with Figure A.27b leads us to the complete graph of $y = \tan x$ shown in Figure A.28.

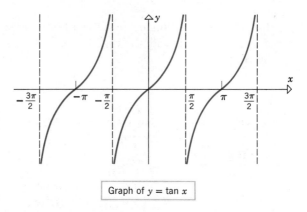

Figure A.28

Graph of $y = \tan x$

REMARK.   Unlike the sine and cosine functions that have domain $(-\infty, +\infty)$. the tangent function is not defined for all x. Since $\tan x = \sin x / \cos x$, the tangent function is undefined whenever $\cos x = 0$; this occurs when x is an odd integer multiple of $\pi/2$, that is, when

$$x = \pm \frac{\pi}{2}, \pm \frac{3\pi}{2}, \pm \frac{5\pi}{2}, \dots$$

DEFINITION 1   A function f is called **periodic** if there is a positive number p such that

$$f(x + p) = f(x) \tag{3}$$

whenever x and $x + p$ lie in the domain of f. We call p a **period** of the function. The smallest positive period is called the **fundamental period** of f or sometimes the **period** of f.

In words, (3) states that the values of f repeat themselves every p units.

▶ Example 1   For the sine and cosine functions, 2π is a period since

$$\sin (x + 2\pi) = \sin x, \quad \cos (x + 2\pi) = \cos x$$

Also, 4π, 6π, 8π, etc., are periods for sine and cosine since

$$\sin (x + 4\pi) = \sin x, \quad \sin (x + 6\pi) = \sin x, \quad \sin (x + 8\pi) = \sin x$$
$$\cos (x + 4\pi) = \cos x, \quad \cos (x + 6\pi) = \cos x, \quad \cos (x + 8\pi) = \cos x$$

and so forth. The fundamental period of sine and cosine is 2π.
For the tangent function, π is a period since

$$\tan (x + \pi) = \tan x$$

Also, 2π, 3π, 4π, etc., are periods; but π is the fundamental period.   ◄

Functions of the form

$$a \sin bx \qquad \text{and} \qquad a \cos bx \tag{4}$$

occur frequently in applications. The graphs of such functions are considered in the following examples.

► **Example 2**   Sketch the graphs of

    (a) $y = 3 \sin x$,   (b) $y = \sin 4x$,   (c) $y = 3 \sin 4x$.

Solution (*a*).   The factor of 3 in

$$y = 3 \sin x$$

has the effect of multiplying each y-coordinate in the graph of $y = \sin x$ by 3, thereby stretching the graph of $y = \sin x$ vertically by a factor of 3 (Figure A.29*a*).

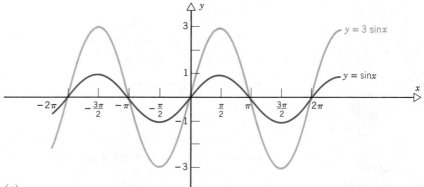

Figure A.29   (*a*)

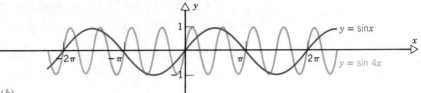

Figure A.29 (b)

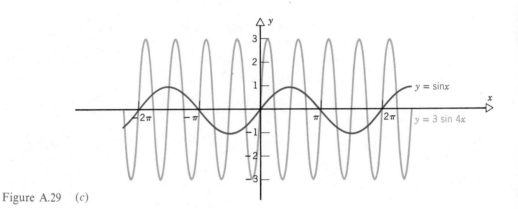

Figure A.29 (c)

Solution (b). The graph of $y = \sin x$ repeats itself when x changes by 2π. Thus, the graph of $y = \sin 4x$ repeats itself whenever $4x$ changes by 2π, or equivalently when x changes by $\frac{1}{4}(2\pi) = \pi/2$. Thus, the factor of 4 in

$$y = \sin 4x$$

has the effect of reducing the fundamental period by a factor of $\frac{1}{4}$. This leads to the graph in Figure A.29*b*.

Solution (c). The graph of

$$y = 3 \sin 4x$$

combines both the stretching effect discussed in (a) and the period effect discussed in (b). The graph is shown in Figure A.29*c*. ◀

The last example illustrates the following general result.

THEOREM 2 *If $a \neq 0$ and $b \neq 0$, then the functions $a \sin bx$ and $a \cos bx$ have fundamental period $2\pi/|b|$ and their graphs oscillate between $-a$ and a.*

We leave the proof as an exercise.

▶ Example 3 The graph of the function $6 \cos 7x$ oscillates between -6 and 6 and has fundamental period $2\pi/7$; the function $\frac{1}{3} \sin \frac{1}{4}x$ oscillates between $-\frac{1}{3}$ and $\frac{1}{3}$ and has fundamental period 8π. ◀

We leave it as an exercise to prove the following result.

THEOREM 3 *If $b \neq 0$, then the function $\tan bx$ has fundamental period $\pi/|b|$.*

▶ Trigonometry Exercises II.

1. Find the fundamental period of
   (a) $\sin 5x$ (b) $\cos \frac{1}{3}x$ (c) $\tan \frac{1}{8}x$
   (d) $\tan 7x$ (e) $\sin \pi x$ (f) $\sec \frac{\pi}{5}x$
   (g) $\cot \pi x$ (h) $\csc \frac{x}{k}$.

As stated in Theorem 2 of this section, the graphs of $a \sin bx$ and $a \cos bx$ oscillate between $-a$ and a. The number $|a|$ is called the **amplitude** of the function. In Exercises 2 and 3, find the amplitudes of the functions.

2. (a) $4 \sin 6x$ (b) $-8 \cos \pi x$
   (c) $\frac{1}{2} \sin x$ (d) $-\frac{1}{8} \cos 2x$.

3. (a) $5 \cos 2x$ (b) $\frac{1}{3} \sin 4x$
   (c) $-\frac{1}{2} \cos x$ (d) $-\sin \pi x$.

In Exercises 4–11, sketch the graph of the equation.

4. $y = \sin 3x$.
5. $y = 2 \sin 3x$.
6. $y = 2 \sin x$.
7. $y = \frac{1}{2} \cos \pi x$.
8. $y = -4 \cos x$.
9. $y = \cos(-4x)$.
10. $y = \sin(2\pi x)$.
11. $y = -2 \sin(-2x)$.
12. (a) How are the graphs of $y = \cos x$ and $y = \cos(-x)$ related?
    (b) How are the graphs of $y = \sin x$ and $y = \sin(-x)$ related?

13. Use the graphs of $\sin x$, $\cos x$, $\tan x$, and the relationships
    $$\cot x = \frac{1}{\tan x}, \quad \sec x = \frac{1}{\cos x}, \quad \csc x = \frac{1}{\sin x}$$
    to help sketch the graphs of
    (a) $\cot x$ (b) $\sec x$ (c) $\csc x$.

In Exercises 14–17, sketch the graph of the equation.

14. $y = 3 \cot 2x$.
15. $y = \frac{1}{4} \sec \pi x$.
16. $y = 2 \csc \frac{x}{3}$.
17. $y = 2 \cot \frac{x}{4}$.

In Exercises 18–32, sketch the graph of the equation.

18. $y = \sin(x + \pi)$.
19. $y = \cos(x - \pi)$.
20. $y = \cos\left(x - \frac{\pi}{4}\right)$.
21. $y = \sin\left(x + \frac{\pi}{2}\right)$.
22. $y = 2 - \cos x$.
23. $y = 1 + \sin x$.
24. $y = 2 \cos\left(4x + \frac{\pi}{2}\right)$.
25. $y = 3 \sin\left(2x + \frac{\pi}{4}\right)$.
26. $y = \tan\left(\pi x - \frac{1}{2}\right)$.
27. $y = 2 \tan(2\pi x + 1)$.
28. $y = |\tan x|$.
29. $y = |\cos x|$.
30. $y = \sin x - \cos x$.
31. $y = \sin x + \cos x$.
32. $y = 2 \sin x + 3 \cos 2x$.

33. Classify the following functions as even, odd, or neither. (See Exercise 22 in Section 2.2.)

(a) $\sin x$ (b) $\cos x$ (c) $\tan x$

(d) $|\sec x|$ (e) $\cot (x^2)$ (f) $3 \sec (-2x)$

(g) $-4 \csc x$ (h) $\dfrac{\sin x}{x}$.

34. Prove: If $0 < x < \pi/2$, then

$$\sin x < x < \tan x$$

[*Hint:* In Figure A.26, compare the areas of triangle OAP, sector OBP, and triangle OBQ. The area of a sector is discussed in Exercise 35 of Unit I of the trigonometry review.]

35. Prove: For all real numbers x,

$$|\sin x| \le |x|$$

[*Hint:* First consider the cases $|x| \ge \pi/2$ and $x = 0$; these are easy. Then use the left-hand inequality in Exercise 34 to complete the proof.]

36. Prove Theorem 2 of this section.

37. Prove Theorem 3 of this section.

appendix 2 supplementary material

I. ONE-SIDED AND INFINITE LIMITS: A RIGOROUS APPROACH

In Section 2.4 we discussed the one-sided limits

$$\lim_{x \to a^-} f(x) = L \qquad \text{and} \qquad \lim_{x \to a^+} f(x) = L$$

from an intuitive viewpoint. We interpreted the first statement to mean that the value of $f(x)$ approaches L as x approaches a from the left side, and we interpreted the second statement to mean that the value of $f(x)$ approaches L as x approaches a from the right side. We can formalize these ideas mathematically as follows:

DEFINITION 1   Let f be defined on some open interval extending to the left from a. We will write

$$\lim_{x \to a^-} f(x) = L$$

if given any number $\epsilon > 0$, we can find a number $\delta > 0$ such that $f(x)$ satisfies

$$|f(x) - L| < \epsilon$$

whenever x satisfies

$$a - \delta < x < a$$

DEFINITION 2   Let f be defined on some open interval extending to the right from a. We will write

$$\lim_{x \to a^+} f(x) = L$$

if given any number $\epsilon > 0$, we can find a number $\delta > 0$ such that $f(x)$ satisfies

$$|f(x) - L| < \epsilon$$

whenever x satisfies

$$a < x < a + \delta$$

Before discussing some examples, let us compare Definitions 1 and 2 to the definition of $\lim_{x \to a} f(x) = L$ as stated in Definition 2.6.1. Given $\epsilon > 0$, the definition of $\lim_{x \to a} f(x) = L$ requires that we be able to find a $\delta > 0$ such that

$$|f(x) - L| < \epsilon \tag{1}$$

whenever x satisfies

$$0 < |x - a| < \delta$$

or equivalently, whenever x lies in the set

$$(a - \delta, a) \cup (a, a + \delta)$$

(See Figure A.30a.)

Figure A.30

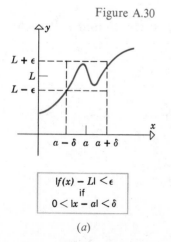

$$|f(x) - L| < \epsilon$$
if
$$0 < |x - a| < \delta$$

(a)

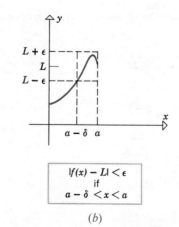

$$|f(x) - L| < \epsilon$$
if
$$a - \delta < x < a$$

(b)

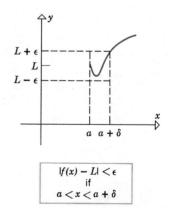

$$|f(x) - L| < \epsilon$$
if
$$a < x < a + \delta$$

(c)

On the other hand, the definition of $\lim\limits_{x \to a^-} f(x) = L$ requires only that (1) hold for x in the set

$$(a - \delta, a)$$

(Figure A.30b), while the definition of $\lim\limits_{x \to a^+} f(x) = L$ requires only that (1) be satisfied on the set

$$(a, a + \delta)$$

(See Figure A.30c.)

▶ **Example 1**   Because the function $f(x) = \sqrt{x}$ is undefined for $x < 0$, we cannot consider the limits

$$\lim_{x \to 0} \sqrt{x} \qquad \text{or} \qquad \lim_{x \to 0^-} \sqrt{x}$$

The expression

$$\lim_{x \to 0^+} \sqrt{x}$$

is the only one that makes sense. Prove that

$$\lim_{x \to 0^+} \sqrt{x} = 0$$

Solution.   Let $\epsilon > 0$ be given. We must find a $\delta > 0$ such that $f(x) = \sqrt{x}$ satisfies

$$|\sqrt{x} - 0| < \epsilon \tag{2}$$

whenever x satisfies

$$0 < x < 0 + \delta$$

or equivalently

$$0 < x < \delta \tag{3}$$

To find δ, we rewrite (2) as

$$\sqrt{x} < \epsilon \tag{4}$$

(since $\sqrt{x} = |\sqrt{x}|$). We must choose δ so that (4) holds whenever (3) does.

We can do this by taking

$$\delta = \epsilon^2$$

To see that this choice of δ works, assume that x satisfies (3). Since we are letting $\delta = \epsilon^2$, it follows from (3) that

$$0 < x < \epsilon^2$$

or, equivalently,

$$0 < \sqrt{x} < \epsilon \qquad (5)$$

Thus, (4) holds since it is just the right-hand inequality in (5). This proves that

$$\lim_{x \to 0^+} \sqrt{x} = 0 \qquad \blacktriangleleft$$

▶ **Example 2**   Let

$$f(x) = \begin{cases} 2x, & x < 1 \\ x^3, & x \geq 1 \end{cases}$$

Prove that

$$\lim_{x \to 1^-} f(x) = 2$$

Solution.   Let $\epsilon > 0$ be given. We must find a $\delta > 0$ such that $f(x)$ satisfies

$$|f(x) - 2| < \epsilon \qquad (6)$$

whenever x satisfies

$$1 - \delta < x < 1 \qquad (7)$$

Because we are considering only x satisfying (7), it follows from the formula for $f(x)$ that

$$f(x) = 2x$$

Thus, we can rewrite (6) as

$$|2x - 2| < \epsilon$$

or

$$|x - 1| < \frac{\epsilon}{2}$$

or

$$1 - \frac{\epsilon}{2} < x < 1 + \frac{\epsilon}{2} \tag{8}$$

We must choose δ so that (8) is satisfied whenever (7) is satisfied. We can do this by taking

$$\delta = \frac{\epsilon}{2}$$

To see that this choice of δ works, assume that x satisfies (7). With $\delta = \epsilon/2$, it follows from (7) that

$$1 - \frac{\epsilon}{2} < x < 1 \tag{9}$$

and this implies that

$$1 - \frac{\epsilon}{2} < x < 1 + \frac{\epsilon}{2}$$

since ϵ is positive.

Thus, we have proved

$$\lim_{x \to 1^-} f(x) = 2$$ ◄

In Section 2.4 we discussed the limits

$$\lim_{x \to -\infty} f(x) = L \qquad \text{and} \qquad \lim_{x \to +\infty} f(x) = L$$

from an intuitive viewpoint. We interpreted the first statement to mean that the value of $f(x)$ approaches L as x moves indefinitely far from the origin along the negative x-axis, and we interpreted the second statement to mean that the value of $f(x)$ approaches L as x moves indefinitely far from the origin along the positive x-axis.

INFINITE LIMITS   The phrase, "x moves indefinitely far from the origin along the positive x-axis" is intended to convey the idea that given *any* positive number N (no matter how large) eventually x is larger than N (Figure A.31a). Similarly, the phrase "x moves indefinitely far from the origin along the negative x-axis" is intended to convey the idea that given any negative number N (no matter how small) eventually x is smaller than N (Figure A.31b).

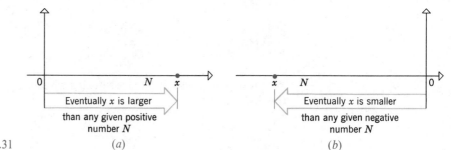

Figure A.31

Eventually x is larger than any given positive number N

(a)

Eventually x is smaller than any given negative number N

(b)

Thus, when we say that $\lim_{x \to +\infty} f(x) = L$, we mean that for any $\epsilon > 0$, we can find a positive number N (sufficiently large) so that $f(x)$ satisfies

$$|f(x) - L| < \epsilon$$

when x is greater than N (Figure A.32). Similarly, when we say that $\lim_{x \to -\infty} f(x) = L$ we mean that for any $\epsilon > 0$, we can find a negative number N so that $f(x)$ satisfies

$$|f(x) - L| < \epsilon$$

when x is smaller than N (Figure A.33).

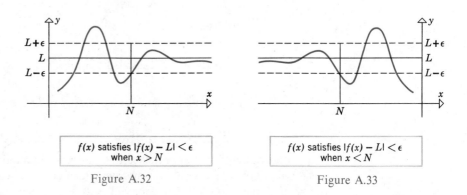

$f(x)$ satisfies $|f(x) - L| < \epsilon$ when $x > N$

Figure A.32

$f(x)$ satisfies $|f(x) - L| < \epsilon$ when $x < N$

Figure A.33

These ideas suggest the following formal definitions.

DEFINITION 3   Let f be a function that is defined on some infinite open interval $(x_0, +\infty)$. We will write

$$\lim_{x \to +\infty} f(x) = L$$

if given any number $\epsilon > 0$, there corresponds a positive number N such that

$$|f(x) - L| < \epsilon$$

whenever x satisfies

$$x > N$$

DEFINITION 4   Let f be a function that is defined on some infinite open interval $(-\infty, x_0)$. We will write

$$\lim_{x \to -\infty} f(x) = L$$

if given any number $\epsilon > 0$, there corresponds a negative number N such that

$$|f(x) - L| < \epsilon$$

whenever x satisfies

$$x < N$$

▶ Example 3   Prove that

$$\lim_{x \to +\infty} \frac{1}{x} = 0$$

Solution.   We must show that for any $\epsilon > 0$ we can find a positive number N such that $f(x) = 1/x$ satisfies

$$\left| \frac{1}{x} - 0 \right| < \epsilon \tag{10}$$

whenever x satisfies

$$x > N \tag{11}$$

To find N, rewrite (10) as

$$\frac{1}{|x|} < \epsilon$$

or

$$|x| > \frac{1}{\epsilon} \tag{12}$$

Thus, we must find a positive number N such that x satisfies (12) whenever x satisfies (11). We can do this by taking

$$N = \frac{1}{\epsilon}$$

To prove that this choice of N works, assume that x satisfies (11). Since we are letting $N = 1/\epsilon$, (11) becomes

$$x > \frac{1}{\epsilon} \qquad (13)$$

But $1/\epsilon$ is positive (since $\epsilon > 0$) so that the x in (13) is positive. Thus, (13) can be written

$$|x| > \frac{1}{\epsilon}$$

which is precisely inequality (12). Thus, we have proved

$$\lim_{x \to +\infty} \frac{1}{x} = 0. \qquad \blacktriangleleft$$

In Section 2.4 we discussed the limits

$$\lim_{x \to a} f(x) = +\infty \qquad \lim_{x \to a} f(x) = -\infty \qquad (14)$$

$$\lim_{x \to a^+} f(x) = +\infty \qquad \lim_{x \to a^+} f(x) = -\infty \qquad (15)$$

$$\lim_{x \to a^-} f(x) = +\infty \qquad \lim_{x \to a^-} f(x) = -\infty \qquad (16)$$

$$\lim_{x \to +\infty} f(x) = +\infty \qquad \lim_{x \to +\infty} f(x) = -\infty \qquad (17)$$

$$\lim_{x \to -\infty} f(x) = +\infty \qquad \lim_{x \to -\infty} f(x) = -\infty \qquad (18)$$

from an intuitive viewpoint. Recall that each of these expressions describes a particular way in which the limit fails to exist. The $+\infty$ to the right of the $=$ sign indicates that the limit fails to exist because $f(x)$ is increasing without bound, and the $-\infty$ indicates that the limit fails to exist because $f(x)$ is decreasing without bound.

To make these ideas mathematically precise, we must clarify the meanings of the phrases "$f(x)$ is increasing without bound" and "$f(x)$ is decreasing without bound." We will limit our discussion to the limits in (14); formal definitions of the remaining limits are discussed in the exercises.

When we say that $f(x)$ increases without bound as x approaches a from either side, we mean that given any positive number N (no matter how large) the value of $f(x)$ eventually exceeds N as x approaches a from either side. Figure A.34 illustrates this idea. For the curve in that figure we have

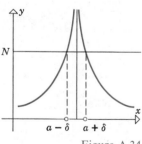

Figure A.34

$$\lim_{x \to a} f(x) = +\infty$$

We picked an arbitrary positive number N and marked it on the y-axis. As x approaches a from either side, x will eventually lie between the points

$$a - \delta \quad \text{and} \quad a + \delta$$

shown in the figure. When this occurs the value of $f(x)$ will satisfy

$$f(x) > N \tag{19}$$

It is evident from the figure that no matter how large we make N, we will always be able to find a sufficiently small positive number δ such that $f(x)$ satisfies (19) when x is between $a - \delta$ and $a + \delta$ (but different from a). This suggests the following definition.

DEFINITION 5   Let f be defined in some open interval containing the number a, except that f need not be defined at a. We will write

$$\lim_{x \to a} f(x) = +\infty$$

if given any positive number N, we can find a number $\delta > 0$ such that $f(x)$ satisfies

$$f(x) > N$$

whenever x satisfies

$$0 < |x - a| < \delta$$

When we say that $f(x)$ decreases without bound as x approaches a from either side, we mean that given any negative number N (no matter how small) the value of $f(x)$ will eventually be smaller than N as x approaches a from either side. Figure A.35 illustrates this idea. For the curve in that figure we have

$$\lim_{x \to a} f(x) = -\infty$$

We picked an arbitrary negative number N and marked it on the y-axis. As x approaches a from either side, x will eventually lie between the points

$$a - \delta \quad \text{and} \quad a + \delta$$

shown in the figure. When this occurs the value of $f(x)$ will satisfy

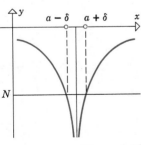

Figure A.35

$$f(x) < N$$

This suggests the following definition.

DEFINITION 6   Let f be defined in some open interval containing the number a, except that f need not be defined at a. We will write

$$\lim_{x \to a} f(x) = -\infty$$

if given any negative number N we can find a number $\delta > 0$ such that $f(x)$ satisfies

$$f(x) < N$$

whenever x satisfies

$$0 < |x - a| < \delta$$

▶ Example 4   Prove that

$$\lim_{x \to 0} \frac{1}{x^2} = +\infty$$

Solution.   We must show that given any positive number N, we can find a positive number δ such that $f(x) = 1/x^2$ satisfies

$$\frac{1}{x^2} > N \tag{20}$$

whenever x satisfies

$$0 < |x - 0| < \delta$$

or equivalently

$$0 < |x| < \delta \tag{21}$$

To find δ we will rewrite (20) as

$$x^2 < \frac{1}{N} \tag{22}$$

Thus, we must find a positive number δ such that x satisfies (22) whenever x satisfies (21). We can do this by taking

$$\delta = \frac{1}{\sqrt{N}}$$

To prove that this choice of δ works, assume that x satisfies (21). Since we are letting $\delta = 1/\sqrt{N}$, (21) becomes

$$0 < |x| < \frac{1}{\sqrt{N}}$$

or equivalently

$$0 < |x|^2 < \frac{1}{N}$$

or since $|x|^2 = x^2$,

$$0 < x^2 < \frac{1}{N} \tag{23}$$

Thus, (22) is satisfied, since it is just the right-hand inequality in (23). This proves that

$$\lim_{x \to +\infty} \frac{1}{x^2} = 0$$ ◄

▶ Exercises

In Exercises 1–6, use Definitions 1 and 2 to prove that the stated limit is correct.

1. $\lim_{x \to 2^+} (x + 1) = 3$.

2. $\lim_{x \to 1^-} (3x + 2) = 5$.

3. $\lim_{x \to 4^+} \sqrt{x - 4} = 0$.

4. $\lim_{x \to 0^-} \sqrt{-x} = 0$.

5. $\lim_{x \to 2^+} f(x) = 2$, where $f(x) = \begin{cases} x, & x > 2 \\ 3x, & x \le 2 \end{cases}$.

6. $\lim_{x \to 2^-} f(x) = 6$, where $f(x) = \begin{cases} x, & x > 2 \\ 3x, & x \le 2 \end{cases}$.

In Exercises 7–14, use Definitions 3 and 4 to prove that the stated limit is correct.

7. $\lim_{x \to +\infty} \frac{1}{x^2} = 0$.

8. $\lim_{x \to -\infty} \frac{1}{x} = 0$.

9. $\lim_{x \to -\infty} \frac{1}{x + 2} = 0$.

10. $\lim_{x \to +\infty} \frac{1}{x + 2} = 0$.

11. $\lim_{x \to +\infty} \frac{x}{x + 1} = 1$.

12. $\lim_{x \to -\infty} \frac{x}{x + 1} = 1$.

13. $\lim_{x \to -\infty} \frac{4x - 1}{2x + 5} = 2$.

14. $\lim_{x \to +\infty} \frac{4x - 1}{2x + 5} = 2$.

In Exercises 15–20, use Definitions 5 and 6 to prove that the stated limit is correct.

15. $\lim_{x \to 3} \frac{1}{(x - 3)^2} = +\infty$.

16. $\lim_{x \to 3} \frac{-1}{(x - 3)^2} = -\infty$.

17. $\lim_{x \to 0} \frac{1}{|x|} = +\infty$.

18. $\lim_{x \to 1} \frac{1}{|x - 1|} = +\infty$.

19. $\lim\limits_{x \to 0} \left(-\dfrac{1}{x^4} \right) = -\infty.$ **20.** $\lim\limits_{x \to 0} \dfrac{1}{x^4} = +\infty.$

21. Define limit statements (15) and (16).

22. Use the definitions in Exercise 21 to prove:

(a) $\lim\limits_{x \to 0^+} \dfrac{1}{x} = +\infty$ (b) $\lim\limits_{x \to 0^-} \dfrac{1}{x} = -\infty.$

23. Use the definitions in Exercise 21 to prove:

(a) $\lim\limits_{x \to 1^+} \dfrac{1}{1 - x} = -\infty$

(b) $\lim\limits_{x \to 1^-} \dfrac{1}{1 - x} = +\infty.$

24. Define limit statements (17) and (18).

25. Use the definitions in Exercise 24 to prove:

(a) $\lim\limits_{x \to +\infty} (x + 1) = +\infty$

(b) $\lim\limits_{x \to -\infty} (x + 1) = -\infty.$

26. Use the definitions in Exercise 24 to prove:

(a) $\lim\limits_{x \to +\infty} (x^2 - 3) = +\infty$

(b) $\lim\limits_{x \to -\infty} (x^3 + 5) = -\infty.$

II. SOME PROOFS OF LIMIT THEOREMS

In this section we will prove some fundamental theorems about limits. Our first result states that a function cannot have two different limits at the same point.

THEOREM 1   *If* $\lim\limits_{x \to a} f(x) = L_1$ *and* $\lim\limits_{x \to a} f(x) = L_2$, *then* $L_1 = L_2$.

REMARK.   In this section we will only consider the cases where a and the limits are finite (that is, not equal to $+\infty$ or $-\infty$.) Other cases are considered in the exercises.

Proof. We shall assume that $L_1 \neq L_2$ and show that this assumption leads to a contradiction. It will then follow that $L_1 = L_2$. Let

$$\epsilon = \frac{|L_1 - L_2|}{2} \tag{1}$$

Since we are assuming $L_1 \neq L_2$, it follows that $\epsilon > 0$.
From the assumption

$$\lim_{x \to a} f(x) = L_1$$

we can find $\delta_1 > 0$ such that

$$|f(x) - L_1| < \epsilon \tag{2}$$

whenever

$$0 < |x - a| < \delta_1 \tag{3}$$

and since

$$\lim_{x \to a} f(x) = L_2 \tag{4}$$

we can find $\delta_2 > 0$ such that

$$|f(x) - L_2| < \epsilon \tag{5}$$

whenever

$$0 < |x - a| < \delta_2 \tag{6}$$

Let δ be the smaller of the numbers δ_1 and δ_2, and choose any x satisfying

$$0 < |x - a| < \delta$$

Since $\delta \le \delta_1$ and $\delta \le \delta_2$, this x satisfies both (3) and (6) and, therefore, $f(x)$ satisfies both (2) and (5). But

$$
\begin{aligned}
2\epsilon &= |L_1 - L_2| && \text{[from (1)]} \\
&= |L_1 - f(x) + f(x) - L_2| \\
&= |(L_1 - f(x)) + (f(x) - L_2)| \\
&\le |L_1 - f(x)| + |f(x) - L_2| && \text{[the triangle inequality]} \\
&< \epsilon + \epsilon && \text{[(2) and (5)]} \\
&= 2\epsilon
\end{aligned}
$$

Thus, we have shown that

$$2\epsilon < 2\epsilon$$

which is a contradiction.   ▌

Next, we prove the result discussed in Example 1 of Section 2.5.

THEOREM 2   *For any constant k,*

$$\lim_{x \to a} k = k$$

Proof. Let $\epsilon > 0$ be given. We must find $\delta > 0$ such that

$$|k - k| < \epsilon \tag{7}$$

whenever

$$0 < |x - a| < \delta \tag{8}$$

But (7) can be written

$$0 < \epsilon$$

which holds for all values of x. Thus, for *any* positive δ whatever, (7) will be satisfied when (8) is satisfied. ∎

Next we will prove part (a) of Theorem 2.5.1.

THEOREM 3   *If* $\lim\limits_{x \to a} f(x) = L_1$ *and* $\lim\limits_{x \to a} g(x) = L_2$, *then*

$$\lim_{x \to a} [f(x) + g(x)] = \lim_{x \to a} f(x) + \lim_{x \to a} g(x) = L_1 + L_2$$

Proof. Let $\epsilon > 0$ be given. To prove that

$$\lim_{x \to a} [f(x) + g(x)] = L_1 + L_2$$

we must show that given $\epsilon > 0$ we can find $\delta > 0$ such that

$$\left|[f(x) + g(x)] - [L_1 + L_2]\right| < \epsilon \tag{9}$$

whenever

$$0 < |x - a| < \delta \tag{10}$$

Since

$$\lim_{x \to a} f(x) = L_1$$

we can find $\delta_1 > 0$ such that

$$|f(x) - L_1| < \frac{\epsilon}{2} \tag{11}$$

whenever

$$0 < |x - a| < \delta_1 \tag{12}$$

and since

$$\lim_{x \to a} g(x) = L_2$$

we can find $\delta_2 > 0$ such that

$$|g(x) - L_2| < \frac{\epsilon}{2} \tag{13}$$

whenever

$$0 < |x - a| < \delta_2 \tag{14}$$

Let δ be the smaller of the numbers δ_1 and δ_2. If x satisfies (10), then x will also satisfy both (12) and (14) since $\delta \le \delta_1$ and $\delta \le \delta_2$. Consequently, $f(x)$ will satisfy both (11) and (13). Therefore,

$$
\begin{aligned}
|[f(x) + g(x)] - [L_1 + L_2]| &= |[f(x) - L_1] + [g(x) - L_2]| \\
&\le |f(x) - L_1| + |g(x) - L_2| \\
&< \frac{\epsilon}{2} + \frac{\epsilon}{2} = \epsilon
\end{aligned}
$$

Thus, for the δ we have selected, (9) is satisfied whenever (10) is satisfied. This completes the proof. ∎

Next we will prove part (c) of Theorem 2.5.1. This proof is a little more complicated than the previous proofs.

THEOREM 4   *If* $\lim\limits_{x \to a} f(x) = L_1$ *and* $\lim\limits_{x \to a} g(x) = L_2$, *then*

$$\lim_{x \to a} [f(x)g(x)] = \lim_{x \to a} f(x) \lim_{x \to a} g(x) = L_1 L_2$$

Proof. Let $\epsilon > 0$ be given. We must find $\delta > 0$ such that

$$|f(x)g(x) - L_1 L_2| < \epsilon \tag{15}$$

whenever

$$0 < |x - a| < \delta \tag{16}$$

To find δ we will express (15) in a different form. We can write

$$f(x) = L_1 + [f(x) - L_1] \qquad \text{and} \qquad g(x) = L_2 + [g(x) - L_2]$$

When we multiply these expressions and subtract $L_1 L_2$, we obtain

$$f(x)g(x) - L_1 L_2 = L_1[g(x) - L_2] + L_2[f(x) - L_1]$$
$$+ [f(x) - L_1][g(x) - L_2]$$

so that (15) can be rewritten

$$|L_1[g(x) - L_2] + L_2[f(x) - L_1] + [f(x) - L_1][g(x) - L_2]| < \epsilon. \quad (17)$$

Thus, we must find $\delta > 0$ such that (17) holds whenever (16) does. Since ϵ is positive, the numbers

$$\sqrt{\epsilon/3}, \qquad \frac{\epsilon}{3(1 + |L_1|)}, \qquad \frac{\epsilon}{3(1 + |L_2|)}$$

are also positive. Therefore, since

$$\lim_{x \to a} f(x) = L_1 \qquad \text{and} \qquad \lim_{x \to a} g(x) = L_2$$

we can find positive numbers $\delta_1, \delta_2, \delta_3,$ and δ_4 such that

$$\left.\begin{array}{llll}
|f(x) - L_1| < \sqrt{\epsilon/3} & \text{when} & 0 < |x - a| < \delta_1 \\[2mm]
|f(x) - L_1| < \dfrac{\epsilon}{3(1 + |L_2|)} & \text{when} & 0 < |x - a| < \delta_2 \\[3mm]
|g(x) - L_2| < \sqrt{\epsilon/3} & \text{when} & 0 < |x - a| < \delta_3 \\[2mm]
|g(x) - L_2| < \dfrac{\epsilon}{3(1 + |L_1|)} & \text{when} & 0 < |x - a| < \delta_4
\end{array}\right\} \quad (18)$$

Let δ be the smallest of the numbers $\delta_1, \delta_2, \delta_3,$ and δ_4. Then $\delta \leq \delta_1, \delta \leq \delta_2,$ $\delta \leq \delta_3,$ and $\delta \leq \delta_4$. Thus, if x satisfies (16), then x will also satisfy the four conditions on the right side of (18). Consequently, $f(x)$ and $g(x)$ will satisfy the four conditions on the left side of (18). Therefore,

$$|L_1[g(x) - L_2] + L_2[f(x) - L_1] + [f(x) - L_1][g(x) - L_2]|$$
$$\leq |L_1[g(x) - L_2]| + |L_2[f(x) - L_1]| + |[f(x) - L_1][g(x) - L_2]|$$
$$= |L_1| |g(x) - L_2| + |L_2| |f(x) - L_1| + |f(x) - L_1| |g(x) - L_2|$$
$$< |L_1| \frac{\epsilon}{3(1 + |L_1|)} + |L_2| \frac{\epsilon}{3(1 + |L_2|)} + \sqrt{\epsilon/3} \sqrt{\epsilon/3} \quad \text{[From (18)]}$$
$$= \frac{\epsilon}{3} \frac{|L_1|}{1 + |L_1|} + \frac{\epsilon}{3} \frac{|L_2|}{1 + |L_2|} + \frac{\epsilon}{3}$$
$$< \frac{\epsilon}{3} + \frac{\epsilon}{3} + \frac{\epsilon}{3}. \quad \left[\text{Since } \frac{|L_1|}{1 + |L_1|} < 1 \text{ and } \frac{|L_2|}{1 + |L_2|} < 1.\right]$$
$$= \epsilon$$

Thus, for the δ we have selected, (17) holds whenever (16) does. This completes the proof. ▮

In Section 2.4 we interpreted the statement

$$\lim_{x \to a} f(x) = L$$

to mean

$$\lim_{x \to a^+} f(x) = L \qquad \text{and} \qquad \lim_{x \to a^-} f(x) = L$$

The following theorem shows that this interpretation is consistent with our formal limit definitions.

THEOREM 5 *If* $\lim_{x \to a^+} f(x) = \lim_{x \to a^-} f(x) = L$, *then* $\lim_{x \to a} f(x) = L$ *and conversely.*

Proof. Assume

$$\lim_{x \to a^+} f(x) = \lim_{x \to a^-} f(x) = L \tag{19}$$

To prove

$$\lim_{x \to a} f(x) = L$$

we must show that given $\epsilon > 0$ there exists a $\delta > 0$ such that

$$|f(x) - L| < \epsilon \tag{20}$$

whenever

$$0 < |x - a| < \delta \tag{21}$$

Because of (19), there exists a number $\delta_1 > 0$ such that (20) is satisfied whenever

$$a < x < a + \delta_1 \tag{22}$$

and there exists a number $\delta_2 > 0$ such that (20) is satisfied whenever

$$a - \delta_2 < x < a \tag{23}$$

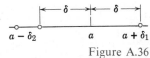

Figure A.36

Let δ be the smaller of δ_1 and δ_2, so that $\delta \le \delta_1$ and $\delta \le \delta_2$. Thus, if x satisfies (21), x will satisfy both (22) and (23) (Figure A.36) and consequently $f(x)$ will satisfy (20). Thus, for the δ we have selected, (20) holds whenever (21) does.

This completes this part of the proof. The proof of the converse is left as an exercise. ▮

▶ Exercises

1. Prove: For any constant k, $\lim\limits_{x \to +\infty} k = k$.

2. Prove: For any constant k, $\lim\limits_{x \to -\infty} k = k$.

3. Prove: If $\lim\limits_{x \to -\infty} f(x) = L_1$ and $\lim\limits_{x \to -\infty} g(x) = L_2$, then

$\lim\limits_{x \to -\infty} [f(x) + g(x)] = L_1 + L_2$.

4. Prove: If $\lim\limits_{x \to +\infty} f(x) = L_1$ and $\lim\limits_{x \to +\infty} g(x) = L_2$, then

$\lim\limits_{x \to +\infty} [f(x) + g(x)] = L_1 + L_2$.

5. Use Theorems 2, 3, and 4 of this section to prove: If $\lim\limits_{x \to a} f(x) = L_1$ and $\lim\limits_{x \to a} g(x) = L_2$, then

$\lim\limits_{x \to a} [f(x) - g(x)] = \lim\limits_{x \to a} f(x) - \lim\limits_{x \to a} g(x)$

$= L_1 - L_2$.

6. Suppose $\lim\limits_{x \to a} f(x) = +\infty$ and $\lim\limits_{x \to a} g(x) = +\infty$.

(a)   Prove: $\lim\limits_{x \to a} [f(x) + g(x)] = +\infty$.

(b)   Is it true that $\lim\limits_{x \to a} [f(x) - g(x)] = 0$?

7. Suppose $\lim\limits_{x \to a} f(x) = -\infty$ and $\lim\limits_{x \to a} g(x) = +\infty$.

(a)   Prove: $\lim\limits_{x \to a} [f(x) - g(x)] = -\infty$

(b)   Is it true that $\lim\limits_{x \to a} [f(x) + g(x)] = 0$?

8. Use Theorems 2 and 4 of this section to prove: If $\lim\limits_{x \to a} f(x) = L$ and k is a constant, then

$\lim\limits_{x \to a} [kf(x)] = kL$.

9. Prove: $\lim\limits_{x \to a} f(x) = L$ if and only if

$\lim\limits_{x \to a} [f(x) - L] = 0$.

10. Prove: If $\lim\limits_{x \to a} f(x) = L$, then $\lim\limits_{x \to a} |f(x)| = |L|$.

11. Finish the proof of Theorem 5 of this section.

III.   PROOF OF THE CHAIN RULE

In this section we will prove Theorem 3.5.2 (the chain rule). We will begin with a preliminary result.

Lemma 1. If f is differentiable at x and if $y = f(x)$, then

$$\Delta y = f'(x) \, \Delta x + \epsilon \Delta x$$

where $\epsilon \to 0$ as $\Delta x \to 0$.

Proof. Define

$$\epsilon = \begin{cases} \dfrac{f(x + \Delta x) - f(x)}{\Delta x} - f'(x) & \text{if } \Delta x \neq 0 \\ 0 & \text{if } \Delta x = 0 \end{cases} \tag{1}$$

If $\Delta x \neq 0$, it follows from (1) that

$$\epsilon \Delta x = [f(x + \Delta x) - f(x)] - f'(x) \Delta x \tag{2}$$

But,

$$\Delta y = f(x + \Delta x) - f(x) \tag{3}$$

so (2) can be written as

$$\epsilon \Delta x = \Delta y - f'(x) \Delta x$$

or

$$\Delta y = f'(x) \Delta x + \epsilon \Delta x \tag{4}$$

If $\Delta x = 0$, then (4) still holds (why?), so (4) is valid for all values of Δx. It remains to show that $\epsilon \to 0$ as $\Delta x \to 0$. But, this follows from the assumption that f is differentiable at x, since

$$\lim_{\Delta x \to 0} \epsilon = \lim_{\Delta x \to 0} \left[\frac{f(x + \Delta x) - f(x)}{\Delta x} - f'(x) \right] = f'(x) - f'(x) = 0 \quad \blacksquare$$

We are now ready to prove the chain rule.

THEOREM 7
The Chain Rule

If g is differentiable at the point x and f is differentiable at the point $g(x)$, then the composition $f \circ g$ is differentiable at the point x. Moreover, if

$$y = f(g(x)) \qquad \text{and} \qquad u = g(x)$$

then

$$\frac{dy}{dx} = \frac{dy}{du} \cdot \frac{du}{dx}$$

Proof. Since g is differentiable at x and $u = g(x)$, it follows from Lemma 1 that

$$\Delta u = g'(x) \Delta x + \epsilon_1 \Delta x \tag{5}$$

where $\epsilon_1 \to 0$ as $\Delta x \to 0$. And since $y = f(g(x)) = f(u)$ is differentiable at $u = g(x)$, it follows from Lemma 1 that

$$\Delta y = f'(u) \Delta u + \epsilon_2 \Delta u \tag{6}$$

where $\epsilon_2 \to 0$ as $\Delta u \to 0$.

Factoring out the Δu in (6) and then substituting (5) yields

$$\Delta y = [f'(u) + \epsilon_2][g'(x)\,\Delta x + \epsilon_1 \Delta x]$$

or

$$\Delta y = [f'(u) + \epsilon_2][g'(x) + \epsilon_1]\,\Delta x$$

or if $\Delta x \neq 0$,

$$\frac{\Delta y}{\Delta x} = [f'(u) + \epsilon_2][g'(x) + \epsilon_1] \tag{7}$$

Since $\epsilon_1 \to 0$ and $\epsilon_2 \to 0$ as $\Delta x \to 0$, it follows from (7) that

$$\lim_{\Delta x \to 0} \frac{\Delta y}{\Delta x} = f'(u)\,g'(x)$$

or

$$\frac{dy}{dx} = f'(u)g'(x) = \frac{dy}{du} \cdot \frac{du}{dx}$$

appendix 3 tables

Table 1
Table of Trigonometric Functions

DEGREES	RADIANS	SIN	COS	TAN	DEGREES	RADIANS	SIN	COS	TAN
0°	0.000	0.000	1.000	0.000					
1°	0.017	0.017	1.000	0.017	21°	0.367	0.358	0.934	0.384
2°	0.035	0.035	0.999	0.035	22°	0.384	0.375	0.927	0.404
3°	0.052	0.052	0.999	0.052	23°	0.401	0.391	0.921	0.424
4°	0.070	0.070	0.998	0.070	24°	0.419	0.407	0.914	0.445
5°	0.087	0.087	0.996	0.087	25°	0.436	0.423	0.906	0.466
6°	0.105	0.105	0.995	0.105	26°	0.454	0.438	0.899	0.488
7°	0.122	0.122	0.993	0.123	27°	0.471	0.454	0.891	0.510
8°	0.140	0.139	0.990	0.141	28°	0.489	0.469	0.883	0.532
9°	0.157	0.156	0.988	0.158	29°	0.506	0.485	0.875	0.554
10°	0.175	0.174	0.985	0.176	30°	0.524	0.500	0.866	0.577
11°	0.192	0.191	0.982	0.194	31°	0.541	0.515	0.857	0.601
12°	0.209	0.208	0.978	0.213	32°	0.559	0.530	0.848	0.625
13°	0.227	0.225	0.974	0.231	33°	0.576	0.545	0.839	0.649
14°	0.244	0.242	0.970	0.249	34°	0.593	0.559	0.829	0.675
15°	0.262	0.259	0.966	0.268	35°	0.611	0.574	0.819	0.700
16°	0.279	0.276	0.961	0.287	36°	0.628	0.588	0.809	0.727
17°	0.297	0.292	0.956	0.306	37°	0.646	0.602	0.799	0.754
18°	0.314	0.309	0.951	0.325	38°	0.663	0.616	0.788	0.781
19°	0.332	0.326	0.946	0.344	39°	0.681	0.629	0.777	0.810
20°	0.349	0.342	0.940	0.364	40°	0.698	0.643	0.766	0.839

Table 1
Table of Trigonometric Functions (Continued)

DEGREES	RADIANS	SIN	COS	TAN	DEGREES	RADIANS	SIN	COS	TAN
41°	0.716	0.656	0.755	0.869	66°	1.152	0.914	0.407	2.246
42°	0.733	0.669	0.743	0.900	67°	1.169	0.921	0.391	2.356
43°	0.750	0.682	0.731	0.933	68°	1.187	0.927	0.375	2.475
44°	0.768	0.695	0.719	0.966	69°	1.204	0.934	0.358	2.605
45°	0.785	0.707	0.707	1.000	70°	1.222	0.940	0.342	2.748
46°	0.803	0.719	0.695	1.036	71°	1.239	0.946	0.326	2.904
47°	0.820	0.731	0.682	1.072	72°	1.257	0.951	0.309	3.078
48°	0.838	0.743	0.669	1.111	73°	1.274	0.956	0.292	3.271
49°	0.855	0.755	0.656	1.150	74°	1.292	0.961	0.276	3.487
50°	0.873	0.766	0.643	1.192	75°	1.309	0.966	0.259	3.732
51°	0.890	0.777	0.629	1.235	76°	1.326	0.970	0.242	4.011
52°	0.908	0.788	0.616	1.280	77°	1.344	0.974	0.225	4.332
53°	0.925	0.799	0.602	1.327	78°	1.361	0.978	0.208	4.705
54°	0.942	0.809	0.588	1.376	79°	1.379	0.982	0.191	5.145
55°	0.960	0.819	0.574	1.428	80°	1.396	0.985	0.174	5.671
56°	0.977	0.829	0.559	1.483	81°	1.414	0.988	0.156	6.314
57°	0.995	0.839	0.545	1.540	82°	1.431	0.990	0.139	7.115
58°	1.012	0.848	0.530	1.600	83°	1.449	0.993	0.122	8.144
59°	1.030	0.857	0.515	1.664	84°	1.466	0.995	0.105	9.514
60°	1.047	0.866	0.500	1.732	85°	1.484	0.996	0.087	11.43
61°	1.065	0.875	0.485	1.804	86°	1.501	0.998	0.070	14.30
62°	1.082	0.883	0.469	1.881	87°	1.518	0.999	0.052	19.08
63°	1.100	0.891	0.454	1.963	88°	1.536	0.999	0.035	28.64
64°	1.117	0.899	0.438	2.050	89°	1.553	1.000	0.017	57.29
65°	1.134	0.906	0.423	2.145	90°	1.571	1.000	0.000	

Table 2
Table of Exponential and Hyperbolic Functions

x	e^x	e^{-x}	SINH x	COSH x	TANH x	x	e^x	e^{-x}	SINH x	COSH x	TANH x
.00	1.0000	1.00000	.0000	1.0000	.00000	.40	1.4918	.67032	.4108	1.0811	.37995
.01	1.0101	.99005	.0100	1.0001	.01000	.41	1.5068	.66365	.4216	1.0852	.38847
.02	1.0202	.98020	.0200	1.0002	.02000	.42	1.5220	.65705	.4325	1.0895	.39693
.03	1.0305	.97045	.0300	1.0005	.02999	.43	1.5373	.65051	.4434	1.0939	.40532
.04	1.0408	.96079	.0400	1.0008	.03998	.44	1.5527	.64404	.4543	1.0984	.41364
.05	1.0513	.95123	.0500	1.0013	.04996	.45	1.5683	.63763	.4653	1.1030	.42190
.06	1.0618	.94176	.0600	1.0018	.05993	.46	1.5841	.63128	.4764	1.1077	.43008
.07	1.0725	.93239	.0701	1.0025	.06989	.47	1.6000	.62500	.4875	1.1125	.43820
.08	1.0833	.92312	.0801	1.0032	.07983	.48	1.6161	.61878	.4986	1.1174	.44624
.09	1.0942	.91393	.0901	1.0041	.08976	.49	1.6323	.61263	.5098	1.1225	.45422
.10	1.1052	.90484	.1002	1.0050	.09967	.50	1.6487	.60653	.5211	1.1276	.46212
.11	1.1163	.89583	.1102	1.0061	.10956	.51	1.6653	.60050	.5324	1.1329	.46995
.12	1.1275	.88692	.1203	1.0072	.11943	.52	1.6820	.59452	.5438	1.1383	.47770
.13	1.1388	.87809	.1304	1.0085	.12927	.53	1.6989	.58860	.5552	1.1438	.48538
.14	1.1503	.86936	.1405	1.0098	.13909	.54	1.7160	.58275	.5666	1.1494	.49299
.15	1.1618	.86071	.1506	1.0113	.14889	.55	1.7333	.57695	.5782	1.1551	.50052
.16	1.1735	.85214	.1607	1.0128	.15865	.56	1.7507	.57121	.5897	1.1609	.50798
.17	1.1853	.84366	.1708	1.0145	.16838	.57	1.7683	.56553	.6014	1.1669	.51536
.18	1.1972	.83527	.1810	1.0162	.17808	.58	1.7860	.55990	.6131	1.1730	.52267
.19	1.2092	.82696	.1911	1.0181	.18775	.59	1.8040	.55433	.6248	1.1792	.52990
.20	1.2214	.81873	.2013	1.0201	.19738	.60	1.8221	.54881	.6367	1.1855	.53705
.21	1.2337	.81058	.2115	1.0221	.20697	.61	1.8404	.54335	.6485	1.1919	.54413
.22	1.2461	.80252	.2218	1.0243	.21652	.62	1.8589	.53794	.6605	1.1984	.55113
.23	1.2586	.79453	.2320	1.0266	.22603	.63	1.8776	.53259	.6725	1.2051	.55805
.24	1.2712	.78663	.2423	1.0289	.23550	.64	1.8965	.52729	.6846	1.2119	.56490
.25	1.2840	.77880	.2526	1.0314	.24492	.65	1.9155	.52205	.6967	1.2188	.57167
.26	1.2969	.77105	.2629	1.0340	.25430	.66	1.9348	.51685	.7090	1.2258	.57836
.27	1.3100	.76338	.2733	1.0367	.26362	.67	1.9542	.51171	.7213	1.2330	.58498
.28	1.3231	.75578	.2837	1.0395	.27291	.68	1.9739	.50662	.7336	1.2402	.59152
.29	1.3364	.74826	.2941	1.0423	.28213	.69	1.9937	.50158	.7461	1.2476	.59798
.30	1.3499	.74082	.3045	1.0453	.29131	.70	2.0138	.49659	.7586	1.2552	.60437
.31	1.3634	.73345	.3150	1.0484	.30044	.71	2.0340	.49164	.7712	1.2628	.61068
.32	1.3771	.72615	.3255	1.0516	.30951	.72	2.0544	.48675	.7838	1.2706	.61691
.33	1.3910	.71892	.3360	1.0549	.31852	.73	2.0751	.48191	.7966	1.2785	.62307
.34	1.4049	.71177	.3466	1.0584	.32748	.74	2.0959	.47711	.8094	1.2865	.62915
.35	1.4191	.70469	.3572	1.0619	.33638	.75	2.1170	.47237	.8223	1.2947	.63515
.36	1.4333	.69768	.3678	1.0655	.34521	.76	2.1383	.46767	.8353	1.3030	.64108
.37	1.4477	.69073	.3785	1.0692	.35399	.77	2.1598	.46301	.8484	1.3114	.64693
.38	1.4623	.68386	.3892	1.0731	.36271	.78	2.1815	.45841	.8615	1.3199	.65271
.39	1.4770	.67706	.4000	1.0770	.37136	.79	2.2034	.45384	.8748	1.3286	.65841

Table 2
Table of Exponential and Hyperbolic Functions (Continued)

x	e^x	e^{-x}	SINH x	COSH x	TANH x	x	e^x	e^{-x}	SINH x	COSH x	TANH x
.80	2.2255	.44933	.8881	1.3374	.66404	3.00	20.086	.04979	10.018	10.068	.99505
.81	2.2479	.44486	.9015	1.3464	.66959	3.10	22.198	.04505	11.076	11.122	.99595
.82	2.2705	.44043	.9150	1.3555	.67507	3.20	24.533	.04076	12.246	12.287	.99668
.83	2.2933	.43605	.9286	1.3647	.68048	3.30	27.113	.03688	13.538	13.575	.99728
.84	2.3164	.43171	.9423	1.3740	.68581	3.40	29.964	.03337	14.965	14.999	.99777
.85	2.3396	.42741	.9561	1.3835	.69107	3.50	33.115	.03020	16.543	16.573	.99818
.86	2.3632	.42316	.9700	1.3932	.69626	3.60	36.598	.02732	18.286	18.313	.99851
.87	2.3869	.41895	.9840	1.4029	.70137	3.70	40.447	.02472	20.211	20.236	.99878
.88	2.4109	.41478	.9981	1.4128	.70642	3.80	44.701	.02237	22.339	22.362	.99900
.89	2.4351	.41066	1.0122	1.4229	.71139	3.90	49.402	.02024	24.691	24.711	.99918
.90	2.4596	.40657	1.0265	1.4331	.71630	4.00	54.598	.01832	27.290	27.308	.99933
.91	2.4843	.40252	1.0409	1.4434	.72113	4.10	60.340	.01657	30.162	30.178	.99945
.92	2.5093	.39852	1.0554	1.4539	.72590	4.20	66.686	.01500	33.336	33.351	.99955
.93	2.5345	.39455	1.0700	1.4645	.73059	4.30	73.700	.01357	36.843	36.857	.99963
.94	2.5600	.39063	1.0847	1.4753	.73522	4.40	81.451	.01227	40.719	40.732	.99970
.95	2.5857	.38674	1.0995	1.4862	.73978	4.50	90.017	.01111	45.003	45.014	.99975
.96	2.6117	.38289	1.1144	1.4973	.74428	4.60	99.484	.01005	49.737	49.747	.99980
.97	2.6379	.37908	1.1294	1.5085	.74870	4.70	109.95	.00910	54.969	54.978	.99983
.98	2.6645	.37531	1.1446	1.5199	.75307	4.80	121.51	.00823	60.751	60.759	.99986
.99	2.6912	.37158	1.1598	1.5314	.75736	4.90	134.29	.00745	67.141	67.149	.99989
1.00	2.7183	.36788	1.1752	1.5431	.76159	5.00	148.41	.00674	74.203	74.210	.99991
1.10	3.0042	.33287	1.3356	1.6685	.80050	5.10	164.02	.00610	82.008	82.014	.99993
1.20	3.3201	.30119	1.5095	1.8107	.83365	5.20	181.27	.00552	90.633	90.639	.99994
1.30	3.6693	.27253	1.6984	1.9709	.86172	5.30	200.34	.00499	100.17	100.17	.99995
1.40	4.0552	.24660	1.9043	2.1509	.88535	5.40	221.41	.00452	110.70	110.71	.99996
1.50	4.4817	.22313	2.1293	2.3524	.90515	5.50	244.69	.00409	122.34	122.35	.99997
1.60	4.9530	.20190	2.3756	2.5775	.92167	5.60	270.43	.00370	135.21	135.22	.99997
1.70	5.4739	.18268	2.6456	2.8283	.93541	5.70	298.87	.00335	149.43	149.44	.99998
1.80	6.0496	.16530	2.9422	3.1075	.94681	5.80	330.30	.00303	165.15	165.15	.99998
1.90	6.6859	.14957	3.2682	3.4177	.95624	5.90	365.04	.00274	182.52	182.52	.99998
2.00	7.3891	.13534	3.6269	3.7622	.96403	6.00	403.43	.00248	201.71	201.72	.99999
2.10	8.1662	.12246	4.0219	4.1443	.97045	6.25	518.01	.00193	259.01	259.01	.99999
2.20	9.0250	.11080	4.4571	4.5679	.97574	6.50	665.14	.00150	332.57	332.57	1.0000
2.30	9.9742	.10026	4.9370	5.0372	.98010	6.75	854.06	.00117	427.03	427.03	1.0000
2.40	11.023	.09072	5.4662	5.5569	.98367	7.00	1096.6	.00091	548.32	548.32	1.0000
2.50	12.182	.08208	6.0502	6.1323	.98661	7.50	1808.0	.00055	904.02	904.02	1.0000
2.60	13.464	.07427	6.6947	6.7690	.98903	8.00	2981.0	.00034	1490.5	1490.5	1.0000
2.70	14.880	.06721	7.4063	7.4735	.99101	8.50	4914.8	.00020	2457.4	2457.4	1.0000
2.80	16.445	.06081	8.1919	8.2527	.99263	9.00	8103.1	.00012	4051.5	4051.5	1.0000
2.90	18.174	.05502	9.0596	9.1146	.99396	9.50	13360.	.00007	6679.9	6679.9	1.0000
						10.00	22026.	.00005	11013.	11013.	1.0000

Table 3
Natural Logarithms

n	$\ln n$	n	$\ln n$	n	$\ln n$
0.0	—	4.5	1.5041	9.0	2.1972
0.1	−2.3026	4.6	1.5261	9.1	2.2083
0.2	−1.6094	4.7	1.5476	9.2	2.2192
0.3	−1.2040	4.8	1.5686	9.3	2.2300
0.4	−0.9163	4.9	1.5892	9.4	2.2407
0.5	−0.6931	5.0	1.6094	9.5	2.2513
0.6	−0.5108	5.1	1.6292	9.6	2.2618
0.7	−0.3567	5.2	1.6487	9.7	2.2721
0.8	−0.2231	5.3	1.6677	9.8	2.2824
0.9	−0.1054	5.4	1.6864	9.9	2.2925
1.0	0.0000	5.5	1.7047	10	2.3026
1.1	0.0953	5.6	1.7228	11	2.3979
1.2	0.1823	5.7	1.7405	12	2.4849
1.3	0.2624	5.8	1.7579	13	2.5649
1.4	0.3365	5.9	1.7750	14	2.6391
1.5	0.4055	6.0	1.7918	15	2.7081
1.6	0.4700	6.1	1.8083	16	2.7726
1.7	0.5306	6.2	1.8245	17	2.8332
1.8	0.5878	6.3	1.8405	18	2.8904
1.9	0.6419	6.4	1.8563	19	2.9444
2.0	0.6931	6.5	1.8718	20	2.9957
2.1	0.7419	6.6	1.8871	25	3.2189
2.2	0.7885	6.7	1.9021	30	3.4012
2.3	0.8329	6.8	1.9169	35	3.5553
2.4	0.8755	6.9	1.9315	40	3.6889
2.5	0.9163	7.0	1.9459	45	3.8067
2.6	0.9555	7.1	1.9601	50	3.9120
2.7	0.9933	7.2	1.9741	55	4.0073
2.8	1.0296	7.3	1.9879	60	4.0943
2.9	1.0647	7.4	2.0015	65	4.1744
3.0	1.0986	7.5	2.0149	70	4.2485
3.1	1.1314	7.6	2.0281	75	4.3175
3.2	1.1632	7.7	2.0412	80	4.3820
3.3	1.1939	7.8	2.0541	85	4.4427
3.4	1.2238	7.9	2.0669	90	4.4998
3.5	1.2528	8.0	2.0794	95	4.5539
3.6	1.2809	8.1	2.0919	100	4.6052
3.7	1.3083	8.2	2.1041	200	5.2983
3.8	1.3350	8.3	2.1163	300	5.7038
3.9	1.3610	8.4	2.1282	400	5.9915
4.0	1.3863	8.5	2.1401	500	6.2146
4.1	1.4110	8.6	2.1518	600	6.3069
4.2	1.4351	8.7	2.1633	700	6.5511
4.3	1.4586	8.8	2.1748	800	6.6846
4.4	1.4816	8.9	2.1861	900	6.8024

▶ Exercise Set 1.1 (Page 14)

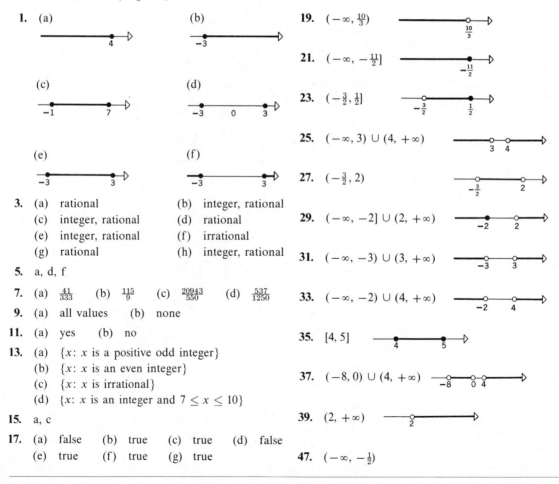

1. (a)

   (b)

   (c)

   (d)

   (e)

   (f)

3. (a) rational   (b) integer, rational
   (c) integer, rational   (d) rational
   (e) integer, rational   (f) irrational
   (g) rational   (h) integer, rational

5. a, d, f

7. (a) $\frac{41}{333}$ (b) $\frac{115}{9}$ (c) $\frac{20943}{550}$ (d) $\frac{537}{1250}$

9. (a) all values   (b) none

11. (a) yes   (b) no

13. (a) $\{x: x$ is a positive odd integer$\}$
    (b) $\{x: x$ is an even integer$\}$
    (c) $\{x: x$ is irrational$\}$
    (d) $\{x: x$ is an integer and $7 \le x \le 10\}$

15. a, c

17. (a) false   (b) true   (c) true   (d) false
    (e) true   (f) true   (g) true

19. $(-\infty, \frac{10}{3})$

21. $(-\infty, -\frac{11}{2}]$

23. $(-\frac{3}{2}, \frac{1}{2}]$

25. $(-\infty, 3) \cup (4, +\infty)$

27. $(-\frac{3}{2}, 2)$

29. $(-\infty, -2] \cup (2, +\infty)$

31. $(-\infty, -3) \cup (3, +\infty)$

33. $(-\infty, -2) \cup (4, +\infty)$

35. $[4, 5]$

37. $(-8, 0) \cup (4, +\infty)$

39. $(2, +\infty)$

47. $(-\infty, -\frac{1}{2})$

▶ Exercise Set 1.2 (Page 21)

1. (a) 7   (b) $\sqrt{2}$ (c) k^2 (d) k^2

5. (a) 2   (b) 1   (c) 14
(d) $3 + \sqrt{2}$ (e) 7   (f) 5

7. (a) -9 (b) 7   (c) 12

9. $-\frac{5}{6}, \frac{3}{2}$ **11.** $\frac{1}{2}, \frac{5}{2}$ **13.** $-\frac{11}{10}, \frac{11}{8}$

15. $1, \frac{17}{5}$ **17.** $(-9, -3)$

19. $(-\infty, \frac{1}{2}] \cup [\frac{9}{2}, +\infty)$ **21.** $(-\infty, \frac{5}{2})$ **23.** $[\frac{7}{10}, \frac{7}{2}]$

25. $(0, +\infty)$ **27.** $(-\infty, -7) \cup (-7, -\frac{3}{2})$

29. $\alpha \in (-\infty, 2] \cup [3, +\infty)$ **31.** $ab \geq 0$ **37.** $\frac{1}{3}$

39. $\frac{13}{11}$ (other answers are possible)

▶ Exercise Set 1.3 (Page 31)

1. 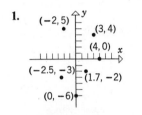   **3.** (a) 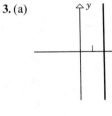   (b) 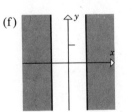   (c)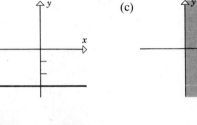

(d) 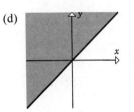   (e) 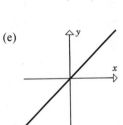   (f)

5. $(6, 9)$

7. (a) vertical; $3 + \sqrt{2}$
(b) vertical; 7
(c) vertical; 5

9. (a) 10   (b) $(4, 5)$

11. (a) $\sqrt{29}$ (b) $(-\frac{9}{2}, -5)$

17. (a) yes   (b) no   (c) yes   (d) yes

19. (a) y-axis (b) origin (c) x-axis, y-axis, origin

21.   **23.**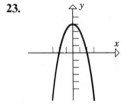

25. 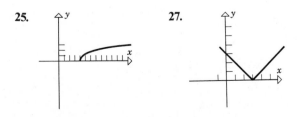   **27.**   **29.**    **31.**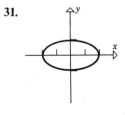

33. The union of the graphs of $x - y = 0$ and $x + y = 0$.

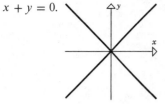

35.

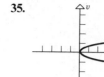

37. $3x - 2y - 5 = 0$

39. 0

▶ Exercise Set 1.4 (Page 39)

1. (a) $\frac{1}{2}$ (b) -1 (c) 0   (d) not defined

3. (a) $\dfrac{1}{\sqrt{3}}$ (b) -1 (c) $\sqrt{3}$

5. (a) $153°$ (b) $45°$ (c) $117°$ (d) $89°$

7.

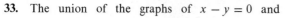

9. (a) parallel   (b) perpendicular   (c) neither

11. (a) 14   (b) $-\frac{1}{3}$

13. 29

15. (a) yes   (b) no

23. (a) $\frac{7}{19}$ (b) $\frac{57}{25}$ (c) $\frac{4}{13}$

25. (a) $20°$ (b) $66°$ (c) $17°$

27. 11.09

▶ Exercise Set 1.5 (Page 45)

1. (a)   (b)   (c)   (d)

3. (a)   (b)   (c)

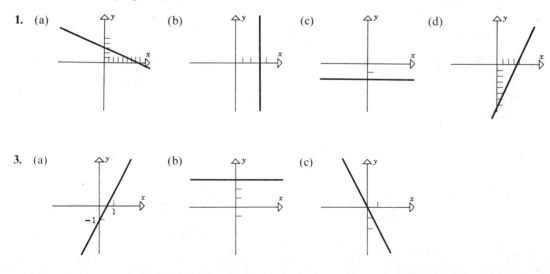

5.

	(a)	(b)	(c)	(d)	(e)
Slope	3	$-1/4$	$-3/5$	0	$-b/a$
y-intercept	2	3	$8/5$	1	b

7. (a) 60° (b) 117°   **9.** $y = -2x + 4$

11. $y = 4x + 7$ **13.** $y = 11x - 18$

15. $y = \dfrac{1}{\sqrt{3}}x - 3$ **17.** $y = \frac{1}{2}x + 2$

19. $y = 1$ **21.** $x = 5$

23. (a) parallel   (b) perpendicular
(c) parallel   (d) perpendicular
(e) neither

25. (a) $(4, -1)$ (b) $(1, -2)$

27. 3   **31.** 4

35. (a) $(\frac{9}{2}, \frac{7}{2})$ (b) $x - 9y + 27 = 0$

37. $(-\frac{29}{8}, -\frac{23}{4})$

39. (a) $F = \frac{9}{5}C + 32$ (b) $\frac{5}{9}$

▶ Exercise Set 1.6 (Page 51)

1. (a) $(0,0)$; 5 (b) $(1,4)$; 4 (c) $(-1,-3)$; $\sqrt{5}$
(d) $(0,-2)$; 1

3. $(x - 3)^2 + (y + 2)^2 = 16$

5. $(x + 4)^2 + (y - 8)^2 = 64$

7. $(x + 3)^2 + (y + 4)^2 = 25$

9. $(x - 1)^2 + (y - 1)^2 = 2$

11. circle; center $(1, 2)$, radius 4

13. circle; center $(-1, 1)$, radius $\sqrt{2}$

15. the point $(-1, -1)$

17. circle; center $(0, 0)$, radius $\frac{1}{3}$

19. no graph

21. circle; center $(-\frac{5}{4}, -\frac{1}{2})$, radius $\frac{3}{2}$

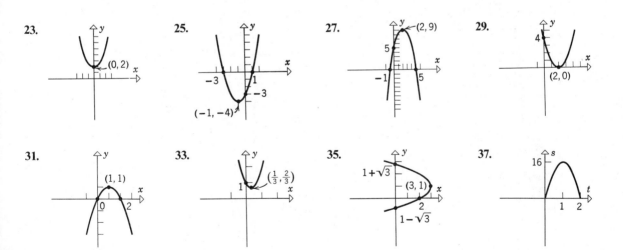

23.   **25.**   **27.**   **29.**

31.   **33.**   **35.**   **37.**

39. (a) equation of path: $2x^2 + 2y^2 - 12x + 8y + 1 = 0$
(b) center $(3, -2)$, radius $\dfrac{5}{\sqrt{2}}$

▶ Chapter 1 Supplementary Exercises (Page 52)

1. (a) $(-3, 5]$ (b) $[-3, 3]$
   (c) $(-\infty, -\frac{1}{2}] \cup [\frac{1}{2}, +\infty)$

3. (a) $(-\infty, -\frac{1}{2}) \cup (3, +\infty)$ (b) $[1, 4]$

5. (a) $(-2, -1] \cup [1, 2)$
   (b) $(-\infty, -5] \cup [-1, +\infty)$

7. (a) Take $a = -2, b = 1$.
   (b) $a + b > 0$

9. Both legs of a right triangle are no longer than the hypotenuse.

11. (a) all points on the y-axis or the line $y = x$
    (b) all points on the line $x = 1$ or the line $y = x + 1$

13. (a) all points on or outside the circle of radius 4 about $(1, 3)$
    (b)

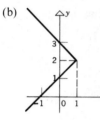

15. at $(2, 4)$ and $(-1, 1)$

17.

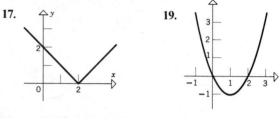

19.

21. $(x - 1)^2 + (y - 2)^2 = 25$

23. four circles: $(x - h)^2 + (y - k)^2 = 25$, where $h = 1$ or 11 and $k = 2$ or 12

25. point $(-2, -1)$ 27. no graph

29. (a) $y = 4x/3$; 10; $(0, 0)$
    (b) $x = 3$; 8; $(3, 0)$
    (c) $y = 4$; 6; $(0, 4)$
    (d) $y = 7 - x$; $\sqrt{2}$; $(\frac{7}{2}, \frac{7}{2})$

31. $(4, -3)$ and $(-4, 3)$

33. (a) $90°$ (b) $0°$ (c) $135°$ (d) $60°$

35. $y = -3$ 37. $y = -\frac{1}{2}x$

39. $L: y - 0 = (-2)(x - 1)$
    $L': y = \frac{1}{2}x - 3;$ $(2, -2)$

41. $L: y - 1 = (\frac{2}{5})(x - 3)$
    $L': 3x + 2y = -8;$ $(-2, -1)$ 43. no

▶ Exercise Set 2.1 (Page 67)

1. (a) 14 (b) 50 (c) 2
   (d) 11 (e) $3t^2 + 2$

3. (a) 4 (b) -8 (c) $\frac{1}{3}$
   (d) $\frac{1}{3.1}$ (e) 5.8

5. (a) 1 (b) 1 (c) -1
   (d) -1 (e) not defined

7. $(-\infty, -\frac{7}{5}) \cup (-\frac{7}{5}, +\infty)$

9. $(-\infty, -2] \cup [1, +\infty)$

11. $(-\infty, +\infty)$ 13. $[5, +\infty)$

15. $(-\infty, -3] \cup [2, +\infty)$ 17. $(-\infty, 3) \cup (3, +\infty)$

19.

21.

23.

25.

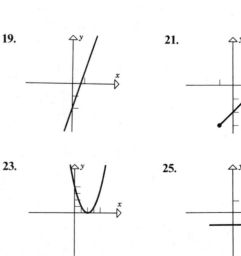

27. **29.** **31.**

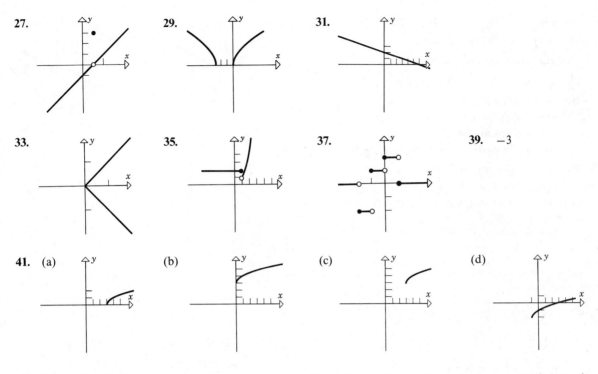

33. **35.** **37.** **39.** -3

41. (a) (b) (c) (d)

43. (a) both (b) y is a function of x. (c) neither (d) both

45. (a) $y = \dfrac{1}{x^2}$ (b) $y = \dfrac{1-x}{1+x}$ (c) $y = -x$

49. $\dfrac{1-(1/x)}{1+(1/x)} = \dfrac{x-1}{x+1}$, $x \neq 0$.

51. (a) $x + 3$, $x \neq 3$ (b) x, $x \neq -3$ or 1 (c) $\sqrt{x} + 1, x > 0$

53. (a) both (b) a function of x (c) a function of y (d) neither

---

▶ Exercise Set 2.2 (Page 74)

1. (a) $t^2 + 1$ (b) $t^2 + 4t + 5$
   (c) $x^2 + 4x + 5$ (d) $\dfrac{1}{x^2} + 1$
   (e) $x^2 + 2hx + h^2 + 1$ (f) $x^2 + 1$
   (g) $x + 1, x \geq 0$ (h) $9x^2 + 1$

3. (a) $x^2 + 2x + 1$ (b) $-x^2 + 2x - 1$
   (c) $2x(x^2 + 1)$ (d) $\dfrac{2x}{x^2 + 1}$
   (e) $2(x^2 + 1)$ (f) $4x^2 + 1$
   (g) $2kx$

5. (a) $\sqrt{x+1} + x - 2$ (b) $\sqrt{x+1} - x + 2$
   (c) $(x-2)\sqrt{x+1}$ (d) $\dfrac{\sqrt{x+1}}{x-2}$
   (e) $\sqrt{x-1}$ (f) $\sqrt{x+1} - 2$
   (g) $k\sqrt{x+1}$

7. (a) $\sqrt{x-2} + \sqrt{x-3}$
   (b) $\sqrt{x-2} - \sqrt{x-3}$
   (c) $\sqrt{x-2}\sqrt{x-3}$ (d) $\dfrac{\sqrt{x-2}}{\sqrt{x-3}}$
   (e) $\sqrt{\sqrt{x-3} - 2}$ (f) $\sqrt{\sqrt{x-2} - 3}$
   (g) $k\sqrt{x-2}$

9. (a) $4x - 15$ (b) $4x^2 - 20x + 25$

11. $g(x) = \sqrt{x}, h(x) = x + 2$

13. $g(x) = x^7, h(x) = x - 5$

15. $g(x) = |x|, h(x) = x^2 - 3x + 5$

19. (a) explicit algebraic
    (b) rational, explicit algebraic
    (c) explicit algebraic
    (d) polynomial, rational, explicit algebraic

21. (a) explicit algebraic
    (b) rational, explicit algebraic
    (c) monomial, polynomial, rational, explicit algebraic
    (d) explicit algebraic

23. (a)   (b)

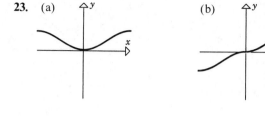

25. yes

27. $\sqrt{\frac{1}{2}x - 3}$

▶ Exercise Set 2.3 (Page 83)

1. (a) $\frac{7}{2}$ (b) $3, y = 3x - \frac{9}{2}$

(c)

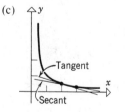

3. (a) $-\frac{1}{6}$ (b) $-\frac{1}{4}, y = -\frac{1}{4}x + 1$

(c)

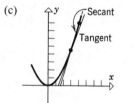

5. (b) $y = 75x - 250$ (c) $y = 3x_0^2 x - 2x_0^3$

7. (a) $2x_0 + 1$ (b) $y = 5x - 4$
   (c) $y = (2x_0 + 1)x - x_0^2$

9. (a) t_0 (b) 0 (c) speeding up
   (d) slowing down

11. It is a straight line with slope equal to the velocity.

13. (a) 320,000 ft (b) 8000 ft/sec
    (c) 45 ft/sec (d) 24,000 ft/sec

15. (a) 720 ft/min (b) 192 ft/min

▶ Exercise Set 2.4 (Page 93)

1. (a) -1 (b) 3
   (c) does not exist (d) 1
   (e) -1 (f) 3

3. (a) 1 (b) 1
   (c) 1 (d) 1
   (e) $-\infty$ (f) $+\infty$

5. (a) 0   (b) 0
   (c) 0   (d) 3
   (e) $+\infty$ (f) $+\infty$

7. (a) $-\infty$ (b) $+\infty$
   (c) does not exist   (d) not defined
   (e) 2   (f) 0

9. (a) $-\infty$ (b) $-\infty$
   (c) $-\infty$ (d) 1
   (e) 2   (f) 2

11. (a) 0   (b) 0
   (c) 0   (d) 0
   (e) does not exist   (f) does not exist

13. all values except -4

▶ Exercise Set 2.5 (Page 106)

1. 7   **3.** π **5.** 36
7. $\sqrt{109}$ **9.** 14   **11.** 0
13. 8   **15.** 4   **17.** $-\frac{4}{5}$
19. $\frac{3}{2}$ **21.** 0   **23.** 0
25. $-\sqrt{5}$ **27.** $\dfrac{1}{\sqrt{6}}$ **29.** $\sqrt{3}$
31. $+\infty$ **33.** does not exist   **35.** $-\infty$

37. $+\infty$ **39.** does not exist   **41.** $+\infty$
43. $-\infty$ **45.** $-\frac{1}{7}$ **47.** -1
49. 6
51. (a) 2   (b) 2   (c) 2
53. 4
57. if $r(a)$ is defined

▶ Exercise Set 2.6 (Page 117)

Note: there are other possible answers for Exercises 1–21, 27.

1. 0.05   **3.** 1/700   **5.** 0.05
7. 1/9000   **9.** 1   **11.** $\delta = \dfrac{\epsilon}{3}$
13. $\delta = \epsilon$ **15.** $\delta = \min\left(\dfrac{\epsilon}{6}, 1\right)$

17. $\delta = \min\left(\dfrac{\epsilon}{36}, \dfrac{1}{4}\right)$ **19.** $\delta = \min(2\epsilon, 4)$
21. $\delta = \epsilon$ **27.** $\delta = \min\left(\dfrac{\epsilon}{8}, 2\right)$

▶ Chapter 2 Supplementary Exercises (Page 118)

1. $|x| \le 2$; $\sqrt{2}, 2, 1$
3. $x \ne -2, 1$; $\frac{1}{2}$, undefined, $\frac{1}{4}$
5. all x; $-1, 3, \sqrt{3}$
7. (a) $(-6 + 6x - 2x^2)/x^2$
   (b) $-9/[x(x + 3)]$
   (c) $(3x^2 - 9x + 3)/(x - 3)$
   (d) $(4x - 3)/(3 - x)$

9. the horizontal line $y = -\pi$; domain: all x; range: $\{-\pi\}$

11. domain: $x \ne -2$; range $y \ne -2$

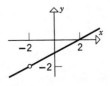

13. domain: $x \geq -\frac{1}{3}$; range: $y \leq 0$

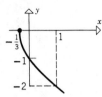

15. domain: $x \neq \pm 2$; range: $y \neq 0, \frac{1}{2}$

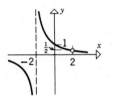

17. Some possible answers are:
  (a) $h(x) = x^3$, $g(x) = x^2 + 3$;
       $h(x) = x^6$, $g(x) = x + 3$
  (b) $h(x) = x^2 + 1$, $g(x) = \sqrt{x}$;
       $h(x) = x^2$, $g(x) = \sqrt{x + 1}$
  (c) $h(x) = 3x + 2$, $g(x) = \sin x$;
       $h(x) = 3x$, $g(x) = \sin (x + 2)$

19. $y = -1$

21. 5

23. (a) -1 (b) does not exist   (c) 1
      (d) 0   (e) $-\infty$ (does not exist)   (f) 0
      (g) 0   (h) $-\infty$ (does not exist)

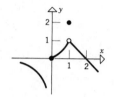

25. 2, 1, 0, does not exist, $+\infty$ (does not exist), does not exist

27. 5, 10, 0, 10, 0, $-\infty$ (does not exist), $+\infty$ (does not exist)

▶ Exercise Set 3.1 (Page 128)

1. $6x$

3. $3x^2$

5. $\dfrac{1}{2\sqrt{x + 1}}$

7. $-\dfrac{1}{x^2}$

9. $2ax$

11. $18, y = 18x - 27$

13. $0, y = 0$

15. $\frac{1}{6}, y = \frac{1}{6}x + \frac{5}{3}$

17. (a) $8x$ (b) 8

19. $8t + 1$

21. $6\lambda - 1$

23. (a) 10   (b) 4

25. (a) 3π (b) 4π

27.

29.

31.

33.

35.

37. $-\dfrac{1}{2x^{3/2}}$

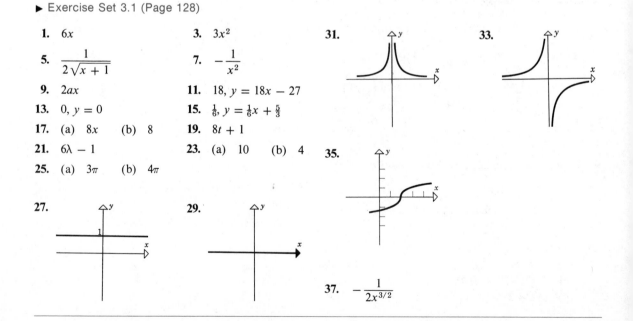

▶ Exercise Set 3.2 (Page 140)

1. $28x^6$

3. $24x^7 + 2$

5. 0

7. $-\frac{1}{3}(7x^6 + 2)$

9. $3ax^2 + 2bx + c$

11. $24x^{-9}$

13. $-3x^{-4} - 7x^{-8}$

15. $18x^2 - \frac{3}{2}x + 12$

17. $-15x^{-2} - 14x^{-3} + 48x^{-4} + 32x^{-5}$

19. $10x^9 + 24x^5 + 8x$

21. $\dfrac{3}{(2x+1)^2}$

23. $\dfrac{(2x^7 + 4x + 2)(24x^7 - 12x) - (3x^8 - 6x^2 + 1)(14x^6 + 4)}{(2x^7 + 4x + 2)^2}$

25. $\left(\dfrac{3x+2}{x}\right)(-5x^{-6}) + (x^{-5} + 1)\left(-\dfrac{2}{x^2}\right)$

27. $\dfrac{1}{4}\left(\dfrac{x}{x+1}\right)(-2x^{-3}) + \dfrac{1}{4}(x^{-2} + 5) \cdot \dfrac{1}{(x+1)^2}$

29. $32t$

31. $3\pi r^2$

33. $\dfrac{7 - 2t^3}{(t^3 + 7)^2}$

35. $-\dfrac{2GmM}{r^3}$

37. (a) $42x - 10$ (b) 24 (c) $\dfrac{2}{x^3}$ (d) $700x^3 - 96x$

39. (a) $-210x^{-8} + 60x^2$ (b) $-6x^{-4}$ (c) $6a$

41. (a) 0 (b) 112 (c) 360

45. $\left(1, \frac{5}{8}\right), \left(2, \frac{2}{3}\right)$

47. $a = 3, b = 2$

51. (a) $2\left(1 + \dfrac{1}{x}\right)(x^{-3} + 7)$

$+ (2x + 1)\left(-\dfrac{1}{x^2}\right)(x^{-3} + 7)$

$+ (2x + 1)\left(1 + \dfrac{1}{x}\right)(-3x^{-4})$

(b) $(-5x^{-6})(x^2 + 2x)(4 - 3x)(2x^9 + 1)$
$+ x^{-5}(2x + 2)(4 - 3x)(2x^9 + 1)$
$+ x^{-5}(x^2 + 2x)(-3)(2x^9 + 1)$
$+ x^{-5}(x^2 + 2x)(4 - 3x)(18x^8)$

(c) $3(7x^6 + 2)(x^7 + 2x - 3)^2$

(d) $100x(x^2 + 1)^{49}$

53. (a) $n(n-1)(n-2) \cdots 1$ (b) 0
(c) $a_n n(n-1)(n-2) \cdots 1$

▶ Exercise Set 3.3 (Page 151)

1. $-2 \sin x - 3 \cos x$

3. $\dfrac{x \cos x - \sin x}{x^2}$

5. $x^3 \cos x + (3x^2 + 5) \sin x$

7. $\sec x \tan x - \sqrt{2} \sec^2 x$

9. $\sec^3 x + \sec x \tan^2 x$

11. $1 + 4 \csc x \cot x - 2 \csc^2 x$

13. $-\dfrac{\csc x}{1 + \csc x}$

15. 0

17. $\dfrac{1}{(1 + x \tan x)^2}$

19. $\sec^3 x + \sec x \tan^2 x$

21. $-x \sin x + 5 \cos x$

23. $-4 \sin x \cos x$

25. 3

27. $\frac{7}{3}$

29. 1

31. 2

33. 0

35. $-\frac{25}{49}$

37. (a) $x = n\pi, n = 0, \pm 1, \pm 2, \ldots$
(b) none
(c) $x = \dfrac{\pi}{2} + n\pi, n = 0, \pm 1, \pm 2, \ldots$

39. (a) $y = x$ (b) $y = 2x - \pi/2 + 1$
(c) $y = 2x + \pi/2 - 1$

41. (a) all x (b) all x
(c) $x \neq \dfrac{\pi}{2} + n\pi, n = 0, \pm 1, \pm 2, \ldots$
(d) $x \neq n\pi, n = 0, \pm 1, \pm 2, \ldots$
(e) $x \neq \dfrac{\pi}{2} + n\pi, n = 0, \pm 1, \pm 2, \ldots$
(f) $x \neq n\pi, n = 0, \pm 1, \pm 2, \ldots$
(g) $x \neq \pi + 2n\pi, n = 0, \pm 1, \pm 2, \ldots$
(h) $x \neq n\pi/2, n = 0, \pm 1, \pm 2, \ldots$
(i) all x

45. (a) 1 (b) 0 (c) 1

47. $\lim\limits_{x \to 0} f(x) = 1$

▶ Exercise Set 3.4 (Page 162)

1. (a) 5   (b) 4   **3.** (a) $-\frac{1}{3}$ (b) -0.5
(c)
(c)

5. $dy = 3x^2\,dx$
$\Delta y = 3x^2\,\Delta x + 3x(\Delta x)^2 + (\Delta x)^3$

7. $dy = (2x - 2)\,dx$
$\Delta y = 2x\,\Delta x + (\Delta x)^2 - 2\,\Delta x$

9. $dy = (12x^2 - 14x + 2)\,dx$

11. $dy = (\cos x - x \sin x)\,dx$ **13.** $2x$

15. -1 **17.** 8.0625

19. 8.9944 **21.** 0.8573

23. (a) $\pm 2\ \text{ft}^2$ (b) side: $\pm 1\%$, area: $\pm 2\%$

25. (a) opposite: $\pm 0.151''$, adjacent: $\pm 0.087''$
(b) opposite: $\pm 3.0\%$, adjacent: $\pm 1.0\%$

27. $\pm 10\%$ **29.** $\pm 6\%$

31. $\pm 0.5\%$ **33.** $0.236\ \text{cm}^3$

▶ Exercise Set 3.5 (Page 170)

1. $37(x^3 + 2x)^{36}(3x^2 + 2)$

3. $-2\left(x^3 - \dfrac{7}{x}\right)^{-3}\left(3x^2 + \dfrac{7}{x^2}\right)$

5. $\dfrac{24(1 - 3x)}{(3x^2 - 2x + 1)^4}$

7. $8x \sec^2 (4x^2)$

9. $-20 \cos^4 x \sin x$

11. $-\dfrac{2}{x^3} \cos\left(\dfrac{1}{x^2}\right)$

13. $28x^6 \sec^2 (x^7) \tan (x^7)$

15. $-3[x + \csc (x^3 + 3)]^{-4}[1 - 3x^2 \csc (x^3 + 3) \cot (x^3 + 3)]$

17. $-x^3 \sec\left(\dfrac{1}{x}\right)\tan\left(\dfrac{1}{x}\right) + 5x^4 \sec\left(\dfrac{1}{x}\right)$

19. $\sin x \sin (\cos x)$

21. $12(5x + 8)^{13}(x^3 + 7x)^{11}(3x^2 + 7) + 65(x^3 + 7x)^{12}(5x + 8)^{12}$

23. $\dfrac{-64x(2x + 1)^{-3}(4x^2 - 1)^{-9} + 6(4x^2 - 1)^{-8}(2x + 1)^{-4}}{(2x + 1)^{-6}}$

25. $5[x \sin 2x + \tan^4 (x^7)]^4[2x \cos 2x + \sin 2x + 28x^6 \tan^3 (x^7) \sec^2 (x^7)]$

27. $-25x \cos (5x) - 10 \sin (5x) - 2 \cos (2x)$

29. $y = -x$ **31.** $y = -1$

33. $\frac{2}{25}(x - 2)$

35. $-\dfrac{9 \sin^2 (1/x) \cos (1/x)}{x^2}$

37. $3 \cot^2 \theta \csc^2 \theta$

39. $\pi(b - a) \sin 2\pi\omega$

41. (b) $\begin{cases} \cos x, & 0 < x < \pi \\ -\cos x, & -\pi < x < 0 \end{cases}$ for both

43. $f'\big(g(h(x))\big)g'(h(x))h'(x)$ **45.** 6

▶ Exercise Set 3.6 (Page 179)

1. $\frac{3}{2}x^{-1/4}$

3. $\dfrac{3x^2}{2\sqrt{1 + x^3}}$

5. $\dfrac{\sec^2 \sqrt{x}}{2\sqrt{x}}$

7. $-\dfrac{1}{7}\left(1 - \dfrac{1}{x^2}\right)\left(x + \dfrac{1}{x}\right)^{-8/7}$

9. $\dfrac{1}{2\sqrt{x}} + (3x)^{-2/3} + \dfrac{5}{4}(5x)^{-3/4}$

11. $-\dfrac{x}{y}$

13. $\dfrac{1 - 2xy - 3y^3}{x^2 + 9xy^2}$

15. $-\dfrac{y^2}{x^2}$

17. $-\dfrac{\sqrt{y}}{\sqrt{x}}$

19. $\dfrac{1 - 70x(x^2 + 3y^2)^{34}}{210y(x^2 + 3y^2)^{34}}$

21. $\dfrac{\frac{3}{2}x^2(x^3 + y^2)^{1/2} - y}{x - y(x^3 + y^2)^{1/2}}$

23. $\dfrac{1 - 2xy^2 \cos(x^2y^2)}{2x^2y \cos(x^2y^2)}$

25. $\dfrac{1 - 3y^2 \tan^2(xy^2 + y) \sec^2(xy^2 + y)}{3(2xy + 1) \tan^2(xy^2 + y) \sec^2(xy^2 + y)}$

27. $\dfrac{3y^2 \sin^2(xy^2) \cos(xy^2)}{2\sqrt{1 + \sin^3(xy^2)} - 6xy \sin^2(xy^2) \cos(xy^2)}$

29. $-\dfrac{14}{13}$

31. 2

33. -2

35. 1

37. $\dfrac{6}{5}$

39. $-2x/y^5$

41. $-3/(y - x)^3$

43. $-\dfrac{\sin 2y + y(\sin^2 y + 1)}{(1 + x \sin y)^3}$

45. $\dfrac{2t^3 + 3a^2}{2a^3 - 6at}$

47. $-\dfrac{b^2\lambda}{a^2\omega}$

49. $-\dfrac{2y^3 + 3t^2y}{(6ty^2 + t^3) \cos t}$

51. $y = \dfrac{\sqrt{3}}{3}x, \; y = -\dfrac{\sqrt{3}}{3}x$

53. (a) $(0, 0), \left(-\dfrac{5}{\sqrt{2}}, 0\right), \left(\dfrac{5}{\sqrt{2}}, 0\right)$

   (b) $9x + 13y = 40$

▶ Exercise Set 3.7 (Page 191)

1. continuous on $(1, 2), [2, 3], (2, 3)$; discontinuous on $[1, 3], (1, 3),$ and $[1, 2]$ at $x = 2$

3. continuous on $(1, 3), (1, 2), (2, 3)$; discontinuous on $[1, 3]$ at $x = 1$ and $x = 3$; on $[1, 2]$ at $x = 1$, and on $[2, 3]$ at $x = 3$

5. none

7. none

9. $x = \pm 4$

11. none

13. $x = n\pi, n = 0, \pm 1, \pm 2, \ldots$

15. $x = n\pi, n = 0, \pm 1, \pm 2, \ldots$

17. none

19. $x = 0, 1$

21. (a) 5   (b) 1

33. not differentiable at $x = 1$

35. (a) $x = \frac{2}{3}$ (b) $x = -2, 2$

39. $-1.65, 1.35$

41. (a) 2.25   (b) 2.235

43. (b) f and all its derivatives up to $f^{(n-1)}(x)$ are continuous on (a, b).

▶ Chapter 3 Supplementary Exercises (Page 193)

1. k

3. $-2/\sqrt{9 - 4x}$ for $x < 9/4$

5. 0

7. $y + 1 = 5(x - 3)$

9. (a) 12   (b) -7 (c) 9
   (d) $-9/4$ (e) 5   (f) 21
   (g) -35 (h) -7 (i) -126
   (j) -12 (k) $3/2$ (l) $-3/2$

11. $2(x - 3)^3(x^2 + 6x + 3)/(x^2 + 2x)^2;\; 3, -3 \pm \sqrt{6}$

13. $-3(3x + 1)^2(3x + 2)/x^7;\; -1/3, -2/3$

15. $(7x^2 + 5x + 3)x^{-1/2}(x^2 + x + 1)^{-2/3}/6;$ none

17. $-2\sqrt{2}/x^3 + 2x^{-2}/5$

19. $\sqrt{3}\; (z = \sin^2 2r)$

21. $2(x - 1)/x^3 \; \left(u = \left(1 - \dfrac{1}{x}\right)^2\right)$

23. 0

25. $2 \; (F(x) = 2x)$

27. ± 1

29. ± 2

31. odd integer multiples of $\pi/4$

33. $\Delta x = \pi/4, \Delta y = 1, dy = \pi/2$

35. (a) $-97/48$ (b) $1 - (\pi/90)$

37. $dh = -\pi/30$ ft

39. (a) 2000 gal/min
   (b) 2500 gal/min

41. $\dfrac{dy}{dx} = \dfrac{y^2 - \cos(x + 2y)}{2\cos(x + 2y) - 2xy}$;

   $y = -\tfrac{1}{2}x$

45. a/b

47. does not exist

49. The sine function is continuous.

▶ Exercise Set 4.1 (Page 202)

1. (a) $\dfrac{dA}{dt} = 2x\,\dfrac{dx}{dt}$ (b) 12 ft^2/min

3. (a) $\dfrac{dV}{dt} = \pi\left(r^2\dfrac{dh}{dt} + 2rh\dfrac{dr}{dt}\right)$
   (b) -20π in.^{3}/sec; decreasing

5. (a) $\dfrac{d\theta}{dt} = \dfrac{\cos^2\theta}{x^2}\left(x\dfrac{dy}{dt} - y\dfrac{dx}{dt}\right)$
   (b) $-\dfrac{5}{16}$ radians/sec; decreasing

7. $1/\sqrt{\pi}$ mph   **9.** $4860\,\pi$ cm^3/min

11. $\tfrac{5}{6}$ ft/sec   **13.** $\tfrac{1}{12}$ radian/sec

15. $\dfrac{9}{20\,\pi}$ ft/min   **17.** $125\,\pi$ ft^3/min

19. 250 mph   **21.** $\tfrac{36}{25}\sqrt{69}$ ft/min

23. $8\pi/5$ km/sec   **25.** $600\sqrt{7}$ mph

27. (a) $-60/7$ units/sec   (b) falling

31. $\dfrac{2}{9\pi}$ cm/sec

▶ Exercise Set 4.2 (Page 209)

1. (a) $(\tfrac{5}{2}, +\infty)$ (b) $(-\infty, \tfrac{5}{2})$
   (c) $(-\infty, +\infty)$ (d) none   (e) none

3. (a) $(-\infty, -2), (-2, +\infty)$ (b) none
   (c) $(-2, +\infty)$ (d) $(-\infty, -2)$ (e) -2

5. (a) $(-\infty, -\tfrac{2}{3}), (\tfrac{2}{3}, +\infty)$ (b) $(-\tfrac{2}{3}, \tfrac{2}{3})$
   (c) $(0, +\infty)$ (d) $(-\infty, 0)$ (e) 0

7. (a) $(1, +\infty)$ (b) $(-\infty, 0), (0, 1)$
   (c) $(-\infty, 0), (\tfrac{2}{3}, +\infty)$ (d) $(0, \tfrac{2}{3})$
   (e) $x = 0, \tfrac{2}{3}$

9. (a) $(\pi, 2\pi)$ (b) $(0, \pi)$
   (c) $(\pi/2, 3\pi/2)$ (d) $(0, \pi/2), (3\pi/2, 2\pi)$
   (e) $\pi/2, 3\pi/2$

11. (a) $(-\pi/2, \pi/2)$ (b) none
   (c) $(0, \pi/2)$ (d) $(-\pi/2, 0)$ (e) 0

13. (a) $(-\infty, -2), (-2, +\infty)$ (b) none
   (c) $(-\infty, -2)$ (d) $(-2, +\infty)$
   (e) -2

15. (a) $(-1, 0), (0, +\infty)$ (b) $(-\infty, -1)$
   (c) $(-\infty, 0), (2, +\infty)$ (d) $(0, 2)$ (e) 0, 2

17. (a)   (b)

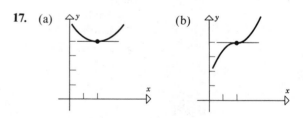

   (c)

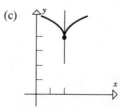

19. (a) 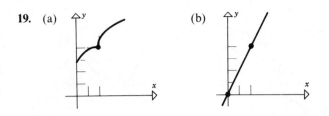   (b)   **21.** none

---

▶ Exercise Set 4.3 (Page 215)

1. $x = \frac{5}{2}$ (stationary)

3. $x = -3, 1$ (stationary)

5. $x = 0, \pm \sqrt{3}$ (stationary)

7. $x = \pm \sqrt{2}$ (stationary)

9. $x = 0$ (not differentiable)

11. $x = n\dfrac{\pi}{3}$, $n = 0, \pm 1, \pm 2, \ldots$ (stationary)

13. $x = n\dfrac{\pi}{4}$, $n = 1, 2, 3, 4, 5, 6, 7$ (stationary)

15. $x = -1$ (stationary); $x = 0$ (not differentiable)

17. relative max of 5 at $x = -2$

19. relative min of 0 at $x = \pi$;
relative max of 1 at $x = \pi/2, 3\pi/2$

21. none

23. relative min of 0 at $x = 1$;
relative max of $\frac{4}{27}$ at $x = \frac{1}{3}$

25. relative min of 0 at $x = 0$;
relative max of 1 at $x = -1, 1$

27. relative min of 0 at $x = 0$

29. relative min of 0 at $x = 0$

31. relative min of 0 at $x = -2, 2$;
relative max of 4 at $x = 0$

33. relative min of 0 at $x = \pm\pi/2, \pm3\pi/2,$
$\pm 5\pi/2, \ldots$;
relative max of 1 at $x = 0, \pm\pi, \pm2\pi, \ldots$

35. relative min of tan (1) at $x = 0$

37. relative min of 0 at $x = \pi/2, \pi, 3\pi/2$;
relative max of 1 at $x = \pi/4, 3\pi/4, 5\pi/4, 7\pi/4$

39. $f(x) = -x^4$ has a relative maximum at $x = 0$,
$f(x) = x^4$ has a relative minimum at $x = 0$,
$f(x) = x^3$ has neither at $x = 0$;
$f'(0) = 0$ for all three functions.

41. (a)    (b)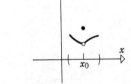

$f(x_0)$ is not an $f(x_0)$ is a relative
extreme value   maximum

(c)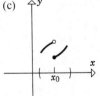

$f(x_0)$ is a relative
minimum

▶ Exercise Set 4.4 (Page 222)

1. 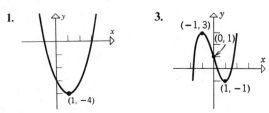   **3.**   **5.** 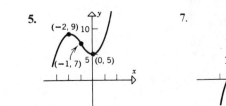   **7.**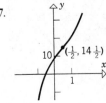

9.

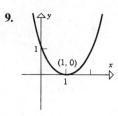

11.

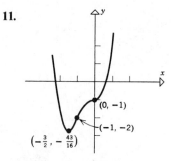

13.

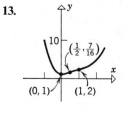

15.

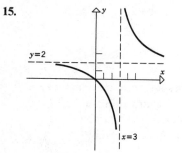

17.

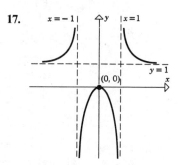

19.

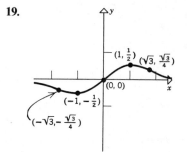

21.

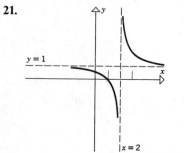

23.

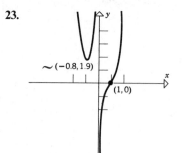

25.

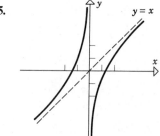

27.

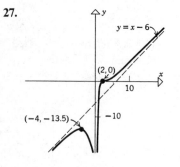

► Exercise Set 4.5 (Page 227)

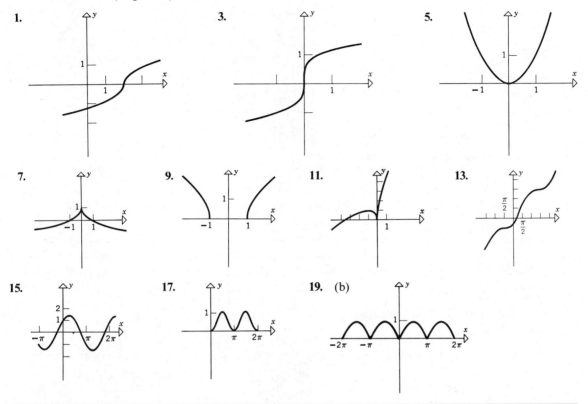

1.   3.   5.

7.   9.   11.   13.

15.   17.   19. (b)

► Exercise Set 4.6 (Page 234)

1. maximum value 1 when $x = 0, 1$;
    minimum value 0 when $x = \frac{1}{2}$

3. maximum value 27 when $x = 4$;
    minimum value -1 when $x = 0$

5. maximum value $3/\sqrt{5}$ when $x = 1$;
    minimum value $-3/\sqrt{5}$ when $x = -1$

7. maximum value 48 when $x = 8$;
    minimum value 0 when $x = 0, 20$

9. maximum value $1 - (\pi/4)$ when $x = -\pi/4$;
    minimum value $(\pi/4) - 1$ when $x = \pi/4$

11. maximum value 2 when $x = 0$;
     minimum value $\sqrt{3}$ when $x = \pi/6$

13. maximum value 17 when $x = -5$;
     minimum value 1 when $x = -3$

15. minimum $-\frac{13}{4}$, no maximum

17. maximum 1, no minimum

19. minimum 0, no maximum

21. no maximum or minimum

23. no maximum or minimum

25. maximum -4, no minimum

27. maximum value $3\sqrt{3}/2$ when $x = (\pi/6) + n\pi$
     $(n = 0, \pm 1, \pm 2, \ldots)$
     minimum value $-3\sqrt{3}/2$ when $x = (5\pi/6) + n\pi$
     $(n = 0, \pm 1, \pm 2, \ldots)$

29. maximum value 2; minimum value $-\frac{1}{4}$

31. (a)   relative min of 0 at $x = a$
     (b)   none

33. (b)   125

35. $a_0 = 9, a_1 = -8, a_2 = 2$

▶ Exercise Set 4.7 (Page 245)

1. $5 + 5$

3. (a) 1 (b) $\frac{1}{2}$

5. 500 ft by 750 ft

7. $10\sqrt{2}$ by $10\sqrt{2}$

9. 5 in. by $\frac{12}{5}$ in.

11. 2 in. square

13. (a) Use all of the wire for the circle.

   (b) $\dfrac{12\pi}{\pi + 4}$ in. for the circle.

15. Each side has length 4.

17. $L/12$ by $L/12$ by $L/12$

19. (a) 7000 (b) yes

21. height $L/\sqrt{3}$, radius $\sqrt{2/3}\, L$

23. height $2\sqrt{(5 - \sqrt{5})/10}\, R$, radius $\sqrt{(5 + \sqrt{5})/10}\, R$

25. $\dfrac{\pi}{3}$

27. $2\pi R^3/(9\sqrt{3})$

29. (a) 24 (b) \$24 (c) \$24.10
    (d) $R'(x) = 10$, $P'(x) = 6 - 0.2x$

33. (a) $\frac{3}{4}$ (b) $\frac{3}{16}$

37. (c) $\frac{1}{4}$ mi downstream from the house

▶ Exercise Set 4.8 (Page 254)

1. (a) 10, 10 (b) no minimum

3. $40\sqrt{2}$ ft (\$1 fencing), $20\sqrt{2}$ ft (\$2 fencing)

5. base: $5\sqrt[3]{4}$ cm square
   height: $10\sqrt[3]{4}$ cm

7. height $=$ radius $= \sqrt[3]{500/\pi}$

9. radius: $\sqrt[6]{\dfrac{450}{\pi^2}}$ cm

   height: $\dfrac{30}{\pi}\sqrt[3]{\dfrac{\pi^2}{450}}$ cm

11. $(-\sqrt{2}, 1)$, $(\sqrt{2}, 1)$

13. (a) no maximum (b) -3

15. closest: $(4, 8)$
    farthest: $(-4, -8)$

17. $\left(-\dfrac{1}{\sqrt{3}}, \dfrac{3}{4}\right)$

19. height: $4R$
    radius: $\sqrt{2}R$

21. $4(1 + 2^{2/3})^{3/2}$ ft

▶ Exercise Set 4.9 (Page 260)

1. 1.4142136

3. 1.8171206

5. -1.6716999

7. 1.2244395

9. 0.5811388

11. -1.4526269

13. 1.8954943

15. 4.4934095

17. (b) 3.1622777

▶ Exercise Set 4.10 (Page 265)

1. 3

3. π

5. 1

7. $\frac{5}{4}$

9. (b) $\tan x$ is not continuous on $[0, \pi]$

▶ Chapter 4 Supplementary Exercises (Page 271)

1. decreasing 39π m^3/sec

3. 60 ft/sec (toward pole)

5. $m = -2 = f(-1)$; $M = 27/256 = f(3/4)$

7. no m; $M = 1/\sqrt{3} = f(\sqrt{3})$

9. $m = -3 = f(3)$; $M = 0 = f(2)$

11. $-2.1149075,\ 0.2541016,\ 1.8608059$

13. $\left(-\frac{1}{\sqrt{3}}, \frac{3}{4}\right)$ $\left(\frac{1}{\sqrt{3}}, \frac{3}{4}\right)$ 15.

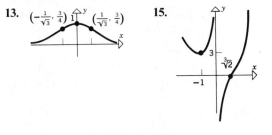

17.   19.

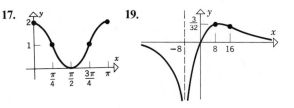

21. at $(1, 2)$, $y = 2x$; at $(2, 3)$, $y = 3$

23. rel. max at $x = \pi/2$ and $3\pi/2$;
    rel. min at $x = 7\pi/6$ and $11\pi/6$

25. rel. min at $x = 9$

27. rel. max at $x = 2\pi/3$ and $4\pi/3$;
    rel. min at $x = \pi$

29. $2\sqrt{8}$ by $3\sqrt{2}$

31. $r = 2P/(3\pi + 8)$ ft.   33. $r/h = \pi/8$

35. satisfied; $c = 0$ 37. satisfied; $c = \sqrt{\pi/2}$

39. satisfied; $c = 1$ 41. satisfied; $c = \frac{1}{2},\ \sqrt{2}$

▶ Exercise Set 5.2 (Page 283)

1. $\frac{1}{9}x^9 + C$ 3. $\frac{7}{12}x^{12/7} + C$

5. $8\sqrt{t} + C$ 7. $\frac{2}{9}x^{9/2} + C$

9. $-\frac{1}{2}x^{-2} + \frac{2}{3}x^{3/2} - \frac{12}{5}x^{5/4} + \frac{1}{3}x^3 + C$

11. $28y^{1/4} - \frac{3}{4}y^{4/3} + \frac{8}{3}y^{3/2} + C$

13. $\frac{1}{2}x^2 + \frac{1}{5}x^5 + C$

15. $3x^{4/3} - \frac{12}{7}x^{7/3} + \frac{3}{10}x^{10/3} + C$

17. $-4\cos x + 2\sin x + C$

19. $\tan x + \sec x + C$ 21. $\sec x + x + C$

23. $\sec x + C$ 25. $\theta - \cos \theta + C$

27. $\frac{1}{2}x^2 - \frac{2}{x} + \frac{1}{3x^3} + C$ 29. $F(x) = \frac{3}{4}x^{4/3} + \frac{5}{4}$

31. $f(x) = \frac{4}{15}x^{5/2} + C_1 x + C_2$

▶ Exercise Set 5.3 (Page 290)

1. (a) $\frac{1}{24}(x^2 + 1)^{24} + C$ (b) $-\frac{1}{4}\cos^4 x + C$
   (c) $-2\cos\sqrt{x} + C$
   (d) $\frac{3}{4}\sqrt{4x^2 + 5} + C$

3. (a) $-\frac{1}{2}\cot^2 x + C$ (b) $\frac{1}{10}(1 + \sin t)^{10} + C$
   (c) $\frac{2}{7}(1 + x)^{7/2} - \frac{4}{5}(1 + x)^{5/2} +$
   $\qquad \frac{2}{3}(1 + x)^{3/2} + C$
   (d) $-\cot(\sin x) + C$

5. $x + \sin x + C$ 7. $\frac{1}{18}(x - 2)^{18} + C$

9. $\frac{5}{\pi}\sin(\pi x - 3) + C$ 11. $\frac{2}{3}\sqrt{3s + 1} + C$

13. $\frac{n}{b(n + 1)}(a + bx)^{(n+1)/n} + C$

15. $\frac{1}{7}\sin^7 x + C$ 17. $\frac{2}{3}\sqrt{x^3 + 1} + C$

19. $\frac{2}{3}(\tan x)^{3/2} + C$ 21. $\frac{1}{3}\tan^3 x + C$

23. $-\cos(\sin \theta) + C$

25. $\frac{1}{b(n + 1)}\sin^{n+1}(a + bx) + C$

27. $\frac{3}{10}(x^2 + 5)^{5/3} + C$

29. $\frac{2}{3}(y + 1)^{3/2} - 2(y + 1)^{1/2} + C$

31. $-\cos x + \frac{1}{3}\cos^3 x + C$

33. $\tan x - x + C$

36. (a) $\frac{1}{2}\sin^2 x + C$; $-\frac{1}{2}\cos^2 x + C$

(b) The two answers differ by a constant. (Subtract the answers and use a trigonometric identity.)

▶ Exercise Set 5.4 (Page 298)

1.

t	s	v	$|v|$	a	direction; motion
1	-5	-9	9	-6	left; speeding up
2	-16	-12	12	0	left; neither
3	-27	-9	9	6	left; slowing down
4	-32	0	0	12	stopped
5	-25	15	15	18	right; speeding up

3. $s(t) = \frac{1}{4}t^4 - \frac{2}{3}t^3 + t + 1$

5. $s(t) = 2t^2 + t$

7. $s(t) = -\cos 2t - t - 2$

9. (a) 12   (b) $t = 2.2$, $s = -24.2$

11.

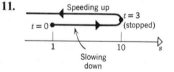

13.

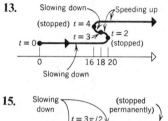

15.

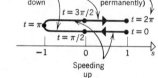

17. (a) $v(3) = 16$ ft/sec, $v(5) = -48$ ft/sec

(b) 196 ft   (c) 112 ft/sec

19. (a) 1 sec   (b) $\frac{1}{2}$ sec

21. (a) $\frac{5}{8}(1 + \sqrt{33})$ sec   (b) $20\sqrt{33}$ ft/sec

23. (a) 5 sec   (b) 272.5 m   (c) 10 sec

(d) -49 m/sec   (e) 12.46 sec   (f) 73.1 m/sec

25. $80\sqrt{10}$ ft/sec

29. (a) negative   (b) negative

(c) speeding up   (d) slowing down

▶ Exercise Set 5.5 (Page 307)

1. (a) 36   (b) 55   (c) 40   (d) 6

3. $\displaystyle\sum_{k=1}^{10} k$

5. $\displaystyle\sum_{k=1}^{49} k(k + 1)$

7. $\displaystyle\sum_{k=1}^{10} 2k$

9. $\displaystyle\sum_{k=1}^{6} (-1)^{k+1}(2k - 1)$

11. $\displaystyle\sum_{k=1}^{5} (-1)^k \frac{1}{k}$

13. $\displaystyle\sum_{k=1}^{4} \sin \frac{(2k - 1)\pi}{8}$

15. $\displaystyle\sum_{k=1}^{5} \frac{k}{k + 1}$

17. (a) $\displaystyle\sum_{k=1}^{5} (-1)^{k+1}a_k$ (b) $\displaystyle\sum_{k=0}^{5} (-1)^{k+1}b_k$

(c) $\displaystyle\sum_{k=0}^{n} a_k x^k$ (d) $\displaystyle\sum_{k=0}^{5} a^{5-k}b^k$

19. 5047

21. 2870

23. 1728

25. 214,365

27. $3^{17} - 3^4$ **29.** $-\frac{399}{400}$ **37.** $\displaystyle\sum_{k=1}^{18} k \sin \frac{\pi}{k}$ **39.** both are valid

31. $a_n - a_0$

33. (a) n^2 (b) -3 (c) $\frac{1}{2}n(n+1)x$ **41.** (a) $\frac{3}{2}(3^{20} - 1)$ (b) $2^{31} - 2^5$ (c) $-\frac{2}{3}\left(1 + \dfrac{1}{2^{101}}\right)$
(d) $(n - m + 1)c$

43. 110

35. (a) $\displaystyle\sum_{k=0}^{14} (k+4)(k+1)$ (b) $\displaystyle\sum_{k=5}^{19} (k-1)(k-4)$

▶ Exercise Set 5.6 (Page 316)

1. (a) 46   (b) 58   **9.** 320   **11.** $\frac{15}{4}$

3. (a) $\dfrac{\sqrt{2}\pi}{4} \approx 1.111$ (b) $\dfrac{(\sqrt{2}+2)\pi}{4} \approx 2.682$ **13.** $\frac{1}{3}$ **15.** (b) $\frac{1}{4}(b^4 - a^4)$

17. $\frac{1}{2}(b^2 - a^2)$

5. $\frac{15}{4}$ **7.** $\frac{1}{3}$

▶ Exercise Set 5.7 (Page 325)

1. (a) $\frac{71}{6}$ (b) 2   **11.** $\displaystyle\lim_{\max \Delta x_k \to 0} \sum_{k=1}^{n} \frac{x_k^*}{x_k^* + 1} \Delta x_k$; $a = 0$, $b = 1$

3. (a) $-\frac{117}{16}$ (b) 3

7. $\displaystyle\int_{-3}^{3} 4x(1 - 3x)\, dx$ **13.** (a) -1 (b) $\frac{5}{2}$ (c) $\frac{21}{2}$ (d) $\frac{1}{2}k^2 - k$

15. $\pi/2$

9. $\displaystyle\lim_{\max \Delta x_k \to 0} \sum_{k=1}^{n} 2x_k^* \Delta x_k$; $a = 1$, $b = 2$

▶ Exercise Set 5.8 (Page 333)

1. $\frac{65}{4}$ **3.** $\frac{81}{10}$ **25.** 0   **27.** $\frac{2}{3}(\sqrt{10} - 2\sqrt{2})$

5. $\frac{22}{3}$ **7.** $\frac{2}{3}$ **29.** $-\frac{5}{3}a^{3/2}$ **31.** 12

9. $-\frac{1}{3}$ **11.** $\frac{52}{3}$ **33.** $\frac{3}{4}(2 - \sqrt{2})$ **35.** $\frac{203}{2}$

13. $\frac{844}{5}$ **15.** $-\frac{55}{3}$ **37.** $\frac{2}{3}$, 3   **39.** 0, 2

17. 0   **19.** 0   **41.** $\frac{9}{4}$, $\frac{11}{4}$ **43.** $-\frac{10}{3}$, $\frac{17}{3}$

21. $\dfrac{\pi^2}{9} + 2\sqrt{3}$ **23.** $\frac{2}{3}$ **45.** $\frac{204}{25}$, $\frac{204}{25}$ **49.** b, c

▶ Exercise Set 5.9 (Page 343)

1. (a) $\displaystyle\int_1^3 u^7\, du$ (b) $-\dfrac{1}{2}\displaystyle\int_7^4 u^{1/2}\, du$ **23.** -4

(c) $\dfrac{1}{\pi}\displaystyle\int_{-\pi}^{\pi} \sin u\, du$ (d) $\displaystyle\int_0^1 u^2\, du$ **25.** (a) $\frac{4}{3}$ (b) $\dfrac{2}{\sqrt{3}}$ (c)

(e) $\dfrac{1}{2}\displaystyle\int_3^4 (u-3)u^{1/2}\, du$

(f) $\displaystyle\int_2^0 (u+5)u^{20}\, du$

3. $2-\dfrac{\sqrt{2}}{2}$ **5.** $\frac{929}{12}$

27. $f_{\text{ave}}=2;\ x^*=4$

7. $\frac{121}{5}$ **9.** 10

29. $f_{\text{ave}}=\frac{1}{2}\alpha(x_0+x_1)+\beta;\ x^*=\frac{1}{2}(x_0+x_1)$ if $\alpha\ne 0$;

11. $\frac{1192}{15}$ **13.** $8-4\sqrt{2}$

x^* is any point in $[x_0, x_1]$ if $\alpha=0$.

15. $-\frac{1}{48}$ **17.** $\frac{106}{405}$

33. 1404π lb   **35.** (b) no

19. 0 **21.** $2(\sqrt{7}-\sqrt{3})$

▶ Exercise Set 5.10 (Page 348)

1. x^3+1 **3.** $\sin\sqrt{x}$ **13.** (a) $\dfrac{3}{x}$ (b) $\dfrac{\cos x}{1+\sin^2 x}$

5. $|x|$ **7.** $\displaystyle\int_2^x \dfrac{1}{t-1}\, dt$ **15.** (a) $3x^2\sin^2(x^3)-2x\sin^2(x^2)$ (b) $\dfrac{2}{1-x^2}$

9. $\displaystyle\int_0^x \dfrac{1}{t-1}\, dt$ **11.** (a) $(0,+\infty)$ (b) $x=1$

▶ Chapter 5 Supplementary Exercises (Page 349)

1. $-\frac{1}{2}x^{-2}+2\sqrt{x}+5\cos x + C$ **17.** $\displaystyle\int_{\pi/2}^{\pi} \dfrac{4}{\pi}\cos u\, du = -\dfrac{4}{\pi}$

3. $\frac{2}{9}(\sqrt{x}+2)^9 + C$

5. $-\frac{1}{2}\cos\sqrt{2x^2-5} + C$ **19.** $17/2$

7. $2x^{3/2}+\frac{6}{17}x^{17/6}+C$ **21.** 128 ft/sec; 256 ft

9. $\frac{1}{5}\tan(\sin 5t)+C$ **23.** (a) 20   (b) 8   (c) $4n$

(d) 8/15   (e) 61/24   (f) 9

11. (a) $\frac{1}{6}y^6+y^4+2y^2+C$ (g) $1+\sqrt{2}$ (h) $3(\sqrt{2}+1)/4$

(b) $\frac{1}{6}(y^2+2)^3+C$

25. (a) $\displaystyle\sum_{k=1}^{9}(-1)^{k-1}\left(\dfrac{k}{k+1}\right)^2 = \sum_{k=2}^{10}(-1)^k\left(\dfrac{k-1}{k}\right)^2$

13. $\displaystyle\int_0^1 u^4\, du = 1/5$

(b) $\displaystyle\sum_{k=1}^{11}\dfrac{(-\pi)^{k+1}}{k} = \sum_{k=2}^{12}(-1)^k\dfrac{\pi^k}{k-1}$

15. $\displaystyle\int_1^4 \dfrac{u-1}{\sqrt{u}}\, du = \dfrac{8}{3}$

27. (a) $64\left[1 - \dfrac{(n+1)(2n+1)}{6n^2}\right]$

  (b) $64\left[1 - \dfrac{(n-1)(2n-1)}{6n^2}\right]$

  (c) $128/3$

29. (a) 12 (b) 12 (c) 12

31. (a) -4 (b) $9\pi/4$
  (c) -2 (d) 0

  (e) 0 (f) 30
  (g) 2

33. (a) 10 (b) 1
  (c) -4 (let $u = -x$)
  (d) 4

35. $f_{\text{ave}} = 7;\ x^* = -\sqrt{7/3}$

37. $f_{\text{ave}} = 13/4;\ x^* = -5/4$

▶ Exercise Set 6.1 (Page 358)

1. $\frac{32}{3}$

3. $\frac{9}{4}$

5. $\frac{49}{192}$

7. $\frac{1}{2}$

9. $\frac{32}{3}$

11. $\pi - 2$

13. $\frac{9}{2}$

15. $\frac{355}{6}$

17. 24

19. $\frac{1}{2}$

21. $4\sqrt{2}$

23. $y = \dfrac{9}{\sqrt[3]{4}}$

▶ Exercise Set 6.2 (Page 365)

1. $\dfrac{32\pi}{5}$

3. $\dfrac{373\pi}{14}$

5. $\dfrac{1296\pi}{5}$

7. $\dfrac{2048\pi}{15}$

9. $\dfrac{\pi}{2}$

11. $\dfrac{\pi}{6}$

13. $\dfrac{3\pi}{5}$

15. 8π

17. 2π

19. $\dfrac{28\pi}{3}$

21. $\dfrac{58\pi}{5}$

23. $\dfrac{72\pi}{5}$

25. $\dfrac{648\pi}{5}$

27. $\dfrac{\pi}{2}$

29. $\frac{4}{3}\pi ab^2$

31. π

33. $\frac{1}{3}\pi r^2 h$

35. $40{,}000\pi$ ft^3

37. $36\sqrt{3}$

39. $\frac{1}{4}(\pi + 2)$

41. $\frac{2}{3}r^3 \tan\theta$

43. $V = 3\pi h^2$ if $0 \le h < 2$, $V = \dfrac{\pi}{3}(12h^2 - h^3 - 4)$
   if $2 \le h \le 4$

▶ Exercise Set 6.3 (Page 372)

1. $2\pi/5$

3. $3\pi\sqrt[3]{4}\,(1 + 3\sqrt[3]{3})$

5. $20\pi/3$

7. $\pi/2$

9. $\pi/5$

11. (a) $7\pi/30$

13. $9\pi/14$

15. $\frac{1}{3}\pi r^2 h$

17. $\dfrac{4\pi}{3}[r^3 - (r^2 - a^2)^{3/2}]$

▶ Exercise Set 6.4 (Page 377)

1. $\sqrt{5}$

3. $\frac{1}{243}(85\sqrt{85} - 8)$

5. $\frac{1}{27}(80\sqrt{10} - 13\sqrt{13})$ **7.** $\frac{17}{6}$

9. (a)

  (b) dy/dx does not exist at $x = 0$
  (c) $\frac{1}{27}(13\sqrt{13} + 80\sqrt{10} - 16)$

▶ Exercise Set 6.5 (Page 382)

1. $35\sqrt{2}\,\pi$ **3.** 8π

5. $\dfrac{16\pi}{9}$ **7.** $40\pi\sqrt{82}$

9. 24π **11.** $\dfrac{16{,}911\pi}{1024}$

▶ Exercise Set 6.6 (Page 387)

1. (a) 210 in.-lb   (b) $\frac{5}{6}$ in.-lb

3. $\frac{9}{5}$ ft-ton

5. 20 lb/ft

7. $900\pi\rho$ ft-lb

9. (a) 926,640 ft-lb   (b) 0.468 horsepower

11. 75,000 ft-lb

13. (a) 96×10^9 (b) 4,800,000 mile-lb

▶ Exercise Set 6.7 (Page 392)

1. (a) 2808 lb   (b) 3600 lb

3. 998.4 lb   **5.** 6988.8 lb

7. 5200 lb

9. $\dfrac{\sqrt{2}}{2}\rho a^3$ **11.** $14{,}976\sqrt{17}$ lb

13. (b) $80\rho_0$ lb/min

▶ Chapter 6 Supplementary Exercises (Page 393)

1. (a) $\displaystyle\int_0^2 (x + 2 - x^2)\,dx$

   (b) $\displaystyle\int_0^2 \sqrt{y}\,dy + \int_2^4 [\sqrt{y} - (y-2)]\,dy$

3. (a) $\displaystyle\int_0^9 2\sqrt{x}\,dx$ (b) $\displaystyle\int_{-3}^3 (9 - y^2)\,dy$

5. (a) $\displaystyle\int_0^2 2\pi x(x + 2 - x^2)\,dx$

   (b) $\displaystyle\int_0^2 \pi y\,dy + \int_2^4 \pi[y - (y-2)^2]\,dy$

7. (a) $\displaystyle\int_0^4 2\pi x[\tfrac{1}{2}x - (2 - \sqrt{4-x})]\,dx$

   (b) $\displaystyle\int_0^2 \pi[(4y - y^2)^2 - 4y^2]\,dy$

9. (a) $\displaystyle\int_0^9 4\pi x^{3/2}\,dx$

   (b) $\displaystyle\int_{-3}^3 \pi(81 - y^4)\,dy$

11. (a) 16/3   (b) 8π

13. 11/4   **15.** $\pi a^2 b/24$

17. (a) $256\pi/15$ (b) $40\pi/3$

19. 61/27   **21.** 779/240

23. $\frac{1}{27}\pi(145^{3/2} - 10^{3/2})$ sq. units

25. $1017\pi/5$ sq. units   **27.** $28\pi\sqrt{3}/5$ sq. units

29. 3 in.-lb   **31.** 10,600 ft-lb

33. 6656π ft-lb   **35.** $28\rho/3$ lb

▶ Exercise Set 7.2 (Page 401)

1. (a) 4   (b) -5

   (c) 1   (d) $\frac{1}{2}$

   (e) -3 (f) 4

   (g) 3   (h) $\frac{1}{2}$

3. 0.01   **5.** e^2

7. 4   **9.** 10^5

11. $\sqrt{\frac{3}{2}}$ **13.** $\dfrac{1}{x}$

15. $\dfrac{2\ln x}{x}$ **17.** $\dfrac{\sec^2 x}{\tan x}$

19. $\dfrac{1 - x^2}{x(1 + x^2)}$

21. $\dfrac{3x^2 - 14x}{x^3 - 7x^2 - 3}$

23. $\dfrac{1}{2x\sqrt{\ln x}}$

25. $-\dfrac{5\cos(5/\ln x)}{x(\ln x)^2}$

27. $4x\ln(x^2 + 1) + 2x[\ln(x^2 + 1)]^2$

29. $\dfrac{x(1 + 2\ln x)}{(1 + \ln x)^2}$

31. $-\dfrac{y}{x(y + 1)}$

33. $\frac{1}{2}\ln|x| + C$

35. $\frac{1}{3}\ln|x^3 - 4| + C$

37. $\ln|\tan x| + C$

39. $-\frac{1}{3}\ln(1 + \cos 3\theta) + C$

41. $\frac{1}{2}x^2 - \frac{1}{2}\ln(x^2 + 1) + C$

43. $\frac{1}{4}(\ln y)^4 + C$

49. (b) $10^{8.2}I_0$ (c) $10{,}000$

▶ Exercise Set 7.3 (Page 407)

1. (a) $r + s$ (b) $s - r$ (c) $-r$
 (d) $2s$ (e) $\frac{1}{5}s$ (f) $-2(r + s)$

5. 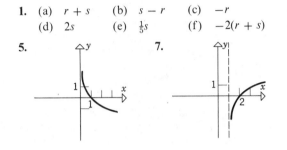 **7.**

9. $\dfrac{1}{2x} + \dfrac{1}{3(x + 3)} + \dfrac{3}{5(3x - 2)}$

11. $\frac{3}{2}(\ln x)^2 + C$

13. $-\frac{1}{2}(\ln x)^2 + C$

15. $x\sqrt[3]{1 + x^2}\left[\dfrac{1}{x} + \dfrac{2x}{3(1 + x^2)}\right]$

17. $\dfrac{(x^2 - 8)^{1/3}\sqrt{x^3 + 1}}{x^6 - 7x + 5}\left[\dfrac{2x}{3(x^2 - 8)} + \dfrac{3x^2}{2(x^3 + 1)}\right.$

$\left. - \dfrac{6x^5 - 7}{x^6 - 7x + 5}\right]$

▶ Exercise Set 7.4 (Page 415)

1. (a) $\dfrac{1}{x}$, $x > 0$

 (b) x^2, $x \neq 0$

 (c) $-x^2$, $-\infty < x < +\infty$

 (d) $-x$, $-\infty < x < +\infty$

 (e) x^3, $x > 0$

 (f) $x + \ln x$, $x > 0$

 (g) $x - \sqrt[3]{x}$, $-\infty < x < +\infty$

 (h) e^x/x, $x > 0$

3. (a) $\sqrt{e}$ (b) $\dfrac{\ln 2}{4\pi}$ (c) 0 (d) $\ln 2$

5. (a) 3.3 (b) 15 (c) 30

7. $-10xe^{-5x^2}$

9. $x^2 e^x(x + 3)$

11. $\dfrac{4}{(e^x + e^{-x})^2}$

13. $(x\sec^2 x + \tan x)e^{x\tan x}$

15. $(1 - 3e^{3x})e^{(x - e^{3x})}$

17. $\dfrac{x - 1}{e^x - x}$

19. $e^{ax}(a\cos bx - b\sin bx)$

21. $3x^2$

23. $-3^{-x}\ln 3$

25. $\pi^{x\tan x}(\ln \pi)(x\sec^2 x + \tan x)$

27. (a) Not of the form a^x, where a is a constant.
 (b) $x^x(1 + \ln x)$

29. $(x^3 - 2x)^{\ln x}\left[\dfrac{3x^2 - 2}{x^3 - 2x}\ln x + \dfrac{1}{x}\ln(x^3 - 2x)\right]$

31. $(\ln x)^{\tan x}\left[\dfrac{\tan x}{x\ln x} + (\sec^2 x)\ln(\ln x)\right]$

33. $x^{(e^x)}\left[\dfrac{e^x}{x} + e^x\ln x\right]$

37. (a) $k^n e^{kx}$ (b) $(-1)^n k^n e^{-kx}$

39. $-\dfrac{1}{\sqrt{2\pi}\sigma^3}(x - \mu)\exp\left[-\dfrac{1}{2}\left(\dfrac{x - \mu}{\sigma}\right)^2\right]$

41. $-\frac{1}{5}e^{-5x} + C$

43. $e^{\sin x} + C$

45. $-\frac{1}{6}e^{-2x^3} + C$

47. $\ln(1 + e^x) + C$

49. $\frac{1}{3}(1 + e^{2t})^{3/2} + C$

51. $\exp(\sin x) + C$

53. $\tan(2 - e^{-x}) + C$

55. $\dfrac{\pi^{\sin x}}{\ln \pi} + C$

57. $\frac{1}{2}x^2\ln 3 - 4\pi e^2\sin x + C$

59. C

61. $2e^{\sqrt{y}} + C$

63. -36

65. $3 + e - e^2$

67. $\ln\left(\frac{21}{13}\right)$

71. $\dfrac{\ln 3}{\ln (2/3)}$

73. exe^{-1}

75. $\dfrac{-qk_0}{2T^2} \exp\left[-(q/2)\left(\dfrac{T - T_0}{T_0 T}\right)\right]$

77. $e^x - 3\ln(e^x + 3) + C$

▶ Exercise Set 7.5 (Page 420)

1. (a) $+\infty$ (b) 0

3. (a) $+\infty$ (b) $+\infty$

5. (a) 1 (b) 1

7.

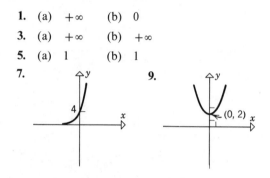

9.

11.

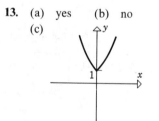

13. (a) yes   (b) no

(c)

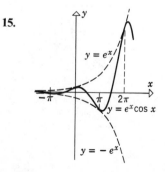

15.

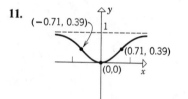

17. -1

19. 0

21. 1

23. 1

25. 1

27. (a) $\displaystyle\lim_{x\to+\infty} xe^x = +\infty, \; \lim_{x\to-\infty} xe^x = 0$

(b)

29. (a) $\displaystyle\lim_{x\to+\infty} \frac{x^2}{e^{2x}} = 0, \; \lim_{x\to-\infty} \frac{x^2}{e^{2x}} = +\infty$

(b)

31. (a)   (b)   (c)

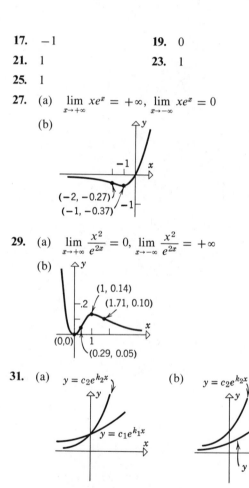

33. (a) 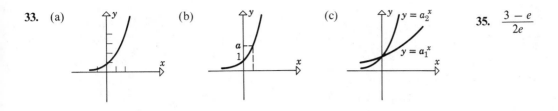 (b) (c) $y = a_2^x$ $y = a_1^x$

35. $\dfrac{3 - e}{2e}$

▶ Exercise Set 7.6 (Page 426)

1.

	$\sinh x_0$	$\cosh x_0$	$\tanh x_0$	$\coth x_0$	$\operatorname{sech} x_0$	$\operatorname{csch} x_0$
(a)	-2	$\sqrt{5}$	$-2/\sqrt{5}$	$-\sqrt{5}/2$	$1/\sqrt{5}$	$-1/2$
(b)	$-3/4$	$5/4$	$-3/5$	$-5/3$	$4/5$	$-4/3$
(c)	$-4/3$	$5/3$	$-4/5$	$-5/4$	$3/5$	$-3/4$
(d)	$1/\sqrt{3}$	$2/\sqrt{3}$	$1/2$	2	$\sqrt{3}/2$	$\sqrt{3}$
(e)	$8/15$	$17/15$	$8/17$	$17/8$	$15/17$	$15/8$
(f)	-1	$\sqrt{2}$	$-1/\sqrt{2}$	$-\sqrt{2}$	$1/\sqrt{2}$	-1

17. $4\cosh(4x - 8)$

19. $-\dfrac{\operatorname{csch}^2(\ln x)}{x}$

21. $\dfrac{\operatorname{csch}\left(\frac{1}{x}\right)\coth\left(\frac{1}{x}\right)}{x^2}$

23. $\frac{1}{7}\sinh^7 x + C$

25. $\frac{2}{3}(\tanh x)^{3/2} + C$

27. $\ln\cosh x + C$

29. $-\frac{1}{3}\operatorname{sech}^3 x + C$

37. $\frac{3}{4}$

▶ Exercise Set 7.7 (Page 440)

1. $y = Cx$

3. $y = Ce^{-\sqrt{1+x^2}} - 1$

5. $y = \ln(\sec x + C)$

7. $y = e^{-2x} + Ce^{-3x}$

9. $y = e^{-x}\sin(e^x) + Ce^{-x}$

11. $y = -\frac{2}{7}x^4 + Cx^{-3}$

13. $y = -1 + 4e^{x^2/2}$

15. $y = 2 - e^{-t}$

17. $y = \sqrt[3]{3t - 3\ln t + 24}$

19. $y = x + \dfrac{1}{x}$

21. $y^2 = -x^2 + 6x - 5$

23. (a) $y = 200 - 175e^{-t/25}$
  (b) 136 lb

25. 25 lb

27. (a) $y = 10e^{-0.005t}$
  (b) 7 mg

29. 196 days

31. 6.8 years

33. (a) 14,400
  (b) 38 years

▶ Chapter 7 Supplementary Exercises (Page 442)

1. (a) $-(2r + s)$ (b) $2s - \frac{3}{2}r$
  (c) $\frac{1}{4}(3r - s)$

3. (a) $\ln 3/(2\ln 5 + \ln 3)$
  (b) $\frac{1}{2}(\ln 5 - \ln 3)$

5. (a) $\sqrt{34}/5$ (b) $-3/\sqrt{34}$
  (c) $-6\sqrt{34}/25$

7. $-1/(2\sqrt{e^x})$

9. $(\ln x - 1)/(\ln x)^2$

11. 0

13. $\ln 10 - \cot x$

15. $x^4 e^{\tan x}\left(\sec^2 x + \dfrac{4}{x}\right)$

17. $(x^2 + a^2)^{-1/2}$

19. $6x\exp(3x^2)$

21. $1/(4x\sqrt{\frac{1}{2}\ln x})$

23. $x^{\pi-1}\pi^x(\pi + x\ln\pi)$

25. $5 \cosh\left(\tanh\left(5x\right)\right) \operatorname{sech}^2\left(5x\right)$

27. $3e^{3x} + 4e^{2x} + e^x$

31. (a) $dy = -e^{-x}\, dx$ (b) $dy = dx/(1+x)$
(c) $dy = 2x\,(\ln 2)\, 2^{x^2}\, dx$

33. (a) $\dfrac{1}{x\sqrt{4+x}}$ (b) $5e^{5x}\sqrt{5x + e^{5x}}$

35. $Y = kt + b,\ b = \ln C$ **37.** $\ln\left(1 + e^x\right) + C$

39. $\dfrac{x^{e+1}}{e+1} + C$ **41.** $4\ln|x| + 3/x + C$

43. $\frac{1}{2}\ln|2\sec x - 1| + C$ **45.** $\frac{1}{2}e^{\sin 2x} + C$

47. $\frac{1}{2}\tanh^2 x + C_1 = -\frac{1}{2}\operatorname{sech}^2 x + C_2$

49. $\ln 2$ **51.** $1/\ln 2$

55. inflection points at $0, 3 - \sqrt{3}$, and $3 + \sqrt{3}$; relative extremum at 3

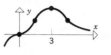

59. $A = \frac{1}{2}(1 - e^{-2b}) = \frac{1}{4}$ when $b = \ln\sqrt{2}$; $A \to \frac{1}{2}$ as $b \to +\infty$

61. $\frac{1}{2}\pi(2 + \sinh 2)$

63. (a) about 352 million
(b) the year 2058

▶ Exercise Set 8.1 (Page 452)

1. (a) yes (b) no (c) yes (d) no

3. yes **5.** no

7. no **9.** yes

11. yes **13.** yes

15. $x^{1/5}$ **17.** $\frac{1}{7}(x + 6)$

19. $\sqrt[3]{\dfrac{x+5}{3}}$ **21.** $\ln\left(x^3 - 1\right)$

23. $\dfrac{1}{\ln x}$ **25.** $\frac{1}{3}e^{1-x}$

27. $\dfrac{1}{15y^2 + 1}$ **29.** $\dfrac{1}{3y^2 e^{1+y^3}}$

31. $\dfrac{1}{10y^4 + 3y^2}$ **33.** (b) $\ln\left(x + \sqrt{x^2 + 1}\right)$

35. (b)

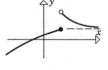

(c) no; $f(g(x)) = x$ for $x > 1$ but the domain of g is $x > 0$

37. $\ln\left(x^2 - 1\right)$ for $x > 1$

39. (b) symmetric about the line $y = x$

41. (b) $1 - \dfrac{\sqrt{3}}{3}$ **43.** (b) $\dfrac{1}{\sqrt[3]{2}}$

45.

▶ Exercise Set 8.2 (Page 460)

1. (a) $-\dfrac{\pi}{2}$ (b) π (c) $-\dfrac{\pi}{4}$
(d) $\dfrac{\pi}{4}$ (e) 0 (f) $\dfrac{\pi}{2}$

3. $\dfrac{1}{2}, -\sqrt{3}, -\dfrac{1}{\sqrt{3}}, 2, -\dfrac{2}{\sqrt{3}}$ **5.** $\frac{4}{5}, \frac{3}{5}, \frac{3}{4}, \frac{5}{3}, \frac{5}{4}$

7. (a) $\dfrac{\pi}{7}$ (b) 0 (c) $\dfrac{2\pi}{7}$

9. (a) $0 \le x \le \pi$ (b) $-1 \le x \le 1$
(c) $-\dfrac{\pi}{2} < x < \dfrac{\pi}{2}$ (d) $-\infty < x < +\infty$
(e) $0 < x \le \dfrac{\pi}{2}, \quad -\pi < x \le -\dfrac{\pi}{2}$ (f) $|x| \ge 1$

11. $\frac{24}{25}$

13. $\frac{\pi}{2}$

15. $-4\sqrt{5}$

19. (a) $\dfrac{1}{\sqrt{1+x^2}}$ (b) $\dfrac{1}{x}$ (c) $\dfrac{\sqrt{x^2-1}}{x}$

      (d) $\sqrt{x^2-1}$

21. (a)   (b)

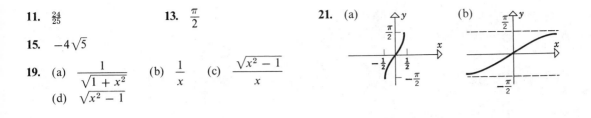

► Exercise Set 8.3 (Page 465)

1. (a) $\dfrac{1}{\sqrt{9-x^2}}$ (b) $-\dfrac{2}{\sqrt{1-(2x+1)^2}}$

3. (a) $\dfrac{7}{x\sqrt{x^{14}-1}}$ (b) $-\dfrac{1}{\sqrt{e^{2x}-1}}$

5. (a) $-\dfrac{1}{|x|\sqrt{x^2-1}}$ (b) $\begin{cases} 1, & \sin x > 0 \\ -1, & \sin x < 0 \end{cases}$

7. (a) $\dfrac{e^x}{x\sqrt{x^2-1}} + e^x \sec^{-1} x$

      (b) $\dfrac{3x^2(\sin^{-1}x)^2}{\sqrt{1-x^2}} + 2x(\sin^{-1}x)^3$

9. (a) $-\dfrac{1}{x^2+1}$

      (b) $10(1 + x\csc^{-1}x)^9\left(-\dfrac{1}{\sqrt{x^2-1}} + \csc^{-1}x\right)$

11. $\dfrac{y\sqrt{1-(x-y)^2} + \sqrt{1-x^2y^2}}{\sqrt{1-x^2y^2} - x\sqrt{1-(x-y)^2}}$

13. $\dfrac{\pi}{2}$ **15.** $-\dfrac{\pi}{12}$

17. $\frac{1}{4}\tan^{-1}4x + C$ **19.** $\tan^{-1}(e^x) + C$

21. $\sin^{-1}(\tan x) + C$ **23.** $\sin^{-1}(\ln x) + C$

25. (a) $\sin^{-1}\left(\dfrac{x}{3}\right) + C$

      (b) $\dfrac{1}{\sqrt{5}}\tan^{-1}\left(\dfrac{x}{\sqrt{5}}\right) + C$

      (c) $\dfrac{1}{\sqrt{\pi}}\sec^{-1}\left(\dfrac{x}{\sqrt{\pi}}\right) + C$

27. $\dfrac{\pi^2}{4}$ **29.** $1 + 2\sqrt{2}$

31. $\dfrac{52\pi}{3}$ mi/min   **33.** $2\sqrt{6}$ ft

► Exercise Set 8.4 (Page 470)

5. (a) $\dfrac{1}{\sqrt{9+x^2}}$ **7.** (a) $-\dfrac{7}{x\sqrt{1-x^{14}}}$

      (b) $\dfrac{2}{\sqrt{(2x+1)^2-1}}$ (b) $-\dfrac{1}{\sqrt{1+e^{2x}}}$

9. (a) $-\dfrac{1}{|x|\sqrt{x^2+1}}$

      (b) $\begin{cases} 1, & x > 0 \\ -1, & x < 0 \end{cases}$

11. (a) $-\dfrac{e^x}{x\sqrt{1-x^2}} + e^x \operatorname{sech}^{-1} x$

      (b) $\dfrac{3x^2(\sinh^{-1}x)^2}{\sqrt{1+x^2}} + 2x(\sinh^{-1}x)^3$

13. (a) $-\dfrac{1}{2x}$

      (b) $10(1 + x\operatorname{csch}^{-1}x)^9\left(-\dfrac{x}{|x|\sqrt{1+x^2}} + \operatorname{csch}^{-1}x\right)$

15. $\cosh^{-1}\left(\dfrac{x}{\sqrt{2}}\right) + C$ **19.** $-\frac{1}{3}\operatorname{csch}^{-1}|x^3| + C$

17. $-\operatorname{sech}^{-1}(e^x) + C$ **21.** -0.2028

23. (a)   (b)

▶ Chapter 8 Supplementary Exercises (Page 471)

1. (a) no   (b) yes   (c) no
  (d) yes   (e) yes

3. does not exist

5. $\frac{1}{2}\ln(x-1)$

7. If $ad - bc \neq 0$, then $f^{-1}(x)$
  $\qquad = (-dx + b)/(cx - a)$

9. (a) $(-\infty, 5/2)$ (b) $(-2, +\infty)$
  (c) $(-\pi/3, 2\pi/3)$

11. $-\dfrac{3}{x^2}$ **13.** $\dfrac{2}{x}$

15. (a) $2\pi/3$ (b) $3/4$
  (c) $3/5$ (d) $3/5$

17. (a) $\pi/4$ (b) $-\pi/4$
  (c) $2\sqrt{6}$ (d) $\pi/3$

19. (a) $33/65$ **21.** $5/4, -3/4, 34/16$
  (b) $56/65$
  (c) 7

23. (a)

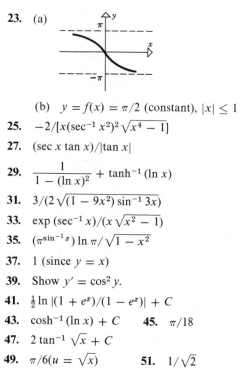

  (b) $y = f(x) = \pi/2$ (constant), $|x| \le 1$

25. $-2/[x(\sec^{-1} x^2)^2 \sqrt{x^4 - 1}]$

27. $(\sec x \tan x)/|\tan x|$

29. $\dfrac{1}{1 - (\ln x)^2} + \tanh^{-1}(\ln x)$

31. $3/(2\sqrt{(1 - 9x^2)} \sin^{-1} 3x)$

33. $\exp(\sec^{-1} x)/(x\sqrt{x^2 - 1})$

35. $(\pi^{\sin^{-1} x}) \ln \pi / \sqrt{1 - x^2}$

37. 1 (since $y = x$)

39. Show $y' = \cos^2 y$.

41. $\frac{1}{2}\ln|(1 + e^x)/(1 - e^x)| + C$

43. $\cosh^{-1}(\ln x) + C$ **45.** $\pi/18$

47. $2 \tan^{-1} \sqrt{x} + C$

49. $\pi/6 (u = \sqrt{x})$ **51.** $1/\sqrt{2}$

▶ Exercise Set 9.2 (Page 483)

1. $-xe^{-x} - e^{-x} + C$

3. $x\ln(2x + 3) - x + \frac{3}{2}\ln(2x + 3) + C$

5. $\frac{1}{4}x^2 \ln x - \frac{1}{8}x^2 + C$

7. $x\cos^{-1}(2x) - \frac{1}{2}\sqrt{1 - 4x^2} + C$

9. $x^2 e^x - 2xe^x + 2e^x + C$

11. $\dfrac{e^x}{2}(\sin x - \cos x) + C$

13. $\dfrac{1}{4\pi}e^{-2\pi x}(\sin 2\pi x - \cos 2\pi x) + C$

15. $\dfrac{e^{ax}}{a^2 + b^2}(a \sin bx - b \cos bx) + C$

17. $x^2 \sin x + 2x \cos x - 2 \sin x + C$

19. $-\frac{1}{3}x \cos(3x + 1) + \frac{1}{9}\sin(3x + 1) + C$

21. $x \tan x + \ln|\cos x| + C$

23. $\frac{1}{2}x[\cos(\ln x) + \sin(\ln x)] + C$

25. $\frac{1}{2}x[\sin(\ln x) - \cos(\ln x)] + C$

27. $\frac{1}{25}(1 - 6e^{-5})$ **29.** $\frac{1}{9}(2e^3 + 1)$

31. $5\ln 5 - 4$ **33.** $\dfrac{5\pi}{6} - \sqrt{3} + 1$

35. $-\dfrac{\pi}{8}$ **37.** $\frac{1}{3}(2 - \sqrt{2})$

39. (a) 1 (b) $\pi(e - 2)$

41. $2\pi^2$

43. (a) $-\frac{1}{3}\sin^2 x \cos x - \frac{2}{3}\cos x + C$
  (b) $\frac{1}{32}(3\pi - 8)$

47. (b) $x^3 e^x - 3x^2 e^x + 6xe^x - 6e^x + C$

▶ Exercise Set 9.3 (Page 490)

1. $-\frac{1}{6}\cos^6 x + C$ **3.** $\frac{1}{2a}\sin^2 ax + C$

5. $\frac{1}{2}\theta - \frac{1}{20}\sin 10\theta + C$

7. $\frac{3}{8}x + \sin\left(\frac{x}{2}\right) + \frac{1}{8}\sin x + C$

9. $\sin\theta - \frac{2}{3}\sin^3\theta + \frac{1}{5}\sin^5\theta + C$

11. $\frac{1}{6}\sin^3 2t - \frac{1}{10}\sin^5 2t + C$

13. $-\frac{1}{5}\cos^5 x + \frac{1}{7}\cos^7 x + C$

15. $-\frac{1}{5}\cos^5\theta + \frac{2}{7}\cos^7\theta - \frac{1}{9}\cos^9\theta + C$

17. $\frac{1}{8}x - \frac{1}{32}\sin 4x + C$

19. $-\frac{1}{6}\cos 3x + \frac{1}{2}\cos x + C$

21. $-\frac{1}{3}\cos\left(\frac{3x}{2}\right) - \cos\left(\frac{x}{2}\right) + C$

23. $\frac{1}{7\cos^7 x} + C$ **25.** $\frac{5\sqrt{2}}{12}$

27. 0 **29.** $\frac{1}{24}$

33. $\frac{\pi}{2}$

35. (a) $\frac{2}{3}$ (b) $\frac{3\pi}{16}$ (c) $\frac{8}{15}$ (d) $\frac{5\pi}{32}$

▶ Exercise Set 9.4 (Page 495)

1. $\frac{1}{3}\tan(3x+1) + C$

3. $\frac{1}{3}\tan^3 x + C$

5. $\frac{1}{16}\tan^4(4x) + \frac{1}{24}\tan^6(4x) + C$

7. $\frac{1}{7}\sec^7 x - \frac{1}{5}\sec^5 x + C$

9. $\frac{1}{4}\sec^3 x \tan x - \frac{5}{8}\sec x \tan x$
$\quad + \frac{3}{8}\ln|\sec x + \tan x| + C$

11. $\frac{1}{6}\sec^3(2t) + C$

13. $\frac{1}{3}\sec^2 x \tan x + \frac{2}{3}\tan x + C$

15. $\frac{1}{5\pi}\sec^4(\pi x)\tan(\pi x) + \frac{4}{15\pi}\sec^2(\pi x)\tan(\pi x)$
$\qquad\qquad + \frac{8}{15\pi}\tan(\pi x) + C$

17. $\frac{1}{3}\tan^3 x - \tan x + x + C$

19. $\frac{1}{6}\tan^3(x^2) + C$

21. $-\frac{1}{5}\csc^5 x + \frac{1}{3}\csc^3 x + C$

23. $-\frac{1}{2}\csc^2 x - \ln|\sin x| + C$

25. $\frac{2}{3}\tan^{3/2} x + \frac{2}{7}\tan^{7/2} x + C$ **27.** $\frac{\sqrt{3}}{2} - \frac{\pi}{6}$

29. $-\frac{1}{2} + \ln 2$ **31.** $\ln(\sqrt{2}+1)$

35. $-\dfrac{1}{\sqrt{a^2+b^2}}\ln|\csc(x+\theta) + \cot(x+\theta)| + C,$
where θ satisfies
$\cos\theta = \dfrac{a}{\sqrt{a^2+b^2}}$ and $\sin\theta = \dfrac{b}{\sqrt{a^2+b^2}}$

▶ Exercise Set 9.5 (Page 501)

1. $\frac{9}{2}\sin^{-1}\left(\frac{x}{3}\right) - \frac{1}{2}x\sqrt{9-x^2} + C$

3. $\frac{1}{16}\tan^{-1}\frac{x}{2} + \frac{x}{8(4+x^2)} + C$

5. $\sqrt{x^2-9} - 3\sec^{-1}\frac{x}{3} + C$

7. $-2\sqrt{2-x^2} + \frac{1}{3}(2-x^2)^{3/2} + C$

9. $\frac{x}{3\sqrt{3+x^2}} + C$ **11.** $\frac{\sqrt{4x^2-9}}{9x} + C$

13. $\frac{x}{\sqrt{1-x^2}} + C$ **15.** $x - \tan^{-1}x + C$

17. $-\frac{\sqrt{9-4x^2}}{9x} + C$ **19.** $-\frac{x}{\sqrt{9x^2-1}} + C$

21. $\frac{1}{2}\sin^{-1}(e^x) + \frac{1}{2}e^x\sqrt{1-e^{2x}} + C$

23. $\frac{2048}{15}$

25. $\frac{1}{2}(\sqrt{3} - \sqrt{2})$

27. $\frac{1}{243}(10\sqrt{3} + 18)$

29. $\frac{1}{2}\ln(x^2 + 4) + C$

31. $\sqrt{5} - \sqrt{2} + \ln\left(\dfrac{2 + 2\sqrt{2}}{1 + \sqrt{5}}\right)$

33. $\frac{\pi}{32}[18\sqrt{5} - \ln(2 + \sqrt{5})]$

35. (a) $\sinh^{-1}\left(\dfrac{x}{3}\right) + C$

    (b) $\ln\left(\dfrac{\sqrt{x^2 + 9}}{3} + \dfrac{x}{3}\right) + C$

▶ Exercise Set 9.6 (Page 505)

1. $\frac{1}{3}\tan^{-1}\left(\dfrac{x - 2}{3}\right) + C$

3. $\sin^{-1}\left(\dfrac{x - 1}{3}\right) + C$

11. $\ln(x^2 + 2x + 5) + \frac{3}{2}\tan^{-1}\left(\dfrac{x + 1}{2}\right) + C$

5. $\ln\left(\sqrt{x^2 - 6x + 10} + x - 3\right) + C$
    or $\sinh^{-1}(x - 3) + C$

13. $\sqrt{x^2 + 2x + 2} + 2\ln\left(\sqrt{x^2 + 2x + 2} + x + 1\right) + C$
    or $\sqrt{x^2 + 2x + 2} + 2\sinh^{-1}(x + 1) + C$

7. $2\sin^{-1}\left(\dfrac{x + 1}{2}\right) + \frac{1}{2}(x + 1)\sqrt{3 - 2x - x^2} + C$

15. $\dfrac{2\pi}{3} - \dfrac{\sqrt{3}}{2}$

9. $\dfrac{1}{\sqrt{10}}\tan^{-1}\dfrac{\sqrt{2}(x + 1)}{\sqrt{5}} + C$

▶ Exercise Set 9.7 (Page 513)

1. $\frac{1}{5}\ln\left|\dfrac{x - 1}{x + 4}\right| + C$

3. $-2\ln|x - 2| + 3\ln|x - 3| + C$

5. $\frac{5}{2}\ln|2x - 1| + 3\ln|x + 4| + C$

7. $-\frac{1}{6}\ln|x - 1| + \frac{1}{15}\ln|x + 2| + \frac{1}{10}\ln|x - 3| + C$

9. $\ln\left|\dfrac{x(x + 3)^2}{x - 3}\right| + C$

11. $\frac{1}{2}x^2 - 2x + 6\ln|x + 2| + C$

13. $3x + 12\ln|x - 2| - \dfrac{2}{x - 2} + C$

15. $\frac{1}{2}x^2 + 3x - \ln|x - 1| + 8\ln|x - 2| + C$

17. $\frac{1}{3}x^3 + x + \ln\left|\dfrac{(x + 1)(x - 1)^2}{x}\right| + C$

19. $3\ln|x| - \ln|x - 1| - \dfrac{5}{x - 1} + C$

21. $\ln\dfrac{(x - 3)^2}{|x + 1|} + \dfrac{1}{x - 3} + C$

23. $\ln|x + 2| + \dfrac{4}{x + 2} - \dfrac{2}{(x + 2)^2} + C$

25. $-\frac{7}{34}\ln|4x - 1| + \frac{6}{17}\ln(x^2 + 1) + \frac{3}{17}\tan^{-1}x + C$

27. $\frac{1}{32}\ln\left|\dfrac{x - 2}{x + 2}\right| - \dfrac{1}{16}\tan^{-1}\dfrac{x}{2} + C$

29. $3\tan^{-1}x + \frac{1}{2}\ln(x^2 + 3) + C$

31. $\frac{1}{2}x^2 - 3x + \frac{1}{2}\ln(x^2 + 1) + C$

33. $\dfrac{1}{\sqrt{2}}\tan^{-1}\left(\dfrac{x + 1}{\sqrt{2}}\right) + \dfrac{1}{x^2 + 2x + 3} + C$

35. $\frac{1}{6}\ln\left|\dfrac{\sin\theta - 1}{\sin\theta + 5}\right| + C$

37. $\ln\dfrac{e^x}{1 + e^x} + C$

39. (a) $\sqrt{2}, -\sqrt{2}$

41. $\pi\left(\frac{19}{5} - \frac{9}{4}\ln 5\right)$

43. $y = \dfrac{3 - 2Ce^x}{1 - Ce^x}$

45. $y = \dfrac{t - 1}{Ct - t + 1}$

47. (a) $(x - 1)(x - 2)(x - 3)$
    (b) $(x - 4)(x^2 + x + 5)$
    (c) $(x - 2)(x - 3)(x^2 + 1)$

49. $\frac{1}{8}\ln|x - 1| - \frac{1}{5}\ln|x - 2|$
    $+ \frac{1}{12}\ln|x - 3| - \frac{1}{120}\ln|x + 3| + C$

▶ Exercise Set 9.8 (Page 520)

1. $2x^{1/2} - 3x^{1/3} + 6x^{1/6} - 6\ln(x^{1/6} + 1) + C$

3. $4\ln\dfrac{v^{1/4}}{|1 - v^{1/4}|} + C$

5. $2t^{1/2} + 3t^{1/3} + 6t^{1/6} + 6\ln|t^{1/6} - 1| + C$

7. $\frac{1}{3}(1 + x^2)^{3/2} - (1 + x^2)^{1/2} + C$

9. $\ln\left|\tan\left(\dfrac{x}{2}\right) + 1\right| + C$

11. 1

13. $\dfrac{4}{\sqrt{3}}\tan^{-1}\left(\sqrt{3}\tan\dfrac{x}{2}\right) - x + C$

17. $\dfrac{2}{\sqrt{3}}\tan^{-1}\left(\dfrac{2\tanh(x/2) + 1}{\sqrt{3}}\right) + C$

▶ Exercise Set 9.9 (Page 528)

1. (a) 0.7366   (b) 0.6532
   (c) 0.6949   (d) 0.6932

3. (a) 1.8961   (b) 1.8961
   (c) 1.8961   (d) 2.0046

5. (a) 1.1261   (b) 0.6352
   (c) 0.8806   (d) 0.8818

7. 3.1416

9. 0.6928

▶ Chapter 9 Supplementary Exercises (Page 529)

1. $\frac{1}{2}x\sin 2x + \frac{1}{4}\cos 2x + C$

3. $\frac{1}{3}\sec^3 x - \sec x + C$

5. $\frac{1}{9}\tan^3 3t + C$

7. $x - \sin x + C$

9. $\frac{1}{6}x^3 + \frac{1}{4}(x^2 - \frac{1}{2})\sin 2x + \frac{1}{4}x\cos 2x + C$

11. $\frac{1}{4}\sec^4 x + C$

13. $-(\sin^3 2x)/48 - (\sin 4x)/64 + x/16 + C$

15. $\frac{1}{4}$

17. $\frac{1}{2}$

19. $\frac{1}{2}$

21. $x/\sqrt{1 + x^2} + C$

23. $\ln|\sec(e^x) + \tan(e^x)| + C$

25. $\sqrt{e^{2x} + 1} + C$

27. $\frac{1}{13}e^{3x}(3\sin 2x - 2\cos 2x) + C$

29. $\frac{5}{6}\pi - \sqrt{3}$

31. $-(\ln x + 1)/x + C$

33. $\sqrt{x^2 - 9} + C$

35. $3\ln(3 + \sqrt{8}) - \sqrt{8}$

37. $\frac{1}{5}\sqrt{2x + 3}(x^2 - 2x + 6) + C$

39. $\frac{1}{2}\sin^{-1}\left(\dfrac{2t + 1}{2}\right) + C$

41. $-\sqrt{a^2 - x^2}/(a^2 x) + C$

43. $\frac{1}{2}[x\sqrt{a^2 - x^2} + a^2\sin^{-1}(x/a)] + C$

45. $-\sqrt{4x - x^2} + C$

47. $\ln|(2x + 1)/(x + 1)| + C$

49. $-\dfrac{\ln|x|}{6} - \dfrac{2}{15}\ln|x + 3| + \dfrac{3}{10}\ln|x - 2| + C$

51. $\frac{1}{3}\ln|x^3 - 3x| + C$

53. $\dfrac{7}{4}\ln\left|\dfrac{x - 1}{x + 1}\right| - \dfrac{3}{2}\tan^{-1}x + C$

55. $\dfrac{1}{26}\left(\ln\left(\dfrac{(x - 3)^2}{x^2 + 4}\right) - 3\tan^{-1}\left(\dfrac{x}{2}\right)\right) + C$

57. $\frac{7}{8}\tan^{-1}\left(\dfrac{x}{2}\right) - \dfrac{24 + 5x}{4(4 + x^2)} + C$

59. $\frac{1}{2}(\ln(x^2 + 2x + 5) - \tan^{-1}(\frac{1}{2}(x + 1)) + C$

61. $\dfrac{1}{\sqrt{2}}\sin^{-1}\sqrt{\frac{2}{3}}x + C$

63. $2(\sqrt{t} - \tan^{-1}\sqrt{t}) + C$

65. $-\frac{7}{4} + 3\ln 2$

67. $\frac{1}{2}(x - \ln|\sin x - \cos x|) + C$

69. $\frac{1}{4}\cot^2(\frac{1}{2}x) + \frac{1}{2}\ln|\tan(\frac{1}{2}x)| + C$

73. (a) $\frac{1}{2}\tan^{-1}(1) = \pi/8$
  (b) $\pi(\pi + 2)/64$ (c) $\pi \ln 2$

75. (a) $e^{2x}(4x^3 - 6x^2 + 6x - 3)/8 + C$
  (b) $(\pi - 2)/125$
  (c) $\frac{1}{6}(\sin x \cos x)^3 + \frac{1}{16}x - \frac{1}{64}\sin(4x) + C$

77. (a) $\frac{1}{4}(\sec^4 \theta - 2\sec^2 \theta) + C_1$
  (b) $\frac{1}{4}\tan^4 \theta + C_2,\ C_2 = C_1 - \frac{1}{4}$

79. (a) 58.9276   (b) 54.7328

81. (a) 1.8277   (b) 1.8279

83. 0.36972

▶ Exercise Set 10.1 (Page 538)

1. 1

3. divergent

5. $\ln \frac{5}{3}$

7. $-\frac{1}{4}$

9. divergent

11. 0

13. divergent

15. $\frac{\pi}{2}$

17. $\frac{9}{2}$

19. 2

21. (b) $\frac{1}{2}$

25. $+\infty$

29. $\dfrac{2\pi NI}{kr}\left(1 - \dfrac{a}{\sqrt{r^2 + a^2}}\right)$

31. (b) 24,000,000

▶ Exercise Set 10.2 (Page 546)

1. 1

3. 1

5. 1

7. $\dfrac{1}{\pi}$

9. -1

11. 0

13. $\frac{1}{2}$

17. $+\infty$

21. (b) 2

25. 2

15. 2

19. $\frac{1}{9}$

23. $k = -1, l = \pm 2\sqrt{2}$

▶ Exercise Set 10.3 (Page 553)

1. 0

3. $-\infty$

5. $+\infty$

7. 0

9. 0

11. π

13. e^2

15. 1

17. 1

19. $-\frac{1}{2}$

21. 0

23. 2

27. (a) 0   (b) $+\infty$
  (c) 0   (d) $-\infty$
  (e) $+\infty$ (f) $+\infty$
  (g) $-\infty$ (h) $-\infty$

29.

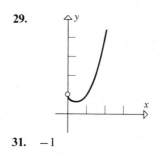

31. -1

▶ Chapter 10 Supplementary Exercises (Page 554)

1. $\pi/2$

3. 1

5. 6

7. $\frac{1}{2}$

9. diverges

11. 1

13. $\pi/4$

15. $n > -1;\ -1/(n+1)^2$

17. $3/2$

19. $+\infty$

21. 0

23. 0

25. $-\dfrac{1}{4\pi^2}$

27. 0

29. 0

31. 1

33. (a) $+\infty$ (b) $3\pi/2$

35. 2

▶ Exercise Set 11.1 (Page 564)

1. $\frac{1}{3}, \frac{2}{4}, \frac{3}{5}, \frac{4}{6}, \frac{5}{7}$; converges to 1

3. 2, 2, 2, 2, 2; converges to 2

5. $\dfrac{\ln 1}{1}, \dfrac{\ln 2}{2}, \dfrac{\ln 3}{3}, \dfrac{\ln 4}{4}, \dfrac{\ln 5}{5}$; converges to 0

7. 0, 2, 0, 2, 0; diverges

9. $-1, \frac{16}{9}, -\frac{54}{28}, \frac{128}{65}, -\frac{250}{126}$; diverges

11. $\frac{6}{2}, \frac{12}{8}, \frac{20}{18}, \frac{30}{32}, \frac{42}{50}$; converges to $\frac{1}{2}$

13. $1, 2^{1/2}, 3^{1/3}, 4^{1/4}, 5^{1/5}$; converges to 1

15. $\left\{\dfrac{2n-1}{2n}\right\}_{n=1}^{+\infty}$; converges to 1

17. $\left\{\dfrac{1}{3^n}\right\}_{n=1}^{+\infty}$; converges to 0

19. $\left\{\dfrac{1}{n} - \dfrac{1}{n+1}\right\}_{n=1}^{+\infty}$; converges to 0

21. $\{\sqrt{n+1} - \sqrt{n+2}\}_{n=1}^{+\infty}$; converges to 0

23. (a) 3, 6, 12, 24, 48, 96, 192, 384
    (b) 1, 1, 2, 3, 5, 8, 13, 21

25. (a) $1, \frac{3}{4}, \frac{2}{3}, \frac{5}{8}$ (b) $\frac{1}{2}$

27. (a) 3   (b) 11   (c) 1001

▶ Exercise Set 11.2 (Page 572)

1. decreasing

3. increasing

5. decreasing

7. increasing

9. nonincreasing

11. not monotone

13. increasing

15. decreasing

17. decreasing

19. converges

21. diverges

23. converges

25. (a) 0   (b) does not exist

▶ Exercise Set 11.3 (Page 580)

1. (a) converges to $\frac{5}{2}$ (b) converges to $\frac{1}{2}$
    (c) diverges

3. $\frac{4}{7}$

5. 6

7. diverges

9. $\frac{1}{3}$

11. $\frac{1}{6}$

13. $\frac{448}{3}$

15. $-\frac{1}{3}$

17. $\frac{4}{9}$

19. $\frac{532}{99}$

21. $\frac{869}{1111}$

23. diverges

29. 1

▶ Exercise Set 11.4 (Page 589)

1. $\frac{4}{3}$

3. $-\frac{1}{36}$

5. (a) converges   (b) diverges
   (c) diverges   (d) diverges
   (e) converges   (f) diverges
   (g) converges   (h) converges

9. diverges

11. diverges

13. converges

15. diverges

17. diverges

19. diverges

21. converges

23. diverges

25. converges

27. converges

29. diverges

35. (a) diverges   (b) diverges
   (c) diverges   (d) converges

▶ Exercise Set 11.5 (Page 597)

1. converges

3. inconclusive

5. diverges

7. diverges

9. converges

11. diverges

13. converges

15. converges

17. converges

19. converges

21. diverges

23. converges

25. converges

27. diverges

29. converges

31. converges

33. converges

▶ Exercise Set 11.6 (Page 604)

13. converges

15. converges

17. diverges

19. converges

21. diverges

23. converges

25. diverges

27. converges

29. diverges

31. converges

33. converges

35. converges

37. $p > 1$

39. converges

▶ Exercise Set 11.7 (Page 616)

1. converges

3. diverges

5. converges

7. converges absolutely

9. diverges

11. converges absolutely

13. conditionally

15. divergent

17. conditionally

19. absolutely

21. conditionally

23. divergent

25. conditionally

27. absolutely

29. conditionally

31. 0.125

33. 0.1

35. 9999

37. 39,999

▶ Exercise Set 11.8 (Page 623)

1. $1, [-1, 1)$

3. $+\infty, (-\infty, +\infty)$

5. $\frac{1}{5}, [-\frac{1}{5}, \frac{1}{5}]$

7. $1, [-1, 1]$

9. $1, (-1, 1]$

11. $+\infty, (-\infty, +\infty)$

13. $+\infty, (-\infty, +\infty)$

15. $1, [-1, 1]$

17. $1, (-2, 0]$

19. $\frac{4}{3}, (-\frac{19}{3}, -\frac{11}{3})$

21. $1, [-2, 0]$

23. $+\infty, (-\infty, +\infty)$

25. $(-\infty, +\infty)$

27. $(a - b, a + b)$

29. $+\infty$

► Exercise Set 11.9 (Page 632)

1. $1 - 2x + 2x^2 - \frac{4}{3}x^3 + \frac{2}{3}x^4$

3. $2x - \frac{4}{3}x^3$　　　**5.** $x + \frac{1}{3}x^3$

7. $x + x^2 + \frac{x^3}{2!} + \frac{x^4}{3!}$　　**9.** $1 + \frac{1}{2}x^2 + \frac{5}{24}x^4$

11. $\ln 3 + \frac{2}{3}x - \frac{2}{9}x^2 + \frac{8}{81}x^3 - \frac{4}{81}x^4$

13. $e + e(x-1) + \frac{e}{2!}(x-1)^2 + \frac{e}{3!}(x-1)^3$

15. $2 + \frac{1}{4}(x-4) - \frac{1}{64}(x-4)^2 + \frac{1}{512}(x-4)^3$

17. $\dfrac{\sqrt{2}}{2} - \dfrac{\sqrt{2}}{2}\left(x - \dfrac{\pi}{4}\right) - \dfrac{\sqrt{2}}{4}\left(x - \dfrac{\pi}{4}\right)^2$
$\qquad + \dfrac{\sqrt{2}}{12}\left(x - \dfrac{\pi}{4}\right)^3$

19. $-\dfrac{\sqrt{3}}{2} + \dfrac{\pi}{2}\left(x + \dfrac{1}{3}\right) + \dfrac{\sqrt{3}\pi^2}{4}\left(x + \dfrac{1}{3}\right)^2$
$\qquad - \dfrac{\pi^3}{12}\left(x + \dfrac{1}{3}\right)^3$

21. $\dfrac{\pi}{4} + \dfrac{1}{2}(x-1) - \dfrac{1}{4}(x-1)^2 + \dfrac{1}{12}(x-1)^3$

23. $\displaystyle\sum_{k=0}^{\infty} (-1)^k \frac{x^k}{k!}$　　**25.** $\displaystyle\sum_{k=0}^{\infty} (-1)^k x^k$

27. $\displaystyle\sum_{k=1}^{\infty} (-1)^{k+1} \frac{x^k}{k}$　　**29.** $\displaystyle\sum_{k=0}^{\infty} (-1)^k \frac{x^{2k}}{4^k (2k)!}$

31. $\displaystyle\sum_{k=0}^{\infty} \frac{x^{2k}}{(2k)!}$　　**33.** $\displaystyle\sum_{k=0}^{\infty} (-1)(x+1)^k$

35. $\displaystyle\sum_{k=1}^{\infty} (-1)^{k+1} \frac{(x-1)^k}{k}$

37. $\displaystyle\sum_{k=0}^{\infty} (-1)^k \frac{\pi^{2k}}{(2k)!}\left(x - \frac{1}{2}\right)^{2k}$

39. $\displaystyle\sum_{k=0}^{\infty} \left(\frac{16 + (-1)^{k+1}}{8}\right)\frac{(x - \ln 4)^k}{k!}$

► Exercise Set 11.10 (Page 643)

1. $\dfrac{2^6 e^{2c}}{6!}x^6$　　　**3.** $-\dfrac{x^5}{(c+1)^6}$

5. $\dfrac{(4+c)e^c}{4!}x^4$　　**7.** $-\dfrac{(1-3c^2)}{3(1+c^2)^3}x^3$

9. $-\dfrac{5}{128c^{7/2}}(x-4)^4$　　**11.** $\dfrac{\cos c}{5!}\left(x - \dfrac{\pi}{6}\right)^5$

13. $\dfrac{7}{(1+c)^8}(x+2)^6$　　**15.** $\dfrac{x^{n+1}}{(1-c)^{n+2}}$

17. $\dfrac{2^{n+1}e^{2c}}{(n+1)!}x^{n+1}$

27. $\displaystyle\sum_{k=0}^{\infty} (-1)^k \frac{2^k x^k}{k!}$, $(-\infty, +\infty)$

29. $\displaystyle\sum_{k=0}^{\infty} (-1)^k x^k$, $(-1, 1)$

31. $\displaystyle\sum_{k=0}^{\infty} (-1)^k 3^k x^{k+2}$, $\left(-\frac{1}{3}, \frac{1}{3}\right)$

33. $1 + \displaystyle\sum_{k=1}^{\infty} (-1)^k \frac{2^{2k-1}}{(2k)!} x^{2k}$, $(-\infty, +\infty)$

35. $\displaystyle\sum_{k=0}^{\infty} (-1)^k \frac{2^{2k+1}}{(2k+1)!} x^{2k+1}$, $(-\infty, +\infty)$

37. $\displaystyle\sum_{k=0}^{\infty} \frac{(-1)^k}{(2k)!} x^{4k}$, $(-\infty, +\infty)$

39. $\displaystyle\sum_{k=0}^{\infty} (-1)^k (x-1)^k$

► Exercise Set 11.11 (Page 652)

1. (a) 9　　(b) 13

3. 1.6487　　　**5.** 0.9877

7. 0.5299　　　**9.** 0.223

11. 0.100　　　**13.** 1.0050

15. $|x| < 0.569$

17. $-0.144 < x < 0.137$

19. (a) 1,999,999　　(b) 8

▶ Exercise Set 11.12 (Page 660)

5. $\displaystyle\sum_{k=1}^{\infty} (-1)^{k+1} k x^{k-1}$

9. $1 - \frac{1}{3}x - \frac{1}{9}x^2 - \frac{5}{81}x^3 + \cdots;\quad 1$

11. $x + \frac{2}{3}x^2 + \frac{8}{9}x^3 + \frac{112}{81}x^4 + \cdots;\quad \frac{1}{2}$

13. 0.764   **15.** 0.494

17. 0.100   **19.** 0.491

21. $1 - \frac{3}{2}x^2 + \frac{25}{24}x^4 - \frac{331}{720}x^6 + \cdots$

23. $x - x^2 + \frac{1}{3}x^3 - \frac{1}{30}x^5 + \cdots$

25. $x^2 - \frac{1}{3}x^4 + \frac{2}{45}x^6 - \frac{1}{315}x^8 + \cdots$

27. $x^2 + \frac{9}{2}x^3 + \frac{79}{8}x^4 + \frac{683}{48}x^5 + \cdots$

29. (a) 0   (b) $-\frac{3}{2}$

31. (a) $x - \frac{1}{6}x^3 + \frac{3}{40}x^5 - \frac{5}{112}x^7$

(b) $x + \displaystyle\sum_{k=1}^{\infty} (-1)^k \frac{1 \cdot 3 \cdot 5 \cdots (2k-1)}{2^k k! (2k+1)} x^{2k+1}$

(c) 1

35. $\displaystyle\sum_{k=0}^{\infty} \binom{m}{k} x^k$ **37.** $\displaystyle\sum_{k=0}^{\infty} (-1)^k \binom{1/3}{k} x^k;\ R = 1$

39. $\displaystyle\sum_{k=0}^{\infty} \binom{2/3}{k} x^{2k};\ R = 1$ **41.** $\displaystyle\sum_{k=0}^{\infty} \binom{-1/2}{k} x^{k+1};\ R = 1$

▶ Chapter 11 Supplementary Exercises (Page 661)

1. $L = 0$ **3.** $L = 0$

5. does not exist

7. (a) 2, 3, 6   (b) 4   (c) 1, 5

9. (a) $|q| > \sqrt{\pi}$ (b) $q > 1/3$

(c) none $(p = 1)$ (d) $q > e$ or $0 < q < 1/e$

11. (a) $1 + 36 \displaystyle\sum_{k=1}^{\infty} (.01)^k$

(b) 15/11

13. (a) 1/4   (b) diverges   (c) 1

15. converges   **17.** converges

19. converges   **21.** diverges (general term

23. converges absolutely   does not approach zero)

25. diverges   **27.** $R = 1;\ 0 \le x \le 2$

29. $R = 2;\ -1 < x < 3$ **31.** $R = 0;\ x = 1$

33. (a) $(x-2) - (x-2)^2/2 + (x-2)^3/3$

(b) $\dfrac{-1}{4(c-1)^4}(x-2)^4$, c between 2 and x

(c) $\dfrac{(\frac{1}{2})^4}{4(\frac{1}{2})^4} = \frac{1}{4}$

35. (a) $1 + (x-1)/2 - (x-1)^2/8$

(b) $\dfrac{(x-1)^3}{16c^{5/2}}$, c between 1 and x

(c) $\dfrac{(5/9)^3}{16(2/3)^5} < 0.0814$

37. $\ln a + \displaystyle\sum_{k=0}^{\infty} (-1)^k \frac{(x/a)^{k+1}}{k+1};\ R = a$

39. $\dfrac{1}{3}\left\{ 1 + \displaystyle\sum_{k=1}^{\infty} (-1)^k \frac{1 \cdot 3 \cdots (2k-1)}{2 \cdot 4 \cdots (2k)} (x/9)^k \right\};$

$R = 9$

41. $1 + x^2/2 + 5x^4/24$

43. $1 - x^2/4 - x^4/96$

45. (a) $\displaystyle\sum_{k=0}^{\infty} \frac{(-1)^k x^{2k}}{(2k+2)!}$ (b) $\displaystyle\sum_{k=0}^{\infty} \frac{(-1)^k x^k}{k+1}$

(c) $\frac{1}{2}$, 1

47. $|x| \le 1.17$

49. 0.444

▶ Exercise Set 12.2 (Page 672)

1.

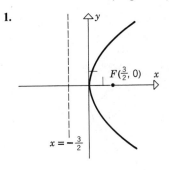

$F(\frac{3}{2}, 0)$

$x = -\frac{3}{2}$

3.

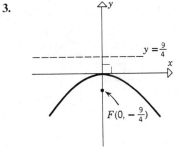

$y = \frac{9}{4}$

$F(0, -\frac{9}{4})$

5.

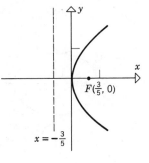

$F(\frac{3}{5}, 0)$

$x = -\frac{3}{5}$

7.

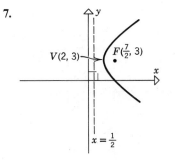

$V(2, 3)$ $F(\frac{7}{2}, 3)$

$x = \frac{1}{2}$

9.

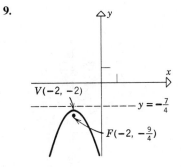

$V(-2, -2)$

$y = -\frac{7}{4}$

$F(-2, -\frac{9}{4})$

11.

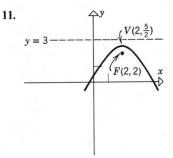

$V(2, \frac{5}{2})$

$y = 3$

$F(2, 2)$

13.

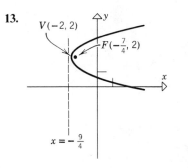

$V(-2, 2)$

$F(-\frac{7}{4}, 2)$

$x = -\frac{9}{4}$

15.

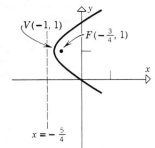

$V(-1, 1)$

$F(-\frac{3}{4}, 1)$

$x = -\frac{5}{4}$

17. $y^2 = 12x$

19. $y^2 = -28x$

21. $y^2 = 2x$

23. $x^2 = -12y$

25. $y^2 = -5(x - \frac{19}{5})$

27. $y^2 = 6(x - \frac{3}{2})$

29. $(x - 1)^2 = 12(y - 1)$

31. $(x - 5)^2 = 2(y + 3)$

33. (a) $y = \frac{7}{6}x^2 - \frac{23}{6}x + 3$

  (b) $x = -\frac{7}{6}y^2 + \frac{17}{6}y + 2$

35. vertex: $\left(-\dfrac{B}{2A}, \dfrac{4AC - B^2}{4A}\right)$

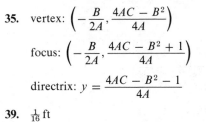

focus: $\left(-\dfrac{B}{2A}, \dfrac{4AC - B^2 + 1}{4A}\right)$

directrix: $y = \dfrac{4AC - B^2 - 1}{4A}$

39. $\frac{1}{16}$ ft

▶ Exercise Set 12.3 (Page 678)

1.
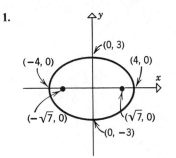

3.
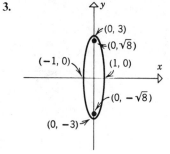

5.
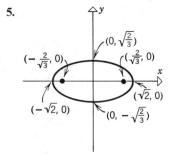

7.
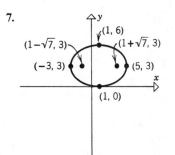

9.
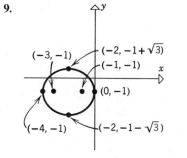

11.
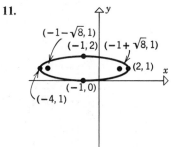

13.
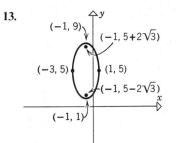

29.
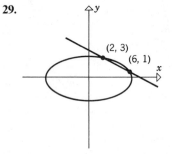

15. $\dfrac{x^2}{9} + \dfrac{y^2}{4} = 1$ 17. $\dfrac{x^2}{169} + \dfrac{y^2}{144} = 1$

19. $\dfrac{x^2}{3} + \dfrac{y^2}{2} = 1$

21. $\dfrac{x^2}{16} + \dfrac{y^2}{4} = 1, \dfrac{x^2}{4} + \dfrac{y^2}{16} = 1$

23. $\dfrac{x^2}{36} + \dfrac{y^2}{81/8} = 1$

25. $(x-1)^2 + \dfrac{(y-3)^2}{2} = 1$

27. $\dfrac{(x-4)^2}{32} + \dfrac{(y-3)^2}{36} = 1$

31.
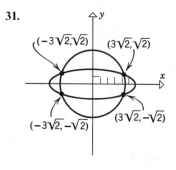

33. $k = -4$ at $(-2, -1)$
    $k = 4$ at $(2, 1)$

35. πab

▶ Exercise Set 12.4 (Page 687)

1.

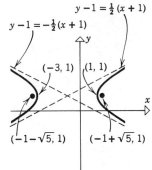

$y = -\frac{1}{2}x$ $y = \frac{1}{2}x$
(−4, 0)   (4, 0)
(−2√5, 0)   (2√5, 0)

3.

(0, 2)   (0, √13)
$y = -\frac{2}{3}x$ $y = \frac{2}{3}x$
(0, −2)   (0, −√13)

5. $y = -2\sqrt{2}x$ $y = 2\sqrt{2}x$

(−1, 0)   (1, 0)
(−3, 0)   (3, 0)

7.

$y = -x$ $y = x$
(−1, 0)   (1, 0)
(−√2, 0)   (√2, 0)

9. $y - 4 = -\frac{2}{3}(x-2)$ $y - 4 = \frac{2}{3}(x-2)$

(5, 4)
(2 − √13, 4)
(−1, 4)   (2 + √13, 4)

11. $y + 3 = -3(x+2)$ $y + 3 = 3(x+2)$

(−2, −3 + 2√10)
(−2, 3)
(−2, −9)   (−2, −3 − 2√10)

13. $y - 1 = \frac{1}{2}(x+1)$
$y - 1 = -\frac{1}{2}(x+1)$

(−3, 1)   (1, 1)
(−1 − √5, 1)   (−1 + √5, 1)

15. $y + 3 = -4(x-1)$
$y + 3 = 4(x-1)$

(−1 − 3)   (3 − 3)
(1 − 2√17, −3)   (1 + 2√17, −3)

17. $\dfrac{x^2}{4} - \dfrac{y^2}{5} = 1$

19. $x^2 - \dfrac{y^2}{4} = 1$

21. $\dfrac{x^2}{64/9} - \dfrac{y^2}{16} = 1,\ \dfrac{y^2}{36} - \dfrac{x^2}{16} = 1$

23. $\dfrac{y^2}{56} - \dfrac{x^2}{56} = 1$

25. $\dfrac{x^2}{4} - \dfrac{y^2}{4/3} = 1$

27. $\dfrac{(x-2)^2}{4} - \dfrac{(y+3)^2}{5} = 1$

29. $\dfrac{(x-6)^2}{16} - \dfrac{(y-4)^2}{9} = 1$

31. $8xy - 4x - 4y + 1 = 0$

33.
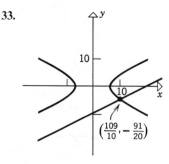

35.
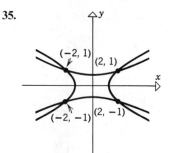

39. $\left(\tfrac{3}{2}\sqrt{13},\,-9\right),\,\left(-\tfrac{3}{2}\sqrt{13},\,-9\right)$

▶ Exercise Set 12.5 (Page 697)

1. (a) $(-1 + 3\sqrt{3},\ \sqrt{3} + 3)$

   (b) $\dfrac{x'^2}{4} - \dfrac{y'^2}{12} = 1$

   (c)
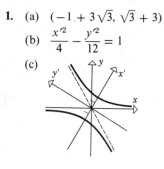

3. $\theta = 45°;\ \dfrac{y'^2}{18} - \dfrac{x'^2}{18} = 1,$
   hyperbola
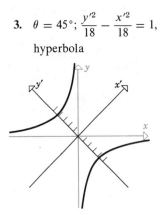

5. $\theta = \tan^{-1}\tfrac{1}{2};$
   $\dfrac{x'^2}{3} - \dfrac{y'^2}{2} = 1,$ hyperbola
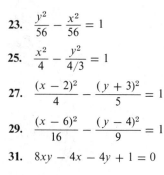

7. $\theta = 60°;\ y' = x'^2$, parabola
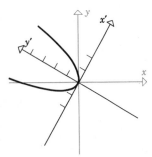

9. $\theta = \tan^{-1}\tfrac{3}{4};$
   $y'^2 = 4(x' - 1)$, parabola
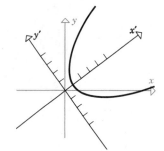

11. $\theta = \tan^{-1}\tfrac{3}{4};$
   $\dfrac{(x' + 1)^2}{4} + y'^2 = 1$, ellipse

15. $x^2 + xy + y^2 = 3$

21. ellipse, point, or no graph

23. parabola, line, pair of parallel lines, or no graph

25. ellipse, point, or no graph

▶ Chapter 12 Supplementary Exercises (Page 698)

1. parabola: $V(-2, 3)$, $F(-5, 3)$, directrix $x = 1$

3. ellipse: $C(-2, 1)$, $F(-2, 1 \pm \sqrt{5})$, axis lengths 6 and 4

5. hyperbola: $C(2, 1)$, $F(2 \pm \sqrt{10}, 1)$, $V(2 \pm 3, 1)$, asymptotes $y - 1 = \pm(x - 2)/3$

7. parabola: $V(3, 1)$, $F(21/8, 1)$, directrix $x = 27/8$

9. $(y - 3)^2 = 16(x - 1)$ **11.** $y^2/9 - x^2/16 = 1$

13. $x^2/25 + y^2/16 = 1$ **15.** $(x - 3)^2 = 4(y - 3)$

17.

19.

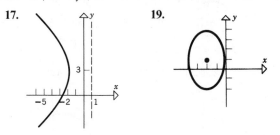

21. $\pi/4$; ellipse; $(x'/\sqrt{2})^2 + (y')^2 = 1$

23. $\pi/6$; hyperbola; $4x'^2 - y'^2 = 1$;

25. $\tan^{-1}(4/3)$; parabola; $y'^2 = 4(x' - 1)$

27.

29.

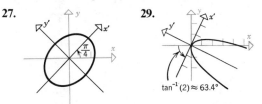

$\tan^{-1}(2) \approx 63.4°$

▶ Exercise Set 13.1 (Page 703)

1.

$(5, \frac{2\pi}{3})$ $(3, \frac{\pi}{4})$ $(1, \frac{\pi}{2})$ $(0, \pi)$ $(4, \frac{7\pi}{6})$ $(2, \frac{4\pi}{3})$

3. (a) $(3\sqrt{3}, 3)$ (b) $\left(-\frac{7}{2}, \frac{7\sqrt{3}}{2}\right)$

(c) $(4\sqrt{2}, 4\sqrt{2})$ (d) $(5, 0)$

(e) $\left(-\frac{7\sqrt{3}}{2}, \frac{7}{2}\right)$ (f) $(0, 0)$

5. (a) $(5, \pi)$ (b) $\left(4, \frac{11\pi}{6}\right)$

(c) $\left(2, \frac{3\pi}{2}\right)$ (d) $\left(8\sqrt{2}, \frac{5\pi}{4}\right)$

(e) $\left(6, \frac{2\pi}{3}\right)$ (f) $\left(\sqrt{2}, \frac{\pi}{4}\right)$

7. (a) $(-5, 0)$ (b) $\left(-4, \frac{5\pi}{6}\right)$

(c) $\left(-2, \frac{\pi}{2}\right)$ (d) $\left(-8\sqrt{2}, \frac{\pi}{4}\right)$

(e) $\left(-6, \frac{5\pi}{3}\right)$ (f) $\left(-\sqrt{2}, \frac{5\pi}{4}\right)$

9. $y = 4$; line

11. $3x^2 + 4y^2 - 12x = 36$; ellipse

13. $x^2 + y^2 + 4x = 0$; circle

15. $r \cos \theta = 7$

17. $r^2 = 16 \cos 2\theta$ **19.** $r = 9 \sec \theta \tan \theta$

21.

23.

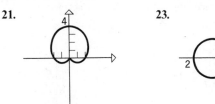

► Exercise Set 13.2 (Page 712)

1.
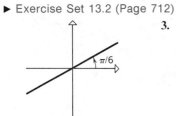
Line

3.
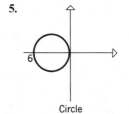
Circle

5.
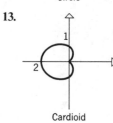
Circle

7.
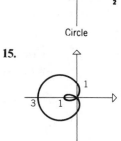
Circle

9.
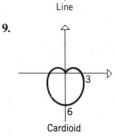
Cardioid

11.
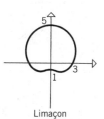
Cardioid

13.
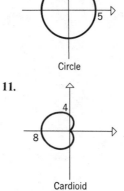
Cardioid

15.
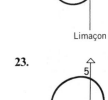
Limaçon

17.
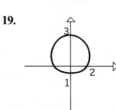
Limaçon

19.
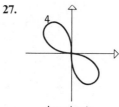
Limaçon

21.

Limaçon

23.
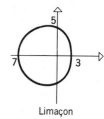
Limaçon

25.

Lemniscate

27.

Lemniscate

29.

Spiral

31.
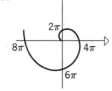
Four-petal rose

33.

Three-petal rose

35.

Eight-petal rose

37.
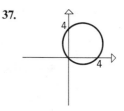

39.
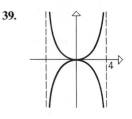

41.
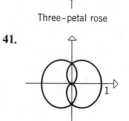

43.
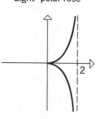

▶ Exercise Set 13.3 (Page 718)

1. $\dfrac{7\pi^3}{1296}$

3. 6π

13. $\dfrac{5\pi}{4}$

15. $100\cos^{-1}\left(\tfrac{3}{5}\right) - 48$

5. $\dfrac{8\pi}{3} + \sqrt{3}$

7. $\dfrac{9\sqrt{3}}{2} - \pi$

17. $\tfrac{4}{3}a^2$

19. $2\sqrt{3} - \dfrac{2\pi}{3}$

9. 1

11. 4π

▶ Exercise Set 13.4 (Page 728)

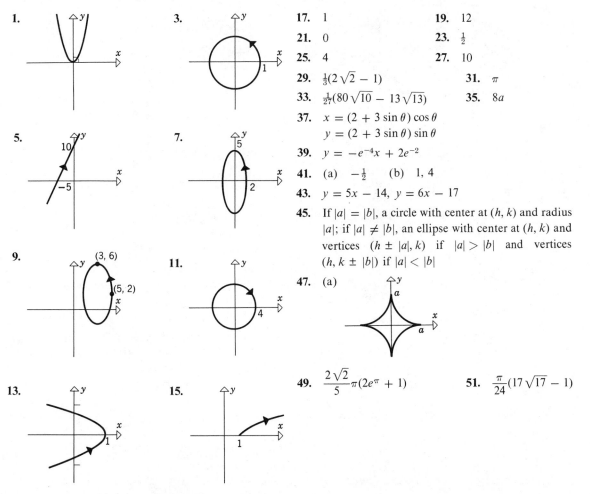

1.

3.

5.

7.

9.

11.

13.

15.

17. 1

19. 12

21. 0

23. $\tfrac{1}{2}$

25. 4

27. 10

29. $\tfrac{1}{3}(2\sqrt{2} - 1)$

31. π

33. $\tfrac{1}{27}(80\sqrt{10} - 13\sqrt{13})$

35. $8a$

37. $x = (2 + 3\sin\theta)\cos\theta$
$y = (2 + 3\sin\theta)\sin\theta$

39. $y = -e^{-4}x + 2e^{-2}$

41. (a) $-\tfrac{1}{2}$ (b) $1, 4$

43. $y = 5x - 14,\ y = 6x - 17$

45. If $|a| = |b|$, a circle with center at (h, k) and radius $|a|$; if $|a| \neq |b|$, an ellipse with center at (h, k) and vertices $(h \pm |a|, k)$ if $|a| > |b|$ and vertices $(h, k \pm |b|)$ if $|a| < |b|$

47. (a)

49. $\dfrac{2\sqrt{2}}{5}\pi(2e^\pi + 1)$

51. $\dfrac{\pi}{24}(17\sqrt{17} - 1)$

▶ Exercise Set 13.5 (Page 736)

1. $\dfrac{1}{\sqrt{3}}$ **3.** $\dfrac{\tan 2 - 2}{2\tan 2 + 1}$ **17.** $\dfrac{a}{3}[(\pi^2 + 4)^{3/2} - 8]$ **19.** $8a$

5. -2 **7.** 1 **21.** $\theta = \dfrac{\pi}{2}, \dfrac{3\pi}{2}, \sin^{-1}\frac{1}{4}, \pi - \sin^{-1}\frac{1}{4}$

9. 0 **11.** $-\dfrac{1}{2\sqrt{3}}$ **23.** $\tan^{-1} 3\sqrt{3}$

13. $\dfrac{\sqrt{10}}{3}(e^6 - 1)$ **15.** $2\pi a$

▶ Chapter 13 Supplementary Exercises (Page 737)

1. (a) $(1, \sqrt{3})$ (b) $(0, -2)$
    (c) $(0, 0)$ (d) $(-1, 1)$
    (e) $(-3, 0)$ (f) $(3/5, -4/5)$

3. (a)   (b)

5. $2xy = 1$ (hyperbola)   **7.** $y = -4$ (line)

9. $y = \sqrt{3}x$ (line)   **11.** $x = y = 0$ (point)

13. $r\cos\theta = -3$ **15.** $\tan\theta = 3$

17.   **19.**

21.   **23.**

25. $(a/2, \pm\pi/6)$, $(a/2, \pm\pi/3)$, $(a/2, \pm 2\pi/3)$,
    $(a/2, \pm 5\pi/6)$

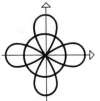

27. (a) $\displaystyle\int_0^{\pi/3} (1 + \cos\theta)^2\, d\theta + \int_{\pi/3}^{\pi/2} (3\cos\theta)^2\, d\theta$

   (b) $\displaystyle 2\int_{\pi/3}^{\pi} \sqrt{2(1 + \cos\theta)}\, d\theta$

29. (a) $\displaystyle\int_{\pi/2}^{\pi} 2[\sin^2\theta - (1 + \cos\theta)^2]\, d\theta$

   (b) $\displaystyle\int_0^{\pi/2} 2\, d\theta$

31. $a^2[\sqrt{3} - \pi/3]$ **33.** $4 - \pi$

35. (a) the ellipse $4(x - 1)^2 + 9(y + 1)^2 = 36$ oriented from $(4, -1)$ counterclockwise to $(-2, -1)$
   (b) $0, -2/9, y = 1$

37. (a) $y = -\ln x$; oriented from $(1, 0)$ to $(e^{-1}, 1)$
   (b) $-2, 4, y = -2(x - \frac{1}{2}) + \ln 2$

39. $\ln(2 + \sqrt{3})$

41. 4 **43.** $\sqrt{2}(e^{2\pi} - 1)$

45. (a) $\left(0, \dfrac{\pi}{6} + \sqrt{3}\right)$ and $\left(0, \dfrac{5\pi}{6} - \sqrt{3}\right)$ (b) $\left(-1, \dfrac{\pi}{2}\right)$

47. $t = 1$

49. slope $= 1$, $\psi = 3\pi/4$

▶ Exercise Set 14.1 (Page 746)

1.

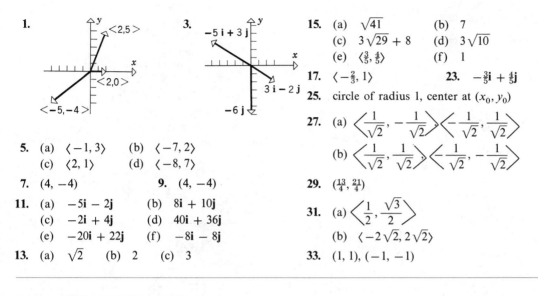

3.

15. (a) $\sqrt{41}$ (b) 7
    (c) $3\sqrt{29} + 8$ (d) $3\sqrt{10}$
    (e) $\langle \frac{3}{5}, \frac{4}{5} \rangle$ (f) 1

17. $\langle -\frac{2}{3}, 1 \rangle$ 23. $-\frac{3}{5}\mathbf{i} + \frac{4}{5}\mathbf{j}$

25. circle of radius 1, center at (x_0, y_0)

27. (a) $\left\langle \frac{1}{\sqrt{2}}, -\frac{1}{\sqrt{2}} \right\rangle \left\langle -\frac{1}{\sqrt{2}}, \frac{1}{\sqrt{2}} \right\rangle$
    (b) $\left\langle \frac{1}{\sqrt{2}}, \frac{1}{\sqrt{2}} \right\rangle \left\langle -\frac{1}{\sqrt{2}}, -\frac{1}{\sqrt{2}} \right\rangle$

5. (a) $\langle -1, 3 \rangle$ (b) $\langle -7, 2 \rangle$
   (c) $\langle 2, 1 \rangle$ (d) $\langle -8, 7 \rangle$

7. $(4, -4)$ 9. $(4, -4)$

29. $(\frac{13}{4}, \frac{21}{4})$

11. (a) $-5\mathbf{i} - 2\mathbf{j}$ (b) $8\mathbf{i} + 10\mathbf{j}$
   (c) $-2\mathbf{i} + 4\mathbf{j}$ (d) $40\mathbf{i} + 36\mathbf{j}$
   (e) $-20\mathbf{i} + 22\mathbf{j}$ (f) $-8\mathbf{i} - 8\mathbf{j}$

31. (a) $\left\langle \frac{1}{2}, \frac{\sqrt{3}}{2} \right\rangle$
    (b) $\langle -2\sqrt{2}, 2\sqrt{2} \rangle$

13. (a) $\sqrt{2}$ (b) 2   (c) 3

33. $(1, 1), (-1, -1)$

▶ Exercise Set 14.2 (Page 753)

1.

3.

11. $5\mathbf{i} + (1 - 2t)\mathbf{j}$ 13. $\left\langle -\frac{1}{t^2}, \sec^2 t \right\rangle$

15.

17.

5.

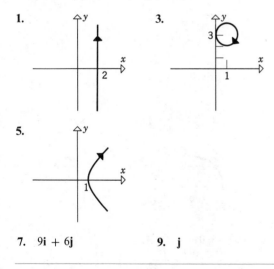

19. $3t\mathbf{i} + 2t^2\mathbf{j} + \mathbf{C}$ 21. $\langle 0, -\frac{2}{3} \rangle$

23. $\frac{52}{3}\mathbf{i} + 4\mathbf{j}$

25. $\langle (t - 1)e^t, t(\ln t - 1) \rangle + \mathbf{C}$

27. $4\mathbf{i} + 8(4w + 1)\mathbf{j}$ 29. $2we^{w^2}\mathbf{i} - 8we^{-w^2}\mathbf{j}$

31. $(1 + \sin t)\mathbf{i} - (\cos t)\mathbf{j}$ 33. $(t^4 + 2)\mathbf{i} - (t^2 + 4)\mathbf{j}$

7. $9\mathbf{i} + 6\mathbf{j}$ 9. $\mathbf{j}$

▶ Exercise Set 14.3 (Page 761)

1.

3.

5.

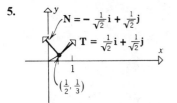

7.

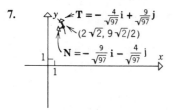

9. $x = \frac{3}{5}s - 2$
$y = \frac{4}{5}s + 3$

11. $x = 3 + \cos x$
$y = 2 + \sin s$, $0 \le s \le 2\pi$

13. $x = \frac{1}{3}[(3s + 1)^{2/3} - 1]^{3/2}$
$y = \frac{1}{2}[(3s + 1)^{2/3} - 1]$ $s \ge 0$

15. $x = \left(\frac{s}{\sqrt{2}} + 1\right) \cos \left[\ln \left(\frac{s}{\sqrt{2}} + 1\right)\right]$

$y = \left(\frac{s}{\sqrt{2}} + 1\right) \sin \left[\ln \left(\frac{s}{\sqrt{2}} + 1\right)\right]$

where $0 \le s \le \sqrt{2}(e^{\pi/2} - 1)$

17. (a) $T = \frac{3}{5}i + \frac{4}{5}j$
$N = -\frac{4}{5}i + \frac{3}{5}j$

(b)

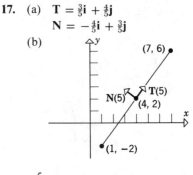

19. $\dfrac{5}{\sqrt{26}}$

▶ Exercise Set 14.4 (Page 769)

1. $\frac{96}{125}$

3. $\dfrac{6}{5\sqrt{10}}$

5. $-\frac{1}{4}$

7. -1

9. $\dfrac{1}{\sqrt{2}}$

11. $\dfrac{4}{5\sqrt{5}}$

13.

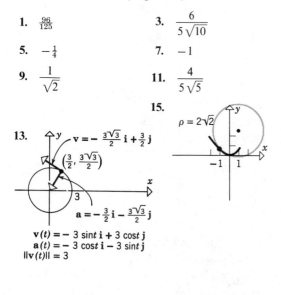

$v(t) = -3 \sin t\, i + 3 \cos t\, j$
$a(t) = -3 \cos t\, i - 3 \sin t\, j$
$\|v(t)\| = 3$

15.

17.

19.

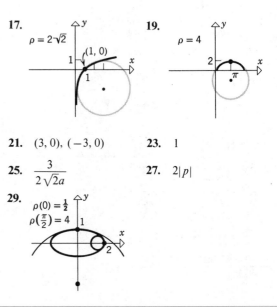

21. $(3, 0), (-3, 0)$

23. 1

25. $\dfrac{3}{2\sqrt{2}a}$

27. $2|p|$

29.

▶ Exercise Set 14.5 (Page 779)

1.

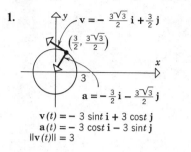

$v(t) = -3\sin t\, i + 3\cos t\, j$
$a(t) = -3\cos t\, i - 3\sin t\, j$
$\|v(t)\| = 3$

3.

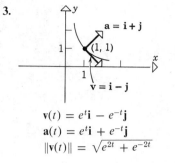

$v(t) = e^t i - e^{-t} j$
$a(t) = e^t i + e^{-t} j$
$\|v(t)\| = \sqrt{e^{2t} + e^{-2t}}$

5.

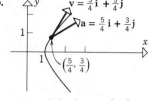

$v(t) = \sinh t\, i + \cosh t\, j$
$a(t) = \cosh t\, i + \sinh t\, j$
$\|v(t)\| = \sqrt{\sinh^2 t + \cosh^2 t}$

7. $r(t) = -16t^2 j$
    $v(t) = -32t j$

9. $r(t) = (t + \cos t - 1)i + (\sin t - t + 1)j$
    $v(t) = (1 - \sin t)i + (\cos t - 1)j$

11. (a) $x = 160t$
         $y = 160\sqrt{3}t - 16t^2,\ t \geq 0$
    (b) 1200 ft
    (c) $1600\sqrt{3}$ ft
    (d) 320 ft/sec

13. $40\sqrt{3}$ ft

15. 800 ft/sec

17. $15°,\ 75°$

19. $|a_T| = 0,\ |a_N| = 2$

21. $|a_T| = 0,\ |a_N| = \sqrt{2}$

23. $|a_T| = 2\sqrt{5},\ |a_N| = 2\sqrt{5}$

29. 9×10^{10} kilometers/sec^2

▶ Chapter 14 Supplementary Exercises (Page 780)

1. (a) $3i - 4j,\ 5$ (b) $-i + 4j,\ \sqrt{17}$

3. $-3i + 4j$ 5. $\frac{3}{5}(4i + 3j)$

7. $-6i + 6\sqrt{3}j$ 9. $i + \frac{1}{2}j$

11. $F_3 = -(F_1 + F_2) = i + 5j$

13. (a) $\frac{1}{2}(t + 4)^{-1/2}i + 2j,\ -\frac{1}{4}(t + 4)^{-3/2}i$
    (b) The graph is the part of the parabola $y = 2(x^2 - 4)$ starting at $(0, -8)$ and directed up and toward the right. At $(1, -6)[t = -3]$, $r' = \frac{1}{2}i + 2j$, $r'' = -\frac{1}{4}i$; at $(2, 0)[t = 0]$, $r' = \frac{1}{4}i + 2j$, $r'' = -1/32i$

15. (a) $\langle 6t^2, 3t^2 \rangle,\ \langle 12t, 6t \rangle$
    (b) The graph is the straight line $x - 2y + 3 = 0$, directed to the right and up; at $(-1, 1)[t = 0]$, $r' = r'' = \langle 0, 0 \rangle$; at $(-5/4, 7/8)[t = -1/2]$, $r' = \langle 3/2, 3/4 \rangle$, $r'' = \langle -6, -3 \rangle$.

17. (a) $t(ki + mj) + C$
    (b) $\langle 4, 4 \rangle$
    (c) 2
    (d) $\sqrt{t^2 + 3}\,i + \ln(\sin t)j + C$

19. (a) $1 + 1/t^2$
    (b) $\langle \sqrt{s^2 + 4},\ 2\ln[\frac{1}{2}(s + \sqrt{s^2 + 4})]\rangle$

21. (a) $(2i - j)/\sqrt{5}$ (b) $(i + 2j)/\sqrt{5}$
    (c) $6/5^{3/2}$

23. (a) j (b) $-i$ (c) -2

25. $-2/25$

27. -1

29. $(x - 1)^2 + (y - \frac{1}{2})^2 = \frac{1}{4}$; at $(1, 0)$, $dy/dx = 0$, $d^2y/dx^2 = 2$

31. $(1/w)\langle e^t, 4e^{2t} \rangle = \langle 1, 4w \rangle$

33. $|a_T| = 1,\ |a_N| = t$

35. (a) $\langle -1, 1 \rangle$, $\langle 1, 1 \rangle$, $\sqrt{2}$

  (b) $(-\mathbf{i} + \mathbf{j})/\sqrt{2}$, $(-\mathbf{i} - \mathbf{j})/\sqrt{2}$, $-1/\sqrt{2}$

  (c) 0, $-\sqrt{2}$

  (d) Trajectory: the branch of the hyperbola $xy = 1$ in the first quadrant, traced so y increases with t.

  (e) $C(2, 2)$

37. $(2t - \sin t)\mathbf{i} + (e^{2t} - 1)\mathbf{j}$

39. (a) $a_T = 0$, $|a_N| = 32$

  (b) $|a_T| = 64/\sqrt{5}$, $|a_N| = 32/\sqrt{5}$

41. $a_T = 0$, $a_N = 8\pi^2 (\text{m/sec}^2)$

43. (a) $\sqrt{5}\,e^t$ (b) $2\sqrt{5}$

▶ Exercise Set 15.1 (Page 790)

1. (a) $\sqrt{14}$; $(1, \frac{1}{2}, \frac{3}{2})$ (b) $\sqrt{11}$; $(\frac{9}{2}, \frac{3}{2}, \frac{9}{2})$

  (c) $\sqrt{30}$; $(\frac{1}{2}, -\frac{1}{2}, 4)$ (d) $\sqrt{42}$; $(\frac{3}{2}, 1, -\frac{5}{2})$

3. $(-6, 2, 1)$, $(-6, 2, -2)$, $(-6, 1, -2)$, $(4, 2, 1)$, $(4, 1, 1)$, $(4, 1, -2)$, $(4, 2, -2)$, $(-6, 1, 1)$

5. (b) $(2, 1, 6)$

  (c) 49

7. distance to x-axis is $\sqrt{y_0^2 + z_0^2}$.

  distance to y-axis is $\sqrt{x_0^2 + z_0^2}$.

9. $(x + 2)^2 + (y - 4)^2 + (z + 1)^2 = 36$

11. $x^2 + (y - 1)^2 + z^2 = 9$

13. (a) $(x - 2)^2 + (y + 1)^2 + (z + 3)^2 = 9$

  (b) $(x - 2)^2 + (y + 1)^2 + (z + 3)^2 = 1$

  (c) $(x - 2)^2 + (y + 1)^2 + (z + 3)^2 = 4$

15. $(x + \frac{1}{2})^2 + (y - 2)^2 + (z - 2)^2 = \frac{5}{4}$

17. $(x - 3)^2 + (y + 2)^2 + (z - 4)^2 = 41$

19. $x^2 + y^2 + z^2 = 30 \pm 2\sqrt{29}$

21. sphere, center $(-5, -2, -1)$, radius 7

23. sphere, center $(\frac{1}{2}, \frac{3}{4}, -\frac{5}{4})$, radius $\frac{3}{4}\sqrt{6}$

25. no graph

27.

(a)
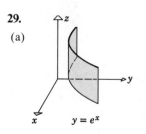

(b)
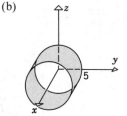

(c)
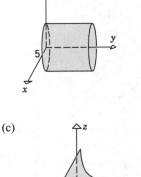

29.

(a)
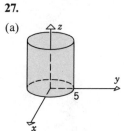
$y = e^x$

(b)
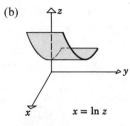
$x = \ln z$

(c)
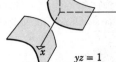
$yz = 1$

▶ Exercise Set 15.2 (Page 796)

1. (a) $\langle -3, 6, 1 \rangle$ (b) $\langle 1, -3, -5 \rangle$
(c) $\langle -1, 6, 1 \rangle$ (d) $\langle 5, 0, 0 \rangle$

3. $6\mathbf{i} - 8\mathbf{j} - 2\mathbf{k}$

5. (a) $\sqrt{3}$ (b) $\sqrt{14}$ (c) $\sqrt{21}$ (d) 3

7. $x = 5 - 3t$ **9.** $x = -t$
  $y = -2 + 6t$ $y = 6t$
  $z = 1 + t$ $z = t$

11. same as for Exercise 7 with $0 \le t \le 1$

13. same as for Exercise 9 with $0 \le t \le 1$

15. $x = -1 + 3t$ **17.** $x = -2 + 2t$
  $y = 2 - 4t$ $y = -t$
  $z = 4 + t$ $z = 5 + 2t$

19. $x = 3 + t$
  $y = 7$
  $z = 0$

21. (a) $(-2, 10, 0)$
  (b) $(-2, 0, -5)$
  (c) does not intersect yz-plane

23. $x = x_1 + at$
  $y = y_1 + bt$
  $z = z_1 + ct$

25. $(1, -1, 2)$

29. (a) no   (b) yes

31. $\left(1, \frac{14}{3}, -\frac{5}{3}\right)$

▶ Exercise Set 15.3 (Page 808)

1. (a) -10 (b) -3 (c) 0   (d) -20

3. (a) obtuse   (b) acute   (c) obtuse
  (d) orthogonal

5. (a) $\frac{14}{13}\mathbf{i} + \frac{21}{13}\mathbf{j}$ (b) $\langle 2, 6 \rangle$
  (c) $-\frac{11}{13}\mathbf{i} + \mathbf{j} + \frac{55}{13}\mathbf{k}$ (d) $\langle -\frac{32}{89}, -\frac{12}{89}, \frac{73}{89} \rangle$

9. (a) 6   (b) 36   (c) $24\sqrt{5}$ (d) $24\sqrt{5}$

11. $\dfrac{1}{5\sqrt{2}}, \dfrac{4}{\sqrt{65}}, \dfrac{9}{\sqrt{130}}$

13. The right angle is at vertex B.

15. (a) $-\dfrac{3}{4}$ (b) $\dfrac{1}{7}$ (c) $\dfrac{48 \pm 25\sqrt{3}}{11}$
  (d) $\dfrac{4}{3}$

19. (a) $\dfrac{2}{5}$ (b) $\dfrac{2}{\sqrt{5}}$ (c) $\dfrac{2}{\sqrt{5}}$

21. -12 foot-pounds

▶ Exercise Set 15.4 (Page 818)

1. $\langle 7, 10, 9 \rangle$

3. $\langle -4, -6, -3 \rangle$

5. (a) $\langle -20, -67, -9 \rangle$ (b) $\langle -78, 52, -26 \rangle$
  (c) $\langle 24, 0, -16 \rangle$ (d) $\langle -12, -22, -8 \rangle$
  (e) $\langle 0, -56, -392 \rangle$ (f) $\langle 0, 56, 392 \rangle$

9. (a) $\dfrac{\sqrt{374}}{2}$ (b) $9\sqrt{13}$

11. ambiguous, needs parentheses

13. 80

15. 1

17. (a) 16   (b) 45

19. (a) 9   (b) $\sqrt{122}$ (c) $\sin^{-1}\left(\frac{9}{14}\right)$

29. (a) $\frac{2}{3}$ (b) $\frac{1}{2}$

▶ Exercise Set 15.5 (Page 829)

1.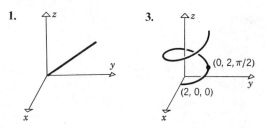

3. (0, 2, π/2), (2, 0, 0)

5.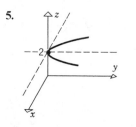

7. $T = -\dfrac{4}{\sqrt{17}}i + \dfrac{1}{\sqrt{17}}k$
$N = -j$
$\kappa = \dfrac{4}{17}$

9. $T = i$
$N = j$
$\kappa = 1$

11. $T = -\dfrac{3}{\sqrt{10}}i + \dfrac{1}{\sqrt{10}}k$
$N = -j$
$\kappa = \dfrac{2}{5}$

13. $T = \dfrac{1}{\sqrt{5}}j + \dfrac{2}{\sqrt{5}}k$
$N = -\dfrac{2}{\sqrt{5}}j + \dfrac{1}{\sqrt{5}}k$
$\kappa = \dfrac{2}{5\sqrt{5}}$

15. $v = i + j + k$
$\|v\| = \sqrt{3}$
$a = j + 2k$

17. $v = -\sqrt{2}i + \sqrt{2}j + k$
$\|v\| = \sqrt{5}$
$a = -\sqrt{2}i - \sqrt{2}j$

19. $v = e^{\pi/2}i - e^{\pi/2}j + k$
$\|v\| = \sqrt{2e^{\pi} + 1}$
$a = -2e^{\pi/2}j$

21. $\sqrt{14}$

23. 28

25. $e - e^{-1}$

27. $\frac{1}{3}t^3 i - t^2 j + \ln|t|k + C$

29. $\frac{1}{2}(e^2 - 1)i + (1 - e^{-1})j + \frac{1}{2}k$

31. (a) $7t^6$
(b) $12t(\sec t)\tan t + 12\sec t - \dfrac{\sin t}{t} - (\cos t)\ln t$

33. (a) $\dfrac{1 + 2t^2}{t}$ (b) $\dfrac{1 + 2t^2}{t}$ (c) $8 + \ln 3$

43. (a) $r(t) = -16t^2 k$
$v(t) = -32t k$
$\|v(t)\| = 32t$
(b) $r(t) = i + 2j + (1 - 16t^2)k$
$v(t) = -32t k$
$\|v(t)\| = 32t$
(c) $r(t) = (3 + t)i - j + (4 + t - 16t^2)k$
$v(t) = i + (1 - 32t)k$
$\|v(t)\| = \sqrt{1 + (1 - 32t)^2}$

▶ Exercise Set 15.6 (Page 837)

1. $x + 4y + 2z = 28$

3. $z = 0$

5. (a) $2y - z = 1$
(b) $x + 9y - 5z = 16$

7. (a) yes   (b) no

9. (a) yes   (b) no

11. (a) $35°$ (b) $79°$

13. (a) $z = 0$ (b) $y = 0$ (c) $x = 0$

15. $4x - 2y + 7z = 0$

17. $4x - 13y + 21z = -14$

19. $x = 5 - 2t$
$y = 5t$
$z = -2 + 11t$

21. $x + y - 3z = 6$

23. $7x + y + 9z = 25$

25. $2x + 4y + 8z = 29$

29. (a) $x = -\frac{11}{7} - 23t$ (b) $x = -5t$
$y = -\frac{12}{7} + t$ $y = -3t$
$z = -7t$ $z = 0$

31. $\frac{5}{3}$

33. $\frac{137}{21}$

35. $\dfrac{5}{3\sqrt{6}}$

37. $\dfrac{25}{\sqrt{126}}$

39. $\dfrac{95}{\sqrt{1817}}$

► Exercise Set 15.7 (Page 845)

1. $x = 1, y = 2$

3. $x_1 = -\frac{7}{32}, x_2 = \frac{5}{16}$

5. $x = 2, y = -1, z = 3$

7. $x_1 = \frac{26}{21}, x_2 = \frac{25}{21}, x_3 = \frac{5}{7}$

9. $x' = x \cos\theta + y \sin\theta$
$y' = -x \sin\theta + y \cos\theta$

► Exercise Set 15.8 (Page 856)

1.

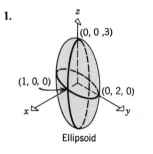

Ellipsoid

3.

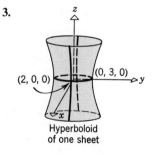

Hyperboloid
of one sheet

5.

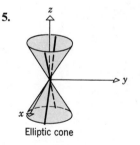

Elliptic cone

7.

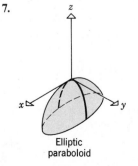

Elliptic
paraboloid

9.

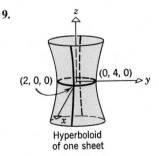

Hyperboloid
of one sheet

11.

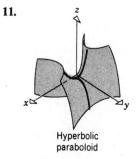

Hyperbolic
paraboloid

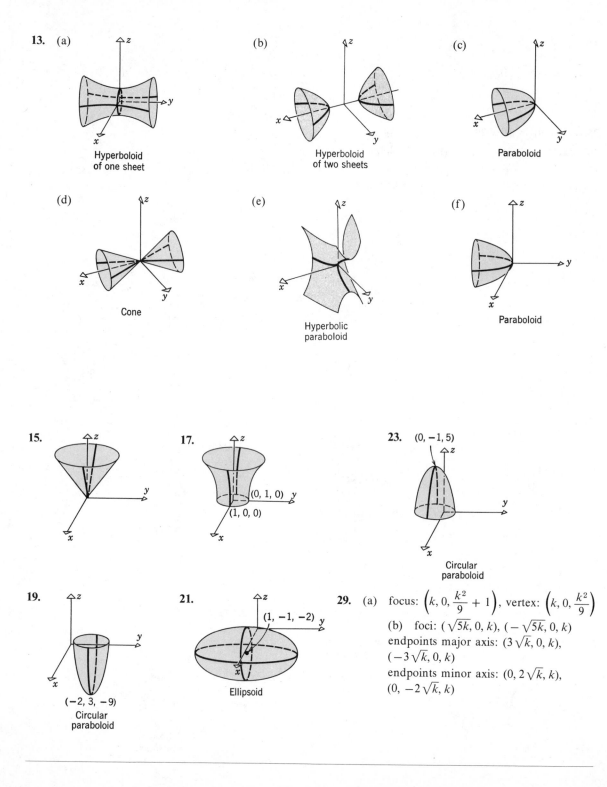

13. (a)

Hyperboloid
of one sheet

(b)

Hyperboloid
of two sheets

(c)

Paraboloid

(d)

Cone

(e)

Hyperbolic
paraboloid

(f)

Paraboloid

15.

17. $(0, 1, 0)$
$(1, 0, 0)$

23. $(0, -1, 5)$

Circular
paraboloid

19. $(-2, 3, -9)$
Circular
paraboloid

21. $(1, -1, -2)$
Ellipsoid

29. (a) focus: $\left(k, 0, \frac{k^2}{9} + 1\right)$, vertex: $\left(k, 0, \frac{k^2}{9}\right)$

(b) foci: $(\sqrt{5k}, 0, k)$, $(-\sqrt{5k}, 0, k)$
endpoints major axis: $(3\sqrt{k}, 0, k)$,
$(-3\sqrt{k}, 0, k)$
endpoints minor axis: $(0, 2\sqrt{k}, k)$,
$(0, -2\sqrt{k}, k)$

► Exercise Set 15.9 (Page 861)

1. (a) $(8, \pi/6, -4)$ (b) $(5\sqrt{2}, 3\pi/4, 6)$
   (c) $(2, \pi/2, 0)$ (d) $(8, 5\pi/3, 6)$
   (e) $(2, 7\pi/4, 1)$ (f) $(0, 0, 1)$

3. (a) $(2\sqrt{2}, \pi/3, 3\pi/4)$ (b) $(2, 7\pi/4, \pi/4)$
   (c) $(6, \pi/2, \pi/3)$ (d) $(10, 5\pi/6, \pi/2)$
   (e) $(8\sqrt{2}, \pi/4, \pi/6)$ (f) $(2\sqrt{2}, 5\pi/3, 3\pi/4)$

5. (a) $(2\sqrt{3}, \pi/6, \pi/6)$ (b) $(\sqrt{2}, \pi/4, 3\pi/4)$
   (c) $(2, 3\pi/4, \pi/2)$ (d) $(4\sqrt{3}, 1, 2\pi/3)$
   (e) $(4\sqrt{2}, 5\pi/6, \pi/4)$ (f) $(2\sqrt{2}, 0, 3\pi/4)$

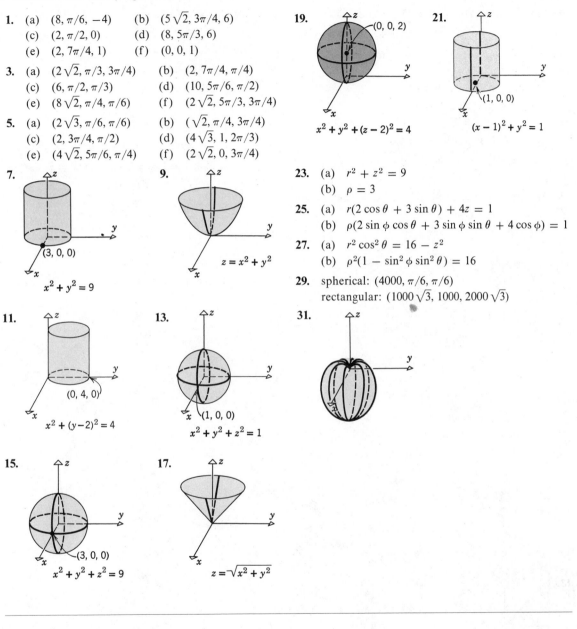

7.

$x^2 + y^2 = 9$

9.

$z = x^2 + y^2$

11.

$(0, 4, 0)$

$x^2 + (y-2)^2 = 4$

13.

$(1, 0, 0)$

$x^2 + y^2 + z^2 = 1$

15.

$(3, 0, 0)$

$x^2 + y^2 + z^2 = 9$

17.

$z = \sqrt{x^2 + y^2}$

19.

$(0, 0, 2)$

$x^2 + y^2 + (z - 2)^2 = 4$

21.

$(1, 0, 0)$

$(x - 1)^2 + y^2 = 1$

23. (a) $r^2 + z^2 = 9$
    (b) $\rho = 3$

25. (a) $r(2\cos\theta + 3\sin\theta) + 4z = 1$
    (b) $\rho(2\sin\phi\cos\theta + 3\sin\phi\sin\theta + 4\cos\phi) = 1$

27. (a) $r^2\cos^2\theta = 16 - z^2$
    (b) $\rho^2(1 - \sin^2\phi\sin^2\theta) = 16$

29. spherical: $(4000, \pi/6, \pi/6)$
    rectangular: $(1000\sqrt{3}, 1000, 2000\sqrt{3})$

31.

► Chapter 15 Supplementary Exercises (Page 863)

1. (a) $\sqrt{6}$ (b) -3
   (c) $5(\mathbf{i} - \mathbf{j} - \mathbf{k})$ (d) $5(-\mathbf{i} + \mathbf{j} + \mathbf{k})$
   (e) $5\sqrt{3}/2$ (f) $\langle -1, 8, -9 \rangle$

3. (a) $2/3$ (b) $2/5$ (c) $\cos^{-1}(-2/15)$
   (d) $3/5, -4/5, 0$
5. Each side reduces to $2\mathbf{i} - 2\mathbf{j} + \mathbf{k}$.

7. $\langle -3/\sqrt{2}, 0, 3/\sqrt{2} \rangle$

9. (a) $-\frac{1}{2}\mathbf{v}$,   (b) $\frac{1}{2}(3\mathbf{i} + 5\mathbf{j} - 4\mathbf{k})$

11. (a) $\|\mathbf{u}\| = \|\mathbf{v}\|$ **13.** $\pm(5\mathbf{i} + 7\mathbf{j} - \mathbf{k})/(5\sqrt{3})$

15. $x - y - z + 4 = 0$ **17.** $x - y - z = -1$

19. (a) $k = -2, l = 12$

   (b) yes, at $(7, 0, 0)$

   (c) $(136/13, -15/13, -60/13)$

21. (a) $x = 2t + 1, y = 3t - 1, z = -3t + 2$

   (b) $x = 1, y = 5t - 3, z = -7t + 4$

23. (a) $x = t, y = -t + 2, z = t - 1$

   (b) $60°$

25. (a) the region inside the elliptic paraboloid $z = 4x^2 + 9y^2$

   (b) the point $(0, 0, 0)$

27. elliptic cone   **29.** ellipsoid

31. hyperboloid of two sheets   **33.** 13 ft-lb

35. The parabola $y = x^2 + 1$ in the plane $z = 1$, traced so x increases with t.

37. $\mathbf{v} = a\mathbf{i}$, $\|\mathbf{v}\| = a$, $\mathbf{a} = -a(\mathbf{j} + \mathbf{k})$, $\mathbf{T} = \mathbf{i}$, $\mathbf{N} = -(\mathbf{j} + \mathbf{k})/\sqrt{2}$, $\kappa = \sqrt{2}/a$

39. $4\pi\sqrt{37}$

41. $x = 1 - t, y = 1 + 2t, z = 1$

43. (a) $\langle 9, 9, -9 \rangle$ (b) $\langle 3 + 4t^3, -4t, 6t^2 \rangle$

45. (a) $x = 3, y = 1, z = 2$ (b) $x = \frac{1}{2}, y = -1$

47. (a) $(1, 1, 1)$ (b) $(\sqrt{3}, \pi/4, \tan^{-1}\sqrt{2})$

49. (a) $z = x^2 - y^2$ (b) $xz = 1$

▶ Exercise Set 16.1 (Page 875)

1. (a) 5   (b) 3

   (c) 1   (d) -2

   (e) $9a^3 + 1$ (f) $a^3b^2 - a^2b^3 + 1$

3. (a) $x^2 - y^2 + 3$ (b) $3x^3y^4 + 3$

5. $x^3 e^{x^3(3y+1)}$

7. (a) $t^2 + 3t^{10}$ (b) 0   (c) 3076

9.

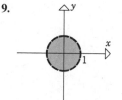

11.

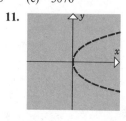

13.

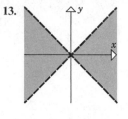

15.

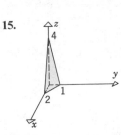

17.

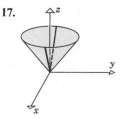

19.

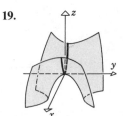

21.

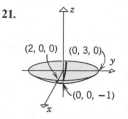

23.

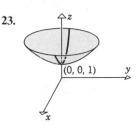

25.

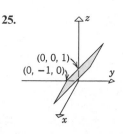

27.

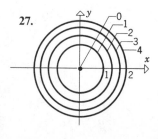

29.

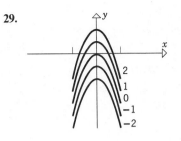

31.

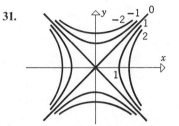

33.

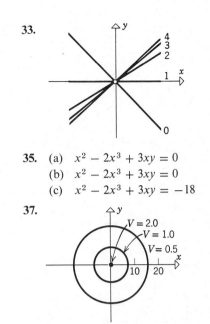

35. (a) $x^2 - 2x^3 + 3xy = 0$
(b) $x^2 - 2x^3 + 3xy = 0$
(c) $x^2 - 2x^3 + 3xy = -18$

37.

► Exercise Set 16.2 (Page 882)

1. $\dfrac{\partial z}{\partial x} = 9x^2y^2$, $\dfrac{\partial z}{\partial y} = 6x^3y$

3. $\dfrac{\partial z}{\partial x} = 8xy^3 e^{x^2y^3}$, $\dfrac{\partial z}{\partial y} = 12x^2y^2 e^{x^2y^3}$

5. $\dfrac{\partial z}{\partial x} = \dfrac{x^3}{y^{3/5} + x} + 3x^2 \ln(1 + xy^{-3/5})$,

$\dfrac{\partial z}{\partial y} = -\dfrac{3x^4}{5(y^{8/5} + xy)}$

7. $f_x(x, y) = \frac{3}{2}x^2y(5x^2 - 7)(3x^5y - 7x^3y)^{-1/2}$,
$f_y(x, y) = \frac{1}{2}x^3(3x^2 - 7)(3x^5y - 7x^3y)^{-1/2}$

9. $f_x(x, y) = \dfrac{y^{-1/2}}{y^2 + x^2}$,

$f_y(x, y) = -\dfrac{xy^{-3/2}}{y^2 + x^2} - \dfrac{3}{2}y^{-5/2}\tan^{-1}\left(\dfrac{x}{y}\right)$

11. $f_x(x, y) = -\frac{4}{3}y^2 \sec^2 x(y^2 \tan x)^{-7/3}$,
$f_y(x, y) = -\frac{8}{3}y \tan x(y^2 \tan x)^{-7/3}$

13. (a) -6 (b) -21

15. (a) $\dfrac{1}{\sqrt{17}}$ (b) $\dfrac{8}{\sqrt{17}}$

17. $\dfrac{\partial z}{\partial x} = -\dfrac{x}{z}$, $\dfrac{\partial z}{\partial y} = -\dfrac{y}{z}$

19. $\dfrac{\partial z}{\partial x} = -\dfrac{2x + yz^2 \cos(xyz)}{xyz \cos(xyz) + \sin(xyz)}$

$\dfrac{\partial z}{\partial y} = -\dfrac{xz^2 \cos(xyz)}{xyz \cos(xyz) + \sin(xyz)}$

21. $f_{xx} = 8$,
$f_{yy} = -96xy^2 + 140y^3$,
$f_{xy} = f_{yx} = -32y^3$

23. $f_{xx} = e^x \cos y$,
$f_{yy} = -e^x \cos y$,
$f_{xy} = f_{yx} = -e^x \sin y$

25. $f_{xx} = -\dfrac{16}{(4x - 5y)^2}$,

$f_{yy} = -\dfrac{25}{(4x - 5y)^2}$,

$f_{xy} = f_{yx} = \dfrac{20}{(4x - 5y)^2}$

27. (a) $30xy^4 - 4$
(b) $60x^2y^3$
(c) $60x^3y^2$

29. (a) -30 (b) -125 (c) 150

31. (a) $\dfrac{\partial^3 f}{\partial x^3}$ (b) $\dfrac{\partial^3 f}{\partial y^2 \partial x}$

     (c) $\dfrac{\partial^4 f}{\partial x^2 \partial y^2}$ (d) $\dfrac{\partial^4 f}{\partial y^3 \partial x}$

35. 6

37. (a) 8   (b) -2

39. (a) $\dfrac{\partial V}{\partial r} = 2\pi r h$ (b) $\dfrac{\partial V}{\partial h} = \pi r^2$

     (c) 48π (d) 64π

41. (a) $\frac{1}{5}$ (b) $-\frac{25}{8}$

43. $\dfrac{\partial z}{\partial x} = -1 - \dfrac{\cos(x - y)}{\cos(x + z)}$,

$\dfrac{\partial z}{\partial y} = \dfrac{\cos(x - y)}{\cos(x + z)}$,

$\dfrac{\partial^2 z}{\partial x \partial y} = -\dfrac{\cos^2(x + z)\sin(x - y) + \cos^2(x - y)\sin(x + z) + \cos(x - y)\sin(x + z)\cos(x + z)}{\cos^3(x + z)}$

45. (a) 4 degrees per centimeter
     (b) 8 degrees per centimeter

▶ Exercise Set 16.3 (Page 894)

1. 0.872

3. $\frac{7}{2}$

5.

7.

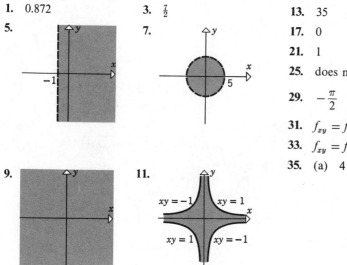

9.

11.

13. 35   **15.** -8

17. 0   **19.** does not exist

21. 1   **23.** 0

25. does not exist

29. $-\dfrac{\pi}{2}$

31. $f_{xy} = f_{yx} = 12x^2 + 6x$

33. $f_{xy} = f_{yx} = -6xy^2 \sin(x^2 + y^3)$

35. (a) 4   (b) 5

▶ Exercise Set 16.4 (Page 901)

1. $42t^{13}$

3. $\dfrac{3\sin(1/t)}{t^2}$

5. $-\frac{10}{3}t^{7/3}e^{1-t^{10/3}}$

7. $\dfrac{\partial z}{\partial u} = 24u^2v^2 - 16uv^3 - 2v + 3$,

     $\dfrac{\partial z}{\partial v} = 16u^3v - 24u^2v^2 - 2u - 3$

9. $\dfrac{\partial z}{\partial u} = -\dfrac{2 \sin u}{3 \sin v}$, $\dfrac{\partial z}{\partial v} = -\dfrac{2 \cos u \cos v}{3 \sin^2 v}$

11. $\dfrac{\partial z}{\partial u} = e^u$, $\dfrac{\partial z}{\partial v} = 0$

13. $\dfrac{\partial z}{\partial u} = \dfrac{2e^{2u}}{1 + e^{4u}}$, $\dfrac{\partial z}{\partial v} = 0$

15. $\dfrac{\partial T}{\partial r} = 3r^2 \sin \theta \cos^2 \theta - 4r^3 \sin^3 \theta \cos \theta$,

$\dfrac{\partial T}{\partial \theta} = -2r^3 \sin^2 \theta \cos \theta + r^4 \sin^4 \theta +$

$\qquad\qquad r^3 \cos^3 \theta - 3r^4 \cos^2 \theta \sin^2 \theta$

17. $\dfrac{\partial t}{\partial x} = \dfrac{x^2 + y^2}{4x^2 y^3}$, $\dfrac{\partial t}{\partial y} = \dfrac{y^2 - 3x^2}{4xy^4}$

19. $-\pi$

21. $\dfrac{\partial z}{\partial r}\bigg|_{r=2,\ \theta=\pi/6} = \sqrt{3}e^{\sqrt{3}}$,

$\dfrac{\partial z}{\partial \theta}\bigg|_{r=2,\ \theta=\pi/6} = (2 - 4\sqrt{3})e^{\sqrt{3}}$

23. $\dfrac{x^2 - y^2}{2xy - y^2}$ **25.** $\dfrac{2\sqrt{xy} - y}{x - 6\sqrt{xy}}$

27. -39 mi/hr

29. $-\frac{7}{36}\sqrt{3}$ radians/sec

35. $\dfrac{\partial z}{\partial x} = \dfrac{x}{\sqrt{x^2 + y^2}}\dfrac{\partial z}{\partial r} - \dfrac{y}{x^2 + y^2}\dfrac{\partial z}{\partial \theta}$,

$\dfrac{\partial z}{\partial y} = \dfrac{y}{\sqrt{x^2 + y^2}}\dfrac{\partial z}{\partial r} + \dfrac{x}{x^2 + y^2}\dfrac{\partial z}{\partial \theta}$

▶ Exercise Set 16.5 (Page 910)

1. $4\mathbf{i} - 8\mathbf{j}$

3. $\dfrac{x}{x^2 + y^2}\mathbf{i} + \dfrac{y}{x^2 + y^2}\mathbf{j}$

5. $-36\mathbf{i} - 12\mathbf{j}$

7. $4\mathbf{i} + 4\mathbf{j}$

9. $6\sqrt{2}$

11. $-\dfrac{3}{\sqrt{10}}$

13. 0

15. $-8\sqrt{2}$

17. $\dfrac{1}{2\sqrt{2}}$

19. $\dfrac{1}{2} + \dfrac{\sqrt{3}}{8}$

21. $2\sqrt{2}$

23.

25.

27. $\dfrac{3}{\sqrt{13}}\mathbf{i} - \dfrac{2}{\sqrt{13}}\mathbf{j}$, $4\sqrt{13}$

29. $\frac{4}{5}\mathbf{i} - \frac{3}{5}\mathbf{j}$, 1

31. $-\dfrac{1}{\sqrt{10}}\mathbf{i} - \dfrac{3}{\sqrt{10}}\mathbf{j}$, $-2\sqrt{10}$

33. $\dfrac{3}{\sqrt{10}}\mathbf{i} - \dfrac{1}{\sqrt{10}}\mathbf{j}$, $-\sqrt{5}$

35. $\dfrac{1}{\sqrt{5}}$

37. $-\frac{3}{2}e$

39. $-\dfrac{4}{\sqrt{17}}\mathbf{i} + \dfrac{1}{\sqrt{17}}\mathbf{j}$

41. (a) 5 (b) 10 (c) $-5\sqrt{5}$

43. $\dfrac{8}{\sqrt{29}}$

45. (a) $2e^{-\pi/2}\mathbf{i}$

▶ Exercise Set 16.6 (Page 919)

1. $48x - 14y - z = 64$, $\quad x = 1 + 48t$
$\qquad\qquad\qquad\qquad\quad y = -2 - 14t$
$\qquad\qquad\qquad\qquad\quad z = 12 - t$

3. $x - y - z = 0$, $\quad x = 1 + t$
$\qquad\qquad\qquad\quad y = -t$
$\qquad\qquad\qquad\quad z = 1 - t$

5. $3y - z = -1$, $x = \pi/6$
$$y = 3t$$
$$z = 1 - t$$

7. $3x - 4z = -25$, $x = -3 + \frac{3}{4}t$
$$y = 0$$
$$z = 4 - t$$

9. $7\,dx - 2\,dy$

11. $\dfrac{y}{1 + x^2y^2}\,dx + \dfrac{x}{1 + x^2y^2}\,dy$

13. 0.10 **15.** 0.03

17. (a) all points on the x-axis and y-axis
      (b) $(0, -2, -4)$

19. $(\frac{1}{2}, -2, -\frac{3}{4})$ **23.** 0.088 cm

25. 8% **27.** $r\%$

29. 2% **31.** 0.004 radians

33. (a) $(-2, 1, 5)$, $(0, 3, 9)$
      (b) $\dfrac{4}{3\sqrt{14}}$ at $(-2, 1, 5)$, $\dfrac{4}{\sqrt{222}}$ at $(0, 3, 9)$

▶ Exercise Set 16.7 (Page 930)

1. (a) 19 (b) -9 (c) 3
      (d) $a^6 + 3$ (e) $-t^8 + 3$
      (f) $(a + b)(a - b)^2 b^3 + 3$

3. $(y + 1)e^{x^2(y+1)z^2}$

5. (a) t^{14} (b) 0 (c) $16{,}384$

7. all points not on the plane $x + y + z = 0$

9. all points within the cylinder $x^2 + y^2 = 1$

11.   **13.**

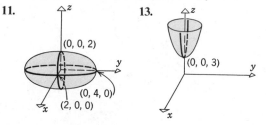

15. $\dfrac{\partial w}{\partial x} = 2xy^4z^3 + y$, $\dfrac{\partial w}{\partial y} = 4x^2y^3z^3 + x$,
$$\dfrac{\partial w}{\partial z} = 3x^2y^4z^2 + 2z$$

17. $\dfrac{\partial w}{\partial x} = \dfrac{2x}{y^2 + z^2}$, $\dfrac{\partial w}{\partial y} = -\dfrac{2y(x^2 + z^2)}{(y^2 + z^2)^2}$,
$$\dfrac{\partial w}{\partial z} = \dfrac{2z(y^2 - x^2)}{(y^2 + z^2)^2}$$

19. $\dfrac{\partial w}{\partial x} = \dfrac{x}{\sqrt{x^2 + y^2 + z^2}}$, $\dfrac{\partial w}{\partial y} = \dfrac{y}{\sqrt{x^2 + y^2 + z^2}}$,
$$\dfrac{\partial w}{\partial z} = \dfrac{z}{\sqrt{x^2 + y^2 + z^2}}$$

21. $f_x = -\dfrac{y^2z^3}{x^2y^4z^6 + 1}$, $f_y = -\dfrac{2xyz^3}{x^2y^4z^6 + 1}$,
$$f_z = -\dfrac{3xy^2z^2}{x^2y^4z^6 + 1}$$

23. $f_x = 4xyz \cosh\sqrt{z}\,\sinh(x^2yz)\cosh(x^2yz)$,
$f_y = 2x^2z \cosh\sqrt{z}\,\sinh(x^2yz)\cosh(x^2yz)$,
$f_z = 2x^2y \cosh\sqrt{z}\,\sinh(x^2yz)\cosh(x^2yz)$
$$+ \dfrac{\sinh\sqrt{z}\,\sinh^2(x^2yz)}{2\sqrt{z}}$$

25. (a) -80 (b) 40 (c) -60

27. (a) $\dfrac{2}{\sqrt{7}}$ (b) $\dfrac{4}{\sqrt{7}}$ (c) $\dfrac{1}{\sqrt{7}}$

29. $\dfrac{\partial w}{\partial x} = -\dfrac{x}{w}$, $\dfrac{\partial w}{\partial y} = -\dfrac{y}{w}$, $\dfrac{\partial w}{\partial z} = -\dfrac{z}{w}$

31. $\dfrac{\partial w}{\partial x} = -\dfrac{yzw\cos(xyz)}{2w + \sin(xyz)}$
$$\dfrac{\partial w}{\partial y} = -\dfrac{xzw\cos(xyz)}{2w + \sin(xyz)}$$
$$\dfrac{\partial w}{\partial z} = -\dfrac{xyw\cos(xyz)}{2w + \sin(xyz)}$$

33. (a) $15x^2y^4z^7 + 2y$ (b) $35x^3y^4z^6 + 3y^2$
      (c) $21x^2y^5z^6$ (d) $42x^3y^5z^5$
      (e) $140x^3y^3z^6 + 6y$ (f) $30xy^4z^7$
      (g) $105x^2y^4z^6$ (h) $210xy^4z^6$

35. $165t^{32}$ **37.** $-2t\cos(t^2)$

39. 3264 **41.** -320

43. $-\frac{314}{741}$

45. $\frac{72}{\sqrt{14}}$

47. $-\frac{8}{63}$

49. $\frac{1}{\sqrt{2}}\mathbf{i} - \frac{1}{\sqrt{2}}\mathbf{j}, \; 3\sqrt{2}$

51. $-\frac{1}{\sqrt{2}}\mathbf{i} + \frac{1}{\sqrt{2}}\mathbf{j}, \; \frac{\sqrt{2}}{2}$

53. $\frac{1}{\sqrt{266}}\mathbf{i} - \frac{11}{\sqrt{266}}\mathbf{j} + \frac{12}{\sqrt{266}}\mathbf{k}, \; -\sqrt{266}$

55. $\frac{3}{\sqrt{11}}$

57. $3x - 2y + 6z = -49, \quad \begin{aligned} x &= -3 + 3t \\ y &= 2 - 2t \\ z &= -6 + 6t \end{aligned}$

59. $x - y - 15z = -17, \quad \begin{aligned} x &= 3 + t \\ y &= 5 - t \\ z &= 1 - 15t \end{aligned}$

63. $(1, \frac{2}{3}, \frac{2}{3}), \; (-1, -\frac{2}{3}, -\frac{2}{3})$

67. $8\,dx - 3\,dy + 4\,dz$

69. $\frac{yz}{1 + (xyz)^2}\,dx + \frac{xz}{1 + (xyz)^2}\,dy + \frac{xy}{1 + (xyz)^2}\,dz$

71. 0.96

73. 2.35 cm^3

75. 39 ft^2

77. 14%

► Exercise Set 16.8 (Page 938)

1. $\dfrac{\partial f}{\partial v} = 8vw^3x^4y^5, \quad \dfrac{\partial f}{\partial w} = 12v^2w^2x^4y^5,$

$\dfrac{\partial f}{\partial x} = 16v^2w^3x^3y^5, \quad \dfrac{\partial f}{\partial y} = 20v^2w^3x^4y^4$

3. $\dfrac{\partial f}{\partial v_1} = \dfrac{2v_1}{v_3^2 + v_4^2}, \quad \dfrac{\partial f}{\partial v_2} = -\dfrac{2v_2}{v_3^2 + v_4^2},$

$\dfrac{\partial f}{\partial v_3} = -\dfrac{2v_3(v_1^2 - v_2^2)}{(v_3^2 + v_4^2)^2}, \quad \dfrac{\partial f}{\partial v_4} = -\dfrac{2v_4(v_1^2 - v_2^2)}{(v_3^2 + v_4^2)^2}$

5. $f_v(1, -2, 4, 8) = 128, \quad f_w(1, -2, 4, 8) = -512,$
$f_x(1, -2, 4, 8) = 32, \quad f_y(1, -2, 4, 8) = \frac{64}{3}$

7. $210t^{29}$

9. $\dfrac{\partial z}{\partial r} = \dfrac{2r\cos^2\theta}{r^2\cos^2\theta + 1}, \; \dfrac{\partial z}{\partial\theta} = -\dfrac{2r^2\sin\theta\cos\theta}{r^2\cos^2\theta + 1}$

11. $\dfrac{\partial w}{\partial\rho} = 2\rho(4\sin^2\phi + \cos^2\phi),$

$\dfrac{\partial w}{\partial\phi} = 6\rho^2\sin\phi\cos\phi,$

$\dfrac{\partial w}{\partial\theta} = 0$

13. $\dfrac{-4\cos 2y\sin 2y + 2y + 1}{2\sqrt{\cos^2 2y + y^2 + y}}$

15. $\sqrt{3} + \dfrac{\pi}{6}$ cm/sec, increasing

17. $\dfrac{\partial z}{\partial x} = -\dfrac{2x + 4z}{4x + 2z - 3y}, \; \dfrac{\partial z}{\partial y} = \dfrac{3z}{4x + 2z - 3y}$

27. (a) $\dfrac{dw}{dt} = \dfrac{\partial w}{\partial x_1}\dfrac{dx_1}{dt} + \dfrac{\partial w}{\partial x_2}\dfrac{dx_2}{dt} + \dfrac{\partial w}{\partial x_3}\dfrac{dx_3}{dt} + \dfrac{\partial w}{\partial x_4}\dfrac{dx_4}{dt}$

(b) $\dfrac{\partial w}{\partial v_1} = \dfrac{\partial w}{\partial x_1}\dfrac{\partial x_1}{\partial v_1} + \dfrac{\partial w}{\partial x_2}\dfrac{\partial x_2}{\partial v_1} + \dfrac{\partial w}{\partial x_3}\dfrac{\partial x_3}{\partial v_1} + \dfrac{\partial w}{\partial x_4}\dfrac{\partial x_4}{\partial v_1},$

$\dfrac{\partial w}{\partial v_2} = \dfrac{\partial w}{\partial x_1}\dfrac{\partial x_1}{\partial v_2} + \dfrac{\partial w}{\partial x_2}\dfrac{\partial x_2}{\partial v_2} + \dfrac{\partial w}{\partial x_3}\dfrac{\partial x_3}{\partial v_2} + \dfrac{\partial w}{\partial x_4}\dfrac{\partial x_4}{\partial v_2},$

$\dfrac{\partial w}{\partial v_3} = \dfrac{\partial w}{\partial x_1}\dfrac{\partial x_1}{\partial v_3} + \dfrac{\partial w}{\partial x_2}\dfrac{\partial x_2}{\partial v_3} + \dfrac{\partial w}{\partial x_3}\dfrac{\partial x_3}{\partial v_3} + \dfrac{\partial w}{\partial x_4}\dfrac{\partial x_4}{\partial v_3}$

► Exercise Set 16.9 (Page 947)

1. $(0, 0)$ relative min

3. $(1, -2)$ saddle

5. $(2, -1)$ relative min

7. $(2, 1), (-2, 1)$ saddle
   $(0, 0)$ relative min

9. $(-1, -1), (1, 1)$ relative min

11. $(0, 0)$ saddle

13. none

15. $(0, 0), (4, 0), (0, -2)$ saddle
    $(\frac{4}{3}, -\frac{2}{3})$ relative min

17. $(-1, 0)$ relative max

19. $\left(\frac{\pi}{2}, \frac{\pi}{2}\right)$ relative max

21. (b) $(0, 0)$ relative min

25. 9, 9, 9

27. $(\sqrt{5}, 0, 0), (-\sqrt{5}, 0, 0)$

29. $\frac{1}{27}$

31. length and width 2 ft, height 4 ft

33. $\frac{3}{2}\sqrt{6}$

35. length and width $\sqrt[3]{2V}$, height $\frac{1}{2}\sqrt[3]{2V}$

► Exercise Set 16.10 (Page 956)

1. max $\sqrt{2}$ at $(\sqrt{2}, 1)$ and $(-\sqrt{2}, -1)$,
   min $-\sqrt{2}$ at $(-\sqrt{2}, 1)$ and $(\sqrt{2}, -1)$

3. max $\sqrt{2}$ at $\left(\frac{1}{\sqrt{2}}, 0\right)$,
   min $-\sqrt{2}$ at $\left(-\frac{1}{\sqrt{2}}, 0\right)$

5. max 6 at $(\frac{4}{3}, \frac{2}{3}, -\frac{4}{3})$,
   min -6 at $(-\frac{4}{3}, -\frac{2}{3}, \frac{4}{3})$

7. max $\frac{1}{3\sqrt{3}}$ at $\left(\frac{1}{\sqrt{3}}, \frac{1}{\sqrt{3}}, \frac{1}{\sqrt{3}}\right)$,
   $\left(\frac{1}{\sqrt{3}}, -\frac{1}{\sqrt{3}}, -\frac{1}{\sqrt{3}}\right), \left(-\frac{1}{\sqrt{3}}, \frac{1}{\sqrt{3}}, -\frac{1}{\sqrt{3}}\right),$
   $\left(-\frac{1}{\sqrt{3}}, -\frac{1}{\sqrt{3}}, \frac{1}{\sqrt{3}}\right);$
   min $-\frac{1}{3\sqrt{3}}$ at $\left(\frac{1}{\sqrt{3}}, \frac{1}{\sqrt{3}}, -\frac{1}{\sqrt{3}}\right)$,
   $\left(\frac{1}{\sqrt{3}}, -\frac{1}{\sqrt{3}}, \frac{1}{\sqrt{3}}\right), \left(-\frac{1}{\sqrt{3}}, \frac{1}{\sqrt{3}}, \frac{1}{\sqrt{3}}\right),$
   $\left(-\frac{1}{\sqrt{3}}, -\frac{1}{\sqrt{3}}, -\frac{1}{\sqrt{3}}\right)$

9. $(\frac{2}{5}, \frac{19}{5})$

11. $(1, -1, 1)$

13. $\frac{1}{2}$

15. $(3, 6)$ closest,
    $(-3, -6)$ farthest

17. 9, 9, 9

19. $(\sqrt{5}, 0, 0), (-\sqrt{5}, 0, 0)$

21. length and width 2 ft, height 4 ft

▶ Chapter 16 Supplementary Exercises (Page 957)

1. (a) 1 (b) yx (c) $e^{r+s} \ln(rs)$

3. (a) the upper nappe of the elliptical cone $z^2 = x^2 + 4y^2$
(b) the plane with x-, y-, z-intercepts a, b, 1

5. $\dfrac{1}{x \sin yz} - \dfrac{3y \ln(xy) \cos yz}{\sin^2 yz}$

7. $\pi/2$; 0; 1, $-\pi^2/4$

9. 2

11. $(2x - 2y + 4r)/(x^2 + y^2 + 2z) = 2/(r + s)$

17. (a) increasing 12 newtons/m²/min
(b) increasing 240 newtons/m²/min

19. (a) $0 = f(0, 0)$ (b) yes

21. (a) $-(6x - 5y + y \sec^2 xy)/(-5x + x \sec^2 xy)$
(b) $-\dfrac{\ln y + \cos(x - y)}{\dfrac{x}{y} - \cos(x - y)}$

23. $\dfrac{dV}{dt} = R\left[\dfrac{dE/dt}{r + R} - \dfrac{dr}{dt}\dfrac{E}{(r + R)^2}\right]$

25. (a) $\langle 6, 45 \rangle$ (b) $-174/\sqrt{17}$

27. (a) $\langle \tfrac{1}{3}, \tfrac{1}{2}, \tfrac{1}{6} \rangle$ (b) $\sqrt{3}/9$

29. (a) $\langle 1, 3, 0 \rangle$ (b) $43/15$

31. (a) $\pm(\mathbf{i} + \mathbf{j})/\sqrt{2}$
(b) $\pm(2\mathbf{i} + \mathbf{j})/\sqrt{5}$

33. $-7/\sqrt{5}$

35. (a) $\mathbf{N} = \langle 8, -6, -5 \rangle$
(b) $8(x - 4) - 6(y + 3) - 5(z - 10) = 0$

37. $(1, 1, 1)$, $(-1, -1, 1)$, $(0, 0, 2)$

39. $(-\tfrac{1}{3}, -\tfrac{1}{2}, 2)$

41. $dV = -\dfrac{1}{3}(.2)\text{m}^3$, $\Delta V = -\dfrac{1}{3}(.218)\text{m}^3$

43. relative min at $(15, -8)$

45. relative min at $(3, 9)$; saddle point at $(0, 0)$

47. $\dfrac{2a}{\sqrt{3}} \times \dfrac{2b}{\sqrt{3}} \times \dfrac{2c}{\sqrt{3}}$

49. $I_1 : I_2 : I_3 = R_1^{-1} : R_2^{-1} : R_3^{-1}$

▶ Exercise Set 17.1 (Page 969)

1. 7

5. 2

9. 3

13. $\tfrac{1}{2}(1 - \ln 2)$

17. $\tfrac{1}{3}$

21.

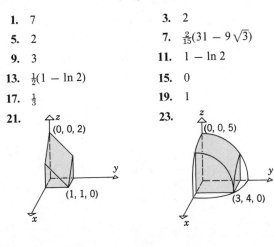

3. 2

7. $\tfrac{2}{15}(31 - 9\sqrt{3})$

11. $1 - \ln 2$

15. 0

19. 1

23.

25.

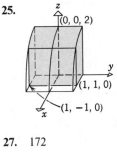

27. 172

29. 8

▶ Exercise Set 17.2 (Page 980)

1. $\frac{1}{40}$

3. 9

37. $\frac{56}{15}$

39. 170

5. $\frac{\pi}{2}$

7. 1

41. $\frac{27}{2}\pi$

43. $\frac{20}{3}$

9. $\frac{2}{3}a^3$

11. $\frac{1}{12}$

45. $\frac{\pi}{2}$

47. $\frac{2000}{3}$

13. 32

15. $-\frac{2}{\pi}$

49. $\int_0^{\sqrt{2}} \int_{y^2}^2 f(x, y)\, dx\, dy$ 51. $\int_1^{e^2} \int_{\ln x}^2 f(x, y)\, dy\, dx$

17. 576

19. $\frac{1}{2}(\sqrt{17} - 1)$

53. $\int_{-1}^1 \int_{-2\sqrt{1-y^2}}^{2\sqrt{1-y^2}} f(x, y)\, dx\, dy$

21. 0

23. $\frac{\pi}{4} - \frac{1}{2}\ln 2$

55. $\int_0^{\pi/2} \int_0^{\sin x} f(x, y)\, dy\, dx$ 57. $\frac{1}{8}(1 - e^{-16})$

25. $\frac{50}{3}$

27. $-\frac{1}{2}$

59. $\frac{1}{3}(1 - \cos 8)$ 61. (a) 0   (b) $-\frac{603}{40}$

29. $\frac{25}{2}$

31. $\sqrt{2} - 1$

33. 32

35. $\frac{125}{12}$

▶ Exercise Set 17.3 (Page 990)

1. $\frac{1}{6}$

3. 0

21. $\frac{\pi}{8}\ln 5$

23. $\frac{\pi}{8}$

5. 0

7. $\frac{3\pi}{2}$

25. $\frac{16}{9}$

27. $\frac{\pi}{2}\left(1 - \frac{1}{\sqrt{1 + a^2}}\right)$

9. $\frac{\pi}{16}$

11. $\frac{4\pi}{3} + 2\sqrt{3}$

29. $\frac{1}{9}(3\pi - 4)a^2 c$

31. $\frac{4\pi}{3} + 2\sqrt{3} - 2$

13. $\frac{4}{3}(27 - 16\sqrt{2})\pi$

15. $\frac{32}{9}$

17. $\frac{5\pi}{32}$

19. $(1 - e^{-1})\pi$

33. (b) $\frac{\pi}{4}$ (c) $\frac{\sqrt{\pi}}{2}$

▶ Exercise Set 17.4 (Page 996)

1. 6π

3. $\frac{\sqrt{5}}{6}$

9. 8π

11. $2(\pi - 2)a^2$

5. $\frac{\pi}{6}(5\sqrt{5} - 1)$

7. $\frac{\pi}{18}(10\sqrt{10} - 1)$

13. 128

15. $\pi a \sqrt{a^2 + h^2}$

▶ Exercise Set 17.5 (Page 1006)

1. 8

3. 7

5. $\frac{81}{5}$

7. $\frac{128}{15}$

9. $\frac{\pi}{2}(\pi - 3)$

11. $\frac{1}{6}$

13. 4

15. $\frac{256}{15}$

17. 9π

19. 2π

21. $\frac{\pi}{6}(8\sqrt{2} - 7)a^3$

23. (a)

(0, 0, 1), (1, 0, 0)

(b)

(0, 9, 9), (3, 9, 0)

25. (a) $\displaystyle\int_0^a \int_0^{b\left(1 - \frac{x}{a}\right)} \int_0^{c\left(1 - \frac{x}{a} - \frac{y}{b}\right)} dz\, dy\, dx,$

$\displaystyle\int_0^b \int_0^{a\left(1 - \frac{y}{b}\right)} \int_0^{c\left(1 - \frac{x}{a} - \frac{y}{b}\right)} dz\, dx\, dy,$

$\displaystyle\int_0^c \int_0^{a\left(1 - \frac{z}{c}\right)} \int_0^{b\left(1 - \frac{x}{a} - \frac{z}{c}\right)} dy\, dx\, dz,$

$\displaystyle\int_0^a \int_0^{c\left(1 - \frac{x}{a}\right)} \int_0^{b\left(1 - \frac{x}{a} - \frac{z}{c}\right)} dy\, dz\, dx,$

$\displaystyle\int_0^c \int_0^{b\left(1 - \frac{z}{c}\right)} \int_0^{a\left(1 - \frac{y}{b} - \frac{z}{c}\right)} dx\, dy\, dz,$

$\displaystyle\int_0^b \int_0^{c\left(1 - \frac{y}{b}\right)} \int_0^{a\left(1 - \frac{y}{b} - \frac{z}{c}\right)} dx\, dz\, dy$

27. $\frac{1}{6}a^3$

29. (a) 0   (b) $\frac{1}{2}(e^2 - 1)$

▶ Exercise Set 17.6 (Page 1021)

1. 10 feet to the right of m_1

3. $\frac{13}{20}, \left(\frac{190}{273}, \frac{6}{13}\right)$

5. $\frac{a^4}{8}, \left(\frac{8a}{15}, \frac{8a}{15}\right)$

7. $\left(\frac{2}{3}, \frac{1}{3}\right)$

9. $\left(-\frac{1}{2}, \frac{2}{5}\right)$

11. $\left(0, \dfrac{4(b^3 - a^3)}{3\pi(b^2 - a^2)}\right)$

13. $\left(0, \frac{8}{3}\right)$

15. $\frac{a^4}{2}, \left(\frac{a}{3}, \frac{a}{2}, \frac{a}{2}\right)$

17. $\frac{1}{6}, \left(0, \frac{16}{35}, \frac{1}{2}\right)$

19. $\left(\frac{1}{4}, \frac{1}{4}, \frac{1}{4}\right)$

21. $\left(\frac{1}{2}, 0, \frac{3}{5}\right)$

23. $\left(\dfrac{3a}{8}, \dfrac{3a}{8}, \dfrac{3a}{8}\right)$

25. $\frac{2}{3}\pi k a^3$

29. $\left(\dfrac{128}{105\pi}, \dfrac{128}{105\pi}\right)$

33. $2\pi^2 k a b$

35. $\left(\dfrac{a}{3}, \dfrac{b}{3}\right)$

▶ Exercise Set 17.7 (Page 1033)

1. $\frac{\pi}{4}$

3. $\frac{\pi}{16}$

5. $\frac{81\pi}{2}$

7. $\frac{8}{3}(10\sqrt{5} - 19)\pi$

9. $\frac{32}{9}(3\pi - 4)$

11. $\frac{2}{3}(\sqrt{3} - 1)\pi$

13. $9\sqrt{2}\pi$

15. $\frac{4}{3}\pi a^3$

17. $\frac{11}{6}\pi$

19. $\frac{1}{4}(2 - \sqrt{2})\pi$

21. $\pi k a^4$

23. $\left(0, 0, \dfrac{7}{16\sqrt{2} - 14}\right)$

25. $\left(\dfrac{3a}{8}, \dfrac{3a}{8}, \dfrac{3a}{8}\right)$

27. $\left(0, \dfrac{195}{152}, 0\right)$

29. $\frac{1}{3}(1 - e^{-1})\pi$

31. 81π

33. $\left(0, 0, \dfrac{2a}{5}\right)$

35. $\left(0, 0, \frac{11}{30}\right)$

37. $\frac{63}{8}\pi$

▶ Chapter 17 Supplementary Exercises (Page 1035)

1. $-\frac{1}{2}\sqrt{2}/\pi$ **3.** $-9/14$

5. $\displaystyle\int_0^1\int_{2y}^2 e^x e^y\,dx\,dy$ **7.** $3/2$

9. (a) (b)

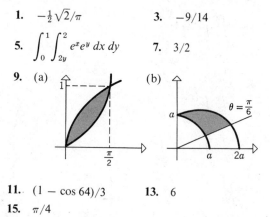

11. $(1-\cos 64)/3$ **13.** 6

15. $\pi/4$

17. $\displaystyle\int_0^{2a}\int_0^{\sqrt{2ay-y^2}}\frac{2xy}{x^2+y^2}\,dx\,dy=a^2$

19. $\pi/4$ **21.** $(37^{3/2}-1)\pi/54$

23. $15\pi\sqrt{2}$ **25.** $512\pi/3$

27. (a) $\displaystyle\int_0^{1/2}\int_0^{1-2y}\int_0^{1-z-2y} z\,dx\,dz\,dy$

(b) $\displaystyle\int_0^4\int_0^{4-y}\int_0^{\sqrt{y}} 3\,dx\,dz\,dy$

29. (a) $\displaystyle\int_0^{2\pi}\int_0^{\pi/3}\int_0^a (\rho^4\sin^3\phi)\,d\rho\,d\phi\,d\theta$

(b) $\displaystyle\int_0^{2\pi}\int_0^{\sqrt{3}a/2}\int_{r/\sqrt{3}}^{\sqrt{a^2-r^2}} r^3\,dz\,dr\,d\theta$

(c) $\displaystyle 4\int_0^{\sqrt{3}a/2}\int_0^{\sqrt{(3a^2/4)-x^2}}\int_{\sqrt{(x^2+y^2)/3}}^{\sqrt{a^2-x^2-y^2}} (x^2+y^2)\,dz\,dy\,dx$

31. $8\pi a^3/3$ **33.** $\pi a^3/9$

35. $\bar{x}=0,\ \bar{y}=5a/6$ **37.** $\bar{x}=0,\ \bar{y}=b/4$

39. $kabc^2/24$

41. $\bar{x}=0,\ \bar{y}=12/7,\ \bar{z}=8/7$

43. $\bar{x}=\bar{y}=0,\ \bar{z}=h/4$

▶ Exercise Set 18.1 (Page 1048)

1. (a) $\frac{4}{3}$ (b) 0 (c) $\frac{4}{3}$

3. $-\frac{8}{3}$ **5.** $1-\pi$

7. 3 **9.** $-1-\frac{\pi}{4}$

11. 0 **13.** $1-e^{-1}$

15. $1-e^3$ **17.** $\frac{23}{6}$

19. $\frac{3}{5}$

21. (a) $\frac{5}{4}-\frac{\pi}{8}+\frac{1}{2}\tan^{-1}2$

(b) $\frac{1}{3}\tan^{-1}2-\frac{2}{3}\tan^{-1}\frac{1}{2}$

(c) $\frac{3}{4}$

23. $-\frac{37}{2}$ **25.** -12

27. 0

▶ Exercise Set 18.2 (Page 1059)

1. $\frac{1}{2}x^2+\frac{1}{2}y^2+K$ **3.** not conservative

5. $x\cos y+y\sin x+K$ **7.** 13

9. -6 **11.** $9e^2$

13. 32 **15.** $-\frac{1}{2}$

17. $\frac{\pi}{4}$

19. (a) $-2a$ (b) $-2a$

(c) $-2a$ (d) 0

▶ Exercise Set 18.3 (Page 1065)

1. 0

3. 0

5. 0

7. 8π

9. -4

11. -1

13. 0

15. πab

17. $\frac{1}{2}ab$

19. $\frac{250}{3}$

21. (b) $\left(0, \frac{4a}{3\pi}\right)$

23. (c) $A = \frac{1}{2}[(x_1 y_2 - x_2 y_1) + (x_2 y_3 - x_3 y_2)$
$+ \cdots + (x_n y_1 - x_1 y_n)]$

(d) 8

▶ Exercise Set 18.4 (Page 1076)

1. $\frac{\pi \delta_0}{6}(5\sqrt{5} - 1)$

3. $\frac{4}{3}\pi \delta_0$

5. $\frac{15\pi}{\sqrt{2}}$

7. $\frac{\pi}{4}$

9. $\frac{3}{\sqrt{2}}$

11. 9

13. $\frac{4\pi}{3}$

15. $\frac{8}{3}\pi a^4$

17. $\frac{1}{4}[37\sqrt{37} - 1]$

19. (a) $\frac{\sqrt{29}}{16} \int_0^6 \int_0^{(12-2x)/3} xy(12 - 2x - 3y)\, dy\, dx$

(b) $\frac{\sqrt{29}}{4} \int_0^3 \int_0^{(12-4z)/3} yz(12 - 3y - 4z)\, dy\, dz$

(c) $\frac{2\sqrt{29}}{9} \int_0^3 \int_0^{6-2z} xz(6 - x - 2z)\, dx\, dz$

21. (a) $\int_0^4 \int_1^2 y^3 z \sqrt{1 + 4y^2}\, dy\, dz$

(b) $\frac{1}{2} \int_0^4 \int_1^4 xz \sqrt{1 + 4x}\, dx\, dz$

▶ Exercise Set 18.5 (Page 1087)

1. $\frac{2}{\sqrt{29}}\mathbf{i} + \frac{3}{\sqrt{29}}\mathbf{j} + \frac{4}{\sqrt{29}}\mathbf{k}$

3. (a) $-\frac{2}{\sqrt{21}}\mathbf{i} - \frac{4}{\sqrt{21}}\mathbf{j} + \frac{1}{\sqrt{21}}\mathbf{k}$

(b) $\frac{3}{5\sqrt{2}}\mathbf{i} - \frac{4}{5\sqrt{2}}\mathbf{j} + \frac{1}{\sqrt{2}}\mathbf{k}$

(c) $\frac{3}{5}\mathbf{i} - \frac{4}{5}\mathbf{k}$

5. 2π

7. $\frac{\pi}{2}$

9. 54π

11. $\frac{14}{3}\pi$

13. 0

15. $4\pi a^3$

19. (a) $\iint\limits_R \mathbf{F} \cdot \left(\mathbf{i} - \frac{\partial x}{\partial y}\mathbf{j} - \frac{\partial x}{\partial z}\mathbf{k}\right) dA$ $\begin{bmatrix} \text{Forward} \\ R = \text{projection on } yz\text{-plane} \end{bmatrix}$

$\iint\limits_R \mathbf{F} \cdot \left(-\mathbf{i} + \frac{\partial x}{\partial y}\mathbf{j} + \frac{\partial x}{\partial z}\mathbf{k}\right) dA$ $\begin{bmatrix} \text{Backward} \\ R = \text{projection on } yz\text{-plane} \end{bmatrix}$

(b) $\iint\limits_R \mathbf{F} \cdot \left(-\frac{\partial y}{\partial x}\mathbf{i} + \mathbf{j} - \frac{\partial y}{\partial z}\mathbf{k}\right) dA$ $\begin{bmatrix} \text{Right} \\ R = \text{projection on } xz\text{-plane} \end{bmatrix}$

$\iint\limits_R \mathbf{F} \cdot \left(\frac{\partial y}{\partial x}\mathbf{i} - \mathbf{j} + \frac{\partial y}{\partial z}\mathbf{k}\right) dA$ $\begin{bmatrix} \text{Left} \\ R = \text{projection on } xz\text{-plane} \end{bmatrix}$

21. 2π

▶ Exercise Set 18.6 (Page 1094)

1. $\operatorname{div} \mathbf{F} = z^3 + 8y^3x^2 + 10zy$

3. $\operatorname{div} \mathbf{F} = ye^{xy} + \sin y + 2\sin z \cos z$

5. $\operatorname{div} \mathbf{F} = \dfrac{1}{x} + xz\, e^{xyz} + \dfrac{x}{x^2 + z^2}$

7. 8

9. $3\pi a^2$

11. 180π

13. $\frac{1}{24}$

15. $\dfrac{\pi}{2}$

17. $\frac{4608}{35}$

19. $4\pi a^3$

23. $\pi a^2 h$

▶ Exercise Set 18.7 (Page 1099)

1. $\operatorname{curl} \mathbf{F} = \mathbf{0}$

3. $\operatorname{curl} \mathbf{F} = -xe^{xy}\,\mathbf{k}$

5. $\operatorname{curl} \mathbf{F} = -xye^{xyz}\mathbf{i} + \dfrac{z}{x^2 + z^2}\mathbf{j} + yz\, e^{xyz}\mathbf{k}$

7. 16π

11. πa^2

15. 0

9. 0

13. $\frac{3}{2}$

17. 2π

▶ Exercise Set 18.8 (Page 1106)

1. 2π

3. 54π

5. $4\pi a^3$

7. 180π

9. No sources or sinks ($\operatorname{div} \mathbf{F} = 0$)

11. Sources at all points except the origin; no sinks.

13. (a) $\frac{3}{2}$ (b) -1 (c) $\mathbf{n} = -\dfrac{1}{\sqrt{2}}\mathbf{j} - \dfrac{1}{\sqrt{2}}\mathbf{k}$

▶ Chapter 18 Supplementary Exercises (Page 1107)

1. 4

3. 6

5. 2

7. not conservative

9. $\dfrac{y^2}{x} + 2y^2 + x^3 + C$

11. $-7\pi^2/16$

13. 3/2

15. 0

17. 32

21. 12/5

25. $(0, 0, \frac{149}{65})$

19. $-45\pi/4$

23. 2

27. $\dfrac{\pi}{2}$

▶ Trigonometry Exercises I (Page A15)

1. (a) $\dfrac{3\pi}{2}$ (b) $\dfrac{13\pi}{6}$ (c) $\dfrac{\pi}{9}$

   (d) $\dfrac{23\pi}{30}$ (e) $\dfrac{\pi^2}{1800}$

3. (a) $12°$ (b) $270°$ (c) $\left(\dfrac{810}{\pi}\right)°$

   (d) $288°$ (e) $900°$

5. (a)   (b)

(c) 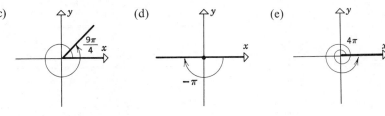 (d) (e)

7.

	θ	$\sin\theta$	$\cos\theta$	$\tan\theta$	$\csc\theta$	$\sec\theta$	$\cot\theta$
(a)	$\dfrac{\pi}{6}$	$\dfrac{1}{2}$	$\dfrac{\sqrt{3}}{2}$	$\dfrac{1}{\sqrt{3}}$	2	$\dfrac{2}{\sqrt{3}}$	$\sqrt{3}$
(b)	$\dfrac{-7\pi}{3}$	$\dfrac{-\sqrt{3}}{2}$	$\dfrac{1}{2}$	$-\sqrt{3}$	$\dfrac{-2}{\sqrt{3}}$	2	$\dfrac{-1}{\sqrt{3}}$
(c)	$\dfrac{-5\pi}{4}$	$\dfrac{1}{\sqrt{2}}$	$\dfrac{-1}{\sqrt{2}}$	-1	$\sqrt{2}$	$-\sqrt{2}$	-1
(d)	-3π	0	-1	0	$-$	-1	$-$
(e)	π	0	-1	0	$-$	-1	$-$

9. (a) $\theta = \pm n\pi,\ n = 0, 1, 2, \ldots$

(b) $\theta = \dfrac{\pi}{2} \pm n\pi,\ n = 0, 1, 2, \ldots$

(c) $\theta = \pm n\pi,\ n = 0, 1, 2, \ldots$

(d) $\theta = \pm n\pi,\ n = 0, 1, 2, \ldots$

(e) $\theta = \dfrac{\pi}{2} \pm n\pi,\ n = 0, 1, 2, \ldots$

(f) $\theta = \pm n\pi,\ n = 0, 1, 2, \ldots$

11. $\theta = \dfrac{5\pi}{4} \pm 2n\pi$ and

$\theta = \dfrac{7\pi}{4} \pm 2n\pi,\ n = 0, 1, 2, \ldots$

13. $\theta = \dfrac{\pi}{3} \pm 2n\pi$ and

$\theta = \dfrac{5\pi}{3} \pm 2n\pi,\ n = 0, 1, 2, \ldots$

15. $\theta = \dfrac{\pi}{3} \pm n\pi,\ n = 0, 1, 2, \ldots$

17. $\theta = \dfrac{4\pi}{3} \pm 2n\pi$ and

$\theta = \dfrac{5\pi}{3} \pm 2n\pi,\ n = 0, 1, 2, \ldots$

19. $\theta = \pi \pm 2n\pi,\ n = 0, 1, 2, \ldots$

21. $\theta = \dfrac{\pi}{6} \pm n\pi,\ n = 0, 1, 2, \ldots$

23. $\theta = \dfrac{7\pi}{6} \pm 2n\pi$ and

$\theta = \dfrac{11\pi}{6} \pm 2n\pi,\ n = 0, 1, 2, \ldots$

25. $\theta = \dfrac{\pi}{6} \pm 2n\pi$ and

$\theta = \dfrac{11\pi}{6} \pm 2n\pi,\ n = 0, 1, 2, \ldots$

27. (a) $\tfrac{1}{4}(\sqrt{2} + \sqrt{6})$

(b) $\tfrac{1}{2}\sqrt{2 - \sqrt{2}}$

(c) $\tfrac{1}{2}\sqrt{2 - \sqrt{2 - \sqrt{3}}}$ or $\sqrt{\dfrac{4 - \sqrt{6 + \sqrt{2}}}{8}}$ are possible answers

29.

	$\sin\theta$	$\cos\theta$	$\tan\theta$	$\csc\theta$	$\sec\theta$	$\cot\theta$
(a)	$\dfrac{a}{3}$	$\dfrac{\sqrt{9-a^2}}{3}$	$\dfrac{a}{\sqrt{9-a^2}}$	$\dfrac{3}{a}$	$\dfrac{3}{\sqrt{9-a^2}}$	$\dfrac{\sqrt{9-a^2}}{a}$
(b)	$\dfrac{a}{\sqrt{a^2+25}}$	$\dfrac{5}{\sqrt{a^2+25}}$	$\dfrac{a}{5}$	$\dfrac{\sqrt{a^2+25}}{a}$	$\dfrac{\sqrt{a^2+25}}{5}$	$\dfrac{5}{a}$
(c)	$\dfrac{\sqrt{a^2-1}}{a}$	$\dfrac{1}{a}$	$\sqrt{a^2-1}$	$\dfrac{a}{\sqrt{a^2-1}}$	a	$\dfrac{1}{\sqrt{a^2-1}}$

31. $\frac{9}{2}\sqrt{2}$, $\frac{9}{2}(1+\sqrt{3})$, $105°$

41. $\sin\left(\dfrac{3\pi}{2}-\theta\right)=-\cos\theta,$

$\cos\left(\dfrac{3\pi}{2}+\theta\right)=\sin\theta$

55. $\sin 3\theta = 3\sin\theta\cos^2\theta - \sin^3\theta,$
$\cos 3\theta = \cos^3\theta - 3\sin^2\theta\cos\theta$

▶ Trigonometry Exercises II (Page A25)

1. (a) $\dfrac{2\pi}{5}$ (b) 6π (c) 8π (d) $\dfrac{\pi}{7}$
  (e) 2 (f) 10 (g) 1 (h) $2\pi k$

3. (a) 5 (b) $\frac{1}{3}$ (c) $\frac{1}{2}$ (d) 1

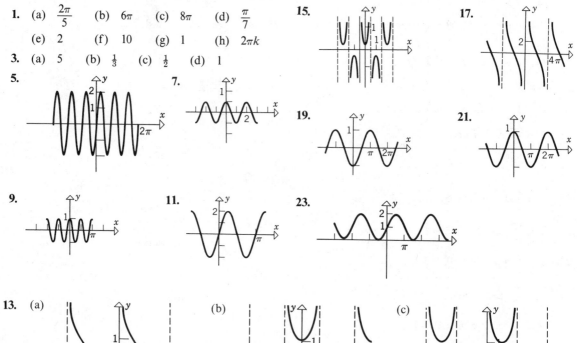

5.

7.

9.

11.

13. (a)

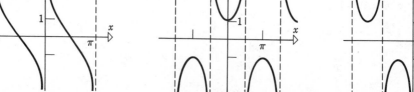

(b)

(c)

15.

17.

19.

21.

23.

25.

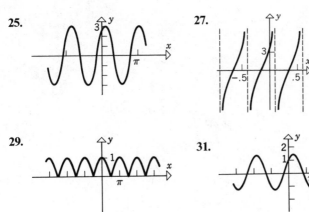

27.

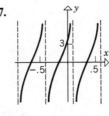

29.

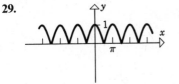

31.

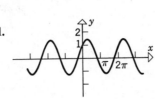

33. (a) odd   (b) even   (c) odd   (d) even
(e) even   (f) even   (g) odd   (h) even

index